Green Design and Manufacturing for Sustainability

Green Design and Manufacturing for Sustainability

Nand K. Jha

CRC Press
Taylor & Francis Group
Boca Raton London New York

CRC Press is an imprint of the
Taylor & Francis Group, an **informa** business

CRC Press
Taylor & Francis Group
6000 Broken Sound Parkway NW, Suite 300
Boca Raton, FL 33487-2742

Printed on acid-free paper
Version Date: 20151109

International Standard Book Number-13: 978-1-4665-0526-1 (Pack - Book and Ebook)

Visit the Taylor & Francis Web site at
http://www.taylorandfrancis.com

and the CRC Press Web site at
http://www.crcpress.com

Contents

Section I Industrial Ecology and Sustainability

Section II Green Design and EOL Treatment

Section IV Design for Environment (DfE)

Section V Environmentally Benign Manufacturing

Section VI Regulations for Sustainable Manufacturing

Preface

Leading institutions in the United States have recognized that sustainable development should have a prominent role in engineering education and practice. The criteria used to accredit engineering education programs have also recognized this need. Universities report that many students are not receiving significant exposure to *sustainability* issues. The World Commission on Environment and Development suggests that "sustainable development should meet the needs of the present without compromising the ability of future generations to meet their own needs."

The importance of sustainable engineering education is evident and engineering educators should recognize the need for sustainability discourse. The engineering education community should include sustainable engineering courses in its curriculum. Unfortunately, few engineering schools have made major updates to their courses and curricula over the past few decades. Only recently have colleges and universities started to include topics in sustainable engineering in their course material, such as life cycle assessment, concepts in renewable energy, and methods of waste minimization. The engineering industry is the highest user of energy and the largest producer of greenhouse gas causing a deterioration in the water quality and soil moisture, and producing toxic air that is unsuitable to breathe. Carbon dioxide, water, and the ecological footprint have become the central foci in sustainable development. Therefore, engineering education has to change to understand the need of society.

The main reason for resistance to change in the curriculum to include sustainable engineering is the lack of suitable textbooks, which could integrate the concept of sustainability in engineering courses such as design and manufacturing. In the typical undergraduate curriculum, the courses on manufacturing and engineering design are the two applied engineering courses most closely related to the real-world practice of engineers.

I have been teaching design and manufacturing courses at graduate and undergraduate levels for close to 40 years and I could not find any textbook that integrates the concept of disassembly and end-of-life considerations, including recycling and reuse in the design of mechanical components such as springs, fasteners, gears, bearings, and so on. We can define green mechanical design as the selection of materials, processes, and geometry that satisfies specified and implied environmental requirements. For sustainability, social, environmental, and economic development must be integrated and balanced. The environmental degradation due to economic activities must be bearable for society in general. The textbook, *Green Design and Manufacturing for Sustainability* attempts to bridge the gap between the need for green design and traditional mechanical design, concentrating only on stress and strain. Green design and manufacturing promises a reduction in materials, energy use, disposal fees, and reduced pollution. It is a big challenge to engineering designers. The topics of ecology, the environmental effect of engineering design and manufacturing, economics of environmental degradation, depletion of energy, depletion of material, and global warming would be of great importance to both engineering students and practicing engineers. The textbook presented here attempts to include and integrate the topics of ecology, economics for sustainability, and optimal design to reduce and minimize the selected material for a mechanical component. In particular, the textbook emphasizes the techniques needed to reduce the use of resources including materials, energy, and water, and so on; to reduce toxic and otherwise harmful emissions to the environment causing global warming; to manage renewable resources to ensure sustainability, quality, and durability; and to design for reuse and recycling.

The National Science Foundation has established the Center for Sustainable Engineering. It has helped faculty become part of a growing network of educators in sustainable engineering. The author has participated in the Center for Sustainable Engineering annual meeting, and general discussion always converges to the lack of response among universities to integrate or include sustainability in their curriculum and a suitable textbook.

The Textbook

The textbook has been organized into six distinct parts. The first chapter is the introduction, which presents sustainability principles and their effect on design and manufacturing. Section I is titled Industrial Ecology and Sustainability, which has three chapters that develop the background and tools that are required for engineers to analyze and respond to environmental constraints.

1. Introduction
2. Industrial ecology; mass and energy balance; mass and energy balance examples
3. Economics of production; mathematical analysis of cash flows; cost estimation
4. Analytical techniques for sustainability analysis; nonlinear optimization; geometric programming; life cycle analysis and life cycle cost analysis

These three chapters are common to green design and sustainable manufacturing. They have been integrated throughout the textbook into the topics of design and manufacturing. Section II presents the concepts of green design and end-of-life treatment including eco-material selection based on the theory developed by Michael Ashby. The end of life and its impact on the environment are presented in detail in the context of design. It has two chapters:

5. Principles of green design and material selection in green design
6. End-of-life treatment and environmental impact

Section III presents the green design of mechanical components including springs, gears, a shaft and its components, bearings, and fasteners:

7. Spring design for sustainability
8. Environmentally conscious gear design
9. Green shaft design
10. Bearing design and sustainability
11. Fastener design

In Section III, the concepts of green design, sustainability, and ecoproperties have been integrated with the traditional mechanical component design. Section IV separately discusses sustainable design concepts including assembly and disassembly and the recyclability and reuse of mechanical components. Recyclability and reuse have to be encouraged to achieve sustainability. The problems associated with recycling and reuse are presented in an easily understood way. It has long been recognized that if components are assembled as snap fit, it would avoid the use of fasteners and welding. The snap fit is used at this stage for plastic products and the theory and practice of snap fitting is discussed in detail. Computer-aided design has found wide acceptance in engineering and several computer-aided design software packages are available for use in engineering practice. The book presents the principles of computer-aided design with finite element analysis. The computer-aided design principles are presented with real-world mechanical component design interwoven with sustainability considerations. The examples also include manufacturing and cost estimations. Most of the computer-aided design examples are undergraduate projects performed by seniors in mechanical engineering.

12. Sustainable design; assembly and disassembly of mechanical systems; recyclability and reuse of mechanical components
13. Snap-fit design

14. Computer-aided design; green design considerations; prototyping (solid modeling) of mechanical components and assembly; finite element analysis
15. Computer-aided design examples

Section V presents the concept and practice of environmentally benign manufacturing. The sustainability principles in manufacturing for minimizing energy are presented first and run through all parts of Section V. Basic metal-removal processes along with the forming and solidification process are integrated with green and sustainability principles.

16. Green manufacturing and metal-removal processes for the minimization of energy
17. Turning
18. Drilling
19. Milling
20. Grinding
21. Cutting tools and sustainability considerations
22. Green forming processes
23. Solidification processes

The concept of life cycle analysis and environmental issues in both net-shape manufacturing and material removal processes is integrated. The demanufacturing and remanufacturing of products are emphasized. In particular, the energy and power consumption during manufacturing processes is presented in detail, including the emission of greenhouse gases and the pollution of the atmosphere. The pollution caused by coolants and lubricants is discussed in detail. Environmentally conscious manufacturing including manufacturing costs and disposal costs is presented. The design considerations for cutting tools are presented in the light of environmental issues. Section VI deals with international regulations for green design and manufacturing. The author thinks that for our graduates to be globally acceptable, they must be aware of the regulations in other countries. The environmental regulations of some selected countries are discussed and presented. The regulations for part materials, design, and manufacturing are more rigorous in the European Union than in the United States. In the United States, the Environmental Protection Agency's regulations are mostly about carbon dioxide emissions while the European Union regulations also deal with material selection, design, assembly, and manufacturing. If components and systems that are designed and manufactured in the United States are to be sold inside European Union countries, these components must conform to the regulations of the European Union or any other country for that matter. The section is straightforward and written more to acquaint engineers who always deal with data, formulae, and equations, with regulations.

24. International regulations for green design and manufacturing; International Organization for Standardization regulations; European Union green engineering regulations; British premises accreditation scheme regulations; U.S. Environmental Protection Agency regulations; India's environmental laws; Chinese environmental regulations; Canadian environmental laws; Brazil's environmental regulations

This textbook is in response to eco-concerns, to help design and manufacture mechanical components using energy and materials. Our objective is to minimize the environmental impact, which may be in conflict with the other noble objective of minimizing cost. It is an attempt to provide perspective and furnish our engineering graduates with tools to integrate environmental concerns in the design and manufacturing of components. The textbook is unique in the sense that it tries to bring environmental issues into the basics of mechanical component design and manufacture. Some principles of environmentally conscious design and sustainable manufacturing are available in a fragmented way in journals and articles; however, they are not presented as a textbook with examples and problems, which can be adapted for undergraduate classes. The range of examples and exercises presented in the textbook makes it easier to adopt it into the engineering curriculum. The intended audience of the textbook is upper-level students in several areas—mechanical, chemical, industrial, manufacturing, and environmental—as well as practicing engineers. The textbook does not include all the data for the mechanical properties of materials, but these days all these data are available online. Furthermore, the author believes that there is enough material in the textbook for two courses: (1) green mechanical component design and (2) sustainable manufacturing.

The first course on green mechanical component design should include Chapters 1 through 15. However, it is a practice in most engineering schools to offer two undergraduate design courses of six credits and Chapters 1 through 15 may be appropriate. Some schools have the practice of offering three credit design courses and an additional three credits for finite element analysis and computer-aided design; this textbook can be adopted for both these courses. However, Chapters 1 through 4 should be common to both the design courses. The textbook is also appropriate for a course at senior level on sustainable manufacturing, including Chapters 1 through 4 and Chapters 16 through 24.

The textbook is suitable for senior-level undergraduates as well as first-year graduates.

Finally, it is always up to the instructors to pick and choose the topics to formulate and cover the course from the wide selection presented in the book. The Instructor's Manual and PowerPoint lectures will be published in due course. The present setup of the manufacturing processes course that the author teaches includes the topics presented in the book and students enjoy the environmental aspect and its integration into manufacturing processes.

Acknowledgments

I wish to thank my wife, Munna, for her patience, understanding, encouragement, and support in writing this textbook. My thanks also go to those at Taylor & Francis who have had to bear with me on every aspect of manuscript preparation. It is they who will have to cope with the marketing and sales of the final product, the textbook.

Author

Nand K. Jha has had an extended and distinguished career in engineering education. Over the past 40 years, he has been involved in mechanical engineering education in various capacities. He is deeply involved in teaching, research, and the development of mechanical engineering curricula at both undergraduate and graduate level. In this extended run of teaching and research, he has contributed significantly to manufacturing science, optimization, robotics and automation, green design, and manufacturing. He has published close to 100 papers and books. In his long tenure in the academic field, several organizations have invited him to deliver lectures on research topics of his interest.

1

Introduction to Sustainability, Principles, and Its Effect on Design and Manufacturing

1.1 Introduction

The World Commission on Environment and Development suggests that sustainable development should meet the needs of the present without compromising the ability of future generations to meet their own needs. This suggests that sustainable development must take a holistic view of the balance between the human need to improve the quality of life and well-being, and the need to preserve natural resources and ecosystems. Sustainable development has become one of the key concepts in the twenty-first century in the present context of global warming, climate change, and population growth. The Intergovernmental Panel on Climate Change (IPCC) stated in 2007 that worsening climate change would create intensification of the hydrological cycle and result in sea-level rise, causing disruption to the lives of millions of people living on shores. The variable and more intense pattern of rainfall will cause floods and drought at the same time in different parts of the earth, which is already evident. Further, greenhouse gas (GHG) emissions in the atmosphere will cause deterioration in water quality and soil moisture and toxic air unsuitable for breathing. Carbon dioxide (CO_2), water, and the ecological footprint have become the central foci in sustainable development. We want a low-carbon economy, if possible with zero CO_2 emission. This would go a long way toward improving the climate and reducing global warming. Figures 1.1 and 1.2 show how these problems are controlled or corrected by the natural environment.

Figure 1.1 shows the recycling of oxygen by photosynthesis from the lithosphere to the biosphere and back. The mechanism of oxygen balancing in nature is unique, and we have not been able to imitate it. Water is an irreplaceable resource on earth, but it is also becoming depleted. However, Figure 1.2 shows how nature has adopted a unique method to recycle this scarce resource.

Nature's unique methods for recycling scarce resources cannot be copied, but efforts must be made to reduce further depletion of scarce resources and to reduce global warming. Sustainable development is the mantra being echoed by the United Nations (UN) and other world organizations. However, for sustainable development, society has to accept the responsibility for cleaner production, reuse, and recycling of industrial products. In general, sustainability interacts with society as depicted in Figure 1.3.

Figure 1.3 shows how the corporate sector plans for production of a product in response to demand from society. However, after coming on to the market, the product may be modified in response to choices by society. If society's opinion of a product changes, product design will change. However, society must face the resulting consequences in terms of environmental impact.

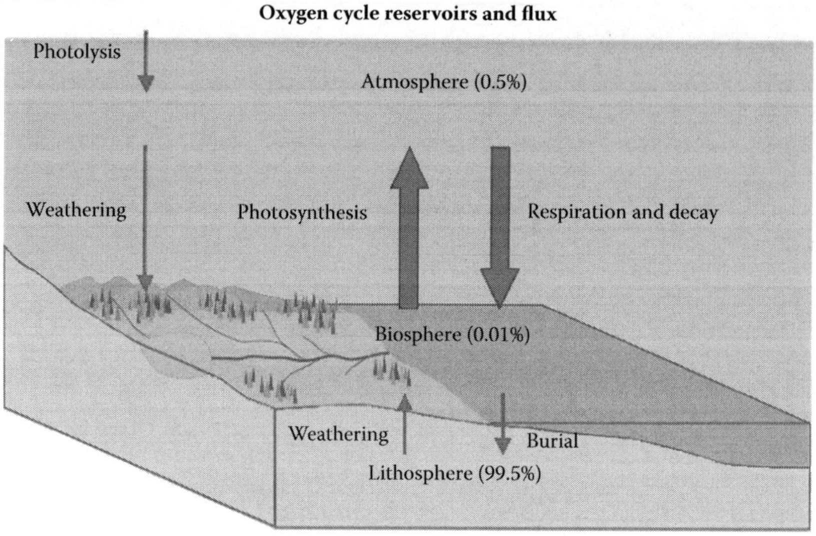

Figure 1.1

Oxygen recycling in the atmosphere.

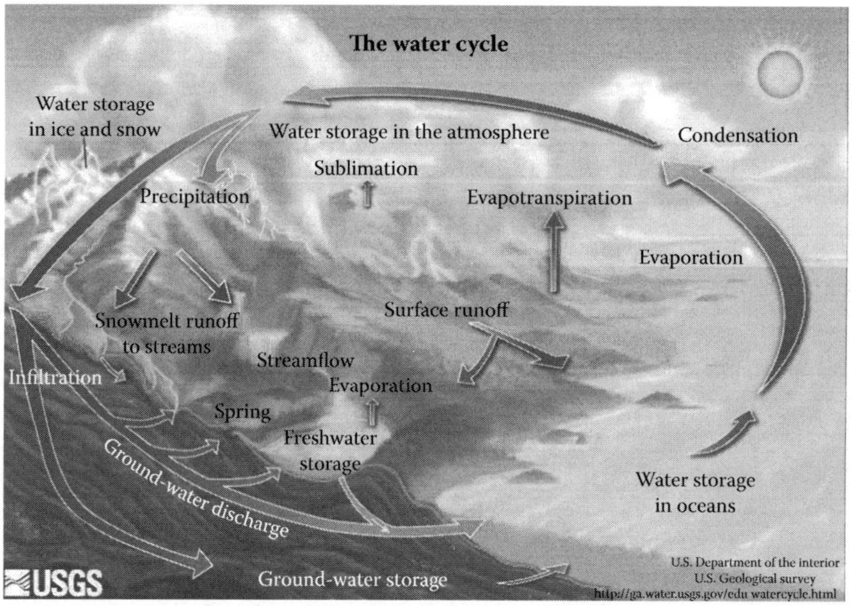

Figure 1.2

Water recycling in nature.

Engineering of products and production is a common endeavor by human beings to develop and feed the population. However, all engineering production needs resources in the form of materials, water, power, and so on, and these will naturally become depleted. Water is the source of life on earth, and it is being depleted faster than it can be replenished.

Water is being drawn from the world's underground stores (aquifers) many times faster than it can be replaced by nature. In manufacturing processes, water is used in electric power generation, to make steam for processing and cleaning, for cooling or controlling process temperature, for rinsing, for food dilution, and in the formulation of products containing water (such as pharmaceuticals and beverages). Although industry consumes only 5% of the water supply, that percentage represents a large volume of freshwater; therefore, industry can play a major role in curbing the rate of water consumption.

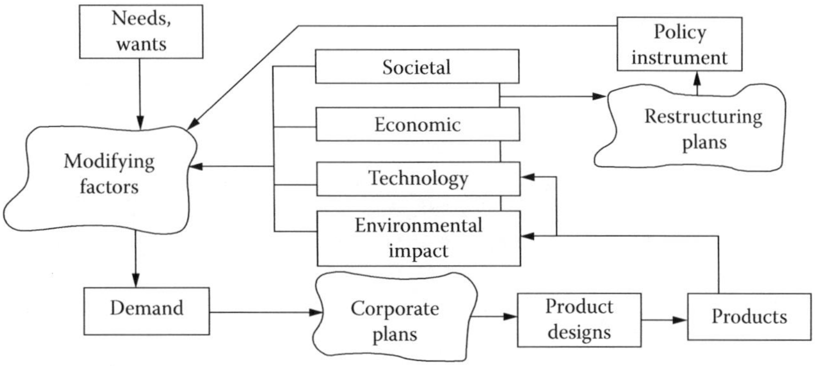

Figure 1.3

Interaction of society, economy, technology, and environment.

1.1.1 Water Has No Substitute

About 70% of the human body is water. At a 1% deficiency, humans sense thirst. At a 5% deficiency, a fever develops, and a 12% deficiency causes death. There are substitutes for oil and most other natural resources, but there is no substitute for water. Water is being drawn from the world's underground stores (aquifers) many times faster than nature can replace it. The rate of net withdrawal is roughly equal to the flow of the Colorado River. About 410,000 million gallons per day (Mgal/d) of water was withdrawn for use in the United States during 2005, according to the US Geological Survey (USGS) Fact Sheet 2009-3098.

Agriculture is the world's largest water consumer, according to the "Global Water Security" report [1], and up to 95% of developing countries' water supply is used for agriculture. In the United States, about 35% of freshwater is consumed by agriculture, and thermoelectric power uses 49% of the water supply, according to the USGS Fact Sheet 2009-3098.

Although industry consumes a smaller share, almost 5% of the water supply, that percentage still represents a large volume of freshwater; therefore, industry can play a major role in curbing the rate of water consumption. It may be surprising just how much industry consumes during processing. It has been estimated that up to 7,060 gal of water may be used to produce 1 ton of milk; 30,000 gal to produce 1 ton of meat; 543,000 gal to produce 1 ton of paper; and 27,500 gal to produce 1 ton of *recycled* paper. These data offer a glimpse of the high ratio of water use per product.

How much water is required to produce a product varies depending on how early in the process measurement begins. These rates would be even higher if they included the water required to grow the crops or access the raw materials. Regardless of the method used to assess the total, clearly, industry consumes a staggering amount of water. Recently, Ford Motor Co. announced that it has set the goal of reducing its water consumption to produce a single vehicle from 2,510 gal in 2009 to 924 gal by 2015.

1.1.2 Sustainable Water Management

Conservation is defined as "keeping and protecting natural resources intact or unimpaired." The industrial war on water consumption starts with designing and redesigning processes to use less water to begin with—doing more with less.

That can be achieved by limiting rinse steps to a meaningful specification and ending the rinse process when it provides no additional benefit to the intermediate or final product. Another way is to specify water concentration to eliminate evaporation (and the corresponding energy consumption) at a later stage. Steps such as repairing leaks of pipes, fittings, and valves; clean-in-place systems, which conserve water by using specially designed spray nozzles; and condensate recovery can go a long way toward reducing depletion of water.

1.1.3 Recycling Technologies

Recycling water is restoring it to its original condition or to new quality specifications through a cycle of changes or treatment. It is important to know the temperature limits of the treatment materials and equipment being deployed to recycle water. For example, the polymeric membranes used in reverse osmosis and anion-exchange resins have a maximum temperature rating as low as 110°F, so the process temperature cannot exceed this limit. It is essential to designing a water conservation, reuse, and recycling program that the final and intermediate product specifications are known and clearly stated.

1.1.4 Getting Started

To get started on your effort to conserve, reuse, and recycle water, you should begin with one of your plant's process streams. From a practical standpoint, you may want to start with a small process to "test the water."

Generally, from a capital investment perspective, it makes sense to start with conservation. The key is knowing how much you can reduce your water consumption without affecting your product quality or plant performance. Next, look for reuse opportunities, and finally, recycling opportunities.

Manufacturing needs lot of water for processing and subsequent cleaning of the chips and other factory waste. Subsequently, all manufacturing operations emit GHGs, causing pollution and global warming. Therefore, for sustainable development, green design and manufacturing is an important problem consideration for engineers.

The two most important aspects of engineering are design and manufacturing. Green engineering addresses environmental problems by changing the underlying composition of a product, process, or system, or by changing the context in which the system operates. Green design encompasses methods of realizing design changes for reduction and prevention of environmental impact. The manufacturing of more environmentally benign materials and products is part of green design. Design for environment (DfE), ecodesign, and design for sustainability (D4S) are all part of the same family. Methods that can be applied during the design stage have a greater effect on a product's overall environmental impact. Tools that can be applied during the conceptual stage often require more detailed information about material and energy flows with respect to a product. New designs must improve safety, minimize waste, and consume less energy. D4S takes a holistic approach, analyzing operational efficiency, safety, functionality, productivity, materials use, ease of operation, and maintenance. Additionally, the use of mechatronics and control in mechanical system design has gained acceptability. Manufacturers across many industries are increasingly emphasizing sustainability.

There are ten basic design principles that help achieve sustainable product development:

1. Eliminate nonessential drive components.

   Simplified designs with fewer components run more efficiently and are less expensive to manufacture. Older mechanical power-transmission and actuator technologies, such as chain, rack-and-pinion, and worm drives, often stand in the way of more efficient designs. These older drives use components that wear over time and demand frequent maintenance, including lubrication and tensioning.

   Designers rely more and more on sophisticated performance and simulation software to help eliminate many unnecessary components, such as line shafts and costly pneumatics and hydraulics. With these mechatronic tools, engineers can analyze energy use, build virtual prototypes, and select the best mechanical design to maximize performance. This approach reduces energy consumption and maintenance costs while improving uptime and reliability.

   Machine manufacturers also use direct-drive motion technology to improve mechanical efficiencies. This technology helps create more accurate, reliable, and energy-efficient machines that are less costly to maintain. For example, a Rockwell Automation customer replaced a motor-worm gearbox with a direct-drive servomotor that boosted the mechanical efficiency of the machine from 29% to 98%. The efficiency realized with direct-drive technology allows designers to use smaller servo drives that, in turn, use less energy.
2. Minimize the mass of essential mechanical components.

   After eliminating all nonessential mechanical components, the next step is to minimize the mass of remaining components. The wide availability of analytical, modeling, and development tools makes the task easier. Such tools can also test a machine's mechanical performance envelope using advanced stress and strain analysis techniques. This allows designers to evaluate potential alternative materials that are lighter and more energy efficient.

   Design techniques widely used in the aerospace industry minimize the weight of structural components without compromising strength. Machine manufacturers can use 3D computer-aided design and finite element analysis tools to optimize the mechanical design, as structural elements can take on more complex shapes. Advanced production equipment can shape and mill contours in mechanical components that eliminate superfluous material. Furthermore, the removal of excess material directly reduces the force needed to move the mass of the structure, allowing designers to choose smaller motors and drives that can consume less energy.

3. Substitute electric actuation for fluid power.

   Wherever and whenever possible, use electric actuation rather than hydraulics and pneumatics. Through hydraulic and pneumatic actuators may have lower initial investment, they are highly expensive to run and maintain. The hidden costs, such as expensive recycling fees for hydraulic fluids and operational energy costs, are tremendous.

   Electric cylinders combine high speed with high accuracy, repeatability, and reliability. Furthermore, their modular design allows them to fit a wide range of machines. Controls of electric cylinders range from simple indexers to more complex programmable, multiaxis motion controllers, giving designers more choices and allowing them to better weigh the costs and benefits of each.

4. Perform a safety audit after mechanical design but before control system design.

   A safety audit helps engineers evaluate and investigate risks early in the development process. It identifies the required safety control system integrity level and helps guide the selection of overall control architecture. By performing the audit before the control system design, safety can be an integral part of the control system rather than a tacked-on afterthought. This saves critical time and helps machine builders to get their equipment to market faster. In addition, end users gain optimized production, thanks to an automation system that helps operate machinery and processes in the most efficient and safest way.

5. Guard or control access to moving parts.

   Hazards cannot be removed during design. Machine manufacturers typically install a fixed physical barrier that protects users from being injured by the moving parts of the machine. Nonfixed barriers such as removable, swinging, or sliding doors find application where there must be frequent access to the dangerous area. While relays and other devices are effective, many safety applications require a level of programming or more sophisticated safety logic that is best met through a safety controller. Safety controllers offer significant benefits in multiple shutdown or ramp-down sequences, such as in transfer line applications, because they provide the necessary logic through software rather than the hard-wired logic of relays. An integrated safety controller works well for any application that needs these advanced functions, such as zone control. Properly designed safety controls and guarding reduce access time and help make the machine safer and more efficient.

6. Integrated safety systems reduce control system complexity.

   The more designers integrate standards and safety control functions, the better the opportunity to eliminate duplicate equipment, boost productivity, and lower costs. Integrated controls reduce the number of special components on the factory floor, which, in turn, reduces crib inventory costs as well as maintenance team training. End users benefit from less waste with fewer parts to maintain and replace throughout the machine life cycle. In addition, integrated control systems give broader access to data about machine operations and status. This reduces the number of accidental shutdowns and prolonged restarts, further improving machine efficiency and productivity.

   There are some control systems available that can reduce speed to a level acceptable for safe human–machine interaction (HMI). This eliminates the need for a separate, dedicated safety controller while providing use across multiple platforms. Safe-speed controllers allow operators to perform maintenance and other tasks while the machine remains in operation. Safe-speed control also helps improve uptime and lowers energy costs, because a machine need not be completely shut down and then restarted.

   Networking and Wi-Fi offer another way to integrate safety and standard controls. The introduction of networking to the plant floors may bring many benefits to manufacturers, including more productivity, less wiring, better diagnostics, and easier access to plant-floor data. Using an existing network to handle safety information creates seamless communication of a complete automation process on one standard set of network hardware and wiring.

7. Distribute control and interface devices near point of use.

   While some control components traditionally reside on the machine, technology advancement makes it possible to move the entire control system close to the application point. Standard automation components—including controllers, motors starters, drives, sensors, contactors, network media, distribution boxes, input/output (I/O), and HMI devices—have become on-machine applications.

On-machine design reduces design and installation time along with the labor needed to assemble the system. This approach also helps eliminate installation mistakes, because there are fewer wire runs and manual connections. In fact, some machine builders have seen their total machine tear-down and reinstallation time drop more than 50%. For end users, the compact nature of on-machine controls frees up the plant floor to reduce overhead and conserve energy.

8. Develop modular control.

   Programming can consume up to 80% of the control system design budget. Modular application development generates and reuses programming code modules as outlined in standardized programming methods and models such as ISA-88. Using the same specification document throughout the process helps cut engineering time and improves the quality and consistency of the design.

   In the same vein, information storage from one program to the next remains consistent, with data found exactly where the engineer expects it to be. The ability to troubleshoot multiple programs in the same way improves uptime. However, customers get up to speed quickly when programs have the same look and feel.

9. Make better use of diagnostics.

   It is now easy to embed data-gathering devices into machine controls. This allows equipment to monitor itself to predict and prevent failures. Moreover, this technology can relay machine status back to the machine manufacturers for value-added monitoring and analysis services without compromising existing resources or hindering profitability. Machines designed with Ethernet and Internet connections permit remote troubleshooting, with obvious implications for maintenance cost. For example, the ability to remotely monitor equipment from a distant location saves fuel, cuts emissions, and virtually eliminates travel time and related expenditures.

10. Designing IT connectivity into machines.

   Machines capable of connecting to an IT infrastructure can provide information for energy efficiency and overall equipment-effectiveness (OOE) calculations that help plant managers reduce waste and optimize productivity.

   IT connectivity also helps maximize the benefits of track-and-trace capabilities. Using advanced information software, manufacturers can track and record relevant data at every step of the process to identify when and where resources were used. This visibility offers end users a wealth of data for waste reduction and other improvement programs. It has been observed in manufacturing industries that about 15%–20% of power is used during standby, and if this portion could be reduced through IT and software it would minimize the power consumption, resulting in savings and improvements in environmental impact. It would make machining more sustainable.

   In addition, these systems also help track product genealogy through the full chain of custody. In so doing, they help document data required by regulators, identify potential product quality issues before they reach the market, and if necessary, respond faster to product recalls.

1.2 Design for Sustainability

The D4S concept outlines methodologies for making sustainable improvements (social, economic, and environmental) to products by applying elements of life cycle thinking. D4S builds on the work of ecodesign to include economic and social concerns, and its methodology includes both incremental and radical innovation. The UN Environment Programme and Delft University of Technology, the Netherlands, in collaboration with key partners, work to support, illustrate, and diffuse targeted D4S demonstration efforts, including the European Commission–funded Clear Production for Better Product project in Vietnam, that are needed to change unsustainable consumption and production patterns.

The ecodesign methodology D4S has developed from general cleaner production methods to focus on products and to include social, economic, and environmental elements of production. D4S is a unique methodology, because it not only encompasses all three pillars of sustainability (people, profit, and planet), but is also applicable to supporting sustainable production capacity in developing countries. For better understanding of the concepts and the players implementing D4S, some contextual and background information is presented in the following subsections.

1.2.1 Sustainable Initiatives

Sustainable innovation and design are not necessarily about new technologies, but about rethinking how to meet the need for growth while at the same time reducing negative environmental and social impacts.

This concept, known as "decoupling," is of great importance, as it seeks to break the link between economic growth and environmental degradation. Decoupling is an idea that could be employed to reduce the environmental impacts associated with production processes. Sustainable innovation allows leapfrogging in production and design processes to make it possible for newly industrializing countries to achieve decoupling in the beginning, not at the end, of the industrialization timeline.

1.2.2 Cleaner Production

Cleaner production (CP) is defined as "the continuous application of integrated preventive environmental strategies to processes and products to reduce risks to humans and environment" [2]. This strategy has been found to be cost effective, is being used in many forms all over the world, and has evolved to cover areas such as energy efficiency, multilateral environmental agreement targets, and sustainable products (including D4S).

1.2.3 Design for Sustainability

D4S is an ecodesign concept that has evolved to include both the social and economic elements of production. D4S can be used in small-and-medium-sized enterprises (SMEs) in developing countries, which have economies that offer few incentives and little support for innovation.

1.3 Objectives for Green Design

There is no universal agreement on the particular goals to be pursued by green design. Some believe that green design and pollution prevention should be the sole criteria in order to reduce costs; others focus on recycling to save raw materials. Another school of thought wants to direct all attention to global warming and environmental problems. Sometimes, pollution prevention may be socially desirable but economically unwise. These days, most people tend to relate green design to a sustainable future in regard to both resources and environmental health. It may be debatable, but this book will advance the following goals for green design for a sustainable future:

- Reduce the use of all kinds of resources, including materials, energy, water, and so on.
- Reduce toxic and otherwise harmful emissions to the environment causing global warming.
- Manage renewable resources to insure sustainability.

The most acceptable objective of green design can be stated clearly: to develop more energy-efficient devices that can reduce consumption of energy and toxic emissions. Sometimes a complete product may be made of recycled materials, use renewable energy, and be biodegradable, but may not be green enough compared with another product that uses fewer resources and emits less toxic gases into the atmosphere. It is rare for one product to be greener in every aspect (resources and energy use, emissions, recyclability, etc.) than other products; there are usually some trade-offs with regard to these characteristics. A fuel-efficient car may mean a lighter body of plastic or aluminum. But it may be revealing to know that both aluminum and plastic require more energy for production than steel. Again, it should be known that some components may require fewer resources during manufacturing but need much more energy during operation and use. Another problem which should be recognized is that some materials, such as carbon fibers, have advantages in terms of material properties but cannot be recycled. However, green products imply more efficient resource use, reduced emissions, and reduced waste and harmful emissions into the atmosphere. Green design and manufacturing promises a reduction in costs of manufacturing and, in turn, more profit to the company. It should result in a reduction in material requirements, energy use, disposal fees, and pollution, and finally, fewer problems for society in general. Green design offers much to the present generation, as well as for sustaining future generations.

International regulations are becoming tough regarding environmental pollution and global warming. European countries in particular have environmental laws applicable to all European Union countries regarding green design and manufacturing for sustainability. Most of the international regulations will be discussed and presented in Chapter 16. These days, products are designed with the aspects of disassembly and recycling in mind. France, Germany, and the Netherlands have special governmental agencies to ensure clean technologies. In the United States, many government purchasing departments specify the use of recycled materials. The US federal government has initiated a number of energy efficiency programs for products and buildings.

1.4 Rubicon of Engineering Design

Design is a complicated and messy process. Designers are faced with numerous competing requirements and constraints. For example, a personal computer must be fast, powerful, and cheap. To be green, it must also be energy efficient and easily recycled. Designers have struggled to achieve the first set of attributes. Achieving environmental goals makes the task more difficult. For most consumers, energy efficiency and recyclability are less important product attributes, which means that designers cannot compromise other product attributes for the sake of becoming green.

Green design and manufacturing requires appropriate knowledge, tools, production methods, and incentives. Aids for green design must be easy and quick to use and understand. Ideally, these design tools will help identify design changes that have lower costs while improving material use and recyclability. For example, using snap fits rather than metal fasteners may add little to the cost at the design stage while significantly increasing recycling potential.

Green design methods and tools are outlined in 1.5 below.

1.5 Some Green Design Methods and Tools

1.5.1 Systems Thinking

A central idea in green design is the notions of system design. It is now being recognized that in a holistic view of a new product, materials supply, manufacture, use, and disposal may all be relevant. It is now being recognized that it may not be a wise decision to structure a society around the automobile. Cities such as Los Angeles and Dallas, to name just two, where a car is required to get to work, shops, schools, and recreation areas may not, after all, be a sustainable living model. A community in which people can walk to work, school, and the shops is inherently greener and more efficient in terms of energy and obviously less polluting.

1.5.2 Environmental Life Cycle Assessment (LCA)

LCA is a technique for tracing all the environmental effects and resource needs of a new product or process, from material suppliers, through manufacture and use, to disposal at the end of its useful life. It is intended to provide a comprehensive assessment of environmental effects. For reasons of lack of data, complexity of analysis, and uncertainty, LCAs often draw arbitrary boundaries (to eliminate all possible suppliers) and stop at the stage of just producing an inventory of emissions and resources rather than proceeding to assessment and improvement.

1.5.3 Material Flows and Cycles

Material flows and cycles is a technique for tracing material use and location over time. For example, steel is routinely recovered from products such as automobiles, melted, and reused in a closed recycle loop. Some materials are disposed of in landfill, although they may at some time in the future be recovered and reused. Materials can move between the anthroposphere (the man-made environment) and the natural environment. In recycling materials, there is a distinction between closed loop (reuse for the same function) and open loop (reuse for a different function, typically with lower quality requirements). In tracing material flows, it is important to be clear about the boundaries of analysis and the uncertainty of mass measurements. It involves tracing materials or energy in and out of an analysis area, such as a manufacturing station, a plant, or a watershed. Ideally, mass balances are based on measurements of inflows, inventories, and outflows (including products, wastes, and emissions). In actuality, it is unusual for all the data needed to be available or even consistent.

1.5.4 Green Indices

How can an analyst compare a pound of mercury dumped into the environment with a pound of dioxin? Green indices or ranking systems attempt to summarize various environmental impacts into a simple scale. The designer or decision-maker can then compare the green scores of alternative materials, processes, and so on, and choose the one with the minimal environmental impact. This would contribute to products with reduced environmental impact. For example, Volvo's environmental priority strategies system (EPS) involves calculating "environmental load units" of alternatives. Although none of these tools is yet capable of incorporating many different types of environmental impact, all provide at least rudimentary guidance to the designer in choosing materials, components, or process alternatives that have reduced environmental impacts.

1.5.5 Design for Disassembly and Recycling Aids

Designing for disassembly and recycling (DFD/R) means making products that can be taken apart easily for subsequent recycling and parts reuse. For example, Kodak's "disposable" cameras snap apart, allowing 87% of parts (by weight) to be reused or recycled. Unfortunately, the economic costs associated with physically taking apart products to get at valuable components and materials often exceed the value of materials. Reducing the time (and thus costs) of disassembly might reverse this balance. Thus, DFD/R acts as a driver for recycling and reuse. DFD/R software tools generally calculate potential disassembly pathways, indicate the fastest pathway, and reveal obstacles to disassembly that can be "designed out."

1.5.6 Risk Analysis

Risk analysis is a means of tracing the chances of different effects occurring. For example, the risk of toxic emission is estimated by estimating the amount and type of emissions, transport in the environment, ecological and human exposure, and probable damage. All these analytical steps are likely to be uncertain. Risk is a useful concept for integrating effects over several media (such as air, water, and land). However, we have difficulty in judging and measuring risk, particularly for low-risk/high-consequence actions.

1.6 Mathematical Representations of Sustainability

Sustainability is defined as conformance with principles of sustainable development, encompassing the environmental, health and safety, social, economic, and ethical aspects of a corporation or other entity. It is desired that a business should be able to anticipate and meet the economic, environmental, and social needs of present and future generations of customers and other stakeholders. Environmental awareness is an all-encompassing consideration these days due to resource depletion and global warming. Finding a path to sustainable development will require worldwide collaboration and innovation in all phases of business: extraction of materials, transportation, manufacturing, delivery to customers, and finally end-of-life considerations, including disposal. There have been attempts to objectively measure sustainability and analyze it quantitatively, but there is very little agreement on the quantitative measurement of sustainability. It has been suggested that sustainability measurement should include the number of people using a technology as some measure of the resource consumption or environmental degradation caused by it. It is also suggested that the role of the technology in the economy should be included in the measurement of sustainability. The following algorithm has been suggested. It assumes that population (P) and per capita gross domestic product (GDP/P) are prime factors in sustainability measurement. The algorithm also includes energy consumption per GDP (E/GDP) as an important factor in this measurement. However, emission of GHG needs to be included in the sustainability equation. Finally, the sustainability equation is represented as

$$S = (\mathrm{P}/\mathrm{GDP}/\mathrm{P}) \times (\mathrm{E}/\mathrm{GDP}) \times (\mathrm{CO_2}/\mathrm{E}) - (\mathrm{CO_2})_{\mathrm{sq}}$$

(the Kaya equation, developed by Yoichi Kaya).

The term (CO_2/E) is defined as CO_2 emitted per unit of energy (E), in BTU. The term $(CO_2)_{sq}$ is defined as the amount of CO_2 sequestered. Another means of CO_2 reduction could be the use of renewable energy (e.g., solar energy, wind energy, or photovoltaic electricity).

1.7 Material Selection and Label Advisors

Any one of several materials can produce a component or product of a particular quality. However, they have different environmental implications. Material selection guidelines attempt to guide designers toward the environmentally preferred material. For example, Graedel and Allenby [3] present the following guiding principles for material selection:

- Choose abundant, nontoxic materials where possible.
- Choose materials familiar to nature (e.g., cellulose) rather than man-made materials (e.g., chlorinated aromatics).
- Minimize the number of materials used in a product process.
- Try to use materials that have an existing recycling infrastructure.
- Use recycled materials where possible.

In addition to these generic guidelines, companies such as IBM and Chrysler have been developing and using specific materials selection guidelines for environmentally sound product development, which describe in detail the materials that should be used for specific applications. Label advisors are generally marks on materials or products that reveal information about the material content relevant to materials handling and waste management. For example, the plastic bottles used in many consumer products usually have a plastic identification symbol that can be used in plastics resorting and recycling. Even labels are subject to green design. Embedding labels into a material is preferable to a separate label material attached with an adhesive. The environmental impact of materials is measured in terms of emission of CO_2 or its equivalent. The choice is always the material that reduces CO_2 emissions as much as possible.

1.8 Green Power Generation

Scientific opinion on climate change has reached a new level of concern. The IPCC has concluded that the earth's climate will change, although we may not agree on by when and by how much. Global mean surface temperature has increased in the range of 0.3°C–0.6°C since the nineteenth century. In addition, global sea levels have risen by 10–25 cm during the same period, much of which may be related to the temperature increase. According to the IPCC, these changes are "unlikely to be entirely natural in origin." The balance of evidence suggests an identifiable human influence on global climate. Global warming is mostly attributed to GHGs, which help solar radiation to penetrate the atmosphere and absorb the infrared radiation reflected back by the earth's surface. This infrared radiation trapped in the atmosphere causes the air temperature to rise. Since the Industrial Revolution, GHGs from human sources (anthropogenic) have been accelerating in proportion to the growing use of fossil fuels. CO_2 is responsible for most global warming, followed by methane (CH_4) and nitrous oxide (N_2O). The concentration of CO_2 in the atmosphere has gone up by 28% during the last 1,000 years. It is estimated that global temperature will rise by 1°C–3.5°C and the sea level will rise by 15–95 cm by the year 2100.

The United States is the world's largest GHG contributor, accounting for 25% of global emissions. The vast majority of GHG emissions in the United States result from energy consumption, of which electricity comprises a significant proportion. Figure 1.4 shows the components of US greenhouse gas emissions in 1998. Electric utilities consume 25% of the primary US energy and are the largest single source of GHG emissions. It is estimated that 40% of US CO_2 emissions are due to combustion of fossil fuels by electric utilities. Therefore, it is obvious that generation of electricity should be diverted as far as possible to renewable sources or green power generation.

1.9 Net Energy Analysis

Net energy analysis (NEA) is a comparison of the useful energy output of a system with the total energy consumed by the system over its life cycle. A life cycle approach includes "upstream" processes such as the mining of raw materials and "downstream" processes such as plant decommissioning. NEA is an important tool for evaluating energy options with consideration of ultimate resource availability.

Determining the output energy is a simple calculation based on the average power output of the plant. Determining the energy inputs is a difficult process for which, in this green power generation study, we employ two basic methods: I/O and process chain analysis (PCA). The I/O method is used to correlate dollar cost to energy use. An I/O model divides the entire economy into distinct sectors, which form the basis for a matrix that distributes the total cost of outputs and total energy inputs of the US economy. The economic input-output life cycle assessment (EIOLA) model developed by Green Design Initiative at Carnegie Mellon University is a useful tool for I/O analysis.

PCA evaluates the material and energy flows for each process within the system life cycle. When possible, actual data for energy expended during a process is used. To determine the energy required for system materials, the mass of material is multiplied by an embodied energy factor in gigajoules (GJ/ton material). This factor accounts for the energy required to mine, mill, and fabricate the raw material.

PCA is generally considered to be more accurate than the I/O method. However, it is difficult to evaluate an entire life cycle using PCA, because data on all the materials used and energy consumed is not readily available. Cost data is frequently available, making the I/O more applicable for many processes. We use PCA as far as possible, then use the I/O method to complete missing portions of the life cycle study of green power generation.

1.10 Greenhouse Gas Emission Rates

The energy requirements for each phase of the life cycle can be used to estimate GHG emissions. The methodology provides a better estimate of GHG impact than simply estimating plant emissions for various green power generation technologies. For this study, we estimate the life cycle emissions normalized in terms of tons CO_2-equivalent emitted per gigawatt-hour electricity produced (ton/GWeh), allowing comparison between alternative technologies of green power generation.

CO_2 is a by-product of fossil fuel combustion. Because the vast majority of US energy is produced by using fossil fuels, each energy input within the life cycle has corresponding CO_2 emissions. CO_2 is the most significant GHG based on total global emissions. Methane (CH_4) and N_2O are actually stronger warming agents but have far lower global emission rates. These less important gases are accounted for in terms of CO_2-equivalents based on their global warming potential.

When averaged over 100 years, CH_4 has a global warming potential 21 times stronger than CO_2, which means that 1 ton of CH_4 emissions can be counted as 21 tons of CO_2-equivalent emissions. Sometimes CH_4 is the main component of input energy emissions.

N_2O has a global warming potential 310 times stronger than CO_2. N_2O is a product of the reaction between nitrogen and oxygen during fuel combustion. High temperature destroys almost all N_2O; therefore, power plant emissions contain a very low concentration of this molecule.

Green power generation is studied through the life cycle assessment of power plants to evaluate net energy and GHG emissions. The life cycle includes input production and transmission, fabrication of equipment and structural materials, and plant construction, operation, and decommissioning. The result of the net energy analysis is an energy payback ratio (EPR), which is the ratio of useful electrical output to the total energy inputs. In the United States, the EPR is 11 for coal, 16 for fusion, 27 for fission, and 23 for wind turbines compared with 4.1 for gas turbines. The GHG emission rate is calculated using the inputs from the NEA. Normalized over its life cycle, a gas turbine emits 464 tons of carbon dioxide equivalent for every gigawatt-hour of electricity produced (GWeh). This emission rate is lower than conventional coal (974 tons/GWeh) but higher than fission (15 tons/GWeh), fusion (9 GWeh), and wind (14 tons/GWeh) technologies.

1.11 Carbon Accounting in Corporations

Carbon accounting can be used as part of *sustainability accounting* by both for-profit and nonprofit organizations. A corporate or organizational "carbon" or GHG emissions assessment is used to quantify the GHGs produced directly and indirectly by the activities of a business or organization within a set of boundaries. It is known as a *carbon footprint*. Accounting for GHG emissions is increasingly framed as a standard requirement for business.

1.11.1 Environmental Full Cost Accounting Methodologies

Many corporations and consumers want to support green products and sustainability but do not know how to make greener decisions. Designers and plant managers are specialists who cannot be expected to be environmental experts capable of estimating the environmental and sustainable implications of their decisions. As a result, a company often incurs high costs by using a material or process that creates environmental problems when an environmentally benign material or process exists. Often consumers purchase products that create environmental problems because they do not know about green alternatives. Companies need management information systems that can reveal the cost to the company of decisions about materials, products, and manufacturing processes. This cost accounting system is called an "environmental *full cost accounting* (EFCA)" system. For example, when an engineer is choosing between protecting a bolt from corrosion by plating it with cadmium and choosing a stainless steel bolt, a full cost accounting system could provide information about the purchase price of the two bolts and the additional cost to the company of choosing a toxic material such as cadmium. In many cases, the choices that designers or consumers make also impose costs on society. For example, choosing a cadmium coating increases the possibility of human exposure to a toxic substance. The designer and consumer might be informed by showing them this social cost of cadmium; that is, the cost of preventing exposure and the potential health risks of exposure. This information might be communicated by having a "social" cost listed on the price tag. A still stronger step would be to actually charge the designer and consumer for the social costs of environmentally damaging materials or products. Thus, the government might add an environmental tax or effluent fee that would account for social damage. A number of companies are beginning to launch preliminary full cost accounting (FCA) efforts in order to spotlight products that result in relatively large environmental costs. Since costs and advantages are usually considered in terms of environmental, economic, and social impacts, full or true cost efforts are collectively called the "triple bottom line." A large number of standards now exist in this area, including FCA, which is now more commonly used in management accounting. The International Organization for Standardization (ISO) has several accredited standards useful in FCA or true cost accounting (TCA), including, for GHGs, the ISO 26000 series for corporate social responsibility coming in 2010, and the ISO 19011 standard for audits, including all these.

FCA embodies several key concepts that distinguish it from standard accounting techniques. The following list highlights the basic tenets of FCA, accounting for

1. Costs rather than outlays: expenditure of cash to acquire or use a resource. A cost is the cash value of the resource as it is used. For example, an outlay is made when a vehicle is purchased, but the cost of the vehicle is incurred over its active life (e.g., 10 years). The cost of the vehicle must be allocated over a period of time, because every year of its use contributes to the *depreciation* of the vehicle's value.
2. Hidden costs and externalities: the value of goods and services is reflected as a cost even if no cash outlay is involved. One community might receive a grant from a state, for example, to purchase equipment. This equipment has value, even though the community did not pay for it in cash. The equipment, therefore, should be valued in an FCA analysis.
3. Overhead and indirect costs: FCA accounts for all overhead and indirect costs, including those that are shared with other public agencies. Overhead and indirect costs might include legal services, administrative support, data processing, billing, and purchasing. Environmental costs as indirect costs include the full range of costs throughout the life cycle of a product, some of which even do not show up in the firm's bottom line. It also contains fixed overheads, fixed administration expenses, and so on.
4. Past and future outlays: past and future cash outlays often do not appear on annual budgets under cash accounting systems. Past (or upfront) costs are initial investments necessary to implement services such as the acquisition of vehicles, equipment, or facilities. Future (or back-end) outlays are costs incurred to complete operations such as facility closure and postclosure care, equipment retirement, and postemployment health and retirement benefits.

Various motives for adoption of FCA/TCA have been identified. The most significant of these tend to involve anticipating market or regulatory problems associated with ignoring the comprehensive outcome of the whole process or event accounted for. In *green economics*, this is the major concern. Corporate decision-makers sometimes call on FCA/TCA measures to decide whether to initiate *recalls* or practice voluntary *product stewardship* (a form of recall at the end of a product's useful life). This may be regarded as a hedge against future liabilities arising from consumers negatively affected by the toxicity of waste products.

Figure 1.4 presents the relationships between FCA and conventional accounting. First, it shows TCA, which does not include the external cost borne by society: environmental cost in terms of degradation of

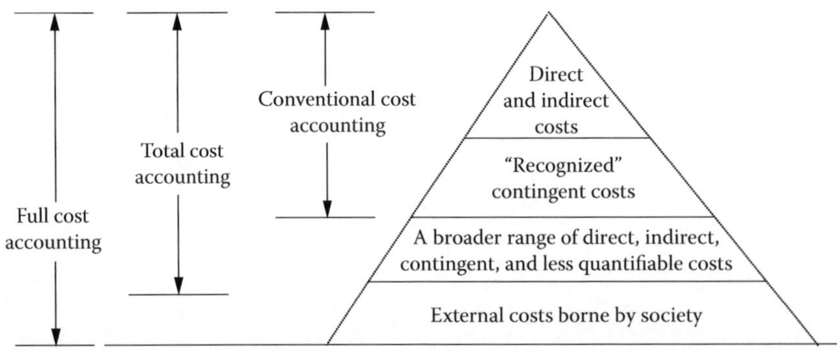

Figure 1.4

Relationship between FCA and conventional accounting.

land, water, and air. Second, allocating monetary value to nature would make its conservation reliant on fluctuating markets. This could lead to devaluation of services that were previously considered financially beneficial. Third, conservation programs for the sake of financial benefit overestimate human ingenuity to invent and replace ecosystem services by artificial means.

References

1. National Intelligence Council. (2012). Global water security: Intelligence community assessment, ICA2012-08. Department of State.
2. United National Industrial Development Organization (UNIDO). (2013). Lima Declaration: Towards inclusive and sustainable industrial development, UNIDO General Conference, Lima, Peru, 2 June.
3. Graedel, T. E. and Allenby, B. R. (1995). *Industrial Ecology.* Prentice Hall, Upper Saddle River, NJ.

Bibliography

Ashby, M. F. (2009). *Materials and the Environment: Eco-Informed Material Choice.* Butterworth Heinemann, Oxford.

Ashby, M. F. (2011). *Materials Selection in Mechanical Design*, 4th edn. Butterworth-Heinemann, Oxford, Chapter 16.

Ashby, M. F., Shercliff, H., and Cebon, D. (2007). *Materials: Engineering, Science, Processing and Design.* Butterworth-Heinemann, Oxford, Chapter 20.

Bock. (2011). Sensible handling of resources. http://bocknorthamerica.com/hp440/Environment.htm.

Crittenden, J. C., National Center for Clean Industrial and Treatment Technologies (CenCITT), EPA Grant # R825370, 1996–2000, Michigan Technological University.

Dahmus, J. and Gutowski, T. (2004). An environmental analysis of machining. ASME International Mechanical Engineering Congress and RD&D Expo, Anaheim, CA.

EPA (Environmental Protection Agency). (2006). Life cycle assessment (LCA): Principles and practice. EPA/6000R-06/060, May.

Fengting, L. (2014). Countdown to UNEA regional perspective. United Nations Environmental Program (UNEP), Bonn Climate Change Conference 4–15 June.

Fiksel, J. (2009). *Design for Environment: A Guide to Sustainable Product Development.* McGraw Hill, New York.

GaBi, PE International. (2008). www.gabi-software.com.

Graedel, T. E. (1998). *Streamlined Life Cycle Assessment.* Prentice Hall, Upper Saddle River, NJ.

Granta Design Limited. (2009). CES EduPack user guide. Granta Design Limited, Cambridge. www.granta-design.com.

Gutowski, T., Jeffrey, D., Thiriez, A. (2006). Electrical energy requirements for manufacturing processes. In *13th CIRP International Conference on Life Cycle Engineering*, Leuven, May 31–June 2 2006.

Ishii, K. and Lee, B. (1996). Reverse fishbone diagram: A tool in aid of design for product retirement. ASME Design Technical Conference, Irvine, CA.

Jan, E. and Bras, B. (1998). ISO 14000 and activity based life cycle assessment in environmentally conscious design and manufacturing: A comparison. 1998 ASME Design Engineering Technical Conference. American Society of Mechanical Engineers, Atlanta, GA.

Keolein, G. A. and Menerey, D. (1994). Sustainable development by design: Review of life cycle design and related approaches. CSS Publication # CSS94-01. *Journal of the Air and Waste Management Association* 44 (5): 645–668.

Kordonowy, D. N. (2001). A power assessment of machining tools. BS Thesis, MIT, Department of Mechanical Engineering, Cambridge, MA.

Lohmann, L. (2009). Toward a different debate in environmental accounting: The cases of carbon and cost-benefit. *Accounting, Organizations and Society* 34: 499–534.

MacKenzie, D. (2009). Making things the same: Gases, emission rights and the politics of carbon markets. *Accounting, Organizations and Society* 34 (3–4): 440–455.

McCarthy, J. J. et al. (2009). *Climate Change Impacts on the United States.* Cambridge University Press, Cambridge, ISBN 978-0-521-14407-0 paperback.

Minnesota Pollution Control Agency (MPCA). (2009).
United Nations. (1997). Kyoto Protocol to the United Nations Framework Convention on Climate Change. Document FCCC/CP 1997/7/ADD.1.
Yong, L. (2014). Briefing by director at United Nations Development Organization (UNIDO), Inclusive and Sustainable Industrial Development (ISID), Vienna, 20 May.

Section I
Industrial Ecology and Sustainability

2

Industrial Ecology and Sustainability

2.1 Introduction

Industrial ecology applies the structure and processes of the natural ecosystem for organizing activities across several domains: materials, energy, environmental quality, and information exchange. To support the increasing consumption of raw materials and other resources and production in developing society, industrial ecology emphasizes the concept of sustainability—which means composite (integrated) development while consuming as few resources as possible and minimizing the pollution of land, water, and air.

Figure 2.1 shows a Venn diagram for sustainable development, integrating environment, economic, and social development. Sustainable development is at the core of all these activities. The figure shows that economic and social development must be equitable; economic development should be viable for the environment; and the impact on the environment should be bearable for society. Economic development should not disturb the environmental balance (Figure 2.2) but should reduce global warming and climate change. Sustainability advocates a systematic and global approach in order to minimize resource consumption, decarbonize energy, minimize greenhouse gas (GHG) emissions, and encourage recyclability and reuse of materials. The emphasis on recycling and reuse in the production process reduces the need for raw material extraction and waste disposal. It strives to encourage the use of renewable energy with low carbon and GHG emissions. Sustainable engineering integrates green technology in design and manufacturing with environmental management. In the past, we were not concerned with environmental management in engineering education, but there is now an emphasis on this part of industrial activity for energy sustainability.

The tools for sustainable engineering include life cycle assessment (LCA), material flow analysis, input/output economic models, and indicators for measuring or assessing sustainability. According to ISO 14040 standards, LCA shall include four phases: goal and scope definition, inventory analysis, impact assessment, and interpretation of the result, as shown in Figure 2.3.

1. *Goal and scope definition*: The stage of goal and scope definition describes the purpose, scope, and boundaries of the study, as well as the functional unit, key assumption to be made, and likely limitations of the work. It is critical for making appropriate comparisons by weight of material produced or by functional unit, such as tons of concrete required, in several projects with similar features.
2. *Inventory analysis*: The inventory analysis constitutes a detailed compilation of all the quantified listings of the energy and raw materials in each stage of the product life cycle, related emissions

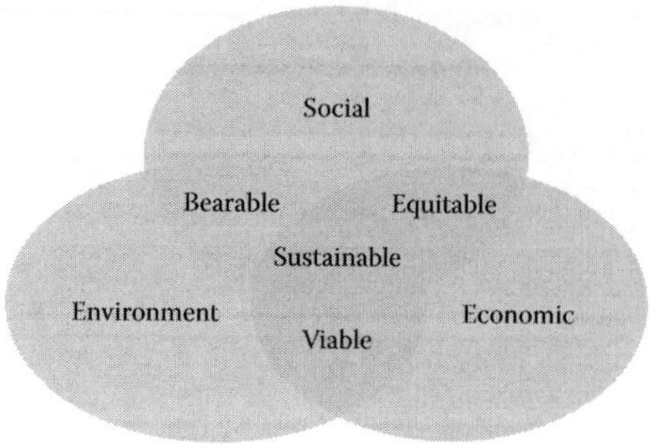

Figure 2.1

(See color insert) Sustainable development.

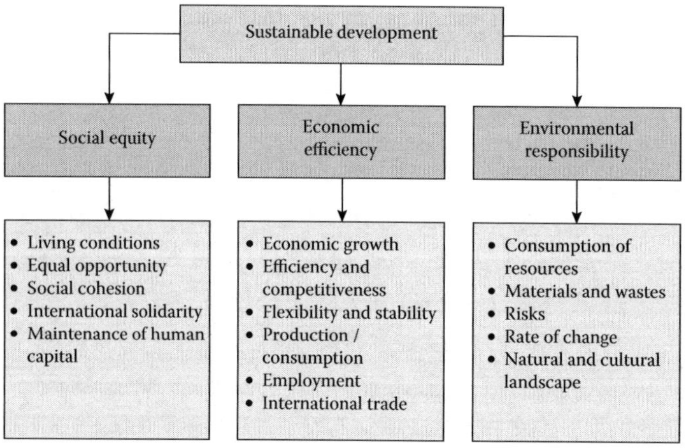

Figure 2.2

Components of sustainable development.

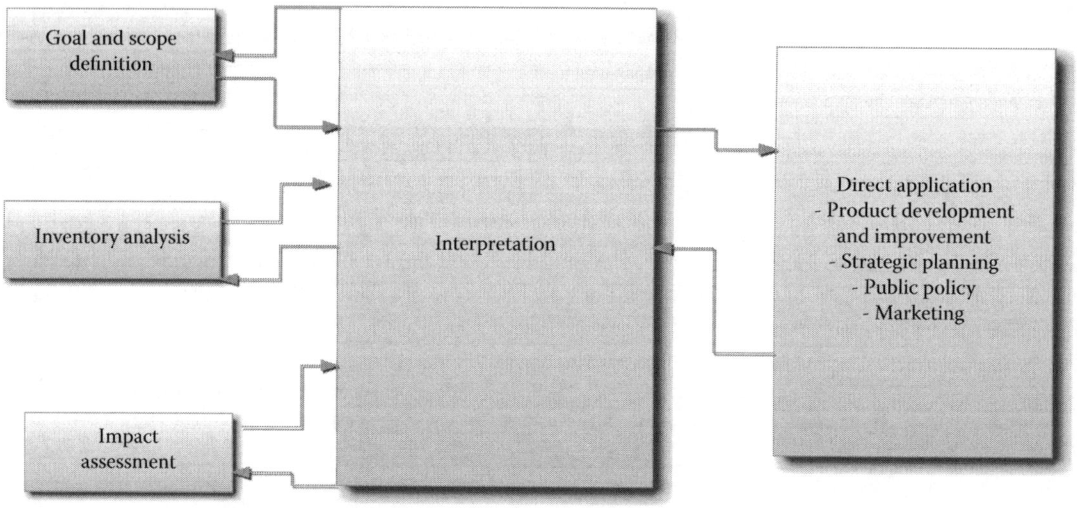

Figure 2.3

Processes required in performing LCA.

to air and water, and solid waste produced in the production, use, recycle, reuse, and disposal of a product.

3. *Impact assessment*: The impact assessment connects inventory data with environmental impact in terms of carbon footprint, GHG emissions, and embodied energy. Impact assessment commonly incorporates three stages: classification, characterization, and weighting. In this regard, global warming potential, depletion of natural resources, acidification potential, eutrophication

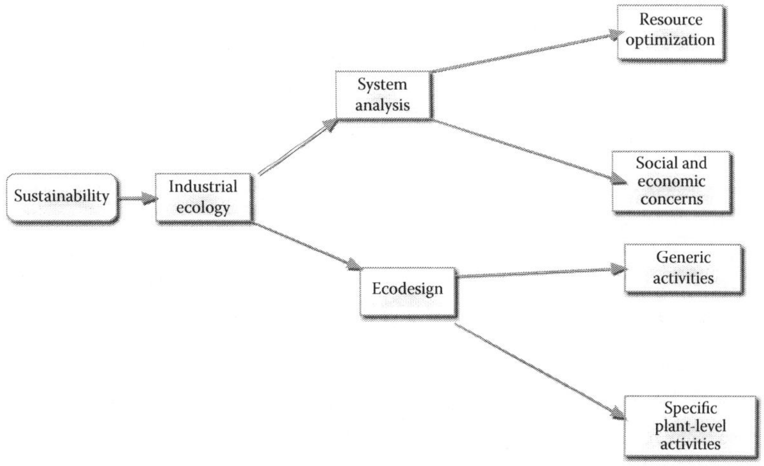

Figure 2.4

Industrial ecology activities.

potential, human toxicity potential, photochemical oxidation, and air odor potential are oftentimes involved.

4. *Interpretation of result*: Interpretation of the result of impact assessment may lead to finding ways to improve or redesign the production process (manufacturing), so as to reduce the costs and the material used.

The goal of industrial ecology is optimization of resource flow, which is the basis of sustainability. It is commonly said that industrial ecology is a subset of sustainability. *Pollution prevention* (P2) or *cleaner production* (CP) is often confused with industrial ecology, which is misleading. Industrial ecology takes a system view that draws the analytical boundary more broadly, including recycling and reuse, whereas P2 or CP is a firm-level approach. It could easily be recognized as falling under the broader rubric of industrial ecology. These elements are tied together in Figure 2.4, showing how conceptual aspects of industrial ecology are linked with more concrete, application-oriented tools and activities.

It is not always possible to recycle the toxic or hazardous materials coming out of industrial processes in an environmentally acceptable manner. One example could be the risk associated with recycling of nuclear power plants at the end of their life cycle, or for that matter nuclear arms, and so on. Industrial ecology takes into consideration risk assessment and environmental economics regarding damage to humans and the ecosystem, and sometimes examines what lies upstream of the sources of pollutants and the human activities that generate them. Therefore, industrial ecology appears to be a group of concepts, tools, and objectives. Industrial ecology is one way of working toward sustainability; it may not be the only way. However, the evolution of industrial ecology has forced us to think of production processes not only as a linear system, in which resources are consumed and wastes are dissipated into the environment, but as a more closed system, like an ecological system (Figure 2.5).

Climate change research is currently a topic of great interest for economic researchers. In particular, environmental input-output analysis is playing an increasing role in measuring the economic and environmental effects of sustainable development policies in Europe. Other approaches also exist, such as econometric modeling, whereby impacts are quantified on statistical grounds and with certain desirable properties (efficient estimates, confidence intervals, hypothesis testing, and so on) that are not found in the input-output approach. Consequently, there is a need to merge both approaches to address the calculation of unbiased and consistent carbon dioxide emission multipliers and their respective confidence intervals. The use of the supply and use system instead of the symmetric input-output table also presents an opportunity to avoid the common problems associated with the construction of technical coefficients (technology assumptions, negatives, and so on). A new policy-relevant application of these multipliers has now been introduced for the quantification of the performance of carbon dioxide emission reductions by industries via external trade.

Industrial ecology evokes powerful images and strong reactions, both positive and negative. Powerful analogies can be made between the evolution of natural ecosystems and the potential evolution of industrial

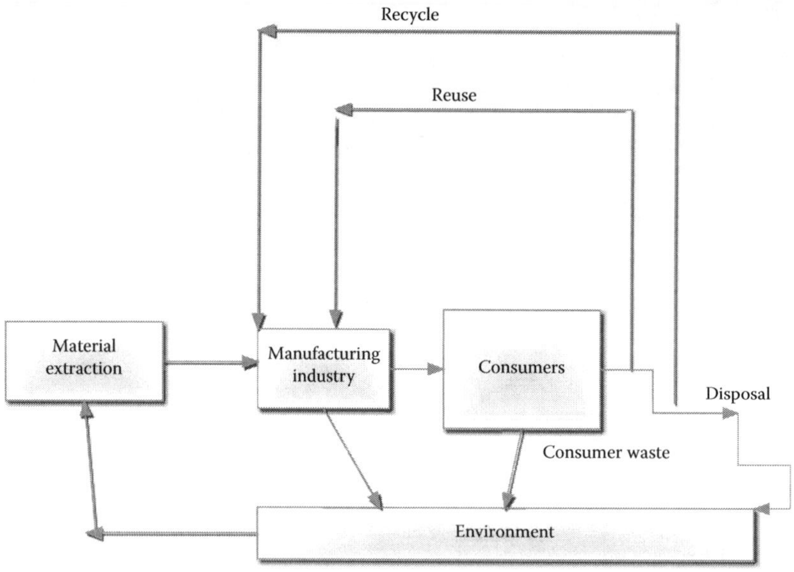

Figure 2.5

Closed form of production process.

systems. Current technology does not bode well for mass conservation of networks that exist in nature. Industrial ecology is much more than a metaphor or a fad; it is a field to which engineers can make significant contributions. At the heart of industrial ecology is the knowledge of how to reuse, modify, and recycle wastes—making waste into raw materials. This will not only save money for materials but also reduce costs, GHG emissions, and energy consumption. This will reduce the depletion of resources and extraction from nature and thereby reduce natural degradation. The goal for engineers should be to take on design tasks such as managing the recycling and reuse of the product at end of life. The target should be to make disposal activities more intricately networked and efficient in design. Not all of the tools needed to accomplish these goals are available yet, but we need to make the effort to provide such tools, which the next generation of engineers will require. The basic principles of industrial ecology may be the same for every major field of engineering, although a little tweaking or modification may be required for different engineering fields. However, it is time for academia to be serious about industrial ecology, either by developing and offering separate courses in industrial ecology or by including concepts of ecology in existing engineering courses. Due to the multidisciplinary nature of environmental problems, the course should be a multidisciplinary offering. The United Nations World Commission on Environment and Development has defined sustainable development as "meeting the needs of the present generation without sacrificing the needs of future generations." This goal is very broad; no part of academia can remain untouched. The complexities of most environmental problems require expertise from a variety of fields—law, economics, business, health, engineering, ecology—to contribute to the development of industrial ecology. Materials and energy are used by every country and every human being, and their consumption causes emission of CO_2, SO_2, NO_x, and low-grade heat along with gaseous, liquid, and solid waste. Collectively known as GHG emissions, and higher dose they are hazardous and damaging to the health of humans and all living beings. The burden on society is tremendous. Global warming is a direct result of GHG emissions. The study of resource consumption, emissions, and their impact is called life cycle assessment (LCA).

The most important tool for industrial ecology is material and energy balancing. Material balances are fundamental to the control of production, particularly in the control of product yields. Volumes of materials as they pass through production systems can be described by the statements of conservation of mass. Material balances are first determined in the exploratory stages of a new process, improved during the planning stage, and then refined and maintained as a control instrument as production continues. With every step in the production process, such as design, material, rate of production, and equipment, the material balances need to be determined again. The increasing cost of energy has caused industry to examine means of reducing energy consumption in production. Energy balances are used in the examination of the various stages of a process, over the whole production process, and even extending over the total production system from the raw material to the finished material and its disposal. Material and energy balances can be very complicated at stages, but the basic approach is straightforward. Experience in working with single-item manufacturing can be extended to more complicated production systems. The increasing use of computers allows mass and energy balance models to be manipulated quite easily, and they can therefore be used in production management to maximize production and minimize costs. We should not forget that the whole concept of industrial ecology is based on the principle of reducing GHG emissions and use of energy and materials.

2.2 Global Warming Potential

The values of global warming potential for GHGs used in calculations will be taken from Table 2.1, which presents data for global warming from different sources. The table shows the global warming potential (GWP) of most industrial materials.

2.2.1 Basic Principles

Figure 2.6 presents the mass and energy balance of a manufacturing system. The mass and energy going into the manufacturing system must balance the mass and energy coming out (Figure 2.7).

The law of conservation of mass leads to the material balance equation

$$\text{Mass In} = \text{Mass Out} + \text{Mass Stored}$$

$$\text{Raw Materials} = \text{Products} + \text{Wastes} + \text{Stored Materials} \sum m_R = \sum m_p + \sum m_w + \sum m_s \tag{2.1}$$

where $\sum$ denotes the sum of all terms.

$$\sum m_R = \sum m_{R1} + \sum m_{R2} + \sum m_{R3} = \text{Total Raw Materials} \tag{2.2}$$

$$\sum m_p = \sum m_{p1} + \sum m_{p2} + \sum m_{p3} = \text{Total Products} \tag{2.3}$$

$$\sum m_w = \sum m_{w1} + \sum m_{w2} + \sum m_{w3} = \text{Total Waste Products} \tag{2.4}$$

$$\sum m_s = \sum m_{s1} + \sum m_{s2} + \sum m_{s3} = \text{Total Stored Products} \tag{2.5}$$

It is possible that in a manufacturing system more than one product, that is, p = 1, 2, 3,…,*n*, is being manufactured. Here, only three products have been shown for illustration. The energy input is the embodied energy for separate products and materials. Waste materials in the manufacturing system (particularly machining) can be considered as the chips being created after machining. We need energy during machining, and this must be included in the energy balance equation.

Material balance follows the law of conservation, that is, energy is conserved in manufacturing operations. The embodied energy into the operation and the energy used for manufacturing must balance the energy coming out and the energy stored.

$$\text{Embodied Energy In} + \text{Energy Used in Manufacturing} = \text{Energy Out (Embodied Energy in the Products)}$$

$$\sum E_E = \sum E_{e1} + \sum E_{e2} + \sum E_{e3} = \text{Total Embodied Energy} \tag{2.6}$$

$$\sum E_m = \sum E_{m1} + \sum E_{m2} + \sum E_{m3} = \text{Total Manufacturing Energy} \tag{2.7}$$

Table 2.1 Recent (Except for CH_4) Global Warming Potentials (GWP) Relative to CO_2

Industrial Designation or Common Name	Chemical Formula	GWP for 100-year Time Horizon
Carbon dioxide	CO_2	1
Methane	CH_4	25
Nitrous oxide	N_2O	298
Substances Controlled by the Montreal Protocol		
CFC-11	CCl_3F	4,750
CFC-12	CCl_2F2	10,900
CFC-13	$CClF_3$	14,400
CFC-113	CCl_2FCClF_2	6,130
CFC-114	$CClF_2CClF_2$	10,000
CFC-115	$CClF_2CF_3$	7,370
Halon-1301	$CBrF_3$	7,140
Halon-1211	$CBrClF_2$	1,890
Halon-2402	$CBrF_2CBrF_2$	1,640
Carbon tetrachloride	CCl_4	1,400
Methyl bromide	CH_3Br	5
Methyl chloroform	CH_3CCl_3	146
HCFC-22	$CHClF_2$	1,810
HCFC-123	$CHCl_2CF_3$	77
HCFC-124	$CHClFCF_3$	609
HCFC-141b	CH_3CCl_2F	725
HCFC-142b	CH_3CClF_2	2,310 (revised value)
HCFC-225ca	$CHCl_2CF_2CF_3$	122
HCFC-225cb	$CHClFCF_2CClF_2$	595
Hydrofluorocarbons		
HFC-23	CHF_3	14,800
HFC-32	CH_2F_2	675
HFC-125	CHF_2CF_3	3,500
HFC-134a	CH_2FCF_3	1,430
HFC-143a	CH_3CF3	4,470
HFC-152a	CH_3CHF_2	124
HFC-227ea	CF_2CHFCF_3	3,220
HFC-236fa	$CF_2CH_2CF_2$	9,810
HFC-245fa	$CF_3CH_2CF_3$	1,030
HFC-365mfc	$CH_3CF_2CH_2CF_3$	794
HFC-43-10mee	$CF_3CHFCHFCF_2CF_3$	1,640
Perfluorinated Compounds		
Sulfur hexafluoride	SF_6	22,800
Nitrogen trifluoride	NF_3	17,200
PFC-14	CF_4	7,390
PFC-116	C_2F_6	12,200
PFC-218	C_3F_8	8,830
PFC-318	$c\text{-}C_4F_8$	10,300
PFC-3-1-10	C_4F_{10}	8,860
PFC-4-1-12	C_5F_{12}	9,160
PFC-5-1-14	C_6F_{14}	9,300
PFC-9-1-18	$C_{10}F_{18}$	>7,500
Trifluoromethyl sulfur pentafluoride	$S_{F5}CF_3$	17,700
Fluorinated Ethers		
HFE-125	CHF_2OCF_3	14,900
HFE-134	CHF_2OCHF_2	6,320

Table 2.1 (Continued) Recent (Except for CH_4) Global Warming Potentials (GWP) Relative to CO_2

Industrial Designation or Common Name	Chemical Formula	GWP for 100-year Time Horizon
HFE-143a	CHF_3 OCF3	756
HCFE-235da2	$CHF_2OCHClCF_2$	350
HEF-245fa2	$CHF_2OCH_2CF_2$	708
HFE-254cb2	$CH_2OCF_2CHF_2$	659
HFE-254fa2	$CHF_2OCH_2CF_3$	359
HFE-347mcc3	$CH_3OCF_2CF_2CF_3$	575
HFE-347pcf2	$CHF_2CF_2OCH_2CF_3$	580
HFE-356pcc3	$CH_3OCF_2CF_2CHF_2$	110
HFE-449s1(HFE-7100)	$C_4F_9OCH_3$	297
HFE-569sf2 (HFE-7200)	$C_4F_9OC_2H_5$	59
HFE-43-10-pccc124 (H-Galden 1040x)	$CHF_2OCF_2OC_2F_4OCHF_2$	1,870
HFE-236ca12(HG-10)	$CH_2OCF_2OCHF_2$	2,800
HFE-338pcc13(HG-01)	$CHF_2OCF_2CF_2OCHF_2$	1,500
Perfluoropolyethers		
PFPMIE	$CF_2OCF(CF_2)$	10,300
	$CF_2OCF_2OCF_3$	10,300
Hydrocarbons and other compounds—direct effects		
Dimethylether	CH_3OCH_3	1
Methylene chloride	CH_2Cl_2	8.7
Methyl chloride	CH_2Cl	13

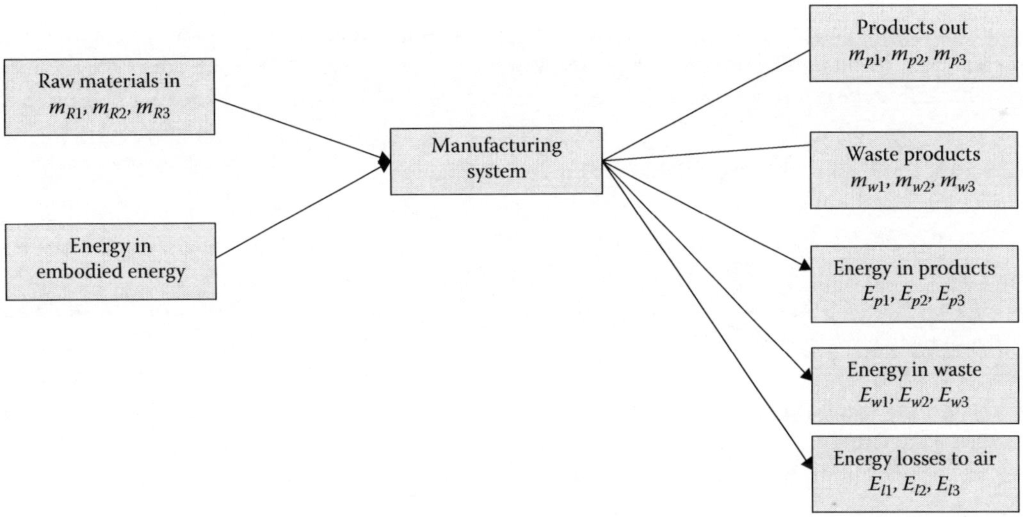

Figure 2.6

Mass and energy balance of manufacturing system.

$$\sum E_w = \sum E_{w1} + \sum E_{w2} + \sum E_{w3} = \text{Total Energy Embodied Wasted in Chips} \tag{2.8}$$

$$\sum E_l = \sum E_{l1} + \sum E_{l2} + \sum E_{l3} = \text{Total Energy Loss to Atmosphere (in Form of Heat)} \tag{2.9}$$

$$\sum E_{eo} = \sum E_{eo1} + \sum E_{eo2} + \sum E_{eo3} = \text{Total Embodied Energy Out} \tag{2.10}$$

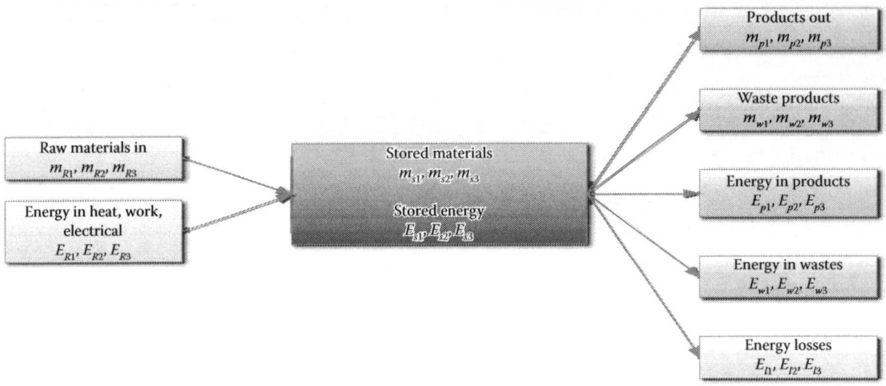

Figure 2.7

Mass and energy balance of reheating system.

For the energy balance, Equations 2.6 and 2.7 should be added on the left side of equality, and Equations 2.8 through 2.10 should be on the right side. We can show the complete energy balance equation as

$$\sum E_E + \sum E_m = \sum E_w + \sum E_l + \sum E_{eo} \tag{2.11}$$

Energy balances are often complicated because forms of energy can be interconnected, for example mechanical energy to heat energy, but overall the quantities must balance.

We can represent the thermodynamic analysis of resources used in manufacturing processes graphically, as shown in Figure 2.8. This figure shows the inputs of raw materials and energy into the manufacturing equipment and the outputs of product and waste. Product is the useful part coming out of the manufacturing system, and waste is scrap, chips, and heat.

Example 2.1: Mass and Energy Balance of Reheating Furnace

Figure 2.7 shows a diagram for a reheating furnace. From this figure, it is clear that exhaust flue gas losses are a key area for priority attention. Since the furnace operates at high temperatures, the exhaust gases leave at high temperatures, resulting in poor efficiency. Hence, a heat recovery device such as an air preheater is a necessary part of the system. The lower the exhaust temperature, the higher the furnace efficiency.

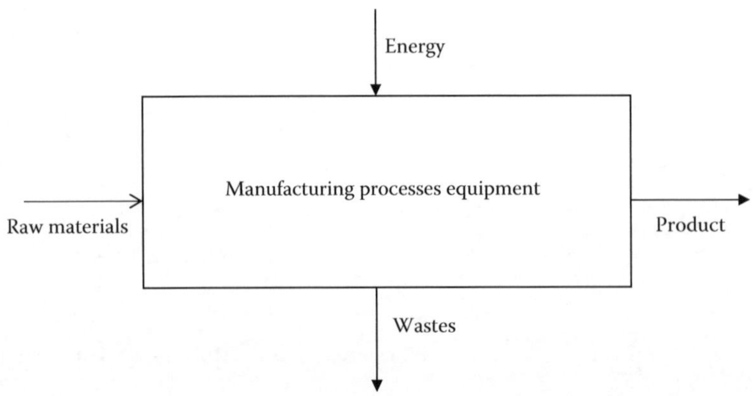

Figure 2.8

Resources analysis in manufacturing processes.

2.3 Material Balances

The first step in material balancing is to look for three basic categories: materials in, materials out, and materials stored. Then, for each category, it must be considered whether the materials are to be treated as a whole, a gross mass balance, or whether various constituents should be treated separately, and if so, what constituents. A simple example might be dry solids as opposed to total material; this really means separating two groups of constituents: nonwater and water. More complete dissection can separate out chemical types such as minerals, or chemical elements such as carbon. The choice and detail depend on the reasons for making the balance and on the information that is required. A major factor in industry is, of course, the value of the materials, and so expensive raw materials are more likely to be considered than cheaper ones, and products more likely than waste materials.

2.4 Bases and Units

Once it has been decided which constituents need consideration, the basis of the calculations has to be selected. This might be some mass of raw material entering the process in a batch system, or some mass per hour in a continuous process. It could be: some mass of a particular predominant constituent (for example, mass balance in a bakery might all be related to 100 kg of flour entering); or some unchanging constituent, such as in combustion calculations with air, in which it is helpful to relate everything to the inert nitrogen component; or carbon added in the nutrients in a fermentation system, because the essential energy relationships of the growing microorganisms are related to the combined carbon in the feed; or the essentially inert nonoil constituents of the oilseeds in an oil-extraction process. Sometimes it is unimportant what basis is chosen; in such cases, a convenient quantity such as the total raw materials into one batch or passed in per hour to a continuous process is often selected. When the basis has been selected, the units can be chosen, such as mass or concentration, which can be by weight or can be molar if reactions are important.

2.5 Total Mass and Composition

Material balances can be based on total mass, mass of dry solids, or mass of particular components, for example protein.

Example 2.2: Constituent Balance

Skim milk is prepared by the removal of some of the fat from whole milk. This skim milk is found to contain 90.5% water, 3.5% protein, 5.1% carbohydrate, 0.1% fat, and 0.85 ash. If the original milk contained 4.5% fat, calculate its composition, assuming that only fat was removed to make skim milk and that there are no losses in processing.

Basis: 100 kg of skim milk

This contains, therefore, 0.1 kg of fat. Let the fat which was removed from it to make skim milk be x kg. The original fat = $(x+0.1)$ kg and total original mass = $(100+x)$ kg. We know the original fat contained in the milk is 4.5%, so $(x + 0.1)/(100 + x) = 0.045$; we can find from this equation that the fat removed from the skim milk is $x = 4.6$ kg.

Therefore, the composition of the whole milk is fat = 4.5%, water = 90.5/104.6 = 86.5%, protein = 3.5/104.6 = 3.3%, carbohydrate = 5.1/104.6 = 4.9%, and ash = 0.8%.

Here, we have followed weight/weight (w/w) calculation of concentration; however, weight/volume, or molar concentration (M), or mole fraction could also be used.

Example 2.3: Concentration of Salt in Water

A solution of common salt in water is prepared by adding 20 kg of salt to 100 kg of water, to make a liquid of density 1323 kg/m^3. Calculate the concentration of salt in this salt solution as (a) weight fraction and (b) weight/volume fraction.

(a) Weight fraction:

20/(100 + 20) = 0.167; % weight/weight = 16.7%.

(b) Weight/volume:

A density of 1323 kg/m^3 means that 1 m^3 of solution weighs 1323 kg, but 1323 kg salt solution contains (20 × 1323)/(100 + 20) = 220.5 kg salt/m^3. Therefore, 1 m^3 solution contains 220.5 kg salt.

Weight/volume fraction = 220.5/1000 = 0.2205.
And so weight/volume = 22.1%.

Example 2.4: Air Composition

If air consists of 77% by weight of nitrogen and 23% by weight of oxygen, calculate (1) the mean molecular weight of air and (2) the mole fraction of oxygen.

1. On the basis of 1000 kg of air, this contains 77/28 moles of N_2 and 23/32 moles of O_2.
   Total number of moles = 2.75 + 0.72 = 3.47 moles.
   So the mean molecular weight of air = 100/3.47 = 28.8.
2. The mole fraction of oxygen = 0.72/(2.75 + 0.72) = 0.21.

Example 2.5: Gas Composition

In the concentration of a soft drink, the total quantity of carbon dioxide required is the equivalent of three volumes of gas to one volume of water at 0°C and atmospheric pressure. Calculate (a) the mass fraction and (b) the mole fraction of the CO_2 in the drink, ignoring all components other than CO_2 and water.

On the basis of 1 m^3 of water = 1000 kg
Volume of carbon dioxide added = 3 m^3
From the gas equation, pV = nRT
$1 \times 3 = n \times 0.08306 \times 273$, so we get
n = 0.134 moles
Molecular weight of carbon dioxide = 44
And so weight of carbon dioxide added = $0.134 \times 44 = 5.9$ kg.

a. Mass fraction of carbon dioxide drink = $5.9/(1000 + 5.9) = 5.9 \times 10^{-3}$
b. Mole fraction of carbon dioxide in the drink = $0.134/(1000/18 + 0.134) = 2.41 \times 10^{-3}$

2.6 Continuous Processes

In continuous processes, time also enters into consideration, and the balances are related to unit time. Thus, in considering a continuous centrifuge separating whole milk into skim milk and cream, if the material contained in the centrifuge is constant both in mass and in composition, the quantities of the components entering and leaving in the different streams in unit time are constant, and a mass balance can be written on this basis. Such an analysis assumes that the process is in a steady state, that is, flows and quantities contained in vessels do not change with time.

Example 2.6: Balance across Equipment in Continuous Centrifugation of Milk

If 35,000 kg of whole milk containing 4% fat is to be separated in a 6-h period into skim milk with 0.45% fat and cream with 45% fat, what are the flow rates of the two output streams from a continuous centrifuge which accomplish this separation?
Basis: 1 h flow of whole milk

Mass in
Total mass = 35,000/6 = 5833 kg/h
Fat = $5833 \times 0.04 = 233$ kg
And so water plus solids-not-fat = 5600 kg
Mass out
Let the mass of cream be x kg, then its total fat content is $0.45x$. The mass of skim milk is $(5833-x)$ and its total fat content is $0.0045(5833-x)$.
Material balance on fat
Fat in = fat out
$5833 \times 0.04 = 0.0045(5833-x)+0.45x$, and so we get

$x = 465$ kg

So that the flow of cream is 465 kg/h and skim milk (5833–465) = 5368 kg/h.

This time unit has to be considered carefully in continuous processes, as normally such processes operate continuously for only part of the total factory time. Usually there are three periods—start-up, continuous processing (so-called steady state), and closedown—and it is important to decide what material balance is being considered. Also, the time interval over which any measurements are taken must be long enough to allow for any slight periodic or chance variation.

2.7 Blending

Another class of situations that can arise are blending problems, in which various ingredients are combined in such proportions as to give a product of some desired composition. Complicated examples, in which an optimum or best achievable composition must be sought, need quite elaborate calculation methods, such as linear programming, but simple examples can be solved by straightforward mass balance.

2.8 Drying

In setting up a material balance for a process, a series of equations can be written for the various individual components and for the process as a whole. In some cases, when groups of materials maintain constant ratios, the equations can include such groups rather than their individual constituents. For example, when drying vegetables the carbohydrates, minerals, protein, and so on can be grouped together as "dry solids," and then only dry solids and water need be taken through the material balance.

Example 2.7: Drying Yield

Potatoes are dried from 14% total solids to 93% total solids. What is the product yield from each 1000 kg of raw potatoes, assuming that 8% by weight of the original potatoes is lost in peeling?

Basis: 1000 kg potato entering

As 8% of potatoes are lost in peeling, potatoes to drying are 920 kg, solids 129 kg.

Mass In (kg)	Mass Out (kg)	
Potato solids: 140 kg	Dried product 92	
Water: 860 kg	Potato solids: 140 × (92/100) = 129 kg	
	Associated water: 10 kg	
	Total product: 139 kg	
	Losses	
	Potato peelings	
	Solids	11 kg
	Water	69 kg
	Water evaporated: 781 kg	
	Total losses: 861 kg	
	Total: 1000 kg	
Potato yield = 139/1000 = 14%		

Often, it is important to be able to follow particular constituents of the raw material through a process. This is just a matter of calculating each constituent.

2.9 Energy Balances

An energy balance can be used to track energy through a system, and is a very useful tool for determining resource use and environmental impacts, using the first and second laws of thermodynamics, to determine how much energy is needed at each point in a system, and in what way this energy is a cost in various environmental issues. The energy accounting system keeps track of energy in, energy out, and nonuseful energy vs. work done and transformations within the system.

Energy takes many forms, such as heat, kinetic energy, chemical energy, and potential energy, but because of interconversions, it is not always easy to isolate the separate constituents of energy balances. However, under some circumstances, certain aspects predominate. There are many heat balances in which other forms of energy are insignificant. In some chemical situations, mechanical energy is insignificant; in some mechanical energy situations, such as fluid flow in pipes, the frictional losses appear as heat, but the details of the heating need not be considered. We are seldom concerned with internal energies.

Therefore, particular applications of energy balances tend to focus on particular dominant aspects, and so a heat balance, for example, can be a useful description of important cost and quality aspects of the process situation. Then, after some preliminary calculations, the important aspects emerge, and other, minor ones can be lumped together or even ignored without introducing substantial errors. With experience, minor aspects can perhaps be left out completely, through this always raises the possibility of error.

Energy balances can be calculated on the basis of external energy used per kilogram of product, or raw material processed, or on dry solids or some key component. The energy consumed in food production includes *direct energy*, which is fuel and electricity used on the farm, in transport and factories, and in storage, selling, and so on; and *indirect energy*, which is used to actually build the machines, to make the packaging, to produce the electricity and the oil, and so on. Food itself is a major energy source, and energy balances can be determined for animal or human feeding; food energy input can be balanced against outputs in heat and mechanical energy and chemical synthesis.

In the SI system there is only energy unit, the Joule. In some heat balances, people still use British Thermal Units (BTU).

2.10 Heat Balances

The most common important energy form is heat energy, and the conservation of this can be illustrated by considering operations such as heating and drying. In these, *enthalpy* (total heat) is conserved, and as with mass balances, so enthalpy balances can be written around the various items of equipment or process stages, or round the whole system (plant), and it is assumed that no appreciable heat is converted to other forms of energy such as work.

Enthalpy (H) is always referred to relative to some reference level or datum. Working out the energy balance, then, is just a matter of considering the various quantities of materials involved, their specific heats, and their changes in temperature or state (since latent heats arising from phase changes are quite frequently encountered). Figure 2.9 illustrates a heat balance.

Heat is absorbed or evolved by some reactions during processing, but usually the quantities are small when compared with the other forms of energy entering into the system, such as sensible heat and latent heat. Latent heat is the heat required to change, at constant temperature, the physical state of materials from solid to liquid, liquid to gas, or solid to gas. Sensible heat is heat that, when added to or subtracted

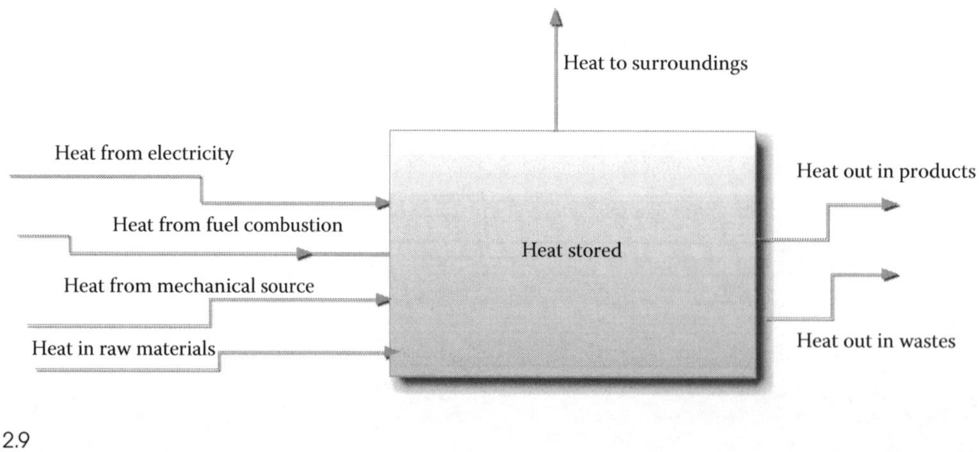

Figure 2.9

Heat balance.

from materials, changes their temperature and thus can be sensed. The units of specific heat are joules per kilogram per degree Kelvin, and sensible heat change is calculated by multiplying the mass by the specific heat by the change in temperature, that is, $m \times c \times \Delta T$. The units of latent heat are joules per kilogram, and total latent heat change is calculated by multiplying the mass of material that changes its phase by the latent heat. Having determined these factors, which are significant in the overall energy balance, the simplified heat balance can then be used with confidence in energy studies. Such calculations can be quite simple and straightforward, but they give a quantitative feeling for the situation and can be of great use in design of equipment and processes.

Example 2.8: Dryer Heat Balance

A textile dryer is found to consume 4 m^3/h of natural gas with a calorific value of 800 KJ/mole. If the throughput of the dryer is 60 kg of wet cloth per hour, drying it from 55% moisture to 10% moisture, estimate the overall thermal efficiency of the dryer, taking into account the latent heat of evaporation only.

60 kg of wet cloth contains
60×0.55 kg water = 33 kg moisture
$60 \times (1-0.55) = 27$ kg dry cloth

As the final product contains 10% moisture, the moisture in the product is $27/9 = 3$ kg.

And so moisture removed/h = $33-3 = 30$ kg/h
Latent heat of evaporation = 2257 kJ/K
Heat necessary to supply = $30 \times 2257 = 6.8 \times 10^4$ kJ/h

Assuming the natural gas to be at standard temperature and pressure, at which 1 mole occupies 22.4 l,

Rate of flow of natural gas = 4 m^3/h = $(4 \times 1000)/22.4 = 179$ moles/h
Heat available from combustion = $179 \times 800 =$ n14.3×10^4 kJ/h
Approximate thermal efficiency of dryer = heat needed/heat used = $6.8 \times 10^4/14.3 \times 10^4 = 48\%$.

To evaluate this efficiency more completely, it would be necessary to take into account the sensible heat of the dry cloth and the moisture, and the changes in temperature and humidity of the combustion air, which would be combined with natural gas. However, as the latent heat of evaporation is the dominant term, the above calculation gives a quick estimate and shows that a simple energy balance can give useful information.

Similarly, an energy balance can be carried out on thermal processing operations and, indeed, any processing operation in which heat or another form of energy is used.

Example 2.9: Autoclave Heat Balance in Canning

An autoclave contains 1000 cans of pea soup. It is heated to an overall temperature of 100°C. If the cans are to be cooled to 40°C before leaving the autoclave, how much cooling water is needed if it enters at 15°C and leaves at 35°C?

The specific heats of the pea soup and the can metal are, respectively, 4.1 kJ/kg°C and 0.50 kJ/kg°C. The weight of each can is 60 g, and it contains 0.45 kg of pea soup. Assume that the heat content of autoclave walls above 40°C is 1.6×10^4 kJ, and there is no heat loss through walls.

Let w = the weight of cooling water required, and the datum temperature = 40°C, the temperature of the cans leaving the autoclave.

Heat Entering

Heat in cans = weight of cans × specific heat × temperature above datum = $1000 \times 0.06 \times 0.5 \times (100-40)$ kJ = 1.8×10^3 kJ

Heat in can contents = weight pea soup × specific heat × temperature above datum = $1000 \times 0.45 \times 4.1 \times (100-40) = 1.1 \times 10^5$ kJ

Heat in water = weight of water × specific heat × temperature above datum = $w \times 4.186 \times (15-40) = -104.6$ w kJ

Heat Leaving

Heat in cans = 1000 × 0.06 × 0.50 × (40–40) (cans leave at datum temperature) = 0
Heat in can contents = 1000 × 0.45 × 4.1 × (40–40) = 0
Heat in water = w × 4.186 × (35–40) = –20.9 w
Heat energy entering = heat energy leaving
  127,800–104.6 × w = (–20.9 w)
  w = 1527 kg
Amount of cooling water required = 1527 kg

2.11 Other Forms of Energy

Motor power is usually derived, in factories, from electrical energy, but it can be produced from steam engines or water power. The electrical energy input can be measured by a suitable wattmeter and the power used in the drive estimated. There are always losses from the motors due to heating, friction, and windage: motor efficiency, which can be obtained from the motor manufacturer, expresses the proportion (usually as a percentage) of the electrical input energy, which emerges at the motor shaft and is available for use.

When considering movement, whether of fluids in pumping, of solids in solid handling, or of foodstuffs in mixers, the energy input is largely mechanical. The flow situations can be analyzed by recognizing the conservation of total energy, whether as energy of motion, or potential energy such as pressure energy, or energy lost in friction. Similarly, chemical energy released in combustion can be calculated from the heats of combustion of the fuels and their rates of consumption. Eventually, energy emerges in the form of heat, and its quantity can be estimated by summing the various sources.

Example 2.10: Refrigeration Load

It is desired to freeze 10,000 loaves of bread each weighing 0.75 kg from an initial room temperature of 18°C to a final temperature of –18°C. The bread-freezing operation is to be carried out in an air-blast freezing tunnel. It is found that the fan motors are rated at a total of 80 horsepower, and measurements suggest that they are operating at around 90% of their rating, under which conditions their manufacturer's data claims a motor efficiency of 86%. If refrigerating 1 ton takes 3.52 kW, estimate the maximum refrigeration load imposed by this freezing installation, assuming that (a) fans and motors are all within the freezing tunnel insulation and (b) the fans but not their motors are in the tunnel. The heat-loss rate from the tunnel to the ambient air has been found to be 6.3 kW.

Extraction rate from freezing bread (maximum) = 104 kW.
Fan rated horsepower = 80.
Now, 0.746 kW = 1 horsepower and the motor is operating at 90% of rating,
and so (fan = motor) power = (80 × 0.9) × 0.746 = 53.7 kW.

a. With motors + fans in tunnel:
Heat load from fans + motors = 53.7 kW.
Heat load from ambient = 6.3 kW.
Total heat load = (104 + 53.7 + 6.3) kW = 164 kW.
As 3.52 kW is equivalent to refrigerating 1 ton, we find total refrigeration = 164/3.52 = 46 tons.

b. With motors outside, the motor inefficiency = (1–0.86) does not impose a load on the refrigeration:
Total heat load = (104+[0.86 × 53.7]+ 6.3) = 156 kW.
= 156/3.52 = 44.5 tons.

In practice, material and energy balances are often combined, as the same stoichiometric information is needed for both.

The objective of a *material and energy (M&E)* balance is to assess the inputs, conversion efficiency, output, and losses. An M&E balance, used in conjunction with diagnosis, is a powerful tool in establishing the basis for improvements and potential savings.

- Material and energy balances can be worked out quantitatively if the amounts of materials entering into the process and the nature of the process are known.
- Material and energy balances take the basic form.

Content of inputs = content of products + waste/losses + changes in the stored materials.

- In continuous processes, a time balance must be established.
- Energy includes heat energy (enthalpy), potential energy (energy of pressure or position), kinetic energy, work energy, and chemical energy. It is the sum over all of these that is conserved.

2.12 Flow Chart

2.12.1 Method for Preparing Process Flow Chart

Identifying and drawing up a unit operation/process is a prerequisite for M&E. The procedure for drawing up the process flow diagrams is illustrative of the production process, involving various input resources, conversion steps and outputs, and recycled streams. The process flow may be constructed stepwise, that is, by identifying the inputs/outputs/wastes at each stage of the process. A graphical representation is provided in Figure 2.10.

Process steps should be sequentially drawn from raw material to finished product. Intermediates and any other by-product should be represented. Operating process parameters such as temperature, pressure, percent concentration, and so on should be represented.

- Inputs of the process should include raw materials, water, steam, and energy (electricity, and so on).
- The flow rate of various streams should also be represented in appropriate units, such as cubic meters per hour or kilograms per hour.
- Wastes/by-products could include solids, water, chemicals, energy, and so on. An energy and mass balance diagram should be drawn for each process step (unit operation) as well as for the entire plant.
- The output of the process is the final product produced in the plant.

2.12.2 Guidance for Material and Energy Balance

M&E balances are important, since they make it possible to identify and quantify previously unknown losses and emissions. These balances are also useful for monitoring the improvements made in an ongoing project while evaluating cost benefits. Raw materials and energy are not only major cost components in any manufacturing activity but also major sources of environmental pollution. Inefficient use of raw materials and energy in production processes is reflected as waste.

- For a complex production stream, it is better to draft the overall material and energy balance first.
- When splitting up the total system, choose simple, discrete subsystems. The process flow diagram could be useful here.

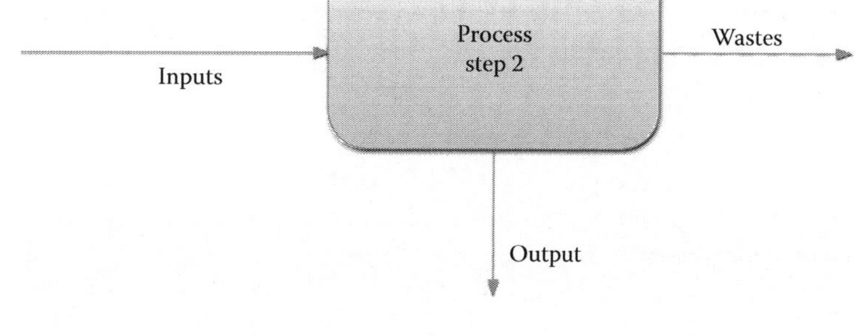

Figure 2.10

Process flow chart.

- Choose the M&E balance envelope such that the number of streams entering and leaving is as small as possible.
- Always choose recycle streams (M&E) within the envelope.
- The measurement units may include time factors or production linkages.
- Consider a full batch as the reference in the case of batch operation.
- It is important to include start-up and cleaning operation consumption of M&E.
- Calculate the gas volumes at standard conditions.
- In the case of shutdown losses, averaging over long periods may be necessary.
- Highlight losses and emissions (M&E) at part-load operation if prevalent.
- For each stream, where applicable, include energy quality (pressure, temperature, enthalpy, kilocalories per hour, kilowatts, amps, volts, and so on).
- During the preparation of M&E balances, analytical data and flow and energy measurements have to be accurate, especially in the case of references over a short time span.

M&E balances need to be developed at various levels according to the above guidelines. They could be developed for the entire production system, or a section of the system, or equipment-wise. The equipment-wise M&E balance could help assess the performance of the equipment, which in turn could help identify and quantify avoidable losses of energy and material, and could be of immense help in equipment design in the event of manufacturing system problems.

2.12.3 Steps in Mass and Energy Balance Calculation

Basically, M&E balance calculation checks whether directly or indirectly measured energy and mass flows are in agreement with the mass and energy conversion principles. In order to use it correctly, a step-by-step approach is advocated:

1. Clearly identify the problem to be studied. In the case of manufacturing system problems, we have to identify material removal processes such as turning, milling, grinding, and so on, as each of them runs on different principles, and the mechanics of material removal are quite different.
2. Define a boundary that encloses the complete system or subsystem or the equipment to be analyzed.
3. Choose the enclosing boundary such that measurement is accurate and easy.
4. Select the appropriate test period depending on the type of process and product. The test period may vary between equipment, as the cycle time or time per product may vary for different material removal processes.
5. Calculate the energy and mass flow based on the measurements in Step 3.
6. Check the M&E balance. If the balances are outside the acceptable limits, repeat the measurements.
7. Energy release or use in endothermic and exothermic processes should be taken into account in the energy balance.

Example 2.11: Heat Balance in Boiler

A heat balance is an attempt to balance the total energy entering the system (e.g., a boiler) with that leaving the system in different forms. Figure 2.11 shows the heat balance and different losses occurring during the generation of steam.

The heat balance in the boiler shows that only 73.85% of the heat is useful, while the remaining 26.15% is waste in various forms. The analysis indicates to the boiler manager that dry flue gas loss is the greatest source of waste of energy, and some device to trap the heat should be used. The heat loss due to hydrogen is 8.15%, suggesting that the fuel needs to be changed.

Example 2.12: Mass Balance in a Cement Plant

The cement process involves gas, liquid, and solid flows with heat and mass transfer, combustion of fuel, reactions of clinker compounds, and undesired chemical reactions that produce, among other things, sulfur, chlorine, and alkalis.*

* Based on figure from Austrian BAT proposal [1], Cembureau for mass balance for production of 1 ton of cement.

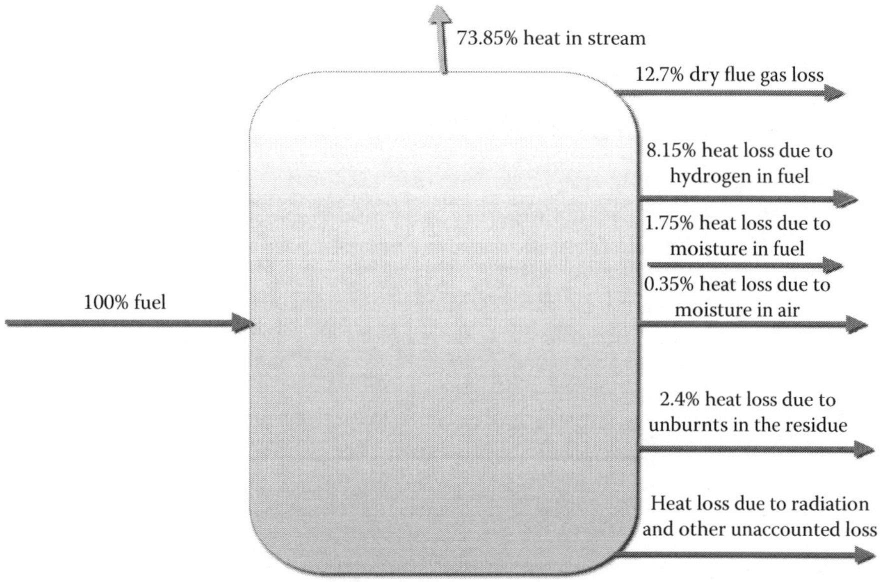

Figure 2.11

Heat balance in boiler.

It is important to understand these processes in order to optimize the operation of the cement kiln, diagnose operational problems, increase production, improve energy consumption, lower emissions, and increase refractory life.

The global cement industry contributes around 20% of all man-made CO_2 emissions, and is consequently responsible for around 10% of man-made global warming (*Global Cement* magazine). Energy consumption by the cement industry is estimated at about 2% of global primary energy consumption, or almost 5% of total global industrial energy consumption. Cement production is currently increasing by about 3%/year [1]. This rate is set to increase as developing nations rapidly become richer and spend proportionately more on cement-intensive infrastructure. There are two sources of CO_2 emissions from the cement plant: first, due to the energy it uses, and second, the evolution of CO_2 as a by-product in the calcination process. The cement plant releases 1 ton of CO_2 for every ton of cement produced, half from the fuel it uses and the other half from the calcination process. Cement manufacturing is highly energy and emissions intensive because of the extreme heat required in the process. Producing a ton of cement requires 4.7 million BTU of energy, equivalent to about 400 lb of coal, and generates nearly a ton of CO_2. Given its high emissions and critical importance to society, cement is an obvious place to look to reduce GHG emissions, which represent around 40% of cement emissions. Finally, the electricity used to power additional plant machinery and the final transportation of cement represents another source of indirect emissions, and accounts for 5%–10% of the industry's emissions.

Indirect emissions from burning fossil fuels to heat the kiln can be reduced by switching to alternative fuels, including natural gas, biomass, and waste-derived fuels such as tires, sewage sludge, and municipal solid wastes. These less carbon-intensive fuels could reduce overall cement emissions by 18%–24% from 2006 levels by 2050. Alternatively, efficiency measures can reduce the demand for fuel by addressing the production process itself (such as switching from inefficient wet kilns to dry ones) or through technical and mechanical improvements (such as preventative maintenance to repair kiln leaks). While some estimate that energy efficiency improvements could achieve emission reductions of up to 40%, some industry analysts suggest that producers may have already exhausted this potential. Finally, CO_2 emissions can be captured after they are produced through carbon capture and storage (CCS). In addition to traditional CCS methods, which are already employed in some power plants, concrete producers can actually use their own product as a sink for CO_2. Through the process of *accelerated carbonation*, CO_2 penetrates concrete and reacts with calcium hydroxide in the presence of water to form calcium carbonate; the result is stable, long-term CO_2 storage. As a mitigation technology, accelerated carbonation can be achieved by exposing freshly mixed concrete to flue gases with high CO_2 concentrations.

Example 2.13: Mass Balance Calculation in Chemical Industries for Air Pollution

This problem shows how a mass balance calculation can be used to check the results of an air pollution monitoring study. A fabric filter (bag filter) is used to remove the dust from the inlet gas stream so that the outlet gas stream meets the required emission standards in the cement, fertilizer, and other chemical industries.

During an air pollution monitoring study, the inlet gas stream to a bag filter is 169,920 m^3/h, and the dust loading is 4577 mg/m^3. The outlet gas stream from the bag filter is 185,040 m^3/h, and the dust loading is 57 mg/m^3.

What is the maximum quantity of ash that will have to be removed per hour from the bag filter hopper based on these test results?

Solution

Based on dust balance

$$\text{Mass}_{\text{in}} = \text{Mass}_{\text{out}}$$

Inlet gas stream dust = outlet gas stream dust + hopper ash

1. Calculate the inlet and outlet dust quantities in kilograms per hour

$$\text{Inlet gas quantity} = 169{,}920\,(\text{m}^3/\text{h}) \times 4577(\text{mg/m}^3) \times \frac{1}{10^6}(\text{kg/mg})$$

$$= 777.7 \text{ kg/h}$$

$$\text{Outlet dust quantity} = 185{,}040\,(\text{m}^3/\text{h}) \times 57\,(\text{mg/m}^3) \times \frac{1}{10^6}\,(\text{kg/mg})$$

$$= 10.6 \text{ kg/h}$$

2. Calculate the quantity of ash that will have to be removed from the hopper per hour

Hopper ash = inlet gas dust quantity − outlet gas dust quantity

$$= (777.7 - 10.6)\text{ kg/h} = 767.1\text{ kg/h}$$

Example 2.14: Energy Balance for Ethanol from Corn

Ecologist David Pimentel [2] finds that producing ethanol and biodiesel from corn and other crops is not worth the energy. Turning plants such as corn, soybeans, and sunflowers into fuel uses much more energy than the resulting ethanol or biodiesel generates.

These strategies are *not sustainable.*

The data for inputs of 1 gallon of ethanol is presented below. The analysis applies to typical US Midwest conditions and practices.

Items	Input of energy	Output of energy	Comments	Comparison with other energy sources
Corn production	55,000 BTU		Production costs include fertilizer and operation of farm machinery	
Fermentation/distillation	74,300 BTU		Typical US Midwest conditions	
Total input of energy/gallon	129,000 BTU			
Output in combustion of 1 gallon of ethanol		76,000 BTU		Gasoline: 120,000/gallon
Output/input ratio (United States)	76,000/129,000 = 0.589 United States: output/input = 0.6			Gasoline: output/input = 7 Brazil: output/input = 3

In terms of energy output compared with energy input for ethanol production, Pimentel found that

- Corn requires 29% more fossil energy than the fuel produced.
- Witch grass requires 45% more fossil energy than the fuel produced.
- Wood biomass requires 57% more fossil energy than the fuel produced.

It is concluded that ethanol production does not provide a net energy balance or gain, and it is not a renewable energy source or an economical fuel. Further, its production and use contribute to air, water, and soil pollution and global warming.

Example 2.15: Approximate Energy Balance for 1000 MW_e Nuclear Power Plant

The energy balance for a nuclear power plant is presented below. The energy balance shows that the output/input ratio is about 14. This is much higher than for biodiesel or gasoline. There is a considerable amount of energy involved in constructing and decommissioning nuclear power plants; even so, the nuclear power plant is a winner in terms of efficiency of energy production.

Energy to construct and decommission a 1000 MW_e plant	4×10^9 kWh_{th}
Energy for fuel (30 years)	7×10^9 kWh_{th}
Energy for operation (30 years)	4×10^9 kWh_{th}
Total **input**	**15×10^9 kWh_{th}**
Electrical energy output (30 years)	**210×10^9 kW_e**
Output/input ratio	14 kWh_e/kWh_{th}

The analysis provided is for a 1000 MW_e pressurized water reactor (PWR) for a 30-year economic life. The ratio of thermal to electrical energy is about 3:1.

The basic principle of M&E balance is to quantify all materials entering into the production process. This will help identify the waste generated as well as how to reduce the waste. This is the basis of sustainable manufacturing. The energy balance helps identify loss of energy due to radiation and conduction. In this age of high energy costs and global warming, it is important to identify and correct sources of energy loss. It is well known that energy consumption and use is a measure of GHG emission and global warming. Regarding global warming, we consider only CO_2 emissions, but other toxic emissions such as SO_2 and NO_2 should not be neglected. These are equally toxic and polluting to the earth, rivers, and the air.

M&E balance equations and networks show quantitatively how much material and energy are lost. Therefore, M&E balance equations are able to pinpoint hitherto unknown losses and emissions. Inefficient use of raw materials and energy in manufacturing processes is identified as waste, and the management can take quick action to reduce waste and improve the financial health of the company. Therefore, it can be stated that M&E balance analysis is at the heart of sustainable, or green, manufacturing. The main objective of M&E balance is to assess input, conversion efficiency, output, and losses, and it is a powerful tool for achieving savings and improvement in production processes.

2.13 Ecological Economics

Ecological economics (EE) is referred to as both a *transdisciplinary* and an *interdisciplinary* field of academic research that aims to address the interdependence and *coevolution* of human *economies* and natural *ecosystems* over time and space. EE is distinct from environmental economics, which is the economic analysis of the environment, in its treatment of the economy as a subsystem of the ecosystem and its emphasis upon preserving natural capital. Green economics is synonymous with EE. Ecological economists have questioned fundamental mainstream economic approaches such as cost-benefit analysis. EE shares many of its perspectives with feminist economics, including the focus on sustainability, nature, justice, and care values.

A simple *circular flow of income* diagram is replaced in EE by a more complex flow diagram reflecting the input of solar energy, which sustains natural inputs and environmental services, which are then used as units of production. Once consumed, natural inputs pass out of the economy as pollution and waste. The potential of an environment to provide services and materials is referred to as an "environment's source function," and this function is depleted as resources are consumed or pollution contaminates the resources. The "sink function" describes an environment's ability to absorb and render harmless waste and pollution: when waste output exceeds the limit of the sink function, long-term damage occurs. The economic value of natural capital and ecosystem services is accepted by mainstream environmental economics, but is emphasized as especially important in EE. Ecological economists may begin by estimating how to maintain a stable environment before assessing the cost in dollar terms.

Global warming is now widely acknowledged as a major issue, with all national scientific academies expressing agreement on its importance. As population growth and energy demand increase, the world faces an energy crisis. Some economists and scientists forecast a global ecological crisis if energy use is not contained.

A primary objective of EE is to ground economic thinking and practice in physical reality, especially in the laws of physics (particularly the laws of thermodynamics) and in knowledge of biological systems. It accepts as a goal the improvement of human well-being through development and seeks to ensure achievement of this through planning for the sustainable development of ecosystems and societies. EE is distinguishable from *neoclassical economics* primarily by its assertion that the economy is embedded within an environmental system. Ecology deals with the energy and matter transactions of life and the earth, and the human economy is by definition contained within this system. Ecological economists argue that neoclassical economics has ignored the environment, at best considering it to be a subset of the human economy.

The neoclassical view ignores much of what the natural sciences have taught us about the contributions of nature to the creation of wealth: for example, the planetary endowment of scarce matter and energy, along with the complex and biologically diverse ecosystems that provide goods and ecosystem services directly to human communities: micro- and macroclimate regulation, water recycling, water purification, storm-water regulation, waste absorption, food and medicine production, pollination, protection from solar and cosmic radiation, the view of a starry night sky, and so on.

EE challenges the conventional approach toward natural resources, claiming that it undervalues natural capital by considering it as interchangeable with human-made capital—labor and technology.

The impending depletion of natural resources and increase of climate-changing GHGs should motivate us to examine how political, economic, and social policies can benefit from alternative energy. Interestingly enough, reducing our dependence on fossil fuels with the specific intention of affecting any one of the abovementioned factors easily benefits at least one of the others. Neoclassical economists tend to maintain that man-made capital can, in principle, replace all types of natural capital. This is known as the *weak sustainability view*: essentially, that every technology can be improved upon or replaced by innovation, and that there are substitutes for any and all scarce materials.

At the other extreme, the *strong sustainability view* argues that the stock of natural resources and ecological functions is irreplaceable. From the premises of strong sustainability, it follows that economic policy has a fiduciary responsibility to the greater ecological world, and that sustainable development must therefore take a different approach to valuing natural resources and ecological functions.

2.14 Energy Economics

A key concept of energy economics is net energy gain, which recognizes that all energy requires energy to produce it. To be useful, the energy return on energy invested has to be greater than 1. The net energy gain from production of coal, oil, and gas has declined over time, as the easiest production sources have been most heavily depleted.

Thermoeconomics is based on the proposition that the role of energy in biological evolution should be defined and understood through the second law of thermodynamics, but also in terms of such economic criteria as productivity, efficiency, and especially the costs and benefits (or profitability) of the various mechanisms for capturing and utilizing available energy to build biomass and do work. As a result, thermoeconomics is often discussed in the field of EE, which itself is related to the fields of sustainability and sustainable development.

Energy analysis is performed in the field of industrial ecology in order to use energy more efficiently. In recent decades, use of energy has spread outside physics and engineering to the fields of industrial ecology, EE, systems ecology, and energetics.

Embodied energy is the sum of all the energy required to produce any goods or services, considered as if that energy were incorporated or "embodied" in the product itself. The concept can be useful in determining the effectiveness of energy-producing or energy-saving devices, or the "real" replacement cost of a building, and, because energy inputs usually entail GHG emissions, in deciding whether a product contributes to or mitigates global warming. One fundamental purpose for measuring this quantity is to compare the amount of energy produced or saved by the product in relation to the amount of energy consumed in producing it.

Embodied energy is an accounting method that aims to find the sum total of energy necessary for an entire product life cycle. Determining what constitutes this life cycle includes assessing the relevance and extent of energy used in raw material extraction, transport, manufacture, assembly, installation, disassembly, deconstruction, and decomposition, as well as human and secondary resources. Different methodologies produce different understandings of the scale and scope of application and the type of energy embodied.

The history of constructing a system of accounts that records the energy flows through an environment can be traced back to the origins of accounting itself. The main methods of embodied energy accounting, as they are used today, grew out of Leontief's input–output model and are called *input–output embodied energy analysis*. The total amount of energy, direct and indirect, for the entire production process is called the embodied energy [3].

2.14.1 Embodied Energy Methodologies

Embodied energy analysis is interested in the energy going to support a consumer, so all energy depreciation is assigned to the final consumer demand. Different methodologies use different scales of data to calculate energy embodied in the products and services of nature and human civilization. International consensus on the appropriateness of data scales and methodologies is still awaited. This difficulty results in a wide range in embodied energy values for any given material. There is no comprehensive global embodied energy public dynamic database. With no stated estimate of the embodied energy error, it is difficult to calibrate the *sustainability index* and hence the value of any given material, process, or service to environmental and human economic processes. Scientists have not yet agreed absolute universal values for embodied energy because there are many variables to take into account, but most agree that products can be compared with each other to determine their relative levels of embodied energy. Comparative lists (for example, see the University of Bath *Embodied Energy & Carbon Material Inventory* [4]) contain average absolute values and explain the factors that have been taken into account when compiling the lists [5].

Typical embodied energy units are MJ/kg (megajoules of energy needed to make a kilogram of product) and tCO_2 (tons of carbon dioxide created by the energy needed to make a kilogram of product). Converting megajoules to tCO_2 is not straightforward, because different types of energy (oil, wind, solar, nuclear, and so on) emit different amounts of carbon dioxide, so the actual amount of carbon dioxide emitted when a product is made will be dependent on the type of energy used in the manufacturing process. For example, the Australian government gives a global average of 0.098 tCO_2 = 1 GJ. This is the same as 1 MJ = 0.098 $kgCO_2$ = 98 gCO_2 or 1 $kgCO_2$ = 10.204 MJ. This is the standard that is commonly used worldwide.

2.14.2 Embodied Energy in Common Materials

Selected data from the Inventory of Carbon and Energy (ICE) prepared by the University of Bath, United Kingdom is presented in the table below.

Material	Energy (MJ/kg)	Carbon (CO_2 kg/kg)	Density (kg/m^3)
Concrete (1:1.5:3)	1.11	0.159	2,400
Bricks (common)	3	0.24	1,700
Concrete block (medium density)	0.67	0.073	1,450
Aerated block	3.5	0.3	750
Limestone block	0.85		2,180
Marble	2	0.116	2,500
Cement mortar (1:3)	1.33	0.208	
Steel (general, average recycled content)	20.1	1.37	7,800
Stainless steel	56.7	6.15	7,850
Timber (general, excludes sequestration)	8.5	0.46	480–720
Glue laminated timber	12	0.87	
Cellulose insulation (loose fill)	0.94–3.3		43

(*Continued*)

Material	Energy (MJ/kg)	Carbon (CO_2 kg/kg)	Density (kg/m³)
Cork insulation	26		160
Glass fiber insulation (glass wool)	28	1.35	12
Flax insulation	39.5	1.7	30
Rockwool (slab)	16.8	1.05	24
Expanded polystyrene insulation	88.6	2.55	15–30
Polyurethane insulation (rigid foam)	101.5	3.48	30
Wool (recycled) insulation	20.9		25
Straw bale	0.91		100–110
Mineral fiber roofing tile	37	2.7	1,850
Slate	0.1–1.0	0.006–0.058	1,600
Clay tile	6.5	0.45	1,900
Aluminum (general and including 33% recycled)	155	8.24	2,700
Bitumen (general)	51	0.38–0.43	
Medium-density fiberboard	11	0.72	680–760
Plywood	15	1.07	540–700
Plasterboard	6.75	0.38	800
Gypsum plaster	1.8	0.12	1,120
Glass	15	0.85	2,500
PVC (general)	77.2	2.41	1,380
Vinyl flooring	65.64	2.92	1,200
Terrazzo tiles	1.4	0.12	1,750
Ceramic tiles	12	0.74	2,000
Wool carpet	106	5.53	
Wallpaper	36.4	1.93	
Vitrified clay pipe (DN 500)	7.9	0.52	
Iron (general)	25	1.91	7,870
Copper (average including 37% recycled)	42	2.6	8,600
Lead (including 61% recycled)	25.21	1.57	11,340
Ceramic sanitary ware	29	1.51	
Paint (water-borne)	59	2.12	
Paint (solvent-borne)	97	3.13	

Photovoltaic (PV) Cell Type	Energy (MJ/m²)	CO_2 (kg/m²)
Monocrystalline (average)	4750	242
Polycrystalline (average)	4070	208
Thin film (average)	1305	67

References

1. European Commission. (2001). Reference document on best available techniques (BAT) in the cement and lime industries. EC.
2. Pimentel, D. and Patzek, T. W. (2005). Ethanol production using corn, swtichgrass, and wood: Biodiesel production using soybean and sunflower. *Natural Resources Research* 14 (1): 65–76.
3. Sudhakar, J. (2012). An input-output analysis to estimate embodied energy of goods. *International Journal of Scientific and Research Publications* 2 (11).
4. Kenneth, B. (2004). General system theory: The skeleton of science. *Management Science* 22: 197–208.
5. Hammond, G. P. and Jones, C. I. (2008). Embodied energy and carbon in construction materials. *Proceedings of the Institution of Civil Engineers-Energy* 161 (2): 87–98.

Bibliography

Allenby, B. (1991). Design for the environment: A tool whose time has come. *SSA Journal* 5–9.

Ayres, R. U. (1989). Industrial metabolism, in *Technology and Environment* (pp. 23–49), National Academy Press, Washington, DC. Ed. Jesse H. Ausubel and Hedy E. Sladovich.

Boulding, K. *General Systems Theory—The Skeleton of Science*, Management Science, 2: 197–208.

Curran, M. A. (1993). Broad-based environmental life cycle assessment. *Environmental Science and Technology* 27 (3): 430–436.

Frosch, R. A. (1992). Industrial ecology: A philosophical introduction. *Proceedings of the National Academy of Sciences, USA* 89: 800–803.

Jackson, T. (ed.). (1993). *Clean Production Strategies: Developing Preventive Environmental Management in the Industrial Economy*, Lewis Publishers, New York.

Keoleian, G. A. and Menerey, D. (1994). Sustainable development by design: Review of life cycle design and related approaches. *Air and Waste: Journal of the Air and Waste Management Association* 44 (55): 645–668.

Odum, E. P. (1993). *Ecology and Our Endangered Life-Support Systems*, 2nd edn. Sinauer Associates, Sunderland, MA.

Society of Environmental Toxicology and Chemistry Foundation for Environmental Education. (1990). *A Technical Framework for Life-Cycle Assessment—SETAC Workshop*, 18 August, Smuggler's Notch, VT.

3

Economics of Sustainable Engineering

3.1 Introduction

In this chapter, we present the mathematical basis of economic analysis for sustainable engineering. We will present engineering economics principles in the simplest terms, as most engineers are exposed to engineering economics principles in detail. In any engineering project, the cost input in the project is most important to analyze the future profitability of the endeavor. Cost terms that are relevant to project economy are fixed cost, variable cost, and incremental costs. Fixed costs are unaffected by increase or decrease of production level. Typical fixed costs include insurance and taxes on facilities, general management and secretarial salaries, and interest on borrowed capital. However, if there is any addition to the existing facilities, fixed costs will change. Variable costs are those that vary with output of production, such as material and labor costs. An incremental cost (or incremental revenue) is the additional cost (or revenue) resulting from increasing the output by one (or more) units. This may be useful in making a decision on whether to increase production by a certain amount, as sometimes incremental cost may prohibit such a decision.

3.2 Total Revenue, Total Costs, and Profit Relations

3.2.1 Total Cost Function

$$TC = C_F + c_v N \tag{3.1}$$

where:

TC = total cost
C_F = fixed costs
c_v = variable cost per unit of production
N = number or volume of production

3.2.2 Total Revenue Function

$$TR = p \cdot N \tag{3.2}$$

where:

TR = total revenue
p = price per unit

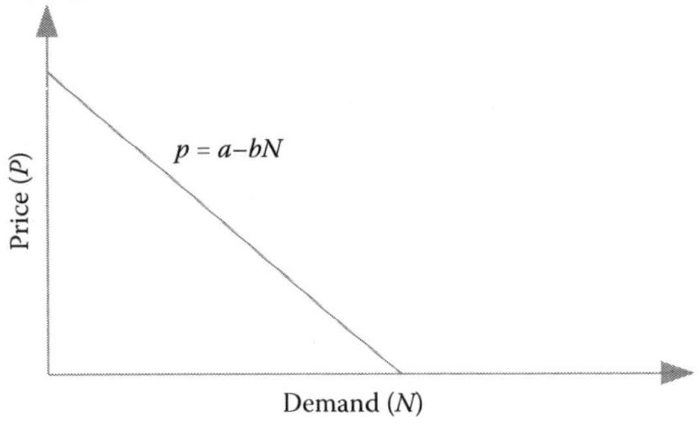

Figure 3.1

Price vs. demand curve.

Normally, there is some relation between the volume sold and price (p) per unit. This may be either linear or nonlinear. Here, we will assume it to be linear:

$$p = a - bN \tag{3.3}$$

where a and b are constants.

This is explained more clearly in Figure 3.1.

We can find the volume (N) of production at which total revenue (TR) will be maximum, which may be the objective of the enterprise. In Figure 3.1, a represents the intercept on the price axis and $-b$ represents the slope of the line.

The total revenue (TR) can be reformulated by replacing the p by $a - b.\ N$ in Equation 3.2:

$$TR = (a - bN)N = aN - bN^2 \quad \text{for } 0 \le N \le \frac{a}{b} \text{ and } a > 0,\ b > 0 \tag{3.4}$$

We can use simple calculus and find the point (volume) of maximum TR by taking the derivative of Equation 3.4 and equating it to zero.

$$\frac{dTR}{dN} = a - 2bN = 0 \tag{3.5}$$

from which we maximize TR, $N^* = a/2b$.

However, the necessary and sufficient condition for maximization of any function is that the second derivative should be negative, and we should check this.

$dTR^2/dN^2 = -2b$, which confirms that $N^* = a/2b$ will maximize the TR function, and N^* represents the demand (sale) volume that will give maximum TR.

3.2.3 Total Profit Analysis

Profit function is developed by deducting TC from the TR function. The profit function is presented in Equation 3.6:

$$\text{Profit} = TR - TC = (aN - bN^2) - (C_F + c_v N) \tag{3.6}$$

To maximize profit and find the sale volume at which we can maximize profit, we can use simple calculus. Taking the derivative of Equation 3.6 and equating to zero, we can find the maximum profit point.

$$\frac{d(\text{Profit})}{dN} = a - 2bN - c_v = 0 \tag{3.7}$$

We get maximum profit point as

$$N^*_{\text{profit}} = \frac{a - c_v}{2b} \tag{3.8}$$

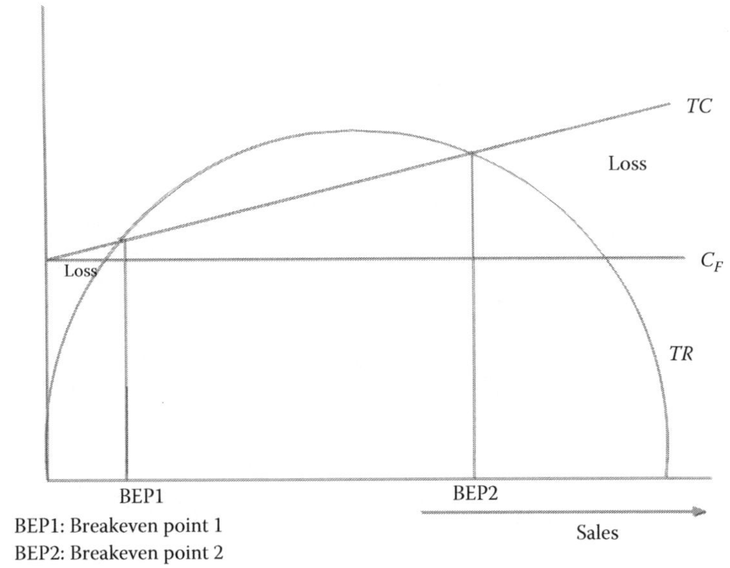

Figure 3.2

Sales vs. break-even point (BEP).

We should also check the sufficiency condition by taking the second derivative of Equation 3.6 as

$$\frac{d^2(\text{profit})}{dN^2} = -2b \tag{3.9}$$

which confirms the sufficiency condition of negativity.

If we examine the N^*_{profit} equation, it shows that to maximize profit, the constant a should be as high as possible, b should be as low as possible, and the variable cost (c_v) should also be as low as possible. These conditions could be very well simulated in a spreadsheet (Figure 3.2).

3.3 Life Cycle Cost Analysis

In engineering practice, we often encounter the life cycle cost. This refers to a summation of all costs related to a product, structure, system, or service during its life span from birth to grave. However, the term *life cycle* has become vague, and we try to present it more clearly here. We are concerned with the *physical life cycle*. The product life cycle is a sequence of transformations in materials and energy, including extraction and processing of materials, product manufacture and assembly, distribution, use, and recycling of product materials. We are concerned with life cycle cost analysis here. Similarly, the process life cycle is defined as a sequence of transformations in materials and energy that includes extraction and processing of materials used for process equipment and supplies, process operation and control, equipment cleaning and maintenance, and waste disposal or recovery. We will discuss both product life cycle and process life cycle of green design and manufacturing for sustainability.

However, there is another term, *business life cycle*, which is normally understood as a sequence of activities including the creation of a product concept, its development, launch, production, maintenance, maturity, revaluation, and renewal in the form of a next generation, which is similar to the product life cycle. Manufacturing a product may involve a variety of different processes, and they must be studied together. Both physical life cycle and business life cycle are presented in Figures 3.3 and 3.4.

The timescale of the business life cycle depends on the rate of technological development and market conditions. We know that in the manufacturing sector new and technologically advanced machine tools are coming onto the market, from manual machine tools, to computer numerically controlled (CNC) machine tools, to energy-saving CNC machine tools, each of which has a very different production rate. Sometimes product obsolescence may also affect the business life cycle. As we have discussed, manufacturers can easily find the maximum revenue point as well as the maximum profit point and make business decisions

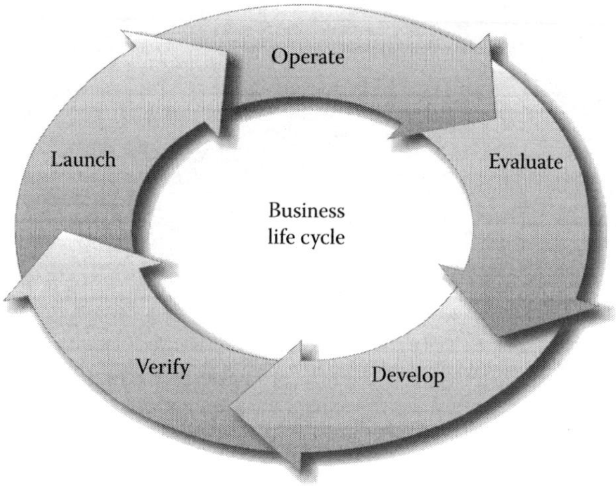

Figure 3.3

Business life cycle.

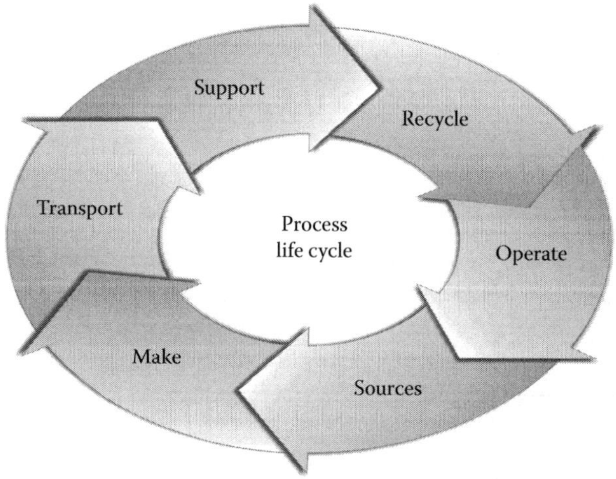

Figure 3.4

Process life cycle.

accordingly. The *sustainability* of the business life cycle depends on the ability of a company to innovate by developing and delivering product extensions or upgrades that serve the needs of its consumers. The sustainability of the physical cycle depends upon the ability of various groups, including the manufacturer, to use energy and materials inputs cost-effectively and to recover or refurbish discarded products. In fact, many new terms have been coined, such as *cradle to grave*, *reincarnation*, and *end of life*.

3.4 Efficiency and Life Cycle Cost Calculation of a Pump

3.4.1 Beyond Purchase Price

Most customers have a good understanding of the total cost of ownership. After purchasing a car or home or appliance, they realize that with the initial purchase price come the added costs of maintenance, taxes or fees, repair of parts, and operating costs (energy). Most equipment purchases should be analyzed on this basis. They need to be installed, sometimes repaired, and always maintained. They also require electricity (or another motive power such as air) to run.

Also, as an example, industrial pumps have a long life (in some cases 50 years or more). This means that a poor decision can haunt a user for many years. For this reason, it is critical to consider a pump's total life cycle cost before making this decision.

3.4.1.1 Calculating Total Life Cycle Cost

The life cycle cost (LCC) for a pump is calculated as follows:

$$LCC = C_{ic} + C_{in} + C_e + C_o + C_m + C_s + C_{env} + C_d \quad (3.10)$$

where:

C_{ic} = initial costs, purchase price (pump, system, pipe, auxiliary services)
C_{in} = installation and commissioning cost (including training)
C_e = energy costs (predicted cost for system operation, including pump driver, controls, and any auxiliary services)
C_o = operation costs (labor cost of normal system supervision)
C_m = maintenance and repair costs (routine and predicted repairs)
C_s = downtime costs (loss of production)
C_{env} = environmental costs (contamination from pumped liquid and auxiliary equipment)
C_d = decommissioning/disposal costs (including restoration of the local environment and disposal of auxiliary services)

Each of these values will vary slightly in significance from application to application. In industries such as the petrochemical or pharmaceutical industry, downtime costs are substantial, which would rule out pumps that require frequent maintenance or are prone to catastrophic failure. In other cases, perhaps the local code requires special fees for the disposal of certain materials. In that case, added costs would be applied for environmental disposal and/or decommissioning.

Unfortunately, many pump users lack the knowledge or foresight to recognize the significance of LCC calculations. All too often, the final decision is dictated by one of the more insignificant costs of total LCC: purchase price.

On the date of purchase, the quoted price for a pump may seem to be the most important matter at hand. In actuality, the initial purchase price is only a fraction of the pump's total LCC. Figure 3.5 represents a pump's total LCC over a period of 7 years. The figure is based on an internal gear pump operating continuously for a period of 7 years and does not account for increases in electricity prices over that period, which would further decrease the significance of the initial purchase price.

Obviously, these percentages vary from pump to pump and from application to application, but the energy cost typically remains the highest single source of cost over a pump's life.

3.4.2 Reduced Energy Costs

Lengthy papers have been devoted to savings on maintenance and parts. While each should be included in LCC calculations, energy costs remain the largest contributor, and yet are often overlooked.

Energy savings can be obtained in many different ways. Specifically to pumps, the three most common means are to alter the system to decrease the input power requirement, to implement controls to maximize pump efficiency, and to replace an inefficient pump with a more efficient model for the application.

3.4.3 System Changes

Unfortunately for system designers, too often a tight budget can trump good foresight. Narrow pipes running through a maze of twists and turns increase the system head required and, in turn, the pump power required. Add to that any changes in viscosity, product buildup, or corrosion on the pipe walls, and the

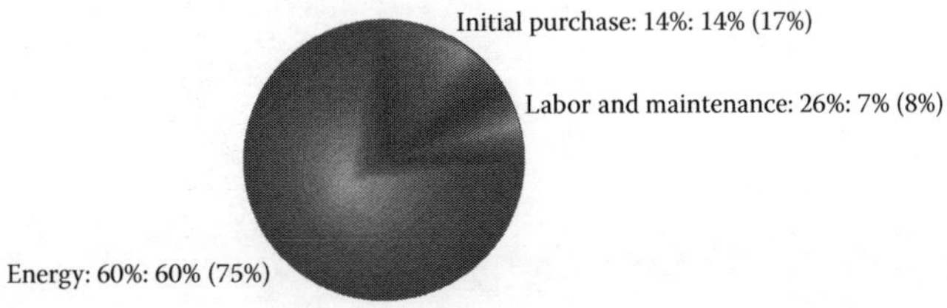

Figure 3.5
Life cycle cost of a pump.

power requirement can climb further. Too often, this is only taken into consideration when looking at the size and initial purchase price of a motor. Motors, like pumps, carry the bulk of their price in energy costs over the life of the motor.

For new installations, a little foresight can save thousands, but for existing installations, this may not be an option.

3.4.4 Energy Savings through Pump Control

Advancements in control technology, decreases in controller prices, and increases in electricity rates have made pump control a popular means of obtaining energy savings. Examples include

- The use of variable-speed devices (variable frequency drives [VFDs]) to control pump flow rate or maintain pressure. VFD use is becoming more and more common for use with centrifugal pumps to keep them running at their best efficiency point (BEP).
- Power monitors are used to track motors' power usage, and can shut them down if power deviates from the desired range. They can be used to keep pumps from running dry or to keep them from running at high pressure.
- Programmable logic controllers (PLCs) or similar logic controls are used to monitor flows, power consumption, temperature, and so on. These are the most complex and costly controls, but allow a few operators to monitor multiple pumps and systems.

Unlike piping and system design, controls are often easy to implement on established pumps. For example, while this may seem to be a simple enough concept, many users of unloading pumps leave the pumps running well after the rail car or tanker is empty. This is an obvious waste of energy, not to mention hard on the pump. Installation of a power monitor takes less than hour and often pays for itself in energy savings within the first year, as well as providing added insurance against a pump failure.

3.4.5 Maximize Pump Efficiency

When selecting a pump, it is critical that all the service conditions are known. Changes such as pressure or viscosity variations can dramatically affect a pump's performance and efficiency. Centrifugals are designed to run on thin liquids at a set operating point. With different impellers and auxiliary controls, they can be configured to operate on moderate-viscosity liquids at varying conditions, but are highly inefficient. They also become inefficient at higher viscosities and low flow/high head conditions.

Pump efficiency is calculated as follows:

$$\text{Mechanical Efficiency} = \frac{\text{Differential Pressure (psi)} \times \text{Capacity (GPM)} \times 100}{\text{Horsepower (bhp)} \times 1715} \quad \text{(U.S. units)} \tag{3.11}$$

The following tables compare two applications. The first is for #2 fuel oil at 100 GPM and 100 ft. The second table is for the same fuel oil, but at higher pressure. The third table is for thicker liquid, in this case lube oil.

Conditions: #2 Fuel Oil (3 cP), 100 GPM, 100 ft

Pump	bhp Required (hp)	Efficiency (%)	Yearly Energy Cost ($)
End suction centrifugal	3.5	63	1418
Rotary PD—internal gear	3.0	72	1215

Conditions: #2 Fuel Oil (3 cP), 100 GPM, 200 ft

Pump	bhp Required (hp)	Efficiency (%)	Yearly Energy Cost ($)
End suction centrifugal	7.8	65	3159
Rotary PD—internal gear	6.8	79	2633

Conditions: 80 W Gear Oil (150 cP), 100 GPM, 200 ft

Pump	bhp Required (hp)	Efficiency (%)	Yearly Energy Cost ($)
End suction centrifugal	13.8	37	5589
Rotary PD—internal gear	7.0	72	2835

As viscosity, flow, and pressure vary, so will power and efficiency. These tables merely illustrate the importance of selecting the appropriate pump for the application and making sure that energy savings are included in the selection process. The most worrying aspect is that application examples like these are all too commonplace, and in a plant with hundreds of pumps, this equates to hundreds of thousands of dollars lost each year due to ignorance or nearsightedness at the date of initial purchase.

3.4.6 LCC Educated Pump User

Here are some steps to take toward reducing pump energy consumption:

1. Run LCC calculations for all new pump installations. Existing pumps should also be examined.
2. Choose "right size" pumps for application. Avoid running centrifugals off their "best efficiency point," and do not misapply them.
3. Install power monitors to track electricity usage and identify poorly performing pumps. This also helps to avoid running dry or against "dead head" conditions.

We can see that LCC can be an effective engineering tool for providing decision support in the design and procurement of major open systems, infrastructure, and so on. LCC overcomes many of the shortcomings of traditional cost accounting, and gives useful insights to management. LCC has emerged as the main engineering analytical tool for environmental or sustainability purposes.

3.5 Mechanics of Time Value of Money

It is common knowledge that a dollar today is worth more than a dollar in the future because of the profit (interest) it can earn. Equity capital earns profit, whereas a borrower has to pay interest on the borrowed capital. Therefore, the value of money changes over time, and the term *time value of money* makes sense. Interest on borrowed capital can also be termed the *cost of money*. When compounded continuously, the present worth of money, P, converts to future worth, F, after n years at the nominal rate r per year: $F=Pe^{nr}$. Similarly, we can calculate present worth from future worth by $F=Pe^{-rn}$.

We have assumed r to be constant, which may not be the case in the real financial world. Uniform series (annuities) A relations for continuous compounding are also developed:

$$F = A\left[\frac{e^{rn}-1}{e^{r}-1}\right], P = \left[\frac{e^{rn}-1}{e^{rn}(e^{r}-1)}\right] \tag{3.12}$$

3.5.1 Relationships for Discrete Compounding

In business and the real world, some cash flows are based on discrete compounding, meaning that the interest is compounded at the end of a finite period, such as a month or a year. The interest rate, i, for discrete compounding converts present worth, P, to future worth, F, as

$$F = P(1+i)^{n}, \text{ and for uniform series (annuities), } A,$$

$$F = A\left[\frac{(1+i)^{n}-1}{i}\right], \quad \text{whereas } P = A\left[\frac{(1+i)^{n}-1}{i(1+i)^{n}}\right] \tag{3.13}$$

In the financial world, we often face the problem of a uniform (arithmetic) gradient of cash flow. This could be explained by some cash flows involving receipts or expenses that are projected to increase or decrease by a uniform amount (G) per period, thus forming an arithmetic gradient series; see the table below.

End of Period	Cash Flows
1	A
2	$A+1(G)$
3	$A+2(G)$
⋮	⋮
$n-1$	$A+(n-2)G$
N	$A+(n-1)G$

This cash flow could also be depicted graphically. We can convert all cash flows to present worth, P, and add them together to formulate the equations

$$P=G\left\{\frac{1}{i}\left[\frac{(1+i)^n-1}{i(1+i)^n}-\frac{n}{(1+i)^n}\right]\right\} \tag{3.14}$$

and

$$F=G\left\{\frac{1}{i}\left[\frac{(1+i)^n-1}{i}-n\right]\right\} \tag{3.15}$$

We can convert cash flows in arithmetic gradient series to uniform series, A:

$$A=G\left[\frac{1}{i}-\frac{n}{(1+i)^n-1}\right] \tag{3.16}$$

In the market, there is also the problem of cash flows of geometric gradient, where each payment or disbursement at end of year increases or decreases by a certain percentage r.

3.5.2 Geometric Gradient: Constant Percentage Increment

Let C = first end-of-period payment of an incrementally increasing series, for example first-year-end salary.

r = rate of growth or decline of the series. It could be positive or negative.

The cash flow in a geometric gradient series can be depicted as in Figure 3.6.

The present worth (P) of all the cash flow in the diagram can be written as

$$P=C\left[\frac{1}{1+i}+\frac{1+r}{(1+i)^2}+\cdots+\frac{(1+r)^{n-2}}{(1+i)^{n-1}}+\frac{(1+r)^{n-1}}{(1+i)^n}\right] \tag{3.17}$$

To write the equation in a more presentable form, it is multiplied by $((1+r)/(1+r))$ on both sides, giving

$$P=\frac{C}{(1+r)}\left[\frac{1+r}{1+i}+\frac{(1+r)^2}{(1+i)^2}+\cdots+\frac{(1+r)^{n-1}}{(1+i)^{n-1}}+\frac{(1+r)^n}{(1+i)^n}\right] \tag{3.18}$$

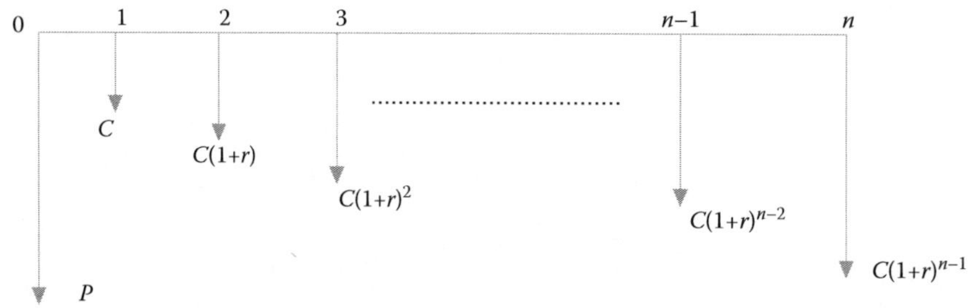

Figure 3.6

Cash-flow diagram of geometric gradient.

There are three possible scenarios to present this equation:

Scenario 3.1

$r > i$; when rate of growth is greater than the interest rate offered by financial institutions on deposits. Let us find a derived variable

$$1+w=\frac{1+r}{1+i} \text{ and substitute it into Equation 3.18 to obtain}$$

$$P=\frac{C}{(1+r)}\left[(1+w)+(1+w)^2+\cdots+(1+w)^{n-1}+(1+w)^n\right] \tag{3.19}$$

Or it can be written as

$$P=\frac{C}{(1+r)}(1+w)\left[1+(1+w)+\cdots+(1+w)^{n-2}+(1+w)^{n-1}\right] \tag{3.20}$$

The summation of the terms in brackets can be obtained using algebra and written as

$$P=\frac{C}{(1+i)}\left[\frac{(1+w)^n-1}{w}\right] \quad \text{for } r>i \text{ and } w=\frac{1+r}{1+i}-1 \tag{3.21}$$

Scenario 3.2

when $r<i$ and $\frac{1}{1+w}=\frac{1+r}{1+i}$

The present worth for scenario 3.2 is

$$P=\frac{C}{(1+r)}\left[\frac{(1+w)^n-1}{w(1+w)^n}\right] \quad \text{for } r<i \tag{3.22}$$

Scenario 3.3

when $r=i$

The present worth value is simplest:

$$P=\frac{Cn}{1+r} \quad \text{for } r=i \tag{3.23}$$

Example 3.1

A municipal power plant in Yonkers is expected to experience an increase in net profit of 10% per year for the next 30 years. The base net profit this year is \$1 million. Interest is 6% on borrowed funds from a local bank.

a. Estimate the present worth of this income to the power plant.
b. If current long-term debt now stands at \$10 million, how much additional borrowing could be secured by the estimated present worth of the net profit if lenders will lend on 100% of that value?

Solution

$n=30$ years, $r=0.1$, and $i=0.06$, $r>I$, therefore $w=1.1/1.06-1=0.0377358$

$$\textbf{a. } P=\frac{C}{1+i}\left[\frac{(1+w)^{20}-1}{w}\right]=\frac{10^6}{1.06}\left[\frac{(1.0377358)^{20}-1}{0.0377358}\right]=\$27,441,693 \tag{3.24}$$

b. Additional borrowing possible = \$27,441,693 − \$10,000,000 = \$17,441,693.

Example 3.2

Assume that the income for an engineering project increases as a geometric gradient, but the costs remain constant. As a result, the net income could begin as a negative value (loss) dominated by high costs, but subsequently rise through zero into the positive range (profit) as income improves. Find the present worth of the net income with $i=10\%$. The table below shows yearly net income.

Year	Net Income (Income Costs (\$))
1	−50,000
2	−35,000
3	−17,750
4	+2090
5	24,900
⋮	⋮
25	2,712,520

Solution

In this problem, base amount of cost (*C*), income, and growth rate (*r*) are unknown. The constants *r* and *C* may be found by the following steps.

1. List the net income for the first three years.
2. Find the change in net income from years 1 to 2, and from years 2 to 3, and list these under Δ Net, as shown in the table below.
3. Find the change in the change and list under ΔΔ net as shown.
4. Growth rate $r=\Delta\Delta/\Delta$.
5. Base rate $C=\Delta/r$. Here, C is the first year's income.

Year	Net Income (\$)	Δ Net	ΔΔ Net
1	−50,000	15,000	2250
2	−35,000	17,250	
3	−17,750		

Then, $r=2250/15,000=0.15$, and $C=15,000/0.15=100,000$.
From this information, the income and cost values can be reconstructed as below.

Year	Income (\$)	Cost (\$)	Net Income (\$)
1	100,000	−150,000	−\$50,000
2	115,000	−150,000	−35,000
3	132,250	−150,000	−17,750
4	152,090	−150,000	+2090
5	174,900	−150,000	+24,900
⋮	⋮	⋮	⋮
25	100,000 × 1.1524	−150,000	+2,712,520

Now, we can find the present worth of the net income with $i=10\%$, $r=15\%$, and $C=100,000$.

$$r>i, \text{ so } w=1.15/1.11=0.04545$$

$$P=\frac{C}{(1+i)}\left\{\frac{(1+w)^n-1}{w}\right\}=\frac{100,000}{1.10}\left[\frac{1.04545^{25}-1}{0.04545}\right]=+4,076,000 \quad (3.25)$$

Present worth of cost is calculated as

$$PC=150,000\left(P/A,10\%,25\right)=1,362,000 \quad (3.26)$$

Hence, at end of year 0, the net income from the project is worth $P+PC=2,714,000$.

3.5.3 Nominal and Effective Interest Rate

Very often the interest period, or time between successive compoundings, is less than 1 year (e.g., daily, weekly, monthly, quarterly, or 6-monthly or semiannually). It has become standard in financial markets to quote interest rates on an annual basis, followed by the compounding period if this differs from 1 year in length For example, if interest is quoted at 6%, it should be understood as 6% per year. However, if it is quoted as 6% semiannually, then 6% is to be understood as the nominal interest rate, and effective interest would be different and greater than 6%. So, effective interest, i_e, is calculated as

$$i_e=(1+i/m)^m-1 \quad (3.27)$$

where m is number of compoundings per year, for example, $m=2$ for semiannually, $m=4$ for quarterly, $m=52$ for weekly, and $m=360$ for daily. We can see from Equation 3.27 that $i_e>i$ when $m>1$.

This introduction to engineering economy has been brief, covering as much as is customary to complete one semester of the full subject.

Example 3.3

State regulation requires that the Indian Point nuclear power plant must deposit certain uniform monies in a sinking fund every year so that it can accumulate $500 million to decommission the plant in 25 years. Find the uniform yearly deposit for decommissioning the plant. Let interest rate, *i*, be 5.5% per year.

Solution

Continuous Compounding:

$$\int_0^{30} A\cdot e^{-0.05t}dt \rightarrow 15.53739679703140342.1A=15.537A \quad (3.28)$$

$$300.e^{1.5}=66.939 \quad (3.29)$$

Discrete Compounding:

$$A=\frac{66.939}{15.537} \quad A=300\cdot\frac{0.05}{(1+0.05)^{30}} \quad A=4.308 \quad (3.30)$$

Example 3.4

In a renewable energy project selection problem, two options for energy-saving devices are available. Both have a useful life of 10 years, and the initial investment is $1 million in both cases. Device X is expected to save $500,000 per year, and device Y is likely to save $650,000 annually. However, device X has $25,000 recyclable value at the end of its useful life, but Y's recyclable value is zero. It is expected that the company will finance the project by borrowing from a bank at a rate of 5%. Which device should be selected?

$$\text{NPVX}(i):=-10^6+500,000\left[\frac{(1+i)^{10}-1}{i\cdot(1+i)^{115}}\right]+25,000(1+i)^{-10} \quad (3.31)$$

$$\text{NPVY}(i) := -10^6 + 650{,}000\left[\frac{(1+i)^{10}-1}{i\cdot(1+i)^{10}}\right] + 0(1+i)^{-10} \tag{3.32}$$

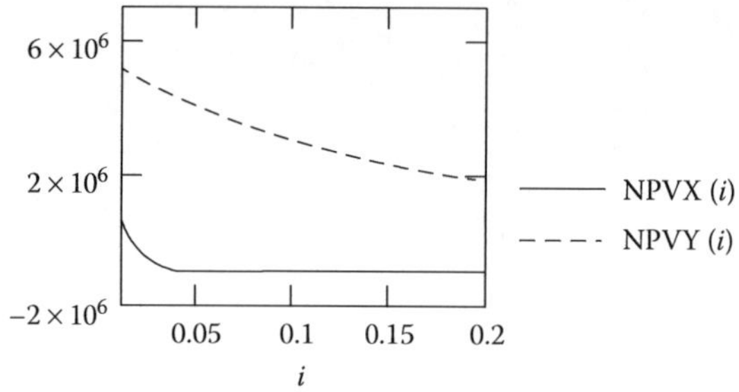

As we see, device Y is always preferable to device X. We should also examine this when the useful life for device X = 15 years and for device Y = 12 years.

$$\text{NPVX}(i) := -10^6 + 500{,}000\left[\frac{(1+i)^{15}-1}{i\cdot(1+i)^{115}}\right] + 25{,}000(1+i)^{-15} \tag{3.33}$$

$$\text{NPVY}(i) := -10^6 + 650{,}000\left[\frac{(1+i)^{12}-1}{i\cdot(1+i)^{12}}\right] + 0(1+i)^{-12} \tag{3.34}$$

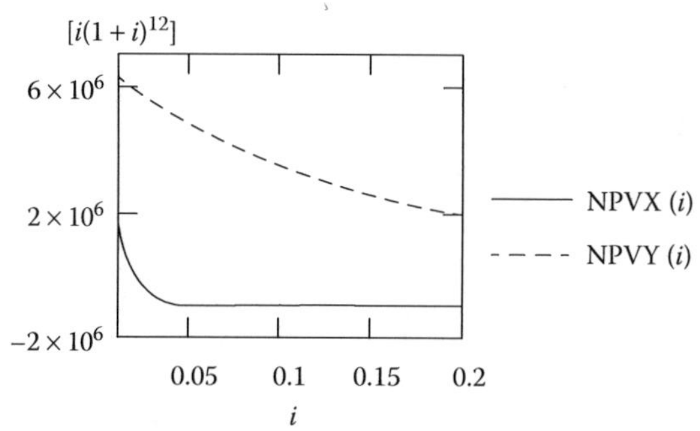

It is obvious again that device Y is preferable at the interest rate considered here.

3.6 Economic Project Selection

Benefit–cost analysis used in engineering projects collects all the relevant costs and benefits, and a ratio of benefit to costs is estimated. In this analysis, the time value of monies must be considered to account for the timing of cash flows (or benefits) occurring after the start of the project. The conventional B/C ratio with present worth (*PW*) is

B = *PW* of benefits
C = *PW* of costs
It is normally written as

$$B/C = \frac{PW(B)}{I - PW(MV) + PW(O\,\&\,M)} \tag{3.35}$$

where:

B = benefits of the proposed project
I = initial capital investment in the proposed project
MV = market value at end of useful life (salvage value)
$O\&M$ = operation and maintenance costs

We can see that $PW(MV)$ in the denominator is a benefit at end of useful life, and a modified B/C ratio with *PW* is presented as

$$B/C = \frac{PW(B) - PW(O\&M)}{I - PW(MV)} \quad (3.36)$$

The numerator now represents the present worth of the benefits minus the present worth of operating and maintenance costs, while the denominator includes only initial value minus salvage value. The above relations of B/C ratio can also be written in terms of annual worth (*AW*) as

$$B/C = \frac{AW \text{ of benefits}}{AW \text{ of costs}} \quad (3.37)$$

$$B/C = \frac{AW(B)}{AW(I) - AW(Salvage + AW(O\&M)} \quad (3.38)$$

The modified B/C ratio in *AW* is presented as

$$B/C = \frac{AW(B) - AW(O\&M)}{AW(I) - AW(Salvage)} \quad (3.39)$$

The importance of B/C lies in assessing overall benefits of the project over costs; that is, for the viability of the proposed project, the B/C ratio must be greater than 1. It includes all the costs and benefits of the proposed project in its life cycle from inception to the end of useful life. It is similar to life cycle analysis, so it may also be termed *life cycle benefit–cost analysis* (LCBCA). Therefore, we can state that LCBCA is useful for *sustainable* development. It emphasizes that for any public project to be successful in terms of sustainable development, its life cycle benefits must outweigh its life cycle costs. It gives a holistic view of costs and benefits, including planning, design, operation, maintenance, inspection, repair, and disposal.

Example 3.5

We will examine Example 3.3, presented above, by the LCBCA method at $i = 5\%$.

For device X: *PW* of cost = $\$10^6$

PW of benefit:

$$\text{NPVXB}(i) := 500{,}000\left[\frac{(1+i)^{10} - 1}{i \cdot (1+i)^{10}}\right] + 25{,}000(1+i)^{-10} \quad (3.40)$$

$$\text{NPVXB}(0.05) = 3.876 \times 10^6 \quad (3.41)$$

represents the *PW* of benefits of device X.

B/C = $3.876 \times 10^6/10^6 = 3.876$ for device X.

We can similarly calculate the B/C ratio for device Y.

$$\text{NPVBY}(i) := 650{,}000\left[\frac{(1+i)^{10} - 1}{i \cdot (1+i)^{10}}\right] + 0(1+i)^{-10} \quad (3.42)$$

$$\text{NPVBY}(0.5) = 5.019 \times 10^6 \quad (3.43)$$

As initial investment is same in both devices, we can calculate the B/C ratio for device Y as

B/C = $5.019 \times 10^6/10^6 = 5.019$

We can see that, even from B/C ratio considerations, device Y is preferable to device X.

We can also draw graphs for the B/C ratio and study the problem as shown below.

$$\text{BCX}(i) = \frac{500{,}000\left[\frac{(1+i)^{10} - 1}{i \cdot (1+i)^{10}}\right] + 25{,}000(1+i)^{-10}}{10^6} \quad (3.44)$$

$$\text{BCY}(i) = \frac{650{,}000\left[\frac{(1+i)^{10}-1}{i\cdot(1+i)^{10}}\right]+0(1+i)^{-10}}{10^6} \tag{3.45}$$

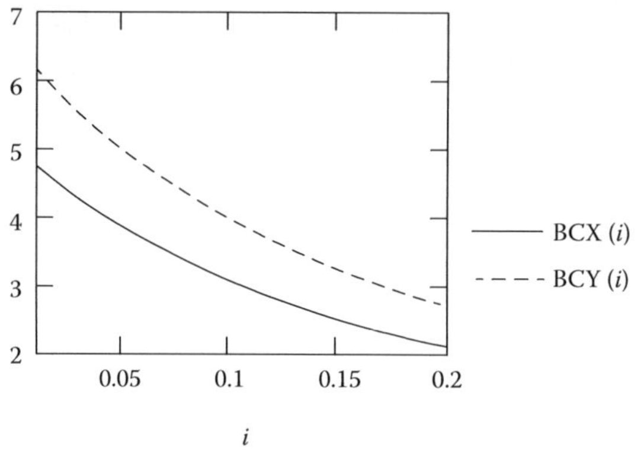

We can see from the B/C ratio graph that device Y remains preferable in the range of borrowing rates presented above.

We have just presented the economic analysis in terms of dollars and cents, but for sustainability, this is no longer a sufficient argument. Greenhouse gas (GHG) emissions are most important for sustainability, as they affect global warming.

3.7 Benefit–Cost Relation for Recycling

The cost function of recycling is defined as

$$C_o = C_d + C_s + C_r + D_c \tag{3.46}$$

where:

C_o = cost of recycling
C_d = cost of disassembly
C_s = cost of shredding
C_r = cost of material recovery
D_c = cost of dumping

The cost of disassembly depends on the kind of fasteners employed in assembly. Naturally, joints will take the same time to disassemble. The cost of disassembly is estimated as

$$C_d = C_{db}\sum_{j=1}^{n}\sum_{i=1}^{m-g} T_{ij} \tag{3.47}$$

where:

C_{db} = hourly cost of disassembly
m = number of fasteners of the same type in the product
n = number of different types of fastener (screws, etc.)
g = number of fasteners that contact parts of the same or compatible materials

The shredding cost is estimated as

$$C_s = C_{sh}(S_t)(W_s) \tag{3.48}$$

where:

C_{sh} = hourly shredding cost
S_t = time to shred 1 ton of material
W_s = weight of shredding materials (tons)

The cost of recovery (C_r) naturally increases with amount and type of material recovered. For manual separation to be carried out, there must be significant value retained in the recycled product in order for the separation to be economically feasible. This scenario is applicable to most other products constructed from subassemblies consisting of a variety of materials. The West European hourly rates and material prices bulletin shows the amount of material (in grams) that has to be removed per minute for recycling to be cost neutral for manual disassembly. Although the numbers are in grams per minute, a reasonable approximation can be made in other units; for example, at least 5.1 g of silver needs to be removed for the process to be economically feasible.

Chen et al. [1] present the cost of material recovery equation as

$$C_r = \sum_{i=1}^{k} R_{ci} . W_{ri} \tag{3.49}$$

where:

R_{ci} = cost of material recovery per ton of type i material (i may be steel, plastics, etc.)
W_{ri} = weight of i material recovered
k = number of different types of material in the product

The cost of dumping (D_c) is estimated as

$$D_c = d_c.(W_d) \tag{3.50}$$

where:

d_c = the cost of dumping 1 ton of material (it may include land, labor, and materials, such as plastic or iron sheet, or any other kinds of chemicals used in dumping)
W_d = weight in tons of dumped material

Cost-benefit analysis is a very useful technique in economic analysis of projects, particularly for government agencies, such as airport, highway, or power generation projects. All projects are studied for maximization of social benefits; however, this is usually decided by other concerns than just economic considerations. The benefit–cost (B/C) ratio should always be greater than 1, and the project with maximum B/C should be selected.

Chen et al. [1] have selected a B/C method for analyzing the disassembly design, termed the *recycling B/C method*:

$$R_{b/c} = \frac{\textit{Total benefit}}{\textit{Total cost}} = \frac{R_p + R_m + E_r}{C_d + C_s + C_r + D_c} \tag{3.51}$$

where:

R_p = revenue from used parts
R_m = revenue from recovered materials
E_r = energy-saving benefit of emission reduction

Chen et al. [1] have presented an equation for the benefit of emission reduction due to using recycled parts or materials. The benefit of energy saving

$$E_r = Es_i \sum_{i=1}^{y} \sum_{j=1}^{h} Em_j \left(Ec_j \right) \tag{3.52}$$

where:

Es_i = energy saving of recycled material of type i
Em_j = type j emission reduction
Ec_j = extemality cost of type j emission
h = number of types of emissions, such as SO_2, particulates, etc.
x = number of reusable parts disassembled in a product
y = number of types of recycled material in a product

The revenue from used parts (R_p) is calculated as

$$R_p = \sum_{i=1}^{x} Pu_i \tag{3.53}$$

where Pu_i = revenue from usable parts

The revenue after selling the recovered material is calculated as

$$Rm = \sum_{i=1}^{y} Pm_i \tag{3.54}$$

where Pm_i = revenue from type i recycled material.

The profit obtained from recycling is calculated as

recycling profit = total recycling revenue − total recycling cost, that is,

$$RP = (R_p + R_m + E_r) - (C_d + C_s + C_r + D_c) \tag{3.55}$$

where RP is recycling profit.

The break-even point (BEP) can also be calculated to find the amount of materials (tons) that must be recycled for total revenue and total cost to be equal, after which profit will result to the recycling company. This is important for the recycling industry to know. Obviously, lower BEP is better from the perspective of business viability. We can also examine RP; as the recycling cost decreases, so profit will rise, and for this to happen, costs of disassembly, shredding, dumping, and material recovery should be as low as possible. It is also possible to increase profit by raising the total recycling revenue, which can happen if there is greater acceptability of recycled parts and materials in industry.

3.8 Selection of Renewable Energy over a Time Period

The energy payback for renewable energy systems is commonly measured in terms of time to break even (or payback time) t_b and energy return on investment (EROI). Renewable energy not only meets our increasing needs for power on the earth but also reduces CO_2 in the air. It is likely to reduce pollution of land, water, and air as well, and in turn reduce global warming. Here, we present the method of calculation of the break-even period for various renewable energy systems, such as wind power, solar power, hydropower, natural gas power generation, and nuclear power. The simple payback period can be calculated as $t_B = -P/A$, where P is the initial investment in dollars and A is the incremental annual savings in dollars per year. The payback time is expressed as number of years. This is fine for dollars, but in international comparisons, it should be adjusted to take account of fluctuating exchange rates. We should also note that due to inflation, the value of the dollar itself may vary.

However, Gutowski [2] has suggested a different analytical tool for such comparison. He defines the terms as:

E_M = the energy investment to manufacture the system

e = annual energy produced (power)

t_L = total useful life of the plant

Payback time (break-even time) $t_B = E_M/e$ and

$$\text{EROI} = \frac{e \cdot t_L}{E_M} = \frac{e \cdot t_L}{e \cdot t_B} = \frac{t_L}{t_B}$$

Both numerator and denominator are in years. Gutowski [2] include a technological development factor in this analysis. Let there be d_0 growth of technology in year 1. The energy cost of for this is $E_M d_0$. In year 1, these systems produce energy ed_0, and an additional rd_0 systems are built. In later years, r is the annual growth rate as a fraction of the cumulative number of systems built.

3.9 Cost-Estimating Technique

The objective of this topic is to present an assortment of methods for estimating important factors in an engineering economy analysis. Whenever an engineering economic analysis is performed for a major capital investment, the cost-estimating effort for that analysis should be an integral part of a comprehensive planning and design process requiring the active participation of not only engineering designers but also personnel from marketing, manufacturing, finance, and top management. Results of cost estimating are used for a variety of purposes, including:

1. Providing information used in setting a selling price for quoting, bidding, or evaluating contracts
2. Determining whether a proposed product can be made and distributed at a profit (for simplicity, price = cost + profit)
3. Evaluating how much capital can be justified for process changes or other improvements
4. Establishing benchmarks for productivity improvement programs

There are two fundamental approaches for cost estimating: the *top-down* approach and the *bottom-up* approach. The top-down approach basically uses historical data from similar engineering projects to estimate the costs, revenues, and other data of the current project, modifying these data for changes in inflation or deflation, activity level, weight, energy consumption, size, and other factors. This approach is best used early in the estimating process, when alternatives are still being developed and refined.

The bottom-up approach is a more detailed method of cost estimating. This method breaks down a project into small, manageable units and estimates their economic consequences. These small unit costs are added, together with other types of costs, to obtain an overall cost estimate. This approach usually works best when the detail concerning the desired output (a product or service) has been defined and clarified.

Example 3.6: Estimating the Cost of a College Degree at Manhattan College

Top-Down Approach

A top-down approach would take the published cost of a 4-year degree at Manhattan College or a similar university and adjust it for inflation and extraordinary items that an incoming student might encounter, such as fraternity/sorority membership, scholarships, or tutoring. For example, suppose that the published cost of attending Manhattan College is \$15,750 for the current year. This figure is anticipated to increase at a rate of 6% per year, and includes full tuition and fees, dorm, and a weekly meal plan. Not included are the costs of books, supplies, and other personal expenses. For our initial estimate, these expenses are assumed to remain constant at \$5000 per year.

Year	Tuition, Fees, Room, and Board (\$)	Other Expenses (\$)	Total Estimated Cost Per Year (\$)
1	15,750 × 1.06 = 16,695	5000	21,695
2	16,695 × 1.06 = 17,697	5000	22,697
3	17,697 × 1.06 = 18,759	5000	23,759
4	18,759 × 1.06 = 19,885	5000	24,885
			Grand total \$93,036

Bottom-Up Approach

In the bottom-up approach, we would first break down anticipated expenses into typical categories, as follows:

Tuition and Fees	Books and Supplies	Living Expenses	Transportation
Tuition	Books	Rent	Fuel
Activities fees	Duplication	Food	Maintenance on car
Membership	Supplies	Clothing	Insurance
Medical insurance	Computer rental	Recreation	Traffic ticket
Lab fee	Software	Utilities	
Software fees			

Tuition and fees can be estimated fairly accurately in each year, as can books and supplies. The other two categories, living expenses and transportation, are probably more dependent on lifestyle. For example, whether you own and operate a car and live in a "high-end" apartment off-campus can dramatically affect the estimated expenses during your college years.

Cost and Revenue Structure

Categories of costs and revenues that are typically needed in an engineering project are: (i) capital investment (fixed and working), (ii) labor costs, (iii) material costs, (iv) maintenance costs, (v) property taxes and insurance, (vi) overhead costs, (vii) revenues or return, (viii) disposal costs, (ix) quality (and scrap) costs, (x) market (or salvage) values, and so on.

3.10 Estimating Techniques

The estimating technique, together with the detailed cost and revenue data, is used to develop individual cash-flow estimates and the overall net cash flow for each alternative. Normally, estimates show future cost projections—they do not produce exact data about the future, which is virtually impossible. Cost and revenue estimates can be classified according to detail, accuracy, and their intended use, as follows:

1. Order-of-magnitude estimates: used in the planning and initial evaluation stage of a project. They typically provide accuracy in the range of ±30% to 50%, and they are developed through semiformal means such as conferences and questionnaires.
2. Semidetailed, or budget, estimates: used in the preliminary or conceptual design stage of project. Budget (semidetailed) estimates are compiled to support the preliminary design effort and decision-making during this project period. Their accuracy usually lies in the range of ±15%.
3. Definitive (detailed) estimates: used in the detailed engineering/construction stage of a project. Detailed estimates are used as a basis for bids and to make detailed design decisions. Their accuracy is ±5%. They are made from specifications, drawings, site surveys, vendor quotations, and in-house records.

As estimates become more detailed, accuracy typically improves, but the cost of making the estimates increases dramatically. There are a number of techniques available in the literature for estimating costs [3]. However, depending on the type of industry and production process, one technique may be preferred over others. In sustainable engineering, the reduction of costs, including material, power, and disposal, is a prime objective. The attempt here is to present them in a fashion that is understandable and usable in sustainable engineering studies.

3.10.1 Sources of Estimating Data

Information sources useful in cost and revenue estimating are too numerous to list completely. The following four are major sources, in order of importance:

1. Accounting records: These are a prime source of information for economic analysis.
2. Other sources within the firm: Engineering, sales, production, quality, purchasing, *and personnel.*
3. Sources outside the firm.
   a. Published information. Technical directory, buyer indexes, U.S. government publications, reference books, and trade journals offer a wealth of information.
   b. Personal contact.
4. Research and development: Sometimes no published data or personal contact can provide adequate information, and the only alternative may be to undertake research and development to generate it.

The Internet can also be a good source of cost-estimating data, though you should ensure that the information is from a reputable source.

www.enr.com	*Engineering News-Record*	Construction and labor costs
www.kbb.com	*Kelley Blue Book Automobile*	Pricing
www.factsonfuel.com	*American Petroleum Institute*	Fuel costs

3.11 Indexes

Cost and prices vary with time for a number of reasons, including (1) technological advances, (2) availability of labor and materials, and (3) inflation. An *index* is a dimensionless number that indicates how a cost or a price has changed with time (typically escalated) with respect to a base year. Indexes provide a convenient means for developing a present and future cost and price estimates of an item from historical data. An estimate of the cost or selling price of an item in year n can be obtained by multiplying the cost or price of an item at an earlier point in time (year k) by the ratio of the index value in year n (I_n) to the index value in year k (I_k), that is,

$$\text{cost}_n = \text{cost}_k\left(\frac{I_n}{I_k}\right) \tag{3.56}$$

where:

k = reference year (e.g., 2000) for which cost or price of an item is known
n = year for which cost or price is to be estimated ($n > k$)
cost_n = estimated cost or price of item in year n
cost_k = cost or price of item in reference year k

Use of this technique allows the cost or potential selling price of an item to be taken from historical data with a specified base year and updated with an index.

Example 3.7: Indexing the Cost of a New Boiler

A certain index for the cost of purchasing and installing a utility boiler is keyed to 2008, where its baseline value was arbitrarily set at 100. Company NKJ installed a 50,000 lb/h boiler for $325,000 in 2010, when the index had a value of 468. This company is to install another boiler of the same size in 2012. The index in 2012 is 742. What is the approximate cost of the new boiler?

Solution

In this example, n is 2012 and k is 2010. From the indexing equation, the approximate cost of the boiler in 2012 is

$$\text{Cost}_{2012} = \$325{,}000(742/468) = \$515{,}278$$

Indexes can be created for a single item or for multiple items. For a single item, the index value is simply the ratio of the cost of the item in the current year to the cost of the same item in the reference year, multiplied by the reference year factor (typically 100).

A composite index is created by averaging the ratios of selected item costs in a particular year to the costs of the same items in a reference year. The developer of an index can assign different weights in the index according to their contribution to the total cost. For example, a general weighted index is given by

$$I_n = \frac{w_1(\text{Cost}_{n1}/\text{Cost}_{k1}) + w_2(\text{Cost}_{n2}/\text{Cost}_{k2}) + \cdots + w_n(\text{Cost}_{nm}/\text{Cost}_{km})}{w_1 + w_2 + \cdots + w_m}\overline{I_k} \tag{3.57}$$

where:

m = total number of items in the index
Cost_{nm} = unit cost (or price) of the mth item in year n
Cost_{km} = unit cost (or price) of the mth item in year k
w_m = weight assigned to the mth item
I_k = composite index value in year k

The weights $w_1, w_2, \ldots w_n$ can sum to any positive number, but typically 1.00 is used. Almost any combination of labor, material, products, services, and so on can be used for a composite cost or price index.

Example 3.8: Weighted Index for Gasoline Cost

Based on the following data, develop a weighted index for the price of a gallon of gasoline in 2006, when 1992 is the reference year, having an index value of 99.2.

Items	Price ($/gallon) in 2008	Price ($/gallon) in 2010	Price ($/gallon) in 2012
Premium	2.25	3.25	3.90
Unleaded plus	1.83	1.88	2.0
Regular unleaded	1.95	2.45	2.75

Solution

Roughly 2.5 times as much regular unleaded is sold compared with premium or unleaded plus, so $w_{premium} = 1$, $w_{unleaded} = 1$, $w_{regular} = 2.5$.

In this example, k is 2008 and n is 2012. From the composite equation of indexing, the value of I_{2012} is

$$I_{2012} = \frac{(1)(3.90/2.25) + (1)(2.0/1.83) + (2.5)(2.75/1.95)}{1+1+2.5}(145) = 205 \quad (3.58)$$

Now, if the index in 2013, for example, is estimated to be 253, it is a simple matter to determine the corresponding 2013 price of gasoline from $I_{2012} = 205$.

Premium: \$3.90/gallon (253/205) = \$4.813/gallon.

Unleaded plus: \$2.00/gallon (253/205) = \$2.47/gallon.

Regular unleaded: \$2.75/gallon (253/205) = \$2.395/gallon.

3.12 Parametric Cost Estimating

Parametric cost estimating is the use of historical cost data and statistical techniques to predict future costs. Statistical techniques are used to develop cost-estimating relationships (CERs) that tie the cost or price of an item (e.g., a product, good, service, or activity) to one or more independent variables (i.e., cost drivers). Table 3.1 shows a variety of items and associated cost drivers.

The output of the parametric model or estimated cost is used to gauge the impact of design decisions on the total cost. This is used in early design stages to get an idea of how much the product will cost on the basis of physical attributes (such as weight, volume, and power). Various statistical and mathematical techniques are used to develop the CERs. For example, simple linear regression and multiple linear regression models, which are standard statistical methods for estimating the value of a dependent variable (the unknown quantity) as a function of one or more independent variables, are often used to develop estimating relationships. However, for sustainability reasons, we should also include GHG emission and recycling parameters. Two commonly used estimating relationships, the power-sizing technique and the learning curve technique, are presented in Sections 3.10.1 and 3.10.2.

3.12.1 Power-Sizing Technique

The *power-sizing* technique is sometimes also referred to as an *exponential model.* It is frequently used for developing capital investment estimates for industrial plants and equipment. We should include the cost involved in cleaning up the land, water, and air after use of the plant or equipment. End-of-life considerations are a must for sustainable engineering This CRE recognizes that cost varies as some power of the change in capacity or size. That is,

Table 3.1 Cost Drivers

Product	Cost Driver (Independent Variables)
Construction	Floor space, roof area, wall surface area, GHG emissions, water usage
Truck	Empty weight, gross weight, horsepower, GHG emissions, recycling and reuse
Passenger car	Curb weight, wheel base, passenger room, horsepower, GHG emissions, recycling and reuse
Turbine engine	Maximum thrust, cruise thrust, specific fuel consumption
Internal combustion engine	Piston displacement, compression ratio, horsepower, recycling and reuse
Aircraft	Empty weight, speed, wing area, GHG emissions
Electric power plant	Kilowatts, GHG emissions, recycling and reuse
Motor	Horsepower, toxic elements, reusability

$$\frac{\text{Cost}_A}{\text{Cost}_B} = \left(\frac{S_A}{S_B}\right)^{C_f} \tag{3.59}$$

where:

S_A = size of plant A
S_B = size of plant B
c_f = cost-capacity factor reflecting economics of scale

c_f will depend on the type of plant or equipment being estimated. Typically, $c_f \cong 2/3$. For example, $c_f = 0.68$ for nuclear generating plants and 0.79 for fossil fuel generating plants. It could safely be assumed that $c_f < 1$ indicates decreasing economies of scale (each additional unit of capacity costs less than the previous unit), $c_f > 1$ indicates increasing economies of scale (each additional unit of capacity costs more than the previous unit), and $c_f = 1$ indicates a linear cost relationship with size.

Example 3.9: Power-Sizing Model for Cost Estimating

Suppose that an aircraft manufacturer desires to make a preliminary estimate of the cost of building a 600 MW fossil fuel plant for the assembly of its new long-distance aircraft. It is known that a 200 MW plant cost \$170 million 20 years ago when the approximate cost index was 400, and the cost index is now 1200. The cost-capacity factor for a fossil fuel power plant is 0.70.

Solution

Before using the power-sizing model to estimate the cost of the 600 MW plant (C_A), we must first use the cost index information to update the known cost of the 200 MW plant 20 years ago to a current cost. Using Equation 3.56, we find that the current cost of a 200 MW plant is

C_B = \$170 million (1200/400) = \$510 million

Now, using Equation 3.59, we obtain the following estimate for the 600 MW plant:

C_A = \$510 million (600 MW/200 MW)$^{0.70}$
C_A = \$1100.41 million

Example 3.10

If a 50 MWe power station cost \$2000/kWe, 20 years before when the cost index was 400, but now the cost index is 1200, find the predicted cost for a 1000 MWe unit.

Solution

Again, we should update the cost 20 years before to a current cost, using Equation 3.56.
Present cost of 2000 kWe:

$$\text{Cost}_{50\text{ MWe}} = 2000\left(\frac{1200}{400}\right) = \$6000\text{ kWe} \tag{3.60}$$

$$\text{Cost}_{1000} = \left(6000\frac{\$}{\text{kWe}}\right)\left(\frac{1000}{50}\right)^{2/3} = \$737/\text{kWe} = \$2242.10/\text{kWe} \tag{3.61}$$

Economy of scale refers to the general proposition that "bigger is cheaper" per unit output. The economy of size principle is a fairly established principle of engineering economy, and is central to the costing of most energy-supply and end-use technologies (and a wide variety of other industrial systems).

3.12.2 Learning Curve Method

A *learning curve* is a mathematical model that explains the phenomenon of increased worker efficiency and improved organizational performance with repetitive production of goods or services. The learning curve method is sometimes called an *experience curve* or a *manufacturing progress function*; fundamentally, it is an estimating relationship. For example, the learning curve can be used in estimating the professional hours expended by engineering staff to accomplish successive detailed designs within a family of products, as well as in estimating the labor hours required to assemble automobiles. The basic concept of learning curves is that some input resources (e.g., energy costs, labor hours, material costs, engineering hours) decrease, on a per-output-unit basis, as the number of units produced increases. For example, if 100 labor hours are required to produce the first unit, and a 90% learning curve is assumed, then 100(0.9) = 90 hours would be needed to produce the second unit. Similarly, $100(0.9)^2 = 81$ labor hours would be needed to produce the third unit.

Equation 3.62 can be used to compute resource requirements assuming a constant percentage reduction in input resources each time the output quantity is doubled.

$$Z_N = K(N)^{le} \tag{3.62}$$

where:

N = output unit number
Z = the number of input resource units needed to produce output unit N
K = the number of input resource units needed to produce the first output unit
s = the learning curve slope parameter expressed as a decimal (s = 0.9 for a 90% learning curve)
$le = \ln(s)/\ln(0.2)$

Example 3.11: Learning Curve for a Mini Baja Car Design

The Mechanical Engineering Department at Manhattan College has a student team that is designing a Mini Baja car for a national competition. The time required for the team to assemble the first is 100 h. Their improvement (learning rate) is 0.8, which means that as output is doubled, their time to assemble a car will be reduced by 20%. Use this information to determine

a. The time it will take the team to assemble the 10th car
b. The total time required to assemble the first 10 cars
c. The estimated cumulative average assembly time for the first 10 cars

Solution

Using Equation 3.62, and assuming a proportional decrease in assembly time for output units between doubled quantities, we have

a. $Z_{10} = 100(10)^{\log 0.8/\log 2} = 100(10)^{-0.322} = 47.6$ h
b. The total time to produce x units, T_x, is given by

$$T_x = \sum Z_u = \sum K(u^n) \tag{3.63}$$

Using Equation 3.63, we get

$$T_{10} = 100\sum u^{-0.332} = 100\left(1^{-0.332} + 2^{-0.332} + \cdots + 10^{-0.332}\right) = 631\text{ h} \tag{3.64}$$

c. The cumulative average time for x units, C_x, is given by

$$C_x = T_x/X = T_{10}/10 = 631/10 = 63.1\text{ h} \tag{3.65}$$

Spreadsheet Solution:

Learning Curve Example 3.10

K	100		
s	0.8		
N	−0.32193		
Units	Unit Time	Cumulative Total	Cumulative Average
1	100	100	100
2	79.44344	179.4434	89.72172
3	69.43777	248.8812	82.9604
4	63.99362	312.8748	78.21871
5	59.55684	372.4317	74.48633
6	55.16375	427.5954	71.2659
7	52.4116	480.007	68.57243
8	50.13882	530.1458	66.26823
9	48.21603	578.3619	64.26243
10	47.6431	626.005	62.6005

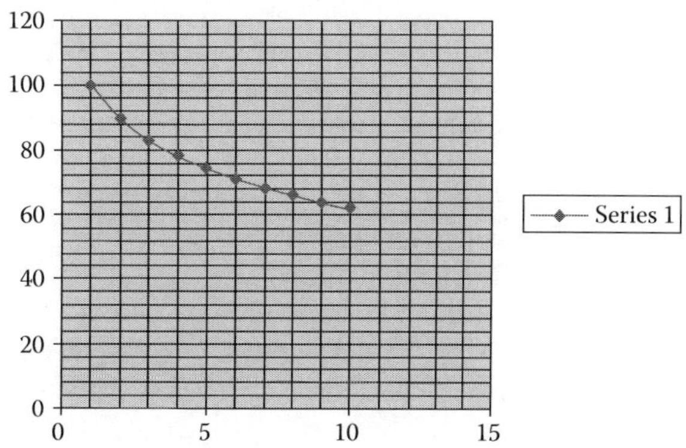

Example 3.12: Photovoltaic Panel Production

Let a photovoltaic panel cost \$200/m² for first-of-a-kind production. Estimate the cost of a 10^4 unit production run. Assume the learning curve slope as 0.85.

Solution

a. Using Equation 3.62, we can estimate the decreasing production time for the photovoltaic panel as

$$Z_{10^4} = K(N)^{le} \tag{3.66}$$

where:

$K = \$200$

$le = \ln(0.85)/\ln(2) = -0.234$

$Z_{10^4} = 200(10^4)^{-0.234} = \$23.17/\text{m}^2$

This proves the principle that learning curves effect savings by sequential mass production of large numbers of identical units (same size). In practice, car engines are a good example of this principle.

b. We can also calculate the total cost to produce x units, Cost_x, given by

$$\text{Cost}_x = \sum Z_u = \sum Z(N1^e) \tag{3.67}$$

Using Equation 3.67, we get

$$\text{Cost}_{10}^{4} = 200\sum Z^{-0.234} = 100\left(1^{-0.234} + 2^{-0.234} + \cdots + 10^{4(-0.234)}\right) = \$3.024\times 10^5 \tag{3.68}$$

c. The cumulative average cost for x units, C_x, is given by

$$\text{Average Cost}_x = \text{Cost}_x / X = \text{Cost}_{10^4} / 10^4 = \$32.34 \tag{3.69}$$

The average cost of production of photovoltaic cell falls to \$32.24/m². This shows the merit of the learning curve. However, we should be careful in applying the learning curve method and keep in mind the following:

1. Larger size normally yields lower reliability, and average unit cost may increase.
2. The learning curve should only be applied to the same product (same size or same design), to the same workforce, and in the same setting (e.g., factory). In practice, the setting may change, or the workforce may alter.

3. The learning curve method does not factor into the principles of sustainability, for example resources depletion, GHG emissions, or technological innovation.

3.13 Developing a Cost-Estimating Relationship (CER)

A CER is a mathematical model that describes the cost of an engineering project as a function of one or more design variables. CERs are useful tools, because they enable a cost estimate to be developed quickly and easily. Furthermore, estimates can be made early in the design process before detailed information is available. As a result, engineers can use CER to make early design decisions that are cost effective in addition to meeting technical requirements. There are three basic steps in developing a CER:

1. *Problem definition*
   The first step in any engineering analysis is to define the problem to be addressed. A well-defined problem is much easier to solve.
2. *Data collection and normalization*
   The collection and normalization of data is the most critical step in the development of a CER. Without reliable data, the cost estimate obtained by using the CER would be meaningless. Data can be obtained from both internal and external sources.
3. *CER equation development*
   The most important step is formulation of an equation that accurately captures the relationship between the selected cost driver(s) and project cost. The table below shows the four general equation types commonly used in CER development.

Typical Equation Forms

Type of Relationship	Generalized Equation
Linear	$\text{Cost} = b_0 + b_1x_1 + b_2x_2 + b_3x_3 + \ldots$
Power	$\text{Cost} = b_0 + b_1x_1b_{11} \cdot x_2b_{12} + \ldots$
Logarithmic	$\text{Cost} = b_0 + \text{b}_1 \cdot \log(x_1) + b_2 \cdot \log(x_2) + \text{b}_3 \cdot \log(x_3) + \ldots$
Exponential	$\text{Cost} = b_0 + \text{b}_1 \cdot \exp(b_{11}x_1) + b_2 \cdot \exp(b_{21}x_2) + \ldots$

In these equations, b_0, b_1, b_2, and b_3 are constants, while x_1, x_2, and x_3 represent design variables. The most common technique to solve for the coefficients' values is the method of least squares. All the equations presented in the table above can easily be transformed into a linear form. The following equations can be used to calculate the values of the coefficients b_0 and b_1 in the simple linear equation $y = b_0 + b_1 \cdot x$:

$$b_1 = \frac{n\sum_{i=1}^{n} x_i y_i - \left(\sum_{i=1}^{n} x_i\right)\left(\sum_{i=1}^{n} y_i\right)}{n\sum_{i=1}^{n} x_i^2 - \left(\sum_{i=1}^{n} x_i\right)^2} \tag{3.70}$$

$$b_0 = \frac{\sum_{i=1}^{n} y_i - b_1 \sum_{i=1}^{n} x_i}{n} \tag{3.71}$$

Note that the variable n in Equations 3.70 and 3.71 is equal to the number of data sets used to estimate the values of b_0 and b_1.

Example 3.13: CER for a Spacecraft

In the early stages of design, it is believed that the cost of a Martian Rover spacecraft is related to its weight. Cost and weight data for six spacecraft, collected and normalized, are shown in the table below. A plot of the data suggests a linear relationship. Use a spreadsheet model to determine the values of the coefficients for the CER.

Spacecraft *i*	Weight (lb) x_i	Cost ($ million) y_i
1	400	278
2	530	414
3	750	557
4	900	689
5	1,130	740
6	1,200	851

The cost equation developed by CER is as follows: $\text{Cost} = 48.283 + 0.66x$

3.14 Sustainability Considerations in Cost Accounting

We have not considered the environmental pollution of land, water, and air due to mass industrial production. As we know, the regulations regarding environmental degradation are becoming increasingly tough in most of the industrialized world. Penalties for atmospheric pollution are severe, and these are add-on costs for the industries. Examples include cooling towers for virtually all Rankine cycle electric-generation stations and stack-gas scrubbers for coal-fired units. Discharge of industrial effluents (as liquid, gas, or solid) into the atmosphere results in severe penalties for the polluters and creates a serious problem for the general public in terms of health and cleaning costs. For example, toxic substances (e.g., sulfur and nitrogen oxides) released by coal-fired generating stations produce the highly injurious acid rain. How we include these factors in a cost estimation is an emerging field in sustainable and industrial ecology. Detrimental effects on infrastructure are also of concern to society at large in terms of depletion of raw materials, water, roads, transport, etc.

A generic life cycle road map for externalities assessment is presented in Figure 3.7. Figure 3.8 shows the cost assignment in sustainable engineering.

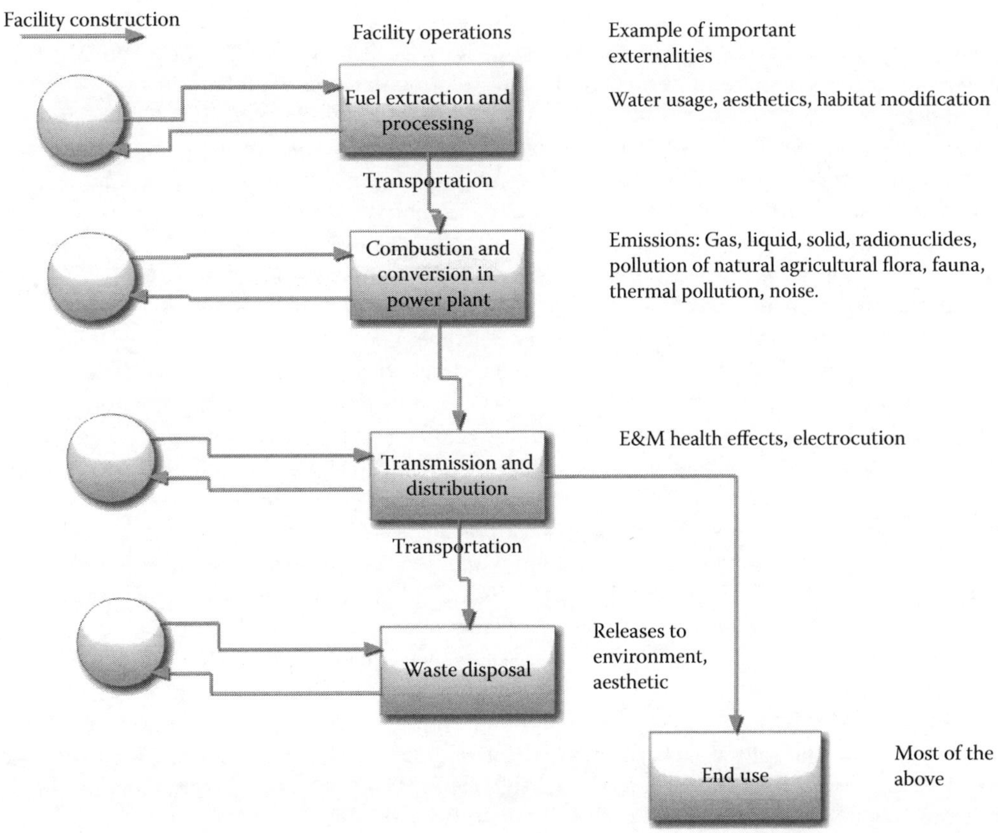

Figure 3.7

Generic life cycle for externalities assessment.

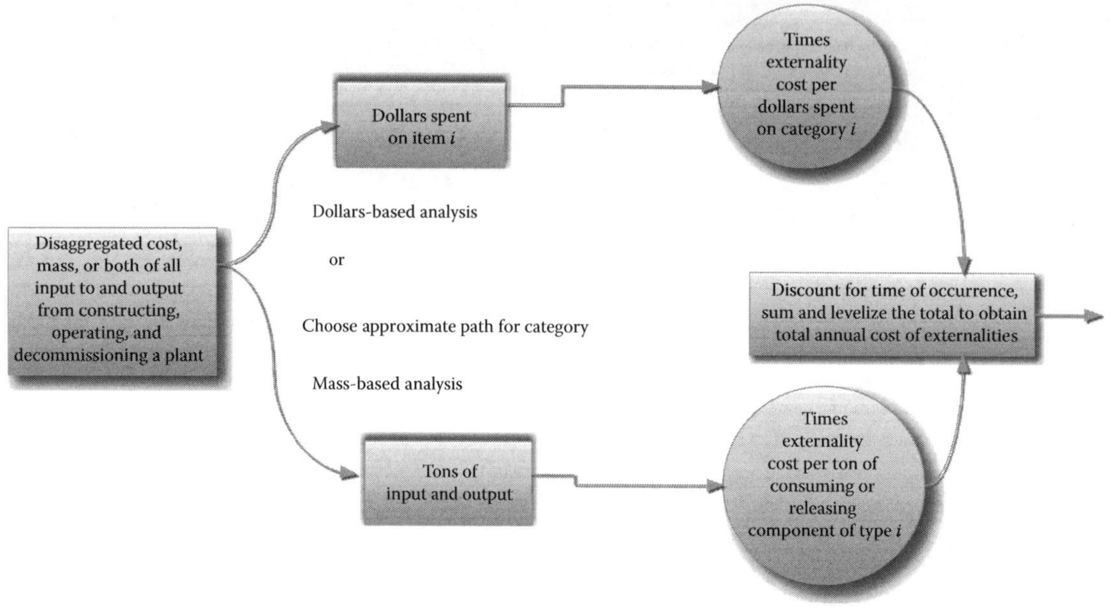

Figure 3.8

Procedure for externalities cost assessment.

3.15 Carbon Dioxide Sequestration and Storage

GHG emissions are believed to be the main reason for global warming, and the Intergovernmental Panel on Climate Change (IPCC) has examined various scenarios to reduce GHG emissions. GHG is thought to be composed of various harmful gases, such as NO_2, CO, CO_2, SO_2, and so on, but carbon dioxide (CO_2) is the main constituent of GHG. It has received wide attention in terms of global warming. Every country has passed legislation to limit and reduce CO_2 emissions. In the United States, the federal Environmental Protection Agency (EPA)'s Clean Air Market Program is called *allowance trading* or *cap-and-trade*. Cap-and-trade is an attempt to control large amounts of emissions from a group of sources, whether stationary or mobile, at costs that are lower than if sources were regulated individually. The approach first sets the overall cap, or maximum amount of emissions per compliance period, that will achieve the desired environmental effects. The total emission allowances provided cannot exceed the cap.

In addition, it has been identified that geologic sequestration (GS) is the most technically viable approach to capture and store CO_2 and thus to control climate change. Energy-related and industrial facilities, such as coal-fired power plants, ethanol and petroleum refineries, natural gas processing plants, and cement and chemical plants, are the major source of CO_2 emissions. GS captures CO_2 from power plants and oil refineries and stores CO_2 in oil and gas reservoirs (i.e., enhanced oil recovery [EOR]), unmineable coal seams, and deep saline formations. There are efficient technologies available to capture CO_2 both postcombustion and precombustion. These technologies, either chemical absorbents (e.g., CO_2 hydrates) or physical absorbents, are applied at sources of emission. An example on carbon sequestration is presented [4].

For example, four plants emitting CO_2 plan to join forces to set up a carbon sequestration facility. These plants are an integrated gasification combined cycle (IGCC) plant, a steel plant, a petrochemical plant, and an oil refinery plant. Annual CO_2 emissions of these plants are 3.5, 4.5, 3.5, and 4.7 million tons, respectively. But the maximum capture rates of CO_2 from these plants vary depending on the technologies employed: 55%, 70%, 60%, and 80%, respectively.

The minimum capture rate of CO_2 is about 30% for all plants. Under the cap-and-trade rules, carbon credit may be earned and sold in the credit market with a price of at least $12.00 per ton if these plants can remove more CO_2 above the minimum level. The CO_2 capture cost varies for different plants based on the technologies adopted. The CO_2 capturing costs are provided in Table 3.2.

For pipeline transportation of CO_2, a cost function with fixed charge and variable cost can be developed based on the existing database. Thus, average cost of CO_2 transportation as a function of the CO_2 mass flow rate (MFR) is expressed as

Table 3.2 Capturing Cost of CO_2

Carbon-Emitting Source	Capturing Cost (\$/ton)	Average (\$/ton)
IGCC plant	$26 \le C_1 \le 74$	$C_{1av} = 50$
Steel plant	$70 \le C_2 \le 90$	$C_{2av} = 80$
	$90 \le C_3 \le 150$	$C_{3av} = 120$
	$150 \le C_4 \le 240$	$C_{4av} = 195$

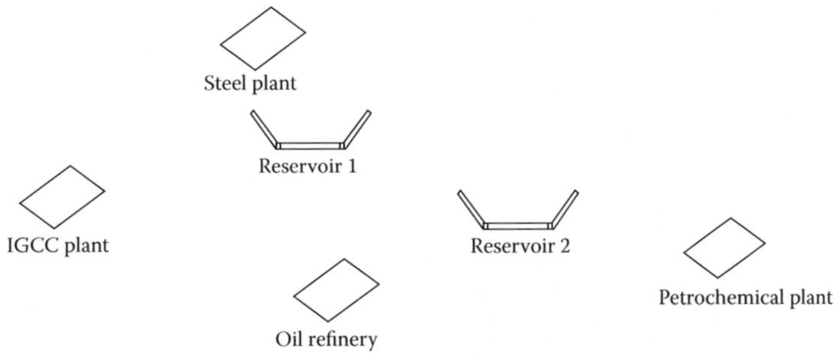

Figure 3.9

Schematic of proposed sequestration plan.

$$C = a(MFR)^b \tag{3.72}$$

where:

C is the average cost (\$/ton/100 km)

MFR is the CO_2 mass flow rate (10^6 tons/year)

both a and b are regression coefficients: $a = 3.287$, $b = 0.619$

Two candidate sites, including a deep saline formation (denoted as reservoir 1) and a depleted oil and gas reservoir (denoted as reservoir 2), can be identified in the region as possible CO_2 storage reservoirs. Reservoirs 1 and 2 can store 5 million and 4 million tons of CO_2 per year, respectively. The locations of these reservoirs are shown in Figure 3.9.

It is estimated that these two reservoirs will be able to store CO_2 for the next 20 years.

The amortized injection cost for pumping stations at two reservoirs is \$20/ton of CO_2.

Table 3.3 shows transportation distances from the plants to the sites.

Formulate the sequestration plan problem as a geometric programming problem, and solve it.

The data in Table 3.4 shows the average cost function for transportation of CO_2.

Table 3.3 Transportation Distances from Plants to Sites

CO_2-Emitting Source	Reservoir 1 (km)	Reservoir 2 (km)
IGCC plant	140	300
Steel plant	200	300
Petrochemical plant	350	25
Oil refinery	200	50

Table 3.4 Average Cost Function for Transportation of CO_2

Mass Flow Rate (MFR), tons CO_2/year	Cost (\$/ton CO_2/100 km)
1	4
5	1.2
10	0.85
20	0.65
30	0.45
40	0.25
50	0.20

Linear regression results:

$$\text{cost} = -0.064 + 2.675(MFR) \tag{3.73}$$

Nonlinear regression:

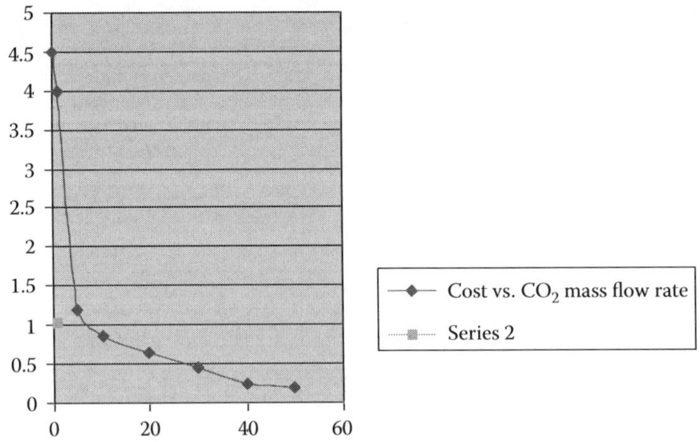

$$\text{cost} = 4.205(MFR)^{-0.719} \tag{3.74}$$

where *MFR* is mass flow rate.

We can see a difference in the curve fit between the linear and nonlinear regression techniques. Nonlinear fitting shows a better result with less error.

3.16 Sustainable Economy Problems

Example 3.14

Engineering economy is all about deciding among competing alternatives.

Consider the situation of replacing a gas guzzler in a family. Susan and Jack are faced with a car replacement situation in which the interest rate can be ignored. Jack's old clunker, which averages 10 miles per gallon (mpg) of gasoline, can be traded in toward a vehicle that gets 15 mpg. Or, as an alternative, Susan's 25 mpg car can be traded in toward a new hybrid vehicle that averages 50 mpg. If they drive both cars 12,000 miles per year, and their goal is to minimize annual gas consumption, which car should be replaced—Jack's or Susan's? They can only afford to upgrade one car at this time.

Solution

Jack's trade-in will save

$$\begin{aligned}&12{,}000 \text{ miles per year}/10 \text{ mpg} - (12{,}000 \text{ miles/year})/15 \text{ mpg}\\&= 1200 \text{ gallons/year} - 800 \text{ gallons/year}\\&= 400 \text{ gallons/year}\end{aligned} \tag{3.75}$$

Susan's trade-in will save

$$\begin{aligned}&12{,}000 \text{ gallons per year}/25 \text{ mpg} - (12{,}000 \text{ miles per year})/50 \text{ mpg}\\&= 480 \text{ gallons/year} - 240 \text{ gallons/year}\\&= 240 \text{ gallons/year}\end{aligned} \tag{3.76}$$

Therefore, Jack should trade in his vehicle to save more gasoline.

Example 3.15: Total Cost in Material Selection: Choosing the Most Economic Material for a Part

Suppose the demand for a part is 100,000 units. The part is produced on a high-speed turret lathe, using 1112 screw-machine steel costing \$0.30 per pound. A study was conducted to determine whether it might be cheaper to use brass screw stock, costing \$1.40 per pound. Because the weight of steel required per piece was 0.0353 pounds and that of brass was 0.0384 pounds, the material cost per piece was \$0.0106 for steel and \$0.0538 for brass. However, when the manufacturing engineering department was consulted, it was found that although 57.1 defect-free parts per hour were being produced by using steel, the output would be 102.9 defect-free parts per hour if brass were used. Which material should be used for this part?

Solution

The machine attendant was paid \$15.00 per hour, and the variable (i.e., traceable) overhead costs for the turret lathe were estimated to be \$10.00 per hour. Thus, the total cost comparison for the two materials was as follows:

	112 Steel	Brass
Material	\$0.30 × 0.0353 = \$0.0106	\$1.40 × 0.0384 = \$0.0538
Labor	\$15.00/57.1 = 0.2627	\$15.00/102.9 = 0.1458
Variable overhead	\$10.00/57.1 = 0.1751	\$10.00/102.9 = 0.0972
Total cost per piece	\$0.4484	\$0.2968
Saving per piece by use of brass	\$0.4484 − \$0.2968 = \$0.1516	

Because 100,000 parts are made each year, revenues are constant across the alternatives. Rule 2 would select brass, and its use will produce savings of \$151.60 per thousand (a total of \$15,160 for the year). It is also clear that costs other than the cost of material were important in the study.

Example 3.16: Trade-Offs in Energy Efficiency Studies

Energy efficiency affects the annual expense of operating an electrical device such as a pump or motor. Typically, more energy-efficient equipment requires a higher capital investment than does less energy-efficient equipment, but extra capital investment usually produces annual savings in electrical power expenses relative to a second pump or motor that is less energy efficient.

Two pumps capable of delivering 100 hp to an agricultural application are being evaluated in a present economy study. The electrical pump will only be used for one year, and it will have no market value at the end of the year. Pertinent data are summarized as follows:

	ABC Pump	XYZ Pump
Purchase price	\$2900	\$6200
Annual maintenance	\$170	\$510
Efficiency	80%	90%

If electric power costs \$0.10 per kWh, and the pump will be operated 4000 h per year, which pump should be selected? Recall 1 hp = 0.746 kW.

Solution

The annual expense of electric power for the ABC pump is

$$(100\,\text{hp}/0.80)(0.746\ \text{kW/hp})(\$0.10/\text{kWh})(4000\ \text{hours/year}) = \$37{,}300 \tag{3.77}$$

For the XYZ pump, the annual expense of electric power is

$$(100\,\text{hp}/0.90)(0.746\ \text{kW/hp})(\$0.10/\text{kWh})(4000\ \text{hours/year}) = \$33{,}156 \tag{3.78}$$

Thus, the total annual cost of owning and operating the ABC pump is $40,370, while the cost of owning and operating the XYZ pump for 1 year is $39,866. Consequently, the more energy-efficient XYZ pump should be selected to minimize total annual cost. Notice the difference in annual energy expense ($4144) that results from a 90% efficient pump relative to an 80% efficient pump. This cost reduction more than balances the extra $3300 in capital investment and $340 in annual maintenance required for the XYZ pump.

Example 3.17: Economics of Demanufacturing of Computers

Let us consider a case study that deals with a timely and important issue concerning environmental economics. What are companies to do with all the old computers that typically accumulate at their facilities (Figure 3.10)?

As a possible solution to this problem, a number of states and companies have jointly established *resources recovery* facilities to process the handling of old electronic equipment, such as computers. Demanufacturing of the computers involves disassembly, refurbishing, and either donating or reselling the units. However, there are some residuals, or remaining components, that cannot be used, are harmful to the environment, and contribute to the cost of demanufacturing.

Let us consider the case of one resource recovery center in the northeast that currently handles approximately 2000 computers per year. Of these computers, approximately 50% are refurbished and donated, 40% are manufactured and resold, and 105 are "residuals." What to do with residuals is a challenging problem.

The facility currently has a contract with an outside source to take the monitors and central processing units (CPUs). This leads to storage requirements in their facility, and there are issues concerning the determination of the optimal truckload facility. The labor costs for the outside contractor are $10.00 per unit, and transportation costs are $1.70 per unit, based on full capacity truck load.

The recovery facility is seeking an alternative solution to the handling of residuals. One alternative under consideration is to establish a method for demanufacturing of the computers that will consider the CPUs only, since there are environmental issues associated with computer monitors.

Figure 3.10

Components of a computer after disassembly.

The recovery facility recently completed a work-measurement study to develop a method for demanufacture of the CPU. This method is as follows:

1. Remove fasteners or screws.
2. Remove metal casting.
3. Remove power and data cables from floppy drives.
4. Unfasten floppy drive.
5. Unfasten supporting bar.
6. Remove support and floppy drive.
7. Remove power and data cables from power supply.
8. Remove motherboard.
9. Unfasten and remove power supply.
10. Unfasten drive housing.
11. Remove rubber feet.

From a predetermined time study, it is determined that the demanufacturing rate is 7.5 min/unit. The facility uses an internal labor rate of $12.00 per hour, and it is estimated that training costs for the demanufacturing method will be 10% of total labor costs. Setup costs and transportation costs are estimated at $155 and $205 of total labor costs, respectively. The facility's goal is to demanufacture 200 units during regular operations. Determine the estimated cost per unit for the proposed demanufacturing method. What percentage reduction is this over the per-unit cost of using the outside contractor?

Solution:

Details of Demanufacturing Costs

The industrial engineer performs the following calculations for labor costs, overhead costs, and transportation costs for 200 units.

The labor capacity is estimated as

$$7.35 \text{ min/unit}(200 \text{ units}) = 1470 \text{ min} = 24.5 \text{ h} \tag{3.79}$$

The industrial engineer develops the following cost template:

	Unit Elements	Factory Estimates	
Demanufacturing Cost Elements	Units Cost/Unit	Factor of Row	Row Total
A: Factory labor	24.5 h		$294.00
B: Training costs	$12.0/h	10% A	$29.40
C: Total labor			$323.40
D: Factory overhead—Setup costs		150% C	$485.10
E: Transportation cost		20% C	$64.68
F: Total direct charge			$55.08
			....
G: Facility rental			$873.48
H: Total demanufacturing cost			$200
I: Quantity—Lot size			
J: Demanufacturing cost per unit			$4.37
Outside cost per unit	$11.70		

The unit cost of the proposed internal demanufacturing method is $4.47, while the per-unit cost of using the outside contractor is $11.70. Should the proposed demanufacturing method be adopted, the estimated per-unit cost savings is $7.73 for a 62.6% reduction over the unit cost for the outside contractor.

The method proposed here not only shows a savings of $62.65 but also minimizes the number of computer residuals entering the waste stream. What makes economic sense is also good for the environment.

References

1. Chen, R. W., Navin-Chandra, D., and Print, F. B. (1993). A cost-benefit analysis model of product design for recyclability and its application. In *IEEE International Symposium on Electronics and the Environment* (pp. 178–183), Arlington, VA.
2. Gutowski, T. G., Gershwin, S. B., and Buonassisi, T. (2010). Energy payback for energy systems ensembles during growth. In *IEEE International Symposium on Sustainable Systems and Technology*, Washington, DC.
3. Sullivan, W. G., Wicks, E., and Patrick, K. (2012). *Engineering Economy*, 15th edn. Prentice Hall, Upper Saddle River.
4. Chang, N. B. (2011). *Sustainable Engineering*. McGraw Hill, New York.

Bibliography

Emblemsvåg, J. (2003). *Life-Cycle Costing*. Wiley, New York.

Newnann, D. and Lavelle, J. P. (2014). *Engineering Economic Analysis*, 12th edn. Oxford University Press, Oxford.

Weber, T. A. (2011). *Optimal Control Theory with Applications in Economics*. MIT Press, Cambridge, MA.

4

Analytical Techniques for Sustainability Analysis

4.1 Nonlinear Optimization

4.1.1 Introduction to Optimization and Green Optimal Design

Engineers want to design new, better, more efficient, and less expensive systems. We can define mechanical design as the *selection* of materials and geometry that satisfy specified and implied functional requirements while remaining within the confines of unavoidable limitations. The successful application of a mathematical programming algorithm to a complex engineering problem requires a careful interfacing of needs and requirements between the optimization operator and the engineering system. For most specific real-world design problems, we could find an infinite number of possible design solutions for the overall objective. In order to obtain a more explicit method of design, the overall objective must be more clearly defined, and all undesirable effects must be avoided [1–3]. For instance, in any specific mechanical element, such effects as stresses, deflections, vibrations, space occupancy, weight, and cost might be undesirable, and the degree of significance for each effect depends on the particular application. Hence, tolerable limits exist for the permissible degree of each undesirable effect. Similarly, there are *desirable effects*, such as power transmission, capability, energy-absorption capability, speed capability, usable length of life, and safety factors.

Engineering optimization is a mathematical technique used for finding and identifying the best candidate from a collection of design alternatives without having to explicitly enumerate and evaluate all possible alternatives. The development of methods for solving the constrained parameter optimization or nonlinear programming problem has been an area of great interest and activity in the applied mathematics and numerical analysis field. Powerful state-of-the-art optimization tools exist, and further developments are expected. The use of these tools in the design of complex systems and processes has paralleled algorithm development. The successful application of optimization methodology to complex system design can result in quite reasonable results in terms of both system efficiency and design flexibility. Unfortunately, the successful application of a nonlinear programming approach to a complex engineering problem requires a careful understanding of the engineering system as well as the algorithm. There are some areas where interface requirements have not been successfully resolved. Issues are identified that require resolution by developers of nonlinear programming algorithms as well as system design engineers. It is these areas of mismatch between theory and practice that constitute the frontiers of optimal design.

4.1.2 Optimization Problem and Model Formulation

In this section we present the optimization problem, its various components, and its formulation as a mathematical programming problem. A brief introduction to the art of modeling is also discussed.

4.1.2.1 Basic Components of an Optimization Problem

An *objective function* expresses the main aim of the model, which is either minimization or maximization. For example, in a manufacturing process, the aim may be to *maximize the profit* or *minimize the cost.* In comparing the data prescribed by a user-defined model with the observed data, the aim is to *minimize the total deviation* of the predictions based on the model from the observed data. In designing a bridge pier, the goal is to *maximize strength* and *minimize size.*

A set of *unknowns* or *variables* control the value of the objective function. In the manufacturing problem, the variables may include the *amounts of different resources used* or the *time spent on each activity.* In the data-fitting problem, the unknowns are the *parameters* of the model. In the pier design problem, the variables are the *shape and dimensions* of the pier.

A set of constraints allow the unknowns to take on certain values but exclude others. In the manufacturing problem, one cannot spend a negative amount of time on any activity, so one constraint is that the "time" variables must be nonnegative. In the pier design problem, one would probably want to limit the breadth of the base and to constrain its size.

The optimization problem is then to find values of the variables that minimize or maximize the objective function while satisfying the constraints.

4.1.3 Objective Function

As already stated, the objective function is the mathematical function one wants to maximize or minimize, subject to certain constraints. Many optimization problems have a single-objective function, and if they do not, they can often be reformulated so that they do. The two exceptions are

- *No Objective Function:* In some cases (e.g., design of integrated circuit layouts), the goal is to find a set of variables that satisfies the constraints of the model. The user does not particularly want to optimize anything, and so there is no reason to define an objective function. This type of problem is usually called a *feasibility problem.*
- *Multiple Objective Functions:* In some cases, the user may wish to optimize a number of different objectives concurrently. For instance, in the optimal design of a panel for a door or window, it would be good to *minimize weight* and *maximize strength* simultaneously. Usually, the different objectives are not compatible; the variables that optimize one objective may be far from optimal for the others. In practice, problems with multiple objectives are reformulated as single-objective problems either by forming a weighted combination of the different objectives or by treating some of the objectives as constraints.

4.1.4 Statement of an Optimization Problem

An optimization or a mathematical programming problem can be stated as follows:

$$\text{To find } \mathrm{X} = \begin{Bmatrix} x_1 \\ x_2 \\ x_3 \\ \cdot \\ \cdot \\ x_v \end{Bmatrix}, \text{ which minimizes } f(\mathbf{X})? \tag{4.1}$$

Subject to the constraints

$$g_i(\mathbf{X})0 \leq, i = 1,2,\ldots,m$$

$$l_j(\mathbf{X})0 =, j = 1,2,\ldots, p(1)$$

where:

$\mathbf{X}$ is an n-dimensional vector called the design vector

$f(\mathbf{X})$ is called the *objective function*

$g_i(\mathbf{X})$ and $l_j(\mathbf{X})$ are known as inequality and equality constraints, respectively

The number of variables n and the number of constraints m or p need not be related in any way. This type of problem is called a *constrained optimization problem.*

4.1.5 Basic Mathematical Model

The basic mathematical problem of interest is stated as follows. Find the variables $X=(x_1,x_2,\ldots,x_n)$ to minimize the objective function

$$f(X)=f(x_1,x_2,\ldots,x_n)$$

subject to the equality constraints

$$c_i(X)=0,\ i=1,2,\ldots,k \tag{4.2a}$$

and inequality constraints

$$c_j(X)\geq 0,\quad j=\left(k+1\right),\ldots,m$$

It is usually assumed that the functions are continuous and differentiable through second order, and that $k\leq$m.

The historical "direct search" techniques have generally been replaced by sophisticated well-structured techniques with strong theoretical foundations. There are many optimization techniques available in the literature. Penalty function and augmented Lagrangian or multiplier methods have been presented by Fiacco and McCormick, Hestenes, Powel, and Betts, among others. Reduced gradient methods have been presented by Rosen, Abadie and Carpentier, Lasdon and Warren, Segent, and Betts. Recent work on "exact" penalty function methods have been reported by Fletcher, Han and Mangasarian, and Mayne and Maratos. Hybrid techniques and methods involving quadratic programming (QP) subproblems, have been presented by Betts, Murray, Biggs, Han, and Powell. The books by Fletcher, Gill and Murray and by Gill, Murray, and Wright [4] provide a fairly complete overview of the field.

4.1.6 Requirements for Application of Optimization Methods to Optimal Design

In order to apply optimization theory to design problems, it is necessary to delineate the boundaries of the engineering system to be optimized [5, 6]:

Formulation of the engineering design problem
Objective to be obtained or the objective function
Restrictions on the parameters
Functional relationships between parameters
Independent and dependent variables

When considering the application of optimization methods to design and operations, we should keep in mind that optimization is only one step in the overall process of arriving at an optimal design or an efficient operation. Optimization theory is also used in other branches of engineering, such as

Design of components or entire systems
Planning and analysis of existing operations
Engineering analysis and data reduction
Control of dynamic systems

4.1.7 Basic Design Problem

It is hoped that the example presented here will inspire interest in optimal design. The example provides an explicit method of synthesis, simultaneously resulting in the most desirable (optimal) design solution. Regardless of the method used, the essential operation of the optimization process is similar. The process always begins with guess values for the design variables, and the objective function and constraints are calculated at these values. Using the computed values of the functions, the algorithm constructs a new estimate of the design variables. This process is repeated until the iteration is terminated by the algorithm. Ideally, the final point x^* is the solution of the stated problem. We will try to explain the optimal engineering design concept as a sustainable engineering tool.

Example 4.1 Design of Tray

Design a plastic tray which must be capable of holding a specified volume of liquid *V*, such that the liquid has a specified depth *H*, and the wall thickness of the tray is to be *T*. The tray is to be manufactured in large numbers.

4.1.8 Optimal Design

The problem does not state what the basic geometrical shape of the tray should be. However, by intuition, we always start with a rectangular shape. Assuming wall thickness *T* to be small compared with the dimensions *b* and *l*, we can write

$$V = blH \tag{4.2b}$$

where:

b = width of tray
l = length of tray
H = height of tray

Therefore, we can arbitrarily select any value for *b* and then calculate the corresponding value of *l*. However, there would be an infinite number of possible solutions to the above problem. In optimal engineering design and design for manufacturing (DFM), we are required to select the manufacturing process along with the materials to be used for the tray. We know from experience that a vacuum-forming process will be good for manufacturing the tray. When selecting the material, we should consider whether the tray will chemically react with the liquid it is holding. Let us select acrylic thermoplastic sheet as the material for the tray. We would just like to stress that in DFM, such factors as manufacturing process and appropriate material for the product must be considered before the design stage. In all optimal engineering design problems, the undesirable effects of stresses, vibrations, deflections, and weight are significant, and these effects must be considered either as constraints or sometimes as objectives. The optimal design for this example is relatively simple, and we will not consider these effects. As we know, the cost of a tray may make or break the tray-manufacturing industry. Hence, in every optimal design, one way or another, we must consider the cost of manufacture. So, in this problem, we consider minimization of cost as our objective function. We have framed our optimal design problem such that our optimum design solution needs to be the best with respect to shape (geometry), material, and manufacturing process for minimization of cost for mass production of the tray. Our first task in optimal design is to develop the objective function, which is cost minimization in this case. The total cost (*TC*) of the tray is developed as:

$$TC = f_o + f_t + f_l + f_m$$

where:

f_0 = overhead cost/tray
f_t = tooling cost/tray
f_l = labor cost/tray
f_m = material cost of plastic/tray

As we have selected vacuum forming as our manufacturing process, the overhead, tooling, and labor costs will be essentially independent of reasonable shapes and feasible plastic materials. These assumptions may not be correct in every optimal design problem. However, in this case, we reasonably conclude that our optimal tray design will be based only on the material cost of the tray. Let us assume *c* as the unit weight cost of the plastic material and *w* as weight density of the plastic material in pounds per cubic inch. If *H* and *T* are specified values in inches, and we wish to find *b* and *l* values of the optimal tray design, we can develop the cost equation as

$$TC = f_m = c \cdot w \cdot \left(bl + 2bH + 2lH)T\right) \tag{4.3}$$

where $(bl + 2bH + 2lH)T$ represents the volume of plastic material used in the manufacture of a rectangular-shaped plastic tray. However, the volume of liquid (V) in tray of thickness (T) and depth (H) is specified by Equation 4.2b, and we can replace the length of the tray as

$$l = \frac{V}{bH} \tag{4.4}$$

Replacing $l = V/(b \cdot H)$, we can rewrite the equation as

$$f_m = cw\left(\frac{V}{H} + 2bH + 2\frac{V}{b}\right)T \tag{4.5}$$

The optimum design is based on the f_m equation. The terms c and w in the equation depends on the material selected. Hence, in order to minimize f_m, we should select from the feasible materials the material having the lowest value for the factor cw. Therefore, the cost per unit volume term ($c \times w$) might be thought of as the material selection (MS) factor. As we can see from the f_m equation, this MS factor is quite independent of the geometrical shape of the tray and of the particular values for V, H, T, b, and l. For simplicity, we assume that only two feasible thermoplastics, acrylic and polystyrene sheets, are available to us at this time. Values of weight density w can be obtained from handbooks, and values of unit cost c of the materials can be obtained from vendors. Hence, we can develop the MS factor in terms of cw for each feasible and available material.

Material	w (lb/in^3)	c ($/lb)	cw ($/in^3)
Acrylic	0.04	2.00	0.08
Polystyrene	0.038	0.60	0.023

Therefore, for this optimal design problem, the optimum available material is polystyrene, because its value of cw is lower. In this particular example, we can immediately select polystyrene because of its considerably lower material unit cost c. However, for many real-world problems, we might have a long list of available feasible materials, and only rigorous optimum MS would be on the basis of the minimum value of the factor cw.

4.1.9 Selection of Optimal Shape

With respect to the selection of the optimum geometrical shape, let us examine the variation of cost f_m with respect to the independent variable b, from the optimum design equation developed earlier. Equation 4.5 has only one independent variable b, and we can find the optimal design by taking the derivative of the cost equation and equating to zero. However, in most real-world optimal design problems, we have to use a constrained nonlinear programming algorithm to find the optimal design, as optimal design problems are inherently nonlinear in nature. For the present example:

$$\partial f_m / \partial b = (cw)(2T)\left[H - V/b^2\right] = 0$$

$$\therefore b_{\text{opt}} = \sqrt{V/H}$$

Entering the value of b_{opt} into the equation $l = V/bH$, we get $l_{\text{opt}} = V/H(b_{\text{opt}})$.

This shows that, from the objective of minimizing cost, the optimum rectangular shape is a square. For each square tray, the minimum cost would be

$$(f_m)_{\text{opt}} = (cw)\left[V/H + 2(\sqrt{V/H})H + 2V/\sqrt{V/H}\right]T$$

Therefore

$$(f_m)_{\text{opt}} = (cw)\left[V/H + 4\sqrt{V/H}\right]T \tag{4.6}$$

In this equation, $(f_m)_{\text{opt}}$ is the material cost of an ideal square tray, (cw) is the MS factor for the optimum material previously selected, and V, H, and T are specified functional requirements from the optimal design problem stated.

In real optimal design problems, there are often space restrictions. Hence, for the intended use of the tray, it might be required to fit between two structures while resting on a shelf of specified length. Therefore, restrictions might exist on permissible values of b and l, as follows:

$$b \le b_{\max}$$

$$l \le l_{\max}$$

In the above equations, $b_{\max}$ and $l_{\max}$ are values of space restrictions existing on feasible values for b and l, respectively. Obviously, particular restrictions on b and l might affect the optimum geometry of the rectangular tray. If $b_{\max}$ and $l_{\max}$ are both equal to or greater than $\sqrt{V/H}$, the ideal square tray is still the optimum design. However, if either $b_{\max}$ or $l_{\max}$ is less than $\sqrt{V/H}$, the optimum design is the tray that just fits within the space restriction. Also, it is possible that no geometrical design satisfying the specifications within the space restriction is feasible. This possibility might be termed *incompatible specifications*, which would be the case if the following relationship existed:

$$b_{\max} \cdot l_{\max} < V/H$$

Here, the rectangular shape was arbitrarily selected by intuition. However, if we designed a circular tray, we might find that the circular shape would minimize the cost (material) function.

4.1.9.1 Circular Tray Design

The material cost function for a circular tray is

$$f_m = (cw)\left[\pi D^2/4 + \pi DH\right]T$$

where D is the diameter of the tray, and all other terms remain the same as for the rectangular tray. For the circular tray, the specified liquid volume V can be expressed as

$$V = (\pi D^2/4)H$$

By eliminating D, we get

$$f_m = (cw)(V/H + 2\sqrt{\pi VH})T$$

Therefore,

$$f_m = (cw)\left[V/H + 3.54\sqrt{VH}\right]T$$

All the terms on the right side are independent of each other. Now, again the optimum material is selected from the list of available materials on the basis of *MS* factor (cw), which is the same as for the rectangular shape. From the liquid volume equation, D can be uniquely determined as

$$D = \sqrt{4V/\pi H} \tag{4.7}$$

And if we compare the circular tray f_m equation with the rectangular tray f_m equation, we find that the optimum cost of the circular tray is lower than for the rectangular tray. For circular tray design, again, space restrictions may apply as they did for the rectangular tray. If space restrictions exist, it turns out that an oval tray would be best for minimizing material cost.

We can even examine the selection of the manufacturing process. Perhaps another available feasible manufacturing process, such as a low-cost molding process, would result in a lower tray cost than vacuum forming. I think it is better to examine the optimum design problem for various possible manufacturing methods from the standpoint of *TC*, including overhead, tooling, labor, and material. This example represents the interaction of *MS*, manufacturing process, and optimal design.

Unfortunately, the optimization process does not always work quite so well in practice, in many cases due to underlying incompatibilities between the optimization algorithm and the functions. In fact, more research is needed: (1) to make engineers and optimizers aware of differing interface considerations, and

(2) to suggest areas where additional attention is needed to make optimization algorithms more tolerant of typical real-world function generators in the field of design and manufacturing.

4.1.9.1.1 Function Generator The purpose of the function generator is to compute the objective function or constraints, or both, for an input set of variables or to indicate that evolution is impossible. The "function error" concept has proven useful for complex systems, preventing breakdown of the computational process, but is not used in lieu of constraints. For example, if the function evaluation requires a numerical quadrature, failure to meet integration error tolerance might negate the validity of the computed functions, and this failure can be indicated by function error. Typically, the function generator models a complex engineering process or system. It is usually formulated and perhaps coded by system design engineers. For many complex systems, the computational expense of a function evaluation is considerably greater than the expense of typical algorithm computations. The computed quantities may be quite nonlinear and may require many computational subprocesses and subroutines. Because the function generator is complex, gradients may be so difficult to obtain that for all practical purposes, they are unavailable. This is not to say that they do not exist—they are just impractical to compute. During the modeling process, the system engineer may also introduce discontinuities in the computed quantities or the derivatives. Regardless of whether the discontinuities are intentionally or inadvertently introduced, the fact remains that they are present in many function generators. Finally, the problem posed may simply be poorly formulated.

4.1.9.1.2 Optimization Algorithm The purpose of the optimization algorithm is to compute a series of estimates for the variables using function information provided by the function generator. Additionally, the optimization algorithm is responsible for terminating the iteration, either at the solution or in some other "reasonable" manner. Typically, the optimization algorithm is developed and coded by a mathematician [5,6]. Considerable effort is expended to develop sophisticated processing to deal with questions of computer storage, computational efficiency, and numerical precision. In order to establish the convergence properties of the algorithm, the algorithm developer must often assume that the functions being optimized have certain mathematical attributes. For example, convexity, continuity, and differentiability are often required to establish convergence. In fact, the necessary conditions for a constrained optimum (the Kuhn–Tucker conditions) involve the gradients of the problem functions. Sufficient conditions require second-order information. Furthermore, the meaningful application of the necessary conditions may require the existence of a solution, the existence of a feasible region, and linear independence of the constraint gradients. Commonly, the algorithm developer merely assumes that all necessary properties are present for a given application. In summary, then, the overall optimization process entails two distinct subprocesses, each developed with different goals and assumptions. What can be done to resolve discrepancies in the overall method is described in Section 4.1.10.

4.1.10 Issues to Be Resolved by the Algorithm Developer

4.1.10.1 Gradient-Free Algorithms

Perhaps the single most important area of research is the development of general nonlinear programming algorithms that do not require derivatives. For applications involving complex engineering systems, it is unrealistic to expect that gradients will be available. Furthermore, the "brute-force" use of finite difference approximations in some algorithms requiring gradients is too costly, inaccurate, and unreliable.

4.1.10.2 Diagnosis of Function Generator Difficulties (Knowledge-Based Optimization)

It is equally important to develop software that can diagnose difficulties in the function generator if at all possible. For example, if an algorithm requires convexity in the problem function, then the algorithm should be able either to detect functions that are not convex or to work acceptably without the convexity assumptions. It is unreasonable to expect the system engineer to guarantee (or perhaps even understand) convexity for complex system functions. It is probably not unreasonable to expect continuity and differentiability of the problem functions. However, it is unreasonable to expect the casual user of an algorithm to diagnose the detrimental effects of discontinuities on an algorithm he may not understand. Good software should detect the presence of discontinuities and notify the user of their presence as clearly and concisely as possible. A precise diagnosis of a difficulty in the function generator is invaluable to the design engineer. Unfortunately, when many current algorithms are applied to function generators with "difficulties," no useful information is provided other than that they "just don't work very well." The situation must be remedied.

4.1.10.3 Robust, Sturdy Software

Because of the nature of real-world problems, it is especially important that software be robust and sturdy. Recognizing the inevitability of poorly posed problems, a good algorithm should be user-friendly, providing useful diagnostic information when possible. Furthermore, an algorithm should die gracefully when given an impossible or poorly formulated problem.

If the projected Hessiam matrix is singular, the algorithm should tell the user that the optimum is not uniquely defined. If a Lagrange multiplier is zero or near zero, the algorithm should tell the user that the active constraint set may be wrong, a constraint is redundant, or the problem is poorly defined. If a Lagrange multiplier is large, the algorithm should indicate constraint dependence or poor scaling. If a feasible region does not exist, the algorithm should terminate in a reasonable fashion and notify the user of inconsistent constraint. It is the responsibility of the algorithm to communicate diagnostic information in terms understandable by the user of the tool, rather than in language useful to the algorithm professional.

4.1.10.4 Noise-Tolerant Algorithms

Recognizing that complex engineering systems may never be noise free, it seems promising to pursue the development of optimization algorithms that are more tolerant of noise in the function generator. To date, little has been done to cast the optimization process in a stochastic framework, thereby using statistical methodology to guide the behavior of an algorithm.

4.1.10.5 Application-Oriented Software Design

With the growth in both size and complexity of current software systems, it is important that the structure of the program be carefully organized. In addition to the fairly obvious need for a modular approach to algorithm procedures, two other items have proven noteworthy in the organization of algorithm structure. First, an algorithm should be able to respond to a function error. While this artifice may be aesthetically displeasing to a purist, it has proven to be invaluable in practice. Second, the so-called "reverse communication" structure suggested by Lawson and Krogh for interfacing numerical integration algorithms is most useful for optimization/system interfaces [7]. For most software implementations, a function evaluation is accomplished by calling a subroutine (the function generator). Using the reverse communication philosophy, a function evaluation is accomplished by returning to the main executive program with a function evaluation flag set. In essence, then, the main program calls both the function generator and the optimization algorithm. This approach permits sophisticated input/output from the function generator and facilitates the use of an overlay structure.

4.2 Single and Multiobjective Optimization: A Brief Overview

In mathematics, optimization is the discipline concerned with finding inputs of a function that minimize or maximize its value, which may be subjected to constraints. Combinatorial optimization is a branch of optimization that is concerned with the optimization of functions with discrete variables. Computational optimization can be defined as the process of designing, implementing, and testing algorithms for solving a large variety of optimization problems. Computational optimization includes the disciplines of mathematics to formulate the model, operations research to model the system, computer science for algorithmic design and analysis, and software engineering to implement the model. Nowadays, researchers can solve real-life problems that in the past were thought to be unsolvable thanks to new technological developments in algorithms and computer hardware. Despite its name, optimization does not necessarily mean finding the optimum solution to a problem, since it may be unfeasible due to the characteristics of the problem, which in many cases is included in the category of NP-hard problems. Yet, for optimization problems that are NP-hard, no polynomial time algorithm exists; that is, the algorithms used might in the worst case need exponential computation time to obtain the optimum, which leads to computation times that are too high for practical purposes. As a result, in recent decades, many authors have proposed approximate methods, including heuristic approaches and artificial neural networks (ANN), to solve these problems instead of using traditional optimization methods such as linear programming (LP), the Nelder–Mead Simplex (NMS) method, Lagrangian relaxation (LR), QP, and so on [8]. Heuristic methods can be seen as simple procedures that rapidly provide satisfactory, but not necessarily optimal, solutions to many instances of complex problems. Metaheuristics are generalizations of heuristics in the sense that they can be applied to a wide set of problems, needing few modifications to be adapted to a specific case. In some cases, the complexity of the problems to be solved is so high that even heuristic and metaheuristic methods are not able to obtain accurate solutions in reasonable runtimes. In these cases, parallel processing becomes an interesting way to obtain good solutions in reduced runtimes.

The most common classification of metaheuristic algorithms is based on trajectory methods vs. population-based methods. Other possible classifications are memory-based vs. memory-less methods, nature-inspired vs. nonnature-inspired, and so on.

- *Trajectory Metaheuristics:* These use a single solution during the search process, and the outcome is also a single optimized solution. Most of them are extensions of simple iterative improvement procedures incorporating techniques that enable the algorithm to escape from local optima. The main trajectory-based metaheuristics include hill climbing (HC), simulated annealing (SA), tabu search (TS), greedy randomized adaptive search procedures (GRASP), variable neighborhood search (VNS), iterated local search (ILS), and so on.
- *Population-Based Metaheuristics:* These use a population of solutions, which evolve during a given number of iterations, also returning a population of solutions when the stop condition is fulfilled. The main population-based metaheuristics include genetic algorithms (GA) and evolutionary algorithms (EA), scatter search (SS), path relinking (PR), memetic algorithms (MA), ant colony optimization (ACO), particle swarm optimization (PSO), estimation of distribution algorithm (EDA), differential evolution (DE), artificial bee colony optimization (ABCO), and so on.

On the other hand, it should be noted that to date, most computational optimization methods have focused on solving single-objective problems, including constraints in some cases. Nevertheless, a large number of applications require the simultaneous optimization of several objectives that are often in conflict, and so some authors have proposed multiobjective algorithms. These multiobjective approaches are often divided into two main categories: aggregate weight functions and Pareto-based optimization methods. Aggregating functions consist of combining all the objectives to be optimized in a single mathematical function, in which the relative importance of each objective is adjusted according to relative weights. Despite its simplicity, this approach has several drawbacks; for example, it is very difficult to adjust the weights of the objectives to be optimized, especially when they have different scales. Further, this approach only returns a single solution as a result of the search process, which becomes an important limitation in the decision-making process, in which the decision-maker must select one solution from several alternatives. The drawbacks of aggregating functions have been solved using Pareto-based multiobjective optimization, which establishes relationships among solutions according to the Pareto-dominance concept. Given a multiobjective optimization problem with $K \geq 2$ objectives to optimize, instead of giving a scalar value to the objective function $f1 \ldots K(s)$ of solution *s*, a partial order is defined according to Pareto-dominance relationships. It is said that solution *s*1 dominates another *s*2 when *s*1 is better than *s*2 in at least one objective, and not worse in the others, and that two solutions, *s*1 and *s*2, are indifferent if neither dominates the other.

The set of nondominated solutions constitutes the so-called Pareto optimal set, which usually contains not one solution, but several. As all the objectives are equally important, the aim of multiobjective optimization is to find this entire set or a representative sample of it. A large number of multiobjective metaheuristics have been presented in recent decades, and they can also be classified into the categories of trajectory methods and population-based methods. Trajectory methods include the Pareto archived evolution strategy (PAES), multiobjective simulated annealing (MOSA), and so on [6]. Population-based metaheuristics include multiobjective tabular search (MOTS), nondominated sorting genetic algorithm (NSGA/NSGA-II), Pareto simulated annealing (PSA), single front genetic algorithm (SFGA), strength Pareto evolutionary algorithm (SPEA/SPEA2), and Pareto envelope-based selection algorithm (PESA/PESA-II). Some authors have also proposed hybrid approaches that combine aspects of two or more methods, such as genetic tabu search (GTS), multiobjective genetic local search (MOGLS), memetic-PAES (M-PAES), multiobjective simulated annealing and tabu search (MOSATS), and so on.

Energy resources are very important from an economic and political perspective for all countries, which is why technological change in energy systems is a very important and inevitable factor that researchers need to deal with. Applications of optimization methods for solving problems found in renewable energy systems seems to be an emerging field. Taking into account the increasing worldwide demand for energy, the expansion of distribution networks has become a problem of primary interest. Due to the important investment costs of creating a renewable energy structure, a primary interest from the point of view of the design and long-term planning of energy systems is to select the best alternative among the different renewable energy systems. Community-scale renewable energy system planning is an important problem, consisting of justifying the allocation patterns of energy resources and services and formulation of local policies regarding energy consumption, economic development, and energy structure.

The model development phase is an iterative process that may require returning to the model definition and formulation phase.

4.2.1 Model Development

Model development includes

- Mathematical description
- Parameter estimation
- Input development
- Software development

The next phase in modeling is validation and evaluation as a whole.

4.2.2 Model Validation and Evaluation

Model validation consists of validation of the assumptions and parameters of the model. The performance of the model is to be evaluated using standard performance measures such as root mean squared error and R^2 value. *Sensitivity analysis* to test the model inputs and parameters is also an iterative process and may require returning to the model definition and formulation phase.

One important aspect of this process is that in most cases, data used in the formulation process should be different from those used in validation.

4.2.3 Modeling Techniques

Different modeling techniques have been developed to meet the requirement of different types of optimization problems. Major categories of modeling approaches include:

- Classical optimization techniques
- Linear programming
- Nonlinear programming
- Geometric programming
- Dynamic programming
- Integer programming
- Stochastic programming
- Evolutionary algorithms, and so on

These approaches will be discussed in Section 4.3. The automatic mathematical modeling of optimal design may be a challenging field of research. This approach will increase the acceptability of optimization techniques in designing and process optimization of engineering problems.

4.2.3.1 Optimization Methods: Introduction and Basic Concepts

If the locus of all points satisfying $f(\mathbf{X}) = \mathbf{c}$, a constant, it is considered that it can form a family of surfaces in the design space, called the *objective function surfaces* [8]. When this is drawn with the constraint surfaces, as shown in Figure 4.1, we can identify the optimum points (maxima). This is possible graphically only when the number of design variables is two. When we have three or more design variables, because of complexity in the objective function surface, we have to solve the problem as a mathematical problem, and this visualization is not possible.

4.2.4 Variables

These are essential. If there are no variables, we cannot define the objective function and the problem constraints. In many practical problems, one cannot choose the design variables arbitrarily. They have to satisfy certain specified functional and other requirements.

4.2.5 Constraints

Constraints are not essential. It has been argued that almost all problems really do have constraints. For example, any variable denoting the "number of objects" in a system can only be useful if it is less than the number of elementary particles in the known universe. In practice, though, answers that make good sense in terms of the underlying physical or economic criteria can often be obtained without putting constraints on the variables.

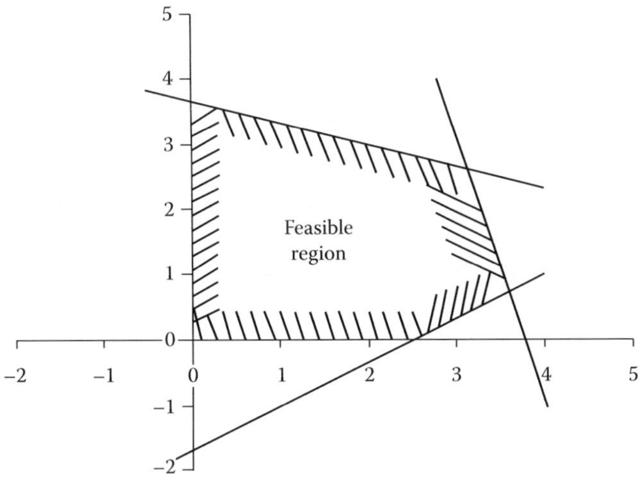

Figure 4.1
Feasible region in optimization.

Design constraints are restrictions that must be satisfied to produce an acceptable design. Constraints can be broadly classified as

1. *Behavioral or Functional Constraints*: These represent limitations on the behavior performance of the system.
2. *Geometric or Side Constraints*: These represent physical limitations on design variables such as availability, fabricability, and transportability.

4.2.6 Constraint Surfaces

Consider the optimization problem presented in Equation 4.1 with only the inequality constraint $g_i(\mathbf{X}) \leq 0$. The set of values of X that satisfy the equation $g_i(\mathbf{X}) \leq 0$ forms a boundary surface in the design space called a *constraint surface.* This will be a $(n-1)$-dimensional subspace, where n is the number of design variables. The constraint surface divides the design space into two regions: one with $g_i(\mathbf{X}) \leq 0$ (feasible region) and the other in which $g_i(\mathbf{X}) > 0$ (infeasible region). The points lying on the hypersurface will satisfy $g_i(\mathbf{X}) = 0$. The collection of all the constraint surfaces $g_i(\mathbf{X}) = 0$, $j = 1,2,\ldots, m$, which defines the acceptable region, is called the *composite constraint surface.*

Figure 4.1 shows a hypothetical two-dimensional design space in which the feasible region is denoted by hatched lines. The two-dimensional design space is bounded by straight lines, as shown in the figure. This is the case when the constraints are linear. However, constraints may be nonlinear as well, and the design space will be bounded by curves in that case. A design point that lies on more than one constraint surface is called a *bound point,* and the associated constraint is called an active constraint [4]. *Free points* are those that do not lie on any constraint surface. The design points that lie in the acceptable or unacceptable regions can be classified as follows:

1. Free and acceptable point
2. Free and unacceptable point
3. Bound and acceptable point
4. Bound and unacceptable point

4.2.7 Formulation of Design Problems as Mathematical Programming Problems

In mathematics, the term *optimization*, or *mathematical programming*, refers to the study of problems in which one seeks to minimize or maximize a real function by systematically choosing the values of real or integer variables from within an allowed set. This problem can be represented in the following way.

Given: A function f: A R from some set A to the real numbers

Sought: An element x_0 in A such that $f(x_0) \leq f(x)$ for all x in A ("minimization") or such that $f(x_0) \geq f(x)$ for all x in A ("maximization").

Such a formulation is called an optimization problem or a *mathematical programming problem* (a term not directly related to computer programming, but still in use, for example, in linear programming). Many real-world and theoretical problems may be modeled in this general framework.

Typically, A is some subset of the Euclidean space R^n, often specified by a set of constraints, equalities, or inequalities that the members of A have to satisfy. The elements of A are called *candidate solutions* or *feasible solutions*. The function f is called an *objective function* or *cost function*. A feasible solution that minimizes (or maximizes, if that is the goal) the objective function is called an *optimal solution*. The domain A of f is called the *search space*.

Generally, when the feasible region or the objective function of the problem does not present convexity, there may be several local minima and maxima, where a *local minimum* x^* is defined as a point for which there exists some $\delta > 0$, so that for all x such that

$$\|x - x^* \leq \delta\|$$

and

$$f(x^*) \leq f(x) \tag{4.8}$$

that is to say, on some region around x^*, all the function values are greater than or equal to the value at that point. Local maxima are defined similarly.

A large number of algorithms proposed for solving nonconvex problems—including the majority of commercially available solvers—are not capable of making a distinction between local optimal solutions and rigorous optimal solutions, and will treat the former as the actual solutions to the original problem. The branch of applied mathematics and numerical analysis that is concerned with the development of deterministic algorithms that are capable of guaranteeing convergence in finite time to the actual optimal solution of a nonconvex problem is called global optimization.

4.2.8 Problem Formulation

Problem formulation is normally the most difficult part of the process. It is the selection of design variables, constraints, objective function(s), and models of the discipline/design.

4.2.9 Selection of Design Variables

A design variable that takes a numeric or binary value is controllable from the point of view of the designer. For instance, the thickness of a structural member can be considered a design variable. Design variables can be continuous (such as the length of a cantilever beam), discrete (such as the number of reinforcement bars used in a beam), or Boolean. Design problems with continuous variables are normally solved more easily. Design variables are often bounded, that is, they have maximum and minimum values. Depending on the adopted method, these bounds can be treated as constraints or separately.

4.2.10 Selection of Constraints

A constraint is a condition that must be satisfied to render the design feasible. An example of a constraint in beam design is that the resistance offered by the beam at points of loading must be equal to or greater than the weight of the structural member and the load supported. In addition to physical laws, constraints can reflect resource limitations, user requirements, or bounds on the validity of the analysis models. Constraints can be used explicitly by the solution algorithm or can be incorporated into the objective by using Lagrange multipliers.

4.2.11 Objectives

An objective is a numerical value that is to be maximized or minimized. For example, a designer may wish to maximize profit or minimize weight. Many solution methods work only with single objectives. When using these methods, the designer normally weights the various objectives and sums them to form a single objective. Other methods allow multiobjective optimization, such as the calculation of a Pareto front.

4.2.12 Models

The designer also has to choose models to relate the constraints and the objectives to the design variables. These models are dependent on the discipline involved. They may be empirical models, such as a regression analysis of aircraft prices; theoretical models, such as from computational fluid dynamics; or reduced-order models of either of these. In choosing the models, the designer must trade off fidelity with the time required for analysis.

The multidisciplinary nature of most design problems complicates model choice and implementation. Often, several iterations are necessary between the disciplines' analyses in order to find the values of the objectives and constraints. As an example, the aerodynamic loads on a bridge affect the structural deformation of the supporting structure. The structural deformation in turn changes the shape of the bridge and hence the aerodynamic loads. Thus, it can be considered a cyclic mechanism. Therefore, in analyzing a bridge, the aerodynamic and structural analyses must be run a number of times in turn until the loads and deformation converge.

4.2.13 Representation in Standard Form

Once the design variables, constraints, and objectives and the relationships between them have been chosen, the problem can be expressed as shown in Equation 4.1.

Maximization problems can be converted to minimization problems by multiplying the objective by −1. Constraints can be reversed in a similar manner. Equality constraints can be replaced by two inequality constraints.

4.2.14 Problem Solution

The problem is normally solved by choosing the appropriate techniques from those available in the field of optimization. These include gradient-based algorithms and population-based algorithms, among others. Very simple problems can sometimes be expressed linearly; in that case, the techniques of linear programming are applicable.

- Population-Based Methods
  - Genetic algorithms [5]
  - Particle swarm optimization
- Other Methods
  - Random search
  - Grid search
  - Simulated annealing

Most of these techniques require a large number of evaluations of the objectives and the constraints. The disciplinary models are often very complex, and a single evaluation can take a significant amount of time. The solution can therefore be extremely time consuming. Many optimization techniques are adaptable to parallel computing. Much of the current research is focused on methods of decreasing the computation time.

The following steps summarize the general procedure used to formulate and solve optimization problems. Some problems may not require that the engineer follow the steps in the exact order, but each of the steps should be considered in the process.

1. Analyze the process itself to identify the process variables and specific characteristics of interest, that is, make a list of all the variables.
2. Determine the criterion for optimization and specify the objective function in terms of the above variables together with coefficients.
3. Develop via mathematical expressions a valid process model that relates the input-output variables of the process and associated coefficients. Include both equality and inequality constraints. Use well-known physical principles such as mass balances, energy balance, empirical relations, implicit concepts, and external restrictions. Identify the independent and dependent variables to obtain the number of degrees of freedom.
4. If the problem formulation is too large in scope:
   a. Break it up into manageable parts, or
   b. Simplify the objective function and the model.

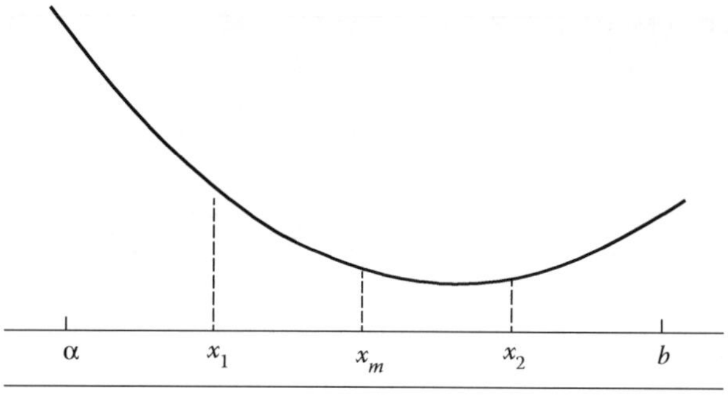

Figure 4.2

Monotonic increasing function.

5. Apply a suitable optimization technique for mathematical statement of the problem.
6. Examine the sensitivity of the result to changes in the values of the parameters in the problem and the assumptions.

4.3 Function of a Single Variable in Optimal Engineering Design

4.3.1 Monotonic Function

A function $f(x)$ is *monotonic* (either increasing or decreasing) if for any two points x_1 and x_2, with $x_1 \leq x_2$, it follows that

$f(x_1) \geq f(x_2)$ (monotonically decreasing) (Figure 4.2)

4.3.1.1 Unimodal Function

A function $f(x)$ is unimodal on the interval $a \leq x \leq b$ if and only if it is monotonic on either side of the single optimal point x^* in the interval.

Minimize $(x) = x^2 + 54/x$ in the interval (0,5) (Figure 4.3)

In other words, if $x^* = 3$ is the single minimum point of $f(x)$ in the range, then $f(x)$ is a unimodal on the interval if and only if for any two points x_1 and x_2,

$$2 \leq x^* \leq 4 \text{ implies that } f(x^*) \leq f(x_1) \leq f(x_2)$$

A function does not have to be continuous to be *unimodal*. This is an extremely important functional property used in optimization.

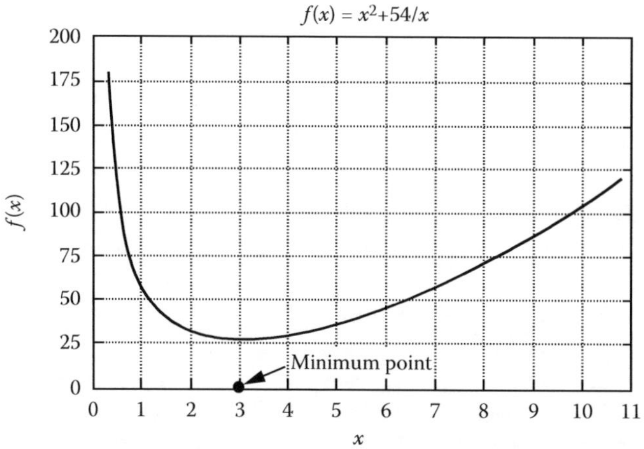

Figure 4.3

Unimodal function.

4.3.1.2 Optimality Criteria

In considering optimization problems, two questions generally must be addressed:

1. How to recognize a variable is optimal
2. If it is not, how to find an optimal solution

For a single variable function $f(x)$, the *necessary* condition for optimality is

$$\frac{df}{dx} = 0\left(\text{at } x = x^*\right) \tag{4.9}$$

The *sufficient* condition for *minimization* is

$$\frac{d^2 f}{dx^2}\bigg|x = x^* \geq 0 \tag{4.10}$$

and for *maximization*, the *sufficient* condition is

$$\frac{d^2 f}{dx^2} \leq 0$$

However, it should be noted that the abovementioned conditions are a guarantee that x^* is a local minimum (maximum).

4.3.1.3 Inflexion Point or Saddle Point

An inflection or saddle point is a stationary point that does not correspond to a local optimum (maximum or minimum). The conditions for inflection points are described below.

Suppose at a point x^* the first derivative is zero and *first nonzero* higher-order derivative is denoted by n.

i. If n is *odd*, x^* is a point of inflection.
ii. If n is *even*, x^* is a local optimum. Moreover,
    a. If that derivative is positive, the point x^* is a local minimum.
    b. If that derivative is negative, the point x^* is a local maximum.

Example 4.2

$$f(x) = 5x^6 - 36x^5 + \frac{165}{2}x^4 - 60x^3 + 36$$

$$\frac{d}{dx}f(x) \rightarrow 30x^5 - 180x^4 + 330x^3 - 180x^2$$

After factoring

$$30x^2(x-1)(x-2)(x-3) = 0$$

and solutions are $x=0,1,2,3$, which produce the optimality criteria.

Optimality Criteria

The function is continuous at all points. But it is not differentiable at the point $x=2$, and the function attains its maximum at $x=2$, which is not a stationary point by our definition.

Example 4.3

Consider the function

$$f(X) = 5x^6 - 36x^5 + \frac{165}{2}x^4 - 60x^3 + 36$$

defined on the real line. The first derivative of this function is

$$f'(x) = 30x^5 - 180x^4 + 330x^3 - 180x^2 = 30x^2(x-1)(x-2)(x-3)$$

Clearly, the first derivative has stationary points at $x = 0,1,2,3$. The second derivative of f is

$$f''(x) = 150x^4 - 720x^3 + 990x^2 - 360x$$

Evaluating this derivative at the four candidate points $x = 0,1,2,3$, we obtain

x	f(x)	f″(x)
0	36	0
1	27.5	60
2	44	120
3	5.5	240

We conclude that $x = 1$, 3 are local minima and $x = 2$ is a local maximum. In order to characterize $x = 0$, the next higher derivative must be evaluated:

$$f'''(x) = 600x^3 - 2160x^2 + 1980x - 360 \text{ . So}$$

$$f'''(0) = -360$$

Since this is an odd-order derivative and is nonzero, the point $x = 0$ is not an optimum point but an inflection point.

Let us check for the sufficiency condition. After finding the second derivative of $f(x)$, we get

$$150x^4 - 720x^3 + 990x^2 - 360x = 0$$

We find the second derivative values at $x = 0,1\ldots3$.

We can conclude that $x = 1,3$ are local minima and $x = 2$ is a local maximum. In order to characterize $x = 0$, the next higher derivative must be evaluated.

$$y(x) = 600x^3 - 2160x^2 + 1990x - 360$$

At

$$x = 0,1..\,3$$
$$y(x) =$$

−360
70
−220
2.37 × 10³

After putting $x = 0$, we get

$$\frac{d^3}{dx^3}f(x) = -360$$

Since this is an odd derivative and is nonzero, the point $x=0$ is not an optimum point but an inflection point.

We can plot the $f(x)$ function and see the optimum points and inflexion points.

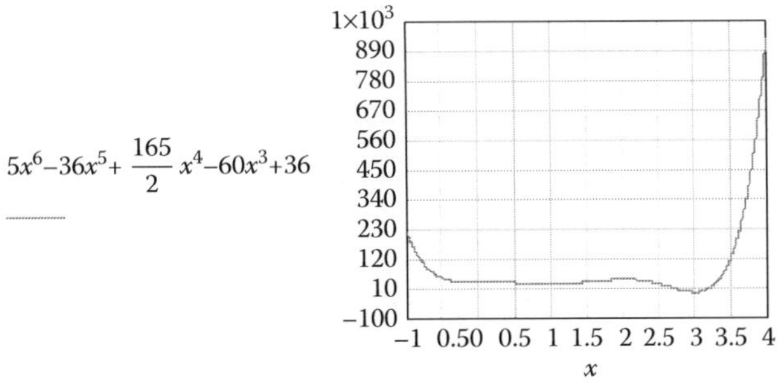

4.3.2 Concept of Global Optimum

The global optimum has to be a local best. The algorithm for global optimum is given as below:

Maximize $f(x)$
Subject to $a \le x \le b$

where a and b are practical limits on the values of variable x.

4.3.2.1 Algorithm

Step 1. Set $df/dx=0$ and compute all stationary points.

Step 2. Select all stationary points in the range a and b. Call them $x_1, x_2, \ldots x_n$. These points, along with a and b, are the only points that can qualify for a local optimum.

Step 3. Find the largest value of $f(x)$ out of $f(a)$, $f(b)$, $f(x_1)$, $f(x_2), \ldots, f(x_n)$. This value becomes the *global maximum point.*

4.3.3 Region-Elimination Methods

Search methods that can locate a single variable optimum by successively eliminating subintervals so as to reduce the remaining interval of search are called *region-elimination methods.*

Concept: Suppose f is strictly unimodal on the interval $a \le x \le b$ with a minimum at x^*. Let x_1 and x_2 be two points in the interval such that $a < x_1 < x_2 < b$. Computing the functional values at x_1 and x_2, we can conclude:

(i) If $f(x_1) > f(x_2)$, then the minimum of $f(x)$ does not lie in the interval (a, x_1). In other words, x^* lies in the range (x_1, b).

(ii) If $f(x_1) < f(x_2)$, the minimum does not lie in the interval (x_2, b), and x^* lies in the interval (a, x_2).

4.3.4 Global Optimum

The next question is how to determine the global maximum or minimum of a function of one variable. Since the global optimum has to be a local optimum, a simple approach is to compute a H local optima and choose the best. An algorithm based on this is presented as:

Maximize $f(x)$
Subject to constraint: $a \le x \le b$

where a and b are practical limits on the values of the variable x.

Once a function is bounded in an interval, in addition to the stationary points, the boundary points can also qualify for the local optimum.

Step 1. Set $df/dx=0$, and compute all stationary points.

Step 2. Select all stationary points that belong to the interval $[a,b]$. Call them $x_1, x_2, .. x_n$. These points, along with a and b, are the only points that can qualify for a local optimum.

Step 3. Find the largest value of $f(x)$ out of $f(a)$, $f(b)$, $f(x_1)$, $f(x_2), \ldots, f(x_n)$. This value becomes the global maximum point.

Note: we did not try to classify the stationary points as local minimum, local maximum, or inflection points, which would have involved the calculation of higher-order derivatives. Instead, it is easier just to compute their functional values and discard them.

Example 4.4

Maximize $f(x) = -x^3 + 3x^2 + 9x + 10$ in the interval $-2 \leq x \leq 4$.
Set $df/dx = -3x^2 + 6x + 9$.

Solving the derivative $f(x)$ equation, we get $x = 3$, and $x = -1$ within the interval. To find the global maximum, evaluate $f(x)$ at $x = 3, -1, -2$, and 4. $f(3) = 37, f(-1) = 5, f(-2) = 12, f(4) = 30$.

Hence, $x = 3$ maximizes f over the interval $(-2, 4)$.

Instead of evaluating at the stationary points and their functional values, we could use certain properties of the function to determine the global optimum more quickly. Earlier, we introduced unimodal functions, for which a local optimum is the global optimum. Unfortunately, the definition of unimodal functions does not suggest a simple test for determining whether or not a function is unimodal. However, there exists an important class of unimodal functions in optimization theory, known as convex and concave functions, which can be identified with some simple tests.

Example 4.5

Let us examine the properties of the function

$$f(x) = (2x + 1)^2(x - 4)$$

$$f'(x) = (2x + 1)(6x - 15)$$

$$f''(x) = 24(x - 1)$$

For all $x \leq 1, f''(x) \leq 0$, and the function is concave in this region. For all $x \geq 1, f''(x) \geq 0$, and the function is convex in the region.

Note that the function has two stationary points, $x = -1/2$ and $x = 5/2$. The function has a local maximum at $x = -1/2$ and a local minimum at $x = 5/2$.

Bounding Phase Method

In the initial phase, starting at some selected trial point, the optimum is roughly bracketed within a finite interval by using the elimination property. Typically, this bounding search is conducted using some heuristic expanding pattern, although extrapolation methods have also been devised. An example of an expanding pattern is in Swann [13], in which the $(k + 1)$st test point is generated using the recursion

$$X_{k+1} = X_k + 2^k\Delta \text{ for } k = 0, 1, 2, \ldots \quad (4.11)$$

where:

X_0 is an arbitrarily selected starting point
Δ is a step-size parameter of suitably chosen magnitude

The sign of Δ is determined by comparing $f(X_0)$, $f(X + [\Delta])$, and $f(X - [\Delta])$. If

$$f(X_0 - \Delta) \geq f(X_0) \geq f(X + \Delta)$$

then, because of the unimodality assumption, the minimum must lie to the right of X_0, and Δ is chosen to be positive. If the inequalities are reversed, Δ is chosen to be negative; if

$$f(X_0 - \Delta) \geq f(X_0) \leq f(X + \Delta)$$

the minimum has been bracketed between $X_0 - \Delta$ and $X_0 + \Delta$, and the bounding search can be terminated. The remaining case is ruled out by the unimodality assumption. However, occurrence of the above condition indicates that the given function is not unimodal.

Example 4.6

Consider the problem of minimizing $f(x) = (100 - x)^2$ given the starting point $x_0 = 30$ and a step size $\Delta = 5$.

The sign of Δ is determined by comparing

$$f(30) = 4900, \quad f(x_0 + \Delta) = f(35) = 4225 \quad \text{and} \quad f(x_0 - \Delta) = f(25) = 5625$$

Since

$$f(x_0 - \Delta) \geq f(x_0) \geq f(x_0 + \Delta)$$

Δ must be positive, and the minimum point x^* must be greater than 30. Then $x_1 = x_0 + \Delta = 35$.

Next, $x_2 = x_1 + 2\Delta = 45$

$$f(45) = 3025 < f(x_1)$$

therefore, $x^* > 35$.

$$x_3 = x_2 + 2^2\Delta = 65$$

$$f(65) = 1225 < f(x_2)$$

therefore, $x^* > 45$.

$$x_4 = x_3 + 2^3\Delta = 105$$

$$f(105) = 25 < f(x_3)$$

therefore, $x^* > 65$.

$$x_5 = 2^4\Delta = 185$$

$$f(185) = 7225 > f(x_4)$$

therefore, $x^* < 185$. Consequently, in six iterations x^* has been bracketed within the interval $65 \leq x^* \leq 185$.

Note that the effectiveness of the bounding search depends directly on the size of Δ. If Δ is large, a poor bracket, that is, a large initial interval, is obtained. On the other hand, if Δ is small, many iterations may be necessary before the bounds can be established.

4.3.5 Interval Refinement Phase

Once a bracket has been established around the optimum, then more sophisticated interval-reduction schemes can be applied to obtain a refined estimate of the optimum point. The portion of subinterval eliminated depends on the location of the trial points x_1 and x_2 within the search interval. Since we have no prior knowledge of the location of the optimum, it is reasonable to expect that the location of the trial points might be above or below the optimum point.

Example 4.7

Minimize $f(x) = (100 - x)^2$ over the interval $60 \leq x \leq 150$. Here $a = 60$ and $b = 150$, and $L = 150 - 60 = 90$, where L presents the range of x values.

$$x_m = \frac{60 + 150}{2} = 105$$

Stage 1

$x_1 = a + L/4 = 60 + 90/4 = 82.5$
$x_2 = b - L/4 = 150 - 90/4 = 127.5$
$f(82.5) = 306.25 > f(105)$
$f(127.5) = 756.25 > f(105)$

Hence, drop the intervals (60, 82.5) and (127.5, 150). The length of the sear interval is reduced from 90 to 45 as $82.5 \leq x \leq 127.5$.

Stage 2

$a = 82.5, b = 127.5, x_m = 105$
$L = 127.5 - 82.5 = 45$
$x_1 = 82.5 + 445 = 93.75$
$x_2 = 127.5 - 41 = 116.25$

We find

$$f(93.5) = 39.06 > f(I\ 05) = 25 \text{ and } f\ (162.5) = 264.06 > f\ (i\ 05)$$

Hence, the interval of uncertainty is (93.75, 116.25).

Stage 3

$a = 93.75, b = 116.25, x = 105$
$L = 116.25 - 93.75 = 22.5$
$x_1 = 99.375$
$x_2 = 110.625$
$f(99.375) = 391 < f(105) = 25$

Golden Section Method

In the golden section method, the interval-reduction strategy is independent of the number of trials. This method deals with the division of an interval into two unequal parts such that the ratio of the smaller to the larger interval is equal to the ratio of the larger to the whole. This is shown graphically in Figure 4.4.

According to the rule stated above, we can write

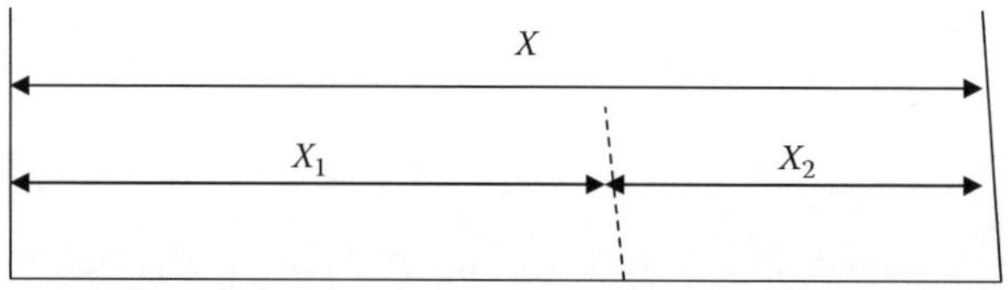

Figure 4.4

Golden section method.

$$\frac{X_2}{X_1} = \frac{X_1}{X} \quad (4.12)$$

We know

$$X = X_1 + X_2 \quad (4.13)$$

Combining Equations 4.12 and 4.13, we can write

$$\left(\frac{x_2}{x_1}\right)^2 + \frac{x_2}{x_1} = 1, \text{ Let } R = X_2 / X_1$$

so the equation is written as

$$R^2 + R - 1 = 0 \quad (4.14)$$

The positive root of Equation 4.14 gives $R = 0.618$. If the golden section rule is applied in a search scheme, the interval uncertainty is reduced by the value of 0.618. By repeated application of the method, the interval of uncertainty after n evaluations is

$$I_n = I_1 (0.618)^{n-1} \quad (4.15)$$

where:

R_n is interval of uncertainty after n evaluations
R_1 is initial interval of uncertainty

Normally the search is carried out over a range of values between the upper bound (B_u) and the lower bound (B_l).

Total range of evaluation can be defined as

$$I_1 = B_u - B_l \quad (4.16)$$

The search scheme is shown graphically in Figure 4.5.

Graphically, it is evident that $X_l = B_u - 0.618I$ and $Xu = B_l + 0.618I$. Both these values are numerically calculated at iteration 1, and the search continues till an optimum is found and interval I is extremely small with a preassumed accuracy of ε.

It would be better to explain the algorithm by an example [10].

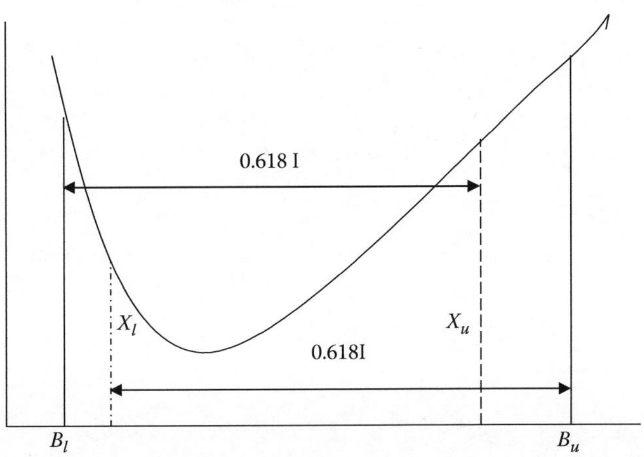

Figure 4.5

Golden section method search.

Example 4.8: Golden Section Method

Find the maximum value of the function $f(x) = x \cos \pi x^2$. The upper (B_u) and lower bounds (B_l) are given as $0.0 \leq x \leq 0.7$, that is, $B_u = 0.7$ and $B_l = 0.0$.

Iteration 1: The interval $I = 0.7 - 0.0 = 0.70$ and we calculate $X_u = 0.0 + 0.618(0.7) = 0.4326$ and $X_l = 0.7 - 0.618(0.7) = 0.2674$. We should calculate $f(X_u) = f(0.4326) = 0.3600$ and $f(X_l) = f(0.2674) = 0.2600$.

Since $f(X_u) > f(X_l)$, the maximum will lie in the range between X_l, that is, 0.2674 and B_u, that is, 0.7. The algorithm now defines $B_l = 0.2674$ and $B_u = 0.7$. The algorithm goes on defining the new interval till I becomes very small or less than the accuracy assumed as ε.

Iteration 2: new $I = 0.7 - 0.2674 = 0.4326$. According to the algorithm, the second interval I may be called $I_2 = 0.4326$. New values of X_{u2} and X_{l2} are calculated. The subscript 2 refers to the second iteration of new lower and upper values of X.

$X_{l2} = 0.7 - 0.618(0.4326) = 0.4326$ and $X_{u2} = 0.2674 + 0.618(0.4326) = 0.53475$.

Now, we can generalize the interval calculation as $I_n = I_1(0.18)^{n-1}$ for the third iteration we calculate.

Iteration 3: $I_3 = 0.7(0.618)^{3-1} = 0.26735$. We can use Equation 4.15 to calculate the interval at any iteration. Now we can calculate the new values of X_{u3} and X_{l3}. The iterations will terminate as follows.

#	BL ($BL_n = XL_{n-1}$)	BU ($BU_n = BU_{n-1}$)	XL ($XL_n = BU_n -$ gold(I_n))	XU ($XU_n = BL_n -$ gold(I_n))	f(XL) $f(X) = x \cos \pi x^2$	f(XU) $f(X) = x \cos \pi x^2$	I(($I_n := I_l$) gold^{n-1})
1	0	0.7	0.267	0.433	0.261	0.36	0.7
2	0.267	0.7	0.433	0.535	0.36	0.33	0.433
	....	....	....	....	....	......	....
19	0.46	0.4589	0.4589	0.4589	0.362	0.363	0.0001

The last iteration shows the $\varepsilon = 0.46 - 0.4589 = 0.0011$, which is a small value, and the accuracy is acceptable. The golden section search scheme can be implemented very easily in MATLAB®, MathCAD or any other software. The golden section method also has a magnificent past, as it was used extensively by Greek architects as well as by Leonardo de Vinci in his paintings of the human body.

4.3.6 Methods Requiring Derivatives

The search methods discussed in the preceding sections required the conditions of unimodality and, in some cases, continuity of the performance being optimized. It seems reasonable that, in addition to continuity, it is assumed that the function is differentiable [4, 11]. When $f(x)$ is a function of third-degree or higher terms involving x, analytic solution of $f'(x) = 0$ would be difficult. Hence, a search method successively approximating the stationary point of $f(x)$ is required. These gradient-based methods are broadly grouped as

- Newton's method
- Steepest descent
- Conjugate gradient
- Sequential quadratic programming

We describe a classical search scheme for finding the root of a nonlinear equation that was originally developed by Newton and later refined by Raphson [11].

4.3.7 Newton–Raphson Method

The Newton–Raphson scheme requires that the function f be twice differentiable. It begins with a point x that is the initial estimate or approximation to zero. In MATLAB, the roots of a polynomial are obtained very easily by writing an M-file as below.

Roots of a polynomial:

```
% Polynomial "5*x^5 + 7*x^4 + 2*x^2 – 6*x + 10"
% Coefficients of the Polynomial
A = [5 7 0 2–6 10]
% roots (A)
X = roots (A)
```

This is based on the Newton–Raphson method. The algorithm calculates the new value of X as follows:

$$X_{n+1} = X_n - \frac{f(x_n)}{f'(x_n)}$$

where n refers to number of iterations. This iteration continues till $|X_{n+1} - X_n| \le \varepsilon$, a very small value or desired accuracy, or even an acceptable error such as 10^{-3}. The Newton–Raphson method should be differentiable. We have to assume an initial value of x for iteration. The selection of initial value will determine the number of iterations needed, but it may not affect the accuracy of the solution.

Example 4.9

$f(x) = 2x^2 + 1 - e^x$

$f^{'}(x) = 4x - e^x$

Assuming initial value $x_0 = 0.7$, $f(0.7) = -0.03375$ and $f'(0.7) = 0.78625$.
We calculate the new value of x_1 from the equation

$$x_1 = x_0 - \frac{f(x_0)}{f^{'}(x_0)} = 0.74293$$

Then, x_2 is calculated as in the first step, and we get $x_2 = 0.74086$.

We calculate $|x_2 - x_1| = 0.00207$. We calculate $x_3 = 0.74085$. Then, $|x_3 - x_2| = 0.00001$, which is below $\varepsilon = 10^{-3}$, and so it is the final solution, $x = 0.741$.

The Newton–Raphson method can be modified further by including the second derivative in the iteration, provided function $f(x)$ is differentiable again or the second derivative exists.

4.3.8 Newton Second-Order Method

The algorithm is presented as follows.

$$X_{n+1} = X_n - \left[\frac{f(x_n)}{f(x_n) - \left\{ \frac{f^{''}(x_n) f(x_n)}{2 f^{'}(x_n)} \right\}} \right]$$

$$f(x) = 2x^2 + 1 - e^x$$

Example 4.10

$$f'(x) = 4x - e^x \text{and } f''(x) = 4x - e^x$$

Initial value $x_0 = 0.7$ (assumed). Values for $f(x_0) = 0.03375$, $f'(x_0) = 0.78625$, and $f''(\text{x}) = 1.98624$. From these values, we find $x_1 = 0.740272$. Similarly, if we iterate, we get $x_2 = 0.74085$ and $x_3 = 0.74085$. This shows no change in the value of x, so we accept the final solution as $x = 0.74087$ with an error of zero.

4.3.9 Constrained Optimization

Most real-world problems in engineering are constrained for design parameters as well as functions: the design dimensions should be constrained within a feasible range, and some functional parameters, such as stresses and strains, should be lower than the property of materials. Most design problems would be limited by the strength of the material being used. Sometimes, manufacturing problems are constrained by the parameters of the dimensions of the machine tools as well as by cutting tools. However, it should be obvious to engineers that solutions for constrained and unconstrained problems are different. These constrained optimization problems of design and manufacturing can be divided into equality-constrained and inequality-constrained problems.

4.3.10 Equality-Constrained Optimization

Equality-constrained problems are converted to unconstrained optimization problems by eliminating or reducing the number of variables by the number of equality constraints in the problem. That is, if there are n variables and k equality constraints, the size of the problem can be reduced to $(n-k)$ by explicitly eliminating k independent variables using equality constraints. As an example, we can present

$$\text{Minimize } f(x) = x_1 x_2 x_3$$

$$\text{Subject to constraint } g_1(x) = x_1 + x_2 + x_3 = 1$$

From the equality constraint given above, we can reduce or eliminate one of the variables in terms of another, for example, $x_1 = 1 - x_2 - x_3$.

Then, the objective function is written as

$$\text{Minimize } f(x) = (1 - x_2 - x_3) x_2 x_3$$

This becomes an unconstrained problem in two variables $\{x_2, x_3\}$, such that the size of the problem is reduced from 3 to 2. In engineering problems, practicing engineers can think of other relationships among variables and reduce the size of the problem further. This variable elimination method is applicable as long as the equality constraint can explicitly be solved for a given set of independent variables. In some cases, it may not be possible to eliminate or reduce the size of the problem due to nonlinearity in the constraint.

4.3.11 Lagrange Multiplier Method

This method converts an equality-constrained problem to an unconstrained problem by using Lagrange multipliers. This creates a set of necessary conditions to find the optimum points [3].

$$\text{Minimize } f(X), \text{ subject to } g_j(X) = 0$$

The Lagrange multiplier method converts this constrained optimization problem to an unconstrained problem:

$$\text{Minimize } L(X, \lambda) = f(X) - \lambda g_j(X)$$

where:

$j = (1, 2, \ldots m)$
$\lambda = (1, 2, \ldots m)$

The unconstrained function $L(X,\lambda)$ is called the Lagrangian function, and λ is the Lagrange multiplier, which may be either positive or negative. These multipliers have both geometrical and physical meaning in optimization. Their values depend on the form of cost and constraint functions.

Example 4.11

$$\text{Minimize } f(x_1, x_2) = x_1^2 + x_2^2$$

$$\text{Subject to constraint } g_1(x_1, x_2) = 2x_1 + x_2 = 6$$

The Lagrangian function can be written by taking the Lagrange multiplier as

$$L(x_1, x_2, \lambda) = x_1^2 + x_2^2 - \lambda(2x_1 + x_2 - 6)$$

Finding the derivatives of the Lagrange function with respect to x_1, x_2, and λ, we get

$$\frac{\partial L}{\partial x_1} = 2x_1 - 2\lambda = 0, \frac{\partial L}{\partial x_2} = 2x_2 - \lambda = 0, \text{ and } \frac{\partial L}{\partial \lambda} = 2x_1 + x_2 - 6 = 0$$

From the first two gradients, we get $x_1^* = \lambda$ and $x_2^* = \lambda/2$. We have to test whether x_1^* and x_2^* are minimizing, and for this purpose the Hessian of the Lagrange function with respect to x_1 and x_2 needs to be written as

$$H(x_1, x_2, \lambda) = \begin{bmatrix} \dfrac{\partial^2 L}{\partial x_1^2} & \dfrac{\partial^2 L}{\partial x_1 \partial x_2} \\ \dfrac{\partial^2 L}{\partial x_2 \partial x_1} & \dfrac{\partial^2 L}{\partial x_2^2} \end{bmatrix} = \begin{bmatrix} 2 & 0 \\ 0 & 2 \end{bmatrix}$$

The Hessian is positive definite as $|H| = 4$. This implies that the function is a convex function for all values of x. We determined $x_1^* = \lambda$ and $x_2^* = \lambda/2$. Entering these values into the constraint equation, we get $\lambda = 12/5$, and therefore $x_1^* = 12/5$ and $x_2^* = 6/5$. The minimum value of $f^*(x_1, x_2) = (12/5)^2 + (6/5)^2 = 18/5$. The Lagrange multiplier method has been extended to more than one equality constraint, as presented below.

Minimize $f(x)$, subject to constraints $g_j(x) = 0$, $j = 1, 2, \ldots, m$.

The Lagrange function can be developed by using m Lagrange multiplier as

$$L(x, \lambda) = f(x) - \sum_{j=1}^{m} \lambda_j g_j(x)$$

By finding the partial derivatives with respect to x and equating them to zero,

$$\frac{\partial L(x, \lambda)}{x_i} = 0, \; i = 1, 2, \ldots n$$

These n partial derivatives along with m constraints or total $(n + m)$ equations are sufficient to solve uniquely for n variables with m Lagrange multipliers.

4.3.12 Kuhn–Tucker Conditions

It was previously established that for both an unconstrained optimization problem and an optimization problem with an equality constraint, the first-order conditions are sufficient for a global optimum when the objective and constraint functions satisfy appropriate concavity/convexity conditions. The same is true for an optimization problem with inequality constraints. Kuhn and Tucker extended the Lagrange multiplier method to both inequality and equality constraints. This is the foundation of general nonlinear programming (NLP) optimization problems. The Kuhn–Tucker (K-T) conditions are both necessary and sufficient if the objective function is concave and each constraint is linear or each constraint function is concave, that is, the problems belong to a class called *convex programming problems*.

Consider the following optimization problem:

$$\text{Minimize } f(X)$$

$$\text{subject to constraints } g_j(x) \geq 0 \text{ for } j = 1, 2, \ldots, J$$

and

$$h_k(x) = 0, \text{ for } k = 1, 2, \ldots, K$$

where $X = [x_1 \, x_2 \ldots x_n]$.

Then, the K-T conditions for $X^* = [x_1^* \, x_2^* \ldots x_n^*]$ to be a local minimum are

$$\nabla f(x) - \sum_{j=1}^{J} u_j \nabla g_j(x) - \sum_{k=1}^{K} v_k \nabla h_k(x) = 0 \tag{4.17}$$

$$g_j(x) \geq 0 \text{ for } j = 1,2,\ldots J \quad (4.18)$$

$$h_k(x) = 0 \text{ for } k = 1,2,\ldots K \quad (4.19)$$

$$u_j g_j(x) = 0 \text{ for } j = 1,2,\ldots,J \quad (4.20)$$

where:

$u_j \geq 0$ for $j = 1,2,\ldots,J$ (note: u_j are restricted to either zero or positive)

v_k is unrestricted in sign

Example 4.12

$$\text{Minimize } f(x) = x_1^2 - x_2$$

$$\text{subject to constraints } x_1 \geq 1, \quad x_1^2 + x_2^2 \leq 26$$

$$x_1 + x_2 = 6$$

Applying the K-T conditions:

$\nabla f(x) = (2x_1, -1)$ (derivative of objective function)
$g_1(x) = x_1 - 1$ $\nabla g_1(x) = (1,0)$ (derivative of first constraint)
$g_2(x) = 26 - x_1^2 - x_2^2$ $\nabla g_2(x) = (-2x_1, -2x_2)$ (derivative of second constraint)
$h_1(x) = x_1 + x_2 - 6$ $\nabla h_1(x) = (1,1)$ (derivative of first equality constraint)

Applying the K-T condition (Equation 4.17), we get

$$2x_1 - u_1 + 2u_2x_2 - v_1 = 0$$

and

$$-1 + 2u_2x_2 - v_1 = 0$$

Applying Equations 4.18 and 4.19, we have

$$x_1 - 1 \geq 0,\ 26 - x_1^2 - x_2^2 \geq 0, \text{ and } x_1 + x_2 - 6 = 0$$

Finally, applying Equation 4.20, we find

$$u_1(x_1 - 1) = 0 \text{ and } u_2(26 - x_1^2 - x_2^2) = 0$$

As indicated earlier, u_1 and u_2 are restricted to being positive or zero, while v_1 is unrestricted in sign. We can write all the K-T conditions together and find the solution.

$$2x_1 - u_1 + 2u_2x_2 - v_1 = 0 \quad (4.21)$$

$$-1 + 2u_2x_2 - v_1 = 0 \quad (4.22)$$

$$x_1 - 1 \geq 0 \quad (4.23)$$

$$26 - x_1^2 - x_2^2 \geq 0 \quad (4.24)$$

$$x_1 + x_2 - 6 = 0 \quad (4.25)$$

$$u_1(x_1 + 1) = 0 \quad (4.26)$$

$$u_2(26 - x_1^2 - x_2^2) = 0 \quad (4.27)$$

where:

$u_j \geq 0$,

$j = 1,2$, and

v_1 is unrestricted

Now, we should test the Hessian matrix of $f(x)$.

$$H(x)=\begin{bmatrix}\dfrac{\partial^2 f(x)}{\partial x_1^{\ 2}} & \dfrac{\partial^2 f(x)}{\partial x_1 \partial x_2}\\ \dfrac{\partial^2 f(x)}{\partial x_2 \partial x_1} & \dfrac{\partial^2 f(x)}{\partial x_2^2}\end{bmatrix}=\begin{bmatrix}2 & 0\\ 0 & 0\end{bmatrix}$$

The $H(x)$ is positive semidefinite, and therefore the objective function $f(x)$ is convex. Similarly, we should also find the Hessian for all the three constraints.

The constraints $g_1(x)$ and $h_1(x)$ are linear. The Hessian of $g_2(x)$ is

$$H(g_2(x))=\begin{bmatrix}2 & 0\\ 0 & 2\end{bmatrix}$$

The determinant $|H(g_2(x))|=0$; hence, the constraint is negative definite, and so the constraint is concave. The sufficiency condition of the K-T theorem is satisfied, as all the constraints are concave, and we know the objective function is convex. Therefore, the solution $(x_1^*, x_2^*, u_1^*, u_2^*, v_1^*)$ satisfying the K-T equations (A) is the optimal solution for the NLP. We find the solution as follows.

Examining Equations 4.23 and 4.25, we can conclude that $x_1^*=1$, $x_2^*=5$. Now, examining Equations 4.26 and 4.27, we can conclude that u_1 and u_2 are not zero. Examining Equations 4.21 and 4.22, we can find $u_1^*=2.2$ and $u_2^*=0.1$ by setting $v_1^*=0$. Then for $v_1^*=0$, it is concluded that $x_1^*=1$, $x_2^*=5$, and $f^*(x)=-4$.

However, it must be recognized that there is another solution for the problem. Let us state very clearly the K-T sufficiency theorem:

If the objective function f(x) is a convex function and the constraints g(x) and h(x) are concave, then the solution (x_1^*, x_2^*) *obtained for the set of equations A is an optimal.*

Example 4.13

$$\text{Minimize } f(x)=x_1^2+x_2^2-3x_1x_2$$

$$\text{subject to constraint } \ g(x)=x_1^2+x_2^2-6\le 0$$

Solution

The Lagrange function for the problem is

$$L=x_1^2+x_2^2-3x_1x_2+u(x_1^2+x_2^2-6+s^2)$$

There is only one constraint, all points of feasible region are regular, and the K-T necessary conditions are applicable. The K-T conditions are

$$\frac{\partial L}{\partial x_1}=2x_1-3x_2+2ux_1=0 \qquad \text{(E4.13.i)}$$

$$\frac{\partial L}{\partial x_2}=2x_2-3x_1+2ux_2=0 \qquad \text{(E4.13.ii)}$$

$$x_1^2+x_2^2-6+s^2=0 \qquad \text{(E4.13.iii)}$$

$$us=0 \qquad \text{(E4.13.iv)}$$

$$u\ge 0 \qquad \text{(E4.13.v)}$$

Equations (E4.13i) through (E4.13iv) have four variables: x_1, x_2, s, and u. These are nonlinear equations and difficult to solve analytically for all the roots. However, the K-T conditions can be analyzed as:

Case 1. $u=0$, then Equations (E4.13i) and (E4.13ii) reduce to

$$2x_1 - 3x_2 = 0$$

$$2x_2 - 3x_1 = 0$$

These are linear equations, and the solution can be obtained as $x_1^* = 0$ and $x_2^* = 0$, which is a trivial solution. Now let us examine the other case.

Case 2. When $s=0$, we can solve Equations (E4.13i) and (E4.13iii) simultaneously for x_1, x_2, and u as follows.

Equation (E4.13i) can be used to solve for u as

$$u = -1 + 1.5\frac{x_2}{x_2},$$

and replacing u in Equation (E4.13ii), we get

$$2x_2 - 3x_1 + 2\left(-1 + 1.5\frac{x_2}{x_1}\right)x_2 = 0$$

which yields

$$x_1^2 = x_2^2$$

Now, solving Equation (E4.13iii) with these values, we get four different solutions:

$$x_1 = x_2 = \sqrt{3}, u = 1/2$$

from Equation (E4.13v)

$$x_1 = x_2 = -\sqrt{3}, \text{ and } u = 1/2$$

$$x_1 = -x_2 = \sqrt{3}, \text{and } u = -\frac{5}{2}$$

and finally

$$x_1 = -x_2 = -\sqrt{3}, \text{and } u = -\frac{5}{2}$$

The last two solutions do not satisfy the K-T conditions, as $u<0$. Therefore, the first two are the solution, that is, $x_1 = x_2 = \sqrt{3}$ and $x_1 = x_2 = -\sqrt{3}$.

Case 3. $u=0$ and $s=0$. With these conditions, solutions from Equations (E4.13i) and (E4.13ii) $x_1 = x_2 = 0$, and substituting these values in Equation (E4.13iii), we get $s^2 = 6 \neq 0$, and the K-T conditions cannot be satisfied.

Hence, the solution set for this optimization problem is $\boldsymbol{X}^* = [0\ 0\ 0]$.

Example 4.14

$$\text{Minimize } f(x) = x_1^2 + x_2^2 - 3x_1x_2$$

$$\text{Subject to constraint } g(x) = x_1^2 + x_2^2 - 6 \leq 0$$

Solution

The Lagrange function for the problem is

$$L = x_1^2 + x_2^2 - 3x_1x_2 + u(x_1^2 + x_2^2 - 6 + s^2)$$

There is only one constraint, all points of feasible region are regular, and the K-T necessary conditions are applicable. The K-T conditions are

$$\frac{\partial L}{\partial x_1} = 2x_1 - 3x_2 + 2ux_1 = 0 \quad \text{(E4.14.i)}$$

$$\frac{\partial L}{\partial x_2} = 2x_2 - 3x_1 + 2ux_2 = 0 \quad \text{(E4.14.ii)}$$

$$x_1^2 + x_2^2 - 6 + s^2 = 0 \quad \text{(E4.14.iii)}$$

$$us = 0 \quad \text{(E4.14.iv)}$$

$$u \geq 0 \quad \text{(E4.14.v)}$$

Equations (E4.14i) through (E4.14iv) have four variables: x_1, x_2, s, and u. These are nonlinear equations and difficult to solve analytically for all the roots. However, the K-T conditions can be analyzed as:

Case 1. $u=0$, then Equations (E4.14i) and (E4.14ii) reduce to

$$2x_1 - 3x_2 = 0$$

$$2x_2 - 3x_1 = 0$$

These are linear equations, and the solution can be obtained as $x_1{}^* = 0$ and $x_2{}^* = 0$, which is a trivial solution. Now let us examine the other case.

Case 2. When $s=0$, we can solve Equations (E4.14i) and (E4.14iii) simultaneously for x_1, x_2, and u as follows.

Equation (E4.14i) can be used to solve for u as

$$u = -1 + 1.5\frac{x_2}{x_1},$$

and replacing u in Equation (E4.14ii), we get

$$2x_2 - 3x_1 + 2\left(-1 + 1.5\frac{x_2}{x_1}\right)x_2 = 0$$

which yields

$$x_1^2 = x_2^2$$

Now, solving Equation (E4.14iii) with these values, we get four different solutions:

$$x_1 = x_2\sqrt{3}, u = 1/2$$

from Equation (E4.14v)

$$x_1 = x_2 = -\sqrt{3}, \text{ and } u = \frac{1}{2}$$

$$x_1 = -x_2 = \sqrt{3}, \text{ and } u = -\frac{5}{2}$$

and finally

$$x_1 = -x_2 = -\sqrt{3}, \text{ and } u = -\frac{5}{2}$$

The last two solutions do not satisfy the K-T conditions, as $u < 0$. Therefore, the first two are the solution, that is, $x_1 = x_2 = \sqrt{3}$ and $x_1 = x_2 = -\sqrt{3}$.

Case 3. $u = 0$ and $s = 0$. With these conditions, solutions from Equations (E4.14i) and (E4.14ii) $x_1 = x_2 = 0$, and substituting these values in Equation (E4.14iii), we get $s^2 = 6 \neq 0$, and all K-T conditions cannot be satisfied. Hence, the solution set for this optimization problem is $\boldsymbol{X}^* = [0\ 0\ 0]$. This solution set satisfies Equations 4.15 through 4.19, which was desired. Therefore, the solution set for this optimization problem is $\boldsymbol{X}^* = [80\ 40]$.

4.4 Engineering Design (Optimal Design) Examples

Example 4.15

ABC Corporation received an order from a well-drilling company to produce the piston rod for an air cylinder that would form part of a braking system. The customer has provided the information that because of space limitations, the relation between the length and the diameter of the rod is $L = 12d^{3.5}$ and $6\,\text{in} < L < 8\,\text{in}$. The thrust load is expected to be 20 kips. The material of the rod is expected to be either cold-drawn or hot-drawn steel.

Solution

Johnson's formula for a short column is

$$P_{cr} = A^* S_y \left(1 - \frac{Q}{4r^2}\right) \tag{4.28}$$

Euler's formula for a long column is

$$P_{cr} = \frac{A^* S_y^* r^2}{Q} \tag{4.29}$$

where:

$$Q = \frac{S_y^* L^2}{\pi^2 C * E} \tag{4.30}$$

A = area of section = $\pi^* d^2/4$
L = length of column
C = coefficient of end conditions
E = modulus of elasticity
r = radius of gyration = $\sqrt{I/A}$
I = moment of inertia

Assuming fixed-pinned condition, let $C = 2$ and factor of safety $(n) = 2.0$.
Replacing the values of A, r, and Q in Equation 4.28, we can write

$$P_{cr} = \frac{\pi d^2 * S_y}{4}\left(1 - \frac{4S_y^* L^2}{C\pi^2 E d^2}\right) \tag{4.31}$$

Replacing $P_{cr} = P_n = 2P$ in Equation 4.31, we write

$$P = \frac{\pi d^2 S_y}{8}\left(1 - \frac{4S_y L^2}{2\pi^2 E d^2}\right) \tag{4.32}$$

There are a wide variety of cold-drawn and hot-rolled steels available. Here, G10450 steel is selected from each group. The material properties are

	Cold-drawn steel		Hot-rolled steel	
	Yield strength	*E*	Yield strength	*E*
G10450	77 kpsi	30 Mpsi	69 kpsi	30 Mpsi

Cold-drawn steel is selected for optimal design.

Replacing yield strength and Young's modulus in Equation 4.32, we get

$$P = 30{,}238d^2\left(1 - \frac{5.201\times10^{-4}L^2}{d^2}\right)$$

subject to constraint

$$L = 12d^{3.5}$$

And $6 < L < 8$ (Figure 4.6).

```
%Design of Air Cylinder Piston
% L=12d^3.5
% 6 in<L<8 in
% Thrust load 20 kips
% Data for Hot Rolled Steel and Cold drawn steel
% Yield strength
YSM=[44*10^3 26*10^3; 60*10^3 32*10^3;67*10^3 45*10^3; 71*10^3 64*10^3;...
77*10^3 69*10^3; 84*10^3 78*10^3]
P=20,000;
C=2;
N=2; PI=22./7.;E=30*10^6;drold=1;
%func=R1C*d^2-R2C*d^7-4*(20,000)*2
%dfunc=2*R1C*d-7*R2C*d^6
% Assuming Short Column and applying Johnson;s formula
```

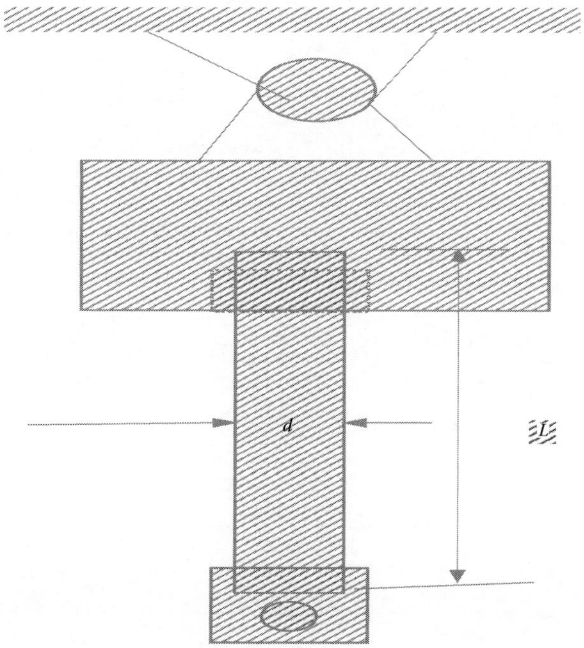

Figure 4.6

Air cylinder design.

```
fprintf(' i R1C R1Hn');
fprintf('.............................n')
for i=1:6
R1C=PI*YSM(:,1)
R2C=576*YSM(:,1).²/(C*PI*E)
dr=drold;
func=R1C*dr²-R2C*dr⁷-4*(20,000)*2
dfunc=2*R1C*dr-7*R2C*dr⁶
dr=dr-func(dr)/dfunc(dr)
end
for j=2:6
R1H=PI*YSM(:,2)
R2H=576*YSM(:,2).²/(C*PI*E)
dr=drold;
func=R1H*dr²-R2H*dr⁷-4*(20,000)*2
dfunc=2*R1H*dr-7*R2H*dr⁶
dr=dr-func(dr)/dfunc(dr)
fprintf(' %5i %8.3f %e %en',i, dr, R1C(i),R1H(j))
end
```

Example 4.16

Design a solid shaft to carry a torsional load of 350 kN and an axial load of 400 kN. We have to find the shaft diameter under the load conditions given. The material of the rod is AISI 1020 CD; yield stress = 420 MPa (Figure 4.7).

Solution

Assume factor of safety (n) = 2. We can apply maximum shear stress theory to design the shaft. We know that $\sigma_x = 4P/\pi d^2$ and $\tau_{xy} = 16T/\pi d^3$.

The maximum shear stress is represented as

$$S_{sy} = \sqrt{\left(\frac{\sigma_x}{2}\right)^2 + \tau_{xy}^2}, \; S_{ssy} = \frac{S_y}{n}$$

Shear yield strength = S_{sy}. We can write

$$\left(\frac{S_y}{2n}\right)^2 = \left(\frac{4P}{\pi d^2}\right)^2 + \left(\frac{16T}{\pi d^3}\right)^2$$

Replacing the values, we can write the above equation as

$$1.103 \times 10^4 = \frac{2.504 \times 10^5}{d^4} + \frac{3.177 \times 10^6}{d^6}$$

Or it can be written in a polynomial form:

$$f(d) = 2.504 \times 10^5 d^6 + 3.177 \times 10^6 d^4 - 1.103 \times 10^4 d^{10}$$

This equation can be solved by the Newton–Raphson method.

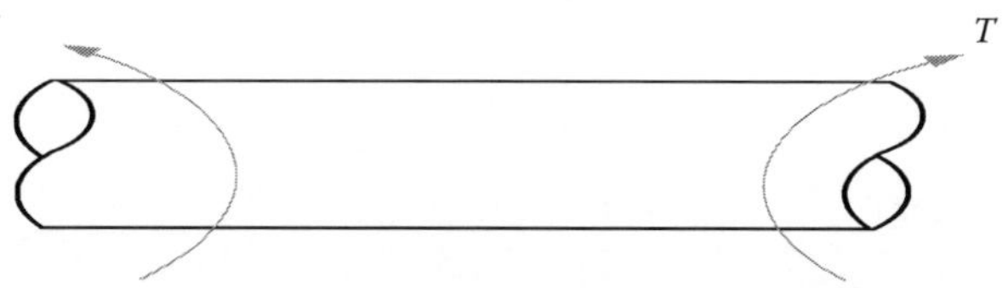

Figure 4.7

Solid shaft for torsional load.

Example 4.17

A hollow steel shaft is to be designed for power transmission. The power source is 30 hp, and the speed of the shaft is 100 rpm. The inside diameter of the shaft is 2 in. If the permissible shearing stress is 11,000 psi, what is the thickness of the shaft t? The objective is to minimize the weight of the shaft. The length of the shaft is given as 5 ft (Figure 4.8).

(Ans. 0.24 in.)

Materials and Properties for Torsion Rod

Material	Density ρ (lbm/in³)	Allowable Shear Stress τ_a (Mpsi)	Elastic Modulus E (Mpsi)	Shear Modulus G (Mpsi)	Poisson Ratio (υ)
4140 Alloy steel	0.283	40	30.45	11.6	0.30
Aluminum alloy 24ST4	0.1	24	10.8	4	0.32
Magnesium alloy A261	0.065	13	6.5	2.4	0.35
Beryllium	0.067	16	43.5	21	0.02
Titanium	0.163	24	16	6.8	0.30

Formulation of the optimal design problem:

The torque on the shaft:

$$T = \frac{63{,}025(\text{hp})}{n} = \frac{63025(30)}{100} = 18{,}908 \text{ lbf.in}$$

The weight of the tubular shaft is

$$f(r_0) = \frac{\pi}{4}(d_0^2 - d_i^2)(60)(0.283) = 13.4(d_0^2 - d_i^2)$$

$$\text{Minimize } f(d_0) = 13.4(d_0^2 - d_i^2)$$

$$\text{Subject to constraint } \frac{18{,}908(d_0+2)/2}{\pi/32(d_0^4 - d_i^4)} \le 11{,}000 \left(\text{shear stress constraint}\right)$$

This can be simplified to write

$$87.547\frac{(d_0 + d_i)}{(d_0^4 - d_i^4)} \le 1$$

It is finally written as

$$1 - 87.547\frac{(d_0 + d_i)}{(d_0^4 - d_i^4)} \ge 0 \left(\text{shear stress constraint}\right)$$

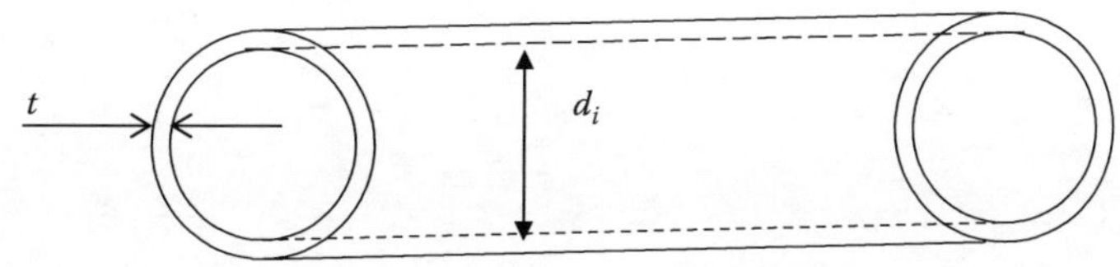

Figure 4.8

Design of hollow shaft.

Let us assume the ratio between the inside and the outside diameter of the transmission shaft to be $0.06 \le d_i/d_0 \le 0.999$. We will assume here that $d_i/d_0 = 0.80$, which is in the middle of the range, and replace $d_i = 0.80\ d_0$ in all the equations above. The reformulated equations are as follows:

$$\text{Minimize } f(d_0) = 4.825 d_0^2$$

$$\text{Subject to constraint } 1 - 263 d_0^{-3} \ge 0$$

Applying geometric programming, we can write

$$\text{Maximize } d(w) = \left(\frac{4.825}{w_{01}}\right)^{w_{01}} \left(\frac{263 w_{11}}{w_{11}}\right)^{w_{11}}$$

$$\text{Subject to constraint } w_{01} = 1$$

$$2w_{01} - 3w_{11} = 0$$

which is solved for $\overset{*}{w_{01}} = 1$, $\overset{*}{w_{11}} = 2/3$, and $d(w)^* = 191$.

Primal variable: $4.825 d_0^2 = \overset{*}{w_{01}} d(w)^* = 191$. We get $d_0 = 6.325$, $d_i = 5$, and $t = 0.63$ in.

This design problem was easy to solve due to the assumed ratio between inside and outside diameters.

Example 4.18

An endless V-belt with a pitch diameter of 500 mm is tightly wrapped around pulleys of pitch radii 600 mm and 40 mm. Determine the central distance of the pulleys.

Let R and r be the pitch radii of the larger and smaller pulleys, respectively, and υ is the half contact angle of the smaller pulley. L is length of pulley, and C is the center distance of the pulley. The relation between these variables is derived as

$$\tan\theta - \theta = \frac{0.5L - \pi R}{R - r}$$

which is calculated from the given data as 3.075.

Writing the above equation as

$$f(\theta) = \tan\theta - \theta - 3.075 \tag{4.33}$$

and

$$C = \frac{R - r}{\cos\theta}$$

Equation 4.33 can be solved using Newton–Raphson.

$$f(0) = \tan(\theta) - \theta - 3.075$$

$$\frac{d}{d\theta} f(\theta) \rightarrow \tan(\theta)^2 \qquad f_1(\theta) = \tan(\theta)^2$$

$$\theta_0 = 0.45\pi$$

$$\theta_1 = \theta_0 - \frac{f(\theta_0)}{f_1(\theta_0)} \qquad \theta_1 = 1.3679345$$

$$\theta_2 = \theta_1 - \frac{f(\theta_1)}{f_1(\theta_1)} \qquad \theta_2 = 1.3502188$$

$$\theta_3 = \theta_2 - \frac{f(\theta_2)}{f_1(\theta_2)} \qquad \theta_3 = 1.34848$$

The final value of θ is 1.34848 in radians. The center distance C can be calculated as 91.1 mm.

Example 4.19

Design of a Crane Hook

A hoist hook is a sturdy, question mark-shaped piece of equipment used to attach heavy objects to ropes or chains in order to lift them. Typically, a hoist hook is made of metal. Some kinds have an open hook, while others latch shut once they have been hooked onto the object they are being used to lift.

The exact shape of a hoist hook may vary depending on its intended use. Grab hooks are designed to hold onto a link of chain, while slip hooks are designed to let the chain slide through them. Grab hooks are generally both lighter and smaller. Slip hooks are often made with a latch to keep the attached object in place.

The end opposite a hoist hook may be designed differently depending on how it will attach to the rope or chain. An eye hook has a circle at the end through which a rope can be threaded. A clevis hook has a U-shaped end with holes at the top of each branch of the U. A small pin can be slid through these holes and secured so that a cable or other piece of equipment can be securely attached. A shank hook ends in a straight bar that can be inserted into a hole in another piece of equipment.

A 250 kN crane hook is to be designed. Hosting speed is assumed to be 4.5 m/min and maximum lift is 10 m.

The cross section of the hook is assumed to be trapezoidal and the factor of safety 1.2 (Figures 4.9 and 4.10).

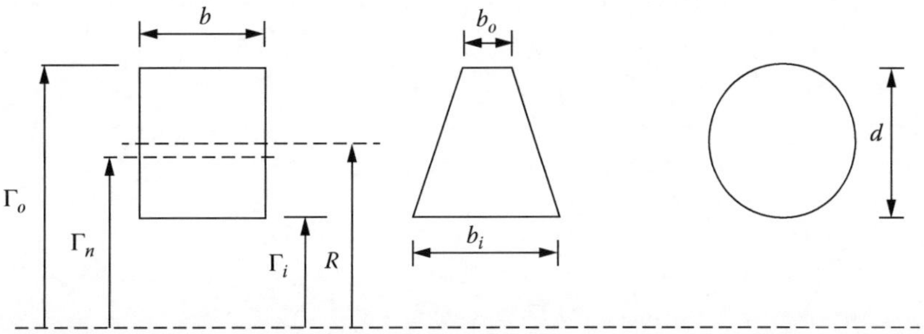

Figure 4.9

Design of hook to lift weight.

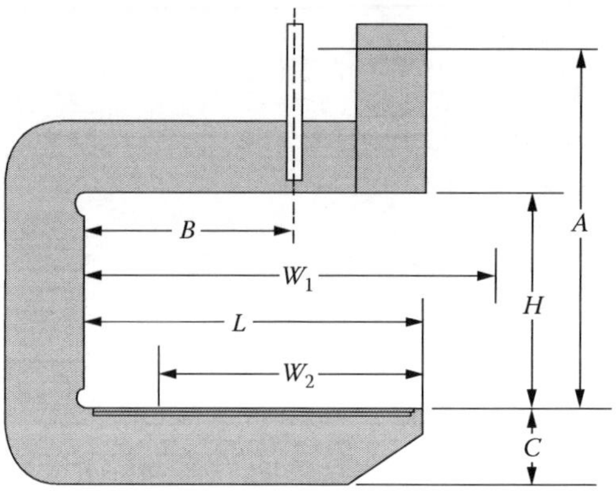

Figure 4.10

Another hook to lift weight.

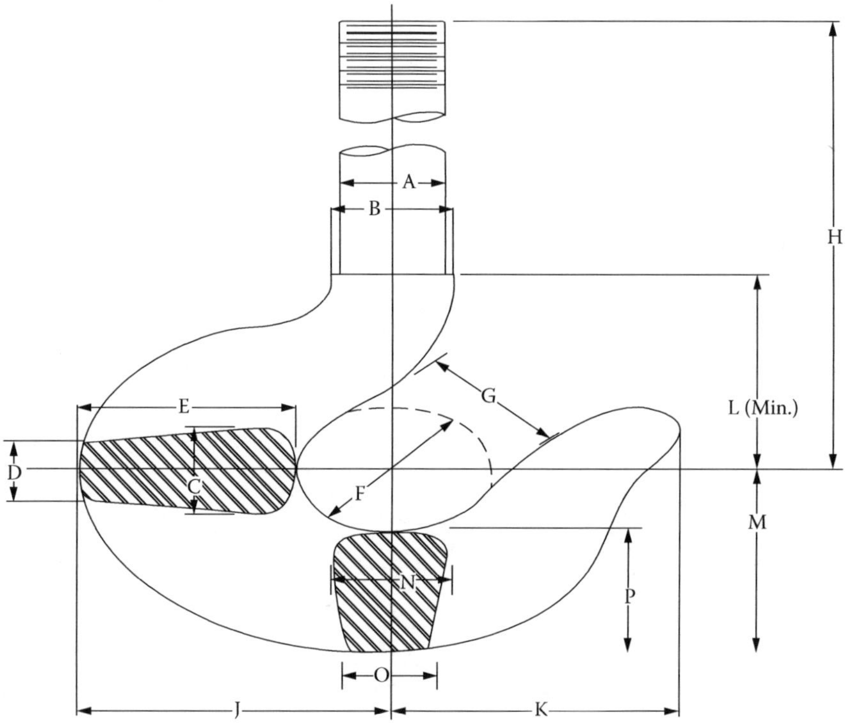

$C = 207$ mm (assumed)

a. Select suitable type and size of wire rope for an expected life of 10 months.
b. Select standard hook, material, and design stresses. Check the induced stresses at the most critical sections.

Solution

Design of Hook

Duty factor = 1.2 for application class II
Safe load $P = 250 \times 1.2 = 300$ KN
$P = 30$ tons
Proof load = 60 tons
Mild steel is selected for hook and $C = 207$ mm.
Various dimensions of hook are provided in terms of C as follows:

A = 2.75, $C = 570$ mm; B = 1.31, $C = 271$ mm; D = 1.44, $C = 298$ mm; E = 1.25, $C = 259$ mm; F = 207 mm; $C = 193$ mm; J = 0.75, $C = 155$ mm; K = 0.92, $C = 190$ mm; L = 0.7, $C = 145$ mm;

M = 0.6, C = 124 mm; N = 1.2, C = 248 mm; mP = 0.5, C = 104 mm; R = 0.5, C = 104 mm; U = 0.3, C = 62 mm; Z = 0.12, C = 25 mm; G = 110 mm, M = 110 mm, H = 193 mm

Design handbooks give these empirical (proportionate) dimensions in terms of C, and this empirical relation has been adopted here.

Checking the hook at the most critical section,

$$\text{Area of trapezium } (A) = \frac{1}{2}(b_i + b_0)h$$

(A) = ½(124 + 50) × 193
(A) = 16.791×10^3 mm^2

Direct tensile stress

σ_t = [P]/A = $300 \times 10^3/16.791 \times 10^3 = 17.89$ N/mm^2
b_i = M = 124 mm
b_0 = 2 Z = 25 × 2 = 50 mm
h = H = 193 mm
$r_i = C/2 = 207/2 = 103.5$ mm
$h = r_0 - r_i$
$r_0 = r_i + h = 103.5 + 193 = 296.5$ mm
r_n = distance from the center of curvature to the neutral axis
R = distance from the center of curvature to the centroidal axis
$h_i = r_n - r_i$
$h_i = 171.51 - 103.5$
$h_i = 68.01$ mm distance of innermost fiber from neutral axis (NA)
$h_0 = r_0 - r_n$
$h_0 = 296.5 - 171.51$
$h_0 = 124.99$ mm distance of outermost fiber from NA.
Bending moment (Mb) = $P \times$ R
$P \times$ R = $300 \times 10^3 \times 104$
= 55.893 × 106 N-mm.

Bending stress maximum at outer fiber PSG 6.2

$$r_n = \frac{H}{\ln(r_0/r_i)} = \frac{193}{\ln(296.5/103.5)} = 183.3 \text{ mm}$$

$$R = r_i + \frac{h}{3}\left(\frac{b_i + 2b_0}{b_i + b_0}\right)$$

At the section when
R = 186.31 mm
Bending moment (Mb) = $P \times R$
$P \times R = 300 \times 10^3 \times 186.31$
= 55.893×10^6 N-mm
e = $R - r_n$
e = 186.31 – 183.3.51
e = 3.10 mm

$$\sigma_m = \frac{F}{A} + \frac{M*y}{A*e(r_n - y)} = \frac{300*10*^3}{16.73*10^3} + \frac{(55.893*10^6)*124.09}{(16.79.10^3)(3.1)(183.3-124.99)} = 2.33*10^3 \text{N/mm}^2$$

MS for hook
Selecting C-45 as hook material,
yield stress = 36 kgf/mm^2
$\sigma_{yt} = 360$ N/mm^2
factor of safety = 2
$[\sigma_t] = 180$ N/mm^2
As $\sigma_t < \sigma_{yt}$,

design is safe in tension.

Example 4.20

Design a tubular column of mean radius r_m and thickness t to minimize the weight subject to stress and buckling constraints. The material of the column is 1030 HR steel of 37.5 kpsi yield strength and mass density 0.28 lb/in^3, and $E = 30$ Mpsi.

The objective function can be written as

$$\text{Minimize weight } W = \rho(AL) = 2\pi\rho L r_m t$$

$$\text{Subject to constraint } \frac{P}{2\pi r_m t} \leq \sigma \text{ (stress constraint)}$$

and

$$P \leq \frac{\pi^3 E r_m^3 t}{4L^2} \text{ (buckling load constraint)}$$

The column is expected to carry a load of 5000 lb, and length of the column (L) is desired to be 15 ft due to functional requirements. Design the tubular column using the K-T conditions (Figure 4.11). Reformulate the problem with data given above.

$$\text{Minimize } f(r_m, t) = 316.673 r_m t$$

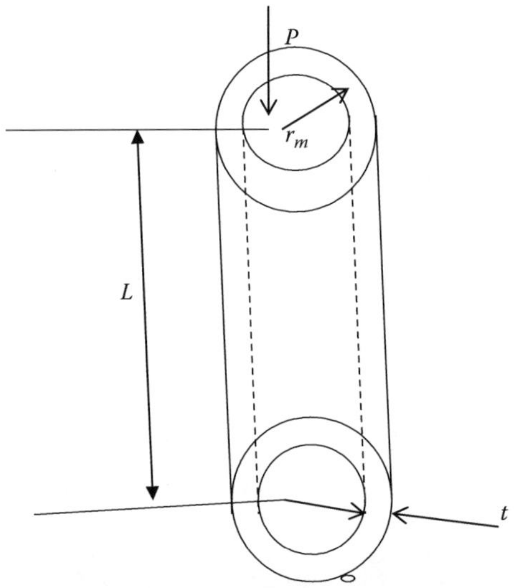

Figure 4.11

A hollow tubular column under buckling.

$$\text{Subject to constraint } 0.021r_m^{-1}t^{-1} \leq 1$$

or

$$g_1(r_m, t) = 1 - 0.021r_m^{-1}t^{-1} \geq 0$$

$$0.617r_m^{-3}t^{-1} \leq 1$$

or

$$g_2(r_m, t) = 1 - 0.617r_m^{-3}t^{-1} \geq 0$$

$$r_m \geq 0, t \geq o$$

Now applying K-T conditions as discussed in Section 4.3.10,

$$\nabla f(r_m, t) = (316.673t, 316.673r_m)$$

$$\nabla g_1 = (0.021r_m^{-2}t^{-1}, 0.021r_m^{-1}t^{-2})$$

$$\nabla g_2 = (1.851r_m^{-4}t^{-1}, 0.617r_m^{-3}t^{-1})$$

The equations generated by K-T conditions are

$$316.617t - u_1(0.021r_m^{-1}t^{-1} - u_2(1.851r_m^{-4}t^{-1}) = 0 \tag{4.34}$$

$$316.673r_m - u_1(0.021r_m^{-1}t^{-3} - u_2(0.617r_m^{-3}t^{-1}) = 0 \tag{4.35}$$

$$u_1(1 - 0.021r_m^{-1}t^{-1}) = 0 \tag{4.36}$$

$$u_2(1 - 0.617r_m^{-3}t^{-1}) = 0 \tag{4.37}$$

$$1 - 0.021r_m^{-1}t^{-1} \geq 0 \tag{4.38}$$

$$1 - 0.617r_m^{-3}t^{-1} \geq 0 \tag{4.39}$$

Writing Equations 4.38 and 4.39 in the original form and dividing Equation 4.39 by 4.38, we can write

$$\frac{0.617r_m^{-3}t^{-1}}{0.021r_m^{-1}t^{-1}} \leq 1$$

or

$$29.4r_m^{-2} \leq 1$$

Solving this, we find $r_m = 5.4$ in. and $t = 0.004$ in.

Plugging these values into Equations 4.34 and 4.35, we get two linear equations in terms of u_1 and u_2:

$$u_1 + 0.544u_2 = 1.27 \text{ and } 60764u_1 + 0.98u_2 = 1710$$

Solving these linear equations, we find $u_1^* = 0.0281$ and $u_2^* = 2.3$.

We can also test Equations 4.37 and 4.38 and see that they are satisfied.

Table 4.1 List of Materials from Environmental Considerations

Material		Yield Stress σ_y (MPa)	Specific Weight ρ (kN/mm³)	Embodied Energy e (MJ/kg)	Carbon Footprint C (kg/kg)	Recycle Percentage
Aluminum	2014–T4	290	26.67	200–240	11–13	33–55
	6061–T6	276	26.85			
Magnesium alloy	AM 100 A	152	17.81	356–394	500–1500	9.2–11.4
	Red Brass	496	85.76			
Steel	AISI 1010 HR	179	77.27	29–35	2.2–2.8	40–44
	AISI 4140 HR	434	77.27			

Usually, there is a ratio between the values of mean radius (r_m) and thickness (t) of the tubular column from the manufacturing constraints. We assume here $r_m/t \leq 50$.

With this assumption, we can modify the value of $t = r_m/50 = 5.4/50 = 0.11$ in.

From practical manufacturing considerations, the preferred sizes are $r_m^* = 5.4$ in and $t^* = 0.12$ in; the minimum weight of the column is $W^* = 316.673(5.4)(0.12) = 206$ lb.

Example 4.21: Rectangular Torque-Carrying Member Ecodesign

An environmentally suitable rectangular, torque-carrying member is to be used in the aircraft. It is concluded after engineering analysis of weight considerations that a thin-walled tube should be used. From their knowledge of engineering materials, engineers prescreen a list of materials in terms of environmental considerations, and these are presented in Table 4.1.

The ecoproperties given in the last two columns are taken from Ashby [13].

We will minimize the total embodied energy (TEE):

$$\text{Minimize } TEE = e(\text{volume})(\text{density}) = e(AL)(\rho) = e\rho\left[2t(b+h) - 4t^2\right] \tag{4.40}$$

$$\text{Subject to } \frac{T}{2bht} \leq \tau_a \leq \frac{\sigma_y}{2n} \tag{4.41}$$

where T is the torque for which the member is to be designed and n is the factor of safety, assumed to be 2. The cross section of the torque-carrying member is shown in Figure 4.12.

From the constraint equation, we can write

$$b = \frac{T \cdot n}{ht\sigma_y}$$

T, n, and σ_y are known. Therefore, b can be replaced in terms of h and t in the objective function. The new objective function is

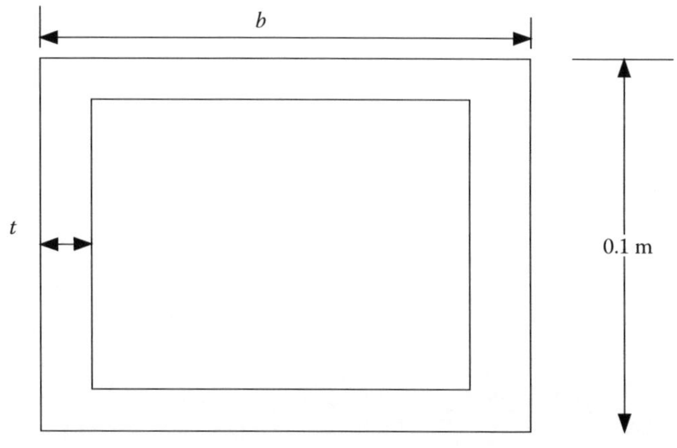

Figure 4.12

Rectangular torque-carrying mechanical component.

$$TEE = e\left[2t\left(\frac{Tn}{ht\sigma_y}+h\right)-4t^2\right](\rho)$$

Applying another constraint on height (h) as $h \le h_{max}$, we can replace h by h_{max} in the objective function and present it as

$$\text{TEE} = e\rho\left[2t\left(\frac{Tn}{h_{max}t\sigma_y}+h_{max}\right)-4t^2\right] \tag{4.42}$$

To minimize *TEE*, we should maximize the material index (*MI*) (Figure 4.12):

$$M_1 = \frac{\sigma_y}{e\rho}$$

Ashby denoted three terms on the right-hand side as functional, geometric, and material parameters [13]. The objective function in more general terms

$$f(TE) = f(F,G,M), \text{ function } f(TE)$$

is called the performance index. However, for MS to minimize the environmental damage, it is required to minimize the term $(e\rho/\sigma_y)$. This is customarily written as maximizing *MI*.

To minimize *TEE*, we should maximize *MI*:

$$M_1 = \frac{\sigma_y}{e\rho} \tag{4.43}$$

Ashby's material performance chart could be used to select the candidate materials for design of the torque-carrying member of an aircraft [13]; however, in this case it is provided in the design. The yield strength vs. energy content MS chart can be generated using Cambridge Engineering Selector V6 software and then selecting the candidate materials.

For minimization of carbon footprint, the M_2 can be developed as

$$M_2 = \frac{\sigma_y}{C\cdot\rho}$$

In engineering design problems, the objectives could be either minimization of embodied energy or minimization of carbon footprint. In Figure 4.13, both calculations are shown.

The *MI* line can be drawn at a slope of 1. This is explained as follows.

Taking the log of Equation 4.43 on both sides of equation, we can write

$$\log(\sigma_y) = \log(e\rho) + \log(M_1) \tag{4.44}$$

Equation 4.44 shows that the slope of the line σ_y vs. $(e\rho)$ should be 1, or 45°. Selecting material for environmentally friendly aircraft parts is based on minimum embodied energy, and this is analyzed in the table below.

Material		Yield Stress σ_y (MPa)	Specific Weight ρ (kN/mm³)	Embodied Energy e (MJ/kg)	Carbon Footprint C (kg/kg)	$M_1 = \sigma_y/e\rho$	$M_2 = \frac{\sigma_y}{C\rho}$
Aluminum	2014–T4	290	26.67	200–240	11–13	49425.64	906,137
Magnesium alloy	AM 100 A	152	17.81	356–394	22.4–24.8	22,759	36,787
Steel	AISI 1010 HR	179	77.27	29–35	2.2–2.8	72,392.26	880,031

The calculated values of maximum values of *MI* for minimization of *TEE* for aluminum, magnesium alloy, and steel are 49425.64, 22759, and 72392.26, respectively. Therefore, steel seems to be the

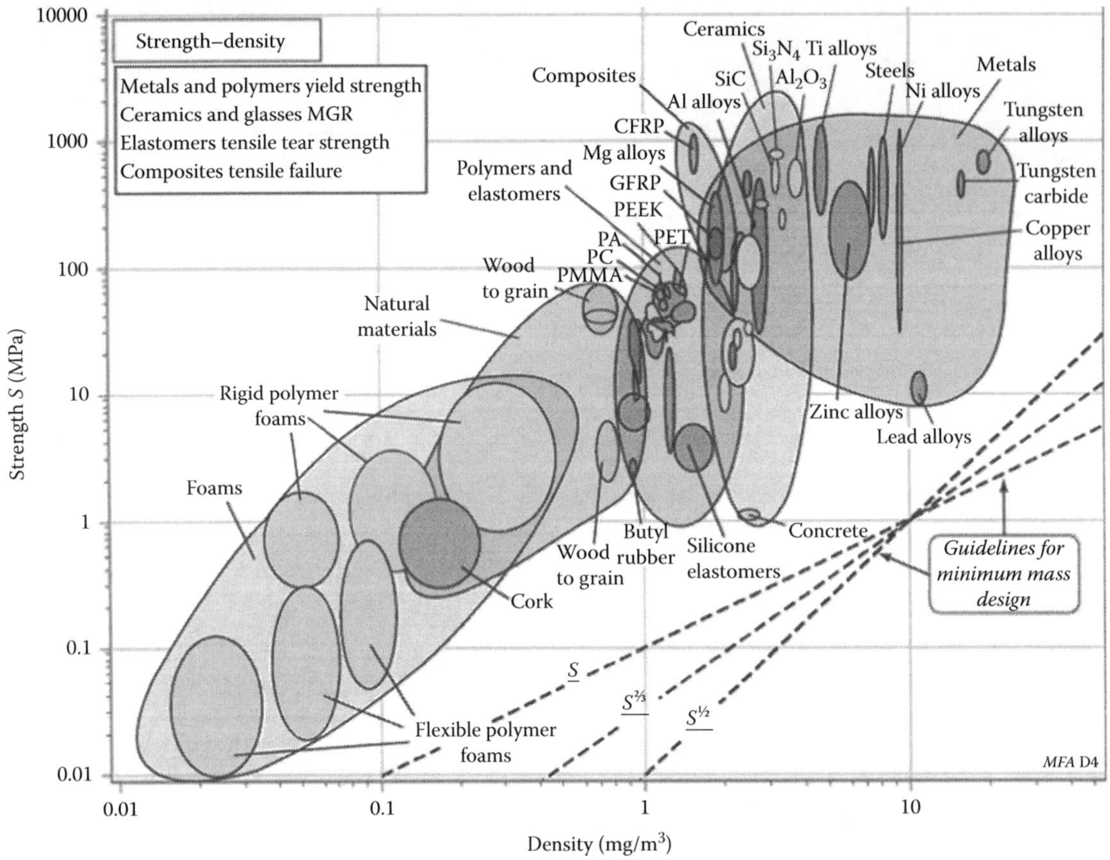

Figure 4.13

Density vs. strength graph. (From Ashby, M. F. *Materials and the Environment*. Elsevier, 2009.).

preferred material for an aircraft torque converter. However, if we examine the problem from multiple objective considerations of maximizing both M_1 and M_2 simultaneously, the material selected could be quite different.

Material		$M_1 = \sigma_y / e\rho$	$M_2 = \sigma_y / C\rho$	Max(M_1, M_2)
Aluminum	2014–T4	49,425.64	906,137	906,137
Magnesium alloy	AM 100 A	22,759	36,787	36787
Steel	AISI 1010 HR	72,392.26	880,031	880,031

According to the maximum (M_1, M_2) values, aluminum is the best material from environmental considerations. Therefore, Aluminum 2014–T4 is the material selected for the eco-friendly design of the aircraft torque-carrying part.

For design of the part, we will apply a simple single-variable optimization principle to Equation 4.42:

$$TEE = e\rho\left[2t\left(\frac{Tn}{h_{\max} t \sigma_y} + h_{\max}\right) - 4t^2\right]$$

By differentiating the equation above with respect to t and equating to zero, we obtain

$$t = \frac{h_{\max}}{4}$$

and therefore

$$b = \frac{4Tn}{h_{\max}{}^{2}.\sigma_y}$$

Now the dimensions of an eco-friendly torque-carrying member of an aircraft are calculated. Let torque $T = 107$ N.m, $h_{\max} = 0.1$ m, and factor of safety $n = 2$.

$$t = \frac{0.1}{4} = 0.25 \text{ m, and } b = 4(107)\frac{2}{(0.1^2 \times 290 \times 10^6)} = 0.285 \text{ m}$$

Example 4.22

A cylindrical pressure vessel is to be designed for minimum cost, as shown in Figure 4.14. The design engineers have decided that candidate material for the pressure vessel is SAE 285 C low-carbon steel. The joint efficiency j is assumed to be known. According to ASME Pressure Vessel standards (Figure 4.14),

$$\text{The thickness of hemispherical heads } (x_4) \geq \frac{px_2}{(2\sigma_y j - 0.2p)} \tag{4.45}$$

$$\text{The thickness of cylindrical shell } (x_1) \geq \frac{px_2}{(\sigma_y j - 0.6p)} \tag{4.46}$$

The pressure vessel will be used to store gas at pressure $p = 300$ psi, and the capacity of the pressure vessel will be 750 ft³. Due to space limitations, the length of the shell (x_2) cannot be more than 20 ft. Shell of such length cannot be rolled, so it is likely to be rolled in two halves and then joined by two longitudinal welds to form a cylinder. The hemispherical ends are to be forged and then welded to the shell. All welds will be single-welded butt joints with backing strip and will be fully radiographed. Let $\sigma_y = 17{,}500$ psi, $j = 1.0$.

The objective function to be formulated is for minimum cost. The cost components are variable costs and fixed costs. Variable costs include material costs and manufacturing costs of welding, forging, and rolling together. These variable costs are a function of the design variables x_1, x_2, x_3, and x_4. Fixed costs such as overhead costs do not depend on the design variables. Therefore, we will only include variable costs in the formulation of the optimal design of the pressure vessel. The variable costs will include cost of shell, cost of heads, and costs of welding the shell and the heads. The costs of the shell and heads can be formulated as

$$\text{Shell cost} = 2\pi\rho C_s x_2 x_1 x_3 \tag{4.47}$$

$$\text{Head cost} = 4\pi\rho C_h x_2^2 x_4 \tag{4.48}$$

where:

ρ is the specific weight of steel (lb/in³)

C_s and C_h represent the costs of material of shell and head, respectively ($/lb)

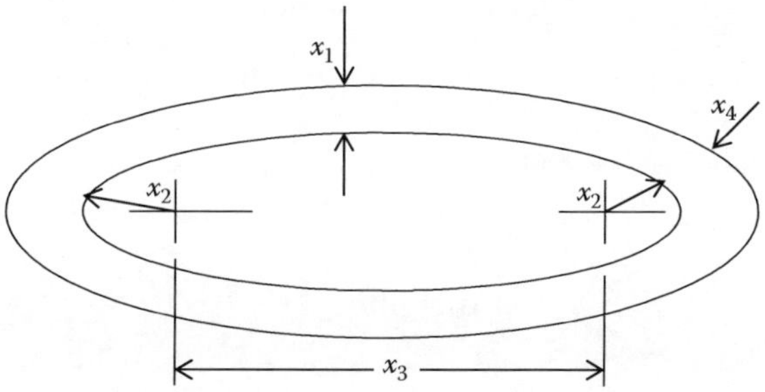

Figure 4.14

Cylindrical pressure vessel.

The welding cost for both shell and head is formulated as

$$\text{Total welding cost} = (\pi C_w \rho / 3\cos^2 30^0)x_1^2 x_3 + (2\pi^2 C_w \rho / 3\cos^2 30)x_2 x_4 \quad (4.49)$$

The sides of the weld are assumed to be at 60° angles. The cost coefficients in the above equations are assumed as $C_s = \$0.45/\text{lb}$, $C_h = \$5.0/\text{lb}$, and $C_w = \$30/\text{lb}$. The material is low-carbon steel, and $\rho = 0.283\,\text{lb/in}^3$. The *TC* objective function is then formulated as

$$TC = \text{shell cost} + \text{head cost} + \text{welding cost}$$

Entering the values given above, we can write the *TC* equation as
Minimize

$$TC = 0.8x_1 x_2 x_3 + 17.8x_2^2 x_4 + 11{,}856x_1^2 x_3 + 74.5x_2 x_4 \quad (4.50)$$

4.5 Geometric Programming

A geometric program (GP) is a type of mathematical optimization problem characterized by objective and constraint functions that have a special form [9, 14]. Recently developed solution methods can solve even large-scale GPs extremely efficiently and reliably; at the same time, a number of practical problems, particularly in circuit design, have been found to be equivalent to (or well approximated by) GPs. Putting these two together, we get effective solutions for practical problems. The basic approach in GP modeling is to attempt to express a practical problem, such as an engineering analysis or design problem, in GP format. In the best case, this formulation is exact; when this is not possible, we settle for an approximate formulation. We start with the basic definitions and facts and some methods used to transform problems into GP format. We show how to recognize functions and problems compatible with GP, and how to approximate functions or data in a form compatible with GP (when this is possible). We give some simple and representative examples and also describe some common extensions of GP, along with methods for solving (or approximately solving) them.

The function $h_m(t)$ is a polynomial function. Duffin et al. [14] have shown how to maximize the dual. Minimize $h_0(t)$, $t_j > 0$, $j = 1,2,\ldots,r$ subject to constraints: $h_m(t) \leq 1$, $m = 1,2,\ldots$, where $h_m(t) = \sum_{k=1}^{I(m)} C_{mt} \Pi_{j=1}^{r} t_j^{a_{mkj}}$, $m = 0,1,2,\ldots$, and $C_{mt} > 0$ and a_{mkj} are real numbers. The coefficients C_{mk}, the exponents a_{mkj}, and I (m) are all known. The dual is formulated as

$$\text{Maximize } V(n) = \Pi_{m=0}^{P} \Pi_{k=1}^{I(m)} \left(\frac{C_{mk} N_m}{\delta_{mk}} \right)^{\delta_{mk}} \quad (4.51)$$

$$\text{Subject to constraints } \delta_{mj} \geq 0, \quad m = 0,1,\ldots,P, \quad j = 1,2\ldots\ldots,I(m)$$

$$\sum_{j=1}^{I(0)} \delta_{0j} = 1 \text{ (normality condition)} \quad (4.52)$$

$$\sum_{m=0}^{P} \sum_{j=1}^{I(m)} a_{mjk} \delta_{mj} = 0 \text{ (orthogonality condition)}, \; \delta_m = \sum_{j=1}^{I(m)} \delta_{mj} \quad m = 0,1,2,\ldots \quad (4.53)$$

4.5.1 Geometric Programming Model

The general primal problem of GP is to

$$\text{Minimize } g_0(x)$$

$$\text{Subject to constraints } g_i(x), i = 1,2,\ldots m,\, x > 0$$

where $x \subset R^m$ and each function $g_i(x)$ is a polynomial with T_i terms, that is,

$$g_i(x) = \sum_{m=1}^{T_i} c_{mt} \prod_{k=1}^{N} x_k^{\,mtk}$$

The exponents a_{mtk} are real numbers, but the coefficients c_{mt} are real positive constants, and decision variables x_n are required to be strictly positive. The degree of difficulty (DOD) is defined as DOD = T–(, + 1). The DOD represents the complexity and difficulty involved in the solution. However in GP there are ways to reduce DOD and make the solution easier by modifying the mathematical model. We will be exploring such tactics during the solution, as the electroactive polymer (EAP) actuator model is highly nonlinear and complex.

4.5.1.1 Examples for Geometric Programming Problems

Example 4.23

A new overhead storage tank is to be designed for storing water. It is proposed that the tank capacity should be about 50,000 ft³. The material cost of the top and bottom of the tank is $60.00 per square foot. However, it is estimated that the sides will cost $50.00 per square foot. From design considerations, the tank should have two supporting beams of 3 in radius to avoid hurricane damage. The cost of the supporting beams is $50.00 per foot. The schematic of the tank is shown here.

Objective function formulation

Total cost function (TC) = cost of sides + cost of top and bottom + cost of supporting beams

Cost of sides = $2\pi R_1 H(\$50.0)$

Cost of top and bottom = $2 \times \pi R_1^2(\$60.00)$

Cost of the beams = $2 \times 50.0(\sqrt{H^2 + 4R_1^2}$

Total cost $TC(R_1, H) = 314.16 R_1 H + 120\pi R_1^2 + 100\sqrt{H^2 + 4R_1^2}$

Replacing $R_1 = x_1$, $H = x_2$, and $\sqrt{H^2 + 4R_1^2} = x_3$

We can write the TC equation as

$$TC(x_1, x_2, x_3) = 314.16 x_1 x_2 + 377 x_1^2 + 100 x_3 \tag{4.54}$$

Now we should develop the constraints. We can write

$$\sqrt{R_1^2 + 4H^2} = x_3$$

or

$$4x_1^2 + x_2^2 = x_3^2$$

This can be written as

$$4x_1^2 x_3^{-2} + x_2^2 x_3^{-2} \le 1 \tag{4.55}$$

Volume constraint should also be formulated

$$\text{Volume} \ge 5000$$

Therefore

$$\text{Volume} = \text{volume of tank} - \text{volume of beams}$$

$$\text{Volume} = \pi R_1^2 H - (2\pi R_2^2)\sqrt{H^2 + 4R_1^2}$$

Writing in X, $R_2 = 3$ in is given in the problem, so

$$\text{Volume} = 3.14 x_1^2 x_2 - 0.393 x_3$$

Volume constraint can be written as

$$3.14x_1^2 - 0.393x_3 \geq 5000$$

Or it can be reformulated as

$$5000 + 0.393x_3^{0.5} \leq 3.14x_1^2 x_2$$

or

$$1592.357x_1^{-2}x_2^{-1} + 0.1252x_1^{-2}x_2^{-1}x_3 \leq 1 \tag{4.56}$$

The complete optimal design mathematical model is shown below:

$$\text{Minimize } TC(x_1, x_2, x_3) = 314.16x_1x_2 + 377x_1^2 + 100x_3 \tag{4.57}$$

$$\text{Subject to constraints } 4x_1^2x_3^{-2} + x_2^2x_3^{-2} \leq 1 \tag{4.58}$$

$$1592.357_1^{-2}x_2^{-1} + 0.1252x_1^{-2}x_2^{-1}x_3 \leq 1 \tag{4.59}$$

$$x_1 \geq 0, x_2 \geq 0, x_3 \geq 0$$

Normally, in the GP format the complete mathematical model for optimal design is written as

$$g_0(X) = 314.16x_1x_2 + 377x_1^2 + 100x_3 \tag{4.60}$$

Subject to constraints

$$g_1(X) = 4x_1^2x_3^{-2} + x_2^2x_3^{-2} \tag{4.61}$$

$$g_2(X) = 1592.357x_1^{-2}x_2^{-1} + 0.1252x_1^{-2}x_2^{-1}x_3 \tag{4.62}$$

$$x_1 \geq 0, x_2 \geq 0, x_3 \geq 0$$

Equations 4.60 through 4.62 are together called the *primal geometric program* (PGP). Now the PGP can be solved. The dual of the PGP can be written as

$$d(w) = \left(\frac{314.16}{w_{01}}\right)^{w_{0101}} \left(\frac{377}{w_{02}}\right)^{w_{02}} \left(\frac{100}{w_{03}}\right)^{w_{03}} x \left[\frac{1592..357w_{10}}{w_{11}}\right]^{w_{11}} \left[\frac{0.1252}{w_{12}}\right]^{w_{12}} \left[\frac{4w_{20}}{w_{21}}\right]^{w_{21}} \left[\frac{w_{20}}{w_{22}}\right]^{w_{22}} \tag{4.63}$$

Subject to constraints

$$w_{01} + w_{02} + w_{03} = 1$$

$$w_{01} + 2w_{02} + 2w_{11} - 2w_{21} - 2w_{22} = 0$$

$$w_{01} + 2w_{12} - w_{21} - w_{22} = 0$$

$$w_{03} - 2w_{11} - 2w_{12} + w_{22} = 0$$

$$w_{10} = w_{11} + w_{12}$$

$$w_{20} = w_{21} + w_{22}$$

Now the *dual geometric program* (DGP) can be solved to find optimum dual values, and finally, optimal design parameters (x) are obtained.

$$w_{01}^* = 0.525, \quad w_{02}^* = 0.00625, \quad w_{03}^* = 0.4688$$

$$w_{11}^* = 0.4522, \quad w_{12}^* = 0.088, \quad w_{10}^* = 0.5402$$

$$w_{21}^* = 0.0073, \quad w_{22}^* = 0.271, \quad w_{20}^* = 0.2783$$

$$d(w^*) = 17.2 = g_0(X^*)\,(\text{minimum cost value})$$

The primal optimal design solution is calculated using the GP method presented earlier:

$$x_1^* = 0.52, \quad x_2^* = 55.72, \quad x_3^* = 80.1$$

That is, radius of tank $R_1 = 0.52$ in; height of tank $H = 55.72$ in; and length of supporting beam $x_3 = 80.1$ in.

GP helps in finding the weightage of different terms of a function even when the coefficients of the terms are not known. However, this is limited to zero-DOD problems.

$$g_0(x) = c_1 x_1^{-3} x_2^{-2} + c_2 x_1^3 x_2 + c_3 x_1^{-3} x_2^3$$

Example 4.24

Let these three terms of the $g_0(x)$ function represent different costs of a *TC* function. The first term represents the set up cost; the second term represents the production costs; and the third term represents the handling costs. We can write dual GP as

$$\text{Maximize } d(w) = \left(\frac{c_1}{w_{01}}\right)^{w_{01}} \left(\frac{c_2}{w_{02}}\right)^{w_{02}} \left(\frac{c_3}{w_{03}}\right)^{w_{03}}$$

$$\text{subject to } w_{01} + w_{02} + w_{03} = 1$$

$$-3w_{01} + 3w_{02} - 3w_{03} = 0$$

$$-2w_{01} + w_{02} + 3w_{03} = 0$$

These three constraints are independent linear equations in three unknown weights. There are exactly enough equations to determine the optimal weights uniquely. This is because the problem is of zero DOD. Any simultaneous equation solution technique, such as Gauss elimination, Gauss–Jordan, or the matrix inversion method, can be used to find $w_{01} = 0.4$, $w_{02} = 0.5$, and $w_{03} = 0.1$.

This illustrates that the *TC* is the summation of 40% set up cost, 50% production cost, and 10% handling cost. However, if we want to know the absolute value of the *TC*, we must know the cost coefficients. Suppose $c_1 = 75$, $c_2 = 80$, and $c_3 = 35$, then

$$g_0^*(x) = \left(\frac{75}{0.4}\right)^{0.4} \left(\frac{80}{0.5}\right)^{0.5} \left(\frac{35}{0.1}\right)^{0.1} = 184.36$$

To find the primal variables x^*, we can write

$$75 x_1^{-3} x_2^{-2} = 0.4(184.36)(\text{first term})$$

$$35 x_1^{-3} x_2^3 = 0.1(184.36)(\text{third term})$$

We can find

$$x_2^* = 0.883 \text{ and } x_1^* = 1.1$$

Minimize

$$g_0(X) = 1000x_1 + 4x10^9 x_1^{-1}x_2^{-1} + 2.5x10^5 x_2$$

subject to constraint $x_1x_2 \geq 9000$, which may be written as

$$g_1(X) = 9000x_1^{-1}x_2^{-1} \leq 1$$

The dual function and constraints are

$$d(w) = \left(\frac{1000}{w_{01}}\right)^{w_{01}} \left(\frac{4x10^9}{w_{02}}\right)^{w_{02}} \left(\frac{2.5x10^5}{w_{03}}\right)^{w_{03}} \left(\frac{9000w_{11}}{w_{11}}\right)^{w_{11}} \tag{4.64}$$

subject to constraints

$$w_{01} + w_{02} + w_{03} = 1$$

$$w_{01} - w_{02} - w_{11} = 0$$

$$-w_{02} + w_{03} - w_{11} = 0$$

All dual variables are supposed to be greater than or equal to zero. The DOD in this problem is T − (N + 1) = 4−3 = 1. The variables w_{01}, w_{02}, and w_{03} can be expressed in terms of w_{11} as

$$w_{01} = w_{03} = \left(\frac{1+w_{11}}{3}\right) \quad \text{and} \quad w_{02} = \left(\frac{1-2w_{11}}{3}\right) \tag{4.65}$$

Replacing w_{01}, w_{02}, and w_{03} into Equation 4.64, we may maximize the dual function subject to the nonnegativity constraints. The optimal solution is

$$d(w)^* = 3.444 \times 10^6$$

and $w_{01}^* = w_{03}^* = 0.435$ and $w_{02}^* = 0.130$, $w_{11}^* = 0.305$.

Now, the primal variables can be computed for $t = 1$ and $t = 3$, as below.

$$1000x_1^* = 0.435 \times 3.444 \times 10^6, \quad x_1^* = 1500$$

$$2.5x \times 10^5 x_2^* = 0.435 \times 3.444 \times 10^6, \quad x_2^* = 6$$

Example 4.25

A beam of rectangular cross section is to be designed. It is to carry a moment of 2160 N.m and maximum shear force of 4800 N. The allowable bending stress is 12 MPa, and the maximum shear stress is 0.8 MPa. Let x_1 be the width and x_2 be the height of the rectangular cross section.

Solution

The allowable bending and shear stresses of the beam in terms of width and height can be written as follows:

$$\sigma = \frac{6M}{x_1x_2^2} \text{ and } \tau = \frac{3V}{2x_1x_2}$$

where M and V are the maximum moment and shear force, respectively. We can write the optimal design problem as

$$\text{Minimize } x_1x_2$$

$$\text{Subject to constraints } \frac{6(2160)}{x_1x_2^2} \le 12(10^6) \text{ and } \frac{3(4800)}{2x_1x_2} \le 0.8(10^6) \text{ (PGP)}$$

Now we write dual constraints of the PGP problem as follows:

$$w_{01} = 1$$

$$w_{01} - w_{10} - w_{20} = 0$$

$$w_{01} - 2w_{10} - w_{20} = 0$$

Example 4.26

Optimal Journal Bearing Design

A sleeve journal bearing consists of a journal and two oil-lubricated journal bearings, one on each end of the journal. A load of 200 lb is shared by the bearings. The half-length l of the bearing and the journal radius r are to be designed optimally. As stated earlier, in optimal design, objective function and constraints must be recognized before design. It can be concluded that there are two objectives in journal bearing design:

(i) By keeping the temperature of the bearing constant, we should design based on the principle that the total weighted sum of frictional moment in the bearing and the angle of twist in the shaft is minimum. This is given by the function

Minimize

$$P = aM(\omega, r, l) + b\vartheta$$

where M is the moment caused by the friction force in the bearing.

$M = 2F \cdot r$ for both bearings. According to lubrication theory, the frictional force is given by

$$F = \frac{\mu\bar{\omega}lr^2}{C}\left[\frac{4\pi}{\left(1-n^2\right)^{0.5}}\right]$$

Therefore

$$M = \frac{8\pi}{\left(\sqrt{1-n^2}\right)}\left[\frac{\mu\bar{\omega}}{C}r^3l\right]$$

where:

M = frictional moment on both bearings (lb.in).
C = radial clearance.
μ = viscosity of the lubricant in the bearing (lb-s/in^2).
ω = angular rotation (rad/s).
$n = e/C$, eccentricity ratio.
e = eccentricity of bearing relative to journal.

The angle of twist is given in mechanics as

$$\theta = TL/GJ$$

where:

T = torque on the journal (in-lb).
L = journal length between bearings.
G = shear modulus of rigidity (psi).
J = polar moment of inertia (in^4).

The angle of twist equation can expressed as

$$\theta = \frac{S_s L}{G \cdot r}$$

where S_s is the shear yield strength for the bearing material

Therefore, the objective function for minimization of weighted moment and angle of twist can be developed for the following data, given for the journal bearing:

$C/r=0.0015$, $n=0.9$, $L=10$ in., S_s (shear yield strength for the bearing material) $=30{,}000$ psi, $G=12\times10^6$ psi, and viscosity of the lubricant $\mu=10^{-6}$ lb-s/in^2.

$$\text{Minimize } P = 0.038\omega r^2 l + 0.025 r^{-1}$$

Weightage values $a=b=1$ are assumed, assigning equal importance to the moment and angle of twist. However, they could vary depending on the importance given to both the objectives.

The constraint equations can be developed as follows. It is desired that the journal rotate at a minimum of 100 rad/s, so the constraint is written as

$$\omega \geq 100$$

and a constraint on load-carrying capacity of the bearing at 11.86:

$$\omega r^{-1} l^3 = 11.6 \text{ (minimum total friction constraint)}$$

In GP, the solution becomes easy for zero DOD, producing a number of equations equal to the number of variables in the dual program. To achieve this, the PGP is written as

$$\text{Minimize } P = 0.45 r^3 l^{-2} + 0.025 r^{-1}$$

$$\text{Subject to constraint } 8.75 r^{-1} l^3 \leq 1$$

There are three terms and only two variables in the design, so it is a zero-difficulty problem, and the DGP can be written as

Maximize

$$f(w) = \left(\frac{0.45}{w_{01}}\right)^{w_{01}} \left(\frac{.025}{w_{02}}\right)^{w_{02}} \left(\frac{8.62 * w_{11}}{w_{11}}\right)^{w_{11}}$$

Subject to constraints

$$w_{01} + w_{02} = 1 \text{ (normality constraint)}$$

$$3w_{01} + w_{02} - w_{11} = 0 \text{ (orthogonality constraint)}$$

$$-2w_{01} + 3w_{11} = 0 \text{ (orthogonality constraint) (DGP)}$$

$$w_{01} \geq 0, w_{02} \geq 0, w_{11} \geq 0 \text{ (these are dual variables)}$$

We can solve the constraint equations as linear equations by any known methods of the Gauss or Gauss–Jordan method. The solution of dual variables is

```
>> A=[1 1 0;3 -1 -1;-2 0 3]
>> B=[1;0;0]
>> AB
ans=
0.3000
0.7000
0.2000
```

which means that $w_{01}=0.3$, $w_{02}=0.7$, and $w_{11}=0.2$.

This dual solution suggests that angle of twist has higher importance in terms of our objective. However, we need to calculate the design variables r and l from the dual variables. We have calculated the numerical value of the objective function as

$$f(w)=\left(\frac{0.45}{0.3}\right)^{0.3}\left(\frac{0.025}{0.7}\right)^{0.7}\left(\frac{8.62(0.2)}{0.2}\right)^{0.2}=0.1686$$

Now we can calculate the primal variables (optimal design dimensions):

$$0.45r^3l^{-2}=0.3(0.1686)=0.051 \tag{4.66}$$

and

$$0.025r^{-1}=0.7(0.1686)=0.12 \tag{4.67}$$

From Equation 4.67, we get $r=0.212$, in and from Equation 4.66, we find $l=0.29$, that is, the radius of the journal bearing and half-length of the sleeve bearing. One important invariance property of these solutions should be noted: the optimal values of dual variables w_{01}^*, w_{02}^*, and w_{11}^* will not change if any of the coefficients 0.45, 0.025, or 8.62 are changed, which tells us that the optimal values will not change if any of the input data are varied. Therefore, it can be concluded that optimal design parameters are not tied only to a specific set of input data. In this design problem, the optimum dual variables show that $C_{01}R^3l^{-2}$ will always be 3/7 of $C_{02}r^{-1}$ for any positive value of C_{01} and C_{02}. We should also recognize that w_{11} is positive, showing that the constraint is always tight for a minimum total friction.

Example 4.27

A better optimal design will result if we include temperature rise in the journal bearing (Figure 4.15). This objective will be an extension of the objective function in (i), and it could be written as

$$\text{Minimize } P=aM(\omega,r,l)+b\theta(r)+cT^*(r,\omega)$$

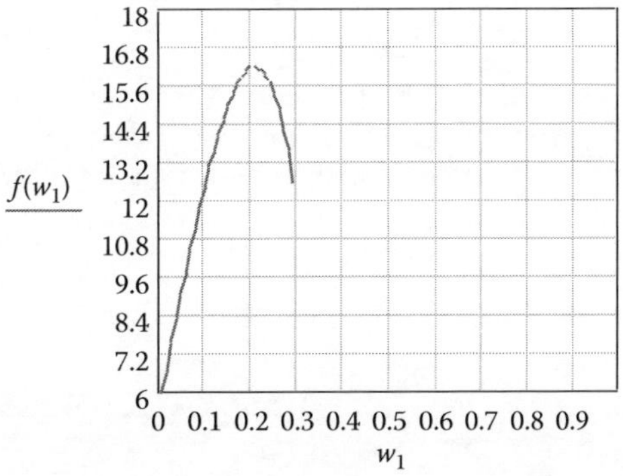

Figure 4.15

Bearing design.

The third term represents the temperature rise as

$$T^* = t_2 - t_1 = \frac{42.4(hp)}{(0.03C_pQ)60}$$

where hp is horsepower loss due to friction in two bearings in a petroleum-based lubricant. We may choose C_p = 0.5 BTU/lb-°F. The oil flow from the ends of both bearings is calculated as $Q=4Cn\omega rl$, and horsepower loss due to friction is calculated as

$$\text{hp} = 3.81\times10^{-3}\frac{\mu\omega^2}{C(1-n^2)^{0.5}}r^3l$$

Putting Q and hp in the temperature rise equation above, we obtain the temperature rise in the bearings as

$$T^* = 0.045\frac{\mu\omega r^2}{C^2n(1-n^2)^{0.5}}$$

The temperature equation variables are r and ω. The constraints are the same as in case (i). Entering the data as given in case (i), the following mathematical model is developed for case (ii).

Minimize

$$p_2 = 0.038\omega r^2 l + 10r - 1 + 0.051\omega$$

subject to constraints

$$\omega r^{-1}l^3 = 11.6$$

and $\omega \geq 100$. This model is further simplified and presented as

Minimize

$$P_2 = 0.44r^3l^{-2} + 10r^{-1} + 0.592rl^{-3}$$

subject to constraint

$$8.62r^{-1}l^3 \leq 1$$

$$r > 0, l > 0$$

Case (ii) has four terms and only two primal design variables: r and l. Therefore, there is one DOD in case (ii) as there are four dual variable in three dual constraints. In mathematical parlance, we can say that these are only necessary, but not sufficient, conditions for optimal solution. We can develop the dual constraints as

$$\omega_{01} + w_{02} + w_{03} = 1\ (\text{normality constraint})$$

$$3\omega_{01} - \omega_{02} - \omega_{03} - w_{11} = 0\ (\text{orthogonal constraint})$$

$$-2\omega_{01} - 3\omega_{03} + 3\omega_{11} = 0\ (\text{orthogonal constraint})$$

All dual variables are strictly positive, that is, ≥0. The dual objective function can be developed as

$$f(\omega)=\left(\frac{C_{01}}{\omega_{01}}\right)^{\omega_{01}}\left(\frac{C_{02}}{\omega_{02}}\right)^{\omega_{02}}\left(\frac{C_{03}}{\omega_{03}}\right)^{\omega_{03}}\left(\frac{C_{11}\cdot\omega_{11}}{\omega_{10}}\right)^{\omega_{11}}$$

$$C_{01}=0.44, C_{02}=10, C_{03}=0.592, \text{ and } C_{11}=8.62$$

From the dual constraint set, we can write any three dual variables in terms of fourth dual variables as

$$\omega_{02}=\frac{7}{3}\omega_{01},\ \omega_{03}=1-\frac{10}{3}\omega_{01},\ \text{and } \omega_{11}=1-\frac{8}{3}\omega_{01}$$

Here, three dual variables are replaced in terms of the fourth dual variable, ω_{01}. On substituting these in the dual objective function, we can write

$$f(\omega_{01})=f(\omega_{01})=\left(\frac{0.44}{\omega_{01}}\right)^{\omega_{01}}\left(\frac{10(3)}{7\omega_{01}}\right)^{7/3\omega_{01}}\left(\frac{0.59(3)}{(3-10\omega_{01})}\right)\left(\frac{(3-10\omega_{01})}{3}\right)\left(\frac{8.62\omega_{11}}{\omega_{11}}\right)^{\left(\frac{(3-8\omega_{01})}{3}\right)}$$

$$f(\omega_1)=\left(\frac{0.44}{\omega_1}\right)^{\omega_1}\left(\frac{30}{7\omega_1}\right)^{\frac{7}{3}\omega_1}\left[\frac{0.593}{3-10\,\omega_1}\right]^{\left(1\frac{10}{3}\omega_1\right)}(8.62)^{\left(1\frac{8}{3}\omega_1\right)}$$

This equation can be solved by finding the positive root of the equation.

The values of all dual variables are: $\omega_{01}^*=0.21, \omega_{02}^*=0.48, \omega_{03}=0.32$, and $\omega_{11}^*=0.45$.

In the case of a single-term constraints, $\omega_{10}=\omega_{11}$ and

$$f^*(\omega)=16.2$$

Now we can find the primal (design) variables as

$$10r^{-1}=0.48(16.2) \text{ and so } r^*=1.29\,\text{in}$$

$$0.59rl^{-3}=0.32(16.2) \text{ and } l^*=0.53\,\text{in}$$

If we compare the optimal design of the two cases, we see a tremendous difference in the design values (Figure 4.15).

Example 4.28

Approximate Solution

GP is a very practical method for nonlinear programming problems. Most design and manufacturing problems are inherently nonlinear, and the GP technique is very suitable. A hatch opening must be covered by a box beam of span length and width shown in Figure 4.16.

The flange thickness (x_1) and the height (x_2) are to be designed to minimize the weight of the cover. The objective function can be written as

Minimize

$$f(x)=120x_1+x_2$$

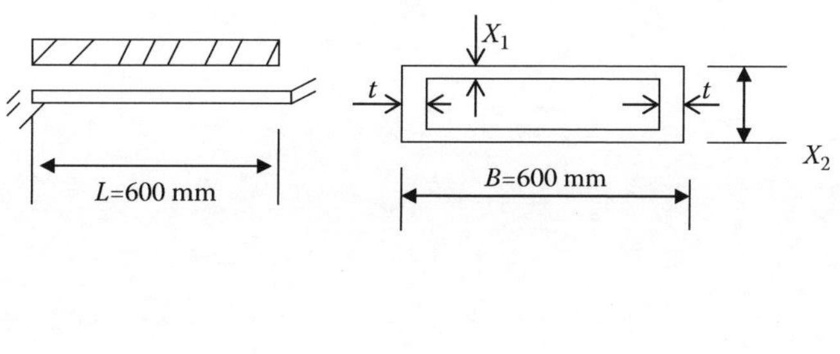

Figure 4.16

Hatch cover example.

$$\text{Subject to constraints } 4x_2^{-1} \le 1 \left(\text{maximum shearing stress}\right)$$

$$\frac{45}{7} x_1^{-1} x_2^{-1} \le 1 \left(\text{maximum bending stress}\right)$$

$$\frac{45}{7} x_1^{-3} x_2^{-1} \le 1 \left(\text{buckling constraint}\right)$$

$$3.125 \times 10^{-3} x_1 x_2 \le 1 \left(\text{deflection constraint}\right)$$

This problem was formulated by Franklin E. Grange of Standard Oil of Indiana. This is a three-DOD problem, and solution would be difficult. However, an approximate solution could be obtained as follows.

To make it a zero-DOD problem, we can start solution with just one of the constraints. Let us examine the problem with only the bending stress constraint, as follows:

$$\text{Minimize } f(x) = 120x_1 + x_2$$

$$\text{Subject to constraint } \frac{45}{7} x_1^{-1} x_2^{-1} \le 1$$

The dual of the problem can be presented as

$$\text{Maximize } d(w) = \left(\frac{120}{w_{01}}\right)^{w_{01}} \left(\frac{1}{w_{02}}\right)^{w_{02}} \left(\frac{45}{7}\right)^{w_{11}}$$

$$\text{Subject to constraints } w_{01} + w_{02} = 1$$

$$w_{01} - w_{11} = 0$$

$$w_{02} - w_{11} = 0$$

The dual solution is $w_{01}^* = w_{02}^* = w_{11}^* = 0.5$. From this, we can find the primal (design) variables as $x_1^* = 0.116\text{cm}$, $x_2^* = 13.90\text{cm}$, and $f(x)^* = 27.775$. We can examine the other constraint with these values. Let us test the buckling constraint.

$$\frac{45}{7} x_1^{-3} x_2^{-1} = 45/7(0.116)^{-3}(13.9)^{-1} = 296 > 1$$

which shows that the buckling constraint is violated. Let us now apply the buckling constraint and solve the problem as

$$\text{Minimize } f(x) = 120x_1 + x_2$$

$$\text{Subject to } \frac{45}{7} x_1^{-3} x_2^{-1} \le 1$$

Write the dual of the primal problem:

$$\text{Maximize } d(w) = \left(\frac{120}{w_{01}}\right)^{w_{01}} \left(\frac{1}{w_{02}}\right)^{w_{02}} \left(\frac{45}{7}\right)^{w_{11}}$$

$$\text{Subject to } w_{01} + w_{02} = 1$$

$$w_{01} - 3w_{11} = 0$$

$$w_{02} - w_{11} = 0$$

Solving the dual, we get $w_{01}^* = 0.75$ and $w_{02}^* = w_{11}^* = 0.25$

Then the new primal (design) variables are $x_1^* = 0.633\text{cm}$ and $x_2^* = 25.326\text{cm}$. Now let test shearing stress constraints.

$4x_1^{-1} = 4(25.326)^{-1} = 0.158 < 1$, that is, the shearing stress constraint is satisfied. Now examine the bending constraint.

$45/7x_1^{-1}x_2^{-1} = 45/7(0.633)^{-1}(25.326)^{-1} = 0.368 < 1$, that is, the bending stress constraint is satisfied. Finally, examine the deflection constraint.

$3.125 \times 10^{-3}x_1x_2 = 3.125 \times 10^{-3}(0.633)(25.326) = 0.05 < 1$, that is, the deflection constraint is also satisfied.

The final design converted to standard dimensions: $x_1^* = 80\text{mm}$ and $x_2^* = 300\text{mm}$.

Now, if we use CVX software [12], the hatch optimization problem gives the following solution:

$x_1 = 0.6332$ mm, $x_2 = 25.3264$ mm, and objective function value is Minimum $f(x) = 101.356$ mm^2. All the constraints are satisfied: these are 0.1579, 0.4009, 1.0, and 0.0501, respectively. This shows that buckling is the most critical constraint.

Example 4.29

A hollow steel shaft is to be designed for power transmission. The power source is 30 hp, and speed of the shaft is 100 rpm. The inside diameter of the shaft is 2 in. If the permissible shearing stress is 11,000 psi, what is the thickness of the shaft *t*? The objective is to minimize the weight of the shaft. The length of the shaft is given as 5 ft.

(Ans. 0.24 in.)

Materials and Properties for Torsion Rod

Material	Density ρ (lbm/in^3)	Allowable Shear Stress τ_a (Mpsi)	Elastic Modulus *E* (Mpsi)	Shear Modulus *G* (Mpsi)	Poisson Ratio (υ)
4140 Alloy steel	0.283	40	30.45	11.6	0.30
Aluminum alloy 24ST4	0.1	24	10.8	4	0.32
Magnesium alloy A261	0.065	13	6.5	2.4	0.35
Beryllium	0.067	16	43.5	21	0.02
Titanium	0.163	24	16	6.8	0.30

Formulation of the optimal design problem:

The torque on the shaft:

$$T = \frac{63{,}025(\text{hp})}{n} = \frac{63025(30)}{100} = 18{,}908 \text{ lbf.in}.$$

The weight of the tubular shaft is

$$f(r_0) = \frac{\pi}{4}(d_0^2 - d_i^2)(60)(0.283) = 13.4(d_0^2 - d_i^2)$$

$$\text{Minimize } f(d_0) = 13.4(d_0^2 - d_i^2)$$

$$\text{Under the constraint } \frac{18{,}908(d_0 + 2)/2}{\pi/32(d_0^4 - d_i^4)} \leq 11{,}000 \left(\text{shear stress constraint}\right)$$

This can be simplified to write

$$87.547\frac{(d_0 + d_i)}{(d_0^4 - d_i^4)} \leq 1$$

It is finally written as

$$1 - 87.547\frac{(d_0 + d_i)}{(d_0^4 - d_i^4)} \geq 0 \left(\text{shear stress constraint}\right)$$

Let us assume the ratio between the inside and the outside diameter of the transmission shaft to be $0.06 \leq d_i/d_0 \leq 0.999$. We will assume here that $d_i/d_0 = 0.80$, which is in the middle of the range, and replace $d_i = 0.80\ d_0$ in all the equations above. The reformulated equations are as follows.

$$\text{Minimize } f(d_0) = 4.825d_0^2$$

$$\text{Under the constraint } 1 - 263d_0^{-3} \geq 0$$

Applying GP, we can write

$$\text{Maximize } d(w) = \left(\frac{4.825}{w_{01}}\right)^{w_{01}} \left(\frac{263w_{11}}{w_{11}}\right)^{w_{11}}$$

$$\text{Under the constraints } w_{01} = 1$$

$$2w_{01} - 3w_{11} = 0$$

which is solved for $w_{01}^* = 1$, $w_{11}^* = 2/3$, and $d(w)^* = 191$.

Primal variable: $4.825d_0^2 = w_{01}^* d(w)^* = 191$. We get $d_0 = 6.325$, $d_i = 5$, and $t = 0.63$ in.

This design problem was easy to solve due to the assumed ratio between inside and outside diameters.

Example 4.30

Design of Two-Bar Truss

Design a two-bar truss subjected to a vertical load of $2P = 66{,}000$ lb for minimum weight. The members have a tubular section with mean diameter d and thickness t. The maximum permissible stress in each member is equal to 60,000 psi. Determine h and d for given values $t = 0.1$ in, $b = 30$ in, and density (ρ) of material = 0.3 lb/in^3 (Figure 4.17).

Problem Formulation

The objective function is given by

$$f(d,h) = 2\rho\pi dt\sqrt{b^2 + h^2} = 0.188d\sqrt{900 + h^2}$$

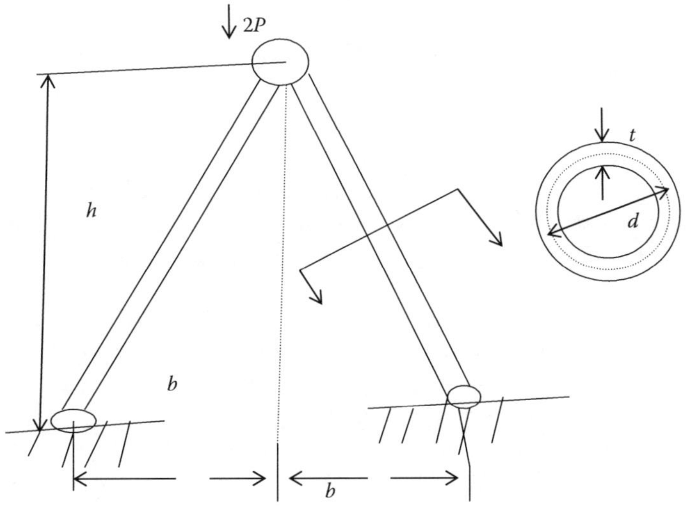

Figure 4.17

Two bar under stress.

$$\text{Subject to constraint } \sigma = \frac{2P}{\pi\, dt}\frac{\sqrt{900+h^2}}{h} = \frac{33{,}000}{\pi d(0.1)}\frac{\sqrt{900+h^2}}{h} \le 60{,}000$$

The constraint could be simplified by entering the given values into

$$1.75\frac{\sqrt{900+h^2}}{dh} \le 1$$

We can observe that the equations are not in the polynomial form. They can be converted to polynomial as follows. Let

$$y = \sqrt{900+h^2} \text{ or } y^2 = 900 + h^2$$

We can add a new constraint: $900y^{-2} + y^{-2}h^2 \le 1$. Now let us write the reformulated problem in polynomial form.

$$\text{Maximize } f(y,d) = 0.188yd$$

$$\text{Subject to constraints } 1.75yh^{-1}d^{-1} \le 1$$

$$900y^{-2} + y^{-2}h^2 \le 1$$

y, h, and d are positive values.

Using CVX, we obtain the solution as $y=42.4264$ in, $h=30$ in, and $d=2.475$ in.

The minimum weight of the two-bar truss is 19.74 lb. All the constraints are satisfied.

Example 4.31

Revisiting Carbon Dioxide Sequestration and Storage

Greenhouse gas (GHG) emissions are believed to be the main reason for global warming, and the Intergovernmental Panel on Climate Change (IPCC) has examined various scenarios to reduce GHG emissions. GHG is thought to be composed of various harmful gases, such as NO_2, CO, CO_2, SO_2, and so on, but carbon dioxide (CO_2) is the main constituent of GHG. It has received wide attention for its role in global warming. Every country has passed legislation to limit and reduce CO_2 emissions. In the United States, the Clean Air Market Program of the federal Environmental Protection Agency (EPA) is called "allowance trading" or "cap-and-trade." Cap-and-trade is an attempt to control large amounts of emissions from a group of sources, whether stationary or mobile, at costs that are lower than they would be if sources were regulated individually. The approach first sets an overall cap, or maximum amount of emissions per compliance period, that will achieve the desired environmental effects. Total emission allowances cannot exceed the cap. In addition, it has been identified that geologic sequestration (GS) is the most technically viable approach to capturing and storing CO_2 and thus controlling climate change. Energy-related and industrial facilities, such as coal-fired power plants, ethanol and petroleum refineries, natural gas processing plants, and cement and chemical plants, are the major source of CO_2 emissions. GS captures CO_2 from power plants and oil refineries and stores it in oil and gas reservoirs (i.e., enhanced oil recovery, EOR), unmineable coal seams, and deep saline formations. Efficient technologies are available to capture CO_2 both postcombustion and precombustion. These technologies are either chemical absorbents, CO_2 hydrates, or physical absorbents, and are applied at appropriate sources of emission. An example of carbon sequestration is presented in [15].

Four plants emitting CO_2 plan to join hands to set up a carbon sequestration facility. These plants are an integrated gasification combined cycle (IGCC) plant, a steel plant, a petrochemical plant, and an oil refinery plant. The annual CO_2 emissions of these plants are 3.5, 4.5, 3.5, and 4.7 million tons, respectively. But the maximum capture rates of CO_2 from these plants vary depending on the technologies employed; they are 55%, 70%, 60%, and 80%, respectively. The minimum capture rate of CO_2 is about 30% for all plants. Under the cap-and-trade rules, carbon credit may be earned and sold in the credit market with a price of at least $12.00 per ton if these plants can remove more CO_2 above the minimum level. The CO_2 capture cost varies for different plants based on the technologies adopted. The range of CO_2 capture costs is shown in Table 4.2.

For pipeline transportation of CO_2, a cost function with fixed charge and variable cost can be developed based on the existing database. Thus, the average cost of CO_2 transportation as a function of the CO_2 mass flow rate (*MFR*) is expressed as $C=a(MFR)^b$, where C is the average cost ($/ton/100 km) and *MFR* is the CO_2 mass flow rate (10^6 tons/year), and a and b are regression coefficients ($a=3.287$, $b=0.619$).

Two candidate sites in the region, a deep saline formation (denoted as reservoir 1) and a depleted oil and gas reservoir (denoted as reservoir 2), can be identified as possible CO_2 storage reservoirs. Reservoirs 1 and 2 can store 5 and 4 million tons of CO_2 per year, respectively. The locations of these reservoirs are shown in Figure 3.9.

It is estimated that these two reservoirs will be able to store CO_2 for next 20 years. The amortized injection cost for pumping stations at two reservoirs is $20/ton of CO_2. Table 4.3 shows transportation distances from the plant to the sites.

Formulate the sequestration plan problem as a GP problem and solve it.

Decision variables:

X_{ij} = the amount of CO_2 transported from plant i to reservoir j annually (tons/year)

i = 1,2,3,4 IGCC plant 1, steel plant 2, petrochemical plant 3, and oil refinery 4

Table 4.2 Capturing Cost of CO_2

Carbon-Emitting Source	Capturing Cost ($/ton)	Average ($/ton)
IGCC plant	$26 \leq C_1 \leq 74$	$C_{1av}=50$
Steel plant	$70 \leq C_2 \leq 90$	$C_{2av}=80$
	$90 \leq C_3 \leq 150$	$C_{3av}=120$
	$150 \leq C_4 \leq 240$	$C_{4av}=195$

Table 4.3 Transportation Distances from Plant to Site

CO_2-Emitting Source	Reservoir 1	Reservoir 2
IGCC plant	140 km	300 km
Steel plant	200 km	300 km
Petrochemical plant	350 km	25 km
Oil refinery	200 km	50 km

j = 1,2
y_i = actual CO_2 capture rate at plant i (%)
E_i = CO_2 emissions at plant i
$Ymax$ = maximum CO_2 capture rate at plant i (%)
$Ymin_i$ = minimum CO_2 capture rate at plant i (%)
C_{iav} = average cost of CO_2 capture at plant i (\$/ton)
d_{ij} = distance between plant i and reservoir j (km)
TC_{ij} = average cost of transportation of CO_2 from plant i to reservoir j (\$/ton/100 km)
Cap_j = capacity of reservoir j (million tons/year)
ic = amortized injection cost for pumping CO_2 into the reservoir (\$/ton of CO_2)
z = carbon credit earning per ton of CO_2

Mathematical model:

The objective is to minimize the total system cost, which includes CO_2 capture cost, transportation cost, and CO_2 injection cost, less carbon credit earnings. Thus, the objective function is to minimize:

Minimize

$$f = \sum_{i=1}^{4} E_{ii} y_i C_{iav} + \sum_{i=1}^{4}\sum_{j=1}^{2} x_{ij} d_{ij} TC_{ij} + \sum_{i=1}^{4}\sum_{j=1}^{2} x_{ij} ic - \sum_{i=1}^{4} zE_i \left(y_i - Y\min_i \right)$$

subject to the following constraints:

1. CO_2 mass balance constraint: this constraint ensures that mass balance is confirmed at the source's side

$$\sum_{j=1}^{2} x_{ij} = E_i y_i, \quad i = 1,2,3,4$$

2. Reservoir capacity constraint: this constraint ensures that reservoir operational capacity is not exceeded annually

$$\sum_{i=1}^{4} x_{ij} \le Cap_j, \quad j = 1,2$$

3. Technology constraint: this constraint ensures that the actual capture rate is lower than the maximum capture rate at all plants

$$y_i = Y\max_i, \quad i = 1\ldots 4$$

4. Policy constraint: this constraint ensures that the actual capture rate is higher than the minimum capture rate at all plants

$$y_i \ge Y\min_i, \quad i = 1\ldots 4$$

5. Negativity constraint: all decision variables should be greater than or equal to zero $x_{ij}, y_i \ge 0$

In all, there are 12 decision variables.

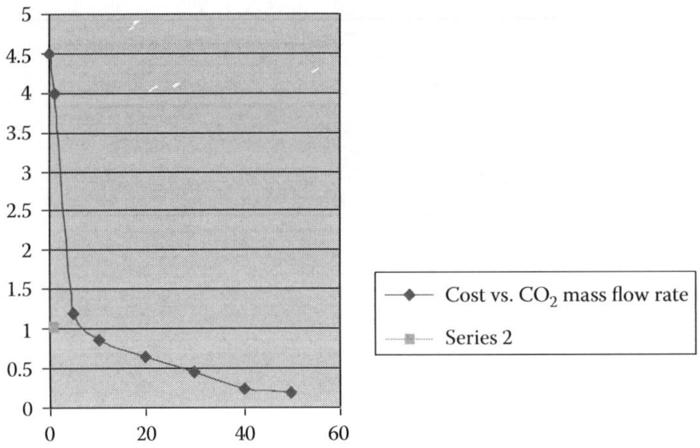

$$\text{Cost} = 4.205(MFR)^{-0.719}$$

Example 4.32

Energy Resources Allocation

Renewable energy sources will play an important role in future energy resources management. To develop an optimal mix of energy technologies, LP models may lead to better management of energy resources. Let us assume that three nuclear power plants have been put into operation in the state, and renewable energy resources such as wind, solar, and biomass will play a very important role in the future. To allocate the environmentally friendly benign energy resources as cost-effectively as possible, we may formulate a compromise programming model to provide a quantitative assessment of energy resources allocation in a region where wind and solar energy resources are abundant. Table 4.4 shows the total energy demand based on the "Annual Energy Outlook" in the United States. Table 4.5 shows a cost matrix that illustrates the list of all possible combinations of energy resources with different end uses. Table 4.6 shows the emission factors associated with seven different energy resources

Table 4.4 Energy Demand Annually (10^9 J/year)

End Uses	Total Energy Demand
Residential	1,339,504
Commercial	1,089,180
Industrial	558,887
Transportation	1,614,301

Table 4.5 Cost Matrix for Different End Users ($/$10^6$ J)

Energy Resources	Residential	Commercial	Industrial	Transportation
Natural gas	19.08	12.10	9.78	11.86
Coal	4.51	3.25	3.25	–
Oil	31.99	17.96	14.60	19.97
Biomass	24.63	3.53	2.49	–
Nuclear	8.20	8.20	8.20	–
Solar	146.50	145.50	146.50	146.50
Wind	20.51	20.51	20.51	–

Table 4.6 Calculated Emissions Associated with Different Forms of Energy Generation (kg/10^6 J)

Energy Generation	Carbon Dioxide	Sulfur Dioxide	Nitrogen Oxides
Natural gas	49.07	–	1.10
Coal	205.00	0.20	0.31
Oil	68.46	2.04	0.82
Biomass	83.18	18.4	–

(i.e., three conventional fossil fuel energy resources vs. four renewable energy resources). The objective is minimization of *TC* and total emissions of CO_2, SO_x, and NO_x (Tables 4.5 and 4.6).

Mathematical Modeling:

Decision variables:

Let X_i = the amount of energy that has been allotted to various end users (10^6 J/year)

Given Data

C_i = unit cost of energy resources *i* that needs to be charged (\$/10^6 J)

E_i = aggregate emission factor for air pollution, including carbon dioxides, sulfur dioxides, and nitrogen oxides, associated with energy resources *i* (kg/10^6 J)

D_j = energy demand of sector *j* in a year (*j* = 1 for residential sector, *j* = 2 for commercial sector, *j* = 3 for industrial sector, and *j* = 4 for transportation sector) (10^6 J/year)

Model Formulation

$$\text{Min } Z_1 = \sum_{1=1}^{7} C_i X_i$$

$$\text{Min } Z_2 = \sum_{i=1}^{7} E_i X_i$$

Subject to Constraints

1. Energy demand constraint: this constraint ensures that the allocation of energy resources *i* for each designated sector is greater than the demand

$$\sum_{j=1}^{7} X_j = D_j,\ j = 1,2,3,4$$

2. Negativity constraint: this constraint ensures that all decision variables are nonnegative

$$X_i > 0$$

Example 4.33

We now revisit the storage tank with CNX programming and compare with the geometric programming results of Example 4.23. A new overhead storage tank is to be designed for storing water. It is proposed that the tank capacity should be about 50,000 ft^3. The material cost of the top and bottom of the tank is \$60.00 per square foot. However, it is estimated that the sides will cost \$50.00 per square foot. From design considerations, the tank should have two supporting beams of 3 in radius to avoid hurricane damage. The cost of the supporting beams is \$50.00 per foot.

Objective function formulation

Total cost function (*TC*) = cost of sides + cost of top and bottom + cost of supporting beams

Cost of sides = $2\pi R_1 H(\$50.0)$

Cost of top and bottom = $2 \times \pi R_1^2(\$60.00)$

Cost of the beams = $2 \times 50.0(\sqrt{H^2 + 4R_1^2}$

Total cost $TC(R_1, H) = 314.16 R_1 H + 120\pi R_1^2 + 100\sqrt{H^2 + 4R_1^2}$

Replacing $R_1 = x_1$, $H = x_2$, and $\sqrt{H^2 + 4R_1^2} = x_3$

We can write the *TC* equation as

$$TC(x_1, x_2, x_3) = 314.16 x_1 x_2 + 377 x_1^2 + 100 x_3 \tag{4.68}$$

Now we should develop the constraints. We can write

$$\sqrt{R_1^2 + 4H^2} = x_3$$

or

$$4x_1^2 + x_2^2 = x_3^2$$

This can be written as

$$4x_1^2 x_3^{-2} + x_2^2 x_3^{-2} \leq 1 \tag{4.69}$$

Volume constraint could also be formulated

$$\text{Volume} \geq 5000$$

Therefore,

$$\text{Volume} = \text{volume of tank} - \text{volume of beams}$$

$$\text{Volume} = \pi R_1^2 H - \left(2\pi R_2^2\right)\sqrt{H^2 + 4R_1^2}$$

Writing in X, $R_2 = 3$ in. is given in the problem, so

$$\text{Volume} = 3.14x_1^2 x_2 - 0.393x_3$$

Volume constraint can be written as

$$3.14x_1^2 - 0.393x_2 \geq 5000$$

Or it can be reformulated as

$$5000 + 0.393x_3^{0.5} \leq 3.14x_1^2 x_2$$

or

$$1592.357x_1^{-2} x_2^{-1} + 0.1252x_1^{-2} x_2^{-1} x_3 \leq 1 \tag{4.70}$$

The complete optimal design mathematical model is shown below:

$$\text{Minimize } TC(x_1, x_2, x_3) = 314.16x_1 x_2 + 377x_1^2 + 100x_3 \tag{4.71}$$

Subject to constraints:

$$4x_1^2 x_3^{-2} + x_2^2 x_3^{-2} \leq 1 \tag{4.72}$$

$$1592.357_1^{-2} x_2^{-1} + 0.1252x_1^{-2} x_2^{-1} x_3 \leq 1 \tag{4.73}$$

$$x_1 \geq 0, x_2 \geq 0, x_3 \geq 0$$

Normally, in the GP format the complete mathematical model for optimal design is written as

$$g_0(X) = 314.16x_1 x_2 + 377x_1^2 + 100x_3 \tag{4.74}$$

Subject to constraints

$$g_1(X) = 4x_1^2x_3^{-2} + x_2^2x_3^{-2} \tag{4.75}$$

$$g_2(X) = 1592.357x_1^{-2}x_2^{-1} + 0.1252x_1^{-2}x_2^{-1}x_3 \tag{4.76}$$

$$x_1 \ge 0, x_2 \ge 0, x_3 \ge 0$$

Equations 4.74 through 4.76 together are also called PGP.

Now the PGP can be solved. The dual of the PGP can be written as

$$d(w) = \left(\frac{314.16}{w_{01}}\right)^{w_{0101}} \left(\frac{377}{w_{02}}\right)^{w_{02}} \left(\frac{100}{w_{03}}\right)^{w_{03}} x \left[\frac{1592..357w_{10}}{w_{11}}\right]^{w_{11}} \left[\frac{0.1252}{w_{12}}\right]^{w_{12}} \left[\frac{4w_{20}}{w_{21}}\right]^{w_{21}} \left[\frac{w_{20}}{w_{22}}\right]^{w_{22}} \tag{4.77}$$

Subject to constraints

$$w_{01} + w_{02} + w_{03} = 1$$

$$w_{01} + 2w_{02} + 2w_{11} - 2w_{21} - 2w_{22} = 0$$

$$w_{01} + 2w_{12} - w_{21} - w_{22} = 0$$

$$w_{03} - 2w_{11} - 2w_{12} + w_{22} = 0$$

$$w_{10} = w_{11} + w_{12}$$

$$w_{20} = w_{21} + w_{22}$$

Now the DGP can be solved to find optimum dual values, and finally optimal design parameters (x) are obtained.

$$w_{01}^* = 0.525, \quad w_{02}^* = 0.00625, \quad w_{03}^* = 0.4688$$

$$w_{11}^* = 0.4522, \quad w_{12}^* = 0.088, \quad w_{10}^* = 0.5402$$

$$w_{21}^* = 0.0073, \quad w_{22}^* = 0.271, \quad w_{20}^* = 0.2783$$

$$d(w^*) = 17.2 = g_0(X^*) \text{ (minimum cost value)}$$

The primal optimal design solution is calculated using the GP method presented earlier:

$$x_1^* = 0.52, \quad x_2^* = 55.72, \quad x_3^* = 80.1$$

That is, radius of tank $R_1 = 0.52$ in.; height of tank $H = 55.72$ in.; and length of supporting beam $x_3 = 80.1$ in.

Solving the problem with the CVX GP solver gives

$$x_1 = 8.8310, \quad x_2 = 20.4620, \quad x_3 = 27.0303$$

where x_1 = radius of tank = 8.831 in., x_2 = height of tank = 20.462 in., and finally x_3 = length of supporting beam = 27.0303 in. The objective function value for minimum cost = \$8.8860e+004. Checking the constraints, and we find that the left-hand side values for constraints 1 and 2 are 0.6214 and 1, respectively. The second constraint, cost of beam, is active.

4.6 Comparing Classical and Advanced Techniques for Optimization

The differences between classical and advanced optimization techniques are discussed below.

4.6.1 Classical Optimization Techniques

Classical optimization techniques are useful in finding the optimum solution or unconstrained maxima or minima of continuous and differentiable functions. These are analytical methods that make use of differential calculus in locating the optimum solution. The classical methods have limited scope in practical applications, as some of them involve objective functions that are not continuous or differentiable. Yet, the study of these classical techniques of optimization forms the basis for developing most of the numerical techniques that have evolved into advanced techniques more suitable to today's practical problems. These methods assume that the function is differentiable twice with respect to the design variables and that the derivatives are continuous. Three main types of problems can be handled by the classical optimization techniques: single-variable functions, multivariable functions with no constraints, and multivariable functions with both equality and inequality constraints. For problems with equality constraints, the Lagrange multiplier method can be used. If the problem has inequality constraints, the Kuhn–Tucker conditions can be used to identify the optimum solution. These methods lead to a set of nonlinear simultaneous equations that may be difficult to solve. These classical methods of optimization are presented below.

The other methods of optimization include

- *Linear Programming* studies the case in which the objective function f is linear and the set A is specified using only linear equalities and inequalities (A is the design variable space).
- *Integer Programming* studies linear programs in which some or all variables are constrained to take on integer values.
- *Quadratic Programming* allows the objective function to have quadratic terms, while the set A must be specified with linear equalities and inequalities.
- *Nonlinear Programming* studies the general case in which the objective function or the constraints or both contain nonlinear parts.
- *Stochastic Programming* studies the case in which some of the constraints depend on random variables.
- *Dynamic Programming* studies the case in which the optimization strategy is based on splitting the problem into smaller subproblems.
- *Combinatorial Optimization* is concerned with problems for which the set of feasible solutions is discrete or can be reduced to a discrete set.
- *Infinite-Dimensional Optimization* studies the case when the set of feasible solutions is a subset of an infinite-dimensional space, such as a space of functions.
- *Constraint Satisfaction* studies the case in which the objective function f is constant (this is used in artificial intelligence, particularly in automated reasoning).

4.6.2 Advanced Optimization Techniques

- *Hill Climbing*: Hill climbing (HC) is a graph search algorithm in which the current path is extended with a successor node that is closer to the solution than the end of the current path.

  In *simple* HC, the first closer node is chosen, whereas in *steepest ascent* HC, all successors are compared and the closest to the solution is chosen. Both forms fail if there is no closer node. This may happen if there are local maxima in the search space that are not solutions. Steepest ascent HC is similar to best first search, but the latter tries all possible extensions of the current path in order, whereas steepest ascent only tries one.

  HC is used widely in artificial intelligence fields for reaching a goal state from a starting node. The choice of next node starting node can be varied to give a number of related algorithms.
- *Simulated Annealing*: The name and inspiration come from the annealing process in metallurgy, a technique involving heating and controlled cooling of a material to increase the size of its crystals and reduce their defects. The heat causes the atoms to become unstuck from their initial positions (a local minimum of the internal energy) and wander randomly through states of higher energy; the slow cooling gives them more chances of finding configurations with lower internal energy than the initial one.

In the simulated annealing (SA) method, each point of the search space is compared to a state of some physical system, and the function to be minimized is interpreted as the internal energy of the system in that state. Therefore, the goal is to bring the system from an arbitrary initial state to a state with minimum energy.

- *Genetic Algorithms*: A genetic algorithm (GA) is a search technique used in computer science to find approximate solutions to optimization and search problems. Specifically, it falls into the category of local search techniques and is therefore generally an incomplete search. GAs are a particular class of EA that use techniques inspired by evolutionary biology, such as inheritance, mutation, selection, and crossover (also called recombination).

  GA are typically implemented as a computer simulation in which a population of abstract representations (called chromosomes) of candidate solutions (called individuals) to an optimization problem evolves toward better solutions. Traditionally, solutions are represented in binary as strings of 0s and 1s, but different encodings are also possible. The evolution starts from a population of completely random individuals and occurs in generations. In each generation, the fitness of the whole population is evaluated, and multiple individuals are stochastically selected from the current population (based on their fitness) and modified (mutated or recombined) to form a new population. The new population is then used in the next iteration of the algorithm.

- *Ant Colony Optimization*: In the real world, ants (initially) wander randomly, and upon finding food, return to their colony while laying down pheromone trails. If other ants find such a path, they are likely not to keep traveling at random, but instead follow the trail laid by earlier ants, returning and reinforcing it, if they eventually find any food.

  Over time, however, the pheromone trail starts to evaporate, thus reducing its attractive strength. The more time it takes for an ant to travel down the path and back again, the more time the pheromones have to evaporate. A short path, by comparison, is marched over faster, and thus the pheromone density remains high, as it is laid on the path as fast as it can evaporate. Pheromone evaporation also has the advantage of avoiding convergence to a local optimal solution. If there were no evaporation at all, the paths chosen by the first ants would tend to be excessively attractive to following ants. In that case, the exploration of the solution space would be constrained.

  Thus, when one ant finds a good (short) path from the colony to a food source, other ants are more likely to follow that path, and such positive feedback eventually leaves all the ants following a single path. The idea of the ant colony algorithm is to mimic this behavior with "simulated ants" walking around the search space representing the problem to be solved.

  Ant colony optimization (ACO) algorithms have been used to produce near-optimal solutions to the traveling salesman problem. They have an advantage over SA and GA approaches when the graph may change dynamically. The ACO algorithm can be run continuously and can adapt to changes in real time. This is of interest in network routing and urban transportation systems.

4.6.3 Optimization Software

A large number of optimizations are available to solve linear and nonlinear mathematical programming problems. All standard software packages have a toolbox for linear, nonlinear, and GP. MATLAB, IMSL, and ACM are very popular. However, some recent software, such as TOMLAB and CVX, is quite useful and easy to use. CVX in particular is free to download. Some optimization software, such as Genesis, NADTRAN, ANSYS, and NSAOPT, is mostly used for solving structural problems and is quite adept and robust in handling large-scale problems.

References

1. Arora, J. S. (1989). *Introduction to Optimum Design*. McGraw-Hill, Inc., New York.
2. Belgundu, A. D. and Chandrupatla, T. R. (1999). *Optimization Concepts and Applications in Engineering*. Prentice Hall, New Jersey.
3. Rao, S. S. (2009). *Engineering Optimization: Theory and Practice*. John Wiley, New Jersey.
4. Gill, P. E., Murray, W., and Wright, M. H. (1981). *Practical Optimization*. Academic Press, London.
5. Goldberg, D. E. (1989). *Genetic Algorithms in Search, Optimization, and Machine Learning*. Addison-Wesley, Reading, MA.
6. Laarhoven van, P. J. and Aarts, E. H. (1987). *Simulated Annealing: Theory and Applications*. Springer Science & Business Media, Springer Netherlands.

7. Lawson, C. L., Hanson, R., Kincaid, D., and Krogh, F. (1979). Basic linear algebra subprograms for Fortran usage. *ACM Trans. Math. Software* 5 (3): 308–323.
8. Ravindran, A., Ragsdell, K. M., and Reklaitis, G. V. (2006). *Engineering Optimization: Methods and Applications.* John Wiley, New Jersey.
9. Wilde, D. J. and Beightler, S. (1967). *Foundation of Optimization*, Prentice-Hall, New Jersey.
10. Swann, W. H. (1964). Report on the development of a direct search method of optimization. Central Instr. Res. Lab. Res., Note 64/3. ICI Ltd., London.
11. Fletcher, R. (1974). *Practical Methods of Optimization.* Wiley-Interscience, New York.
12. Boyd, S. (2015). CVX: Matlab software for disciplined convex programming. Stanford University.
13. Ashby, M. F. (2009). *Materials and the Environment.* Elsevier.
14. Duffin, R. J., Peterson, E., and Zener, C. (1967). *Geometric Programming: Theory and Applications.* John Wiley, New York.
15. Chang, N. B. (2011). *System Analysis for Sustainable Engineering: Theory & Applications.* McGraw Hill, New York.

Bibliography

Ni Bin Chang. (2011). *Sustainable Engineering*, McGraw Hill.

Tora, E. A. and El-Halwagi, M. M. (2009). Optimal design and integration of solar systems and fossil fuels for sustainable and stable power outlet. *Clean Technologies & Environmental Policy* 11 (4): 401–407.

Section II
Green Design and EOL Treatment

5

Principles of Green Design

Traditionally, in mechanical engineering we do not consider the practical application of pollution control, resource conservation, reuse, and recyclability of a product. We are not concerned with pollution and resources but deal mainly with stress, strain, and strength in designing a product. The time has come to include the concept of materials preservations due to depletion of resources, energy preservation, and pollution. *Sustainability* is the practical application of these principles. Sustainability involves a broad range of disciplines, and engineers should be sufficiently informed about the vitality of sustainability. Sometimes, the concept of sustainability is applied in system design, but never in component design. Sustainability considerations will reduce the depletion of resources and emission of greenhouse gases (GHG) into the atmosphere. It will help to conserve resources and the earth itself for future generations.

Green design is intended to develop more environmentally benign products and processes.

The application of green design involves a particular framework for the consideration of environmental issues, the application of relevant analysis and synthesis methods, and a challenge to traditional design and manufacturing procedures. The environmental aspects of mechanical component design are generally ignored. The American Society of Mechanical Engineers (ASME) has a general broad definition of Designing for the Environment, popularly known as DfE. ASME defines DfE as the "front-end planning discipline which simultaneously takes into account impacts of design, manufacturing life cycle, use and disposal of products on the environment." In the engineering curriculum, we must discuss and include the concepts of the health and safety of society at large, including toxicity, recycling of products, and the status of components at the end of their lives, reuse of components, and minimization of energy usage and waste, along with its disposal. As has been established, energy usage directly affects the pollution of land, water, and air. Less energy usage means less pollution. Mechanical engineers use components and systems which can be easily recycled and reused. The concept of recycling and reuse must be incorporated in the initial design itself, and down the chain during manufacturing this concept should be reinforced. Traditional mechanical engineering courses should be reformulated to include sustainability considerations in design and manufacturing. Energy requirements for both products and processes must be reduced. Even the nanotechnologies should have suitable guidelines for sustainability. Sustainable engineering transcends the various engineering disciplines such as environmental engineering, chemical engineering, civil engineering, and electrical and computer engineering. Countries are increasingly developing regulations and guidelines for sustainable engineering practices for components and systems. If we want our engineers to compete globally, or find employment globally, they must be trained in the principles and practices of sustainability.

Pollution prevention is a bundle of activities that reduce the amount of pollution generated by the manufacturing process. It encourages us to create strategies for the reduction of material consumption, and adopts manufacturing processes that reduce waste creation, including pollution. The term cleaner production (CP) also espouses a similar concept of conservation of materials and energies, leading to reduction in toxic raw materials and GHG emissions. CP is coined by the United Nations to mean "sustainable business," and aims to reduce the environmental impacts of manufacturing and services by adopting green technology and environmental management strategies and methods.

In particular, the use of fossil fuels to generate energy is supposed to be the root cause of emissions of toxic and hazardous gases, including CO_2, and NO_x. These GHG emissions lead to ozone depletion, global warming, acidification, and eutrophication, among others.

5.1 Design for the Environment

DfE is a design methodology. It is one of the DFX methodologies, where X may stand for one of the aims: X = C(ost); X=M(anufacturability); X=T(estability), and so on. Manufacturers have realized that end-of-pipe pollution control has its limits. The most common DfE practices are: design for recycling (DFR); design for remanufacture; design for disassembly; design for energy efficiency and hazardous material minimization, or elimination.

Product design for the environment (DfE) means minimizing the negative impacts on the environment during the life cycle of the product from material extraction, manufacturing, consumption, including recycling and reuse, and finally, its disposal to landfill. The earlier concept of pollution prevention was limited to subdisciplines like green chemistry and green design (also known as DfE). At the design stage itself, we should have knowledge and understanding of design decisions on environmental loads. There are, however, different environmental goals for different products. For example, designing a car for lower fuel consumption may lead to using structural composite materials for weight reduction, whereas designing for recyclability would most likely lead to the use of metals for the structural components. DfE is a set of guidelines to help designers meet design goals. Two of the most prominent guidelines for environmental protection are reduced hydrocarbon fuel consumption and increased recyclability. A full list of guidelines would be: reduction in GHG emissions and toxins from use of heavy metals, electric and direct fossil fuel usage, sulfur and NO_x emissions, and materials usage and waste. Graedel [1] suggested a simplified list of guidelines for life cycle analysis (LCA): (1) materials choice, (2) energy use, (3) solid residues, (4) liquid residues, and (5) gaseous residues. He called it streamlined life cycle analysis (SCLA). Graedel [1] suggested assigning a numerical value from 0, being the worst, to 4, being the best. In general, a score of 0 is assigned to a very poor or uninformed practice that raises significant environmental concern, while a score of 4 indicates excellent environmental performance with no known serious concerns. A perfect product would thus obtain a score of 100.

There is no widespread consensus or agreement on the particular goals to be pursued by green design. Some argue that green design and pollution prevention should be practiced solely to reduce costs. For them, any waste from a process or product is an opportunity. Others focus on particular strategies, such as recycling to save raw materials, and develop goals specifically for these strategies. Another approach is to direct all attention to a particular environmental problem, such as global warming, or a particular medium, such as air pollution, and ignore other environmental effects. However, there is some agreement on the broad goals of green design. These are as follows:

- Reduce or minimize the use of nonrenewable resources
- Manage renewable resources to insure sustainability; and
- Reduce, with the ultimate goal of elimination, toxic and otherwise harmful emissions to the environment, including emissions contributing to global warming

LCA is a study to evaluate all resources' consumption and all emissions in the product life cycle. LCA usually consists of three stages:

- Life cycle inventory—a quantitative measure of the inputs and outputs of each stage of the life cycle of a product
- Impact assessment—a value judgment made as to the relative importance of the findings, and
- Improvement plan—an analysis of the changes that are needed to bring about an environmental consciousness in product or process.

Knowledge gained during LCA needs to be transferred into the initial design of a new product. It is actually possible to focus on a specific stage of the product's life such that the environmental impact is minimized in that stage as well as emphasizing the entire life of the product. Researchers have analyzed different stages of a product's life and developed techniques and logistics to improve the design of the product from an environmental perspective. Fiksel [2] presents an excellent overview of DfE concepts and practices. Fiksel defines DfE as: "… a systematic consideration of design performance with respect to environmental, health and safety objectives over the full product and process life cycle." According to the author, DfE can be broken down into many stages, including manufacturing, consumer use, and the end of life of the product. Throughout these stages, different forms of design strategies can be envisioned as the pieces of DfE. For example, in order to minimize the effect of the product on the environment at the manufacturing stage, design objectives may include design for energy conservation to reduce the energy use in production and to be able to use renewable forms of energy and design for minimizing the discharge of hazardous by-products during production. Similar concerns are also valid during the distribution of the product. Finally, during the end-of-life stage of the product, there are design objectives to increase the output of product recovery. These include design for material and product recovery, disassembly, and waste minimization. DFR suggests making better choices for material selection (MS) such that the processes of material separation and material recovery become more efficient. Some general characteristics of DFR are as follows:

- Long product life with minimized use of raw materials (source reduction).
- Easy separation of different materials.
- Fewer number of different materials in a single product while maintaining compatibility with the existing manufacturing infrastructure.
- Fewer components within a given material in an engineered system, increased awareness of life cycle balances, and reprocessing expenses.
- Increased number of parts or subsystems that are easily disassembled and reused without refurbishing.
- More adaptable materials for multiple product applications and fewer secondary operations, reducing the amount of scrap and simplifying the recovery process. On the other hand, design for remanufacturing (or part recovery) suggests the use of reusable parts and packaging.

Disassembly is used both in recycling and remanufacturing to increase the recovery rate by allowing selective separation of parts and materials. Thus, designing for disassembly (DFD) is important and, therefore, it has been given special attention. DFD initiatives lead to the correct identification of design specifications to minimize the complexity of the structure of the product by minimizing the number of parts, increasing the use of common materials, and choosing the fastener and joint types which are easily removable. DFD is often used both in recycling and remanufacturing to increase the recovery rate by allowing selective separation of parts and materials. Thus, DFD is important and therefore it has been given special attention. DFD is often carried out using software due to the complexity of the problem. Lee et al. [3] present a computer-based method that supports the design of a product in terms of disassembability. In order to identify the disassembability of the product, the authors propose an assessment method. Kroll and Hanft [4] propose a rating scheme that allows the designers to translate properties of a design into quantitative scores and thus provide a means of identifying weaknesses in the design and comparing alternatives. Lee et al. [3] present a benchmark of the software available for DFD.

DFD is just one of the aspects of DfE. However, DfE comes with more than one task. There are three components of DfE: the challenges faced by product designers and environment specialists, the green design challenges from a company's point of view, and the external factors and forces that drive the design decisions. The ultimate goal of green design is to reduce the overall environmental damage when producing goods and providing services [5]. To achieve this goal, in addition to this cooperative work, availability of guidelines, checklists, and software-based DfE tools also play a key role. Azzone and Noci [6] introduce an integrated measuring method for the greenness of new products. In this integrated method, at every stage of the product development, required decision-making initiatives are identified. DfE aims to reduce time to market, lower cost and increase quality of the product. However, the production of the designed product also adds to pollution and global warming. The designer should be familiar with the manufacturing processes and their characteristics as well as with the measurement methods of the *impact*, because, during LCA analysis, it is essential to identify a functional unit and establish a quantitative measurement method so that sensible comparisons between alternate products or processes can be made. Therefore, environmentally conscious production (ECP) is the next important consideration in sustainable development.

5.2 Environmentally Conscious Production

In addition to environmentally friendly product designs resulting from DfE initiatives, issues involving production must also be addressed to have a complete concept of environmentally conscious manufacturing. These issues include selecting energy sources necessary for production, designing cooling systems, and handling hazardous by-products. Currently, numerous production techniques, material handling systems, and energy sources are available. Utilizing some sort of assessment tool to select between them may be valuable financially, as well as improving the environmental features of the production system. Chen et al. [7] develops a tool to come up with a good material and process combination. Similar models have been developed to analyze how the selection of different manufacturing processes affect the environment. Many companies monitor their waste generation as a result of their manufacturing processes. Several techniques have been proposed for such a monitoring process. For example, weighting methods were proposed to measure the chemical and toxic discharges of different manufacturing methods.

Major developments to date are waste-stream models for machining techniques such as turning, tapping, milling, drilling, and nontraditional machining techniques such as electrical discharge machining (EDM) processes. The waste streams identified with their associated process performance measures for traditional and EDM processes are depicted in the table below.

		Turning, Tapping, Milling, Drilling	Electro Discharge Machining (EDM)
Water Stream	Gas/aerosol	• Cutting fluid mist • Dust (dry machining)	• Electrolyte vapor • Gaseous products of chemical reactions
	Solid	• Chips, worn tools	• Sludge
	Liquid	• Spent cutting fluids	• Used electrolyte
Process Performance Measures		• Material removal rates • Chip morphology • Power consumption • Heat generated • Mist size and mass distribution • Tool life • Cutting fluid life	• Material removal rates • Surface quality

The table above shows the waste stream of gas, liquid, and solid from traditional machining processes. There is considerable research on environmentally benign manufacturing but it has not yet been incorporated into the engineering curriculum.

Models have been developed to understand the effect of cutting fluids on chip morphology in machining processes, with special focus on drilling operations. These efforts have been supported by Ford, NSF, NSF-EPA, Caterpillar, Setco, Kennametal, M.A. Ford, Cheboygan, Heldun, Chesterton, and Valenite.

5.3 Green Design Tools

The objective of green design is to pursue these goals for a sustainable future. Some green design terms and tools are explained below.

5.3.1 Mass Balance Analysis

Mass balance analysis involves tracing the materials or energy in and out of an analysis area such as a manufacturing station, a plant, or a watershed. Ideally, mass balances are based on measurements of inflows, inventories, and outflows (including products, wastes, and emissions).

In actuality, all the data needed is rarely available or even consistent.

5.3.2 Green Indices

Green indices or ranking systems attempt to summarize various environmental impacts into a simple scale. The designer or decision-maker can then compare the green score of alternatives (materials, processes, etc.) and choose the one with the most minimal environmental impact. This would contribute to products with reduced environmental impacts. Some of the most common environmental impact assessment tools are carbon footprints, eco-indicators, and embodied energy content.

5.3.3 Energy Content

Primary production of metals is energy intensive. Many among them, such as aluminum, magnesium, and titanium, require at least twice as much energy per unit weight (or five times per unit volume) as compared to polymers. However, most metals can be recycled efficiently and the energy required for recycling is much less than that required for primary extraction and production. Therefore, the energy content of a material is the energy used in producing it from the natural ore and manufacturing it to make it usable. For example, the energy content of aluminum is the electric power absorbed in its extraction from bauxite ore. Obviously, the energy content of a polymer would be the energy contained in the oil itself and that required for subsequent processing. This is mostly called "embodied energy."

5.3.4 Eco-Indicators

The *eco-indicator* of a material or process is a number that indicates the damage to the ecosystem or life. The eco-indicators of engineering materials include greenhouse (GHG) effects, which really mean an increasing concentration of gases such as carbon dioxide (CO_2), methane (CH_4), and carbon monoxide (CO) causing global warming. There is also high-altitude decomposition of the ozone layer, increasing ultraviolet radiation on Earth. Due to the increases in CO_2, CH_4, and CO in the atmosphere, acid rain causes degradation in forests; numerous such cases have been reported from China, Germany, and other countries. There are some components manufactured which contain heavy metals such as cadmium (Cd), mercury (Hg), and lead (Pb); these are highly toxic and pollute land, water, and air. Some countries are introducing legislation and regulations preventing the use of heavy metals in any product; in particular, European countries are at the forefront of this through the waste electrical and electronic equipment (WEEE) regulations. Some substances found in various materials used for engineering production cause cancer in people. Due to high emissions of dust, along with sulfur dioxide (SO_2) in the atmosphere, winter and summer smog is reported from many countries like China, India, and so on. For each environmental effect, a score is calculated and is normalized by comparing it with the current level of environmental effects per person per year.

The eco-indicator is expressed in millipoints (mPts) per functional unit and is calculated according to the Eco-indicator 99 method (Pre-Consultants, Netherlands, www.pre.nl).

A life cycle inventory (LCI) is completed to calculate the eco-indicator. Sometimes, these life cycle inventories or the environmental effects parameters are multiplied by a weighting factor determined by the seriousness of an effect; for example, if GHG emissions are more serious they should be multiplied by a larger value compared to, say, ozone layer depletion. These weightings are still subjective and no algorithm has yet been developed to quantify them exactly or scientifically. Finally, the weighted normalized effects are summed to provide the eco-indicator value. The eco-indicators are used at the design stage to compare the environmental effects of products and the lowest eco-indicator material is selected.

Some metals are ductile and are shaped by rolling, forging, drawing, and extrusion; these are easy to machine with precision and they can be joined in many different ways. However, some metals like steel, aluminum, and so on can also be shaped by metal-removal processes such as turning, milling, shaping, and grinding. However, there are other metals like cast iron that can be cast into different shapes by foundry processes. For reasons of flexibility of design, at this time polymers are increasingly being used wherever feasible.

5.4 Footprint Indicators

Mostly, a footprint is understood as the total burden it places on the environment. However, the most important of these footprints is the carbon footprint. There are other footprint indicators, for example, energy use, solid waste, and air emissions. For green design and manufacturing, the ecoproperties of materials need to be known, as without these the design and manufacturing parameters cannot be calculated.

However, as has been indicated earlier, there are no standardized methods to measure embodied energies and carbon footprints of engineering materials.

5.4.1 Design for Disassembly and Recycling Aids

Design for disassembly and recycling (DFD/R) refers to making products that can be taken apart easily for subsequent recycling and parts reuse. For example, Kodak's "disposable" cameras snap apart, allowing 87% of the parts (by weight) to be reused or recycled. Unfortunately, the economic costs associated with physically taking apart products to get at valuable components and materials often exceed the value of the materials. Reducing the time (and thus cost) of disassembly might reverse this balance. Thus, DFD/R acts as a driver for recycling and reuse. Later, the techniques for disassembly, resuse, recycling, and landfill are presented in the mechanical component design. Designing and manufacturing green products requires appropriate knowledge, tools, production methods, and incentives. Ideally, these design tools will help identify design changes that have lower costs while improving materials use and recyclability. For example, using snap fits rather than metal fasteners may have little additional cost burden at the design stage but may significantly increase recycling potential.

5.4.2 Recycling

It is now a well-established principle in mechanical design theory that designers should consider the end-of-life scenario before the final design. The disposal of the product after its useful life must be thought out at design stage: How we are going to recycle the component or system, what percentage of materials are going to be reused, and finally, how the remainder would be disposed of in landfill, incinerated, and so forth. So, recycling has become a new parameter in the design of parts or components. Still, green design is supposed to be a system problem, but it need not be. It should be considered and included at the component stage. Recycling is becoming a serious issue in some societies. The reasons for this could be many, such as depletion of resources, environmental degradation, global warming, and so on. Recycling is an attempt to reduce the use of new material and its extraction from nature. As mentioned earlier, extraction of ore for material requires more energy compared to recycling parts.

Most materials are recyclable and reusable. Recycling can lead to substantial energy savings, thereby reducing atmospheric pollution. Among metals, aluminum alloys are a good example of recycling, as a low ratio of energy is required to refine recycled aluminum. Ashby [8] has indicated that 33%–55% of aluminum is recycled. However, the quality of the material tends to decrease after successive recycling. It has been reported that components made of alloys are difficult to recycle. Examples include bimetallic strips used in sensors and galvanized steel cans. Thermoplastics are easier to recycle. Table 5.1 shows the recycling code adopted in the United States with their numbers and associated polymers.

In general, crystalline ceramics are not recycled, whereas amorphous glass containers are recycled extensively. This use of recycled glass provides reduced emissions of harmful and hazardous gases in the atmosphere. However, glass containers with polymeric coatings are difficult to recycle due to dissimilar material combinations. There have been great strides in the field of composite materials such as metal matrix composites (aluminum matrices with carbon-fiber reinforcement), polymer matrix composites (carbon- or glass-fiber–reinforced epoxy composites), and ceramic matrix composites (carbon and carbon composites) due to increased use of composites in industry. However, they are difficult to recycle due to fine-scale combination of different materials. For example, microelectronic circuit boards are manufactured of composites and are difficult to recycle. In the United States, such components are under strict watch by regulatory agencies before disposal.

Table 5.1 Identification Codes for Polymer Recycling

Recycle Code	Associated Polymer
1	Polyethylene terephthalate (PET or PETE)
2	High density polyethylene (HDPE)
3	Polyvinyl chloride (PVC)
4	Low-density polyethylene (LDPE)
5	Polypropylene (PP)
6	Polystyrene (PS)

5.5 Factors in Material Selection for Environmentally Conscious Design

The performance of a product depends on the material selected at design stage. For an environmentally sustainable design, we should look at the ecoproperties of materials' carbon footprints and embodied energies. However, we should know that these ecoproperties are not exact. We know that engineering properties of materials, mechanical, thermal and electrical attributes, are well characterized. There are standard well-established methods and equipment that can provide exact values for physical, chemical, mechanical, and dimensional properties, but there is no sophisticated test equipment to measure ecoproperties. Therefore, the ecoproperties cited in the literature are subjective and contribute to error. Some of the ecoproperties reported vary by a standard deviation of ±10%, which suggests they may differ by as much as 20%. It is also noted that some of the data related to recycling and reuse content vary widely from country to country. Metals used in product design are invariably alloys. Steel, which is widely used, is an alloy of iron with carbon and a myriad of other alloying elements to make them harder, tougher, or more corrosion resistant. Metals such as aluminum alloys, magnesium alloys, titanium alloys, copper alloys, nickel-chrome alloys, low-carbon steel, low-alloy steel, stainless steel, and cast iron are widely used in industrial production. Parts subjected to fatigue load require high endurance strength and, for short struts, high compressive strength. However, high buckling strength is needed for slender columns. For green design, we should include ecoproperties such as energy content, eco-indicators, and so on.

The challenge of green design is to alter conventional design and manufacturing procedures to incorporate environmental considerations systematically and effectively. This requires changes in the existing procedures. Change to any existing process is difficult. Changing design procedures is particularly difficult because designers face many conflicting objectives, uncertainties, and a work environment demanding speed and cost-effectiveness.

5.6 Green Designing and Machining

We know that various materials have different levels of manufacturability. Commonly employed manufacturing processes are described below. From the environmental perspective, it is desirable to produce components to near net shape so that the wastage of materials (scrap) can be minimized. The manufacturing selection depends on the material, component geometry (shape as well as size), and required dimensional tolerances. As we know, sand casting tolerance is much lower than milling or turning and powder consolidation gives a much better tolerance than any metal-removal process. For green manufacturing, we should use manufacturing processes which consume minimum energy and creates minimum environmental damage in terms of atmospheric emissions, liquid and solid waste. In manufacturing engineering, we should lay stress on manufacturing processes with the least specific energy consideration to minimize the consumption of energy.

Many material-processing technologies liberate various toxic by-products. Under the U.S. Clean Air Act, various product manufacturers have to follow the National Emissions Standards for Hazardous Air Pollutants (NESHAPS). The release of chemicals such as asbestos, benzene, mercury, beryllium, lead, inorganic arsenic, copper, nickel, and radionuclides should be minimized. Also, airborne particles from aerosols are increasingly focused upon by legislators.

Environmental concerns must be introduced in practical and meaningful fashions into these complicated design processes. For the future, we wish to prepare consumers and designers to think proactively about the environment.

Several materials can produce a particular quality component or product. However, they have different environmental implications. MS guidelines attempt to direct designers toward environmentally preferred materials. For example, Graedel and Allenby [9] present the following guiding principles for MS:

- Choose abundant, nontoxic materials where possible.
- Choose materials familiar to nature (e.g., cellulose), rather than man-made materials (e.g., chlorinated aromatics).
- Minimize the number of materials used in a production process.
- Try to use materials that have an existing recycling infrastructure.
- Use recycled materials where possible.

In addition to these generic guidelines, companies such as IBM and Chrysler have been developing and using specific material selection guidelines for environmentally sound product development, which describe in detail the materials that should be used for specific applications. Label advisors are generally markings on materials or products that reveal information about the material content relevant to materials handling and waste management. For example, the plastic bottles used in many consumer products usually have a plastic identification symbol (see Table 5.1) that can be used in plastics resorting and recycling efforts. Even labels are subject to green design. Embedding a label into a material is preferable to a separate label material attached with an adhesive. The material data collected in *Materials and Environment* (2009) is most comprehensive. A sample of data is presented here for use in green design and manufacturing for sustainability problems. International standard ISO 14040 has provided some guidelines and procedures but it is very difficult to apply to engineering problems. Here, we present the relevant eco-material properties with regard to green mechanical component design. Synopses of engineering materials along with its typical use are presented below.

- *Aluminum alloys*; Composition; Al plus alloying elements, for example, Mg, Mn, Zn, Zr, Li.
  Typical usage: Aerospace engineering, automotive engineering-pistons, clutches, housing, exhaust manifolds; sports equipment such as golf clubs and bicycles; die-cast chassis for household and electronic products; sidings for buildings; reflective coatings for mirrors; foil for containers and packaging beverage cans; electrical and thermal conductors.
- *Magnesium alloys*; composition: Mg plus alloying elements, for example, Al, Mn, Si, Zn, Cu, Li, rare earth elements.
  Typical usage: Aerospace; automotives; sports goods such as bicycles; nuclear fuel cans; vibration damping and shielding of machine tools; engine case castings; crank cases; transmission housings; automotive wheels, ladders; housings for electronic equipment, particularly cell phone and portable computer chassis; camera bodies; office equipment; marine hardware and lawnmowers.
- *Titanium alloys*; composition: Ti plus alloying elements, for example, Al, Zr, Cr, Mo, Si, Sn, Ni, Fe, V.
  Typical usage: Aircraft turbine blades; general aerospace applications, chemical engineering; pressure vessels; high-performance automotive parts such as connecting rods; heat exchangers; bioengineering; medical; missile fuel tanks; compressors; valve bodies; light springs; surgical implants; marine hardware; paper-pulp equipment; sports equipment such as golf clubs and bicycles.
- *Copper alloys*; composition: Cu with up to 40% Zn or 30% Sn, Al, or Ni.
  Typical usage: Electrical wiring, cables, bus bars, high-strength, high-conductivity wires and sections, overhead lines, contact wires, resistance-welding electrodes, terminals, high-conductivity items for use at raised temperatures, heat exchangers, coinage, pans, kettles and boilers, plates for etching and engraving, roofing and architecture, cast sculptures, pumps, marine propellers.
- *Lead alloys*; composition: Pb plus 0%–25% Sb or 0%–60% Sn, sometimes with some Ca.
  Typical usage: Roofs, wall cladding, pipe work, window seals, and flooring in buildings; sculpture and tableware such as pewter; solder for electrical circuits and for mechanical joining, bearings, painting type, ammunition, pigments, X-ray shielding, corrosion-resistant material in the chemical industry and electrodes for lead-acid batteries.
- *Zinc die-casting alloys*; composition: Zn plus 3%–30% Al, typically, often with up to 30% Cu.
  Typical usage: Die-castings; automotive parts and tools; gears; household goods; office equipment; building hardware; padlocks; toys; business machines; sound reproduction equipment; hydraulic valves; pneumatic valves; soldering; handles; automotive components.
- *Nickel-chromium alloys*; composition: Ni plus 10%–30% Cr plus 0%–10% Fe.
  Typical usage: Heating elements and furnace windings; bimetallic strips; thermocouples; springs; food-processing equipment; chemical engineering equipment.
- *Nickel-based superalloys*; composition: Ni plus 10%–25% Cr plus Ti, Al, Co, Mn, Zr, B, and Fe in varying proportions.
  Typical usages: Blades, disks, and combustion chambers in turbines and jet engines, rocket engines; general structural aerospace applications; light springs, high-temperature chemical engineering equipment; bioengineering and medical.
- *Low-carbon steel*; composition: Fe plus 0.2%–0.3% C.
  Typical usage: Reinforcement of concrete, steel sections for construction, sheet for roofing, car body panels, cans, and pressed-sheet products. Low-carbon steel is widely used in engineering and no list can be complete.
- *Low-alloy steel*; composition: Fe plus 1.0% C, 2.5% Ni, 2.5% Mo, 2.5% V.

Typical usage: Springs, tools, ball bearings, rollers, crankshafts, gears, connecting rods, knives and scissors, pressure vessels.

- *Stainless steel*; composition: Fe plus 0.25% C, 16%–30% Cr, 3.5%–3.7% Ni, 10% Mn plus Si, P, S (plus N for 200 series).
  Typical usage: Railroad cars, trucks; trailers, food-processing equipment, sinks, stoves, cooking utensils, cutlery, flatware, scissors and knives, architectural metalwork, laundry equipment, chemical processing equipment, jet-engine parts, surgical tools, furnace and boiler components, oil-burner parts, petroleum-processing equipment, dairy equipment, heat-treating equipment, automotive trim, structural uses in corrosive environments; for example, nuclear plants, ships, offshore installations, underwater cables, and pipes.
- *Cast iron, ductile (nodular)*; composition: Fe plus 3.2–4.1% C, 1.8–2.8% Si, 0.85% Mn, 0.15% P, 0.03% S.
  Typical usage: Brake disks and drums; bearings; camshafts; cylinder liners; piston rings; machine tools, structural parts; engine blocks, gears, crankshafts, heavy-duty gear cases; pipe joints; pump castings; components in rock crushers.

Now, we will present the ecoproperties of the above mentioned materials. The physical and mechanical properties are not presented here as they are available in mechanical design texts.

The ecoproperties of materials for processing and recycling are shown in Table 5.2.

5.7 Steps in Eco-Material Selection for Mechanical Components

Intuitively, mechanical design engineers *screen* the materials for functional requirements before finding the dimensions. Some requirements are critical to a component, such as weight for aerospace products and elastic energy capacity for springs; electric power transmission lines must be good conductors, and building beams should be able to resist buckling.

After screening the materials for functional requirements, the materials are ranked using the *material performance index*. This index is a function of material properties, which is used to optimize the design performance under a set of real-world constraints. For green design, the energy content and eco-impact (eco-indicator) should be minimized under constraints of stress, stiffness, and so on. We will be developing the performance index for the design of different components. We can also minimize cost, weight, or volume under constraints of yielding, stiffness, buckling, and so forth. The life cycle cost is usually minimized in MS for green design; the cost includes that of materials and manufacturing along with recycling, demanufacturing, and disposal.

Finally, designers short-list the materials based on failure behavior and environmental impact. No doubt factors such as cost, availability, and past usage will guide the designer to select the optimum material for the product to be designed. However, the objective is invariably the minimization of the material in the product and it should have no adverse environmental effect. The designer engineers would naturally consider all possible new materials and technology for the component so that the overall (life cycle) cost is minimized. Designers realize that cost is highly dependent on the mass of material and that, in turn, affects the environmental impact. We will present some useful MS methods used by mechanical design engineers.

5.7.1 Selection of Material by Cost/Unit Volume Consideration

The method is presented through an example of selecting material for a tray to hold a specified volume of liquid.

Example 5.1

Design a plastic tray which must be capable of holding a specified volume of liquid V, such that the liquid has a specified depth H, and the wall thickness of the tray T. The tray is to be manufactured in large quantities (Figure 5.1).

Optimal Design

As can be seen, the problem does not say what the basic geometrical shape of the tray should be. However, by intuition, we always start with a rectangular shape. Assuming wall thickness T to be small compared to the dimensions b and l, we can write

$$V = blH \tag{5.1}$$

Table 5.2 Ecoproperties of Some Selected Materials

	Ecoproperties: Materials				Ecoproperties: Processing				Recycling		
Material	Embodied Energy (MJ/kg)	CO_2 Footprint (kg/kg)	Water Usage (l/kg)	Eco-Indicator (millipoints/kg)	Casting Energy (MJ/kg)	Casting CO_2 footprint (kg/kg)	Deformation Processing Energy (MJ/kg)	Deformation processing CO_2 footprint (kg/kg)	Embodied Energy Recycling (Mj/kg)	CO_2 Footprint, Recycling (kg/kg)	Recycle Fraction (%)
Aluminum alloys	200–240	11–13	125–375	740–820	2.4–2.9	0.14–0.17	2.4–2.9	0.19–0.23	18–21	1.1–1.2	33–55
Magnesium alloys	356–394	22.4–24.8	500–1500	-	2.1–2.6	0.13–0.16	3.8–4.6	0.3–0.37	8.4–9.3	1.8–1.95	9.2–11.4
Titanium alloys	600–740	38–44	470–1410	-	5.02–5.77	0.301–0.346	4.71–5.7	0.377–0.456	2.28–2.81	14.4–16.7	20–24
Copper alloys	68–74	4.9–5.6	150–450	1300–1500	2.5–3.8	0.15–0.17	1.8–2.2	0.14–0.16	17–18.5	1.2–1.4	40–60
Lead alloys	53–58	3.3–3.7	175–525	610–670	0.4–0.49	0.024–0.03	1.6–2.7	0.13–0.14	8.4–9.2	0.46–0.51	70–75
Zinc die-casting alloys	70–75	3.7–4	160–521	3000–3400	1.09–1.32	0.065–0.08	-	-	12.6–13.5	0.66–0.72	20–25
Nickel-chromium alloys	127–140	7.89–8.82	134–512	4900–5500	3.68–4.06	0.221–0.243	2.52–2.78	0.202–0.222	31.8–35	1.97–2.21	22–26
Nickel-based superalloys	135–150	7.89–9.2	134–484	4900–5500	3.38–4.00	0.20–0.24	3.26–3.95 *11.3–13.6	0.26–0.31* 0.9–1.09	33.8–37.5	1.97–2.3	22–26
Low-carbon steel	29–35	2.2–2.8	23–69	79–87	4.2–4.6	0.25–0.28	2.2–2.6	0.18–0.21	8.1–9.8	0.6–0.8	40–44
Low-alloy steel	32–38	1.95–2.3	37–111	100–120	3.8–4.6	0.23–0.28	3.2–3.9	0.26–0.31	9.0–11	0.55–0.64	40–44
Stainless steel	77–85	4.7–5.4	112–336	860–960	3.9–4.4	0.24–0.29	3.0–3.7	0.26–0.31	22–24	1.4–1.5	35–40
Cast iron, ductile (nodular)	16–18	1.0–1.1	13–39	38–42	3.2–3.7	0.19–0.22	2.5–3.1	0.2–0.24	4.2–5.5	0.29–0.32	60–80

Source: All the materials data has been obtained from one source, but the exhaustive database is available in the Cambridge Engineering Selector (CES), developed by Granta Design, Cambridge, United Kingdom. The ASM Handbook is also a good source for ecoproperties of materials.

- Data not available.

* Powder forming.

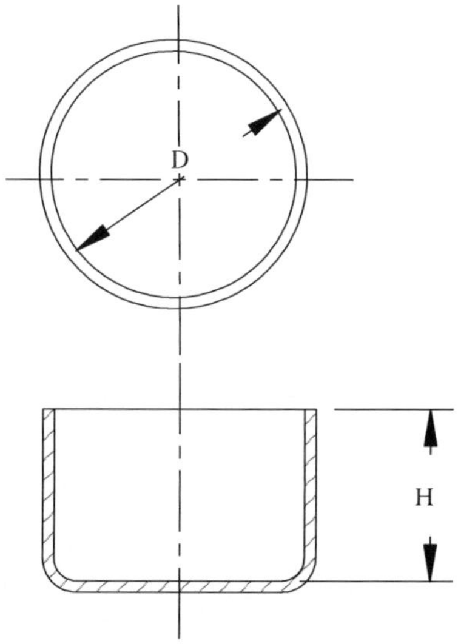

Figure 5.1
Green tray design.

Therefore, we can arbitrarily select any value for b and then calculate the corresponding value of l. However, there would be an infinite number of possible solutions to the above problem. In this DfE problem, we are required to select a material for the tray that is environmentally suitable. We know from our experience that a vacuum-forming process will be suitable for the manufacture of the tray. When selecting the material, we should consider whether the tray will react chemically with the liquid it is holding. Let us select acrylic thermoplastic sheet as the material for the tray. In all optimal engineering design problems, the undesirable effects of stresses, vibrations, deflections, and weight are significant, and these effects must be considered either as constraints or sometimes as objectives. The optimal design for this example is a relatively simple one and we will not consider these effects. As we know, the cost of the tray may make or break the tray-manufacturing industry. Hence, in every optimal design, we must consider the cost of manufacture in one way or another. So, in this problem, we consider the minimization of cost as our objective function. We have framed our optimal design problem such that for the minimization of cost for the mass production of the tray to consider whether our optimum design solution is the best with respect to the shape (geometry), material, and manufacturing process. Our first task in optimal design is to develop the objective function, which in this case is cost minimization. The total cost (TC) of the tray is developed as:

$$TC = f_o + f_t + f_l + f_m \tag{5.2}$$

where:

f_0 = overhead cost/tray
f_t = tooling cost/tray
f_l = labor cost/tray
f_m = material cost of plastic/tray

As we have selected vacuum forming as our manufacturing process, the overheads, tooling, and labor costs will essentially be independent of reasonable shapes and feasible plastic materials. These assumptions may not be correct in every optimal design problem. However, in this case we may reasonably conclude that our optimal tray design will be based only on the material cost of the tray. Let us assume c to be the unit weight cost of the plastic material and w to be the weight density of the plastic material in pounds per cubic inch. If H and T are specified values in inches and we are supposed to find the values of b and l of the optimal tray design, we can develop the cost equation as

$$TC = f_m = cw\left(bl + 2bH + 2lH\right)T \tag{5.3}$$

where $(bl + 2bH + 2lH)T$ represents the volume of plastic material used in the manufacture of a rectangular-shaped plastic tray. However, it has been specified that the volume of liquid (V) in the tray of thickness T and depth H are derived from Equation 5.1 and we can replace the length of the tray as

$$l = \frac{V}{bH} \tag{5.4}$$

Replacing l by the above in Equation 5.3, we can rewrite the equation as

$$f_m = cw\left(\frac{V}{H} + 2bH + 2\frac{V}{b}\right)T \tag{5.5}$$

The optimum design is based on the f_m equation. The terms c and w in the equation depend on the material selected. Hence, to minimize f_m we should select the material having the lowest value for the factor cw from the list of feasible materials. Therefore, the cost per unit volume term (cw) may be thought as the MS factor. As we can see from the f_m equation, this factor is quite independent of the geometrical shape of the tray and of the particular values for V, H, T, b, and l. For simplicity of illustration, we will assume that only two feasible thermoplastics, such as acrylic and polystyrene sheets, are available to us at this time. The values of weight density w may be obtained from handbooks, and values of unit cost c of the materials can be obtained from vendors. Hence, we might develop the MS factor in terms of cw for each feasible and available material (Table 5.3).

Therefore, for this optimal design problem, the optimum available material is polystyrene, because of its lower cw value. In this particular example, we can select *polystyrene* directly because of its considerably lower material unit cost c. However, if we examine the environmental effects, such as embodied energy and eco-indicators, these also support our selection of polystyrene. We have tried to impose environmental consideration over cost consideration and both support the same MS.

For many real-world problems, we may have a long list of available feasible materials, and only rigorous optimum MS would be on the basis of minimum value of the factor cw.

Determination of Optimal Shape

To determine the optimum geometrical shape, let us examine the variation of cost f_m with respect to the independent variable b from the optimum design equation developed earlier. This equation has only one independent variable b and we can find the optimal design by taking the derivative of the cost equation and making it equal to zero. However, in most real-world optimal design problems, we have to use a constrained nonlinear programming algorithm to find the optimal design, as optimal design problems are inherently nonlinear in nature. For the present example, we can say

$$\partial f_m / \partial b = (cw)(2T)\left[H - V / b^2\right] = 0$$

$$\therefore b_{opt} = \sqrt{\frac{V}{H}} \tag{5.6}$$

Plugging the value of b_{opt} into the equation $l = V/bH$, we get

Table 5.3 Cost-Weight-Based Material Selection

Materials	w,lb/in³	c, $/lb	Embodied Energy (MJ/kg)	Eco-indicator (millipoints/kg)	(*cw*), $/in³
Acrylic	0.04	2.00	91–102	380–420	0.08
Polystyrene	0.038	0.60	86–99	340–380	0.023

$$l_{opt} = \frac{V}{H(b_{opt})} \tag{5.7}$$

which shows that from the objective of minimizing cost, the optimum rectangular shape is a *square*. For each square tray, the minimum cost would be

$$(f_m)_{opt} = (cw)\left[V/H + 2(\sqrt{V/H})H + 2V/\sqrt{V/H}\right]T$$

Therefore

$$(f_m)_{opt} = (cw)\left[V/H + 4\sqrt{V/H}\right]T \tag{5.8}$$

In this equation, $(f_m)_{opt}$ is the material cost of an ideal square tray, *cw* is the MS factor for the optimum material previously selected, and *V*, *H*, and *T* are specified functional requirements from the optimal design problem stated.

Often, space restrictions exist in real optimal design problems. Hence, for the intended usage of the tray, it might be required to fit between two structures while resting on a shelf of specified length. Therefore, restrictions might exist on permissible values of *b* and *l* as follows:

$$b \le b_{max} \tag{5.9}$$

$$l \le l_{max} \tag{5.10}$$

In these equations, b_{max} and l_{max} are the values of space restrictions that exist for feasible values of *b* and *l*, respectively. Obviously, particular restrictions on *b* and *l* might affect the optimum geometry of the rectangular tray. If b_{max} and l_{max} are both equal to or greater than $\sqrt{V/H}$, the ideal square tray is still the optimum design. However, if either b_{max} or l_{max} is less than $\sqrt{V/H}$, the optimum design is the tray which barely fits within the space restriction. For the case where b_{max} is less than $\sqrt{V/H}$, the optimum design is depicted in Figure 5.1. Also, it is possible that no geometrical design satisfying the specifications within the space restriction is feasible. This possibility might be termed a case of *incompatible specifications*, which would exist if the following relationship existed:

$$b_{max} \cdot l_{max} < V/H \tag{5.11}$$

Here, a rectangular tray was arbitrarily selected using our intuition. However, if we design a circular tray we might find that the circular shape would minimize the cost (material) function.

Circular Tray Design

The material cost function for a circular tray is

$$f_m = (cw)\left[\pi D^2/4 + \pi DH\right]T \tag{5.12}$$

where *D* is the diameter of the tray; all other terms remain the same as for the rectangular tray. For the circular tray, the specified liquid volume *V* can be expressed as

$$V = \left(\frac{\pi D^2}{4}\right)H$$

By eliminating *D*, we get

$$f_m = (cw)(V/H + 2\sqrt{\pi VH})T$$

Therefore

$$f_m = (cw)\left[V/H + 3.54\sqrt{VH}\right]T \tag{5.13}$$

All the terms on the right-hand side are independent of each other. Now again, the optimum material is selected from the list of available materials on the basis of the MS factor (cw), which is the same as for the rectangular shape. From the liquid volume equation, D could be determined uniquely as

$$D = \sqrt{\frac{4V}{\pi H}} \tag{5.14}$$

And if we compare the circular tray f_m equation with the rectangular tray f_m equation, we find that the optimum cost of the circular tray is lower than that of the rectangular tray. For the circular tray design, space restrictions may again apply, as they did for the rectangular tray. If space restrictions exist, it turns out that an *oval* tray would be the best possible for minimizing the material cost.

We can even examine the selection of the manufacturing process. Perhaps another available feasible manufacturing process, such as a low-cost molding process, would result in a lower tray cost than vacuum forming. It is better to examine the MS problem from the standpoint of the life cycle cost, including recycling and disposal.

5.7.2 Merit Index Method

The second MS method is called the *merit index* (MI) or weighted factor method. It is widely used in mechanical design. The example of the tray presented in Example 5.1 will be revisited and analyzed through the MI method.

Example 5.2

We will apply weightings to cost, embodied energy, and eco-indicators. The sum of these weightings (ω_i) must add up to 1. These weighting assignments are subjective; however, they depend on the choice and expertise of designers. If the environmental impacts are more important, then obviously embodied energy and eco-indicators should be assigned more weight compared to cost, or vice-versa. Here, we assign the weightings 0.25, 0.45, and 0.30 to cost, embodied energy, and eco-indicator, respectively. Design engineers should use their good judgment or past experience in choosing appropriate values of ω_i. Usually, the most important parameter is given the highest value of ω_i, and so on. Here, we may notice that values for embodied energy and eco-indicators are given in a range, so we will use the average value given in parentheses for our calculation (see Table 5.4):

$$\text{Acrylic:} \quad \frac{0.45(2) + 0.30(96.5) + 0.30(400)}{400} = 0.375$$

$$\text{Polystyrene:} \quad \frac{0.45(0.6) + 0.30(92.5) + 0.25(360)}{360} = 0.378$$

If we look at the MI column, both values are essentially the same. However, the balance is tipped slightly in favor of polystyrene. However, this supports the solution obtained in Example 5.1.

Table 5.4 Merit Index Material Selection

Materials	w (lb/in^3)	c (\$/lb)	Embodied Energy (MJ/kg)*	Eco-indicator (millipoints/kg)*	MI
Acrylic	0.04	2.00	91–102 (96.5)	380–420 (400)	0.375
Polystyrene	0.038	0.60	86–99 (92.5)	340–380 (360)	0.378

*Data from Table 5.3.

Example 5.3

A short list of materials has been made for an environmentally conscious leaf spring design. The materials are 304 stainless steel, 6061 Al, Ti-6 AI-4V, and carbon-fiber-reinforced composite (CFRP). The leaf spring is to be designed for the maximum storage of elastic energy. The elastic energy per unit volume then becomes the objective function. If weight is considered, the elastic energy (U) is given by

$$U = \frac{\sigma_y^{\ 2}}{\rho E} \tag{5.15}$$

where:

σ_y is the yield strength of the spring material
ρ is the density of the spring material in megagrams per cubic meter
E is Young's modulus of the material in gigapascals (see Table 5.5)

Applying weights of 0.45, 0.30, and 0.25 to elastic energy, embodied energy, and eco-indicator, respectively, we can calculate MI as

$$\text{304 stainless steel:} \quad \frac{0.45(65.22)+0.30(81)+0.25(910)}{910} = 0.309$$

$$\text{6061 aluminum:} \quad \frac{0.45(76.19)+0.30(220)+0.25(780)}{780} = 0.379$$

$$\text{Ti6-AI-4V titanium alloy:} \quad \frac{0.45(1700.28)+0.30(670)+0.25(90)}{1700.28} = 0.581$$

$$\text{CFRP:} \quad \frac{0.45(7407.41)+0.30(272.5)+0.30(22)}{7407.41} = 0.462$$

All the values calculated above are normalized as they are divided by the maximum value in the row. For this example problem, we find that the titanium alloy has the highest MI value, followed by CFRP. Thus, it is the most suitable material for leaf spring design. However, it is very expensive compared to CFRP. Due to this reason, CFRP is the most widely used material for truck springs.

5.7.3 Performance Index Method

Ashby has developed a systematic method for selection of material for engineering design [10]. The MI method discussed in Section 5.7.2 depends on subjective human judgment in applying the weightings. Asbhy has created a database in the Cambridge Material Selector (CMS), created by Granta Design [10]. For specialized design problems, a more rigorous mathematical approach based on the objective function and real-world constraints is advocated. The aim is to identify the materials for which accurate design calculation will

Table 5.5 Merit Index for Leaf Spring

Material	Density (ρ) (Mg/m³)	Young's Modulus (E), (GPa)	Yield Strength (σ_y), (MPa)	Elastic Energy, $\sigma_y^2/(E\rho)$	Embodied Energy (MJ/kg)	Eco-indicators (mp/kg)	MI
304 Stainless steel	7.85	200	320	65.22	77–85 (81)*	860–960 (910)*	0.309
6061 Aluminum	2.70	70	120	76.19	200–240 (220)*	740–820 (780)*	0.3791
Ti-6AI-4V	4.4	120	950	1700.28	600–740 (670)*	80–100 (90)*	0.581
CFRP	1.35	100	1000	7407.41	259–286 (272.5)*	22	0.462

*Average value used in this computation.

be performed later. The requirements for mathematical model development is a usual optimization model such as

1. Identify the decision variables or design variables along with fixed (imposed) variables.
2. Develop the objective function for the design, such as minimize weight, maximize strength, minimize costs, and so forth. This is also called the performance function.
3. Develop the design constraints, such as yield constraint, stiffness constraint, and so on.

This is one of the most important concepts of MS. It is an algebraic formula expressing a compromise between two material properties. In its simplest form, an MI can be the property itself (or the inverse of it), but in most cases it is written as a fraction, where the numerator is the quantity one wishes to maximize and the denominator the quantity to be minimized. In other words, an MI is a coupled equation and its use allows a quick comparison of alternatives without having to resort to mechanical design calculation.

For instance, taking strength as a design criteria and low weight as a critical MS requirement, it can be shown that the resulting MS is a function of the loading mode, and the most suitable candidate material is the one in which that ratio is maximized. Alternatively, if two properties have to be simultaneously maximized, then the MI value is expressed by their product. Therefore, for each case it is possible to identify or deduce one (or more) suitable merit indices. This methodology is quite simple in some cases but can be difficult in others, mainly when multiple objectives are involved.

5.8 Materials Properties Chart

There are bidimensional graphic representations in which the axes are assimilated to materials properties plotted on logarithmic scales. The range of the axes is chosen to include all materials, but regions of interest can, of course, be enlarged at will. These maps were introduced by Ashby in a pioneering paper and constitute one of the most powerful tools for MS [8]. Recently, materials properties charts (MPC) were integrated with a number of databases forming the CES4.1 software commercialized by Granta Design. Integration of MPC with MI is a very useful tool, and a number of recent papers have used this software.

Let us demonstrate these steps through an example, which is taken from Ashby [8].

A case of tie-rod design is demonstrated through the performance index method. A tie-rod usually carries a uniaxial tensile load and the length of the tie-rod is usually known. It is desired in the design to find the cross-sectional area (A), or we can say that it is the design variable or free variable. Let us apply a constraint on the design dimension that the tie-rod does not yield while in use or loaded. The objective function for the DfE problem should be to minimize the environmental impact, such as minimizing the embodied energy content of the tie-rod, or the carbon footprint, or the eco-indicator. Any one or more than one objective could be used for DfE. Here we choose to minimize the embodied energy content of the tie-rod. We can develop the objective function as

$$\text{minimize total energy } (TE) = em$$

where e is the energy content of the material in kilojoules per kilogram, and m is the mass of the tie-rod.

$$\text{Mass, } m = \text{volume} \times \text{density, or}$$

$$TE = e(AL)(\rho) \tag{5.16}$$

where:

TE = total embodied energy
L = length of tie-rod
A = cross-sectional area of tie-rod
ρ = density

The total embodied energy TE needs to be minimized subject to the design constraint of the yield strength. The objective is that the tie-rod should not yield (fail) under the load P. We can write the yield strength constraint as

$$\frac{P}{A} \le \sigma_y \quad \text{or} \quad A = \frac{P}{\sigma_y} \tag{5.17}$$

Replacing A from Equation 5.17 in Equation 5.16, we get

$$TE = (PL)\frac{e\rho}{\sigma_y} \tag{5.18}$$

Examining Equation 5.18, we can say that the material that will minimize the total energy content of the tie-rod may be the one that maximizes the performance index M_1:

$$M_1 = \frac{\sigma_y}{e\rho} \tag{5.19}$$

The performance index is essentially the specific strength divided by the energy content e multiplied by the density ρ of the material. If we further examine the performance index, it can either maximize the yield strength σ_y of the material or minimize the energy content of the material. Materials with the same performance index M_1 should have same energy content and, therefore, the same environmental impact.

Let us examine the problem by replacing the yield constraint by a stiffness constraint k, which means that the tie-rod should not elongate more than a prescribed elongation δ under the applied load P. The new MS for this DfE problem is explained below. We know from mechanics that

$$k = \frac{P}{\delta}$$

and the stiffness defined by the tie-rod geometry is

$$k = \frac{AE}{L}$$

where E is the Young's modulus of the material and L is the length of the tie-rod, which is known in advance.

We can write

$$A = \frac{P}{\delta}\frac{L}{E} \quad \text{or} \quad A = kL\frac{1}{E} \tag{5.20}$$

We can write the objective function from Equation 5.16 as

$$TE = kL^2\left(\frac{e\rho}{E}\right) \tag{5.21}$$

As k and L are fixed variables, the performance index in this case is

$$M_2 = \frac{E}{e\rho} \tag{5.22}$$

which indicates that to minimize the total embodied energy *TE*, we need either to maximize the tensile strength *E* or minimize the embodied energy and density of the material. Again, materials with the same value of M_2 should have the same environmental effect.

We see that the performance index is essentially a measure of efficiency of materials. Material efficiency in industrial production focuses on the amount of a particular material needed to produce a particular product; in this case, a tie-rod. Mathematically, it is the ratio of the product divided by the raw material used, so the number is never smaller than one. Material efficiency can be improved either by the reducing the amount of material contained in the final product ("light weighting"), or by reducing the amount of material that enters the production process but ends up in the waste stream. In a slightly broader sense, taking into account the industrial production–consumption cycle, material efficiency can refer to the amount of virgin natural resources required for the production of a given amount of product, with recycling of postconsumption waste material back into production contributing to material efficiency. Therefore, we can identify the most important components of material efficiency, as (1) light weighting in the production process, (2) waste reduction in the production process, (3) recycling of material in the production–consumption cycle.

Ashby's materials properties charts (MPC) combined with the MI is very useful in eco-material selection and therefore very useful for the DfE. Some of the useful charts with eco-material properties and merit indices are presented in Figures 5.2 through 5.8.

Ashby has developed a powerful systematic method for *material selection charts* according to the importance of each property through the graphs and bubble charts in Figures 5.2 through 5.8. He has advocated the use of the performance index in selection of materials for design. The performance metric *P* of a structural element depends on (1) the functional requirements, (2) the geometry, and (3) the material properties of the structure. That is,

P _ [(functional requirements, *F*), (geometry, G), (material properties, M)]

$$P = f\left(F, G, M\right) \tag{5.23}$$

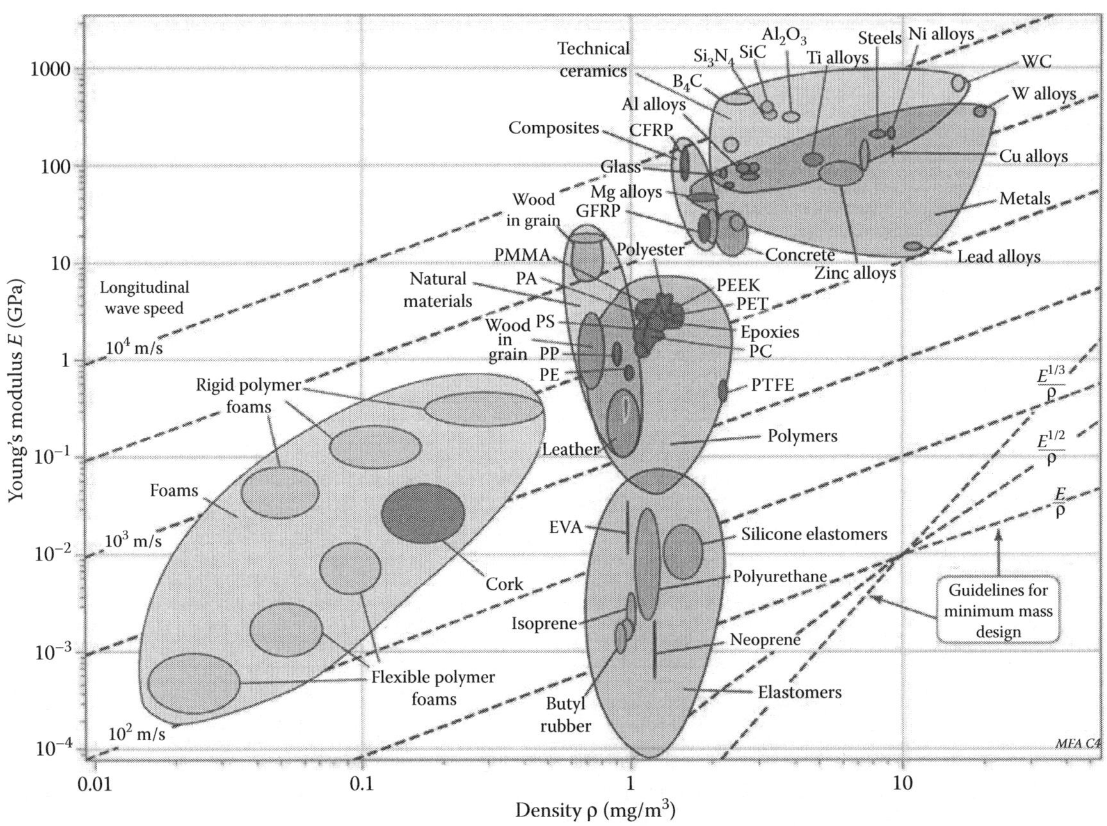

Figure 5.2

(See color insert) Density vs. Young's modulus of materials. (Copyright © The McGraw-Hill Companies, Inc. Permission required for reproduction or display.)

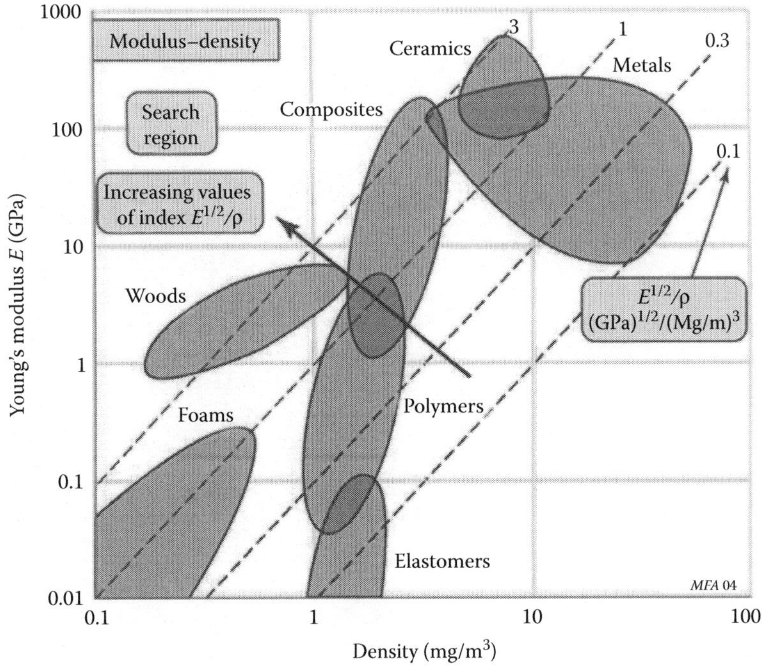

Figure 5.3

Density vs. Young's modulus for increasing values of $E^{1/2}/\rho$. (Copyright © The McGraw-Hill Companies, Inc. Permission required for reproduction or display.)

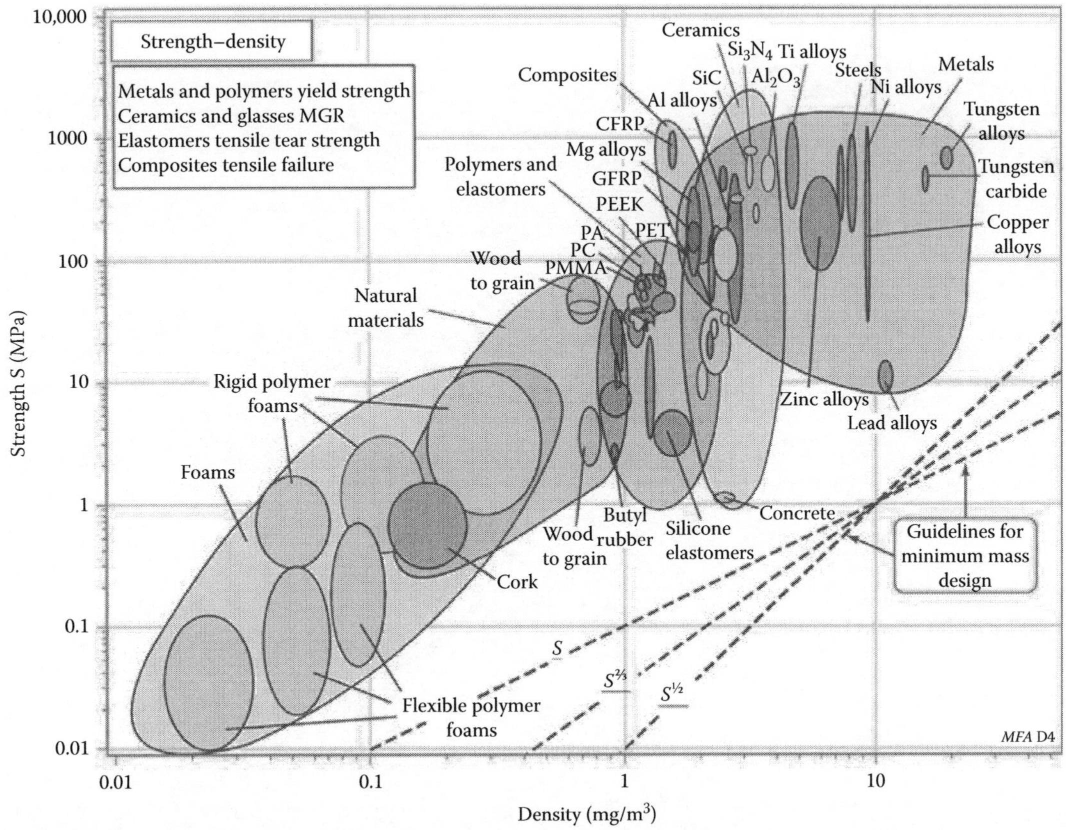

Figure 5.4

(See color insert) Density vs. strength of materials. (Copyright © The McGraw-Hill Companies, Inc. Permission required for reproduction or display.)

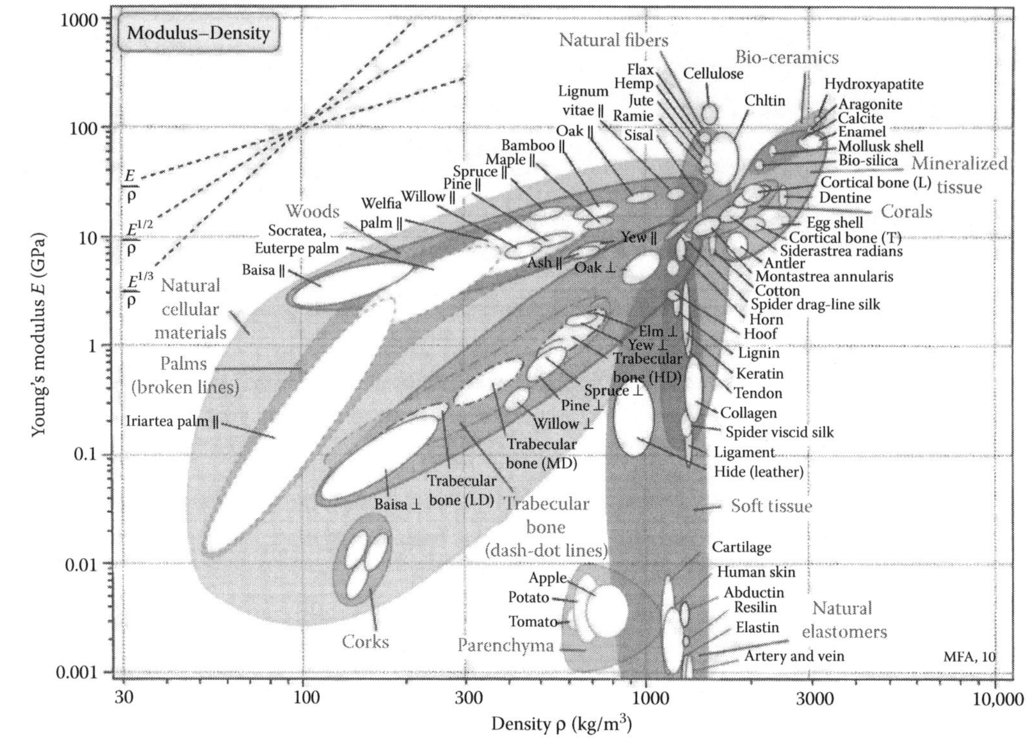

Figure 5.5

Density vs. Young's modulus for several constraints on material selection.

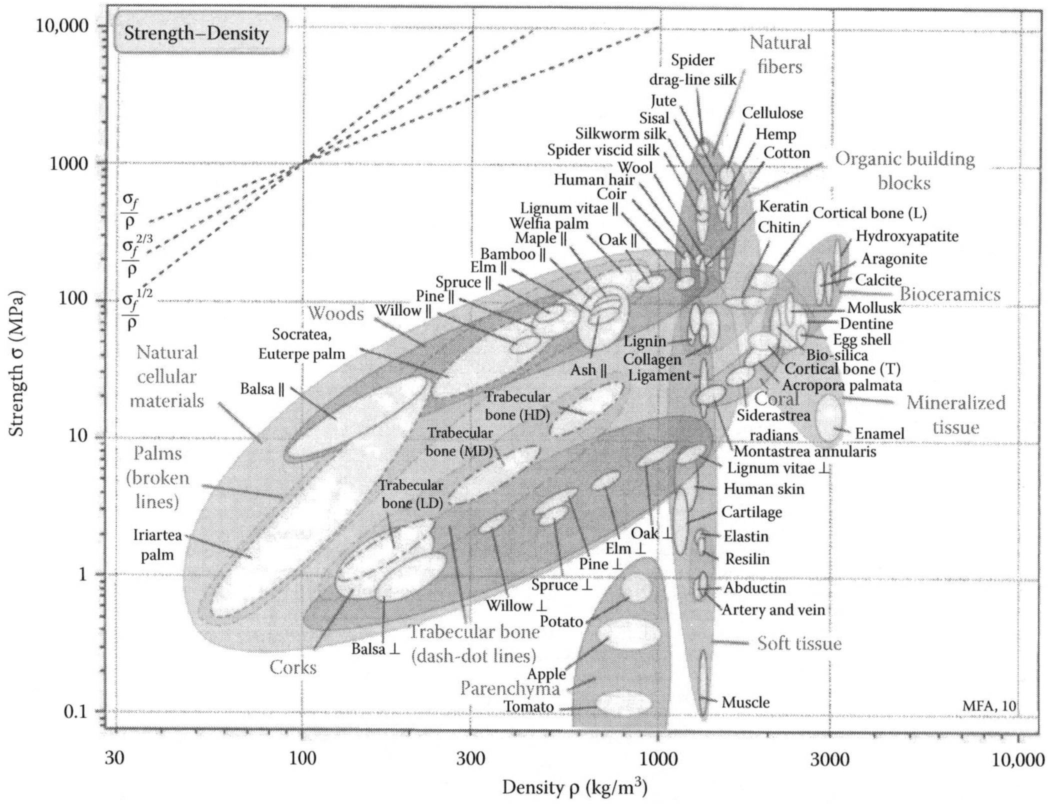

Figure 5.6

(See color insert) Density vs. strength under various constraints.

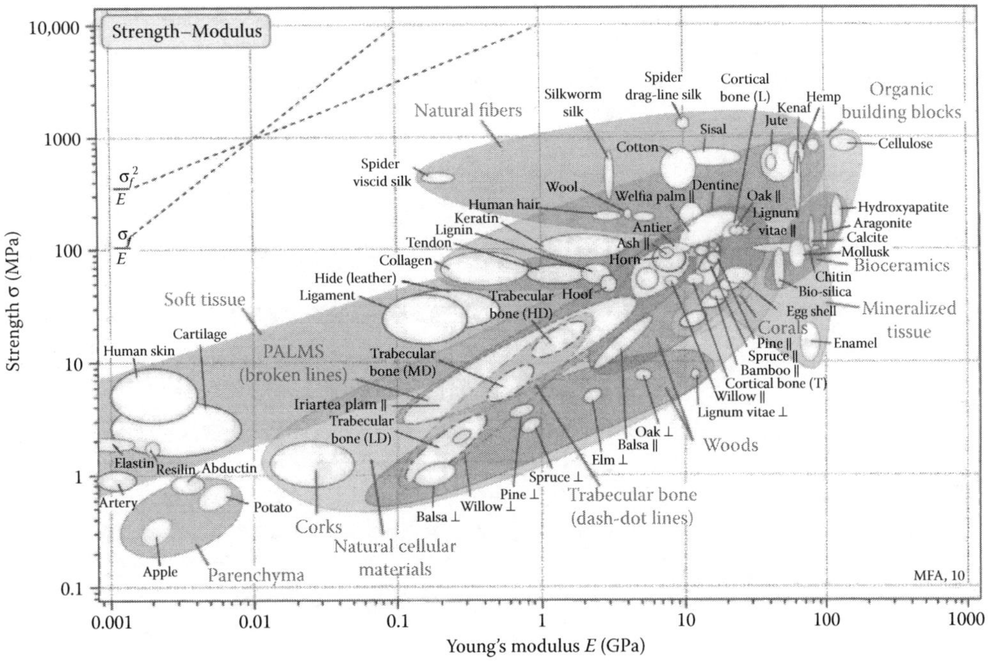

Figure 5.7

Young's modulus vs. strength under two constraints.

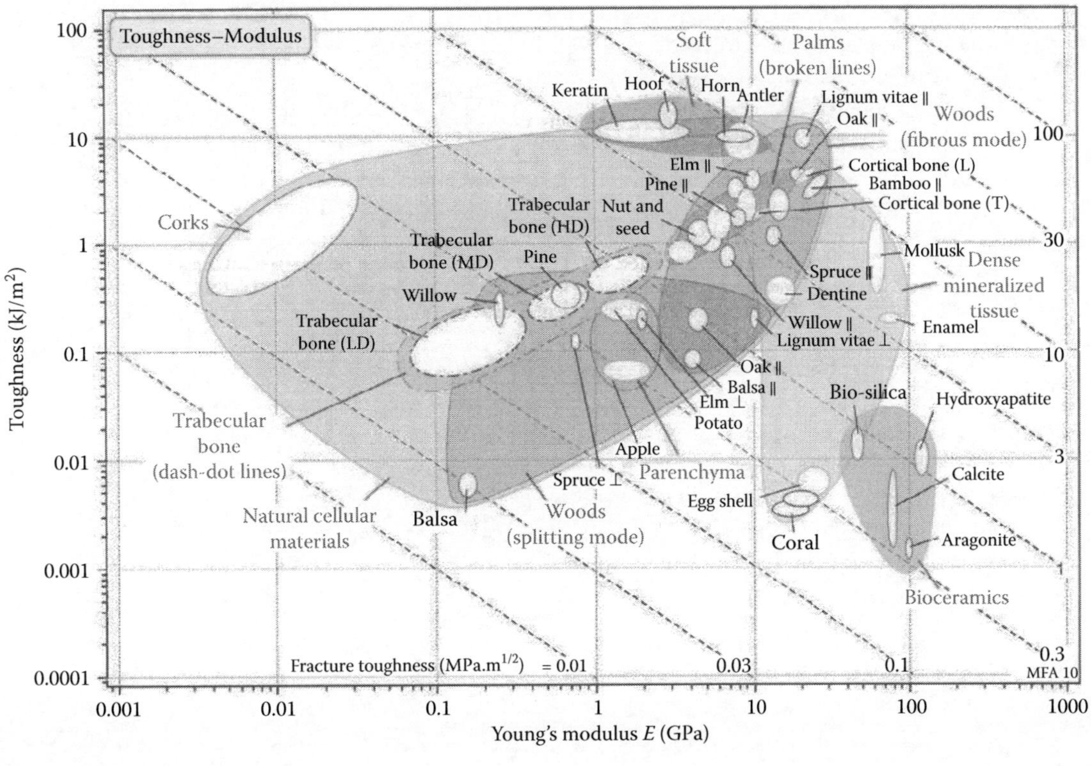

Figure 5.8

Young's modulus vs. toughness.

By using the charts of Ashby [8] for the performance indices M_1 and M_2, we can identify the materials useful for the tie-rod to meet objectives stated above. Intuitively, we can see that materials lying in the left-upper hand portion of the figures are the best materials for energy conscious design or DfE.

The short list of materials in the left-upper hand corner of Figures 5.8 are presented in the table below.

Materials	Density(ρ), kg/m³	Young's Modulus (E), GPa	Yield Strength, (σ_y), MPa	Embodied Energy (e), MJ/kg	Cost, $/ kg	$M_1 = \frac{\sigma_y}{e\rho}$	$M_2 = \frac{E}{e\rho}$	Weighted Performance Index
Silicon carbide	3–3.2	280–310	1000–5250	70–78	15–21	13.623	1.286	
Low-alloy steel	7800–7900	205–317	400–1500	32–38	1.0–1.1	.00346	.00095	
Cast iron	7050–7250	165–180	250–680	16–18	0.7–0.8	.0038	.00142	
Tungsten carbide	440-	120(GPa)	505	312–346	19–29	.0038	.000842	
Zinc alloys	4900–7000	68–100	80–450	70–75	0.9–3.40	.006	.00021	
Titanium alloys	4400–4800	110–120 (GPa)	750–1200 (GPa)	740	96.7–107	.0003	.000034	

References

1. Graedel, T. E. (1998). *Streamlined Life Cycle Assessment.* Prentice Hall, Upper Saddle River, NJ.
2. Fiksel, J. (2012). *Design for Environment.* McGraw-Hill, New York.
3. Lee, S. G., Lye, S. W., and Khoo, M. K. (2001). A multi-objective methodology for evaluating product end-of-life options and disassembly. *International Journal of Advanced Manufacturing Technology* 18: 148–156.
4. Kroll, E. and Hanft, T. A. (1998). Quantitative evaluation of product disassembly for recycling. *Research in Engineering Design-Theory, Applications, and Concurrent Engineering* 10: 1–14.
5. Emblemsvag, J. (2003). *Life-Cycle Costing: Using Activity-Based Costing and Monte Carlo Methods to Manage Future Costs and Risks.* Hoboken, NJ: Wiley.
6. Azzone, G., Manzini, R., and Noci, G. (1996). Environmental trends in environmental reporting. *Business Strategy and the Environment* 5 (4): 219–230.
7. Chen, R. W., Navin-Chandra, D., and Print, F. B. (1993). A cost-benefit analysis model of product design for recyclability and its application. *IEEE International Symposium on Electronics and the Environment* (pp. 178–183), Arlington, VA.
8. Ashby, M. F. (2009). *Materials and the Environment: Eco-informed Material Choice.* Elsevier, New York.
9. Graedel, T. E. and Allenby, T. E. (1995). *Industrial Ecology.* Prentice Hall, Upper Saddle River, NJ.
10. Ashby, M. F. (2011). *Materials Selection in Mechanical Design*, 4th edn. Elsevier, New York.

Bibliography

Botthroyd, G., Dewhurst, P., and Knight, W. (2002). *Product Design for Manufacture and Assembly*, 2nd edn. Marcel Dreker, New York.

ISO 14044:2010. (2010). *Environmental Management-Life Cycle Assessment, 2010.*

United States Environmental Protection Agency (EPA). (2010). *Life Cycle Assessment, 2010.*

Whitney, D. E. (2004). *Mechanical Assembly: Their Design, Manufacture, and Role in Product Development.* New York: Oxford University Press.

6

End-of-Life Treatment and Environmental Impact Scenario

The economic feasibility of the end-of-life (EOL) scenario of a mechanical component depends on the disassembly cost, the revenue from disassembled components, and the regulatory requirements on components and materials. Regulatory requirements are quite diverse among countries and even between different states in the United States. While designers have to obey the regulations of the country or state, mostly economic considerations outweigh everything else. If we look at the disassembly cost, which depends on the number of fasteners in the assembly, it is clear that the disassembly of a product will be disadvantaged by a large number of fasteners in the assembly. To disassemble such a component will involve locating the fasteners, gripping them in difficult-to-reach places, and handling them in a confined space. They need to be reached and removed in an optimal sequence; otherwise, disassembly will take more time and cost more money. Therefore, it is desirable for designers to look into the assembly process and EOL treatment of a product. In reality, this means that designers should design components with fewer fasteners in the assembly, such that disassembly, recycling, and reuse of components are easy. Sometimes this is called *design for disassembly* (DFD) or *product-embedded disassembly*. Since disassembly cost is a major driver of recycling and reuse, the time component of the disassembly should be carefully estimated. Sometimes, the design of fasteners or fastener configuration should be examined carefully. There are number of objectives we want to achieve at the EOL, most importantly,

1. Maximize profit, as no business will attempt to disassemble, recycle, or reuse a component if it is not profitable. So, we can understand that a recycling or reuse business will not pick up unless it is profitable. We can write the objective function for profit as total return minus total costs of disassembly. This can be presented as Equation 6.1:

$$f(x_i)=\sum_{i=1}^{n}p_i(x_i)-\sum_{i=1}^{n}c_i(x_i) \tag{6.1}$$

where:

$i = 1,2,\ldots n$ = number of components in the assembly
p_i = sales price of each component
c_i = the disassembly and associated costs

2. Minimize the environmental impact of the EOL scenario:

$$f(z_i) = \sum_{i=1}^{n} e_i(z_i) \tag{6.2}$$

where $e_i(z_i)$ is the environmental impact of the *i*th component at EOL. The environmental impact can be measured in different ways, such as minimization of greenhouse gases (GHG), minimization of the energy embodiment, minimization of eco-indicators, or all of them together. These values can be determined by life cycle analysis (LCA).

6.1 LCA Analysis at EOL

All the materials and products we use in our daily lives have an impact on the environment. Life cycle assessment (LCA) provides a methodology for quantifying a holistic view of the environmental impact of a product at all stages of its life (*cradle to grave*). LCA is just one of several systems analysis tools, which are used in conjunction with other tools such as risk analysis, full cost accounting, pollution prevention, and product stewardship to perform holistic assessment. According to the ISO 14040 standard, LCA must include four essential phases: (1) goal and scope definition, (2) inventory analysis, (3) impact assessment, and obviously (4) interpretation of result. A brief description of these phases is given below.

1. Goal and scope definition: This stage presents the purpose, scope, and boundaries of the project, including the functional unit. It must include the key assumptions and likely limitations of the work.
2. Inventory analysis: This entails the detailed compilation of all the quantified listings of the energy and raw materials at each stage of the product life cycle and GHG emissions to air and water. It should also indicate recycling, reuse, and disposal at the EOL. ISO standards have very clearly specified the protocol in this regard.
3. Impact assessment: The impact assessment links the inventory data with environmental impact and assesses its negative impact (i.e., global warming, depletion of resources, acidification of river water, human toxicity, air pollution, etc.). No standard methodology has been developed or accepted for impact assessment. However, a functional unit accounting system may be used to normalize energy carriers (e.g., coal, hydro, natural gas, wind farm, solar energy, fusion energy, etc.), material resources, and emissions, including disposal.
4. Interpretation of result: Interpretation of the impact assessment result may suggest improvements in the design and manufacturing process (remanufacturing) to reduce the costs and materials used.

LCA software packages have been developed with varying emphasis. Some LCA software packages widely used in industries are shown in the table below.

Software Package	Website
1. EIO-LCA (economic input–output life cycle assessment)	http://www.eiolca.net, Green Design Initiative, CMU
2. SimaPro	http://www.pre.nl/default.html
3. LCAPIX	http://www.kmlmtd.com/pas/index.html
4. TEAM 5.2	http://www.ecobilan.com/uk_team06.php
5. CMLCA	http://www.cmlca.eu/
6. Umbert	http://www.umberto.de/en/
7. BEES 4.0	http://www.bfr.nist.gov/oae/software/bees
8. GaBi	http://gabi-software.com/

We will look into the prospects of recycling and reuse of different components disassembled from an assembly. EOL treatment of components can be grouped into three categories: (1) reuse, (2) recycling, and (3) landfill. First, we try to reuse a component after disassembly, if possible. As an example, some gears from a gearbox can be reused after being disassembled; some can be sent for recycling; and some have to be sent to landfill.

Reuse of a component should be the preferred alternative; however, the market for reused components fluctuates, and hence revenue is difficult to estimate *a priori*. Revenue from recycling can be calculated from the material composition of the component.

For $i = 1,2,\ldots,n$ components of an assembly, let components 1 to 5 be reused, 6 to $n-2$ be recycled, and the last two sent to their graveyard in the landfill. We can write the profit equation (Equation 6.3) as

$$f(x_i) = \sum_{i=1}^{5} Sr_i + \sum_{i=6}^{n-2} Sc_i - \sum_{i=1}^{5}\left(C_i^{trans} + C_i^{refurb}\right)_{reuse}$$

$$-\sum_{j=6}^{n-2}\left(C_j^{trans} + C_j^{sherd}\right)_{recycle} - \sum_{k=n-3}^{n}\left(\left(C_k^{trans} + C_k^{landfill}\right)_{landfill}\right) \tag{6.3}$$

where:

Sr_i and Sc_i refer to the sales price for the reused and recycled components, respectively, new item C_i^{trans}, C_i^{refurb}, C_j^{shred}, and $C_k^{landfill}$ are the cost of transportation, refurbishment, shredding, and landfill, respectively

If data is available for all eight terms in Equation 6.3, we can find the total profit for the recycling and reuse of components after disassembly of a product. If we can minimize the costs of transportation, shredding, and landfill, the profits of the recycling business will increase. We know landfill costs are rising due to unavailability and cost of municipal land. Due to the regulatory requirements of municipalities, states, and countries, landfill is becoming difficult, and the recycling business is not picking up. The sale price of recycled and reused components could be increased to achieve a higher profit, but there is a ceiling on sales price; a reused automobile cannot be sold on a par with a new car. Revenue from recycling is based on the material composition of the recyclable components.

The cost data for refurbishment, shredding, and landfill are calculated from Boothroyd et al. [1]. Refurbishment cost data is very difficult to find; Sato and Iwasa have assumed the costs to be [2] has assumed the costs to be

$$C_i^{refurb} = 0.5.reuse_i \tag{6.4}$$

Transportation cost estimation by Boothroyd et al. [1] is

$$C_i^{trans} = \left(\frac{\$}{\text{kg}} \cdot \text{km}\right) \cdot D_i \cdot W_i \tag{6.5}$$

where:

W_i is weight of the component
D_i is the distance transported
\$/kg.km is normally known

The transportation charge is $2.07e^{-4}$ \$/kg.km [1]. The cost of shredding and landfill could be written as

$$C_i^{shred} = \Delta C_i^{shred} \cdot m_i \tag{6.6}$$

$$\Delta C_i^{shred} = 0.12\,[\$/\text{kg}]$$

and landfill cost can be calculated as

$$C_i^{landfill} = \Delta C_i^{landfill} \cdot m_i \tag{6.7}$$

where $\Delta C_i^{landfill} = 0.02\,[\$/\text{kg}]$. The landfill cost varies largely according to the toxicity of the component and the regulations in the respective area.

Example 6.1

A system was disassembled into 12 components. The total weight of the equipment is 5525 kg. A list of components, along with their material composition, weights, and EOL decisions, is given in Table 6.1. It can be seen from Table 6.1 that there are five items for reuse and three for recycling, while four are going straight to landfill. Total revenue from recycling can be calculated as follows:

$$f(x_i) = \sum_{i=1}^{5} Sr_i + \sum_{i=6}^{8} Sc_i - \sum_{i=1}^{5}\left(C_i^{trans} + C_i^{refurb}\right)_{reuse}$$

$$-\sum_{i=6}^{8}\left(C_j^{trans} + C_j^{shred}\right)_{recycle} - \sum_{9}^{12}\left(C_k^{trans} + C_k^{landfill}\right)_{landfill}$$

The sales prices for reuse and recycling are estimated below. The reused items are of high-tech quality and difficult to manufacture, so every item has a very high reuse sales price:

Item	Reuse Sales Price ($)
1	3500
3	4200
4	4000
9	3500
10	7500

Although Sato and Iwasa [2] have included a refurbishing cost of 0.5 × reuse cost, it is assumed here that all items except item 10 are reused as they are, and only item 10 is refurbished at a cost of 0.5 × $7500 = $3750.

All the items for reuse are to be transported about 3500 km from New York to Seattle, and so the total transportation cost of the reused items = $\$2.7 \times 10^{-4} \times 2155 \times 3500 = \$2037$.

Recycling Cost: Recycled items' sale prices are given in the table below.

Item	Recycled Sale Price
2	$0.7/kg × 1500 = $1050
5	$1.7/kg × 1200 = $2040
6	$3.5/kg × 125 = $437.5

The shredding and refurbishing of recycled items is calculated as below. The recycled items 2, 5, and 6 are shredded and recycled into the furnace without refurbishing. However, shredding and recycling costs are calculated as $0.12/kg × 2825 = $ 339.00.

Table 6.1 List of Components from Disassembled Equipment

Part #	Material	Weight	EOL Decision
1	Cast iron	500 kg	Reuse
2	Low-carbon steel	1500 kg	Recycle
3	Low-alloy steel	1200 kg	Reuse
4	Stainless steel	250 kg	Reuse
5	Aluminum	1200 kg	Recycle
6	Copper alloy	125 kg	Recycle
7	PVC	270 kg	Landfill
8	Glass ceramic	200 kg	Landfill
9	Tungsten carbide	50 kg	Reuse
10	CFRP	155 kg	Reuse
11	Silicon elastomers	25 kg	Landfill
12	Butyl rubber	50 kg	Landfill

The transportation costs of recycled items are calculated based on the assumption that these items are being transported 1000 km for shredding and recycling, so the transportation cost for recycled items = $2.7 × 10^{-4} × 1000 × 2825 = $762.75.

Now we need to calculate the landfill costs of items 7, 8, 11, and 12. First, these items need to be transported a distance of 100 km to landfill, and so

$$\text{Transportation cost of landfill items} = \$2.7 \times 10^{-4} \times 100 \times (270 + 200 + 25 + 50) = \$147.15$$

$$\text{Landfill cost} = \$0.02\,/\,\text{kg} \times 545 = \$109.00$$

Now, the total return after calculating the reuse, recycle, transportation, and landfill cost is calculated as

$$\begin{aligned} \text{Total return}(f(x)) &= (3500 + 4200 + 4000 + 3500 + 7500)_{reusesale} \\ &+ (1050 + 2040 + 437.50)_{salesrecycled} - (3750 + 2037)_{reusetrans+refur} \\ &- (\$339 + 762.50)_{recycledshredtrans} - (\$147.15 + 109)_{translandfill} \\ &= (22{,}700 + 3527.50) - (5787 + 1101.5 + 256.15) = \$19{,}082.85 \end{aligned}$$

Therefore, the total return from disassembly of the system, after reusing and recycling, is approximately $19,082.85.

We need to consider the environmental impact of a recycling business. If the environmental impact is on a par with that of a newly manufactured product, it does not save the earth from environmental pollution. We can formulate the environmental impact assessment from three different angles: (i) embodied energy, (ii) carbon emissions or (GHG) emissions, and (iii) eco-indicators. These can all be minimized, either separately or jointly.

6.2 Minimization of Embodied Energy

All embodied energy estimations have been taken from Table 6.3.

We will start with minimization of embodied energy formulation. Equation 2 is formulated for $i = 1,2,\ldots,n$ components of a disassembled product.

$$f(z_i) = \sum_{i=1}^{5} (ereuse_i + etrans_i + erefurb_i)_{reuse} + \sum_{j=6}^{n-2} (erecycle_j + etrans_j + eshred_j)_{recycle} + \sum_{k=n-3}^{n} (etrans_k + elandfill_k)_{landfill} \quad (6.8)$$

where all these terms refer to energy consumption of reuse, recycling, and landfill. The energy consumption of reuse is the inverse of the energy recovered from reusing component i:

$$ereuse_i = -\sum me_i^{\text{intense}} \cdot W_{ij} \quad (6.9)$$

where:

me_i^{intense} is the energy intensity of material i

W_{ij} is the weight of material j in component i

It is obvious that if reuse is possible, it is the best option at EOL. It is useful to calculate the energy (negative) for reuse items 1, 3, 4, 9, and 10.

$$\text{Reuse energy} = -[(18\text{ MJ/kg} \times 500\text{ kg})_{cast\ iron} + (38\text{ MJ/kg} \times 1200)_{lowalloysteel}$$

$$+ (85\text{ MJ/kg} \times 250)_{stainlesssteel} + (91\text{ MJ/kg} \times 50)_{tungstencarbide} + (286\text{ MJ/kg} \times 155)_{CFRP}$$

$$= \text{Reuse energy savings} = -1.25 \times 10^5\text{ MJ}$$

The energy required to recycle items 2, 5, and 6 is calculated as follows:

$$\text{Recycling energy consumption} = (9.8\text{ MJ/kg} \times 1500\text{ kg})_{lowcarbonsteel}$$

$$+ (2.1\text{ MJ/kg} \times 1200\text{ kg})_{aluminum} + (18.5\text{ MJ/kg} \times 125)_{copper} = 19{,}532.5\text{ MJ}$$

$$erecycle_j = -\sum me_j^{recovery} \cdot W_{ij} \tag{6.10}$$

where $me_j^{recovery}$ is the recovered energy of material j.

It is difficult to find data for energy consumption for refurbishment; however, the EOL computer disposal is available as

$$erefurb_j = 1.106.W_j \tag{6.11}$$

where W_j is the weight of the jth component. In this example, only item 10 (CFRP) is refurbished, and energy consumption for refurbishing is calculated as below:

$$\text{Refurbishing energy} = 1.106 \times 155 = 164.3\text{ MJ}$$

The energy consumption of transportation is estimated as

$$etrans_j = \Delta e_j^{trans} \cdot D_j \cdot W_j \tag{6.12}$$

where $\Delta e_j^{trans} = 1.17e - 3[\text{MJ/kg.km}]$ [1].

Therefore, energy consumption during transportation is calculated as

$$\text{Transportation energy consumption} = 1.17 \times 10^{-3} \times (2155 \times 3500 + 2825 \times 1000 + 545 \times 100)$$

$$= 12{,}193.74\text{ MJ}$$

The energy consumption for shredding and landfill, respectively, is estimated as

$$eshred_j = \Delta eshread_j \cdot W_j \tag{6.13}$$

and

$$elandfill_k = \Delta elandfill_k \cdot W_k \tag{6.14}$$

The cited values are $\Delta eshredj = 1.0\,[\text{MJ/kg.km}]$ and $\Delta elandfill_k = 20{,}000\,[\text{MJ/kg.km}]$ [1].

The energy consumption for shredding and landfill, respectively, is calculated as

$$\text{Shredding and landfill energy consumption}$$

$$= 10\text{ MJ/km} \times 1000 \times 2825 + 20{,}000\text{ MJ/kg.km} \times 100 \times 545$$

$$= 1118.25 \times 10^6\text{ MJ}$$

Therefore, total net energy consumption = $-1.25 \times 10^5 + 19{,}532.5 + 164.5 + 12{,}193.74 + 1118.25 \times 10^6 =$ 1118.16×10^6 MJ.

6.3 Minimization of Carbon Footprint

The amount of energy used in manufacturing an engineering component is directly linked to the amount of carbon dioxide (CO_2) emitted into the air. Some countries have already formulated regulations requiring GHG labeling on the product, so that consumers can compare the amount of GHG emitted into the atmosphere during the making of similar products and decide accordingly. It is possible that a lower-GHG emitting product may be more expensive than a higher-GHG emitting product. It is also possible that a lower-GHG emitting product may need more power during operation further along the chain and may thus emit more GHG during use. Therefore, we can conclude that carbon emission is due to energy consumption. However, we should consider the types of energy available:

- Green energy, no carbon emission
- Carbon-based energy, with carbon emission

Primary energy sources can be divided into two groups: green energy and fossil fuel energy. Our goal is to eliminate carbon emissions, and energy sources such as hydro, solar, or wind are a response to our need. The energy available on the electrical grid from these sources is clean energy, emitting no carbon. The energy generated by fossil fuels such as coal, oil, and natural gas is CO_2-emitting primary energy and should be avoided as far as possible. This is easier said than done, due to the requirement for energy by the increasing global population and industrial development everywhere; we cannot avoid power generation from fossil fuels. However, there are a number of sources of green energy that have not been exploited in full. There are several reasons, for instance lack of technology, requirement for a large initial investment, and a long period of development before power can be generated. Vanloon and Duffy [] have tabulated the carbon content of fossil fuels, as shown in Table 6.2 [3]. This table shows that coal is the greatest polluter, contributing most GHG to the atmosphere. Biomass generates nearly the same amount of carbon per gigajoule of heat. However, biomass is considered to be a renewable resource, because the same amount of CO_2 is sequestered during photosynthesis as is released in combustion. We can see from Table 6.2 that every primary source of energy listed inevitably emits carbon as a by-product. Jeswiet and Kara have used the CO_2 emission in the last column to calculate the carbon emission signature (CES) [4].

The CO_2 emission due to reuse and recycling of parts in this example can be estimated as follows:

$$\text{Negative CO}_2 \text{ emission due to reuse of items 1, 3, 4, 9, and 10}$$

$$= -(500 \times 13 + 1200 \times 2.3 + 250 \times 5.4 + 50 \times 4.9 + 155 \times 18.3)$$

$$= -13{,}692 \text{ kg CO}_2$$

This is the saving in energy use due to the reuse of components 1, 3, 4, 9, and 10.

$$\text{CO}_2 \text{ emission during recycling of items 2, 5, and 6}$$

$$= (1500 \times 2.8 + 1200 \times 13 + 125 \times 5.6)$$

$$= 20{,}500 \text{ kg CO}_2 \text{ emitted during recycling}$$

The data for CO_2 emission during transportation (reuse + recycling + landfill) are not available; therefore, it is calculated based on the data for energy consumption during transportation for recycle, reuse, and landfill.

Table 6.2 Energy-producing fuels, heat, and CO_2 released

Type of Fossil Fuel	1 GJ of Heat Produced Releases	ΔH (kJ)	CO_2 (kg)
Coal	$C + O_2 = CO_2$	−394	112
Heavy oil	$C_{20}H_{42} + 30\ O_2 = 20\ CO_2 + 21\ H_2O$	−13,300	66
Natural gas	$CH_4 + 2\ O_2 = CO_2 + 2\ H_2O$	−890	49
Biomass	$CH_2O + O_2 = CO_2 + H_2O$	−440	100

Note: ΔH = enthalpy; heat content; thermodynamic potential.

CO_2 emission during transportation: reuse, recycle, landfills, and refurbishing

$$=\left(12{,}193.74+1118.25\times10^{6}\right)\times112/10^{3}=126\text{ kg }CO_2$$

Therefore, the total CO_2 emission during reuse, recycling, and landfill = 13,692 + 20,500+126 = 34,318 kg CO_2 emitted into the air due to system disassembly, reuse, recycling, refurbishing, and landfill. I have used the data for coal. Significantly, about 53.5% of energy in the US grid comes from coal.

Products are perceived to be one of the greatest sources of GHG. The design and development of low GHG emission (i.e., low-carbon) products are crucial for mitigating climate change. Therefore, the industry needs to take action to reduce GHG emissions from its products and services under mounting pressure from the implementation of the Kyoto and Copenhagen protocols. Designers have to design products which emit less GHG than has been common practice in recent decades. The product's carbon footprint plays a major role as the quantitative foundation of low-carbon product design. Reduction of GHG emission in the entire life cycle of a product and the life cycle cost of a product are chosen as the global warming performance and the economic performance of the product, respectively. Remanufacturing is an effective way to maintain products in a closed loop, reducing both the environmental impact and the cost of the manufacturing processes. However, the calculation of a product's GHG emissions is a very difficult and complex problem when data collection is involved.

6.4 Product's GHG Flow

A product system can be defined as an organic combination of materials and components that meet certain functions and all activities over the whole life cycle of the product, including raw material collection and refinement; design and development; manufacture, assembly and packaging; distribution and storage; use and maintenance; and the disassembly and recovery stage. Over the whole life cycle, there are energy flows, material flows, and discharge flows between the product system and the environmental system. Energy and natural resources are essential inputs into the system. One part of the output is the product and its function; the other is useless emissions, including gas (e.g., GHG), liquid, and solid waste. During the raw material stage, the inputs are energy and natural resources, and the major outputs are materials or workpiece. The GHG emissions during this stage come from the refinement and transport of raw materials. The GHG emissions are bound up with product attributes such as geometric topological characteristics, structure, weight, size, number, material, process route and processing methods, and so on. During the assembly and packaging stage, parts are assembled into finished products and packed. The GHG emissions come from assembly and transport processes.

The amount of GHG emissions during the use and maintenance stage is affected by time of operation in use, lifetime of product, energy consumption, and so on. During the EOL stage, only the emissions from product decomposition and recycling are considered; these are affected by dismantling methods (i.e., destructive disassembly and nondestructive disassembly), recycling rates (i.e., product reuse, parts reuse, parts manufacturing, materials recovery, and energy recovery, in descending order), weight of parts and materials, and so on.

6.5 Low-Carbon Design

The GHG emissions of a product throughout its life cycle need to be estimated, with problematic parts and materials identified to reduce GHG emissions by the product. The amount of GHG emissions from a product depends on the GHG emissions of the individual components. The embedded GHG emissions of a part represent the GHG emissions of the part for its entire life cycle, which is defined as the sum of two sets of GHG emissions: the first from the use of raw materials, manufacturing, and transport of the part itself, and the second from the emissions attributable to the part during the product's manufacturing, distribution, use, and EOL stages. A low-carbon product design system consists of several steps.

In order for companies to better confront GHG emission issues, accurate estimation of GHG emissions over the entire life cycle of a product is necessary [5]. GHG emissions data can be used to identify significant issues that require improvement. Significant issues refer to problematic components, materials, or processes that make a large contribution to the GHG emissions. When a product is redesigned, the design should seek to reduce the GHG emissions throughout the entire life cycle of the product by improving these significant issues. Ecodesign is a process that integrates significant environmental aspects into product design and development [6]. When significant environmental aspects are related to GHG emissions only, a focal point of ecodesign would be a reduction in the GHG emissions of a product over its entire life cycle. The typical outcome of ecodesign is the generation of improvement strategies and ideas for the reduction of GHG emissions throughout the entire life cycle of a product [7]. The embodiment design stage includes determining the performance and quality specifications of a product in order to realize the intended function the product will fulfill. Once the specification is fixed, parts suitable for the specification are chosen to form the bill of materials (BOM) of the product. Thus, a conventional ecodesign method (ISO, 2002; Wimmer et al., 2004; IEC, 2008) cannot provide specific guidance as to the choice of parts capable of reducing GHG emissions during the embodiment design stage. Normally, part information consists of parameters such as weight, material composition, size, thickness, and color, among others. If the GHG emission data of the part is available, it can be combined with the part information. With the part information, including the GHG emission data, a product designer can formulate a BOM using parts with lower GHG emissions. The outcome of this design is a product with low GHG emissions or a low-carbon product. The essential element of a low-carbon product design system is the GHG emission data of a part over its entire life cycle. The GHG emission data of a part over its entire life cycle is also termed the *embedded GHG emission* of the part. The normal practice when setting the system boundary of a part in a conventional environmental assessment method, such as an LCA, is to include only the use of the raw materials and manufacturing stages of the part, also known as the *cradle to gate*, in the system boundary. However, transport of the part to the manufacturer should be included in this system boundary. The rationale behind this practice is that once the part is integrated into the product, the environmental impact of the product beyond the product manufacturing (mainly the assembly of parts) stage depends on the distribution, use, and EOL stage of the product. Thus, it is an attempt to estimate the environmental impact accrued by the part from cradle to grave. There is a potential danger of using global warming as the only impact category in assessing the environmental impact of a product. Other impact categories, such as acidification and toxicity, can be significant sources of impact caused by the product. However, the toxicity and hazardousness of a product, for instance, can be minimized or eliminated using a routine screening procedure during the manufacturing of a product. In contrast, global warming is the single most important impact category in today's world, and most stakeholders, including customers and environmental nongovernmental organizations (NGOs), are demanding improvements in the impact of products on global warming. Therefore, there is a need to develop products tailored toward minimizing impact on global warming resulting from GHG emission.

6.6 Estimation of GHG Emission of a Product

British Standards Institute (BSI) Publicly Available Specification (PAS) 2050 deals with the quantification of GHG emissions over the entire life cycle of a product or the GHG emissions calculation method. An international standard on product carbon footprint (PCF) is under development by ISO/TC 207, which will become ISO 14067 when completed. Figure 6.1 shows the procedure for estimating the GHG emissions of a product and for developing an embedded GHG emission database as part of the proposed low-carbon product design system. As shown in Figure 6.1, the GHG emissions of a *product* must first be estimated before estimating the embedded GHG emissions of a part. The following sections describe the step-by-step procedure of developing an embedded GHG emission database for a part based on the procedure outlined in Figure 6.1.

As shown in Figure 6.1, the initial step in developing the embedded GHG emission database of a part involves defining the functional unit of a product and then setting up the system boundary of the product. Normally, the functional unit of a product is one piece of the product. Calculation of the GHG emissions of a product in each life cycle stage follows the method described in Section 6.7.1.

6.6.1 Raw Materials Stage

This stage consists of the acquisition of natural resources, the processing of resources, including raw and ancillary materials, the manufacturing of parts, and the associated transport of the materials. The major outputs of this stage are parts and materials. Thus, the GHG emissions from this stage can be quantified by gathering GHG emission data from the upstream processes, including the part-manufacturing process itself.

6.6.2 Manufacturing Stage

GHG emissions from the identified emission sources are measured directly or calculated indirectly.

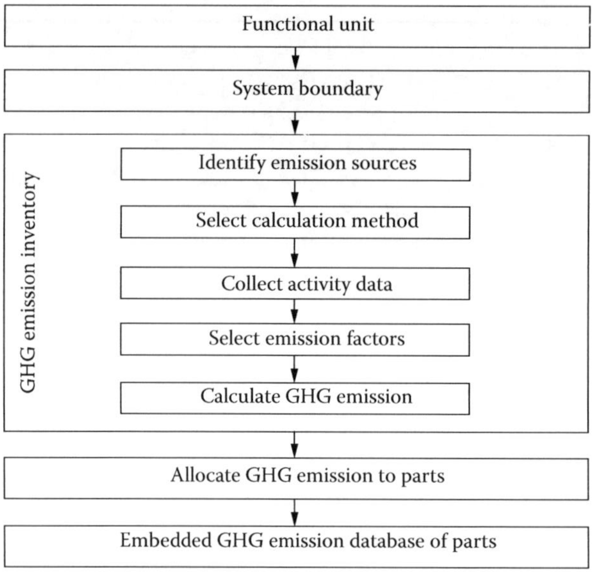

Figure 6.1

Process for developing an embedded GHG emission database for parts within the low-carbon product design system.

6.6.3 Distribution Stage

GHG emissions mainly stem from fuel combustion during transportation of material or product. A distribution scenario based on the product weight, mode of transport, transport distance, and number of products transported is used for the calculation of the GHG emissions during this stage.

6.6.4 Use Stage

The average energy consumption during the use of the product should be estimated by taking into account the product lifetime and consumed electricity (on, off, and standby mode). The EuP directive (Deutsche Umwelthilfe and Fraunhofer IZM, 2007) is mostly used in calculating average power. We can formulate the use scenario of a 10-year product lifetime, with 4 h of use per day, 365 days per year. The total amount of operating time is multiplied by the power consumption of the product, and this value is then converted into GHG emissions by multiplying by the emission factor for electricity. Here, the factor used was 0.424 kg CO_2-e/kWh (KEMCO, 2006).

6.7 End-of-Life Scenario

The recycling methods of the recyclable materials and parts and the disposal methods for the remaining waste depend on the type and weight of the materials of each part of a product. As such, an EOL scenario based on the type of materials and weight of each part of the product is made for the calculation of the GHG emissions during this stage.

6.7.1 Estimation of Embedded GHG Emission of a Part

The embedded GHG emission of a part can be calculated using Equation 6.15. As mentioned earlier, a product is assembled from more than one component, so we need to estimate the GHG emission of each component separately.

$$GHGE_i = GHGRaw_i + GHGUse_i + GHGManufR_i + GHGTrans_i + GHGEMfg_i + GHGEDistr_i + GHGEUse_i + GHGEOL_i \tag{6.15}$$

where:

$GHGE_i$ = embedded GHG emission of the ith component of the product

$GHGERaw_i$ = the amount of GHG emissions from the use of raw materials for component i (kg CO_2)

$GHGEManufR_i$ = the amount of GHG emissions from manufacturing component i (kg CO_2)
$GHGETrans_i$ = the amount of GHG emissions from the transportation of material for component i (kg CO_2-e)
$GHGEMfg_i$ = the amount of allocated GHG emission from the GHG emissions of the product's manufacturing stage
$GHGEDistr_i$ = the amount of allocated GHG emission from the GHG emissions of the product's distribution stage
$GHGEU_{sei}$ = the amount of GHG emissions in product's use stage
$GHGEOL_i$ = the amount of allocated GHG emission from the GHG emissions of the product's EOL stage

The part's GHG emission data (the first three terms in Equation 6.15), the use of raw materials, and the manufacturing and transport of the part itself can be collected directly from the field. The GHG emission data allocated to parts by the suppliers were collected during the use of raw materials and manufacturing stages. For the transport of the part, GHG emissions from the transportation method and distance were allocated to the part on a mass basis. The part's GHG emissions from the other terms in Equation 6.16 (product manufacturing, distribution, use, and EOL stages) all require allocation of the product's GHG emissions to each part. Allocation of the GHG emissions occurring during the product's manufacturing and distribution stages was based on the mass fraction of the part to the product. Separate allocation methods are applied to a "component" and a "part." A component is composed of several lower-level parts.

There are two types of component: those that require power (electricity) for operation and those that support the structure of the product without requiring power. In the allocation, parts using electricity directly were included, while those not using electricity directly, but only indirectly affecting the electricity consumption of the product, were excluded. A part using electricity directly is defined as requiring electricity for its operation, with the amount of electricity required specified by the part supplier. Examples of parts affecting the electricity consumption of a product indirectly include cables, insulators, and cooling fans, among others. For instance, in the case of a washing machine, the energy consumption of each part differs depending on the washing mode (e.g., standard, wool, washing, etc.), although the machine is in the "on mode." It is not effective to design a product by considering all the product function and use patterns. When the energy consumption of the parts is reduced, the energy consumption of the product will decrease not only in the "on mode" but also in other modes. Thus, the focus should be on the development of a low-carbon product design method by choosing parts with low GHG emissions. In general, the power consumption of a part is difficult to measure; thus, the power consumption of a component is allocated to parts based on the rated power consumption in the part specification. Equation 6.16 can be used to determine this allocation. Multiplying the emission factor of power (electricity) by the amount of power consumed by the part gives the GHG emission of that part.

$$PP_i = MP_C \left(\frac{SP_i}{\sum_{i=1}^{n} SP_{P_i}} \right) \tag{6.16}$$

where:

PPi = power consumption of the ith part
C = component
MP_C = measured power consumption of the component
SP_{Pi} = typical power consumption in the specification
i = the ith part of the component

When a component does not require power, no allocation is made to the part, because it does not use power during the use stage of the product.

The calculation of the GHG emissions occurring during the EOL stage is done using the aforementioned EOL scenario for the component. For each part, the component portion of the GHG emissions is allocated based on its mass fraction. The allocation criteria of a part are shown in Table 6.2.

6.7.2 Concept of Low-Carbon Product System

Figure 6.2 shows the process of the low-carbon product design system. The key elements of this design system include: establishment of the GHG emission target, establishment of the BOM structure, formation of the g-BOM, estimation of the GHG emission of the product, identification of problematic parts, selection of alternative parts, and evaluation of the newly designed product's GHG emissions. The use of the g-BOM during the embodiment design stage allows a quick calculation of the GHG emissions of a product. Hence, a designer can easily and quickly evaluate alternative parts in the design of the low-carbon product.

6.7.3 Establishment of GHG Emission Target

The GHG emission target should be established per product, as a clear target for each product has to be furnished to the product designer. The target should be established by considering the GHG emissions of the product generated over its entire life cycle. Furthermore, a target during the use stage should also be set, because the power consumption of a product during use is separately regulated by power consumption regulations (EC, 2005; USEPA, 2008). Thus, two different GHG emission targets should be established, based on the GHG emissions during the entire life cycle of the product and during the use stage only.

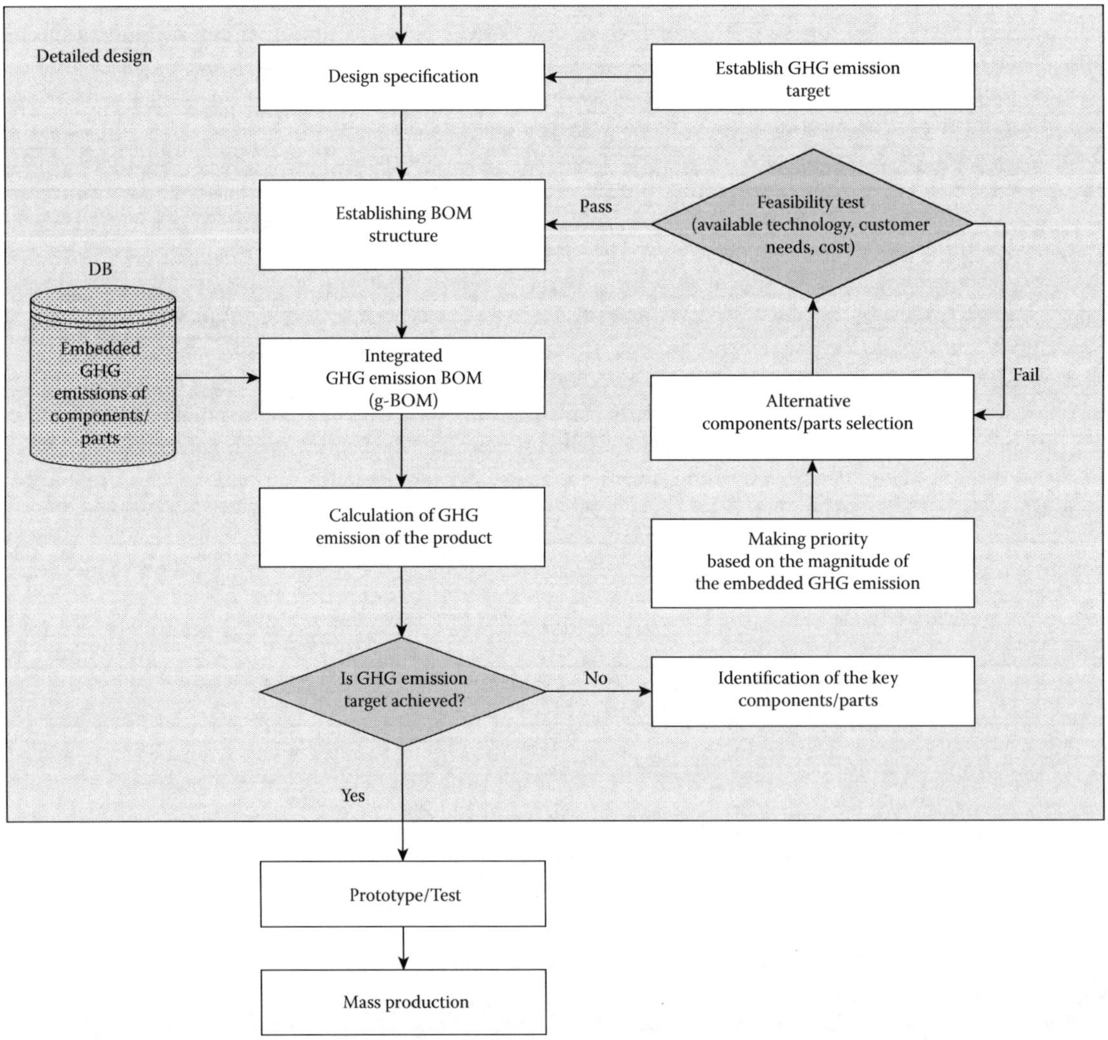

Figure 6.2

Process of the low-carbon product design system.

6.7.4 Establishment of BOM Structure

The embodiment design includes the establishment of the design specification. The design specification determines the product specifications, such as the function, performance, and appearance and quality of the design. Once the design specification is complete, the BOM of the product is established. The BOM contains the product structural information and the part information. During the formation of the BOM, the performance, appearance, and quality of the product are determined by selecting specific parts relevant to the design specification. In other words, the BOM indicates which parts will be used, from which suppliers, and where they will be connected within the product. The BOM represents the hierarchy with respect to the parts in a product. Parts are represented as levels 0, 1, 2, 3, and so on, depending on the level of assembly, where level 0 means the product, level 1 a part comprising the product, level 2 parts comprising level 1 parts, and level 3 parts comprising level 2 parts. Thus, a single part can be level 1, while a component can be level 3. Therefore, there is no relationship between the part/component and the part level.

6.7.5 Formation of g-BOM

Integration of the embedded GHG emission data of the parts in the BOM leads to the formation of the g-BOM, which involves the integration of the embedded GHG emission data of the part into the corresponding part data (i.e., the part information) of the BOM allocation criteria of components and parts per life cycle stage (Table 6.3).

6.7.6 Estimation of GHG Emissions Using g-BOM of the Product

Figure 6.2 shows an example of the g-BOM. Adding GHG emissions along the column (each life cycle stage) results in the GHG emissions of the product per life cycle. Adding GHG emissions along the row (each part) shows the GHG emissions of a part over its entire life cycle stage.

6.7.7 Identification of Problematic Parts with High GHG Emissions

The parts with higher GHG emissions are identified as the problematic parts. The embedded GHG emission data of each part is divided by the total GHG emissions of all parts comprising the product. Any part of the fractional contribution exceeding certain value (e.g., 1%) is identified as a problematic part. A problematic part becomes a target for improvement in the form of replacement by an alternative part.

6.7.8 Selection of Alternative Parts to Replace Problematic Parts

Alternative parts that replace the identified problematic parts or materials should be chosen first. Alternative components or parts are chosen from those whose embedded GHG emissions are already known. For this purpose, the embedded GHG emissions of the new components or parts are calculated through allocation of a typical product that will require those components or parts. The embedded GHG emissions of a part are estimated following the calculation method of the part's GHG emissions in clause 2.0. The choice of alternative parts must be assessed with respect to four different criteria: technological feasibility, customer requirements, cost, and potential reduction of GHG emissions:

- **a.** Technical feasibility: Is it technically feasible to apply different technology to replace the problematic part with a different parts? Is the alternative part exchangeable?
- **b.** Customer requirements: Does the alternative part meet the customer's requirements?
- **c.** Cost: This is related to the cost difference between the original and the replacement part.
- **d.** GHG emission reduction potential: This pertains to the reduction in the GHG emissions between the original and the replacement part. To assess alternative parts with respect to the four criteria,

Table 6.3 Integration Criteria of GHG Emission

Life Cycle	Scenario	Allocation Criteria	
		Component	Part
Manufacturing		Weight	Weight
Distribution	Distance, transportation mode, type of fuel	Weight	Weight
Use	Usage hours, electricity consumption	Contribution of power, consumption measured	Contribution of power, consumption measured
End	Reuse, recycling, incineration, landfill	Field data using GHG database	Weight

the Must-Want matrix in the six sigma method of SEC (2006) can be used. The Must-Want matrix first screens alternative parts meeting the "must" conditions. Those meeting the "must" conditions are then assessed with respect to the "want" condition for the selection of the alternative to the problematic part. The "must" conditions are normally technical feasibility and customer requirements. The "want" conditions used are cost and potential GHG emission reduction.

References

1. Boothroyd, G., Dewhurst, P., and Knight, W. (2002). *Product Design for Manufacture and Assembly*, 2nd edn. Marcel Dekker, New York.
2. Sato, K. and Iwasa, Y. (2012). *Groundwater Updates*. Springer.
3. vanLoon, G. and Duffy, S. (2005). *Environmental Chemistry: A Global Perspective*. Oxford University Press, Oxford.
4. Jeswiet, J. and Kara, S. (2008). Carbon emissions and CES™ in manufacturing. *CIRP Annals—Manufacturing Technology 57*, 17–20.
5. Minx, J., Baiocchi, G., Wiedmann, T., Barrett, J., Creutzig, F., Feng, K., Forster, M., Pichler, P., Weisz, H., and Hubacek, K. (2013). Carbon footprints of cities and other human settlement in the UK. *Environmental Research Letters* 8 (3).
6. ISO 20022. (2015). Universal financial industry. ISO.
7. Schischke, K., Nissen, N.F., Sherry, J., O'Rafferty, S., O'Connor, F., Sitek, J., Pamminger, R., and Wimmer, W. (2012). Life cycle thinking in small and medium sized enterprises: Status quo and strategic needs in the electronic sector. *Electronics Goes Green*, September, Berlin. IEEE.

Bibliography

Ashby, M. F. (2009). *Materials and the Environment: Eco-informed Material Choice*. Elsevier, New York.

Ashby, M. F. (2011). *Materials Selection in Mechanical Design*, 4th edn. Elsevier, New York.

Chen, R. W., Navin-Chandra, D., and Print, F. B. (1993). A cost-benefit analysis model of product design for recyclability and its application. *IEEE International Symposium on Electronics and the Environment* (pp. 178–183), Arlington, VA.

Emblemsvag, J. (2003). *Life-Cycle Costing: Using Activity-Based Costing and Monte Carlo Methods to Manage Future Costs and Risks*. Hoboken, NJ: Wiley.

ISO 14044:2010. (2010). *Environmental Management-Life Cycle Assessment, 2010*.

Kroll, E. and Hanft, T. A. (1998). Quantitative evaluation of product disassembly for recycling. *Research in Engineering Design—Theory, Applications, and Concurrent Engineering* 10: 1–14.

Lee, S. G., Lye, S. W., and Khoo, M. K. (2001). A multi-objective methodology for evaluating product end-of-life options and disassembly. *International Journal of Advanced Manufacturing Technology* 18: 148–156.

United States Environmental Protection Agency (EPA). (2010). *Life Cycle Assessment, 2010*.

Whitney, D. E. (2004). *Mechanical Assembly: Their Design, Manufacture, and Role in Product Development*. New York: Oxford University Press.

SECTION III
Green Design of Mechanical Components

7

Spring Design for Sustainability

7.1 Green Spring Design

Coiled springs are useful in many practical applications. Springs exhibit flexibility, with either a linear or a nonlinear relationship between deflection and load. Springs have been extensively studied by mechanical designers. They are mass produced (low cost) and can be classified as wire springs, flat springs, and specially shaped springs. Springs are designed to store energy. The overall configuration of the spring family is shown in Figure 7.1.

Wire springs under tension or compression undergo twisting. The design expressions for wire springs are presented below.

The most common design variables for spring are: d = wire diameter, D_m = mean coil diameter, and N_a = number of active coils. The equations are presented below.

Load deflection equation:

$$P = K\delta$$

where:

P is applied load

δ is deflection along the axis of the spring

Spring constant:

$$K = \frac{d^4 G}{8 D_a^3 N_a}$$

where G is the shear modulus of the spring material

Shear stress:

$$\tau = \frac{8CPD_a}{\pi d^3} \tag{7.1}$$

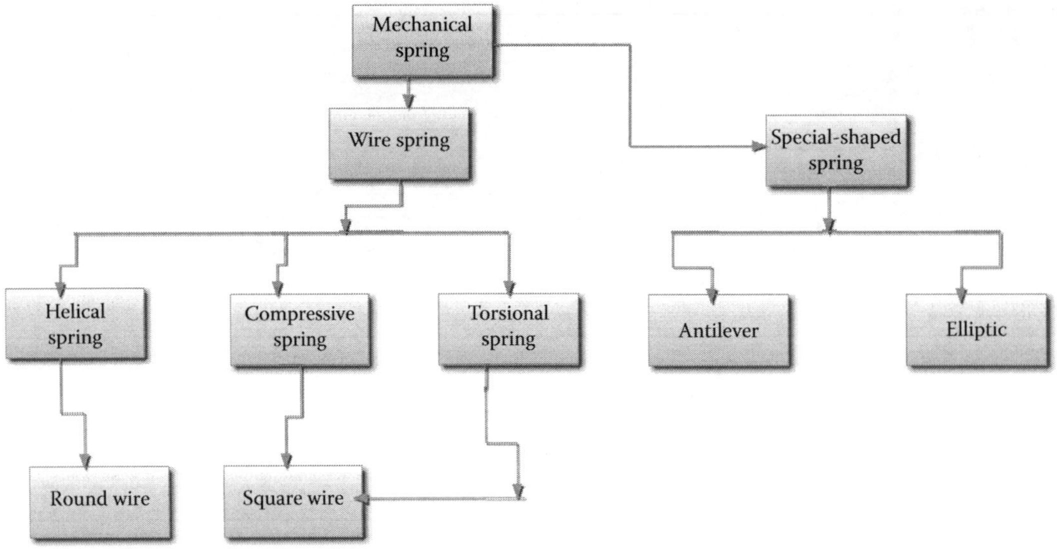

Figure 7.1

Families of mechanical springs.

Wahl stress concentration factor:

$$C = \frac{(4D_a - d)}{(4D_a - 4d)} + \frac{0.615d}{D_a} \tag{7.2}$$

Frequency of surge waves:

$$\varpi = \frac{d}{2\pi D_a^2 N_a}\sqrt{\frac{G}{2\rho}}$$

where ρ is the mass density of spring material.

Castigliano's theorem is used to develop the total strain energy (stored energy) of a helical spring as

$$U = \frac{4P^2 D_m^3 N_a}{d^4 G} + \frac{2P^2 D_m N_a}{d^4 G} \tag{7.3}$$

The total deflection of the spring is given as

$$\delta = \frac{8PD_m^3 N_a}{d^4 G} \tag{7.4}$$

It is obvious that if the free length of the spring is large, it is likely to buckle, and care should be taken in the design of the spring. Certain empirical relations are provided in Budynas and Nisbett regarding this consideration (Table 7.1) [1].

The expression for the Wahl stress concentration factor C accounts for usually high stresses at certain points in the spring.

The critical deflection for stability is given by

$$\delta_{cr} = l_f C_1' \left[1 - \left(1 - \frac{C_2'}{\lambda_{cr}^2} \right)^{0.5} \right]$$

Table 7.1 End Conditions in Spring Design

Design Parameters	Plain Ground	Plain and Ground	Squared or Closed	Squared and Ground
	Types of Spring Ends			
End coils	0	1	2	2
Total coils, N_t	N_a	N_a+1	N_a+2	N_a+2
Free length, l_f	pN_a+d	$p(N_a+1)$	pN_a+3d	pN_a+2d
Solid length, l_s	$d(N_t+1)$	dN_t	$d(N_t+1)$	dN_t
Pitch, p	$(l_f-d)/N_a$	$l_f/(N_a+1)$	$(l_f-3d)/N_a$	$(l_f-2d)/N_a$

Table 7.2 End Condition Constants

End Condition	α
Spring supported between flat parallel surfaces (fixed ends)	0.5
One end supported by flat surface perpendicular to spring axis (fixed) and other end pivoted (hinged)	0.707
Both ends pivoted (hinged)	1
One end clamped, other end free	2

λ_{cr} is the *effective slenderness ratio* and is given as $\lambda_{cr}=\alpha l_f/D$, where l_f is free length.
C_1 and C_2 are elastic constants:

$$C_1' = \frac{E}{2(E-G)}, \quad C_2' = \frac{2\pi^2(E-G)}{(2G+E)}$$

α is the *end-condition constant* and varies with end conditions. Budynas and Nisbett have provided a set of conditions' end-condition constant (Table 7.2) [1].

The condition of absolute stability, that is, the free length of spring that will not buckle under the force P, is given by

$$l_f < \frac{\pi D_m}{\alpha}\left[\frac{2(E-G)}{2G+E}\right]^{0.5}$$

As we can see, the free length that resists buckling depends not on the force applied but on the properties of the spring material. For steel, the free length that resists buckling becomes $l_f=2.63D/\alpha$, and if $\alpha=0.5$, for squared and ground ends, $l_{fs}=5.26D_a$. l_{fs} is the free length of the spring coil. This condition emphasizes that the calculated l_f should always be less than the value calculated to resist buckling.

Example 7.1

Design for environment (DfE) of a compression spring for a static load over a known deflection of 0.75 in. The range of force P is $100<P\leq 150$ lbf.

DfE of Helical Compression Spring

The short-listed spring materials, with other parameters, are given in Table 7.3.

7.2 Helical Compression Spring Design

Springs come in many shapes and have many purposes; helical springs, leaf springs, spiral springs, and torsion springs. Figure 7.2 shows a round-wire helical compression spring. It is loaded by an axial force F, D is the mean coil diameter, and d is the wire diameter.

If the spring is cut at a point and a portion removed, then, from equilibrium of the cut portion, a torque equation can be written as

$$T = \frac{FD}{2}$$

Table 7.3 Short-Listed Spring Materials with Other Parameters

Materials	Density (ρ), lb/in³	Young's Modulus (E), Mpsi	Modulus of Rigidity (G), Mpsi	Yield Strength (σ_y), kpsi	Embodied Energy (e), Btu	Cost, \$/kg	$M_1 = (\sigma_y / e\rho)$	$M_2 = (E / e\rho)$	Weighted Performance Index
Ti alloys	0.6	16.5	6.2	143–750	66.36×10^3	15–21			
Music wire	0.28	28.0–29.5	11.6–12.0	57.25–217.4	$30.336 \times 10^{3-}$ 36.024×10^3	1.0–1.1			
HD spring	0.282	28.5–28.8	11.4–11.7	36.28–98.7	$15.168 \times 10^{3-}$ 17.04×10^3	0.7–0.8			
Valve spring	440	29.5	11.2	80	312–346	19–29			
Chrome–silicon	18.1–25.8	29.5	11.2	11.61–65.31	70–75	0.9–3.40			
304 Stainless	16.22–17.744	28.0	10	174.17	72.996×10^3 $- 80.580 \times 10^3$	6.77–10.910			

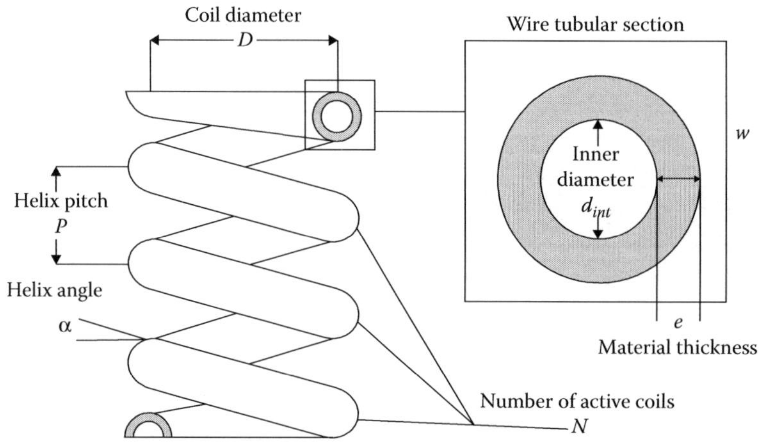

Figure 7.2

Geometric and material-related design parameters.

The maximum stress in the spring wire may be computed by superposition of the direct shear stress in the wire and the torsional shear stress:

$$\tau_{max} = \frac{T_r}{J} + \frac{F}{A}$$

inside the spring fiber.

Replacing $\tau_{max} = \tau$, $T = FD/2$, $r = d/2$, $J = \pi d^4/32$, and $A = \pi d^2/4$ gives

$$\tau_{max} = \frac{8FD}{\pi d^3} + 4\frac{F}{\pi d^2}$$

It is usual to write the shear stress equation as

$$\tau_{max} = k_s \frac{8FD}{\pi d^3} \leq S_{sy} = \frac{\sigma_y}{2}$$

Normally, shear yield strength S_{sy} can be replaced by $\sigma_y/2$.

7.3 Principles of Green Spring Design

7.3.1 Minimize Total Embodied Energy

The best material for a spring, regardless of its shape or the way it is loaded, can be selected according to the following principles. The total embodied energy (*TE*) of a spring can be written as

$$TE = e\left(\text{volume}\right)\left(\text{density}\right) = e\left(AL\right)(\rho)$$

The cross-sectional area of a spring wire of diameter d

$$A = \frac{\pi}{4}d^2$$

The length of the spring wire in a helical compression spring of coil diameter D is $L = \pi DN$, where N is number of coils.

The TE equation is

$$TE = \left(\frac{\pi^2}{4}d^2 DN\right)e\rho$$

where e is the embodied energy of spring material in megajoules per kilogram.

We need to minimize *TE*, as this minimizes not only the mass of material used but also the CO_2 footprint. The minimization of material selection model (MSE) should be achieved under the shear stress formulated above.

$$\text{Minimized } TE = \left(\frac{\pi^2}{4}d^2 DN\right)\left(e\rho\right) \tag{7.5}$$

Subject to the constraint

$$k_s \frac{8FD}{\pi d^3} \le \frac{\sigma_y}{2} \tag{7.6}$$

Now, let us define the known and unknown decision variables in the MSE. F, D, and N are known, and only d is unknown or a free variable. From the constraint in Equation 7.6, d is derived as

$$d = \left(\frac{16k_s FD}{\pi\sigma_y}\right)^{1/3}$$

and, entering the d value into Equation 7.5, we minimize *TE*.

Or, the *TE* equation can be written in simplified form:

$$TE = \left(\frac{\pi^2}{4}\left(\frac{16k_s}{\pi}\right)^{2/3}\right)\left(F^{2/3}D^{5/3}N\right)\left(\frac{e\rho}{\sigma_y^{2/3}}\right) \tag{7.7}$$

Ashby denoted the three terms on the right-hand side as functional, geometric, and material parameters [2]. The objective function in more general terms, $f(TE) = f(F, G, M)$, function $f(TE)$, is called the *performance index*. However, for material selection to minimize environmental damage, it is necessary to minimize the term

$$\left(\frac{e\rho}{\sigma_y^{2/3}}\right)$$

It is customarily written as maximizing the material index.

To maximize:

$$M_1 = \frac{\sigma_y^{2/3}}{e\rho} \tag{7.8}$$

Ashby provides a chart for the best materials are those with high values of σ and low values of *E*: in the upper left corner [2]. Spring materials are those with large values of *M*.

$$M = \frac{\sigma^2}{E}$$

while elastic hinge materials are those with large values of $M = \sigma / E$.

What is the best natural material for springs? What is best for elastic hinge applications? Materials with high values of toughness are at the top of the chart (antler, bamboo, and bone); they are good for impact loading.

7.3.2 Deflection of Spring

Another useful constraint in spring design relates to the deflection of the spring. It defines the free length, the solid length, and the length that is able to resist buckling. The deflection δ of a spring is also limited by the nature of its application. The deflection of a spring is

$$\delta = \frac{8FD^3N}{d^4G} \tag{7.9}$$

The wire diameter (*d*) under deflection constraint can be calculated as

$$d = \left(\frac{8FD^3N}{G\delta}\right)^{1/4} \tag{7.10}$$

Now, *TE* can be rewritten with the *d* value under deflection constraint as

$$TE = \left(\frac{2.6\sqrt{8}\pi^2}{4\delta^{1/2}}\right)\left(F^{0.5}D^{2.5}N^{0.5}\right)\left(\frac{e\rho}{E^{0.5}}\right) \tag{7.11}$$

To minimize under deflection constraint, we should minimize the material function, or maximize

$$M_2 = \frac{E^{0.5}}{e\rho} \tag{7.12}$$

We have two constraints—shear stress and deflection—and for material selection, it is recommended that material related to the dominant constraint should be selected. If the spring is to meet both requirements or constraints, its *TE* has to be the greater of TE_1 and TE_2. We can write this as a min–max optimization model:

$$TE_{\text{selected}} = \max(TE_1, TE_2) \tag{7.13}$$

Table 7.4 Analytical Selection of Material under Multiple Constraints

Material	σy (MPa)	E (GPa)	e(MJ/kg)	ρ (kg/m₃)	TE_1	TE_2	TE_s
Ti alloys	750–1200	110–120	600–746	4400–4800	7.698×10^4	2.418×10^3	7.698×10^4
Carbon-fiber-reinforced plastic (CFRP)	550–1050	60–150	259–286	1550	1.204×10^4	339.553	1.204×10^4
Spring steel (low alloy)	400–1500	460–1200	32–38	7800–7900	6.782×10^3	80.236	6.782×10^3
Music wire (low-carbon steel)	250–395	200–215	29–35	7800–7900	1.312×10^4	146.54	1.312×10^4
HD spring (high-carbon steel)	400–1155	200–215	29–35	7800	1.08×10^4	145.647	1.08×10^4
Valve spring (medium)	305–900	200–216	29–35	7800	8.58×10^3	45.607	8.58×10^3
Stainless steel	170–1000	189–210	77–85	7600–8100	2.23×10^4	378.75	2.23×10^4
Nickel–chromium	365–460	200–230	127–140	8300–8500	4.98×10^4	645.408	4.98×10^4
Nickel-based superalloys	300–1900	150–245	135–150	7750–8650	2.69×10^4	645.408	2.69×10^4

We select the material with the higher of the two *TE*s. An analytical method is presented below. Min–max models are normally used for continuous function, but in our case the *TE* function is discrete, and an analytical method would work better. We will use charts generated by CSE between the material indexes M_1 and M_2 to short-list materials useful for the spring and then use the analytical method. The short-listed spring materials and calculations for TE_1 and TE_2 are presented in Table 7.4.

Examining the TE_s values, we find that *spring steel* is the best material, meeting both constraints and minimizing the *TE*. In order of preference, the materials are

1. Spring steel (low-alloy steel)
2. Valve spring (medium-carbon steel)
3. HD spring (high-carbon steel)
4. Carbon-fiber-reinforced polymer (CFRP)
5. Music wire (low-carbon steel)
6. Stainless steel
7. Nickel-based superalloys
8. Nickel–chromium
9. Ti alloys

Spring steel minimizes the embodied energy (*TE*) and meets both constraints. Therefore, spring steel (low-alloy steel) is selected for design of the complete spring.

As the shear stress constraint is dominant in environmentally conscious material selection, we find the spring wire diameter

$$d = \left(\frac{k_s(16FD)}{\pi\sigma_y} \right)$$

C is normally in the range $4 \leq C \leq 12$ for all practical design. Assuming $C=6$, $K_B=(4C+2)/(4C-3)=1.238$, and $K_B=k_S$ (where K_B and k_S are stress concentration factors). $F=9$ kg. An average value of $\sigma_y=950MPa$ is used to calculate $d=1.434$ mm and mean coil diameter $D_m=Cd=8.6$ mm. The outer coil diameter $D_o=D_m+d=9.944$ mm, and inner coil diameter $D_i=D_m-d=7.166$ mm. Stability considerations are very important in spring coil design to avoid buckling under compressive load. It is assumed in this case that the space limitation between two flat surfaces $l_f=100$ mm. From stability considerations, it is necessary that l_f be less than the critical free length

$$\left(l_f\right)cr = \frac{\pi D}{\alpha}\left[\frac{2(E-G)}{2G+E}\right]^{1/2}$$

For squared and ground ends, $\alpha=0.5$. Taking the average E and G values from Table 7.4, for spring steel, $(l_f)cr=2.63D/\alpha=126.25$ mm. $l_f<(l_f)cr$, and so the spring coil is stable (not likely to buckle). We can calculate the number of actual turns in the spring coil; $l_f=pN_a+d$, where p is the pitch of the spring coil and may depend on manufacturing considerations; here it is assumed $p=5$ mm. Now, $N_a=19.77$ turns, and total number of turns for squared and ground spring $N_t=N_a+2=21.77$ turns.

Therefore, the complete environmental design of the helical compression spring is

$$d=1.434\text{ mm}, D_m=8.6\text{ mm}, D_o=9.944\text{ mm}, D_i=7.166\text{ mm}, N_a=19.77\text{ turns, and } N_t=21.77\text{ turns}$$

Example 7.2: Life Cycle Cost (LCC) Calculation of Helical Spring

Several life cycle analysis (LCA) techniques have evolved lately that can evaluate cradle-to-grave inventory and impact assessment of all mechanical components and systems. These LCA techniques lead to environmentally more sensible design at a time when greenhouse gas (GHG) emissions, global warming, and UN Intergovernmental Panel on Climate Change (IPCC) issues have notably appeared as important topics of discussion.

Springs are mass-produced items, and the life cycle cost of a large number of springs should be calculated together. Most consumers have a good understanding of the total cost of a product. The life cycle cost of springs can be calculated as follows:

$$LCC_{\text{spring}}=C_{\text{ic}}+C_{\text{installation}}+C_{\text{m}}+C_{\text{dis}}+C_{\text{d}}-C_{\text{recycling}}-C_{\text{reuse}}$$

where:

C_{ic} = initial costs, purchase price
$C_{\text{installation}}$ = installation cost
C_{m} = maintenance cost
C_{d} = disassembly cost
$C_{\text{recycling}}$ = recycling cost
C_{reuse} = reuse cost

Example 7.3: End-of-Life Considerations; Reuse and Recycle

Mattresses and box springs become problematic once they reach the end of their useful life. There is no easy way to dispose of mattresses; not only are they difficult to landfill, but they also pose challenges for some incinerators due to their inability to be easily compressed and crushed. However, mattress recycling has been increasing as an alternative to landfilling or incineration. A new company in Nashville removes cotton, peels away layers of foam, and shoves the steel springs into a baler, where they are compacted into a manageable clump. It has been estimated that the process takes about 7.62 min for a mattress and about 5.03 min for a box spring. It is estimated that a mattress or box contains about 25 lb of steel, which will go to landfill if it is not recycled. Sometimes these mattresses and boxes are refurbished and reused. These new products are made from old springs with a new cover, to be sold in discount stores. The Connecticut Department of Consumer Protection issues licenses for proper refurbishment, reuse, and recycling.

We know that springs are mass produced. For the above-designed springs, let us calculate the weight of one spring:

$$W=AL\rho=\frac{\pi d^2}{4}(\pi DN)(\rho)=\frac{\pi^2 d^2 DN\rho}{4}$$

We get

$$W=\frac{\pi^2(1.434/1000)^2}{4}(9.44/1000)(21.44)(7900)=0.032\text{ kg}$$

If the weight of each spring is 0.032 kg, and we manufacture 10^6 springs/day, the total weight of the springs manufactured is 3.2×10^4 kg. These springs can be used in a number of engineering applications, such as

- Medical devices
  - Diagnostic instruments
  - Syringes
- Aerospace
  - Landing gear
  - Cockpit controls
- Specialty commercial
  - Security fencing wire forms
  - Factory hand tools

or in mattresses and spring boxes as described earlier. We will consider recycling from mattresses and boxes here as an example. Normally, one mattress and spring box contains about 1000 helical springs, so that these 10^6 springs are sufficient for 1000 mattresses and boxes. If all of them are recycled and reused, we would like to find out how much money, energy, and environmental pollution are saved.

Cost Saving from Recycling

As mentioned earlier, each mattress and spring box takes about 7.62 and 5.03 min, respectively, to demanufacture (disassemble). The cost of demanufacturing is

$$Tme_{\text{dis}} = 500\times(7.62+5.03) = 6.325\times10^3 \text{ min}$$

The time for disassembly is about 106 h. If the company pays \$10.00/h to the workers, the total cost of demanufacturing or disassembly is \$1060. Now, we will calculate how much money is saved by recycling all the 3.2×10^4 kg of recycled springs. The spring is made from low-alloy steel, and the going rate for such steel in the market is \$1.1/kg.

Therefore, if we used new (virgin) steel for springs, the material cost would be \$35,420. This illustrates the importance of recycling; there is a substantial saving of \$34,360 on 10^6 springs.

Energy Saving from Recycling

TE values for primary production (extraction) from the mines and processing are provided by Ashby [2], and we can use these to estimate the energy content of 3.2×10^4 kg of steel.

$$TE \text{ of virgin material} = 3.2\times10^4.\frac{(32+38)^*}{2} = 1.12\times10^6\ MJ$$

$$\text{Recycling } TE = 3.2\times10^4.\frac{(9+11)^*}{2} = 3.2\times10^5 \text{ MJ}$$

$$\text{Recycling } TE = 3.2\times10^5 \text{ MJ}$$

$$\text{Energy saved by recycling} = 1.12\times10^6 - 3.2\times10^5 = 800{,}000 \text{ MJ}$$

Here, we have not considered the energy needed to manufacture spring through deep drawing of wire; the energy savings due to recycling would be tremendous.

Reduction in Emission of CO_2

$$CO_2 \text{ footprint, primary production} = 3.2\times10^4.\frac{(1.95+2.3)}{2} = 68{,}000 \text{ kg}$$

$$\text{CO}_2 \text{ footprint of recycled spring} = 3.2\times 10^4 . \frac{(0.55+0.64)}{2} = 19{,}040 \text{ kg}$$

$$\text{Reduction in emission of CO}_2 = (68{,}000 - 19{,}040) = 48{,}960 \text{ kg}$$

We can also calculate the water usage in manufacturing steel. Water is also considered a scarce, rapidly decreasing resource.

Water Saving

Depending on available data for water usage in production of low-alloy steel [2] (l/kg), we can estimate the quantity of water used in manufacturing 3.2×10^4 kg of steel:

$$3.2\times(10^4).\frac{(37+111)^*}{2} = 2{,}368{,}000\ \text{l}$$

The recycled spring will not need any water, so the total water saving is 2,368,000 L.

We can see benefits from recycling in terms of depletion of resources (material and water), money, energy, and carbon footprint.

Equations 7.3, 7.4, 7.7, and 7.8 can be redeveloped for minimization of carbon footprint, minimization of cost, or simply minimization of mass, as shown in Table 7.5.

Mass of spring = 0.391 lbf

Spring mass = 0.391 × 0.454 = 0.178 kg

Table 7.5 Examples of Performance Indices to Minimize Environmental Impact

Objective	Objective Function	Constraint	Material Indices
Minimization of embodied energy	$TE=\left(\frac{\pi^2}{4}\left(\frac{16k_s}{\pi}\right)^{2/3}\right)\left(F^{2/3}D^{5/3}N\right)\left(\frac{e\rho}{\sigma_y^{2/3}}\right)$	Shear stress	$M_1=\frac{\sigma_y^{2/3}}{e\rho}$
	$TE=\left(\frac{2.6\sqrt{8}\pi^2}{4\delta^{1/2}}\right)\left(F^{0.5}D^{2.5}N^{0.5}\right)\left(\frac{e\rho}{E^{0.5}}\right)$	Deflection	$M_2=\frac{E^{0.5}}{e\rho}$
Minimization of carbon footprint	$TCO_2=\left(\frac{\pi^2}{4}\left(\frac{16k_s}{\pi}\right)^{2/3}\right)\left(F^{2/3}D^{5/3}N\right)\left(\frac{CO_2\rho}{\sigma_y^{2/3}}\right)$	Shear stress	$M_1=\frac{\sigma_y^{2/3}}{CO_2\rho}$
	$TCO_2=\left(\frac{2.6\sqrt{8}\pi^2}{4\delta^{1/2}}\right)\left(F^{0.5}D^{2.5}N^{0.5}\right)\left(\frac{CO_2\rho}{E^{0.5}}\right)$	Deflection	$M_2=\frac{E^{0.5}}{CO_2\rho}$
Minimization of cost	$Tp=\left(\frac{\pi^2}{4}\left(\frac{16k_s}{\pi}\right)^{2/3}\right)\left(F^{2/3}D^{5/3}N\right)\left(\frac{p\rho}{\sigma_y^{2/3}}\right)$	Shear stress	$M_1=\frac{\sigma_y^{2/3}}{p\rho}$
	$Tp=\left(\frac{2.6\sqrt{8}\pi^2}{4\delta^{1/2}}\right)\left(F^{0.5}D^{2.5}N^{0.5}\right)\left(\frac{p\rho}{E^{0.5}}\right)$	Deflection	$M_2=\frac{E^{0.5}}{p\rho}$
Minimization of mass	$Tm=\left(\frac{\pi^2}{4}\left(\frac{16k_s}{\pi}\right)^{2/3}\right)\left(F^{2/3}D^{5/3}N\right)\left(\frac{\rho}{\sigma_y^{2/3}}\right)$	Shear stress	$M_1=\frac{\sigma_y^{2/3}}{\rho}$
	$Tm=\left(\frac{2.6\sqrt{8}\pi^2}{4\delta^{1/2}}\right)\left(F^{0.5}D^{2.5}N^{0.5}\right)\left(\frac{\rho}{E^{0.5}}\right)$	Deflection	$M_2=\frac{E^{0.5}}{\rho}$

Source: Data from Kutz, M. *Environmentally Conscious Mechanical Design*, Wiley, 2007 [3].

Note: e = embodied energy (MJ/kg); CO_2 = carbon footprint (kg/kg); p = price ($/kg); m = mass (kg).

Embodied energy lower = $32 \times 0.178 = 5.7$ MJ
Embodied energy higher = $38 \times 0.178 = 6.764$ MJ
Carbon footprint lower = $1.95 \times 0.178 = 0.347$ kg
Carbon footprint higher = $2.3 \times 0.178 = 0.41$ kg
Thus, embodied energy and carbon footprint both have upper and lower limits:

$$\text{Embodied energy: } 5.7 \text{ MJ} \leq \text{Embodied energy} \leq 6.764 \text{MJ}$$

$$\text{Carbon footprint: } 0.347 \text{ kg} \leq \text{Carbon footprint} \leq 0.41 \text{ kg}$$

We can redesign the spring to lower the embodied energy as well as the carbon footprint.

References

1. Budynas, R. G. and Nisbett, J. K. (2008). *Shigley's Mechanical Engineering Design*, 8th edn. McGraw Hill, New York.
2. Ashby, M. F. (2009). *Materials and the Environment: Eco-Informed Material Choice*. Elsevier, Canada.
3. Kutz, M. (2007). *Environmentally Conscious Mechanical Design*. Wiley, Hoboken, NJ.

Bibliography

Ashby, M. F. (2011). *Material Selection in Mechanical Design*, 4th edn. Elsevier, New York.

Botthroyd, G., Dewhurst, P., and Knight, W. (2002). *Product Design for Manufacture and Assembly*, 2nd edn. Marcel Dekker, New York.

Chen, R. W., Chandra, D. N., and Print, F. B. (1993). A cost-benefit analysis model of product design for recyclability and its application. In *IEEE International Symposium on Electronics and the Environment* (pp. 178–183), Arlington, VA.

Desai, A. and Mital, A. (2005). Incorporating work factors in design for disassembly in product design. *Journal of Manufacturing Technology Management* 16: 712–732.

Dowie, T. and Kelly, P. (1994). Estimation of disassembly times, Technical Report DDR/TR15, Manchester Metropolitan University, UK.

ISO 14044:2010, Environmental management—Life Cycle Assessment, 2010.

Kroll, E. and Carver, B. S. (1999). Disassembly analysis through time estimation and other metrics. *Robotics and Computer-Integrated Manufacturing* 15: 191–200.

Kroll, E. and Hanft, T. A. (1998). Quantitative evaluation of product disassembly for recycling. *Research in Engineering Design—Theory, Applications, and Concurrent Engineering* 10: 1–14.

Lee, S. G., Lye, S. W., and Khoo, M. K. (2001). A multi-objective methodology for evaluating product end-of-life options and disassembly. *International Journal of Advanced Manufacturing Technology* 18: 148–156.

Lidwell, W., Holden, K., and Butler, J. (2003). *Universal Principles of Design*. Rockport Publishers, Beverly, MA.

McGlothlin, S. and Kroll, E. (1995). Systematic estimation of disassembly difficulties: Application to computer monitors. In *Proceedings of the 1995 IEEE International Symposium on Electronics & the Environment* (pp. 83–88), Scottsdale, AZ, USA.

Mott, R. (2004). *Machine Elements in Mechanical Design*, 4th edn. Prentice Hall, Upper Saddle River, New Jersey.

Segur, A. B. Predetermined Motion Time System (PMTS).

Skakoon, J. G. (2000). *Detailed Mechanical Design: A Practical Guide*. ASME Press, New York.

Skakoon, J. G. (2008). *The Elements of Mechanical Design*. ASME Press, New York.

Slocum, A. H. (1992). *Precision Machine Design*. Prentice Hall, NJ.

Soemers, H. M. J. R. (2010). *Design Principles for Precision Mechanisms*. Herman Soemers Publications.

Suga, T., Saneshige, K., and Fujimoto, J. (1996). Quantitative disassembly evaluation. In *Proceedings of the 1996 IEEE International Symposium on Electronics & the Environment* (pp. 19–24), Arlington, VA, USA.

United States Environmental Protection Agency. (2010). *Life Cycle Assessment*. United States Environmental Protection Agency (EPA).

Whitney, D. E. (2004). *Mechanical Assembly: Their Design, Manufacture, and Role in Product Development*. Oxford University Press, New York.

8

Environmentally Conscious Design of Gears

8.1 Introduction

Gears are used to transmit power between shafts: they may be parallel, at an angle, or perpendicular to one another. Power transmission is necessary for multiple reasons, including providing torque multiplication at low speeds; reducing engine revolutions per minute (rpm) at highway speeds; allowing the engine to operate within its most efficient rpm range; and allowing the engine to be disengaged from the rear wheels while the vehicle is not moving (torque converter and clutch). There are three basic purposes of transmission: (1) to alter shaft rpm; (2) to multiply torque; and (3) to reverse the direction of shaft rotation.

8.2 Gear Types

The gears used for changing the rpm and the torque of the power moving through the transmission as well as the direction of rotation are

- Planetary gears
- Helical gears
- Spur gears

8.2.1 Spur Gears

These are the simplest gears with straight-cut teeth parallel to the axis of the transmitting shaft. They are, however, noisy in operation (Figure 8.1).

8.2.2 Helical Gears

These are set to transmit power between parallel shafts as well as between perpendicular shafts. Helical gears to transmit power at an angle are called spiral gears. Helical gears have teeth inclined to the axis of rotation, with at least two teeth in mesh at any time. This distributes the tooth load, resulting in quieter operation. Helical gears can be used in the same applications as spur gears, but the inclined teeth develop thrust load and bending couples, which are not present in spur gearing (Figure 8.2).

Advantages of helical gears:

- Noiseless motion even at higher speeds.
- Smooth transmission of small gears with fewer teeth at a large transmission ratio (up to 15:1).

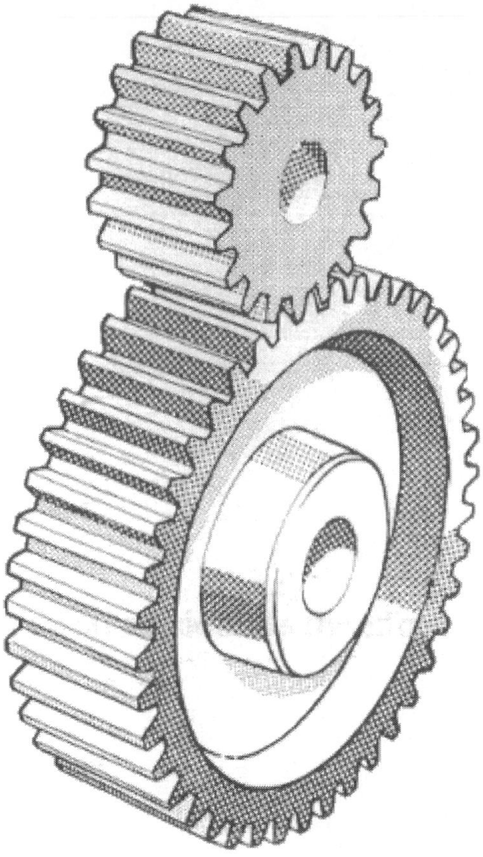

Figure 8.1

Spur gear set.

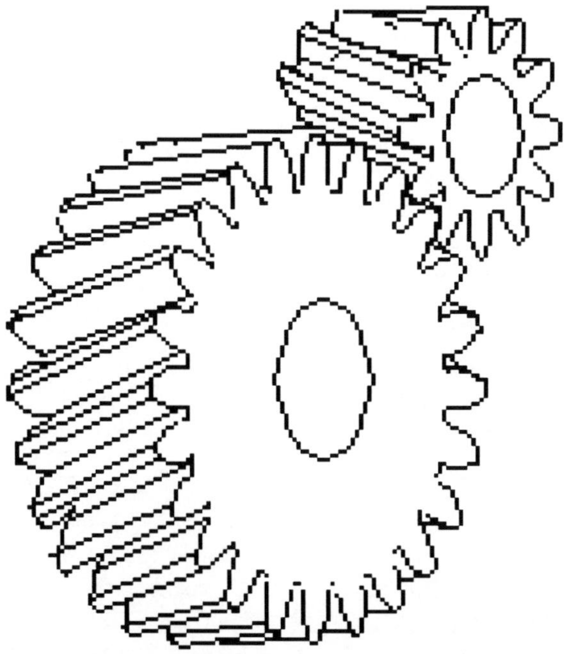

Figure 8.2

Helical gear set.

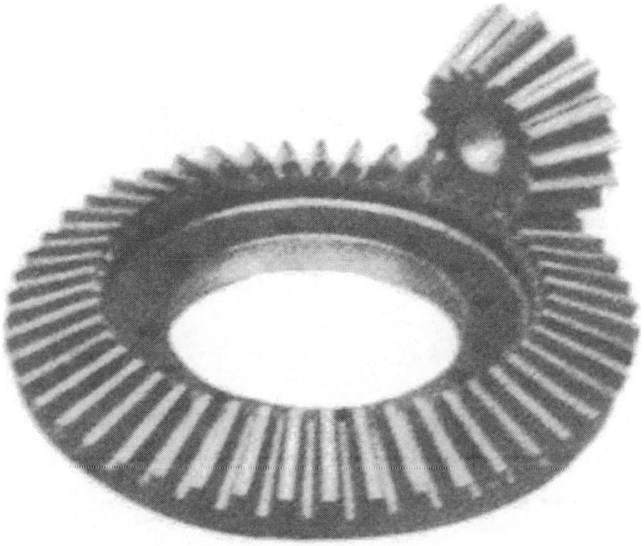

Figure 8.3

Straight bevel gear set.

- These gears must be rigidly installed on the shaft. They result in an axial force in one direction, depending upon the direction of rotation, and are used for transmitting low power.

8.2.3 Bevel Gears

These have teeth cut on conical surfaces and are used mostly for transmitting motion between intersecting shafts. There are straight-tooth bevel gears and spiral bevel gears (Figure 8.3).

8.2.4 Hypoid Gears

These are quite similar to spiral bevel gears, except that the shafts are offset and nonintersecting. Bevel gear transmission is employed for transmitting motion from one direction to the other at any angle. In general, 90° transmission is used.

8.2.5 Worms and Worm Gears

These are the fourth basic gear type. The worm resembles a screw.

Figure 8.4

Rack and pinion set.

8.2.6 Planetary Gears

This is the most complex type of gear and is used in all automatic transmissions. It contains three parts: spur gear, planet gear, and internal gear (ring gear).

8.2.7 Rack and Pinion

The rotary motion of the pinion (1) (small gear) is transformed into the linear motion of the rack (2). For transmitting high power, worm and worm rack transmission is employed (Figure 8.4).

8.3 Common Terms in Gear Design: Power, Torque, Horsepower, and Gear Ratio

Torque is a measure of twisting force, and power is a measure of how quickly work can be done. Power is dependent on torque and rpm and is calculated as torque multiplied by revolutions per minute. When two gears are in mesh, a gear ratio exists. For example, if the driver gear has 14 teeth and the driven gear has 28 teeth, the gear ratio is 28/14 = 2:1. This simply means that the drive gear must rotate twice to make the driven gear rotate once. It is obvious that when two gears are in mesh, one will spin in the opposite direction to the other. Idle gears are used to reverse direction. The smaller of the two meshing gears is called a *pinion* and the larger a gear; the combination is called a *gear and pinion set*. The change in rpm from the input gear to the output gear is directly proportional to the gear ratio. For example, for a 3:1 gear ratio, if the input gear rotates at 900 rpm, the output gear will rotate at 300 rpm. Similarly, for the same gear-pinion set, if input gear turns at 50 lb/ft of force, then the output gear will turn the driveshaft with 150 lb/ft force.

In some power transmission, we can use multiple sets of gear and pinion. In such cases, individual gear ratios can be multiplied to calculate a total gear ratio.

8.4 Gear Defects

- Wear of some or all of the teeth
- One or more teeth broken or twisted
- Burrs on the bore or the key way
- Burrs on the internal splines of the gear bore or the tooth surface
- Crack or damage on the rim or the bore of the gear wheel

8.5 Limits of Wear for Discarding

With open gear transmission and surface speed up to 2 m/s, the tooth thickness (t) reduces due to wear by (0.5 m) mm, where m is the module, or about 33% of the original tooth thickness. With closed gear

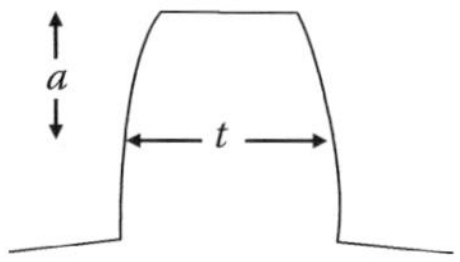

Figure 8.5

Tooth thickness (t) and addendum (a).

transmission (gear boxes) for surface speeds over 2 m/s, the tooth thickness (t) reduces due to wear by 0.3 mm (Figure 8.5).

8.6 Lubrication of Gears

- Proper lubrication with timely addition/replacement plays a vital role in maintaining the efficiency of the gear box and increasing its working life.
- The oil level should be checked in all the gear box's oil indicators/dipsticks.
- The oil level should be in between the minimum and maximum limits of the dipstick (or indicator).
- The lubricating oils need to be tested periodically. The following tests are to be done:
  1. Dirt contamination
  2. Moisture
  3. Volatile materials (benzene, kerosene, spirit, etc.)
  4. Viscosity
  5. Acidity
  6. Alkalinity due to soda (Na_2CO_3), for example (Table 8.1)

8.7 Steps in Spur Gear Drive Design

Spur gears are used to transmit power and rotary motion between parallel shafts. Figure 8.6a and b show the nomenclature of a spur gear. During the design of spur gear, we need to define all the nomenclature. The circular pitch, p, of spur gear is defined as the distance, on the pitch circle, from a point on a tooth to the corresponding point of the adjacent tooth. The diametral pitch, p_d, is defined as the number of teeth of the gear divided by the pitch circle diameter. They are calculated in Equations 8.1 and 8.2

Table 8.1 Module vs. Addendum and Thickness Spur Gear

Module (m)	1	1.5	2	2.5	3	3.5	4	5	6	7	8	9	10
$a = 0.7485\ m$	0.75	1.12	1.5	1.87	2.24	2.62	2.99	3.74	4.49	5.24	5.98	6.73	7.48
$t = 1.387\ m$	1.39	2.08	2.77	3.47	4.16	4.85	5.56	6.94	8.32	9.71	11.1	12.48	13.87

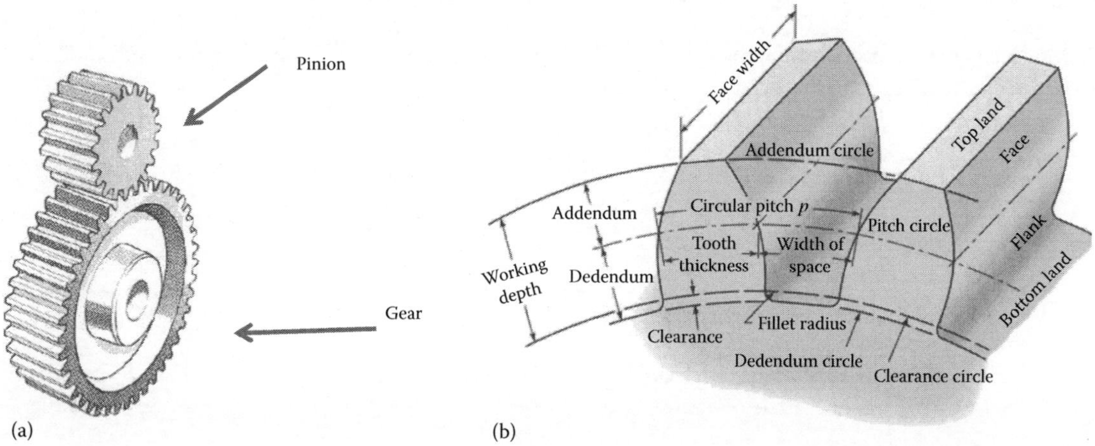

Figure 8.6

(a) Spur gear set. (b) Nomenclature of spur gears.

$$p = \frac{\pi d}{N_t} \tag{8.1}$$

and

$$p_d = \frac{N_t}{d} \tag{8.2}$$

The multiplication of p and p_d is a constant, $p_p = \pi$.

When two gears are in mesh, the center distance, c, of two gears is

$$c = \frac{d_p + d_g}{2} \tag{8.3}$$

where d_p and d_g are the diameter of pinion and gear, respectively. Velocity ratio is calculated as

$$r_v = \frac{\omega_2}{\omega_1} = \frac{n_2}{n_1} = \frac{N_{t_1}}{N_{t_2}} = \frac{d_1}{d_2}$$

where:

ω = angular velocity, rad/s
n = rpm
N_t = number of teeth
d = pitch diameter

Subscripts 1 and 2 refer to the driver and driven gear, respectively.

The American Gear Manufacturers Association (AGMA) have standardized the proportions of these dimensions for various pressure angles, and only the most commonly used are presented below.

	20°	22.5°	25°
Addendum	$1/p_d$	$1/p_d$	$1/p_d$
Dedendum	$1.25/p_d$	$1.25/p_d$	$1.25/p_d$
Clearance	$0.25/p_d$	$0.25/p_d$	$0.25/p_d$
Working depth	$2/p_d$	$2/p_d$	$2/p_d$
Whole depth	$2.25/p_d$	$2.25/p_d$	$2.25/p_d$

Module is the inverse of diametral pitch in MKS units, $m = d/N$ (mm). Diametral pitch and module are provided in standard sizes by AGMA:

Module: 1, 1.25, 1.5, 2.0, 2.5, 3, 4, 5, 6, 8, 10, 12, 16, 20, 25, 32, 40, 50
Diametral pitch (p_d): 2, 2.25, 2.5, 3, 4, 6, 8, 10, 12, 16

Steps in design of spur gear:

1. From design requirements, identify speed of pinion, *nP*, desired output speed of gear, *nG*, and power to be transmitted, *P*.
2. Choose type of material for the gears (steel, cast iron, bronze, etc.).
3. Select a trial value for the diametral pitch, P_d.

Note that diametral pitch must be a standard size (see list above).

As P_d decreases, tooth size increases, but as P_d increases, the number of teeth increases, the gear train runs more smoothly and quietly, and the drive becomes smaller as well.

Example 8.1

A gear pair is to be a part of the drive for a milling machine requiring 20 hp with the pinion speed to be 550 rpm and the gear speed to be between 180 and 190 rpm.

Driven = milling machine, power = 20 hp, pinion speed = 550 rpm, $N_p = 24$ teeth, and $N_g = 71$ teeth. Output speed = 180–190 rpm ≈ 185 rpm, continuous use = 30,000 h.

Find:

- Compact gear design
- Trial size

$P_d = 6$ T/in, so $p = (\pi/p_d) = (\pi/6)$ and from velocity ratio, we can write: $(185/550) = (N_p/N_g) = (d_p/d_g)$ The circular pitch, $p = (\pi d_p/N_p) = (\pi d_p/N_p) = (\pi/6)$ and $d_p = N_p/6 = 24/6 = 4$ in and $d_g = 71/6 = 11.833$ in. $D_p = 4.80$ in, $D_g = 14.2$ in, $N_p = 24$ teeth, $N_g = 71$ teeth.

Central distance between gear and pinion:

$$C = \left(\frac{Dp + D_g}{2}\right) = \frac{4. + 11.833}{2} = 7.92 \text{ in}$$

Velocity of transmission:

$$V_t = \frac{\pi D_p n_p}{12} = \frac{\pi.4.550}{12} = 575.96 \text{ ft/min}$$

Force transmitted:

$$W_t = \frac{33000.(HP)}{V_t} = \frac{33000(20)}{575.96} = 1145.913\text{lbf}$$

Face width constraint is usually $8/P_d \le F \le 16/P_d$; we select average value to get nominal face width: $F\,12/P_d = 12/6 = 2$ in.

Rim thickness factor

Addendum = $1/P_d = 1/6 = 0.167$ and C = clearance = $0.25/6 = 0.041667$.
Rim thickness = $h_t = 2a + C = 0.45$. If the gear blank is not solid, then, $h_t = 1.2\ (0.45) = 0.54$ in.

Dynamic factor (K_V)

The dynamic factor K_v is used to account for the dynamic effects of gear transmission. Normally dynamic factors depend on the pitch-line velocity for a particular material, which is related to the gear manufacturing process. Most of the dynamic factors have been developed by AGMA. They are as follows:

$$K_v = \frac{600 + V}{600} \text{ (cast iron, cast profile)}$$

$$K_v = \frac{1200 + V}{1200} \text{ (cut, milled profile)}$$

where V is the pitch-line velocity in feet per minute. The above dynamic factors are termed the *Bath equation*. Later, AGMA added the following equations for dynamic effects:

$$K_v = \frac{50 + \sqrt{V}}{50} \text{ hobbed or shaped profile)}$$

$$K_v = \sqrt{\frac{78 + \sqrt{V}}{78}} \text{ (shaved or ground profile)}$$

In SI units, all the above dynamic effects are modified as

$$K_v = \frac{3.05 + V}{3.05} \text{ (cast iron, cast profile)}$$

$$K_v = \frac{6.1 + V}{6.1} \text{ (cut or milled profile)}$$

$$K_v = \frac{3.56 + \sqrt{V}}{3.56} \text{ (hobbed or shaped profile)}$$

$$K_v = \sqrt{\frac{5.56 + \sqrt{V}}{5.56}} \text{ (shaved or ground profile)}$$

where:

V is in meters per second (m/s)

K_v can be calculated from the above equations

8.7.1 Guidelines for Adjustments in Successive Iterations

1. Decreasing the numerical value of the diametral pitch results in larger teeth and generally lower stresses. Also, the lower value of the pitch usually means a larger face width, which decreases stress and increases surface durability.
2. Increasing the diameter of the pinion decreases the transmitted load, generally lowers the stresses, and improves surface durability.
3. Increasing the face width lowers the stress and improves surface durability, but has less impact than either the pitch or the pitch diameter.
4. Gears with more and smaller teeth tend to run more smoothly and quietly than gears with fewer and larger teeth.
5. Standard values of diametral pitch should be used for ease of manufacture and lower cost.
6. Use of high-alloy steels with high surface hardness results in the most compact system, but the cost is higher.
7. Use of gears with high-quality number Q_v adds cost but lowers the load distribution factor Km.
8. The number of teeth in the pinion should be as small as possible to make the system compact. But the possibility of interference is greater with fewer teeth. Check Table 8.6 to ensure no interference will occur.

8.8 Helical Gear Design

High-efficiency gears are environmentally friendly and save both energy and money. Some helical gears are virtually maintenance free. Tooth profiles are so precise that there is practically no wear and no backlash. Helical gear units are sealed, and oil-filled housings may require no oil changes, which is a blessing for the environment as well as for the bottom line. Power transmission can be achieved in many other ways, such as belts and pulley, or chains and sprockets, but they are high-wear and high-maintenance parts compared with helical gears. Helical gear reducers also come in many shapes, sizes, and configurations, and machine designers prefer to use these gears whenever possible.

Although worm and spur gears are less expensive, they have a much shorter life span than helical gears. In the case of fluctuating stresses, the number of cycles-to-failure for helical gears is much larger than for other forms of gears. It is estimated that spur and worm gears may require replacement every 3 years. This frequent replacement requires more capital and investments and drives up maintenance, downtime, and waste-disposal costs. All this adds up to depletion of resources, maintenance and disposal costs, and degradation of the environment, which are all contrary to design for the environment. Therefore, it can be concluded that both spur and worm gears are not environmentally sustainable.

In machine design, we strive for components creating less noise and low vibration, which is reflected in the precision and life of the component. Rotating machines that are noisy and vibrate can have any number of quality issues: improperly balanced rotating components; mating components with excessive tolerances; and components that are not rigid enough, fluctuating under load and leading to misalignment. If we compare the noise generated by helical gears with that of spur gears, it is about 10–12 dB lower. In human terms, this means that 16 helical gear units generate as much noise as eight worm gears or a single spur gear unit.

As design engineers, we look into the heat generated by the machine during its operation. As we know, any heat loss is wasted energy and most deplorable in the days of green design. It is also a sign of low efficiency of the system. Heat generation in rotating machines can also be an indication of misalignment, incorrect fits, and seal drag; overall, an inefficient design. Inefficient design of gear reducers leads to excessive energy consumption, and these gearboxes typically get so hot that one cannot touch them for more than a second or two. Here, again, helical gear reducers are a winner. Helical gearing is 20%–30% more efficient than a single-stage worm gearing, leading to cooler operation and less noise, including vibration. It has also been estimated that with a 2 hp motor, helical gears can save up to 4000 kWh annually, which points to lower gear reducer losses. It can also help engineers to opt for a smaller motor, and this saves on investment.

Engineers can visualize the tremendous loss of energy in the operation of a machine, and we can avoid this by better design and selection of components. Low-quality gear reducers, along with low-quality motors, are major contributors to excessive energy consumption and emissions of harmful gases into the atmosphere. There are endless miles of conveyors and thousands of pieces of equipment being driven by inefficient motors and even less efficient drives. Energy audits performed in assembly lines show that many factories and processing plants are operating at or about 60%–70% efficiency, and there is ample room for improvement. It can be honestly said that highly efficient equipment saves money, energy, materials, and time, and so is more productive. Therefore, green design will lead U.S. industries to be highly competitive in the global market. Going green is no longer a fad or a feel-good choice but economic. It is sound business sense and our opportunity to regain a competitive edge in the world market.

In spur gears, the teeth are parallel to the axis, whereas in helical gears, the teeth are inclined to the axis. Both the gears are transmitting power between two parallel shafts.

A helical gear can be thought of as an ordinary spur gear machined from a stack of thin shim stock, each limitation of which is rotated slightly with respect to its neighbors. When power is transmitted, both shafts are subjected to thrust load on the shaft.

A herringbone or double helical gear, shown in Figure 8.7, consists of two helical gears with opposing helix angle stacked together. As a result, two opposing thrust loads cancel, and the shafts are not acted on by any thrust load. The advantages of elimination of thrust load in herringbone gears are obliterated by considerably higher machining and mounting costs. This limits their applications to very heavy power transmission.

8.8.1 Kinematics

When two helical gears are engaged as in Figure 8.7, the helix angle has to be the same on each gear, but one gear must have a right-hand helix and the other a left-hand helix.

8.8.2 Helical Gears, Geometry, and Nomenclature

The helix angle ψ is always measured on the cylindrical pitch surface (Figure 8.8). The ψ value is not standardized. It ranges between 15° and 45°. Commonly used values are 15°, 23°, 30°, and 45°. Lower values give less end thrust. Higher values result in smoother operation and more end thrust. Above 45° is not recommended.

The circular pitch (p) and pressure angle (ϕ) are measured in the plane of rotation, as in spur gears. These quantities in normal plane are denoted by suffix n (p_n, ϕ_n), as shown in Figure 8.9.

From geometry, we have normal pitch as

$$p_n = p_t \cos(\phi) \tag{8.4}$$

where p_t is called *transverse circular pitch*.

Figure 8.7

Helical gear.

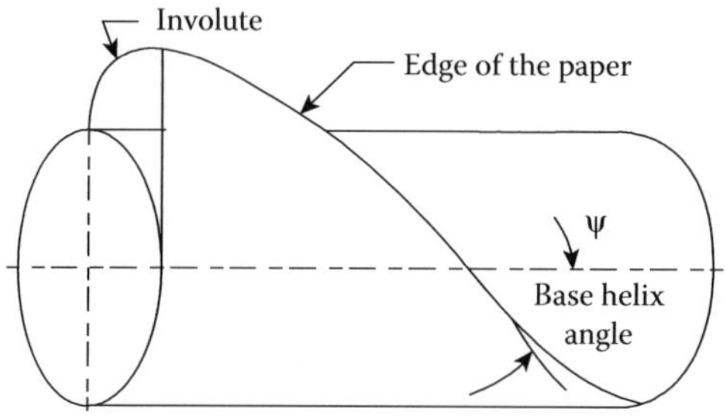

Figure 8.8

Illustration of helical gear tooth formation.

Normal module m_n is

$$m_n = m \cos \psi \tag{8.5}$$

m_n is used for hob selection. The pitch diameter (d) of the helical gear is

$$d = Zm = Zm_n / \cos \psi \tag{8.6}$$

The axial pitch (p_a) is

$$p_a = p / \tan \psi \tag{8.7}$$

For axial overlap of adjacent teeth,

$$b \geq pa \tag{8.8}$$

In practice, $b = (1.15\text{–}2)\ p_a$ is used.

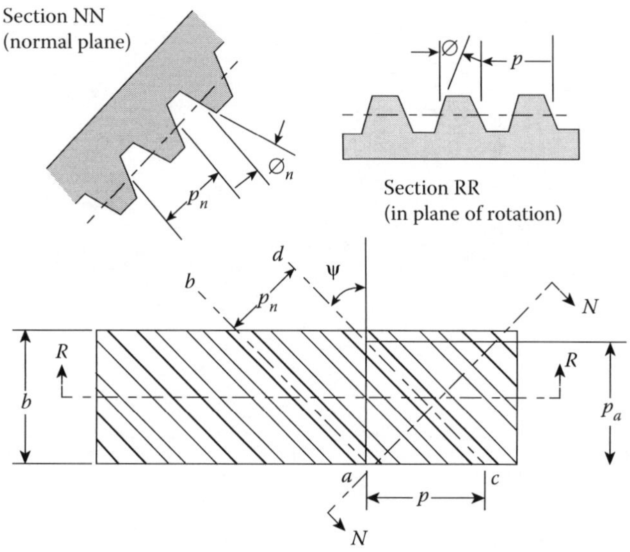

Figure 8.9

Portion of helical gear.

The relation between normal and transverse pressure angles is

$$\tan\phi n = \tan\phi \cdot \cos\psi \tag{8.9}$$

In the case of a helical gear, the resultant load between mating teeth is always perpendicular to the tooth surface. Hence, bending stresses are computed in the normal plane, and the strength of the tooth as a cantilever beam depends on its profile in the normal plane. Figure 8.9 shows the view of a helical gear in normal and transverse plane.

Figure 8.9 shows the pitch cylinder and one tooth of a helical gear. The normal plane intersects the pitch cylinder in an ellipse.

The shape of the tooth in the normal plane is nearly the same as the shape of a spur gear tooth having a pitch radius equal to radius R_e of the ellipse.

$$\mathrm{R_e} = d/(2\cos^2\psi) \tag{8.10}$$

The equivalent number of teeth (also called virtual number of teeth), Z_v, is defined as the number of teeth in a gear of radius R_e:

$$Z_v = \frac{R_e}{m_n} = \frac{d}{m_n \cos^2\varphi} \tag{8.11}$$

Substituting $m_n = m \cos\psi$, and $d = Z\,m$, we get

$$Z_v = \frac{Z}{\cos^2\varphi} \tag{8.12}$$

When we compute the bending strength of helical teeth, values of the Lewis form factor Y are the same as for spur gears having the same number of teeth as the virtual number of teeth (Z_v) in the helical gear and a pressure angle equal to ϕ_n.

Determination of the geometry factor J is also based on the virtual number of teeth. These values are plotted in Figure 8.10.

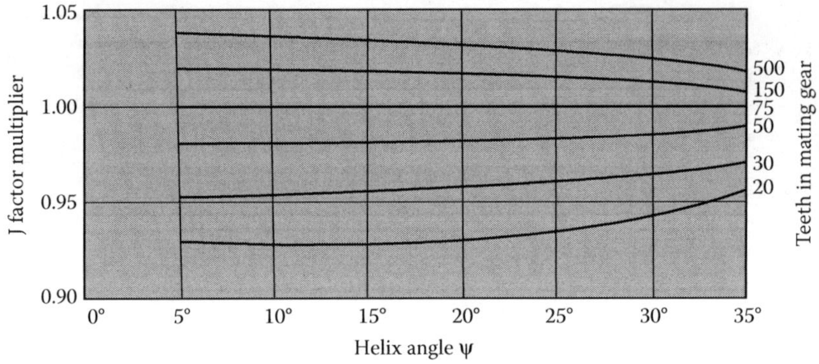

Figure 8.10

Helical gear J factor multipliers to be used with mating gears with other than 75 teeth.

8.8.3 Helical Gears: Force Analysis

A three-dimensional view of the forces acting on a helical gear tooth is shown in Figure 8.11.

$$F_r = F_n \sin \phi_n \tag{8.13}$$

$$F_t = F_n \cos \phi_n \cos \psi \tag{8.14}$$

$$F_a = F_n \cos \phi_n \sin \psi \tag{8.15}$$

$$F_r = F_t \tan \phi \tag{8.16}$$

and finally

$$F_n = \frac{F_t}{\cos(\varphi)\cos(\phi_n)} \tag{8.17}$$

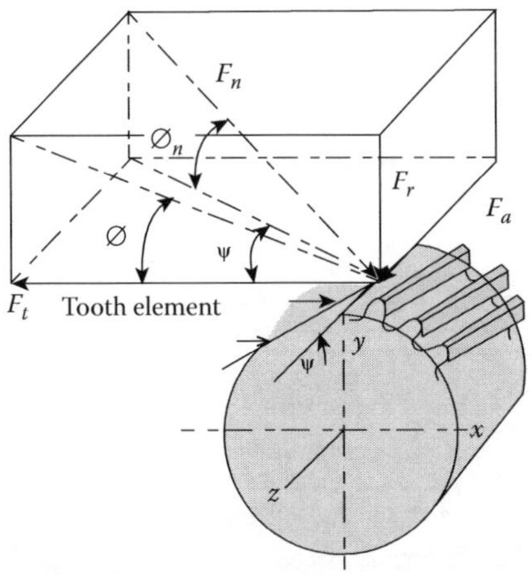

Figure 8.11

Tooth force acting on a right-hand helical gear.

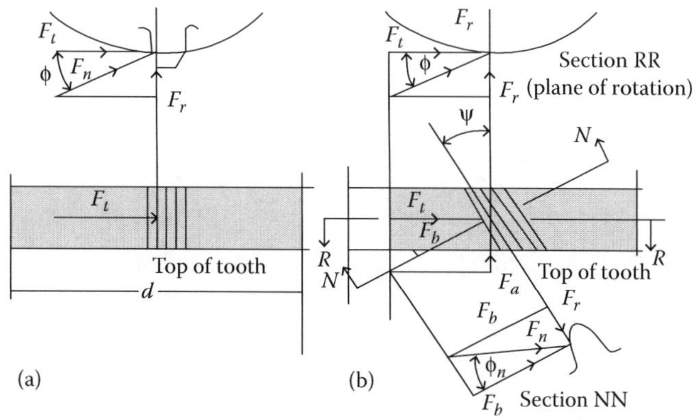

Figure 8.12

Comparison of force components on (a) spur and (b) helical gears.

Figure 8.12 illustrates the tooth forces acting on spur and helical gears. For spur gears, the total tooth force consists of tangential F_t and radial F_r forces. For helical gears, component F_a is added, and normal section *NN* is needed to show a true view of total tooth force F_n.

The vector sum F_t and F_a is labeled F_b; the subscript *b* is chosen because F_b is the bending force on the helical tooth (just as F_t is the bending force on the spur tooth).

The force component associated with power transmission is only F_t.

$$F_t = \frac{1000W}{V} \tag{8.18}$$

where:

- F_t is in Newtons
- W is in kilowatts
- V is velocity in meters per second

$$F_b = F_t / \cos(\varphi) \tag{8.19}$$

$$F_r = F_b \tan(\phi_n) \tag{8.20}$$

$$F_r = F_t \tan(\phi) \tag{8.21}$$

Combining Equations 8.15, 8.20 and 8.21,

$$\tan(\phi_n) = \tan(\phi).\cos(\varphi) \tag{8.22}$$

8.8.3.1 Helical Gears Force Analysis: Bending Stress

The bending stress equation for helical gear teeth is given as

$$\sigma = \frac{F_t}{bmJ} K_v K_o (0.93 K_m) \tag{8.23}$$

The introduction of constant 0.93 with the mounting factor reflects the slightly lower sensitivity of helical gears to mounting conditions. The J factor can be determined from Figures 8.13 and 8.14.

The velocity factor K_v is calculated from Equation 8.21 and Figure 8.15.

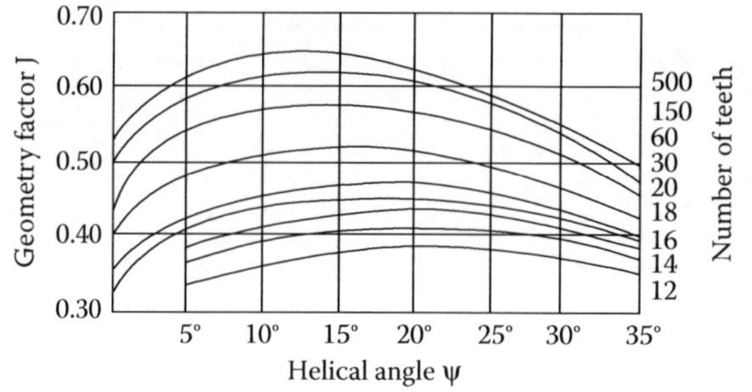

Figure 8.13

Geometry factor J for helical gear with 03C6n=20° and mating with 75 tooth gear.

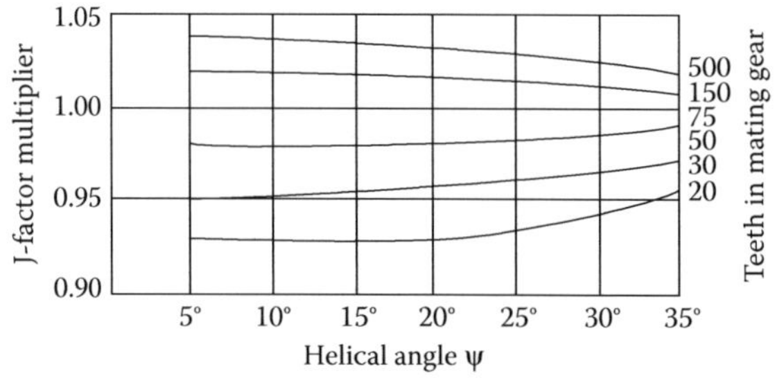

Figure 8.14

J-factor multiplier when the mating gear has more or fewer than 75 teeth.

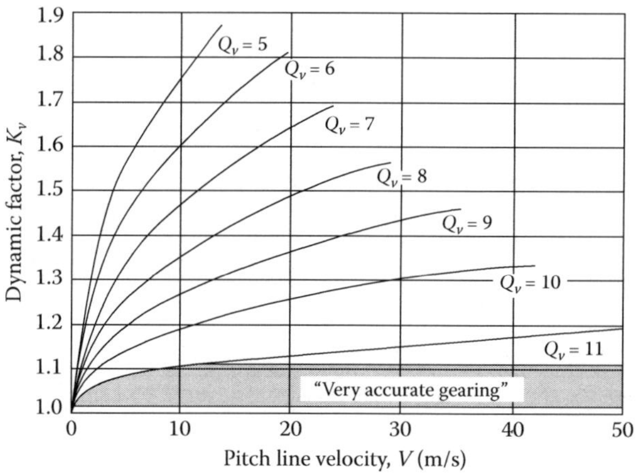

Figure 8.15

Dynamic factors for the helical gear tooth.

$$K_v = \left[\frac{78 + (200V)^{0.5}}{78}\right]^{0.5} \tag{8.24}$$

K_o is the overload factor, which reflects the degree of nonuniformity of driving and load torques. It is given in Tables 8.2 and 8.3.

8.8.3.1.1 Helical Gear: Permissible Tooth Bending Stress (AGMA)

Fatigue strength (σ_e) of the material is given by

$$\sigma_e = \sigma_e' \, k_L \, k_V \, k_S \, k_r \, k_T \, k_M \tag{8.25}$$

Table 8.2 Overload Factor K_o

Source of Power	Driven Machinery: Uniform	Moderate Shock	Heavy Shock
Uniform	1.00	1.25	1.75
Light shock	1.25	1.50	2.00
Medium shock	1.50	1.75	2.25

Table 8.3 Load Distribution Factor K_m

Characteristics of Support	Face Width (mm): 0–50	150	225	400 up
Accurate mountings, small bearing clearances, minimum deflection, precision gears	1.2	1.3	1.4	1.7
Less rigid mountings, less accurate gears, contact across the full face	1.5	1.6	1.7	2.0
Accuracy and mounting such that less than full-face contact exists	Over 2.0	Over 2.0	Over 2.0	Over 2.0

Table 8.4 Reliability Factor k_r

Reliability factor R	0.50	0.90	0.95	0.99	0.999	0.9999
Factor k_r	1.000	0.897	0.868	0.814	0.753	0.702

where:

σ'_e = Endurance limit of rotating-beam specimen

k_L = Load factor = 1.0 for bending loads

k_V = Size factor = 1.0 for $m < 5$ mm

k_M = 0.85 for $m > 5$ mm

k_S = Surface factor is based on the ultimate strength of the material and for cut, shaved, and ground gears

k_r = Reliability factor, given in Table 8.4

k_L = Temperature factor = 1 for $T \leq 120°C$; for >120°C, an attempt should be made to lower the temperature below 120°C through heat exchanges more than 120°C

$k_T < 1$ to be taken from AGMA standards

k_M = Factor for miscellaneous effects (Figure 8.16). For idler gears subjected to two-way bending, = 1. For other gears subjected to one-way bending, the value is taken from Figure 8.13. Use $k_M = 1.33$ for $\sigma_{out} < 1.4$ GPa

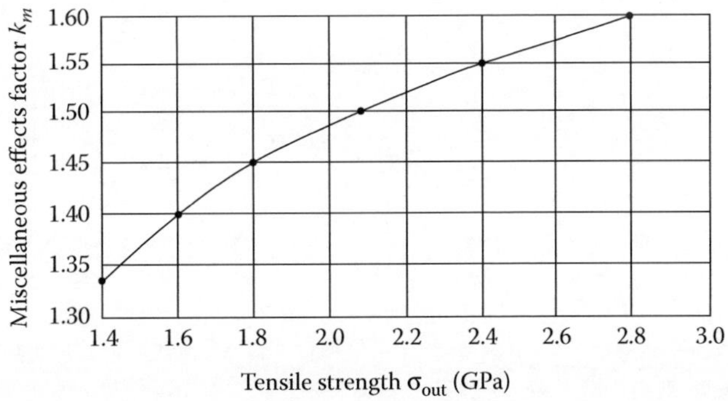

Figure 8.16

Miscellaneous effects factor k_M.

Permissible bending stress

$$\sigma = \frac{\sigma_e}{S} \tag{8.26}$$

Hence, the design equation from bending considerations is

$$\sigma \le [\sigma] \tag{8.27}$$

8.8.3.2 Helical Gear: Contact Stress

In the case of spur gears of contact ratio (*CR*) less than 2, the theoretical length of tooth contact is 1.0. With helical gears, the length of contact per tooth is $b/\cos\psi$, and the helical action causes the total length of tooth contact to be approximately $b/\cos\psi$ times the *CR* at all times. The AGMA recommends that 95% of this value be taken as the length of contact when computing contact stress.

The contact stress equation is given as

$$\sigma_c = C_p \sqrt{\frac{W^t}{FdI}\left(\frac{\cos(\psi)}{0.95CR}\right) K_V K_o (0.93 K_m)} \tag{8.28}$$

The elastic coefficient factor C_p is given by (Table 8.5)

$$C_p = 0.564 \sqrt{\frac{1}{\frac{1-\mu_p^2}{E_p} + \frac{1-\mu_g^2}{E_g}}} \tag{8.29}$$

μ and E are Poisson ratio and Young's modulus, respectively. Subscripts p and g refer to pinion and gears, respectively.

The geometry factor I is given by

$$I = \frac{\sin(\phi)\cos(\phi)}{2} \frac{i}{i+1} \tag{8.30}$$

where:

Speed ratio $i = n_1/n_2 = d_2/d_1$

ϕ is the transverse contact angle

k_V, k_o, and K_M as taken for bending stress calculation.

The *CR* is given by

$$CR = \left(\frac{\sqrt{(r_p + a)^2 - r_{bp}^2 + (r_g + a)^2 - r_{bg}^2} - (r_p + r_g)^2 \sin\phi}{\pi.m.\cos(\phi)} \right) \tag{8.31}$$

Table 8.5 Elastic Coefficients *Cp* for Spur Gears and Helical Gears ($\sqrt{\text{MPa}}$)

	Gear Materials			
Pinion Materials	Steel	Cast Iron	Al Bronze	Tin Bronze
Steel, $E = 207$ GPa	191	166	162	158
Cast iron, $E = 131$ GPa	166	149	149	145
Al bronze, $E = 121$ GPa	162	149	145	141
Tin bronze, $E = 110$ GPa	158	145	141	137

Table 8.6 Surface Fatigue Strength σ'_{st} (MPa) for Metallic Spur Gears

Material	σ'_{st} MPa
Steel	2.8(BHN)—69 MPa
Nodular iron	0.95(2.8[BHN]−69MPa)
Cast iron grade 20	379
Cast iron grade 30	482
Cast iron grade 40	551
Tin bronze AGMA 2C (11% Sn)	207
Aluminum bronze (ASTM B 148-12) (Alloy 9C-H.T.)	448

Note: 107 life cycles with 99% reliability and temperature <120°C

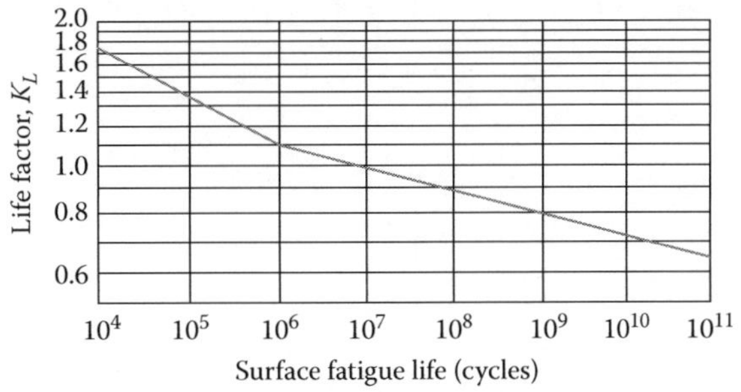

Figure 8.17

Life factor K_L.

where:

- R is the pitch circle radius
- r_b is the base circle radius, suffix p for pinion and g for gear
- a is the addendum
- ϕ is the transverse pressure angle

8.8.4 Helical Gear: Surface Fatigue Strength

Due to fluctuation of load and direction of forces, fatigue failure is common in helical gears. Surface fatigue strength of the material is given by

$$\sigma_{st} = \sigma'_{st} K_L K_H K_R K_T \tag{8.32}$$

where:

- σ_{st} = surface fatigue strength of the material, given in Table 8.6
- K_L = life factor, given in Figure 8.17.
- K_H = hardness ratio factor; may be taken as 1.0
- K = ratio of Brinell hardness of the pinion to Brinell hardness of the gear: K_H = 1.0 for $K < 1.2$
- K_R = reliability factor, given in Table 8.7

The generation of heat during the operation of gear is a common phenomenon and is normally estimated based on lubricant temperature. K_T = temperature factor = 1 for $T \leq 120°C$, based on lubricant temperature. Above 120°C, it is less than 1, to be taken from AGMA standards.

8.8.5 Helical Gear: Allowable Surface Fatigue Stress (AGMA)

Allowable surface fatigue stress for design is given by

Table 8.7 Reliability Factor

Reliability (%)	50	99	99.9
k_R	1.25	1.00	0.80

$$\sigma_H = \frac{\sigma_{st}}{n} \tag{8.33}$$

where n is a factor of safety in the range 1.1–1.5. These values can sometimes change for a particular gear operation and use. The value should be adjusted according to engineers' experience and expertise.

Example 8.2: Helical Gear Design

Design a two-stage helical gear box to transmit 10 KW power from an electric motor rotating at 1440 rpm to a machine with approximately overall reduction ratio of 10. Design a spur gear pair based on bending and wear strength considerations. Check the gears for dynamic load using Buckingham's equation. Sketch the gear with all constructional features.

Design Calculation

- Given data: Two-stage gear box
- Design based on bending and wear strength
- Check the gears for dynamic load
- Power: $p = 10$ KW
- Speed: $N = 1440$ rpm
- Transmission ratio: $i' = 10$

In a multistage gear box consisting of two or three stages, the velocity ratio at each stage should not exceed 6:1. In this case, the intermediate speeds are arranged in geometric progression. If i' is the total transmission ratio, then the speed reduction at each stage (i) is obtained in the following way:

1. Calculation of velocity ratio in each stage
   For two stages, $i = \sqrt{i'} = \sqrt{10} = 3.16$ (velocity ratio)
2. Selection of standard speed reduction ratio (PSG 8.12)
   Velocity ratio $(i) = 3.16$
3. Selection of standard tooth profile
   200-Full depth involute system
4. Selection of number of teeth on pinion
   $Z_p = 18$ (in order to avoid interference)
5. Calculation of number of teeth on gear
   We know that
   $$Z_g = 57 \text{ (one hunting tooth)}$$
6. Selection of material for pinion and gear (PSG 8.4)
   When speed ratio less than 4,
   Pinion: 40Ni2Cr1Mo28
   Gear: 15Ni2Cr1Mo15
7. Selection of design bending stress and design surface stress, kgf/cm^2
   By considering module (m) up to 6,
   $$[\sigma_b]p = 4000 \text{ kgf/cm}^2 [\sigma_c]p = 11{,}000 \text{ kgf/cm}^2$$
   $$[\sigma_b]g = 3200 \text{kgf/cm}^2 [\sigma_c]g = 9500 \text{kgf/cm}^2$$
8. Selection of Lewis form factor for pinion y_p and for gear y_g [1]
   $Z_p = 18, y_p = 0.377$
   $$Z_g = 57, y_g = 0.480 \text{ (by using interpolation formula)}$$

9. Decide the weaker element
Beam strength of gear tooth (Lewis equation): $Sb = m.\ b.\ \sigma_b.\ y$
Beam strength for pinion: $[\sigma_b\ x\ y]p = 4000 \times 0.377 = 1507$ kgf/cm^2
Beam strength for pinion: $[\sigma_b\ x\ y]g = 3200 \times 0.480 = 1536$ kgf/cm^2
From the above calculation, the pinion is weaker than the gear
10. Calculation of nominal twisting moment (M_t), kgf.cm
$M_t = 97{,}420\ \text{KW}/n = 97{,}420 \times 10/1440 = 676.52$ kgf.cm
11. Calculation of design twisting moment (M_t), kgf.cm
$(M_t) = M_t.\ Kd.K$ for symmetric scheme $Kd.K = 1.3 = 676.52 \times 1.3 = 879.48$ kgf.cm
12. Selection of $\vartheta_m = b/m$ (face width to module ratio) = 10
13. Calculation of module (m)
Formula based on beam (bending) strength of pinion tooth

$$m > 1.263\sqrt[3]{\frac{[M_t]}{y.[\sigma_b].\psi_m Z_p}}$$

$$m > 1.263\sqrt[3]{\frac{[879.4]}{0.377.[3200].10.18}}$$

$m = 2.008$ mm
14. Calculation of equivalent Young's modulus (E_{equ})
For steel, $E_{equ} = 2.15 \times 106$ kgf/cm^2
15. Calculation of center distance (a)
16. $$a = m\left(\frac{Z_p + Z_g}{2}\right) = 2\left(\frac{18+57}{2}\right) = 75\ \text{mm}$$

Calculation of face width (b)
$b = 10\ m = 10 \times 2 = 20$ mm
17. Calculation of surface (contact, compressive) strength (Mischke)
Check the module for wear

$$\sigma_c = 0.74\frac{i+1}{a}\sqrt{\frac{i+1}{i.b}E_{\text{equ}}[M_t]}$$

$$\sigma_c = 0.74\frac{3.16+1}{7.5}\sqrt{\frac{3.16+1}{3.16.20}2.15.10^6[879.14]} = 14483.48\ \text{kgf/cm}^2$$

$[\sigma_c]_p = 11{,}000$ kgf/cm^2

From the above calculation, it is observed that the design surface stress is higher than the actual surface stress, so that gear tooth failure occurs due to wear. Therefore, it becomes necessary to increase the module of the gear tooth in order to avoid failure of the tooth.

8.9 Crossed Helical Gear

Crossed helical gear is shown in Figure 8.18.

1. Crossed helical gears are identical to other helical gears but are mounted on nonparallel shafts. They are basically nonenveloping worm gears, since the gear blanks have a cylindrical form.
2. The relationship between the shaft angle and the helix angles of mating gears is

$$\sigma = \psi_1 \pm \psi_2 \tag{8.34}$$

where:
σ = the shaft angle
+ sign is used when the gears have the same hand
− sign when they have the opposite hand

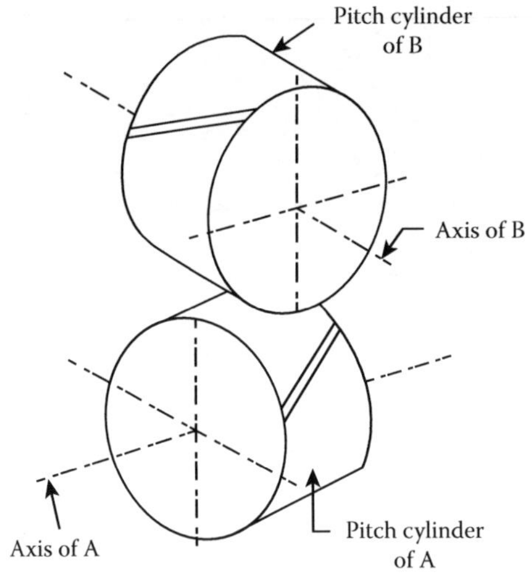

Figure 8.18

View of the pitch cylinder of a pair of crossed helical gears.

3. Opposite hand-crossed helical gears are used when the shaft angle is small.
4. The most common shaft angle is 90°, which results in mating gears with complementary helix angles of the same hand.
5. The action of the crossed helical gears differs fundamentally from that of parallel helical gears in that the mating teeth slide across each other as they rotate.
6. The sliding velocity increases with increasing shaft angle.
7. For a given shaft angle, the sliding velocity is lowest when the two helix angles are the same.
8. Mating crossed helical gears must have the same p_n and ϕ_n but not necessarily the same p and ϕ.

   The pitch diameter d is

$$d = mZ = \frac{m_n Z}{\cos(\varphi)} \tag{8.35}$$

9. Furthermore, the velocity ratio is not necessarily the ratio of pitch diameters; it must be calculated as the ratio of the numbers of teeth.

8.9.1 Crossed Helical Gear: Design Tips

Due to the different nature and direction of forces in crossed helical gear applications, the designer should note the following suggestions:

1. Crossed helical gears have very low load-carrying capacities—usually less than a resultant tooth load of 400 N.
2. The limitation is one of surface deterioration, not bending strength.
3. Since they have point contact, to increase the load capacity, *CR*s of 2 or more are usually used.
4. Low values of pressure angle and relatively large values of tooth depth are commonly specified to increase the *CR*.
5. There are no standards for crossed helical gear tooth proportions. Many different proportions give good tooth action.

8.10 Worm Gear Design

Advantages of worm gear drives and self-locking properties:

- The most important characteristic of worm gear drives is their high-speed reduction. A speed reduction as high as 100:1 can be obtained with a single pair of worm wheels.
- Worm gear drives are compact with small overall dimensions, compared with equivalent spur or helical gear drives having the same speed reduction.

- The operation is smooth and silent.
- Provision can be made for self-locking operation, whereby the motion is transmitted only from the worm to the worm wheel. This is advantageous in applications such as cranes and lifting devices.

8.10.1 Materials Used for Worm and Worm Wheel

Worms are made of case-hardened steel with a surface hardness of 60 HRC and a case depth of 0.75–4.5 mm. The following varieties of steel are used for the worm:

Normalized carbon steels: 40C8, 55C8
Case-hardened carbon steels: 10C4, 14C8
Case-hardened alloy steels
Nickel-chromium steels

Worm wheel material should be soft and confirmable. Phosphor bronze with a surface hardness of 90–120 BHN is widely used for the worm wheel.

Example 8.3: Worm Gear Design

A two-tooth right-hand worm transmits 2 kW at 2950 rpm to a 32-tooth worm gear. The worm gear is of 4 mm module, 20° pressure and a face width of 30 mm. Pitch diameter of the worm is 50 mm with a face width of 65 mm. The worm is made of steel case that is carburized, oil quenched, tempered, and ground. The worm gear is made of phosphor bronze.

1. Find the center distance, the lead, and the lead angle.
2. Find the bearing reactions on the worm gear and worm shaft and the torque output.
3. Find the efficiency.

The general arrangement and isometric views are shown in Figures 8.19 and 8.20.

Given that

$W = 2\,\text{kW}, n_1 = 2950\text{ rpm}, Z_1 = 2, Z_2 = 32, m = 4\text{ mm}, \varphi = 20°, d_1 = 50\text{ mm}, b_1 = 65\text{ mm}, b_2 = 28\text{ mm}$

Pinion material: case carburized steel; gear material: phosphor bronze
Find: (a) $C=?$, $L=?$, $\lambda=?$, Q- (b) bearing reactions? $T_2=?$ and Q (c) $\eta=?$

Solution: (a)

$d_2 = m\,Z_2 = 4\times 32 = 128$ mm
Center distance $C = 0.5(d_1 + d_2) = 0.5(50 + 128) = 79$ mm
Axial pitch: $p = \pi.m = 3.14\times 4 = 12.56$ mm
Lead: $L = p\,Z_1 = 12.56\times 2 = 25.12$ mm
Lead angle: $\lambda = \tan^{-1}(L/\pi\,d_1) = \tan^{-1}(25.12/\pi\times 50) = 9.09°$

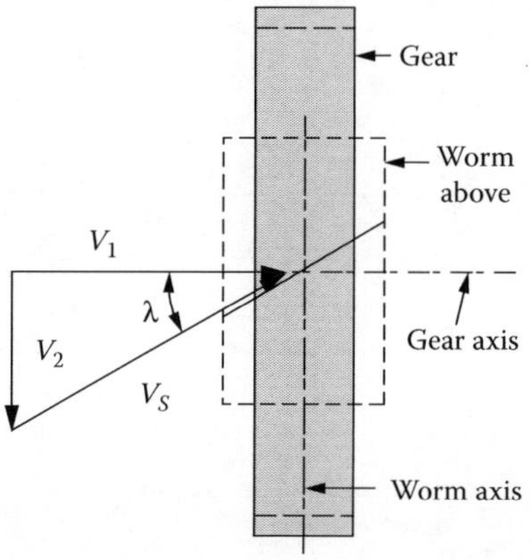

Figure 8.19

General arrangement of the worm drive.

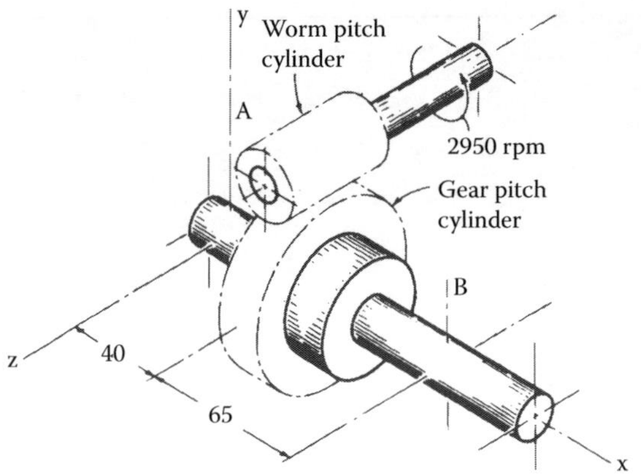

Figure 8.20

Isometric view of the worm gears in mesh. (From Budynas, R. G., et al. *Mechanical Engineering Design*, 8th edn., McGraw Hill, 2008.)

Solution: (b)

$$V_1 = V_m = \frac{\pi.d_1.n_1}{6000} = \pi\times 50\times 2950 / 60{,}000 = 7.72\,\text{m/s}$$

$$n_2 = n_1 / i = \{n_1(Z_2/Z_1)\} = 2950/(32/2) = 184.38 \text{ rpm}$$

$$V_2 = (\pi d_2 n_2 / 60{,}000) = \pi\times 128\times 184.38/60{,}000 = 1.24\,\text{m/s}$$

$$V_S = V_1/\cos\lambda = 7.72/\cos 9.09° = 7.82\,\text{m/s}$$

For sliding velocity V_S=7.82 m/s and the given materials, the coefficient of friction f=0.024 (Figure 8.21).

Since the helix angle of the gear is the same as the lead angle of the worm,

$$\Phi_n = \tan^{-1}(\tan\varphi.\ \cos\psi) = \tan^{-1}(\tan 20°\ \cos 9.09°) = 19.77°$$

$$F_{t1} = W/V_1 = 2000/7.72 = 259\ N$$

$$F_n = \frac{F_t}{\cos\varphi\cdot\sin\lambda + f\cdot\cos\lambda} = \frac{259}{\cos\ 19.79\cdot\sin\ 9.09 + 0.024\cdot\cos\ 9.09} = 1503\ N$$

$$F_{r1} = F^y = F_n\ \sin\varphi_n = 1503\ \sin 19.77° = 508\ N$$

$$F_{a1} = F^z = F_n(\cos\varphi_n\cos - f\sin\lambda) = 1503\ (\cos\ 19.77°\ \cos 9.09 - 0.024\ \sin\ 9.09°) = 1391\ N$$

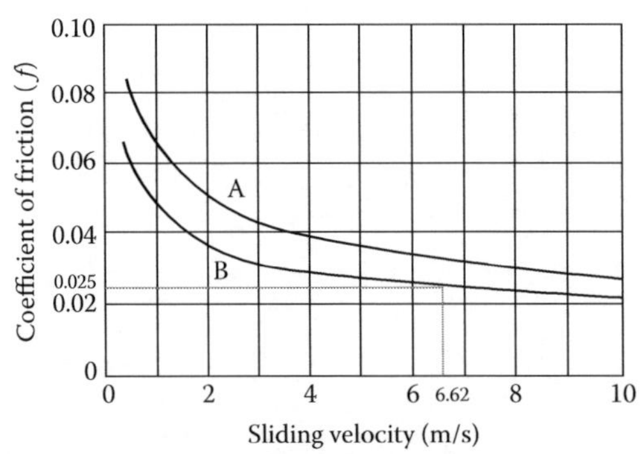

Figure 8.21

Sliding velocity vs. coefficient of friction. Friction of well-lubricated worm gears, A for cast iron worm and gear and B for case-hardened steel worm and phosphor bronze worm gear.

8.10.2 Worm Gear Force Analysis

Referring to Figure 8.22, we can now write down the forces acting on the worm gear tooth.

$$F_{t2} = F_{a1} = 1391\,N, \text{and } F_{a1} = F_{t2} = 259\,N\, F_{r2} = F_{r1} = 508\,N$$

Since Bearing B takes the entire thrust load (Figures 8.23 and 8.24),

$$F_B^X = F_{a2} = 259\text{N}$$

Taking moment about the z-axis through A, we get $F_B^y = 351\,N$ and then we can find $F_a^Y = 508 - 351 = 157$ N
By taking moment about y-axis through A, we have

$$F_{t2} \times 40 - F_B^z \times 105 = 0$$

$$F_B^z = 530\text{ N, from which } FA^Z = 1391 - 530 = 861\text{ N}$$

$$T = F_{t2} \times r_2 = 1391 \times 64 \times 10^{-3} = 89.02\text{ Nm}$$

Since the bearing at C takes the entire thrust (Figure 8.25),

$$F_C^z = F_{a1} = 1391\text{ N}$$

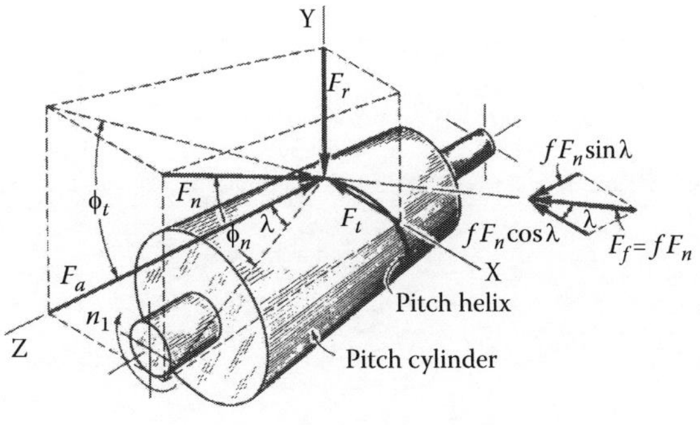

Figure 8.22

Forces on the worm gear tooth on the pitch cylinder. (From Budynas, R. G., et al. *Mechanical Engineering Design*, 8th edn., McGraw Hill, 2008.)

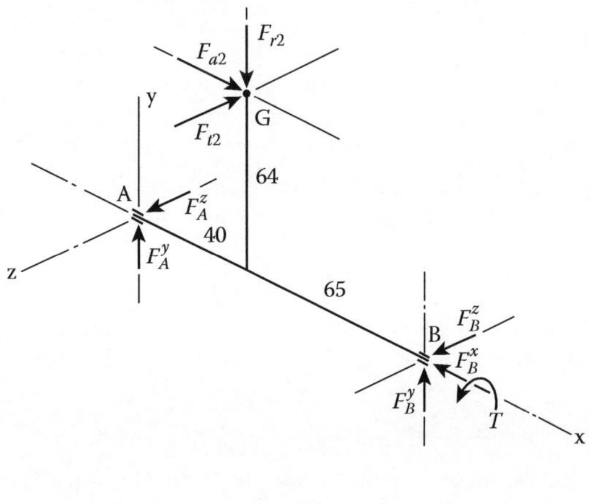

Figure 8.23

Sketch showing the forces acting on the worm gear shaft.

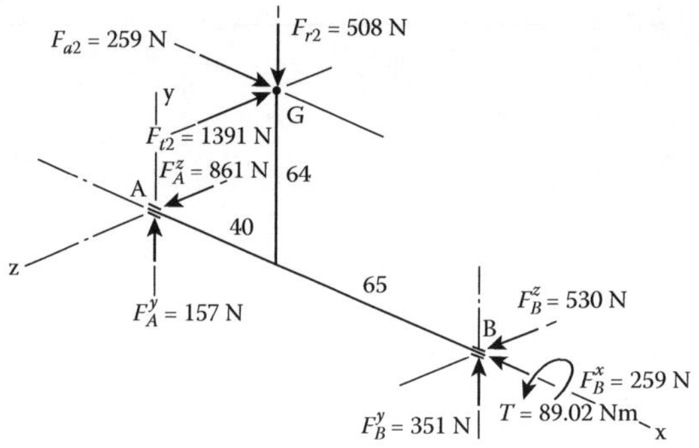

Figure 8.24

Sketch showing the calculated value of forces acting on the worm gear shaft.

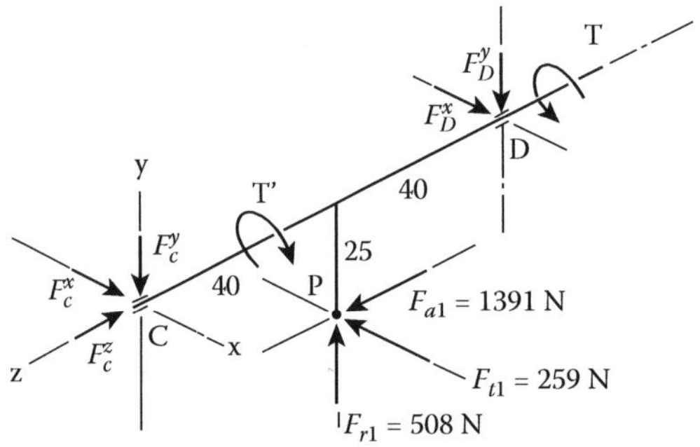

Figure 8.25

Forces acting on the pinion shaft and bearing reactions.

and taking moment about y (vertical) axis through D, we get

$$F_c^x = 129.5 \text{ N and } F_D^x = 129.5 \text{ N}$$

Taking moment about x (horizontal) axis through D,

$$F_c^y \times 80 - F_{a1} \times 25 - F_{r1} \times 40 = 0$$

$$80\ F_c^y - 1391 \times 25 - 508 \times 40 = 0$$

and so we get $F_C^y = 689\,\text{N}$ from which we can find $F_D^y = -181\,\text{N}$.

Solution: (c) Efficiency of the Gearbox

The efficiency of the gearbox is given by η (Figure 8.26)

$$\eta = \frac{\cos\ \phi_n - f \cdot \tan\lambda}{\cos\ \phi_n + f \cos\ \lambda} = \frac{\cos\ 19.77 - 0.024\ \tan 9.09}{\cos\ 19.77 + 0.024 \cot 9.09} = 0.859$$

Example 8.4

Design a worm gear set to deliver 12 kW from a shaft rotating at 1500 rpm to another rotating at 75 rpm.

Solution

A 20° normal pressure angle worm gear is assumed, for which the lead angle should not exceed 25° (Table 8.8) and Z_2 minimum is 21 (Table 8.9). Allowing 6° lead per thread of the worm, the worm could have four or fewer teeth. $Z_1 = 4$ or quadruple-threaded worm is assumed.

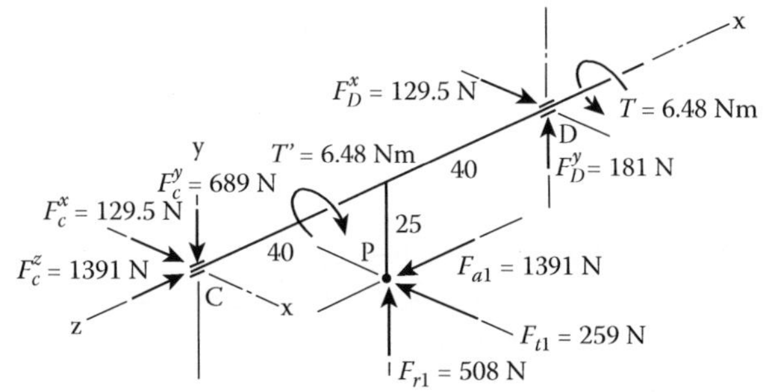

Figure 8.26

Calculated values of forces acting on the pinion shaft and bearing reactions. (From Budynas, R. G., et al. *Mechanical Engineering Design,* 8th edn., McGraw Hill, 2008.)

Table 8.8 Maximum Worm Lead Angle and Worm Gear Lewis Form Factor for Various Pressure Angles

Pressure Angle, ϕ_n (°)	Maximum Lead Angle, λ (°)	Lewis Form Factor, y	Modified Lewis Form Factor, Y
14.5	15	0.1	0.314
20	25	0.125	0.393
25	35	0.15	0.473
30	45	0.175	0.55

Table 8.9 Minimum Number of Teeth in the Worm Gear

Pressure angle, ϕ_n	14.5	17.5	20	22.5	25	27.5	30
Minimum Z_2	40	27	21	17	14	12	10

From the worm gear design guidelines we have

$i = n_1/n_2 = 1500/75 = 20 = Z_2/Z_1$

$\omega_1 = 2\pi n_1/60 = 157$ rad / s, $Z_2 = i \times Z_1 = 20 \times 4 = 80$

A center distance of 250 mm (as per R10 series) is assumed.

$$d_1 \geq C^{0.875}/3 = 250^{0.875}/3 \geq 42\,\text{mm}$$

and

$$\frac{C^{0.875}}{3} \leq d_1 \leq \frac{C0.875}{1.75}$$

As we can calculate for C = 250 mm, $42 \leq d_1 \leq 75$, we can assume any value of d_1 in the range. We assume $d_1 = 72$ mm. As we can find circular pitch (p_2) as $d_1 \approx 4p_2$, we get $p_2 = 18$ mm. We can now find module m as $m = p_2/\pi = 18/3.14 = 5.73$ mm and take standard module of 6 mm.

Hence, $d_2 = m\,Z_2 = 6 \times 80 = 480$ mm and actual center distance $C = 0.5\,(d_1 + d_2) = 0.5(72 + 480) = 276$ mm.

Check for $d_1 \leq C^{0.875}/1.7 \leq 80.4$ mm, $d_1 = 80$ mm is taken.

$C = 0.5(d_1 + d_2) = 0.5(80 + 480) = 280$ mm

Lead $= N_{tw} \times p_a = 4 \times 18.84 = 75.36$ mm, and so

$\tan \lambda = L/\pi\, d_1 = 75.36/3.14 \times 72 = 0.3333$

$\lambda = 18.43° = \psi$

$\omega_2 = (2\pi.n_2/60) = (2 \times 3.14 \times 75/60) = 7.85$ rad/s

$V_2 = \omega_2\, r_2 = 7.85 \times (0.5 \times 480) \times 10^{-3} = 1.884$ m/s

$F_t = 1000\ W/V = 1000 \times 12/1.884 = 6370$ N

$b \leq 0.5\ d_{a1}$, $b \leq 0.5(d_1 + 2\ m) \leq 0.5 \times (80 + 2 \times 6) \leq 46$, so $b = 45$ is assumed

$Y = 0.393$ from Table 8.8 for $\varphi_n = 20°$

$$F_d = F_{2t}\left(\frac{6.1 + V_2}{6.1}\right) = 8133\text{N}$$

Choosing phosphor bronze for the gear and heat-treated C45 steel for the ground worm, $[\sigma_b] = 80$ MPa from Table 8.10.

Beam strength of the worm gear $F_b = \sigma_b \cdot b \cdot m \cdot Y = 8489$N.

Worm gear bending and surface fatigue strengths are given in Table 8.10.

F_b (8489) > F_d (8133). Hence, the design is safe from bending fatigue considerations.

Check for the wear strength.

$F_w = d_2.b.K_w = 480.(45).(0.518) = 11{,}189$ N

=

$K_w = 0.518$ for steel worm vs. bronze worm gear with $\lambda < 25°$ from Table 8.11.

F_w (11,189) > F_d (8133). Hence, the design is safe from wear strength consideration also.

AGMA recommendation for the axial length of the worm is

$$L_W = p_a\left(4.5 + \frac{Z_2}{50}\right) = 8.84\left(4.5 + \frac{80}{50}\right) = 115 \text{ mm}$$

Worm velocity $V_1 = \omega_1\, r_1 = 157 \times 0.04 = 6.28$ m/s

$$V_s = \frac{V_1}{\cos\lambda} = \frac{6.28}{\cos 18.43} = 6.62 \text{ m/s}$$

From the worm gear friction characteristics shown in Figure 8.21, for $V_s = 6.62$ m/s and $f = 0.025$,

$$F_n = \frac{F_{t2}}{\cos\phi_n . \cos\lambda} = \frac{6370}{\cos 20^0 . \cos 18.43^0} = 7145\text{N}$$

$\eta = 94.8\%$

$$\eta = \frac{\cos\phi_n - f.\tan\lambda}{\cos\phi_n + f\cos\lambda} = \frac{\cos 20 - 0.025\tan\ 18.43}{\cos 20 + 0.025\cot\ 18.43} = 0.948$$

$\eta = 94.8\%$

Heat generated during operation: $H_g = (1-\eta)W = (1-0.918)12{,}000 = 984$ Nm.

Surface area A for conventional housing designs may be roughly estimated from this equation:

$$A = 14.75\ C^{1.75}$$

where:

- A is in m^2
- C (the distance between the shafts) is in m
- Therefore, $A = 14.75(0.28)^{1.75} = 1.694$ m^2.

Table 8.10 Permissible Stress in Bending Fatigue

Materials of the Gear	σ_b (MPa)
Centrifugally cast Cu–Sn bronze	23.5
Phosphor bronze	80
Aluminum alloy, Al–Si	11.3
Zn alloy	7.5
Cast iron	11.8

Table 8.11 Worm Gear Wear Factors k_w

Material		kW (MPa)		
Worm	Gear	$\lambda<10$	$\lambda<25$	$\lambda>25$
Steel, 250 BHN	Bronze	0.414	0.518	0.621
Hardened steel (Surface	Bronze	0.552	0.690	0.828
500 BHN)	Chill-cast bronze	0.828	1.036	1.243
Cast Iron	Bronze	1.036	1.277	1.553

8.11 Bevel Gear Design

8.11.1 Nomenclature of Bevel Gears (Figure 8.27)

- Back angle, H. Face angle, P. Pitch angle
- Back cone angle, I. Face width, Q. Pitch cone apex
- Back cone distance, J. Front angle, R. Pitch cone apex to crown
- Clearance, K. Mean cone distance, S. Pitch diameter
- Crown point, L. Midface, T. Root angle
- Crown to back, M. Mounting distance, U. Shaft angle
- Dedendum angle, N. Outer cone distance, V. Equivalent pitch radius
- Outside diameter

Example of typical bevel gear dimensions by Miter:

Gears	Miter
Number of teeth (Z_0)	31
Module	3.4031 mm
Spiral angle	35°0′
Hand of spiral	Left hand
Pitch cone angle η	45°0′
Mean cone distance R_m	91.07 mm

Figure 8.28 shows two views of a bevel gear with dimensions, and Figure 8.29 shows orthographic views of a bevel pinion.

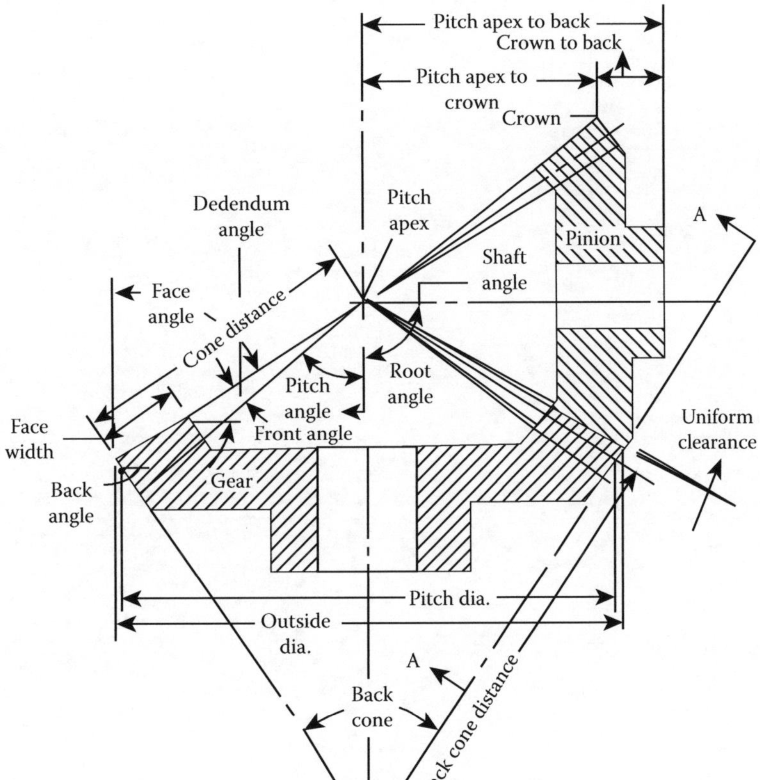

Figure 8.27

Nomenclature of bevel gear.

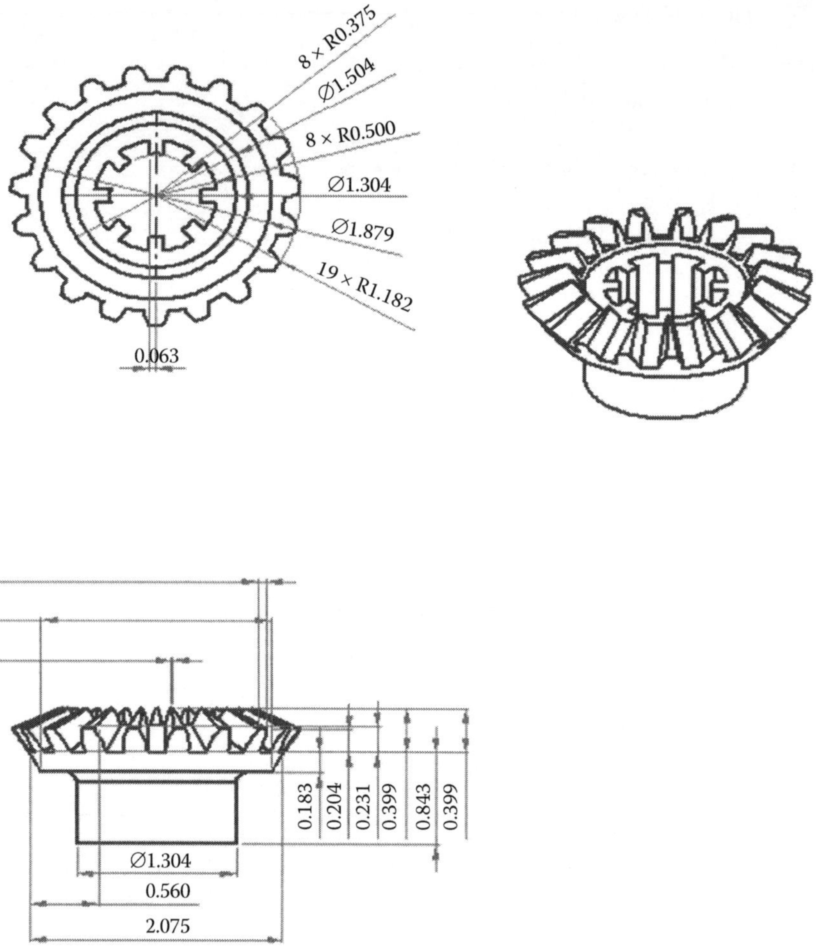

Figure 8.28

Bevel gear orthographic views.

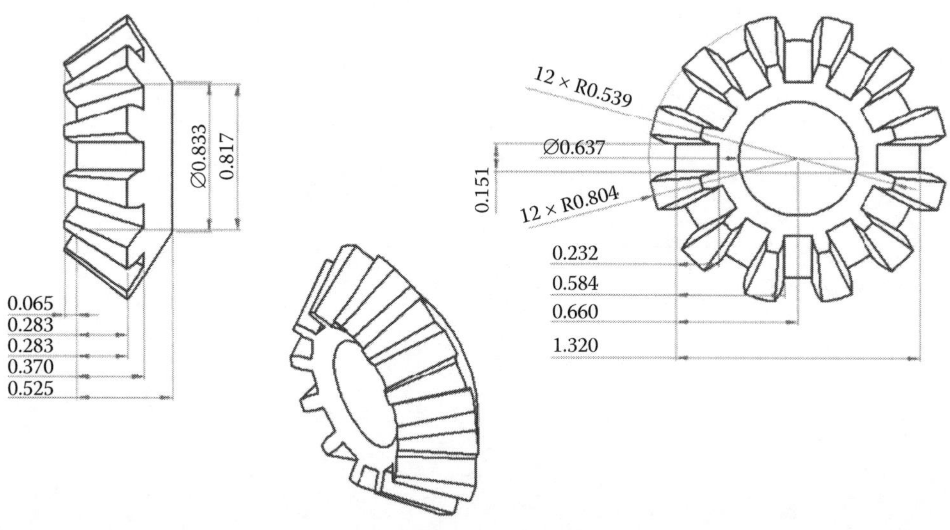

Figure 8.29

Orthographic views of bevel pinion.

Example 8.5 Gearbox Design

A complete design of gear is presented below. It is presented in step-by-step fashion and every detail of the gearbox is calculated. Shaft design, selection of environmentally friendly lubricant, and estimated heat losses are included.

1. Design of gears is based on beam strength, pitting, and scuffing (high-speed gears) considerations. The minimum pitch diameter of the pinion should be

   $d_{1\min} = 2 \times \text{bore} + 0.25\ m$

   where:

   d is the bore diameter
   m is the module expressed in mm.
2. As per Nuttall Works of the Westinghouse Electric and Manufacturing company, the minimum thickness of metal $t_{\min}$ between the keyway and the root circle shall be

   $$t_{\min} = m\sqrt{\frac{Z}{5}}$$
3. The outside diameter of the hubs in larger gears should be 1.8 times the bore for steel, 2 times for custom industrial (CI), and 1.65 times for forged steel. The hub length should be at least 1.25 times the bore and never less than the width of the gear.
4. Design of the shaft is based on fatigue strength and rigidity considerations. (At the contact region, the deflection of the shaft should be less than 0.01 module, the slope of the shaft at the radial bearing should be less than 0.008 radians, and for self-aligning bearings it should be less than 0.05 radians.)
5. Bearings selection is based on 90% reliability for the following life:

   8 h operation per day life = 20,000–30,000 h.
   8–16 h operations per day life = 40,000–50,000 h.
   16–24 h operations per day life = 50,000–60,000 h.
6. Selection of lubricant is based on peripheral velocity, load, type of application, and operating temperature, and so on. SAE 30, 40, 50, 60, 80, 90 are recommended. For low-friction and high-temperature operation, synthetic oils are used. However, for green engineering considerations, we should select lubricants that are environmentally suitable.
7. Selection of method of lubrication is based on the peripheral speed. Up to a peripheral speed of 15 m/s, oil bath (oil immersion/splash) lubrication is used. Greater depth of immersion is recommended for low-speed application. The maximum depth should not exceed 100 mm. At higher speed, the depth of immersion is reduced to cut churning losses: 0.7 tooth height or a minimum of 1 module is taken. However, the depth of immersion should not be less than 10 mm. The generally recommended depth of immersion is 3 to 4 times the module, to a maximum of 6 modules. Above a peripheral velocity of 15 m/s, stream (jet) lubrication is used. Oil is delivered by a pump through a filter and, if necessary, through a cooler directly to the teeth of the gears as they are coming out of the mesh.
8. The quantity of oil required is shown below by rule to thumb (it differs widely from application to application).

   For splash lubrication, $Q = (2.5 - 8)\ L_t$ (L) or

   $$Q = (0.35 - 0.7)\ W\ \text{(L)}$$

   For spray lubrication, rate of supply should be

   $$Q_e = 30\ (L_t/\Delta T)\ \text{(L/m)}$$

   where:

   L_t = loss of power at the teeth contact (kW)
   W = power transmitted (kW)
   ΔT is the difference in oil temperature between the outlet and inlet in °C

   Oil circulation time $t' = \dfrac{Q}{Q_e}$ min

Table 8.12 Alternate Approach for Wall Thickness *s* for Gearboxes

	Noncase-Hardened Gears (mm)	Case-Hardened Gears (mm)
CI castings	$0.007\,L+6$	$0.010\,L+6$
Steel castings	$0.005\,L+4$	$0.007\,L+4$
Welded construction	$0.004\,L+4$	$0.005\,L+4$

Note: L is the largest dimension of the housing

$t'=1$–2.5 for splash lubrication with no external circulation, where space is lacking in a compact design
$t'=4$–30 for oil circulation with cooling or reservoir

9. Gear box housing design is based on rule of thumb and thermal consideration. The wall thickness of the CI housing can be found from the empirical formula (Table 8.12):

$$s=2\ (0.1\ T)\ 0.25\geq 6\ \text{mm}$$

and that of the cover from $s_c=0.9$ s, where T is torque on the low-speed shaft in Nm.

The diameter d of the bolts for securing the cover should be $d=(T)^{1/3}\geq 10$ mm, and that of foundation bolts, $d_f=(2T)^{1/3}\geq 12$ mm. The thickness of the foundation flange should be $s_{ff}\geq 1.5\ d_f$. The width of the flanges at the base and at the two halves of the housing should be $W_f=(2.1$–$2.5)\ d$.

Wall thickness of the load-carrying upper half is taken as 0.8 *s*, and that of the nonload-carrying upper half of the housing as 0.5 *s*. Flange thickness is taken as 1.5 *s* for castings and 2 *s* for welded construction. Outside dimension of the bearing housing is kept at 1.2 times the outside diameter of the bearing. Flange screw/bolt diameter = 1.2 *s* for casting and 1.5 *s* for welded construction. Bolt spacing is 6–10 times the bolt diameter.

In designing reducing gear housing, simple geometric shapes are to be preferred, with the outside as plain as possible. In order to reduce air draft noise, the gap between the gear and the side wall should be at least 15 mm.

8.12 Losses in Gearboxes

From green design considerations, the reduction of power loss in gearboxes is very important. Besides the power losses outlined below, the heat loss due to radiation needs to be considered.

$$\text{Total power loss } L = L_t + L_{ch} + L_b$$

where:

L_t = Power loss at tooth engagement
L_{ch} = Churning power losses
L_b = Bearing power losses

The power loss due to tooth engagement is calculated as

$$L_t=2.3f\left(\frac{1}{z_1}+\frac{1}{z_2}\right)W\ \text{kW}$$

or

$$L_t=\left(\frac{1}{z_1\cos\varphi}+\frac{0.3}{V+2}\right)W\ \text{kW}$$

where:

f is the coefficient of friction (Table 8.13)
Z_1 and Z_2 are the number of teeth on the pinion and the gear, respectively
V is peripheral velocity (m/s)
ϕ is the helix angle

Table 8.13 Coefficient of Friction for 20° Pressure Angle Gears

V (m/s)	0.7	7	14	21	42
f	0.08	0.06	0.05	0.04	0.025

Note: V is peripheral speed

The power loss due to churning is estimated as

$$L_{CH} = c.b.V\left(\frac{200V\mu}{Z_1 + Z_2}\right)^{0.5} 10^{-3}(kW)$$

where:

- V = peripheral speed (m/s)
- b = face width of the gear (mm)
- c = factor equal to 0.009 for splash lubrication, 0.006 for stream lubrication
- μ = viscosity of oil at the operating temperature (cP)

The bearing power loss is calculated as

$$L_b = 5.23 \times 10^{-8}\, F f_b\, d\, n\; kW$$

where:

- F = radial load on the bearing (N)
- f_b = coefficient of friction at the bearing reduced to the shaft diameter 0.005-0.01
- d = shaft diameter (mm)
- n = shaft speed (rpm)

The heat generated by the total power loss will raise the temperature of the oil and the housing. The housing will dissipate heat H by radiation, convection, and conduction through the foundation plate or the frame. When equilibrium conditions are reached, the heat generated and the heat dissipated will be the same. This equilibrium temperature should be less than the maximum operating temperature for the oil; otherwise, the oil will become oxidized. If the temperature exceeds this value, additional heat has to be dissipated by a separate cooling arrangement.

$$H = k_t\left(T_o - T_a\right)A\left(1 + U\right)\left(\text{kcal/h}\right)$$

where:

- A = free surface of the housing from which heat is removed to cool the drive (including 50% of the surface of the fins) (m²)
- T_o and T_a = temperature of the oil and the surrounding air (°C)
- k_t = heat transfer coefficient, equal to 10–16 kcal/(m².°C.h); larger values are used under favorable conditions of air circulation; in new standard reducing gears, the value adopted here is

$$k_t = 14\ \text{kcal}\ (\text{m}^2.^\circ\text{C.h})$$

The U factor takes into account heat transfer to the foundation plate or frame of the machine and amounts to up to 0.3 when the housing seating surface is large.

$$\text{Heat generated per hour: } H_g = 3{,}600{,}000\,L\,/\,J$$

where:

- J is the mechanical equivalent of heat
- J = 4270 Nm/kcal
- L is the total power loss in kW.
- H_g = 843 L, $H_d \geq H_g$, and $T_o \leq 93$°C; otherwise, provide a heat exchanger for cooling the oil. Thus, the design is acceptable in terms of thermal considerations.

8.13 End-of-Life Considerations: Reuse, Recycling of Gears, and Sustainability Considerations

Steel is a unique material because it always contains recycled steel. Each year, millions of tons of pre- and postconsumer steel products, including used steel cans, appliances, automobiles, and construction materials, are recycled by North American steel mills into every ton of new steel produced.

In fact, with the exception of the earliest steelmaking methods, recycling has always been an integral part of the steelmaking process. What is more, all new steel products made from recycled steel can be recycled again at the end of their useful lives. It is estimated that 40% of steels are recycled and about 60% of cast iron is recycled. Used steel cans are recycled into part of a guard rail, which may one day be recycled into an appliance. And an infrastructure of ferrous scrap processors is widely available to prepare all types of steel products for recycling. Processors prepare and ship steel scrap to steel mills and foundries for remelting into new steel. The North American steel industry uses one of two types of furnaces to make new steel. Both furnaces recycle old steel products into new steel, but each is used to create different products for varied applications. The basic oxygen furnace uses an average of 25% steel scrap to make new steel. The other type of steelmaking furnace, the electric arc furnace, melts more than 90% steel scrap to make new steel. Iron foundries also melt steel scrap to make new iron products, which contain about 75% scrap.

Many states have instituted "buy recycled" mandates that require purchasing agents to buy products with recycled content when possible. As purchasing agents look for products with recycled content, they should recognize the recycled content and recyclability of steel. Bridges, guard rails, utility poles, fire hydrants, residential and commercial construction materials, appliances, cars, trucks, back hoes, desks, file cabinets, and cans are some of the many steel products that contain recycled steel and, just as importantly, are recyclable again at the end of their useful lives. When purchasing agents and consumers buy products made with recycled materials, they complete the recycling loop and help develop and maintain markets for recyclables. An ordinary magnet, which sticks to steel, can help determine whether a product is made from recycled, recyclable steel.

8.14 Environmental Benefits of Gear and Steel Recycling

Recycling steel saves energy and natural resources. It also reduces the carbon footprint. In a year, the steel industry saves enough energy to power about 18 million households for a year. When one ton of steel is recycled, 2500 pounds of iron ore, 1400 pounds of coal and 120 pounds of limestone are conserved. It saves about 56,000 MJ of energy and reduces CO_2 emission into the air by 4200 kg. By making recycling integral to steelmaking, the steel industry leads the "buy recycled" effort. Each year, more than 40%–70% of the steel the domestic industry produces is recycled.

8.15 Innovative Gearboxes for the Waste and Recycling Industry

Engineers should pay more attention to green product manufacturing based on environmental management systems, energy conservation, development of cleaner products, and waste and pollution reduction. At the design stage, engineers should consider minimization of materials use by designing parts and reduce the use of materials by designing the gearbox for minimization of volume or weight. It is possible to disassemble the gearboxes at end of life, refurbish the pinion and gear, and reuse them, if possible. At least, it is always possible to recycle the gear materials. From the environmental impact viewpoint, recycling itself can reduce the carbon footprint and energy content of the component. The eco-indicators or indices can be reduced as far as possible. Compact innovative design of gearboxes can also save a lot of materials and thereby reduce the embodied energy and carbon footprint. Some gearbox manufacturers (e.g., Brevini) have developed very innovative combined helical and bevel-helical units that offer maximum torque and very high reduction ratios in an extremely compact unit. These gearboxes are used in crushers, shredders, conveyors, and many more waste and recycling industries. Normally, for such operations, the requirement is for high torque and low speed. In such industries, output torque requirement may be of the order of 9000 N.m and output speed 4.0 rpm. Such high reduction ratios allow the drive systems to be powered by small hydraulic motors and small power packs, which keeps running costs to a minimum. Another advantage is the long lifetime of such gearboxes, with the strength, capacity, and durability of bevel-helical gears.

Innovative gear boxes for the waste and recycling industry.

It has been reported that some automotive parts recyclers remove the contaminated fluid, which is clearly an environmental hazard, from the gearbox or rack and pinion. They report that they need two people to complete the job. The steps described below illustrate the disassembly process for gearboxes.

8.15.1 Flush Procedure for Gear/Rack Replacement

Two people required.

1. Replace gear in car.
2. Do not connect pressure or return hose to gear. Place end of pressure hose into bucket and close off return line on reservoir.
3. Fill pump with new fluid.
4. Start and stop engine four times to push old fluid out of pump. Repeat this until fluid is clean.
5. Connect lines to gearbox or rack and pinion.
6. Check fluid level. *Keep it full.*
7. Start engine, turn steering wheel right and left.

Using the right gearboxes, drives, couplings, and motors in waste handling and recycling can significantly reduce long-term operating costs.

Waste handling and recycling creates one of the toughest operating environments for all types of motion and control equipment. In particular, gearboxes, drive motors, couplings, and chains have to withstand extreme loads, shock, and contamination from dust, debris, and glass, plus widely varying temperatures and, in many instances, infrequent maintenance. Despite this, it is important for waste management and recycling companies to maximize machine uptime while minimizing maintenance and long-term operating costs.

Aggressive materials such as glass are typically found in waste and recycling debris, and yet many gearboxes, drive motors, couplings, and chains are not equipped for the challenge. Couplings, for example, often wear out rapidly in recycling plants due to a lack of tolerance in their connections and the swift development of misalignment resulting from incorrect installation. However, the use of improved couplings can reduce the likelihood of failure. While the specification of improved components offers one method of increasing efficiency, the addition of one particular component offers another. The variable-speed drive will not only save energy but also reduce shock loading on belts, with features such as "soft start" and programmable ramp times, minimizing both maintenance and operating costs.

To combat the aggressive conditions within waste and recycling—moisture, dust, abrasive particles—it is also necessary to specify rolling bearings that offer a longer life span. NSK Super TF and Hi-TF bearings can offer up to seven times the life span of alternatives made with ordinary carburized steel in applications such as waste and recycling, where lubricant is contaminated. There will never be a shortage of waste and recycling: the challenge is to process these materials in the most cost-effective way, and that means relying on the most up-to-date methods of production, state-of-the-art products, and new technologies. Plant reliability is vitally important in minimizing downtime and maximizing uptime—even a couple of hours' downtime can result in huge losses—but a range of components are now available that can reinforce that reliability.

Let us consider the business of recycling and composting. It is likely that the sectors are set to double over the next few years as landfill space gets ever scarcer. We are never likely, as a society, to recycle 100% of our waste. True, the market for recyclable materials handling systems is a much "younger" market than that for normal compaction refuse collection vehicles (RCVs)—even if history shows us that "recycling" and separating reusable resources such as newspaper and cardboard is hardly a new idea. The worrying fact today is that, though it is easy enough to justify the cost of a high-specification, purpose-built RCV against a cheaper alternative, the actual utilization of some specialist recycling units is still very low.

The RCV should be fitted with an automatic gearbox. This might add to the initial cost by as much as $10,000, but will more than justify itself by performing well over an extended period with little downtime.

Interestingly, there is another factor to consider. In the past, and certainly in the U.S. market, there was a distinct difference between widespread autobox provision in lighter-duty vans and light trucks, RCVs, and trucks used for urban distribution, and manual transmission in heavy-duty line-haul trucks.

Reference

1. Budynas, R. G., Shigley, J. E., and Nisbett, J. K. (2008). *Mechanical Engineering Design*, 8th edn. McGraw Hill.

Bibliography

Ashby, M. F. (2009). *Materials and the Environment: Eco-Informed Material Choice*. Elsevier, Boston.

Ashby, M. F. (2011). *Material Selection in Mechanical Design*, 4th edn. Elsevier, New York.

Botthroyd, G., Dewhurst, P., and Knight, W. A. (2002). *Product Design for Manufacture and Assembly*, 2nd edn. Marcel Dekker, New York.

Chen, R. W., Navin-Chandra, D., and Print, F. B. (1993). A cost-benefit analysis model of product design for recyclability and its application. In *IEEE International Symposium on Electronics and the Environment* (pp. 178–183). Arlington, VA.

Desai, A. and Mital, A. (2005). Incorporating work factors in design for disassembly in product design. *Journal of Manufacturing Technology Management* 16: 712–732.

Dowie, T. and Kelly, P. (1994). Estimation of disassembly times, Technical Report DDR/TR15, Manchester Metropolitan University, UK.

ISO 14044:2010. (2010). *Environmental Management—Life Cycle Assessment, 2010*. Leonardo academy.

Kroll, E. and Carver, B. S. (1999). Disassembly analysis through time estimation and other metrics. *Robotics and Computer-Integrated Manufacturing* 15: 191–200.

Kroll, E. and Hanft, T. A. (1998). Quantitative evaluation of product disassembly for recycling. *Research in Engineering Design—Theory, Applications, and Concurrent Engineering* 10: 1–14.

Lee, S. G., Lye, S. W., and Khoo, M. K. (2001). A multi-objective methodology for evaluating product end-of-life options and disassembly. *International Journal of Advanced Manufacturing Technology* 18: 148–156.

Lidwell, W., Holden, K., and Butler, J. (2003). *Universal Principles of Design*. Rockport, Gloucester, MA.

McGlothlin, S. and Kroll, E. (1995). Systematic estimation of disassembly difficulties: Application to computer monitors. In *IEEE International Symposium on Electronics & the Environment* (pp. 83–88), Orlando, FL. *Jurnal Mekanikal*, December 2008.

Mott, R. (2004). *Machine Elements in Mechanical Design*, 4th edn. Prentice Hall.

Segur, A. B. Predetermined Motion Time System (PMTS) mtm-international.org.

Shigley, J. E. and Mischke, C. R. (2001). *Mechanical Engineering Design*, 6th edn. McGraw Hill.

Skakoon, J. (2000). *Detailed Mechanical Design: A Practical Guide*, Technology & Engineering. ASME Press, New York.

Skakoon, J. G. (2008). *The Elements of Mechanical Design*. ASME Press, New York.

Slocum, A. H. (2007). *Fundamentals of Design*. Massachusetts Institute of Technology, Cambridge, MA.

Slocum, A. H. (1992). *Precision Machine Design*, Technology & Engineering. Prentice Hall, NJ 07632.

Soemers, H. (2010). *Design Principles for Precision Mechanisms*, ISBN 90365 31039.

Suga, T., Saneshige, K., and Fujimoto, J. (1996). Quantitative disassembly evaluation. In *Proceedings of the 1996 IEEE International Symposium on Electronics & the Environment* (pp. 19–24). Dallas, TX.

United States Environmental Protection Agency (EPA). (2010). *Life Cycle Assessment, 2010*, Science & Technology Law & Regulation.

Whitney, D. E. (2004). *Mechanical Assembly: Their Design, Manufacture, and Role in Product Development*. Oxford University Press, New York.

9

Shaft Design

9.1 Introduction and Applications of Shaft

Shafts are important and very common machine elements. We come across some examples of shafts in the industrial world; for example, gears must be mounted on some sort of shaft, and a gate valve or globe valve is opened and closed by a handwheel turning another type of shaft often referred to as a spindle. It is normal practice to buy items such as gate or globe valves, gears, pulleys, and many other items as standard components "off the shelf." In other words, a manufacturer makes large quantities of industry-standard items which are sold for many different purposes. Generally, shafts do not fall into this category. A shaft is usually designed to perform a specific task in a specific machine. There may of course be thousands of similar machines produced, each using a shaft of that design, and the manufacturer may provide extra shafts as "spare parts," but that shaft design generally has no use outside the machine for which it was designed. It follows that designers will frequently be called upon to design shafts and it is therefore important for them to understand the requirements of shafts and the design features needed to fulfill these requirements.

Basic requirements are (i) good fatigue strength, (ii) high stiffness, (iii) low cost, and (iv) good wear resistance.

9.1.1 Definitions of Some Common Terms

Since shafts take on several different configurations and are used for many different purposes, several different definitions are in common use, as listed below. The nomenclature is not always clear cut and there is often an overlap of function and therefore of definition.

Shaft In general, a *rotating* member used for the transmission of power.

Axle Generally, a *stationary* member used as a support for rotating members such as bearings, wheels, idler gears, and so on.

Spindle A short shaft, usually of small diameter, usually rotating, for example, valve spindle for gate valve, but consider also the *headstock spindle* of a lathe, which is quite large and usually has a hole right through its center.

Stub shaft A shaft which is integral with an engine, motor or prime mover and is of suitable size, shape, and projection to allow its easy connection to other shafts.

Line shaft (or *power transmission shaft*) A shaft connected to a prime mover which transmits power to a number of machines—now mostly superseded by machines having individual motors.

Jack shaft A short shaft used to connect a prime mover to a machine or another shaft. May also be a short shaft placed as an intermediate shaft between a prime mover and a driven machine.

Flexible shaft Permits the transmission of power between two shafts (e.g., motor shaft and machine shaft) whose rotational axes are at an angle or where the angle between the shafts may change.

Some examples of different types of shafts are presented here. Figure 9.1 shows a preliminary drawing of a shaft drawn for design purposes. The locations of two support *bearings* are shown schematically. There are two *keyways*, suggesting something like a pulley on the right-hand end and perhaps a gear between the two bearings, with *shoulders* for axial location of the gear and the two bearings.

Figure 9.2 shows a *sector shaft*. Although it oscillates only through 50°–60° in order to steer the front wheels on a Mazda truck, this component from the steering box is described as a sector shaft. It features part of a gear (a sector) which gives the shaft its name, and tapered splines onto which a lever called the *Pitman arm* is fitted tightly by tightening a nut on the right-hand threaded end.

Figure 9.3 shows a large component, referred to as a shaft, in a wind turbine. A notable feature is the large flange which appears to have been formed by forging. The shaft has been machined all over after forging. It appears that this shaft acts as a pivot to allow the blade and generator assembly to turn to face the wind.

Figure 9.4 shows an exploded view showing the components which are assembled onto the *stub axle* (on left of diagram) on the front axle of a Toyota light truck. The *stub axle assembly* is able to rotate, say, 35° about a near-vertical *king pin* (not visible) to allow the front wheels to be steered, but the stub axle does not rotate about its own axis. The front wheel rotates on bearings mounted on the stub axle. See also Figure 9.5, which is a sectioned view of a stub axle and hub.

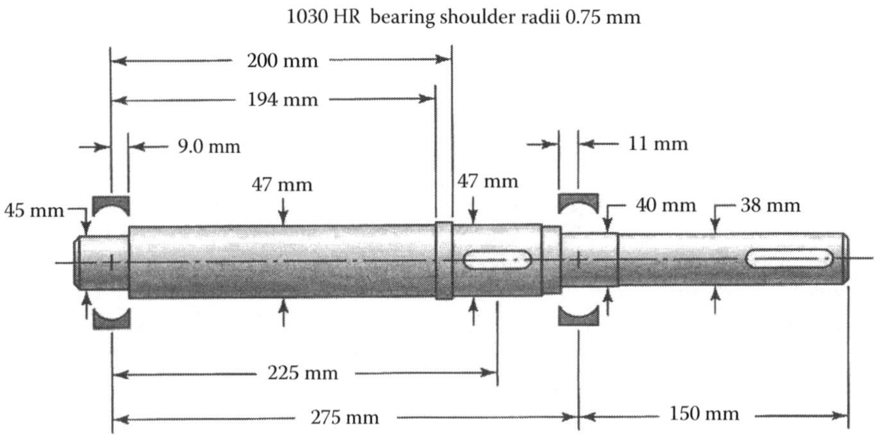

Figure 9.1

A preliminary design of shaft. (From Shigley, J. E., Mischke, C. R., *Mechanical Engineering Design*, 6th edn., McGraw Hill, 2001.)

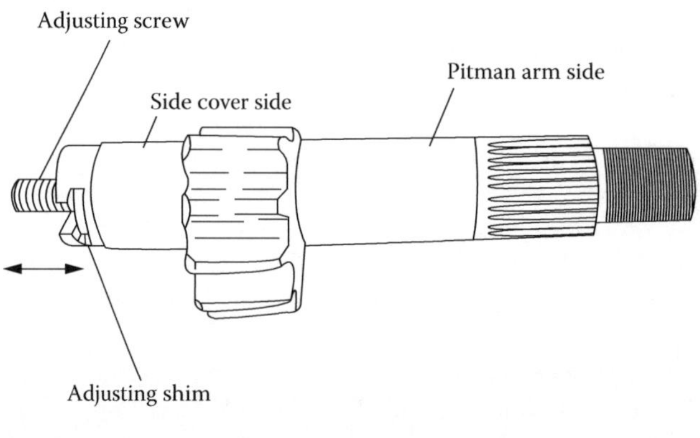

Figure 9.2

Sector shaft. (From Mazda T2600 Workshop Manual 1989, pp. N-20.)

Figure 9.3

Wind turbine shaft.

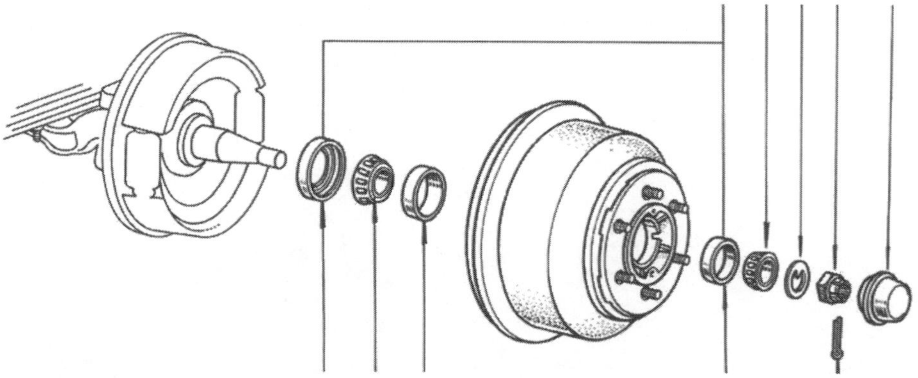

Figure 9.4

Stub axle.

Figure 9.5

Headstock assembly.

Figure 9.5 shows a sketch of a *headstock spindle* in a lathe. The spindle rotates and has a threaded end onto which can be screwed a three- or four-jaw chuck. The work to be "turned" is clamped in the chuck.

Figure 9.6 shows a fully assembled *gate valve* (a) and a sectioned view of a gate valve, showing the spindle used to raise or lower the *gate* to allow or block flow through the valve (b).

Figure 9.7 shows a Japanese-style backpack mower with flexible shaft between the red power pack and the usual hand-held mower assembly. In general, flexible shafts are not used to transmit high power or high speed.

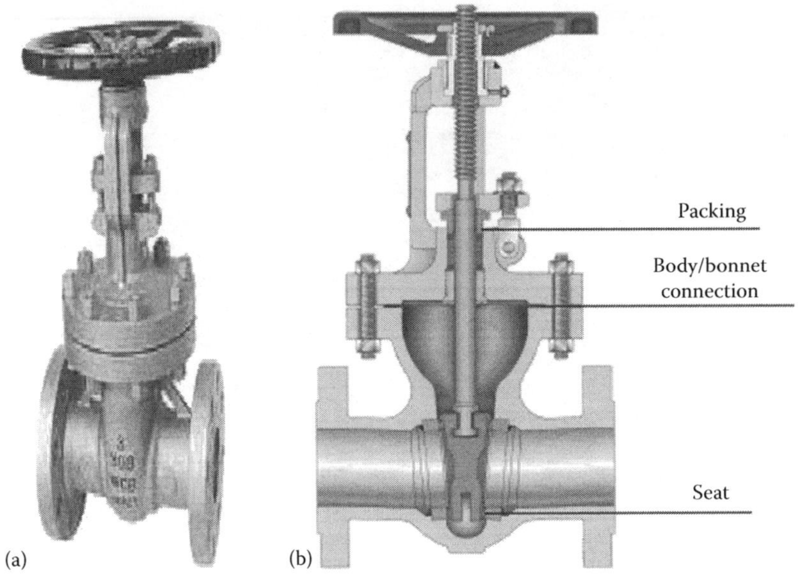

Figure 9.6

Gate valve. (From http://www.valvediagnostics.com/media/pictures/gate.gif.)

Figure 9.7

Flexible shaft. (From http://www.szwholesale.com/uploadFiles/Japanese%20style%20backpack%20mower%20with%20flexible%20shaft_600.jpg.)

9.1.2 Shaft Materials

Most shafts are made from steel, either low or medium carbon. However, high-quality alloy steel, usually heat treated, may be chosen for critical applications.

Examples of alloy steel are: Low- or medium-carbon steel, AISI 1010, 1018, 1020, 1035; low-alloy steel, AISI 4140, 4340 or 8640; stainless steel, titanium, or Inconel. Other metals, for example, brass, stainless steel, or aluminum, may be used where corrosion is a problem or lightness is required. Small, light-duty shafts, for example, in household appliances, may be injection molded in a plastic material such as nylon or Delrin.

9.1.3 Functions of a Shaft

As previously noted, shafts transmit power. Since

$$P = T\omega \tag{9.1}$$

this implies both rotation (ω) and the transmission of torque (T).

Shafts locate members such as gears and pulleys in their correct relative positions.

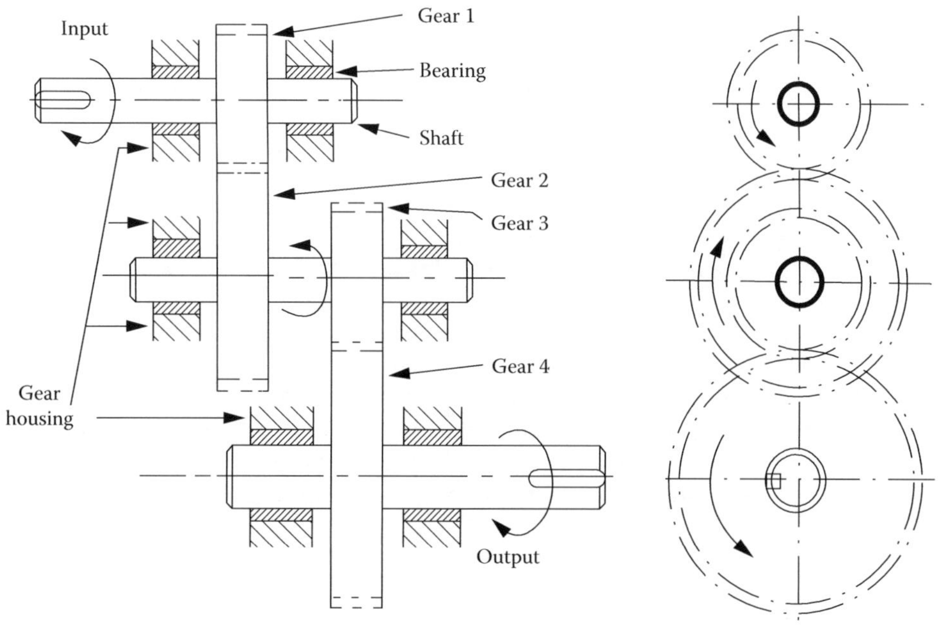

Figure 9.8

Double reduction gear mounted on shaft. (Reproduced from J. Carvill, *The Student Engineer's Companion*, Butterworths, 1980.)

Shafts connect members such as gears and transmit torque from one member to another. In Figures 9.8 and 9.9, the intermediate shaft of the double reduction gear train transmits torque from one gear to the other on the same shaft. The input and output shafts transmit torque between the single gear on each of these shafts and some other component (such as a belt pulley or chain wheel) which is mounted on the projecting end of the shaft. Note that these are simplified diagrams.

Figure 9.8 is an example of the use of shafts to mount the gears used in a double reduction (or compound) gear train. Note that while this simplified sketch shows the shafts, the bearings which locate them, and the housing into which the bearings fit, there is no provision for axial location of the shafts. Note also the increased diameter of the output shaft, needed to transmit the increased torque created by the reduction in speed. With two stages of gear reduction, the input and output shafts rotate in the same direction. Figure 9.9 shows the principles of a compound gear train.

9.1.4 Shaft Bearings

Shafts rotate and they must therefore be mounted in bearings which

- Allow rotation
- Reduce friction to a low level
- Reduce wear of both shaft and bearings to an acceptable level

Figure 9.9

Compound gear train. (From http://www.brighthub.com/engineering/mechanical/articles/66020.aspx?image=68312.)

Since shafts are often used to locate items such as gears in precise locations, bearings must

- Have minimal clearance (or slack) between bearing and shaft
- Locate the shaft both radially and axially

The different types of bearings available will be described later in this project.

9.1.5 Transmitting Torque to and from Shafts

In many machine designs, torque must be transmitted between shafts and gears, pulleys, chain-wheels, and other hubs. In almost all cases, this must be done without allowing slip between the two parts. The state of stress at a selected critical point on the surface of a rotating power transmission shaft may involve fluctuating torsional shearing stress, transverse shearing stress, bending stress, or axial stress components. Because of the simultaneous occurrence of torsional shear and normal stresses due to bending, the stress analysis of a shaft virtually always involves the use of a combined stress approach. The recommended approach for shaft design and analysis is the distortion energy theory of failure. In general, shaft design equations must be based on multiaxial states of stress produced by fluctuating loads. As shown below, there are a number of ways in which this torque transfer may be achieved.

9.1.6 Keys

Perhaps the most widely used method of torque transfer is by the use of *keys*. A longitudinal groove called a *keyway* is machined into the shaft and a corresponding groove into the bore of the hub. The key fits simultaneously into both grooves, locking them together. The key is subjected to direct shear (Figure 9.10).

9.1.7 Key Data

The keys are made either straight or tapered. A straight key transmits torque by means of shear and compression, but it cannot transmit heavy torque. It is fitted on the sides. However, tapered keys transmit torque through friction. A tapered key has play between the hub and the shaft and it has a tendency to throw the pulleys or gears out of alignment. It can transmit higher torque. Figure 9.11 shows the dimensions of standard straight keys. The keys are square up to $w = 1.5$ in; in larger sizes h is often made smaller than w. In all sizes, w is made about equal to $d/4$, where d is the diameter of the shaft. Square and flat key standards are shown in Figure 9.11.

Figure 9.11 shows examples of types of keys in general use, including the square key already illustrated in Figure 9.10. Note that the flat key (Figure 9.11b) is incorrectly drawn—it should have a rectangular cross section with the larger dimension resisting direct shear as shown in Figure 9.12, presenting details of feather keys and the profiled keyways into which they fit.

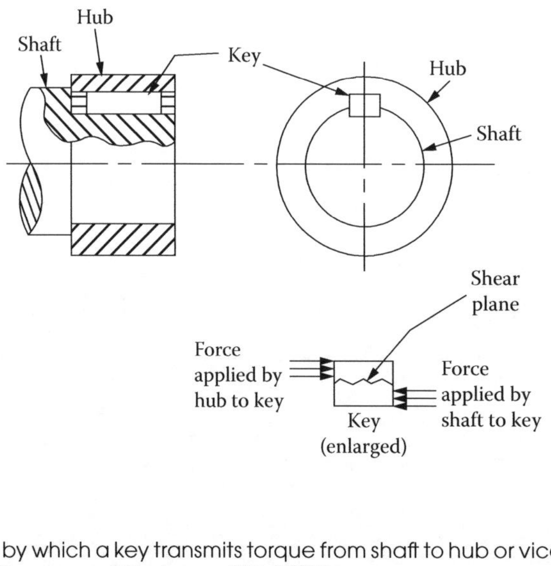

Figure 9.10

The principle of direct shear by which a key transmits torque from shaft to hub or vice versa. (From Juvinall, R. C., *Fundamentals of Machine Component Design*, p. 537, 1983.)

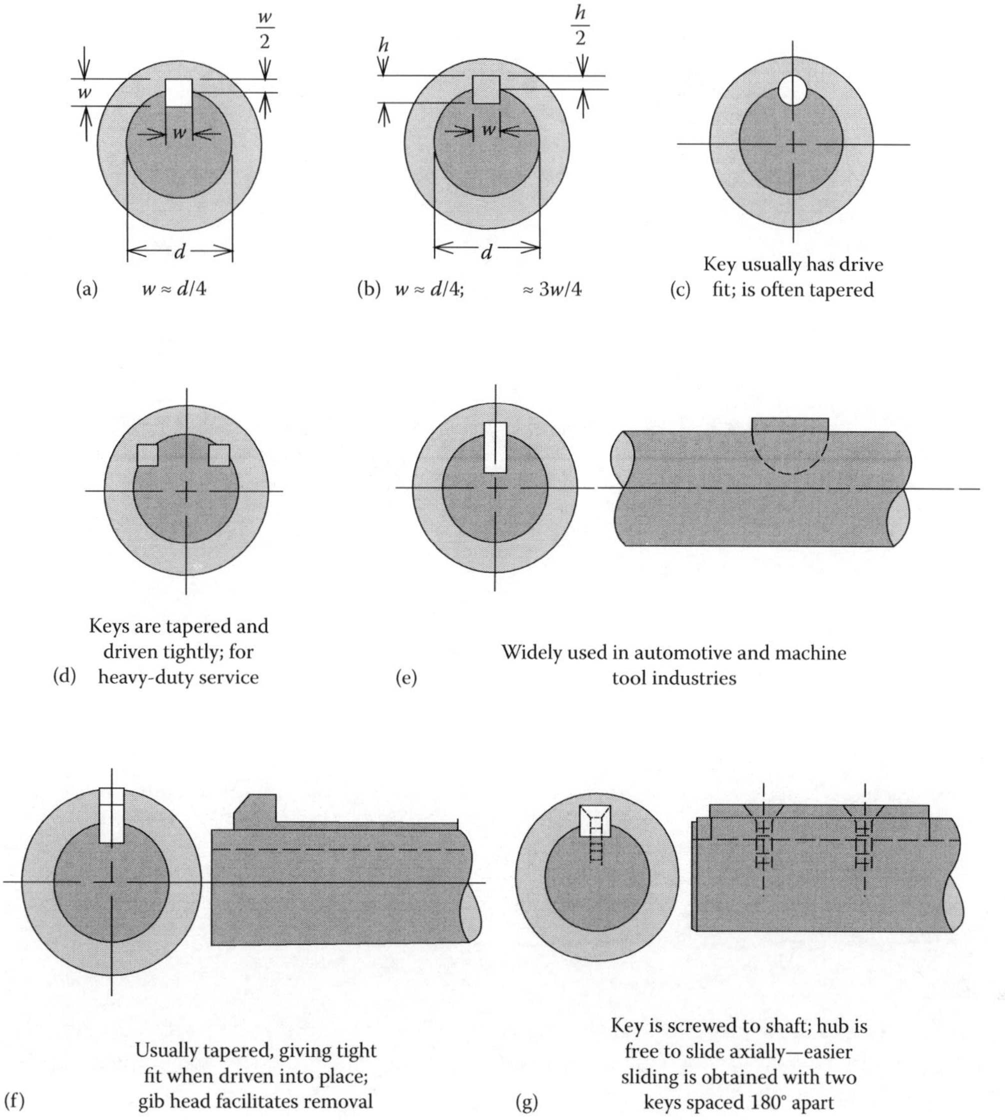

Figure 9.11

Various keys in general engineering use. (a) Square key, (b) flat key, (c) round key, (d) kennedy keys, (e) woodruff key, (f) gib-head key, (g) feather key.

Note that, in general, keys do not provide axial location of the hub. In fact, feather keys are designed to allow the "hub" (in whatever form it is) to slide along the shaft. One exception is the gib-head key, which is tapered and provides reasonable axial location when the key is hammered into place.

Example 9.1: Design of Square Key for Power Transmission

Design the dimensions of a square fitted key (Figure 9.11a), for a 3(7/16) in steel shaft to transmit 95 hp at 200 rpm. The torque fluctuates and we assume the key is well fitting.

From hp given, torque is calculated as

$$T = \frac{63{,}030 \times 95}{200} = 29{,}900 \text{ lb-in}.$$

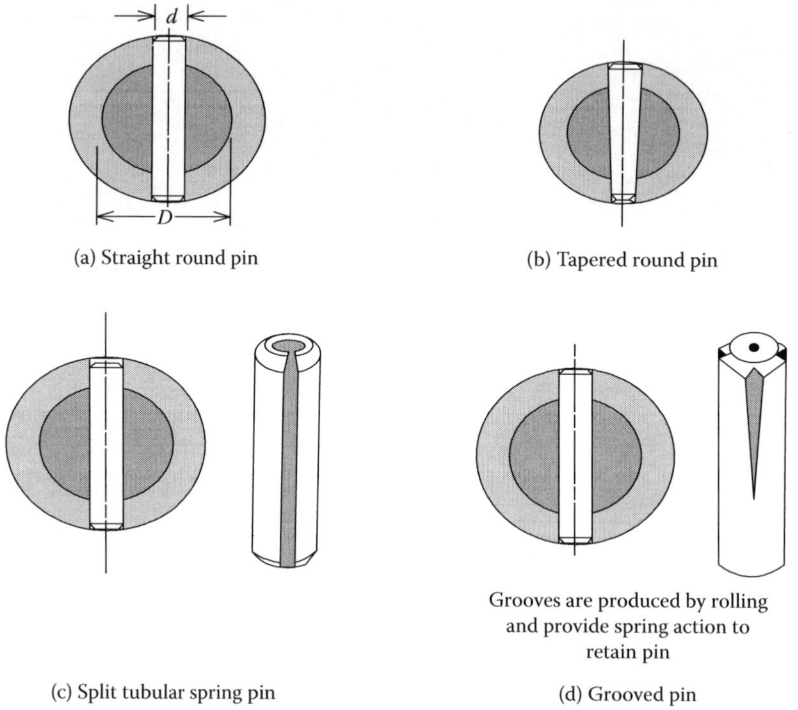

Figure 9.12

Round pins of various types are sometimes used in place of keys. Pins are generally regarded as suitable only for light duty. (From Juvinall, R. C., *Fundamentals of Machine Component Design*, p. 538, 1983.)

The key dimensions from Figure 9.11 are calculated as below.
$w = d/4 = 0.859375$ in, $h = w$ (square key).

9.1.8 Grub Screws and Set Screws

These are regarded as light-duty attachments. Sometimes the end of the screw merely bears against the surface of the shaft. In other cases (dog-end or cone-end screws) the end of the screw may enter a drilled hole in the shaft, as in Figures 9.13 and 9.14.

9.1.9 Splines

Splines are essentially axial grooves or recesses which are machined into the shaft, very like a series of keyways. These splines consists of multiple integral keys with parallel sides milled on the outside diameter of the shaft. Splines are an integral part of the shaft (cf. keys, which are loose parts). Corresponding grooves are cut (*broached*) into the bore of the hub so that the shaft/hub assembly forms a series of interlocking projections.

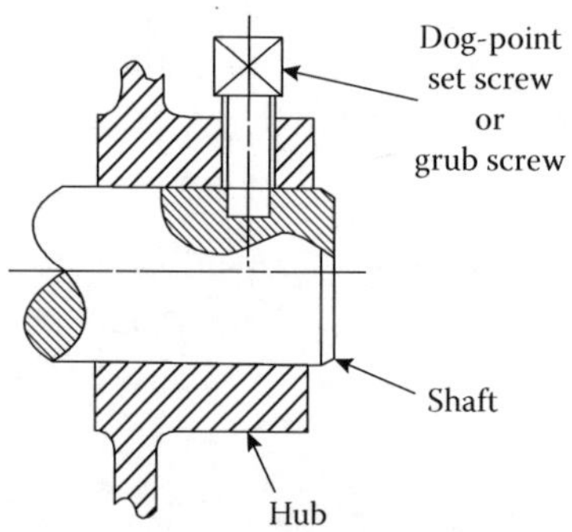

Figure 9.13

Transmission of torque by means of a set screw or grub screw. The end of the screw is subjected to direct shear.

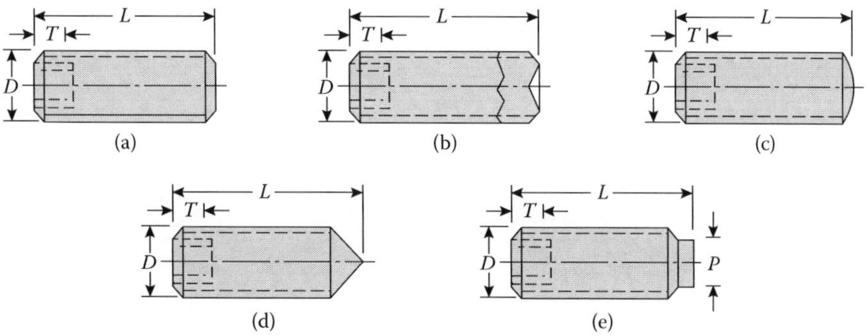

Figure 9.14

Drawings of different types of set screws or grub screws. In general, set screws or grub screws may have the internal hexagon shown in this figure. (From Shigley, J. E. and Mischke, C. R., *Mechanical Engineering Design*, 6th edn., p. 1141, McGraw Hill, 2001.)

The resulting connection is stronger than a keyed joint and is used in heavy-duty applications. Spline profiles may be square, involute, or triangular. Splines are often designed to allow axial movement of a gear or hub whilst continuing to transmit torque. For axial sliding to occur satisfactorily, the bearing pressure on the faces of the spline must be low and good lubrication must be provided.

All dimensions of splines are expressed as functions of the shaft diameter d; the height h of the spline is between $0.075d \leq h \leq 0.125d$ and the width of the spline w is in the range $0.156d \leq w \leq 0.25d$. The torque that an integral multiple shaft can transmit safely by pressure between the sides of the spline is expressed as

$$T = \frac{1}{2} phnl(d - h)$$

where:

p = allowable pressure in pounds per square inch
n = number of splines
l = engaged length of each spline, in inches

To transmit a required torque, the length of the spline may be calculated for a selected pressure. The pressure p must be much lower than the elastic limit S_y in compression, to prevent wear of the sliding surfaces. Since p may be considered as bearing pressure, a very low pressure of 1000 psi is recommended for a safe fit.

It can rise to 3000 psi depending on the type of fit from loose to tight.

9.1.10 Shaft Failure Modes

In general, shafts can be subjected to various combinations of axial, transverse (bending), and torsional loads, and these loads may be static or fluctuating. The primary potential failure modes to be considered when designing a power transmission shaft are: Fatigue failure; force-induced elastic deformation failure; wear failure. If plain-bearing journals, gear teeth, splines, or cam lobes are integral features of a shaft, wear must be also considered as a potential failure mode.

In general, a shaft may fail by

- Excessive lateral deflection, which causes items such as gears to move laterally from their proper location, resulting in incorrect meshing.
- Torsional deflection, which destroys the precise angular relationship or "timing" between sections of a mechanism.
- Wear, which may take place on bearing surfaces (*journals*) or other contact areas, such as cams.
- Fracture. Unless the shaft was grossly under-designed, fracture usually occurs by *fatigue cracking*. This is discussed below.

9.2 Stress Concentration and Fatigue Cracking

9.2.1 Mechanism of Fatigue Cracking

Fatigue cracking occurs under conditions of *repeated loading* or *cyclic loading*. It is not the magnitude of the load *in itself* which causes the failure, but the *cumulative* effect of many thousands, often millions, of repetitions of loading, or *load cycles*, which results in *cumulative damage*.

9.2.2 Load Cycles

Figure 9.15 is a schematic representation of cyclic loading of a rotating shaft subjected to a bending moment. The shaft is mounted in two bearings which allow it to rotate with angular velocity ω, driven by the drive pulley at the left-hand end. A mass carrier at the right-hand end of the shaft is mounted on the shaft by another bearing. This bearing ensures that, as the shaft rotates, the mass carrier always hangs vertically downwards. It should be clear that the force due to the mass carrier is $F = mg$. This force applies a bending moment to the right-hand section of the shaft and the bending moment due to F reaches a maximum at Shaft Bearing 2. Consider a point A at the top of the shaft. Since the force F is tending to bend the shaft downwards, the shaft material at A must be in *tension*.

Similarly, the shaft material at a point B at the bottom of the shaft must be loaded in *compression*. Now let the shaft rotate through 180° so that point A moves to the bottom and B to the top. The shaft material at A is now in compression and the material at B is in tension. After a further shaft rotation of 180°, A is again at the top and in tension, while B is again at the bottom and is in compression. The total of 360° of shaft rotation results in one complete load cycle for points A and B and, indeed, for all points on the shaft. Figure 9.16 shows a graph of the load cycles for points A and B.

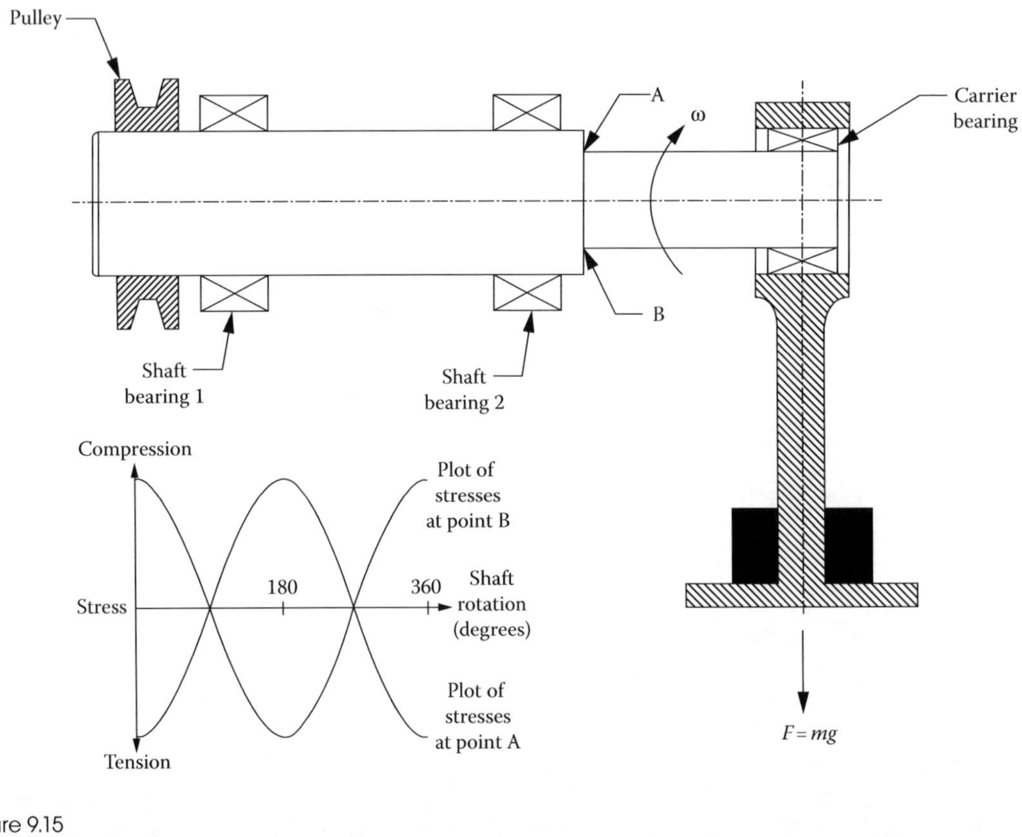

Figure 9.15

Schematic representation of cyclic loading of a rotating shaft subjected to a bending moment.

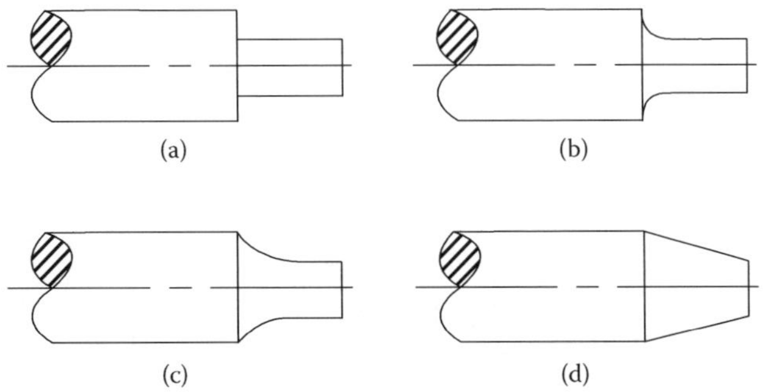

Figure 9.16

Methods of reducing stress concentration at a change of shaft diameter. Stress concentration effects *decrease* from (a) to (d).

9.2.3 Fatigue Crack Propagation

Within certain limits, the application of large numbers of load cycles may cause tiny cracks to occur at highly stressed sections of the component. Subsequent load cycles cause this microscopic crack to increase in size, slowly at first, but with increasing rapidity. If not detected or checked in some way, the crack will eventually become so large, and will so weaken the component that it will fracture completely. The final failure may be sudden and may have catastrophic consequences; nevertheless, fatigue cracking is a *progressive failure* which has been occurring for some time prior to the final failure.

Due to the insidious nature of fatigue cracking and its potential to cause catastrophic failure, designers always need to be on guard against design features which may lead to fatigue failure. Items such as shafts which may undergo a very large number of rotations during their lifetime always need to be designed with fatigue failure in mind. It is generally accepted that fatigue cracking is caused essentially by *tensile* stresses. However, you need to be aware of the complexity of the stress fields in many engineering components. Shafts in pure torsional shear, for example, can fail by fatigue cracking, presumably due to normal (i.e., tensile) stresses acting along planes at 45° to the shaft axis.

9.3 Stress Concentrations

It is observed in practice that fatigue cracks frequently occur where there is a sharp or abrupt change in the shape of the component. Theoretical analysis and experimental techniques both show that stresses within the material rise sharply at abrupt changes of section. Such features are referred to as *stress concentrations* or *stress raisers.*

Referring to Figure 9.17, it is seen that there is an abrupt change of shaft diameter close to Bearing 2 called *shoulder* on the shaft. Such a shoulder causes a significant stress concentration. The location of the shoulder is usually in the region of high bending moment and therefore of high tensile and compressive stresses. The combination of these two factors makes this section of the shaft particularly vulnerable to fatigue failure. Changes of cross section other than shoulders causing stress concentration in shafts include:

- Keyways
- Splines
- Circlip grooves
- Threads
- Oil holes
- Undercuts

9.3.1 Stress Concentration Factors

The stress analysis process for fatigue is highly dependent on stress concentrations.

Stress concentrations for shoulders and keyways are dependent on size specifications that are not known the first time through the process. Fortunately, since these elements are usually of standard proportions, it is possible to estimate the stress concentration factors for the initial design of the shaft. These stress concentrations will be fine-tuned in successive iterations, once the details are known.

Shoulders for bearing and gear support should match the catalog recommendation for the specific bearing or gear. A look through bearing catalogs shows that a typical bearing calls for the ratio of D/d to be between 1.2 and 1.5. For a first approximation, the worst case of 1.5 can be assumed. Similarly, the fillet radius at the shoulder needs to be sized to avoid interference with the fillet radius of the mating component [1]. Mott has also provided a table for preliminary estimates of stress concentration factors (Table 9.1) [2].

As we may notice in the table, the stress concentration factor (K_t) is highest in the axial direction and in the case of the retaining ring groove. It gives a hint to where the failure is most likely to happen.

9.3.2 Reducing Stress Concentrations

As designers, we need to understand that the potential for stress concentrations to produce fatigue cracking can be reduced in two ways:

- Reduce the stress concentration effect by making the change of shape more gradual
- Relocate the stress concentration or change of shape to an area subjected to lower stresses

Figure 9.16 shows four ways of designing a change of shaft diameter. The sharp shoulder in (a) produces a very high stress concentration. The shoulder with *fillet radius* (b) is significantly better; the larger fillet radius (c) is better again and the long taper (d) has a very low stress concentration.

Figure 9.17 shows how the effect of a stress concentration on shaft fatigue life depends on its location in the shaft. The circlip grooves on either side of the central pulley are in a location of high stress and seriously weaken the shaft. The circlip groove on the right-hand bearing is in an unstressed location and will have no effect on shaft life.

In Figure 9.17, the center pulley is mounted midway between the bearings. Due to the belt force on the pulley, the central section of the shaft is subjected to high bending moments causing high bending stresses. The left-hand section of the shaft also transmits torque and is subjected to torsional stresses. In this figure, the pulley is located axially by means of two circlips. The circlip grooves which must be machined into the shaft cause abrupt changes of section and are known to cause high stress concentrations. The keyway which must be used to transmit the torque between the pulley and the shaft also forms a severe stress concentration.

Furthermore, the circlips and the keyway significantly reduce the cross section of the shaft available to carry bending and torsion. This combination leads to a high risk of fatigue cracking and the design should be changed. On the other hand, the right-hand shaft bearing has been retained axially by a circlip and this circlip groove is in an unstressed region at the end of the shaft (no bending, no torque) so that the circlip may be used without the risk of causing fatigue cracking. It should be recognized that a shaft is rotational component and its design must take fatigue into consideration.

The fatigue failures are based on the following theories: (1) Modified Goodman theory, (2) Gerber theory, (3) ASME elliptic theory, (4) Soderberg theory, and (5) Langer theory.

Table 9.1 First Estimate of Stress Concentration Factor K_t

	Bending	Torsion	Axial
Shoulder fillet sharp ($r/d = 0.02$)	2.7	2.2	3.0
Shoulder fillet-well rounded ($r/d = 0.01$)	1.7	1.5	1.9
End mill (keyseat) ($r/d = 0.02$)	2.2	3.0	Not available
Sled keyseat	1.7	Not available	Not available
Retaining ring groove	5.0	3.0	5.0

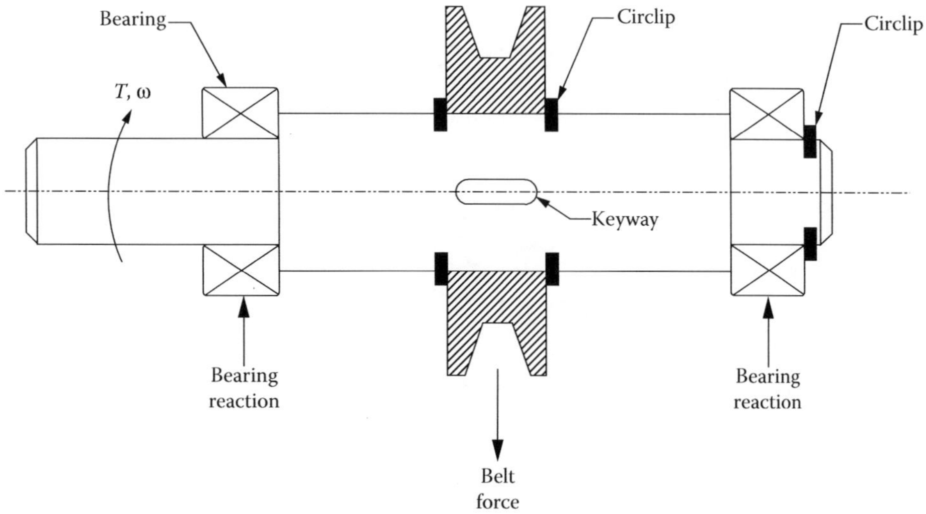

Figure 9.17

The effect of a stress concentration on shaft fatigue life depends on its location on the shaft.

1. Modified Goodman equation for fatigue failure:

$$\frac{S_a}{S_e} + \frac{S_m}{S_{ut}} = 1 \tag{9.2}$$

where:
S_a = alternating strength
S_m = mean strength
S_e = endurance strength
S_{ut} = ultimate tensile strength

Endurance strength of steel is calculated as:

$S'_e =$	$0.5S_{ut}$	$S_{ut} \leq 200$ kpsi (1400 MPa)
	100 kpsi	$S_{ut} > 200$ kpsi
	700 MPa	$S_{ut} > 1400$ MPa

where S_{ut} is the *minimum* tensile strength. The prime mark on S'_e in this equation refers to the *rotating-beam specimen* itself. The unprimed symbol S_e is reserved for the endurance limit of any particular machine element subjected to any kind of loading.

The S'_e value is modified to find the value of S_e as

$$S_e = k_a k_b k_c k_d k_e k_f S'_e$$

where:
k_a = surface condition modification factor, which depends on the quality of finish of the actual part
k_b = size modification factor, which depends on the design dimension of the part. The dimension needs to be converted to effective dimension d_e by equating the volume of material to the equivalent cylindrical dimension
k_c = load modification factor, which depends on the type of load; for a bending load it is 1
k_d = temperature medication factor, which depends on the operating temperature compared to room temperature

k_e= reliability factor, which depends on the scatter of the data

k_f= miscellaneous effect factor, which depends on the residual stresses, corrosion, and so on

For more comprehensive treatment of these modifications, see [1–4].

The fatigue factor of safety n can be included in this equation and written as

$$\frac{\sigma_a}{S_e} + \frac{\sigma_m}{S_{ut}} = \frac{1}{n} \tag{9.3}$$

where:

σ_a = alternating stress

σ_m = mean stress in fluctuating stress

2. The Gerber equation for fluctuating stress is presented as

$$\frac{n\sigma_a}{S_e} + \frac{(n\sigma_m)^2}{S_{ut}^2} = 1 \tag{9.4}$$

3. The ASME elliptic equation for fatigue failure is given by

$$\left(\frac{n \cdot \sigma_a}{S_e}\right)^2 + \left(\frac{n\sigma_m}{S_y}\right)^2 = 1 \tag{9.5}$$

4. Soderberg theory:

$$\frac{\sigma_a}{S_e} + \frac{\sigma_m}{S_y} = \frac{1}{n} \tag{9.6}$$

5. Langer theory:

$$\sigma_a + \sigma_m = \frac{S_y}{n} \tag{9.7}$$

This is called Langer's static equation.

It is advised that in the cases of modified Goodman and Gerber fatigue failure, the Langer equation for static failure must be applied; whichever factor of safety is lower should be accepted as the final factor of safety. In the cases of ASME elliptic and Soderberg, there is no need to check for Langer's factor of safety as the yield strength S_y is included in these equations.

Figure 9.18 shows some stress–time relations, Figure 9.19 is a fatigue diagram showing various criteria of failure, and Figure 9.20 shows sensitivity curves for materials in reversed torsion. In Figures 9.18 and 9.19, σ_{min} = minimum stress, σ_m = midrange component, σ_{max} = maximum stress, σ_r = range of stress, σ_a = amplitude component, σ_S = static or steady stress.

$$\sigma_m = \frac{\sigma_{max}}{2} + \frac{\sigma_{min}}{2} \tag{9.8}$$

and

$$\sigma_a = \frac{\sigma_{max}}{2} - \frac{\sigma_{min}}{2} \tag{9.9}$$

Based on the fatigue equations from Equations 9.2 through 9.7, the diameter of the shaft equations are developed below.

A shaft is subjected to

1. Shear stress because of *torsional load*
2. Bending stress because of weight of the elements such as gears, pulleys, and so on, in addition to the weight of the shaft itself

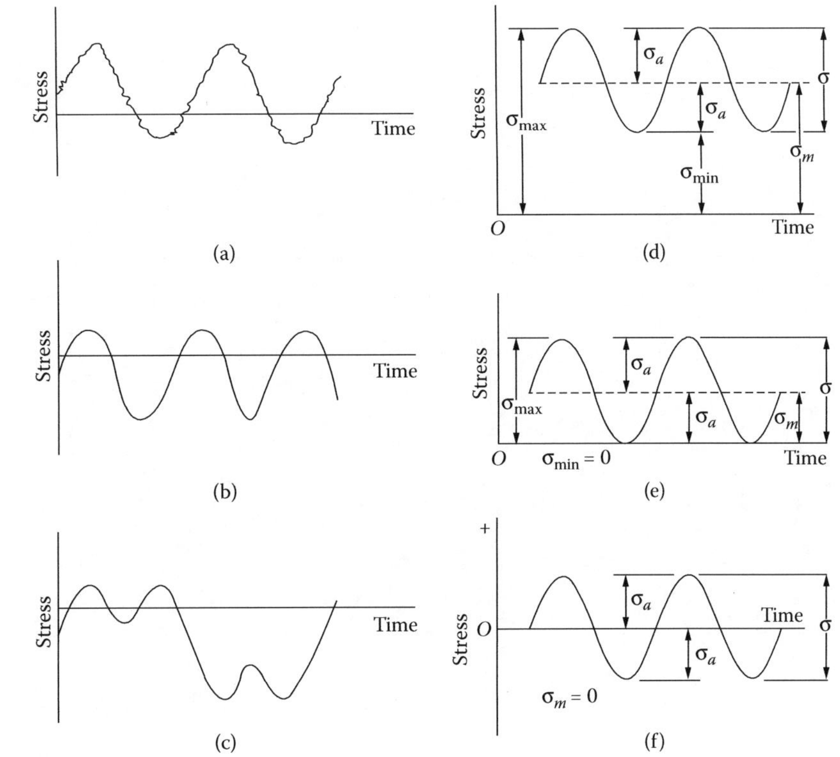

Figure 9.18

Some stress-time relations. (a) fluctuating stress with high frequency ripple; (b and c), nonsinusoidal fluctuating, stress; (d) sinusoidal fluctuating stress; (e) repeated stress; (f) completely reversed sinusoidal stress.

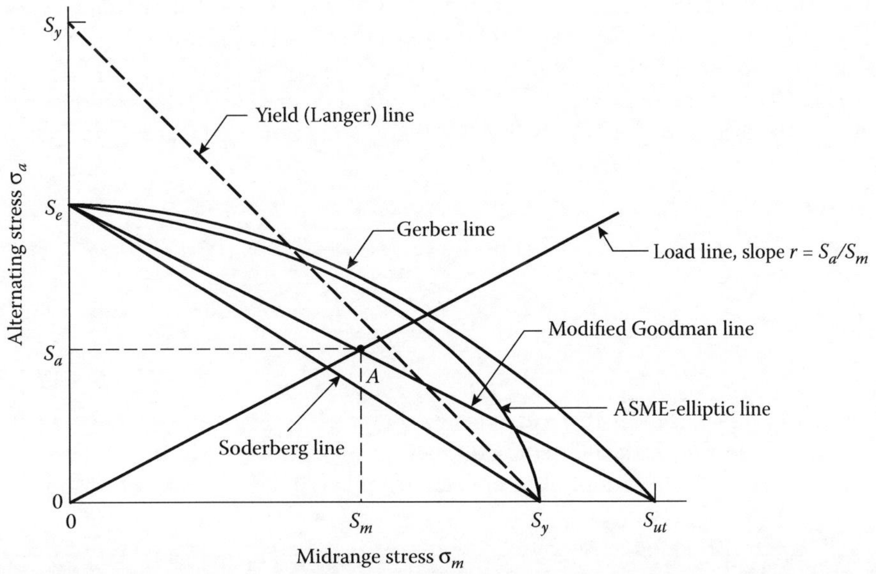

Figure 9.19

Fatigue diagram showing various criteria of failure. For each criterion, points on or "above" the respective line indicate failure.

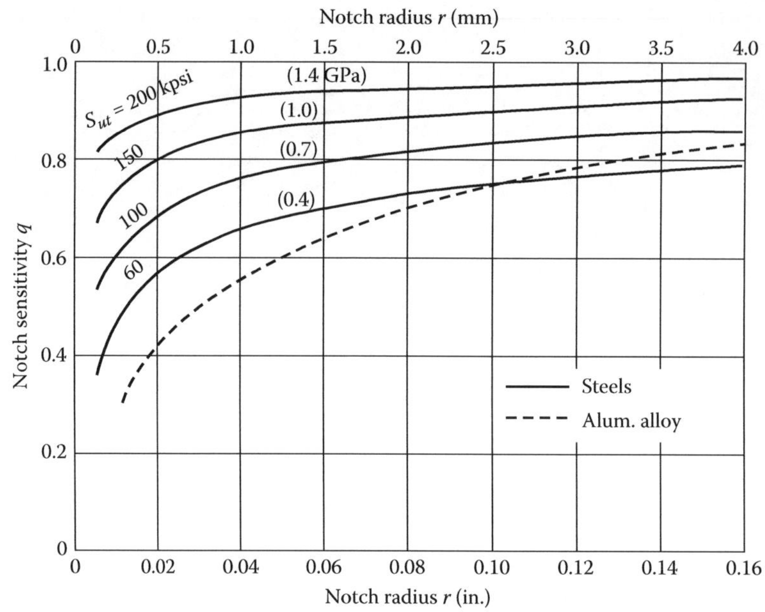

Figure 9.20

Sensitivity curves for materials in reversed torsion. For larger notch radii, use the values of q_{shear} corresponding to r=0.16 in (4 mm).

Shafts can be subjected to different kinds of loads such as

- Shafts subjected to *twisting loads/torque*
- Shafts subjected to *bending only*
- Shafts subjected to *torque and bending*
- Shafts subjected to *axial loads in addition to torque and bending*

9.4 Shaft Stresses and Design

Bending, torsion, and axial stresses may be present in both midrange and alternating components. For analysis, it is simple enough to combine the different types of stresses into alternating and midrange von Mises stresses, as shown in Section 6.14. It is sometimes convenient to customize the equations specifically for shaft applications. Axial loads are usually comparatively very small at critical locations where bending and torsion dominate, so they will be left out of the following equations. The fluctuating stresses due to bending and torsion are given by

$$\sigma_a = K_f \frac{M_a C}{I} \quad \text{and} \quad \sigma_m = K_f \frac{M_m C}{I}$$

$$\tau_a = K_{fs} \frac{T_a r}{J} \quad \text{and} \quad \tau_m = K_{fs} \frac{T_m r}{J} \tag{9.10}$$

where:
M_m and M_a are the midrange and alternating bending moments
T_m and T_a are the midrange and alternating torques
K_f and K_{fs} are the fatigue stress concentration factors for bending and torsion, respectively

Assuming a solid shaft with round cross section, appropriate geometry terms can be introduced for c, I, and J resulting in

$$\sigma_a = K_f \frac{32M_a}{\pi d^3} \quad \text{and} \quad \sigma_m = K_f \frac{32M_m}{\pi d^3}$$

Similarly, we can write

$$\tau_a = K_{fs}\frac{16T_a}{\pi d^3} \quad \text{and} \quad \tau_m = K_{fs}\frac{16T_m}{\pi d^3} \tag{9.11}$$

Combining these stresses in accordance with the distortion energy failure theory, the von Mises stresses for rotating round, solid shafts, neglecting axial loads, are given by

$$\sigma'_a = \sqrt{\sigma_a^2 + 3\tau_a^2} = \left[\left(K_f\frac{32M_a}{\pi d^3}\right)^2 + 3\left(K_{fs}\frac{16T_a}{\pi d^3}\right)^2\right]^{0.5}$$

and

$$\sigma'_m = \sqrt{\sigma_m^2 + 3\tau_m^2} = \left[\left(K_f\frac{32M_m}{\pi d^3}\right)^2 + 3\left(K_{fs}\frac{16T_m}{\pi d^3}\right)^2\right]^{0.5} \tag{9.12}$$

Now, the design of shaft that is the diameter of the shaft based on different fatigue failure theories can be found. For the modified Goodman fatigue failure, the diameter of the shaft d may be established by replacing σ'_a and σ'_m from Equation 9.12 into the following equation:

$$\frac{1}{n} = \frac{\sigma'_a}{S_e} + \frac{\sigma'_m}{S_{ut}}$$

The diameter d of the shaft is

$$d = \left(\frac{16n}{\pi}\left[\frac{1}{S_e}\left\{4(K_f M_a)^2 + 3(K_{fs}\tau_a)^2\right\}^{0.5} + \frac{1}{S_{ut}}\left\{4(K_f M_m)^2 + 3(K_{fs}T_m)^2\right\}^{0.5}\right]\right)^{1/3} \tag{9.13}$$

It is standard practice to identify the significant failure theory followed by the fatigue failure locus name. For example, DE-Gerber indicates the stresses are combined using the distortion energy (DE) theory, and Gerber criteria are used for fatigue failure.

DE-Gerber

$$d = \left(\frac{8nA}{\pi S_e}\left\{1 + \left[1 + \left(\frac{2BS_e}{AS_{ut}}\right)^2\right]^{0.5}\right\}\right)^{0.5} \tag{9.14}$$

where:

$$A = \sqrt{4(K_f M_a)^2 + 3(K_{fs}T_a)^2}$$

and

$$B = \sqrt{4(K_f M_m)^2 + 3(K_{fs}T_m)^2}$$

DE-ASME elliptic

$$d = \left\{\frac{16n}{\pi}\left[4\left(\frac{K_f M_a}{S_e}\right)^2 + 3\left(\frac{K_{fs}T_a}{S_e}\right)^2 + 4\left(\frac{K_f M_m}{S_y}\right)^2 + 3\left(\frac{K_{fs}T_m}{S_y}\right)^2\right]^{0.5}\right\}^{1/3} \tag{9.15}$$

DE-Soderberg

$$d = \left(\frac{16n}{\pi}\left\{\frac{1}{S_e}\left[4(K_f M_a)^2 + 3(K_{fs}T_a)^2\right]^{0.5} + \frac{1}{S_{yt}}\left[4(K_f M_m)^2 + 3(K_{fs}T_m)^2\right]^{0.5}\right\}\right)^{1/3} \tag{9.16}$$

DE-Goodman

$$d=\left(\frac{16n}{\pi}\left\{\frac{1}{S_e}\left[4\left(K_f M_a\right)^2+3\left(K_{fs}T_a\right)^2\right]^{0.5}+\frac{1}{S_{ut}}\left[4\left(K_f M_m\right)^2+3\left(K_{fs}T_m\right)^2\right]^{0.5}\right\}\right)^{1/3} \tag{9.17}$$

In Equations 9.14 through 9.17, n is the factor of safety, and some of the loads M_a, M_m, T_a, or T_m may be absent due to the nature of the load and such terms should be replaced by zero in the relevant equation, such as in components with constant bending and torsion, and so $M_m = 0$, and $T_a = 0$.

References

1. Budynas, R. G., Shigley, J. E., and Nisbett, J. K. (2008). *Mechanical Engineering Design*, 8th edn. McGraw Hill, New York.
2. Mott, R. (2004). *Machine Elements in Mechanical Design*, 4th edn. Prentice Hall.
3. Boothroyd, G., Dewhurst, P., and Knight, W. (2002). *Product Design for Manufacture and Assembly*, 2nd edn. Marcel Dekker, New York.
4. Whitney, D. E. (2004). *Mechanical Assembly: Their Design, Manufacture, and Role in Product Development*. Oxford University Press, New York.

Bibliography

Ashby, M. F. (2009). *Materials and the Environment: Eco-Informed Material Choice*. Elsevier, Boston.

Ashby, M. F. (2011). *Material Selection in Mechanical Design*, 4th edn. Elsevier, New York.

Chen, R. W., Navin-Chandra, D., and Print, F. B. (1993). A cost-benefit analysis model of product design for recyclability and its application. In *IEEE International Symposium on Electronics and the Environment* (pp. 178–183). Arlington, VA.

Desai, A. and Mital, A. (2005). Incorporating Work Factors in Design for Disassembly in Product Design. *Journal of Manufacturing Technology Management* 16: 712–732.

Dowie, T. and Kelly, P. (1994). Estimation of disassembly times, Technical Report DDR/TR15, Manchester Metropolitan University, UK.

ISO 14044:2010. (2010). *Environmental Management-Life Cycle Assessment, 2010.*

Kroll, E. and Carver, B. S. (1999). Disassembly analysis through time estimation and other metrics. *Robotics and Computer-Integrated Manufacturing*: 191–200.

Kroll, E. and Hanft, T. A. (1998). Quantitative evaluation of product disassembly for recycling. *Research in Engineering Design-Theory, Applications, and Concurrent Engineering* 10: 1–14.

Lee, S. G., Lye, S. W., and Khoo, M. K. (2001). A multi-objective methodology for evaluating product end-of-life options and disassembly, *International Journal of Advanced Manufacturing Technology* 18: 148–156.

Lidwell, W., Holden, K., and Butler, J. (2003).*Universal Principles of Design*. Rockport, Gloucester, MA.

McGlothlin, S. and Kroll, E. (1995). Systematic estimation of disassembly difficulties: application to computer monitors. In *IEEE International Symposium on Electronics & the Environment* (pp. 83–88), Orlando, FL. *Jurnal Mekanikal,* December 2008.

Segur, A. B. Predetermined Motion Time System (PMTS). mtm_international.org

Skakoon, J. (2000). *Detailed Mechanical Design: A Practical Guide, Technology & Engineering*. ASME Press, New York.

Skakoon, J. G. (2008). *The Elements of Mechanical Design*. ASME Press, New York.

Slocum, A. H. (2007). *Fundamentals of Design*, Massachusetts Institute of Technology, Cambridge, MA.

Slocum, A. H. (1992). *Precision Machine Design, Technology & Engineering*. Prentice Hall.

Soemers, H. (2010). *Design Principles for Precision Mechanisms,* ISBN 9036531039.

Suga, T., Saneshige, K., and Fujimoto, J. (1996). Quantitative disassembly Evaluation. In *Proceedings of the 1996 IEEE International Symposium on Electronics & the Environment* (pp. 19–24). Dallas, TX.

United States Environmental Protection Agency (EPA). (2010). *Life Cycle Assessment*, 2010, Science & Technology.

10

Bearings and Green Engineering

10.1 Shaft Bearings

A major problem in dynamic machinery occurs due to friction, which causes loss of energy. From the viewpoint of sustainability considerations, this is a major challenge. It has been estimated that from one-third (1/3) to one-half (1/2) of all the energy produced in the world is wasted in friction. Friction may be defined as the opposing force that is developed when two surfaces move relative to each other. Lubrication reduces the friction developed, but it has its own pitfalls; lubricants, which may be in the liquid, solid, or gaseous state, can be toxic and hazardous to health. Plain bearings, as distinguished from rolling-element bearings, are usually classified as plain journal or sleeve, thrust, spherical, pivot, or shoe-type thrust. Sometimes, bearings are classified based on the type of lubricant used. A hydrodynamically lubricated bearing is one that uses a fluid lubricant (liquid or gas) to separate the moving surfaces completely. When this ideal condition is attained, it is commonly called a *thick film*. Other commonly used terms for this type of lubrication are *fluid film*, *stable*, or *perfect*. Bearings are essential components of practically all systems, ranging from automobiles to CD players to industrial production machinery. Virtually any application involving moving elements contains one or more bearings.

Environmental impacts exist at each step of the bearing life cycle. These include energy consumption during production (embodied energy), energy loss caused by friction during use, and the production of waste, for example through lubricant leakage or the disposal of obsolete bearings. Due to the omnipresence of bearings, the environmental problems they cause are present everywhere, from land to sea to air. We will first present the basics of bearing technology and then discuss our objective of bearings design with a smaller environmental footprint in terms of reduced energy and lubrication consumption.

Some of the requirements of bearings were set out earlier when shaft location was discussed in Chapter 9. Bearings may be classified in a number of different ways. However, for our purposes, it is sufficient to use two groups:

- Sliding contact
- Rolling contact

Before moving on to consider these two groups, it is worth pointing out the necessary conditions for stable and adequately constrained mounting of a shaft. While being free to rotate, the whole shaft must generally be constrained against *radial* and *axial* movement. In general, adequate radial constraint requires two bearings, relatively widely spaced along the length of the shaft. Only in the case of *very* short shafts should the use of only one bearing be considered. Again, in general, the shaft will need axial restraint in two directions. Sometimes, both axial restraints are applied by one bearing; in other cases, each of two bearings may provide restraint in one axial direction.

Long shafts usually require more than two bearings, especially if lateral rigidity (small deflection under load) is required. Especially for long shafts, questions of axial expansion due to heating must be considered. Potential changes in length of the housing or the shaft (or both) due to heating while in use usually require both axial constraints to be provided by the same bearing.

10.2 Sliding-Contact Bearings

The simplest type of these bearings is known as a journal, plain, or sleeve bearing. A bearing of this type locates a shaft radially. There is no provision for axial location. Figure 10.1 shows two examples of this type of bearing and the different constructions that are available.

When the journal and the bearings are properly lubricated, that is, there is a film of clean, noncorrosive lubricant between them, separating the two surfaces in contact, the only requirement of the bearing material is that it should have sufficient strength and rigidity. Materials used for sliding-contact bearings should possess the following properties: compressive strength, fatigue strength, confirmability, corrosion resistance, and so on.

The materials commonly used for sliding-contact bearings are

Babbit metal: tin base babbit, lead base babbit

Bronzes: gunmetal, phosphor bronze, cast iron, silver, nonmetallic bearings (Figures 10.2 and 10.3)

10.2.1 Materials for Plain Bearings

Where rubbing contact occurs between two machine parts, it is usual to make the parts from dissimilar materials. In the case of journal bearings, the shaft to be supported is almost always made from carbon steel, so bearings are seldom made of steel. Frequently used bearing materials are

- Bronze, usually in the form of a bush
- White metal, a tin/antimony/copper alloy that is often bonded to a steel shell
- Copper/lead/indium, often used for automotive engine bearings, usually bonded to a steel shell
- Cast iron, with a shaft running directly in a machined bore in the cast component
- Various plastics such as PTFE, nylon, Delrin

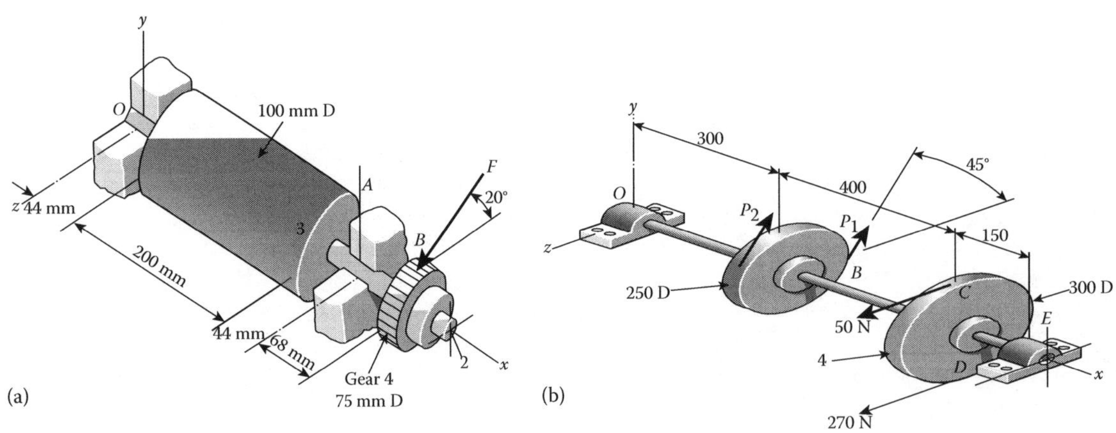

Figure 10.1

Examples of journal bearings. (a) The shaft simply runs in holes in the structural members of the machine, which may be steel or cast iron. (b) The shaft runs in separate components of the general style known as plummer blocks. Plummer blocks are usually fitted with bushes made from a material which has good wearing properties, such as bronze. The advantage of using bushes is that the bush can be replaced when it becomes worn. (From Shigley, J. E. and Mischke, C. R., *Mechanical Engineering Design*, 6th edn. McGraw Hill, 2001 (1). With permission.)

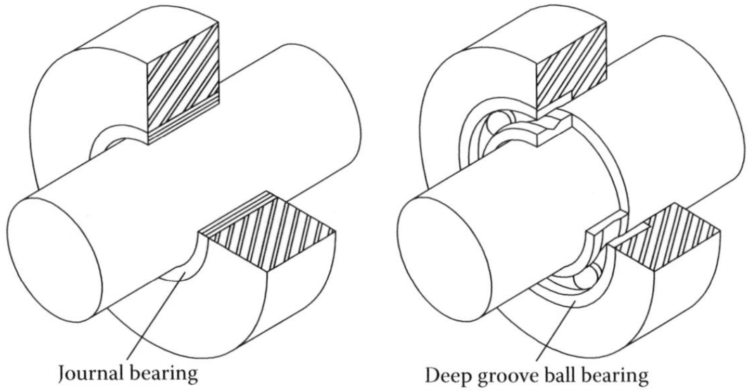

Figure 10.2

Comparison of the configuration of a bush-type bearing with a rolling-contact bearing.

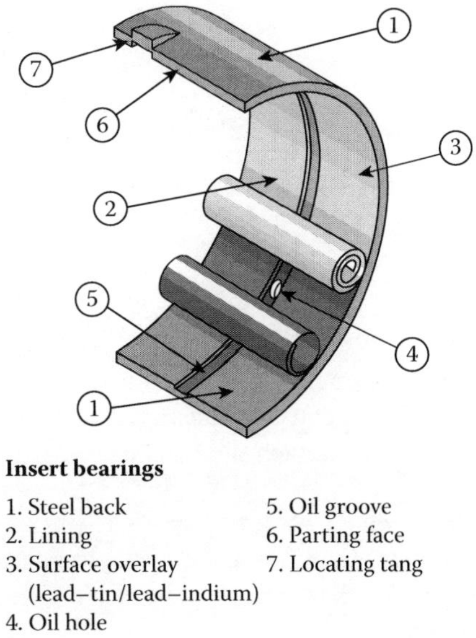

Figure 10.3

An important class of plain bearings is the *slipper bearing* or *insert bearing*, almost universally used for automotive engine bearings. Two semicircular insert bearings are used to form the complete bearing. Note the steel backing or shell, the lining that forms the main load-carrying section of the bearing, and the thin surface overlay of softer material for reduction of friction and quick "bedding-in" of the bearing.

10.2.2 Lubrication of Plain Bearings

Plain bearings are usually lubricated by grease or oil, supplied by an oil-drip lubricator or a ring oiler or by periodic application of an oilcan. Alternatively, it is possible to design a plain bearing as an air- or gas-lubricated bearing, which requires a constant supply of gas under pressure. Gas-lubricated bearings are beyond the scope of these notes.

Plain bearings in heavy-duty applications require lubrication by a continuous flow of oil under pressure. The pressure helps to create a film of oil between the two metal surfaces. Also, by conducting away heat, the oil flow provides an important cooling function to prevent overheating of the bearing. One well-known example of this type of lubrication is the crankshaft bearings of virtually all car engines (Figure 10.4). These bearings fail very rapidly if the oil supply fails for any reason.

10.2.3 Characteristics of Plain Bearings

- Friction is higher than for rolling-contact bearings. Starting friction of plain bearings is significantly higher than running friction.
- Well-designed plain bearings can have an extremely long life. However, they can fail without warning.
- Plain bearings run more quietly than rolling-contact bearings.

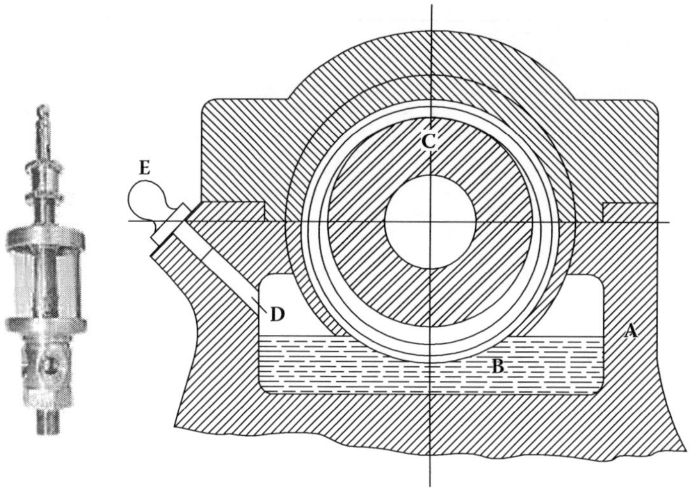

(a) (b)

Figure 10.4

(a) An *oil lubricator* in which the bowl is filled with oil and is set to drip oil at a predetermined rate into the bearing. (b) Schematic of a *loose oil ring oiler.* The lower part of the *loose ring* (B) dips into the oil bath (A) and drags oil to the top of the bearing (C). Oil is replenished when necessary through (D). (E) Oil lubricator.

Pillow blocks, also known as *plummer blocks*, are a useful off-the-shelf component that can simplify shaft mounting in some cases. They are normally available with a range of different rolling-contact bearings fitted within the housing: ball or roller, single or double row, self-aligning or deep-groove type, depending on the application and the probability of misalignment. The housing also makes provision for seals to retain lubricant, while the mounting feet make it easy to attach the housing to a base plate or frame.

10.2.3.1 Design of Plain Bearing

The design procedure used in this example makes use of dimensionless groups plotted against the Sommerfeld number on a series of charts to determine other values for the bearing. These charts are reproduced in many texts on mechanical engineering design.

Example 10.1

A plain bearing with hydrodynamic lubrication is to be designed to carry a load of 2500 N at a shaft speed of 30 revolutions per second.

Initial assumptions are that SAE grade 40 oil will be used, and the mean oil temperature in the bearing is 68.3°C (which corresponds to 155°F). At this temperature, the oil has an absolute viscosity of 4 microreyn (i.e., 27.5 mPa.s or 0.0275 Pa.s), as shown in Figure 10.5. The density is assumed to be 850 kg/m^3 and the specific heat 1800 J/kg °C.

As a starting point, it will be assumed that the length/diameter ratio = 1, and for a first calculation, $l = d = 2r = 40$ mm will be used.

It is also assumed that the bearing/shaft is between a "free running fit" (H9/d9) and a "close running fit" (H8/f7) with a radial clearance of 0.04 mm.

First check that the "unit load" (load/projected area of bearing) is acceptable.

Unit load = load/(length × diameter) = 2500/(0.04 × 0.04) = 1.56 MPa. This is perhaps on the high side for machinery (although low for internal combustion engines), but this value will be accepted for this example.

Bearing designs are commonly based on charts using the "bearing characteristic number" or Sommerfeld number. This is defined (nondimensionally) as

$$S = (r/c)^2(\mu N/P)$$

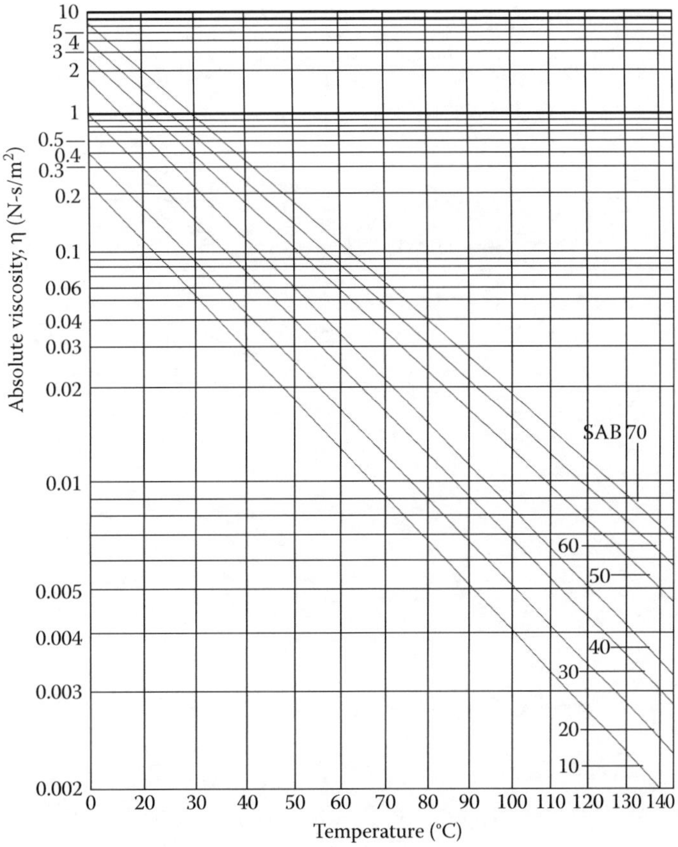

Figure 10.5

Temperature vs. viscosity of oil.

where:

- r is radius
- c is the radial clearance
- μ is the absolute viscosity
- N is shaft speed in revolutions per second
- P is the unit load

As there is no information given about rotation of the bush or load vector, it is assumed that these are not rotating and that $S = S'$. So

$$S = (0.02/0.00004)^2 (0.0275 \times 30/1{,}563{,}000)$$

$$= 0.132$$

The first chart in the series plots the minimum film thickness variable h_o/c, where h_o is the minimum film thickness, against the Sommerfeld number for a variety of l/d ratios (Figure 10.6). From this chart, $h_o/c = 0.41$, so the minimum film thickness is

$$h_o = 0.41 \times 0.00004 = 0.0000164 \text{ m}$$

It is important that the minimum film thickness is significantly greater than the maximum surface roughness. Keep in mind the fact that surface roughness is usually given as an average roughness amplitude (*Ra*) and that the extreme peak-to-valley depths may well be three to six times greater than the *Ra*.

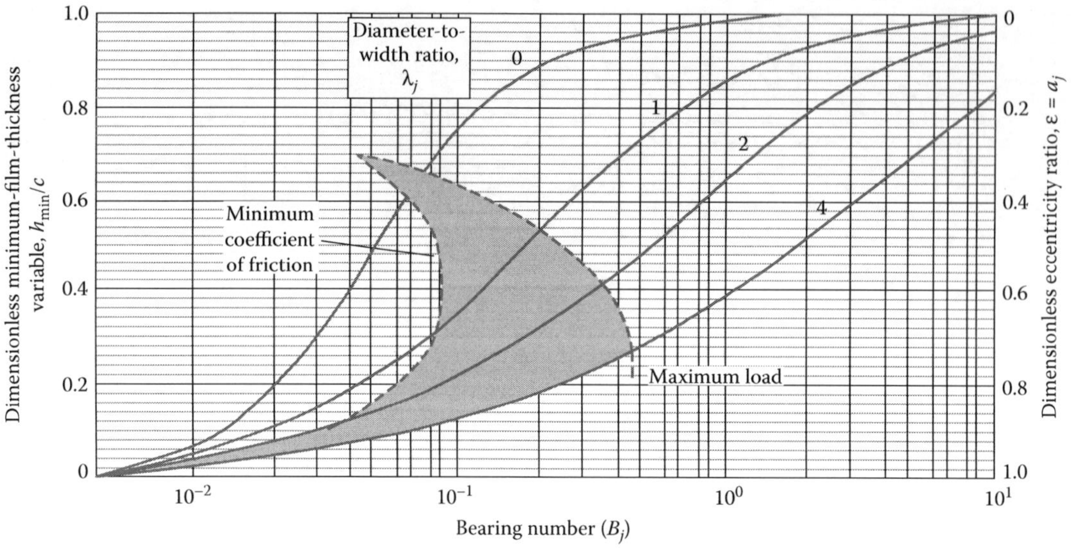

Figure 10.6

Bearing number vs. minimum film thickness.

Minimum film thickness = radial clearance − eccentricity

$h_o = c - \varepsilon$, divide both sides by c
$h_o/c = 1 - \varepsilon$, where ε is the eccentricity ratio

The eccentricity of this design is $e = c - h_o$:

$$e = 0.00004 - 0.0000164 = 0.0000236\ \text{m}$$

Two lines indicating optima, maximum load, and minimum power loss are also marked on the first chart, and the zone between these lines is normally considered to be a recommended operating region (Figure 10.6).

The second chart needed plots the coefficient of friction variable $f\,r/c$, where f is the coefficient of friction, against the Sommerfeld number (Figure 10.7). From this chart, the value of $f\,r/c$ is 3.3, so

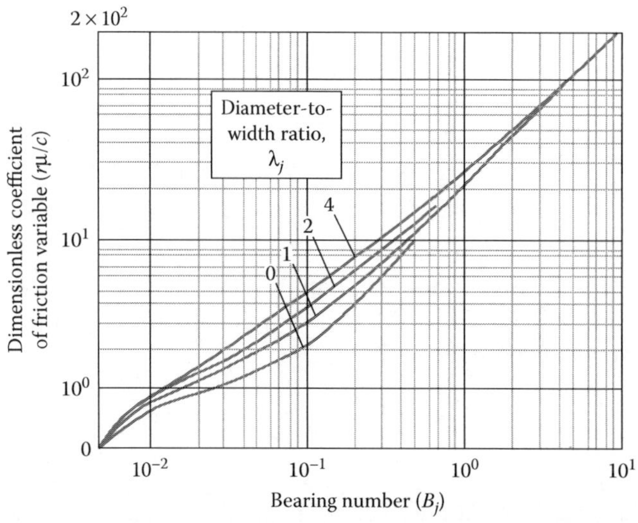

Figure 10.7

Bearing number vs. friction coefficient variable.

$$f = 3.3 \times 0.00004 / 0.02 = 0.0066$$

Torque to overcome the friction = coefficient of friction × load × radius, substituting values

$$T = 0.0066 \times 2500 \times 0.02 = 0.33 \text{ Nm}$$

Power lost is torque × angular velocity, substituting values

Power lost = 0.33 × 2 × 3.14159 × 30 = 62.2 W

The third chart plots the flow variable $Q/rcNl$ against S. From this chart, $Q/(rcNl) = 4.3$, substituting values (Figure 10.8). Total flow $Q = 4.3 \times 0.02 \times 0.00004 \times 30 \times 0.04 = 4.128\ e^{-6}\ m^3/s$.

The fourth chart plots the (side) flow ratio (leakage ratio) against S, and from this chart the (side) flow ratio $Q_s/Q = 0.66$, substituting values (Figure 10.9). The side leakage is $Q_s = 0.66 \times 4.128\ e^{-6} = 2.724\ e^{-6}\ m^3/s$.

10.2.3.1.1 Rise in Oil Temperature It is assumed that the oil that is retained in the bearing rises in temperature by ΔT, and the oil that leaks out of the side of the bearing rises in temperature by $\Delta T/2$. First, the volume flow rates need to be converted to kilograms per second.

Total oil mass flow rate = $4.128\ e^{-6} \times 850 = 0.003509$ kg/s
Side leakage mass flow rate = $0.66 \times 0.003509 = 0.002316$ kg/s
Mass flow rate remaining in bearing = $0.34 \times 0.003509 = 0.001193$ kg/s

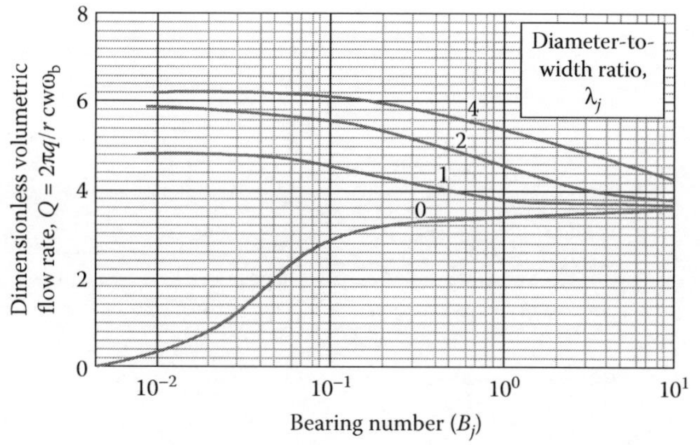

Figure 10.8

Bearing characteristic number vs. flow rate.

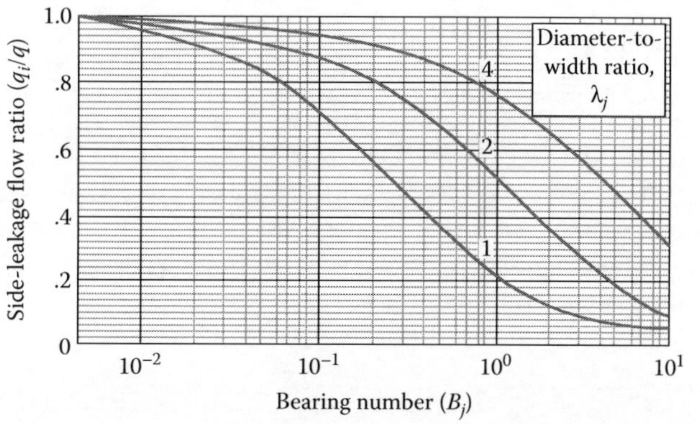

Figure 10.9

Bearing number vs. side-leakage ratio.

Carrying out an energy balance, assuming that all the energy put into the oil raises only the oil temperature,

$62.2 = 1800\ (0.002316\ \Delta T/2 + 0.001193\ \Delta T)$
Rise in oil temperature $\Delta T = 14.7°C$, assume 15°C

10.2.3.1.2 Oil cooling As it was assumed at the outset that the mean oil temperature was 68.3°C, this means that the inlet oil temperature would be 68.3–7.5 = 60.8°C and the outlet temperature 75.8°C. Provided there is sufficient cooling available to ensure that the oil from the bearing outlet is cooled to 60°C or less, this design may well be acceptable. As the viscosity of most oils changes very rapidly (typically halving for a 10°C–15°C temperature rise), it is important to check that operation will still be satisfactory outside these limits—for example, when starting from cold and during short periods of overloading. As well as Figures 5.5 through 5.9, there are also charts to provide additional information such as the maximum film pressure, the angular position where the oil film thickness is a minimum, and where it terminates.

Example 10.2

Design of Hydrodynamic Journal Bearing

Shaft diameter (D) = 40 mm
Bearing length (L) = 40 mm
Diametric clearance (ΔD) = 40×10^{-3} mm
Rotational speed (n) = 300 rpm
Minimum film thickness (h_0) = 3×10^{-3} mm
Viscosity (μ) = 0.04 Pas
Density (ρ) = 0.85×10^3 kg/m^3
Specific heat capacity (c) = 1.8×10^3 J/(kg.K)

Calculated Values

Bearing length number (L/D) = 1
Clearance number ($D/\Delta D$) = 1000
Eccentricity ratio $\left(\varepsilon = \dfrac{\Delta D/2 - h_0}{\Delta D/2}\right) = 0.85$
Sommerfeld number (S) = 9.116
Attitude angle (β) = 37.4°
Flow number (Cq)(C_q) = 0.68
Friction number $\left(f\dfrac{R}{\Delta R}\right) = 1.199$
Coefficient of friction (f) = 0.0012
Load capacity (F) = 9.164×10^3 N
Stiffness $\left(S = -\dfrac{dF}{dh}\right) = 3701 \times 10^6$ N/m
Power loss (N) = 6.9 W
Side leakage (Q) = 0.342×10^{-6} m^3/s
Frictional heating (ΔT) = 13.2 K

Hydrodynamic journal bearings become unstable with shaft eccentricity. For this reason, high-speed or low-loaded hydrodynamic bearings need to be designed with less stiffness. Frictional heating is calculated assuming adiabatic boundary conditions for the bearing, that is, power loss N is drained with the fluid flow Q through the bearing. Maximum pressure occurs in the converging part of the film, while cavitation pressure manifests in the diverging part of the film.

10.3 Rolling-Contact Bearings

These bearings use elements such as balls or rollers to avoid sliding contact. The usual practice is to provide two more or less cylindrical housings or races, the inner of which fits onto the shaft and rotates with it, while the outer race fits into the fixed or nonrotating component. The rolling members (balls or rollers) run freely between the two races. In practice, it is impossible to achieve pure rolling, and there will always be some sliding movement within the bearing. Despite this, such bearings run very freely and are often referred to as *anti-friction bearings.*

In general, cylindrical roller bearings do not provide axial restraint—the rollers in their cage are free to slide axially along the outer race. While this feature may be useful to allow for the change of shaft length due to heating, these bearings cannot be used unless there is provision elsewhere to provide axial location and, if necessary, carry axial loading or thrust. Some types of cylindrical roller bearings do provide axial restraint in one direction by means of a shoulder in the outer race. Note that any axial or thrust loading carried by a roller bearing must be resisted by the ends of the rollers contacting the shoulders in the inner and outer races. Since this will result in sliding motion rather than rolling, axial loading must be kept low to avoid scoring, overheating, and seizure.

However, bearings must

1. Allow rotation
2. Reduce friction to a low level
3. Reduce wear of both shaft and bearings to an acceptable level

Since shafts are often used to locate items such as gears in precise locations, bearings must

4. Have minimal clearance (or slack) between bearing and shaft
5. Locate the shaft both radially and axially

Bearings are one of the key components that determine the service interval for rotating machinery; they can even limit the life of the machine. Therefore, knowledge and understanding of bearings are important to both designers and users. In selecting the bearing to be used, one must consider a variety of factors:

- Loads, maximum and minimum, static and dynamic
- Speed and running pattern
- Stiffness
- Temperature levels and gradients, heat conduction
- Lubrication: viscosity, stiffness, durability
- Degree of contamination, shielding
- Mounting
- Maintenance
- Lifetime and service time

Bearing selection or analysis is futile without rotor dynamic data from the shaft and other rotating parts. A correctly dimensioned shaft is fundamental, and there are many criteria that need to be fulfilled:

- Magnetic behavior, especially for two-pole motors
- Fatigue durability
- Limited deflection for impeller, seal, and rotor positions
- No natural frequencies of the rotating system close to rotational speed
- Low impact on the surroundings regarding vibrations
- Limited angular displacement at bearing positions
- Temperature conduction to ensure suitable bearing temperature

Types of rolling-contact bearing:

1. Deep-groove ball bearing
2. Cylindrical roller bearing
3. Angular contact bearing
4. Self-aligning bearing
5. Taper roller bearing
6. Thrust ball bearing

10.3.1 Characteristics of Rolling-Contact Bearings

- They have very low friction, particularly when starting.
- They require larger radial sizes than plain bearings but need shorter axial length.
- They are generally more costly than plain bearings.

Figure 10.10

An example of a cylindrical roller bearing.

- They have a finite life. The rolling elements and races eventually fail by flaking of the contact surfaces, a form of fatigue failure.
- Rolling-contact bearings are not as quiet in operation as plain bearings. As progressive surface flaking occurs, the bearings become increasingly noisy in operation and provide ample warning of impending failure. Compare with plain bearings, which can fail very rapidly and almost without warning.

A roller bearing does not last forever; sooner or later, fatigue or wear, or lubricant deterioration, will ultimately destroy the bearing's ability to function properly. In general, cylindrical roller bearings do not provide axial restraint—the rollers in their cage are free to slide axially along the outer race (Figure 10.10).

10.3.2 Tapered Roller Bearing

Tapered roller bearings have very high radial-load capacity and high axial-load capacity in one direction. For this reason, tapered roller bearings are often used in pairs, mounted in opposite directions, to cater for high radial loads and axial loads in either direction. Tapered roller bearings are often used, as in Figure 10.11, in the wheel bearings of cars and heavy trucks. Note that each roller is actually a frustrum of a cone, so that true rolling contact is maintained over the whole length of the roller.

Figure 10.11

A double-row tapered roller bearing, used to increase the load-carrying capacity of the bearing and to provide axial restraint in two directions.

10.3.3 Roller Thrust Bearing

Roller thrust bearings are also tapered rollers to maintain true rolling over the length of the roller. Roller thrust bearings have a greater load capacity than ball thrust bearings (Figure 10.12).

10.3.4 Needle Roller Bearings

One of the main advantages of needle roller bearings is their small diameter, which allows them to be fitted into small spaces. They usually do not have a separate inner race, but run directly onto the shaft, which must be suitably hard and with a precision-ground surface. When these bearings fail, they often damage the shaft surface, and the shaft needs to be replaced, unlike ball and roller bearings, which have an inner race, so that bearing damage is normally confined to the bearing races, and the shaft is undamaged (Figure 10.13).

10.3.5 Ball Bearings

In ball bearings in general, the rolling elements are spheres of high-quality, hardened, polished alloy steel, rolling in hardened alloy steel inner and outer races. Figure 10.14 shows a deep-groove ball bearing and the associated terminology. The balls are uniformly spaced around the races by a cage, which is riveted into place as the last step in the assembly process. As well as spacing the balls, the cage performs the essential function of preventing the balls from contacting each other, which would cause scuffing and rapid wear.

Figure 10.12

Roller thrust bearings.

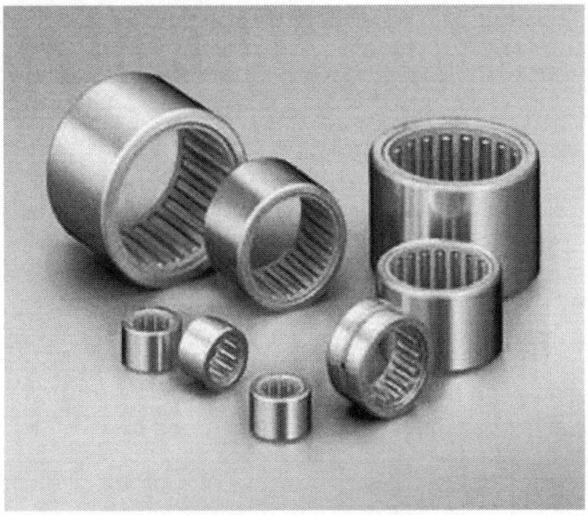

Figure 10.13

Examples of needle roller bearings.

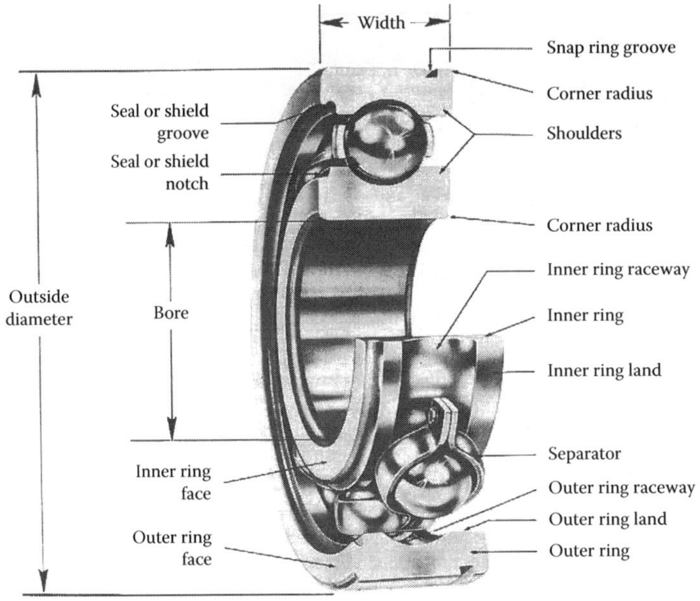

Figure 10.14

Ball bearing and terminology.

However, there remains some sliding contact and consequent friction between the balls and the cage, so lubrication is essential for long life.

In ball bearings in general, the rolling elements are spheres of high-quality, hardened, polished alloy steel, rolling in hardened alloy steel inner and outer rings or races.

Figure 10.16 shows how the commonly used deep-groove ball bearing is assembled. After the maximum number of balls has been inserted between the inner and outer races, the balls are uniformly spaced around the races by a cage, which is riveted into place as the last step in the assembly process. As well as spacing the balls, the cage performs the essential function of preventing the balls from contacting each other, which would cause scuffing and rapid wear. However, there remains some sliding contact and consequent friction between the balls and the cage, so lubrication is essential for long life.

An alternative to deep-groove ball bearings is the *filling notch ball bearing*, in which small "notches" are cut out of one side of the inner and outer races. When the notches are aligned, the required number of balls may be slipped into the raceways before being spaced by the cage, as for the deep-groove bearing. However, the notches create a slight discontinuity in the path of the balls, tending to decrease the life of the bearing. Since in deep-groove bearings the balls run in shallow grooves in the races (Figures 10.15 and 10.16), these bearings are able to support some axial loading as well as radial.

Figure 10.15

Double-row spherical self-aligning ball bearings.

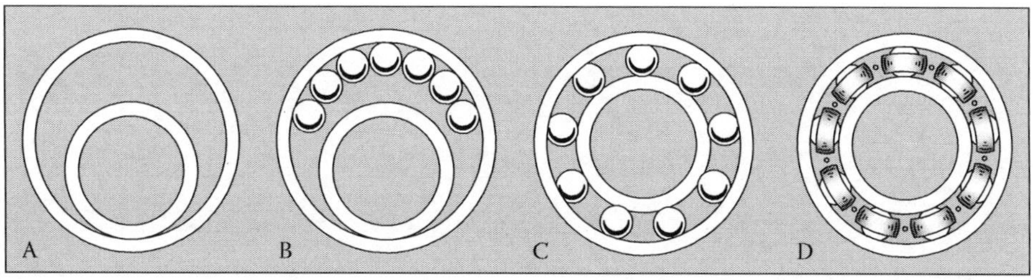

(a)

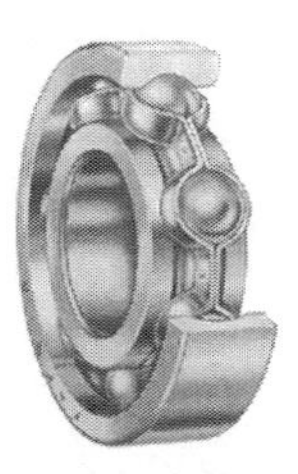

(b)

Figure 10.16

(a) Steps in the assembly of the deep groove-type ball bearing. (b) Sectioned view of an assembled deep-groove ball bearing. (Courtesy of The Timken Company, at http://science.howstuffworks.com/bearing3.htm.)

10.3.5.1 Ball Bearing Load Ratings and Life Calculations: General Notes about Bearing Life

Bearing life means how long the ball bearing will last under standard operating conditions. There will be a number of factors involved in the life of a bearing, including the amount of bearing load the ball bearing will be expected to handle. It is important to know the bearing life of ball bearings so that one can plan ahead to replace the bearing. Bearing life is calculated in number of revolutions. We need to establish how much time a revolution takes, and for what percentage of time the ball bearing is in continuous revolution in the application, in order to determine bearing life. The bearing life statistic is a measure of the amount of time, in revolutions, by which 90% of the ball bearings can be expected to have survived. This does not mean that every bearing will fail as soon as this number has been exceeded, of course. The median life for ball bearings, also referred to as the mean time before failure (MTBF), is about five times the basic life number for the bearing. This means that by five times the basic life rating revolutions (see Section 10.3.5.2), about half the ball bearings can be expected to have failed. It is important to remember that there may be some variation in results. Factors such as proper lubrication, care and handling of the bearing, and stress on the bearing can result in very significant fluctuations in the life of the ball bearing.

10.3.5.2 Basic Life Rating

The basic life rating (L_{10}) is defined in the manufacturer's specification as "dynamic load ratings and rating life for rolling bearings" as follows:

The basic life rating is the life obtained with 90% reliability, when an individual bearing or an identical group of bearings are manufactured with common materials, common manufacturing processes and quality, and operate under the same conventional conditions. L_{10} life is the accumulated rotation where 90% of survive without material flaking when they are operated under fixed conditions, of a population of bearings.

10.3.6 Calculation of Bearing Life

Bearing life calculation refers to the life of an individual bearing, in terms of revolutions, that can be assumed under continuous operating conditions designed to specific system operating conditions.

$$L_{10} = \left(10^6/60n\right)\times\left(C/P\right)^3$$

where:

L_{10} = life of an individual bearing under continuous run (hours)
C = dynamic load (Newtons)
n = no. of rotations per minute
P = applied load (effective)
P_{0r} = $X_0F_r + Y_0F_a$ (3 for ball bearings and 10/3 for roller bearings)
(X_0 = 1 and $Y_0 = 0$ always)

X_0 and Y_0 are factors required for the calculation of equivalent dynamic bearing load for single- and double-row deep-groove ball bearings, dependent upon axial load (F_a) and radial load (F_r).

Therefore, if $P_0 < F_r$, $P = P_0 = F_r$, P_0 = axial load and is equal to

$$P_o = 0.6F_r + 0.5F_a$$

where:

F_r is the radial force
F_a is the axial force

10.3.6.1 Calculation of Maximum Permissible Axial Load in Newtons

$$F_{ap} = 3Bd$$

where:

B = bearing width
d = bearing bore diameter
F_{ap} = maximum permissible load

10.3.6.2 Calculation of Minimum Loads

Calculation of minimum loads in a bearing under operating conditions means calculation of the loads acting over a bearing under static or dynamic conditions:

$$F_{am} = A(n/1000)^2$$

where:

F_{am} = minimum loads (Newtons)
n = number of rotations per minute
A = cross-sectional area of bore

10.3.6.3 Calculation of System Bearing Life

Let us understand what system bearing life means. This can be clarified as the criteria that apply to the bearing "system" and not to the individual bearings alone, which should be general practice in the industry. This means that the individual life of one bearing may be 100,000 revolutions, while another may last for only 26,500 revolutions. But, if the value is plugged in, the system bearing life becomes 25,000 revolutions. This means that, as an effect of their combined life, all the bearings employed in a system should perform at least 25,000 revolutions at continuous run.

According to American Petroleum Institute (API) Standards, the value for the system bearing life should be less than the life of the individual bearing with the shortest life span. The formula for system bearing life is

$$L_{10h} = \left[\left(1/L_{10hA}\right) + \left(1/L_{10hB}\right) + \left(1/L_{10hB}\right) + \cdots\cdots + \left(1/L_{10hN}\right)\right]$$

where:

L_{10hA} is the basic rating life, L_{10h}, per ISO 281 for bearing A
L_{10hB} is the basic rating life, L_{10h}, per ISO 281 for bearing B
L_{10hN} is the basic rating life, L_{10h}, per ISO 281 for bearing N
N is the number of bearings

10.3.7 General Notes about Bearing Load

There are two types of bearing load to consider with a ball bearing: radial load, which represents loads perpendicular to the shaft, and axial, or thrust, load, which represents loads parallel to the shaft. A ball bearing can handle both these kinds of loads, but different loads affect bearings in different ways, so multiple bearing rating calculations are required. The load-bearing calculations are outlined by the Japanese Industrial Standards (JIS) system, which provides standards not only for ball bearings but also for a wide variety of industrial activities requiring accurate measurements. JIS measurements are widely accepted standards throughout the world. Ball bearing standards are found under JIS B, the classification regarding mechanical engineering, which includes bearings (other classifications include A for civil engineering and C for electrical engineering). Use the load ratings to determine how many of each type of ball bearing will be useful, and which type of bearing will be appropriate to your needs, so that the bearing will have a long, effective life.

Table 10.1 Axial Load Ratio vs. X and Y Values

Axial Load Ratio		$\frac{F_a}{F_r} < e$		$\frac{F_a}{F_r} > e$		
Units						
N	kgf	X	Y	X	Y	e
$\frac{F_a}{iZD_w^2}$						
0.172	0.0175				2.30	0.19
0.345	0.0352				1.99	0.22
0.689	0.0703				1.71	0.26
1.03	0.105	1	0	0.56	1.55	0.28
1.38	0.143				1.45	0.30
2.07	0.211				1.31	0.34
3.45	0.352				1.15	0.38
5.17	0.527				1.04	0.42
6.89	0.703				1.00	0.44

Note: i, no. of rows; Z, no. of balls; D_w, Ball diameter (mm). Values for X and Y that are not in Table 10.1 can be calculated by linear interpolation.

The static and dynamic capacity of a bearing are important considerations in roller-bearing design. The static load-carrying capacity of a bearing is defined as the static load that corresponds to a total, permanent deformation of balls and races, at the most heavily stressed point of contact, equal to 0.0001 of the ball diameter.

Dynamic load-carrying capacity of a bearing is defined as the radial load in radial bearings (or thrust load in thrust bearing) that can be carried for a minimum life of 1 million revolutions.

10.3.8 Dynamic Equivalent Radial Load Factor (P_r)

The dynamic equivalent radial load factor (P_r) is defined as "the direction and magnitude to the bearing, which is able to obtain the same life under the actual load and rotation conditions." From the calculation formula and Table 10.1, the axial and radial loads are replaced by P_r.

$$P_r = XV\,F_r + Y\,F_a$$

where:

X and Y are taken from Table 10.1
F_r = radial load (N or kgf)
V = 1 and 1.2 for rotation of inner and outer ring, respectively
F_a = axial load (N or kgf)

10.3.9 Life of a Ball Bearing

The required life of a ball bearing, and its definition, depend on the application and requirements of the equipment. Because there are many different applications for equipment, definitions of life also vary. Some different types of life include rating life, noise life, lubricant life, and performance life.

When noise becomes louder than the setup level, it is the end of noise life. When lubricants lose their function due to degradation, it is the end of lubricant life. When revolutions per minute and runout go beyond their setup points, it is the end of performance life.

In this section, the rating life of a single-row deep-groove ball bearing, which is specified in JIS B 1518, is explained. Rating life is a predicted life calculated on the basis of the basic dynamic radial load. In general, if bearings are made of high-quality steel with high-level production skills, the load rating and rating life can be calculated based on the specification defined in JIS and ISO.

Example 10.3: Calculations of Life for Ball Bearing

Under the conditions below, the L_{10} life of R-830ZZ is calculated.

$$\left.\begin{array}{l}\text{Basic dynamic radial load rating}\left(C_r\right)=553\text{ N}\\ \text{Ball diameter}\left(D_w\right)=1.5875\text{ mm}\\ \text{Number of balls}\left(Z\right)=6\\ \text{RPM}\left(n\right)=3600/\text{min}\\ \text{Radial load}\left(F_r\right)=6\text{ N}\\ \text{Axial load}\left(F_a\right)=8\text{ N}\end{array}\right]$$

1. Calculate the ratio of axial load

$$\frac{F_a}{ZD_w{}^2}=\frac{8}{6\times\left(1.5875\right)^2}=0.529$$

2. Calculate e value relative to the ratio of axial load

$$e=0.22+\frac{\left(0.529-0.345\right)}{\left(0.689-0.345\right)}\times\left(0.26-0.22\right)=0.24$$

3. Calculate the ratio of radial and axial loads

$$\frac{F_a}{F_r}=\frac{8}{6}=1.33$$

4. Compare load ratio and e value

$$\frac{F_a}{F_r}=1.33,\ \mathrm{e}=0.24\ \therefore\ \frac{F_a}{F_r}>\mathrm{e}$$

5. Determine X and Y

$$X=0.56$$

$$Y=1.99-\frac{\left(0.529-0.345\right)}{\left(0.689-0.345\right)}\left(1.99-1.71\right)=1.84$$

6. Calculate dynamic equivalent load

$$P_r=0.56\times6+1.84\times8=18.08$$

7. Calculate life hours

$$L_{10}=\left[\frac{10^6}{60\times3600}\right]\times\left[\frac{533}{18.08}\right]^3=132{,}473\text{ h}$$

Basic rating life: 132,472 h

10.3.10 Basic Static Radial Load Rating

This is defined as "The basic static load rating is the amount of static radial load that equates to a stress level of 4200MPa on the most heavily stressed ball/raceway contact area (the center)." A total permanent deformation (ball and raceway) caused by the contact stress is about 0.0001 times the ball diameter. The basic static radial load ratings can be found in the dimension tables.

Example 10.4

A deep-groove ball bearing having SKF # 6207 is subjected to load cycles as given in the table below. Under each phase (cycle) loads occur with high shocks. Determine expected life of bearing in hours with probability of survival of 95%.

Phase	Radial Load F_r (kN)	Axial Load F_a (kN)	N (rpm)	Percentage
I	3	2	600	15
II	3.5	2	800	25
III	5	2	900	30
IV	0.5	2	1500	40

Calculation of permissible mean load in Newtons at 95% probability:
First, we calculate the average N:

$$N = \frac{600\times15+800\times25+900\times30+1500\times40}{100} = 1160 \text{ rpm}$$

The mean load (P_m), also called cubic mean load, is estimated as

$$P_m = \sqrt[3]{\frac{P_{r_1}N_1^3 + P_{r_2}N_2^3 + P_{r_3}N_3^3 + P_{r_4}N_4^3}{N}}$$

P_r, P_{r1}, P_{r2}, P_{r3}, and P_{r4} are calculated using the equation given above as

$$P_r = X\,F_r + Y\,F_a$$

The X and Y values can be selected from Table 10.1. For the SKF 6207 ball bearing, inner diameter $D_w = 35$ mm.

To use Table 10.1, we need to find axial load ratio (r):

$$r = \frac{F_a}{iZD_w^2}$$

where:

i = number of rows of ball bearings
Z = number of balls
D_w = diameter of ball

These values can be obtained from manufacturers' specifications. Here $i=1$, $Z=6$, and $D_w=$ 1.5875 mm. The manufacturers also specify the basic dynamic load $(C_r) = 553$ N.

We can calculate r values for all phases as

$$r_1 = r_2 = r_3 = r_4 = 0.272$$

and from Table 10.1 we get

$$e_j = 0.19 + \frac{0.272-0.134}{0.314-0.134}(0.22-0.19) = 0.213,\ j = 1,2,3,4$$

$F_a/F_r = 2/3 = 0.66 > e$ for first phase
$F_a/F_r = 2/3.5 = 0.571 > e$ for second phase
$F_a/F_r = 2/5 = 0.4 > e$ for third phase
$F_a/F_r = 2/0.5 > e$ for fourth phase

Therefore, $X = 0.56$. However, Y values need to be interpolated:

$$Y = 2.30 + \left(\frac{(0.272-0.134)}{(0.314-0.134)}\right)(0.22-0.19) = 2.34$$

Now, dynamic equivalent loads for all phases can be calculated as

$P_r = 0.56(3000) + 2.34(2000) = 6360$ N for first phase
$P_r = 0.56(3500) + 2.34(2000) = 6640$ N for second phase
$P_r = 0.56(5000) + 2.34(2000) = 7480$ N for third phase
$P_r = 0.56\ (500) + 2.34(2000) = 4960$ N for fourth phase

Now, average load can be calculated as

$P_m = 27.359$ N
Cubic mean load $(P_m) = 27.359$ kgf

Now, the life of the SKF 6207 bearing can be estimated:

$$L_{10} = \left(\frac{10^6}{60N}\right) \times \left(\frac{C}{P}\right)^3$$

C is the dynamic load rating obtained from the specification of the bearing. $C = 553$ N. Therefore

$$L_{10} = \left(\frac{10^6}{60 \times 1160}\right)\left(\frac{553}{27.359}\right)^3 = 118{,}650 \text{ h}$$

Life of bearing with 95% probability is calculated as

$$\frac{L_{95}}{L_{10}} = \left[\frac{\ln(1/0.95)}{\ln(1/0.90)}\right]^{1/b}$$

where $b = 1.34$ [2].

$$\frac{L_{95}}{L_{10}} = \left[\frac{\ln(1/0.95)}{\ln(1/0.90)}\right]^{1/b} = 0.3646$$

Finally, the life of the SKF 6207 bearing with 95% probability is $L_{95} = 43{,}255$ h.

To ensure the long life of a product, many factors have to be taken into account. A poor installation with unfavorable inlet conditions, unfavorable duty points, bad anchoring, and so on can cause structural disturbances that may be detrimental, while natural frequencies in the structure from poorly designed supports of the pump, pipes, valves, and so on are often the cause of high vibration levels. The advent of high-speed engines, automatic production machinery, high-speed milling, and so on provided the impetus for extensive research and development of rolling bearings (also called anti-friction bearings). The Anti-friction Bearing Manufacturers Association (AFBMA) standardized bearing dimensions and the basis for their selection. Thus, it is possible for a designer to select a bearing from the catalog of one manufacturer and replace it from the catalog of another manufacturer at end of life.

10.4 Use of Bearing and End-of-Life Considerations

10.4.1 Bearing Failure

Bearing failure causes are listed below in order of likelihood.

10.4.1.1 Penetrating Fluid or Particles

Cleanliness is essential for good performance. Particles generate high stresses in the bearing components and thereby create premature fatigue failure. Particles also generate wear, which shortens life. Particles, especially light alloy particles such as zinc, may act as catalysts for the grease and cause premature aging. If a fluid enters the bearing and is overrolled, the fluid acts as a jet and forces the grease out from the raceway. If the fluid is oil, and contains just a tiny amount of water (0.1%), the bearing's lifetime will be ruined even if it is flooded with that oil. Furthermore, the fluid may also cause corrosion.

10.4.1.2 High Temperature or Gradient

High temperatures due to bad cooling or too much heat generated can be suspected (if the bearing is not totally ruined) if the grease has darkened or if it feels as though carbon particles are dissolved in the oil. If synthetic oil is used, it may have polymerized and created a lacquer layer on the surface, mainly seen in oil lubrication. A bearing that has been overfilled with grease may also cause too high a temperature. If the temperature is only slightly higher than allowed, the signs are not so obvious; however, the lifetime will be reduced due to lower viscosity and faster degradation of the lubricant. Cages with plastic material will also age faster at increased temperatures. The normal temperature limit for a Flygt product is 90°C, although peak temperatures up to 120°C are allowed. If too much heat is generated, perhaps from overload, but with adequate cooling, the gradient over the bearing may be too high, generating a preload that can destroy the bearing. A correct internal bearing clearance is vital.

10.4.1.3 Incorrect Mounting

Improper mounting out in the field causes a large number of bearing failures. Precautions that must be taken include

- Keep everything clean and corrosion inhibiting during mounting and storage.
- Do not overheat. The maximum temperature is 150°C for very short periods. The bearing is stabilized to a certain temperature, and temperatures above this may alter the internal structure of the material, although it takes some time. The cage may also be damaged if subjected to overheating.
- Use suitable tools. Avoid mounting in such a way that the mounting force goes through the rolling element, which may cause indentations and destroy the bearing.
- Use the correct grease or oil, and do not overfill. Bearing lifetime is calculated for a specific lubricant. Note that different greases and oils must not be mixed, as the result may be disastrous.
- Use the right bearing as specified by Flygt. Standard bearings may have improper lubricants, clearances, or angles or may be unsuitable in other ways.

There are many ways to go wrong and only a few to do it right. Always follow the guidelines supplied by manufacturers.

10.4.1.4 Shock Loads from Handling

Shock load may result from mounting of other parts, that is, the impeller, or during transport and installation.

10.4.1.5 Vibration at Standstill

During operation, the bearing can withstand rather high vibration. Normally, 25 mm/s root mean square (rms) does not have a significant effect on the lifetime, although the grease may be "shaken" away, depending on installation and grease type. At standstill, the allowable levels are much lower; false brinelling may occur at level as low rods come into contact with the machine and your car. Be aware, though, that all bearings create some noise, especially during start-up, while the clearance is still at its nominal value. When the machine is warm, the noise normally decreases.

- *Vibration*: Vibrations caused by loose parts contain many frequencies and can be hard to separate from the cluttering noise of cavitation and debris in pumped media. Vibration caused by a defective bearing can only be detected with sophisticated tools, and even then, it is hard unless you have the vibration pattern of a good operating condition with the actual bearing. Condition monitoring (with acoustic emission, see-technique, and so on) is a good tool if you are able to measure the vibration close to the bearing, and if you set up individual limits for different running conditions. Vibration from a damaged bearing close to failure or a loose part may be spotted by simpler vibration-measuring devices. This can be used to stop the machine before total breakdown.
- *Temperature*: Temperature measurements are unable to detect defective bearings. In some cases, you may see a temperature rise just before breakdown. Trend measurements of temperatures may sometimes give information that indicates a malfunction, in rare cases due to increased heat generation in the bearing.

10.4.2 Bearing Alignment

Ball and roller bearings are precision components and are manufactured with very small internal clearances. Therefore, they provide very accurate positioning of the shaft. However, due to the small clearances, they do not perform well if they are misaligned. If there are doubts concerning accurate alignment of the housing into which the bearings will fit (or the amount of deflection of the shaft on which they are mounted), *self-aligning bearings* may be specified, as shown in Figure 10.15.

In these bearings, self-alignment is achieved by allowing the plane of the balls or rollers to tilt relative to the spherical race in the outer housing. Figure 10.15 shows a double-row ball bearing (which has greater radial load capacity than a single-row bearing) allowing the outer race to tilt within a spherical seating. An alternative construction (not shown) has an outer race with an external spherical surface, enclosed within an additional casing that has an internal spherical surface.

Pillow blocks, also known as plummer blocks, are a useful off-the-shelf component that can simplify shaft mounting in some cases. They are normally available with a range of different rolling-contact bearings fitted within the housing—ball or roller, single or double row, self-aligning or deep-groove type, depending on the application and the probability of misalignment. The housing also makes provision for seals to retain lubricant, while the mounting feet make it easy to attach the housing to a base plate or frame.

10.4.3 Bearing Preload

Since all engineering components deflect to some extent under load, shaft bearings that have been set up with zero clearance when there is no load on the shaft will almost certainly have some clearance or slackness when the shaft has a load applied. This is true whether the load on the shaft is radial, axial, or a combination of the two. This clearance when running, which results in unnecessary wear of the bearings, can in many cases be overcome by preloading the bearings.

The outer races of the two tapered roller bearings (Figure 10.11) are located in the bearing housing. They cannot come closer together because each race is against a shoulder in the housing. The bearing inner races are mounted on the shaft. The right-hand inner race bears against a shoulder on the shaft. The left-hand inner race is held axially by the P5-42 washer and nut on the end of the shaft. Tightening the nut will therefore push the two inner races closer together, squeezing them into the tapered outer races. Initial tightening of the nut will take up all the clearance in the bearing assembly. Further tightening of the nut will provide bearing preload. Preload may be regarded as negative clearance. Ideally, the preload should be sufficient to ensure that the bearings do not run slack (i.e., with clearance) under normal operating conditions, since clearance is known to increase the rate of wear of tapered roller bearings. However, too much preload may result in excessive loading and overheating of the bearing. In practice, the adjusting nut of a slotted nut and a split pin are used, and this method is common practice in industry.

10.4.4 Ball Bearings: Guide to Selection, Applications, and Calculations

The discussion in this section should help mechanical engineers understand the selection of ball bearings depending upon the application and load conditions. Primarily, it helps to calculate the minimum and maximum load-bearing capacity of the bearing and, most importantly, the life of the bearing as an individual. Here, the designation number of the ball bearing provides information on the design, working nature, and physical properties of the bearings.

The calculation accuracy, operating temperatures, and relubrication features, along with applications, utility, and system of operations, recommended by the American Society of Mechanical Engineers (ASME) are presented below. It also obeys the rules of the American National Standards Institute (ANSI/API). It gives significant relevance to bearing operations from a maintenance point of view.

Each shaft shall be supported by two radial bearings and one double-acting thrust bearing, which may or may not be combined with one of the radial bearings. This means that the thrust bearing may or may not be in the same axis as the rest of the two radial bearings. In other words, the thrust bearing can be arranged in such an order that it may or may not be a part of the system bearing calculation, despite supporting the shaft.

Here, only three systems of arrangements are possible:

1. Rolling-element radial and thrust
2. Hydrodynamic radial and rolling-element thrust
3. Hydrodynamic radial and thrust

10.4.5 Understanding the Bearing Specification Provided by Manufacturers

For any designated number in any bearing company, the bearing I.D. is equal to five times the last two digits. For example, 6208 = I.D. 40 mm, and 2313 = I.D. 65 mm.

Additional information is given by suffixes added to the designation number of the bearings.

Example 10.5

-626z: Here, 6 mm is the I.D., and 6 mm is the thickness, with single shield.
627-2z: Here, 7 mm is the I.D., and 7 mm is the thickness, with double shield.

Note: For the selection of the bearing, C is important. If it is equivalent to $C/P \times F_r$, in terms of Newtons, then a bearing matching the load-bearing capacity will be suitable.

Supplementary Designations:

1. K: Roller and cage assembly of a cylindrical roller thrust bearing. Example: KNU07
2. L: Removable inner and outer ring of a separate ring. Example: LNU207
3. R: Separate bearing without removable inner and outer ring

Suffixes:

1. A, B, C, D, E: Differing or modified internal design

External Design:

Ref	Description
X	Boundary dimensions altered to conform to ISO standards
RS and LS	Rubbing seals at one side of the bearing
2RS and 2LS	Rubbing seals at both sides of the bearing
Z	Shield (nonrubbing seals) at one end of the bearing
2Z	Shield (nonrubbing seals) at both sides of the bearing
K	Tapered bore, 1:12 on diameter
K30	Tapered bore, 1:30 on diameter
N	Snap ring groove in outer ring
NR	Snap ring groove in outer ring, with snap ring
ZN	Shield at one side of the bearing and snap ring groove in outer ring at other side
ZNR	As ZN, with snap ring
N2	Two locating slots in outer ring
G	Single-row angular contact ball bearing for paired, back-to-back, or face-to-face mounting

Cage:

Ref	Description
J	Pressed steel cage
Y	Pressed brass cage
M	Machined brass cage
F	Machined steel or spheroidal graphite cast iron cage
L	Machined light alloy cage
P	Injection-molded cage of glass-fiber-reinforced plastic
TH	Snap type of fabric-reinforced phenolic plastic
IH	Injection-molded cage of plastic
V	Full component bearing
VH	Full component bearing with nonseparable roller component

10.4.6 Other Bearing Features

10.4.6.1 Accuracy

a. P6: ISO clearance class 6
b. P5: ISO clearance class 5
c. CLN: ISO clearance of class 6X for taper roller bearings

10.4.7 Relubrication Feature

Classification by number of holes provided in the bearings for lubrication:

- **a.** W 20: three holes in outer ring
- **b.** W 26: six holes in inner ring
- **c.** W 33: lubrication groove and three holes in the outer ring
- **d.** W 33X: lubrication groove and six holes in the outer ring
- **e.** W 5/3: this class has combined features of class W 26 and W 33
- **f.** W 5/8: this class has combined features of class W 20 and W 26
- **g.** Q E5: Special electric motor quality, dimensional and running accuracy to P6 for high demands of silent running
- **h.** Q E6: Normal electric motor quality, for quiet running application
- **i.** Q 05: vibration peaks extra low
- **j.** Q 06: vibration peaks lower than normal
- **k.** Q 5: vibration level extra low (supersedes C6)
- **l.** Q 6: vibration lower level than normal (supersedes C6)
- **m.** Q 55: this class has combined features of class Q 5 and Q 05
- **n.** W 66: this class has combined features of class Q 6 and Q 06

10.4.8 Classification according to Temperature

- **a.** MT: grease for medium temperature (−30°C–110°C)
- **b.** LT: grease for medium temperature (−50°C–80°C)
- **c.** HT: grease for medium temperature (−20°C–130°C)

10.4.9 Stabilization Feature

Bearing rings are dimensionally stabilized for operating temperature.

- **a.** S0: up to +150°C
- **b.** S1: up to +200°C
- **c.** S3: up to +250°C
- **d.** S4: up to +300°C
- **e.** S5: up to +350°C

10.5 Green Bearing Design

There are many sound reasons why environmental issues should be considered in the design and product development process. They help us meet customer and legislative requirements. They put us in a stronger competitive and commercial position. They play a major part in ensuring we maintain innovative credibility. But, most of all, they help us achieve our ambition of reducing the overall environmental impact of our products across their life cycle. Here, we want to demonstrate the environmental impacts of bearings in terms of embodied energy (energy consumed during production), carbon footprint, and eco-indicators of bearing range for a variety of applications. New bearings designed according to the concept of green design include seal technologies (hard seal coatings and surface design), lightweight parts (polymers), and "lube-for-life" thin film lubrication, which makes relubrication unnecessary throughout a bearing's life span. An estimated 50 billion bearings are in use at any time, and reduction in energy consumption by 30%–70%, depending on the type of use, would result in tremendous savings in energy; reduce carbon dioxide emissions into the atmosphere, as well as saving resources and money; and lead to an enormous reduction in lubricant disposal. Ships trading in the world's oceans and seas can eliminate both operational and accidental stem tube oil pollution, while also reducing their operating costs.

10.5.1 Reducing Environmental Footprint by Reusing and Recycling Bearings

As mentioned in the previous section, nearly 50 billion bearings are in use in the world every day, and if we can reuse and recycle a fraction of them, we will save a large amount of materials and embodied energy, as well as reducing the carbon footprint.

The reduced material and energy requirements resulting from bearing remanufacture represent a reduction in environmental impact.

Remanufacturing is a major contributing element to reducing the life cycle cost of industrial equipment, giving overall benefits such as

- Extending operating life of the component
- Reducing maintenance cost
- Enhancing reliability for production up-time and productivity
- Reducing likelihood of failures
- Maintaining the condition of replacement stock
- Improving overall asset reliability
- Providing input for reliability improvements

Reducing negative environmental impact is an increasingly important issue for our society, and remanufacturing can be a significant contribution to improved sustainability, by reducing natural resource and energy use.

Certain application conditions—such as contamination or sporadic metal-to-metal contact in the rolling contact zone—can cause all sorts of damage to the bearings. As a result, the service life of a bearing is often shorter than the calculated rating life. The alternative is to apply a controlled remanufacturing process before any major damage or bearing failure occurs. This can substantially prolong the service life of the bearing in question, reducing costs and lead times. And, since it requires less energy than manufacturing a new bearing, it is better for the environment as well.

The typical candidates for industrial bearing remanufacturing are

- Large-size bearings
- Caster bearings
- Backing bearings
- Slewing bearings

10.5.2 Lubrication Management

10.5.2.1 Lubrication of Bearings

Lubrication of rolling contact bearings is important because it

- Prevents corrosion
- Greatly reduces the effects of the sliding friction present in all bearings, particularly roller bearings
- Carries heat away from heavily loaded bearings

Lubrication may be achieved by

- Drip, splash or mist (oil only).
- Lubricators.
- Oil circulation within a gearbox by entrainment and splashing, and so on. Grease is a suitable lubricant provided rotational speed is not too high, and mineral oils are good general purpose bearing lubricants, although many modern applications demand special high-pressure lubricants.

10.5.2.2 Sealing

Seals are divided into two main classes: static and dynamic. Static seals are used to prevent the loss of fluid in a pressure vessel. Dynamic seals are used to prevent loss of fluid in a sliding or rotating joint. Numerous types of materials are used, such as organic and mineral fibers, natural and synthetic rubber, cork, paper, and soft metals.

For static seals, a flat rigid surface and a thin gasket are preferable, and a minimum amount of packing surface should be exposed to the fluid. Sometimes, sealing is created by friction. Sometimes, the gasket is confined to a chamber and distorted to affect a seal by jamming the packing across the leakage path. For a dynamic seal, it is important that sliding members have a very smooth, hard surface. Mechanical seals are very effective for sealing a rotating shaft against gas or liquid under pressure. The sealing member may be either stationary or rotating.

Good sealing is essential. Seals should

- Keep dirt out, thereby preventing premature wear of the bearing
- Keep lubricant in, ensuring that the bearing (and possibly other components) will not run short of lubricant

A lubrication management program can be defined as the sum of all the activities performed in a given facility to help ensure the right lubricant is provided in the right quantity to the right point at the right time with the right method.

The benefits of a lubrication program are related to cost savings in

- Downtime
- Spares
- Labor
- Overtime labor
- Energy consumption

Problems are mainly caused by

- Lubricant contamination
- Lubricant chemical degradation
- Wrong lubricant selection
- Insufficient or misleading knowledge

10.5.3 Recycling of Lubricants

Lubricating oils are viscous liquids used for lubricating moving parts of engines and machines. Since lubricating oils are obtained from petroleum—a finite product—and with dwindling production from world oil reserves, there is an increasing need to recycle used lubricating oils. Accordingly, recycling effort focuses on a comparative study of four methods of recycling used lubrication oils: acid/clay, distillation/clay, acid, and activated charcoal/clay treatment methods. Tests carried out on the recycled lubrication oil include flash point, pour point, specific gravity, metal content, viscosity, and sulfur content. The results from the tests showed that the viscosity of used lube oil increased from 25.5 cSt to 86.2 with distillation, 89.10 with acid/clay treatment, and 80.5 with activated/clay treatment. This is compared with 92.8 cSt for fresh lube oil. Other results from the different tests showed varied degrees of improvement, with the best results obtained using the acid/clay treatment.

Lubricating oils from petroleum consist essentially of complex mixtures of hydrocarbon molecules. They are mostly composed of isoalkanes, having slightly longer branches, and monocycloalkanes and monoaromatics, which have several short branches on the ring. These hydrocarbon molecules generally range from low-viscosity oils having molecular weights as low as 250 to very viscous lubricants with molecular weights as high as 1000 [3]. The number of carbon atoms ranges from 20 to 34. Lubricating oils are viscous liquids and are used for lubricating moving parts of engines and machines. Grease, which is a semisolid, also belongs to this group. There are three major classes of lubricating oils: lubricating greases, automotive oils, and industrial lubricating oil. When lubricating oils are used in service, they help to protect rubbing surfaces and promote easier motion of connected parts. In the process, they serve as a medium to reduce the buildup of temperature on the moving surfaces. Further buildup of temperature degrades the lubricating oils, leading to reduction in properties such as viscosity, specific gravity, and so on. Dirt and metal parts worn from the surfaces are deposited into the lubricating oils. With increased length of use, the lubricating oil loses its lubricating properties as a result of overreduction of desired properties, and thus must be evacuated and replaced with fresh oil. With the large amount of engine oils used, the disposal of lubricating oils has now become a major problem. Many nations are now addressing the problem of environmental pollution posed by waste or used lubricating oils in their countries [4]. In the United States, for example, about 2 billion gallons of oils are generated annually. This has led industries and governments to look for satisfactory solutions that will reduce the contribution of used lubrication oil to pollution and also recover this valuable hydrocarbon resource. Many people dispose of used oil by using it for dust prevention. This method of disposal is in many ways unsatisfactory, as the lead-bearing dust and runoff constitute air and water pollution. Another method for disposal of used oil is incineration. This method represents another poor use of such a valuable product, and the attendant emission of probably carcinogenic products, contributing to environmental pollution. Recycling of used lubricants is now attracting more attention. This is partly because of the fear of dwindling world oil reserves, but more as a result of the environmental concern posed by used lubricant. The following three reasons explain interest in the recycling of waste lubricating oils:

1. Needing to conserve crude reserves
2. Minimizing unemployment through the building/construction of used lubricating oil recycling plants
3. Eliminating the environmental pollution risk of used lubricant

Environmental considerations regarding the conservation of resources have maintained interest in the concept of recycling up to the present day. The reclamation of spent crankcase oils is now a subject of pressing national interest in some countries. On the other hand, used lubrication oils are now recognized to account for more pollution than all oil spills at sea and offshore put together. Some countries have petroleum storage, and spent oils represent a precious commodity, which must not be wasted. For this reason, the conservation of petroleum resources has been declared a national policy for several countries, and the benefits of wise resource management are obvious. The recent impetus toward waste recovery has led to renewed interest in redistilling to convert this used lubricating oil into useful original lubricating oils. Redistilling or refining is the use of distilling or refining processes on used lubrication oil to produce high-quality base stock for lubricants or other petroleum products. The use of this method has increased tremendously in developed countries, in some cases reaching up to 50% of the countries' needs for lubricating oil (Thrash, 1991), and different methods have been developed by Western countries for refining used lubricating oil for reuse. The basic principle remains the same and uses many of the following basic steps:

- Removal of water and solid particles by settling
- Sulfuric acid treatment to remove gums, greases, and so on
- Alkaline treatment to neutralize acid
- Water washing to remove "soap"
- Clay contacting to bleach the oil and absorb impurities
- Stripping to drive off moisture and volatile oils
- Filtering to remove clay and other solids
- Blending to specification

It is essential to recognize that all used oils should be collected for controlled disposal. Some products, such as transformer oils and hydraulic oils, can be readily collected from large industrial concerns, regenerated to a recognized standard, and returned to the original source. Oil from automotive sources will include monograde and multigrade crankcase oils from petrol and diesel engines, and industrial lubricants that have been inadequately segregated may also be included. Lubrication oil is used to provide a film between the moving parts of machines and engines to prevent wear with little or no loss of power. The conventional steps in lubricating oil manufacture are pretreatment of the crude oil charge, followed by distillation of the crude in two steps (an atmospheric tower and a vacuum tower), deasphalting (as required by the nature of the crude oil charge), dewaxing, solvent extraction, filtering, and blending, including mixing various additives with the final lubricating oil. The prime objective in the production of lubricating oil is the separation of wax distillate and cylinder stock without any decomposition or cracking of the lubrication oil fractions; thus, a vacuum distillation unit is used to separate the wax distillate and the bottom stock at a lower temperature. The properties that make the high-boiling paraffin hydrocarbon suitable for lubricating manufacture include stability at high temperature, fluidity at low temperature, only a moderate change in viscosity over a broad temperature range, and sufficient adhesiveness to keep it in place under high shear forces. The desired fractions for the manufacture of lubricating oil have high boiling points, and separation into various boiling point-range cuts must be accomplished under reduced pressure. The vacuum tower produces some fuel oil overhead, which is sold as a separate product or sent to another area of the refinery for further processing and blending. The two main products from the vacuum tower are wax distillate and cylinder stock, which is the bottom product. Both streams contain desirable lubricating oil constituents as well as by-products. The wax distillate is charged directly to the dewaxing unit. The vacuum tower bottoms, or cylinder stock, are charged to the deasphalting unit. These two fractions form the basic stock for lubricating oil manufacture.

General characteristics of lubricating oils:

All lubricants are characterized by highly specific properties, including

1. Viscosity index, cloud point, flash point, pour point
2. Total base number (TBN), ash content, water content
3. Corrosive properties, relative density, solubility, total acid number (TAN), and so on

10.5.3.1 Viscosity

Viscosity is defined as the force acting on a unit area where the velocity gradient is equal at a given density of the fluid. Viscosity is strongly dependent on temperature and must be stated for a certain temperature. The most important fluid characteristic of a lubricant is its viscosity under the operation conditions to which it is subjected in the unit. The relationship between a shearing stress and the viscosity gradient it produces in a liquid is characteristic of that liquid. Lubrication oils are identified by Society of Automotive Engineers (SAE) number. The SAE viscosity numbers are used by most automotive equipment manufacturers to describe the viscosity of the oil they recommend for use in their products. The higher the SAE viscosity number, the more viscous or heavier is the lubricating oil. Viscosity numbers are given in terms of Saybolt universal second (SSU). Certain additives are used for the improvement of viscosity–temperature characteristics.

10.5.4 Specific Gravity (Density)

From ordinary theory, we know that the density of a substance is equal to the mass of a substance divided by the volume of the substance, that is

$$\text{Density } (\rho) = \text{Mass } (m) / \text{Volume } (V)$$

Specific gravity is the ratio of the density of the material to the density of an equal volume of water. The temperature at which the density has been measured must be known, since density changes with temperature.

10.5.5 Cloud Point and Pour Point

Cloud and pour points are listed in ASTM D97-47. Cloud point is the temperature at which paraffinic wax or other oils are cooled under given conditions. Pour point is a measure of the temperature at which the oil ceases to flow under the service conditions of a specific system. It is very important to users of lubricants in a low-temperature environment.

10.5.6 Water Content

The presence of water in lubricating oil during operation depends on where the automobile is being used. In almost all systems, traces of water in the lubricant are unavoidable, arising from such sources as leaking oil coolers, engine cooling system leaks, and atmospheric condensation in all types of machinery. Accordingly, the water content must not exceed the "action" level (more than 0.5) recommended for the grade of oil and application. In places where there are bad roads and drainage systems, one is bound to see water as a contaminant of oil. The water in the radiator may also contribute to the presence of water in lubricating oils during use. The presence of excessive water contamination will affect the viscosity of the oil, which may give rise to emulsion formation, and can also lead to gear tooth and bearing problems.

10.5.7 Flash Point

The flash point test gives an indication of the presence of volatile compounds in oil, and the flash point is the temperature to which the oil must be heated under specific conditions to give off sufficient vapor to form a flammable mixture with air. There are various methods of determining the flash point of oils, as included in ASTM [29]. Flash point, open cup is the temperature at which a flash appears on the surface of the sample when a small flame of specified size is passed across the cup at regular temperature intervals while the oil in the cup is heated at a specified rate.

10.5.8 Fire Point

Fire point and flash point are significant in cases where high-temperature operations are encountered. Fire point, open cup is the temperature at which the oil ignites and continues to burn for at least 5 s. Fire point is obtained as a continuation of the flash point test. Fire and flash points by Cleveland open cup are included in ASTM D92–56.

10.5.9 Methodology for Recovering Used Lubricants

The following materials are used in the recovery of used lubricating oil: concentrated sulfuric acid, caustic soda, clay, activated charcoal, and ethyl acetate. The following apparatus and equipment is used:

Buchner funnel
Filtering flask

- Separating funnels
- Distillation column
- Flash point tester
- Thermometer (360°C)
- Erlenmeyer flask (250 mL)
- Beaker
- Measuring cylinder
- Vacuum pump
- Sulfur content analyzer
- Atomic absorption spectroscopy (AAS)
- Coiled condenser
- Ubbelohde viscometer
- Erlenmeyer flask
- Round bottom flask

The methods used in the re-refining of used lubricating oils include filtration to remove impurities, acid treatment, acid/clay treatment, distillation/clay treatment, and activated charcoal/clay treatment.

10.5.10 Recycling of Bearings at End of Life

Bearings are recycled as follows:

1. Disassemble and clean the components returned from the customers.
2. Check for damage to the parts: balls, inner races, and outer races.
3. Remove the worn portions and adjust the shape by rough machining.
4. Mask the areas that must not be sprayed.
5. Roughen the surfaces to be sprayed by shot blasting.
6. Form a coating up to the necessary thickness by spraying to restore the dimensions.
7. Finish by turning or polishing.
8. Inspect the coating and dimensions.
9. Assemble each component.

The parts to be recycled vary widely in materials and properties. We investigated whether the bond strength of the coating varies with the substrate material. It is known that the bond strength of the coating mainly depends on the anchor effect, and the surface conditions after shot blasting, performed as a pretreatment before the spray, have an impact on the bond strength. The surface condition is indicated by (1) roughness, (2) hardness, (3) existence of graphite (cast iron and steel), (4) cleanliness, and so on.

One of the main potential approaches to reducing the environmental impact of bearings is the replacement of common bearing materials such as steel, brass, bronze, and so on with plastic. This is likely to reduce the use of precious materials and the energy needed for extraction and manufacture of parts. In this regard, plastic seems to be a good candidate for replacing common bearing materials. It is widely used for a large number of applications.

10.6 Plastic Bearing Design

Greene and Tweed offer three standard configurations of plastic bearings: sleeve bearings, flange bearings, and thrust bearings [30]. Sleeve and flange bearings are available in both solid and split variations. Because plastic materials tend to expand as temperature increases and contract as temperature decreases, dimensional changes caused by thermal expansion will affect the performance of plastic bearings. At high temperatures, the bearing cross section may grow to the point of interference with its fit, causing excessive friction and wear. At cryogenic temperatures, the bearing will try to shrink, tightening around the shaft and causing increased friction. The gap on the split bearings is designed for ease of bearing installation and to minimize the undesirable effects of thermal expansion and contraction. Split bearings are therefore recommended for all rod and piston applications, other than instances where the bearing must be press-fitted for retention. However, the following aspects must be predecided before selecting a plastic bearing:

- Define the bearing diameter needed.
- Select the bearing cross section needed.

- Define the length of the bearing.
- Calculate the unit loading, surface velocity, and pressure–velocity that the bearing will experience.

Bearing performance is affected by many different factors. It is important to keep in mind the following factors when designing a bearing.

10.6.1 Bearing Loading

It is important to maintain the lowest possible loading on a bearing. Lower unit loading on the bearing can be accomplished by increasing the bearing length. PTFE-based bearings perform best at low load and high velocity. Bearing pressure is the amount of load that the bearing has to support for a given bearing area. Bearing pressure is calculated as follows:

$$P = \frac{F}{DL}$$

where:

P = bearing pressure (psi)
F = total load on the bearing (lb)
D = bearing I.D. (in.)
L = length of the bearing (in.)

10.6.2 Running Velocity

Bearing surface speed is calculated as follows:

For rotary applications: $V = (D \times \pi \times \text{rpm})/12$
For reciprocating applications: $V = (LS \times C \times 2)/12$
For thrust bearings: $V = (0.52 \times \text{rpm})(0.6R_1 + 0.4R_2)$

where:

V = velocity (ft/min)
D = dynamic diameter (in)
LS = length of stroke (inches)
C = cycles per minute
R_1 = radius of the thrust bearing I.D.
R_2 = radius of the thrust bearing O.D.

10.6.3 Bearing PV Limits

$$\text{PV} = \text{Pressure (psi)} \times \text{Velocity (ft/min)}$$

PV limits are affected by variations in temperature, speed, loading, lubrication, and surface finish. Exceeding PV limits will result in accelerated wear and premature bearing failure. PV limits are provided by manufacturers.

10.6.4 Bearing Media

PTFE bearings have almost unlimited chemical compatibility. Thermoplastic and composite bearings have wide chemical compatibility, except for use with acids and strong oxidizing agents.

10.6.5 Plastic Bearing Lubrication

Plastic bearings are designed to operate maintenance free, thus eliminating the need for periodic lubrication. Lubricated bearings, however, will exhibit a lower coefficient of friction and longer service life. Therefore, the application of grease or lubricant at the time of installation is recommended whenever possible.

10.6.6 Metal Surface Hardness

Some of the fillers in PTFE and the reinforcements in the composites are abrasive to soft metals. Therefore, hardened running surfaces are recommended. Moreover, surface hardness of $>Rc$ 40 is further recommended.

10.6.7 Metal Surface Finish

A good surface finish on the dynamic surfaces will reduce the friction and the wear rate on bearings. It is recommended to maintain 8–32 *Ra* surface finish for best bearing performance.

Bearing materials and applications: the following table shows the materials code and its applications in bearing design.
PTFE compound material codes are 015, 064, 348, 357, 379, 392
Thermoplastic compound material codes are 036, 039, 372, 913
Composite compound material codes are 94, 915

Material Code	Application
015	Tough, abrasion resistant. Recommended for abrasive environments and water applications. FDA approved.
064	Low deformation, good temperature dissipation. Recommended for numerous general applications.
348	FDA compliant, excellent wear properties, compatible with LOX, MMH, hydrazine, and oxidizers. Not recommended for water or steam applications.
357	Good wear, temperature, and deformation properties. Nonabrasive against soft mating surfaces. Ideal for semiconductor manufacturing applications.
379	Excellent wear and deformation characteristics. Recommended for general applications.
392	Excellent wear and deformation properties. May be abrasive against mating surfaces. Recommended for water and steam applications.
036	High strength, low deformation under load, high-temperature capabilities, good chemical resistance. Recommended for all general applications.
039	Low coefficient of friction, low deformation, good chemical compatibility. Recommended for steam and general applications.
372	High temperature resistance, high strength, excellent chemical compatibility. Recommended for high-temperature applications. Not recommended for water or steam applications.
913	High impact resistance, thermal shock resistant, good chemical resistance. Not recommended for abrasive media.
914	High impact resistance, low deformation under load, low thermal expansion, good chemical compatibility, excellent wear resistance. Not recommended for abrasive media.
915	Good wear resistance, excellent dimensional stability, and good chemical compatibility. Recommended for hydraulic cylinders and underwater applications.

Note: FDA, Food and Drug Administration; LOX, liquid oxygen; MMH, monomethyl hydrazine.

10.6.8 Typical Plastic Bearing Applications (4–6)

Plastic bearing applications are now found in a wide variety of equipment and instruments. Possible applications include

1. Automotive
2. Construction equipment
3. Conveyors
4. Cryogenic equipment
5. Dryer oven conveyors
6. Exercise equipment
7. Farm equipment
8. Hydraulic actuators
9. Hydraulic presses
10. Intensifiers
11. Machine tools
12. Marine
13. Medical equipment
14. Mining
15. Oil-field equipment
16. Packaging equipment
17. Photographic processing equipment

18. Pumps
19. Robotics
20. Vacuum metalizing equipment

References

1. Shigley, J. E. and Mischke, C. R. (2001). *Mechanical Engineering Design*, 6th edn. McGraw Hill.
2. Budynas, R. G. and Nisbett, J. K. (2008). *Mechanical Engineering Design*, 8th edn. McGraw Hill, Boston.
3. Concawe. (1985). The collection, disposal and regeneration of waste oils and related materials. Concawe Report No. 85/53. Concawe, Brussels.
4. Thurman, F. M. (ed). (1982). *Organic Chemistry of Natural Waters*, Kluwer Academic, Dordrecht.
5. ASTM (American Society for Testing and Materials). (2010). ASTM F1194-99. Standard guide for documenting the results of chemical permeation testing of materials used in protective clothing. Standard volume 1103. ASTM, West Conshohocken, PA.
6. Greene Tweed. Plastic bearings. Greene Tweed, Kulpsville, PA.

Bibliography

Ashby, M. F. (2009). *Materials and the Environment: Eco-Informed Material Choice*, Elsevier, Canada.

Ashby, M. F. (2011). *Material Selection in Mechanical Design*, 4th edn. Elsevier, New York.

Botthroyd, G., Dewhurst, P., and Knight, W. (2002). *Product Design for Manufacture and Assembly*, 2nd edn. Marcel Dekker, New York.

Chen, R. W., Navin-Chandra, D., and Print, F. B. (1993) A cost-benefit analysis model of product design for recyclability and its application. In *IEEE International Symposium on Electronics and the Environment* (pp. 178–183), Arlington, VA.

Desai, A. and Mital, A. (2005). Incorporating work factors in design for disassembly in product design. *Journal of Manufacturing Technology Management* 16: 712–732.

Dowie, T. and Kelly, P. (1994). Estimation of disassembly times, Technical Report DDR/TR15, Manchester Metropolitan University, UK.

http://www.nskamericas.com/cps/rde/xchg/na_en/hs.xsl/needle-roller-bearings.html

ISO 14044. (2010). *Environmental Management-Life Cycle Assessment—Requirements and Guidelines.*

Kroll, E. and Carver, B. S. (1999). Disassembly analysis through time estimation and other metrics. *Robotics and Computer-Integrated Manufacturing* 15: 191–200.

Kroll, E. and Hanft, T. A. (1998). Quantitative evaluation of product disassembly for recycling. *Research in Engineering Design—Theory, Applications, and Concurrent Engineering* 10: 1–14.

Lee, S. G., Lye, S. W., and Khoo, M. K. (2001). A multi-objective methodology for evaluating product end-of-life options and disassembly. *International Journal of Advanced Manufacturing Technology* 18: 148–156.

Lidwell, W., Holden, K., and Butler, J. (2003). *Universal Principles of Design*. Rockport Publishers, Gloucester, MA.

McGlothlin, S. and Kroll, E. (1995). Systematic estimation of disassembly difficulties: Application to computer monitors. In *Proceedings of the IEEE International Symposium on Electronics & the Environment* (pp. 83–88), IEEE Xplore Press, Orlando, FL.

Mott, R. (2004). *Machine Elements in Mechanical Design*, 4th edn. Pearson, Prentice Hall, Upper Saddle River.

Raimondi, A. and Boyd, J. A solution for the finite journal bearing and its application to analysis and design, parts I, II and III. *Trans. ASME* 1(1): 159–209.

Segur, A. B. *Predetermined Motion Time System (PMTS)*, mtm_international.org

Skakoon, J. (2000). *Detailed Mechanical Design: A Practical Guide*. ASME Press, New York.

Skakoon, J. (2008). *The Elements of Mechanical Design*. ASME Press, New York.

Slocum, A. (1992). *Precision Machine Design*, Prentice Hall, NY.

Soemers, H. (2010). *Design Principles for Precision Mechanisms*. T-Point Print VOF.

Suga, T., Saneshige, K., and Fujimoto, J. (1996). Quantitative disassembly evaluation. In *Proceedings of the 1996 IEEE International Symposium on Electronics & the Environment* (pp. 19–24), Dallas, TX.

United States Environmental Protection Agency (EPA), 2010, *Life Cycle Assessment*, 2010.

Whitney, D. E. (2004). *Mechanical Assembly: Their Design, Manufacture, and Role in Product Development.* Oxford University Press, New York.

11

Fastener Design

11.1 Introduction to Fasteners

Mechanical assembly is a term that is sometimes used to describe the process of putting together components on an *assembly line*. As a product travels down the assembly line, a part is added at each work station. In some lines, the conveyor may stop at each station, while in other factory line setups, the conveyor may move continuously. As the part passes the work station, the worker or an *industrial robot* performs a task that contributes to the assembly of the completed product. *Just-in-sequence* is an approach to manufacturing where components arrive at an *assembly line* in a specific order at the precise moment they are needed, and not before. It may be a component of just-in-time manufacturing, where companies minimize the amount of supplies they have in storage to cut costs and streamline operations. Using these approaches makes a business more flexible and more efficient. They can increase the bottom line without forcing the company to compromise the quality and reliability of its products. The automotive industry uses just-in-sequence manufacturing most widely. In car manufacturing, components from a variety of sources must be brought together to create a complete car. If the company maintains a parts *inventory* for the assembly line, it must sink capital into the purchase of parts and needs to create room to store and maintain them. This may consume substantial production resources and may also create situations where the company has a pressing need for parts and must wait for them because the shipper is not used to rapid delivery. With just-in-sequence, the car company works with the shipper and supplier to get deliveries of the right parts for manufacturing, just as they are needed. Parts come in a specific order, and workers on the assembly line can unpack them directly from shipping containers and install them, without a stay in storage or the need for sorting. The manufacturer can specify the order of the parts down to the color and special features so workers will always have the right part when they need it. One advantage of the just-in-sequence approach is tremendous flexibility. Up until the order is pulled by the supplier, the company can change it to add elements, reorder the sequence, and so forth. The familiarity with rapid order pulling and delivery also allows the supplier and shipper to meet unexpected needs quickly. If the company's production needs change, it can quickly meet them with minimal lag time, since it is already operating on a just-in-sequence system.

Mechanical assembly carries the connotation of putting component parts together to make a complete product or perform a function. Mechanical assembly requires special engineering techniques to ensure cost-effective and on-time assembly of products. To design products for mechanical assembly, engineers must essentially build the product in the design and then disassemble it. By doing so, the engineer can decide the steps necessary to build the product in a logical order using assembly line technology. This careful engineering is important because, if the steps are taken out of order, the result could be missing components. If this happens, the item would have to be disassembled to put parts in place. By creating a complete concept of the assembly process during the engineering phase, products can go into assembly line production in multiple plants using the plan and parts from a variety of sources. In many cases, parts not specifically engineered for a certain product, such as nuts and bolts, can be purchased from a low bidder, thus saving production costs. One of the key objectives of design for manufacturing (DFM) is to reduce the number of fasteners. However, the need for fasteners in assembly cannot be entirely eliminated.

Products assembled in this manner can be easily disassembled and reassembled in remote locations. Production costs are often lowered as component parts can be created off-site and assembled in the factory. These products can also be repaired easily with in-stock replacement parts. With the ability to easily take apart mechanical assemblies and reassemble them, these products can be upgraded easily.

The choice of the sequence in which parts or subassemblies are put together in the mechanical assembly of a product can drastically affect the efficiency of the assembly process. For example, one sequence may require less fixturing, less changing of tools, and include simpler and more reliable operations than others. The choice of the assembly sequence is usually made by a human expert. In the case of manufacturing, the choice is typically made by an industrial engineer. In the case of repair, the choice is made by the maintenance personnel. No clear systematic procedure seems to be followed in either case. Humans seem to use common sense and past experience blended in a fuzzy, sometimes inconsistent, and not well understood way. There is a growing need to systematize and to computerize the generation of assembly sequences for several reasons:

- Although many experienced industrial engineers have a knack for devising efficient ways to assemble a given product, systematic procedures are needed to guarantee that no good assembly sequence has been overlooked. For complex products, the number of feasible assembly sequences may be so large that even skillful engineers may fail to notice many possibilities. The availability of a systematic procedure that is proven correct and complete will guarantee that all feasible sequences and only the feasible ones will be generated.
- The planning and programming chores in manufacturing are time consuming and error prone. For small batches of production, the cost of planning and programming can weigh heavily in the total production cost. Moreover, the time spent in planning and programming may excessively delay the actual production. The automation of these chores will expedite their execution, reduce their cost, and improve their quality. Systematic procedures are needed in order to facilitate the automation of planning and programming of assembly systems. For some products, small changes in the design can have a large impact on the assembly alternatives. Autonomous systems for applications such as space or deep-sea exploration will need the ability to generate assembly or disassembly sequences that fit the particular situation they encounter. It is virtually impossible to preprogram all possible situations those systems might face, particularly if execution errors can occur and the systems are expected to recover autonomously.
- In less structured, more dynamic manufacturing systems or facilities, there is a need to adapt the assembly process to different machines. The need to produce different products in the same shop may lead to the choice of an assembly sequence for a product that may not be the most efficient but uses the idle equipment in the shop. Knowledge of all assembly sequence options of each product is required in order to optimize the overall use of machines and tools. Similarly, when the same product is assembled in different shops, the knowledge of all assembly sequences is needed in the selection of the sequence most suitable to the equipment available in each shop.

Manufacturing processes, in which single parts are combined to form an assembly, are referred to as manufacturing processes for joining and assembling. These processes can be divided into three major classes:

1. Processes for nonpermanent combining, which allow for multiple disassembly and assembly of single parts or subassemblies.
2. Processes for permanent combining of single parts or subassemblies. Eventual disassembly would result in severe damages to the components in the assembly and the subsequent assembly if attempted would not be possible any more. Processes for permanent assembly by welding joining

processes, in which two (or more) parts or subassemblies are joined together to form a permanent assembly. Examples are welding, adhesive bonding, brazing and soldering.

   a. Oxyfuel gas welding
   b. Arc welding with consumable electrodes
   c. Arc welding with nonconsumable electrodes
   d. Resistance welding
   e. Solid-state welding
      i. Forge welding
      ii. Cold roll welding
      iii. Diffusion welding
      iv. Explosion welding
      v. Friction welding

3. Mechanical assembly, which involves the use of various fastening methods to mechanically attach two (or more) parts or subassemblies together. This group includes processes for permanent (riveting, press or shrink fitting) or nonpermanent (assembly with threaded fasteners) assembly.

Threaded fasteners are components that have external or internal threads for assembly of parts. The common threaded fastener types are screws, bolts, studs, and nuts.

A bolt is an externally threaded fastener that is inserted through holes in the parts and screwed into a nut on the opposite side; a screw is an externally threaded fastener that is generally assembled into a blind threaded hole and no nut is required; a stud is an externally threaded fastener, but without the usual head possessed by a bolt. Studs can also be used to assemble two parts using a nut. They are available with threads on one end or both; a nut is an internally threaded fastener having standard threads. The typical assemblies that result from the use of screws, bolts, studs, and nuts are shown in Figure 11.1.

Threaded fasteners come in a variety of sizes, threads, and shapes. Also, numerous head styles are available on bolts and screws, some of which are illustrated in the figure. The geometries of these heads, as well as the variety of sizes available, require different hand tools for the operator. Various head styles are available on screws and bolts. In addition to screws, bolts, studs, and nuts, other types of threaded fasteners and related hardware are also available. These include screw thread inserts, captive threaded fasteners, and washers. Whether a threaded fastener serves its purpose depends to a large degree on the amount of torque applied to tighten it. Once the threaded fastener has been rotated until it is seated

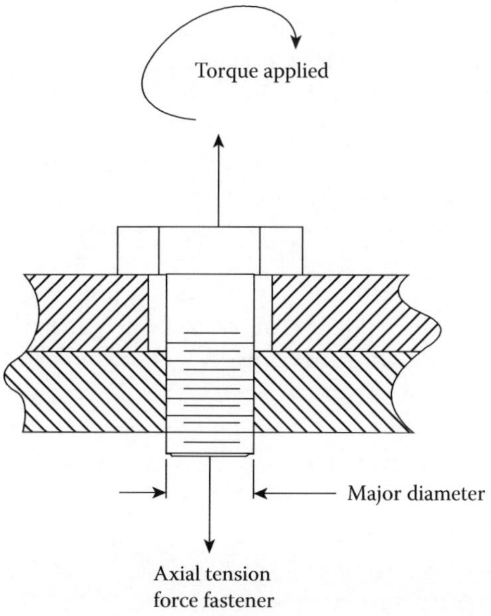

Figure 11.1

Tightening of threaded fasteners.

against the part surface, additional tightening will increase the amount of tension in the fastener (and, simultaneously, the amount of compression in the parts being held together) and an increasing torque will resist the tightening. Various methods are employed to apply the required torque. These are torque wrenches, powered wrenches designed to stall when the required torque is reached, and torque-turn tightening, in which the fastener is initially tightened to a low torque level and then rotated a specified additional amount.

11.2 Riveting

A rivet is an unthreaded, headed pin used to join two (or more) parts by passing the pin through holes in the parts and then forming (upsetting) a second head in the pin on the opposite side. The deforming operation can be performed hot or cold and by hammering or steady pressing. Once the rivet has been deformed, it cannot be removed except by breaking one of the heads. Riveting is a fastening method that offers high production rates, simplicity, dependability, and low cost. Despite these apparent advantages, its applications have declined in recent decades in favor of threaded fasteners, welding, and adhesive bonding. It could be stated that rivet fasteners are not sustainable or not appropriate to design for environment (DfE). Riveting is used as one of the primary fastening processes in the aircraft and aerospace industries for joining skins to channels and other structural members. Much of the equipment used in riveting is portable and manually operated. Automatic drilling and riveting machines are available for drilling the holes and then inserting and upsetting the rivets.

11.2.1 Press and Shrink Fits

There are several assembly methods that are based on mechanical interference between the two mating parts being joined. The methods discussed here include press fitting and shrink fitting.

- A press-fit assembly is one in which the two components have an interference fit between them. The typical case is when a pin of a diameter *dp* is pressed into a hole of a slightly smaller diameter d_h. Applications of press fitting include locating and locking the components such as the assembly of collars, gears, pulleys, and similar components on to shafts. The major limitations of press fitting include the necessity of a substantial press force, and the possible damage to the surfaces of components during the process of press fitting. These limitations are overcome in the process of shrink fitting.
- To assemble by shrink fitting, the external part is heated to enlarge by thermal expansion, and the internal part either remains at room temperature or is cooled to contract its size. The parts are then assembled and brought back to room temperature so that the external part shrinks and, if previously cooled, the internal part expands to form a strong interference fit. It could be argued that there is loss of energy in shrink fitting and obviously more CO_2 emission.
- A modification of the shrink fitting method is the so-called expansion fit, which occurs when only the internal part is cooled to contract it for assembly. Once inserted into the mating component, it warms to room temperature, expanding to create the interference assembly. Various methods are used to accomplish the heating or cooling of the work parts. Suitable heating equipment includes torches, furnaces, electric resistance heaters, and electric induction heaters. Cooling methods include conventional refrigeration, packing in dry ice, and immersion in cold liquids, including liquid nitrogen. The change in diameter that results from heating or cooling a cylindrical workpiece depends on the coefficient of thermal expansion and the temperature difference that is applied to the parts.
- The shrink fitting method is used to fit gears, pulleys, sleeves, and other components onto solid and hollow shafts, but the most popular application is to fit bearings on to shafts.

11.2.2 Benefits of Cold Heading

- *Savings in Material*: Cold heading scrap loss is as low as 1% compared to turning, shaving or forging where up to 75% (or more) scrap loss is common.
- *Higher Strength*: Cold heading creates continuous, unbroken material grain flow instead of cutting the grains of the fastener.

- *Longer Life Part*: Properly designed cold headed parts are resistant to impact, fatigue, and shear failure.
- *Eliminate Costly Secondary Operations*: Often a part can be formed complete in the header, eliminating the need for expensive secondary forming and handling.
- *Production Rates*: Cold heading can produce up to 450–500 pieces per minute. This sounds unrealistic, but it is true.

Solid rivets were commonly used from the 1800s for all sorts of steelwork (bridges, ships, tanks and pressure vessels, truck chassis, machinery) until being mostly superseded by welding in the mid-1900s.

Rivets are still used in aircraft—and can be used in combination with adhesives which give higher strength and ensure a sealed joint. Rivets are specified by their length, diameter, head, and type. Solid aluminum rivets used are in aircraft (Figure 11.2).

Countersunk head of flush aircraft rivet showing inner swaging dimensions.

In structures, rivets are made of tough and ductile low-carbon steel or nickel steel. According to the ASME Boiler Construction Code, the rivet material must have a tensile strength between 45,000 and 55,000 psi and a minimum yield point equal to one-half the tensile strength. Brass rivets are used in small sizes.

Today, the most common type of rivet is the pop-rivet—where the swaging is done from the outside by a self-breaking pin through the center of the (hollow) rivet (Figures 11.3 and 11.4). Solid rivets are more difficult to fit but are stronger and more reliably sealed than the hollow pop-rivet.

11.2.3 Common Industry Guidelines for Semitubular Rivets

The clinch allowance for *semitubular rivets* is an important parameter in rivet design. The allowance is normally up to a diameter of 1/2 in. or 50%–70% of the shank diameter of the rivet, dependent on application.

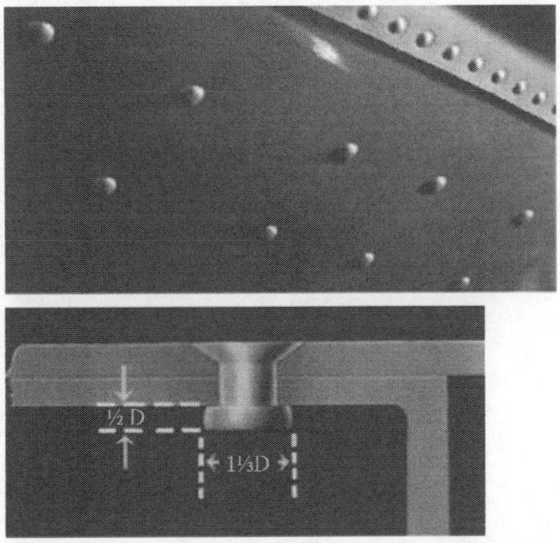

Figure 11.2

Rivets in use.

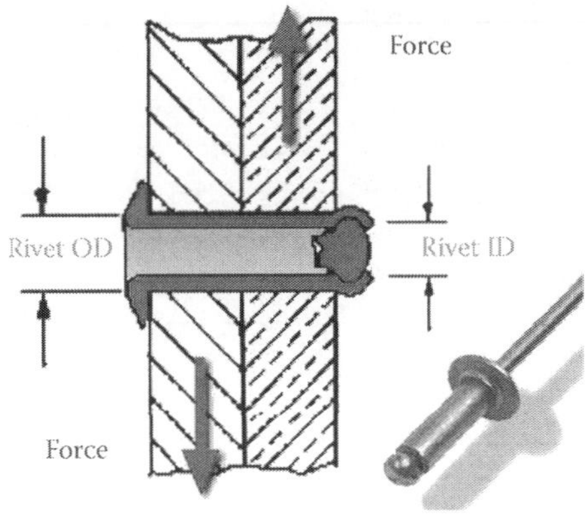

Figure 11.3

Pop-rivet after pulling and breaking (popping) the pin, forming a clamping fastener between the two sheets. Pop-rivets are limited to light duty, usually thin sheet metal and some (nonbrittle) plastics. They are often aluminum (soft), but can also be made from steel, stainless steel, and brasses.

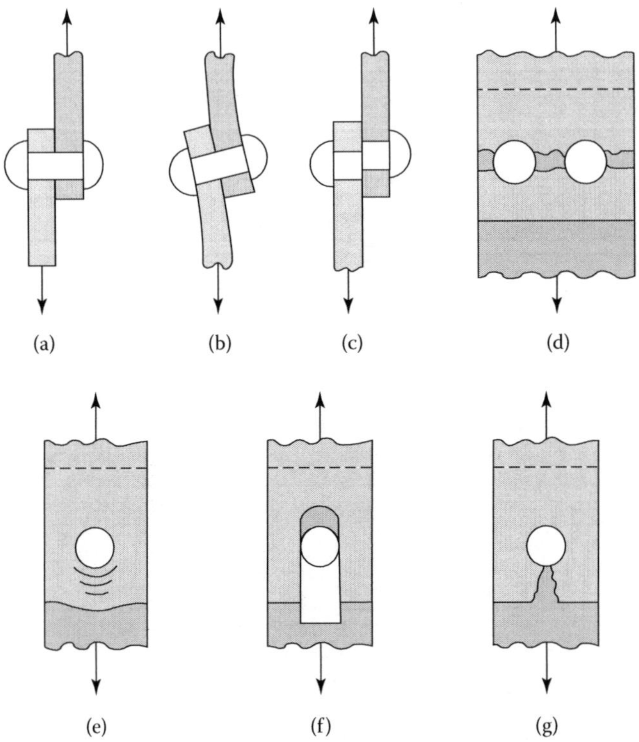

Figure 11.4

(a–g) Forces on rivets.

Clinch allowance = 50%–70% of shank body diameter of the rivet
Shank length of the rivet = combined workpiece thickness plus the clinch allowance of the rivet

The workpiece hole diameter has to be greater than the shank diameter of the rivet. It is usually about 107% of that value. It could also be stated that the workpiece hole diameter is about 1/16 in. larger than the shank diameter of the rivet. If multiple workpiece holes are used, larger hole(s) to accommodate stack-up tolerances are suggested (Figure 11.5).

The maximum force F needed to punch a hole in a plate may be estimated as follows:

$$F = 1.25\pi dtS_{su}$$

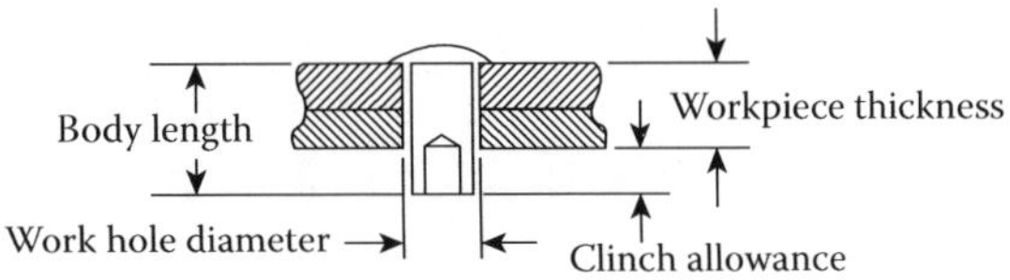

Figure 11.5

Some important parameters in rivet design.

where:

d is the workpiece hole diameter in inches

t is the thickness of the plate in inches

S_{su} is the ultimate strength in shear of the plate material in pounds per square inch

Normally, the ultimate shear strength is about 0.75 of the ultimate tensile strength of the material. The work of punching a hole is calculated as follows:

$$W = 0.68Ft$$

11.2.4 Types of Joint

Two arrangements common in joining plates are *lap joints* and *butt joints.*

Lap joints consist of overlapping plates held together by one or more rows of rivets. A single riveted lap joint is shown in Figure 11.5. In double or triple riveted joints, two or three rows of rivets are arranged either in staggered or in chain form.

Butt joints are formed by butting the plates against each other and joining them by overlapping straps or cover plates. A butt joint may have one strap on the outside, but most butt joints have one on the outside and another on the inside.

11.2.5 Stresses in Joints

Rivets should always be placed at 90° to the acting force and the maximum stress induced is either *shear* or crushing. The length of rivets between the heads should not exceed four or five times their diameter, as if it is longer, there is a danger of the head snapping when the rivet cools.

Failure of a riveted joint: According to the conventional theory, the failure of joints may occur in one of the following ways:

1. Shearing of rivets (Figure 11.4c)
2. Rupturing of the plate by tension (Figure 11.4d)
3. Crushing of the plate or rivets (Figure 11.4e)
4. Shearing of the margin (Figure 11.4f)
5. Tearing of the margin (Figure 11.4g)

Figures 11.6 through 11.9 show different arrangements of lap and butt jount of rivets. There are various types of rivets in use, but standard dimensions are shown in Figure 11.10.

Example 11.1: Design of Rivet

Two 15 × 1/2 in. plates are joined by lap joints with rivets. The rivets have a diameter of 7/8 in. and the working stresses are as follows: shear, 13,500 psi; bearing, 27,000 psi; and tension, 18,000 psi. Design the riveted connection.

The riveted joint should be designed for maximum efficiency, and usually it is assumed that the row of rivets immediately adjacent to the load carries the full load; the maximum tensile load carried

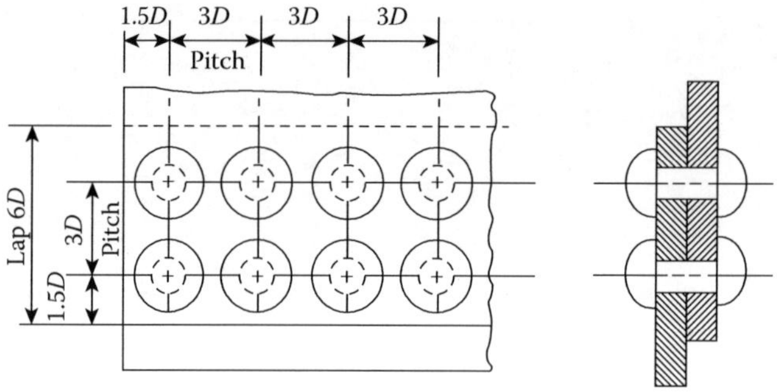

Figure 11.6

Double-riveted lap joint (chain). From design considerations and failure of riveted joints, the pitch and margin dimensions are shown relative to the diameter of the workpiece hole. A lap itself of *6D* is suggested.

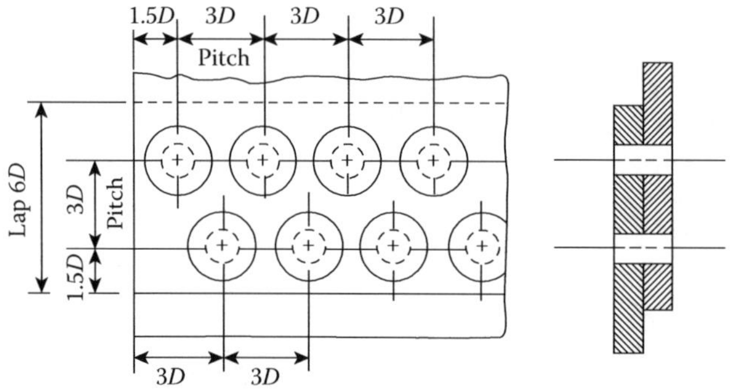

Figure 11.7

Double-riveted lap joint (zigzag). In this joint, all dimensions are very similar to the double riveted lap joint (chain).

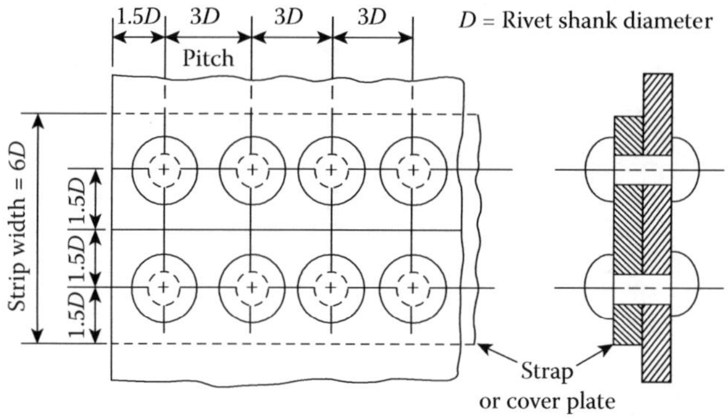

Figure 11.8

Single-strap chain-riveted butt joint (single row). In this joint, the horizontal and vertical pitch differs considerably.

by such a row occurs when only one rivet is in that row. The *strength of the plate with one rivet* is calculated as

$$F_{t1} = S_t(w - D)t$$

D is the hole diameter, and it should be = 7/8 + 1/16 = 15/16 in.

$$F_{t1} = 18000\left(15 - \frac{15}{16}\right)(1/2) = 126{,}563 \text{ lb}$$

The strength of joint in *bearing* and *shear* depends on *the number of rivets.*

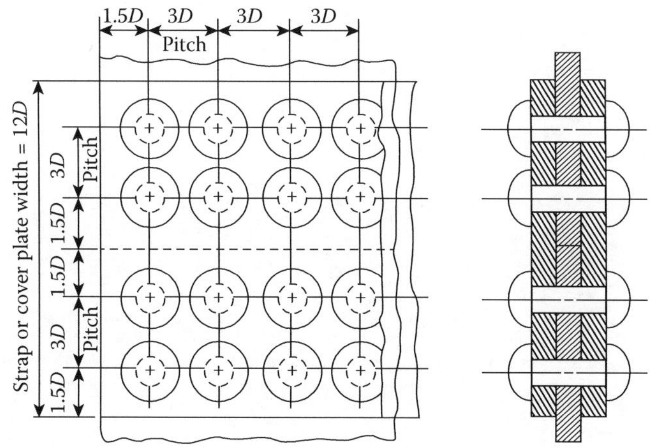

Figure 11.9

Double-strap chain-riveted butt joint (double row).

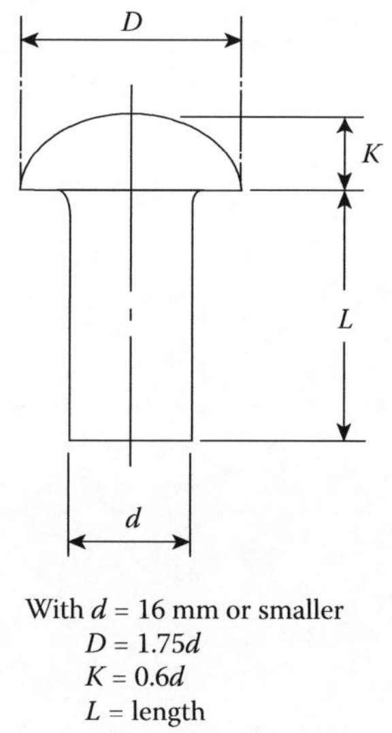

Figure 11.10

Standard dimensions of rivets.

Shear Strength

Shear strength of one rivet (F_s)

$$F_s = S_s \frac{\pi}{4}\left(\frac{7}{8}\right)^2 = 13{,}500\frac{\pi}{4}\left(\frac{7}{8}\right)^2 = 8{,}110 \text{ lb}$$

Bearing Strength

$$F_b = S_b t.d = 27{,}000(1/2)(7/8) = 11{,}800 \text{ lb}$$

Therefore, the rivets are weakest in shear and the number of rivets necessary is

$$n = \frac{126{,}563}{8{,}110} = 15.6 \approx 16$$

that is, 16 rivets are necessary to permit 126,563 lb of load to be applied.

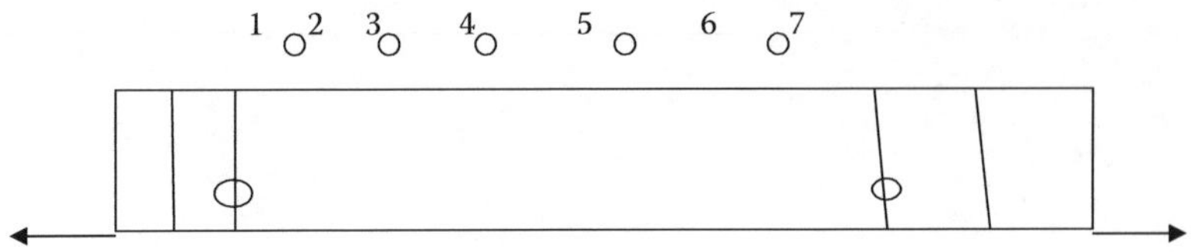

Figure 11.11

Arrangement of rivets in lap joint.

Now, the distribution of 16 rivets is to be determined. In doing this, it should be decided that for lap joints having equal plate thickness, the last row of the rivets (farthest from the load) in the upper plate is the first row (the row closest to the load of the lower plate). Therefore, the first and last row will have one rivet as depicted in Figure 11.11.

The two plates are joined as a lap joint and 16 rivets are arranged in chain. The outer two rows are shown with one rivet each and the remaining 14 rivets are arranged in Rows 2–6 in the following manner.

The portion of load carried by an interior row is determined by

$$P_{t(\text{actual})} = \frac{n-n'}{n} P$$

where:

$P_{t(\text{actual})}$ = portion of externally applied load P acting on a particular row
n = total number of rivets in the lap joint
n' = total number of rivets in the rows between the row being checked in tension and external load

$$P_{t(\text{actual})} = \frac{(16-1)}{16}(126{,}563) = 118{,}653 \text{ lb}$$

Strength of plate in Row 2:

$$P_{t(\text{allow})} = 18{,}500\big(15 - 2(15/16)\big)(1/2) = 121{,}406 \text{ lb}$$

$P_{t(\text{allow})}$ is greater than $P_{t(\text{actual})}$ and, therefore, the joint is safe. Therefore, Rows 2 and 6 will have two rivets each. Let us next examine Row 3 if three rivets are placed in it.

Load on Row 3:

$$P_{t(\text{actual})} = \frac{16-3}{16}(126{,}563) = 102{,}832.45 \text{ lb}$$

Strength of plate in Row 3:

$$P_{t(\text{allow})} = 18{,}000(15 - 3(15/16))(1/2) = 109{,}688 \text{ lb}$$

Here, again, it is found that $P_{t(\text{allow})} > P_{t(\text{actual})}$, and Rows 3 and 5, with three rivets each, are safe. We have so far accounted for 12 rivets in Rows 1, 2, 3, 5, 6, and 7. We only have Row 4 left and we can assign four rivets to it.

$$P_{t(\text{actual})} = \frac{16-6}{16}(126{,}563) = 79{,}102 \text{ lb}$$

$$P_{t(\text{allow})} = 18{,}000(16 - 4(15/16))(1/2) = 101{,}250 \text{ lb}$$

$P_{t(\text{allow})} > P_{t(\text{actual})}$; therefore, Row 4 with four rivets is safe.

It is advisable to check the spacing requirement for Row 4. It is shown in Figure 11.7 that the center-to-center distance is $3D$ and the margin is $1.5D$. The required width is $3(3\times 7/8)+2(1.5\times 7/8)=10.50$ in.

Since this is less than the width of the plate (15 in.), the spacing requirement is met. The final rivet pattern is thus satisfactory. Finally, the efficiency of the joint is calculated:

$$\text{Efficiency} = \frac{P_{\text{allow(max)}}}{\text{tensile strength of plate}}$$

$$\eta = \frac{126{,}563}{18000\times 15\times 1/2} = 0.9375$$

Therefore, the efficiency of the joint is 93.75%, which is very good.

11.3 Design of Bolts and Screws

An example of an automotive bolt in a connecting rod is presented to demonstrate the importance of bolts in engineering industries. The connecting rod is a vital component of an automobile, which will not function without the bolt connecting the two halves of the big end.

Example 11.2: Design Procedure for Automotive Bolt in Connecting Rod

The design of automotive bolts is a complex process, involving a multitude of factors. These include the determination of operating loads and the establishment of geometric configuration. The process for connecting rod bolts is described in the following paragraphs as an example.

11.3.1 Fastener Load

The first step in the process of designing a connecting rod bolt is to determine the load that it must carry. This is accomplished by calculating the dynamic force caused by the oscillating piston and connecting rod. This force is determined from the classical concept that force equals mass times acceleration. The mass includes the mass of the piston plus a portion of the mass of the rod. This mass undergoes oscillating motion as the crankshaft rotates. The resulting acceleration, which is at its maximum value when the piston is at top dead center and bottom dead center, is proportional to the stroke and the square of the engine speed. The oscillating force is sometimes called the reciprocating weight.

It is seen that the design load, the reciprocating weight, depends on the square of the rpm speed. This means that if the speed is doubled, for example, the design load is increased by a factor of four.

11.3.2 Reciprocating Weight

A typical value for this reciprocating weight is in the vicinity of 20,000 lb. For purposes of bolt design, a rule of thumb is to size the bolts and select the material for this application such that each of the two rod bolts has a strength of approximately 20,000 lb (corresponding to the total reciprocating weight). This essentially builds in a nominal safety factor of two. The stress is calculated according to the following formula, so that the root diameter of the thread can be calculated from the formula

$$\text{Stress} = \frac{\text{Force}}{\text{Area}} = \frac{\text{Recip.Wt.}}{\frac{\pi}{4}D^2}$$

$$D = \sqrt{\frac{4\times \text{Recip.Wt.}}{\pi \times \text{Allowable Stress}}} \tag{11.1}$$

This formula shows that the thread size can be smaller if a stronger material is used. Or, for a given thread size, a stronger material will permit a greater reciprocating weight. The graph shows the relationship between thread size and material strength.

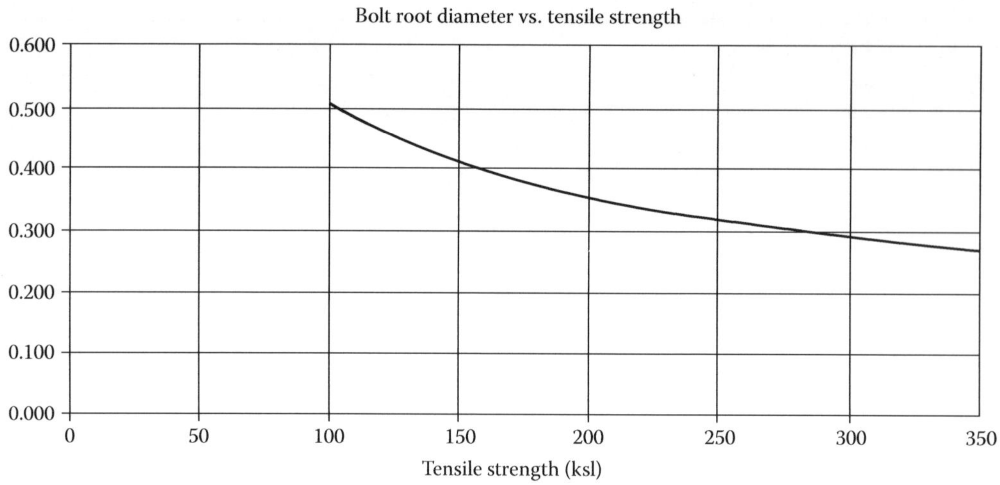

Figure 11.12

Tensile strength vs. bolt root diameter.

11.3.3 Other Stresses

It must be realized that the direct reciprocating load is not the only source of stresses in bolts. A secondary effect arises because of the flexibility of the journal end of the connecting rod. The reciprocating load causes bending deformation of the bolted joint (even steel deforms under load). This deformation causes bending stresses in the bolt as well as in the rod itself. These bending stresses fluctuate from zero to their maximum level during each revolution of the crankshaft (Figure 11.12).

11.3.4 Geometric Configuration

The next step is to establish the details of the geometric configuration. Here the major consideration is fatigue, the fracture that could occur due to frequent repetition of high stresses, such as the bending stresses described above. Several factors must be considered in the prevention of fatigue; attention to design details is essential.

11.3.5 Stress Riser Elimination

Fatigue failure is frequently caused by localized stress risers, such as sharp corners. In bolts, this would correspond to the notch effect associated with the thread form. It is well known that the maximum stress in an engaged bolt occurs in the last engaged thread. By removing the remaining, nonengaged threads, the local notch effect can be reduced. This leads to the standard configuration used in most automotive racing products (ARP) rod bolts: a reduced-diameter shank and full engagement for the remaining threads. Providing a local fillet radius at the location of the maximum stress further reduces the local notch effect. Thus, this configuration represents the optimum with respect to fatigue strength.

The reduced-diameter shank is helpful in another sense. It reduces the bending stiffness of the bolt. Therefore, when the bolt bends due to deformation of the connecting rod, the bending stresses are reduced below what they would otherwise be. This further increases the fatigue resistance of the bolt. A typical bolt configuration is shown in Figure 11.13.

With respect to bolt fatigue strength, thread rolling is a major consideration. Threads are rolled after heat treating. This process, which deforms the metal, produces a beneficial compressive stress in the root of the thread. It is beneficial because it counteracts the fluctuating tensile stresses that can cause fatigue cracking. If heat treatment were to occur after rolling, the compressive stresses would be eliminated. This would therefore reduce the fatigue resistance of the bolt.

11.3.6 Engaged Thread

An additional factor must be taken into account in defining the bolt configuration: the length of engaged thread. If too few threads are engaged, the threads will shear at loads that are lower than the strength of the bolt. As a practical matter, the thread length is always selected so that the thread shear strength is significantly

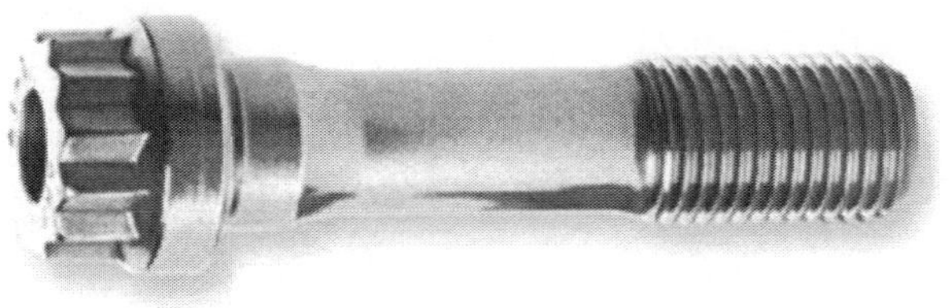

Figure 11.13

Typical bolt used in connecting rod.

greater than the bolt tension strength. This problem is especially important in bolts used in aluminum rods because of the fact that the shear strength of aluminum is much lower than the shear strength of steel.

11.3.7 Preload in Connecting Rod

Finally, although not a design parameter, the subject of bolt installation preload must be addressed. It is a fundamental engineering concept that the force in a bolt in an ideal preloaded joint will remain equal to the preload until the externally applied force exceeds it. Then, the force in the bolt will be equal to the external force. This means that fluctuating external forces will not cause fluctuating forces in a preloaded bolt as long as the preload exceeds the external force. The result is that fatigue failure will not occur.

In a nonideal joint, such as in a connecting rod, the bolt will feel fluctuating stresses due to fluctuating rod distortions. These are additive to the preload, so that fatigue could result. In connecting rods, precise preloads are required, because if they are too low, the external forces (the reciprocating weights) will exceed the preloads, thus causing fatigue. If they are too high, they provide a high mean stress that combines with the fluctuating stresses due to rod distortion, and fatigue is promoted. The objective, then, is to preload a bolt so that it just exceeds the external load, and no higher. To sum up: Both insufficient preloads and excessive preloads can lead to fatigue failures.

These preloads can be attained in a connecting rod by applying proper torque using a torque wrench or by measuring the amount of stretch in the bolt using a stretch gauge (it is known that a bolt stretches in proportion to the tension in it). The torque method is sometimes inaccurate because of the uncertainty in the coefficient of friction at the interface between the bolt and the rod. This inaccuracy can be minimized by using the lubricant.

11.3.8 Screw Thread Design: Screw Thread Fundamentals

Screw threads serve three basic functions in mechanical systems: (1) to provide a clamping force, (2) to restrict or control motion, and (3) to transmit power. Geometrically, a screw thread is a helical inclined plane. A helix is the curve defined by moving a point with uniform angular and linear velocity around an axis. The distance the point moves linearly (parallel to the axis) in one revolution is referred to as the pitch or lead. The term "internal threads" refers to threads cut into the sidewall of an existing hole.

External threads refer to threads cut or rolled into the external cylindrical surface of a fastener or stud. The size most commonly associated with screw threads is the nominal diameter. Nominal diameter is a more of a label than a size. For example, a bolt and nut may be described as having a diameter of 1/2 in. But, neither the external threads of the bolt nor the internal threads of the nut have a diameter of exactly 0.500 in. In fact, the bolt diameter is a little smaller and the nut diameter a little larger. However, it is easier to specify the components by a single size designation since the bolt and nut are mating components. Some common terms used in bolt and screws are

Major Diameter: The largest diameter of the thread
Minor Diameter: The smallest diameter of the thread
Crest: The peak of the thread for external threads; the valley of the thread for internal threads
Pitch Diameter: Nominally, the mean of the major and minor diameters
Thread Angle: The included angle between two adjacent thread walls

11.3.9 Effect of Thread Angle on Strength

The lower the value of the thread angle, the greater the load-carrying capability of the thread. The force (F) of mating threads is normal to the surface of the thread. The components of the force F, transverse and

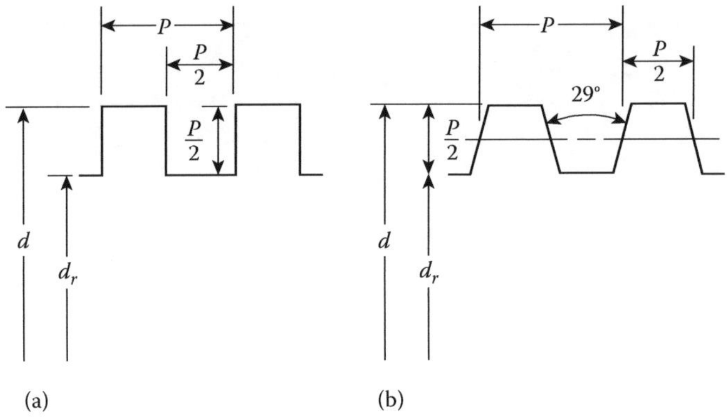

Figure 11.14

Pitch and configuration of a thread. (From The McGraw-Hill Companies, Inc. Permission required for reproduction or display.)

parallel to the axis are shown as F_t and F_a. The component of force typically responsible for failure is that applied transverse to the axis of the thread.

A screw thread is defined as a ridge of uniform section in the form of a helix on either the external or internal surface of a cylinder. Internal threads refer to those on nuts and tapped holes, while external threads are those on bolts, studs, or screws.

The thread form is the configuration of the thread in an axial plane; or, more simply, it is the profile of the thread, composed of the crest, root, and flanks. At the top of the threads are the crests, at the bottom the roots, and joining them are the flanks. The triangle formed when the thread profile is extended to a point at both crests and roots is the fundamental triangle. The height of the fundamental triangle is the distance, radially measured, between the sharp crest and sharp root diameters.

The distance measured parallel to the thread axis, between corresponding points on adjacent threads, is the *thread pitch* (Figure 11.14). Unified screw threads are designated in threads per inch. This is the number of complete threads occurring in 1 in. of threaded length. Metric thread pitch is designated as the distance between threads (the pitch) in millimeters.

On an internal thread, the minor diameter occurs at the crests and the major diameter occurs at the roots (Figure 11.15). On an external thread, the major diameter is at the thread crests, and the minor diameter is at the thread roots. The flank angle is the angle between a flank and the perpendicular thread axis. *Flank*

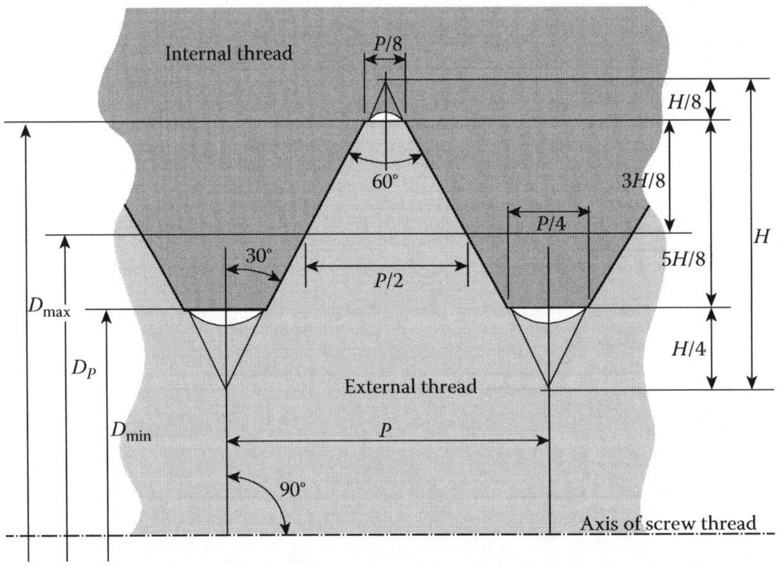

Figure 11.15

Internal thread specifications.

angles are sometimes termed "half-angle" of the thread, but this is only true when neighboring flanks have identical angles; that is, the threads are symmetrical. Unified screw threads have a 30° flank angle and are symmetrical. This is why they are commonly referred to as 60° threads.

Pitch diameter is the diameter of a theoretical cylinder that passes through the threads in such a way that the distance between the thread crests and thread roots is equal. The number of threads per inch cracking in internal threads, especially under cyclic loads, is called the *pitch*. Internal threads are more susceptible, since they are typically cut, and cutting operations in metals produce surface irregularities that can contribute to crack growth. External threads are typically rolled onto a fastener and, therefore, lack the surface flaws of cut threads. As the thread angle decreases, the transverse component F_t gets smaller. This is why square and buttress threads are usually used for power transfer applications. In an ideal product, these widths would each equal one-half of the thread pitch.

An intentional clearance is created between mating threads when the nut and bolt are manufactured. This clearance is known as the allowance. Having an allowance ensures that when the threads are manufactured, there will be a positive space between them. For fasteners, the allowance is generally applied to the external thread. Tolerances are specified amounts by which dimensions are permitted to vary for convenience of manufacturing. The tolerance is the difference between the maximum and minimum permitted limits.

11.3.10 Thread Series

There are three standard thread series in the unified screw thread system that is highly important for fasteners: UNC (coarse), UNF (fine), and 8-UN (8 thread). A chart listing the standard sizes and thread pitches with their respective thread stress areas is listed in the Fastenal technical reference guide, along with a special series designated UNS.

Below are some of the aspects of fine and coarse threads.

11.3.11 Fine Thread

- Since they have larger stress areas, the bolts are stronger in tension.
- Their larger minor diameters develop higher torsional and transverse shear strengths.
- They can tap better in thin-walled members.
- With their smaller helix angle, they permit closer adjustment accuracy.

11.3.12 Coarse Thread

- Stripping strengths are greater for the same length of engagement.
- Improved fatigue resistance.
- Less likely to cross thread.
- Quicker assembly and disassembly.
- Tap better in brittle materials.
- Larger thread allowances allow for thicker platings and coatings. Numerous arguments have been made for using either fine or coarse threads; however, with the increase in automated assembly processes, bias toward the coarse-thread series has developed.

11.3.13 UNR Threads

The UNR thread is a modified version of a standard UN thread. The single difference is a mandatory root radius with limits of 0.108–0.144 times the thread pitch. When first introduced, decades ago, it was necessary to specify UNR (rounded root) threads. Today, all fasteners that are roll threaded should have a UNR thread because thread rolling dies with rounded crests are now the standard method for manufacturing most threads.

11.3.14 UNJ Threads

The UNJ thread is a thread form having root radius limits of 0.150–0.180 times the thread pitch. With these enlarged radii, minor diameters of external thread increase and intrude beyond the basic profile of the UN and UNR thread forms. Consequently, to offset the possibility of interference between mating threads, the minor diameters of the UNJ internal threads had to be increased.

The standard thread tolerances for UNJ threads are 3A/3B. UNJ threads are now the standard for aerospace fasteners and have some usage in highly specialized industrial applications. UNJ bolts are similar to UNR, but the curve of the thread root is gentler, which requires that it be shallower. In fact, the thread root is so shallow that the bolt thread cannot mate with a UN nut, so there is a UNJ nut specification as well.

11.3.15 Thread Strength

Two fundamentals must be considered when designing a threaded connection:

1. Ensure that the threaded fasteners were manufactured to a current ASTM, ANSI, DIN, ISO or other recognized standard.
2. Ensure that the design promotes bolts to break in tension prior to the female and/or male threads stripping. A broken bolt is an obvious failure. However, when the threads strip prior to the bolt breaking, the failure may go unnoticed until after the fastener is put into service. The strength of bolts loaded in tension can be easily determined by the ultimate tensile strength. To determine the amount of force required to break a bolt, multiply its ultimate tensile strength by its tensile-stress area, determining the strength of the threads is more complicated. Since the male threads pull past the female threads, or vice versa, the threads fail due to shear and not in tension. Therefore, the stripping strength of an assembly depends on the shear strength of the nut and bolt materials. To determine the force required to strip the threads, multiply the shear strength by the cross-sectional area being sheared. The difficulty lies in determining the cross-sectional area in which the shear will occur.

Here are three possible scenarios for this type of failure:

1. The nut material is stronger than the bolt material. In this example, the nut threads will shear out the bolt threads. The failure will occur at the root of the bolt threads.
2. The bolt material is stronger than the nut material. In this scenario, the bolt threads will shear out the nut threads. The failure will occur at the root of the nut threads.
3. The nut and bolt are the same strength. In this scenario, both threads will strip simultaneously. This failure will occur at the pitch line.

The tensile strength of most fasteners is usually specified, whereas shear strength is not. In order to avoid shearing the threads, ensure that the length of engagement between the internal and external threads is long enough to provide adequate cross-sectional thread area.

Failure scenarios 1 and 3 can typically be avoided by ensuring proper thread engagement. With proper engagement, those scenarios would result in a tensile failure of the bolt rather than thread stripping.

Taking proper precautions during the design phase is vital to avoid thread failure. Once the first engaged thread begins to shear, the threads behind it will also shear in rapid succession. Generally, the hardness and the actual material strength of a nut is less than that of the bolt. For example, if we compare the hardness of an SAE J995 Grade 8 nut (HRC 24–32, up to a diameter of 5/8 in.), it is likely to be less than the SAE J429 Grade 8 bolt (HRC 33–39). This is designed to yield the nut threads to ensure the load is not carried solely by the first thread. As the thread yields, the load is further distributed to the next five threads. Even with the load distribution, the first engaged thread still takes the majority of the load. In a typical situation, the first engaged thread carries 34% of the load. Using internally threaded materials with higher strengths and hardness can often result in fatigue or loosening.

The strength capacities of standard nuts are listed as the nut's *proof stress*. This should not be confused with the proof strength of the bolts. Proof stress is the ultimate load the nut can support without thread failure. For design purposes, the most important aspect of choosing the appropriate nut is to select a nut with a proof stress equal to or greater than the ultimate tensile strength of the bolt. For carbon-steel fasteners (including tapped holes) the length of engagement would be limited to approximately one nominal diameter (approximately 1-1/2 times the diameter for aluminum). After that, there is no appreciable increase in strength. Once the applied load has exceeded the first thread's capacity, it will fail and subsequently cause the remaining threads to fail in succession. Fundamentally, in thread connection design, the nut or tapped hole should support more load than the bolt. Thus, the design criteria for threaded connections also leads to nut selection criteria which help the designer to ensure functionality in the joint. The following are the basic rules:

1. Ensure that the nut adheres to a specification which is compatible with the specification of the bolt (ASTM A193 and ASTM A194, SAE J429 and SAE J995, etc.)
2. Ensure that the selected nut has a proof stress greater than or equal to the tensile strength of the bolt. If the nut proof stress does not exceed the proof strength of the bolt, stripping failure is very likely.

11.3.16 Standardization of Threads (Standard Inch Units)

To facilitate their use, screw threads have been standardized. This is the unified thread series and consists of specifications for Unified Coarse (UNC), Unified Fine (UNF) and Unified Extra-Fine (UNEF) threads. Metric threads are also standardized. Metric thread specification is given through ISO standards.

11.3.16.1 Unified Coarse

UNC is the most commonly used thread on general purpose fasteners. Coarse threads are deeper than fine threads and are easier to assemble without cross threading. UNC threads are normally easier to remove when corroded, owing to their sloppy fit. A UNC fastener can be procured with a Class 3 (tighter) fit if needed (classes of fits are discussed in Section 11.3.17).

11.3.16.2 Unified Fine

UNF threads have a larger minor diameter than UNC threads, which gives UNF fasteners slightly higher load-carrying capacity (in shear) and better torque locking capabilities than UNC fasteners of the same material and outside diameter. The fine threads have tighter manufacturing tolerances than UNC threads, and the smaller lead angle allows for finer tension adjustment. UNF threads are frequently used in cases where thread engagement is minimized due to thinner wall thickness.

11.3.17 Class Fit: Thread Fit

Thread fit is a combination of allowances and tolerances and a measure of tightness or looseness between them. A clearance fit is one that provides a free-running assembly and an interference fit is one that has a positive interference, thus requiring tools for the initial rundown of the nut.

For unified inch screw threads, there are six standard classes of fit: 1B, 2B, and 3B for internal threads; and 1A, 2A, and 3A for external threads. All are considered clearance fits; that is to say, they assemble without interference. The higher the class number, the tighter the fit. The "A" designates an external thread, and "B" designates an internal thread.

- Classes 1A and 1B are considered an extremely loose tolerance thread fit. This class is suited for quick and easy assembly and disassembly. Outside of low-carbon threaded rod or machine screws, this thread fit is rarely specified.
- Classes 2A and 2B offer optimum thread fit that balances fastener performance, manufacturing, economy, and convenience. Nearly 90% of all commercial and industrial fasteners use this class of thread fit.
- Classes 3A and 3B are suited for close tolerance fasteners. These fasteners are intended for service where safety is a critical design consideration. This class of fit has restrictive tolerances and no allowance. Socket products generally have a 3A thread fit.

The axial distance through which the fully formed threads of both the nut and bolt are in contact is called the *length of thread engagement*. The *depth of thread engagement* is the distance the threads overlap in a radial direction. The length of thread engagement is one of the key strength aspects and one of the few which the designer may be able to control.

Per the acceptance requirements of ASME B1.3, System 21, the allowance specified for the Class 2A external threads is used to accommodate the plating thickness. The plain finished parts (or plated parts prior to plating) would be tested for adherence to these tolerances with a 2A go/no-go thread gauge. The 2A go gauge would ensure the pitch diameter falls below the maximum requirement; the no-go gauge would verify that the pitch diameter is above the minimum requirement.

Several assembly methods are based on mechanical interference between two mating partsbeing joined. The methods include press fitting, shrink fitting, expansion fits, snap fits, and retaining rings.

Class fit is a specification of how tightly mating external and internal threads will mesh.

It is based upon the difference in the values of the respective pitch diameters. These differences are in the thousandths of an inch. For the unified thread form, the classes of fit are

Class 1: Loose fit: Threads may be assembled easily by hand. It is used in cases where frequent assembly/disassembly is required. Typically requires use of locking devices such as lock washers, locking nuts, jam nuts, and so on. Class 1 fits are common for bolts and nuts.

Class 2: Standard fit: Threads may be assembled partly by hand. It is the most common fit in use and is used in semipermanent assemblies.
Class 3: Tight fit: This may be started by hand, but requires assistance (tools) to advance threads. It is common for set screws and mostly used in permanent assemblies.

An additional designation is made for external (A) vs. internal (B) threads and is included as a postscript to the numerical designation.

Examples: Standard inch unit thread specification

0.4375–20UNF – 2A, LH (external left-handed thread of 0.4375 in. diameter and 20 threads per in)
0.500–13UNC – 1A (external thread)
0.375–24UNEF – 2B (internal thread)

11.3.18 Specification of Metric Threads

Metric threads are defined in the standards document ISO 965-1. Metric thread specifications always begin with the thread series designation (e.g., M or MJ), followed by the fastener's nominal diameter and thread pitch (both in units of millimeters) separated by the symbol "×."

Examples: Metric thread specification

MJ6 × 1–4H5H (metric thread of 6 mm nominal dia and 1 mm pitch and followed by class of fit)
M8 × 1.25–4h6h LH
M10 × 1.5–4h5h 7

11.3.19 Metric Thread Series

There exist multiple metric thread series used for special applications. The standard is the M series. The MJ series is one of the most common of the special application threads.

M Series: Standard metric thread profile
MJ Series: Modified series in which crest and root radii are specified

11.3.20 Metric Thread Fits

A fit between metric threads is indicated by the internal thread class fit followed by the external thread tolerance class separated by a slash; for example, M10 × 1.5–6H/6g. The class fit is specified by tolerance grade (numeral) and by tolerance position (letter).

11.3.20.1 General Purpose Fit

6g (external), 6H (internal)

11.3.20.2 Close Fit

5g6g (external), 6H (internal)

If the thread fit designation (e.g., "–6g") is omitted (e.g., M10 × 1.5), it specifies a "medium" fit, which is 6H/6g. The 6H/6g fit is the standard ISO tolerance class for general use.

The English unit internal and external thread class fit 2B/2A is essentially equivalent to the ISO thread class fit 6H/6g. The English unit class fit 3B/3A is approximately equivalent to the ISO class fit 4H5H/4h6h.

11.3.20.2.1 Default Metric Fastener Thread Pitch If metric thread pitch designation (e.g., "× 1.5") is omitted, it specifies coarse pitch threads. For example, M10 or M10-6g, by default, specifies M10 × 1.5. The standard metric fastener thread series for general purpose threaded components is the M thread profile and the coarse pitch thread series (Figure 11.16).

11.3.20.2.2 Metric Fastener Threads Series Compatibility Metric fastener thread series M is the common thread profile. Thread series MJ designates that the external thread has an increased root radius (shallower root relative to external M thread profile), thereby having higher fatigue strength (due to reduced stress concentrations), but requires a truncated crest height of the MJ internal thread to prevent interference at the external MJ thread root. M external threads are compatible with both M and MJ internal threads.

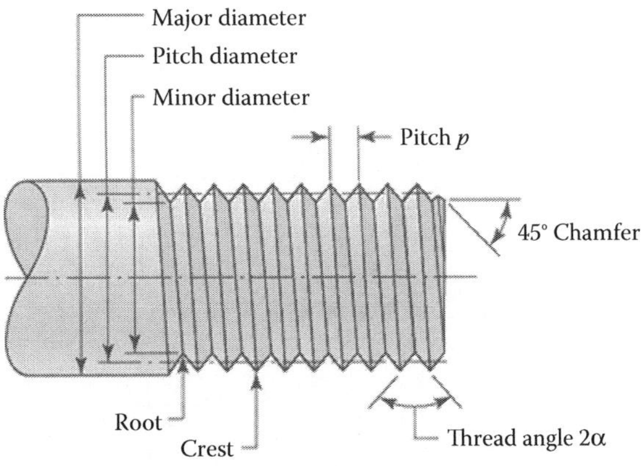

Figure 11.16

Threaded fastener. (From The McGraw-Hill Companies, Inc. Permission required for reproduction or display.)

11.3.21 Left-Hand Threads

Unless otherwise specified, screw threads are assumed to be right-handed. This means that the direction of the thread helix is such that a clockwise rotation of the thread will cause it to advance along its axis. Left-handed threads advance when rotated counterclockwise. Left-handed threads are often used in situations where rotation loads would cause right-hand threads to loosen during service. A common example is the bicycle.

The pedals of a bicycle are attached to the crank arm using screw threads. The pedal on one side of the bicycle uses a right-hand thread and the other uses a left-hand one. This prevents the motion of pedals and crank from unscrewing the pedal and having it fall off during use. Left-hand threads must be indicated in the thread specification. This is accomplished by appending "LH" to the end of the specification.

11.3.22 Writing Notes for Threaded Holes

The *note* for a threaded hole is a specification of all information required for the creation of the hole. This includes: (1) The diameter (and depth if blind) of the pilot hole drilled prior to thread creation; (2) the specification of the internal threads for the hole. Again, a depth is given if the hole is blind.

The creation of the internal threads is a metal-cutting process referred to as "tapping." It should be apparent that to cut metal, the diameter of the pilot hole must be smaller than the major diameter of the threads. This difference in diameters is very important. If the pilot hole diameter is too small, too much material will have to be cut and the thread cutting tool (tap), being very hard (and therefore brittle), will break. If the pilot-hole diameter is too large, the thread height will be too small and the load-carrying capability of the threads will be compromised. In practice, the diameter of the pilot hole will set the minor diameter of the internal threads. Typically, the thread height for internal threads is approximately 75% of the mating external threads (it may be as low as 50% for materials such as steel). This means that a gap will exist between the crest of the external thread and the root of the internal. For this reason, threads may not be considered a seal in and of themselves.

The diameter of the pilot hole is specific for each thread series and form. This unique diameter is determined by referencing the thread series and form within a standard table (Table 11.6). Typically, this value is referred to in the table as the tap drill diameter. The following table also provides the values of threads per inch for specific nominal diameters and thread series. Notice that in Figure 11.17, the tap drill diameter is given in fractions, letters, and numbers.

11.4 Design of Statically Loaded Bolted Joints

The principal steps in selecting a threaded fastener to sustain a steady load are discussed below. Like any design procedure, it cannot cover every case and eventuality, nor can it be followed blindly, and the user must exercise thought and judgment during the process. It may be necessary to iterate some of the steps before a satisfactory solution is obtained.

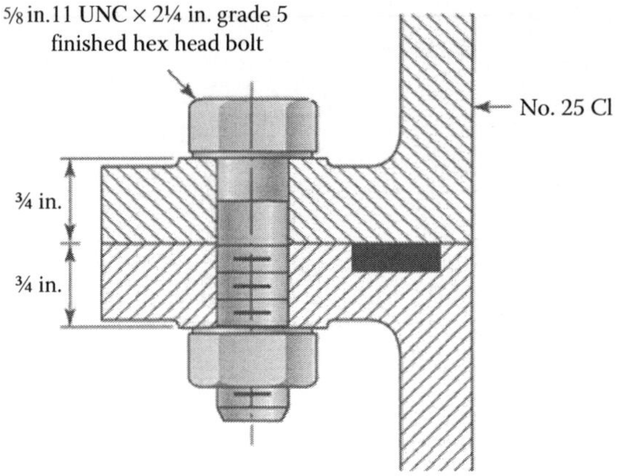

Figure 11.17

Information regarding bolt specifications inscribed.

11.5 Estimation of Stiffness of Bolt and Joint Members

The stiffness k of any axially loaded member is given by the equation

$$k = \frac{AE}{l}$$

where:

A = loaded cross-sectional area
E = Young's modulus
l = loaded length

Thus, in order to achieve the desired high ratio of joint stiffness to bolt stiffness, the following ratios should be maximized:

$$\frac{A_j}{A_b}, \frac{E_j}{E_b}, \frac{l_b}{l_j}$$

These ratios are described below.

11.5.1 Length Ratio

In the case of through-drilled joints clamped by a nut and bolt, the clamped length of the joint and the grip length of the bolt are clearly identical. However, since the tensile stress in the bolt does not fall immediately from its full value to zero at the loaded faces of the bolt head and nut, it is usual to consider the effective length of the bolt to be slightly greater than the actual grip length. This has the beneficial effect of slightly reducing the bolt stiffness. Where the bolt is screwed into a tapped hole in one of the joint members, the latter is clearly not wholly in compression. This tends to reduce the stiffness of the joint compared to the bolts. The following effective lengths are suggested:

An addition of 0.4 times the local shank diameter at each end is suitable—see Figure 11.18.

Here, the effective joint length l_{je} is greater than the effective bolt length, resulting in a less favorable stiffness ratio than for the case of a bolt in a through-drilled hole (Table 11.1).

11.5.2 Area Ratio

The bolt area will be determined by strength considerations. With a standard bolt, the critical section is usually the thread root diameter, which is less than the nominal bolt diameter, and is also subject to stress concentrations. The bolt shank may therefore be wasted down or drilled axially to give a cross-sectional area

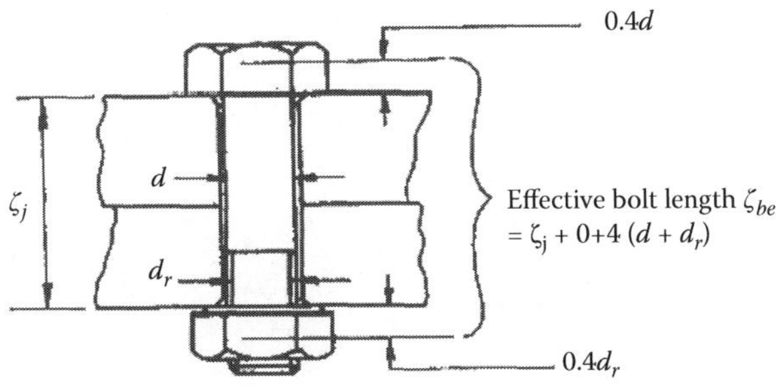

Figure 11.18

Effective grip length of bolt with nut.

Table 11.1 Bolt Loads and Torques

Thread Size (mm)	Maximum Axial Force (kN) for Bolts/Screws of Grade			Torque to Produce This Force (Nm)		
	8.8	10.9	12.9	8.8	10.9	12.9
M3	2.3	3.25	3.9	1.3	1.8	2.1
M4	4.0	5.65	6.75	2.7	3.8	4.6
M5	6.55	9.2	11.1	5.5	8.0	9.5
M6	9.25	13.0	15.6	9.5	13.0	16.0
M8	17.0	23.9	28.7	23.0	32.0	39.0
M10	27.1	38.0	45.7	46.0	64.0	77.0
M12	39.5	55.5	66.7	80.0	110	135
M14	54.0	76.0	91.3	125	180	215
M16	75	105	126	195	275	330
M20	117	164	197	385	540	650
M24	169	237	284	60	930	1100
M27	221	311	374	980	1740	1650
M30	269	379	454	1350	1850	2250
M36	393	553	664	2280	3210	3850

somewhat less than the thread root area, without loss of strength. A small length of the full shank diameter may have to be retained near the joint interface for lateral location purposes, but the resulting bolt will be more resilient than a standard one, and will therefore attract a smaller proportion of any load applied to the joint. It is also possible to increase the underhead radius, and thus relieve the stress concentration in this region. Both measures increase the effective *fatigue strength* of the bolt, which is important when the joint load has a dynamic component (Figure 11.19).

The joint area should always be as large as possible, provided the joint members can be made stiff enough to ensure reasonable parity of pressure over the whole joint interface. It is pointless to employ large, thin flanges which will distort under the clamping load and result in a concentration of load around the bolts, but little elsewhere (Figure 11.20).

For design purposes, we must try to estimate an effective area for use in calculating the joint stiffness. If the diameter of the joint faces D_j does not exceed the bolt head or washer diameter and bolt hole diameter D_h, the compressive load in the joint may be assumed to be uniform, so that

$$\text{effective joint area}\,(A_j) = \frac{\pi}{4}\left(D_j^2 - D_h^2\right)$$

For irregularly shaped joints, the joint diameter D_j may be estimated by drawing lines at 45° to the bolt axis from the collar diameter of the bolt or nut to the joint interface (see Figure 11.21).

To ensure effective clamping of the joint by the bolts in a heavy-duty connection, the size and number of bolts should be chosen so that the joint diameters of adjacent bolts touch or overlap slightly.

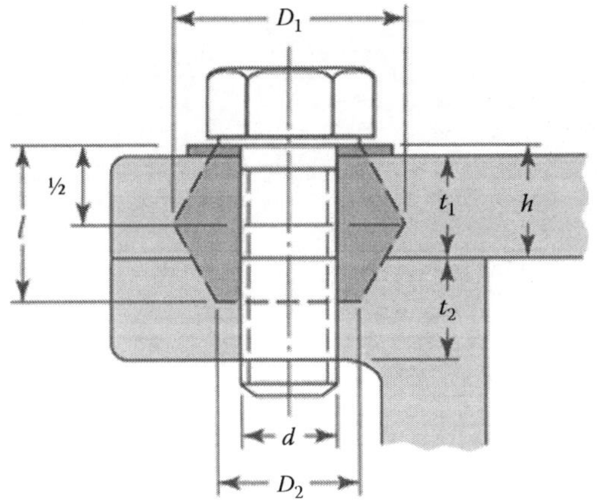

Figure 11.19

Bolt and cone for estimation of stiffness.

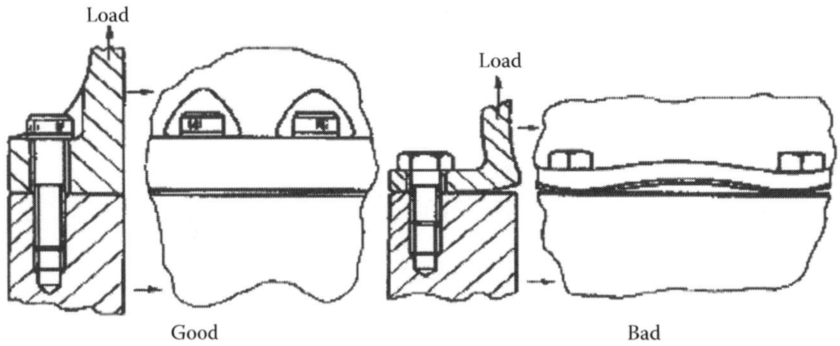

Figure 11.20

Good and bad proportions of bolting flange.

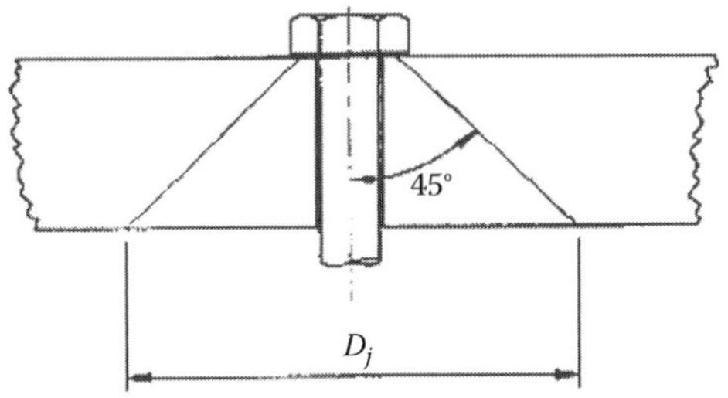

Figure 11.21

Estimation of joint diameter (D_j).

11.5.2.1 Sizing and Strength: How Big of a Screw Is Needed?

The selection of a screw is not a simple matter. For example, if a screw can withstand 100 lb of force before yielding, it is recommended for a number of reasons (discussed in Section 11.5.3) that it be tightened to about 80 lb of tension/clamping force for installation alone. It turns out that only a portion of the external load is seen by the bolt; a rough estimate is about 1/3, but this depends on lots of things. It seems reasonable to think that the tension on a screw/bolt is the sum of the initial clamping load plus any external loads, but this is not so.

Let us consider the following picture and imagine that the screw (or spring) is clamping the box against the ground with 10 lb of force. If we attempt to pull up on the box with 5 lb of force, it would only reduce the pressure between the red box and the ground; it would not stretch the screw at all. In fact, the screw would not feel any change in force whatsoever until the upward force exceeded 10 lb and the red box lifts. The force/tension in a spring depends only on its stretch force ($F = k\delta$), and if the bolt head does not move, the tension does not change.

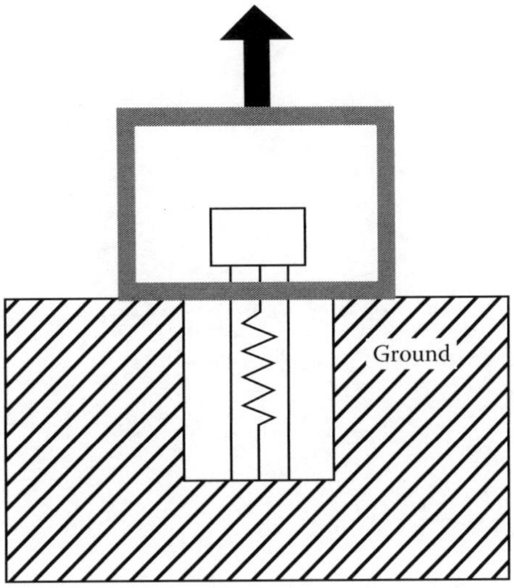

In most situations, however, the ground would act as a spring as well (in compression). As an external load is applied, the ground would decompress as the clamping force lowered, so the bolt head *would* indeed move slightly. In the end, a screw/bolt feels some fraction of an external load, and the amount it feels depends on how the spring rate of the bolt compares to the spring rate of the joint ("ground" in the picture). The forces each feels for a given preload (F_p) and external load (F_e) are

$$F_b = F_p + \frac{k_b}{k_b + k_j} F_e \quad \text{and} \quad F_j = F_p - \frac{k_j}{k_b + k_j} F_e$$

where:

F_b = bolt load
F_p = preload from tightening
F_e = externally applied load
F_j = joint clamping load
k_b = bolt spring constant (AE/L)
k_j = joint spring constant

These loads are proportional to their stiffness. The joint spring constant or the spring constant of the material under the bolt head is difficult to calculate; however, it is given as follows:

$$\frac{k_j}{ED} = Ae^{BD/l}$$

where:

D = nominal diameter of bolt
E = tensile strength of joint material
l = nearly half of the grip length
A and B are constants (as given in Table 11.2)

The last row of general expression should be used in most cases and k_j will be largely based on the E value of the joint material.

Table 11.2 Values of A and B Constants for *K* Calculation

Joint Material	A	B
Steel	0.78715	0.62873
Aluminum	0.79670	0.63816
Copper	0.79568	0.63553
Gray cast iron	0.77871	0.61616
General expression	**0.78952**	**0.62914**

Source: Budynas, R. G. and Nisbett, J. K. *Mechanical Engineering Design*, 8th edn. McGraw Hill, Boston, 2008 [1].

Example 11.3

A Grade 5 bolt is used to clamp a 1/2 in. thick plate of $E=30(10^6)$ psi. Find k_j if the grip length of the bolt is 1.1 in. and the nominal diameter of the bolt is 0.5 in.

From the above equation,

$$\frac{k_j}{30(10^6)(0.5)} = 0.78952e^{(0.62914*0.5/.55)}$$

$$K_j = 20.98(10^6)\ \text{lb / in.}$$

11.5.3 Rough Guide for Selection of Bolt or Screw

Start with the load that needs to be held in tension; let us call this *F*. If there is a shear (sideways) load, it should be so designed that friction or dowel pins will bear the load, not the bolt, but if this is not an option note that the shear strength is 60% of tensile strength in many steels.

Let us assume a safety factor of 2.5, so the design load is now 2.5*F*. We now need to select a screw with enough strength so that it can withstand the combined external load and preload from tightening. It may be safely assumed that 80% of the bolt's proof strength is being used up in preload, which leaves 20% to handle 1/3 of the external load. Or, in other words, we are looking for a bolt where 60% of its proof strength is greater than the load.

Example: 11.4

What size Grade 2 bolt is necessary to hold 100 lb? The proof strength of Grade 2 bolts between 0.25 and 0.75 in. is 55 kpsi, and 60% of this is 34.2 kpsi. So, the bolt tensile-stress area is 0.442 in.^{2} greater than our load (2.5×100 lb) divided by 33 kpsi, or 0.0076 in^2. A #6 UNC 9 (united screw thread) should work. For perspective, the diameter of a #6 screw is 0.138 in. We should keep in mind that the ultimate strength (breaking strength) of a Grade 2 bolt is 74 kpsi, so a #6 screw could theoretically hold 672 lb in pure tension. If you are wondering why bolts you see in cars and weight machines are so large, it is partly to guard against loosening and fatigue failure in addition to safety factors. It is therefore desirable, in the interests of overall joint stiffness, to employ confined gaskets or O-rings for sealing and allow the jointed components to abut directly, in preference to using interposed gaskets (see Figure 11.22).

11.5.4 Preloading of Bolted Joints: Behavior of a Bolted Joint under a Varying Axial Load

The three diagrams of Figure 11.23 show a bolted joint: Firstly, when the nut has been hand tightened so that the components are snug but before significant load is applied; secondly, when the nut has been tightened to produce a preload or pretension *P* in the bolt; and thirdly, under the subsequent influence of an external load F_e applied to the joint. When the preload is applied, the bolt stretches and the joint members compress under the influence of forces applied at the nut and bolt faces. The compressive load applied to the joint is clearly equal to the bolt tension *P*, so that the extension of the bolt relative to the amount by which the joint members compress will be in inverse proportion to their relative stiffnesses. Since the joint members are usually stiffer than the bolt, the amount of joint compression, δ_j, is usually less than bolt stretch; that is, $\delta_j<\delta_b$.

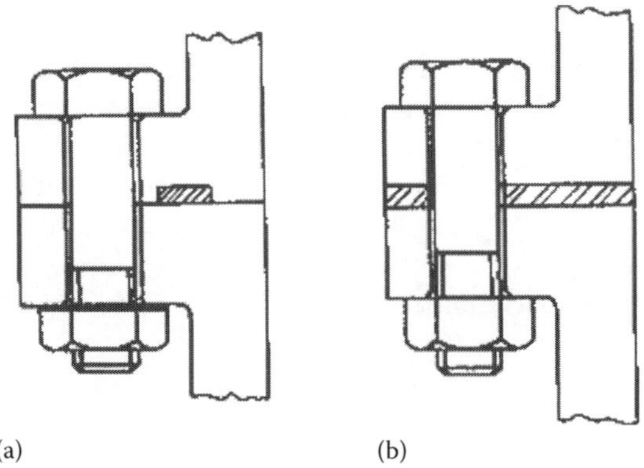

Figure 11.22

Confined and interposed gaskets: (a) confined gasket, giving stiff joint; (b) interposed gasket, giving low joint stiffness.

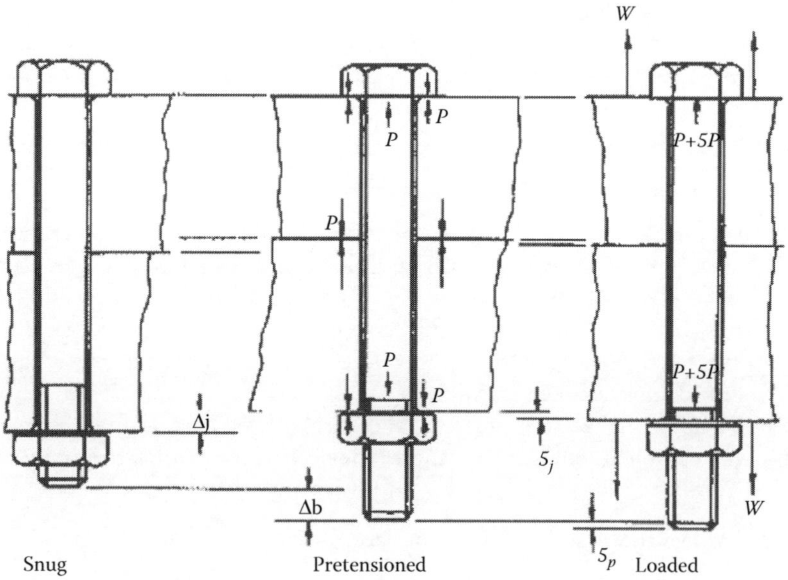

Figure 11.23

Behavior of bolted joint under preload.

It is assumed that the extension of the bolt and the compression of the joint materials are equal, to guard against joint separation; that is, $\delta_b = \delta_j$, and the relations for loads shared by bolt and joint are shown above, however, it is shown below.

External load F_e is shared between an additional load over the preload on the bolt and a *reduction* in compression of the joint, in proportion to their stiffnesses:

$$F_b = F_p + \frac{k_b}{k_b + k_j} F_e \ldots\ldots F_j = F_p - \frac{k_j}{k_b + k_j} F_e$$

Thus, by designing the joint members to be very stiff and the bolt to be resilient, and applying a high preload to the bolt so that the joint faces do not separate when an external load is applied, a large external load can be seen to cause a small increase in the bolt load but a much larger decrease in the joint load. This will have important implications later, when fatigue is considered in joints with fluctuating loads but, in the context of steady loads applied and removed relatively infrequently, it is important to note the need for a preload to prevent joint faces separating and also that the increase in the bolt load is likely to be very small.

11.5.5 Choice of Preload

Having established that it is desirable to preload a bolted joint that is subsequently subjected to a tensile load, we must now decide how much preload to apply. On the one hand, we must ensure that the joint faces will not separate under the application of an external load to the joint, and on the other we must not apply such a high preload that the bolt will fail when we tighten it up. Let us define the minimum preload to avoid separation of the joint members as P_{min} and the maximum preload which can be applied without causing failure of the bolt in service as P_{max}.

The safety factor on separation of the joint faces may be estimated. If F_0 is the load at which joint separation is likely to occur, then

$$n = \frac{F_0}{F_j}$$

In order to ensure that the joint members do not separate, it is prudent to apply a high safety factor to the condition of joint parting. A factor of at least two is suggested. The minimum preload P_{min} can also be restricted to be $P_{min} = 2F_j$.

The upper limit of the preload must be such that the bolt will not fail, either during tightening or under subsequent external load application. It is suggested that a maximum value of preload, P_{max}, is made equivalent to a nominal tensile stress at the minimum bolt cross section of 85%–90% of the bolt material yield strength.

Methods to achieve a specified preload in threaded fasteners are given in Table 11.3.

11.5.6 Control of Preload

A permissible range of preload, $F_{p(min)}$ to $F_{p(max)}$, has now been established, based on criteria of joint parting and bolt strength. The tightening of the nut must be controlled and monitored to an accuracy which will result in a preload somewhere within this safe range. If a preload equal to the mean of $F_{p(max)}$ and $F_{p(min)}$ is specified, the permissible error in preload control is $\pm(F_{p(max)} - F_{p(min)})/(F_{p(max)} + F_{p(min)})$. The choice of the method used to control the tightening of the nut therefore influences the required strength of the bolt, and indeed the design of the complete bolted assembly.

Table 11.3 gives a guide to the accuracy of preload control offered by a range of commercially available methods, together with comments on their limitations. In selecting the method to be employed, the designer

Table 11.3 Methods to Achieve a Specified Preload in Threaded Fasteners

Method	Accuracy	Remarks
Yield control	± 5%	Yield is detected during tightening by continuous monitoring of torque (T) and angle of turn of the nut (θ) and noting the reduction in $dT/d\theta$ which occurs at yield. Complex equipment is required. A similar accuracy may be achieved using a proprietary load-indicating bolt, known as the "Rotabolt."
Angle control (turn of the nut)	± 10% at best	The nut is tightened hand-tight and then turned through a further specified angle, calculated to produce the required strain in the bolt. The hand-tight condition is difficult to control accurately unless the joint members are stiff and flat.
Direct measurement of bolt stretch	± 10% at best	The extension of the bolt on tightening is measured and compared to a specified value. Both ends of the bolt must be accessible for measurement, which must be accurately carried out since extension is normally small. (Typical strain[a] 0.1%.)
Torque wrench		
Strain calibrated	± 10–15%	The nut is tightened to a specified torque. The resulting tension produced in the bolt varies considerably with thread and collar friction. A torque wrench setting is therefore worthless unless the strain is accompanied by an indication of whether the assembly should be dry or lubricated. Additional errors will result from lack of consistency of the wrench and any inaccuracy in calibration.
Torque calibrated	± 15–25%	
Uncalibrated	± 25–50%	
Air-stall power wrench		
Calibrated	± 25–40%	Air-operated power wrenches in which the spanner: socket carrier is designed to stall at a preset torque. Used for automated assembly in mass production. Maintenance of accuracy depends upon frequent recalibration against a known standard.
Uncalibrated	± 30–100%	

Table 11.4 Typical Embedding at Interfaces of a Bolted Joint (Steel)

Type of Surface	Machined Thread	Formed Thread	Machined Surface	Ground Surface
Amount of embedding at each interface (mm)	0.005	0.003	0.004	0.002

should be sure that his intentions are within the capabilities of the service environment, as well as his assembly shop, if the fasteners concerned will require periodic retightening in service.

11.5.7 Loss of Preload

The initial preload applied to the bolt tends to be degraded by subsequent application and removal of the external load on the joint. This is due to the embedding of the contacting surfaces and the flattening of surface roughness, and tends to occur rapidly with the first few load applications, and more slowly thereafter. The effect is most pronounced with soft materials and poorly finished surfaces. For design purposes, the amounts of embedding shown in Table 11.4 may be taken as representative for pairs of steel surfaces.

For aluminum surfaces, the figures quoted should be trebled. For soft gasket materials (which may also bed down due to heat) the amount of embedding may be much larger. The problem may be minimized by using large joint face areas, placing hardened washers under the nut and bolt faces, and by employing long, resilient bolts to reduce the loss of preload for a given amount of embedding. The significance of loss of preload on the safety of the joint members against parting should be checked during design of bolt or fasteners. The final bolt extension or deflection (δ_t) may be calculated from the initial extension as

$$\delta_t = \delta_b - \delta_c\left(\frac{k_j}{k_b + k_j}\right)$$

where:

δ_b is the initial bolt extension

δ_c is the total amount of embedding obtained by summing the embedding at each surface

Example 11.5

From Table 11.4, the embedding can be estimated as shown.

Bolt head/upper joint member	0.004
Interface-upper/lower members (ground surfaces)	0.002
Nut face/lower joint member	0.004
Threads (formed)	0.003
$\delta_c = 0.013$	

The final residual preload is given by

$$\text{Residual preload} = F_{p(\text{mean})}\left(\frac{\delta_t}{\delta_b}\right)$$

where

$$F_{p(mean)} = \frac{F_{p(\max)} + F_{p(\min)}}{2}$$

The maximum and minimum preloads are normally estimated from proof load as

$$0.75A_tS_p \le F_p \le 0.90A_tS_p$$

where A_t is the tensile-stress area of the bolt and S_p is the proof strength of the bolt material. The proof strength of the material is about $0.85S_y$, where S_y is yield strength of material.

11.5.8 Design of Bolt Assembly

Generalized torque coefficients and a universal torque table give engineers an easier path to accurate bolt preloading. Engineers must frequently calculate the bolt preload torque needed to securely hold parts together without overstressing joints. They mainly rely on torque equations and torque tables, time-consuming tools that were developed more than a half-century ago. Here is a user-friendly alternative.

The bolt preload torque T consists of three components that share a common factor.

$$T = (C_L + C_t + C_c)(D \times F_p)$$

where the torque coefficients for lifting (C_L), thread friction (C_t), and collar friction (C_c), respectively, are as follows:

11.5.8.1 Torque Coefficient for Lifting

$$C_L = \frac{L}{2\pi D}$$

where:

- D is the nominal diameter of bolt
- L is the grip length of bolt

11.5.8.2 Torque Coefficient for Thread

$$C_t = \left[\frac{\mu_t}{\sin\alpha(\cos\theta - \mu_t \sin\alpha)}\right]\frac{L}{2\pi D}$$

where:

- α is the helix angle
- θ is the thread angle of the thread in use
- μ_t is the coefficient of the threads

11.5.8.3 Torque Coefficient for Collar

$$C_c = \mu_c \frac{D_c}{2D}$$

where:

- μ_c is the collar friction and
- D_c is the collar diameter (about $1.25D$)

Denoting the overall torque coefficient as $C = C_L + C_t + C_c$, the equation reduces to

$$T = CDFp$$

where:

- D is the major diameter of the bolt
- L is the grip length of bolt
- F_p is the preload

This equation is similar to the classical approximation $T = 0.2DF_p$ that applies to the coefficient of friction (c.o.f.), $\mu_c = \mu_t = 0.15$, for small screws. The torque coefficient C however, is generalized to embrace all screw sizes and c.o.f.s. Only two factors affect the equations—bolt size and friction—provided $\mu_c = \mu_t$ and $D_c = 1.25D$. D_c is the washer face diameter, usually between 1.25 and 1.5 times the diameter of the bolt D. The generalized torque-coefficients table lists values for C for selected screw sizes and c.o.f.s. The data directly link torque and clamping force, thus eliminating the need to toil over the torque equation.

11.5.8.4 Determining Torque

The general applied torque equation is presented as

$$T = KF_iD$$

where K is torque coefficient. K is a function of the coefficients of friction and collar friction in the joint, and these are based on variables such as surface finish, coatings, and so on. Accepted coefficient values for both types of friction for steel-on-steel bolts are 0.15. With a coefficient of friction = 0.15, $K \approx 0.20$ for all ranges of bolt sizes in both coarse and fine threads. For materials other than clean steel on steel, determine appropriate values from similar applications or by testing actual bolt and joint configurations. Typical K values are shown in the table below.

	K Values
Nonplated, black finish	0.3 (most common) (0.2–0.3)
Zinc plated	0.20 (most common) (0.17–0.22)
Lubricated	0.18 (most common) (0.12–0.18)
Cadmium plated	0.15 (most common) (0.11–0.15)
With Bowman antiseize	0.12
With Bowman grip nuts	0.09

Source: Budynas, R. G. and Nisbett, J. K. *Mechanical Engineering Design*, 8th edn. McGraw Hill, Boston, 2008.

When calculating loads and torques, the female thread always has a greater shear area than that of the bolt. This means there is a range where the base material yield strength is lower than the bolt's, but the bolt can still be torqued to the maximum. In fact, with a length of engagement of 1.5D, the range where full torque can be applied to lower-strength base material is quite large. F_i is the preload calculated earlier. Specialized torque charts are a better option for taking advantage of this range without going through a series of calculations for each application. They also determine the correct torque and bolt load under rated conditions. The torque wrench is a relatively inexpensive means of giving a rough estimate of the bolt's preload. For most practical applications, the torque wrench will provide the degree of accuracy desired. However, even perfect input torque can give a variation of preload by as much as 25%–30%. The following example will demonstrate some complications that can be involved in calculating torque.

Example 11.6

Let us determine the clamping torque produced from 70 to 80 lbf-in. torque on a 0.250–28 UNF screw. Assume a coefficient of friction (c.o.f.) of 0.10–0.15 for both threads and nut collar.

From the table, torque coefficients for a 0.250 in. diameter screw are $C_{max} = 0.195915$ at maximum friction $\mu_{max} = 0.15$; and $C_{min} = 0.138035$ at minimum friction $\mu_{min} = 0.10$.

C is the overall torque coefficient, including lifting, thread friction, and collar friction.

$$F_{p(\min)} = \frac{T_{\min}}{C_{\max}D} = 1429 \text{ lbf}$$

$$F_{p(\max)} = \frac{T_{\max}}{C_{\min} D} = 2318 \text{ lbf}$$

The classical equation $T = 0.2\ DF_i$ is reasonably accurate for c.o.f. = 0.15 and size 0.25 in. and smaller screws; that is, $C \approx 0.2$. The bolt size has a moderate effect on C but the effect of the c.o.f. is substantial. Average torque coefficients for all sizes are close to a straight line: $C = 0.02 + 1.17\mu$. Therefore, engineers can use linear interpolation to determine C across c.o.f.s for a specific bolt size. Typical values of c.o.f. are about 0.1–0.2.

Example 11.7: Torque Calculation for Hex Cap Screw 1/2-13 Zinc-Plated Grade 5

Using torque $= K \times d \times F$, $K = 0.20$ (safe assumption from table above) and $d = 0.50$ in. It is safe to assume that the preload F_p applied is in the range 0.75(proofload) $\leq F_p \leq$ 0.90(proofload)

Tensile-stress area $(A_t) = 0.1419$ in^2.
F_p = (75% of the proofload) = 0.75 × 85,000 psi × 0.1419 in.^{2} = 9045 lb
T = (0.20)(0.5 in.)(9045 lb) = 905 in.-lb = 75 ft-lb

In this example, we selected $K = 0.20$. This is a common K factor for zinc-plated fasteners. However, if the fastener is installed with excessive lubrication, then it should not be installed with the same torque. In such conditions, the value of K may drop to a value of 0.10.

If $T = K \times d \times F_p$, then we can also determine the preload, F_p, in the same way.

$$F_p = \frac{T}{(K \times d)}$$

$$F_p = \frac{905 \text{ in.-lb}}{(0.10 \times 0.50 \text{ in.})} = 18{,}100 \text{ lb}$$

The value obtained for our preload, 18,100 lb, would exceed the tensile strength of a 1/2-13 Grade 5 bolt. More than likely the bolt in question would have failed. The exact opposite may have also happened, resulting in a lower clamp load than may be required, causing a field failure.

11.5.8.5 Universal Torque Table

Applied torque (T) concurrently produces both *tensile* and *shear* stresses in a bolt. Shear stress originates from the bolt twisting torque T_b where

$$T_b = C_b \frac{T}{C}$$

and the torque coefficient (C_b) is given as

$$C_b = C - 0.625\mu$$

$$\text{Tensile stress } \sigma_t = \frac{F_i}{A_t} = \left(\frac{C_b}{C}\right)\left(\frac{T}{DA_t}\right)$$

$$\text{Shear stress } \tau = \left(\frac{16}{\pi D_r^3}\right)\left(\frac{C_b}{C}\right)T$$

where D_r is the root diameter or minor diameter of the bolt.

Conventional methods combine tension and shear stresses into an equivalent stress and compare it with material strength. A general notion is that if this equivalent stress reaches the material strength, the part fails. The maximum distortion energy criterion of failure stipulates that inelastic failure occurs when the von Mises stress approaches the material yield stress.

For most bolt materials, von Mises stress criteria apply. The von Mises stress during preloading is

$$\sigma_v = (\sigma_t^2 + 3\tau^2)^{0.5} \left(\text{von Mises stress}\right)$$

This is one of the most taxing tasks in bolt-stress analysis. It happens to be important because engineers are often challenged to determine a torque that uses every bit of allowable stress based on material strength and the safety factor.

The following embedding technique circumvents the hurdle. First, embed τ into σ_t.

$$\sigma_v = \sigma_t \sqrt{1 + 3\left(\frac{\tau}{\sigma_t}\right)^{2}} = \xi_v \sigma_t,$$

$$\xi_v = \sqrt{1 + 3\left(\frac{16 C_b D A_t}{\pi d_r^{\,3}}\right)^{2}} \tag{11.2}$$

The proportional constant ξ_v is a multiplier to convert σ_t to σ_v, while τ is embedded into σ_t. Second, substitute

$$T = \left(\frac{CDA_t}{\xi_v}\right)\sigma_v = K_v \sigma_v$$

The proportional constant K_v depends on bolt size and coefficient of friction (c.o.f), and could be tabulated in the same way as torque coefficients. Alternatively, universal torque values T_u can be tabulated for a fixed value of von Mises stress σ_v based on the relation $T_u = K_v \sigma_v$.

The universal torque and torque coefficient link preload torque, stress, and clamping force. Given any one, determine the other two through simple math.

11.5.9 Torque Tables

Traditional torque tables typically list bolt-preload torques for a specific bolt-material strength and surface condition. It is not uncommon for a project to require more than a dozen torque tables. Those tables are suitable for assembly shops but ineffective for stress analysis because they convey neither the von Mises stress produced from the torque nor the c.o.f. Engineers require the former to determine the margin of safety at assembly, and the latter is a key parameter to calculate stress, clamping force, and so on. The information must have been used to develop the torque tables but is not passed along with the tables. A universal torque table, used in conjunction with torque coefficients, makes all essential information and processes transparent. Further, one universal torque table applies to all materials and surface conditions, because preload control relates torque and von Mises stress irrespective of the material strength, and the c.o.f. is the core parameter in preload control, irrespective of surface conditions (Table 11.5).

11.5.10 Fastener Property Classes

The metric system relies on property classes for strength requirements. The numbering system is very simple; the number prior to the decimal indicates, the approximate minimum tensile strength in megapascals when multiplied by 100; the number following the decimal indicates yield as a percentage of the tensile strength.

Example 11.8: Property Class 10.9

The first number (10) indicates a tensile strength of approximately 1000 MPa. ".9" indicates that 90% of the tensile strength is approximately the yield strength, or approximately 940 MPa. Most ASTM

Table 11.5 Clamping Force and Applied Torque: Bolts and Nuts (ISO Metric Fasteneres)

	Clamping Force (kN)				Applied Torque (Nm)									
	Property Class				PC 8.8/8		PC 9.8/9		PC 10.9/10		PC 12.9/12			
	8.8/8	9.8/9	10.9/10	12.9/12	Min	Max	Min	Max	Min	Max	Min	Max		
M3	2.2	–	3.1	3.7	1.1	1.6	–	–	1.4	2.2	1.7	2.6		
M4	3.8	–	5.5	6.4	2.5	3.7	–	–	3.7	5.3	4.0	5.7		
M5	6.2	6.9	8.9	10.4	5.0	7.4	5.6	8.2	7.2	106	80	11.6		
M6	8.7	9.8	12.5	14.6	8.4	12.4	9.5	14.0	12.1	17.8	13.5	19.5		
M8	15.9	17.8	22.8	26.6	20	30	23	34	29	43	33	47		
M10	25.3	28.3	36.1	42.2	41	60	46	68	59	85	65	94		
M12	36.7	41.1	52.5	61.4	71	105	80	118	102	150	114	164		
M14	50	56.1	71.6	84	112	168	127	187	161	240	182	265		
M16	68.2	76.5	97.5	114	175	260	198	292	250	371	282	406		
M18	86.2	–	119	140	255	372	–	–	353	513	390	560		
M20	110	–	152	178	355	520	–	–	491	718	550	795		
M22	136	–	189	220	500	705	–	–	675	989	750	1080		
M24	159	–	220	256	620	928	–	–	867	1283	950	1370		

specifications allow for both metric series and inch series fasteners to be made. For example, ASTM A490 & ASTM A490M. The ASTM A490M is the metric A490 structural bolt. The clamping forces specified are equal to 75% of the proof load of the property classes of bolts, as given in ISO 898/1.

11.5.11 Margins of Safety

It is good practice to limit preload to provide an adequate margin of safety.

Design for preload in three steps:

1. Choose preload stress $\sigma_v \leq \sigma_y / FOS$ (*FOS* represents factor of safety).
2. Scale torque using the universal torque table.
3. Calculate clamping force using the torque coefficient.

A bolt (or stud) is called an externally threaded fastener and a nut is called an internally threaded fastener. The thread is an example of the inclined plane, one of the basic mechanical principles. The use of torque has long been associated with methods of tightening and inspecting bolts. The object of tightening bolts is to achieve a minimum tension, not torque, in the bolts. There is a relationship between torque and tension, but it is highly variable and must be used with caution. The variables in the torque–tension relationship include lubrication, thread fit, the use of a washer and the tension in the bolt. The torque used to tighten a bolt is consumed by overcoming the friction between nut and washer (about 60%), overcoming the friction between bolt thread and nut threads (about 30%), and providing energy to elongate the bolt (about 10%). Tests have indicated that torque–tension relationships for structural bolts easily vary by as much as 40%.

Most engineers who deal with bolted joints are familiar with torque tables for SAE-grade bolts, but these are all too often used without regard for the yield strength of internal threads or the bolt's length of engagement. In some instances that can be a serious mistake. That is because in static-bolting applications—assuming sufficient thread engagement—there are three cases to consider in determining maximum torque.

The first is when the yield strength of the internal thread *equals* or exceeds the proof strength of the bolt, $S_i \approx S_t$, in which case the bolt may be fully torqued.

The second is when the yield strength of the internal thread is *less* than the bolt proof strength, $S_i < S_t$, but because of the greater shear area of the internal thread the bolt can still be fully torqued.

The final case has an internal-thread yield strength less than the bolt proof strength, $S_i < S_t$, and even with the added shear area the internal threads have insufficient strength to support full torque.

In the first case, when the internal thread is as strong as the bolt and the length of engagement is at least 1.5 times the nominal bolt diameter, designers can use standard torque tables. For the other two cases, where the internal thread material has a lower yield strength than the bolt's proof strength, maximum allowable torque must be based on the length of engagement, bolt proof strength, and yield strength of the internal thread.

11.5.12 Torque Tension Performance Data

Data for guidance is given in the graphs in Section 11.5.13, but it should be noted that the actual component that is coated, its dimensions, and the coating dimensions of mating parts can have a significant effect on the actual tightening performance. Most figures have been obtained from testing parts that conform to the ISO specification for coarse series of bolts and, in many cases, M10 flange fasteners have been used. However, it seems quite useful in practice as it shows the bands of specifications.

11.5.13 Typical Torque Tension Specification Requirements

Band A

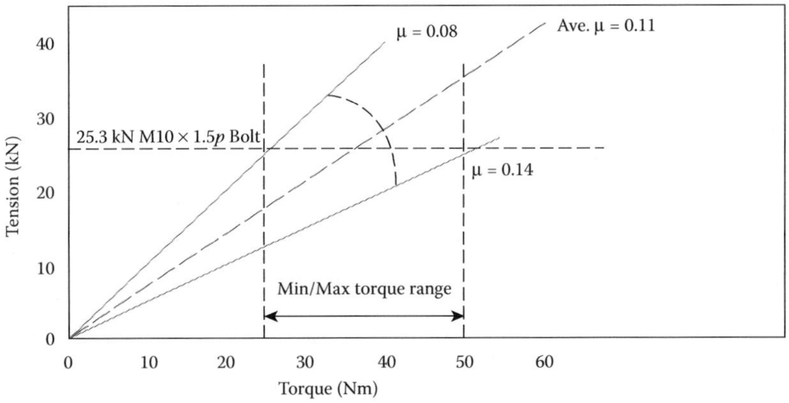

Specifications that fall within Band A

VW TL 246
Opel GME 00255
Mercedes
BMW GS 90010
GM 3359
Honda D2008-5

Materials:
DELTAPROTEKR®+VH 301GZ
GEOMET®321 +VL

Band B

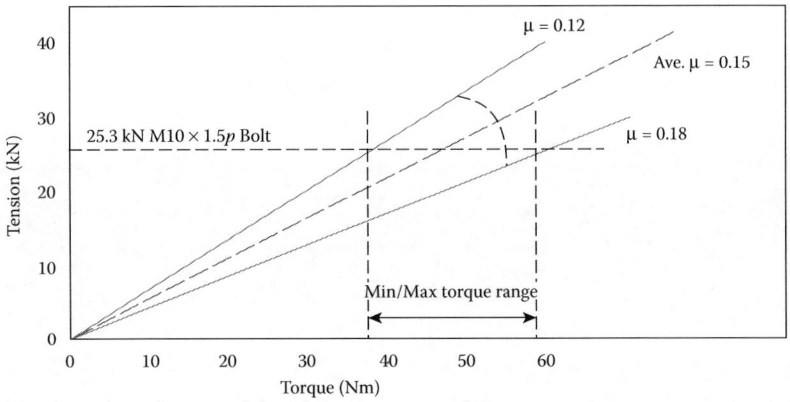

Specifications that fall within Band B

Ford WZ100 (μ= 0.12 – 0.18) (S438 S447, S448)
Ford WZ101 (μ= 0.11 – 0.17) (S439, S440, S442)
Renault 01-71-002/-Q
Peugeot/Citroen B152230

Materials:
DELTAPROTEKT®+ VH 302GZ
GEOMET®321
GEOMET®500
GEOBLACK®
MAGNI 565 and 560

Band C

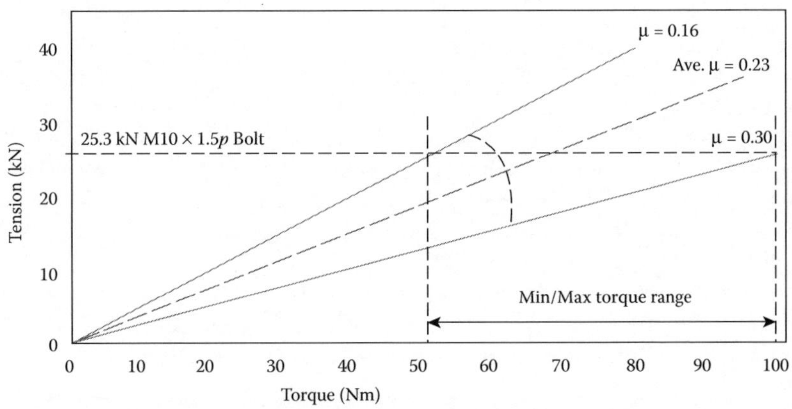

Specifications that fall within Band C

Honda
Ford S4

Materials:
Unlubricated zinc
Uncoated unlubricated bolt
Zinc nickel

11.5.14 Failure Modes

For typical static bolted joint, engineers should evaluate three failure modes to determine the highest load that the bolted joint can withstand. These are

- Tensile failure of the bolt body
- Shear failure of the bolt threads
- Shear failure of the internal threads

11.5.14.1 Tensile Load Calculation

The lowest load that leads to failure determines the limiting torque. In the first instance, calculate the tensile-stress area of a bolt from ASME B1.1–1989 Appendix B.

11.5.14.2 Metric System

$$D_p = D - 0.649519p$$

where

D is the nominal diameter of the bolt

p is pitch known. The root diameter $D_{\min} = D - 1.226869p$.

11.5.14.3 US System

$$D_p = D - 0.649519p \text{ and } D_{\min} = D - 1.299038p$$

The mean of D_p and $D_{\min}$ is used to calculate the tensile-stress area (A_t). Therefore, $A_t = \pi/16(D_p + D_{\min})$.

However, there are tables available in the literature where the tensile-stress area for different conditions is given. The calculated value A_t is not based on the minimum root diameter. This value is slightly larger than the area defined by the root diameter to account for the added strength of the actual thread in the bolt cross section. A more conservative value would be the minimum root diameter of the bolt, which disregards the area of the spiraling thread. For hollow threaded parts, use the minimum root diameter.

The allowable tensile load in the bolt based only on the tensile area is determined from

$$P_t = S_t \frac{A_t}{n}$$

For a typical static application, the factor of safety $n = 1.25$. However, factor of safety values may vary depending on the practical situation.

11.5.15 Evaluate Shear Stress in the Bolt

Unlike the tensile area, length of engagement affects the bolt's shear area. The shear area of the bolt is the effective area at the maximum minor diameter of the internal thread multiplied by the number of engaged threads. This is calculated as

$$P_s = 0.5 S_t \frac{A_t}{n}$$

An option is to calculate shear area using a 1 in. length of engagement. This gives shear area per unit length, and minimizes recalculation if the length of engagement changes. Multiplying the shear area per unit length by the length of engagement gives the actual shear area.

11.5.16 Shear Stress in Internal Thread

The final stress to evaluate is shear stress in the internal thread. Shear stress area (A_s) in a bolt is a function of the bolt's length of engagement. The shear area of the internal thread is the effective area at the minimum major diameter of the external thread multiplied by the number of engaged threads. This is found from

$$A_s = 0.5\pi D L_e$$

where L_e is thread engagement length. Normally it could be assumed that four threads are in engagement and it could be estimated that $L_e = 2p$.

The shear stress in screw and nut threads can be calculated as

$$\tau_{(\text{nut})} = \frac{P}{A_s}$$

$$\tau_{(\text{screw})} = \frac{P}{\pi D_r (L_e/2)}$$

and the bearing stress between threads is calculated as

$$\sigma_b = \frac{P}{(\pi/4)(D^2 - D_r^{\ 2})(L_e/p)}$$

where:

D_r = root diameter of screw

D = nominal diameter of screw

P is the load to be supported by the joint. The direct tension or compression stress can be calculated as $\sigma = P/A$, where A is the cross-sectional area based on the diameter of the screw

The maximum tensile load in the bolt, based on the shear area of the internal threads, is calculated as $P_i = 0.5S_i A_i/n$. The lowest calculated allowable tensile load of P_t, P_s, and P_i should be used to determine maximum torque (Table 11.6).

Table 11.6 Diameters and Tensile-Stress Areas

		Coarse Series—UNC			Fine Series—UNF		
Size Designation	Nominal Major Diameter, in.	Threads per in., N	Tensile-Stress Area A, in.^{2}	Minor-Diameter Area A_r, in.^{2}	Threads per in., N	Tensile-Stress Area A_t, in.^{2}	Minor-Diameter Area A_r, in.^{2}
0	0.0600				80	0.00180	0.00151
1	0.0730	64	0.00263	0.00218	72	0.00278	0.00237
2	0.0860	56	0.00370	0.00310	64	0.00394	0.00339
3	0.0990	48	0.00487	0.00406	56	0.00523	0.00451
4	0.1120	40	0.00604	0.00496	48	0.00661	0.00566
5	0.1250	40	0.00796	0.00672	44	0.00880	0.00716
6	0.1380	32	0.00909	0.00745	40	0.01015	0.00874
8	0.1640	32	0.0140	0.01196	36	0.01474	0.01285
10	0.1900	24	0.0175	0.01450	32	0.0200	0.0175
12	0.2160	24	0.0242	0.0206	28	0.0258	0.0226
1/4	0.2500	20	0.0318	0.0269	28	0.0364	0.0326
5/16	0.3125	18	0.0524	0.0454	24	0.0580	0.0524
3/8	0.3750	16	0.0775	0.0678	24	0.0878	0.0809
7/16	0.4375	14	0.1063	0.0933	20	0.1187	0.1090
1/2	0.5000	13	0.1419	0.1257	20	0.1599	0.1486
9/16	0.5625	12	0.182	0.162	18	0.203	0.189
5/8	0.6250	11	0.226	0.202	18	0.256	0.240
3/4	0.7500	10	0.334	0.302	16	0.373	0.351
7/8	0.8750	9	0.462	0.419	14	0.509	0.480
1	1.0000	8	0.606	0.551	12	0.663	0.625
11/4	1.2500	7	0.969	0.890	12	1.073	1.024
1	1.5000	6	1.405	1.294	12	1.581	1.521

This table was compiled from ANSI Bl.1-1974. The minor diameter was found from the equation $d_r = d - 1.299038p$, and the pitch diameter from $d_p = d - 0.649519p$. The mean of the pitch diameter and the minor diameter was used to compute the tensile-stress area.

Example 11.8: Examining the Safety of Bolt in a Joint

The bolt in the joint shown in Figure 11.18 is 5/8-in. course threaded. The parts it connects are subjected to a load P that fluctuates from 1000 to 8000 lb. The bolt and parts are made of the same material. Let us assume that the material has a yield strength of 70,000 psi and an endurance limit of 30,000 psi, and that the stress concentration factor for the thread is 3.8.

1. Is the bolt safe for continuous action?
2. If the preload is 10,000 lb is it safe for continuous action?
3. If the preload is 16000 lb is it safe for continuous action?

Solution

i. $P_m = (1000 + 8000)/2 = 4500$ lb.
ii. $P_a = (8000 - 1000)/2 = 3500$ lb.
iii. P_m and P_a are steady and alternating components of force.

In computing the stress in the screws in bending or torsion, it is safer to base the calculations on the minor diameter.

The stress area A_t can be taken from table above or calculated as $A_t = 0.226$ in^2. Therefore,

$$\sigma_m = \frac{4{,}500}{0.226} = 19{,}000\,\text{psi}$$

$$\sigma_a = \frac{3{,}500(3.8)}{0.226} = 58{,}000\ \text{psi}$$

σ_m and σ_a are known as the midrange and amplitude components of stresses. The location of Case 1's solution can be plotted on a Soderberg fatigue diagram and judged for its safety.

ii. The bolt is only threaded just enough to permit the nut to engage; therefore, the gross area of bolt should be used to calculate the midrange and amplitude components of stresses. However, in this case, the stiffness constant must be determined.

$$k_b = \frac{A_b E_b}{l_b}\left(\text{stiffness of bolt}\right)$$

$$A_b = \frac{\pi}{h}\left(\frac{5}{8}\right)^2 = 0.31\,\text{in.}^2$$

and therefore,

$$k_b = \frac{0.31E}{l}$$

The area of the joint is rather difficult; however, a reasonable assumption is that the outside diameter of an annular object is $3D$ and the inside diameter is D. Hence, the joint area is

$$A_f = \frac{\pi}{4}\left[\left(3\times\left(\frac{5}{8}\right)\right)^2 - \left(\frac{5}{8}\right)^2\right] = 2.47\,\text{in.}^2\left(\text{area of joint}\right)$$

$$k_j = \frac{A_j E}{l}\left(\text{joint stiffness or spring index}\right)$$

$$F_m = \frac{k_b F_m}{k_b + k_j} + F_p = \frac{(0.31E/l)4{,}500}{(0.31E/l) + (2.47E/l)} + 10{,}000 = 10{,}502 \text{ lb}$$

$$\sigma_m = 10{,}502/0.26 = 46{,}500 \text{ psi}$$

$$F_a = \frac{k_b F_a}{k_b + k_j} = \frac{0.31(3500)}{0.31 + 2.47} = 390 \text{ lb}$$

$$\sigma_a = \frac{3.8(3900)}{0.226} = 6550 \text{ psi}$$

Again, both values would be plotted on a Soderberg fatigue curve.

$$\text{iii. } F_m = \frac{k_b}{k_b + k_j}(4{,}500) + 16{,}000 = 16{,}502 \text{ lb}$$

and

$$\sigma_m = 16{,}502/0.226 = 73{,}000 \text{ psi}$$

$$\sigma_a = 6550 \text{ psi as in section (ii)}$$

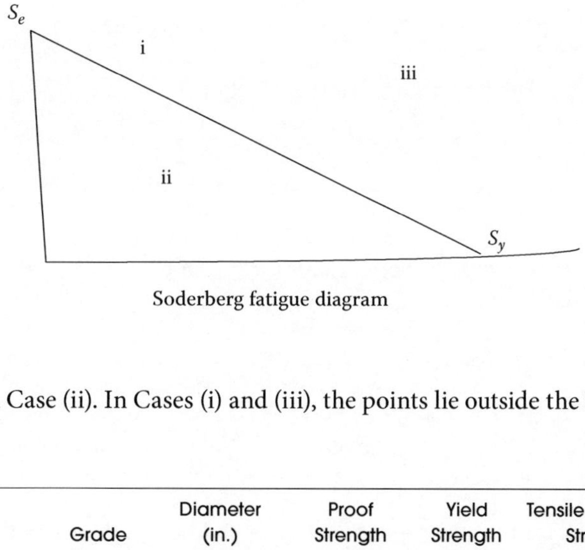

Soderberg fatigue diagram

The bolt is safe only in Case (ii). In Cases (i) and (iii), the points lie outside the Soderberg line and are unsafe.

Head Marking	Grade	Diameter (in.)	Proof Strength	Yield Strength	Tensile (Ultimate) Strength
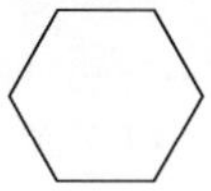	0.2	1/4–3/4	55,000 psi	57000 psi	74,000 psi
		3/4–1 1/2	33	36	60
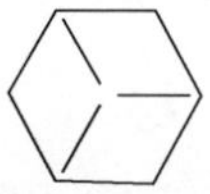	0.5	1/4–1	85	92	120
		1–1 1/2	74	81	105
	0.8	1/4–1 1/2	120	130	150

The combined clamped member compliance is the sum of compliance of two flanges and one gasket:

1. In addition to the indicated grade marking, all grades, except A563 grades O, A, and B, must be marked for manufacturer identification.
2. The markings shown for all grades of A194 nuts are for cold formed and hot forged nuts. When nuts are machined from bar stock the nut must be additionally marked with the letter B.
3. Nuts are not required to be marked unless specified by the purchaser. When marked, the identification marking shall be the grade letter O, A, or B.
4. Properties shown are those of nonplated or noncoated coarse-thread hex nuts.
5. Properties shown are those of coarse-thread heavy hex nuts.
6. Properties shown are those of coarse-thread heavy hex nuts.
7. Properties shown are those of coarse 8-pitch thread heavy hex nuts.
8. Hardness is shown by Brinell hardness numbers.
9. The nut manufacturer, at his option, may add other markings to indicate the use of atmospheric corrosion-resistant steel.
10. Specifications: ASTM A563—Carbon and Alloy Steel Nuts.

11.5.17 Preloading Mechanism Compliance

In some cases, the nut rotation angle may be too small for this technique to be practical. It is not unusual to only require about 10° of nut rotation for preloading. For a given preload requirement, choosing an extra-fine thread over a coarse thread and a larger compliance for the preloading components will increase the nut rotation angle.

When specifying torque for preloading, the corresponding nut rotation angle is usually not calculated. As a result, there is no clue whether the specified torque is difficult to work with because the corresponding wrench rotation angle is small.

The nut position method is not exactly an alternate method for preloading bolted joints. Rather, it is an alternate way of controlling the magnitude of preloading. Whether specifying torque or nut rotation angle, in the end, a torque will be applied to tighten the nut. Two aspects govern the torque process: force to pull the wrench and distance to pull it. Torque specifications reflect the former; nut rotation reflects the latter.

The nut rotation angle is related to torque T by the equation

$$T = C \times D \times F_p$$

where:

C = the torque coefficient (a function of bolt size and coefficient of friction)
D = the nominal bolt diameter

Calculate the nut angle in degrees with

$$\theta = 360\left(\frac{\mathrm{N}}{\mathrm{ki}}\right)\left(\frac{\mathrm{T}}{\mathrm{CD}}\right)$$

Therefore, nut angle and torque are proportional, and the force on the wrench and the distance to rotate the nut are two indicators of the same thing.

Giving the preloading loop more compliance is an effective way to increase the preloading nut angle. Perhaps more fundamentally, it is best to avoid grossly oversized bolts. Some of the C values for different grip lengths are shown in Table 11.7.

11.5.18 Thermal Aspects in Bolt Design

Most fastener materials are temperature sensitive. This means that their properties are influenced by a change in temperature. A metallic fastener's strength properties decline with an elevation in temperature. Once the temperature increases significantly, other problems also begin to occur. Some of these problems include, but are not limited to, a breakdown in the coatings, high-temperature corrosion,

Table 11.7 Computation of Bolt and Member Stiffness for Grip Length

Bolt Grip, in.	Stiffnesses, M lbf/in. k_b	k_m	C	$1-C$
2	2.57	12.69	0.168	0.832
3	1.79	11.33	0.136	0.864
4	1.37	10.63	0.114	0.886

Note: Steel members clamped using a 1/2 in.-13 NC steel bolt. $C = (k_b)/(k_b + k_m)$.

differential thermal expansion coefficients between the fastener and the joint, and creep and stress relaxation.

The strength of most fasteners will decrease as temperatures rise. Any type of plating or coating will also alter the results; for example, zinc-plated fasteners are usually not recommended to be used above a temperature of 250°F.

One of the most celebrated examples of temperature effects on bolts is with Grade 8.2 (low-carbon martensite) bolts. At room temperatures, Grade 8.2 and SAE J429 Grade 8 (medium-carbon alloy) bolts have similar properties. However, Grade 8.2 bolts are made of low-carbon boron steel. The Grade 8 fastener is a medium-carbon alloy steel and recommended for temperatures between 650°F and 800°F and is not intended for use at higher temperatures.

Mild steels with ultimate tensile strengths (S_u) of less than 64,000 psi should not be used for bolts whose temperatures may exceed 570°F. Harder steels can be used for temperatures up to 840°F. Special steels should be used for temperatures above 840°F. The influence of temperatures above 392°F on the elastic limit of steel may be taken into account by decreasing the elastic limit at room temperature by 1% for every 9°F above 392°F, up to 932°F.

Every bolting material has a temperature above which it would be unsafe to use. Oftentimes this is referred to as the fasteners' high-temperature service limit. Although we saw that the fastener loses strength as the temperature increases, the service limit is usually determined by an occurrence known as stress relaxation. A fastener is tightened in the joint. This action places the bolt under significant stress. The length of the bolt does not increase. The joint and the nut will determine the length. Once exposed to a higher temperature, the bolt begins to relieve itself of a significant amount of the stress. Since the stress and the preload are related, this implies that the clamping force with which the bolt holds the joint together will be significantly reduced.

One of the most problematic temperature effects that must be taken into account when designing a joint is differential thermal expansion between the bolt and joint members. As the temperature rises, all bolt and joint materials expand, but not all at the same rate.

As an example, aluminum will expand about twice as much as some carbon-steel fasteners. If using a SAE J429 Grade 8 fastener to clamp an aluminum joint, we would expect to see a significant increase in the tension of the bolts, which would increase the clamping force as the temperature increases. This reaction could damage the joint or gasket material or even break the bolt. If the bolt would expand more than the joint, the joint could lose preload and clamping force. It should be recognized that differential expansion problems could occur even if the fastener and the joint are made of the same material. If the bolt and the joint heat up at different rates, the corresponding thermal expansion will also cause the bolt and the joint to expand differently. There are various other temperature-related effects, which must be considered when designing a bolted joint. Two of the more common occurrences that are very closely related are creep and stress relaxation. If a constant load is applied to a fastener and we raise the service temperature and the temperature places the bolt in its "creep range," the bolt will begin to stretch even if the load is well within the fastener's mechanical limits. Eventually, the bolt may stretch to a point where it may not be able to support the load and will fail. Creep is defined as the slow increase in length of a material under a constant, heavy load. Stress relaxation is very similar. However, if the bolt is exposed to a high temperature, the bolt begins to relieve itself of some of the stress and we can lose our preload. The behavior of the bolted joint will depend, to a large extent, on the clamping force on the joint in service.

11.5.19 Nuts and Bolts of Ceramic Fasteners

Few fasteners can beat threaded ceramic nuts and bolts when it comes to withstanding heat, pressure, and corrosive environments. Just do not torque them too tightly. So it seems there is a definite market for reliable fasteners that are electrically nonconductive, nonmagnetic, and capable of withstanding temperatures to 1600°C (2900°F). Many engineers are inclined to reject the idea of ceramic fasteners on the grounds that they must be impossible to torque and, therefore, impossible to install. Ceramics do have little resistance to torque, and they do not degrade or break down gracefully. Overtorque them by just a hair, drop them, or strike an installed nut with a hammer, and they shatter. And if even a small fracture starts, one can be sure the fastener will soon be in pieces. But this just makes alumina and zirconia fasteners tricky to install, not impossible. Alumina and zirconia also have lower coefficients of thermal expansion than do metals. And with engineers mixing and matching, using ceramic nuts on metal bolts and vice versa, they must take into account operating temperatures and thermal expansion. So, if you plan to use an alumina nut on a threaded, metal rod, be aware it could split if the metal expands more than the alumina at higher temperatures. What overcomes engineers' reluctance to use such brittle materials for nuts and screws is usually a challenging project. They have to attach a critical component to a device or another component, but the fastener cannot conduct electricity and provide a path for a short circuit. Or, the nuts and bolts must ignore high magnetic fields in an MRI machine, for example, where magnetic fields could push and pull metal fasteners. Ceramics are also inert and practically immune to corrosion. The only acid that affects them, for example, is hydrofluoric acid. This allows them to be used in plating operations.

11.5.20 Working with Alumina and Zirconia

In general, ceramics encompass all materials that are neither organic nor metals. The four basic classes include oxides (MgO, Al_2O_3, and mullite), nonoxides (carbides, nitrides, and several combinations of C, Si, and N), glasses (SiO_2 and B_2O_3), and various salts (chlorides, sulfates, and nitrates). For now, however, commercial fasteners are only made out of alumina (Al_2O_3, which is also known as corundum or polycrystalline sapphire) and zirconia (ZrO_2) (Table 11.8).

11.6 Screw Fasteners

Screw fasteners are based on metric or imperial (inch) thread forms. The size of the thread is the nominal (rounded-off) outside diameter. For example M12 means metric with 12 mm nominal outside diameter.

11.6.1 Screw Fasteners Have the Following Advantages

- Removable
- Simple to install with common tools
- Relatively cheap (because they are so common)

Table 11.8 Properties of Ceramic Fasteners

Property	Alumina (A99.9%)	Zirconia (PSZ)
Composition by weight	99.9% Al_2O_3	5.4% Y_2O_3
Density (gm/cm^3)	3.89	6.0
Porosity (volume %)	Impervious	Impervious
Color	Ivory	Off-white
Hardness (GPa Knoop)	11.1	11.7
Four-point bend strength (MPa)	330	760
Maximum temperature (°C)	1,650	2,200
Thermal expansion coefficient (from 25 to 700°C $\times 10^6$ K^{-1})	7.5×10^{-6}	11.2×10^{-6}
Thermal conductivity (J/s mm °C)	35	2
(Btu/h in. °F)	242	14
Dielectric strength (kV/mm)	10–35	51
Dielectric constant	10.0	28
Volume resistivity (ohm.cm)	Greater than 10^{14}	Greater than 10^{13}

11.6.1.1 Disadvantages

- Critical applications must not loosen (friction nuts, lock washers, torque, Loctite, deformed thread, wiring, pinning)
- Can be damaged during installation by overtightening (requires torque wrench)
- Requires a hole (not potentially as strong as welding/brazing/gluing in some applications)
- Can be bulky in some applications (e.g., sheet metal)

11.6.2 Difference between a Screw and a Bolt?

A bolt is designed to fit with a nut (Figure 11.24). A screw is designed to fit into a tapped hole (Figure 11.25). Most screw fasteners can do both, except for a few head types that cannot be torqued—like round head, plow bolt, and so on. These are always called bolts (Figure 11.26).

The term "bolt" is also used when the fastener is designed to be used with a nut (e.g., a structural bolt), even though it can act as a screw.

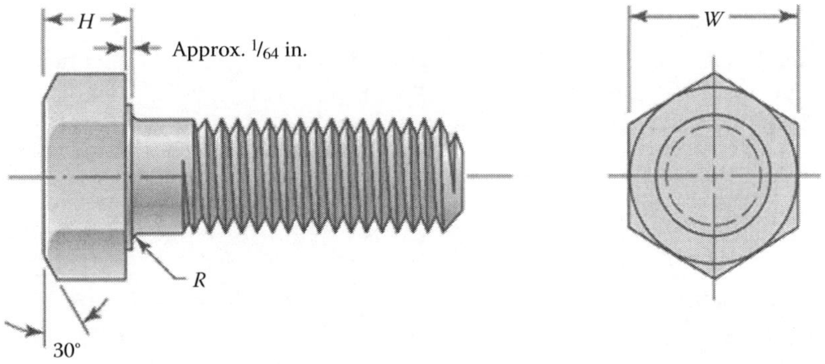

Figure 11.24

Typical bolt. (From The McGraw-Hill Companies, Inc. Permission required for reproduction or display.)

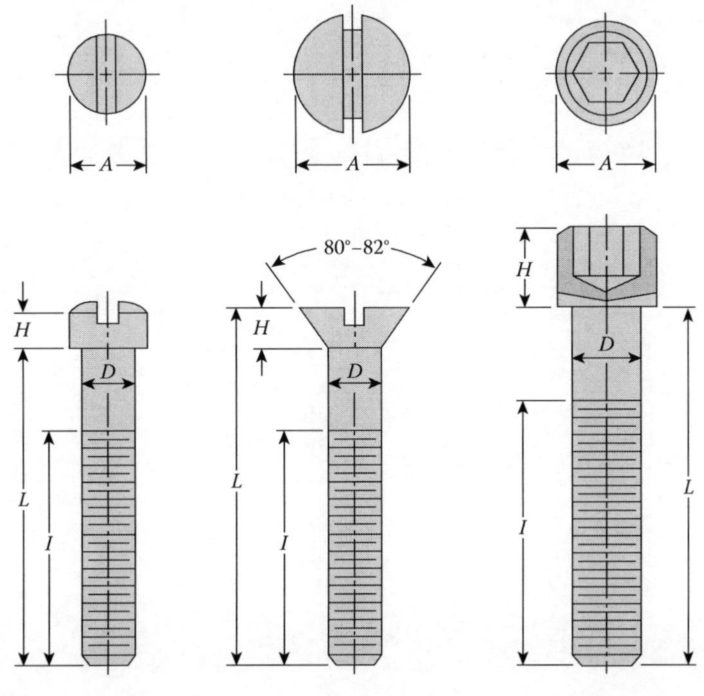

Figure 11.25

Screw dimensions.

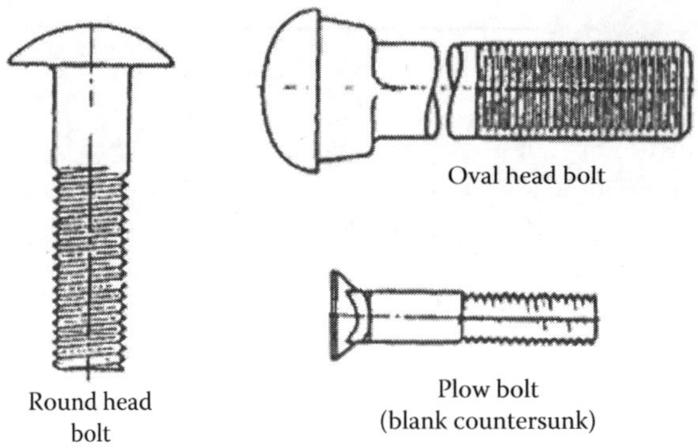

Figure 11.26

Bolt heads that cannot be torqued.

A *set screw* is used to secure an object within another object. It is generally fully threaded. A *grub screw* (or blind screw) has no head, so must be torqued by an internal hex, Allen, slot, Torx, star, Phillips, or straight driver. The common use of a set screw is locking a hub onto a shaft, usually against a flat on the shaft or a key (Figures 11.27 and 11.28).

11.6.3 A Selection of Various Screw Fastener Heads (Figure 11.29)

11.6.3.1 Classification by Fastener Type

- Screw: (sometimes called bolt) A, B, C, D, E, F, G, H, I, J, K, L, M, N, O, P, Q, R, S, T, U, Y, AA, BB, CC
- Set screw: X
- Grub screw: V, W
- Stud: Z

11.6.3.2 Classification by Driver Type; Required for Fasteners

- Straight slotted screwdriver: A, D, G, H, K, O, V
- Philips head screwdriver: B, E, I, L, P
- Posidrive head screwdriver: C, F, J, M
- Allen key (socket head): S, T, U, W
- Hex head (hexagonal spanner or socket driver): N, O, P, Q, R
- Square Head (spanner): X, Y
- Wing: (hand tightened) AA; No head: (stud) Z, (dome) CC; Eye: (attach cable) BB

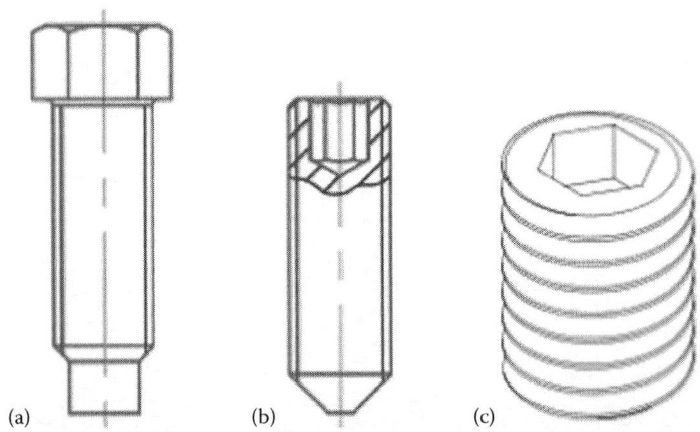

Figure 11.27

Set screws: (a) set screw, (b) grub screw, and (c) hex socket grub screw.

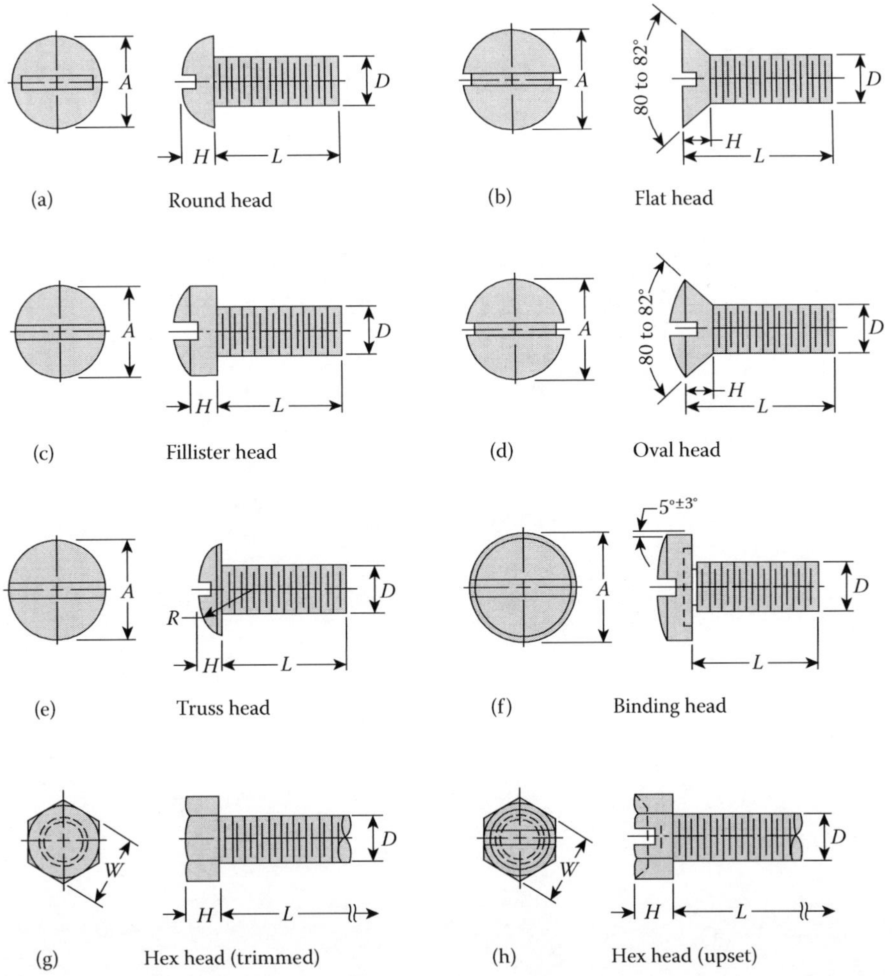

Figure 11.28

Screw heads with dimensions: (a) round head, (b) flat head, (c) fillister head, (d) oval head, (e) truss head, (f) binding head, (g) hex head (trimmed), and (h) hex head (upset).

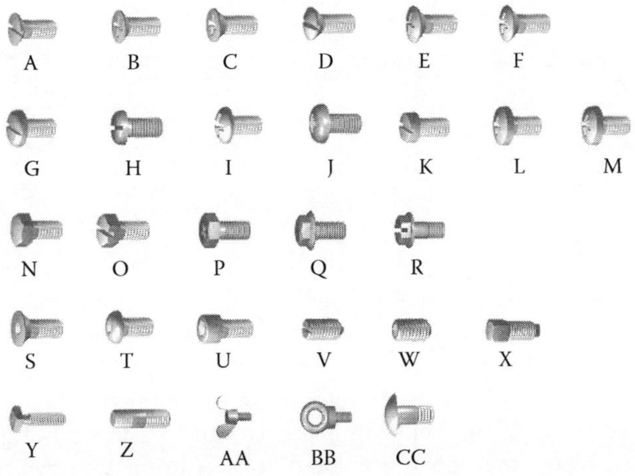

Figure 11.29

Screw fastener heads.

11.6.4 Driver Type Summary

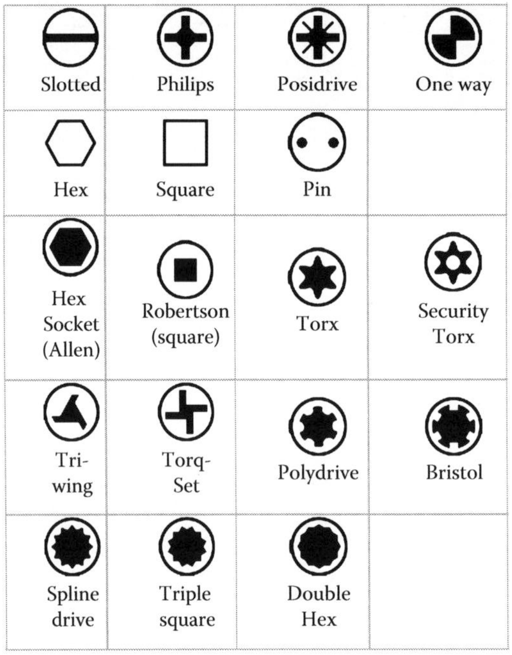

11.6.4.1 Classification by Head Style

- Countersunk: A, B, C, S
- Raised countersunk: D, E, F
- Pan head: G, H, I, J
- Cheese head: K, L, M
- Hex head: N, O, P, Q, R
- Flanged: Q, R
- Cap screw: U
- Dome head: CC/Wing: AA/Eye: BB

11.6.4.2 Head Styles

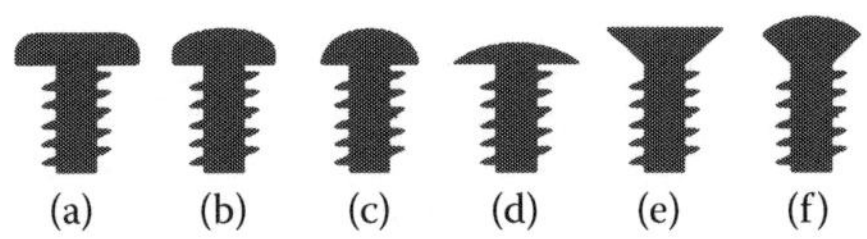

(a) Pan, (b) button, (c) round, (d) truss, (e) countersunk, (f) oval

11.6.4.3 Screw Fastener Materials

- Steel (mild)
- High tensile steel (bolt grades 6.8, 8.8, 10.9, 12.9, and so on)
- Stainless steel (most types available)
- Brass (limited range)
- Aluminum/titanium (rare—special use)

11.6.4.4 Plating of Steel Fasteners

- None (will rust except in very dry environment like air conditioning).
- Zinc electroplated (thin coating with limited corrosion resistance but good finish). Zinc-plated steel almost always has a bright, silver-colored finish. Chromate conversion coatings are applied for increased corrosion resistance. The chromate conversion coatings are available in blue, yellow (gold), olive drab, black, and so on. Chromate conversion is also called *passivated*.
- Hot-dip galvanized (larger, coarse threads typically 10 mm and above, high corrosion resistance).

- Many other metal platings: chrome, nickel, copper, and so on.
- Black (oil) finish: usually high tensile bolts. Slight rust prevention—dry or oily environment only.

11.6.5 Fastener Head Proportions

These are typical proportions of screw fasteners, but it is better to refer to manufacturer's data for exact dimensions.

The grip length of various metric bolts is shown in Figure 11.30.

Thread size vs. spanner (wrench) size: It may be useful to select wrench size for assembly and disassembly.

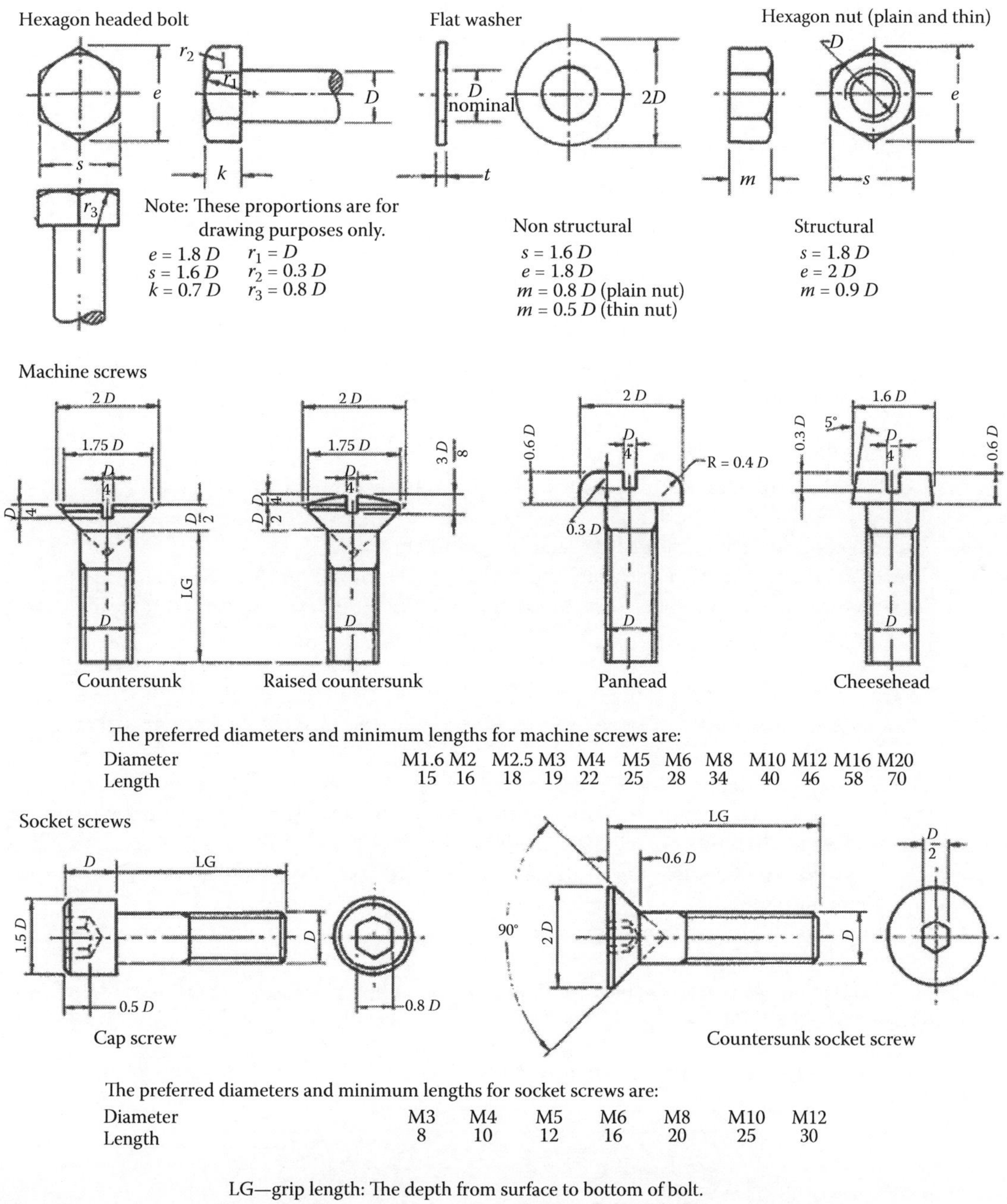

Figure 11.30

Proportions of bolt, screw, and nuts.

ISO metric thread	M1.6	M2	M2.5	M3	M4	M5	M6	M8	M10	M12	M16	M20	M24	M30	M36	M42	M48	M56	M64
Wrench size (mm)	3.2	4	5	5.5	7	8	10	13	17	19	24	30	36	46	55	65	75	85	95

In addition, the following nonpreferred intermediate sizes are specified:

ISO metric thread	M7	M14	M18	M22	M27	M33	M39	M45	M52	M60	M68
Wrench size (mm)	11	22	27	32	41	50	60	70	80	90	100

11.6.6 Specifying a Screw Fastener

Screw fasteners are usually specified in a certain sequence:

M | thread diameter |×| pitch (optional) |×| length LG | driver type | head type | plating | material example: M12 × 50 long Socket Head Cap Screw Zinc Plate. In this case, no pitch is specified, so the standard M12 pitch is used. When material is not specified, steel is assumed.

11.6.6.1 Thread Types: Most Common Types and When to Use Which

The most common thread types are the inch-based Unified Coarse/Fine (UNC/UNF) and metric coarse/fine. Other types and their purposes are described at the end of this section.

11.6.6.2 Coarse or Fine?

To oversimplify, use a coarse thread unless you are tapping into sheet metal. The differences are

- Coarse threads have fewer threads per inch than fine threads.
- Coarse threads are more common, and more shops will have coarse taps.
- Coarse threads are less likely to cross thread or jam because the screw is inserted at an angle. They're also faster to install.
- Screws with fine threads are slightly stronger. This is because the smaller fine threads take up less of the available area. See the loading charts to see the typical differences between the tensile strength of a fine vs. a coarse thread (Table 11.9). A 1/4 UNF thread is about 14% stronger than its UNC counterpart.
- Coarse threads are slightly stronger (against stripping) per length of engagement than finer threads. This may be surprising given the almost universal recommendation that fine threads be used in sheet metal and other thin-walled materials.
- Coarse threads are more tolerant to slight damage or corrosion than fine threads since they have more room for error.
- Fine threads provide finer adjustment since they advance less per rotation than coarse threads.

The metric coarse thread is actually between UN coarse and fine threads, and the metric fine thread is finer than the UNF threads (Table 11.9).

11.6.7 Fine vs. Coarse Threads

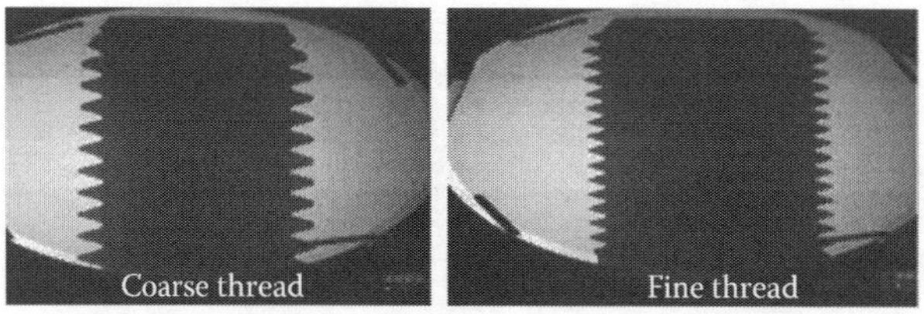

Table 11.9 Standard Metric Thread Sizes

Nominal Diameter *D* (mm)		Pitch *p* (mm)	
1st Choice	2nd Choice	Coarse	Fine
1		0.25	
1.2		0.25	
	1.4	0.3	
1.6		0.35	
	1.8	0.35	
2		0.4	
2.5		0.45	
3		0.5	
	3.5	0.6	
4		0.7	
5		0.8	
6		1	
	7	1	
8		1.25	1
10		1.5	1.25 or 1
12		1.75	1.5 or 1.25
	14	2	1.5
16		2	1.5
	18	2.5	2 or 1.5
20		2.5	2 or 1.5
	22	2.5	2 or 1.5
24		3	2
	27	3	2
30		3.5	2
	33	3.5	2
36		4	3
	39	4	3
42		4.5	3
	45	4.5	3
48		5	3
	52	5	4
56		5.5	4
	60	5.5	4
64		6	4

Fine threads have a smaller pitch for the same diameter. This gives a larger cross section for the bolt—which means it is stronger. They are also better in fatigue because there is a less sudden change in diameter. Fine threads give a higher clamping force for the same tightening torque, which has the added benefit of not coming apart as easily under vibration. The main disadvantage of fine threads is the need for higher accuracy, which means they are more sensitive to coatings (they need to be thin), and dirt (easier to get jammed if rusty, dirty, and so on).

11.6.8 Rolled vs. Cut (Machined) Threads

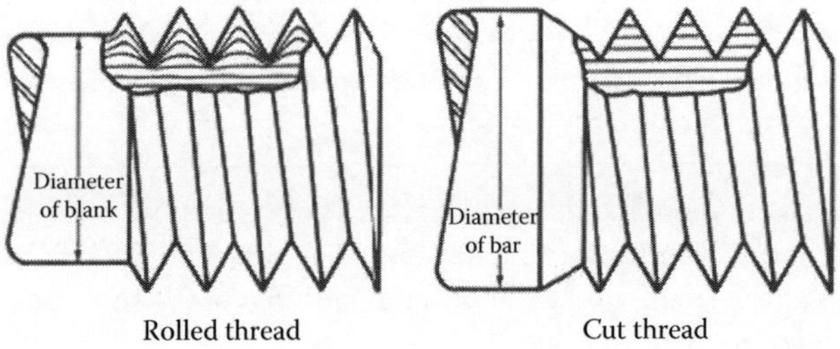

11.6.9 Sustainable Fastener Manufacturing

11.6.9.1 Thread Production

Threads can be produced either by cutting or rolling operations. The shank of a blank designed for cut threading will be full size from the fillet under the head to the end of the bolt. Producing cut threads involves removing the material from a bolt blank with a cutting die or lathe in order to produce the thread. This interrupts the grain flow of the material. Rolled threads are formed by rolling the reduced-diameter (approximately equal to the pitch diameter) portion of the shank between two reciprocating serrated dies. The dies apply pressure, compressing the minor diameter (thread roots) and forcing that material up to form the major diameter (thread crests). Imagine squeezing a balloon with your hand; you compress with your fingers to form the valley, while allowing part of the balloon to expand between your fingers. This is the concept behind roll threading. Rolled threads have several advantages: more accurate and uniform thread dimension, smoother thread surface, and generally, greater thread strength (particularly fatigue and shear strength). Thread cutting requires the least amount of tooling costs. It is generally only used for large diameter or nonstandard externally threaded fasteners. Thread cutting is still the most commonly used method for internal threads.

11.6.10 Substantial Material Saving

When a thread is produced by rolling, material savings are inevitable. A rolled thread requires a smaller blank. The finished part's major thread diameter is produced by the outward flow of displaced material. On the other hand, a cut thread's blank diameter and thread diameter are equal to one another. Thread rolling produces no waste material. The table below clearly shows the tremendous savings inherent in the thread rolling process. It could be stated that thread rolling is good for the environment.

Thread Size	Material Savings (%)	Energy Savings
0.128–32*	24	
1/4–20	25	
3/8–16	27	
1/2–13	19	
5/8–11	19	
3/4–10*	16	
1–8	18	
1 1/4–7	16	
11/2–6	16	
2–4 1/2	15	

*Shows the threads/in. data.

Thread rolling is used to form threads on cylindrical parts by rolling them between two dies. It is the most important process for mass production of bolts and screws in industry. The production rate is roughly about eight bolts and screws. This is much higher than thread cutting on a lathe or CNC. There are many other advantages, as follows:

1. Greater accuracy and uniformity
   The thread rolling dies of today are ground and polished to exacting specifications. In fact, it is possible to maintain a pitch diameter tolerance of 0.001 in. However, the major advantage thread rolling has over all other types of thread production is the ability to remain extremely accurate over very long production runs. The key is in the die. A thread rolling die does not erode over time because the contact point is not concentrated on a sharp cutting edge, but is instead distributed over a broad surface. Also, the thread rolling process itself is relatively friction free, so adhesion does not case problems either. Therefore, if each of the blank's dimensions are uniform, the first part produced will be the same as the one-thousandth part produced. No other thread producing method can touch rolling in the speed department. Whether a production run involves small or large quantities, the labor savings are tremendous.
2. Other cost-saving features
   - Thread rolling dies do not require sharpening. There is no downtime for sharpening or resetting.
   - Thread rolling's uniformity saves money on inspection labor.
   - Rolling can be performed on the collect end behind a part's shoulder, often saving a secondary threading operation.

- Smoother surfaces.
- Better fatigue resistance due to compressive stresses introduced by rolling.

There is a hardened rolling die on each side of the job (which is the bolt being rolled). The rolls apply a large pressure against the workpiece which deforms (forges) the thread onto the job. The rolled thread ends up larger in diameter than the shank of the bolt, which is usually a good thing for fatigue. A rolled thread is stronger than machined threads because of grain flow, making it more difficult to "strip the thread."

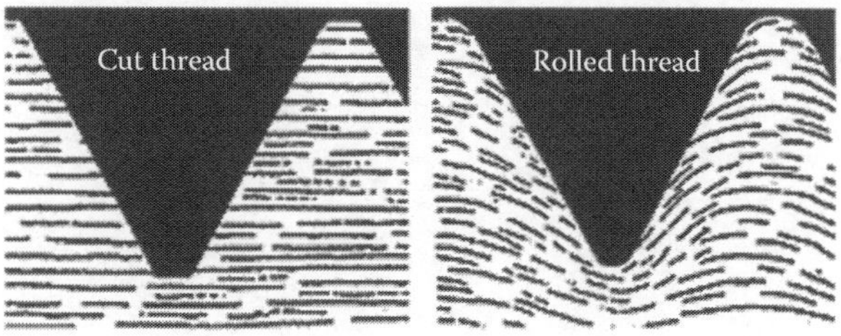

11.6.10.1 UNJ or MJ

These threads are used in situations where fasteners must withstand high fatigue stress, predominantly the aerospace industry. The basic difference between UNJ and UN is a larger root radius. Avoidance of sharp corners is critical for fatigue resistance. The root is given a large enough radius that it could potentially interfere with a typical UN internal thread, so there are both external and internal UNJ threads (and MJs). It is statistically highly likely that an external UNJ thread will fit a regular internal UN thread.

11.6.10.2 Standards

In general, the geometry is defined by ANSI, ASME, and ISO, while material strength properties are defined by ASTM, IFI, SAE, and ISO.

11.6.10.3 Constant Pitch Series

This refers to the many series of threads where the pitch does not increase with diameter. The 8-UN (8 threads per inch) series is apparently very popular above 1 in. diameter fasteners. Typically these are used for adjustment devices and not fasteners.

11.6.10.4 Hole Detailing

Holes are drilled for screw heads in the following manners:

- Countersink (for a countersink head)
- Counterbore (for a cap head)
- Spotface (Figure 11.31)

11.6.11 Washers

Washers are always advised to be used in assembly to seal and insulate.

A selection of washers:

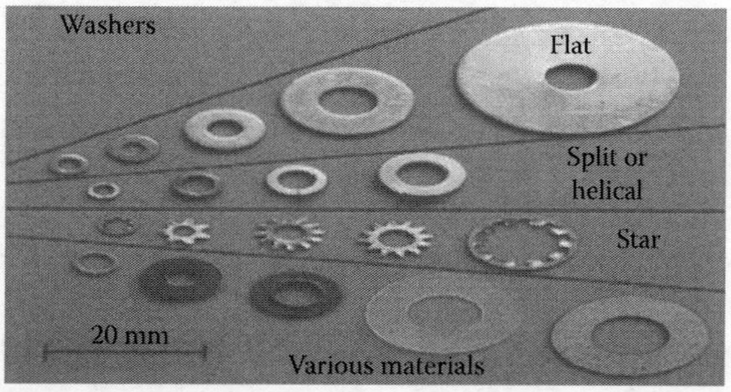

The role of a washer can be:

- *Protection*: To prevent scoring of the surface by the nut or bolt head, or spread the load over a wider area.

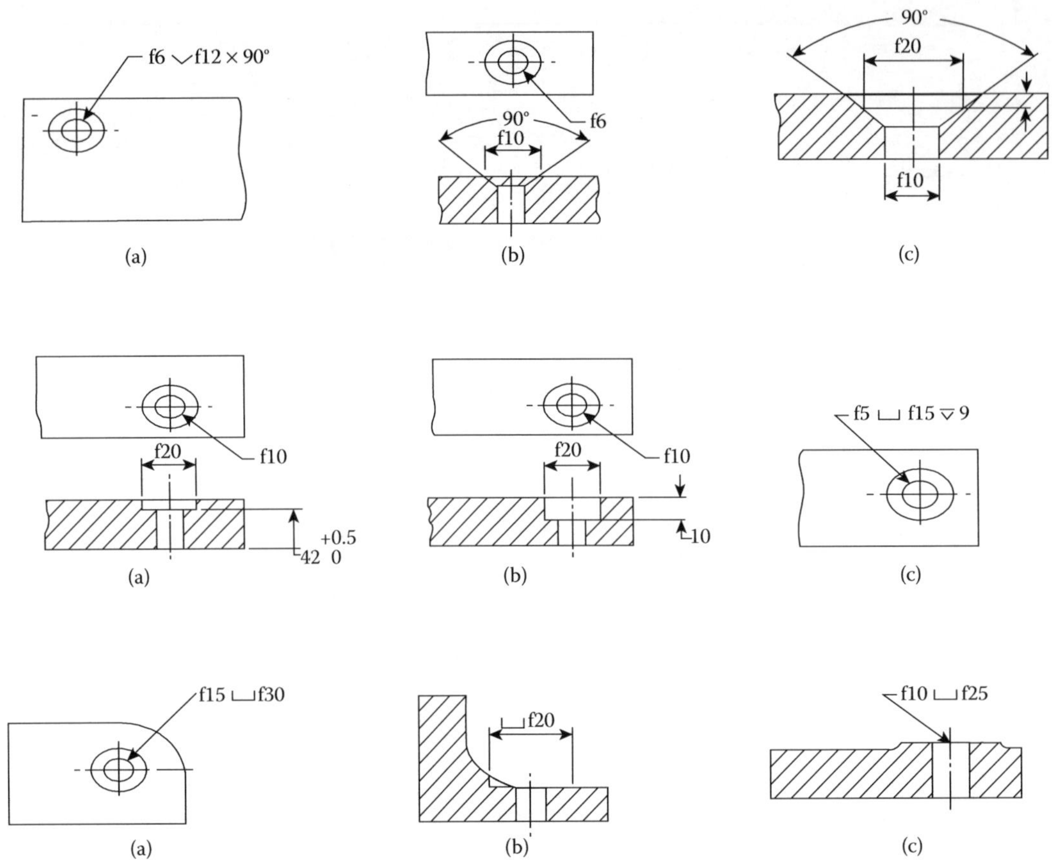

Figure 11.31

Methods of dimensioning countersinks (a), counterbores (b), and spotfaces (c).

- *Locking*: To prevent the nut/bolt from coming loose (helical/spring/split washer, star washer).
- *Insulation*: For thermal, electrical, corrosion reasons.
- *Sealing*: To prevent leakage (e.g., copper washer used for sump plug).
- *Alignment*: Tapered washer to give a level surface on an angled face.

11.6.12 Nuts

These lock the clamped material against the bolt head. In applications where vibration or rotation may work a nut loose, various locking mechanisms may be employed: adhesives (Loctite), *split pins* or lockwire, nylon inserts, or slightly oval-shaped threads.

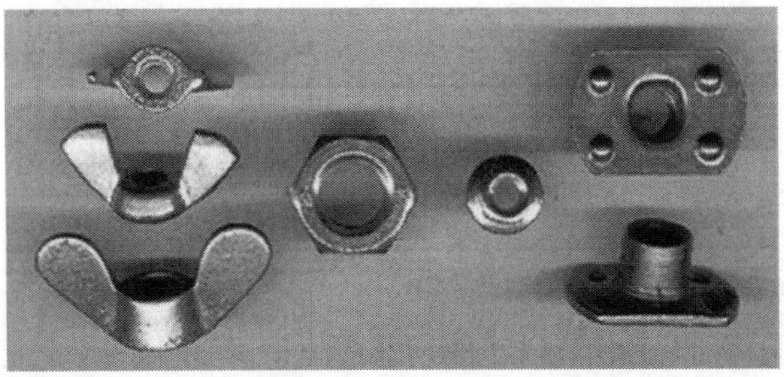

Left to right: wing, hex, hex flange, and slab weld nuts

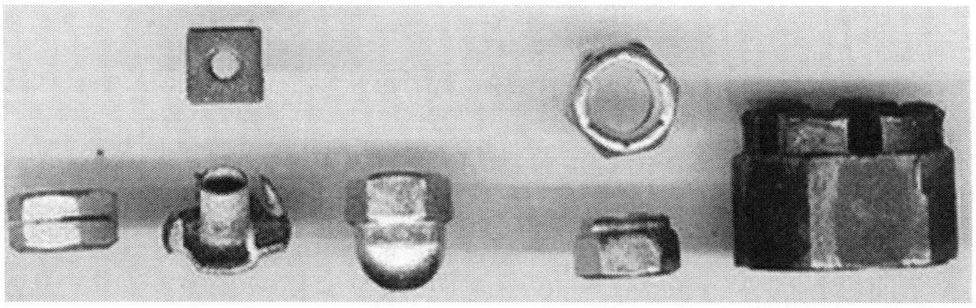

Left to right: slotted, square, T, cap (or acorn), nylon locking, and castellated nuts

11.6.13 Pins

Pins are mostly used to locate two objects together.

Dowel pins are the most accurate, machined to tight tolerances and pressed or free fitting into reamed holes. In a bolted joint, the play would be typically 0.2 mm for a 10 mm (3/8 in.) bolt. When dowels are used, the accuracy would typically be 0.02 mm. They are often used when precise mating alignment is required, such as in differential gear casings, engines, injection molding tools, press tools, and so on.

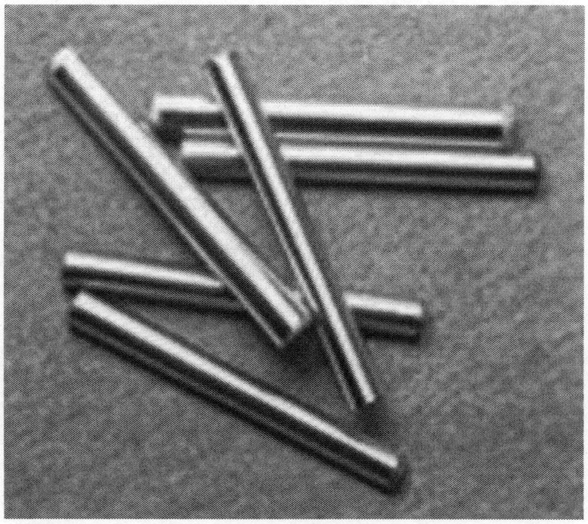

Dowel pins

11.6.14 Spring Pin (or Roll Pin)

These represent the same idea as dowel pins, but do not require accurate holes or pins because the pin "squeezes" into the hole. The friction is reliable and easy to extract but more as a way of holding parts together than accurate location.

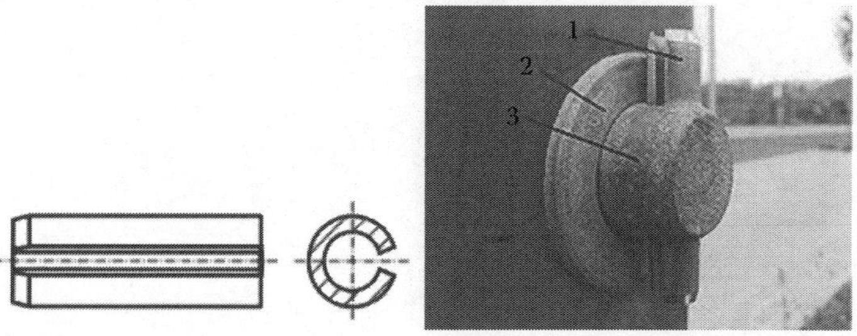

1. Spring pin

11.6.15 Split Pin (or Cotter Pin)

This represents the next level down from the roll pin in terms of accuracy. No real accuracy is needed on the hole and it can even tolerate an incorrect pin diameter to a limited extent. The "R" pin is a similar concept but is more readily removed and reinserted multiple times.

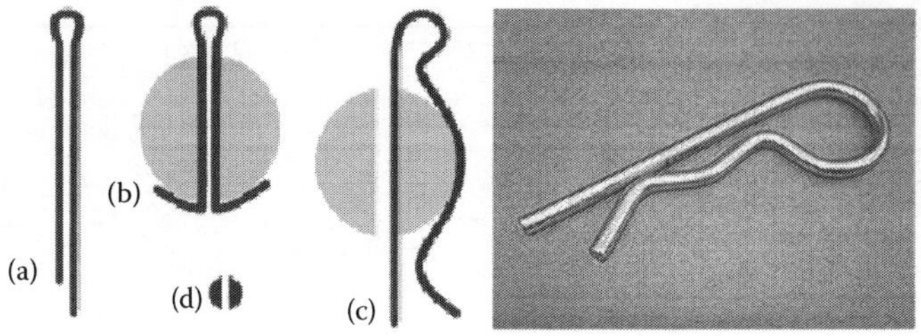

Split Pins (a, b, d) and the "R" pin (c).

11.6.16 Cotter Pin

Cotter pins are shown above. These fasteners are formed from half-round wire into a single two-stem pin as depicted above. They range between 0.8 mm (0.031 in.) and 19 mm (3/4 in.) in diameter, and are in point style. Cotter pins are inserted into holes in the mating parts and their legs are split to lock the assembly. They are used to secure nuts on to shafts as shown in Figure 11.32.

11.6.17 Keys and Keyways

Keys are used to connect a shaft to a hub (which could be a gear, pulley, link arm, and so on).

Parallel keys are the most widely used. They are sized according to the shaft diameter (square keys for smaller shafts, rectangular for shaft diameters over 170 mm. Set screws may be used to lock the mating parts into place so they do not move. The keyway is a longitudinal slot in both the shaft and mating hub.

Woodruff keys (half-moon keys) are semicircular shaped keys that, when installed, leave a protruding tab. The keyway in the shaft is a semicircular pocket and the mating part has a longitudinal slot. They are used to improve the concentricity of the shaft and the mating part, which is critical for high speed operations. The main advantage of the Woodruff key is that it avoids the milling of a keyway near shaft shoulders, which already have stress concentrations. Common applications include machine tools, automotive applications, snow blowers and marine propellers. It can also be used in combination with a tapered (conical) fit.

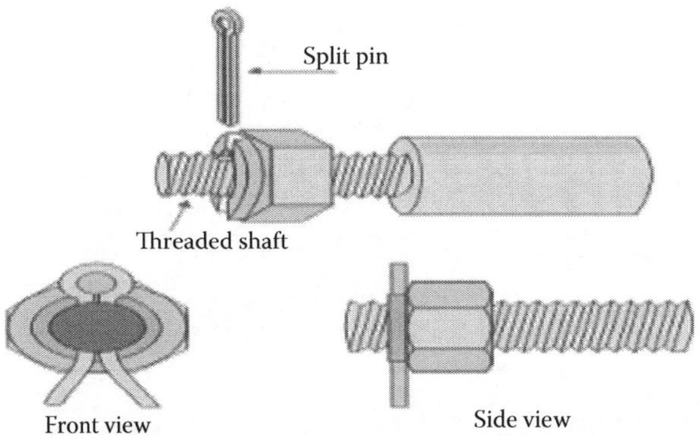

Figure 11.32

Cotter pin.

A Woodruff key and keyway Gear G is held on shaft S by Woodruff key N

The hub length of the gear is usually greater than the shaft diameter, for stability. If the key in this example is made equal in length to hub, it would have enough strength.

Example 11.9

Normally, keys are made of cold-drawn steel. Let there be 6000 lb force on the shaft surface and a gear is to be mounted on the shaft. The shear strength of the key material (S_{sy}) is 37.5 kpsi. The factor of safety is assumed to be 3. A 3/8 in. square is to be designed.

$$\frac{S_{sy}}{3.0} = \frac{F}{tl} \text{ or } \frac{37.5 \times 10^3}{3} = \frac{6000}{(3/8)l} \text{ or length of key } (l) = 1.28 \text{ in}$$

It may be approximated as 17/16 in.

Tapered keys use a wedge action to fit the key. Some taper keys have a gib, or tab, for easier removal during disassembly. Excessive wedge force or a loose fitting shaft can push the hub off-center. A gib-head key is shown below.

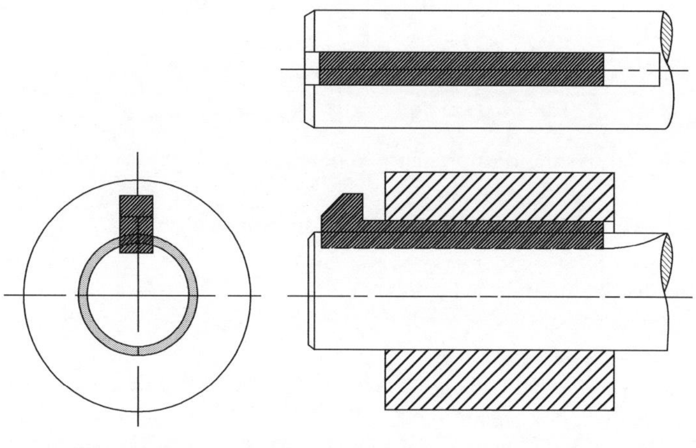

A Scotch key is simply a hole drilled axially into the part and the shaft, so that a round key can be used. It works, but it is difficult to match a replacement part. The circular cross section gives low stress concentration.

11.6.18 Circlips

Circlips (snap rings, retaining rings) prevent axial movement on a shaft, but normally with some play. Common for smaller equipment—rollers, bearings, and so on.

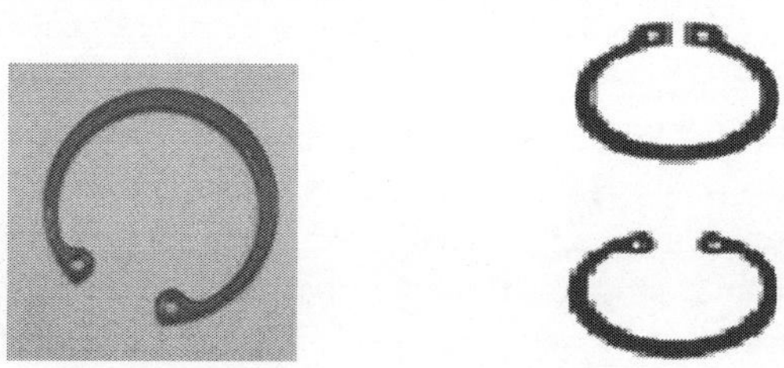

External (right) and internal (left) circlips in the standard design

E clip: can be pushed on from the side of the shaft without the need to sketch over the end of the shaft. Less secure.

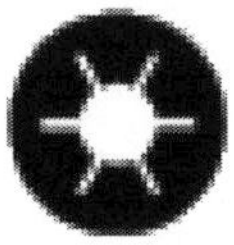

Push-on clip. No groove required.

Most fasteners use one of four designs: Friction-based fasteners are inexpensive and easy. Prevailing-torque nuts typically have a deformed shape or distorted threads and are sometimes referred to as top lock nuts.

Nylon-insert nuts come with a nylon ring inside the threads and are available in a variety of types and colors. Jam nuts are double nuts consisting of a jam nut and a regular nut.

Spring-based fasteners are inexpensive but ineffective against vibrations. They include splitting washers or cupped spring washers, sometimes referred to as Belleville washers.

Interference-fit versions are effective, although extremely time consuming to install and, therefore, not cost efficient. They include different assembly techniques such as safety wiring, cotter pins, peening, welding, and tabs, in which bent metal is forced against the bolt to keep it from turning.

Tension-based, two-piece washer systems rely on cams and serrated friction surfaces. They are simple, effective, and reusable.

11.6.19 Installation Checklist

Below are some guidelines that should improve the reliability of bolted joints:

- Thoroughly clean bolt holes and threads with a solvent to eliminate residual adhesives, lubricants, and particulate matter.
- Make sure an effective moly-based (molybdenum disulfide) or graphite-based lubricant is applied to thread-mating surfaces.
- Torque to proper bolt specifications.
- Make sure new bolts of the correct grade are used in critical applications. (Used bolts may be damaged and visual inspections are not reliable.)
- Consult the fastener manufacturer when special conditions such as high temperatures are present.
- Ensure the service and calibration on your torque wrench is current.
- Tighten flange bolts in torque steps and in a star pattern or other pattern to exert a uniform clamp load where applicable.

11.6.19.1 Fastener Problems

Most modern equipment uses one or more of fasteners. During vibrations, localized slip at the bolt/nut contact surfaces can loosen threaded fasteners. Friction-based, spring-based, and interference-fit fasteners do not handle these conditions well because of their natural tendency to follow the threads and make the bolt move away from the joint. A nut has a preferred direction of rotation when subjected to vibrations and thread friction has been overcome, and only a small amount of vibration-induced fastener movement can significantly reduce clamp load. Once some clamp load is lost, the bolt moves even more, creating more clearance in the shaft, and weakening bolts can shear due to wear and tear.

These designs face other problems as well; many fasteners such as split washers and stover nuts are not reusable, and fasteners such as nylon-insert nuts are temperature sensitive. At high temperatures, nylon inserts melt or degrade. Interference-fit designs can be difficult, expensive, and take twice as long to install due to the extra effort needed to thread the length of the bolt with resistance. Additional skill may be required for wiring, which is also time consuming to install and remove.

Tests can determine a fastener's resistance to vibration and loosening. One, the Junker vibration test (DIN 65151), measures clamp load in joints while they vibrate. It introduces vibrations that cause inertial forces in the tightened bolted joint to be overcome friction in the threads and cause loosening. With most fasteners, a significant or complete loss of clamp load is evident within a few seconds. NASM-1312-7 (MIL-STD1312-7) is the U.S. standard for vibration testing.

11.7 End-of-Life Considerations in Fasteners: Reuse of Fasteners and Bolts, DfE of Fasteners

The reuse and recycling of fasteners is essential to consider sustainable design or for design for environment (DfE). Interest is being shown in the torque tension performance of fasteners when they are reused and our equipment has been used to investigate the change in performance during 10 reuses/cycles. In these cases, the integral lubricants show a vast improvement over the "surface" type of torque tension control. A bolt that has already been used in a given application may or may not be reused, depending on the grade and application.

The bolt may be reused provided it is absolutely certain that the bolt never exceeded proof load and yielded. As the fastened joint is put into use, it will encounter all types of various loads, including tension loads, shear loads, cyclic loads, prying loads, and loads which may be a combination of these and other possibilities. Events such as pressure changes in a pipeline, vibration from an engine, or an impact of a hydraulic ram may produce these loads. These external loads add to or subtract from the initial load of the fastener and in extreme cases may yield the fastener. Torque is only an indirect indication of tension. Improper installation may yield the fastener. One important factor to remember is that, typically, one will not know whether or not the fastener has been yielded. Especially in critical situations, one should never reuse a fastener unless certain the fastener has never been yielded.

There are a tremendous number of conflicting opinions on this subject, but the only definitive published information source is in regard to ASTM A325 and A490 structural bolts: ASTM A490 bolts and galvanized ASTM A325 bolts shall not be reused. Touching up or retightening bolts that may have been loosened by the installation of adjacent bolts shall not be considered to be a reuse. Pretensioned installation involves the inelastic elongation of the portion of the threaded length between the nut and the thread run-out. Black ASTM A325 bolts, however, possess sufficient ductility to undergo more than one pretensioned installation. As a simple rule of thumb, a black ASTM A325 bolt is suitable for reuse if the nut can be run up the threads by hand.

Heat will lower the yield value of the fastener. The yield strengths are determined at room temperature. ASTM A193 B7 has a yield strength of 75–105 kpsi (75 kpsi for diameters over 4 in. and 105 kpsi for material in diameters up to 2-1/2 in.) at 70°, and drops to approximately 53–74 kpsi at approximately 800°. So, if we install a B7 fastener at room temperature expecting each fastener to support a tensile load of 85 kpsi, which would not yield the fastener of diameters of 2-1/2 in. and smaller. Now, if we were to introduce heat into our fastener application, the fastener would begin to yield at a much lower load.

The basic function of the nut is to produce and maintain the clamp load on the assembly. As the fastener elongates, it starts to apply a compressive load to the nut. The first engaged thread of the bolt experiences an enormous amount of tension from the entire body. Some of this load is transferred to the nut, causing less tension in the bolt at the second engaged thread. Then, this thread transfers part of this load to the third thread, and so on. The threads of the bolt will stretch. At the same time, the compressive forces acting on the bearing surface of the nut squeeze the bottom threads of the nut together. Due to this deformation causing an uneven load distribution, the first few threads may plastically deform (yield). The first few threads of the nut will support the majority of the load. In some cases involving UNC threaded nuts, research has shown that the first thread will have to support nearly 35% of the load. The second thread will support about 25% of the load and the third thread about 18%. To best allow the load to distribute to the mating threads, the nut is slightly softer than the bolt to allow its threads to distort a small amount to match the expanding contours of the bolt threads. If a nut were reused, there would no longer be a "perfect" thread match. This will create more friction between the threads during installation, which will significantly alter the installation torque. On a recent demonstration with a 1/2-13 zinc-plated SAE J429 Grade 5 hex cap screw and zinc-plated SAE J995 Grade 5 hex nut, we used an installation torque of 70 ft-lb to obtain a clamp load of 9000 lb (without any added lubrication). On the second installation, this torque had increased to 95 ft-lb to obtain 9000 lb. By the fourth installation, we required 145 ft-lb to reach a clamp load of 9000 lb.

If the bolts have been tensioned beyond their yield point, they enter the "plastic zone" (where they elongate and do not contract once the load is removed), which means they may be subject to premature failure. Since it is virtually impossible to determine visually if a specific fastener has entered its plastic zone when previously used, the decision to reuse a fastener will be determined by the price to replace it vs. the potential cost or liability of that fastener failing.

11.7.1 Lubricants and Concerns for the Environment

The lubricants are essential for smooth functioning of fasteners; they are also damaging to the environment. They can flow into rivers and streams and pollute water resources, and even the soil could be damaged beyond repair.

However, lubricants are critical to healthy bolted joints. When fasteners are not properly lubricated, elevated friction levels can reduce clamp loads while increasing clamp-load deviation. More friction adds resistance during tightening, making bolts harder to install. More friction also increases the torque needed for a joint load because installers must first overcome this additional friction before getting a good clamp load. (Meeting proper torque specifications is a prerequisite for a good clamp load.) It should be realized that approximately 90% of the torque applied to a bolt is used to overcome friction. Only 10% is used to clamp the load.

While crankcase oil from a car may be good for engine bearings, it is not very effective on bolt threads. The latest research indicates graphite or moly-based (molybdenum disulfide) lubricants are best at reducing friction and limiting torsional stress. This stress arises from bolts deformed by a torque or twisting movements during tightening. Depending on the application, experts generally recommend graphite lubricants, as opposed to petroleum-based solvents which cause galling with stainless steels.

Assembly and dismantling of threaded fasteners should require little effort and leave the screw and its tapping undamaged; this is possible with lubricants.

They not only facilitate assembly and dismantling but also reduce friction, increasing the preload force. It will save maintenance, repair, and purchasing costs. Furthermore, lubricants have to work at high temperatures to ensure nondestructive dismantling and prevent expensive repairs. The most suitable lubricant for a particular application depends on the service temperature, media influences, and material combinations.

Using speciality lubricants for screws ensures that no damage occurs during assembly and dismantling and increases preload forces without requiring higher tightening torques. Furthermore, they reduce friction, thus permitting smaller screw and nut dimensions, which cuts costs. Bolt failure, loose bolts, leaking joints, damaged gaskets, and uneven flange alignment are the result of improper bolt preload. Even one poorly assembled critical joint in a refinery, nuclear plant, or paper mill can cause a costly shutdown or even the loss of human life.

11.7.2 Lubricant Benefits

1. *Controls Friction*: Friction during torque can cause galling, seizing and fastener wear. A properly selected lubricant will prevent these problems from occurring by reducing the overall amount of friction in the bolt assembly process.
2. *Assists with Proper Tension*: Even when following prescribed torque, unexpected nut factors can severely reduce the expected fastener preload.

3. *Reduces Corrosion*: Corrosion can wreak havoc on fasteners. With the right protective lubricant, we can reduce and even prevent corrosion from moisture and chemicals thanks to the barrier it provides.
4. *Facilitates Break-out*: It has been shown that a bolt assembled with a lubricant containing a high amount of solids is easier to break out, even years later. This is due to lubricant residue remaining after fastener make-up. By taking a proactive approach to bolt loosening and using a lubricant during initial tightening, one can avoid galling associated with bolt break-out.
5. *Reduces Damage*: When a bolt fails in a critical joint, there is never a good outcome. Loss of production, hazardous emissions, financial damage, dangerous material leaks, and worker injury or death may result. These tragic events can be reduced or even eliminated by adding the proper lubricant to the bolting procedure, since it helps a bolt in make-up preload.

11.7.3 Finding the Right Lubricant for Fastener Assembly (2,3)

It is important to find thread lubricants with a higher percentage of high-pressure grease and solids, since they leave a mechanical barrier in place. Even when the liquid portion of the lubricant evaporates, the solid materials will remain and assist with bolt release.

To find the right lubricant for one's application, it is important to remember the following:

- Environmental conditions play a significant role in how a lubricant works. Extreme temperatures, salt water, humidity, and many other conditions should be considered when finding the right lubricant for one's job.
- A lubricant's chemistry must be compatible with the fastener's material to avoid damage from the lubricant. Otherwise, fasteners can corrode or break down faster than anticipated.
- The lubricant must meet the standards required by the industry and local environmental regulation. The speed at which the moving surfaces move relative to one another can make a difference in which lubricant one should choose. Threaded connections are an example of slow-moving surfaces, and require lubricants that are properly formulated for these slow-moving, high-pressure applications.
- Lubricants with a low-friction factor will result in a greater bolt preload.
- Lubricants with a high-friction factor will cause a decreased bolt preload.

Moly-based (molybdenum disulfide) lubricants are the most suitable for nuts' and bolts' assembly and disassembly. Black Moly and Blue Moly are described below.

1. Black Moly

   Black Moly lubricant is designed for industrial applications requiring nondestructive assembly and disassembly of components. This multipurpose lubricant contains a high concentration of molybdenum disulfide, making it the most effective choice for extreme-pressure applications, such as under heavy loads and slow speeds which can result in boundary lubrication failure. Black Moly withstands pressures to 500,000 psi and temperatures up to 750°F.

   Benefits:

   - Protects and lubricates metal parts in high-friction, high-pressure environments
   - A simple coating provides smooth break-in of new and rebuilt equipment

2. Mariner's Choice

   Never-Seez Mariner's Choice prevents rust, galling, and galvanic corrosion in salt water, salt air, and high-moisture environments. It is especially formulated for wet applications.

   This product was developed from years of experience working with marine engineers to provide superior parts' protection in applications exposed to salt water. The upper temperature limit is 2450°F and the torque reduction value is 41%.

   Benefits:

   - Excellent protection from salt water and fresh water
   - Excellent resistance to water washout
   - Inhibits galvanic corrosion
   - Compatible with soft metals

3. Blue Moly

   Blue Moly is designed for applications requiring nondestructive assembly and disassembly where combinations of high loads, high temperatures and low speeds are encountered to reduce friction during the assembly of press-fit components. Blue Moly withstands pressures of 300,000 psi. The exclusive combination of molybdenum disulfide and pure nickel flake suspended in a premium-grade grease works in a wide range of applications. Blue Moly is especially recommended for those applications in which graphite, zinc, lead, and copper are prohibited or undesirable. The addition of special corrosion inhibitors makes Blue Moly an effective choice of use in harsh environments.

The upper temperature limit is 1500°F and the torque reduction value is 19%.

Benefits:

- Excellent protection against extreme pressure
- Unique blue color allows for easy visual inspection

Industrial usage:

- Heavy construction equipment
- Aircraft components
- Stainless steel fasteners and slow-moving parts
- Automotive engine bolts and body fasteners
- Stainless steel flanges and pump bolts
- Chemical processing machinery
- Gasket dressing
- Military equipment

4. Thread Adhesives

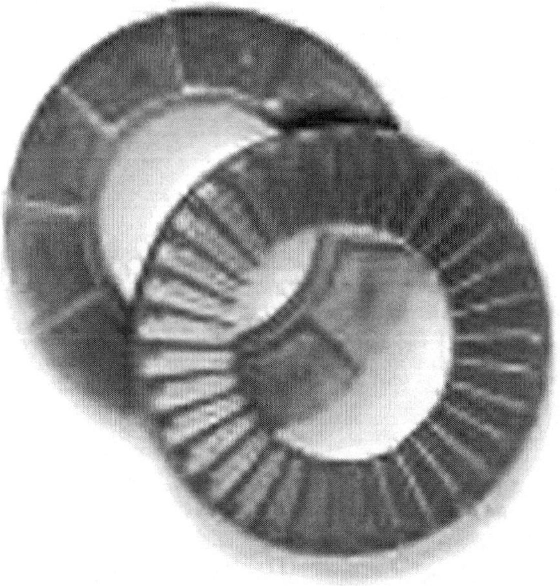

Although adhesives sometimes prevent bolts from loosening, they have several disadvantages. For example, many of them introduce friction which can create a torsion load on the bolt.

Other disadvantages of adhesives besides potential increased friction include

- Difficult to disassemble once cured.
- High and low temperatures can create inconsistent results.
- Limited storage life of the adhesive (needs to be replaced over time).
- Potential health hazards, including (allergic reactions and skin rashes).

Benefits of lubricated fasteners include:

- Improved reusability
- Reduced friction and deviation
- Facilitated assembly and disassembly
- Reduced torsion stress due to minimized thread friction
- Avoids galling and thread seizure
- Additional protection against corrosion

11.7.4 Life Cycle Cost and Profitability

Over the operational life cycle, lubricants give increased operational reliability and lower maintenance costs while significantly reducing the risks of production stops, accidents, and warranty claims. They can help increase profitability by considering the complete life cycle cost for securing bolts.

11.7.5 Removing Rust from Nuts and Bolts

Rusted mechanical elements can be a serious detriment to any manufacturing shop or industrial product user. Unless it is specially treated with rust-resistant materials, nearly every iron-based component is at risk of rusting, particularly under specific environmental conditions that accelerate this type of deterioration. Although not every form of rust is visibly detectable, any significant accumulation will cause a component to malfunction and eventually fail over time.

Rust traditionally refers only to the formation of iron oxides in iron-based metals and alloys, although other types of metal face similar kinds of oxide corrosion. The combination of water and oxygen causes these iron oxides to form a stiff, nonreactive coating along the metal's surface. As rusted parts of the metal component began to flake away, more surface area is exposed and the rusting continues until all the material's oxygen, water, and carbon dioxide are expended, or the material is dissolved entirely. In the case of fasteners, such as bolts and screws, rust corrosion can lead to a larger component or entire machine failing, making it all the more important to address rust problems promptly.

11.7.5.1 Removing a Rusted Fastener

If bolts and screws are exposed to moisture for a protracted length of time, they will begin to rust very quickly. Fasteners in automobile components, ships, and even home appliances used outdoors are subject to high levels of moisture that can degrade the function of the devices in which the fasteners are housed. Removing these parts can be difficult, especially if the threads have become rusted or rust has degraded the metal head, making them hard to grip with a wrench or screwdriver. There are several options for helping to helping to remove rusted components, including

- *Lubrication*: Commercial and industrial lubricants can often reach a fastener's threads, which may be difficult for tools to grasp. Applying lubricant to the bolt, then tapping its head with a hammer, can help the lubricant penetrate the fitting and loosen the bolt sufficiently to be removed.
- *Cutting*: If a fastener has rusted too severely to be salvaged, cutting it out may be the only alternative, even though it will render the fastener unusable. A cut-off wheel, nut splitter, or even industrial shears can be used to permanently extricate a rusted fastener. Similarly, clamping a bolt with a vise-grip and twisting its head off can also free it from its slot.
- *Torching*: Using a moderate to low heat torch is another viable method for removing a rusted bolt. Applying adequate heat may allow the nut or the bolt to expand enough to be loosened. However, care must be taken when relying on torch removal, as some lubricants are flammable and lubricant-treated fasteners may ignite if exposed to a flame.

11.7.5.2 Rust Removal Methods

While many people simply remove and replace rusted fasteners, in some cases, older or product-specific fasteners cannot be substituted with replacements. Restoring bolts and screws via rust removal, cleaning, and reinstallation can be a vital and cost-efficient way of keeping a device functional. There are numerous ways to remove rust, each with its own degree of effectiveness. Some of the more common techniques include:

- *Sanding*: After removing a rusted fastener, rubbing it with a fine-grade sandpaper or steel wool is a straightforward way to scrap off the accumulated rust. This is a relatively simple and inexpensive method, but can be time consuming and often requires a lubricant to make the fastener reusable.
- *Hydrogen peroxide*: This may be applied to fasteners and left to stand while it dissolves some of the surface rust. After the fastener has been removed, it typically needs to be scoured with a hard brush in order to clean the threads and the shank. Hydrogen peroxide is also a relatively inexpensive method, but it may leave residual spots of rust behind.
- *Repainting*: Coatings may be used to limit the spread of rust. After removing rust from a particular area of a metal component, painting over that surface may help to shield the metal from environmental effects, such as moisture, that lead to rust. Likewise, an oily or wax-based coating may have similar effects.
- *Rust cleaner*: There are a variety of different rust cleaner brands available from numerous manufacturers. Most of them include a dissolving agent, such as oxalic acid, that chemically reacts with iron oxide to break it away from the base material. These rust cleaners are particularly useful for small metal parts, such as fasteners, which can be dipped, soaked, or sprayed in the solution to remove rust. As always, the manufacturer's safety recommendations should be closely consulted, as some of these cleaners may be harmful if misused.

References

1. Budynas, R. G. and Nisbett, J. K. (2008). *Mechanical Engineering Design*, 8th edn. McGraw Hill, Boston.
2. Ashby, M. F. (2009). *Materials and the Environment: Eco-Informed Material Choice*, Elsevier, Canada.
3. Ashby, M. F. (2011). *Material Selection in Mechanical Design*, 4th edn. Elsevier, New York.

Bibliography

Botthroyd, G., Dewhurst, P., Knight, W. (2002). *Product Design for Manufacture and Assembly*, 2nd edn. Marcel Dekker, New York.

Chen, R. W., Navin-Chandra, D., and Print, F. B. (1993). A cost-benefit analysis model of product design for recyclability and its application. In *IEEE International Symposium on Electronics and the Environment* (pp. 178–183), Arlington, VA.

Desai, A. and Mital, A. (2005). Incorporating work factors in design for disassembly in product design, *Journal of Manufacturing Technology Management* 16: 712–732.

Dowie, T. and Kelly, P. (1994). Estimation of disassembly times, Technical Report DDR/TR15, Manchester Metropolitan University, UK.

ISO 14044. (2010). *Environmental Management-Life Cycle Assessment—Requirements and Guidelines.*

Kroll, E. and Carver, B. S. (1999). Disassembly analysis through time estimation and other metrics, *Robotics and Computer-Integrated Manufacturing* 15: 191–200.

Kroll, E. and Hanft, T. A. (1998). Quantitative evaluation of product disassembly for recycling, *Research in Engineering Design—Theory, Applications, and Concurrent Engineering* 10: 1–14.

Lee, S. G., Lye, S. W., and Khoo, M. K. (2001). A multi-objective methodology for evaluating product end-of-life options and disassembly. *International Journal of Advanced Manufacturing Technology* 18: 148–156.

Lidwell, W., Holden, K., and Butler, J. (2003). *Universal Principles of Design*, Rockport Publishers, Gloucester, MA.

McGlothlin, S. and Kroll, E. (1995). Systematic estimation of disassembly difficulties: Application to computer monitors. In *Proceedings of the IEEE International Symposium on Electronics & the Environment* (pp. 83–88), IEEE Xplore Press, Orlando, FL.

Mott, R. (2004). *Machine Elements in Mechanical Design*, 4th edn. Pearson, Prentice Hall, Upper Saddle River.

Segur, A. B. *Predetermined Motion Time System (PMTS)*, mtm_international.org.
Skakoon, J. (2000). *Detailed Mechanical Design: A Practical Guide*, ASME Press, New York.
Skakoon, J. (2008). *The Elements of Mechanical Design*, ASME Press, New York.
Slocum, A. (1992). *Precision Machine Design*, SME.
Soemers, H. (2010). *Design Principles for Precision Mechanisms*, T-Point Print VOF.
Suga, T., Saneshige, K., and Fujimoto, J. (1996). Quantitative disassembly evaluation. In *Proceedings of the 1996 IEEE International Symposium on Electronics & the Environment* (pp. 19–24), Dallas, TX.
United States Environmental Protection Agency (EPA). (2010), *Life Cycle Assessment*, 2010, U.S./gov.
Whitney, D. E. (2004). *Mechanical Assembly: Their Design, Manufacture, and Role in Product Development*, Oxford University Press, New York.

SECTION IV
Design for Environment (DfE)

12

Design for Environment

12.1 Introduction to Design for Environment

The most significant benefits of design for environments (DfE) can only be obtained if the product's entire life cycle, starting from need recognition, design development, production, distribution, use, and disposal are all considered together. The concepts underlying industrial ecology require that the all these activities should be within the parameters of a global ecosystem, which includes the biosphere (i.e., all living organisms) and the geosphere (all lands and waters). With these premises, environmental analysis is oriented toward life cycle consideration of a product associated with its physical dimension, focusing on the interaction between the ecosphere and all the activities involved in the product life, from conceptual stage to disposal. Frosch and Gallopoulos [1] have even hypothesized that a product becomes "a transient embodiment of material and energy occurring in the course of material and energy process flows of the industrial system," and the life cycle is understood as a set of processes of transformation, each generating an output of flows of by-products and emissions. The term "green design" involves environmental analysis in DfE. The purpose of the methods is to help designers understand the implications of alternative design choices in terms of their expected effects on ecological resources. However, these methods have limitations. There are very few analytical relationships among ecological considerations and design of components. There are two methods:; carbon footprint indicators and life cycle assessment for DfE. For a complete analysis aimed at evaluation and reduction of the environmental impact, therefore, it is necessary to take into account not only the manufacturing phases of production and machining, but also the activities preproduction of materials and postproduction phases of use, recycling and disposal. This complete product system includes the product, as integral with its life cycle, within the environmental, social, and technological context in which life cycle evolves.

Life cycle design is characterized by three main aspects: the perspective broadened to include the entire life cycle; assumption that the most effective interventions are those made in the first phase of design; and the simultaneity of the operations of analysis and synthesis on the various aspects of the design problems. This last characteristic merges the life cycle design with design techniques oriented toward the control and compression of the times and costs of product development. The selection of design alternatives must be guided by considering the main factors of product success, which define the design targets, in relation to all phases of life cycle: resource utilization (optimizing materials and energy use); manufacturing planning (optimization of production processes); life cycle cost (LCC) (optimization of the total cost of the life cycle);

product properties (harmonizing a wide range of required product properties, such as ease of production, functionality, safety, quality, reliability, aesthetics); company policies (respect for the company objectives); and environmental protection (control and minimization of environmental impacts).

Green design or DfE has the main aim of reducing the waste of material and greenhouse gas (GHG) emissions into the atmosphere (Figure 12.1). It has been noticed that close to 200 million tons of municipal solid waste are sent to landfills. We can understand the enormous cost of materials going to landfills and its transportation costs, including the landfill costs. Many of these waste materials may be usable with proper design and recycling. This is aside from the fact that some of the waste materials are toxic and a health hazard arising from the pollution of land, water and air product development.

Although many factors influence product development (e.g., manufacturability, reliability, and cost considerations), designers have the greatest influence on the product's composition and structure. As a result, the burden of creating products that are easy to dismantle and recycle falls largely upon the designers' shoulders. Design for disassembly (DfD) must be incorporated in the early stages of product design, when the structure of the product is determined. Design tools that enhance the designer's understanding of the disassembly process and aid in the evaluation of alternative designs. It is increasingly recognized by the engineering world that some better method of design and manufacturing needs to be developed. US industries have come to conclude that there is a need for environmentally friendly designs that can reduce waste, encourage recycling and reuse of the components at the end of life (EOL), and reduce emissions of GHG. There have been notable initiatives like the formation of the Vehicle Recycling Partnership (VRP) among the three top automobile manufacturing companies, General Motors, Ford, and Chrysler. It has developed methods to recycle, recover, and reuse of large percentage of scrap metals from cars. However, it must be recognized that separation of parts and components from an assembly and subassembly is a difficult task and sometimes may not be economically feasible or profitable. One can understand the gravity of the problem of the disassembly as most of the assemblies are made so as not to come apart under static and dynamic forces. It tells us that the problem needs to be tackled at assembly stage if we want to encourage disassembly and encourage the reuse of parts and components and sometimes the subassembly itself. Further study of these problems leads us to design it as if parts are designed in a certain way it would be difficult to disassemble. As we know, if we design two components that need to be assembled by nuts and bolts, then the disassembled components may not be reusable. However, if we design such that its assembly is based on snap fit, disassembly may be easier and disassembled components are reusable.

There are two other aspects in green design that we neglect in the traditional design-embodied energy and CO_2 footprint. Embodied energy refers to the energy invested in a component during material extraction and manufacturing. If a component goes to landfill, the embodied energy of the component is also lost. CO_2 footprint of a part refers to the emissions of CO_2 during its extraction, manufacturing, and transportation.

These days the regulations by governments in accepting a product and its final disposal affects design decisions. The science and engineering of DfE or green design has become the foundation of engineering design. In economic terms, these implications refer to component reuse, recycling, and disposal. Designers

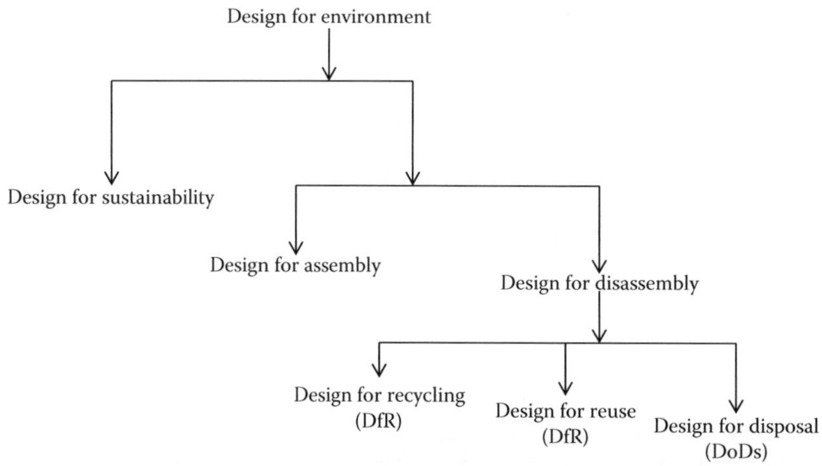

Figure 12.1

Design for environment (DfE).

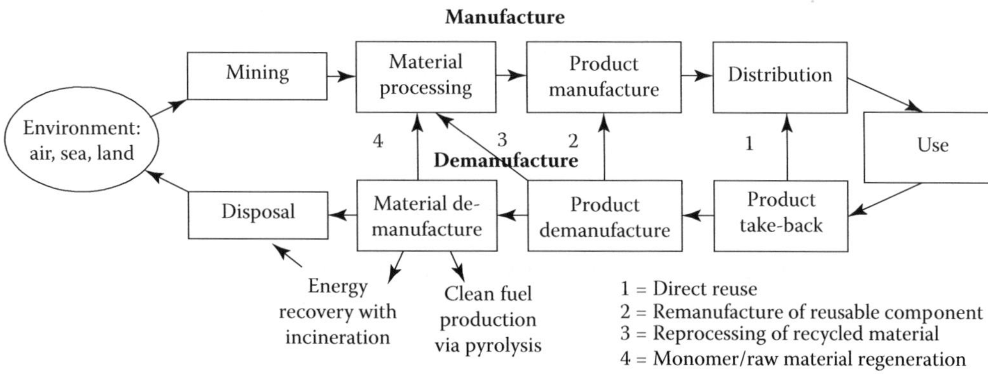

Figure 12.2

Flow diagram of design for environment.

have to account for the total life cycle of the product from grave to grave or at least from grave to gate. DfE is a systematic consideration of design issues related to environmental and human health over the life cycle of the product.

Below are seven aspects that should be included in DfE:

- Manufacture without producing hazardous waste
- Use clean technologies
- Reduce product chemical emissions
- Reduce product energy consumption
- Use nonhazardous recyclable materials
- Use recycled material and reused components
- Design for ease of disassembly product reuse or recycling at the EOL

This concept can be represented in a flow diagram as shown in Figure 12.2.

The term *demanufacture* is appearing more and more, especially in the electronics industry (DEC, Motorola, IBM) to characterize the process opposite to manufacturing involved in recycling materials and products. Companies (such as CinTech E-Waste) have sprung to corner the demanufacturing market.

12.1.1 Growing Importance of Design for Recycling

> Design for recycling is a method that implies the following requirements of a product: easy to dismantle, easy to obtain "clean" material-fractions that can be recycled (e.g., iron and copper should be easy to separate), easy to remove parts/components that must be treated separately, use as few different materials as possible, mark the materials/polymers in order to sort them correct, avoid surface treatment in order to keep the materials "clean."

The importance of design for recycling (DfR) is dictated by reasons of: dwindling resources; landfill space, especially in (over) crowded Europe; and raw material (lesser short-term importance). The social and political climate is changing in the US and the US Congress and president are being forced to follow Europe's lead with regard to environmental regulations. Every township and municipality is feeling the pinch as 80% of electronics are being sent to landfills at EOL. The US National Renewable Energy Laboratory estimates that 30 billion lb. of plastics end up in landfills each year, and only 1% of plastic waste is recycled.

12.1.1.1 Recovery Priorities

Recovery priorities include reuse, recycled materials, and energy recovery.

1. Reuse: This is the series of activities, including collection, separation, and in some cases processing, by which products are recovered from the waste stream for new use in their original intended manner. The remanufactured or refurbished components fall under the reuse category.
2. Recycle: A series of activities, including collection, separation, and processing, by which products or other materials are recovered from or otherwise diverted from the solid waste stream for use in the form of raw materials in the manufacture of new products. However, some materials either cannot be recycled or should be recycled by certain amounts to make the recycling cost effective.

The problem is not a complex assembly of materials, but a mixture of materials that, if separated, can be individually recycled. These require nondestructive disassembly. Everywhere, priority should be given to environmental laws. However, material recycling is most common and even destructive disassembly, as in rivets, is welcome. Even valuable materials such as gold in electronics can be recovered by this process.

3. Energy recovery:
   a. Only energy embodied in materials is preserved through incineration or pyrolysis.

Recycling is one of the best environmental success stories of the late 20th century. In 2003, US recycling and composting efforts diverted more than 72 million tons of material away from landfills and incinerators and instead turned those materials into valuable resources. Collecting recyclable materials is just the first step in a series of actions that generate a host of financial, environmental, and societal returns. There are several key benefits to recycling.

Recycling:

- Protects and expands US manufacturing jobs and increases US competitiveness in the global marketplace, creating 1.1 million jobs
- Saves energy and prevents pollution caused by the extraction and processing of virgin materials and the manufacture of products using virgin materials
- Decreases emissions of GHGs that contribute to global climate change
- Conserves natural resources such as timber, water, and minerals
- Helps sustain the environment for future generations

Recycling not only makes sense from an environmental standpoint, but it also makes solid economic sense. The need for recycled materials has grown steadily as well. Your old cans, bottles, and paper products are in high demand and the Environmental Protection Agency (EPA) is encouraging local communities to increase their recycling collection totals through the "resource conservation challenge." Manufacturing through recycling is good for the development of jobs and an improvement in environmental health.

Reducing the amount of waste generation and reusing all that is possible will go a long way. Waste that cannot be reduced or recycled can be incinerated or disposed through landfills, as long as it is done in accordance with proper regulations. All trash buried in landfill does not decompose into dirt. Highly combustible methane gas is produced from decomposed trash. Recycling is beneficial for your local economy in a broader sense too. According to a report released by the National Recycling Coalition, more than 56,000 recycling and reuse establishments in the United States employ approximately 1.1 million people, generate an annual payroll of $37 billion, and gross $236 billion in annual revenues. According to the report, the number of workers in the recycling industry is comparable to the automobile and truck manufacturing industry and is significantly larger than mining and waste management and disposal industries. In addition, wages for workers in the recycling industry are notably higher than the national average for all industries, according to the report.

12.1.2 Materials Most Commonly Recycled in the United States through Collection Programs

US recycling rates for commonly recycled consumer goods in 2003 are listed below:

- Newspapers: 82.4%
- Corrugated cardboard boxes: 71.3%
- Steel cans: 60.0%
- Yard trimmings: 56.3%
- Aluminum beer and soft drink cans: 43.9%
- Scrap tires: 35.6%
- Magazines: 33.0%
- Plastic milk and water bottles: 31.9%
- Plastic soft drink bottles: 25.2%
- Glass containers: 22.0%

12.1.3 Waste Prevention and Recycling Effects on Global Warming

Reducing waste is good for the environment because it conserves natural resources. The manufacture, distribution, and use of products, as well as management of the resulting waste, all produce GHG emissions.

GHGs, which trap heat in the upper atmosphere, occur naturally and help create climates that sustain life on our planet. Increased concentrations of these gases can contribute to rising global temperatures, sea-level changes, and other climate changes.

12.1.3.1 Reduce Emissions from Energy Consumption

Recycling saves energy. Manufacturing goods from recycled materials typically requires less energy than producing goods from virgin materials. When people reuse goods or when products are made with less material, less energy is needed to extract, transport, and process raw materials and to manufacture products. When energy demand decreases, fewer fossil fuels are burned and less carbon dioxide is emitted into the atmosphere.

12.1.3.2 Reduce Emissions from Incinerators

Recycling and waste prevention divert materials from incinerators and thus reduces GHG emissions from waste combustion.

12.1.3.3 Reduce Methane Emissions from Landfills

Waste prevention and recycling (including composting) diverts organic wastes from landfills, thus reducing the methane that would be released if these materials broke down in a landfill.

12.1.3.4 Increase Storage of Carbon in Forests

Trees absorb carbon dioxide from the atmosphere and store it in wood in a process called *carbon sequestration*. Waste prevention and recycling paper products allows more trees to remain standing in the forest, where they can continue to remove carbon dioxide from the atmosphere.

12.2 Design for Assembly

Historically, all assembly was manual until about 50 years ago but now many assembly operations have been mechanized and several are performed by robots. For mass production to meet needs the world over, interchangeability of parts is required. Interchangeable parts enable use of low-skill assemblers and also save assembly costs.

It has long been recognized that the manufacturing, assembly, and recycling costs can be minimized if all these aspects are considered during the design stage; that is, if we want to reduce manufacturing and material cost, proper product design is most important. To improve the percentage of recycling and reuse, appropriate steps must be taken at the design stage. Reducing the number of parts or components in the assembly makes it easier to manufacture and assemble. The material selection should be the most important aspect of the assembly cost. Although slightly different in detail, they are very similar and all focus on the following issues:

- Material: For example, reduce overall material diversity, avoid the use of laminates or make them out of compatible materials which can be recycled as a mixture.
- Fastener: For example, reduce fastener count and diversity, avoid incompatible adhesives which degrade recyclability of materials, use snap fits where appropriate.
- Component design issues: For example, avoid paints and laminates, build in planes for easy separation and access.

Boothroyd et al. [2] advocated that some of functions of one component may be integrated with other components or eliminated altogether to reduce cost. Consequently, it would reduce the material cost, manufacturing cost, and assembly as well as disassembly cost. It would also encourage recycling and reuse. He has also suggested the following considerations for design for assembly:

1. Determine how a component can be grasped or manipulated for assembly to other component. He called it handling requirements. It should be determined if it needs reorientation before its assembly with other components.
2. Examine which assembly processes, like joining, welding, or snap fitting, are appropriate for different assemblies.

12.2.1 Cost of Assembled Product

An assembled product or system is made of several components. Obviously, mass production is facilitated by robot or some other kind of automated system. However, products made in less than several thousands

Table 12.1 Standard Disassembly Tools with Their Letter Codes

Unscrewing and Screwing	Gripping and Fixturing	Cutting and Breaking
Phillips screwdriver: PS	Vise: VS	Knife: KN
Flathead screwdriver: FS	Pliers: PL	Wire cutter: WC
Nut driver: ND		Handheld shears: SH
Fixed: end wrench: FW		Drill: DR
Adjustable wrench: AW		Handheld power grinder: PG
Socket with ratchet: SR		Grinding wheel: GW
Allen key: AK		Hacksaw: HS
Power wrench: PW		Power saber saw: BS
		Hammer: HM
		Chisel: CH
		Prybar: PR

per year are manually assembled. For cost estimation, the fixed cost, variable cost, and overhead cost is needed. Fixed costs are those incurred in a predetermined amount, regardless of number of products. The variable costs, such as material, labor, and power depend on the number of parts produced. Manual assembly costs are estimated by multiplying the total operation time by hourly labor cost. The assembly operation is helped by tools used. The list of tools for assembly and disassembly of a product is presented in Table 12.1.

A popular method for estimating assembly times has been developed by Boothroyd Dewhurst Inc. and is available as a software tool. They proposed the concept of *assembly efficiency* through design for assembly index (DFA). The expression for the DFA index is presented as

$$DFA_{\text{index}} = \frac{N_{\min} \times 3s}{T_{es}} \tag{12.1}$$

where:

$N_{\min}$ = minimum number of parts in assembly
T_{es} = estimated total assembly time

The “3 seconds” in the numerator is the theoretical minimum time required to handle and insert a part that is appropriate for assembly. The DFA index gives the insight that drives the cost of assembly. To determine theoretical minimum number of parts ($N_{\min}$), the following condition or conditions must be satisfied:

- Does the part need to move relative to the rest of assembly? Small compliance-related movement does not count.
- Must the part be made of different material from the rest of assembly for fundamental physical reasons? Several metals get contaminated after being assembled with other metals. The list below shows the kind of contamination and it affects recycling, reuse, and disposal costs (Table 12.2).
- Does the part hinder accessibility during replacement or repair?

Therefore, it is better to use parts of same material as far as possible and prevent inaccessibility if it is possible. The fundamental operations for assembly, as well as disassembly, are grasping, orienting, and inserting a part depend on the part geometry and the required trajectory of part insertion.

Boothroyd et al. [2] present some ideal characteristics of parts for ease of assembly. They are

Table 12.2 Examples of Contaminants for Metals Recycling

Considered Metal	Contaminated during Recycling
Steel	Copper, tin, zinc, lead, aluminum
Aluminum	Cast iron, steel, chrome, zinc, lead, copper, magnesium
Zinc	Cast iron, steel, lead, tin, cadmium

1. The *z-axis* assembly, which simply means the part is inserted from the top of the assembly utilizing the gravity and the assembler can visually identify the location of assembly.
2. Precise *self-aligning* so that precise motor control is not needed as in case of chamfer. A chamfer can be implemented as a tapered lead on the end of a peg, or a conical widening on the opening of a hole. Parts that require fine positioning require precise motor control and are slow in assembly.
3. Parts requiring no special orientation like a screw, which is slow to assemble. Parts like a sphere do not need any orientation. The parts in reverse order of the need of orientation starting from most to least orientations are: capped and keyed cylinder, capped cylinder, cylinder, and sphere.
4. Parts requiring only one hand for assembly, which means the effort required to manipulate the part is less compared to the part requiring both hands. This also indicates the assembly operation with both hands will take more time compared with the assembly with one hand. The other factor that can dictate the use of both hands is the weight of the part.
5. Parts requiring tools as presented in Table 12.1 above will generally take more time to assemble.
6. Parts assembled by linear motion take much less time than parts requiring circular motion as in assembly with a screw. Pushing in a pin requires less time than a screw.
7. Parts do not need any extra attachment to make it secure. Some parts may require a subsequent securing operation, such as tightening, curing, or addition of other part.

Boothroyd et al. [2] present an equation for the probability of a defective product, containing one or more assembly errors for a product requiring n assembly operations:

$$D_a = 1 - \left[1 - 0.0001\left(t_i - 3.0\right)\right]^n \tag{12.2}$$

t_i is average time per operation.

Alternatively, the expected number of assembly errors in one of these products is given by

$$N_a = 0.0001\left(t_i - 3.0\right)n \tag{12.3}$$

These equations become useful in early stages of design to compare the likely assembly error rates of alternative design concepts for small products assembled in large quantities. The faulty assembly steps may be a major cause of defective products rather than the defective component in the assembly.

The time estimated for assembly of a system is based on the predetermined by the methods-time measurements (MTM). These MTM time standards are estimated times for basic motions involving hand and arms movements, although elements are also provided for eye, leg, foot, and body actions. These standard times fairly well known and documented and are available in a manual by Boothroyd Dewhurst Inc. [2]. There are a number of methods that can be used to calculate the time needed to carry out a job or a task. Time standards can be established by time study, by work sampling, or by the use of predetermined times. A predetermined time system consists of a set of time data and a systematic procedure that analyzes and subdivides any manual or human task into motions, body movements, or other elements of human performance and assigns to each the appropriate time value. Predetermined time systems can have different classifications. They can be classified according to accuracy level, time required for application, and the extent of method description. There are numerous specialized systems available.

The first predetermined time system was developed by A.B. Segur. Segur called it methods-time analysis (MTA) [3].

12.2.1.1 Methods: Time Measurement for Disassembly and Assembly

MTM can be defined as a procedure that analyzes any manual operation or method into the basic motions required to perform it and assigns to each motion a predetermined time standard which is determined by the nature of the motion and the conditions under which it is made. The basic MTM system is MTM-1. In 1963, a new member of the MTM system called General Purpose Data (MTM-GPD) was introduced. MTM systems include MTM-1, MTM-GPD, MTM-2, MTM-3, MTM-V, MTM-M, MTM-C, and 4M DATA.

12.2.1.1.1 MTM-1 MTM-1 is the most detailed system. Motions are broken down into ten categories:

Reach, Move, Turn, Apply Pressure, Grasp, Position, Release, Disengage, Body (leg-foot, horizontal, and vertical) motions, Eye motions. The time for each basic element is given in units of time measurement units (TMU).
1 TMU = 0.000010 h = 0.00060 min = 0.036 s, or conversely 1 s = 27.78 TMU
1 min = 1667 TMU, 1 h = 100,000 TMU. The following is a summarized description of the 10 MTM-1 categories:

12.2.1.1.1.1 Reach Reach is when the hand or finger is moved to a destination. Reach is usually a movement with an empty hand or finger. The time for performing a Reach depends on the following factors:

1. Condition (nature of destination)
2. Length of the motion
3. Type of reach

The time to perform a Reach is affected by the nature of the object toward which the reach is made. The following is a description of these five classes:

A: Reach to object in fixed location, or to object in other hand or on which other hand rests
B: Reach to single object in location, which may vary slightly from cycle to cycle
C: Reach to object jumbled with other objects in a group so that search and select occur
D: Reach to a very small object or where accurate grasp is required
E: Reach to indefinite location to get hand in position for body balance or next motion or out of way.
MTM-1 Reach is in Table 12.3.

Table 12.3 MTM-1 Reach

Distance Moved in Inches	Time TMU				Hand in Motion	
	A	B	C or D	E	A	B
¾ or less	2.0	2.0	2.0	2.0	1.6	1.6
1	2.5	2.5	3.6	2.4	2.3	2.3
2	4.0	4.0	5.9	3.8	3.5	2.7
3	5.3	5.3	7.3	5.3	4.5	3.6
4	6.1	6.4	8.4	6.8	4.9	4.3
5	6.5	7.8	9.4	7.4	5.3	5.0
6	7.0	8.6	10.1	8.0	5.7	5.7
7	7.4	9.3	10.8	8.7	6.1	6.5
8	7.9	10.1	11.5	9.3	6.5	7.2
9	8.3	10.8	12.2	9.9	6.9	7.9
10	8.7	11.5	12.9	10.5	7.3	8.6
12	9.6	12.9	14.2	11.8	8.1	10.1
14	10.5	14.4	15.6	13.0	8.9	11.5
16	11.4	15.8	17.0	14.2	9.7	12.9
18	12.3	17.2	18.4	15.5	10.5	14.4
20	13.1	18.6	19.8	16.7	11.3	15.8
22	14.0	20.1	21.2	18.0	12.1	17.3
24	14.9	21.5	22.5	19.2	12.9	18.8
26	15.8	22.9	23.9	20.4	13.7	20.2
28	16.7	24.4	25.3	21.7	14.5	21.7
30	17.5	25.8	26.7	22.9	15.3	23.2
Additional per inch over 30 in.	0.4	0.7	0.7	0.6		

12.2.1.1.1.2 Move The difference between Move and Reach is that in move the hand or finger is holding something. In Move, the predominant purpose is to transport an object to a destination. The time of Move is affected by its length, nature of destination, type of move, and weight factor (static and dynamic). There are three classes of Move:

Class A Move: Object to other hand or against stop
Class B Move: Object to approximate or indefinite location
Class C Move: Object to exact location.
MTM-1 Move is in Table 12.4.

Eye Travel Time = 15.2xT/D TMU, with a maximum value of 20 TMU
Where T = the distance between points from and to which the eye travel
D = the perpendicular distance from the eye to the line of travel T
Eye Focus Time = 7.3 TMU
Supplementary Information
Area of Normal Vision = Circle 4 in Diameter 16 from eyes

12.2.1.1.1.3 Turn Turn is a movement that rotates the hand, either empty or loaded. The movement rotates the hand, wrist, and forearm about the long axis of the forearm. The time for Turn depends on the following two factors:

1. The number of degrees turned
2. The weight of the object or the resistance against which the turn is made (weight factor)

MTM-1 Turn is in Table 12.5.

Apply Pressure
Apply Pressure is the application of force without resultant movements.

Table 12.4 MTM-1 Move

	Time TMU				Weight Allowance		
Distance Moved in Inches	A	B	C	Hand in Motion B	Weight (lb) Up To	Dynamic Factor	Static Constant TMU
¾ or less	2.0	2.0	2.0	1.7		2.5 1.00 0	
1	2.5	2.9	3.4	2.3			
2	3.6	4.6	5.2	2.9	7.5	1.06	2.2
3	4.9	5.7	6.7	3.6			
4	6.1	6.9	8.0	4.3	12.5	1.11	3.9
5	7.3	8.0	9.2	5.0			
6	8.1	8.9	10.3	5.7	17.5	1.17	5.6
7	8.9	9.7	11.1	6.5			
8	9.7	10.6	11.8	7.2	22.5	1.22	7.4
9	10.5	11.5	12.7	7.90			
10	11.3	12.2	13.5	8.6	27.5	1.28	9.1
12	12.9	13.4	15.2	10.0			
14	14.4	14.6	16.9	11.4	32.5	1.33	10.8
16	16.0	15.8	18.7	12.8			
18	17.6	17.0	20.4	14.2	37.5	1.39	12.5
20	19.2	18.2	22.1	15.6			
22	20.8	19.4	23.8	17.0	42.5	1.44	14.3
24	22.4	20.6	25.5	18.4			
26	24.0	21.8	27.3	19.8	47.5	1.50	16.0
28	25.5	23.1	29.0	21.2			
30	27.1	24.3	30.7	22.7			
Additional	0.8	0.6	0.85	TMU per in over 30 in.			

Table 12.5 MTM-1 Turn

	Time TMU for Degree Turned										
Weight (lb)	30	45	60	75	90	105	120	135	150	165	180
Small 0 to 2	2.8	3.5	4.1	4.8	5.4	6.1	6.8	7.4	8.1	8.7	9.4
Medium 2.1 to 10	4.4	5.5	6.5	7.5	8.5	9.6	10.6	11.6	12.7	13.7	14.8
Large 10 to 35	8.4	10.5	12.3	14.4	16.2	18.3	20.4	22.2	24.3	26.1	28.2

MTM-1 Apply Pressure is in Table 12.6.
Grasp

Grasp is the motion used when the purpose is to gain control of one or more objects. Grasp is almost always followed by Move. There are five types of Grasp (Pickup, Regrasp, Transfer, Select, Contact). MTM-1 Grasp is in Table 12.7.

12.2.1.1.1.4 Position Position is the basic element employed to align, orient, and engage one object with another object, where the motions are so minor that they do not justify classification as other basic elements. The time of Position depends on the following:

1. Class of fit (loose, close, and exact)
2. Symmetry (symmetrical, nonsymmetrical, semisymmetrical)
3. Ease of handling

MTM-1 Position is in Table 12.8.

Release

Release is the relinquishing of control of an object by the hand or fingers. There are two classes of Release:

1. Normal release performed by opening of fingers
2. Contact release: the release begins and is completed at the instant the following Reach begins (no time allowed)

Table 12.6 MTM-1 Apply Pressure

Full Cycle			Components		
Symbol	TMU	Description	Symbol	TMU	Description
APA	10.6	AF + DM+RLF	AF	3.4	Apply force
			DM	4.2	Dwell minimum
APB	16.2	APA+G2			
			RLF	3.0	Release force

Source: Barnes 1980; Konz 1995.

Table 12.7 MTM-1 Grasp

Type of Grasp	Case	Time TMU	Description	
Pick-up	1A	2.0	Any size object by itself, easily grasped	
	1B	3.5	Object very small or lying close against flat surface	
	1C1	7.3	Diameter larger than 1/2″	Interference with grasp on
	1C2	8.7	Diameter 1/4″ to 1/2″	bottom and one side of
	1C3	10.8	Diameter less than 1/4″	nearly cyndrical object
Regrasp	2	5.6	Change grasp without relinquishing	
Transfer	3	5.6	Control transferred from one hand to the other	
Select	4a	7.3	Larger than 1 × 1 × 1	Object jumbled with other
	4B	9.1	1/4 × 1/4 × 1/8	objects so that search
	4C	12.9	Smaller than 1/4 × 1/4 × 1/8	and select occur
Contact	5	0	Contact grasp, hook, or sliding grasp	

Table 12.8 MTM-1 Position

Class of Fit	Symmetry	Easy to Handle	Difficult to Handle
1 Loose	S	5.6	11.2
	SS	9.1	14.7
	NS	10.4	16.0
2 Close	S	16.2	21.8
	SS	19.7	25.3
	NS	21.0	26.6
3 Exact	S	43.0	48.6
	SS	46.5	52.1
	NS	47.8	53.4
Supplementary rule for surface alignment			
P1SE per alignment: >1/16 1/4	P2SE per alignment: 1/16		

MTM-1 Release is in Table 12.9.

12.2.1.1.1.5 Disengage Disengage is the basic element used to break contact between one object and another. The time required for Disengage is affected by the following:

1. Class of fit (loose, close, tight)
2. Ease of handling (easy and difficult)
3. Care of handling

MTM-1 Disengage is in Table 12.10.
Class of Fit: Height of Recoil, Easy to Handle, Difficult to Handle

1. Loose (very slight effort blends with subsequent move)
2. Close (normal effort, slight recoil)
3. Tight (considerable effort, hand recoils markedly)

12.2.1.1.1.6 Body, Leg, and Foot Motions The previous motions were motions of the hand and arm. Body motions are motions of the leg-foot, horizontal torso motions, and vertical torso motions. A look through the tables of Body, Leg, and Foot Motions published by the MTM Association gives a reasonable explanation of these motions. MTM-1 Body, Leg, and Foot Motions are in Table 12.11.

Table 12.9 MTM-1 Release

Case	Time TMU	Description
1	2.0	Normal release performed by opening fingers as independent motion
2	0	Contact release

Table 12.10 MTM-1 Disengage

Class Fit	Height of Recoil	Easy to Handle	Difficult to Handle
1 Loose very slight effort, bends with subsequent move	Up to 1	4.0	5.7
2 Close normal effort, slight recoil	Over 1 to 5	7.5	11.8
Tight considerable effort, hand recoils markedly	Over 5 to 12	22.9	34.7

Table 12.11 MTM-1 Body, Leg, and Foot Motions

Type		Symbol	TMU	Distance	Description
Leg-foot motion		FM	8.5	To 4	Held at ankle
		FMP	19.1	To 4	With heavy pressure
		LM	7.1	To 6	Hinged at knee or hip in any direction
			1,2	Each additional inch	
Horizontal motion	Side step	SSC1	*	<12	*Use Reach or move time when less than 12. Complete when leading leg contacts floor
			17.0	12	
			0.6	Each additional inch	
		SSC2	34.1	12	Lagging leg must contact floor before next motion can be made
			1.1	Each additional inch	
	Turn body	TBC1	18.4		Complete when leading leg contacts floor
		TBC2	37.2		Lagging leg must contact floor before next motion can be made
	Walk	W_FT	5.3	Per foot	Unobstructed
		W_P	15.0	Per foot	Unobstructed
		W_PO	17.0	Per foot	When obstructed or with weight
Vertical motion		SIT	34.7		From standing position
		STD	43.4		From sitting position
		B,S,KOK	29.0		Bend, stoop, kneel on one knee
		AB,AS,AKOK	31.9		Arise from bend, stoop, knee on one knee
		KBK	69.4		Kneel on both knees
		ARBK	76.7		Arise from knee on both knees

12.2.1.1.1.7 Eye Motions Eye time must be considered when the eyes do direct the hands or body movements. There are two types of Eye time: eye focus time (time required to focus the eyes on an object and look at it long enough to determine certain distinguishable characteristics within the area which may be seen without shifting the eyes; and eye travel time (affected by the distance between points from and to which the eye travels, and the perpendicular distance from the eye to the line of travel). MTM-1 Eye Travel (ET) and Eye Focus are in Table 12.12.

12.2.1.1.2 MTM-2 The most accurate predetermined time system of the family of MTM is MTM-1. MTM-1 provides the most detailed description of methods but on the other hand it requires longer time for analysis. In MTM-1, it takes about 250 times the cycle time to analyze the task while in MTM-2 it takes about 100 times the cycle time, and in MTM-3 about 35 times the cycle time. For these reasons, MTM-2 and MTM-3 were developed. MTM-2 was developed by constructing motion combinations from basic motions of MTM-1. If the number of distance range is smaller and the cases of control are fewer than MTM-1, MTM-2 is best used for work that is not highly repetitive and for elements that are not less than one minute long. MTM-2 consists of nine manual motion categories. These nine motion categories and their symbols are as follows:

Motion Symbols:

1- Get G, 2- Put P, 3- Apply Pressure A, 4- Regrasp R, 5-Eye Action E, 6- Crank C, 7- Step S, 8- Foot Motion F, 9- Bend and Arise B.

The two key motion categories are Get and Put. Get and Put are the only two variable categories and only 37 time standards appear in the MTM-2 table (Table 12.12). The following is a brief explanation of the nine categories of MTM-2:

Table 12.12 Fastening Guidelines (MTM-2)

Guidelines	Justification
Minimize the number of fasteners	Most disassembly time is spent on fastener removal
Minimize the fastener removal tools required	Changing tools take time
Fasteners should be easy to remove	Saves time during disassembly
Fastener should be easy to access	Awkward movement slows down disassembly process
Snap fits should be obviously located and able to be opened using standard tools	Special tools may not be available
Try to use fasteners of material compatible with parts being joined	Allow disassembly operations to be avoided
If parts cannot be produced of compatible materials, make them easy to disassembly	The parts must be separated in order to recycle
Eliminate adhesives unless compatible with material of both parts being joined	Many adhesive cause contamination of parts for material recycling
Minimize the number and length of wires and cable used	Flexible elements are slow to remove, copper contaminates steel in recycling
Connections can be designed to break as an alternative to removing fasteners	Fracture is fast disassembly operation

12.2.1.1.2.1 Get Get is reaching with hand or fingers to an object, grasping object, and subsequently releasing it. Get is combining Reach, Grasp, and Release.

The variables of Get are:

1. Action employed: GA-no grasping motion; GB-closing fingers to gain control; GC-complex grasping.
2. The distance reached: 0–2, over 2–6, over 6–12, over 12–18, and over 18 in.
3. Object weight, or its resistance to motion: GW-1 add 1 TMU per 2 pounds moved.

12.2.1.1.2.2 Put Put is moving an object to a destination with the hand or fingers. Put is combining Move and Position. The variables of Put are:

1. The number of correcting motions or no correction required: PA-continuous smooth motion; PB-one correction; and PC-more than one correction.
2. The distance moved (same as for Get).
3. Object weight, or its resistance to motion (same as for Get): PW-1 add 1 TMU per 10 pounds moved.

12.2.1.1.2.3 Apply Pressure An action with the purpose of exerting muscular force on an object.

12.2.1.1.2.4 Regrasp An action performed by the hand with the purpose of changing the grasp on an object.

12.2.1.1.2.5 Eye Action An action with the purpose of recognizing a readily distinguishable characteristic of an object or shifting the aim of the axis of vision to a new viewing area.

12.2.1.1.2.6 Crank A motion with the purpose of moving an object in a circular path of more than half a revolution with the hand or fingers.

12.2.1.1.2.7 Step Either a leg motion with the purpose of moving the body or a leg to motion more than 12 inches.

12.2.1.1.2.8 Foot Motion A short foot or leg motion when the purpose is not to move the body.

12.2.1.1.2.9 Bend and Arise A bend, stoop, or kneel on one knee, and the subsequent rise.

MTM in –2 is Table 12.11.

12.2.1.1.3 MTM-3 The simplest MTM system is MTM-3. This system is used with long-cycle short-run operations. MTM-3 consists of four motion categories. Speed of the analysis with MTM-3 is seven times as fast as with MTM-1 and three times as fast as with MTM-2. The following is the four motion categories of MTM-3 and their symbols:

12.2.1.1.3.1 Motion Symbol 1, Handle H; 2, Transport T; 3, Step and Foot Motion SF; 4, Bend and Arise B.

The two key motion categories are Handle and Transport. These two categories are the only categories of the MTM-3 four categories, which have variable values. Only ten time standards appear in the MTM-3.

The following is a brief explanation of the four motion categories of MTM-3:

Handle

It is getting control over an object with the hand or fingers and placing the object in a new location. Handle is defined as gaining control of an object and placing it in a new location.

Transport

It is placing an object in a new location with the hand or fingers. Transport is defined as placing an object in a new location when the hand already has control of the object.

Step and Foot Motion

Step and Foot combines the S and F categories of MTM-2.

Bend and Arise

A bend, stoop, or kneel on one knee, and the subsequent rise. It is the same as MTM-2.

MTM-3 is in Table 12.12.

However, these standard times are altered by the weight of the component and whether one or both hands are used during the assembly of the components. Boothroyd et al. [2] presents some findings in terms of empirical equations to account for variations from the standard motions during assembly insertions.

Effect of Weight on Handling Time

The effect of increasing weight on grasping and controlling by one hand is found to be proportional increase of the basic MTM time. The total adjustment t_{hw} to handling time is given by

$$t_{hw} = (0.0125 + 0.011xt_h)W \tag{12.4}$$

where W is weight of present component in lb and t_h is predetermined basic time for handling a "light" component when no orientation is required for insertion. An average value for t_h is 1.13, and therefore the total time penalty due to weight would be approximately 0.0249W. For example, if the weight of the component is 10 lb, the penalty for the weight is $0.249 \approx 0.25$ s, however, this is for a very small movement of the part; if the movement is fairly large it has to be modified further. If the part requires both hands a penalty factor of 1.5 should be applied in those cases.

12.3 Design for Disassembly

The work in design for disassembly (DfD) is growing, but it is less extensive than efforts in DfA. In general, DfD efforts are intended to reduce the time and effort required to remove components or materials from a product. For example, complete nesting can slow disassembly by not providing a location for the disassembler to reach, grasp, or otherwise handle. Automobile manufacturers have also established vehicle recycling guidelines. Chrysler, Ford and General Motors have their own recycling guidelines.

However, existing green design for excellence (DfX) approaches mainly focuses on materials, manufacturing and assembly, and the use phases. The EOL phase is often overlooked or neglected due to lack of data and/or high degree of complexity/uncertainty. Integrating EOL aspects in the design phase can facilitate a more complete life cycle analysis (LCA) and life cycle cost (LCC) analysis.

The most common design approach for reducing EOL impacts is DfD. DfD offers potentially the greatest environmental and cost benefits.

One can notice that difficulties of disassembly are sown during assembly itself and we can say during product design stage as well. However, we must recognize that disassembly may not be economically feasible or justifiable but in the context of the green design, it is useful in reducing transportation and landfill costs. However, if regulations regarding recycling and reuse exist, as it does in Europe, companies have to attempt this. It may cost more to start designing for disassembly but later it can be beneficial. It has been seen in European countries that due to strict regulations regarding recycling and reuse, most of the electronic and electrical components/systems manufacturing companies have recycled thermoplastic and saved millions of dollars. The plastic casings of computers are easily recycled and reused. This DfD not only saved money but reduced materials wastage, greenhouse gas emissions, and landfill costs, and above all it improved the environment. Therefore, it is proper to think of environmentally friendly components at the design stage itself. It should start right from material selection so that it is easier for assembly and disassembly, as well as being environmentally friendly.

There are two types of plastics—thermoplastics and thermoset—that can be used in automobiles. Thermoplastics could be useful for interior parts such as dashboards, consoles, etc. At the EOL of the automobile these thermoplastics can be recycled and used to manufacture several components.

The problem of disassembly could be made easier by proper assembly. The components can be assembled either by joining or welding. But when parts are welded for assembly its separation or disassembly is difficult, time-consuming, and expensive. If it is joined by some fasteners, sometimes the disassembled component may not be usable or it will involve money to refurbish it to make it usable. However, a solution to this problem is snap-fit assembly, which can be disassembled easily. There is a problem with snap-fit assembly in that it may not be able to withstand certain kinds of forces, such as axial or torque forces.

If an assembly has a large number of parts, it would take more time and several tools to disassemble them, so fewer parts allow faster disassembly. This in turn could reduce the labor cost, tool cost, and energy costs.

12.3.1 Design for Disassembly Guidelines

Changing fashion trends and rapidly advancing technology has led to the shortening of life spans for many of today's products. Cellular (mobile) telephones, cars, and personal computers are just some of the products which are constantly re-engineered, restyled, and remarketed to meet our growing demands for better products. These demands, however, are placing a heavy burden on our natural and physical resources, particularly during manufacture and when the products reach their EOL. It is becoming clear that we may not be able to sustain such a fast rate of product life turnover without considering both the environmental and economic impacts this has on the planet.

Designers are becoming steadily aware of the problem and are employing techniques that allow them to design with greater responsibility; DfD is one such technique. It involves designing a product to be disassembled for easier maintenance, repair, recovery, and reuse of components/materials. As part of DfE and sustainable product design, DfD is becoming increasingly recognized as an effective tool by designers, manufacturers, and legislative boards alike.

Some countries have passed laws that bind the manufacturers to the principles of sustainable design. However, it is just not legal constraint but also an economically viable process. Reducing waste in the manufacturing and recovery processes using DfD techniques can significantly reduce production costs and allow for greater technical efficiency. Modular design principles within DfD techniques allow for greater flexibility during product development, shorter development timescales and reduced development costs. Implementing DfD into a design specification allows the product and its components to be better suited for reuse or recycling at its end of life, thus reducing resources required creating new products. When designing products with disassembly in mind, there are three important factors that must be considered by the designer:

- The selection and use of materials
- The design of components and product architecture
- The selection and use of fasteners

Sometimes a limiting factor in recycling is the economics of separation into pure material. The separation could be either manual or mechanical. For manual separation to be carried out, there must be a significant value retained in the recycled product in order for the separation to be economically feasible. This scenario is applicable to most other products constructed from subassemblies consisting of a variety of materials. Different guidelines apply, depending on whether manual or mechanical separation is carried out. As a general rule, however, products with a low material removal rate (MRR)—less than 2.26kg/minute for plastics—benefit from mechanical disassembly, whereas it is more economical to manually disassemble products with a high MMR (approx. 4.5 kg/min). The amount of material (grams) that has to be removed per minute for recycling to be cost neutral for manual disassembly (based on West European hourly rates and material prices, September 1995):

Precious Metals		Metals		Plastics		Glass
Gold	0.05*	Aluminum	700	PC, PM	350	Glass 6000
Palladium	0.4	Iron	50,000	PEE	250	
Silver	5.1	Copper	30	ABS	800	
				PS	1000	
				PVC	4000	

* Data shows that if materials are removed more than shown in table above, then it is economically feasible.

12.3.1.1 The Design Problem for DfD

At the conceptual stage, it is the endeavor of the designer to produce a "good design" or even attempt to deliver the "best" design. As we know, this means design must be marketable, have "manufacturability," be easy to assemble and disassemble, be recyclable and reusable, and it has to be environmentally friendly. If we start with the concept of DfE, the material selected should be biodegradable, easier to manufacture and machine, and recyclable. If we look at list of biodegradable materials useful for engineering components, the list is nonexistent. There are materials which are easier to machine compared with others but cost may be higher and even the equipment and machine tool may differ. Most common DfE practices are design for remanufacture, design for energy efficiency, and design for minimization of carbon footprint. The goals for environmental requirements are to minimize the use of natural resources, energy consumption, waste generation, health and safety risks, and ecological degradation.

However, DfD is not a panacea for all problems but applications of DfD principles in mechanical components offers a glimmer of hope that shows us that sophisticated environmental planning is possible. DfD is a subset of the emerging environmental redesign movement that assumes that pollution is a function of design. In the United States, the notion that design has anything to do with pollution is not immediately obvious. Global warming is the product of poor planning and design by geologists, mining engineers, shippers, civil engineers, automotive designers, or all design engineers. The ozone hole is courtesy of refrigeration engineers who were merely trying to give the world a safe way to preserve food and medical products with refrigeration. In fact, virtually everything that can be considered pollution is the product of intense planning and design, down to the last bubble-pack and plastic milk carton clogging our waste dumps.

When it is understood that pollution is a function of planning and design, a whole host of options opens to the design engineers that never existed before. It also exposes the shortcomings of traditional engineering design curriculum where we never considered the problem of recycling, reuse, and, if an item is nonrecyclable, sending it to landfill or incineration.

This requires quite different thinking in a modern engineering education. Trying to organize industrialization on the cheap may be the root cause of our environmental dilemmas. Education has its vital role in recycling but without industrial redesign, many efforts merely result in waste piling up in warehouses rather than landfills.

To understand the DfD regulations, the importance of design, manufacturing issues and environmental issues should be integrated in education. It is well known that German cars already conform to DfD regulations, and auto makers have established sophisticated recycling facilities, while in the US, DfD is still an essentially unknown concept. Industrial-environmental redesign is rapidly becoming the environmental strategy of choice in Europe and, increasingly, Japan.

The design has emerged as the dominant economic environmental factor and good design decisions lead to prosperity. Because energy cost is a design problem, the costs of premium energy-saving materials became a small fraction of the total costs. The materials requiring less energy during processing would bring much bigger energy efficiencies during the life cycle of the product.

DfD is nothing more than the planned process for recycling; therefore, everything that is manufactured must have a plan for its disposal when it is designed. This is the true concept of DfD. Of course, what makes DfD so effective is that environmental problems are placed in the hands of the product designers themselves. It is a design process that allows for the easy recovery of products, parts and materials when a product is disassembled to maximize economic value and minimize environmental impacts. This can be done through reuse, repair, remanufacture and recycling of the product or its parts. DfD should enable the separation of product parts without damage.

Already, basic DfD strategies have emerged: eliminate adhesives (things that are glued together are hard to separate), eliminate coatings (same reason), reduce the number of materials, use easily recycled materials, employ reversible fasteners, and reduce parts count. A product that is easy to disassemble is easy to repair or modify. Waste can be reduced if a large product does not have to be scrapped because a small part fails or becomes obsolete. The time required for separation of parts was identified as a key contributor to the overall cost of recycling.

12.3.2 Origin of the Concept of DfD

Disposability was always a concern to maintain a throughput economy. However, with changing regulations in European countries, particularly with the Waste Electrical and Electronic Equipment (WEEE) directive in Europe (12.1) there was a shift toward extended producer responsibility. However, with depletion of resources and increase in price of essential materials, industries have been forced to think in the design for life cycle of a product and consider recycling and reuse. This tectonic shift has been recognized as a sign of things to come by global manufacturers, driving interest in the DfD strategy.

DfD has become a growing phenomenon within manufacturing industries as greater attention is devoted to the management of the EOL of products to facilitate its disposal or reuse. The effort is driven by the increasing waste of large amounts of consumer goods, and the resulting pollution and the loss of materials and energy that these products contain.

The designer is now confronted with the matter of creating products that have parts or components that are easy to take apart, so that the product components can be recycled or reused conveniently. This feature also lends itself to easier maintenance. Such products have these common features:

1. They are designed so that all components can be easily separated, handled, and cleaned to permit economical recycling.
2. They include two-way snap fits, or have break points on the snap fits, which simplifies both assembly and disassembly.
3. They clearly identify separation points, which also eases disassembly.
4. They reflect tight tolerance design principles so that parts interlock, thereby substantially reducing the disassembly time.

Although most products can be disassembled eventually, lengthy disassembly and the cost of disassembly is likely to be greater than the revenue gained through recycling the parts and materials from the product. It is for this reason that designing products for easy disassembly has increased in popularity, enabling more of the product to be recycled economically.

However, the implementation of the DfD concept is not easy. It requires that engineers should think of the life span of all components in the system or subsystem along with their maintenance requirement and reassembly after disassembly.

In general, designers should attempt the following:

- Use detachable joints such as snap, screw or bayonet instead of welded, glued or soldered connections.
- Use standardized joints so that the product can be dismantled with a few universal tools, for example, one type and size of screw.
- Position joints so that the product does not need to be turned or moved.
- Dismantling guidelines on the product should show how it should be opened nondestructively, for example, where and how to apply leverage with a screwdriver to open snap connections.
- Place parts that are likely to wear out at the same time in close proximity so they can be easily replaced simultaneously.
- Indicate on the product which parts must be cleaned or maintained in a specific way, for example, color-coded lubricating points.

In addition to the techniques given above, authors Dowie and Kelly [4] provided fastening and connection guidelines shown in Table 12.12 which may be employed in DfD.

Two areas which may be considered subsets of DfD and which should also be areas of design concern are DfE and DfR. DfE is a systematic consideration of design issues related to environmental and human health over the life cycle of the product.

DfR deals with the reclamation of product value from materials at the end of their product life, or EOL. Below are listed a number of areas associated with DfR:

- Designing for ease of disassembly, to enable the removal of parts without damage
- Designing for ease of purifying to ensure that the purifying process does not damage the environment
- Designing for ease of testing and classifying to make it clear as to the condition of parts that can be reused and to enable easy classification of parts through proper markings
- Designing for ease of reconditioning; this supports the reprocessing of parts by providing additional material as well as gripping and adjusting features
- Designing for ease of reassembly to provide easy assembly for reconditioned and new parts

There are a number of benefits of achieving efficient disassembly of products as opposed to recycling a product by shredding, which include the following:

- Components that are of adequate quality can be refurbished or reused.
- Metallic parts can be separated easily into categories that increase their recycling value.
- Disassembled plastic parts can be easily removed and recycled.
- Parts made from other material such as glass or hazardous material can easily be separated and reprocessed.

The first requirement for recyclability of a product is that there must be a demand for the material. Therefore, one should design with materials that have a high demand such as metals and plastics. With regard to plastics, currently markets exist for HDPE, PET, LDPE, PS, PP, and PVC. The designer should examine current plastic recycling figures for demand and usage. There are essentially three kinds of plastics:

Thermoplastics: It becomes soft when heated and solidify when cooled. They are molded or shaped with pressure in plastic state. All thermoplastics are recyclable.

Thermosets: Thermosets have no melting point and burn before melting. They can be softened and molded or set when heated. They are branched polymers with three-dimensional, cross-linked networks and they are not recyclable.

Elastometers: They are rubber-like materials and not recyclable.

Types of Plastic by Resin	
PET (polyethylene terephthalate)	For example, soda bottle, water bottle, and food containers
HDPE (high-density polyethylene)	For example, milk and water jugs, detergent bottles
PVC (polyvinyl chloride)	For example, shampoo and window cleaner bottles
LDPE (low-density polyethylene)	For example, trash bags, six-pack rings, flexible lids
PP (polypropylene)	For example, lids, closure caps, snack food wrap
PS (polystyrene)	For example, Styrofoam, clear brittle cups

The second major concern of DfR is a measure of how easy it is to remove the materials from the product. A rating system can be applied to help the designer assess the disassembly and recyclability characteristics of product designs. Examples of such rating systems are given in Tables 12.13 and 12.14. The examples are taken from the automotive industry, one of the current leaders in DfR.

Ashby [5] developed the recycle fraction chart for most of engineering material as shown in Figure 12.3.

Therefore, the design for dismantling (also known as DfR, or DfD) includes the practices to optimize the way a product will be treated at EOL, and to optimize the separation of components and materials for their recovery (repair, recycling, energy recovery). The main objective of design for dismantling is to integrate issues related to the EOL of a product in the design stage. It is actually more effective to adapt

Table 12.13 Separability Ratings

Rating	Description	Example
1	May be disassembled easily by manual techniques, less than 1 min	Pull apart parts
2	May be disassembled with effort manually, less than 3 min	Instrument cluster, radio
3	May be disassembled with effort and some mechanical separation or shredding to separate. The process has been fully proven.	Engine, sheet metal, uncorroded screws
4	May be disassembled with effort and some mechanical separation or shredding to separate. The process is under development	Instrument panels, corroded screws, adhesive
5	Cannot be disassembled, no known process of separation.	Heated backlight

Table 12.14 Recyclability Ratings

Rating	Description	Example
1	Part is remanufacturable	Starter motor, alternator
2	Material in a part is recyclable with clearly defined technology and infrastructure	Most metals, PET, HDPD
3	Material is technically feasible to recycle; infrastructure to support recycling is not available	Most thermoplastics, glass, thermosets
4	Material is technically feasible to recycle with further process or material development required	Armrest, airbag modules, single metal with single thermosets
5	Material is organic, can be used for energy recovery but cannot be recycled	Multithermoplastics, wood products
6	Material is inorganic no known technology for recycling	Heated glass, fiberglass

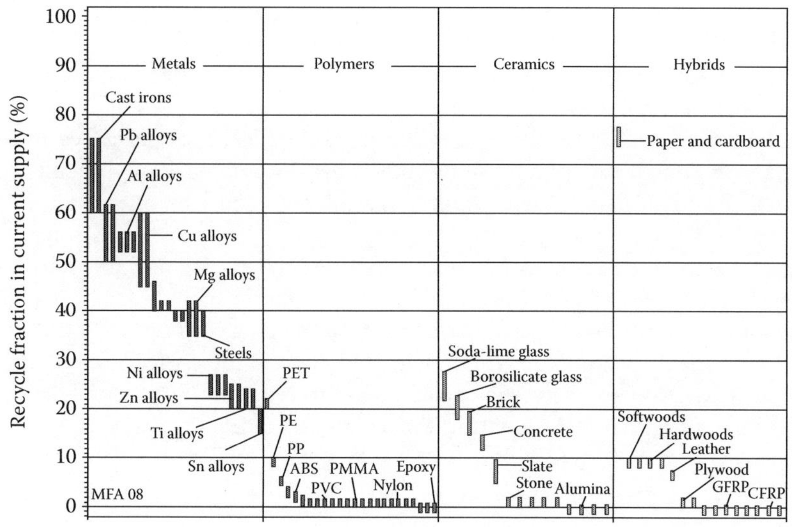

Figure 12.3

Recycle fraction chart. (Michael F. Ashby, Materials and the Environment)

a product based on a chain of dismantling steps (recycling) that might be applied to treat it at EOL than the other way around. It is not possible for economic or technical reasons to separate and recycle all of the components of a product so it is therefore appropriate that the choice of materials and assembly should be best suited for dismantling. However, an upgrade/restoration could mean the product is not at the EOL. It can be upgraded, for example, if we change one or more components of a computer processor, graphics card, etc. and continue using it. It can be refurbished; this is the case of a return of a product during its warranty period. In case of reuse and re-employment, the product has reached the end of its first life (e.g., the owner wants to discard it) as it no longer works, but it can be repaired. After repair, the product can be reused. However, recycling means the product does not work and cannot be repaired. In this case, the product will be recycled.

12.3.3 Some Existing Separation Techniques

- Riddle: Small-size particles pass through a grid while larger particles do not.
- Often used with multiple riddles in a cascade fashion.
- Wind sifter: Solids are separated by gravity in a wind stream according to density, form and dimension; light particles are blown away while heavier particles settle.
- Gas cyclone: Separates dust from airflow by means of a centrifugal force.
- Hydroclone: Same as gas cyclone but used to remove particles from water.
- Float-sink: Solid particles are separated according to density by immersion in a fluid of intermediate density.
- Magnetism: Ferromagnetic elements are lifted away by means of a magnetic field.
- Eddy current: An eddy current is passed through the mixture of metal particles in the presence of a fluctuating magnetic field. Conducting particles are deflected from the flow path while nonconductive ones are not.
- Electrostatic precipitator: Small particles are electrically charged and then removed by an electric field.

The natural tendency of things is to mix rather than unmix (second law of thermodynamics) and separation can be performed at cost of energy. There is a theoretical minimum amount of energy, E_{min}, required to separate an element in concentration c in a containing material (Figure 12.4):

$$E_{min} = -\frac{RT}{c}\left[c\ln c + (1-c)\ln(1-c)\right]$$

where RT is room temperature.

12.3.4 Distinction between Manual and Mechanical Separation

As was seen in the examination of the different vehicles, it is possible to design for manual separation without facilitating mechanical separation. The best example of the opposite situation, designing for mechanical but not manual separation, is metal recovery for the vehicle as a whole. The steel can easily be separated once the vehicle has been shredded, but accomplishing this by hand would be quite difficult. These distinctions occur because manual and mechanical separations have different requirements:

- Manual separation requires more effort in improving the disassembly and sorting process for the component or assembly, since the primary limiting factor for manual separation is the (labor) time required for this separation. If the materials in the part being considered require significant time to separate and identify, manual separation will not be economically feasible. Guidelines appropriate for manual separation are those that suggest ways to reduce the (manual) disassembly effort and those that make visual material identification easier and quicker.

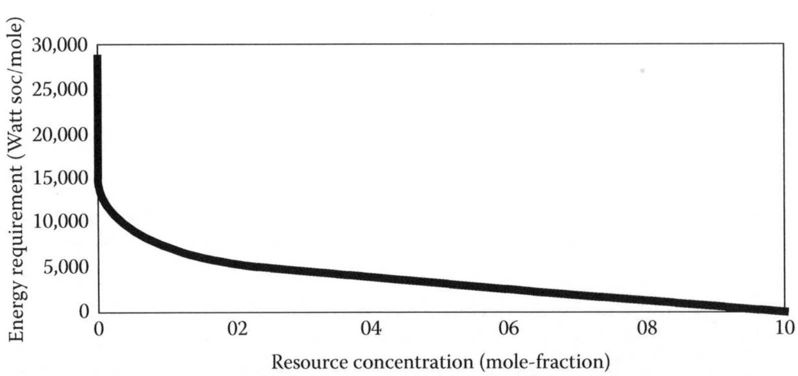

Figure 12.4

Relationship between energy requirement and resource concentration.

- Mechanical separation, on the other hand, requires more effort in creating an assembly or component that can be separated quickly and easily into pure streams of materials based on material properties. If the materials used do not have distinctive properties that can be used for separation, economical recycling will not be possible.

Disassembly effort and visual identification are not important for mechanical separation, but material selection is critical to this separation effort. Many design guidelines, such as reduction of the number of materials used, standardization of material types, and the use of recyclable materials, are applicable for either type of separation. There are, however, a significant number of possible design techniques which are only useful for one type of separation.

12.3.4.1 Fastener Selection

A number of different techniques can be used when designing for manual separation, all intended to reduce the amount of time it takes to dismantle the components. Specifically, these include reducing the number of fasteners, communizing the fastener types, using snap fits, and avoiding nonremovable fasteners. When considering mechanical disassembly, the only concern is the separability of the fastener material from other materials in the component because the fasteners will be shredded with the component. Accordingly, the number and type of fastener used is not important. Instead, integral fasteners and material-compatible fasteners are greatly preferred. If this is not possible, ferrous fasteners are preferred in plastic assemblies because they allow for easy magnetic separation. Some of common screws in assembly of engineering systems are shown in Figure 11.28.

12.3.5 Screw Disassembly Procedure

The screws like round head, flat head, fillister head, oval head, truss head, binding head, as well as hexagonal head, need different sizes and types of screwdrivers during disassembly. To disassemble them from some assembled system will need pressure and then some torque. Referring to MTM-1 (Tables 12.3 through 12.11), we can see the time elements for Reach screw driver, Grasp the screw driver, Move to screw, Position the screwdriver over the screw, Apply pressure, Turn, Disengage, and finally remove the screw. Time from MTM-1 table in TMU:

Motion	TMU	Reference
Reach to screwdriver	10.1	Flathead screws, jumbled-up screws, Table 12.3
Grasp screwdriver	12.9	Object less than 1/4 in, search and select, Table 12.7
Move to screw disassembly	30.7	Move 30 in, Table 12.4
Position the screw driver	47.8	Table 12.8, exact symmetry
Apply pressure	16.2	Table 12.6, apply force, full circle
Turn	16.5	Turn 180°, Table 12.5
Disengage	11.8	Close normal effort, Table 12.8
Total time to disassemble one thread	146.0	

Example 12.1: Design for Disassembly

The time to disassemble the screw will depend on the length of the screw, which depends upon the thickness of two parts it is supposed to assemble, say it is 1 in. If the nominal diameter of the screw is 0.125 in. then there are 40 threads/in. (UNC) threads or 44 threads/in. (UNF) threads. We assume UNC threads for this type of thread, the time to unscrew the screw is 44(140) = 1960TMU = 70.56 s. Finally, the screw needs to be removed and it would depend on the kind of fit, loose or tight. The tight fit and difficult to handle example is considered here for 34.7 TMU. If we assume there 15 screws in this assembly of two steel plates, we can calculate the total disassembly time for 15 screws; disassembly time = 15(70.56) + 15 (34.7)(0.036) s = 1077.138 s = 17.9523 min. Now let us look at the plates to be disengaged and moved to safer place, where they may be recycled or reused.

We assume here plates of 24 in. × 24 in. × 1/2 in. steel plates of weight 81.2 lb weight. These two plates are heavy items and to move those to recycle or reuse bins may require help from one more person and both hands have to be used. It is getting control over an object with the hand or fingers and placing the object in a new location. Handle is defined as gaining control of an object and placing it in a new location. In MTM-1, these steps are: Reach, Grasp, and Move. If we take predetermined times from Tables 12.3, 12.7, 12.13; time for moving one plate is = (23.2 + 2.0 + 22.7) = 48TMU for moving the plate by 30 in. However, these plates are heavy at 81.2 lb, so the time elements modified for weight = 1.5 × 48 × 81.2 = 5846.4 TMU = 210.4704 s = 3.5 min. For both the plates, the time to move them would be 7 min. Therefore, the total time for disassembly would be

$$T_{\text{disassembly}} = 18 + 7 = 25 \text{ min}$$

which seems to be reasonable time to unscrew all 15 screws with flathead screw driver and move the plates about 30 in from the workbench. The screws may or may not be reused. It normally depends on the life of screws, the nature of it assembly, the corrosion, etc. For example, we assume only 25% of the screws are reusable and the rest have to be recycled. The Table 12.15 shows the recyclability, detaching behavior, joining behavior, including the load-carrying capability of nuts, bolts, and screws. All these characteristics are average, except good load-carrying capacity and both static and fatigue strength. Table 12.15 shows the degree of difficulty in assembly as well as disassembly. The degree of difficulty shows the relative degree of disassembly of the nuts and screws, which is normally in the range from 1.5 to 2.0. The time of disassembly should be multiplied by the degree of difficulty (DoD) factor to accurately and appropriately the disassembly time. As we have assumed only 25% screws are reused, the rest, or 75%, are recycled after shredding. The Table 12.16 shows the shredding behavior of fasteners and materials used in the product assembly. We can now estimate the benefit of disassembly, reuse, and recycling. But before we estimate the benefit in dollars, let us assign this disassembly process a separating rating (SR) from Table 12.12 and recyclability rating (RR) from Table 12.13. We can comfortably assign SR as 2 and RR also as 2 from Tables 12.12 and 12.13. It only shows how easy or difficult it has been to disassemble the parts and recycle them. As SR and RR numbers increase, so does difficulty in disassembly and recycling of the parts and it would affect the cost of disassembly and decrease the benefit or profit of disassembly. The following steps can be followed in estimating the EOL or the recycling and reuse benefits:

$$\text{Reuse value} = \text{Cost of component}\left(\$\right) - \text{Miscellaneous cost}\left(\$\right)$$

In this hypothetical example, 25% of crews are reused and so only three screws are reused. The miscellaneous cost is composed of collection plus reprocessing costs. In this particular example, no extra effort is needed in collecting the parts nor is any money spent in reprocessing the collected parts, so it is zero (Table 12.16).

Table 12.15 Fastener Selection Guidelines

Factors Affecting the Dismantling	Improvement Opportunities
The type and number of fasteners	✓ Select fasteners easy to disassemble
	✓ Provide access to linking elements
	✓ Reduce the number of fasteners
	✓ Use of standard fasteners
	✓ Prefer snap-fit fasteners
	✓ Reduce the number and type of fasteners in an assembly
	✓ If metal fasteners are used, prefer ferromagnetic metals to allow a magnetically sorting during dismantling (but beware of corrosion that could make disassembly difficult)
The visibility of fasteners	✓ Facilitate access to fasteners (e.g. through holes)
	✓ Mark not visible links
	✓ Avoid using hidden links
The conditions of disassembly	✓ Permit an automated disassembly
	✓ Favour the use of standard tools
	✓ The use of materials with shape memory allows an active disassembly (e.g. depending on temperature)
	✓ The presence of fluid or gas can disrupt the operations of disassembly

Table 12.16 The Standard Disassembly Operations with Their Letter Codes

Unscrew-Un	Turn-Tu	Wedge/Pry—We	Cut—Cu
Remove-Re	Flip—Fl	Deform -De	Push/Pull- Pu
Hold/grip-Ho	Saw-Sa	Drill—Dr	Hammer—Ha
Peel—Pe	Clean-Cl	Grind—Gr	Impact - Ia

Therefore, the reuse value is

$$\text{Reuse value} = \$ \text{ per part reused} \times \# \text{ of parts reused}$$

$$= \$3.00 \left(\begin{array}{l} \text{here for illustration \$ per part is assumed} \\ \text{as \$1.00, which is not true in market} \end{array} \right)$$

Remanufacture value represents the value of component after disassembly the parts are reprocessed or refurbished before reusing them. The remanufacturing cost sometimes may involve machining, cleaning, removing paint, or cleaning for any corrosion on the part. Just cleaning the part with brush and maybe putting some lubricant all around the part may cost money in terms of labor and material costs. However, in this example, we assume that the three screws are cleaned and lubricated and let the remanufacture value be

$$\begin{aligned} &\text{Remanufacture value} \\ &\quad = \$ \text{ per component} \times \# \text{ of components} \\ &\qquad - \text{remanufacture cost/component} \times \# \text{ of components} \\ &\quad = \$3.00\$ - 0.10 \times 3 = \$2.70 \end{aligned}$$

Primary recycle value here consists of 12 screws and two steel plates. Here the miscellaneous cost of collection and reprocessing is again assumed to be zero. The primary recycles value:

$$\begin{aligned} &\text{Primary recycle value} \\ &\quad = \# \text{ of screws} \times \text{weight/screw} \times \text{cost of recycled material} \\ &\qquad + \#\text{of steel plates disassembled} \times \text{weight per plate} \\ &\qquad \times \text{cost of per pound of material reused} \\ &\quad = 12 \times 0.0625 \,\text{lb/screw} \times \$0.50/\text{lb} + 2 \times 81.2 \,\text{lb/plate} \times \$\ 0.20/\text{lb} \\ &\quad = \$\ 32.9 \end{aligned}$$

Here, screws are of alloy steel and plates are of cast iron. The shredding cost (C_s) needs to be accounted for in the EOL cost as well.

$$\begin{aligned} C_S &= C_{Sh}(S_t)(W_S) \\ &= \text{hourly shredding cost/h} \\ &\qquad \times \# \text{ of hours used to shred one lb of material} \\ &\qquad \times \text{weight of shredded material} \end{aligned}$$

Table 12.17 shows the value of shredded material. Here again assuming some values to show the shredding cost.

$$C_s = \$30/\text{h} \times 1/2 \text{ hr/lb} \times 163.15 = \$2447.25$$

Table 12.17 Degree of Difficulty of Disassembly

Separate Fastener			
Fastener	Degree of difficulty	Integrated fastener	Degree of difficulty
Nail with head	1.5	Cylindrical snap fit	1.6
Nail w/o head	1.8	Cantilever snap fit	1.3
Screw/bolt standard head	1.4	Seam/crimp joint	1.6
Screw/bolt special head	2.2	Interferance fit	1.8
Nut and bolt	2.1	Integrally threaded part	2.2
Rivets/staples	2.0	Socket/plug	1.2
Retaining rings/circlip	2.5		
Tapes	1.7		
Adhesive	2.1		
Welded	4.0		
Valero/zipper	1.0		
Releasable clips	1.8		

The shredding cost seems very high; however, if it is manual shredding it may be acceptable. One aspect can help explain the high cost of shredding is that plates are of cast iron, whereas screws are of alloy steel and they need to be separated. As we know, if the mixture of materials after shredding is of, say, plastics and steel, magnetic process may be used to separate steel and plastics can be picked up later. Therefore, it is quite clear here that manual separation is not economically viable or feasible. This may put the recycling process itself on the backburner.

In this case, the shredded material is not going for incineration; it is just reprocessed and sent to the entry point of the production process.

12.3.6 Embodied Energy in Recycling

However, we will show the value of embodied energy in the recycling. The total material of alloy steel and cast iron are 0.75 and 162.4 lb respectively. The embodied energy of fresh material if used in a production or manufacturing process would be

$$\begin{aligned}
&\text{Embodied energy of fresh material}\\
&\quad = (0.72\ \text{lb} \times 0.454\ \text{kg/lb} \times 35\ \text{MJ/kg})\\
&\qquad + (162.4\ \text{lb} \times 0.454 \times 17\ \text{MJ/kg}) \times 948\ \text{Btu}\\
&\quad = (11.5 + 1253.5) \times 948 = 1.2 \times 10^6\ \text{Btu}
\end{aligned}$$

The calculation above shows that we will need 1.2×10^6 Btu to produce the materials for screws and the plate. However, if we recycle the materials, we can calculate the saving in energy. In this case, these materials will have to go steel processing before using them to manufacture parts.

The energy required to reprocess recycled materials is

$$\begin{aligned}
&\text{Embodied energy for recycled material}\\
&\quad = (0.72 \times 0.454 \times 10\ \text{MJ/kg} + 162.4 \times 0.454 \times 5.2\ \text{MJ/kg}) \times 948\\
&\quad = (3.3 + 383.5) \times 948\\
&\quad = 0.37 \times 10^6\ \text{Btu}
\end{aligned}$$

We can see the big savings in terms of embodied energy $= (1.2 - 0.37) \times 10^6 = 0.83 \times 10^6$ Btu. We can also estimate the savings in terms of dollars.

Normally, energy drawn from an electric grid, coal, heavy oil, and so on differ in price but every one has a different environmental impact in terms of carbon footprint, which is supposed to be the main cause of global warning. The energy price used here is \$0.035/MJ and so the savings in terms of dollars $= (0.83 \times 10^6/9.48) \times$

$ 0.035/MJ = $31.00. All the embodied energy data are taken from Ashby's book. We will estimate the carbon footprint before and after recycling.

12.3.7 Carbon Footprint Estimation

Carbon footprint of fresh material $= (0.72 \times 0.454 \times (2.125\ \text{kg/kg}) + 162.4 \times 0.454 \times (1.1\ \text{kg/kg})) = (0.7 + 81.1) =$ 81.8–82 kg of CO_2 emitted in atmosphere when we use fresh materials in this process. However, if we use recycled materials, the CO_2 emitted in atmosphere is

$$\text{Carbon footprint of recycled materials}$$

$$= (0.72 \times 0.6\ \text{kg/kg} + 162.4 \times 0.31\ \text{kg/kg}) \times 0.454 = 23\ \text{kg}$$

This is the amount of CO_2 emitted in the atmosphere if we use recycled materials. This shows that there is reduction of (82 − 23) kg = 59 kg of CO_2. For this small disassembly process, the reduction in terms carbon footprint is very significant or quantifiable. The design for disassembly, design for recyclability or, simply, design for environment, are successful tools for estimating carbon footprint, energy savings, and so on.

It is interesting to note that during incineration the energy produced can be tapped and used for some useful purposes. If the parts or components are likely to go to grave (landfill), this may require special preparation for a different kind of landfill, as for hazardous or nonhazardous material. Obviously, the landfill preparation for hazardous material will cost more compared to nonhazardous material. The landfill area will depend on the volume or weight of the material multiplied by the landfill cost. Sometimes, there are special handling costs involved for a special material for the safety of workers involved in the landfill operations. We should look into the collection costs, which are composed of handling, transportation, and storage cost. In this example, this cost is not considered.

12.3.7.1 Material Selection

Perhaps the most interesting distinction in the material selection guidelines is that between component and assembly. If an assembly containing two polymers is being considered for manual separation, the designer should attempt to create components made of one material or the other, so that the components do not need to be disassembled as well. However, if the same assembly is being designed for mechanical separation, it does not matter whether the individual components of the assembly are mixed materials or not as the entire assembly will be shredded anyway. For manual separation, large masses of a single material are important. For mechanical separation, reducing the total number of different materials in the assembly is more important. In addition, it is extremely important to note the specific material properties to be used in the mechanical separation and to make sure there is a sufficient distinction to allow easy and accurate separation. A metal plate riveted to a plastic component would be extremely difficult to disassemble manually, yet it can easily be separated using mechanical means.

The technical choices made during the design phase of a product are key to the "performance" of a product during its dismantling. It is estimated that only 10%–20% of the costs and benefits of recycling are due to the optimization of the recycling process, while 80%–90% of these costs are determined by design [6].

More generally, the challenges of designing for dismantling are

- Preserve the raw material resources
- Reduce the amount of nonrecyclable waste sent to landfill
- Facilitate the achievement of recycling regulations requirements

12.3.8 Principles of Design for Dismantling

The design for dismantling revolves based on three key principles:

- Reuse
- Re-employ
- Recycle

12.3.8.1 Use and Materials Selection

Considering the amount of recyclable material in a product and the redemption price of recycled material. Recycling of a product will only be relevant if it is possible to separate it (manually or mechanically) in sufficiently pure material flow. The choice of materials is conditioned in particular by the type of treatment that will be applied at the product at the EOL. Depending on the redemption price of recycled materials, the choice for manual or mechanical removal will be more appropriate. This choice will depend on the amount of the different materials that can be extracted from a product, but also the possible rate of extraction of these materials.

The table below gives an order of magnitude for some materials, of the quantity (in grams per minute) of material to be extracted from a waste stream to make the activity economically viable.

Precious Metals		Metals		Plastics		Glass	
Gold	0.05*	Aluminum	700	PC, PM	350	Glass	6000
Palladium	0.4	Iron	50,000	PEE	250		
Silver	5.1	Copper	30	ABS	800		
				PS	1000		
				PVC	4000		

Example 12.2: Recycling Profitability

In order to make ABS recycling profitable, it must be possible to sort 800 g of ABS in a minute from a waste stream. In contrast, to make extraction and recycling of gold content in waste profitable, just 50 mg (0.05 g) should be extracted in the same period of time.

Parts Marking (Plastic)

Concerning plastic parts, one of the key design measures for dismantling is the marking of parts. This marking is defined by ISO 11469 and ISO 1043-1 to ISO 1043-4.

12.3.8.2 Marking Can

Permit the identification and allow quickly (hand) sorting of a part depending on the material type.

To be practically effective, marking must remain visible throughout the life of the part that wears it; this is why it is recommended that the marking is carried out during molding of the part, or by etching. Labels are especially to be avoided because they can cause contamination during recycling. Painting may disappear during use as well.

The ISO standards do not define a specific location for the marking of plastic parts. However, to be relevant/usable, it should be visible at any time of the product life cycle and especially during the phases of manufacturing, repair, or EOL of the product. It is therefore possible to produce several markings in various locations. Obviously, this should not compromise the aesthetic considerations of the product.

12.3.8.3 Choose Materials with Physical Properties That Can Be Separated

Regarding the choice of plastic, to allow separation by densimetric sorting of different polymers, the polymers should have a difference in density of at least 0.15 g/cm^3.

Moreover, laminated materials are to be avoided because of the difficulty in separating the materials used. Finally, in cases where the materials cannot be separated, they have to be compatible for recycling (see sheet design plastic parts).

12.3.8.4 Use Surface Treatments Wisely

For metals, the design rules for dismantling are to avoid the use of surface coatings, especially when they reduce the recyclability of materials (reduced performance of the recycled material). Moreover, the alloys bring additional elements into the recycled material which may be undesirable; it is therefore necessary to limit their use (e.g., selective surface treatments).

12.3.9 Parts Design and Structure of Products

The rules for the structure of a product adapted to disassembly are

- Create a hierarchical structure.
- Create a modular design.

- Gather wear parts for their simultaneous replacement.
- Reduce the number and type of operations required for the dismantling the product.
- Standardize/reduce the number of components.
- Reduce the number of materials in an assembly.
- Divide the product into functional modules and subassemblies (replaceable, repairable).
- Do not paint plastic parts (identification problems or contamination during recycling).
- Avoid as much as possible the use of parts with laminated materials.

12.3.10 Selection and Use of Fasteners

The third area of design for dismantling is the choice and use of fasteners and connection types. The choice of fasteners relates both to the ability of a product to be repaired (time and associated cost) in its ability to be dismantled or crushed at EOL (separation and release of different materials) (Table 12.15).

Whatever the fasteners used, the disassembled parts must have the following characteristics: good accessibility, low volume, low weight, toughness, not dangerous.

Fasteners must take into account the end use of the parts in order to choose the most appropriate ones and whether they are re-employed/reused, refurbishment, recycled, and incinerated (Figure 12.5).

The following compares different types of fasteners in terms of their behavior against recyclability, disassembly, and accuracy of the link or loading capacity. The basic operations of disassembly are unscrew, remove, hold/grip, turn, flip, saw, clean, wedge/pry, deform, drill, grind, cut, push/pull, hammer, and, of course, inspect. It is obvious all these operations are not required in every disassembly. Each operation is a cost driver. Whitney [7] presents an equation for calculating time in seconds for disassembly operations based on the degree of difficulty score. The degree of difficulty score ranges from 1 to 10. The Table 12.16 presents the degree of difficulty each operation required for disassembly.

Each task is assigned quantitative difficulty scores for five aspects of task performance. The scores are based on a scale of 1 (easy) to 10 (difficult), and are obtained from charts that include ratings for doing each task under various conditions. Since the difficulty ratings are based on task performance time (see

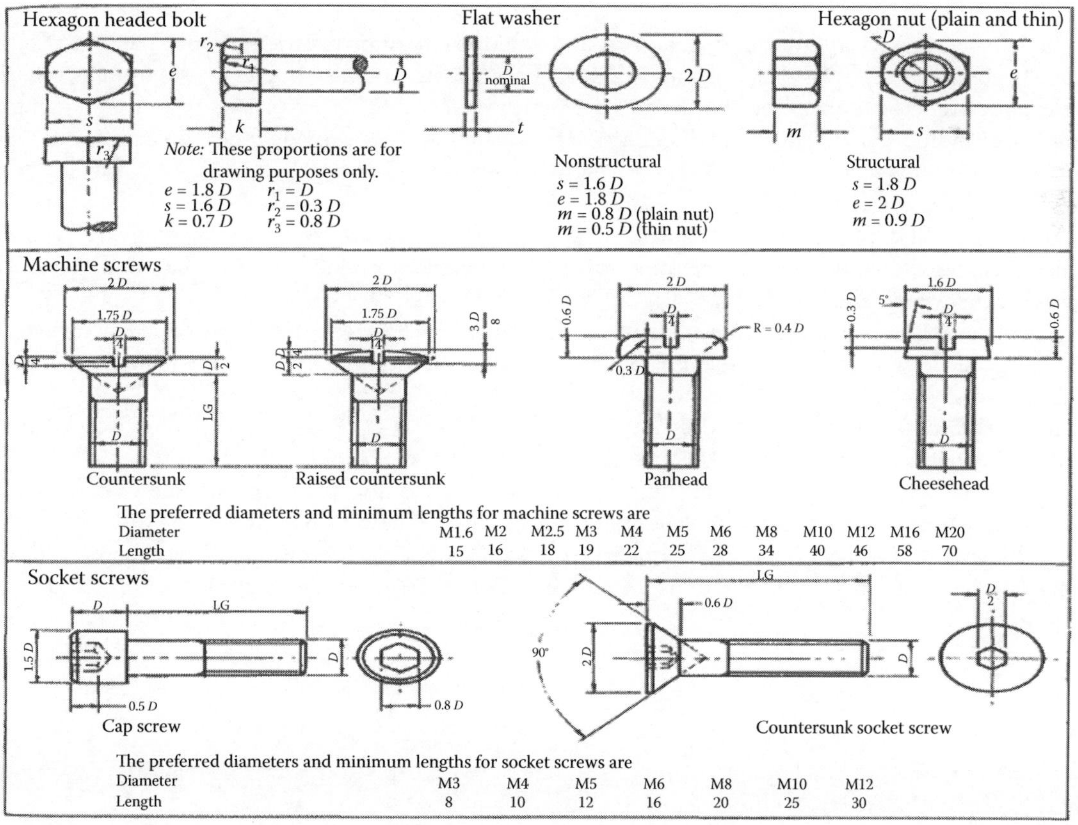

Figure 12.5

Bolts, nuts, and screws for engineering assembly.

Section 12.4 and Kroll and Hanft [8]), scores higher than 10 are possible. The four categories of task performance are

- Accessibility: a measure of the ease with which a part can be accessed or reached by the tool or hand. This is an indication of whether or not adequate clearance exists.
- Positioning: the degree of precision required to position the tool or hand. For example, a higher degree of precision is required to engage a screw with a screwdriver than to simply grasp a loose part.
- Force: a measure of the amount of force required to do a task. For example, less force is required to remove a loose part than to free a part glued to the assembly.
- Base time: the time required to do the basic task movements without difficulty. This category excludes any time spent positioning the tool or overcoming resistance. The basic ease of doing a task is indicated by this score. For example, an unscrew operation will have a higher base time score than a simple flip operation.

The estimated time in second for disassembly is

$$T_{disassembly} = 1.04(dds - 1) + 0.9(NHT)$$

where:

dds = degree of difficulty score
NHT = number of hand and tools employed

The disassembly process, as we know, is invariably focused on the fasteners.

Efficient disassembly requires that a dissembler know how the different parts are fastened together. It is usually not necessary that the dissembler would know the material composition of a fastener, except in the following cases: (i) when the material is known to hinder recycling efficiency, and (ii) when the relative mass of the fasteners is great. Two parts are considered to be mating if they are in physical contact with each other. There is a large variety in fastener types and Table 12.1 provides a list of the different types from a disassembly perspective. From our survey of disassembly practice we found these broad class types to represent the differences in unfastening effort and difficulties. The fastener types are divided into separate and integral fasteners to facilitate identification. The relative unfastening difficulty or U-rating of each fastener type is also listed in the table.

Desai and Mital [6] present a multifactor model to compute the disassembly effort index (DEI) score, which is representative of the total operating cost to disassemble a product. The DEI score is derived from seven factors, which are (i) time, (ii) tools, (iii) fixture, (iv) access, (v) instruct, (vi) hazard, and (vii) force requirements. The relative unfastening difficulty or U-rating of each fastener type is also listed in Table 12.16. Here it is called degree of difficulty of disassembly and presented in the table in a form somewhat more simplified than Desai and Mital [6].

As we can see from the degree of difficulty in disassembly Table 12.17, the nail with head is less difficult to disassemble than without head; similar conclusion can be drawn about bolt with or without head. It is apparent that welded joint is most difficult to disassemble compared with other fastenings. The disassembly time equation [7] also takes into account number of hands and tools required in the disassembly that will affect the disassembly time. For example, if we want to disassemble two plates assembled with two bolts with head and this requires both hands and only a screwdriver for disassembly, we can estimate the disassembly time:

$$T_{disassembly} = 1.04(1.4 - 1) + 0.99(2) = 2.396s \approx 2.4 \text{ s}$$

Sometimes, if disassembly requires more than one tool, we can see this will increase the disassembly time. However, the estimate of disassembly time for one bolt and two bolts disassembly would be 4.8 s.

Normally the tools of disassembly are: powered screwdriver, manual screwdriver, wrench, pliers, spanners, chisel, scissors, hammer, tie-rod, pins, and hacksaw. Each one of these being used involves a different degree of difficulty in disassembly.

The primary objective of a disassembly operation is to generate valuable output: In a typical disassembly facility there are three classes of outputs: (i) retrieved parts or subsystems; (ii) recyclable material mass; and (iii) waste material. There are usually at least two waste material destinations; one contains waste that is directly landfilled, while the second contains waste that will be incinerated. The recyclable material mass is typically collected, then transferred to a recycling facility. The dissembler will contract with several materials-recycling vendors.

12.3.11 Synthesis of Rules of Design for Dismantling

Figure 12.6 presents a synthesis of rules for design for dismantling. The design for dismantling mainly consists to integrate EOL issues at the early stage of design of the product to adapt products to technical and economic limits of dismantling systems. The main rules are summarized as follows:

- Use of recyclable materials/compatible (if not recycled, and avoiding materials prohibited by regulations)
- Streamline the number of materials
- Using materials with characteristics that allow their separation easily during recycling
- Reduce component number by integrating many functions into one component
- Reduce the number and standardize the fasteners used
- Design of separable components; choose materials with different properties (magnetic/nonmagnetic, heavy/light) for easy sorting; keep a gap of 0.15 g/cm^3 between the densities of polymers
- Facilitate access to components/fasteners (if possible to have only one way to access all components)
- Avoid painting (especially plastic parts) and coatings/surface treatments

Table 12.18 indicates the behavior of different types of joint for recyclability, detaching behavior, joint behavior, and carrying capacity. It may be useful to help select a particular joint for a particular application.

Table 12.19 shows liberation behavior after a shredding operation on some connection types (see also Tables 12.20 and 12.21).

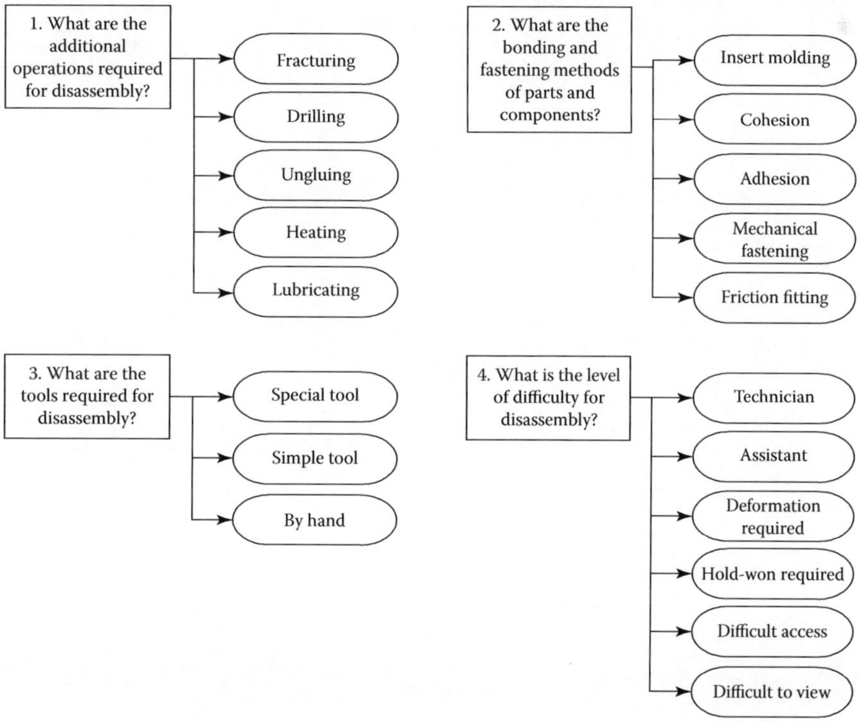

Figure 12.6

Synthesis of rules for design of dismantling.

Table 12.18 Ability of Several Types of Connections Applying to Different Criteria (Recyclability, Disassembly, Junction, Charging)

		Recyclability		Detaching Behavior		Joining Behavior		Carrying Capacity	
		Material Recycling	Product Recycling	Destructive Detaching Expenditure	Detaching Expenditure	Guidance Expenditure	Joining Expenditure	Fatigue Strength	Static Strength
Material connection	Adhesive bonding	Average	Poor	Average	Poor	Poor	Average	Average	Average
	Welding	Good	Poor	Average	Poor	Poor	Average	Good	Good
Fictional connection	Magnetic	Average	Average	n/a	Good	Average	Good	Average	Average
	Velcro	Average	Average	n/a	Good	Good	Good	Poor	Poor
	Nut & bolt (metal)	Average	Average	Average	Average	Average	Average	Good	Good
	Nut & bolt (plastic)	Good	Good	Good	Average	Average	Average	Average	Average
	Spring connection	Good	Good	n/a	Good	Good	Good	Poor	Average
Positive connection	Snap	Good	Poor	Good	Poor	Good	Good	Average	Good
	Bent lever	Good	Good	Average	Good	Average	Good	Average	Good
	1/4 turn	Good	Good	Average	Good	Good	Good	Average	Good
	Press-turn	Average	Good	Average	Good	Good	Good	Average	Average
	Press-press	Average	Good	Average	Good	Good	Good	Poor	Average

Source: VDI Standards. VDI 2243: Recycling-oriented product development. www.vdi.eu/engineering/vdi-standards/.

Table 12.19 Characteristics of Connection Types Related to Their Specific Degree and Nonrandomness of Liberation Behavior after a Shredding Operation (Continuing from Van Schaik and Reuter, 2007) with Examples for Different Connection Complexities, Properties of Connected Materials, Homo/Heterogeneity of Connection, etc.

Connection Types	Before Shredding	After Shredding	After Shredding	Liberation Behavior
Bolting/riveting		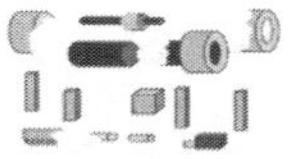		• High liberation • High randomness
Gluing	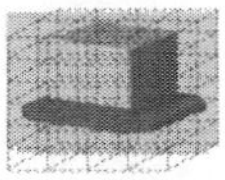	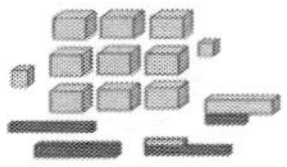		• Medium liberation • Medium randomness
Coating/painting	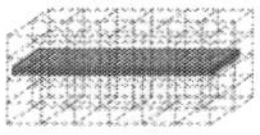			• Low liberation • Low randomness of liberation
Foaming	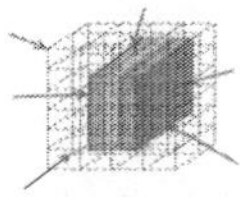	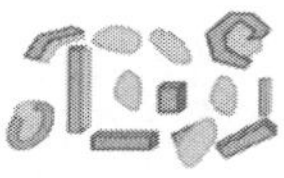		• Medium liberation • Medium randomness
Examples (including effect of characteristics of connections)				
Connected materials/ components of different levels of complexity	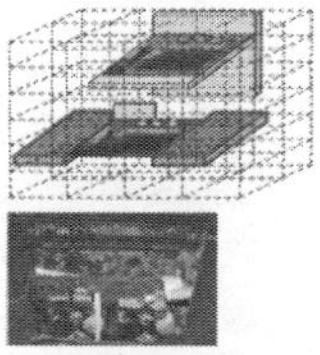	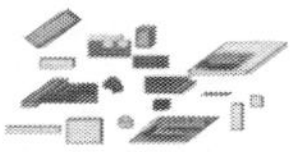	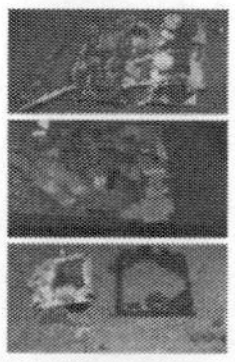	• High/medium liberation from structure • High/medium randomness (both depending on joint type and connected materials)
Heterogeneous/high number of connections per surface area				• Low liberation on component • Low randomness
Material properties, low complexity, low number of joints	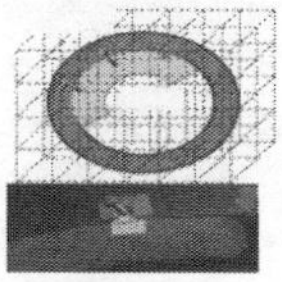			• High liberation • High-medium randomness (depending on joint type and connected materials)

Source: Worrell, E. and Reuter, M. (eds). *Handbook of Recycling: State-of-the-Art for Practicitioners, Analysts, and Scientists.* Elsevier, Waltham, MA, 2014 [11].

Table 12.20 Standard Disassembly Tools with Their Letter Codes

Unscrewing & Screwing	Gripping & Fixturing	Cutting & Breaking
Phillips screwdriver: PS	Vise: VS	Knife: KN
Flathead screwdriver: FS	Pliers: PL	Wire cutter: WC
Nut driver: ND		Handheld shears: SH
Fixed-end wrench: FW		Drill: DR
Adjustable wrench: AW		Handheld power grinder: PG
Socket with ratchet: SR		Grinding wheel: GW
Allen key: AK		Hacksaw: HS
Power wrench: PW		Power saber saw: BS
		Hammer: HM
		Chisel: CH
		Prybar: PR

Table 12.21 Degree of Difficulty of Disassembly

Separate fastener	Degree of difficulty	Integrated fastener	Degree of difficulty
Nail with head	1.5	Cylindrical snap fit	1.6
Nail w/o head	1.8	Cantilever snap fit	1.3
Screw/bolt standard head	1.4	Seam/crimp joint	1.6
Screw/bolt special head	2.2	Interferance fit	1.8
Nut and bolt	2.1	Integrally threaded part	2.2
Rivets/staples	2.0	Socket/plug	1.2
Retaining rings/circlip	2.5		
Tapes	1.7		
Adhesive	2.1		
Welded	4.0		
Valero/Zipper	1.0		
Releasable clips	1.8		

Source: McGlothlin, S. and Kroll, E. Systematic estimation of disassembly difficulties: Application to computer monitors. *IEEE International Symposium on Electronics & the Environment* (pp. 83–88), Orlando, FL, 1995.

12.4 Analytical Methods in Studies of Science for DfD

12.4.1 Learning Curve Method of Assembly and Disassembly

The learning curve method has been applied in engineering production for a long time, and particularly in aerospace industries. The Wight model showed that, in repetitive tasks, operators improve their performance, although with diminishing return [9].

A learning curve is a mathematical model that explains the phenomenon of increased worker efficiency and improved organizational performance with repetitive production of goods or services. The learning curve method is sometimes called an experience curve or a manufacturing progress function; fundamentally, it is an estimating relationship. For example, the learning curve can be used in estimating the professional hours expended by an engineering staff to accomplish successive detailed designs within a family of products, as well as in estimating the labor hours required to assemble automobiles. The basic concept of learning curves is that some input resources (e.g., energy costs, labor hours, material costs, engineering hours) decrease on a per-output-unit basis as the number of units produced increases. For example, if 100 labor hours are required to produce the first unit and a 90% learning curves is assumed, then 100(0.9) = 90 h would be needed to produce the second unit. Similarly, $100(0.9)^2 = 81$ labor hours would be needed to produce third unit.

In particular, Wight developed an equation to compute resource requirements, assuming a constant percentage reduction in input resources each time the output quantity is doubled.

$$Z_N = K(N)^{le} \quad (12.5)$$

where:

- N = output unit number
- Z = the number of input resource units needed to produce output unit N
- K = the number of input resource units needed to produce the first output unit
- s = the learning curve slope parameter expressed as a decimal (s = 0.9 for a 90% learning curve)
- le = ln(s)/ln(0.2)

Example 12.3: Learning Curve for a Mini Baja Car Design

The Mechanical Engineering Department at Manhattan College has a student team that is designing a Mini Baja car for national competition. The time required for the team to assemble the first is

100 h. Their improvement (learning rate) is 0.8, which means that as output is doubled, their time to assemble a car be reduced by 20%. Use this information to determine

a. The time it will take the team to assemble the 10th car
b. The total time required to assemble the first 10 cars
c. The estimated cumulative averages assembly time for the first 10 cars

Solution

Using Equation 12.4, and assuming a proportional decrease in assembly time for output units between doubled quantities, we have

a. $Z_{10} = 100(10)^{\log 0.8/\log 2} = 100(10)^{-0.322} = 47.6$ h
b. The total time to produce × units, T_x, is given by

$$T_x = \sum Z_u = \sum K\left(u^n\right) \tag{12.6}$$

Using Equation 12.6, we get

$$T_{10} = 100\sum u^{-.332} = 100\left(1^{-.332} + 2^{-.332} + \cdots + 10^{-0.332}\right) = 631\,\text{h}$$

c. The cumulative average time for x units, C_x, is given by

$$C_x = T_x / X = T_{10} / 10 = 631/10 = 63.1\,\text{h}$$

Spreadsheet Solution:

K	100
S	0.8
N	−0.32193

Units	Unit time	Cumulative Total	Cumulative Average
1	100	100	100
2	79.44344	179.4434	89.72172
3	69.43777	248.8812	82.9604
4	63.99362	312.8748	78.21871
5	59.55684	372.4317	74.48633
6	55.16375	427.5954	71.2659
7	52.4116	480.007	68.57243
8	50.13882	530.1458	66.26823
9	48.21603	578.3619	64.26243
10	47.6431	626.005	62.6005

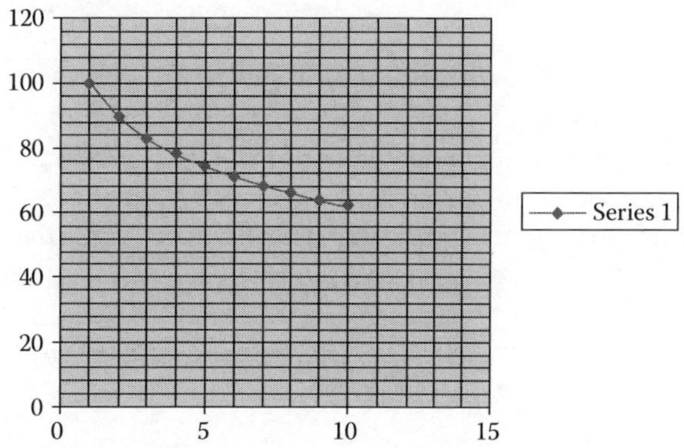

Example 12.4: Photovoltaic Panel Production

Let a photovoltaic panel cost \$200/m^2 for the first-of-a-kind production and estimate the cost of 10^4 unit production run. Assume the learning curve slope as 0.85.

Solution

Using the Equation 12.4, we can estimate decreasing production time for photovoltaic panel as

$$Z_{10^4} = K(N)^{le}$$

where:

K = \$200

le = ln(0.85)/ln(2) = −0.234

$$Z_{10^4} = 200(10^4)^{-0.234} = \$23.17/\text{m}^2$$

This proved the principle that learning curves effect savings by sequential mass production of large number of identical units (same size). In practice, car engines are good example of such principle.

We can also calculate the total cost to produce x units, Cost_x, is given by

$$\text{Cost}_x = \sum Z_u = \sum K\left(N^{1e}\right) \tag{12.7}$$

Using Equation 12.6, we get

$$\text{Cost}_{10}{}^4 = 200\sum u^{-0.234} = 100\left(1^{-0.234} + 2^{-0.234} + \cdots + 10^{4(-0.234)}\right) = \$3.024\times10^5$$

c. The cumulative average cost for x units, C_x, is given by

$$\text{Average Cost}_x = \text{Cost}_x/x = \text{cost}_{10}{}^4/10^4 = \$32.24$$

The average cost of production of the photovoltaic cells falls to \$32.24/m². This shows the merit of the learning curve. However, we should be careful in applying the learning curve method and keep in mind the following:

1. Larger size normally yields lower reliability and average unit cost may increase.
2. Learning curves should only be applied to the same products (same size or same design), same workforce, and in the same setting (e.g., factory). In practice, the setting may change or workforce may alter.
3. The learning curve method does not factor into the principles of sustainability, for example, resources depletion, GHG emissions or technological innovation.

12.4.2 Evaluation Method of Disassemble-ability

Various methodologies have been developed to evaluate the disassembability of a product. The methodologies vary from the methodologies that use the spreadsheet-like chart, energy use, entropy for disassembly, time, to the EOL value of a product. There are also several computerized tools that have developed to help designers in measuring the disassembability of the product at the design stage. The methodologies may be applied not only for mechanical products such as automotive components but also for electrical products.

12.4.2.1 Spreadsheet-Like Chart

Early disassembability methodologies used the spreadsheet-like chart to measure the disassembability. McGlothlin and Kroll [9] designed the spreadsheet-like chart to measure the ease of disassembly of a product. The authors categorized the disassembly difficulties into five subjective categories: accessibility, a measure of the ease with which a part can be accessed; positioning, a measure of how precisely the tool or hand needs to be positioned and oriented in order to perform the task; force, a measure of the amount of force required performing the task; additional time, while each of the previous difficulty sources is related to time, this category has to do with additional time penalties; and special, which is a provision to note special problems encountered that do not fit in any of the other categories (Table 12.22).

Table 12.22 Standard Disassembly Operations with Their Letter Codes

Unscrew: Un	Turn: Tu	Wedge/Pry: We	Cut: Cu
Remove: Re	Flip: Fl	Deform: De	Push/Pull: Pu
Hold/grip: Ho	Saw: Sa	Drill: Dr	Hammer: Ha
Peel: Pe	Clean: Cl	Grind: Gr	Impact: Ia

$$\text{Overall efficiency} = \frac{5x\sum \text{column.2}}{\sum \text{column.13}} \tag{12.8}$$

Equation 12.1 is used to estimate the disassembly time

$$\text{Disassembly time } = \left(\sum col.13 - 5x\sum \text{column.5}\right)x1.04 + \left(\#NH\right)x0.9$$

where NH is the number of hand and tool manipulations based on Whitney's equation [[7].

The authors use the scoring system (1: easy to 4: difficult) to measure the above categories. By using the chart, shown in Table 12.23, the disassembly overall efficiency can be calculated by using the Equation 12.1. Kroll and Hanft [8] extended McGlothin and Kroll's work. They derived the disassembly scores from the work-measurement analyses of standard disassembly tasks. The scores used in this method are 1–10. The authors analyzed the disassembly task by using the Maynard operation sequence technique (MOST) time measurement system and comparing their estimated performance times. The authors also introduced the standard disassembly tasks and the standard disassembly tools with their letter code. By using the chart, disassembly time is calculated, Equation (12.3). Kroll and Hanft [8] did not use the term "design efficiency" but "effectiveness of design" because in the engineering context efficiency means the comparison between output and input. Kroll and Carver presented a method for estimating disassembly time to allow the designer to identify weakness in the design and improve it accordingly [12]. Kroll and Carver stated that the disassembly efficiency or effectiveness is an unrealistic metric because it was derived from the assembly efficiency and there are fundamental differences between assembly and disassembly operations.

Desai and Mital [6] presented a methodology in a spreadsheet-like chart to design products for disassembly. This would facilitate EOL product disassembly with a view to maximizing material usage in the supply chain at a low cost to the environment. The numeric evaluation is made based on the methods-time measurement (MTM) system. The authors also considered the posture allowance for the need of specialized postural requirements and designed a table that is fully equipped with scores for unnatural postural requirements (Table 12.22).

12.4.2.2 Energy and Entropy for Disassembly

Suga et al. [13] presented the quantitative method to evaluate the disassembly evaluation by introducing two parameters, energy for disassembly and entropy for disassembly. Energy for disassembly is energy required to disconnect of an interconnection.

Equations 12.8 and 12.9 are used to calculate the energy for disassembly of a screw.

$$E_n = T_f \times 0.8 \times \theta \tag{12.9}$$

where

T_f = clamping torque

d = diameter of screw

θ = rotational angle producing the axial tension

The screw connections are found everywhere in mechanical assembly. It is accomplished by tightening the screw. Normally, the tightening process starts by applying a clamping force F_f to screw down. The torque created to clamp the screw is T_f and is calculated in Equation 12.12:

$$T_f = \left[\left(\frac{d_m}{2d}\right)\left(\frac{\tan\alpha + \mu\sec\alpha}{1-\mu\tan\lambda\sec\alpha}\right) + 0.625\mu_w\right]F_f d \tag{12.10}$$

Table 12.23 Disassembly Evaluation Chart for Hand Drill (Prepared by Nand Jha, 2/15/2013)

1	2	3	4	5	6	7	8	9	10	11	12	13	14
Part Number	Quantity	Min. # of Parts	Task Type	# of Repetition	Repositioning	Accessibility	Required Tool	Force	Time	Special	Subtotal	Total	Comment
1	8	0	Un	6	2	1	PS	1	3	1	8	48	
2	2	0	Un	2	2	2	PS	1	3	1	9	18	
3	1	1	We	2	1	2	PB	1	2	1	7	21	
			Fl	1	1	1		1	2	1	6	6	
			Cl	1	1	2	Ro	1	2	1	7	7	
4			We	2	1	1	PB	1	2	1	6	12	
			Re	1	1	2		1	1	1	6	6	
5	2	2	Re	2	1	1		1	1	1	5	10	
6	1	1	Ho	1	1	1		1	2	1	6	6	

Source: From Whitney, D. E. *Mechanical Assembly: Their Design, Manufacture, and Role in Product Development.* Oxford University Press, New York, 2004. With permission.

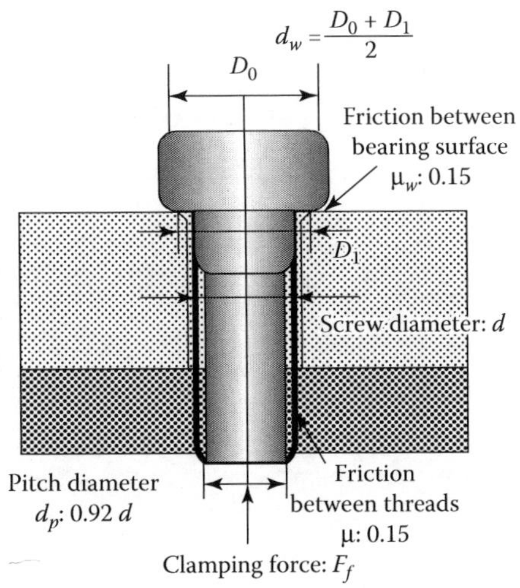

Figure 12.7

Parameter of a screw joint.

Mean screw diameter (d_m) is calculated as $d_m = d - p/2$; p is pitch of screw. The screw pitch diameter (d_p) is calculated as $d_p = d - p$ (Figure 12.7).

α and λ are thread angle and lead angle, respectively. The thread friction (μ) and collar friction (μ_w) are normally equal to 0.15. Equation 12.12 can also be written as

$$T_f = KF_f d \tag{12.11}$$

The interesting fact about Equation 12.13 is that $K = 0.2$ for $\mu = \mu_w = 0.15$ and so

$$T_f = 0.2F_f d \tag{12.12}$$

no matter what size of screw or bolts are used and no matter whether the threads are fine or coarse. Therefore, the tightening torque or moment can always be calculated by using Equation 12.14. The tightening force F_f and the diameter of screw is deciding factor in the clamping toque. Now the Equation 12.11 estimates the loosening energy required for disassembly of a screw joint. It is obvious that in disassembly of the screw joint, the force required is about 80% of the clamping or tightening force because it is in the loosening side of the screw. The energy for disassembly is 80% of the clamping energy $T_f\theta$, where θ is revolving angle in radians for disassembly (E_n).

Example 12.5: Disassembly Time Calculation

A square-thread power screw is to be disassembled. It has a major diameter 36 mm and a pitch of 6 mm. The loosening force (E_f) is 65 kN per screw. Find the loosening torque and the energy required for disassembly of the screw.

The mean diameter of the screw: $d_m = 36 - 3 = 33$ mm and the loosening torque is

$$T_f = 0.2(65)(33) = 448.8 \text{ Nm}$$

The energy for disassembly is $E_f = 0.8T_f\theta$. The revolving angle (θ) can be less than 360°, but here it is assumed as equal to full revolution 2π. Therefore, the energy required for disassembly is

$$E_f = 0.8(448.8)(2\pi) = 2256 \text{ N}\cdot\text{m} = 2256 \text{ J per screw}$$

The total energy required should be calculated based on the number of threads present and it will depend on the length of the screw, that is, number of threads = length of screw/(lead=pitch), if it single threaded screw. For multiple threaded screws the necessary correction should be made. In this example, we have not considered the nut and that can also change the energy calculation. If we consider a

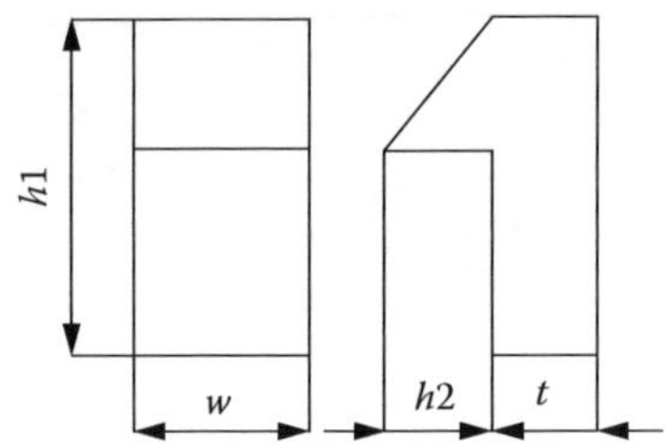

Figure 12.8

Parameter of a simple model of snap fit. (From Whitney, D. E. *Mechanical Assembly: Their Design, Manufacture, and Role in Product Development.* Oxford University Press, New York, 2004. With permission.)

mechanized disassembly, we can calculate the power required for the disassembly and time for disassembly if the angular speed is known.

12.4.2.3 Entropy Calculation

Entropy is the measure of disorder of the fluid. The greater the entropy, the less energy is recoverable as work. The entropy of disassembly is calculated as

$$S_s = \sum_k \left[\ln\left(\frac{N_{ik}!}{\prod_j n_i!} \right) + \sum_i \ln\left(\frac{n_i!}{\prod_j n_{ij}} \right) \right]$$

Entropy for disassembly is used to calculate the degree of disassembly difficulty. It measures the randomness of the total number of connections in subassembly, type of interconnection methods, and the interconnection made by each method for the direction $+x$, $-x$, $+y$, $-y$, $+z$, or $-z$, where N_{ik} is the total number of interconnections belonging to assembly k, n_i is the number of interconnections made by method i (for instance, i = 1 is the screw, and i = 2 is snap fit).

In the example above, there is only one type of joint, the screw joint in one assembly, that is, $k = 1$ and $N = 1$, therefore, $S_s = \ln(1/1) = 0$, which indicates that for this disassembly the degree of disassembly difficulty is zero, which in thermodynamic terms means the disorder in the system is zero and all the energy is converted to useful work of disassembly. For example, we assume $k = 1$, for screw joint (i), $N_{11} = 4$, for snap joint ($j = 2$), $N_{12} = 5$ (Figure 12.8).

We can now calculate the entropy of disassembly or the degree of disassembly difficulty as

$$S_s = \sum \left[\ln\left(\frac{4|}{1x2} \right) + \ln\left(\frac{5|}{1x2} \right) \right] = \ln(12) + \ln(60) = 4.28 + 4.1 = 8.4 \cong 9$$

It shows there are nine degrees of difficulty and randomness or disorder in the disassembly.

12.4.2.4 EOL Value for a Mechanical Component

Lee et al. proposed a guideline for determining feasible EOL options for manufactured products, set against the conflicting objectives of minimizing environmental impact and minimizing deficit and how to determine the EOL value of a product [14]. The guidelines and equation to calculate the EOL value are as follows:

1. If component is made of metal without any other alloy, primary recycling is recommended. If alloys are present, they alter the mechanical properties of the parent metal, so secondary recycling or landfill is more appropriate.
2. If component is polymeric, primary recycling is recommended; otherwise, consider secondary recycling or incineration to recover its energy content.
3. If component is made of ceramic, secondary recycling or landfill is recommended.
4. If component is made of an elastomer or is a composite material, secondary recycling or incineration is recommended, otherwise landfill is recommended.
5. If component contains toxic or hazardous material, special handling is required.

To calculate the EOL value, we propose these equations steps:

Reuse value = Cost of component ($) – Miscellaneous cost ($)
Remanufacture value = Cost of component ($) – Remanufacture cost ($) – Miscellaneous cost ($)
Primary recycle value = Weight of component (kg) ×
Market value of material ($/kg) – Miscellaneous cost ($)
Secondary recycle value = Weight of component (kg) ×
Scrap value of material ($/kg) – Miscellaneous cost ($)
Incinerate value = Energy produced (Watt) ×
Unit cost of energy ($/Watt) – Miscellaneous cost ($)
Landfill cost = –(Weight of component (kg) × Cost of landfill ($/kg)) – Miscellaneous cost ($)
Special handling cost = –(Weight of component (kg) × Cost of special handling ($/kg)) – Miscellaneous cost ($)
Miscellaneous cost = Collection cost ($) + Processing cost ($)
Collection cost = Handling + transportation + storage

For product redesign, Lee et al. [14] proposed the following guidelines:

1. Positioning the component which generates the greatest surplus at the top of the product structure, so as to realize this surplus as soon as possible
2. Positioning the component which generates the greatest deficit at the bottom of the product structure, so it can be disposed into landfill without the need for disassembly
3. Reducing the disassembly time by simplifying the joining technique
4. Avoiding the using of toxic materials, as they have to be disassembled for special handling

12.4.3 Reduction of EOL Impacts through Design for Disassembly

The estimated life cost consists of the manufacture, assembly, maintenance, remanufacture and recycling costs as determined by the choice of fastening or joining method. Each cost includes only expenses resulting directly from the choice of fastening or joining method. For example, the maintenance cost includes the expenses associated with joint disassembly and reassembly necessary for a maintenance task, and not the cost of other activities associated with the maintenance task. Similarly, the recycling cost represents the expense of material separation and not material reprocessing. The assembly and disassembly costs are estimated using time required for disassembly and assembly of various fastening and joining methods from [6]. For maintenance and remanufacture, the assumption is that the joint must be disassembled to enable further tasks.

The first cost consists of the manufacture and first assembly cost as determined by the fastening or joining method. It is assumed that the part-manufacture cost can be separated into a basic part-manufacturing cost that remains constant for different connecting methods, and the additional manufacturing cost of modifying a part to implement a particular fastening method. For example, if the fastening method involves threaded fasteners, the additional manufacturing effort could include drilling holes in the part. The additional cost may also be due to a more complicated mold to achieve molded holes or snap fits. The first cost includes only the portion of the manufacturing cost determined by the connecting method, and not the basic part-manufacturing cost. The first cost also includes the cost of assembly as determined by the type and amount of fasteners or joining compound necessary to achieve the designer-specified joint requirements.

12.4.4 Recycling Cost

The recycling expense includes the cost of extracting material introduced by the fastening method that is not recycling-compatible with the part material, or the cost of separating parts made of different materials. The cost of reprocessing the material of neither the part nor the fastening method is included. It is assumed that the fastening method will not affect the reprocessing expense of the parts if incompatible materials introduced by the fastening method can be removed.

$$C_r = \sum_{i=1}^{k} R_{ci}.W_{ri}$$

where C_r is cost of material recovery equivalent to the product of cost of material recovery (Rc_i) and weight of material recovered (Wr_i), and k represents types of material recovered, like copper, steel, and so on.

12.4.5 Repair and Maintenance Cost

The repair and maintenance cost consists of disassembly and reassembly expenses, which represent time required for disassembly and reassembly at field labor rate, and the expected cost of part and fastener replacement due to damage incurred during disassembly and assembly.

$$C_{RM} = Cda \sum_{j=1}^{n} \sum_{i=1}^{m-g} T_{ij}$$

where

C_{da} is the hourly cost of assembly and disassembly
m is the number of the same type of joint
n is # of different types of joints or fasteners
g is # of fasteners that contact the same type of material

12.4.6 Remanufacture Cost

The remanufacture cost imposed by the fastening method also consists of expenses related to disassembly, reassembly and the probability of part and fastening method failure.

In general, the remanufacture cost is modeled as follows:

$$C_{rm} = ((T_d + T_a)h + p_f.C_f + (p_{da} + p_f \cdot p_{pe} - p_{da} p_f \cdot p_{pe})C_p$$

where:

C_{rm} = remanufacture cost
T_d = disassembly time
T_a = assembly time
h = labor rate ($/h),
p_f = probability of fastener failure in disassembly and assembly
C_f = cost of fastener failure
p_{da} = probability of part failure in disassembly/assembly
p_{pe} = probability of part failure in fastening-method extraction
C_p = cost of part failure

If the fastening method must be destroyed for disassembly, such as the case with rivets, the resulting damage to the part is categorized as part damage during method extraction. The probability of part damage during disassembly is defined to be zero, that is, $p_{pd}=0$. The probability of fastener damage in disassembly, that is, $p_f=1$ and the general remanufacture cost reduces to

$$C_{rm} = (T_d + T_a)h + C_f + p_{pe} C_p$$

There is a further possibility that the parts will be damaged during rivet removal along with the fasteners. If the part cannot be repaired, the consequential cost is part replacement cost. In such cases, $p_f=0$ and $p_{pe}=0$. The new cost equation for remanufacturing is presented below.

The general remanufacture cost then reduces to

$$C_{rm} = (T_d + T_a)h + p_{pd} C_p$$

If failure of both the fastening method during disassembly and the part during fastening-method extraction is unavoidable, the remanufacture cost will include disassembly and the consequential cost of part and fastener failure. For $Pf=1$ and $Ppe=1$, the general remanufacture cost reduces to

$$C_{rm} = (T_d + T_a)h + p_{pd} C_p$$

12.4.6.1 Cost-Benefit Analysis of Recycling

The cost-benefit analysis for recycled parts and material has been presented by Chen et al. [15] and seems to be useful where recycling should be attempted.

The cost function of recycling is defined as

$$C_o = C_d + C_s + C_r + D_c$$

where:

C_o = cost of recycling
C_d = cost of disassembly
C_s = cost of shredding
C_r = cost of material recovery
D_c = cost of dumping

The cost of disassembly depends on the kind of fasteners employed in assembly. Naturally and similarly, joints will take same time in disassembly. The cost of disassembly is estimated as

$$C_d = C_{db} \sum_{j=1}^{n} \sum_{i=1}^{m-g} T_{ij}$$

where:

C_{db} = hourly cost of disassembly
m = the number of the same type of fasteners in the product
n = the number of fasteners of a different type, such as screw and so on
g = the number of fasteners that contact parts of the same or compatible materials

The shredding cost is estimated as

$$C_s = C_{sh}(S_t)(W_s)$$

where

C_{sh} = hourly shredding cost
S_t = time to shred 1 ton of material
W_s = weight of shredding materials in tons

The cost of recovery (C_r) naturally increases with amount and type of material recovered. For manual separation to be carried out there must be a significant value retained in the recycled product in order for the separation to be economically feasible. This scenario is applicable to most other products constructed from subassemblies consisting of a variety of materials. West European hourly rates and material prices bulletins show the amount of material (grams) that has to be removed per minute for recycling to be cost-neutral for manual disassembly. No doubt they present in g/min, but a reasonable approximation can be made in other units, as it is presented that silver should be removed at a rate of at least 5.1 g/min to be economically feasible.

Chen et al. [15] presents the cost of material recovery equation as

$$C_r = \sum_{i=1}^{k} R_{ci} \cdot W_{ri}$$

where

R_{ci} = cost of material recovery per ton of type *i* material (*i* may be steel, plastics, etc.)
W_{ri} = weight of i material recovered
k_i = number of different types of materials in the product

The cost of dumping (D_c) is estimated as

$$D_c = d_c.(W_d)$$

where d_c is cost of dumping of one ton of material; it may include the land, labor, and materials, like plastic sheet, iron sheet, or any other kind of chemicals used in dumpint. W_d is the weight. in tons of dumped material.

The cost-benefit analysis is a very useful technique in economic analysis of projects most of the time in government agencies, like airport, highway, or power generation projects. All the projects are envisaged for maximization of social benefits; however, most of the time it is decided by concerns other than just economic considerations. The benefit-cost (B-C) ratio should be always greater than 1 and the project with maximum B-C should be selected.

Chen et al. [15] has selected B-C the method for analyzing the disassembly design. It has been defined as Recycling B-C method and presented as

$$R_{b/c} = \frac{\text{Total Benefit}}{\text{Total Cost}} = \frac{R_p + R_m + E_r}{C_d + C_s + C_r + D_c}$$

where

R_p = revenue from used parts
R_m = the revenue from recovered materials
E_r = energy-saving benefit of emission reduction

Chen et al. [15] have presented an equation for the benefit of emission reduction due to using recycled parts or materials. The benefit of energy saving (E_r)

$$E_r = \sum_{i=1}^{y} Es_i \sum_{j=1}^{h} Em_j(Ec_j)$$

where:

Es_i = energy saving of recycled material of type i
Em_j = type j emission reduction
Ec_j = externality cost of type j emission
h = number of types of emissions such as SO_2, particulates, and so on
x = number of reusable parts disassembled in a product
y = number of types of recycled material in a product

The revenue from used parts (R_p) is calculated as

$$R_p = \sum_{i=1}^{x} Pu_i$$

where Pu_i = revenue from usable parts.

The revenue after selling the recovered material is calculated as

$$Rm = \sum_{i=1}^{y} Pm_i$$

where Pm_i = revenue from type i recycled material.

The profit obtained from recycling is calculated as Recycling profit = Total recycling revenue – Total recycling cost, that is,

$$RP = (R_p + R_m + E_r) - (C_d + C_s + C_r + D_c)$$

The break-even point (BEP) can also be calculated to find amount of materials (tons) that must be recycled at which total revenue and total cost will be equal and after which profit will occur to the recycling company. It is important for the recycling industry to know. Obviously, lower BEP is better for business viability. We can also examine the recycling profit (RP) and state that as the recycling cost decreases, so the profit will rise; for this to happen, cost of disassembly, shredding, dumping, and material recovery should be as low as possible. It is also possible to increase profit by raising the total recycling revenue, which can happen if there is greater acceptability of recycled parts and materials in industry.

Example 12.6: Learning Curve for Mini-Baja Car Design

A hypothetical example for the recycling of materials from disassembly of a system is presented. The system parts are made essentially of copper and steel, although some other materials are there but they have to be dumped as they could not be recycled. Total weight of the system is about 24 tons. There are three ($m=3$) similar joints but only two ($n=2$) different types of fasteners. However, among the three similar joints, two fasteners are in contact with same type of materials ($g=2$). We need different types of tools to disassemble the components of the system and times taken to disassemble the components are

	$j=1$	$j=2$
$i=1$	$T_{11}=7$	$T_{12}=5$
$i=2$	$T_{21}=1.5$	...

All times are given in hours. Materials that are shredded are mostly dumped and that is about 10 tons. The shredding equipment is quite efficient and takes only 20 h to shred it. The shredded components are dumped and the weight of dumped material is also 10 tons. The cost of materials recovery depends upon the type of material, and for steel recovery some kind of magnetic system is used. The cost of steel recovery is \$15/ton and the cost of copper recovery is \$100/ton. Steel recovered is 12 tons and copper is about 2 tons. The dumping cost depends on the land price, and other special materials, chemicals, cover etc. used in dumping. These costs vary significantly from place to place and country to country. The weight of material dumped is the same as shredded material, equal to 10 tons. The cost of dumping is \$1000.00/ton.

The benefits depend on many aspects, including revenue from used parts and recycled materials. However, we should also consider the benefits resulting from reduction in emissions and energy in our analysis of benefits of recycling. We will present the benefits from these aspects separately as follows:

$$E_r = \sum_{i=1}^{y} Es_i + \sum_{j=1}^{x} Em_j Ec_j$$

where Es_i represents the energy savings from using used steel and copper and Em_j are the reduction in emissions of hazardous gases such as GHG into the atmosphere. Here, only two types of emissions are considered CO_2 and SO_2. Here, only steel and copper are recycled, at 12 and 2 tons, respectively. Low-alloy steel is used here which has an average of 35 MJ/kg as embodied energy for primary production and embodied energy for recycled material is 10 MJ/kg. Therefore, there is savings of 25 MJ/kg from using recycled steel. Similarly for copper, the saving in energy is (71 − 17.75) = 53.25 MJ/kg. The reduction in carbon dioxide emissions by using steel is (2.125 − 0.6) = 1.52 kg/kg and the reduction in emissions due to copper recycling is (5.35 − 1.3) = 4.05 kg/kg. The SO_2 data could not be obtained, so it is not taken into this example (source: Ashby). The external cost of emissions should be considered as the cost of containing the hazards due to CO_2 emissions and if the emissions are reduced due to recycling, it should be considered as savings or benefits of recycling. Let us consider the savings due to reductions in emissions as \$5/ton of reduction in CO_2 emissions. The total benefit from reduction in CO_2 is

$$\begin{aligned} ES = E_s + E_c &= \{(25\ \text{MJ/kg} \times 12\ \text{tons} \times 1000\ \text{kg/ton}) \\ &\quad + (53.25\ \text{MJ/kg} \times 2\ \text{tons} \times 1000\ \text{kg/ton})\} \times \$0.031/\text{MJ} \\ &= (3\times 10^5 + 1.065\times 10^5)\times 0.031 \\ &= \$12601.5 \end{aligned}$$

This is the total energy saving due to recycling of steel and copper. The energy cost of \$0.031/MJ is for the energy from grid. The benefits due to reduction in emissions is

$$E_m = \{(1.52\text{kg/kg} \times 12\text{tons} \times 1000\text{kg/ton}) + (4.05 \times 2\text{tons} \times 1000\text{kg/ton})\} \times \$5/\text{ton}$$

$$= (6240 + 8100) \times 5 = \$71700$$

Total benefits due to recycling on energy savings and reduction in emissions is \$84301.5.
The benefit from recycling of both steel and copper (R_m) is calculated next.

$$R_m = R_{ms} + R_{mc} = (12\text{ tons} \times 1000 \times \$0.85/\text{kg}) + (2\text{tons}/1000 \times \$7.5/\text{kg})$$

$$= (10{,}200 + 15{,}000) = \$25{,}200$$

The revenue from reuse of parts has to be estimated as many parts are used again. In the real world, some of the disassembled components are never used again, like shafts and so on. However, some components are easily reused, like gears, flywheels, springs, and so on. Here, it is assumed that five gears after disassembly are reused and we can find the revenue from reusing the gears. The reused parts are normally sold at lower value than the primary or virgin parts. For gear, we use the sale price of reused gears as \$400.00/gear. Hence, the revenue occurring from reusing of gears is

$$R_p = 5 \times \$400 = \$2000$$

We can estimate the total revenue (TR) from recycling as the following:

$$TR = \$(84{,}301.5 + 25{,}200 + 2000) = \$111{,}501.5$$

Now, we calculate the total cost (TC) of recycling.

$$TC = C_d + C_s + C_r + D_c$$

Disassembly cost (C_s) is

$$C_d = C_{db} \sum_{j=1}^{2} \sum_{i=1}^{3-2} T_{ij} = \$30(7 + 5 + 1.5) = \$405.00$$

Shredding cost (C_s) = $C_{SH}(S_t)(W_S) = \$35(20)(10) = \$7{,}000.00$. The cost of material recovery is:

$$(C_r) = \sum_{i=1}^{2} Rc_i Wr_i = (15 \times 12 + 100(2) = \$380.00$$

Dumping cost (D_C) = $d_c x W_d = \$1000(10) = \$10{,}000.00$
Total cost (TC) = \$405.00 + \$7000 + \$380.00 + \$10,000 = \$17,785.00.
The net profit from recycling = \$(111,501.5 – 17785) = \$93,715.5
The benefit-cost ratio

$$(R_{B-C}) = \frac{111501.5}{17785} = 6.23$$

The benefit-cost ratio is very high and greater than 1 is always acceptable. The total recycling profit (R_P) = \$93,715.5.

Example 12.7: Disassembly of Gearbox

An example of an 165-horsepower gearbox disassembly is taken from the Brown Manufacturing [16] web site for to illustrate the disassembly process.

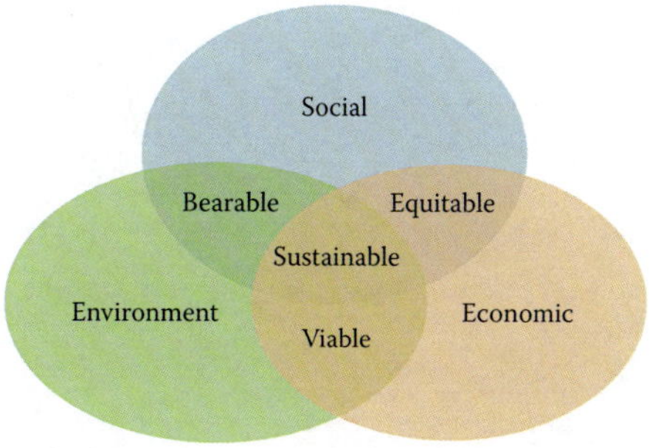

Figure 2.1

Sustainable development.

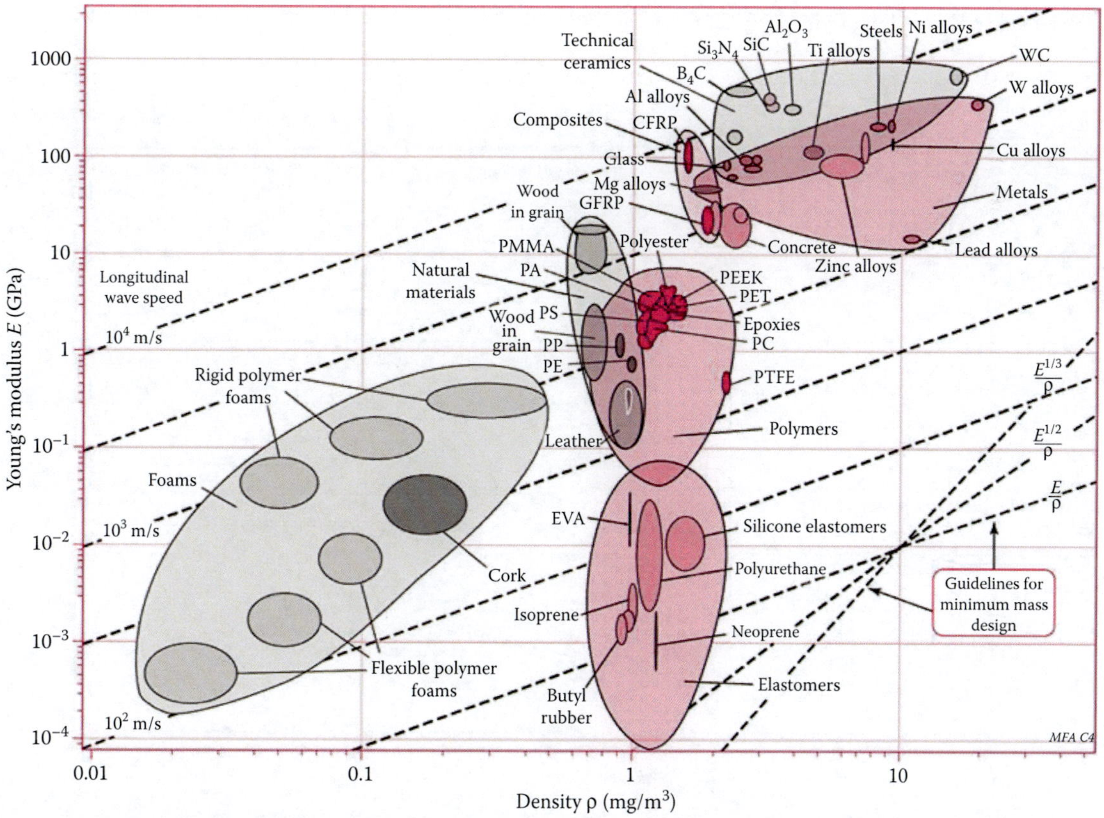

Figure 5.2

Density vs. Young's modulus of materials. (Copyright © The McGraw-Hill Companies, Inc. Permission required for reproduction or display.)

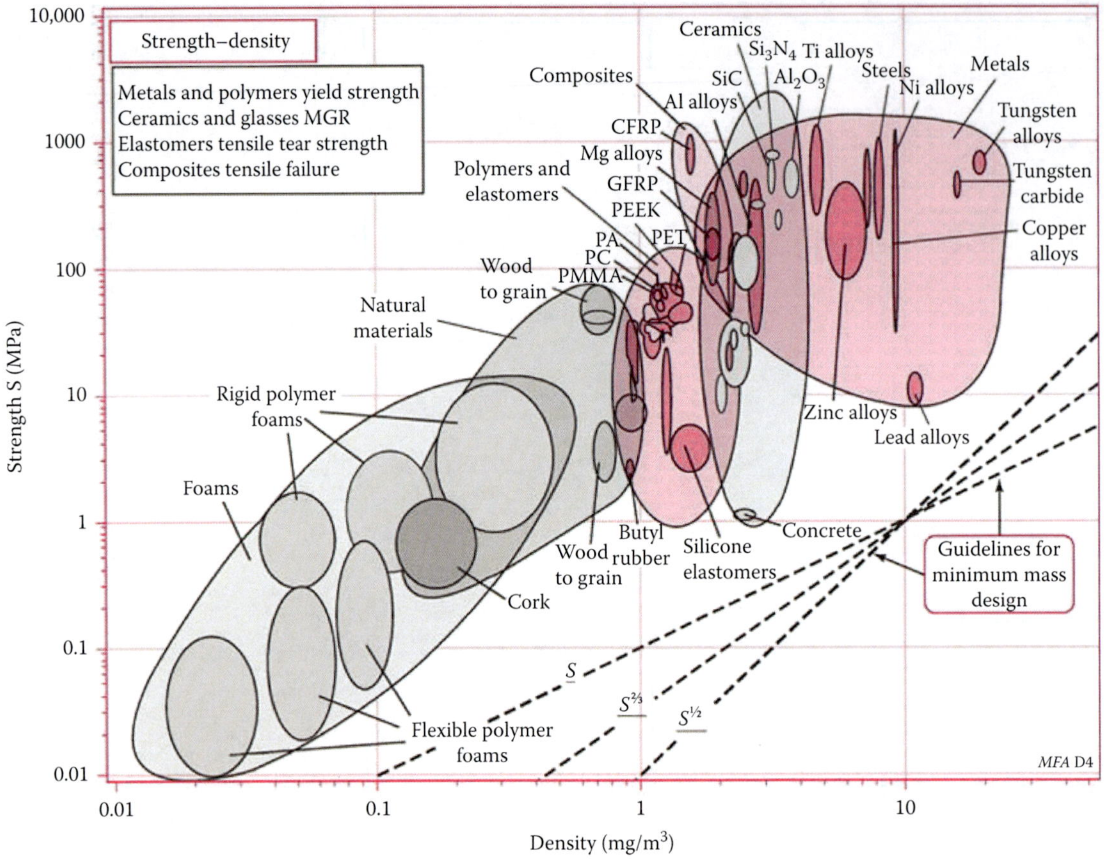

Figure 5.4

Density vs. strength of materials. (Copyright © The McGraw-Hill Companies, Inc. Permission required for reproduction or display.)

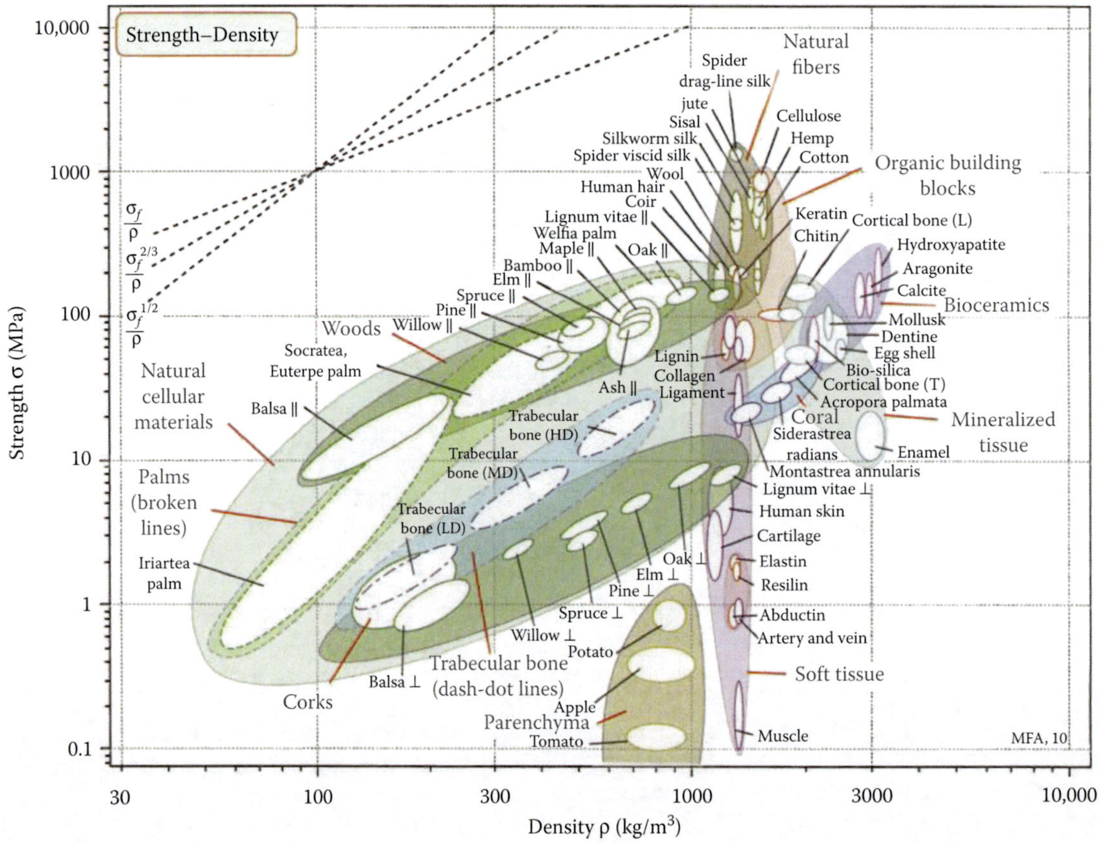

Figure 5.6

Density vs. strength under various constraints.

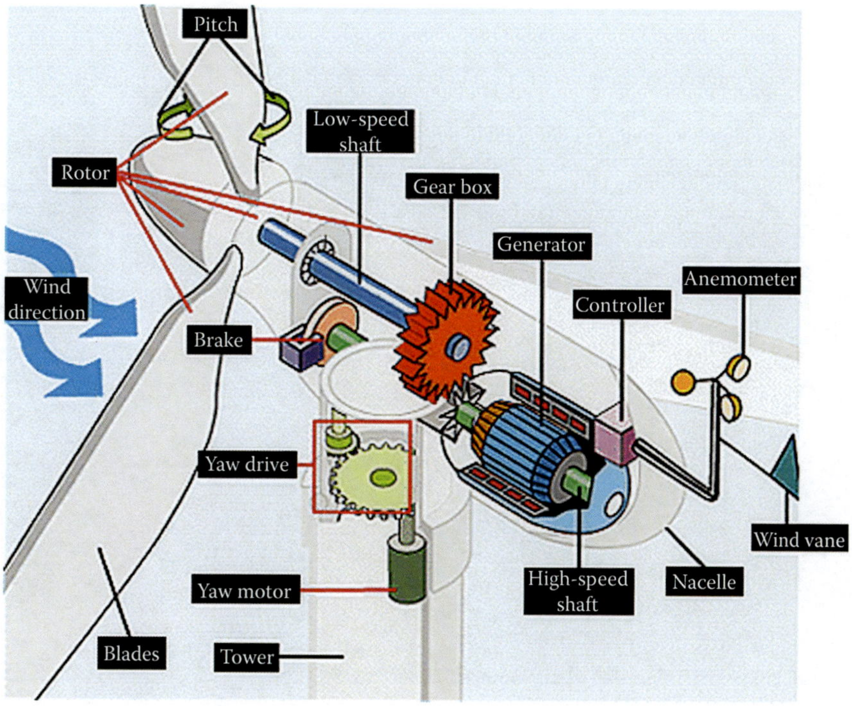

Figure 15.2

Wind turbine system.

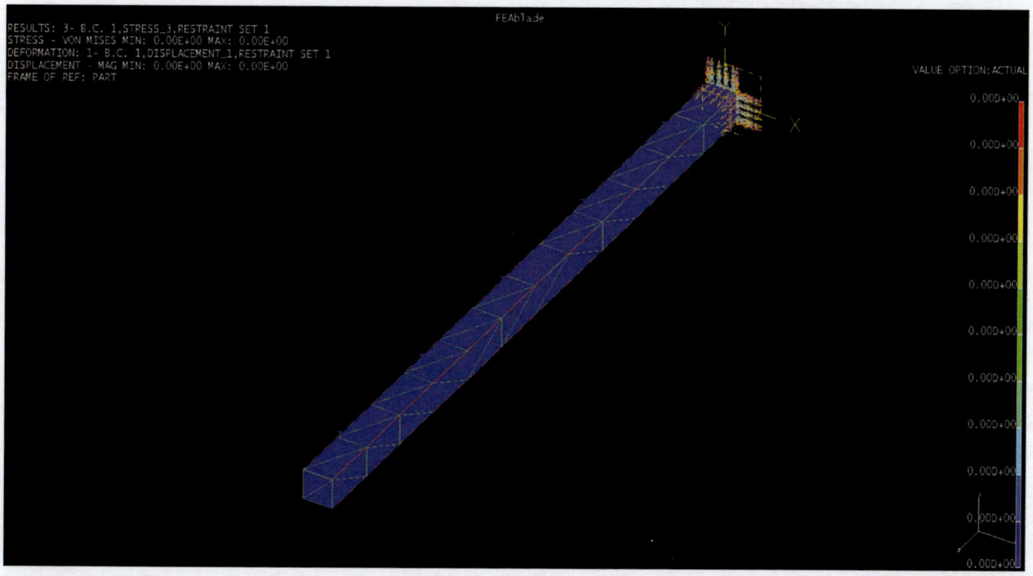

Figure 15.10

FE analysis blade.

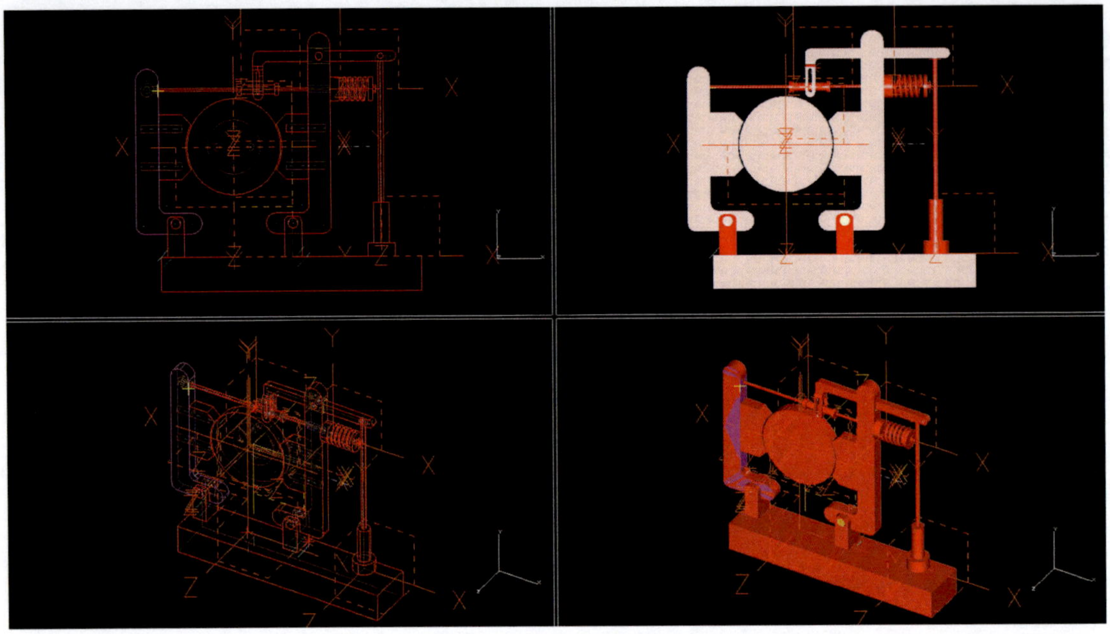

Figure 15.33

Brake assembly.

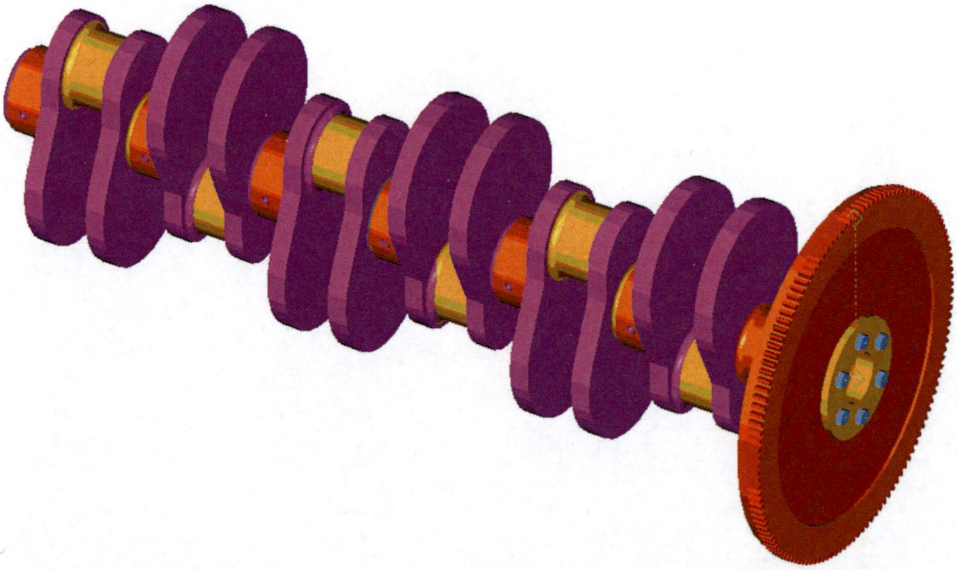

Figure 15.40

Complete solid modeling.

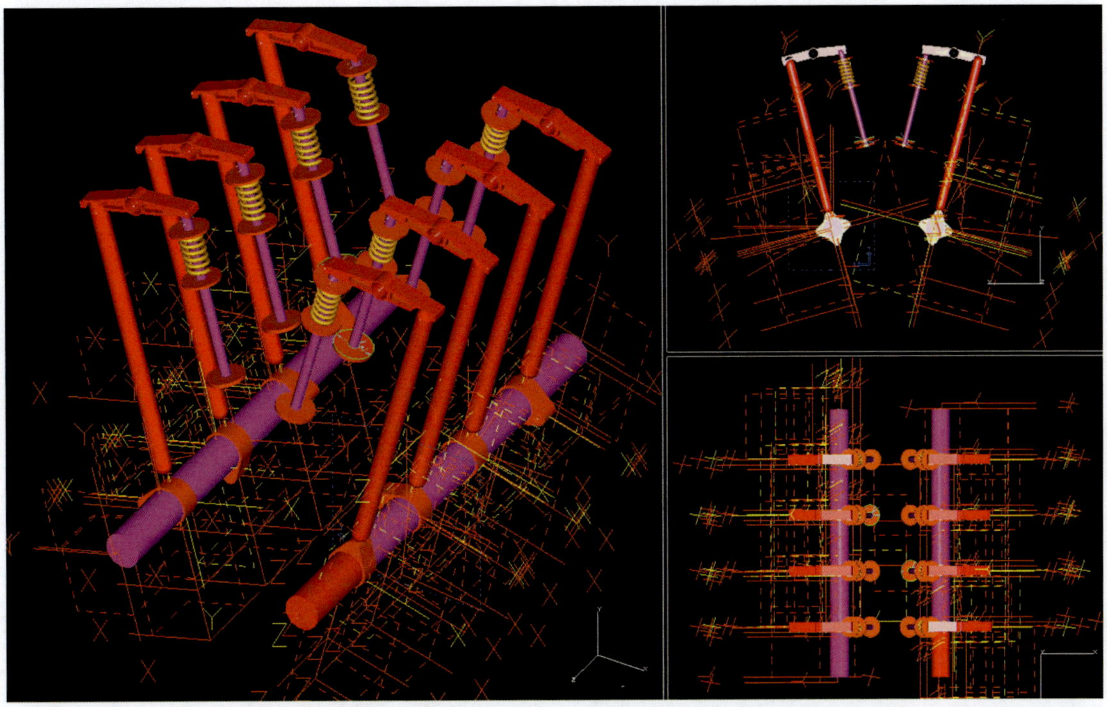

Figure 15.54

Final assembly.

Figure 15.70

Orthogonal view of assembly, inside view.

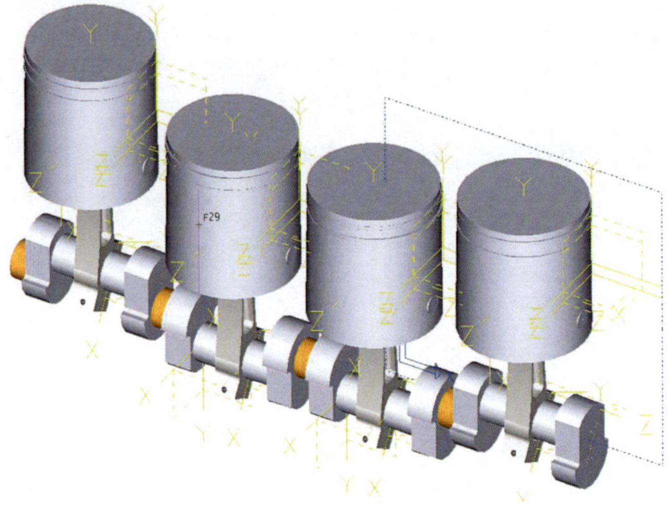

Figure 15.93

Assembled crankshaft.

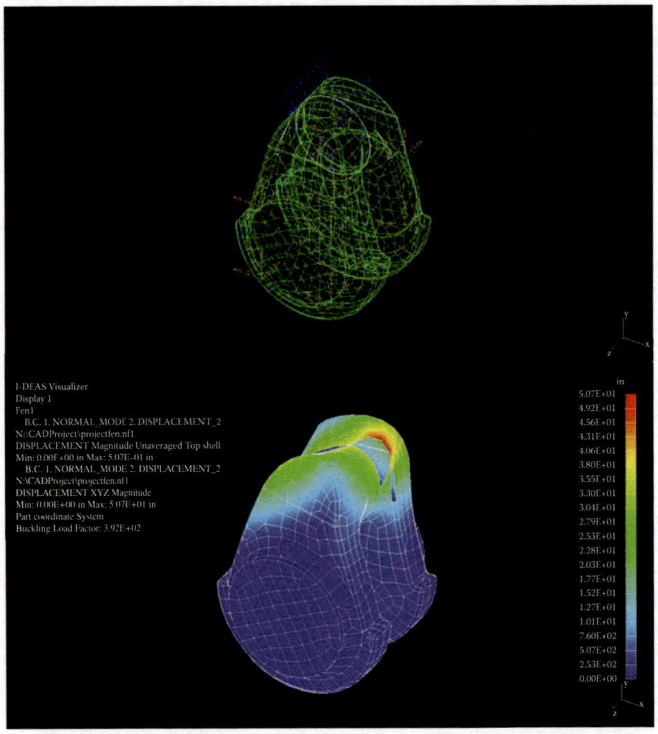

Figure 15.116

Nodal mesh for a crank throw.

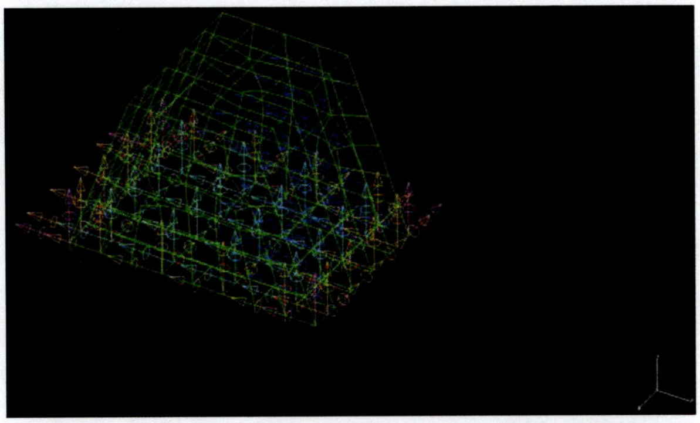

Figure 15.117

Nodal mesh for a flywheel tooth.

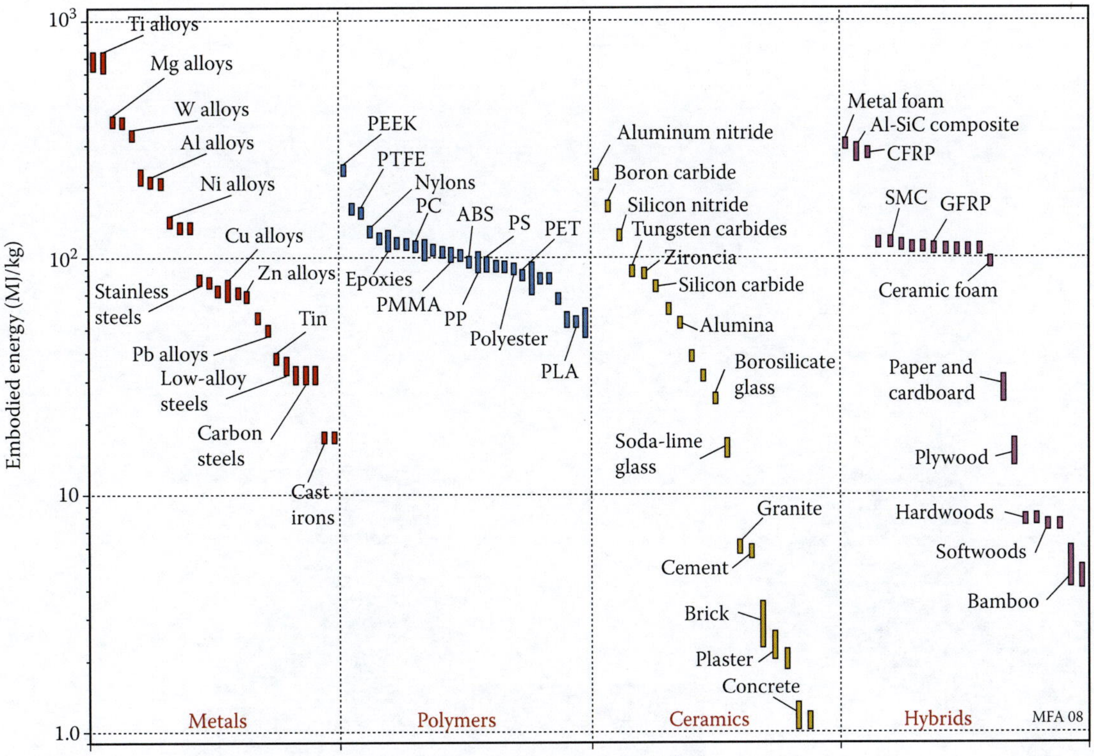

Figure 16.7

A bar chart of the embedded energies of materials per unit mass.

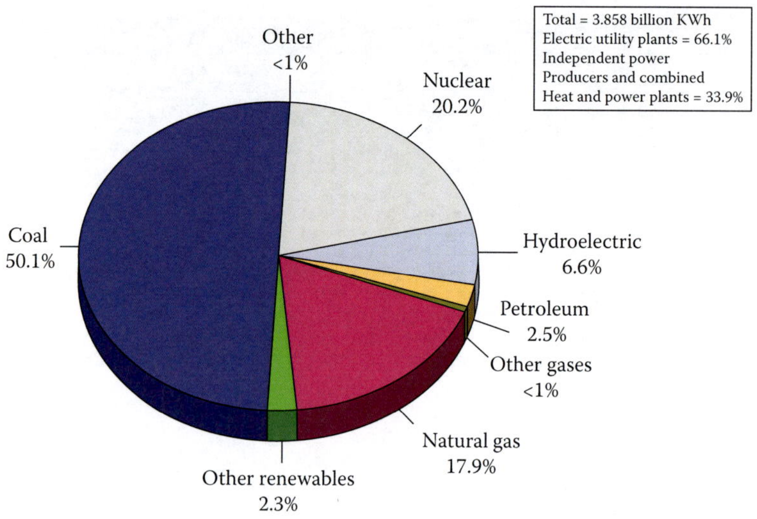

Figure 16.10

Components of U.S. net electric power generation, 2002. (Data from U.S. Energy Information Administration, DOE-STD-1151-2002, DOE, May 2002 (30).)

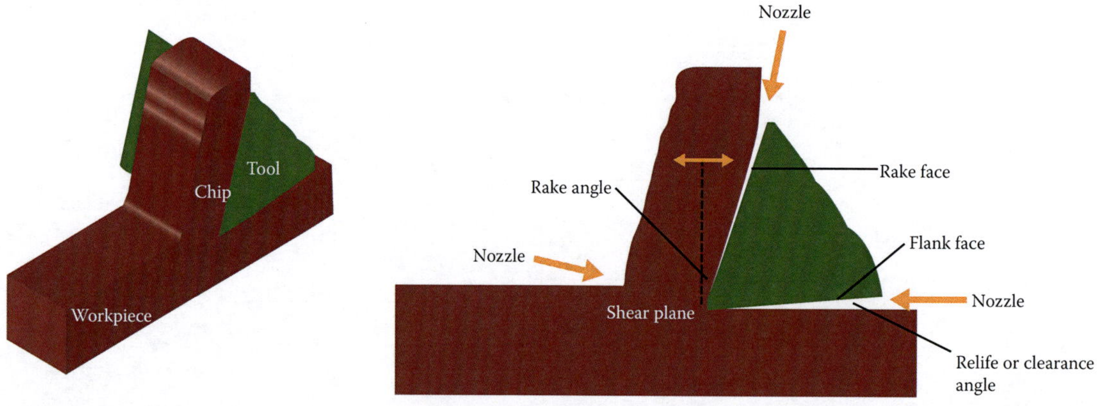

Figure 21.3

Application of cutting fluid.

Disassembly

1. Remove the cotter pin (T5-052), M42x3 nut (T5-054A), output seal guard (T5-048), and the output oil seal (T5-065) from the output shaft.

The cotter pin of 42 mm and 3 mm pitch are removed with the help of an adjustable wrench (AW) and nut driver from Table 12.1. With the predetermined time estimated in Tables 12.3 through 12.10; we get per Reach + Move + Apply pressure + Position + Turn + Grasp + Disengage + Release = (26.7 + 30.7 + 52.1 + 16.2 + 9.4 + 7.3 + 34.7 + 2.0) × 3 = 537.3 TMU. If we consider the degree of difficulty of separation of nut and bolt from Table 12.16 of 2.1 and apply for this disassembly:

Disassembly time of nut and bolt = 537.3 (2.2) = 1128.33 TMU

Time to disassemble the output seal guard: From Table 12.15, we can use the Wedge/Pry tool to dislodge the seal guard. This will need the following operations: Reach + Move + Position + Pull + Move = (26.7 + 30.7 + 53.4 + 9.4 + 30.7) = 150.9 TMU.

Remove the oil seal; this takes the same time as above. It is possible that a screwdriver or wedge pry is used to remove oil seal = 150.9 TMU.

Hence, the total time for the first disassembly operation = (1128.3 + 150.9 + 150.9) = 1430.1 TMU.

2. Remove the top cover (T5-035). Clean sealant from the surface.

Removing the top cover and cleaning the sealant will need the following tools: screwdriver + knife + grinder + brush. Refer to Table 12.1.

The operations required are: Unscrew + Remove + Peel + Clean + Push/Pull.

Time estimates for operation: (9.4 + 30.7 + 53.4 + 34.7 + 28.2) = 156.6 TMU.

3. Remove cover opposite to the input shaft (T5-037). Remove any shims found. Remove any remaining gasket material.

The time estimates for Operation 3 will be approximately as Operation 1. The estimated time for operation 3 = 1430.1 TMU.

4. Tap the input shaft to remove the tapered roller bearing (T5-060) on the opposite side. Remove the input shaft (T5-039 1¾ in × 6 spline or T5-039-20 1¾ in × 20 spline), the crown gear, and the inner race of the tapered roller bearing (T5-064).

Tools required are hammer, screwdriver, chisel, and prying bar. Refer to Table 12.1.

The operations are as follows: tap, remove roller bearing, pull spline shaft out, disengage gear and pull out, remove inner races. Refer to Table 12.15.

Time for operation 4 = Reach to grab 1 lb hammer + Move it near the shaft + tap + move the hammer to safe place + Remove tapered bearing (grasp + apply pressure)+ Remove input shaft (grasp + apply pressure + pull it out) + Use chisel to disengage the gear + remove gear + move gear to the disassembly bin + remove the inner races of the beating = (23.2 + (30.7 + 16.0) + 9.4 + (30.7 + 16.0) + (10.8 + 48.6 + 16.2 + 47.5) + (26.7 + 53.4 + 12.0 + 22.9)) = 364.1 TMU.

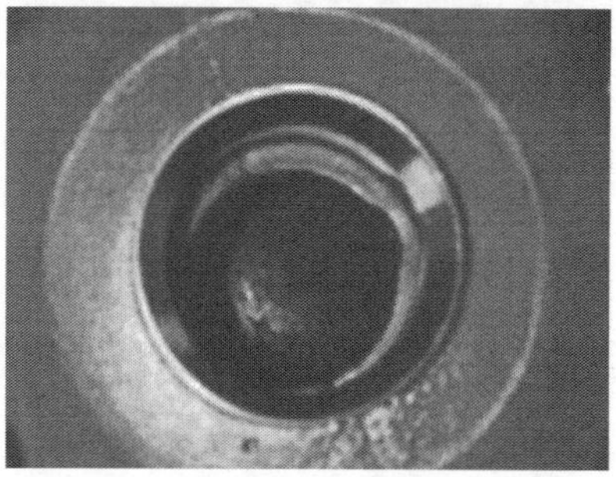

In the estimation of time factors such as weight, difficulty of reaching, reaching for tools, grasp time, moving the weight, and so on, Tables 12.3 through 12.10 are used. We should consider the degree of difficulty factor from Table 12.16. It looks that degree of difficulty should be applied to each element separately; however, here it is being applied together at 2.5. Finally, the time for Operation 3 = (364.2 × 2.5) = 910.25 TMU.

5. Remove the other cover (T5-036) and the oil seal (T5-066).

Time for Operation 5 is the same as that for Operation 3.

Therefore, time for operation 5 = 1430.1 TMU.

6. Remove indentations from locknut (T5-044A).

The indentation removal from the locknut requires the following operation: reaching for tool, applying pressure, removing the indentation. The time estimates for operation 6 = 235.7 TMU.

7. Place an appropriate spanner socket on the lock nut. Be sure the pins are completely engaged.

   The tasks are reaching for spanner socket, grasping, moving to locknut, positioning, applying pressure on spanner, and turning the spanner.

   Time estimates for Operation 7 = (26.7 + 12.9 + (30.7 + 16.0)+53.4 + 16.2 + 28.2) × 2.5 = 460.25 TMU. A DoD of 2.5 has been applied.

8. Unscrew the locking ring using an impact wrench. The ring can be difficult to loosen because of the thread-locking compound (Locktite).

   The tasks are: reach for impact wrench, grasp the wrench, move to locking ring, apply pressure, turn the wrench 10 times, move the locking ring to reuse bin, and move wrench to tool box.

   Time estimate for operation 8 = (26.7 + 12.0 + (30.7 + 47.5) + 16.2 + 10 × 14.8 + (30.7 + 16) + (30.7 + 1.5)) × 2.5 = 900 TMU.

   Here, 10 threads are assumed to be unlocked and degree of difficulty of operation is assumed as 2.5.

9. Remove the spacer (T5-045 on 540 rpm, T5076A on 1000 rpm).

   Tools used are chisel or flathead screwdriver. Tasks are: reach for the tool, grasp it, move to the spacer, position it exactly, apply pressure, and move it or pull it; additionally, move tool to toolbox and remove and release spacer in the recyclable bin.

   Time estimates for operation 9 = (26.7 + 12.9+(30.7 + 1.5) + 53.4 + 16.2+ (15.2 + 9.1) + 30.7 + 15. 2 + 9.1) = 220.7 TMU.

10. Preload the bearing. The correct preload is achieved when rotating torque of output shaft is 20–26 kg-cm (17–23 lb-in).

    Preloading requires applying torque or turn to the bearing and it would depend on the number of turns needed. It is assumed here that the rotating torque of the output shaft is 25 lb-in, hence the time for operation 10 = 28.2 × 25 = 705 TMU. Refer to Table 12.5

11. Remove the output shaft (T5-040), along with the inner races of the tapered roller bearing sets.

    The time of operation 11 is equivalent operation 4 = 364.5 TMU.

12. Remove the outer races of the tapered roller bearing sets from the casting.

    The time required for operation 12 is equivalent to operation 4 = 364.7 TMU.

    All the above time elements are assumed depending on the difficulty of disassembly, tools required, and the motions during disassembly.

    Finally, the total time for disassembly = (1430.1 + 156.6 + 1430.1 + 364.1 + 1430.1 + 235.1 + 460 .25 + 900 + 220.7 + 705 + 364.5 + 364.5) = 8061.05 TMU.

    Time for Disassembly = 8061.05 × 0.036 s = 292 s = 5 min.

    The repair and maintenance crew in an automotive garage normally do not agree with these time estimates; their complaints refer to unavailability of tools at one place, tools available at longer distances (here it was assumed to be 30 in distance), different types of tools, so it takes more time to search and grasp, and tools that do not work as they should meaning a person has to reach for another tool, etc. However, for exact operation, very little difference was reported. However, to account for the problems referred to above, the time estimate is approximated to Time for disassembly = 30 min.

Example 12.8: Eco-economic Analysis of the Disassembly Operation for the 175 HP Gearbox

The parts disassembled here are cotter pin, nut, output seal guard, the output oil seal, the top cover, cover opposite to the input shaft, shims, gasket material, the tapered roller bearing on the opposite side, input shaft, the crown gear, the inner race of the tapered roller bearing, other cover and the oil seal, lock nut, locking ring, spacer, the output shaft, the inner races of the tapered roller bearing sets, the outer races of the tapered roller bearing sets. The parts such as shims, spacers, outer and inner seal guard, output oil seals, cotter pin, locking ring, gasket materials, etc., cannot be reused. The design of shaft is such that they are useful for just one kind of operation; that is, if the input and output shaft for some reason got damaged during the disassembly, they may not reused. A number of reasons are present in the design like presence of splines, the tight fit, life of the gear box, etc., which may bring about damage to the input or output shaft. In all likelihood, it may not happen. There is always some probability of damage and the part not being reused. In such a case, the shaft will be sent for recycling. Crown gears, along with tapered roller bearings, and inner and outer races, are most likely to be reused. Therefore, the list of reusable and recycled parts of the gearbox after disassembly is the following.

Reused: Crown gear, tapered roller bearing, races.

Recycled parts: Shims, spacer, cotter ring, seal guards, oil seals, locking gear, input and output shafts.

Using equations from Chen et al. [15], the cost and benefits are calculated.

Reuse value = Cost of component ($) − Miscellaneous cost ($)

Cost of components = Cost of crown gear + Cost of tapered roller bearings + Cost of races (inner + outer races) = 2 × $65.00 + 2 × $150 − Miscellaneous cost

Miscellaneous cost is calculated as: Miscellaneous cost = Collection cost ($) + Processing cost ($).

In this case, we will assume this cost as zero. Therefore, the reuse value (benefit) is Reuse value (benefit) = $430.00.

The remanufacturing value is also very minimal as no manufacturing operation is involved. There may be operations like cleaning with brush and washing away some old lubricant from gears and bearing required.

Recycle value = Weight of component (kg) × Market value of material ($/kg) − Miscellaneous cost ($)

Recycle values are classified into primary and secondary recycle value. Secondary recycle value refers to the scrap value. In this example, all recycled components are taken as scrap value. Again, the miscellaneous cost is assumed as zero. Therefore, the recycle value is estimated as

Recycle value (benefit) = weight of components (shims, spacer, cotter ring, seal guards, oil seals, locking gear, input and output shafts) × scrap value of materials

It could safely be assumed that oil seals are not recyclable and therefore removed from recyclable list. The number of shims can vary sharply from application to application and here it is assumed that weight of shims = ½ lb. There is usually one spacer on each side but this can vary. In this case, two spacers are assumed and the weight of spacer = 11.5 lb. The major component weights are of the two shafts. The shaft weights are estimated as:

Input and output shaft weight = $2 \times \pi/4(1.75)^2(06) \times 0.28 = 82$ lb

The recycle value (benefit) = (1/2 + 1.5 + 82) × 0.454 × $ 1.5/kg = $57.25

Here, the materials for shims, spacers, and shafts are assumed to be the same as low-alloy steel, which may not be true. The dollar value of scrap steel also varies from country to country, and zones to zones within the United States. The scrap value of $1.5/kg is taken from published scrap values.

To sum up, the reuse and recycle value of the gearbox after disassembly = $430 + $57.25 = $487.25.

If we use the fresh materials and manufactured gears, bearings, input and output shafts, it could safely be said there would be a very high difference in cost; there would be high savings of dollars, as well as materials, in reuse. Think of depletion of material resources. We can also note that the input and output shaft manufacturing involves a lot of dollars and if we can use them, it would bring a tremendous saving.

12.4.6.2 Ecoproperties Calculation

Embodied energy of scrap materials = (84 × 0.454) × 10 MJ/kg = 381.36 MJ

If we use fresh material (instead of recycled material) = (84 × 0.454 × 35) MJ = 1334.76 MJ

The embodied energy savings = 953.4 MJ = 953.4 × 948 = 1×10^6 Btu.

Carbon footprint calculation:

Recycled material carbon footprint = 84 × 0.454 × 0.6 kg/kg = 23 kg of CO_2 emission in the air.

The carbon footprint of fresh carbon alloy steel = 84 × 0.454 × 2.125 kg/kg = 81 kg of CO_2 emission in the air. The savings of carbon footprint by using recycled material = (81–23) = 58 kg of CO_2. These calculations show us how we can save material and limit the carbon dioxide emission in the atmosphere. This estimation helps in sustainability of design and manufacturing and helps reduce global warming. We must not forget that this is the effect of disassembly and recycling.

12.5 Design for Sustainability

The design for sustainability (D4S) concept outlines methodologies for making sustainable improvements (social, economic and environmental) to products by applying elements of life cycle thinking. D4S builds on the work of ecodesign to include economic and social concerns, and its methodology includes both incremental and radical innovation. In the 1990s, concepts such as ecodesign and green product design were introduced as strategies that companies could employ to reduce the environmental impact associated with their production processes. Sustainable innovation and design is not necessarily about new technologies, but about rethinking how to meet the need of growth while at the same time reducing negative environmental and social impacts. Industry is vital for economic growth of all nations, but developing countries operate with much different circumstances and needs that should be taken into account. Sustainable innovation allows for leapfrogging in design and production processes to make it possible for these newly industrializing countries to achieve decoupling in the beginning, not the end, of the industrialization timeline. More production efficiency increases does not, of course, translate into absolute environmental benefit, but cleaner production processes are an essential component of a widespread movement of sustainable design that could challenge entrenched consumption and production patterns.

D4S is an ecodesign concept that has evolved to include both social and economic elements of production. It integrates the three pillars of sustainability, which are people, profit, and the planet, but it goes beyond simply "greening" products to introduce how to meet consumer needs in more holistic, sustainable way. Industries incorporating D4S into long-term product innovation strategies strive to alleviate the negative environmental, social, and economic impacts along a product's supply chain and through its life cycle, or the "cradle-to-grave" mentality. The newly designed products and services offer increased functionality and ease of use, longer life spans, easy disassembly or recyclability, lower environmental impacts that can save the company money and improved materials sourcing and production which can positively affect communities. In other words, sustainability offers added value through better quality and lower price, which are the two motivators for most consumer decisions.

The D4S research program includes the systematic development, testing, and international diffusion of methods and tools for the design of products with superior life cycles, improved eco-efficiency and effectiveness (via intelligent materials and energy applications, integration of emerging product technologies, and economic optimization). The focus is on modeling, actual application, and demonstration of sustainable products, departing from simply an engineering/life cycle design perspective.

The challenges of D4S are to generate knowledge supporting the innovation and design engineering of products and service systems with superior sustainability and to make optimal use of networking and entrepreneurship as success factors for implementation. The key D4S approaches are: redesign and benchmarking (incremental design), and new product design.

Redesign

The goal for D4S redesign is to sustainably redesign an existing product for which the specific market and manufacturing conditions are already known, taking into account its primary function and the associated services provided. A product's improvement potential can be determined relatively easily as the product already exists, so market and manufacturing information is readily available. D4S redesign is of particular interest to developing countries as this more incremental type of innovation requires smaller risks and investment and follows a structured and predictable path, but it can be just as important as a more radical approach in terms of economic and environmental impacts.

12.5.1 Benchmarking

Benchmarking and other approaches based on replication are still the predominant way in which products are designed worldwide, in developing and developed countries alike. The goal of D4S benchmarking is to learn from the best practice of competitors by comparing one's product to those of competitors to determine how to make that product sustainable. The methodology is a structured approach to which improvement options are generated by looking at the environmental, social, and economic aspects of a particular product.

Benchmarking differs from D4S redesign in that it starts with comparing existing products in the market before moving into the design phase.

12.5.2 New Product Design

The D4S new product development approach applies "out-of-the-box," or radical innovation strategies, which can lead to more sustainable impacts while providing the breakthroughs necessary to ensure a company's continued competitiveness. New product design involves a higher level of technical, market, and organizational uncertainty than redesign but can be an inventive and iterative process where new ideas on how to meet needs are converted to products and services. Eco-friendly materials, sustainable development practices, and innovative information and communication technology are all concepts that can help inspire new product design. As consumer needs and expectations evolve, new products and services offer opportunities to enhance product portfolio sustainability, including addressing increasingly important social concerns.

The stages and processes involved with new product design can be viewed as threefold: policy formulation, idea generation, and product development. Policy formulation addresses the company's goal and strategies; idea generation allows the company to brainstorm and develop ideas for new products, taking into account the ability to harness developing technologies, materials and consumer needs; and finally, product development involve debating and testing concepts against the decisions in the idea finding phase. The key challenge with respect to new product design is market demand. Without consumer need, even the most sustainable product will fail.

Today there is a significant and growing trend of recognizing what happens to a product once its user phase is finished or after EOL. Legislative pressures and consumer awareness are driving businesses to develop sustainable product design strategies. Engineers need to have an understanding of production methods, but all processes are not necessarily relevant to all materials, situations or components. Therefore, it is also necessary to have an understanding of how processes influence products and systems. Traditionally, design for manufacture is about optimizing design to aid manufacturing. Today, there is a significant and growing trend of recognizing what happens to product once its use phase has finished. Postconsumer processes are now an important consideration during the initial stages of design. In the past, the focus was limited to design for manufacture and assembly. Now the pressure is on for postconsumer design. EOL processing and DfE are therefore areas of growing interest.

If material is recycled and reused by a manufacturer, then money may be saved as less virgin material will have to be bought. If a product or part of a product has to go to landfill then it has a negative value associated with it, as someone will have to pay for that disposal. These considerations must be factored into the initial design as they alter the costing over the whole product life cycle. For manufacturers required to take back their products at EOL, these issues will directly affect their profitability.

The constraints of an EOL processing technique should be accounted for in the early design phase. Constraints and standards are something that a designer needs to know about before beginning to develop a new component or product. For example, to design a molded plastic part, it is necessary to incorporate draft angles within the design so that it can be ejected easily from the mold. Failure to incorporate draft angles could result in a higher cost or greater difficulty to manufacture. However, DfD and EOL processing are not so well established or understood. There is a need for practical advice and information about the implications of design decisions and material choices.

12.5.3 Life Cycle Assessment for EOL Analysis

Life cycle assessment (LCA), also known as cradle-to-grave analysis, is a technique to assess environmental impacts associated with all stages of a product's life from cradle to grave (i.e., from raw material extraction through material processing, manufacture, distribution, use, repair and maintenance, and disposal or recycling). LCAs can help avoid a narrow outlook on environmental concerns by

- Compiling an inventory of relevant energy and materials inputs and environmental releases
- Evaluating the potential impacts associated with identified inputs and releases
- Interpreting the results to help you make a more informed decision

Life cycle engineering/design is the discipline that incorporates environmental issues and parameters across the life cycle of a product into product development. By using the LCA tool, the weaknesses of a product in term of environmental consequences can be detected.

12.5.3.1 Life Cycle Design (LCD)

Life cycle design is characterized by three main aspects: the perspective broadened to include the entire life cycle; an assumption that the most effective interventions are those made in the first phase of design; and the simultaneity of the operations of analysis and synthesis on the various aspects of the design problems. This last characteristic merges with the life cycle design with design techniques oriented toward the control and compression of the times and costs of product development. The selection of design alternatives must be guided by considering the main factors of product success, which define the design targets, in relation to all phases of life cycle; these are resources utilization (optimizing materials and energy use); manufacturing planning (optimization of production processes); life cycle cost (optimization of the total cost of the life cycle); product properties (harmonizing a wide range of required product properties, such as ease of production, functionality, safety, quality, reliability, aesthetics); company policies (respect for the company objectives); and the environmental protection (control and minimization of environmental impacts).

In the course of the life cycle design directed at environmental improvement of the product, the evaluation of results must be continuous and distributed throughout all the phases of the design process. To evaluate design alternatives and identify the best alternative satisfying the environmental constraints, it is necessary to make use of a proper tool to quantify the environmental performance of the product under development. The most common and complete tool is LCA. LCA allows a complete environmental analysis of the whole production system, giving an evaluation of the performance in relation to the entire life cycle. LCA is the tool that visualizes the environmental and resources consequences of design choices. It compiles the relevant inputs and outputs of the product system; evaluates the potential environmental impacts associated with those inputs and outputs; and interprets the results of the inventory analysis and impact assessment phases in relation to the objectives of the study (Table 12.23).

Traditionally, the manufacturing companies attempt to control the costs of resources, production and distribution; however, that is incomplete in the present global economy. Today the companies have to include costs of the entire life cycle among the parameters of the decision-making process. LCC analysis, or life cycle costing, is a methodology directed to the calculation of all costs associated with an activity or a product over its entire life cycle, making it the most appropriate economic analysis tool in a life cycle design. The importance of estimating and controlling costs during the design phase is the most important factor in controlling the cost of the product. It relates not only to the product cost, but also those corresponding to the phase of use and disposal, which are greatly affected by the design choices. Therefore, life cost analysis is the methodology used to evaluate all the costs associated with a product over the entire life cycle. We can represent the LCA graphically as shown in Figure 12.9.

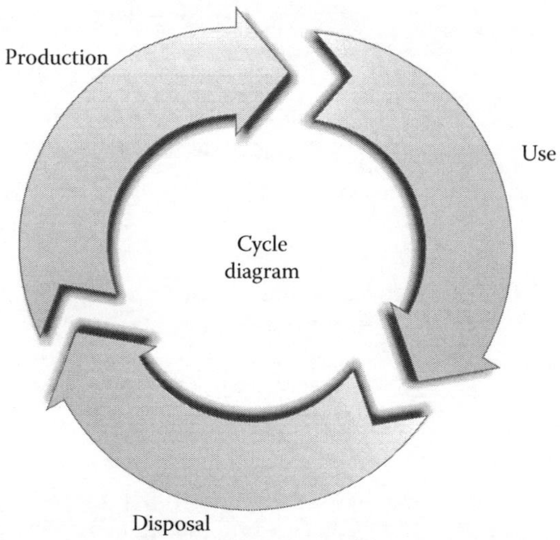

Figure 12.9

Life cycle analysis for the end of life.

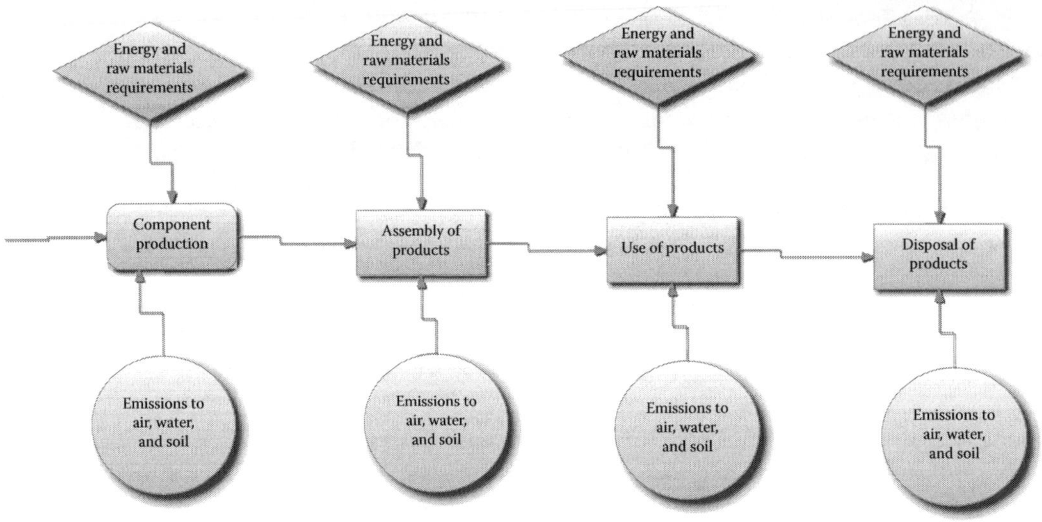

Figure 12.10

LCA diagram.

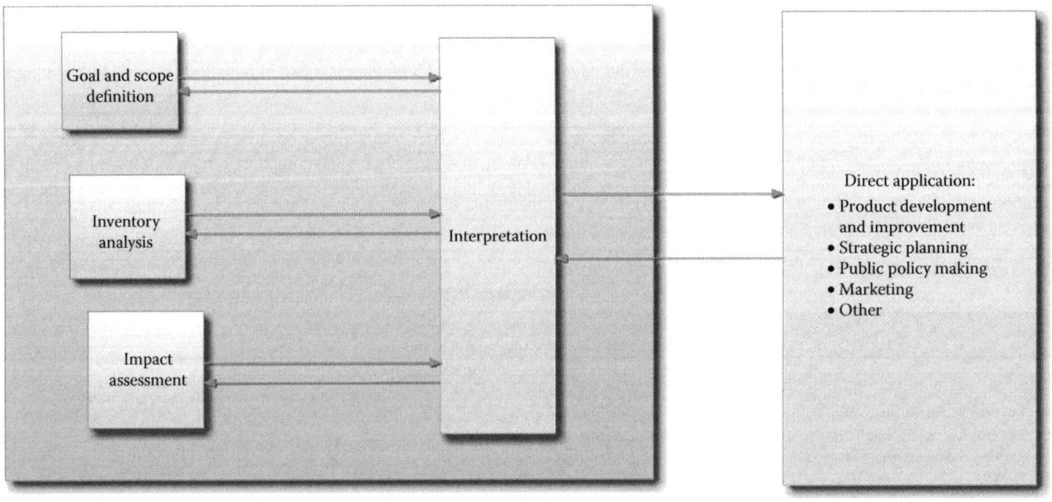

Figure 12.11

Phases of LCA.

To help minimize the environmental pollution and resource use, the application of LCA principles in production, use, and disposal are essential. It is also helpful in improving the profitability of the company. ISO produced a series of standards in 1997/98, which were revised in 2006. ISO 14040 published in 2006, outlined LCA principles and a framework. It has also published 14044 in 2006, which outlined the requirements and guidelines for environmental management including, LCA. Graphically, all the factors considered in an LCA are shown in Figure 12.10.

It should be recognized that LCA is a method for determining the environmental impact of a product or system. However, it is difficult to consider all social, economic, and environmental issues in a short period. Overall impact of production on air, water, and soil is not even documented and LCA software does not have the complete database for all these impacts. Impact assessment framework according to ISO 14040 is detailed below. Four different phases are distinguished by ISO 14040 as shown in Figure 12.11.

LCA software is available that gathers and analyzes all the data collected in a large spreadsheet. Data collection can depend on sources from the specific industry, published literature or a database of the software. In most cases it could be all three. However, the databases in different software are insufficient with regard to the sources of energy conversions. It is very time-consuming to collect the data from different sources. The impact assessment depends on the database created earlier. It is always helpful to have all the materials and related emissions in the inventory (database). However, there are three stages of life cycle impact assessment; classification, characterization, and valuation. It seems that some parts of impact assessment are subjective.

12.5.3.1.1 Classification At this stage of impact assessment, it is advised to group the materials and emissions in different categories of environmental impact. For example, all CO_2 emissions should be in a separate category than the ozone-depleting gases like CFCs and HCFCs. ISO 14040 advocates the consideration of large number of environmental issues. These are: the GHGs, ozone-depletion gases, acidification, eutrophication, summer smog, winter smog, heavy metals, carcinogenic substances, waste, respiratory effects, ionizing radiation, eco-toxic substances, land use, and raw materials. It is difficult to get the complete data for all the above and most of the LCA software does not possess all these data completely.

12.5.3.1.2 Characterization All the emissions that contribute to environmental impact will not necessarily have the same effect. Characterization means quantifying the environmental impacts of different environmental issues. For example, some characterization data for carcinogens are tabulated below.

Carcinogens	kg
As	0.0044
Benzene	0.000011
Benzon[a]pyrene	1
Cr(6+)	0.44
C_xH_y aromatic	0.000011
Ethylbenzene	0.000011
Fluranthene	1
Ni	0.44
PAH	1
Tar	0.000011

The Environmental Protection Agency (EPA) has developed an equation for the normalization of emissions.

The interpretation stage of LCA identifies areas with the potential for improvement within a system. Interpretation means to compare their notes with experts in the industry or practitioner in the real world. This exercise can help improve the calculation of environmental effects of the system.

Example 12.9: LCA: Green Power Generation from Hydro

A reservoir hydro power station works by harnessing the energy from falling water held in the reservoir. The water held above the power station in a lake or reservoir (dam) is channeled through penstocks to the turbine. The turbine extracts the energy from the water, turning it into mechanical energy that spins the generator rotor. This in turn generates electricity in the stator windings. For any given volume of water, the amount of energy that can be extracted is determined by the height from which the water falls from reservoir. This is known as the head of power station. After passing through the turbine, water exits through a draught tube, back to the reservoir or lake. The physical structure of a reservoir hydroelectric generation plant is shown in the figure below.

12.5.3.2 Embodied Energy Analysis for Reservoir Power Plant

The life cycle of a reservoir hydro power station consists of preliminary investigations and river diversion, power plant construction and civil works, operation and maintenance, and plant decommissioning and land reclamation. For the power station, the power plant construction and civil works include the hydro dam, the spillway, the intake and penstocks, and the power house and switchyard. The life cycle of a hydro power station, with energy flows, is shown in the graphic below.

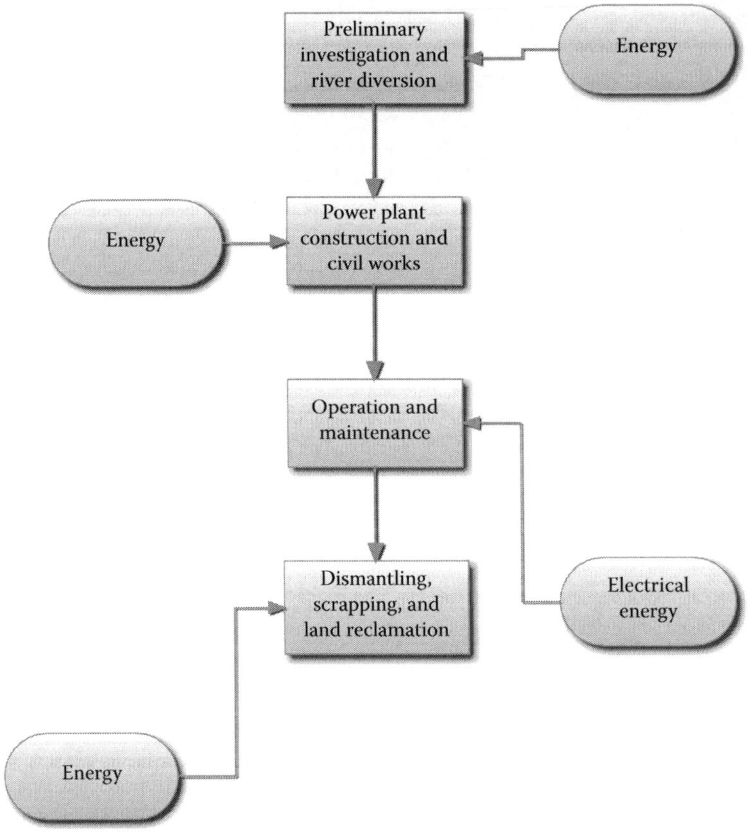

12.5.3.3 Electrical Energy Output: Example

The estimated life cycle electrical energy output of the Benmore power station is estimated by multiplying the average annual energy output to the national system by the estimated life of the Benmore power station. The average annual electrical energy output of a power station is 2215 GWh.

The total life of the power station is assumed to be 200 years, as it has already been in operation for 45 years and Meridian Energy has predicted it will be in operation for more than another 145 years. The total life cycle electrical output was estimated to be 1,595,000,000 GJ. Detailed calculations and explanations are presented below.

Annual energy output: 2,215 GWh
Calendar year lifetime: 200 years
Total life cycle electrical output: 443,000 GWh
Total life cycle electrical output: 1,595,000,000 GJ

12.5.3.4 Preliminary Investigations and River Diversion

The embodied energy of the preliminary investigations is calculated using I/O analysis. The cost of preliminary investigation is \$2004. The embodied energy of the river diversion was calculated using both PCA and I/O analysis.

12.5.3.5 Energy Embodied in Preliminary Investigations and River Diversion

Description	Embodied Energy (GJ)
Preliminary investigations	7,000
Construction materials of river diversion	186,000
Other processes of river diversion	89,000
Total embodied energy	282,000

12.5.3.6 *Major Civil Works*

The major civil works of the power station include the dam, spillway, intake and penstocks, and power house and switchyard. The embodied energy associated with the construction materials of the major civil works is presented below.

The construction materials of the dam, spillway, intake, and penstocks in concrete amounted to 80,000 cubic yards, which is approximately equivalent to 141,910 tons.

12.5.4 Energy Embodied in Construction Materials of Major Civil Works

Material	Weight	Embodied Energy (GJ)
Earth: hydro dam	21,807,000	94.77
Concrete: spillway, intake, and penstocks	489,000	2.12
Concrete: power house and switchyard	142,000	0.62
Concrete: total	631,000	2.74
Steel: spillway	10,000	0.04
Steel: powerhouse and switchyard	182,000	0.79
Steel: total	192,000	0.83
Aluminum: power house and switchyard	1,000	0.01
Chromium: power house and switchyard	124	0.0005
Copper: power house and switchyard	2,000	0.008
Iron: power house and switchyard	8,000	0.04
Manganese: power house and switchyard	4,000	0.02
Molybdenum: power house and switchyard	300	0.001
Plastic: power house and switchyard	4,000	0.02
Silicon: power house and switchyard	3,000	0.01
Vanadium: power house and switchyard	9,000	0.04
Total	22,660,000	54

Earth used in the construction of the hydro dam accounts for nearly 95% of the embodied energy of the construction materials of the major civil works in the power station. Concrete and steel, which are the second and the third highest contributors, account for 2.74% and 0.83%, respectively. The other construction materials only account for about 0.15% of the embodied energy.

The embodied energy associated with the construction of major civil works that are not related to construction materials is calculated using I/O analysis. This includes excavation and disposal of earth, dewatering, engineering and construction, administration, transportation of equipment, and powerhouse and switchyard equipment.

Embodied energy associated with the construction of major civil works not related to construction materials:

Description	Embodied Energy (GJ)
Hydro dam construction	296,000
Spillway, intake, and penstocks construction	358,000
Powerhouse and switchyard construction	250,000
Powerhouse and switchyard equipment	426,000

12.5.5 Ancillary Construction Works

The embodied energy of ancillary construction works include the permanent village and other establishments, and other miscellaneous expenses such as public relations, village maintenance, and flood damage.

12.5.5.1 *Embodied Energy Associated with Ancillary Construction Works*

Description	Embodied Energy (GJ)
Permanent village and other establishments	304,000
Other miscellaneous construction works	86,000
Total embodied energy of ancillary construction works	390,000

12.5.6 Plant Operation and Maintenance

The energy associated with the operation and maintenance of the power station was calculated using I/O analysis. The annual operation and maintenance costs of Benmore power station is approximately \$1,000,000. The power station is undergoing refurbishment after nearly 50 years of operation. This is going to cost approximately \$12,000,000 per annum over five years. A repeat of this type of capital investment is not expected to be made for another 50 years. Therefore, taking into consideration that the assumed lifetime of the power station is 200 years, it is estimated that the annual cost of refurbishing is \$12,000,000.

12.5.6.1 Difficulties Associated with LCA in EOL

As has been indicated earlier, the database for LCA analysis is always incomplete or insufficient. The insufficiency affects both quality and quantity. It has been observed that differing data used in the characterization stage will produce different results after LCA analysis. Use of alternative methods or software may yield incomparable results. So the data reliability is a big problem in LCA. It is strongly advised that improvement of data and results must be checked and compared with the real-world data. However, ISO 14040/44 is a globally accepted standard for life-cycle–based environmental assessment. There is growing use of ISO 14044/44 all over the world.

12.5.6.2 Footprint Indicators

Mostly footprint is understood as the total burden placed on the environment. However, most important of these footprints is carbon footprint. There are other footprint indicators for example, such as that for energy use, solid waste, and air emissions.

For green design and manufacturing, the ecoproperties of materials need to be known, as without these the design and manufacturing parameters cannot be calculated. However, there are no standardized methods to measure embodied energies and carbon footprints of engineering materials. A sample of data is presented here for use in green design and manufacturing for sustainability problems. International standard ISO 14040 has provided some guidelines and procedures but it is very difficult to apply that to engineering problems. Here only relevant eco-material properties data are presented along with its typical use.

- Aluminum alloys. Composition: Al + alloying elements, for example, Mg, Mn, Zn, Zr, Li.

  Typical usage: aerospace engineering, automotive engineering-pistons, clutch housing, exhaust manifolds, sports equipment such as golf clubs and bicycles, die-cast chassis for household and electronic products, sidings for buildings, reflecting coatings for mirrors, foil for containers and packaging beverage cans, electrical and thermal conductors.
- Magnesium alloys. Composition: Mg + alloying elements, for example, Al, Mn, Si, Zn, Cu, Li, rare earth elements.

  Typical usage: aerospace, automotives, sports goods such as bicycles, nuclear fuel cans, vibration damping and shielding of machine tools, engine case castings, crank cases, transmission housings, automotive wheels, ladders, housing for electronic equipment and particularly mobile phone and portable computer chassis, camera bodies, office equipment, marine hardware, and lawnmowers.
- Titanium alloys. Composition: Ti + alloying elements, for example, Al, Zr, Cr, Mo, Si, Sn, Ni, Fe, V.

  Typical usage: aircraft turbine blades, general aerospace applications, chemical engineering, pressure vessels, high-performance automotive parts such as connecting rods, heat exchangers, bioengineering, medical, missile fuel tanks, compressors, valve bodies, light springs, surgical implants, marine hardware, paper-pulp equipment, sports equipment such as golf clubs and bicycles.
- Copper alloys. Composition: Cu with up to 40% Zn or 30% Sn, Al, or Ni.

  Typical usage: electrical wiring, cables, bus bars, high-strength and high-conductivity wires and sections, overhead lines, contact wires, resistance-welding electrodes, terminals, high-conductivity items for use at raised temperatures, heat exchangers, coinage, pans, kettles and boilers, plates for etching and engraving, roofing and architecture, cast sculptures, pumps, marine propellers.
- Lead alloys. Composition: Pb + 0%–25% Sb or 0%–60% Sn, sometimes with some Ca.

  Typical usage: roofs, wall cladding, pipe work, window seals, and flooring in buildings; sculpture and table wear as pewter; solder for electrical circuits and for mechanical joining; bearings; painting type; ammunition; pigments; x-ray shielding; corrosion-resistant material in chemical industry; and electrodes for lead-acid batteries.

- Zinc die-casting alloys. Composition: Zn + 0%–30% Al, typically, often with up to 30% Cu.

  Typical usage: die-castings, automotive parts and tools, gears, household goods, office equipment, building hardware, padlocks, toys, business machines, sound reproduction equipment, hydraulic valves, pneumatic valves, soldering, handles, automotive components.
- Nickel-chromium alloys. Composition: Ni + 10%–30% Cr + 0%–10% Fe.

  Typical usage: heating elements and furnace windings, bimetallic strips, thermocouples, springs, food-processing equipment, chemical engineering equipment.
- Nickel-based superalloys. Composition: Ni + 10%–25% Cr + Ti, Al, Co, Mn, Zr, B, and Fe in varying proportions.

  Typical usage: blades, disks, and combustion chambers in turbines and jet engines; rocket engines; general structural aerospace applications; light springs; high-temperature chemical engineering equipment; bioengineering and medical.
- Low-carbon steel. Composition: Fe/0.2–0.3C.

  Typical usage: reinforcement of concrete, steel sections for construction, sheet for roofing, car body panels, cans, and pressed-sheet products. Low-carbon steel is widely used in engineering and no list can be complete.
- Low-alloy steel. Composition: Fe/,1.0, C/,2.5, Ni/,2.5, Mo/,2.5, V.

  Typical usage: springs, tools, ball bearings, rollers, crankshafts, gears, connecting rods, knives and scissors, pressure vessels.
- Stainless steel. Composition: Fe/,0.25C/16–30 Cr/3.5–3.7 Ni/, 10 Mn + Si, P, S (+N for 200 series).

  Typical usage: railway cars, trucks, trailers, food-processing equipment, sinks, stoves, cooking utensils, cutlery, flatware, scissors and knives, architectural metalwork, laundry equipment, chemical processing equipment, jet-engine parts, surgical tools, furnace and boiler components, oil-burner parts, petroleum-processing equipment, dairy equipment, heat-treating equipment, automotive trim, structural uses in corrosive environments, for example, nuclear plants, ships, offshore installations, underwater cables, and pipes.
- Cast iron, ductile (nodular). Composition: Fe/3.2-4.1%, C/1.8–2.8%, Si/,0.85, Mn/,0.15, P/,0.03%, S

  Typical usage: brake discs and drums, bearings, camshafts, cylinder liners, piston rings, machine tools, structural parts, engine blocks, gears, crankshafts, heavy-duty gear cases, pipe joints, pump castings, components in rock crushers.

12.5.6.3 Goals and Purpose of LCA

The goal of LCA is to compare the full range of environmental effects assignable to products and services in order to improve processes, support policy and provide a sound basis for informed decisions. The term life cycle refers to the notion that a fair, holistic assessment requires the assessment of raw material production, manufacture, distribution, use, and disposal, including all intervening transportation steps necessary or caused by a product's existence. The procedures of LCA are part of the ISO 14000 environmental management standards: in ISO 14040:2006 and 14044:2006. (ISO 14044 replaced earlier versions of 14041 to ISO 14043.) The proposed framework for integrating the assembly and disassembly in life cycle management consists of two steps. The first step is cost analysis for finding the best EOL option. The next step is improving the assembly and disassembly design based on EOL option and design weaknesses. Cost analysis proposed in this research is an optimistic method. In the proposed framework, the Boothroyd et al. [2] design evaluation worksheet is applied to find the assembly cost and Anoop-Mital method is used to get the disassembly cost.

Figure 12.12 shows the proposed worksheet for cost analysis. EOL options are reuse, remanufacturing, and recycling. Figure 12.11 shows a Cost Analysis Worksheet Reuse, associated with disassembly cost and cleaning cost. Remanufacturing is associated with disassembly cost, cleaning cost, redesign cost, and remanufacturing cost. Recycling is associated with disassembly cost, cleaning cost, and recycling cost. Costs are defined as the following:

- Assembly costs (Ac): costs associated with assembling the component in the product structure
- Disassembly costs (Dc): costs associated with taking the component apart from the assembly
- Cleaning costs (Cc): costs associated with cleaning of the component
- Recycling costs (Rec): costs associated with recycling of the component
- Redesign costs (Dec): costs associated with remanufacturing the component

			EOL option												Rate of interest						
			Reuse			Remanufacturing					Recycling										
Part ID	Part name	Photo	Disassembly cost	Cleaning cost	Reuse EOL revenue	Disassembly cost	Cleaning cost	Redesign cost	Remanufacturing cost	Remanufacturing EOL revenue	Disassembly cost	Cleaning cost	Recycling cost	Recycling EOL revenue	Assembly cost	Estimated life	Reuse net present worth	Remanufacturing net present worth	Recycling net present worth		

Figure 12.12

Worksheet for cost analysis under LCA.

There are also some revenues at the EOL of product that can be taken from recycling, reuse, or remanufacturing the product.

- Recycling revenue (Rr): revenue accrued from recycling the component
- Remanufacturing revenue (Dr): revenue accrued from selling the remanufactured component
- Reuse revenue (Rer): Revenue accrued from reusing the component

All of these revenues and costs are brought into account using a net present worth cost analysis. In order to conduct this cost analysis, two more data sets are required. The first is the rate of interest, which is related to the area of manufacturing and using the product, and the next one is product life. Product life shows the estimated years that a product can be used before throwing it away. After conducting this cost analysis, net present worth of reuse, recycling, and remanufacturing are calculated. If there is no limitation on EOL option, the one with highest net present worth should be selected. Otherwise, desired EOL option is selected.

12.5.6.4 Evaluation Methods of Disassemble-ability for Automotive Components

Automotive vehicles have become most common form of transport world over. Automobiles are manufactured in large numbers in every country, consume lots of gas and emit hazardous gases in the atmosphere. The automobile components could be recycled and reused. Automotive manufacturers are facing the impact that vehicles have on the environment. In some European countries, Japan, USA, and Australia, laws require automotive manufacturers to take back their products at the end of their useful life and recycle them. Most of the developed countries also have set new legislation planned to force automotive manufacturers to recover and recycle their products at the end of their life.

Disassembly is the first step to recycle automotive components. In general, the designers do not have the experience in disassembly and recycling to determine the impact of various design aspects on the disassembability at the disassembly stage. Based on that, many researchers have proposed several methods in evaluating disassembability. In particular, the implication of EOL vehicles on current product design practice needs to be explored. At present, approximately 75%–80% are being recycled by weight. However, the remaining 20%–25% in weight, consisting mainly of heterogeneous mix of materials such as resins, rubber, glass, textile, and so on, is still being sent to landfill or incinerator. According to an EU directive [17], the disposal of vehicles is a major source of hazardous waste and toxic emissions. About 25% of a vehicle's weight is classified as hazardous waste. The world is facing up to the impact vehicles have on consumption of energy and resources, waste generation, GHGs, hazardous substance emissions, and disposal issues. In order to enhance the recycling rate of the vehicle, it is important to disassemble every part of the vehicle, depending on the materials. However, it may be impossible as the cost may be too much or the goal is unrealistic. Because of that, the disassembly effort should be determined in the vehicle development phase, where its properties are fixed. Based on that, the disassembability, or the degree of easiness of disassembly, becomes important. The environmental impact of products can be minimized only if the products can be disassembled and recycled easily and the cost-effectiveness of recycling will be increased if disassembly is made easier.

12.6 EOL Vehicle Dismantling Process

In general, the designers do not have the experience in disassembly and recycling to determine the impact of various design aspects on the disassembability at the disassembly stage. It is therefore important that support systems for disassembability are available that encourage designers to incorporate the disassembability issue in order to fulfill the legislation that will be fully implemented in 2015. The system should provide feedback regarding possible improvement to the designer.

The system should provide assistance in making decisions at the early stage of the vehicle design and development process in order to avoid the cost and time consumed through later design. There needs to be fast and correct decision-making in order to remain competitive in the world market. Based on the study of several papers published in the field of EOL vehicles, integrating the constraints from EOL strategies into the early design is one of the important aspects that needs to be considered.

From EOL vehicles, dismantling companies first remove the oil, engine, transmission, tires, battery, catalytic converter, and other parts that are commonly recycled or reused. The dismantling process is the process or procedure of breaking down an EOL vehicle into its individual components and material-specific elements [15]. Shredding companies then sort out the ferrous and non ferrous metals and resin from the remaining vehicle bodies. At shredder facilities, hulks are inspected prior to shredding to ensure that potentially hazardous components such as batteries, gas tanks, and fluids have been removed.

Hulks (and other collected materials) are then shredded into fist-sized pieces using large hammer mills. The post-shredder process separates the stream of materials into the basic streams of ferrous material and nonferrous material. By using air separation, the nonferrous materials are separated into the metal (containing aluminum, brass, bronze, copper, lead, magnesium, nickel, stainless steel, and zinc) and nonmetal known as ASR (containing plastic, glass, rubber, carpet, foam, textile, etc.) fraction. The ferrous metal fraction is sent to the steel smelters for recycling. The metal fraction is typically sent to another specialized facility to separate the stream into its individual metals by a variety of means. The ASR is disposed into landfill [15]. ASR often contains hazardous substances such as lead, cadmium, and printed circuit board (PCB). Therefore, some countries have classified ASR as hazardous waste and have established legislative controls.

References

1. Frosch, R. A. and Gallopoulos, N. E. (1989). Strategies for manufacturing. *Scientific American* 26 (3): 144–152.
2. Boothroyd, G., Dewhurst, P., and Knight, W. (2002). *Product Design for Manufacture and Assembly*, 2nd edn. Marcel Dekker, New York.
3. Segur, A. B. *Predetermined Motion Time System (PMTS)*.
4. Dowie, T. and Kelly, P. (1994). Estimation of disassembly times. Technical Report DDR/TR15, Manchester Metropolitan University, UK.
5. Ashby, M. F. (2009). *Materials and the Environment: Eco-informed Material Choice*. Elsevier, Toronto.
6. Desai, A. and Mital, A. (2005). Incorporating work factors in design for disassembly in product design. *Journal of Manufacturing Technology Management* 16: 712–732.
7. Whitney, D. E. (2004). *Mechanical Assembly: Their Design, Manufacture, and Role in Product Development*. Oxford University Press, New York.

8. Kroll, E. and Hanft, T.A., 1998. Quantitative evaluation of product disassembly for recycling. *Research in Engineering Design-Theory, Applications, and Concurrent Engineering* 10: 1–14.
9. McGlothlin, S. and Kroll, E. (1995). Systematic estimation of disassembly difficulties: Application to computer monitors. *IEEE International Symposium on Electronics & the Environment* (pp. 83–88), Orlando, FL.
10. VDI Standards. VDI recycling-oriented product development. www.vdi.eu/engineering/vdi-standards/.
11. Worrell, E. and Reuter, M. (eds). (2014). *Handbook of Recycling: State-of-the-Art for Practicitioners, Analysts, and Scientists.* Elsevier, Waltham, MA.
12. Kroll, E. and Carver, B. S. (1999). Disassembly analysis through time estimation and other metrics. *Robotics and Computer-Integrated Manufacturing* 15: 191–200.
13. Suga, T., Saneshige, K., and Fujimoto, J. (1996). Quantitative disassembly evaluation. In *Proceedings of the 1996 IEEE International Symposium on Electronics & the Environment* (pp. 19–24), Dallas, TX.
14. Lee, S. G., Lye, S. W., and Khoo, M. K. (2001). A multi-objective methodology for evaluating product end-of-life options and disassembly. *International Journal of Advanced Manufacturing Technology* 18: 148–156.
15. Chen, R. W., Navin-Chandra, D., and Print, F. B. (1993). A cost-benefit analysis model of product design for recyclability and its application. In *IEEE International Symposium on Electronics and the Environment* (pp. 178–183). Arlington, VA.
16. David Brown Gear Systems Ltd., Park Road, Lockwood, Huddersfield HD4 5DD, UK.
17. ISO 14044:2010. (2010). Environmental Management-Life Cycle Assessment, 2010.
18. United States Environmental Protection Agency (EPA). (2010). Life Cycle Assessment, 2010.

13

Snap-Fit Design

13.1 Introduction to Snap-Fit: General Discussion

A snap-fit joint is an economical and quick method of joining plastic parts. A snap-fit joint can be designed either to be easily separated or to be inseparable without breaking one of its components. The strength of the snap-fit joint depends on the material used, its geometry, and the forces acting on the joint. Most snap-fit joint designs share the common design features of a protruding ledge and a snap foot. Whether snap joints are cantilever or cylindrical fit, they function similarly. When snap-fit joints are being designed, it is important to know the mechanical stresses to be applied to the snap beams after assembly, the required mechanical stresses or strains on the snap beams during assembly, the number of times the snap joint will be engaged and disengaged, and the mechanical limits of the material(s) to be used in the design.

A snap fit is an effective method for incorporating the fastening system into the product design itself. A snap fit can be designed either to allow parts to be permanently fastened (or predetermined to be broken off) or for frequent assembly and disassembly. In combination with O-rings or proper seals, even gas and fluid connections can be made. Designing a snap fit is complex due to a combination of factors:

- The functional requirements of the product
- The assembly requirements
- The mechanical properties of the thermoplastics
- The design of the mold and the individual part

Some snap fits can be found in a wide variety of shapes. Examples of typical snap-fit geometries are the cantilever beam type, the cylindrical type, and the spherical type.

Snap fits are commonly used as an assembly method for injection-molded parts. Although they have been around for many years, they have recently become more important in an attempt to simplify assembly and its associated costs. Snap fits are very useful, because they eliminate screws, clips, adhesives, and other joining methods. The snaps are molded into the product, so additional parts are not needed to join them together. Additionally, if designed correctly, they can be disassembled and reassembled several times without any problems.

13.1.1 Reasons to Use Snap Fits

- They reduce assembly costs.
- They are typically designed for ease of assembly, and are often easily automated.
- They replace screws, nuts, and washers.
- They are molded as an integral component of the plastic part.
- No welding or adhesives are required.
- They can be engaged and disengaged.

A snap fit can be designed as either a permanent snap or a multiple snap. Permanent fits are used in disposable parts that are never meant to be disassembled. Multiple snaps are used in most designs in which disassembly for service is expected.

There are complications associated with snap fits. A lot more engineering is required to design a proper snap fit than to design something for screw assembly. Snap fits can also make the injection mold that produces the parts significantly more complicated and expensive. Generally, the savings in assembly costs more than make up for the added cost and complexity of the mold.

13.1.2 Things to Be Aware of When Using Snap Fits

- Some designs require higher tooling cost.
- They are susceptible to breakage due to mishandling and abuse prior to assembly.
- Snap fits that are assembled under stress will creep.
- It is difficult to design snap fits with hermetic seals. If the beam or the ledge relaxes, it may decrease the effectiveness of the seal.

There are three main types of snap fits: annular, cantilever, and torsional.

13.2 Annular Snap Fits

Figures 13.1 through 13.3 show examples of annular snap fits. Because there is so much hoop strain when the parts are assembled, usually only materials with a high elongation at yield are suitable for use. The calculation for the maximum deflections of the fit will be shown in Section 13.3.

13.3 Cantilever Snap Fits

Cantilever snap fits are the most widely used type. A considerable amount of calculation and engineering goes into designing a good snap fit; this section will describe different design techniques. This section presents calculations to find the specific dimensions of the cantilever.

Figure 13.4 shows four different methods of designing snap fits for disassembly. Figure 13.4a shows a snap fit with a 90° hook and a 90° recess; this permanent fit cannot be disassembled. Figure 13.4b uses an angled surface for both the assembly and disassembly portions of the cantilever and the recess. The cap can be removed with the same force as used in assembly. Figure 13.4c has the same 90° angles as Figure 13.4a, but

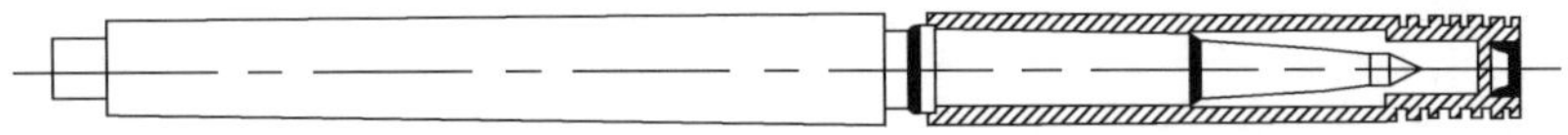

Figure 13.1

A pen sometimes uses an annular snap fit to retain the cap.

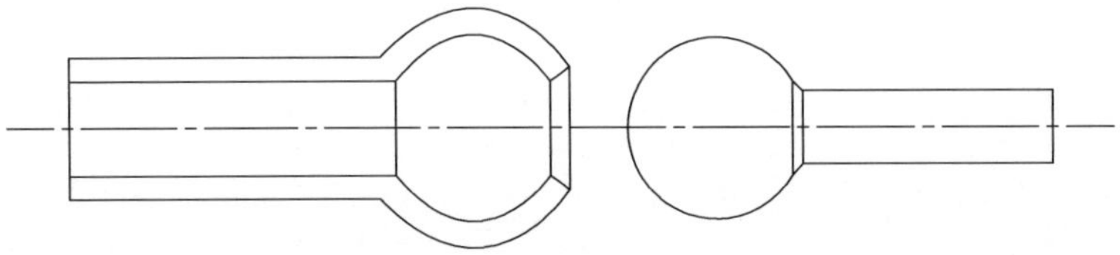

Figure 13.2

A ball and socket joint is a kind of annular snap fit.

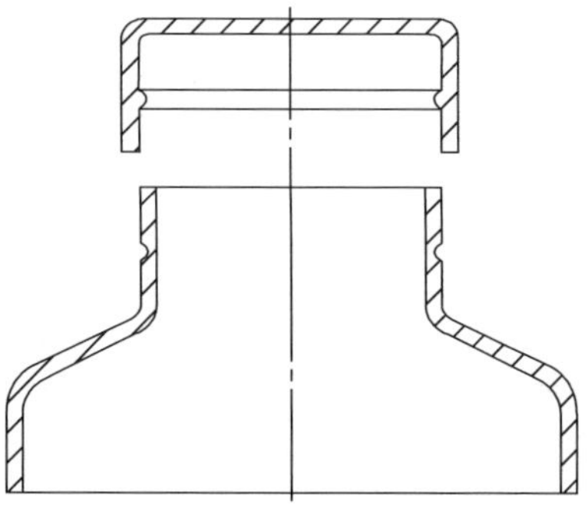

Figure 13.3

This bottle cap uses an annular snap fit.

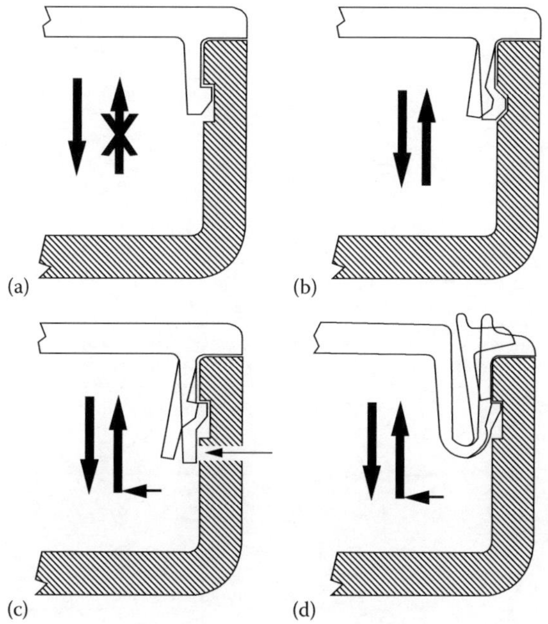

Figure 13.4

(a–d) These four snap-fit designs allow different types of disassembly.

the designer has put a window in the side to allow the snap fit to be disengaged for disassembly. Figure 13.4d uses a U-type beam to disengage the snap from the outside.

A potential problem with the design shown in Figure 13.4c is that the snap fit can be pushed too far; there is no stop. If a snap fit breaks, it is usually impossible to repair it, so designers will commonly put a stop behind it to prevent it from being overstrained. The design shown in Figure 13.4 has a stop incorporated (the cantilever can only be pushed so far). Figure 13.5 shows the mechanism of a cantilever snap fit, and how both assembly and separation can be achieved by using angled surfaces on the snap and the engaging piece of the recess.

U-shaped cantilever snap fits are commonly used in such products as battery doors and covers. Figure 13.6 shows how this type of snap fit works. With this type of snap fit, the plastic does not experience much strain, so multiple flexes are possible without damaging the plastic beam. It also has a built-in stop, so the beam cannot be damaged due to being flexed too much.

One problem with joining two plastic pieces together is holding a high tolerance on the snap fits so that a tight fit is achieved. With snap fits, when two pieces are joined together as shown in Figure 13.6, there is no way to get any preload on the cantilever snap. This is a major disadvantage of using snap fits over screws. One way to get around this is to sandwich an elastomeric material between the mating pieces, as shown in Figure 13.7. The plastic pieces are pressed together, compressing the elastomeric component and engaging the snap fit. The compressed component acts as a spring and gives preload to the snap, resulting in a tight assembly. The elastomeric component can also act as a gasket to keep out dust, water, and other contaminants.

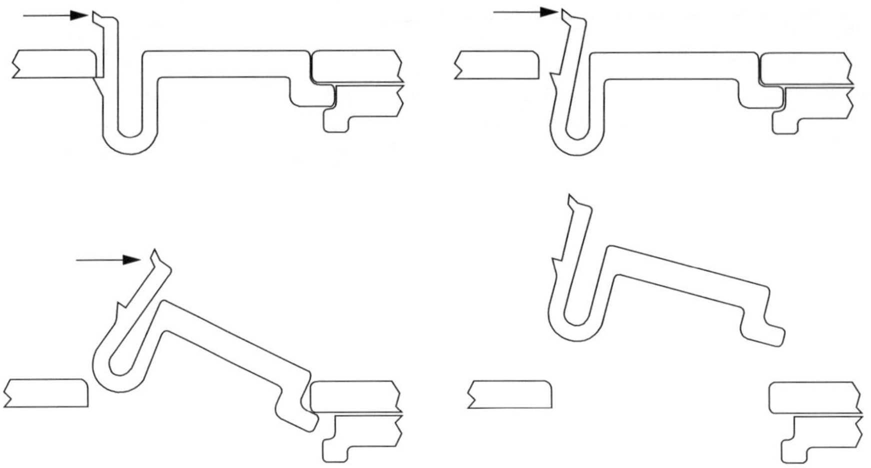

Figure 13.5

The stages of a U-shaped cantilever snap fit.

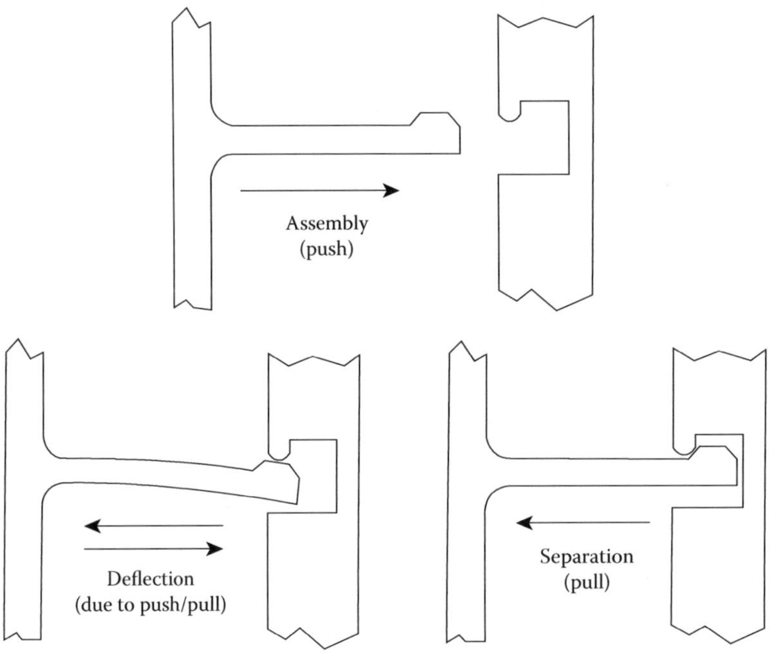

Figure 13.6

Cantilever snap fit.

The design engineer's job is to find a balance between integrity of the assembly and strength of the cantilever beam. While a cantilever beam with a deep overhang can make the unit secure, it also puts more strain on the beam during assembly and disassembly. A typical snap-fit assembly consists of a cantilever beam with an overhang at the end of the beam (see Figure 13.6). The depth of the overhang defines the amount of deflection during assembly. The overhang typically has a gentle ramp on the entrance side and a sharper angle on the retraction side. The small angle at the entrance side (α) helps to reduce the assembly effort. Both the assembly and disassembly force can be optimized by modifying the angles mentioned above. The main design considerations of a snap fit are integrity of the assembly and strength of the beam. The integrity of the assembly is controlled by the stiffness (k) of the beam and the amount of deflection required for assembly or disassembly. Rigidity can be increased either by using a higher-modulus material (E) or by increasing the cross-sectional moment of inertia (I) of the beam. The product of these two parameters (EI) will determine the total rigidity of a given beam length (Figures 13.8 and 13.9).

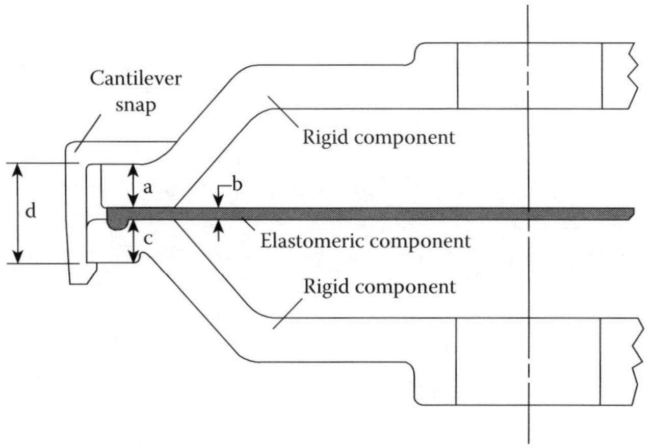

Figure 13.7

An elastomeric material can be used to seal and tighten the fit between mating parts.

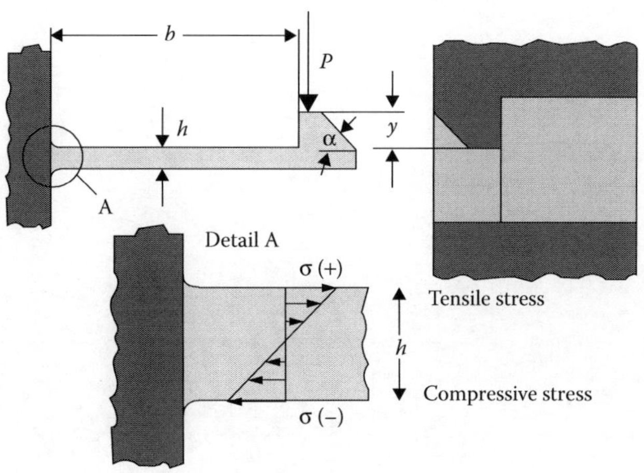

Figure 13.8

Simple snap-fitting hook.

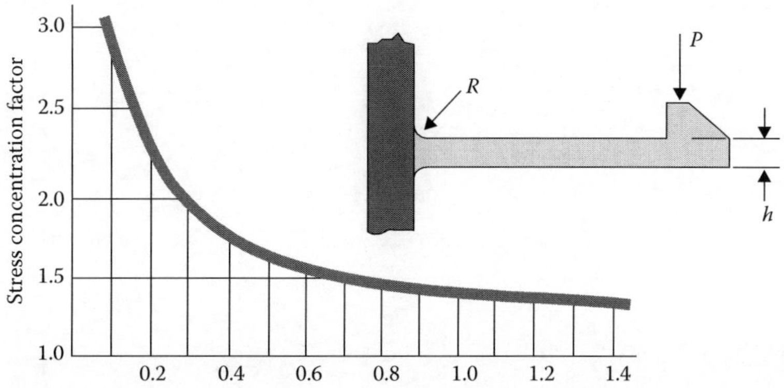

Figure 13.9

Effects of fillet radius on stress concentration.

13.3.1 Design Aids

A large proportion of snap joints are basically simple cantilever snaps (Figure 13.5), which may be either rectangular or of a geometrically more complex cross section (see Figure 13.6). It is suggested that the finger should be designed so that either its thickness (h) or its width (b) tapers from the root to the hook; in this way, the load-bearing cross section at any point bears a more appropriate relation to the local load. The maximum strain on the material can therefore be reduced, and less material is needed. Good results have been obtained by reducing h linearly so that its value at the end of the hook is equal to one-half the value at the root; alternatively, the finger width may be reduced to one-quarter of the base value (see Figure 13.7, designs 2 and 3). With the designs illustrated in Figure 13.7, the vulnerable cross section is always at the root (see also Figure 13.8, Detail A). Special attention must therefore be given to this area to avoid stress concentration. Figure 13.8 graphically represents the effect of the root radius on stress concentration. At first glance, it seems that an optimum reduction in stress concentration is obtained using an R/h ratio of 0.6, since only a marginal reduction occurs after this point. However, using R/h 0.6 would result in a thick area at the intersection of the snap-fit arm and its base. Thick sections will usually result in sinks or voids, which are signs of high residual stress. For this reason, the designer should reach a compromise between a large radius to reduce stress concentration and a small radius to reduce the potential for residual stresses due to the creation of a thick section adjacent to a thin section.

Internal testing shows that the radius should not be less than 0.015 in. in any instance (Figure 13.10), Definitions:

- y = (permissible) deflection or undercut
- ε = (permissible) strain in the outer fiber at the root; in formulae
- l = length of arm
- h = thickness at root
- b = width at root
- c = distance between outer fiber and neutral fiber (center of gravity)
- Z = section modulus = I/C
- I = moment of inertia
- E_s = secant modulus (see Figure 13.10)
- P = (permissible) deflection force
- K = geometric factor (see Figure 13.10)

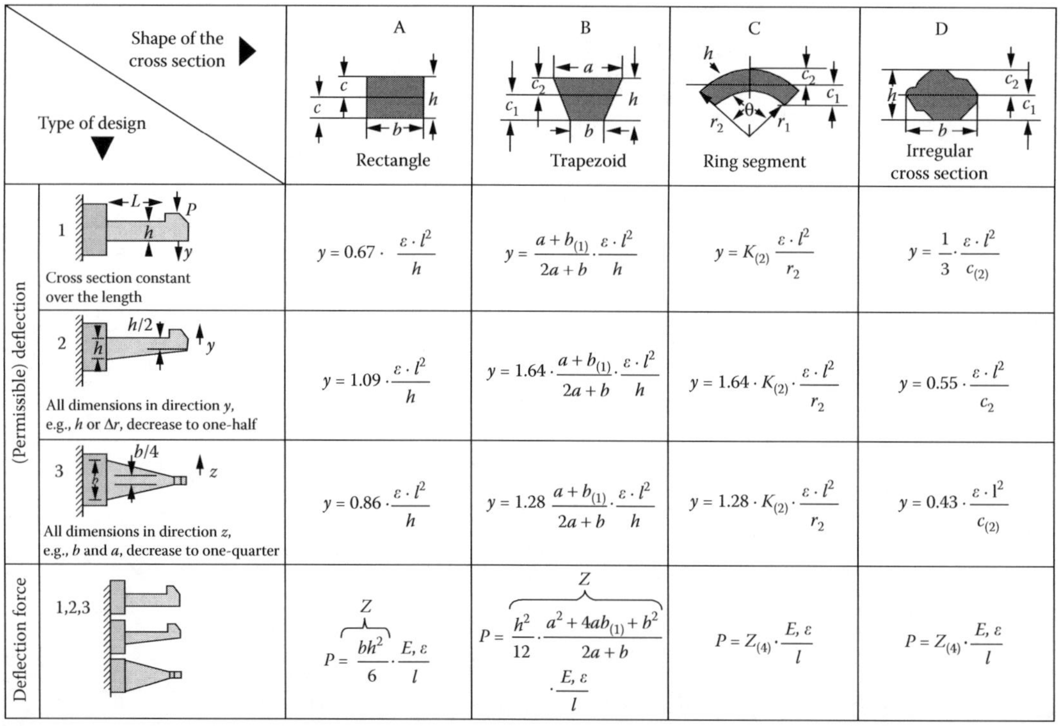

Figure 13.10

Equations for dimensioning cantilevers.

13.3.2 Explanation

1. These formulae apply when the tensile stress is in the small surface area *b*. If it occurs in the larger surface area *a*, however, *a* and *b* must be interchanged.
2. If the tensile stress occurs in the convex surface, use *K*2 (see Figure 13.8); if it occurs in the concave surface, use *K*1.
3. *c* is the distance between the outer fiber and the center of gravity (neutral axis) in the surface subject to tensile stress.
4. The section modulus should be determined for the surface subject to tensile stress. Section moduli for cross-section shape type C are given in Figure 13.11. Permissible stresses are usually more affected than the associated strains by temperature. It is preferable to determine the strain associated with the permissible stress at room temperature. As a first approximation, the computation may be based on this value regardless of the temperature.

Although the equations in Figure 13.12 may appear unfamiliar, they are simple manipulations of the conventional engineering equations to express the analysis in terms of permissible strain levels.

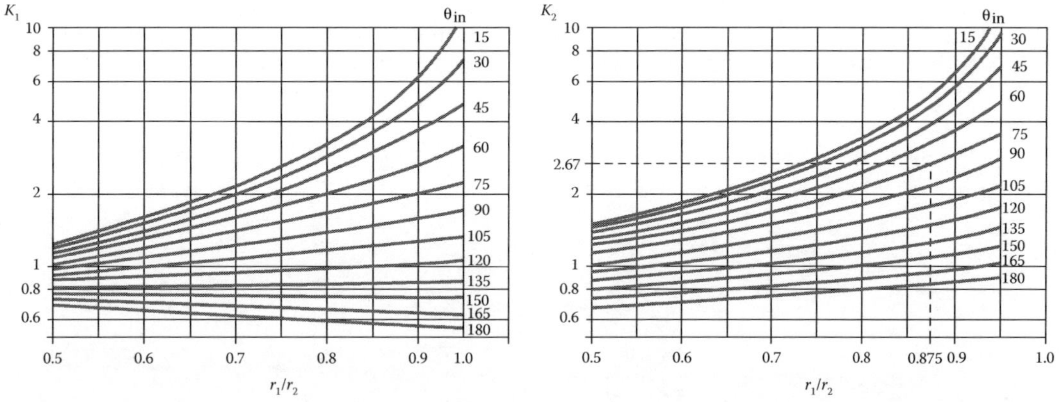

Figure 13.11

Diagrams for determining K_1 and K_2 for cross-sectional shape type C in Figure 13.17 (K_1: concave side under tensile load, K_2: convex side under tensile load).

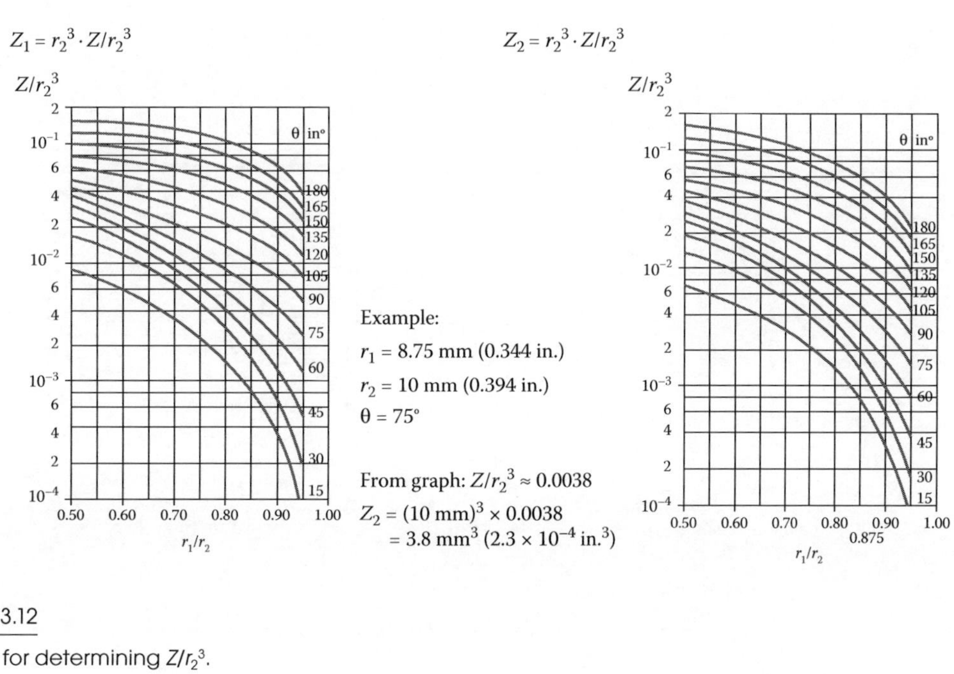

Figure 13.12

Graphs for determining Z/r_2^3.

13.3.3 Geometric Factors *K* and *Z* for Ring Segment

13.3.3.1 Cantilever Snap Joints

Figure 13.13 shows the undercut effect in cantilever as well as annular snap joints. These undercuts are proportional to the maximum yield permissible on the component and will be mathematically estimated in Example 13.1.

13.3.4 Permissible Undercut

The deflection y occurring during the joining operation is equal to the undercut (Figure 13.13). The permissible deflection y (permissible undercut) depends not only on the shape but also on the permissible strain E for the material used. In general, during a single, brief snap-fitting operation, partially crystalline materials may be stressed almost to the yield point and amorphous materials up to about 70% of the yield strain. Glass-fiber–reinforced molding compounds do not normally have a distinct yield point. The permissible strain for these materials in the case of snap joints is about half the elongation at break (see Figures 13.14 and 13.15).

13.3.5 Deflection Force

Using the equations given in Figure 13.10, y can be calculated easily even for cross sections of complex shapes. The procedure is explained with the aid of the following example. A particularly favorable form of snap-fitting arm is design 2 in Figure 13.19, with the thickness of the arm decreasing linearly to half its initial value. This version increases y by more than 60% compared with a snap-fitting arm of constant cross section (design 1). Complex designs such as that shown in Figure 13.13 may be used in applications to increase the effective length. The deflection force P required to bend the finger can be calculated using the equations in the bottom row of Figure 13.10 for cross sections of various shapes. E_s is the strain-dependent modulus

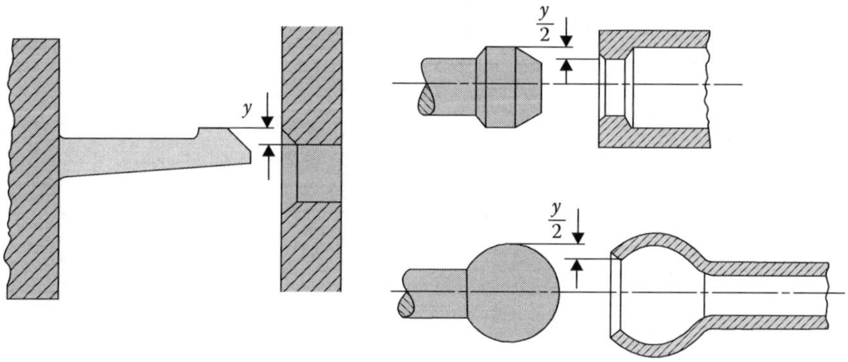

Figure 13.13

Undercut for snap joints.

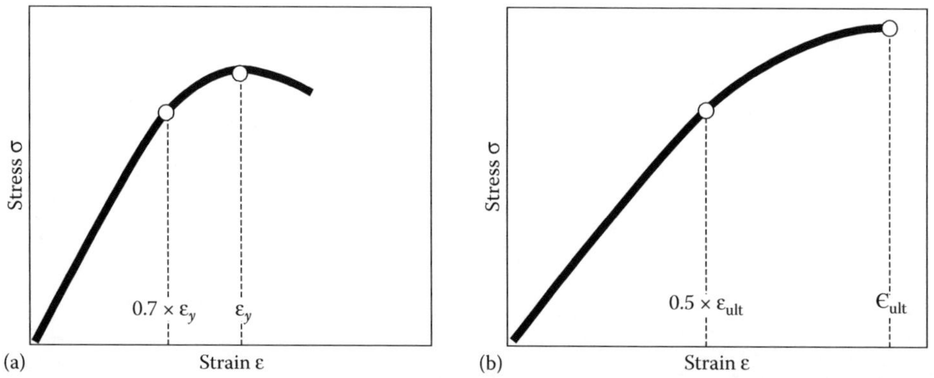

Figure 13.14

Determination of the permissible strain for the joining operation ((a) material with distinct yield point; (b) glass-fiber–reinforced material without yield point).

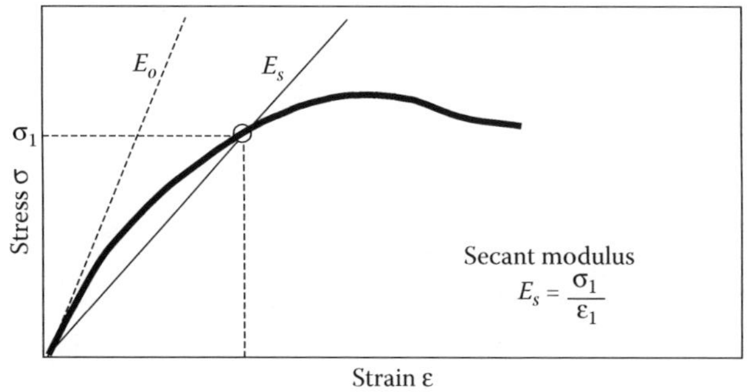

Figure 13.15

Determination of the secant modulus.

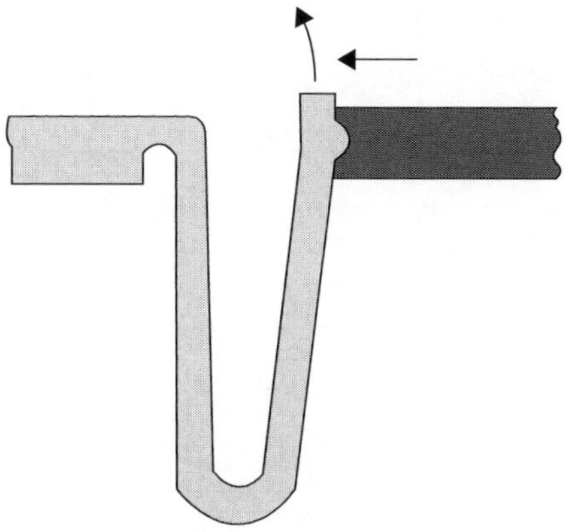

Figure 13.16

U-shaped snap-fitting arm for a lid fastening.

of elasticity or *secant modulus* (see Figure 13.15). Values for E_s for various plastics can be determined from Figure 13.15. The strain value used should always be that on which the dimensioning of the undercut was based (Figures 13.16 and 13.17).

The relationship presented in Figure 13.18 is based on the simple concept of static equilibrium of forces acting at a point. The mating force or the force needed for a snap joint can be calculated as follows.

During the assembly operation, *P* and the friction force *F* have to be overcome. The mating force is given by Equation 13.1:

$$W = P\tan(\alpha + p) = P\frac{\mu + \tan\alpha}{1 - \mu\tan\alpha} \tag{13.1}$$

where:

- P is friction angle
- μ is $\tan P$
- α is the ramp angle at the end of the cantilever

The value for $(\mu + \tan\alpha)/(1 - \mu\tan\alpha)$ can be taken directly from Figure 13.19. Friction coefficients for various materials are given in Table 13.1. In the case of separable joints, the separation force can be determined in the same way as the mating force by using the above equation. The angle of inclination to be used here is the angle α.

The figures depend on the relative speed of the mating parts, the pressure applied, and the surface quality. Friction between two different plastic materials gives values equal to or slightly lower than those shown in Table 13.1. With two components of the same plastic material, the friction coefficient is generally higher. Where the factor is known, it has been indicated in parentheses. The coefficient of friction in Table 13.1 is for plastic on steel. Generally, it can be observed from the data in Table 13.1 that the coefficient of friction is higher by a factor of 1.2–1.5 for plastic on plastic.

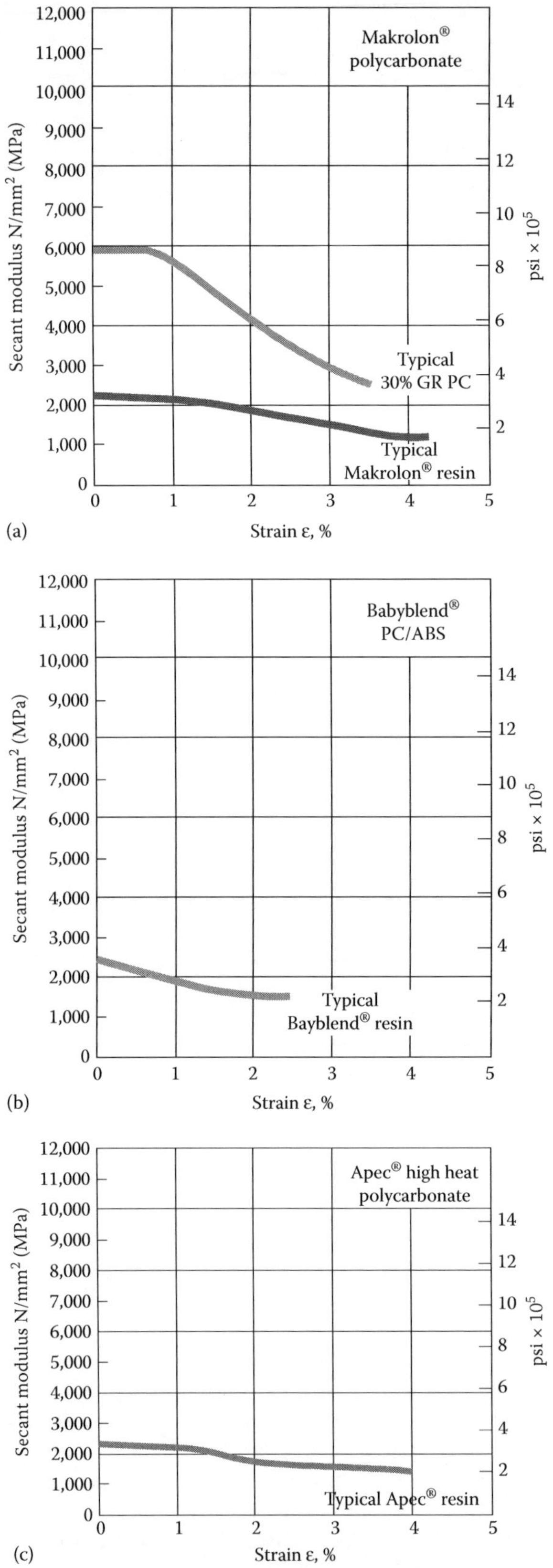

Figure 13.17

(a–c) Secant modulus for Bayer engineering plastics at 23°C (73°F) (Continued).

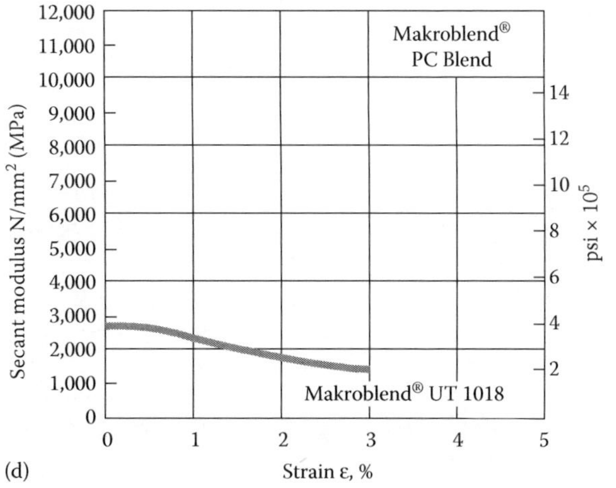

Figure 13.17 (Continued)

(d) Secant modulus for Bayer engineering plastics at 23°C (73°F).

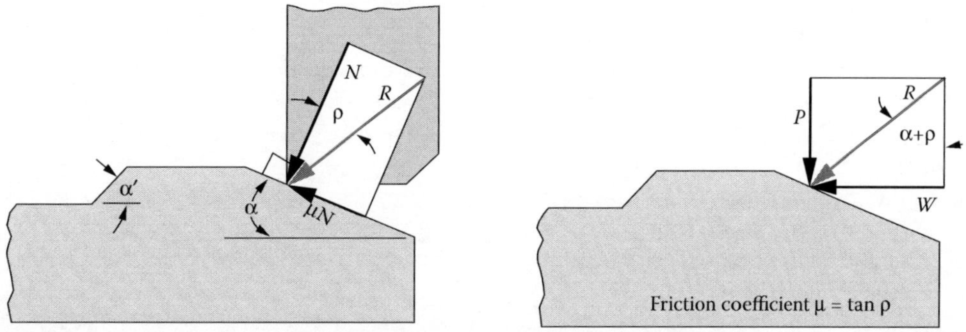

Figure 13.18

Relationship between deflection force and mating force.

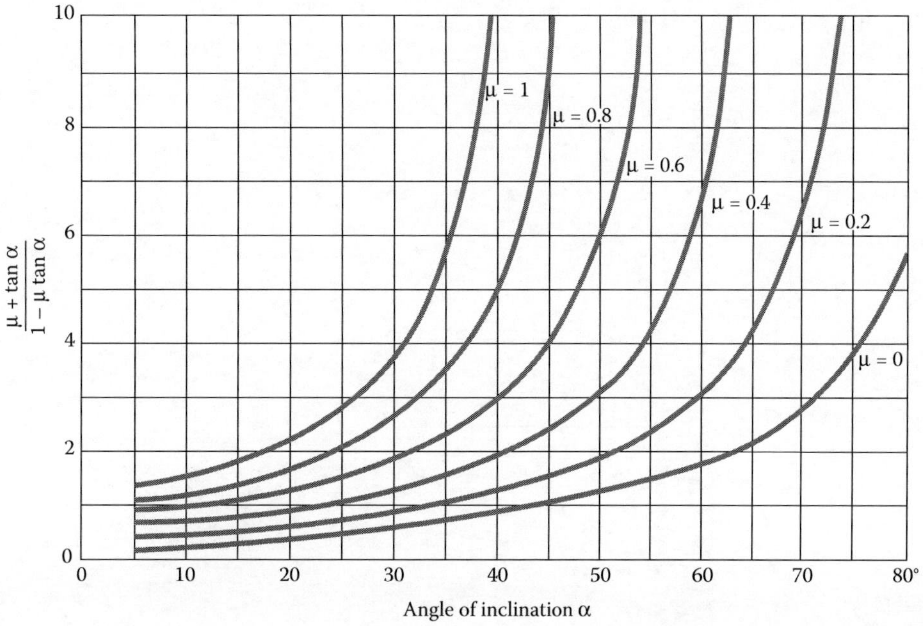

Figure 13.19

Diagram for determining $(\mu + \tan\alpha)/(1 - \mu\tan\alpha)$.

Table 13.1 Coefficient of Friction

Plastic Material	Friction
Polytetrafluoroethylene	0.12–0.22
Polyethylene, rigid	0.20–0.25 (×2.0)
Polypropylene	0.25–0.30 (×1.5)
Polyoxymethylene	0.20–0.35 (×1.5)
Polyamide	0.30–0.40 (×1.5)
Polybutylene	0.35–0.40
Polystyrene	0.40–0.50
Polycarbonate	0.45–0.55 (×1.2)
Styrene acrylonitrile	0.45–0.55
Poly(methyl methacrylate)	0.50–0.60 (×1.2)
Acrylonitrile butadiene styrene	0.50–0.65 (×1.2)
Polyethylene, flexible	0.55–0.60 (×1.2)
Polyvinyl chloride	0.55–0.60 (×1.0)

Example 13.1: Snap-Fitting Hook

This calculation is for a snap-fitting hook of rectangular cross section with a constant decrease in thickness from h at the root to $h/2$ at the end of the hook (see Figure 13.20). This is an example of design type 2 in Figure 13.7 and should be used whenever possible to permit greater deformation and to save material.

Given that

- **a.** Material = Makrolon polycarbonate
- **b.** Length (l) = 19 mm (0.75 in.)
- **c.** Width (b) = 9.5 mm (0.37 in.)
- **d.** Undercut (y) = 2.4 mm (0.094 in.)
- **e.** Angle of inclination (α) = 30°

Find (i) thickness h at which full deflection y will cause a strain of one-half the permissible strain, (ii) deflection force P, (iii) mating force W.

Solution:

- **i.** Determine wall thickness h:

$$y = 1.09\frac{\varepsilon l^2}{h}$$

Permissible strain from Figure 13.24: $\varepsilon_{min} = 4\%$. However, strain required here: $\varepsilon = 2\%$.

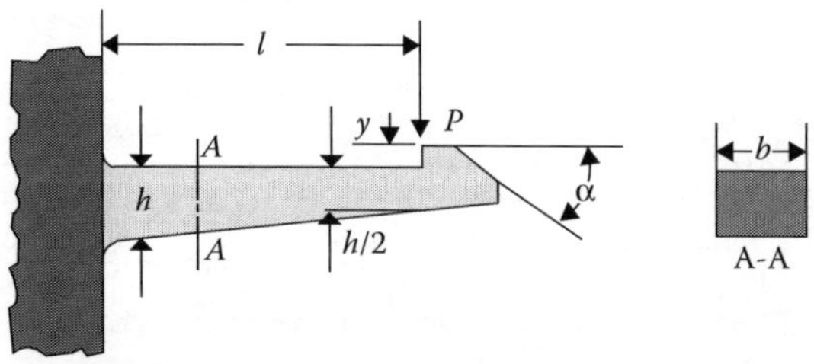

Figure 13.20

Snap-fitting hook, design type 2, shape A.

Using the deflection equation from Figure 13.17, type 2, shape A:

$$h = \frac{1.09 \times \varepsilon \times l^2}{y} = \frac{1.09 \times 0.02 \times 19^2}{2.4} = 3.28 \text{ mm}$$

ii. Determine deflection force P
Deflection force equation from Figure 13.19, cross section A:

$$P = \frac{bh^2}{6l} E_s \varepsilon$$

From Figure 13.24 at $\varepsilon = 2.0\%$, $E_s = 1815$ N/mm^2

$$P = \frac{9.5(3.28)^2}{6(19)}(1815)(0.02) = 32.5 \text{ N}$$

iii. Calculate mating force W
Friction coefficient from friction table above (PC on PC): $\mu = 0.50 \times 1.2 = 0.6$

$$W = P\tan(\alpha + p) = P\frac{\mu + \tan\alpha}{1 - \mu\tan\alpha} = 32.5\frac{(0.5 \times 12) + \tan(30)}{1 - (0.5 \times 12)\tan(30)}$$

$$= (88.5\text{N}) \text{ (very high compared with graph value)}$$

From Figure 13.28,

$$\frac{\mu + \tan\alpha}{1 - \mu\tan\alpha} = 1.8$$

For $\mu = 0.6$ and $\alpha = 30°$
$W = 32.5 \text{ N} \times 1.8 = 58.5 \text{ N}$

Example 13.2: Snap-Fitting Hook

This calculation example is for a snap-fitting hook with a segmented ring cross section decreasing in thickness from h at the root to $h/2$ at the end of the hook (see Figure 13.21). This is design type 2, shape C in Figure 13.7. This taper ratio should be used when possible to distribute stresses evenly during arm deflection. It also reduces material use.

Given: Material = Bayblend® PC/ABS

Length (l) = 25.4 mm (1.0 in), angle of arc (θ) = 75°, outer radius (r_2) = 20 mm (0.787 in), inner radius (r_1) = 17.5 mm (0.689 in), and thickness (h) = 2.5 mm (0.1 in), $\phi = \theta/2 = 37.5°$.

Find the maximum allowable deflection for a snap-fit design which will be assembled and unassembled frequently, snap-fitting hook, design type 2, shape C.

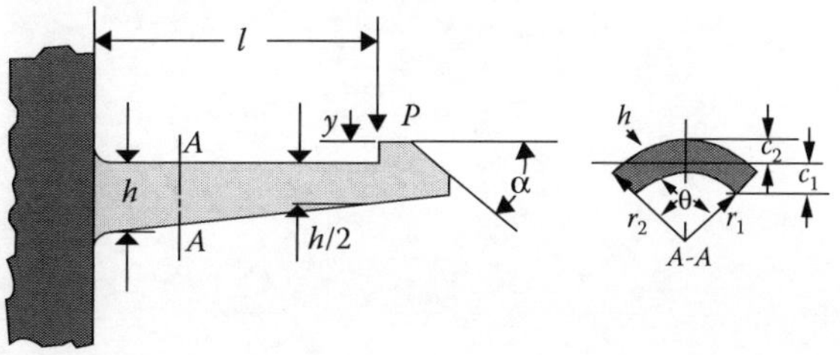

Figure 13.21

Design considerations.

Solution

The permissible strain for a one-time snap-fit assembly in Bayblend® resin is 2.5%. Since the design is for frequent separation and rejoining, 60% of this value should be used or $\varepsilon_{pm} = (0.6)\ (2.5\%) = 1.5\%$.

Deflection equation from Figure 13.21 (type 2, shape C):

$$y = 1.64K(2)\frac{\varepsilon l^2}{r_2}$$

The variable for $K2$ can be obtained from the curves in Figure 13.18. It should be noted that if the member is deflected so that the tensile stress occurs in the convex surface, the curve for $K1$ should be used; if it occurs in the concave surface, $K2$ should be used. In this case, the tensile stress will occur in the convex surface; therefore the curve for $K2$ should be used.

$$r_1/r_2 = 0.875 \text{ and } \theta = 75°, \text{ so from Figure 13.19, } K2 = 2.67$$

$$y = 1.64(2.67)\frac{0.05(25.4)^2}{20} = 2.11\ \text{mm}$$

13.3.6 Deflection

In the case of torsion snap joints, the deflection is not the result of a flexural load as with cantilever snaps but is due to a torsional deformation of the fulcrum. The torsion bar (Figure 13.22) is subject to shear.

The following relationship exists between the total angle of twist and the deflections y_1 or y_2 (Figure 13.22):

$$\sin\phi = \frac{y_1 = y_2}{l_1 l_2}$$

where:

- ϕ is angle of twist
- y_1, y_2 are deflections
- l_1, l_2 are length of lever arm

The maximum permissible angle ϕ_{pm} is limited by the permissible shear strain y_{pm}:

$$\phi_{pm} = \frac{180}{\pi}\frac{\gamma_{pm} \cdot l}{r}$$

where:

- ϕ_{pm} = permissible total angle of twist
- γ_{pm} = permissible shear strain
- l = length of torsion bar
- r = radius of torsion bar

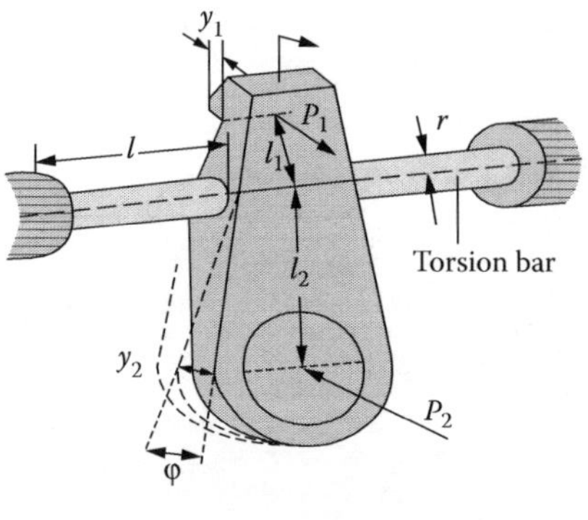

Figure 13.22

Snap-fitting arm with torsion bar.

The maximum permissible shear strain γ_{pm} for plastics is approximately equal to

$$\gamma_{pm} = (1+\nu)\varepsilon_{pm} \text{ or } \gamma_{pm} = 1.35\varepsilon_{pm}$$

13.3.7 Deflection Force

A force P is required to deflect the lever arm by the amount y_1 (or y_2). The deflection force can act at point 1 or 2. For example, see Figure 13.21. In this case,

$$P_1 l_1 = P_2 l_2 = \frac{\gamma G I_p}{r}(x2^*)$$

where:

G = shearing modulus of elasticity

γ = shear strain

$I_p = \frac{\pi r^4}{4}$ = polar moment of inertia for a solid circular cross section

Note: The factor 2 only applies where there are two torsion bars, as in Figure 13.19.

The shear modulus G can be determined fairly accurately from the secant modulus as follows:

$$G = \frac{E_s}{2(1+\upsilon)}$$

where:

E_S = secant modulus (see Figure 13.9)

υ = Poisson's ratio

13.3.8 Annular Snap Joints D: Permissible Undercut

The annular snap joint is a convenient form of joint between two rotationally symmetric parts. Here, too, a largely stress-free, positive joint is normally obtained. The joint can be either detachable (Figures 13.29a and 13.30), difficult to disassemble, or inseparable (Figure 13.23), depending on the dimension of the bead and the return angle. Inseparable designs should be avoided in view of the complex tooling required (split cavity mold).

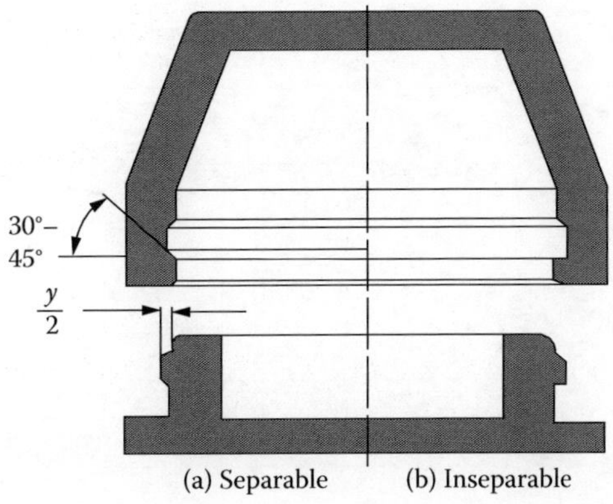

Figure 13.23

Annular snap joint.

The allowable deformation should not be exceeded either during the ejection of the part from the mold or during the joining operation. The permissible undercut as shown in Figure 13.13 is limited by the maximum permissible strain

$$Y_{pm} = \varepsilon_{pm} d$$

where ε_{pm} is absolute value.

This is based on the assumption that one of the mating parts remains rigid. If this is not the case, the actual load on the material is correspondingly smaller (with components of equal flexibility, the strain is halved, that is, the undercut can be twice as large) (Figures 13.23 through 13.25). The definitions used in these figures are as follows: W = mating force, y = undercut, α = lead angle, α' = return angle, t = wall thickness, and d = diameter at the joint.

13.3.9 Deflection Force and Mating Force

The determination of the mating force W is somewhat more complicated for annular snap joints. This is because the snap-fitting bead on the shaft expands a relatively large portion of the tube (Figure 13.26). Accordingly, the stress is also distributed over a large area of the material surrounding the bead.

Experimentally proven answers to this problem are based on the "theory of a beam of infinite length resting on a resilient foundation." Two extreme cases are depicted in Figure 13.27.

1. The force P is applied at the end of the beam. (This corresponds to a snap joint with the groove at the end of the tube.)
2. The force P is applied a long distance (co) from the end of the beam. (This is equivalent to an annular snap joint with the groove remote from the end of the tube.)

A somewhat simplified version of the theory may be expressed as follows for joints near the end of the tube:

$$P = y \cdot d \cdot E_s X$$

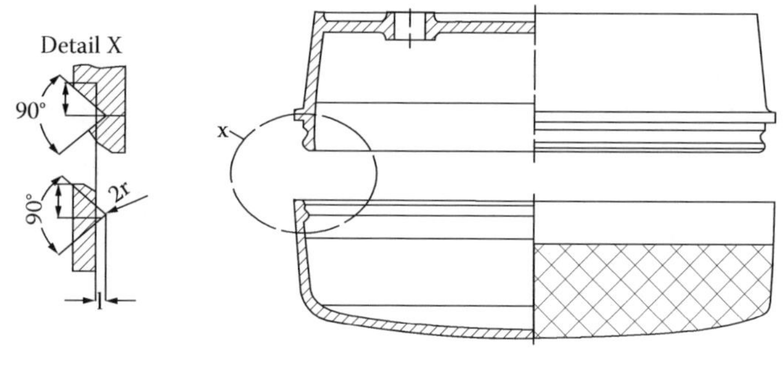

Figure 13.24

Annular snap joint on a lamp housing.

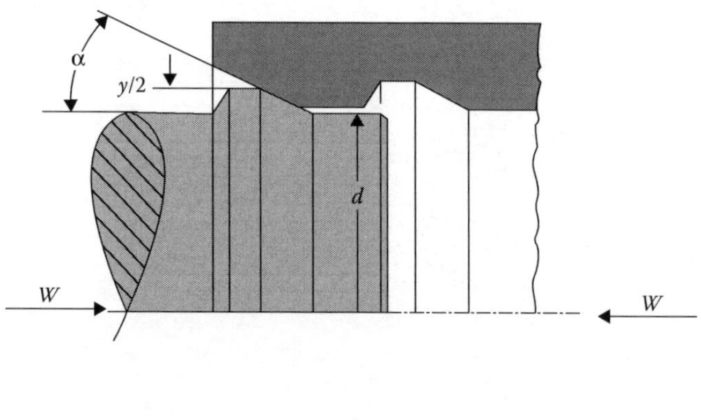

Figure 13.25

Annular snap joint.

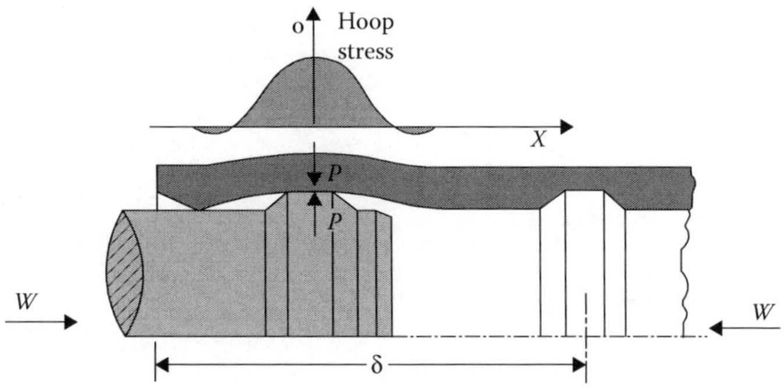

Figure 13.26

Stress distribution during joining operation.

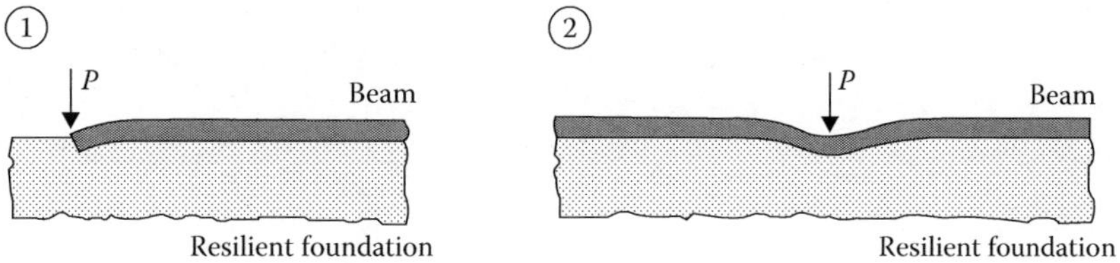

Figure 13.27

Beam resting on a resilient foundation.

where:

- P = transverse force
- y = undercut
- d = diameter at the joint
- E_S = secant modulus
- X = geometric factor

The geometric factor X takes into account the geometric rigidity. As far as the mating force is concerned, friction conditions and joint angles must be taken into account.

$$P = W\frac{\mu + \tan\alpha}{1 - \mu\tan\alpha}$$

where:

- μ = friction coefficient
- α = lead angle

The geometric factor, assuming that the shaft is rigid and the outer tube (hub) is elastic, is

$$X_N = 0.62\frac{\sqrt{(d_o/d - 1) + (d_o/d + 1)}}{\left[(d_o/d)^2 + 1\right]/\left[(d_o/d)^2 - 1\right] + \upsilon}$$

where:

- d_0 = external diameter of the tube
- d = diameter at the joint
- υ = Poisson's ratio

If the tube is rigid and the hollow shaft elastic,

$$X_N = 0.62 \frac{\sqrt{(d/d_i - 1) + (d/d_i + 1)}}{\left[(d/d_i)^2 + 1\right] / \left[(d/d_i)^2 - 1\right] - \upsilon}$$

where:

d = diameter at the joint

d_i = internal diameter of the hollow shaft

The geometric factors X_N and X_W can be found in Figure 13.26.

$$\delta = 1.8\sqrt{dt}$$

where:

d = joint diameter

t = wall thickness

In this case, the transverse force P and mating force W are theoretically four times as great as when the joint is near the end of the tube. However, tests have shown that the actual mating forces rarely exceed a factor of three (Figure 13.28).

The snap joint is considered "remote" if the distance from the end of the tube is at least

$$P_{\text{remote}} = 3P_{\text{near}} \text{ and } W_{\text{remote}} = 3W_{\text{near}}$$

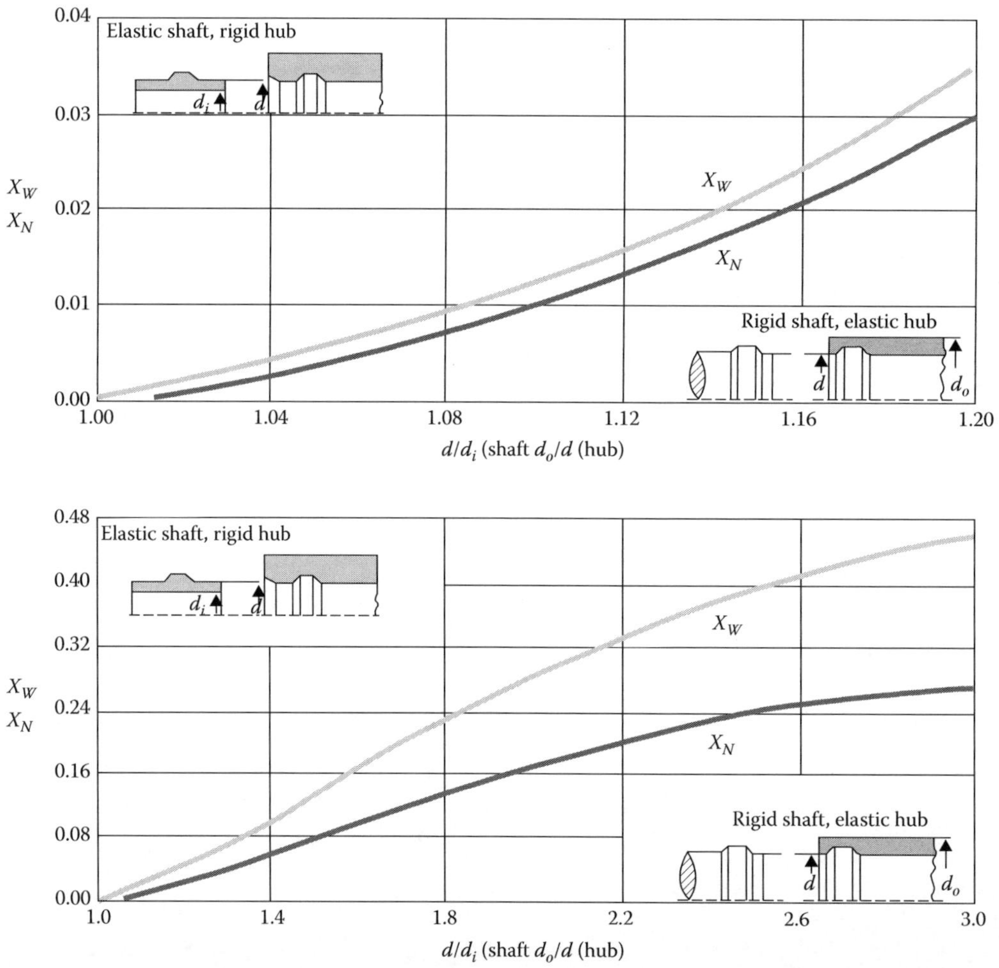

Figure 13.28

Diagrams for determining the geometric factor X for annular snap joints.

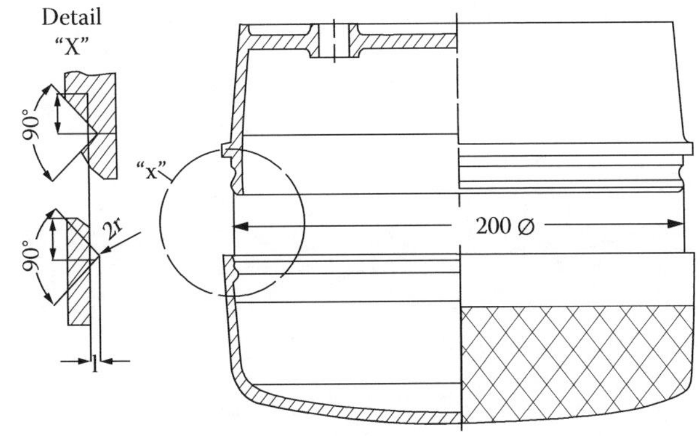

Figure 13.29

Lamp housing with cover.

This means that if the joint lies between 0 and δ, the factor is between 1 and 3. The secant modulus E_s must be determined as a function of the strain ε from Figure 13.24. For the sake of simplicity, it may be assumed here that the strain

$$\varepsilon = \frac{Y}{d} * 100\%$$

where:

y = undercut

d = diameter over the entire wall thickness (In fact, it varies at different points and in different directions on the wall cross section; Figure 13.29.)

Example 13.3: Calculation Example Annular Snap Joint [1]

Given

Lamp cover and housing made of Makrolon® polycarbonate. Snap-fitting groove near the end. Dimensions: $d = 200$ mm (7.87 in), $t = 2.5$ mm (0.098 in) for both mating parts, $y = 2 \times 1$ mm (0.039 in), $\alpha = 45°$, lead and return angle = 45°. The edge is rounded, however, and the effective angle may be assumed to be $\alpha = 30°$.

Required

(1) Occurring strain ε, (2) deflection force P, (3) mating force W

Solution

Strain ε:

$$\varepsilon = \frac{y}{d} = \frac{1}{200} \times 100\% = 0.5\%$$

Since both mating parts have approximately equal stiffness, the deflection for each part is approximately half the undercut. This strain is permissible for polycarbonate according to Figure 13.24. Transverse force P:

$$P = y \times d\, E_s \times X$$

As the mating parts are of approximately equal stiffness, the calculation may be performed for either component. In this case, the lamp cover (hub) has been chosen. X_N from Figure 13.24 with

$$\frac{d_o}{d} = \frac{200 + 2 * 2.5}{200} = 1.025$$

From Figure 13.24, $X_N = 0.0017 = 1.7 \times 10^{-3}$

$$X_N = 0.0017 = 1.7 \times 10^{-3}$$

From Figure 13.16, $E_S = 2200$ MPa (320,000 psi)

$$P = 1\text{ mm} \times 200\text{ mm} \times 2200\text{ MPa} \times 1.7 \times 10^{-3} = 748\text{ N}$$

Mating force $\mu = 0.6$ from Table 13.1:

$$W = P\frac{\mu + \tan\alpha}{1 - \mu\tan\alpha} = 1346\text{ N}$$

The mating and opening force W in this case is considerable. It should be remembered, however, that such a force would only occur if mating parts in true axial alignment were joined by machine. In manual assembly, the greater part of the bead is first introduced into the groove at an angle, and only the remaining portion is pressed or knocked into position. The mating forces occurring under these circumstances are much smaller, as only part of the bead is deformed.

Example 13.4: Disassembly of Snap Joint

Figure 13.30 shows a snap connection. The disassembly energy of the snap connection

$$E = \frac{1}{8}E_s wt^3 \frac{h_2^3}{h_1^3}$$

where:

E_s = modulus elasticity of the material
h_1, h_2, w, t are defined in Figure 13.30

The disassembly energy is similar to strain energy, which changes the height of the pin fitting in the sleeve. The snap joint is considered as a cantilever and E_s is the modulus of elasticity. This calculation is for a snap-fitting hook of rectangular cross design type 1, shape A (see Figure 13.27). This is an example of design type 1 from Figure 13.27 and should be used whenever possible to permit greater deformation and to save material.

Given data: $h_1 = 19$ mm, $W = 9.5$ mm, $h_2 = 2.4$ mm, angle of inclination $(\alpha) = 30°$
Material: Makrolon polycarbonate

Find the disassembly energy E. To find the disassembly energy, we have to find the t of the snap-fit hook joint. Referring to Figure 13.7, design type 1, shape A, we find the deflection or undercut $h_2 = 0.67x\varepsilon x h_1^2 / t$ or

$$t = \frac{0.67x\varepsilon x h_1^2}{h_2}$$

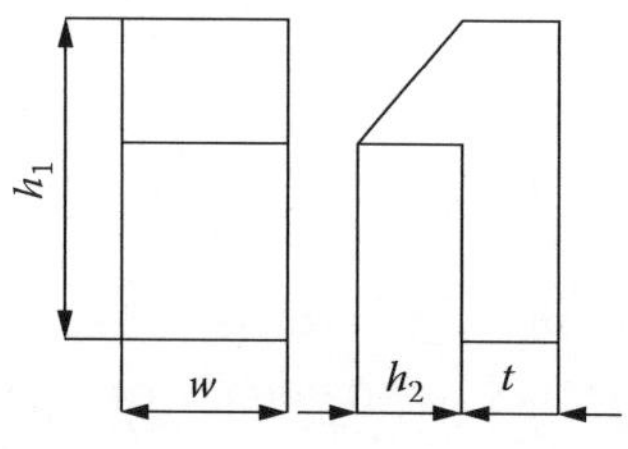

Figure 13.30

Parameters of a simple model of snap fit.

From Figure 13.24, we can find that for polycarbonate, minimum strain $\varepsilon_{min} = 4\%$. It is possible that this strain is less than the ε_{min} value at undercut or deflection h_2; however, this value is used in the calculation of disassembly energy.

t is calculated as

$$t = \frac{0.67 \times \varepsilon \times h_1^2}{h_2} = \frac{0.67 \times 0.04 \times 19^2}{2.4} = 4 \text{ mm}$$

Finally, the disassembly energy (E) is calculated by reading from Figure 13.24, $E_s = 1100$ MPa

$$E = \frac{1}{8} E_S w t^3 \frac{h_2^2}{h_1^3} = \frac{1}{8}(1100)(9.5)(4)^3 \frac{2.4^2}{19^3} = 72 \text{ MJ}$$

To calculate the disassembly entropy, Equation 13.2 is used:

$$S_S = \sum_k \left[\ln\left(\frac{N_{ik}!}{\prod_j n_i!} \right) + \sum_i \ln \frac{n_i!}{\prod_j n_{ij}!} \right] \tag{13.2}$$

where:

N_{ik} is the total number of interconnections belonging to assembly k

n_i is the number of interconnections made by method i (for instance, $i = 1$ is screw, and $i = 2$ is snap fit)

n_{ij} is the number of interconnections for the direction j ($+x$, $-x$, $+y$, $-y$, $+z$, and $-z$)

The disassembly entropy can also be calculated as zero for a single snap joint, which shows that all the disassembly energy is converted to useful work in disassembly.

The time for disassembly can also be estimated by the method proposed by Gungor and Gupta [2]. The authors use total time for disassembly (TTD) as the parameter to measure a given disassembly sequence of a product:

$$TTD = \sum_{i=1}^{n} \left(T_{x_{i-1}, x_i} \right) \tag{13.3}$$

$$T_{x_{i-1}}, x_i = t_i (1 + \alpha + \beta), i = 1, 2, \ldots, n \tag{13.4}$$

T_{x_{i-1}, x_i} = adjusted time for disassembly of X_i being disassembled after X_{i-1}

when $i = 1$, then $T_{x_{i-1}, x_i} = t_i$

α, β = direction factors

α = 0.1–1.0 if the direction changes from X_i to X_{i-1}, otherwise $\alpha = 0$

β = 0.1–1.0 if the joint type of X_{ii} is different from the joint type of X_{i-1}, otherwise $\beta = 0$

For our example of a snap joint, there is only one joint ($I = 1$), and direction factor (α) may be assumed as zero. β has to be equal to zero, as there is only one joint. We can estimate the disassembly time $T_{x..x_i} = t_i$ for the snap joint from the MTM-1 data tables in Chapter 12. We can analyze the disassembly process and establish the following motions: reach, grasp, position, apply pressure, turn, move, and disengage. The class of fit in the snap joint is exact, which must be taken into consideration in the disassembly of the snap joint.

Time estimates from MTM-1 tables = (26.7 + 7.3 + 53.4 + 4.2 + 9.4 + 30.7 + 34.7) TMU = 166.4 TMU = 6.012 s. A snap-fit assembly can be disassembled easily, and so its separability rating is good.

One problem with a snap-fit assembly is that it may not be able to withstand certain kinds of forces, such as axial or torque. In this disassembly, slight axial force is needed. However, the disassembled parts can be reused easily. The degree of difficulty here is related to manual use.

References

1. Bayer Material Science LLC. (2012). 100 Bayes Road, Pittsburgh, PA, 2012.
2. Gungor, A. and Gupta, S. M. (1999). Issues in environmentally conscious manufacturing and product recovery: A survey. *Computers & Industrial Engineering* 36 (4): 811–863.

14

Finite Element Analysis (FEA) and Computer-Aided Design (CAD)

14.1 Introduction to Computer-Aided Design

Computer-aided design is the blending of human and machine, working together to achieve the optimal design and manufacture of a product. It has simplified many design processes. The computer's graphics capability and computing power allow designers to fashion and test their ideas interactively in real time without having to create real prototypes as in conventional approaches to design. A typical CAD system involves both design and manufacturing operations. Complex products are designed and analyzed, and their manufacturing plans are produced. Although sometimes students think of CAD as an electronic drafting board, its functions stretch beyond drawing pictures. An analysis of objects drawn with CAD can be made interactively on the screen, where physical information can be extracted. In engineering, finite element analysis, heat transfer analysis, stress analysis, dynamic simulation of mechanisms, and fluid dynamic analysis are common operations performed with CAD.

14.1.1 Objectives of CAD Programs

The computer's *memory capacity* and *processor speed* are the two top considerations when selecting a CAD software package. Desirable features of CAD software packages include "user-friendly" communication. The criteria for such a system might include

- That it allows the user to spend more time on the design than on commands.
- That the number of steps necessary to activate any command should be minimized.
- That any command affecting the database should have protection against accidental deletion or reorganization.
- That the menu should be arranged for easy interaction with the user.
- That a set of nested commands should be accessible in the midst of drawing, editing, or dimensioning without affecting the progression of the current command. That is, a user can view options during the current command.
- That the overall appearance of the program reflects a simple and logical layout.

Virtually all CAD software is capable of drawing lines, circles, arcs, rectangles, polygons, points, splines, closed curves, and double lines. CAD programs offer various other tools, including on-screen feedback of absolute, relative, or polar coordinates; rubber band lines; grid and object snaps; and parallels, perpendiculars and tangent to the existing entities.

For mechanical design, features such as automatic fillets and chamfers, automated crosshatching, the ability to construct circular or rectangular arrays of selected objects, automatic calculation of areas, and section properties must be included. Other features are the ability to perform simple calculations such as areas, volumes, centroids, moments of inertia, and centers of mass, geometry creation (2D and 3D), dimensioning and labeling, and editing ability.

All CAD programs have editing features. The editing features command usually includes function such as erase, trim, extend, scale, rotate, stretch, undo, move, insert, copy, break, mirror, mirror copy, group, ungroup, and change entity characteristics. Additional features such as zoom/window are used to magnify specific areas for more accurate editing. CAD packages also include macro functions, which allow the user to execute a set of subroutines as one command.

14.1.2 Components of CAD

The CAD system consists of hardware and software. Hardware consists of the computer system, including

- The host computers
- Display devices
- Interactive devices
- Output devices

Software is a set of written instructions, procedures, and rules that directs the operation of the computer. The human factor concerns the training of the designer or the engineer. The designer behind the system is involved with three aspects of computer-aided design:

1. Analysis
2. Visualization
3. Synthesis

Figure 14.1 shows all the components and functions of CAD.

Computer-aided design software can turn 2D and 3D designs into useful models and prototypes for a variety of fields. From floor plans and landscaping to kitchen remodels and 3D printing, CAD software has the tools to help us build and customize designs from the ground up. CAD software program should be tailored to our needs and interests.

CAD programs range from professional applications like AutoCAD to consumer-based packages such as TurboCAD Deluxe, DesignCAD 3D Max and SketchUp Pro. AutoCAD is the best and most recognized CAD software for professional users, but it has steep learning curve and costly licensing.

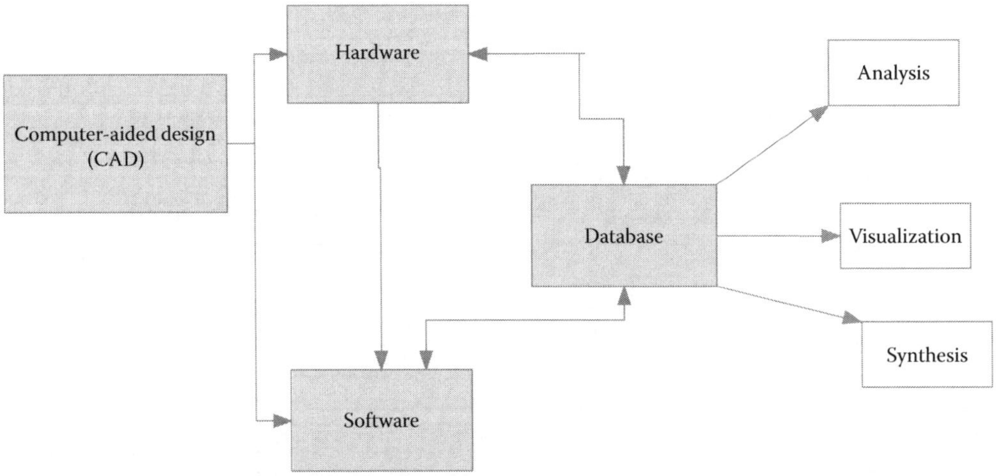

Figure 14.1

Components of computer-aided design.

There are large numbers of vendors, who provide different CAD software for different usage. The list includes (1) AllyCAD (2) ArchiCAD (3) AutoCAD (4) AutoCAD Architecture (5) Autodesk Inventor (6) Cadwork (7) gCAD3D (8) I-DEAS (9) IntelliCAD (10) MicroStation (11) NX (12) ProE (13) Autodesk Revit (14) Sketchpad, and (15) SolidWorks.

However, we are mainly interested in CAD software for mechanical engineering, industrial and product design, manufacturing CAD and computer-aided manufacturing, and graphics and illustration. Software like AutoCAD, Autodesk, NX, and SolidWorks are quite good and easy to use.

14.2 CAD Software: What to Look For

When shopping for computer-aided design software programs, a few features and tools stand out as key indicators of good software.

14.2.1 Design Tools

Designing 2D and 3D models is the essence of any good CAD program, so our choice must have all the right tools. Some programs are only for 2D drawings or 3D modeling, but the best software has a range of feature for both. For architectural designs, look for a wall tool and house wizard to do some of the work for you. Photorealistic rendering is another important feature that enables the user to see what finished designs will really look like.

14.2.2 Interface Features

Easy CAD software might sound like a contradiction, but in truth many programs have resources and features to make the software itself easier to use. The best CAD programs have customizable tool palettes, a command line and the ability to import existing designs. Also, be sure to look for a setup manager and macro recording to save even more time.

14.2.3 Editing Tools

CAD packages should come with many different editing tools. Point markers and layer managers can keep the designs organized. Text and color editing help make notes and distinctions between different elements and pieces of the CAD project. A snap tool is absolutely necessary so you can "snap" to specific locations on solid models. Many programs also come with 2D and 3D symbols and include models so it is not necessary to create each small element individually from scratch.

14.2.4 Help and Support

Both 3D and 2D CAD software can be hard to learn, so it is nice to have a support network to walk the first-time user through the process. E-mail is the main source of direct contact with many CAD software manufacturers, but the best companies will offer phone support as well. Video tutorials, an in-program manual, and community forums are also convenient sources for additional help.

14.2.5 Compatibility

File compatibility is one aspect of CAD drawing software that should not be overlooked. DWG, DXF, DWF and DGN are the most important formats to look for if working with AutoCAD. For 3D printing, an STL export feature is handy. For printing or e-mailing read-only files, look for PDF and various image file formats for easy transfer. The best 3D CAD software not only has a wide range of features but an easy-to-use interface. The special features of the NX software package are described below.

14.3 NX 9.0 CAD Software Package (1)

NX is an advanced version of I-DEAS (Integrated Design Engineering Analysis Software) Master Series and an integrated package of mechanical computer-aided engineering software tools developed by Structural Dynamics Research Corporation (SDRC). NX Master Series is a suite of programs that are used to facilitate a concurrent engineering approach to design, analysis and manufacturing of mechanical engineering products. NX is composed of a number of software modules called "Applications," each subdivided into "Tasks," and all executed from a common user interface and sharing common database.

14.3.1 NX Master Series Applications

1. Design
2. Drafting
3. Simulation
4. Test
5. Manufacturing
6. Management
7. Data Translator/PCB

Under the various Applications, several Tasks are grouped.

14.3.2 I-DEAS Master Series Design Tasks

1. Master Modeler: All solid modeling tasks are performed under this task. It is a flexible feature-based modeling system. The solid part model is used for many downstream uses such as interference studies, mass properties calculation, kinematic analysis, stress analysis, dynamics, manufacturing, testing, drafting or other uses. At this stage a database is created which is used for all downstream applications.

   Solid models are constructed in two ways: with primitives or with boundary definition. Both of methods develop complex geometries from successive combinations of simple geometric operations.

   The primitives approach allows elementary shapes such as blocks and cylinders to be combined in building block fashion. The user positions these primitives as required and then creates a new shape with the proper Boolean logic command.

   In the boundary condition approach, two-dimensional surfaces are swept through space to trace out volume. A linear sweep translates the surface in a straight line to produce an extruded volume, a rotational sweep produces a part with axial symmetry, and a compound sweep moves a surface through a specified curve to generate a more complex solid.

   Most mechanical components with planer, cylindrical, and other simple shapes can be modeled with primitives, but complex contoured components such as automobile exhaust manifolds and turbine blades are easier to model using the boundary definition method. I-DEAS present all these methods and works very well.
2. Drafting: In I-DEAS, drafting is a downstream function rather than the starting point. After solid modeling, especially when making isometric and section views, drawing can be made more quickly. Annotations that describe the manufacturing intent of a part such as geometric dimensioning and tolerancing and surface finish symbols is traditionally placed on 2D drawings. All three views of orthographic projections are created right from the solid model.
3. Simulation: The purpose of the simulation application is to check for possible failure modes of parts. A variety of solution methods are available. The most common solution is the linear statics to calculate deflection and stresses. Other failure modes such as fatigue and buckling are also available. If the applied forces change rapidly, dynamic analysis should be used to calculate the response due to transient loads. If deflections are high, nonlinear analysis should be used. The importance of simulation is not just to check that the design is good enough but to use parameter studies and optimization to help the designers improve the design.

   NX 9.0 Simulation Application has the following Tasks:

   i. Preprocessing: Three steps of preprocessing include (a) modeler, (b) meshing, and (c) boundary conditions.

      Normally you should have your solid model already created and then you should start simulation. Before the solution starts meshing, boundary conditions are applied in that order.

   ii. Solution: Model Solution is really the finite element analysis solution. An analysis will be performed using a "solution set," which contains a "boundary conditions" set of "restraints" and "loads."

      Analysis results are stored in the model file as "analysis datasets." Each dataset contains one set of displacements or stresses that can be displayed. Stresses are stored in a dataset as the entire stress tensor, so any stress characteristic can be computed and displayed within the postprocessing task.

   iii. Postprocessing
4. Manufacturing: The fundamental reason for designing a product is so that it can eventually be manufactured. The idea of concurrent engineering is to integrate all of the different disciplines into

the design phase. Using integrated tools at this stage in design, engineers can pose questions such as which process is best to produce a certain component. Are geometric changes required to make the product easier to manufacture? These questions can be answered using the same part geometry.

Normally the solid model can be converted to IGES file and taken to a computer numerical control machine to manufacture the part.

5. Test: In a traditional engineering process, testing was done mainly to qualify the design after manufacturing. But I-DEAS test applications offer various analytical tools, including fatigue analysis and model testing to extract natural frequency, and mode shapes.
6. Model File: The model file is a database used to store your work-in-progress when you use the save command. With this file saved, you can open this file later and continue where you left. The model file actually consists of two physical files stored on the computer disk. These files have a file name you give, plus the extension .MF1 and .MF2.

14.3.3 Icon Panel of NX (1)

The first section on the top of the icon panel contains icons that are specific to the task you are in. As you change tasks within an application, these icons will change.

The second section contains icons that pertain to the application. These icons will generally remain fixed as you change tasks within an application. They include functions to modify, move, measure, delete, and manage.

The third, or bottom, section contains icons that deal with display operations that are consistent across all applications in I-DEAS. These icons control the display, such as to generate line, hidden line, and shaded image display. They also control the view scale and viewing angles. See Appendix A for a summary of the icons from each application.

14.3.4 Menu Commands

14.3.4.1 File Menu

The file menu lets you save your work in your model file. This menu also contains the exit command.

14.3.4.2 Option Menu

Two important commands in the option menu are Units and Preferences.

Units: A model file has an associated system of units currently being used. All data is actually stored internally in SI units. For example, when you enter a number while you are using inch units, it will be converted to SI units internally. When this number is printed out, it will be scaled back into inch units.

Preferences: These preferences include options on how icons, menus, forms, and the graphics display appear.

14.3.4.3 View and Display

The icons at the bottom of the icon panel control view direction and display mode (line, hidden line, or shaded displays).

14.3.4.3.1 3D Dynamic Viewing Functions Dynamic viewing functions are controlled with function keys F1 through F6.

14.3.4.3.2 Data Management Concepts There are two primary areas where your parts are stored. Initially, parts are created in your model file. You may store it in libraries and catalogs either to share with others or for your own data organization.

Parts that are visible on the screen are described as being on your workbench. Parts in the model file can also be temporarily "put away" into "bins" in the model file, much like storing your work in a drawer into your desk.

In each application of I-DEAS, there are menus to "manage" the storage and retrieval of different entities in the model file. For example, in the Modeler task, there is a "Manage Bins icon. This command gives you a directory of the parts you have stored in the bins. It also allows you to rename, copy, and delete these parts.

14.4 Finite Element Analysis

Finite element analysis (FEA) is a computerized method for predicting how a product reacts to real-world forces, vibration, heat, fluid flow, and other physical effects. Finite element analysis shows whether a product will break, wear out, or work the way it was designed. It is called analysis, but in the product development process, it is used to predict what is going to happen when the product is used. FEA works by breaking down a real object into a large number (thousands to hundreds of thousands) of finite elements, such as little cubes. Mathematical equations help predict the behavior of each element. A computer then adds up all the individual behaviors to predict the behavior of the actual object. Finite element analysis helps predict the behavior of products affected by many physical effects, including

- Mechanical stress
- Mechanical vibration
- Fatigue
- Motion
- Heat transfer
- Fluid flow
- Electrostatics
- Plastic injection molding

Many engineering analyses involve physical quantities, such as stress (or strain) everywhere in a solid cause by applied forces (or pressure), temperature at ever point in a solid caused by heat sources (or sinks), velocities at every location in a fluid due to difference in heads or pressure, and amplitudes of vibration at different parts of a solid caused by sources of excitation. FEA is the modeling of products and systems in a virtual environment, for the purpose of finding and solving potential (or existing) structural or performance issues. FEA is the practical application of the finite element method (FEM), which is used by engineers and scientist to mathematically model and numerically solve very complex structural and fluid problems. FEA software can be utilized in a wide range of industries, but it is most commonly used in the aeronautical, biomechanical and automotive industries.

A finite element (FE) model comprises a system of points, called "nodes," which form the shape of the design [1–3]. Connected to these nodes are the finite elements themselves, which form the finite element mesh and contain the material and structural properties of the model, defining how it will react to certain conditions. The density of the finite element mesh may vary throughout the material, depending on the anticipated change in stress levels of a particular area. Regions that experience high changes in stress usually require a higher mesh density than those that experience little or no stress variation. Points of interest may include fracture points of previously tested material, fillets, corners, complex detail, and high-stress areas.

Typical results calculated by the solver include

1. Nodal displacements
2. Velocities and accelerations
3. Elemental forces, strains, and stresses

14.4.1 FEA Software

The following are examples of FEA software applications.

NX Nastran is a finite element solver that analyzes stress, vibration, structural failure/durability, heat transfer, noise/acoustics and flutter/aeroelasticity.

Femap is a CAD-independent, Windows-native pre- and post-processor for advanced engineering FEA. It provides engineers and analysts with an FEA modeling solution to handle even the most complex tasks easily, accurately, and affordably.

Solid Edge Simulation is a built-in FEA tool for design engineers to validate part and assembly designs digitally within the Solid Edge environment. Based on proven Femap finite element modeling technology, Solid Edge Simulation significantly reduces the need for physical prototypes, thereby reducing material and testing costs, while saving design time.

Parasolid is 3D geometric modeling component software, enabling users of Parasolid-based products to model complex parts and assemblies. It is used as the geometry engine in hundreds of different CAD, CAM and CAE applications.

14.4.2 Benefits of FEA

FEA can be used in new product design or to refine an existing product to ensure that the design will be able to perform to specifications prior to manufacturing. With FEA you can

- Predict and improve product performance and reliability
- Reduce physical prototyping and testing
- Evaluate different designs and materials
- Optimize designs and reduce material usage

We should note that the substances (either solids or fluids) being considered are assumed to be continuous or continua. Therefore, we are really asked to solve problems involving an infinite number of solution points, which implies infinite degree of freedom (DOF).

A degree of freedom can be regarded as one unknown quantity. Common sense suggests that the degree of difficulty of the solution to a problem is proportional to the number of DOF involved.

The FEM is based on the concept of discretization of continua for converting a problem involving infinite DOF to one with a finite number of DOF and is obviously a vital alternative way of solving such a problem.

The use of the discretization concept in engineering and science is not new. Finite difference method is regularly used in numerical differentiation. Discretization procedures used in FEA starts with the subdivision of a continuum into a finite number of subdomains called elements. The elements are connected to corners, or in some cases at selected points on edges, called nodes, as shown in Figure 14.2.

The original solid is subdivided into a finite number of subdomains (elements) with certain shapes (triangular and quadrilateral), and these elements are interconnected at the nodes. After such discretization, the original continuum is no longer a continuous body but contains a number of discrete pieces connected at the corners and/or edges in some cases. Another noticeable change is that the originally continuous curved boundary has now become segments of straight lines. The degree of similarity of two bodies, of course, depends on the number of elements used in discretized body (model). The more elements used in the model, the closer it gets to the original geometry, and thus the result is closer to the exact solution.

The use of FEM has been demonstrated in fields other than stress analysis. These include elastic-plastic stress analysis, heat and mass transfer, thermal stress analysis, dynamics and vibrations, fluid mechanics, diffusion and mass transport, solid mechanics, concrete engineering, aeromechanics, biomechanics and bioengineering, material science, mechanisms and linkages, and so on. However, in no way this list is complete.

If the minimum topological requirements are easily satisfied, the question remains as to how fine to subdivide the major components. The question is particularly relevant for elastic continua, such as slabs and unreinforced shells. In general, as the mesh density increases, you can expect the results to become more accurate. The mesh density required can be a function of many factors. Among them are the stress gradients, the type of loadings, the boundary conditions, the element types used, the element shapes, and the degree of accuracy desired.

14.4.3 Steps in Finite Element Method (1,3)

A general guideline most finite element analyses follow are given below.

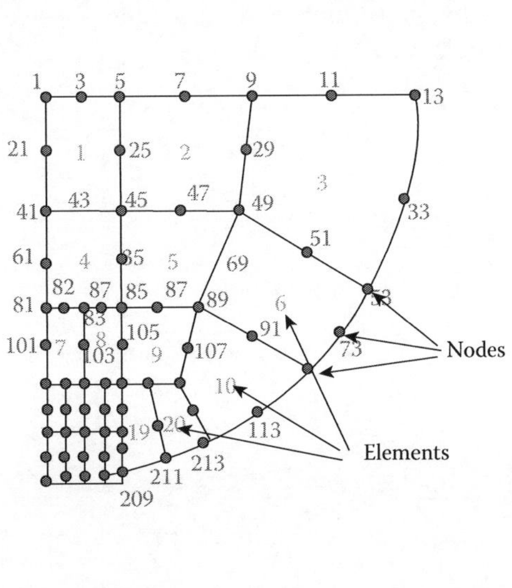

Figure 14.2

Discretization of a solid with nodes and elements.

14.4.3.1 Step 1: Discretize of Real Structure

Depending on the nature of structure, a variety of types of elements are available.

Commonly used elements are (i) bar elements, (ii) triangular plate elements, (iii) quadrilateral plate elements, (iv) triangular torus elements, (v) quadrilateral torus elements, (vi) tetrahedron elements, and (vii) hexahedron elements. Modern finite element computer programs allow various combinations of different element types.

The bar elements are frequently used in modeling trusses and frames, and the plate elements are suitable for structures having planar geometries or when variations of the physical quantities are limited to a plane (i.e., the plane stress or plane strain case).

To simulate the effects of real-world working environments in FEA, various load types can be applied to the FE model, including

- Nodal: Forces, moments, displacements, velocities, accelerations, temperature and heat flux
- Elemental: Distributed loading, pressure, temperature and heat flux
- Acceleration body loads (gravity)

Types of analysis include

- Linear statics: Linear analysis with applied loads and constraints that are static
- Nonlinear statics and dynamics: Effects due to contact (where one part of the model comes into contact with another), nonlinear material definitions (plasticity, elasticity, and so on) and large displacement (strains that exceed small displacement theory that limits a linear analysis approach)
- Normal modes: Natural frequencies of vibration
- Dynamic response: Loads or motions that vary with time and frequency
- Buckling: Critical loads at which a structure becomes unstable
- Heat transfer: Conduction, radiation, and phase change

Finally, for structures with extremely complicated geometry, when a 3D model becomes a necessity, the tetrahedron or hexahedron elements are commonly used. Many other types (e.g., beam and shell elements) are developed for special applications.

14.4.3.2 Step 2: Identify the Primary Unknown Quantities and Appropriate Interpolation Function

Depending on the nature of problem, the primary unknown quantity involved in the solution varies from case to case.

One may use displacement component as the primary unknown quantity in a stress analysis problem, temperature in heat conduction analysis, and velocity in a fluid flow problem.

Example 14.1

For the bar element shown below two nodes were assigned at $x=x_1$ and at $x=x_2$ for Node 1 and Node 2, respectively. We plan to develop a relation between force and displacement for the nodes.

We can write equation for stiffness of the bar as

$$k=\frac{EA}{L}$$

where:

- E is Young's modulus of the material of bar
- A is the cross-sectional area
- L is length of bar $L=x_2-x_1$

Forces at Node 1 and 2 are shown in Figure 14.2. From statistics, we know

$$F_1+F_2=0 \text{ and } F_2=k(x_2-x_1)$$

We can also write $F_1=-k(x_2-x_1)$. By principle of superposition, we can write

$$\begin{bmatrix} F_1 \\ F_2 \end{bmatrix}=\begin{bmatrix} k & -k \\ -k & k \end{bmatrix}\begin{bmatrix} x_1 \\ x_2 \end{bmatrix}=\frac{EA}{L}\begin{bmatrix} 1 & -1 \\ -1 & 1 \end{bmatrix}\begin{bmatrix} x_1 \\ x_2 \end{bmatrix}$$

or

$$F = K \cdot X$$

where:

K is the stiffness matrix

X is displacement vector

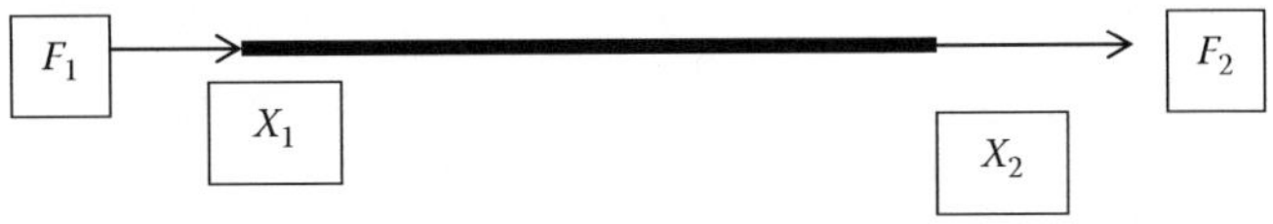

Example 14.2

Consider the assemblage of three of these two-force member elements. (Motion is restricted to one dimension, along the X-axis.)

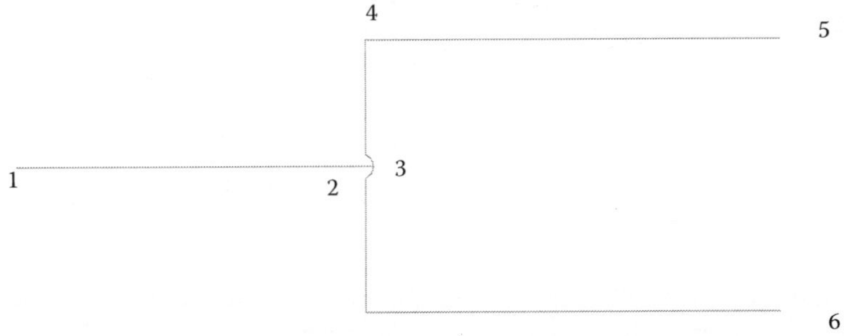

The assemblage has three elements and four nodes. Node 3 is at the intersection of all the three elements. For three elements we can write the equations as:

Element 1:

$$\begin{bmatrix} F_1 \\ F_{21} \end{bmatrix} = \begin{bmatrix} k_1 & -k_1 \\ -k_1 & k_1 \end{bmatrix} \begin{bmatrix} x_1 \\ x_2 \end{bmatrix}$$

Element 2:

$$\begin{bmatrix} F_{22} \\ F_3 \end{bmatrix} = \begin{bmatrix} k_2 & -k_2 \\ -k_2 & k_2 \end{bmatrix} \begin{bmatrix} x_2 \\ x_3 \end{bmatrix}$$

Element 3:

$$\begin{bmatrix} F_{23} \\ F_4 \end{bmatrix} = \begin{bmatrix} k_3 & -k_3 \\ -k_3 & k_3 \end{bmatrix} \begin{bmatrix} x_2 \\ x_4 \end{bmatrix}$$

14.4.3.3 Step 3

Derive general overall (global) force-displacement matrix (containing all nodal displacements)

Element 1:

$$\begin{bmatrix} F_1 \\ F_{21} \\ 0 \\ 0 \end{bmatrix} = \begin{bmatrix} k_1 & -k_1 & 0 & 0 \\ -k_1 & k_1 & 0 & 0 \\ 0 & 0 & 0 & 0 \\ 0 & 0 & 0 & 0 \end{bmatrix} \begin{bmatrix} x_1 \\ x_2 \\ x_3 \\ x_4 \end{bmatrix}$$

Element 2:

$$\begin{bmatrix} 0 \\ F_{22} \\ F_3 \\ 0 \end{bmatrix} = \begin{bmatrix} 0 & 0 & 0 & 0 \\ 0 & k_2 & -k_2 & 0 \\ 0 & -k_2 & k_2 & 0 \\ 0 & 0 & 0 & 0 \end{bmatrix} \begin{bmatrix} x_1 \\ x_2 \\ x_3 \\ x_4 \end{bmatrix}$$

Element 3:

$$\begin{bmatrix} 0 \\ F_{23} \\ 0 \\ F_4 \end{bmatrix} = \begin{bmatrix} 0 & 0 & 0 & 0 \\ 0 & k_3 & 0 & -k_3 \\ 0 & 0 & 0 & 0 \\ 0 & -k_3 & 0 & k_3 \end{bmatrix} \begin{bmatrix} x_1 \\ x_2 \\ x_3 \\ x_4 \end{bmatrix}$$

Summing the three sets of general equation, knowing $F_2 = F_{21} + F_{22} + F_{23}$, the overall global matrix can be formulated as

$$\begin{bmatrix} F_1 \\ F_2 \\ F_3 \\ F_4 \end{bmatrix} = \begin{bmatrix} k_1 & -k_1 & 0 & 0 \\ -k_1 & (k_1 + k_2 + k_3) & -k_2 & -k_3 \\ 0 & -k_2 & k_2 & 0 \\ 0 & -k_3 & 0 & k_3 \end{bmatrix} \begin{bmatrix} x_1 \\ x_2 \\ x_3 \\ x_4 \end{bmatrix}$$

Once the overall global matrix is developed for the structure (component), the next step is to apply the boundary conditions and solve for the unknown displacements.

At every node in the structure, either the externally applied load or the nodal displacement is needed as a boundary condition.

14.4.3.4 Step 4: Solve for Unknown Displacements

This step will be demonstrated with the following example.

Let $F_1 = 40$ lb, $k_1 = 50$ lb/in., $k_2 = 30$ lb/in., $k_3 = 70$ lb/in.

From the assemblage diagram, it is clear that $x_3 = x_4 = 0$. These are the boundary conditions to be applied to the global matrix. Forces F_3 and F_4 are reactions at Nodes 3 and 4.

Hence, in the global matrix, x_1 and x_2 are unknown displacements at Nodes 1 and 2. The reactions at Nodes 2 and 3 are also unknowns.

So, overall there are four unknowns and four equations, so the unique values can be found.

Plugging the boundary conditions, the global matrix is

$$\begin{bmatrix} 40 \\ F_2 \\ F_3 \\ F_4 \end{bmatrix} = \begin{bmatrix} 50 & -50 & 0 & 0 \\ -50 & (50+30+70) & -30 & -70 \\ 0 & -30 & 30 & 0 \\ 0 & -70 & 0 & 70 \end{bmatrix} \begin{bmatrix} x_1 \\ x_2 \\ 0 \\ 0 \end{bmatrix}$$

We can write all four equations as below.

$$40 = 50x_1 - 50x_2 \tag{14.1}$$

$$0 = -50x_1 + 150x_2 \tag{14.2}$$

$$F_3 = 0 \cdot x_1 - 30x_2 \tag{14.3}$$

$$F_4 = 0 \cdot x_1 - 70x_2 \tag{14.4}$$

From Equations 14.1 and 14.2, we get, $x_1 = 1.2$ in., $x_2 = 0.4$ in. Then from Equations 14.3 and 14.4, we get $F_1 = -12$ lb, $F_4 = -28$ lb.

The above steps illustrated the fundamental operation of the finite element analysis method. As can be seen, the formulation of the global force displacement relation equation is based on the general force displacement equation of a single element. The two-force member element (one-dimensional linear spring) is the simplest type of element used in FEA.

14.4.3.5 Step 5: Interpret the Result

Several general purpose FE programs are available. Commercial codes such as ANSYS, NASTRAN, and MARC analysis are very good and able to solve almost all machine component design problems, having extremely complex geometries and boundary conditions. It has become an important component in CAD technology.

Some commercial CAD packages have already integrated the FEM and solid modeling with a common interface and common database for both disciplines. NX I-9.0 is one of those packages.

The proprietary nature of these commercial packages allows the engineers to use them only as "black boxes" and often the important physical sense of this powerful tool is lost.

Example 14.3

A machine component is shown in the figure below. A load of $P = 180$ lb acts on the variable cross-sectional bar in which one end is fixed and a compressive load acts at the other end. Dimensions of the bar are shown in the figure. Calculate the deflection at various points along its length. Calculate the reaction force at the fixed end and compute stresses in each element.

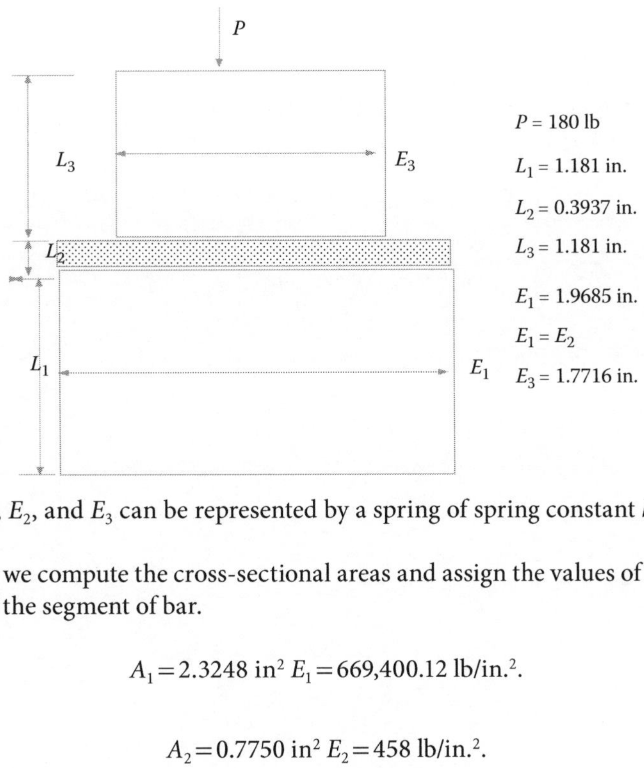

Each element E_1, E_2, and E_3 can be represented by a spring of spring constant k_1, k_2, and k_3, as calculated below.

From the figure, we compute the cross-sectional areas and assign the values of Young's modulus of elasticity to each of the segment of bar.

$$A_1 = 2.3248 \text{ in}^2 \; E_1 = 669{,}400.12 \text{ lb/in.}^2.$$

$$A_2 = 0.7750 \text{ in}^2 \; E_2 = 458 \text{ lb/in.}^2.$$

$$A_3 = 2.0922 \text{ in}^2 \; E_3 = 669{,}400.12 \text{ lb/in.}^2.$$

$$P = 180 \text{ lb}$$

All three segments of bar can be idealized as three different springs of three different lengths: $L_1 = 1.181$ in., $L_2 = 0.3937$ in., and $L_3 = 1.181$ in.

As we can imagine, there are four nodes for the three elements (springs), and their deflections in the vertical direction are y_1, y_2, y_3, and y_4.

Spring constants are also calculated as

$$K_1 = A_1E_1/L_1 = 1{,}317{,}714.986 \text{ lb/in.}$$

$$K_2 = A_2.E_2/L_2 = 901.575 \text{ lb/in.}$$

$$K_3 = A_3.E_3/L_3 = 1{,}185{,}909.253 \text{ lb/in.}$$

Using the spring representation of the bar segments, we can draw a free body diagram of the forces at each node and apply the law of mechanics where $\sum F = 0$, then at each node we have the following:

Node 1	$R_1 - k_1(y_2 - y_1) = 0$
Node 2	$k_1(y_2 - y_1) - k_2(y_3 - y_2) = 0$
Node 3	$k_2(y_3 - y_2) - k_3(y_4 - y_3) = 0$
Node 4	$k_3(y_4 - y_3) - P = 0$

$$\begin{bmatrix} k_1 & -k_2 & 0 & 0 \\ -k_1 & k_1 + k_2 & -k_2 & 0 \\ 0 & -k_2 & k_2 + k_3 & -k_3 \\ 0 & 0 & -k_3 & k_3 \end{bmatrix} \begin{bmatrix} y_1 \\ y_2 \\ y_3 \\ y_4 \end{bmatrix} = \begin{bmatrix} -R_1 \\ 0 \\ 0 \\ P \end{bmatrix}$$

As $y_1 = 0$, which is equivalent to eliminating Row 1 and Column 4 of the stiffness matrix. Hence, the above matrix reduces to

$$\begin{bmatrix} k_1 + k_2 & -k_2 & 0 \\ -k_2 & k_2 + k_3 & -k_3 \\ 0 & -k_3 & k_3 \end{bmatrix} \begin{bmatrix} y_2 \\ y_3 \\ y_4 \end{bmatrix} = \begin{bmatrix} 0 \\ 0 \\ P \end{bmatrix}$$

Plugging the values of all the k's and substituting the value of P, we obtain

$$\begin{bmatrix} 13{,}18{,}616.566 & -901.57 & 0 \\ -901.57 & 11{,}86{,}810.823 & -11{,}85{,}909.253 \\ 0 & -11{,}85{,}909.253 & 11{,}85{,}909.253 \end{bmatrix} \begin{bmatrix} y_2 \\ y_3 \\ y_4 \end{bmatrix} = \begin{bmatrix} 0 \\ 0 \\ 180 \end{bmatrix}$$

The solution of the above matrix will yield

$$y_2 = 0.0001366 \text{ in.}$$

$$y_3 = 0.199788 \text{ in.}$$

$$y_4 = 0.199940 \text{ in.}$$

Plugging $y_1 = 0$ in first equation, we get $R_1 = -k_1.y_2 = 180$ lb. Stresses are obtained using the stress–strain relationship for each segment of the bar, that is,

$$\sigma_1 = \frac{E_1(y_2 - y_1)}{L_1} = 77.4259 \text{ lb/in.}^2$$

$$\sigma_1 = \frac{E_2(y_3 - y_2)}{L_2} = 232.2589 \text{ lb/in.}^2$$

$$\sigma_3 = \frac{E_3(y_4 - y_3)}{L_{31}} = 113.351\,\text{lb/in}^2$$

Example 14.4

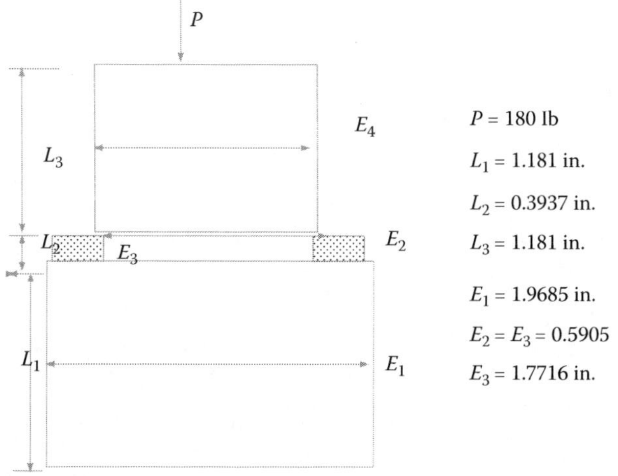

This is similar to Example 14.3; however, Element 2 is replaced with two identical supports, which form Elements 2 and 3.

$$A_2 = A_3 = 0.235\text{ in.}^2\; P = 180\text{ lb}$$

$$A_1 = 3.04\text{ in.}^2$$

$$A_4 = 2.456\text{ in.}^2$$

Spring constants for all four elements are calculated as K values:

$K_1 = E_1.A_1/L_1 = 1{,}724{,}887.221$ lb/in (Here E's are Young's modulus values)

$$K_2 = k_3 = E_2.A_2/L_2 = 318.64\text{ lb/in.}$$

$$K_4 = E_4.A_4/L_4 = 1{,}397{,}158.65\text{ lb/in.}$$

Writing the relation for each node:

Node 1	$R_1 - k_1(y_2 - y_1) = 0$
Node 2	$k_1(y_2 - y_1) - (k_2 + k_3)(y_3 - y_2) = 0$ (two springs stiffness are added)
Node 3	$(k_2 + k_3)(y_3 - y_2) - k_4)(y_4 - y_3) = 0$
Node 4	$k_4(y_4 - y_3) - P = 0$

Writing into matrix form as

$$\begin{bmatrix} k_1 & -k_1 & 0 & 0 \\ -k_1 & k_1 + k_2 + k_3 & -(k_2 + k_3) & 0 \\ 0 & -(k_2 + k_3) & k_2 + k_3 + k_4 & -k_4 \\ 0 & 0 & -k_4 & k_4 \end{bmatrix} \begin{bmatrix} y_1 \\ y_2 \\ y_3 \\ y_4 \end{bmatrix} = \begin{bmatrix} -R_1 \\ 0 \\ 0 \\ P \end{bmatrix}$$

Because $y_1 = 0$, the above matrix can be reduced to

$$\begin{bmatrix} k_1 + k_2 + k_3 & -(k_2 + k_3) & 0 \\ -(k_2 + k_3) & k_2 + k_3 + k_4 & -k_4 \\ 0 & -k_4 & k_4 \end{bmatrix} \begin{bmatrix} y_2 \\ y_3 \\ y_4 \end{bmatrix} = \begin{bmatrix} 0 \\ 0 \\ P \end{bmatrix}$$

Substituting the k's, we get

$$\begin{bmatrix} 1725524.507 & -637.2856 & 0 \\ -637.2856 & 1397795.936 & -1397158.65 \\ 0 & -1397158.65 & 1397158.65 \end{bmatrix} \begin{bmatrix} y_2 \\ y_3 \\ y_4 \end{bmatrix} = \begin{bmatrix} 0 \\ 0 \\ 180 \end{bmatrix}$$

The solution of which is found to be

$y_2 = 0.0001$ in.
$y_3 = 0.2826$ in.
$y_4 = 0.2827$ in.

We can also get $R_1 = 172.3$ lb. And then we can calculate the stresses as

$$\sigma_1 = \frac{E_1(y_2 - y_1)}{L_1} = 56.675\,\text{lb/in.}^2$$

$$\sigma_2 = \frac{E_2(y_3 - y_{21})}{L_2} = 328.63\,\text{lb/in.}^2$$

$$\sigma_4 = \frac{E_4(y_4 - y_3)}{L_3} = 56.675\,\text{lb/in.}^2$$

Example 14.5

A tapering round bar is fixed at one end and a tensile load $P = 180$ lb is applied at the other end. The maximum and minimum radius of the bar is 20 and 10 in., respectively. The bar's modulus of elasticity $E = 669{,}400.12$ lb/in^2. Consider the bar as set of four elements of equal length and uniformly increasing diameter. Find the global stiffness matrix and displacements at each node and reaction force.

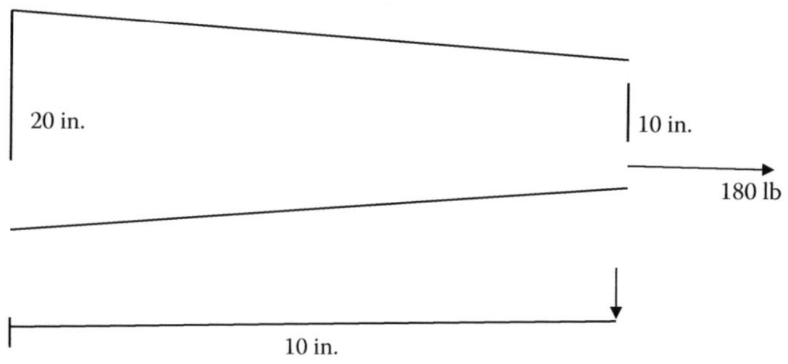

Let the tapered round bar be represented by a four element model as shown below.

The dimensions of the four segments are $D_1 = 10$ in., $l_1 = 2.5.$ in., $D_2 = 16.66$ in., $l_2 = 2.5$ in., $D_3 = 13.33$ in., $l_3 = 2.5$ in., $D_4 = 20$ in., and $l_4 = 2.5$ in.

As we can see, there are four elements of equal length and five nodes in all. At the top, the taper bar is fixed, and a tensile load of $p = 180$ lb. is applied at the minimum diameter. We can calculate $a_1 = 78.5398$ in.$^{-2}$, $a_2 = 139.626$ in.$^{-2}$, $A_3 = 218.166$ in.$^{-2}$, and $A_4 = 314.159$ in.$^{-2}$. We can also calculate $k_1 = A_1.E_1/l_1 = 2{,}102{,}982.5$, $k_2 = 3{,}738{,}635.52$, $k_3 = 58416180.5$, $k_4 = 784119300$. We can now write force displacement relations at each node:

At Node 1	$k_1(y_2 - y_1) - P = 0$
Node 2	$k_2(y_3 - y_2) - k_1(y_2 - y_1) = 0$
Node 3	$k_3(y_4 - y_3) - k_2(y_3 - y_2) = 0$
Node 4	$k_4(y_5 - y_4) - k_3(y_4 - y_3) = 0$
Node 5	$R - k_4(y_5 - y_4) = 0$

Writing the above equations in a compact matrix form as

$$\begin{bmatrix} k_1 & -k_1 & 0 & 0 & 0 \\ -k_1 & k_1+k_2 & k_2 & 0 & 0 \\ 0 & -k_2 & k_2+k_3 & -k_3 & 0 \\ 0 & 0 & -k_3 & k_3+k_4 & -k_4 \\ 0 & 0 & 0 & -k_4 & k_4 \end{bmatrix} \begin{bmatrix} y_1 \\ y_2 \\ y_3 \\ y_4 \\ y_5 \end{bmatrix} = \begin{bmatrix} P \\ 0 \\ 0 \\ 0 \\ -R \end{bmatrix}$$

Applying boundary condition, that is $y_5 = 0$, we reduce the above matrix to four equations:

$$\begin{bmatrix} k_1 & -k_1 & 0 & 0 & 0 \\ -k_1 & k_1+k_2 & k_2 & 0 & 0 \\ 0 & -k_2 & k_2+k_3 & -k_3 & 0 \\ 0 & 0 & -k_3 & k_3+k_4 & -k_4 \end{bmatrix} \begin{bmatrix} y_1 \\ y_2 \\ y_3 \\ y_4 \end{bmatrix} = \begin{bmatrix} P \\ 0 \\ 0 \\ 0 \end{bmatrix}$$

After plugging the k_s and solving, we find

$$y_1 = 0.185 \times 10^{-4} \text{ in.}$$

$$y_2 = 0.1004 \times 10^{-4} \text{ in.}$$

$$y_3 = 0.0522 \times 10^{-4} \text{ in.}$$

$$y_4 = 0.0214 \times 10^{-4} \text{ in. and}$$

$$R = 180.0 \text{ lb}$$

References

1. Irons, B. and Shrive, N. (1983). *Finite Element Primer.* Wiley, Hoboken, NJ.
2. Chandrupatla, T. R. and Belegundu, A. D. (2011). *Introduction to Finite Elements in Engineering.* Prentice Hall, Upper Saddle River, NJ.
3. Baran, N. M. (1988). *Finite Element Analysis on Microcomputers.* McGraw-Hill, New York.

Bibliography

Kirsch, U. (1981). *Optimum Structural Design.* McGraw Hill, New York.
NX 9.0. (2013). Siemens, PLM. Plano, TX.

15

Computer-Aided Design Examples

Solid Modeling, Assembly, Finite Element Analysis, and Cost Estimation

The computer-aided design (CAD) examples presented here consist of all components of design, solid modeling, assembly, finite element analysis (FEA), and cost analysis, as well as tolerance and codes wherever possible. Design engineers should design components within the codes and standards, including tolerance and allowances for the components designed. The design materials selected must be possible to manufacture, and so manufacturing methods of some components are also discussed.

All solid modeling, assembly, and FEA are done using NX 9.0. During material selection, due thought has been given to environmental impact factors such as embodied energy carbon footprint, including cost. The first project on CAD relates to green power generation, and all aspects of CAD, material selection, and life cycle analysis (LCA) for wind power generation are presented in detail.

15.1 CAD Example 1: Wind Power Generation

15.1.1 Introduction

The biggest advantage of wind energy is without a doubt the fact that it is a *renewable green energy source*. Wind energy is green energy, because during the generation of power, the environment is not polluted. Wind energy is renewable energy simply because the "fuel" used to generate electricity, that is, the wind, is practically unlimited.

There is significant variation in wind turbine size depending on purpose. Wind turbines can be used in applications as small as charging batteries or, as has been seen more recently, for producing commercial electric power. Smaller turbines are generally used to power a single household and have a capacity below 100 kW—most commonly around 2 kW. Commercially sized turbines have a capacity of up to 5 MW. In order to convert wind energy into electricity, an average wind speed of 14 mph is required. Wind turbine blade length and height are the main differences between commercial and residential turbines. Residential turbines generally stand at around 10 m tall, while commercial turbines are anywhere from 30 to 100 m tall.

Wind turbines or wind generators are used to convert the natural kinetic energy produced by the wind into useful mechanical energy. The main purpose of the wind turbine is to turn the generator located within the main housing just behind the three large blades; this main housing is called the *nacelle*. The generator will then use the mechanical energy to create electrical power. The power is transmitted within the nacelle through a series of shafts connected by a gearbox. The lower-speed shaft is connected to the rotor which

Figure 15.1

Typical wind turbine.

holds the three large blades, as seen in Figure 15.1. As mentioned above, a wind turbine consists of four main parts: the rotor, the nacelle, the tower, and the foundation. We will demonstrate below complete design and FEA of all main mechanical components. The rotor blades intersect the wind and capture the energy it contains, which causes them to rotate in a vertical plane about the shaft axis. The slow rotation of the shaft is normally increased by use of a gearbox, by which the rotational motion is delivered to a generator. The electrical output from the generator is then transferred through cables, down the turbine tower, to a substation, where the power is eventually fed into the electricity grid. The mechanical components housed at the top of the turbine tower—the rotor, the gearbox, and the generator—are all mounted in the nacelle, which can pivot, or yaw, about the vertical axis, so that the rotor shaft is always aligned with the wind direction.

Figure 15.1 shows the completed solid model, and the different components of a wind turbine system are shown in Figure 15.2. The specialized shape of the blades allows the movement of the entire system. Certain points along the cross section of the blades are shaped like an airfoil, and as with the cross section of an airplane wing, this shape creates a force perpendicular to the length of the blade or tangential to the swept area of the blades. Figure 15.3 shows the shape and forces involved with the blades.

Most wind turbines can produce a velocity tangential to the swept area at the blade tip (V_t) of between 40 and 60 m/s, giving the low-speed shaft an angular velocity of only 4–6 radians per second (40–60 rpm) for a 21 m diameter swept area. The generator, however, requires much higher speeds to produce electric power; therefore, the low-speed shaft is equipped with a large gear with many teeth, which is in contact with a much smaller pinion gear that is then attached to the high-speed shaft, which is directly connected to the generator. The effect of the gearbox is that the high-speed shaft will move at a higher velocity directly proportional to the number of teeth of the larger gear over the number of teeth of the smaller gear. Figure 15.4 shows the final gearbox design with the low-speed shaft on top. This gearbox above has a ratio of only 1:5, which is low for the typical ratio for a turbine, between 1:25 and 1:50. Low-speed generators do exist; however, they are very expensive.

Stresses play a large part in the design of any mechanical system, and in this case, can be broken into two categories: static and dynamic. The base of the turbine, called the tower, is a long and hollow cylindrical piece, which acts as a column to support the blades and nacelle. The tower must be designed to avoid buckling under the weight on top of it and to avoid bending possibly produced by the wind forces hitting the front of the wind turbine. The wind also creates stresses in the blades, which must also support a bending moment. The dynamic stresses are all due to the turning of the blades and both shafts. The shafts must be sized appropriately to avoid deformation due to torque. The gear teeth carry loads proportional to their respective torque over the gear radius. The teeth must be made strong enough to endure repeated stresses.

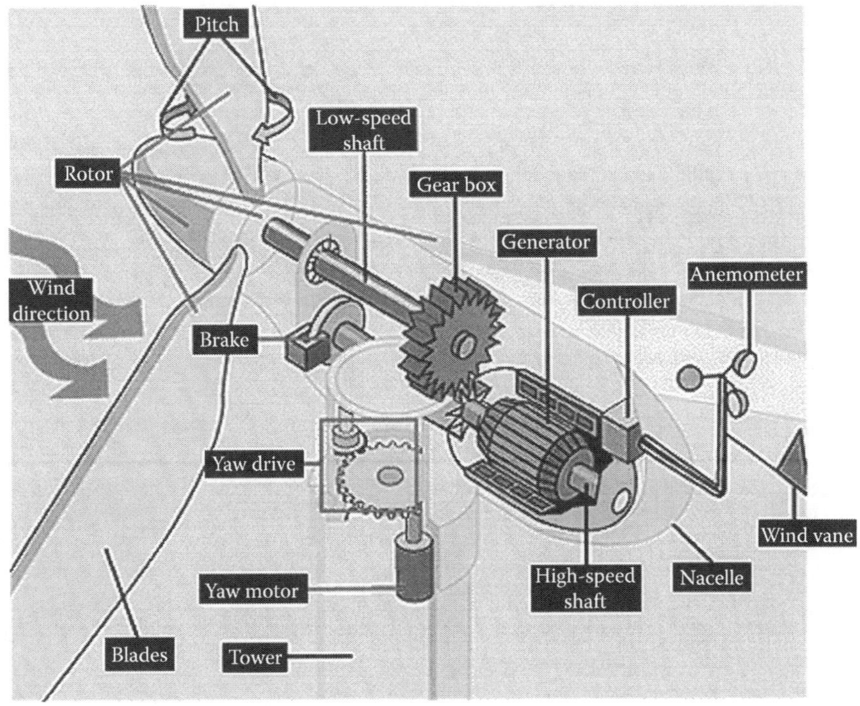

Figure 15.2

(See color insert) Wind turbine system.

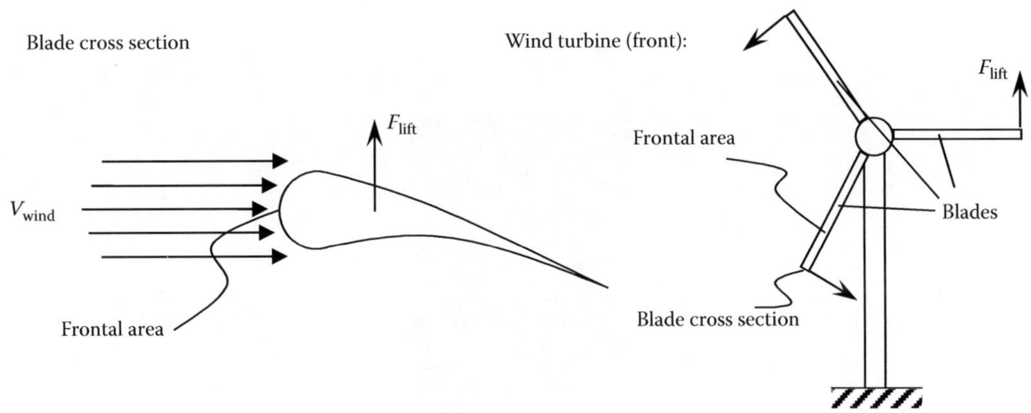

Figure 15.3

Blade design.

Making wind turbines a more significant source of energy in the future depends on the arrival of better and lower-cost materials which can be produced in high volumes and still be reliable. There is trend toward lighter-weight materials, but very sophisticated materials are often too expensive to justify their use. The turbine tower can represent up to 65% of the weight of the entire turbine. Low-cost materials such as reinforced concrete are now being used for the tower.

15.1.2 Mechanical Components Design

15.1.2.1 Analysis and Design

The first component to be analyzed is the tower, which as mentioned before, makes up most of the entire weight. The tower has to support the weight of the nacelle and blades, which for most wind turbines is approximately 73 tons (716 kN). Assuming a 50 m tower made of steel, which is fixed at the base, and using the Euler column formula,

Figure 15.4

Gearbox.

$$F_{cr} = \frac{C\pi^2 EI}{L^2} \Rightarrow$$

$$I = \frac{F_{cr}L^2}{C\pi^2 E} = \frac{(716{,}000\ \text{N})(50\ \text{m})^2}{\frac{1}{4}\pi^2(190\times10^9\ \text{Pa})} = 0.0038\ \text{m}^4 \tag{15.1}$$

a simple hollow cylinder was created in NX-9.0 by first drawing two circles with the same center point in the same plane. The larger circle had a diameter of 2200 mm and the smaller 2000 mm, and then both were extruded together for a distance of 50 m. Figure 15.5 shows the end product.

The second moment of area for this tower would come to

$$I = \frac{\pi}{64}\left(d_0^4 - d_i^4\right) = \frac{\pi}{64}\left(2.2^4 - 2^4\right)\ \text{m}^4 = 0.365\ \text{m}^4 \tag{15.2}$$

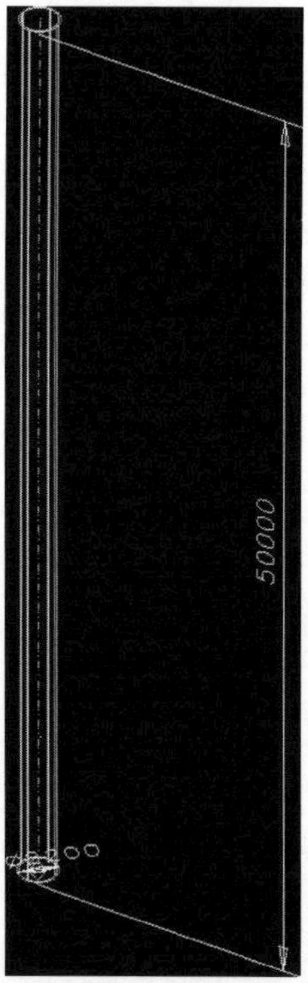

Figure 15.5

Tower with dimensions (mm).

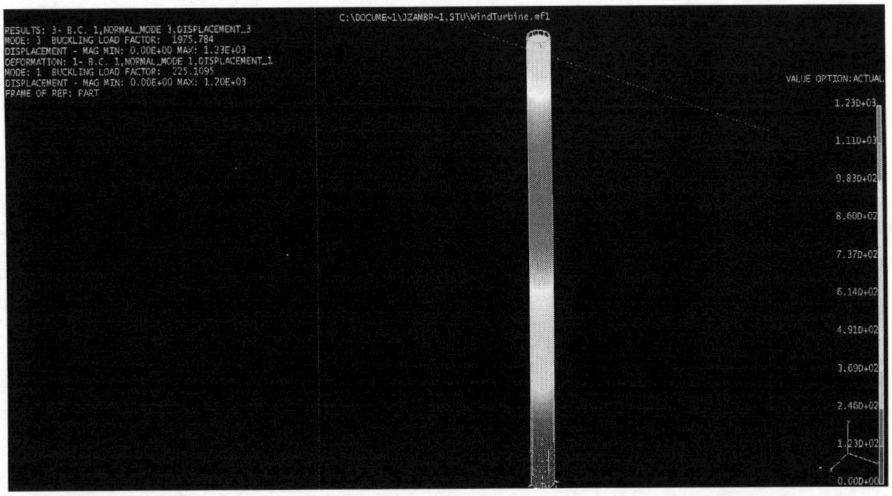

Figure 15.6

FE analysis tower.

The resulting safety factor for the tower is 94, making it far safer than necessary. Material use could be further reduced by redesigning the tower to be wider at the base and smoothly narrow toward the top, making it more stable. An FEA was performed in NX-Ideas to test the tower for buckling. The results showed a buckling load factor of 225, as shown in Figure 15.6.

Another concern is that the tower might bend due to strong winds creating a pressure over the entire frontal area of the turbine. The frontal area would be approximately equal to the areas of the three blades, the nacelle, and the tower added together. The combined force on the turbine is calculated below, using a safe wind speed of 15 m/s at sea level:

$$A_{\text{turbine}} = 3A_{\text{blade}} + A_{\text{tower}} + A_{\text{nacelle}} = 3\left(7.125\ \text{m}^2\right) + \left(30\ \text{m}\right)\left(2.2\text{m}\right) + \pi\left(2.3\ \text{m}\right)^2 = 107\ \text{m}^2 \tag{15.3}$$

$$F_m = \rho_{\text{air}} V_{\max}^2 A_{\text{turbine}} = \left(1.23\frac{\text{kg}}{\text{m}^3}\right)\left(15\frac{\text{m}}{\text{s}}\right)^2\left(104\ \text{m}^2\right) = 28.78\ \text{kN} \tag{15.4}$$

The normal bending stress in the tower due to the wind loading can be calculated and checked against the yield strength of the tower material. Figure 15.7 shows the dimensions that will be used in the stress formula for this and other applications.

The material chosen is SAE 1006 hot-rolled steel with yield strength of 170 MPa. The normal stress equation is shown here:

$$\sigma = \frac{Mc}{I} = \frac{FLc}{\dfrac{\pi\left(d_o^4 - d_i^4\right)}{64}} = \frac{\left(28.88\times10^4\ \text{N}\right)\left(30\ \text{m}\right)\left(1.1\ \text{m}\right)}{0.365\ \text{m}^4} = 23.03\ \text{MPa} \tag{15.5}$$

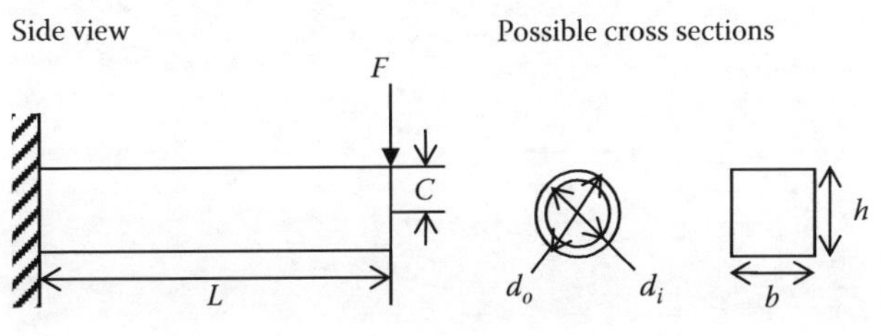

Figure 15.7

Beam dimensions.

$$n = \frac{S_y}{\sigma} = \frac{170}{123.03} = 7.296 \tag{15.6}$$

Wind turbines can be made with any number of blades, but *three* has been found to be the most efficient number when considering aerodynamic efficiency and cost. Aerodynamic efficiency increases by 6% from one to two blades and by 3% from two to three blades, but above three blades the increase is lower. Low material costs are desired; therefore, three is a good compromise. Selecting a blade length is directly related to the desired power output of the wind turbine. A certain amount of power is available in the wind, depending on its speed and density. The wind turbine can capture that power, but the actual power output is reduced by the turbine efficiency, which is generally about 35%. Therefore, the power required from the wind for a 10 hp output is equivalent to 7.45 kW.

$$P_{\text{wind}} = \text{ real power/turbine efficiency} = \frac{7.45\text{ kW}}{0.35} = 21.286\text{ kW} \tag{15.7}$$

The wind power formula based on blade diameter can then be written (Equation 15.8). The radius of the swept area, which approximates to the blade length, is then found to be

$$P_{\text{wind}} = \frac{1}{2}\rho_{\text{air}} A_{\text{swept}} V_{\text{wind}}^3 \tag{15.8}$$

where:

$$A_{\text{swept}} = r^2\pi$$

$$r = \sqrt{\frac{2P_{\text{wind}}}{\rho_{\text{air}} V_{\text{wind}}^3 \pi}} = \sqrt{\frac{2(213{,}057\text{ W})}{\left(1.23\frac{\text{kg}}{\text{m}^3}\right)\left(6\frac{\text{m}}{\text{s}}\right)^3}} = 7.142\text{ m}$$

The blades created in NX-Ideas were made to be 7.25 m in length, as shown in Figure 15.8.

A blade was created by drawing five different cross sections, four airfoils and one circle, initially in the same plane. Each cross section was then moved in the z-direction in increments that would add up to 7.25 m. The circle preceded the largest airfoil, which was followed by the smaller airfoils, with the smallest one on the very end. Once the cross sections were in the desired locations, a loft feature was used to create a solid blade based on the cross sections. Figure 15.9 shows the blade cross sections from two views.

The circular end of the blade was then extruded another 3 m to create a cylinder for insertion into the rotor. The first blade was positioned horizontally into the rotor in the appropriate location and then copied in two other locations at 120° using the center of the rotor as an axis. Each blade would experience a moment due to wind, just as the tower would. The blade was approximated by a square cross section. The following formulas show the acquisition of the force and normal stress associated with the wind and blade:

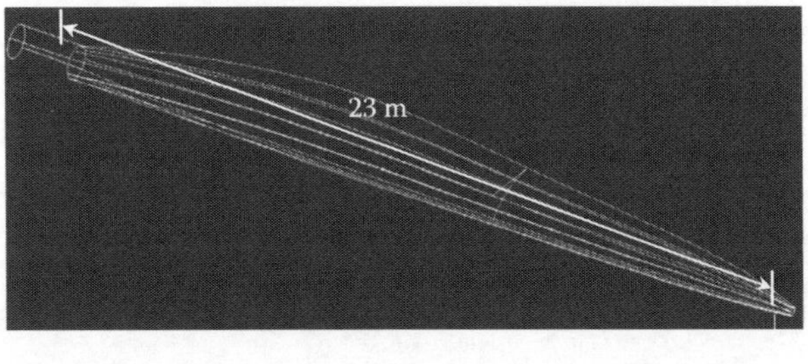

Figure 15.8

Blade solid model.

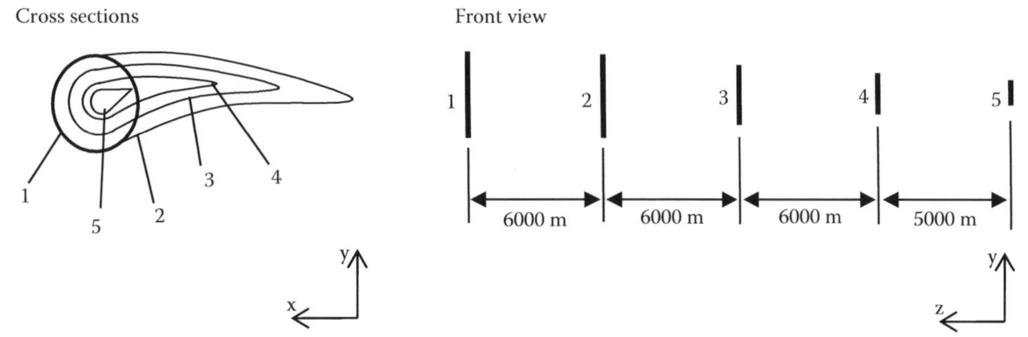

Figure 15.9

Blade design detail.

$$A_{\text{blade}} = rd = (7.125\text{ m})(1\text{ m}) = 7.125\text{ m}^2 \tag{15.9}$$

$$F_{\text{blade}} = \rho_{\text{air}} V_{\text{max}}^2 A_{\text{blade}} = \left(1.23\frac{\text{kg}}{\text{m}^3}\right)\left(20\frac{\text{m}}{\text{s}}\right)^2 (7.125\text{ m}^2) = 3.506\text{ kN} \tag{15.10}$$

$$\sigma = \frac{Mc}{I} = \frac{FLc}{\frac{b^3h}{12}} = \frac{(11{,}316\text{ N})(23\text{ m})(0.5\text{ m})}{\frac{(1)^3(1)}{12}} = 1.6\text{ MPa} \tag{15.11}$$

The yield strength of fiberglass, of which most blades are composed, was found to be 69 MPa, giving a safety factor of 43 for the blades designed. An FEA was also modeled in Ideas and showed no major structural concerns, as shown in Figure 15.10.

The rotor was created by drawing the shape shown in Figure 15.11 and then revolving the shape around an axis of revolution. The rotor was then rotated so that the round part became part of the face of the turbine. The round shape creates an ideal aerodynamic shape to minimize drag. Figure 15.12 shows the completed solid model.

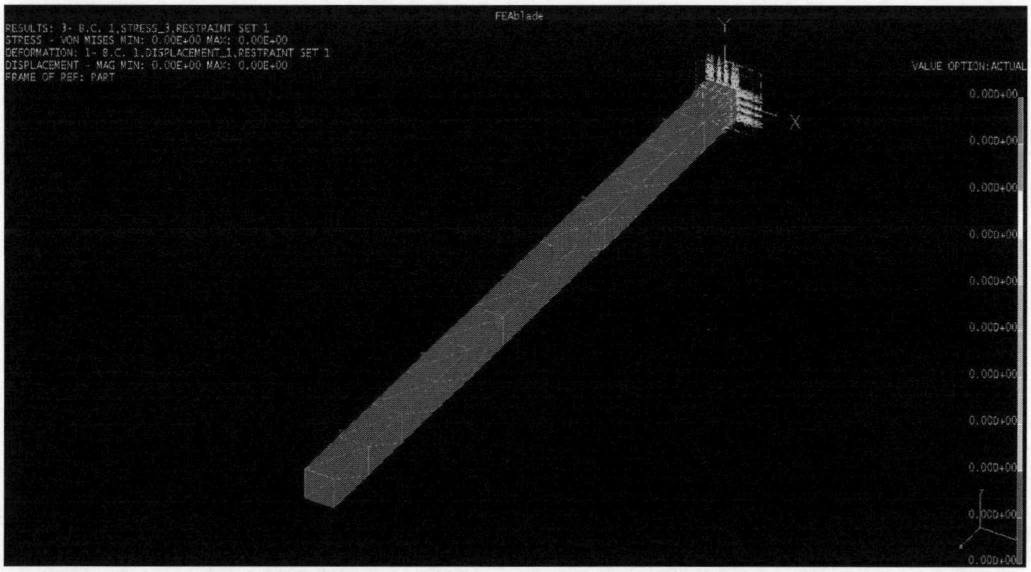

Figure 15.10

(See color insert) FE analysis blade.

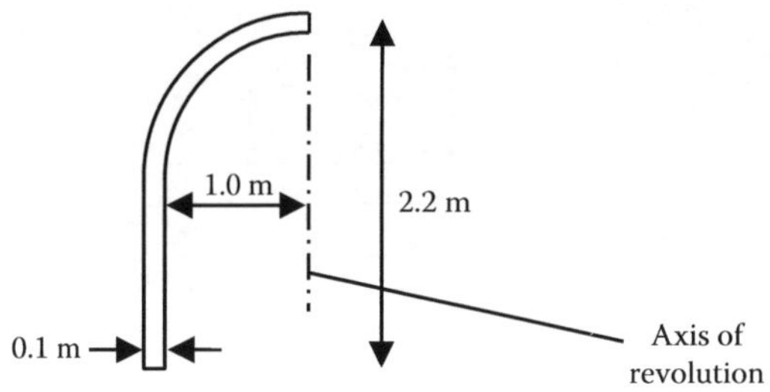

Figure 15.11

Rotor design.

Figure 15.12

Rotor solid model.

The gear teeth experience stresses due to the torque created by the turning of the blades. Wind turbine blades generally move with a tangential velocity at the ends (V_t) of 50 m/s. Higher values are used to make gears and pinion. Using this value, an angular velocity (ω) can be obtained for the low-speed shaft, and then a torque based on the power expected from the wind. The force on a tooth on the first gear is then found using Equation 15.14:

$$\omega = \frac{V_t}{r} = \frac{50\dfrac{\text{m}}{\text{s}}}{10.5\,\text{m}} = 4.76\frac{\text{rad}}{\text{s}} \tag{15.12}$$

$$T = \frac{P}{\omega} = \frac{213{,}057\text{ W}}{4.76\dfrac{\text{rad}}{\text{s}}} = 44760\text{ N}\cdot\text{m} \tag{15.13}$$

$$W_t = \frac{2T}{d_{\text{gear}}} = \frac{2(44{,}760\text{ N}\cdot\text{m})}{2700\,\text{m}} = 33.16\text{ N} \tag{15.14}$$

The stress on one gear tooth can then be found using the same method as with the blade. The gear tooth is also approximated with a rectangular shape.

$$\sigma = \frac{Mc}{I} = \frac{FLc}{\dfrac{b^3h}{12}} = \frac{(33.16\text{ N})(0.2\text{ m})(0.0375\text{ m})}{\dfrac{(0.8)^3(0.075)}{12}} = 77.7\text{ Pa} \tag{15.15}$$

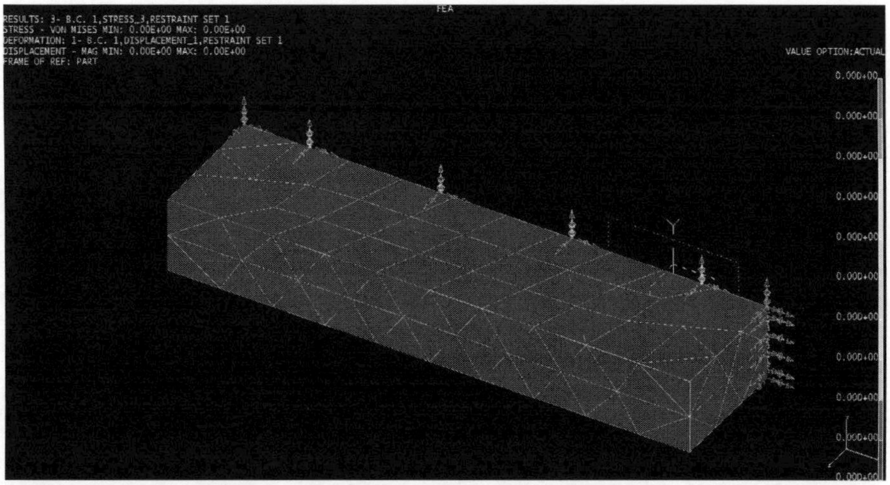

Figure 15.13

FE analysis gear tooth.

The gear is also made out of steel, so the stress is nowhere near the yield stress. An FEA of the gear tooth also showed no problems (Figure 15.13).

The nacelle was created in a similar fashion to the blade, in that two circular cross sections were created and then lofted together. Figure 15.14 shows the two cross sections and the solid model product. This shape, like the rotor, is aerodynamically effective (Figure 15.15).

15.1.2.2 Solid Modeling

A. Tower (Figure 15.16)
   1. Circle-center edge in X–Y plane, diameter = 2200 mm (center at the origin)
   2. Circle-center edge in X–Y plane, diameter = 2000 mm (same center as previous circle)
   3. Extrude, pick both circles, new part, distance = 50,000 mm
   4. Rotate part, pivot point (origin), about X, angle (–90)

B. Rotor (Figure 15.17)
   Draw shape X–Y plane
   1. Polyline, square U-shape
   2. Circle-center edge, quarter perimeter circles
   3. Polyline, axis of revolution (Figure 15.17)
   4. Revolve, choose enclosed shape, axis of revolution, 360°
   5. Rotate part, pivot point (origin), about X, angle (90)

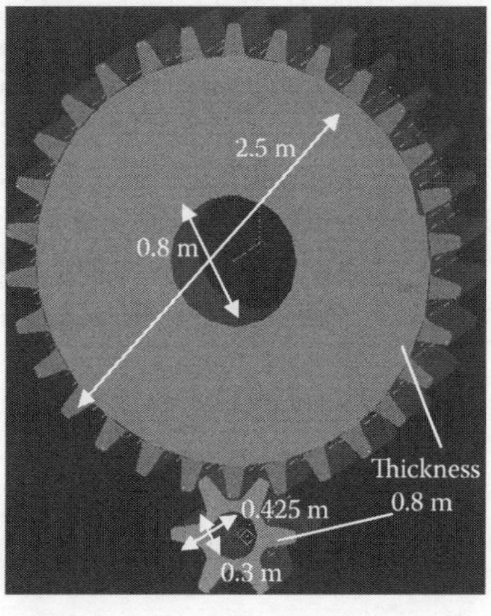

Figure 15.14

Solid model gears.

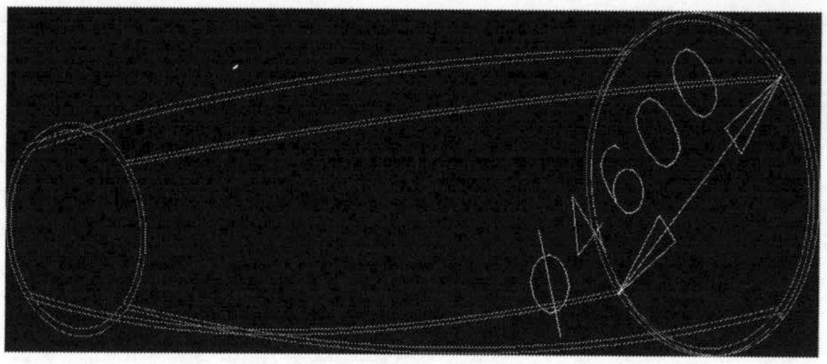

Figure 15.15

Solid model nacelle with dimensions (mm).

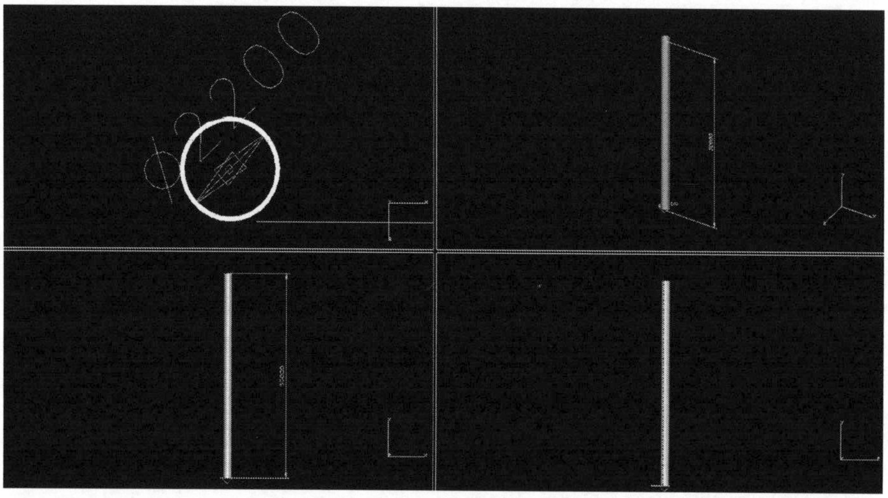

Figure 15.16

Tower.

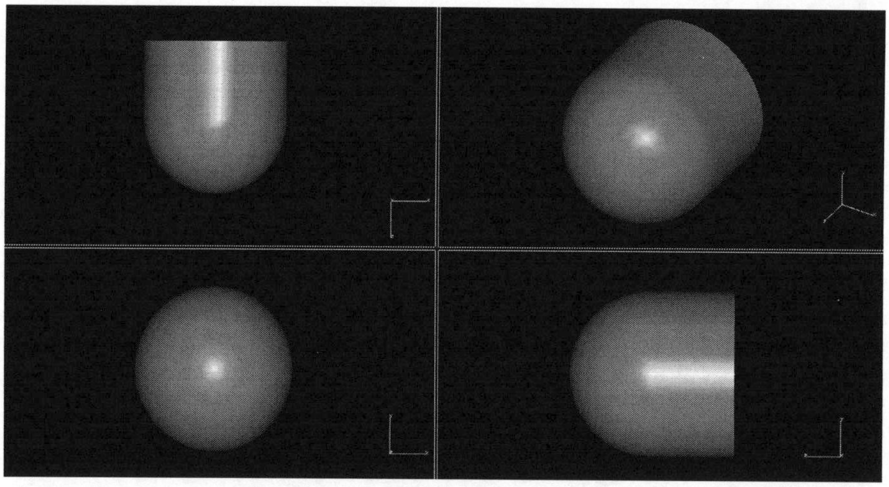

Figure 15.17

Rotor.

- **C.** Blades (Figure 15.18)
  Draw aerofoils X–Y plane
  1. Semicircle, C-shape
  2. Polyline, aerofoil top
  3. Three-point curve, aerofoil bottom
  4. Move aerofoils in Z-direction at equal spaces (Figure 15.18)

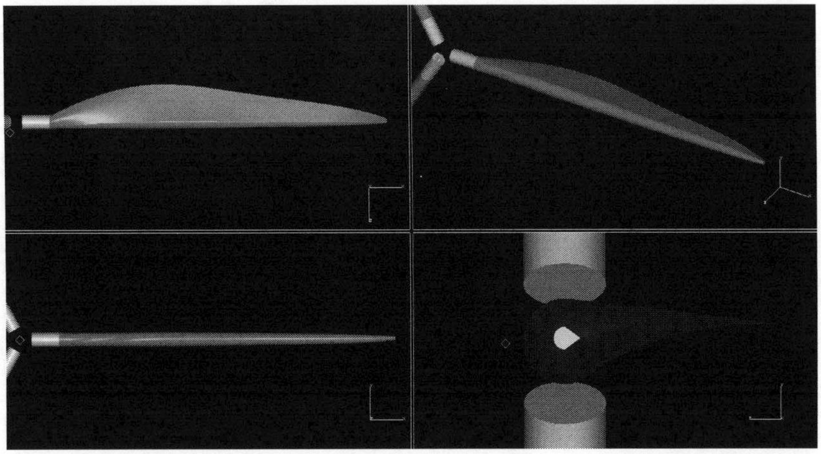

Figure 15.18

Blade.

5. Loft, choose each aerofoil cross section
6. Rotate, pivot point (origin), about Y, angle (90)
7. Move, position within rotor
8. Circular pattern, axis of revolution (same as rotor), 3 copies, 360°

D. Low-speed shaft (Figure 15.19)
1. Circle-center edge in X–Y plane, diameter = 800 mm
2. Extrude, distance = 6000 mm
3. Move, place within rotor between blades

E. Low-speed gear (Figure 15.20)
1. Circle-center edge in X–Y plane, diameter = 2500 mm
2. Extrude, distance = 800 mm
3. Sketch in place, flat circular surface of cylinder
4. Polylines, draw one tooth on circle perimeter
5. Dimension, three sides of tooth and angle between heights and width of tooth
6. Modify, set angles to 100°, heights 200 mm, widths 75 mm
7. Fillet, top corners and base of tooth diameter = 20 mm
8. Extrude, pick tooth, new part, distance = 800 mm
9. Circular pattern, axis of revolution (cylinder center), 30 copies, 360°
10. Move, align with main shaft
11. Cut, use shaft to cut through gear

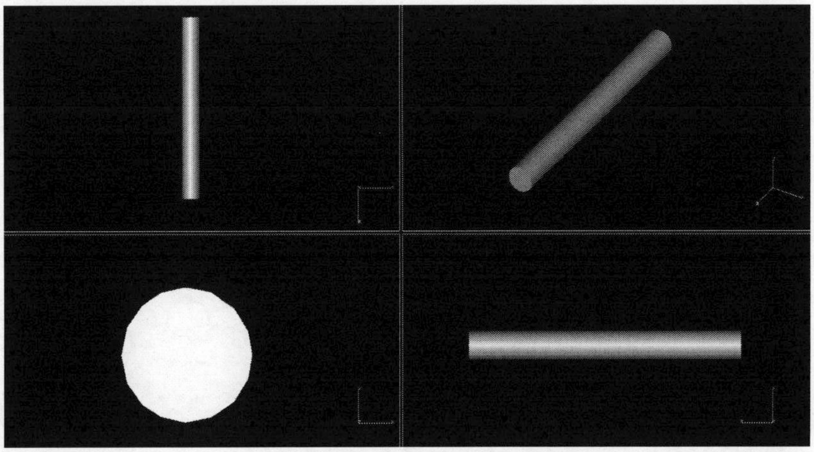

Figure 15.19

Low-speed shaft.

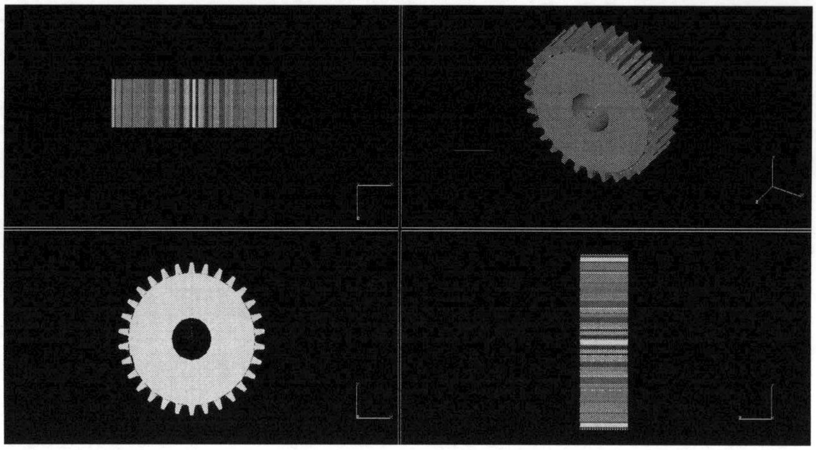

Figure 15.20

Main gear.

F. Pinion gear (Figure 15.21)
   1. Circle-center edge in X–Y plane, diameter = 425 mm
   2. Extrude, distance = 800 mm
   3. Sketch in place, flat circular surface of cylinder
   4. Polylines, draw one tooth on circle perimeter
   5. Dimension, three sides of tooth and angle between heights and width of tooth
   6. Modify, set angles to 100°, heights 200 mm, widths 75 mm
   7. Fillet, top corners and base of tooth diameter = 20 mm
   8. Extrude, pick tooth, new part, distance = 800 mm
   9. Circular pattern, axis of revolution (cylinder center), 6 copies, 360°
   10. Move, align with gear teeth
   11. Cut, use high-speed shaft to cut through gear
G. High-speed shaft (Figure 15.22)
   1. Circle-center edge in X–Y plane, diameter = 300 mm
   2. Extrude, distance = 6000 mm
   3. Move, align with pinion gear
H. Nacelle (Figure 15.23)
   1. Sketch in place, rotor open end
   2. Circle-center edge diameter = 4600 mm
   3. Circle-center edge diameter = 4500 mm

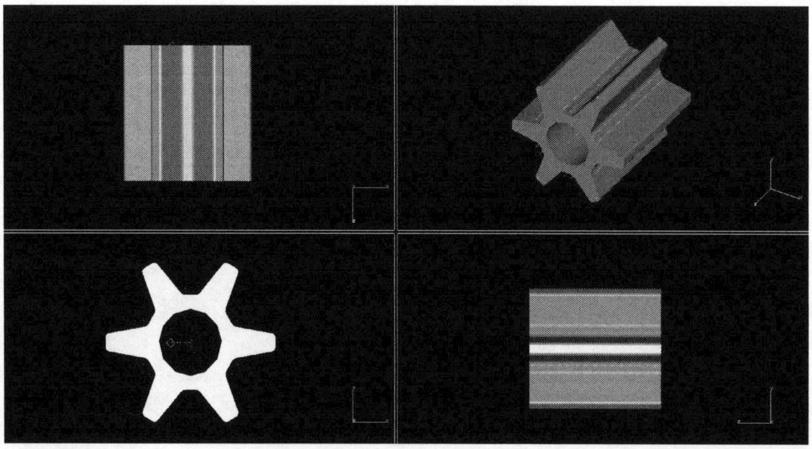

Figure 15.21

Pinion gear.

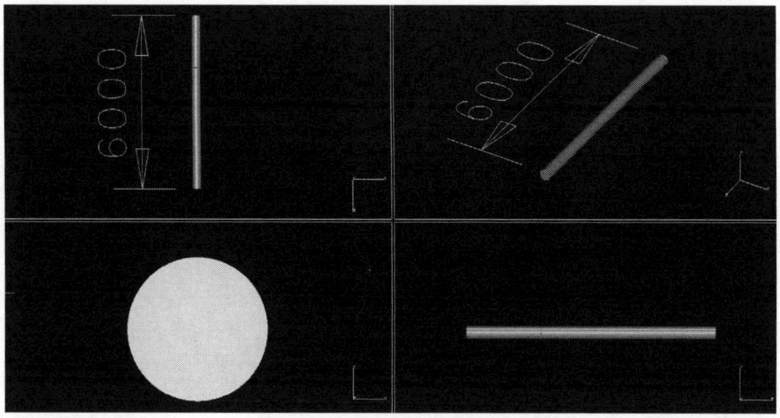

Figure 15.22

High-speed shaft.

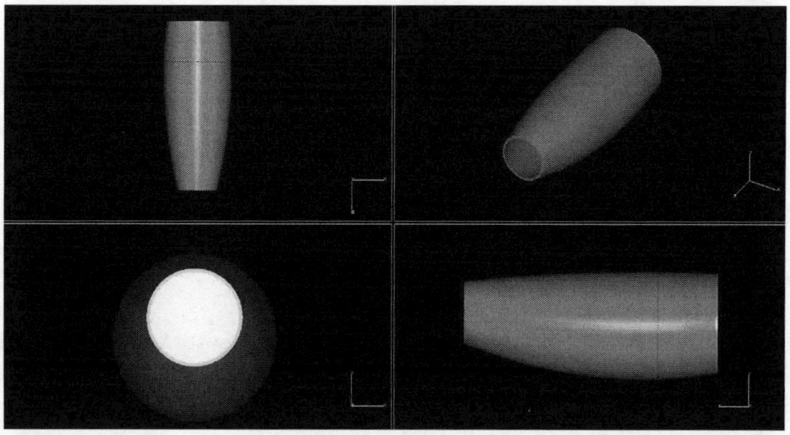

Figure 15.23

Nacelle.

4. Move, pick both circles, (0, −500, −15,000), to form back end of nacelle
5. Loft, pick rotor open end surface and created circles
6. Nacelle automatically in place

15.1.3 Materials (Table 15.1)

15.1.3.1 Cost

There are several parts made of steel: tower, gears, and shafts. The density of steel is 7.85 g/cm^3, and the cost is approximately \$800/ton. The volume calculations are as follows:

$$V_{\text{tower}} = \frac{\pi\left(2.2^2 - 2^2\right)\text{m}^2}{4} \times 50\ \text{m} = 33\ \text{m}^3$$

$$V_{\text{gears}} = \left(\frac{\pi(2.9)^2}{4} + \frac{\pi(0.425)^2}{4}\right)\text{m}^2 \times 0.8\ \text{m} = 5.4\ \text{m}^3$$

$$V_{\text{shafts}} = 5.2\ \text{m}\left(\frac{\pi(0.3)^2}{4}\right)\text{m}^2 + 7.2\ \text{m}\left(\frac{\pi(0.8)^2}{4}\right)\text{m}^2 = 4\ \text{m}^3$$

Table 15.1 Percentage of Materials for Large (Small) Turbines

Component/ Materials	Permanent Magnetic Materials	Prestressed Concrete	Steel	Aluminum	Copper	Glass-Reinforced Plastic	Wood Epoxy	Carbon Filament-Reinforced Plastic
Rotor								
Hub			(95)–100	(5)				
Blades			5			95	(95)	(95)
Nacelle	(17)		(65)–80	3–4	14	1–(2)		
Gearbox			98–(100)	2	2			
Generator	(50)		(20)–65		(30)–35			
Frame, machinery and shell			85–(74)	9–(50)	4–(12)	3–(5)		
Tower		2	98	(2)				

Therefore, the cost of steel will be

$$\$=\left(7.85\frac{\text{g}}{\text{cm}^3}\times\frac{1\times10^6\,\text{cm}^3}{\text{m}^3}\right)\left(800\frac{\$}{\text{ton}}\times\frac{\text{ton}}{1\times10^6\ \text{g}}\right)(33+5.4+4)\,\text{m}^3=\$266{,}272$$

Now, we will calculate the ecoproperties of the materials used in the wind turbine system.

15.1.4 Life Cycle Analysis of Wind Power Generation

15.1.4.1 Embodied Energy Analysis of Wind Power Generation

The life cycle of a wind farm consists of turbine production; turbine transportation to the site; site construction (which includes wind farm fixed assets); wind farm operation and maintenance; and dismantling, scrapping, and land reclamation. The various phases of the life cycle with energy inputs and electrical energy output, which is the only useful product, are illustrated in Figure 15.24.

A sample of data is presented in Table 15.2 for life cycle analysis.

Table 15.2 shows the operational characteristics and the estimated life cycle electrical energy output of the wind power generation system presented earlier. The useful life has been assumed to be 20 years, and with availability factor 90%, a full power lifetime of 18 years is calculated. Normally, the turbine only operates when wind speeds are in the range 4–25 m/s. Within this range, it is assumed that the turbine can only generate electricity at full generation capacity at a nominal speed of 15 m/s. The full power life of the wind farm is calculated as $20\times0.90=18$. The net power output of the wind farm is calculated by multiplying the capacity of the wind farm by the annual load factor. The life cycle electrical energy output of 604,017 W was calculated by multiplying the full power lifetime by net power output.

15.1.5 Turbine Manufacturing and Dismantling

The embodied energy of turbine manufacturing and dismantling is calculated as shown below. Tower, shaft, and gears are made of steel.

15.1.5.1 Steel

The volume of the steel used in turbine manufacture is

$$V_{\text{tower}}=\frac{\pi\left(2.2^2-2^2\right)\text{m}^2}{4}\times30\ \text{m}=19\ \text{m}^3$$

$$V_{\text{gears}}=\left(\frac{\pi(2.9)^2}{4}+\frac{\pi(0.425)^2}{4}\right)\text{m}^2\times0.8\ \text{m}=5.4\ \text{m}^3$$

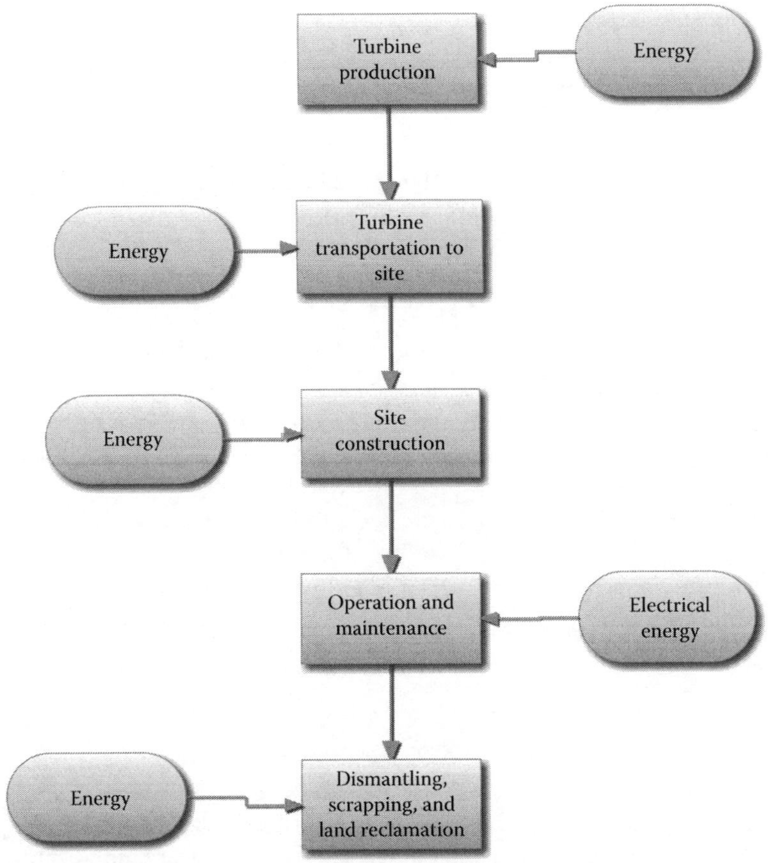

Figure 15.24

Life cycle energy flows of a wind power farm.

Table 15.2 Wind Farm Useful Electrical Energy Data

Description	Amount
Wind power capacity	10 hp (7.6 kW)
Lifetime	20 years
Availability factor	90%
Annual load factor (turbine efficiency)	35%
Full power lifetime	18
Net power output	33,556.5 W
Lifetime net power output	604,017 W

$$V_{\text{shafts}} = 5.2\text{ m}\left(\frac{\pi(0.3)^2}{4}\right)\text{m}^2 + 7.2\text{ m}\left(\frac{\pi(0.8)^2}{4}\right)\text{m}^2 = 4\text{ m}^3$$

Nacelle volume $= \pi/4 \times 4.6^2 \times 15 = 249.3\text{ m}^3$

Frame volume: The length of complete frame and shell over generator, gearbox, and cone is assumed to be three times the length of the slow-speed shaft. The slow-speed shaft is 6000 mm long, and hence the frame length is about 6000 mm. It is likely to be of oval shape, and the largest diameter is assumed to be 1000 mm (twice the size of the main gear). Now, we can calculate the volume of material of the frame and shell:

Frame volume $= \pi/4 \times 1^2 \times 6 = 18.846\text{ m}^3$

Total steel used in manufacture of turbine $= (19 + 5.4 + 4)\text{ m}^3 \times 7860\text{ kg/m}^3 = 334 \times 10^6\text{ g} = 223$ tons.

15.1.5.2 Glass Fiber

Volume of blade = 7.125 × 3 = 21.375 m^3

The blade is made of glass ceramic, and its mass = 21.375 m^3 × 2440 kg/m^3 = 52 tons.

15.1.5.3 Aluminum

We can see from Table 15.1 that aluminum is used in the hub (2%), the nacelle (4%), the gearbox (2%), frames and shells (50%), and the tower (2%). We can calculate the mass of Al as follows.

Total Al = 0.02 × 33 m^3 + 0.04 × 249.3 + 0.02 × 5.4 + 0.5 × 18.846 + 0.02 × 33 = 718.9 m^3

So mass of Al = 718 m^3 × 2710 kg/m^3 = 116 tons.

Copper volume = 0.14 × 249.3 + 0.02 × 9.4 + 0.35 × 100 (volume of generator assumed) + 0.17 × 18.846 = 73.3 m^3

Total mass of copper used = 73.3 × 8940 kg/m^3 = 65 tons.
Permanent magnetic materials are used in the generator (50%) and the nacelle (17%).
The volume of magnetic materials used in the wind turbine is calculated in Table 15.3.
Magnetic material volume = 0.17 × 249.3 + 0.5 × 100 = 42.931 m^3
Total mass of permanent magnet = 7.7 g/cm^3 × 42.931 = 322 tons.
(Density of the permanent magnet is taken from the website of Hunghon Dunben Magnetic Co. Ltd, China.)

In addition to the embodied energy calculated in Table 15.3, we should also calculate the embodied energy in the production of the turbine along with its transportation to the wind farm site. These data were not readily available.

15.1.6 Site Preparation and Construction

The embodied energy of site preparation and construction includes site construction building materials, nonmaterial-related site preparation and construction processes, and fixed assets and equipment (excluding the wind turbine). The site construction building materials include the wind turbine foundations, cable trenches and cables, paths and roads, and site office (if there is one). The results are presented in Table 15.4.

15.1.7 Wind Farm Operation and Maintenance

The embodied energy of wind farm operation and maintenance is calculated as follows.

15.1.7.1 Steel

Total steel used in the wind farm = 223 + 30 + 0.136 + 5 = 258.136 tons.

The cost of steel will be

$$258.136 \text{ tons} \times \$1000/\text{ton} = \$258.136$$

$$\$ = \left(7.85\frac{\text{g}}{\text{cm}^3} \times \frac{1\times10^6\,\text{cm}^3}{\text{m}^3}\right)\left(800\frac{\$}{\text{ton}} \times \frac{\text{ton}}{1\times10^6\,\text{g}}\right)(33+5.4+4)\text{m}^3 = \$266{,}272$$

15.1.7.2 Copper

Total copper used = 65 tons
Copper cost = 65 × 7000 = $455,000

Table 15.3 Embodied Energy of Wind Turbine Construction Material

Material	Mass (Ton)	Energy Intensity (GJ/ton)[a]	Embodied Energy (GJ)	Percentage of Total (%)
Aluminum	116	200–240	2.32×10^4–2.784×10^4	52–53.4
Copper	65	68–74	4.42×10^3–4.81×10^3	9.9–9.2[b]
Steel	223	29–35	6.467×10^3–7.805×10^3	14.5–15.0
Glass fiber	52	23.8–26.3	1.238×10^3–1.368×10^3	2.8–2.6[b]
Permanent magnet	332	28–31	9.296×10^3–1.029×10^4	20.8–19.7[b]
		Total embodied energy	4.462×10^4–5.211×10^4	

[a] Energy intensity data is taken from Ashby, M. F. *Materials and Environment*. Butterworth & Heinemann, Boston, MA, 2011 [1].
[b] It is noticed that an increase in the embodied energy does not result in an increase in the percentage.

Table 15.4 Embodied Energy of Site Preparation and Construction

Description	Amount of Materials (ton)	Embodied Energy (GJ/ton)	Embodied Energy (GJ)	Percentage (%)
Turbine foundation construction materials (steel + concrete)[a]	Concrete: 720	1.1–1.2	792–864	32.7–30.2
	Steel: 30	29–35 (low-carbon steel)	870–1050	35.9–36.7
Cable trenches and cable construction materials[b]	Corrosion resistance steel: 0.136	32–38 (low-alloy steel)	4.352–5.168	0.01797–0.01806
Path and road construction materials[c]	Soil: 50	2.2–2.3	110–115	4.5–40
	Stone: 100	4.9–6.4[b]	490–640	20.2–22.4
Site office construction materials[d]	Concrete: 10	1.1–1.2	11–12	0.4542–0.4194
	Steel: 5	29–35[b]	145–175	
Total embodied energy			2422–2861	

[a] The amount of concrete in the foundation of a turbine is 300 m^3. It is assumed to have a density of 2400 kg/m^3. Hence, the mass of concrete for each foundation is calculated as 720 tons. The amount of steel is about 30 tons.

[b] The tower height is 50 m, and the length of cable is assumed to be 55 m. The material of the cable is corrosion-resistant steel with cast steel as sheave material. The standard wire diameter selected is 3/8 in, and weight per foot is $1.7d^2$ lbf. (Data from Budynas, R. and Nisbett, J. *Mechanical Engineering Design*, 8th edn. McGraw-Hill, New York, 2008.) From this, we can calculate the weight of 55 m cable = $50/0.305 \times 1.7(3/8)^2 \times 12$ = 470 lb or 136 kg.

[c] The path and road construction materials are assumed to be stone and soil. The mass of soil and stone will change with the length of roads and path. It is assumed here that 5 km road was built, and total mass of soil = 50 tons and stone = 100 tons.

[d] The material used in site office construction is, again, concrete and steel. The mass of materials will depend upon the number of rooms at the site; however, it could just be the main transformer room. The masses of steel and concrete materials used are assumed to be steel = 5 tons and concrete = 10 tons.

15.1.7.3 Aluminum

Total aluminum = 116 tons
Cost of Aluminum = 116 tons × \$3000/ton = \$348,000

15.1.7.4 Glass Fiber

Total glass fiber used = 52 tons
Cost of glass fiber = 52 tons × \$6500/ton = \$338,000

15.1.7.5 Permanent Magnet

Total permanent magnet used = 332 tons
A 10 kW permanent magnet for a wind farm can be bought for about \$14,000 from Anhui Hummer Dynamo Co. Ltd, China

15.1.7.6 Concrete

Total concrete used = 750 tons
Total concrete cost = 750 tons × \$65/ton = \$48,750

15.1.7.7 Stone

Total stone used = 100 tons
Stone cost = 100 tons × \$1000/ton = \$100,000.
Total material cost = \$266,272 + \$455,000 + \$348,000 + \$338,000 + \$14,000 + \$48,750 + \$100,000 = $\$1.67 \times 10^6$
The total cost is calculated as follows:

Total cost = material cost + overhead cost
Overhead cost = 25% of material cost
So, total cost = $\$1.67 \times 10^6 + 0.25(1.67 \times 10^6) = \$\ 2.0875 \times 10^6$

Some of the cost data should be converted to 2012 dollars, as these are 2005 data.

The average proportions of cost associated with different activities related to the operation and maintenance of a wind farm are as follows.

Category	Proportion (%)
Administration	21
Insurance	13
Land rent	18
Miscellaneous	17
Power from grid	5
Service and spare parts	26

From the total cost, proportionate costs for different categories can be calculated easily. The assumed industry for each of the categories is listed below:

Administration: Electricity generation and distribution
Insurance: Insurance
Miscellaneous: Electricity generation and distribution
Service and spare parts: Machinery and equipment maintenance
Power from grid: Electricity generation and distribution
Life cycle operation and maintenance (O & M) costs = total cost × energy intensity (GJ/$) × plant life (20 years).
The energy intensity coefficients from [2] were used to calculate the embodied energy.

15.1.8 Wind Farm Decommissioning and Land Reclamation

The energy embodied in wind farm decommissioning and land reclamation was assumed to be zero. The energy associated with wind turbine dismantling was already included in the calculation for turbine production and dismantling. It is also stated in [3] that the recycling and scrapping of the wind turbine has a negative embodied energy. This represents the energy gain because of the lower energy embodied in recycling the materials when compared with the extraction of the materials in their raw state from the earth. It was assumed that the energy embodied in the other stages of plant decommissioning (e.g., plant equipment dismantling) is canceled out by the energy gain in recycling of turbine materials. The land reclamation embodied energy is assumed to be nil.

15.1.9 Discussion

The proportion of energy embodied in the various phases of the wind farm's life cycle is shown below. The wind farm's life cycle is divided into four main phases: turbine production, turbine transportation, site construction and fixed assets, and operation and maintenance. The wind farm decommissioning and reclamation embodied energy is assumed to be zero.

The highest portion of the wind farm's life cycle energy is embodied in turbine production and dismantling, accounting for nearly 43% of the wind farm's total embodied energy. Turbine transportation and site construction and fixed assets contribute about 30% and 20%, respectively, to the total. Wind farm operation and maintenance contributes only 7.38% to the total embodied energy. The direct energy of the wind farm corresponds to operation and maintenance, while indirect energy includes all the other phases of the life cycle.

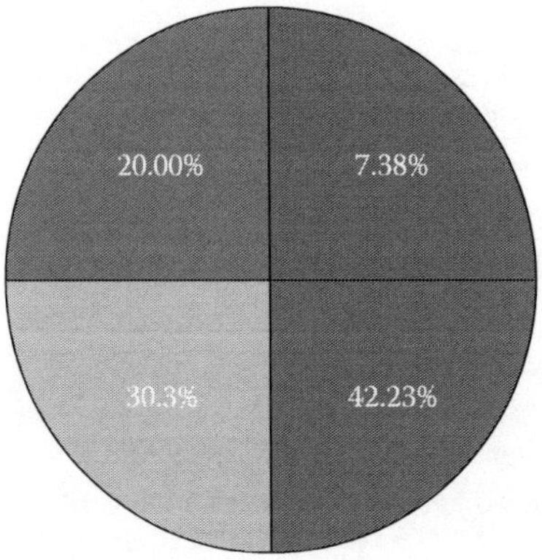

15.2 CAD Example 2: Drum Brake

15.2.1 Introduction

Brakes are a widely used tool for reduction of speeds. They can be used in many different ways; however, they are most commonly seen in systems of transportation. Brakes can be found in trains, buses, ambulances, planes, cars, trucks, wheelchairs, and bicycles, and even on the rollerblades of those pesky neighbors of yours. The majority of rear drum brakes today are internally expanding. However, the drum brake analyzed in this design is actually an externally contracting brake system [4].

A brake system powered by a solenoid actuating arm was analyzed in this project. This brake system consists of a solenoid cylinder joined to connecting rods. The connecting rods are then attached to the collar, which is between two rocking arms that hold cast-iron brakes. The rocking arms are held in place by the force created by the compression spring on the outside of the right rocking arm. The cast-iron drum can be found between the cast-iron brake linings.

The solenoid is activated, which moves the connecting rods. The connecting rods then rotate the collar, which pulls each rocking arm inward in order to allow the brake lining to contact the drum. Once the brake lining is in contact with the drum, the force of the brakes as well as frictional forces will slow the rotation of the drum. When the solenoid moves back to its original position, the compression spring returns the rocking arms, collar, and connecting rods back to their original positions until they are needed again.

Possible manufacturing complications include the necessity for the parts to be manufactured to a high standard. The collar in particular needs to be machined so that there are no issues with the teeth while rotating. Another manufacturing complication could be the spring. The spring is purchased from an outside vendor, who might have different tolerance levels.

15.2.2 Analysis and Design

The materials chosen for the parts that were analyzed are shown in Table 15.5. These materials were shortlisted based on environmental impact considerations.

15.2.2.1 Design of Spring

When designing the spring, it is important to consider the material. The most common material used for springs is *music wire*. However, in this design problem, gauge 24 chrome–silicon is used. “This alloy is an excellent material for highly stressed springs that require long life and are subjected to shock loading” (Budynas, 506). Equation 15.16 is used when estimating the ultimate tensile strength for the spring:

$$S_{ut} = \frac{A}{d^m} \tag{15.16}$$

where:

d is the wire diameter of the spring

constants A and m can be found in table 10-4 in *Shigley’s Mechanical Engineering Design* [4].

The maximum allowable torsional stress is equivalent to the estimate of the yield strength in shear, which is based on distortion energy theory. The estimated yield strength in shear is

$$S_{sy} = 0.45 \cdot S_{ut} \tag{15.17}$$

The axial spring force that is applied to the spring corresponding to this yield strength in shear is given by

Table 15.5 Materials Used

Part	Material	Ultimate Tensile Strength S_t (MPa)
Spring	Chrome–silicon	1475.205
Solenoid	AISI 4140 Q&T	1770
Drum	Cast iron	200
Brake lining	Cast iron	200

$$F = \frac{\pi d^3 S_{sy}}{8K_B D} \tag{15.18}$$

where:

D is the mean coil diameter
K_B is the Bergstrasser factor

The deflection observed by the spring is given as

$$U = \frac{8FD^3N}{d^4G} \tag{15.19}$$

where:

N is the total number of coils in the spring
G is the modulus of rigidity of the spring

The cost of the material is another important factor when considering which material to choose for any given application. The chrome–silicon chosen for this application has a relatively high cost when compared with some of the other options. Of course, to compensate for this, the spring will have a relatively higher elastic limit.

The results of calculations for the spring can be seen in Table 15.6.

15.2.2.2 Design of Solenoid

The solenoid in this design will provide the actuating force that will cause the brakes to close on the drum. It is important to be able to analyze the loading that the solenoid can withstand so that the critical loading of

Table 15.6 Spring Calculation Results

Chrome–Silicon		
Constant	m	0.108
Wire diameter	*d*, mm	14.8336
Constant	*A*	1974
Mean coil diameter	*D*, mm	76.2
Total coils	*N.t*	5
Constant	alpha	0.5
Number of active coils	N_a	3
Modulus of rigidity	*G*	77,200 MPa
Scale of the spring	*k*, N/mm	351.9876 ± 0.05
Deflection	*y*, mm	24.68889 ± 0.05
Solid length	*L.s*, mm	89.0016 ± 0.05
Free length	*L.o*, mm	113.6905 ± 0.05
To avoid buckling	*L.o* <	400.812 ± 0.05
Pitch of the body coil	*p*, mm	23.06323 ± 0.05
Ultimate strength (S_{ut})	MPa	1475.205 ± 0.05
Yield strength (S_{sy})	MPa	663.8425 ± 0.05
Spring index	C	5.136986 ± 0.05
Bergstrasser factor	K.B	1.284934 ± 0.05
Static load (*F*)	Newtons	8690.183 ± 0.05
Cost	$	50.49132 ± 0.05

the column will not be reached. If the critical loading is reached, the solenoid will fail and experience what is called *buckling*. The critical load that a column observes is given by

$$P_{cr} = \frac{C\pi^2 E}{(l/k)^2} \times A \tag{15.20}$$

where:

- C is based on the end conditions
- E is the modulus of elasticity
- A is the cross-sectional area of the column
- l/k is the slenderness ratio

The end conditions for this design problem for the solenoid are one end rounded and pivoted with one end fixed, which gives a value of C equal to 2. The modulus of elasticity is based on the material used. The material chosen for the given design is AISI 4140 quenched and tempered carbon steel. The slenderness ratio for the given design problem is

$$\frac{l}{k} = \sqrt{\frac{2\pi^2 CE}{S_y}} \tag{15.21}$$

where S_y is the tensile yield strength of the material.

Before the cross-sectional area is known, the critical unit load can be found, and is given as

$$\frac{P_{cr}}{A} = \frac{C\pi^2 E}{(l/k)^2} \tag{15.22}$$

The results of the calculations for the buckling of the solenoid can be seen in Table 15.7.

15.2.2.3 Design of Drum

The drum and brake lining are analyzed in order to find the braking power of the system. The drum chosen for this design problem has a diameter of 10.09 in with a friction surface (where the brake contacts the drum) of 2 in. The weight of the drum is given as 11.3 lbf. The braking power of the system is equivalent to the torque observed by the drum, and is given as

Table 15.7 Results for Calculations on the Solenoid

AISI 4140 Q&T	S_t, MPa	1770		
	S_y, Pa	1,640,000,000		
	S.y, kpsi	238		
One end rounded and pivoted, and one end fixed				
	C	2		
	P_{cr}/A		820	MPa/m²
Carbon steel	E, Mpsi	30		
	E, Pa	2.07×10^{11}		
	l/k	70.5899895 ± 0.05		
	A, m²	0.00012668 ± 0.05		
	d, m	0.0127		
	P_{cr}, Pa	103,875.033 ± 0.05		
	P_{cr}, kPa	103.875033 ± .05		

$$T = a \cdot f \cdot dN \tag{15.23}$$

where:

- a is the distance to the pivot point
- f is the friction factor
- dN is the normal force on the drum from the brakes

The friction factor is dependent on the material used, which for this design problem is cast iron for both the drum and the brake lining. The friction factor for this design problem is 0.35. The distance to the pivot point is given as

$$a = \frac{4r\sin(\theta_2)}{2\theta_2 + \sin(2\theta_2)} \tag{15.24}$$

where:

- r is the radius of the drum
- θ2 is the angle above the horizontal at which the brake contacts the drum

The normal force observed by the drum and brake lining is defined as

$$dN = P_{\max} br\big(\sin(\theta_2) - \sin(\theta_1)\big) \tag{15.25}$$

where:

- $P_{\max}$ is the maximum pressure of the drum
- b is the face width of the brakes exposed to the drum
- θ1 is the angle below the horizontal at which the brake contacts the drum

An important analytical concept for brakes is thermal analysis, since brakes convert kinetic energy into heat. The rate of heat loss is given as

$$\text{Heat loss} = \left(h_{\text{rad}} + f_{\text{vent}} \cdot h_{\text{conv}}\right) \cdot A \cdot \Delta T \tag{15.26}$$

where:

- h_{rad} is the radiation component of the coefficient of heat transfer
- h_{conv} is the convective component of the coefficient of heat transfer
- f_{vent} is the ventilation factor (1 for this design problem)
- ΔT is the change in temperature of the disk to ambient air

The radiation and convective components can be found in figure 16-24 in *Shigley's Mechanical Engineering Design*. The calculated results for the drum and brake capacity can be seen in Table 15.8. The thermal analysis calculations can be seen in Table 15.9.

15.2.3 Solid Modeling

The solid modeling for the base and support of the rocking arms can be seen in Figure 15.25. On the left is the solid view of the solid model, while the wire view is on the right. The base and support remain stationary throughout the entire process.

The left rocking arm and brake lining are shown in Figure 15.26. On the left is the solid view, while the wire view of the solid model is on the right. The rocking arm rotates positively along the *z*-axis.

The right rocking arm is shown in Figure 15.27. On the left is the solid view, while the wireframe view of the solid model is on the right. The rocking arm rotates negatively along the *z*-axis.

The drum of the system can be seen in Figure 15.28. On the left is the solid view, while the wireframe view of the solid model is on the right. The drum rotates in a stationary position along the *z*-axis.

Table 15.8 Drum and Brake Lining Calculations

		θ_2	θ_1
		0.523598776 ± .05	−0.5236 ± 0.05
	A	0	±0.05
	B	0.090586074	±0.05
Max pressure (drum)	Pa	700,000	±0.05
Friction coefficient	*f*	0.35	±0.05
Face width	*b*, m	0.0508	±0.05
Drum diameter	*D*, m	0.256286	±0.05
Drum radius	*r*, m	0.128143	±0.05
	a, m	0.133955115	±0.05
	dN, N	4556.76508	±0.05
Torque (brake capacity)	*T*, N-m	213.6406958	±0.05
	Rx	−4556.76508	±0.05
	Ry	−1594.867778	±0.05

Table 15.9 Thermal Analysis

Diameter	*D*, m	0.256286
Radius	*r*, m	0.128143
Arc length	m	0.067096 ± 0.05
Assume:	*T*, °F	200
Convection	*hc*	7.4 ± 0.05
Radiation	*hr*	16.1 ± 0.05
Force vel	assume	12 m/s
Material area	*A*, m²	0.134191 ± 0.05
Rate of energy loss	*H*.loss, W	630.6979 ± 0.05

Figure 15.29 shows the rod and spring configuration. On the left is the solid view, while the wireframe view is on the right. The configuration moves laterally along the *x*-axis.

Figure 15.30 shows the connecting slide. On the left is the solid view, while the wireframe view is on the right. The connecting slide translates laterally along the *x*-axis.

The connecting rod is shown in Figure 15.31. The solid view of the solid model is on the left, while the wireframe view is on the right. The connecting rod rotates about the *z*-axis in the middle due to the tension and compression forces from the solenoid on the right end.

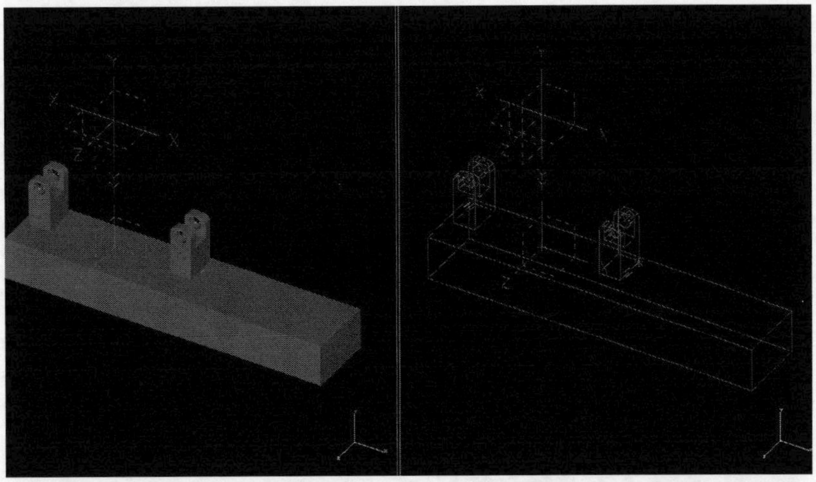

Figure 15.25

Base and support.

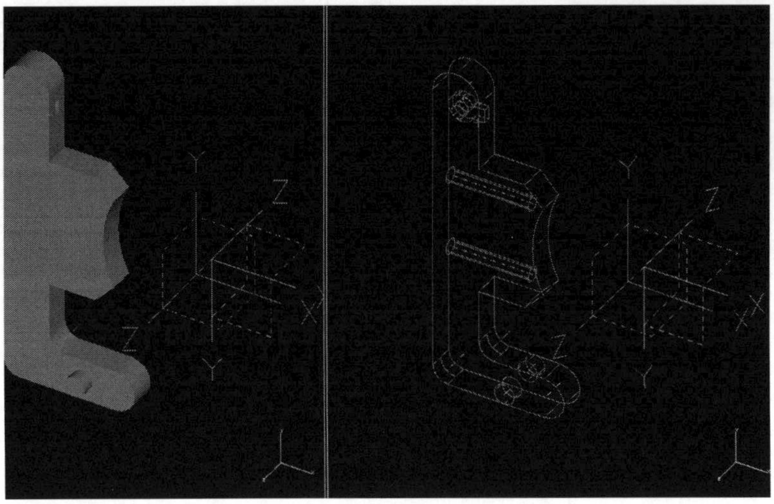

Figure 15.26

Left rocking arm and brake lining.

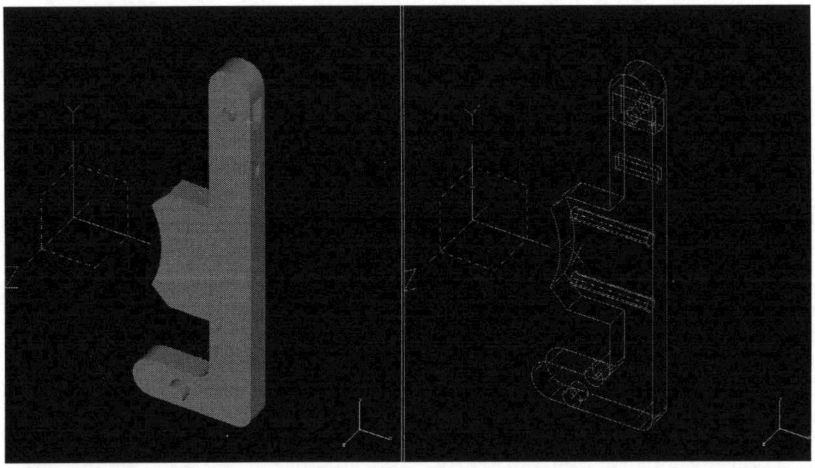

Figure 15.27

Right rocking arm and brake lining.

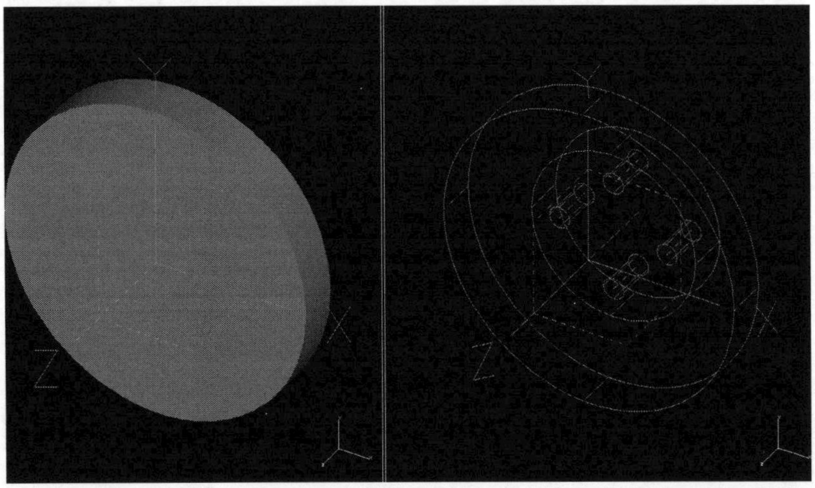

Figure 15.28

Drum.

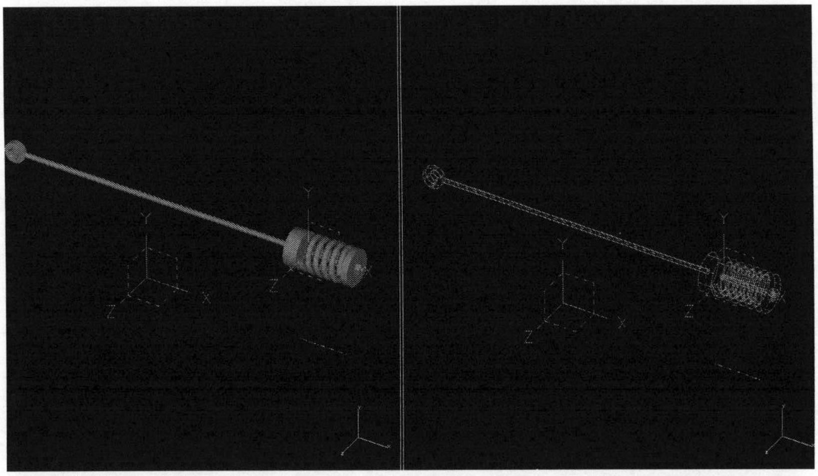

Figure 15.29

Rod and spring configuration.

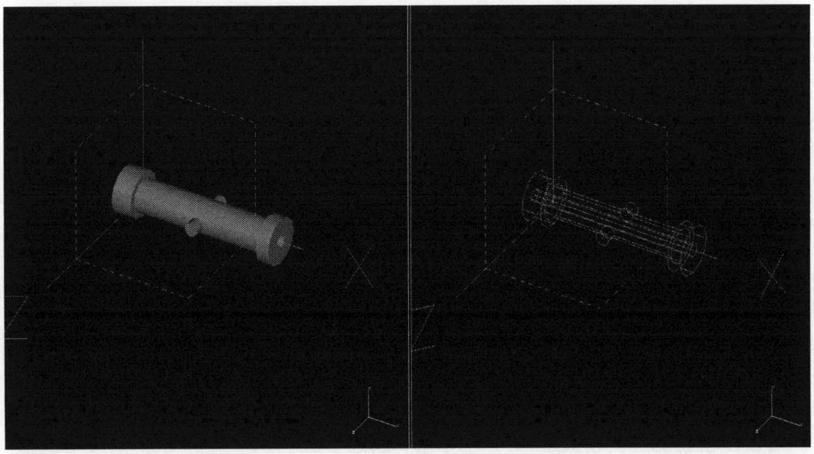

Figure 15.30

Connecting slide.

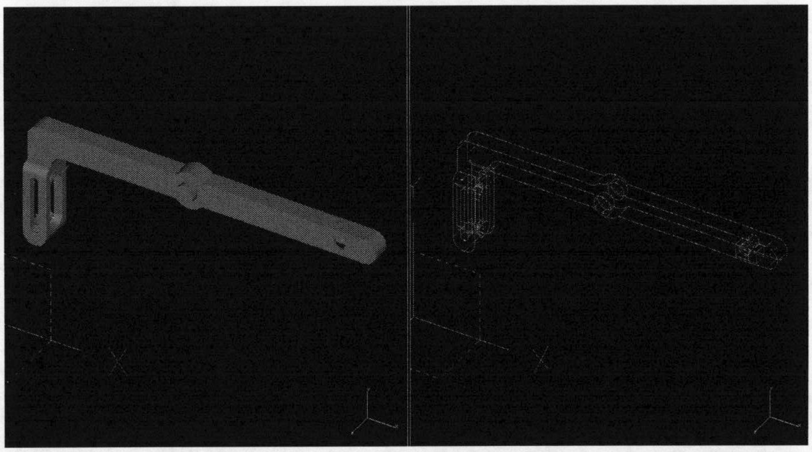

Figure 15.31

Connecting rod.

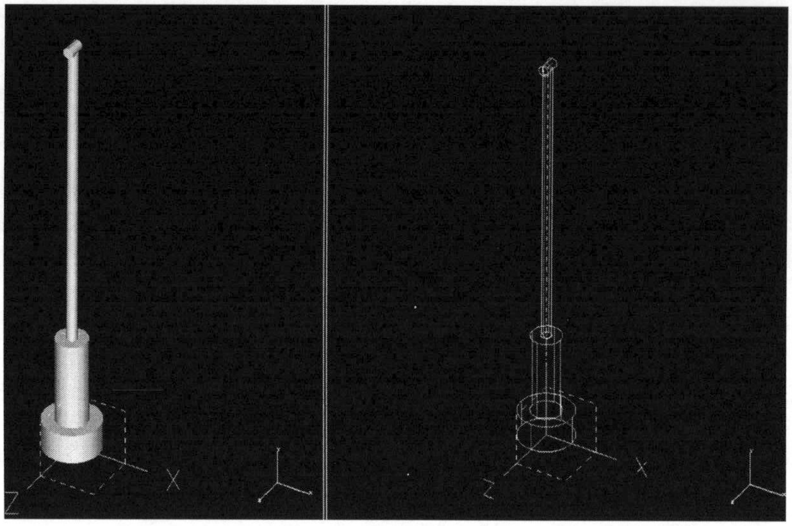

Figure 15.32

Solenoid cylinder.

The solenoid cylinder (actuating rod) is shown in Figure 15.32. The solid view of the solid model is on the left, while the wireframe view is on the right. The actuator translates in the *y*-axis to create the force on the connecting rod.

15.2.4 Assembly

The entire assembly can be visualized for the brake system. All the individual parts work in unison to move the brakes to create contact with the drum. The solenoid pulls on the connecting rod, which in turn moves the connecting slide. The connecting slide moves the rod, which is observing a force from the spring. The rod moves the rocking arms, which are connected to the brakes. The brakes close around the drum and create friction. This friction is then converted into heat and slows the drum down, and if held long enough, will eventually stop the drum entirely (Figure 15.33).

15.2.5 Finite Element Analysis

A thermal FEA was performed on the drum in the system. A thermal analysis is important when analyzing a brake system, since a brake system converts kinetic energy into heat. It is important to know the maximum temperature that can be observed. In this thermal analysis, some restraints were involved. The inner surface of the drum is exposed to 20°F. The outside surface is exposed to ambient air that is 200° higher, at 220°F. The convection coefficient used is 7.4 W/(m^{2}C). The highest temperature is observed at the outer surface, while the lowest temperature is observed at the inner surface. This is to be expected, since the drum is exposed to the friction from the brakes on the outer surface.

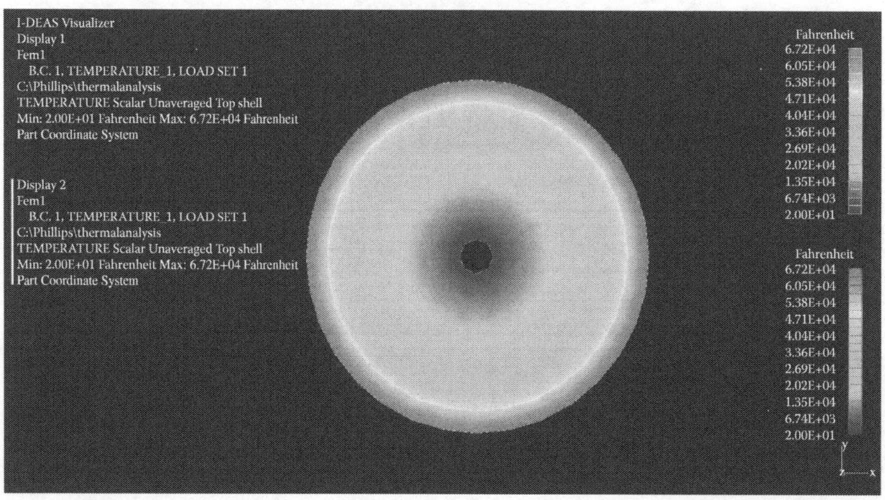

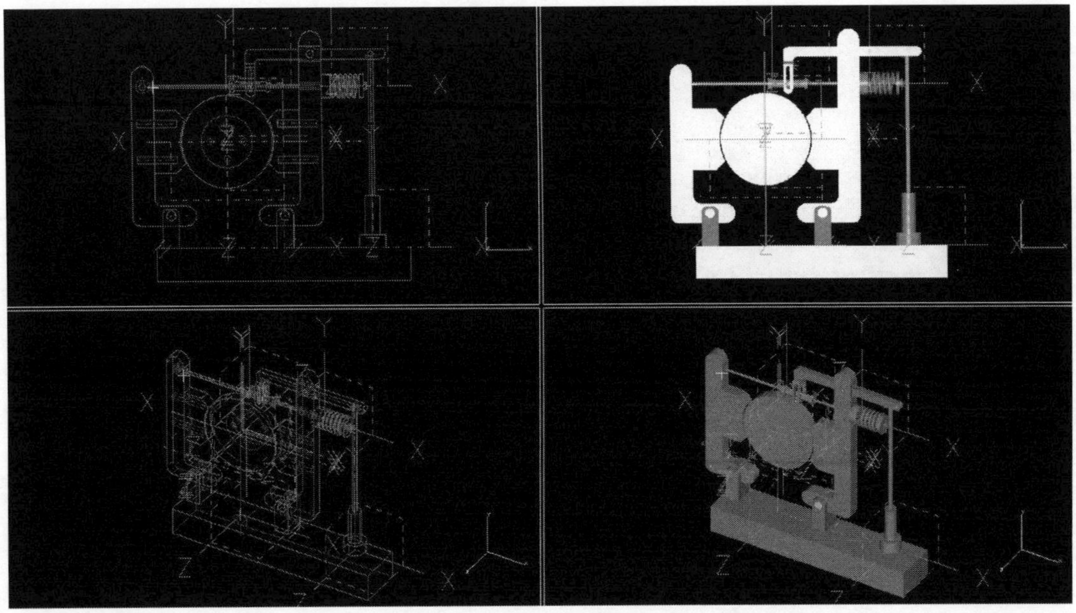

Figure 15.33

(See color insert) Brake assembly.

Table 15.10 Cost Estimation

Component		$ per Part	% of Total
Collar	Material	20.00	1.5
	Machining	100.00	7.7
Connecting rod	Material	300.00	23.0
	Forging	5.00	0.4
Compression spring	Material	50.50	3.9
Drum	Material	600.00	46.0
	Casting	25.00	1.9
Brake lining	Material	70.00	5.4
	Casting	10.00	0.8
Rocking arms	Material	25.00	1.9
	Forging	5.00	0.4
Solenoid actuator	Material	80.00	6.1
	Casting	15.00	1.1
Total		1305.5	100

15.2.6 Cost Estimation

The cost of each part was researched via the Internet. The cost of material was chosen from the lower end of the range that was found. The only part that required calculations for the cost was the spring. This was also the only part that would be purchased from a store without specific machining. The collar (or connecting slide) has a high machining cost because precision machining is needed in order for the part to work properly and without failure. The connecting rod and rocking arms were both forged at a cost shown in Table 15.10. The drum, brake/brake lining, and solenoid actuator were formed by casting, each at a different cost. The most expensive portion of the cost estimation was the drum itself, at 46%, followed by the connecting rod at 23%.

15.3 CAD Example 3: Crankshaft and Flywheel

Today, complicated systems can be designed and simulated through the use of computers. NX is a computer-aided design program that is used to create solid models and perform an FEA to ensure engineering stability. The objective of this project is to design a crankshaft along with a flywheel for use in a 600 hp six-cylinder engine. A mathematical analysis and FEA will be performed to ensure that the design constructed will be feasible.

15.3.1 Introduction

In a combustion engine, a flywheel is coupled to a crankshaft (Figure 15.34). This eliminates friction between the flywheel and crankshaft, which occurs under conditions of torque fluctuation due to clearance between the meshing teeth, by providing a deformable member to reduce this clearance. The deformable member includes a bore with teeth that mesh with the crankshaft teeth [4].

A crankshaft is a shaft with offset throws or journals to which piston assemblies are connected by means of connecting rods. When a crankshaft is rotated, the offset crank throws convert the linear, back-and-forth movement of the pistons into rotary motion at the crankshaft. This works in the same way as pedals function on a bicycle crankshaft. Crankshafts are supported by bearings at main journals. These main bearings require pressurized lubrication any time the engine is running; in this way, shaft main journal to bearing bore contact is avoided.

Bending stress between the main journals occurs between power strokes. Crankshafts are designed to withstand the bending forces that result from the compression and combustion pressures developed in the cylinder. The normal bending stress takes place between the main journals whenever the crank throw between them is loaded by cylinder pressure.

Torsional stresses are the high-speed twisting vibrations that occur in a crankshaft. Crankshaft torsional vibration occurs because when a crank throw is under compression, it slows to a speed slightly less than the average crank speed. The crank throw receiving the power stroke forces accelerates to a speed higher than the average crank speed. Twisting vibrations take place at high frequencies, and crankshaft design, materials, and hardening methods must be taken into account.

The flywheel is a device used to control cyclic fluctuation in engine speed. A flywheel of suitable dimensions attached to the crankshaft makes the moment of inertia of the rotating parts quite high, and thus acts as a reservoir of energy. During the period when the flywheel absorbs energy, its speed increases, and when the flywheel releases energy, its speed decreases.

15.3.2 Analysis/Design

To begin designing this system, it is necessary to choose an appropriate material for the manufacturing process. Selecting a material is a crucial step, because the characteristics of the material should meet the

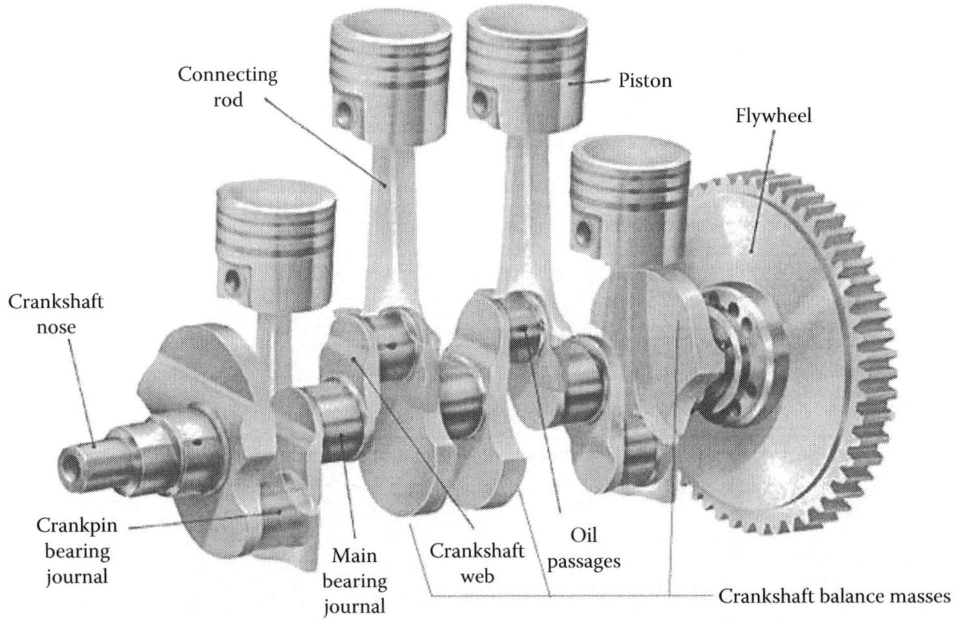

Figure 15.34

Schematic of in-line four-cylinder engine.

Table 15.11 Material Properties

Material	Density ρ (kg/m^3)	Specific strength σ (MNm/kg)
Steel	7800	0.22
Aluminum alloy	2700	0.22
Titanium	4500	0.27
Glass-fiber-reinforced polymer	2000	0.80
Carbon-fiber-reinforced polymer	1500	1.60

requirements for performance and manufacturing. A list of common crankshaft materials can be seen in Table 15.11.

The crankshaft's performance should have a high strength and stiffness to withstand the extra loads in modern engines and to offer opportunities for downsizing and weight reduction. In addition, it should be resistant to fatigue in torsion and bending, and to wear in bending areas.

The manufacturing route for forged steel crankshafts is usually: hot forging, heat treatment, machining, and surface treatment. Controlled air cooling after forging costs less than the traditional quench and temper. For efficient and cost-effective processing, the process should provide a consistent hardening response and good machinability in the hardened condition.

Steel was found to be the ideal choice from both environmental considerations and manufacturing considerations for a crankshaft. Steel has the highest density. The cost of steel is approximately 50 cents/lb. Also, steel is a very easy material to forge and machine.

The first part designed was the crank pin. This is the part of the crankshaft where the connecting rod is located. The piston forces (pressure) cause torque on the crankshaft. It is necessary to size the diameter of the pin based on the shear forces acting between the pin and the webs. Since the power of the engine is known, the maximum torque can be found using Equation 15.27

$$F = \frac{hp \cdot 5252}{d \cdot n} \tag{15.27}$$

where:

- hp is the horsepower of the engine
- d is half of the stroke length
- n is revolutions per minute

The stroke length was an average length taken from crankshafts of a similar size. Now that the maximum torque is known, the minimum diameter size for the pin can be calculated using Equation 15.28

$$D = \frac{F}{\tau_{max} \cdot A} \tag{15.28}$$

where:

- τ_{max} is the maximum allowable stress
- A is the cross-sectional area of the bearing

The maximum stress was found by choosing a factor of safety and dividing the yield strength by this factor. The diameter of the crank pin was found to be 2 in.

The next part to be designed is the journal bearing. The journal bearing has a similar concept as the crank pin except that a bending force is added to the pin. Using the combined loading equations and maximum forces, it is possible to solve for the diameter of the bearing. Equations 15.29 and 15.30 demonstrate how the bending forces and torsional stresses were found:

$$\tau_{xy} = \frac{T \cdot r}{j} \tag{15.29}$$

$$\sigma_x = \frac{M \cdot c}{I} \tag{15.30}$$

where:

- T is the torque that was calculated earlier
- r is the radius of the journal bearing
- j is the polar moment of inertia
- M is the moment acting on the pin
- c is the radius of the pin
- I is the second moment of inertia

To calculate the total stress, Equation 15.31 is used:

$$\sigma = \sqrt{\sigma_x + \tau_{xy}} \tag{15.31}$$

It was found that the diameter of the journal bearing was 2 in.

The web was the next part designed. To solve for the thickness of the web, the diameter of the crank journal pin was multiplied by 0.4. The rest of the geometry, such as the counterweight and fillets, is not important, because the design of a crankshaft varies from company to company.

In order to attach the flywheel to the end of the crankshaft, a mounting hub must be made. The size of the diameter has to be greater than that of the crank pin, because otherwise the crankshaft would move horizontally; the hub keeps the crankshaft constrained. The flywheel mount has an array of six holes for the bolts, spread out evenly at a radius of 1 in from the center. The size of the bolts must be determined before the holes are made in the mount. By using the shear stress equation for a bolt, the minimum diameter was found to be 0.2 in with a factor of safety of 5. However, the preferred size is a nominal diameter of 0.25 in. This would make the manufacturing easier and cheaper.

Finally, the flywheel is the last part to be designed. The center of the flywheel is designed first, remembering that along with the plate, holes have to be designed to fit the hub that was built. Once this is designed, a diameter is chosen so that it may be able to deliver the necessary inertia. Using the kinetic energy equations, the mass can be found. Furthermore, the volume of the material can be found:

$$E = \frac{I}{r^2} V \left(V_{\max} - V_{\min} \right) \tag{15.32}$$

where:

- E is the energy stored in the flywheel
- V is the average operating speed of the flywheel
- $V_{\max}$ and $V_{\min}$ are the maximum and minimum operating speeds in feet per second

The energy stored in the flywheel is taken to be 85% of the power that the engine produces; this gives a flywheel diameter of 11 in.

Once the flywheel diameter has been found, the weight and thickness can be calculated by using Equations 15.33 and 15.34:

$$W = \frac{2g \cdot E}{V \left(V_{\max} - V_{\min} \right)} \tag{15.33}$$

$$t = \frac{4W}{\pi \cdot D^4 \cdot \rho} \tag{15.34}$$

where:

- W is the weight of the flywheel
- g is the acceleration due to gravity

t is the thickness of the flywheel
ρ is the density of the material

The thickness is found to be 0.4 in. These dimensions will satisfy the amount of energy transmitted.

$$F = \frac{\frac{h_p}{2} \times 5252}{df} = 3.601 \times 10^3 \text{ lb}$$

Design:

Assuming Crankshaft is SAE Cold-Drawn Steel

$$S_u = 76{,}000 \text{ psi}, \; S_y = 64{,}000 \text{ psi}, \; n = 5, \; T = 450.171 \text{ lbf ft}, \; s_e = 0.5, \quad S_u = 3.8 \times 10^4 \text{ psi}$$

$$f = 3500$$

$$\tau(d) = \frac{F}{\frac{\pi d^2}{4}}$$

$$J(d_{cp1}) := \frac{\pi d_{cp1}{}^4}{32}$$

$$c(d_{cpl}) = \frac{d_{cpl}}{2}$$

$$d = \frac{1.5}{12} = 0.125$$

This will be a 600 hp six-cylinder engine with an ideal engine speed of 3500 rpm and a standard stroke length of 3 in.

A. Sizing the crank pin:

$$T = F \cdot d = 450.171$$

$$z(d_{cpl}) = \frac{d_{cpl}}{2}$$

$$h_p = 60$$

$$d = 2 \text{ in.}$$

B. Thickness of the webs:

$$\sigma(d_{cpl}) := \sqrt{\sigma_x(d_{cpl})^2 + \tau_{xy}(d_{cpl})^2}$$

$$M = F \cdot 1$$

$$\sigma_x\left(d_{\text{cpl}}\right)=\frac{M\cdot c\left(d_{\text{cpl}}\right)}{I\left(d_{\text{cpl}}\right)}$$

$$d_{\text{web}}=\text{root}\left(\sigma\left(d_{\text{cpl}}\right)-\frac{S_y}{5},d_{\text{cpl}}\right)$$

$$\tau_{\text{xy}}\left(d_{\text{cpl}}\right)=T\cdot\frac{z\left(d_{\text{cpl}}\right)}{J\left(d_{\text{cpl}}\right)}$$

$$I\left(d_{\text{cpl}}\right)=\frac{\pi\cdot d_{\text{cpl}}{}^4}{64}$$

$$d_{\text{web}}=1.756\text{ in.}$$

C. Sizing the Journal Bearing Diameter:

$$l=1\text{ in.}+h$$

$$d_{\text{cp}}:=\text{root}\left(\tau(d)-\tau_{\text{maxallow}},d\right)$$

$$d_{cp1}=d_{cp}+1.24\text{ in}=1.839\text{ in.}$$

$$h=0.4\,d_{\text{cpl}}=0.735\text{ in.}$$

$$\tau_{\text{maxallow}}=\frac{S_y}{5}=1.28\times10^4\,\text{psi}$$

$$T=450.171\text{ lbf}\cdot\text{ft}$$

$$d=0.1\text{ in.}$$

$$d_{\text{cpl}}=1\text{ in.}$$

$$T=450.171\text{ lbf}\cdot\text{ft}$$

$$S_p=50{,}000\text{ psi}$$

$$n_{\text{d}}=5$$

$$\tau\left(d_{\text{bolt}}\right)=\frac{F_{\text{bolt}}}{6\dfrac{\pi\cdot d_{\text{bolt}}^2}{4}}$$

The minimum diameter for the stresses acting on the journal bearing is 1.756. This was rounded up to a standard size of 2 in.

D. Sizing the Bolts That Attach the Flywheel to the Crankshaft:

$$F_{\text{bolt}} := \frac{T}{1\,\text{in}} = 5.402 \times 10^{3}\,\text{lbf}$$

$$\tau_{\text{allow}} = \frac{S_p}{n_{\text{d}}} = 1 \times 10^{4}\,\text{psi}$$

$$d_{\text{bolt}} = 0.25\,\text{in.}$$

$$d_{\text{bolt1}} = \text{root}\left(\tau\left(d_{\text{bolt}}\right) - \tau_{\text{allow}}, d_{\text{bolt}}\right)$$

$$V_{\text{min}}\left(d_{\text{fly}}\right) = V_{\text{minrpm}} \cdot \pi \cdot d_{\text{fly}}$$

$$d_{\text{bolt1}} = 0.339\,\text{in.}$$

A diameter of 0.339 in would be satisfactory for six bolts about a radius of an inch from the center of the shaft. With a factor of safety, the diameter will be 0.50 in.

E. Designing the Flywheel:

$$h_{\text{pf}} = 600\,\text{hp}$$

$$h_{\text{pfly}} = 0.85$$

$$h_{\text{pf}} = 3.803 \times 10^{5}\,\text{W}$$

$$V_{\text{maxrpm}} = \frac{6500}{60\,\text{s}}$$

$$r\left(d_{\text{fly}}\right) = \frac{d_{\text{fly}}}{2}$$

$$V_{\text{minrpm}} = \frac{1}{60\,\text{s}}$$

$$V_{\text{rpm}} = \frac{f}{60\,\text{s}}$$

$$V_{\text{max}}\left(d_{\text{fly}}\right) = V_{\text{maxrpm}} \cdot \pi \cdot d_{\text{fly}}$$

$$V_{\text{m}}\left(d_{\text{fly}}\right) = V_{\text{rpm}}\pi \cdot d_{\text{fly}}$$

$$I\left(d_{\text{fly}}\right) = \frac{\pi \cdot d_{\text{fly}}^{\ 4}}{64}$$

$$E\left(d_{\text{fly}}\right) = \frac{I\left(d_{\text{fly}}\right)}{r\left(d_{\text{fly}}\right)^2} \cdot V_{\text{m}}\left(d_{\text{fly}}\right) \cdot \left(V_{\text{max}}\left(d_{\text{fly}}\right) - V_{\text{min}}\left(d_{\text{fly}}\right)\right) - h_{\text{pfly}}$$

$$d_{fly} = 2 \text{ (guess value)}$$

$$d_{flys} = \text{root}\left(E\left(d_{fly}\right), d_{fly}\right)$$

$$d_{flys} = 10.72 \text{ in.}$$

$$d_{fly1} = 11 \text{ in.}$$

$$W1 = \frac{2 \cdot g \cdot h_{pfly}}{\left(V_{max}\left(d_{fly1}\right)^2 - V_{min}\left(d_{fly1}\right)^2\right)}$$

$$\rho_{steel} = 7850 \frac{\text{kg}}{\text{m}^3}$$

$$t = \frac{4W1}{\pi \cdot d_{fly1}{}^2 \cdot \rho_{steel}}$$

$$t = 67.478 \frac{\text{m}}{\text{s}^3} \cdot \text{in.}$$

$$t_1 := \frac{t}{45.84} \cdot \frac{\text{s}^3}{\text{m}}$$

$$t_1 = 1.472 \text{ in.}$$

F. Bending within the Web:

$$F = 3.601 \times 10^3 \text{ lbf}$$

$$I_{bw} = \frac{b \cdot h_1{}^3}{12} = 0.022 \text{ in.}^4$$

$$\sigma_{bending} = \frac{M_{bw} \cdot y}{I_{bw}} = 7.299 \times 10^4 \text{ psi}$$

$$\sigma_{max} = \sigma_{bending}$$

$$\sigma_{avg} = \frac{(\sigma_{max} + \sigma_{min})}{2} = 0 \text{ psi}$$

Fatigue stress calculation

$$L_2 = \frac{7}{8} \text{ in.}$$

$$L_3 = 0.75 \text{ in.}$$

$$b = \frac{5}{8}\text{ in.}$$

$$h_2 = L_3$$

$$y = \frac{L_3}{2} = 0.375\text{ in.}$$

$$M_{bw} = FL_1 = \frac{F}{2} \cdot L_2 = 483.194\text{ J}$$

$$\sigma_{\text{alt}} = \frac{(\sigma_{\max} + \sigma_{\min})}{2} = 7.299 \times 10^4\text{ psi}$$

Correction values are

$$K_t = 1.5$$

$$q = 0.7$$

$$K_f = 1 + q \cdot (K_t - 1) = 1.35$$

$$\sigma_{\text{alt1}} = \sigma_{\text{alt}} \cdot K_f = 9.853 \times 10^4\text{ psi}$$

$$\text{fos} = \frac{s_e}{\sigma_{\text{alt1}}} = 0.386$$

G. Torque applied to bearings:

$$F = 3.601 \times 10^3\text{ lbf}$$

$$L_4 = 2.5\text{ in.}$$

$$d_t = 2\text{ in.}$$

$$r_t = \frac{d_t}{2}$$

$$T_t = 2 \cdot F \cdot L_4 = 1.801 \times 10^4\text{ lbf} \cdot \text{in.}$$

$$T_t = 1.501 \times 10^3\text{ lbf} \cdot \text{ft}$$

$$J_t := \frac{\pi \cdot d_t^{\ 4}}{32} = 1.571\text{ in.}^4$$

$$\tau_t = \frac{T_t \cdot r_t}{J_t} = 1.146 \times 10^4\text{ psi}$$

$$\text{fos}_t = \frac{S_y}{\tau_t} = 5.583$$

15.3.3 Solid Modeling

To perform this modeling, I will split both parts individually, focusing on one part at a time, using NX 9.0. The initial part will be the crankshaft. This crankshaft will be used in a 600 hp six-cylinder engine and resembles Figure 15.34.

To start modeling this design, a new part was created under the name of "CrankshaftSection1." The units for this design were changed from the default to inches. A 2 in circle was created on the *y–z*-axis and extruded 1.5 in. Then, sketching on the extruded plane, the shape for the web was created using three circles, a 2"×4" rectangle, and two lines. This was then extruded by 0.62 in. The crank pin was the next step in this design. This is where the force will be applied because of the connecting rods.

When a complete crank throw had been created, through the use of the "Manage Bins," "CrankshaftSection1" was copied three times. With the help of the Move-To, Rotate and Join functions, a complete crankshaft has been created.

The second part that needs to be created is the flywheel, which will be a fairly simple design, starting off with a cylinder then adding a disk with teeth to resemble a realistic flywheel. Once both are completed, I will join them and animate them to rotate 360° about a fixed base.

The steps in the animation process are as follows:

1. In Master Assembly, add a parent in hierarchy called Crank-Fly.
2. Click the "add to" and then click on the part that will rotate.
3. Add another parent, called "bearings," and repeat step 2 for the bearings.
4. Click on "manage configurations," make a copy of the existing configuration, and name new configuration "angle 22.5."
5. Select the new configuration that was just made as active.
6. Close all windows and click the "rotate" button.
7. Right click and select hierarchy, then select the Crank-Fly parent.
8. Pick a pivot point and rotate it 22.5°.
9. Repeat steps 4–8 every 22.5° until the part has rotated 360°.
10. Create a sequence by using the "manage sequences" icon, and select all configurations.
11. Click on "animate" button, and play sequence.

15.3.4 Manufacturing Processes

- Forging inspection:

  The appropriate crankshaft material in forged form will be received from stores, visually inspected as per the drawing, and placed in the workplace area.
- Crank web milling, journal milling, and crank pin milling:

  There is a special type of milling cutter used for journal and crank pin milling. This cutter is of a special design, with cartridges mounted on the periphery and sides for simultaneous milling of the pin and web of a crankshaft [2]. The spindle is driven through a gearbox by a 37 kW motor. The spindle head is mounted on slide units, which are driven by servomotors through ball and lead screws. While one of the servomotors controls the feed movement of the cutter, another one is used to index the milling head in various positions. The spindle head slide is provided with a hydraulic counterbalance system.
- Drilling:

  Drilling of flywheel flange

  Drilling of crankpin

  Drilling of crossed oil hole
- Grinding:

  Surface grinding is done for journal, main journal, and crank pin. One of the critical steps in the manufacturing of forged steel crankshafts is the grinding of the sidewalls.

- Crack detection:
  Since the material is forged, there is a possibility that the material may have internal flaws, which will be exposed during machining. Hence, before material is sent for further operations, cracks are detected by the magnaflux method.
- Nitriding:
  The case-hardening process depends on the absorption of nitrogen into the steel. A much higher surface hardness is obtainable with case-hardening steels; they are extremely resistant to abrasion and have a high fatigue strength.
- Dynamic balancing: lapping and superfinish:
  The crankshaft surface is ground to a superfinish so as to avoid cyclic fatigue failure of the crankshaft due to any surface cracks and burrs. Generally, the bearings are selected as per the shaft/crankpin diameter. If the shaft diameter is slightly lower than the recommended bearing size, the next bearings are selected, and the diameter is changed according to bearing size.

15.3.5 Cost Estimation (Tables 15.12 and 15.13)

15.3.6 Assembly (Figures 15.35 through 15.41)

15.3.7 Finite Element Analysis

This is a numerical technique for finding approximate solutions of partial differential equations (PDE) as well as of integral equations. This was done by placing nodes in order to resemble the shape of the six-cylinder crankshaft. As seen in Figure 15.42, there are stable constraints at both ends of the crankshaft. There are 1000 pounds of force acting on the crank pins, where the connecting rods would be placed. Results show that this crankshaft design is optimized (Figure 15.40).

Table 15.12a Breakdown of Cost

	Cast	Forged	μ Alloy	ADI
Variable costs	$26.79	$28.08	$26.56	$31.09
Fixed costs	$12.04	$17.40	$11.60	$18.03
Total costs	$38.83	$45.48	$41.15	$49.13
Forming cost	$23.26	$23.74	$22.83	$25.87
Machining cost	$15.56	$21.74	$18.32	$23.26
Total costs	$38.83	$45.48	$41.15	$49.13
Materials	$17.44	$13.99	$15.06	$16.02
Tooling	$2.76	$3.54	$2.90	$3.84
Labor	$6.69	$12.54	$10.33	$12.09
Capital	$7.33	$10.34	$8.73	$11.32
Maintenance	$1.72	$2.27	$1.91	$2.56
Other	$2.88	$2.89	$2.22	$3.29
Total	$38.83	$45.48	$41.15	$49.13

Table 15.12b Breakdown of Cost

Component	$ per Part	% of Total
Building	1.90	2.7
Raw material	17.53	24.8
Cleaning	1.00	1.4
Shearing	0.69	1.0
Heating	1.38	2.0
Preforming	1.30	0.8
Forging	4.99	7.1
Trimming	0.63	0.9
Coining	0.88	1.2
Inspection	1.56	2.2
Heat treat	2.69	3.8
Shot clean	0.21	0.3
Machining	35.94	50.8
Total	70.70	100.0

Table 15.13 Cost Estimation

Steel Code	Annual Volume × 1000	Material ($)	Forging ($)	Heat Treat ($)	Machining ($)	Total Cost ($)
CS	100	17.50	14.30	2.90	35.90	70.60
CS	200	17.50	14.30	2.90	25.50	60.20
CS	300	17.50	14.20	2.90	22.90	57.50
CS-HS	100	18.50	14.30	2.90	35.00	70.70
CS-HS	200	18.50	14.30	2.90	24.50	60.20
CS-HS	300	18.50	14.20	2.90	22.00	57.60
MA	100	18.30	14.30	0.00	33.20	65.80
MA	200	18.30	14.30	0.00	21.70	54.30
MA	300	18.30	14.20	0.00	18.50	51.00

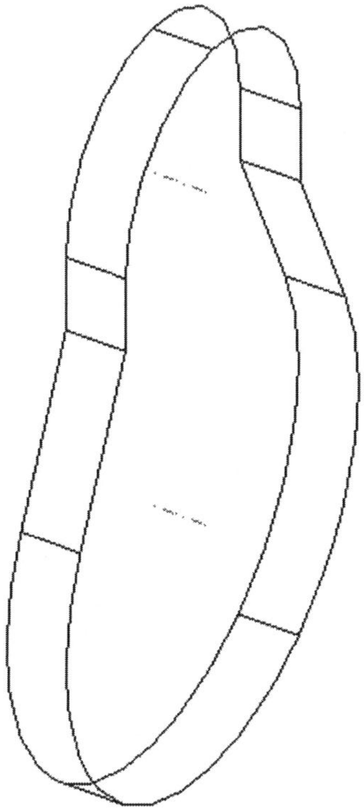

Figure 15.35

Crankshaft web.

15.4 CAD Example 4: Design of Cam, Camshaft, and Rocker Arm

15.4.1 Design a Cam, Camshaft, Rocker Arm, and Assembly for a V-6 200 hp

A cam, camshaft, and rocker arm assembly is a mechanical mechanism used in internal combustion engines. Its purpose is to open and close the intake and exhaust valves on a cylinder at exactly the correct time. If the engine is to run properly, the timing of these valves is critical. If the timing is not correct, the engine may misfire or not run at all.

The energy used to drive the camshaft is taken from the crankshaft. Once the camshaft is in motion, its asymmetrically shaped cams will trigger the pushrod to accelerate upwards. The pushrod will cause the rocker arm to rotate about its shaft and displace the spring holding the valve closed. As soon as the surface of the cam loses contact with the pushrod, the valve returns back to its closed position (Figure 15.43).

This mechanism is comprised of several components. Each of these components is separately designed to properly handle cyclical stress. At the same time, they will be correctly dimensioned (with tolerances) so that

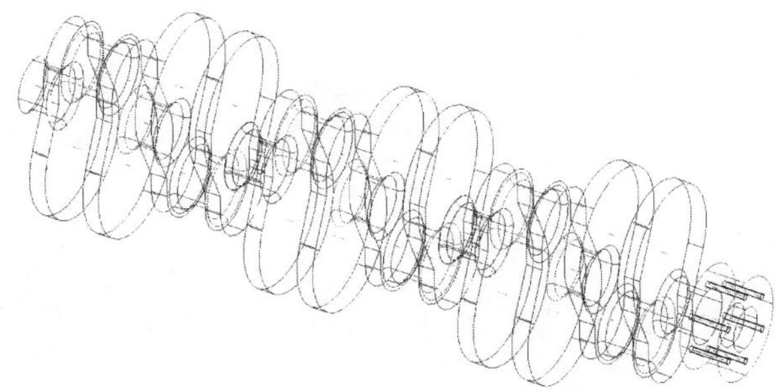

Figure 15.36

Crankshaft.

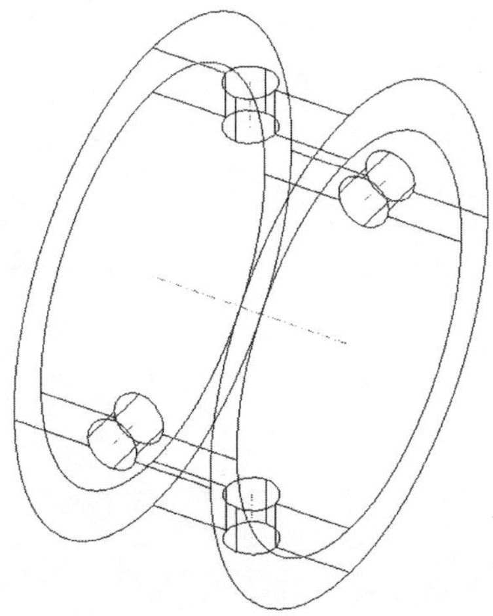

Figure 15.37

Crank pin.

they fit together and move properly. A solid model of each component, as well as the entire assembly, will be completed using NX 9.0. In addition, FEA will be conducted using NX 9.0 on some selected components.

15.4.2 Design

It is important to analyze the forces on each component of the system, and design to adequately handle the stresses. Some specifications, such as horsepower, revolutions per minute, and torque are vital to these stress calculations. A list of specifications follows.

Given design data:

$$\sqrt{\mathrm{HP} = 335\ \mathrm{hp}}$$

$$\sqrt{\mathrm{rev} = \frac{2\pi \cdot \mathrm{rad}}{60}}$$

$$\sqrt{n_{\mathrm{crank}} = 5200 \frac{\mathrm{rev}}{\mathrm{min}}}$$

Torque on camshaft:

$$T_{\mathrm{cam}} = \frac{HP}{n_{\mathrm{cam}}} = 1.015 \times 10^4\ \mathrm{ft} \cdot \mathrm{lbf}$$

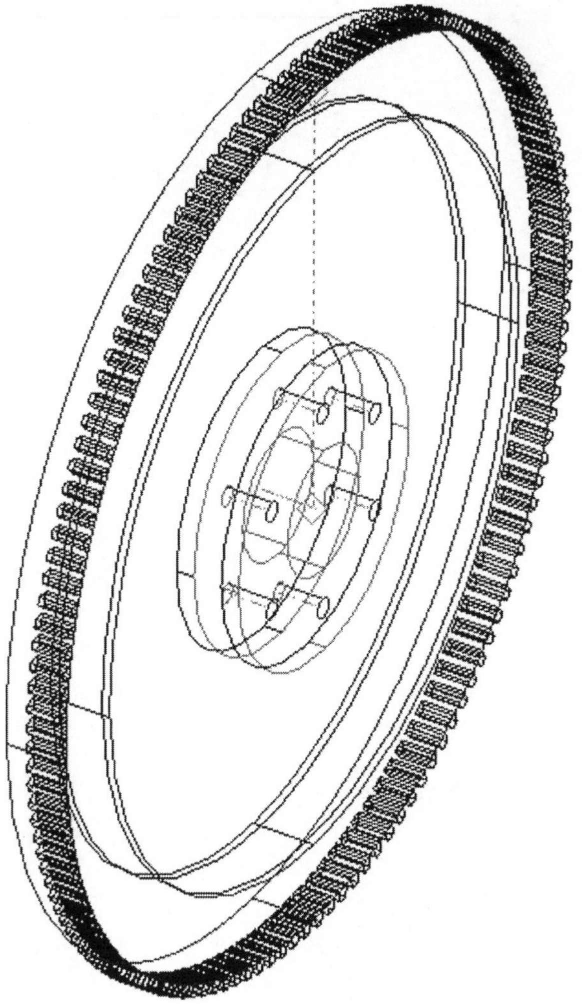

Figure 15.38

Flywheel.

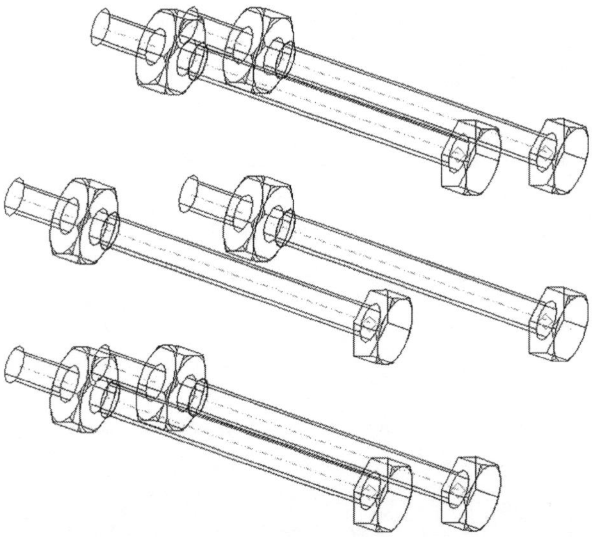

Figure 15.39

Bolts.

Figure 15.40

(See color insert) Complete solid modeling.

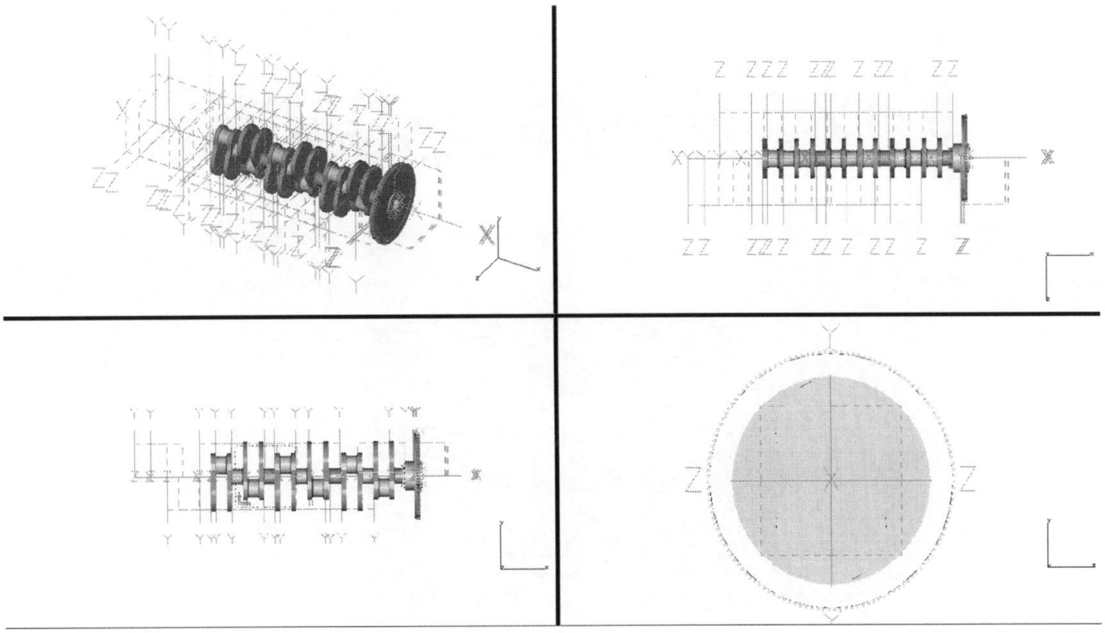

Figure 15.41

Complete solid modelling of crankshaft.

Speed of camshaft in terms of speed of crankshaft:

$$n_{cam} = 2 \cdot n_{crank} = 1.04 \times 10^4 \frac{\text{rev}}{\text{min}}$$

This is the center of the shaft. The forces of the cams are shifted to the center for a more simple analysis (Figure 15.44). The shaft is assumed to be 25 in. Then, the following calculations indicate whether it can handle the stress.

$$\sqrt{F_{min} = 85\ \text{lbf}}$$

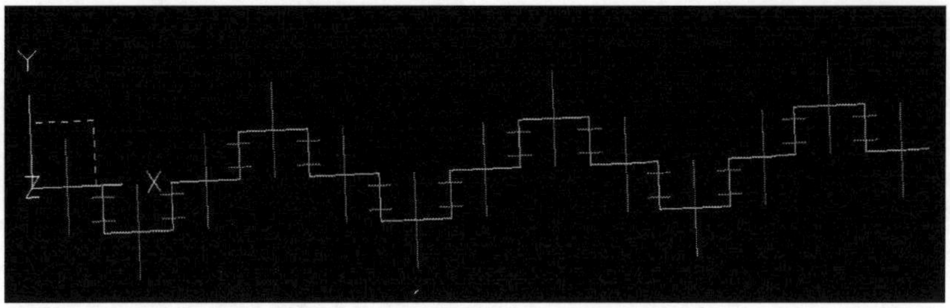

Figure 15.42

Node locations.

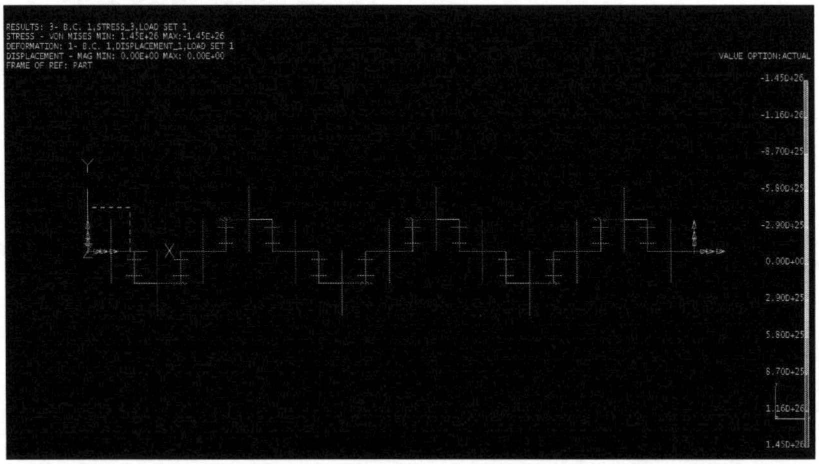

Figure 15.43

Forces applied to crankshaft.

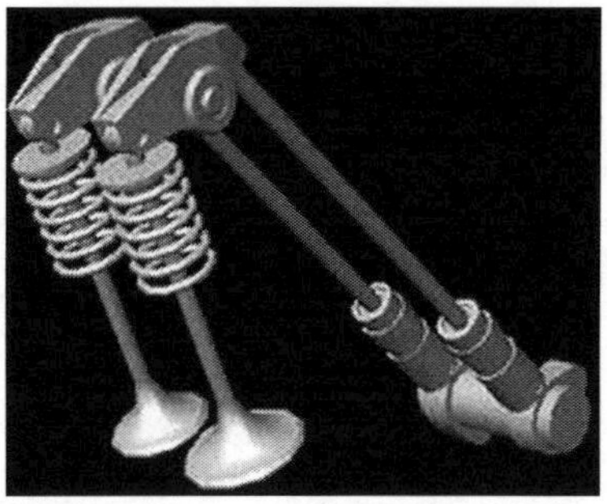

Figure 15.44

Cam, camshaft, and rocker arm assembly.

$$\sqrt{l = 12.5\ \text{in.}}$$

This is the spring preload, and it provides the minimum force on the cam.

$$M = 2 \cdot F_{\min} \cdot l = 177.083\ \text{ft} \cdot \text{lbf}$$

This is the maximum bending moment at the center of the shaft.

The camshaft will be designed to handle the static load applied by the cams contacting the pushrod (Figure 15.45). Then, the design will be checked to ensure that it will not fail due to fatigue stresses.

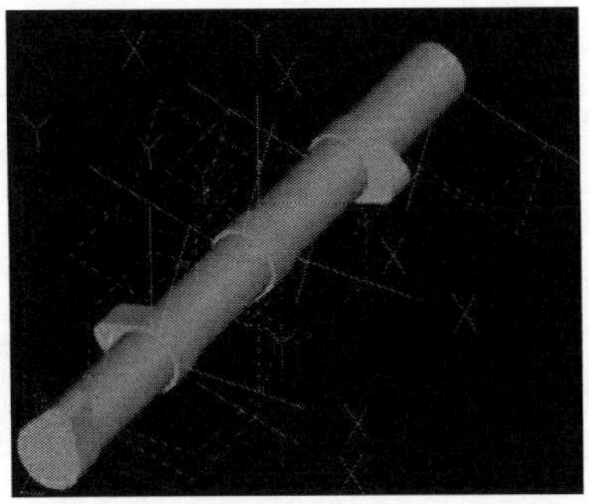

Figure 15.45

Camshaft.

15.4.3 Design of Camshaft—AISI 1020 Steel

Using Mohr's circle to determine σ and τ,

$$\sigma = \frac{S_y}{n} = 1.19 \times 10^5 \text{ psi}$$

Principal stresses σ_1 and σ_2:

$$\sqrt{n = 2}$$

Assumed factor of safety:

$$\sqrt{S_y = 238 \times 10^3 \text{ psi}}$$

A-21 4140 Q&T steel at 400°C

Diameters are calculated from the stress equation. These stress values are then compared with the values of allowable stress. The calculated stress must be less than the allowable stress.

$$\sqrt{d_1 = 1.75 \text{ in.}}$$

$$\sigma_1 = 32\frac{\frac{M}{\pi d_1^3}}{2} + \left[\left(32\frac{\frac{M}{\pi d_1^3}}{2}\right)^2 + \left(16\frac{T_{cam}}{\pi d_1^3}\right)^2\right]^{\frac{1}{2}} = 1.178 \times 10^5 \text{ psi}$$

Equations for principal stresses on the shaft:

$$\sqrt{d_2 = 1.75 \text{ in.}}$$

$$\sigma_2 = 32\frac{\frac{M}{\pi d_2^3}}{2} - \left[\left(32\frac{\frac{M}{\pi d_2^3}}{2}\right)^2 + \left(16\frac{T_{cam}}{\pi d_2^3}\right)^2\right]^{\frac{1}{2}} = -1.138 \times 10^5 \text{ psi}$$

$$\sigma_2 = \sigma_1$$

Both diameters chosen above will be adequate for the design. However, it is better to choose the larger of the two diameters because it will handle the stress better.

Now the dynamic check can be done, based on fatigue caused by repeated load on the spring.

15.4.3.1 Dynamic Check

$$\sqrt{L = 3.98 \text{ in.}}$$

Cylinder length

$$\sqrt{\text{Displacement} = 428 \text{ in.}^3}$$

Engine displacement

$$A = \frac{\text{Displacement}}{L} = 107.538 \text{ in.}$$

Bore area

$$\sqrt{\eta = 0.95}$$

Mechanical efficiency

$$\sqrt{\text{Cyl} = 4}$$

Camshaft rotates twice as fast as crankshaft

$$N = 2 \cdot n_{\text{crank}} = 1.04 \times 10^4 \frac{\text{rev}}{\text{min}}$$

Number of cylinders

$$\text{Power} = \frac{HP}{\text{Cyl} \cdot \eta} = 6.574 \times 10^4 \text{ W}$$

Power produced by engine

$$\text{Pressure} = \frac{\text{Power}}{\text{L} \cdot A \cdot \text{N}} = 74.895 \text{ psi}$$

Pressure in each cylinder

$$F_{\max} = \text{pressure} \cdot A = 8.054 \times 10^3 \text{ lbf}$$

Force on each cylinder

$$\sigma_a = \frac{F_{max} - F_{min}}{2A} = 37.052$$

$$\sigma_m = \frac{F_{max} - F_{min}}{2A} = 37.843$$

Mean and average stresses

$$\sqrt{S_{ut} = 257.10^3 \text{ psi}}$$

Ultimate tensile strength of steel is selected from design book

$$\sqrt{S_{e.prime} = 100.10^3 \text{ psi}}$$

Choose larger diameter calculated before entering into k_b.

$$d_1 = 1.75 \text{ in}$$

Surface factor

$$b = -0.265$$

$$a = 2.7$$

$$k_a = a \cdot 257{,}000^b = 0.099$$

Size factor

$$k_b = \left(\frac{1.482}{0.3}\right)^{-0.1133} = 0.8$$

Load factor

$$k_c = 0.67$$

The camshaft is loaded in torsion, and therefore this load factor must be used in the endurance limit equation.

Temperature factor

$$k_d = 1$$

Endurance limit

$$k_e = 1$$

$$S_e = k_a \cdot k_b \cdot k_c \cdot k_d \cdot k_e \cdot S_{e.\text{prime}} = 5.562 \times 10^3 \text{ psi}$$

Rearranging the Goodman equation for factor of safety n and solving,

The fatigue factor of safety $(n) = 2.45$.

The design of the spring must be strong enough to withstand the tendency to move due to fluctuating pressure in the cylinder. At the same time, it must not exert too much stress on other components of the system, or failure in fatigue could be an issue.

15.4.3.2 Design of Spring—Stainless Steel

$$F_{\min} = 85 \text{ lbf}$$

$$F_{\max} = 8.054 \times 10^3 \text{ lbf}$$

$$\text{height}_{\text{open}} = 1.82 \text{ in.}$$

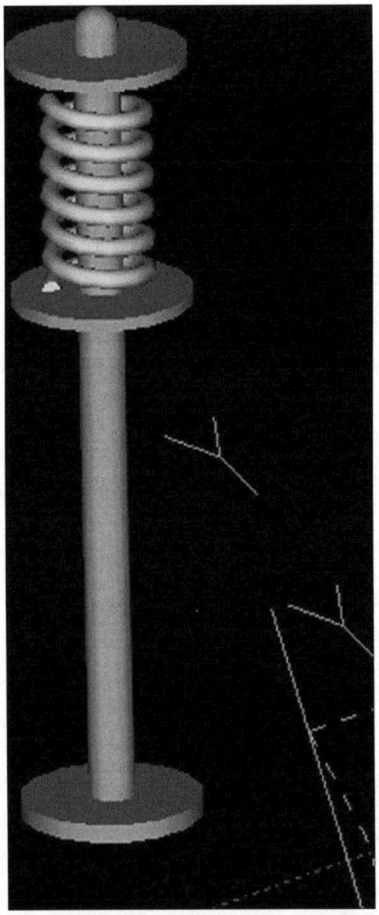

Figure 15.46

Valve and spring (shown with retainers).

Heights of spring when valve is fully open and closed (Figure 15.46)

$$\text{height}_{\text{closed}} = 1.38 \text{ in.}$$

Modulus of rigidity $$G = 11.5 \times 10^6 \text{ psi}$$

Number of available coils $$N_a = 6$$

$$N_t = N_a + 1 = 7$$

Wire diameter $$d = \frac{\text{height}_{\text{closed}}}{N_t} = 0.19$$

Spring pitch $$\text{pitch} = \frac{\text{height}_{\text{open}}}{N_t} = 0.26 \text{ in.}$$

Deflection $$y_s = \text{height}_{\text{open}} - \text{height}_{\text{closed}} = 0.44 \text{ in.}$$

Spring constant $$k = \frac{F_{\max}}{y_s} = 1.83 \times 10^4$$

Coil diameter $$D = \left(\frac{d^4 \cdot G}{k \cdot 8 \cdot N_a} \right)^{\frac{1}{3}}$$

15.4.3.3 Design of Pushrod—Material Selected: AISI 1020 Steel

Designing the pushrod is very important, because it is liable to fail in both shear and buckling. The length and diameter must have the proper ratio, while remaining within the proper size for the system (Figure 15.47).

$$L = 3.98 \text{ in.}$$

$$\text{Stroke length} = \sqrt{L_S = 1.99 \text{ in.}}$$

$$R = \frac{L}{2} = 1.99 \text{ in.}$$

L can range from 7.5 in to 9 in.
Length to bore ratio is defined as $L/\text{Bore} = 1.3$.
Length of the spring as specified by Ford Motors

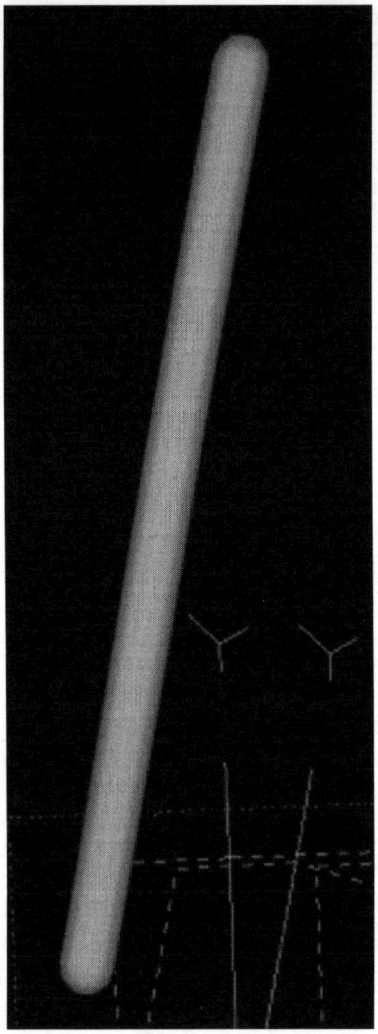

Figure 15.47

Pushrod.

$$L_{pr} = L_{cr} + L_p + L_s = 13.72 \text{ in.}$$

$$n_d = 4$$

$$F_{max} = 8.054 \times 10^3 \text{ lbf}$$

Factor of safety

$$P = F_{max}$$

P. crit to initiate failure $P_{cr} = n_d \cdot P = 3.222 \times 10^4 \text{ lbf}$

End-condition constant $C = 1$

Modulus of elasticity of steel $E = 27.6 \times 10^6$ psi

$$d_{\text{pushrod}} = \left(\frac{64 \cdot P_{cr} \cdot L_{pr}{}^2}{\pi^3 \cdot C \cdot E} \right)^{\frac{1}{4}}$$

diameter of the pushrod with a factor of safety as listed above

Miscellaneous calculations for following parts: material AISI 1020 steel rocker arm (Figure 15.48)

$$\text{Rocker}_{\text{arm}\times\text{ratio}} = 1.73$$

Valve

$$\text{Ratio} = 0.48$$

Diameter of valve face

$$D_{\text{valve}} = \text{bore} \cdot \text{ratio} = 1.982 \text{ in.}$$

Pushrod

$$\text{Bore} = 4.13 \text{ in.}$$

$$L_{pr} = 13.72 \text{ in.}$$

$$L_{\text{head}} = L_{\text{valve}} - L_{\text{stem}} = 1.715 \text{ in.}$$

$$L_{\text{stem}} = \frac{5}{6} \cdot L_{\text{valve}} = 8.575 \text{ in.}$$

$$D_{\text{valve}\times\text{stem}} = 0.37 \text{ in.}$$

$$L_{\text{valve}} = 0.75 \cdot L_{pr} = 10.29 \text{ in.}$$

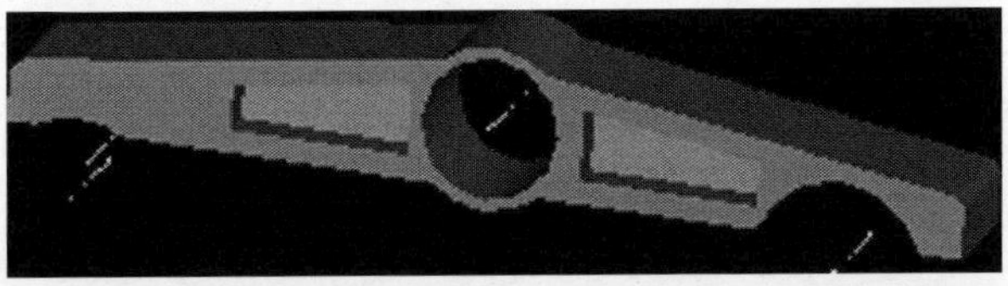

Figure 15.48

Rocker arm.

15.4.4 Tolerance Analysis

Part	Diameter (in)	Length (in)
Pushrod	0.821 ± 0.025	13.27 ± 0.075
Camshaft	1.75 ± 0.03	25 ± 0.075
Rocker arm	1 ± 0.025	3.75 ± 0.04
Valve face	1.982 ± 0.03	10.25 ± 0.075
Valve stem	0.371 ± 0.025	8.47 ± 0.04
Spring	0.197 ± 0.025	1.82 ± 0.025

15.4.5 Manufacturing

Since the camshaft and pushrod are symmetrical shapes, it would be best if they were drawn from some steel alloy. The cams, retainers, valves, and rocker arms would best be made by die-casting. Die-casting involves directing a flow of molten metal into a die. It is important to keep the casting chamber pressurized to make sure the part takes form properly. This is also an ideal manufacturing method due to the parts' irregular shape and small size, and the need for precision.

Machining and forging costs will average \$50 per hour for all parts.

15.4.6 Cost Analysis

15.4.6.1 Materials

Since the cam, camshaft, and rocker arm are all parts that are constantly in motion, they are considered part of the reciprocating weight of the engine. This means that the design of these parts must be as light as possible; otherwise, it will limit the operating speed of the engine. This is why many engineers choose steel as the material for this assembly. Steel offers an ideal balance of strength, durability, and weight.

The cost of material can be found by calculating the total volume of parts to be manufactured and then multiplying by the cost of steel. The following calculations show the cost of material alone.

$$\rho_{\text{steel}} = 487\frac{\text{lbm}}{\text{ft}^3}$$

$$\text{price} = 5\frac{\text{dollars}}{\text{lbm}}$$

$$V_{\text{camshaft}} = \frac{\pi}{4}\cdot d_1^2 \cdot 1 = 0.017\text{ ft}$$

$$V_{\text{pushrod}} := \frac{\pi}{4}\cdot d_{\text{pushrod}}^2 \cdot L_{pr} = 4.2\times10^{-3}\text{ ft}$$

$$d_{\text{rockershaft}} = 0.75\text{ in.}$$

$$V_{\text{rockerarm}} = \left[(w\cdot h + 0.5\cdot f\cdot h)\cdot t\right] - \frac{\pi}{4}\cdot d_{\text{rockershaft}}^2 \cdot t = 1.046\times10^{-3}\text{ ft}$$

$$w = 2\text{ in},\quad t = 1\text{ in},\quad f = 0.5\text{ in},\quad h = 1\text{ in.}$$

$$V_{\text{spring}} := \frac{\pi}{4}\cdot d^2\cdot(N_a\cdot\pi\cdot D) = 9.004\times10^{-5}\text{ ft}$$

$$V_{valve} = \frac{\pi}{4}\left(D^2_{valve\times stem} \cdot L_{stem} + D^2_{valve} \cdot L_{head}\right) = 3.6\times 10^{-3} \text{ ft}$$

$$d_{retainer} = 1.25 \text{ in.}$$

$$h_{vt} = 0.25 \text{ in.}$$

$$V_{valveretainer} := \frac{p}{4} \times d^2_{retainer} \times h_{vt} = 1.775\times 10^{-4} \text{ ft}$$

$$V_{total} = 2\times V_{camshaft} + 8\times\left(V_{pushrod} + V_{rockerarm} + V_{spring} + V_{valve} + 2\times V_{valveretainer} + V_{cam}\right) = 0.116 \text{ ft}^3$$

$$d_{cam} = 1.25 \text{ in.}$$

$$t_{cam} = 1 \text{ in.}$$

$$V_{cam} = \frac{p}{4} \times d^2_{cam} \times t_{cam} \times 1.25 = 8.877\times 10^{-4} \text{ ft}$$

$$m_{total} = V_{total} \times r_{steel} = 56.604 \text{ lbm}$$

$$\text{price}_{total} = m_{total} \times \text{price} = 283.018 \text{ dollars}$$

15.4.7 Overhead Costs

Overhead costs are unforeseen costs that must be accounted for in advance. They are usually estimated to be approximately 15% of overall costs.

15.4.7.1 Solid Modeling

Each of the components designed will be shown individually with dimensions and tolerances.

Pushrod: The pushrod assumes the basic shape of a cylinder. It was created by drawing a circle equal in diameter to that of the pushrod. This circle was then extruded a distance equal to the pushrod's length. The top of the pushrod has rounded ends. These were created by drawing a quarter section of a circle on the end of each rod. This circular section was then revolved about the pushrod's centerline (Figure 15.49).

Camshaft: The camshaft is nothing more than a cylinder, just like the pushrod. However, there are cams located on the shaft at regular intervals. The cam profile was created first. The first step in doing this was to draw a circular section. Then, a triangle was drawn on the circular section by using two poly lines tangent to the perimeter of the circle. The sharp edges were filleted, and the section was extruded by 1 in. Now that the cam is created, cylinders can be attached to it on either side. This will be the first section of the shaft. This section will be copied and joined to the original, thus extending the camshaft to two sections. This is repeated until the shaft has four full sections (Figure 15.50).

Spring: The first step in creating the spring is to draw a circular cross section. This will be the cross section of the spring's wire. Then, this section needs to be revolved in a helical fashion. This means that, as the cross section makes a complete revolution, it is translated some distance upwards. The distance over which the cross section is translated per rotation is referred to as the pitch (Figure 15.51).

Valve: The valve face and stem were designed as a single part. The valve face is nothing more than a short cylinder. A circular cross section was extruded a small distance. This creates the disk-like shape at the valve face. Next, the valve stem was created. The valve stem is a cylinder, created in a way nearly identical to the pushrod. A circular cross section was drawn on the face of the valve and then extruded outwards (Figure 15.52).

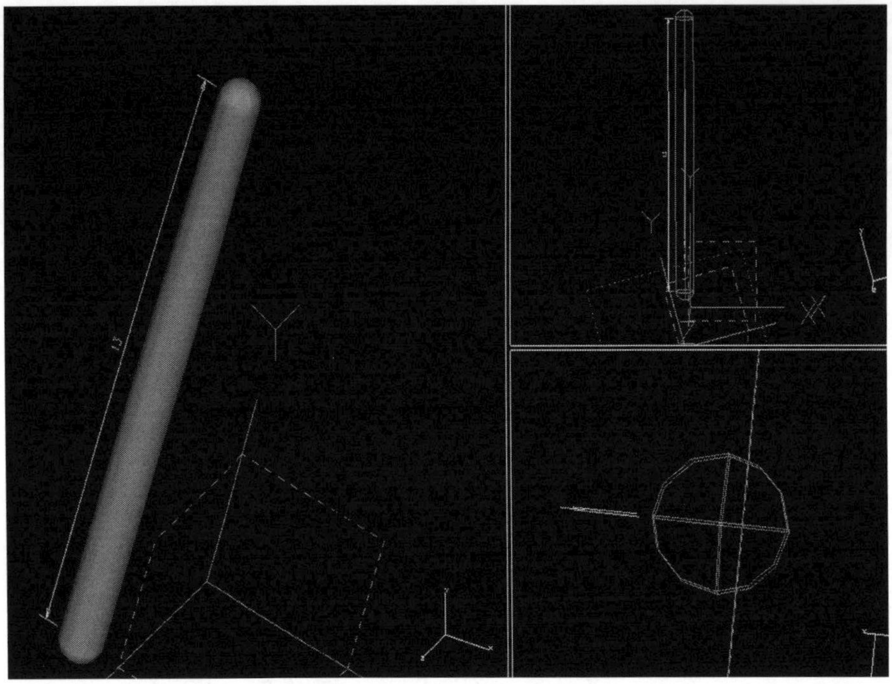

Figure 15.49

Dimensioned pushrod.

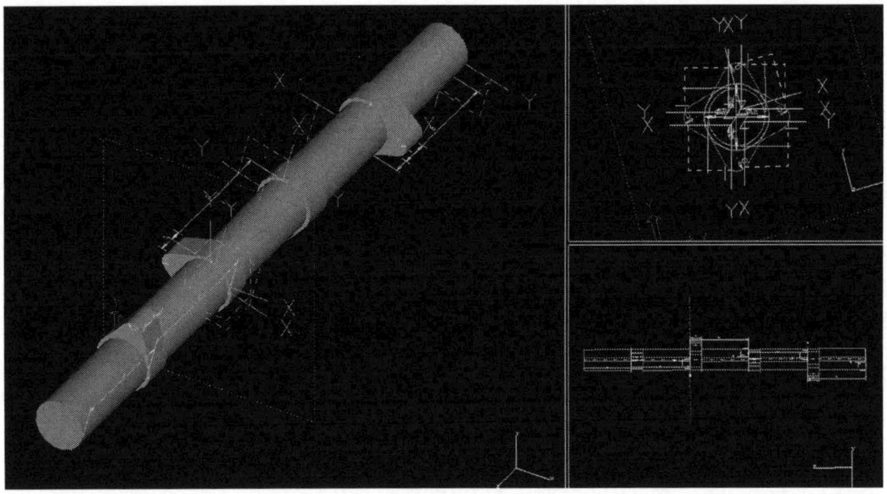

Figure 15.50

Dimensioned pushrod.

Rocker Arm: The rocker arm was a little more complicated to design than the rest of the components. First, the cross section of the part was drawn using poly lines. Then, the sharp edges were filleted out. It is important to make sure that the lines create a symmetrical shape with all the dimensions needed. Next, this shape was extruded 1 in. Now that the general form of the rocker arm has been created, it can be detailed. The cutting function in Ideas was used to cut out sections of the rocker arm's general form. A center hole for the rocker shaft was cut by drawing a circle on the center of the rocker arm. This circle was used as a cutting tool to remove a volume of material. The same thing was done to take out the trapezoidal sections on either side of this hole. These trapezoidal cutouts decrease the weight of the rocker arm, which decreases manufacturing costs as well as the reciprocating weight of the engine.

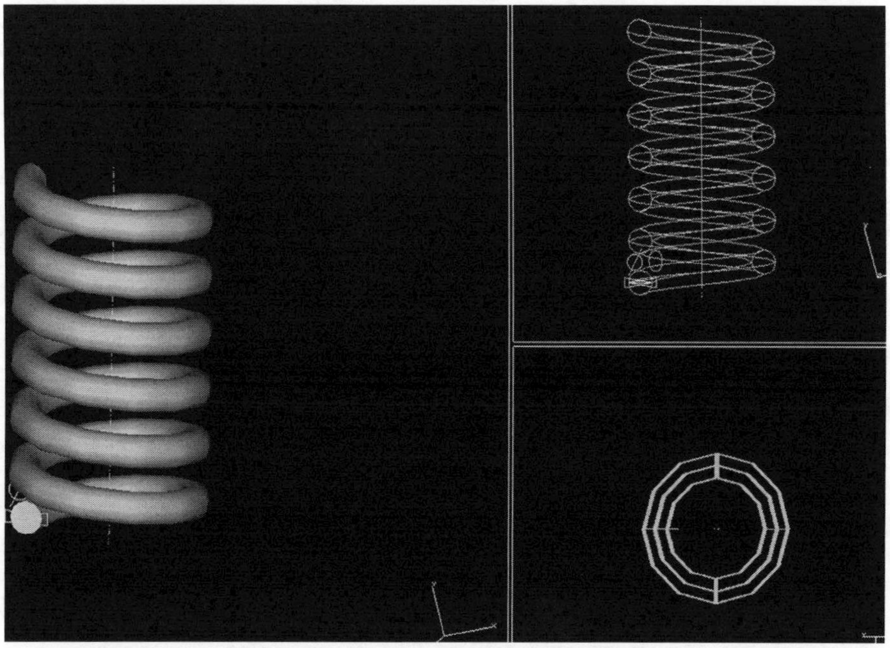

Figure 15.51

Dimensioned spring.

Figure 15.52

Dimensioned valve.

Assembly: Now that each of the parts above is modeled separately, they can be assembled. Once the assembly is complete, it can be copied as a whole and pasted where needed to complete the rest of the assembly (Figures 15.53 and 15.54).

15.4.8 Finite Element Analysis

FEA is a method of analyzing stress in a mechanical member. It is done by dividing the mechanical member into some finite number of small elements. NX-9 measured the stress and deflection in each of these elements and calculated a resultant stress and deflection.

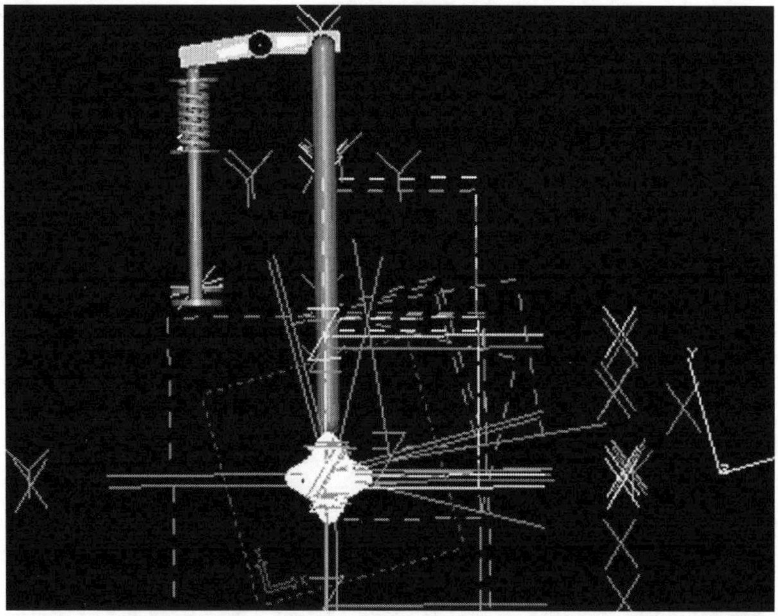

Figure 15.53

First assembly.

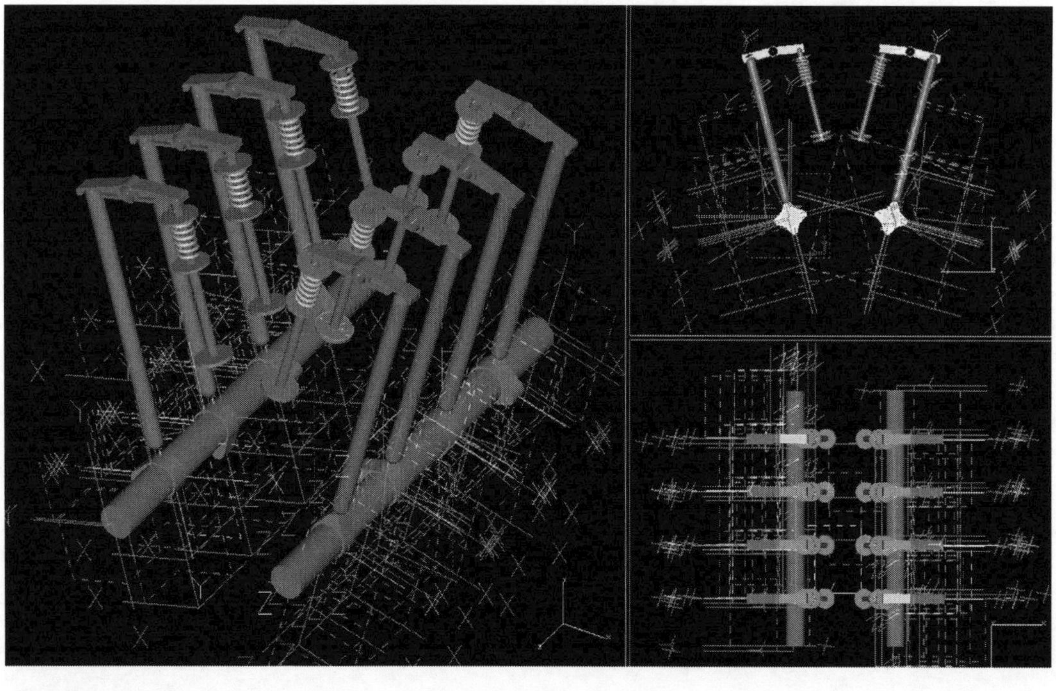

Figure 15.54

(See color insert) Final assembly.

15.4.8.1 FEA: Pushrod

Results from the FEA on the pushrod show that there is no deflection or stress occurring that would indicate failure in either static loading or fatigue.

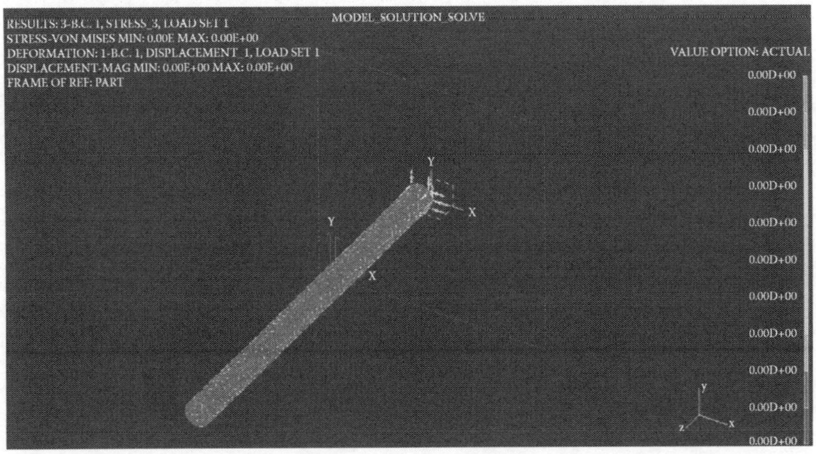

15.4.8.2 FEA: Camshaft

Results from the FEA on the camshaft show that there is no deflection or stress occurring that would indicate failure in either static loading or fatigue.

Most of the stress and deflection occurring is due to torsion, and therefore angular deflections are to be expected. Naturally, the highest stress is at the farthest point on the shaft from its point of application.

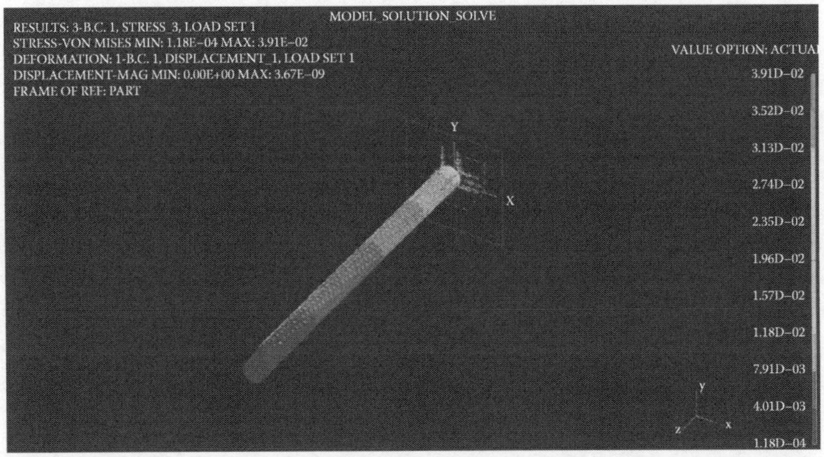

15.4.8.3 FEA: Rocker Arm

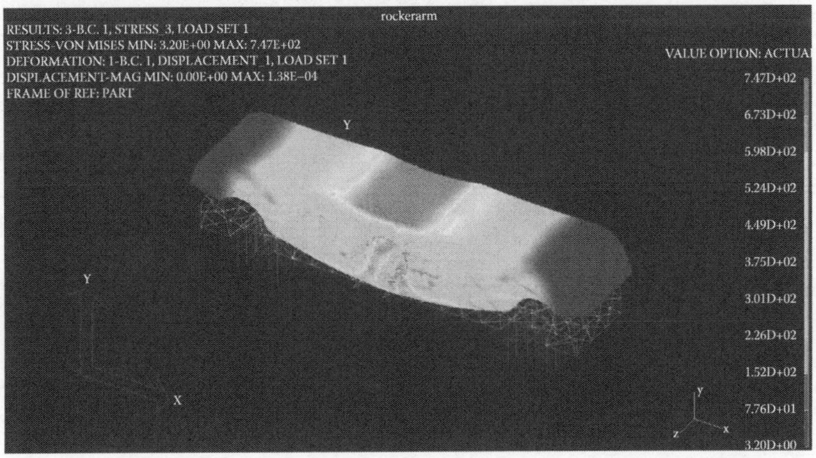

15.5 CAD Example 5: Computer-Aided Design of Central Differential of an Automobile

Design a center differential for a possible SAE Mini Baja vehicle, for which the complete drive train will consist of a 10 hp motor, transmission with a gear ratio of 6.72:1, and chain and sprocket, which have a ratio of 3.23:1, that transmits power to the center differential. It is calculated that 32 lbf × ft of torque will be produced at 2400 rpm, without taking into consideration internal losses of the system components. The safety factor should be at least 2 to account for any unforeseen loading scenarios.

An industry benchmark self-contained differential will be designed and tested in the computer. The goal is to come up with a design by which all components can be subjected to extreme forces that might be encountered in real life and provide long-lasting reliable service with minimal cost. Worst-case scenarios will be designed, and these forces will be applied to the various parts of the model. Once weak points are determined, measures will be taken to strengthen those parts by analyzing various geometries and materials.

A manufacturer is currently producing a self-contained differential for the mass market. One of the major faults of the design is that it is likely to break under long-term use. Also, its applications are limited to light vehicles that do not produce much output torque.

To solve the various problems, an understanding of what is involved is important. The basic differential is used in most vehicles to allow two wheels in the same rotational axis to spin at two different speeds. This is important for the vehicle, because when a vehicle makes a turn, the inside tire will rotate more slowly than the outer tire. If both wheels rotated at the same speed all the time, the vehicle would need new tires more often. A differential also allows the turning radius to be decreased for the same reason. In all, the vehicle becomes more manageable on tarmac or concrete. In Figure 15.55, the differential is composed of a driveshaft that rotates a pinion gear. This pinion gear in turn transmits power to a ring gear. The ring gear is attached to a housing in which spider gears are held in place by a shaft. The spider gears and side gears allow the right and left axle shafts to rotate at different speeds when the vehicle goes round a corner.

The initial conceptual design only calls for spider gears (they will be referred to as pinion gears from now on), side gears (they will be referred to as axle bevel gears from now on), axles, stud, and housing. The ring gear will be substituted by a 40-tooth sprocket that will bolt onto the housing with eight grade 8 3/8–16 bolts with nylon nuts. The pinion gears shown in Figure 15.55 are not needed, since there is no driveshaft in the proposed design.

Most of the computer-aided design will be done in Solidedge due to the ease of gear and sprocket production. Testing will be done in SolidWorks and NX-9.0.

15.5.1 Analysis and Design

This section will detail the format, part name, calculations and theory, and FEA analysis results.

15.5.1.1 Bevel Gear Calculations

Gear characteristics: both bevel gears are the same, 20° straight bevel gears with 16 teeth

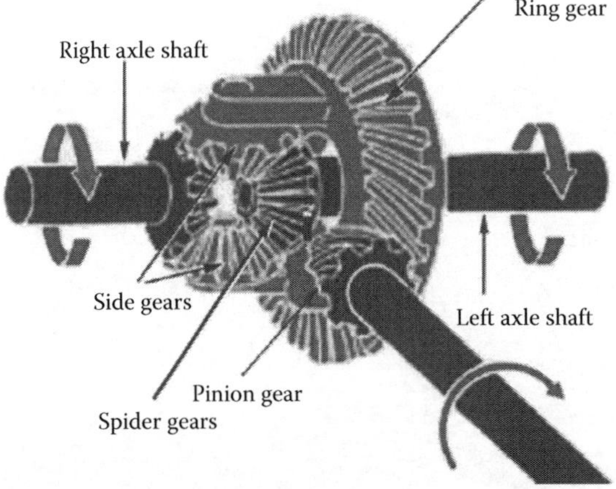

Figure 15.55

Diagram of differential.

$$p_{\text{gear}} = 16\frac{1}{\text{in}}$$

(Assumed)

The pitch of the gear is important because it will dictate the thickness of the teeth, which will in turn affect the strength of each individual tooth.

$$F = \frac{10}{p_{\text{gear}}} = 0.625\ \text{in.}$$

15.5.1.2 Working Depth

This is important, since the teeth need to be robust for long-lasting service.

$$h_k := \frac{2.0}{p_{\text{gear}}} = 0.125\ \text{in.}$$

The clearance for gears is important to ensure easy operation.

15.5.1.3 Addendum of Gear

$$c = \left(\frac{0.188}{p_{\text{gear}}}\right) + 0.0028\ \text{in.} = 0.015\ \text{in.}$$

When $\Gamma = 90$, therefore

15.5.1.4 Gear Ratio

$$m_g = 1$$

$$m_{90} = m_g = 1$$

$$a_{\text{gear}} = \frac{0.54}{p_{\text{gear}}} + \frac{0.460}{p_{\text{gear}} \times (m_{90})^2} = 0.063\ \text{in.}$$

The goal is to design a gear with minimal volume to keep the overall assembly compact.

15.5.2 Outboard Axle Shaft Calculations

15.5.2.1 Shaft Material

4130 chromoly normalized at 870°C (1600°F)

$$r_{\text{axle}} = 0.25\ \text{in.}$$

15.5.2.2 Dimensions of Axle Shaft

$$L_{\text{axle}} = 25\ \text{in.}$$

$$G_{\text{axle}} = 80 \times 10^9\ \frac{\text{lbf}}{\text{in.}^2}$$

$$S_{ut} = 633{,}008\ \frac{\text{lbf}}{\text{in.}^2}$$

15.5.2.3 Maximum Shear Stress, Moment, and Angle of Twist for Axle Shaft

$$J_{\text{axle}} = 0.5 \times p \times r_{\text{axle}}^4 = 6.136 \times 10^{-3}\ \text{in.}^4$$

$$t_{\text{max\_axle}} := \frac{s_{\text{outboard\_axle}} \times J_{\text{axle}}}{r_{\text{axle}}} = 1.383 \times 10^3\ \text{lbf} \cdot \text{in.}$$

The minimum diameter for the shaft is calculated as follows:

$$\sigma_{\text{outboard\_axle}} := T_{\text{max}} \times \frac{r_{\text{axle}}}{J_{\text{axle}}} = 5.636 \times 10^4\ \frac{\text{lbf}}{\text{in.}^2}$$

$$\theta = \frac{(T_{\text{max}} \cdot L_{\text{axle}})}{G_{\text{axle}} \cdot J_{\text{axle}}} = 7.046 \times 10^{-5}$$

$$D = 1.72 \left(\frac{t_{\text{max\_axle}}}{s_{\text{outboard\_axle}}} \right)^{\frac{1}{3}} = 0.5\ \text{in.}$$

Thus, the minimum diameter for the axle would be 0.5 in. By choosing a 1 in diameter axle for the design, any unforeseen loading could be handled. This is also a standard size for rods.

The design uses a few parts that are mass produced as standard, such as bolts and nuts. The bolt chosen is grade 8 3/8–16, 1.5 in long, and the pillow blocks are available in the market as standard.

15.5.3 Solid Modeling

The following format will be used. The part creation within NX 9.0 will be described step by step, and images of the parts created will be shown.

Setup:

1. Set units to inches (pounds).
2. Set grid to desired dimensions.

Axle Bevel Gear:

1. Draw single tooth profile with base and a center line that will be used as an axis of revolution later on.
2. Extrude profile in two directions, keeping the middle on the x–y plane.
3. Chamfer sides of the teeth to make more realistic.
4. Create circular array by choosing extrusion and line drawn earlier as center (Figures 15.56 through 15.59).
5. Fill in gaps within pattern.
6. Fillet sharp edges to reduce stresses within part.
7. Name part and add to library.

Pinion:

1. Check procedure for axle bevel gear (Figures 15.60 through 15.62).
2. Do not create splines on inside; it is a straight bore.
3. Do not create extension for gear.

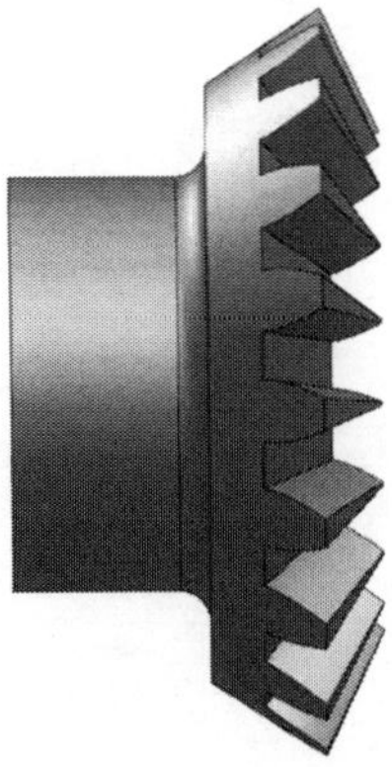

Figure 15.56

Axle bevel gear side view.

Figure 15.57

Axle bevel gear front view.

Figure 15.58

Axle bevel gear orthogonal view.

Axle:

1. Draw circle with diameter 1 in (Figures 15.63 and 15.64).
2. Extrude 5 in.
3. Draw on top of one end single spline profile, 0.125×0.125.
4. Cut out single spline profile by using the extrusion command but switching it to cut.
5. Circular array around center of rod to complete the splines.
6. Name part and add to library.

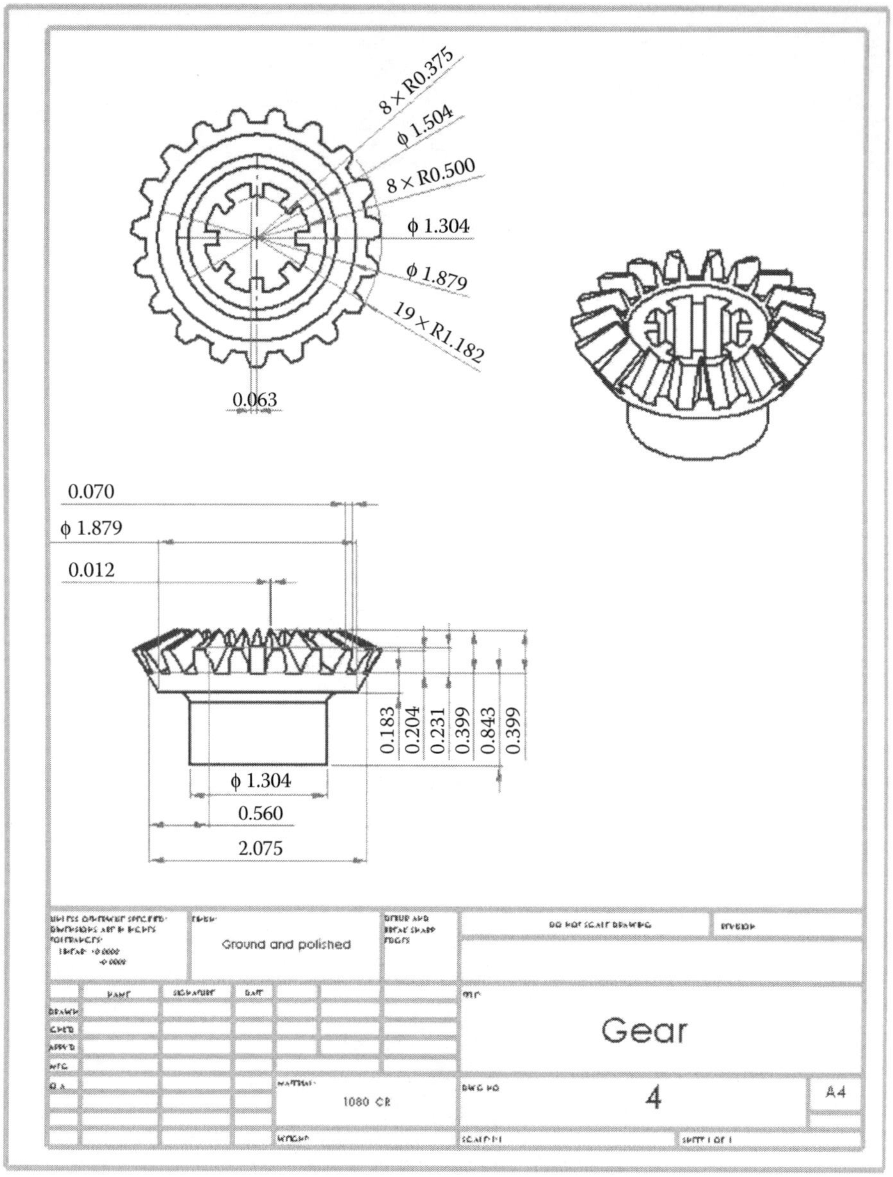

Figure 15.59

Technical drawing of gear.

Figure 15.60

Front view of pinion.

Figure 15.61

Orthogonal view of pinion.

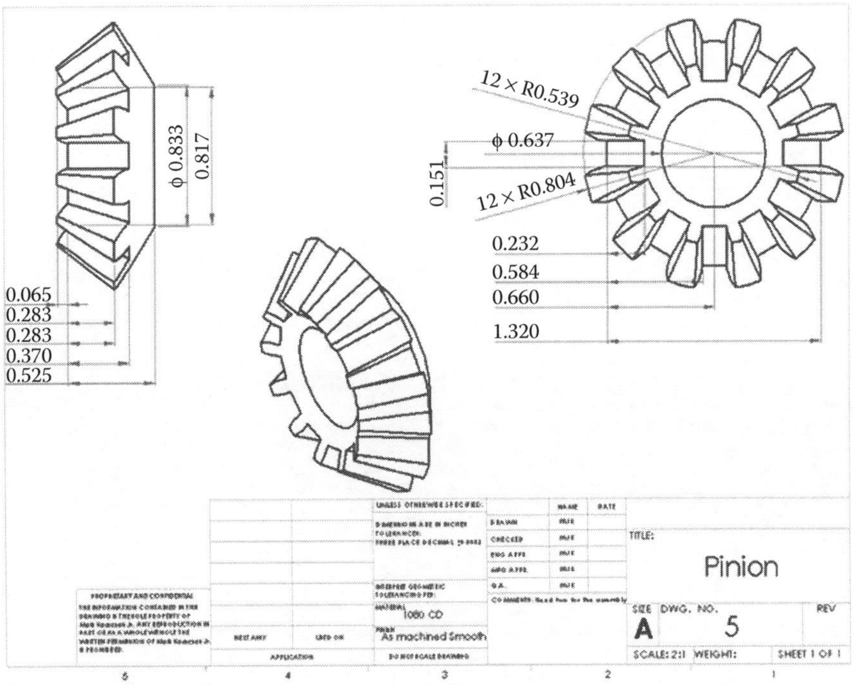

Figure 15.62

Technical drawing of pinion.

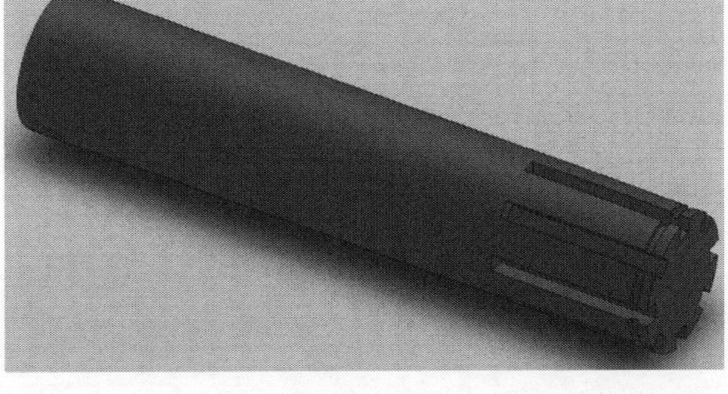

Figure 15.63

Axle orthogonal view.

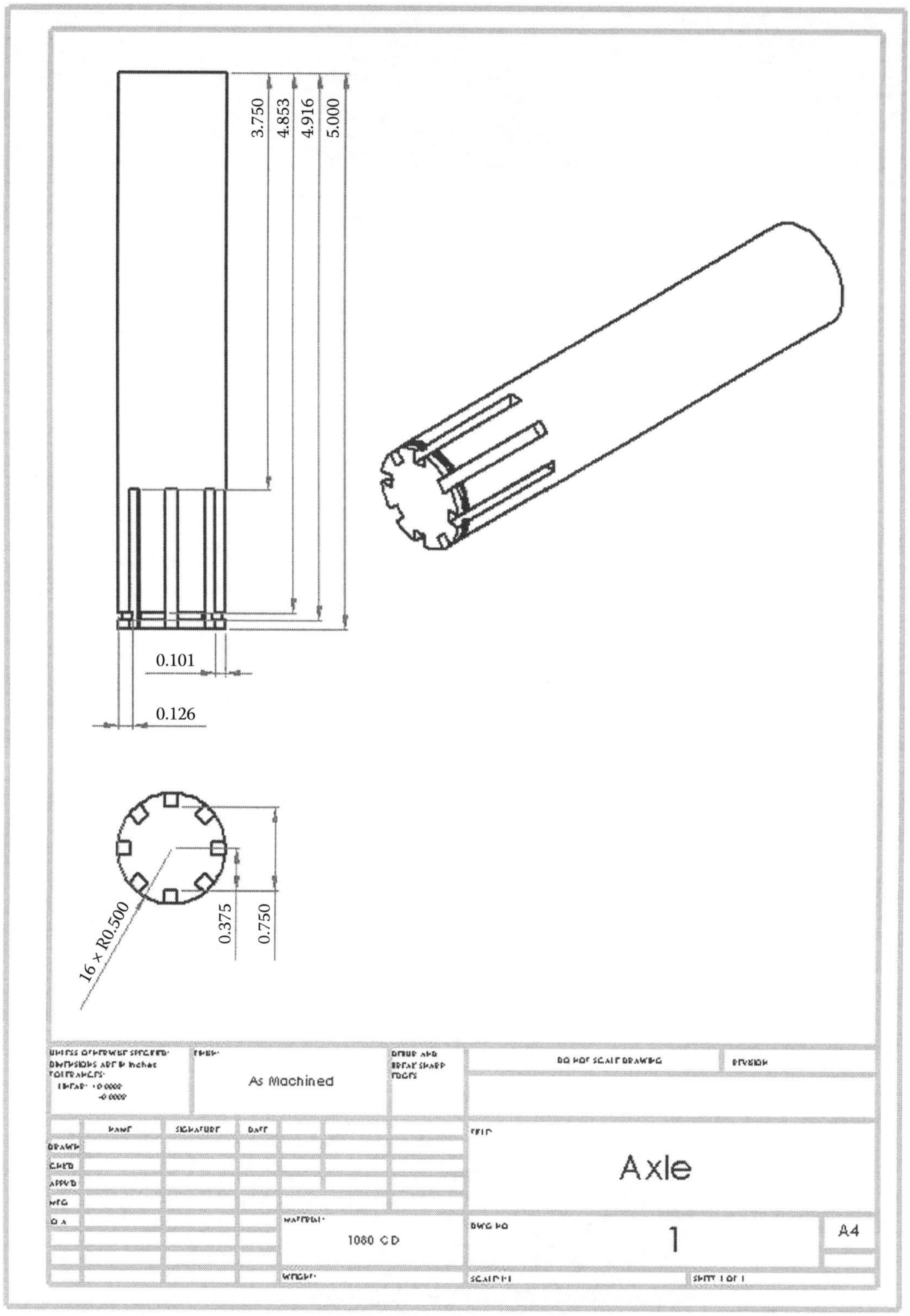

Figure 15.64

Technical drawing of axle with splines.

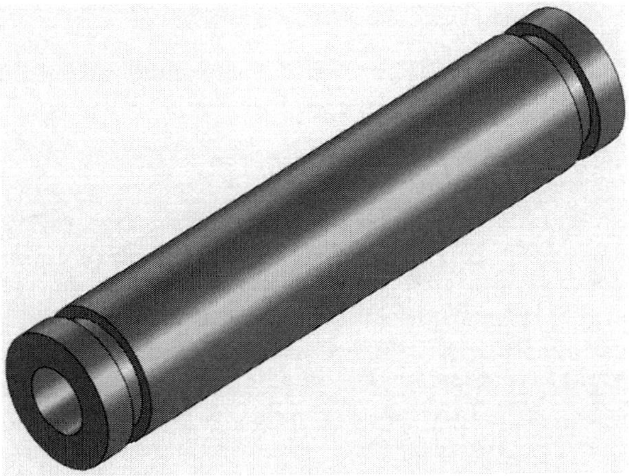

Figure 15.65

Center pin orthogonal view.

Center pin:

1. Draw side profile of pin with notches in it (Figures 15.65 and 15.66).
2. Revolve around center.
3. Name part and add to library.

Assembly:

1. Pull parts out and create mates to configure assembly.
2. Create half of the assembly with one axle, one bearing, and one bevel gear.
3. Mirror the assembly to create the other side of it.

15.5.4 Manufacturing Process

Forming the gears is a multistep process [2]. First, blanks are created on a lathe that have the rough dimensions of the final bevel gear shape minus any material removal between the teeth. The internal splines are cut into the blank by using a broaching process. From there, the teeth are cut out on a vertical mill using a slot mill that has the required geometry. Once the teeth are cut into the blanks, they are sent for heat treatment to be hardened and then surface ground to get rid of any distortion from heat treatment.

The axles are relatively simple to produce with a CNC Lathe/Mill Combo. Raw material 1/16 in greater in diameter is bought in long sections and placed in the machine. The outer dimension of the raw material is turned down to 1.000 inch as per the design, and the snap ring slot is machined into the blank. Once a section is done, the mill activates and uses a 1/8 in diameter end mill to cut the eight splines into the axle. Usually, splines would be cut using a slot cutter, but with the current design it will not be feasible. The long rods are then cut down to around the final length of 5 in plus 0.0625 in to allow for facing the ends of the axle on another lathe. From here, the axles are done and ready for assembly in a different section of the plant.

The center pin is a simple piece to produce. Using a similar process as for the axles, a CNC lathe would be used to turn down a raw rod to the final outer diameter, then cut in the grooves that hold the pin in place within the housing, then drill out the center, and finally, cut it to length plus 0.065 in. Then the cut side would be machined true with the rest of the piece.

The housing is placed on a similar CNC lathe and turned down to the final shape. A blank log of aluminum with an outer diameter of 6.25 in would be ideal.

The sprocket is bought from an aftermarket supplier, so manufacturing the piece will not be a problem for the company.

15.5.5 Cost Estimation

This section will cover the assembly part by part, looking at popular materials for mechanical components. The material costs will be taken from the cost report on SAE Mini Baja, which uses mild steel, alloy steel,

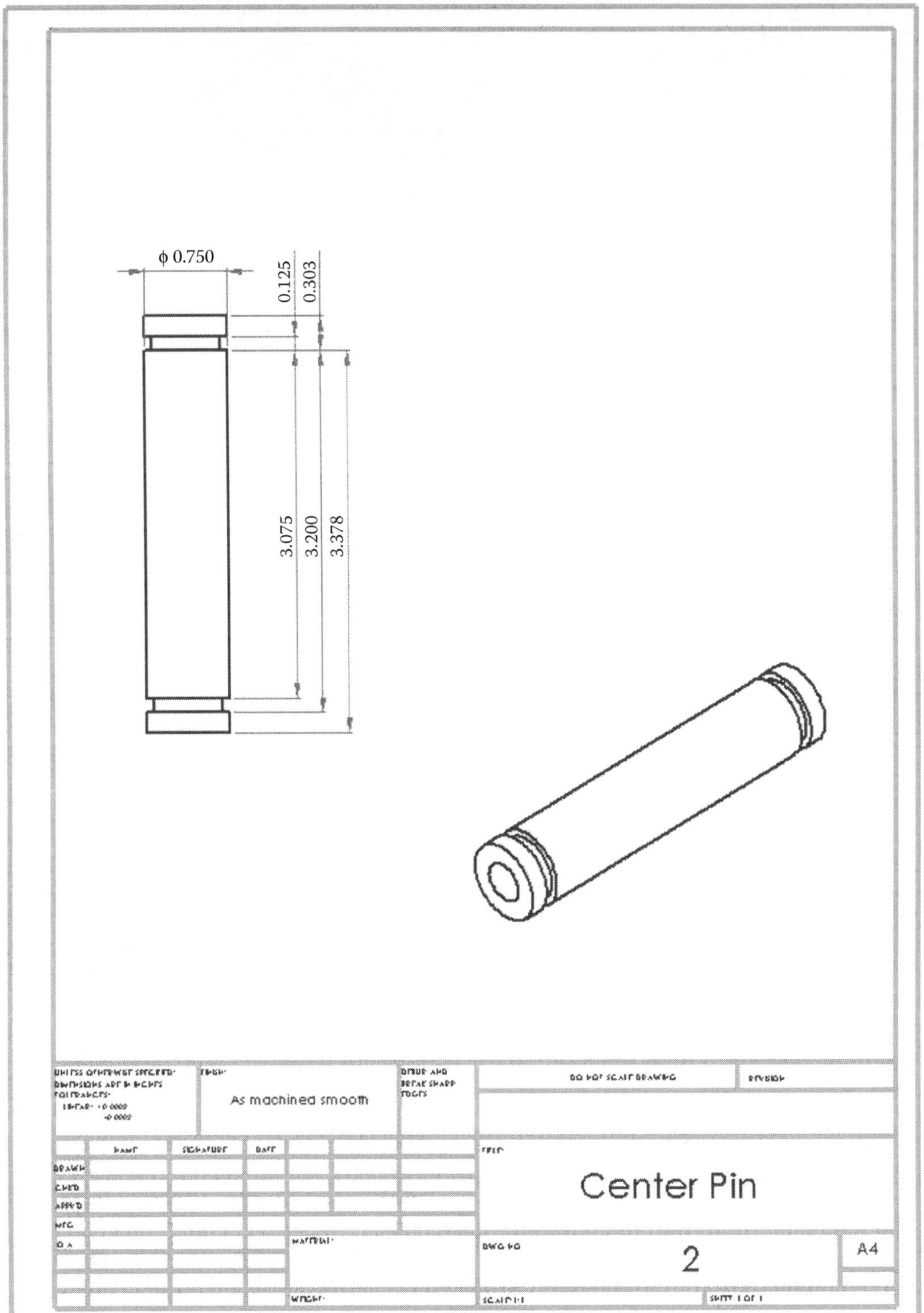

Figure 15.66

Technical drawing of pin.

aluminum, magnesium (Mg), brass, and copper. Since the prices are determined by weight, the volume of each part needed to be determined (Table 15.14). This was done by using the mass properties feature in Solidedge. The average densities of each possible material were also tabulated to be used for cost estimation (Table 15.15).

The cost of the component will be the cost of material plus manufacturing plus overhead. Parametric costs are outlined in Tables 15.16 through 15.22.

Table 15.14 Volume of Each Part

	Volume (in^3)	Volume (ft^3)
Axle	3.77	0.002181713
Pin	1.18	0.00068287
Housing	14.54	0.008414352
Gear	1.48	0.000856481
Pinion	0.37	0.00021412
Snap ring	0.03224	1.86574×10^{-5}

Table 15.15 Tabulated Costs per Pound and Densities

Material	Cost per Pound	Density (lbm/ft^3)
Mild Steel, e.g., 1010, 1025	$1.00	490
Alloy Steel, e.g., 4130, Chromoly	$2.00	490
Aluminum	$5.00	168
Mg	$9.00	108
Brass	$10.00	524
Copper	$10.00	558

Table 15.16 Parameters for Costs

Item	Processing Cost ($)
Material removal cost	150
Foundry cost	125
Forging cost	135
Overhead	150

Table 15.17 Costs for Axle

Material	Total Cost for Part	Material Removal Cost	Foundry Cost	Manufacturing Cost	Overhead Cost	Total Cost
Mild steel, e.g., 1010, 1025	$0.00	$0.00	$0.00	$0.01	$0.00	$0.01
Alloy steel, e.g., 4130, ChroMoly	$0.00	$0.01	$0.01	$0.01	$0.01	$0.02
Aluminum	$0.01	$0.02	$0.01	$0.03	$0.02	$0.06
Mg	$0.02	$0.03	$0.03	$0.06	$0.03	$0.11
Brass	$0.02	$0.03	$0.03	$0.06	$0.03	$0.12
Copper	$0.02	$0.03	$0.03	$0.06	$0.03	$0.12

Table 15.18 Costs for Pin

Material	Total Cost for Part	Material Removal Cost	Foundry Cost	Manufacturing Cost	Overhead Cost	Total Cost
Mild steel, e.g., 1010, 1025	$0.00	$0.00	$0.00	$0.00	$0.00	$0.00
Alloy steel, e.g., 4130, ChroMoly	$0.00	$0.00	$0.00	$0.00	$0.00	$0.01
Aluminum	$0.00	$0.01	$0.00	$0.01	$0.01	$0.02
Mg	$0.01	$0.01	$0.01	$0.02	$0.01	$0.03
Brass	$0.01	$0.01	$0.01	$0.02	$0.01	$0.04
Copper	$0.01	$0.01	$0.01	$0.02	$0.01	$0.04

Table 15.19 Costs for Housing

Material	Total Cost for Part	Material Removal Cost	Foundry Cost	Manufacturing Cost	Overhead Cost	Total Cost
Mild steel, e.g., 1010, 1025	$0.01	$0.01	$0.01	$0.02	$0.01	$0.05
Alloy steel, e.g., 4130, ChroMoly	$0.02	$0.03	$0.02	$0.05	$0.03	$0.09
Aluminum	$0.04	$0.06	$0.06	$0.12	$0.06	$0.23
Mg	$0.08	$0.11	$0.10	$0.22	$0.11	$0.41
Brass	$0.08	$0.13	$0.11	$0.24	$0.13	$0.45
Copper	$0.08	$0.13	$0.11	$0.24	$0.13	$0.45

Table 15.20 Costs for Bevel Gear

Material	Total Cost for Part	Material Removal Cost	Foundry Cost	Manufacturing Cost	Overhead Cost	Total Cost
Mild steel, e.g., 1010, 1025	$0.00	$0.00	$0.00	$0.00	$0.00	$0.00
Alloy steel, e.g., 4130, ChroMoly	$0.00	$0.00	$0.00	$0.00	$0.00	$0.01
Aluminum	$0.00	$0.01	$0.01	$0.01	$0.01	$0.02
Mg	$0.01	$0.01	$0.01	$0.02	$0.01	$0.04
Brass	$0.01	$0.01	$0.01	$0.02	$0.01	$0.05
Copper	$0.01	$0.01	$0.01	$0.02	$0.01	$0.05

Table 15.21 Costs for Pinion

Material	Total Cost for Part	Material Removal Cost	Foundry Cost	Manufacturing Cost	Overhead Cost	Total Cost
Mild steel, e.g., 1010, 1025	$0.00	$0.00	$0.00	$0.00	$0.00	$0.00
Alloy steel, e.g., 4130, ChroMoly	$0.00	$0.00	$0.00	$0.00	$0.00	$0.00
Aluminum	$0.00	$0.00	$0.00	$0.00	$0.00	$0.01
Mg	$0.00	$0.00	$0.00	$0.01	$0.00	$0.01
Brass	$0.00	$0.00	$0.00	$0.01	$0.00	$0.01
Copper	$0.00	$0.00	$0.00	$0.01	$0.00	$0.01

Table 15.22 Costs for Snap Ring

Material	Total Cost for Part	Material Removal Cost	Foundry Cost	Manufacturing Cost	Overhead Cost	Total Cost
Mild steel, e.g., 1010, 1025	$0.00	$0.00	$0.00	$0.00	$0.00	$0.00
Alloy steel, e.g., 4130, ChroMoly	$0.00	$0.00	$0.00	$0.00	$0.00	$0.00
Aluminum	$0.00	$0.00	$0.00	$0.00	$0.00	$0.00
Mg	$0.00	$0.00	$0.00	$0.00	$0.00	$0.00
Brass	$0.00	$0.00	$0.00	$0.00	$0.00	$0.00
Copper	$0.00	$0.00	$0.00	$0.00	$0.00	$0.00

The total cost of the assembly would be roughly \$0.55 without taking into account the cost of bolts, nuts, and other miscellaneous pieces. This is not a realistic figure using the parameters provided.

15.5.6 Codes and Standards

The design of each part used a minimum safety factor of 2, which exceeds most design requirements. Since this assembly will only be released to a small number of customers, risk is minimized for the company. With small production amounts, quality can be increased, thus increasing reliability even more. The design followed the design requirements from SAE for their Baja Series competition.

15.5.7 Assembly

The entire assembly was designed to that it could be taken out of the vehicle by taking out at most four bolts. The rest of the axle was not taken into consideration due to complexity. The four bolts hold the pillow blocks to the frame. From there, the eight bolts holding the two housing pieces are taken off, and the gears inside are accessible. Everything is assembled with a slip fit to make it easier to repair in case of need where hydraulic presses are not available. On the inside of the assembly, there are two snap rings that hold the bevel gear to the axle. These are important for ensuring that the axles do not slip out of the housing (Figures 15.67 through 15.70).

15.5.8 FEA

The most important parts were determined to be the gear teeth, spline, axle shaft, and pinion pin. The required FEA was conducted for each part. For the teeth of the gears, it was determined that the best results would be obtained from a 3-D analysis of a single tooth plus its base (Figures 15.71 and 15.72).

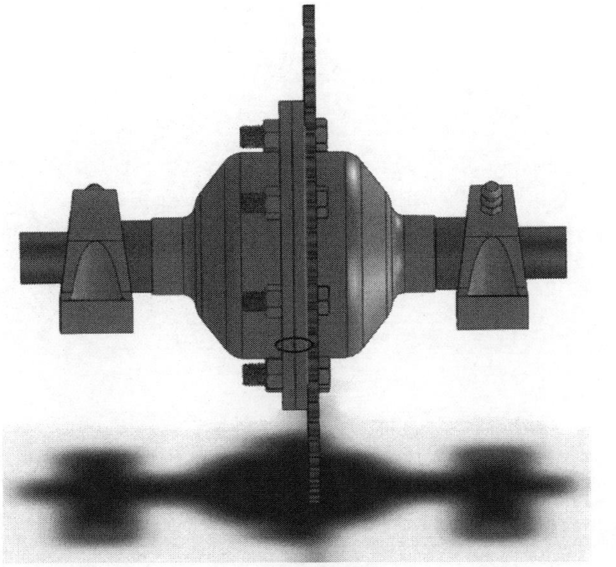

Figure 15.67

Front view of assembly complete.

Figure 15.68

Front view of assembly, inside view.

Figure 15.69

Orthogonal view of assembly complete.

Figure 15.70

(See color insert) Orthogonal view of assembly, inside view.

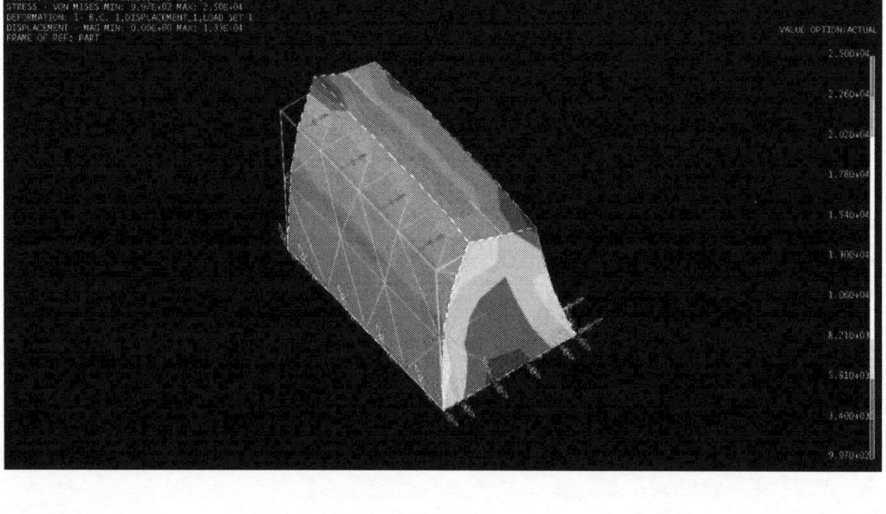

Figure 15.71

Initial tooth stress analysis.

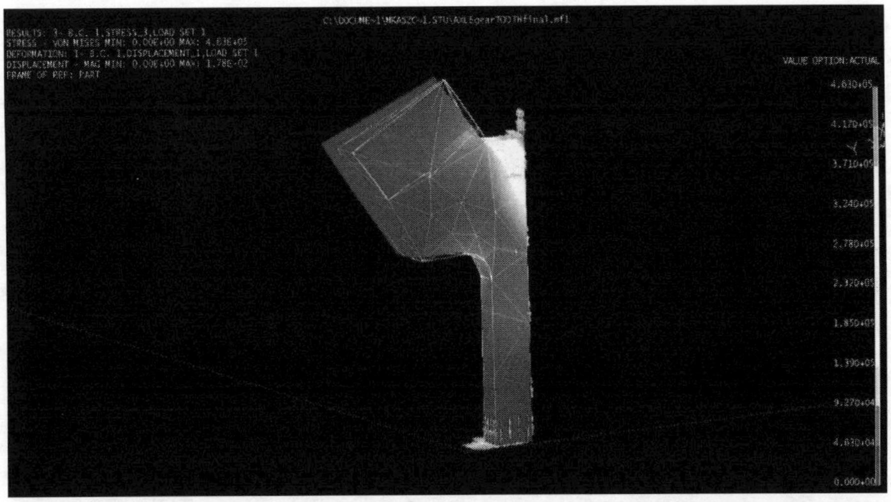

Figure 15.72

Final tooth stress analysis.

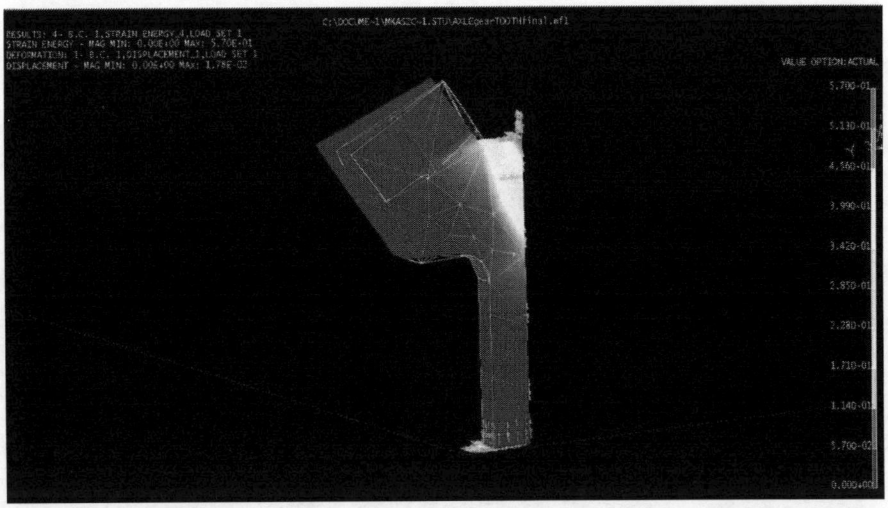

Figure 15.73

Final tooth strain analysis.

It can be seen from these figures that the maximum stress occurs at the right-hand top edge.

The design yielded a maximum stress of 463,000 psi, which is below the yielding strength of the selected material (4140). The safety factor is around 2.1 for the tooth (Figures 15.73 and 15.74).

The spline profile was tested under extreme conditions and found to be safe. The maximum von Mises stress was 6450 psi, which is well below the allowable calculated von Mises stress for 4130 chromoly (Figures 15.75 and 15.76).

15.6 CAD Example 6: Design of Universal Joint in a Car

A universal joint is to be designed to link two driveshafts in series on a 2004 Toyota Tacoma with an engine that produces 150 hp.

A universal joint is a simple mechanical system that is used in a very wide variety of applications. Universal joints, or U-joints, can be seen in many tools and machinery, as well as in the drive train of every automobile. This report will focus on the design and analysis of an automotive universal joint which links two driveshafts in a rear-wheel-drive vehicle.

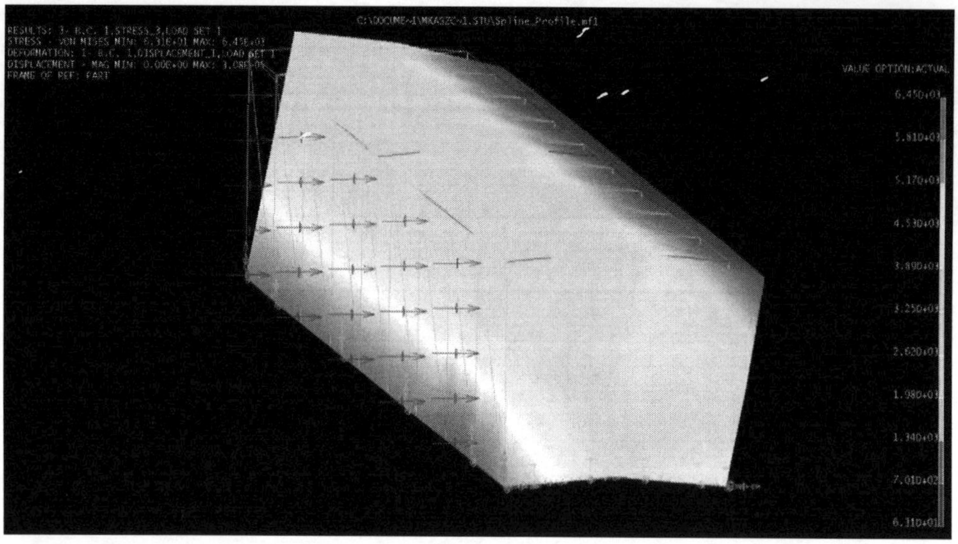

Figure 15.74

Initial spline stress analysis.

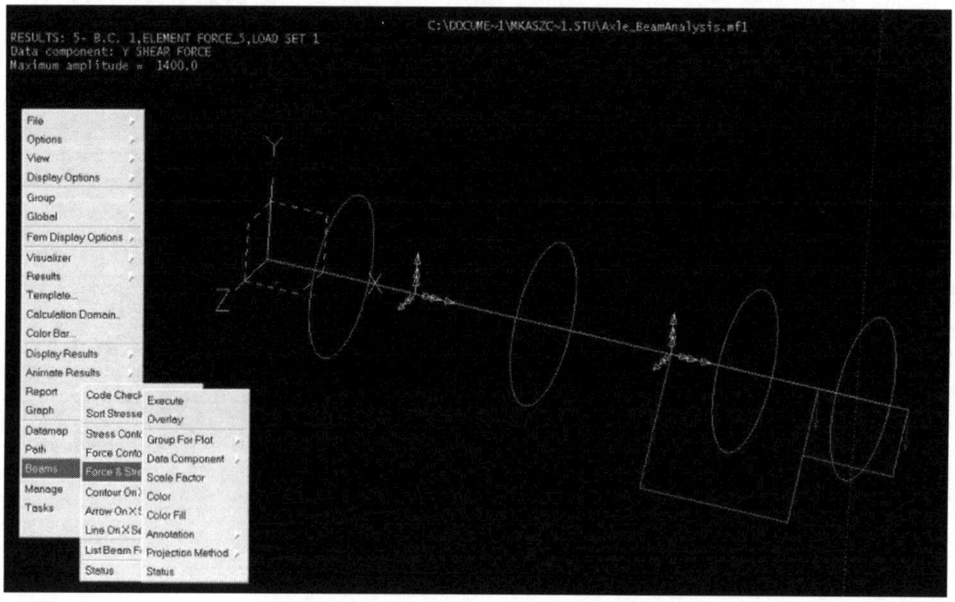

Figure 15.75

Axle shear stress in y.

Figure 15.77 shows a basic sketch of the function and movement of the universal joint to be designed. The input driveshaft is connected directly to the vehicle's transmission, which is spun by the engine. As the engine runs, the input shaft spins. The universal joint transmits power from the input shaft to the output shaft and allows the two to spin together at an angle to each other. The output shaft is connected to the axle gears, which spin the tires and move the vehicle.

As the system operates, several stresses will be experienced by each component. Each driveshaft will experience stress from torsion as it spins. The "yolk" or "ears" of the driveshaft, where the U-joint mounts, will experience shear stress. And the joint itself will experience normal stresses of tension and compression as it transmits power between shafts. These stresses will be analyzed in the next section, and a suitable universal joint will be designed for our application.

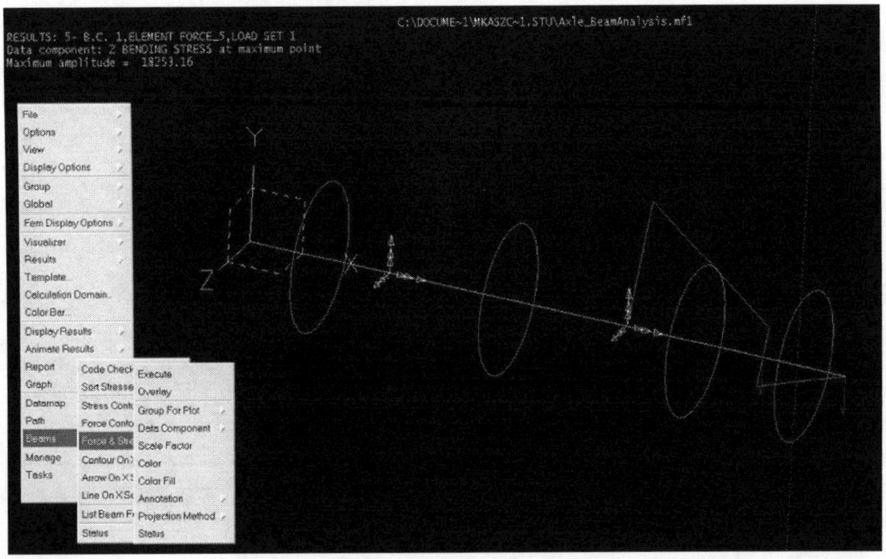

Figure 15.76

Bending stress about z.

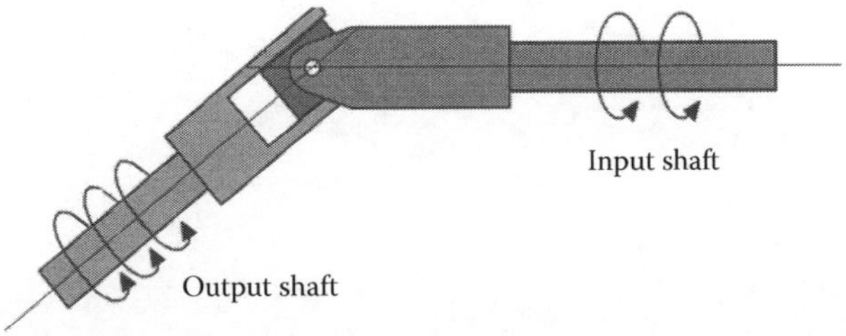

Figure 15.77

Basic sketch of universal joint operation.

15.6.1 Analysis and Design

The goal is to design a universal joint transferring power between a 150 hp motor and a drive axle with driveshafts at 10° to each other at 2000 rpm. Figure 15.78 shows a basic schematic of the universal joint connecting two rotating driveshafts. The input shaft is connected to the transmission, and the output shaft is connected to the axle differential.

Angular Displacement Relationship

$$\tan\theta_2 = \tan\theta_2 \cos\delta$$

Angular Velocity Relationship

$$\frac{\omega_3}{\omega_2} = \frac{\cos\delta}{1-\sin^2\theta_2\sin^2\delta}$$

Angular Acceleration Relationship

$$\alpha_3 = \omega_2^2 \frac{\cos\delta\sin^2\delta\sin\left(2\cdot\theta_2\right)}{\left(1-\sin^2\theta_2\sin\delta\right)}$$

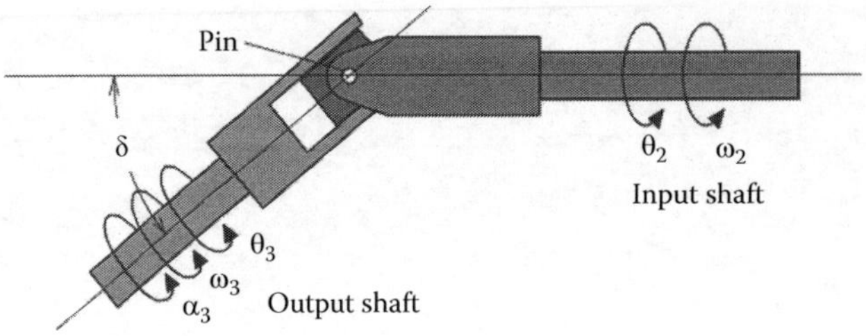

Figure 15.78

Universal joint diagram.

The maximum angular acceleration ratio is [4]:

$$\frac{\alpha_{max}}{\omega^2} = 0.0306 \ \text{for} \ \beta = 10°$$

$$\omega = \frac{(2000)(2\pi)}{60} = \frac{209\,\text{rad}}{s}$$

$$\alpha_{max} = (0.0306)\left(\frac{209\ \text{rad}}{s}\right)^2 = \frac{1336\ \text{rad}}{s^2}$$

$$\text{Weight} = \pi r^2 t \rho_{steel} = 2\ \text{lb}$$

Inertia Torque

$$I = \frac{Wr^2}{2g} = \frac{(2\ \text{lb})(3\ \text{in})^2}{2\left(32.2\frac{ft}{s^2}\right)} = 0.0233\ \text{in}\cdot\text{lb}\cdot\text{s}^2$$

Universal Joint Selection

$$\text{Torque} = \frac{(150\ \text{hp})(63{,}025)}{2000\ \text{rpm}} = 4726\ \text{in}\cdot\text{lb}$$

For continuous operation, a use factor of 68 is used [4].

$$\text{Required torque} = (68)(4726) = 1360\ \text{in}\cdot\text{lb}$$

15.6.2 Design of Driveshaft

The radius of the driveshaft is calculated using this equation. The S_{sy} of the steel alloy 4340 shaft is 230 psi with a factor of safety of $n = 2.5$.

$$J = \frac{\pi d^4}{32} = \frac{\pi 3.75^4}{32} = 19.41$$

$$r = \frac{\tau J}{T} = \frac{\left(\frac{S_{sy}}{n}\right)J}{T} = \frac{\left(\frac{230\text{psi}}{2.5}\right)19.41}{1360\ \text{in.}\cdot\text{lb}} = 1.875\ \text{in.}$$

The driveshaft radius is 1.875 in, and the diameter will equal 3.75 in. The force on the universal joint is divided in half and applied to each end of the "+" shape.

$$F = \frac{T}{r} = \frac{1360\text{ in.}\cdot\text{lb}}{\left(\frac{3.75\text{ in.}}{2}\right)} = 725\text{ lbf}$$

15.6.3 Design of Universal Joint

The stress on the universal joint is defined as follows. The U-joint is made of a stainless steel alloy 440a with an S_y of 400 psi and a factor of safety of $n = 2.5$.

$$M = \left(\frac{F}{2}\right)(x) = \left(\frac{725\text{ lbf}}{2}\right)\left(\frac{3.75\text{ in.}}{2}\right) = 679\text{ lbf}\cdot\text{in.}$$

$$I = \frac{\pi d^4}{64} = \frac{\pi 1\text{ in.}^4}{64} = 0.049$$

$$c = \frac{\sigma I}{M} = \frac{\left(\frac{S_y}{n}\right)I}{M} = \frac{\left(\frac{400\text{ psi}}{2.5}\right)0.049}{679\text{ lbf}\cdot\text{in.}} = 0.467\text{ in.}$$

The calculated universal joint radius is 0.467 in. A preferred size of 0.5 in will be chosen.

The stress on each driveshaft arm is calculated using the following equations. The length of the arm must be larger than half the length of the universal joint to allow free movement, so length b is fixed at 2.5 in. The thickness of the arm, c, is unknown.

$$M = \left(\frac{F}{2}\right)(x) = \left(\frac{725\text{ lbf}}{2}\right)(2.5\text{ in.}) = 906\text{ lbf}\cdot\text{in.}$$

$$I = \frac{bh^3}{12} = \frac{(2.5\text{ in.})(0.5)^3}{12} = 0.026$$

$$c = \frac{\sigma I}{M} = \frac{\left(\frac{Sy}{n}\right)I}{M} = \frac{\left(\frac{230\ psi}{2.5}\right)0.026}{906\text{ lbf}\cdot\text{in.}} = 0.25\text{ in.}$$

Based on the required torque calculated, a 508-0500 series solid shaft universal joint was chosen from a catalog. It is made out of steel, and has end-diameters of 1.00 in and length and width of 3.75 in. This is the joint to be created.

15.6.4 Solid Modeling

Now that the universal joint system has been designed and the dimensions have been chosen, computer-aided modeling can be used to bring the system to life. SolidWorks 2010 was used to model each component and then assemble them into one working assembly.

Each driveshaft was modeled as a hollow tube 24 in. long and 3.75 in. diameter, with a thickness of 1/4 in. The ears and the shaft are one solid piece, with holes to fit the U-joint.

The model was begun by creating a 24 in long solid cylinder. The cylinder was hollowed out to give walls of 1/4 in. The ears were sketched in side profile on the y–z plane of the cylinder. The shape of the ears was cut out, and the corners were rounded using the fillet command. The 1 in hole was drawn and cut out in the same manner (Figure 15.79).

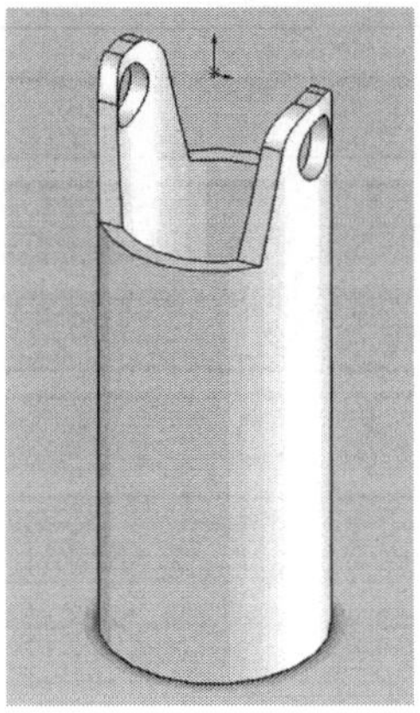

Figure 15.79

Modeled drive shaft.

The universal joint itself was made up of three main components: the cast body, the axle pins, and the rotating bushing. The cast body is shaped like an "X" and has holes for each pin to slide into. The axle pins are 1/2 in in diameter and 1.5 in long. One pin slides into each of the four holes in the cast body. Lastly, the bushings are made of steel and fit over each pin like a cap. These are 1 in in diameter, and have a 1/2 in hole to slide over the pins. These bushings are designed to spin on the axle pin, allowing the universal joint to function.

To begin modeling the U-joint, an X shape was created using two 1 in diameter, 3.75 in long cylinders. They were both centered at the origin, one running along the x-axis and the other running along the z-axis, creating the cross shape. A 1/2 in hole was drawn on each of the four faces and cut out for the pins to be mounted into.

The pins were simply 1/2 in diameter, 2 in long cylinders that would fit perfectly into the holes of the U-joint. The bushings were also simply created, starting with a 1.5 in wide and 1 in tall cylinder. One of the edges has a 0.10 in chamfer. A 1/2 hold is drawn and cut 0.75 in deep on the chamfered end of the bushing, where it will slide onto the pin (Figure 15.80).

15.6.5 Tolerance Analysis

While designing and modeling each component of the universal joint system, tolerances had to be established for some of the parts. A press fit is required in two areas of the design: where the axle pins mount in the cast body, and where the universal joint bushing mounts in the driveshaft ears. In designing for these fits, dimensions must have some intentional interference. The diameter of the pins must be the same or a couple of thousandths of an inch larger in order to achieve a press fit. The same tolerances apply to the

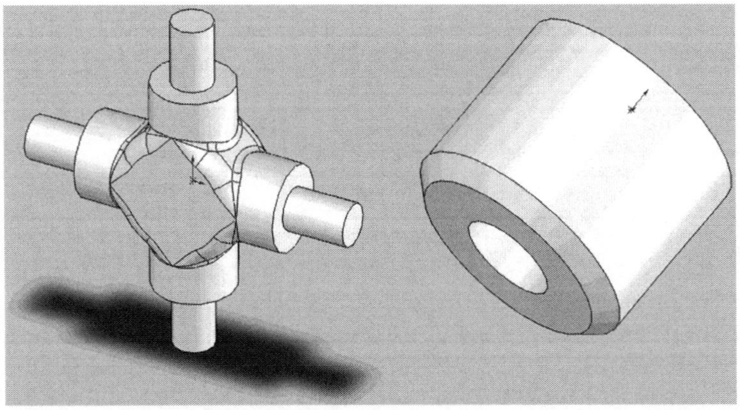

Figure 15.80

Modeled universal joint and bushing.

bushing-in-ear press fit. For each of these fits, there is a manufacturing dimensional tolerance of +0.003 in on the shafts and −0.003 in on the holes.

15.6.6 Manufacturing Process

In order to manufacture the U-joint body, the piece is cast out of steel. A prototype must first be made, and then the die is created. 4140 Steel is melted in an arc furnace and liquefied for pouring. The molten steel is poured into the cast, and the piece is allowed to cool. After it is removed from the cast, any sharp edges are filed off and the part is ready for assembly.

The bushings are machined from a 1 in diameter solid 4340 steel shaft. Each section is cut to length. A metal lathe is used to chamfer each edge. Lastly, the bushing is hollowed by drilling a 1/2 in hole inside, for mounting on the pin.

The driveshafts begin as a piece of 4140 steel tube of outside diameter 3.75 in and thickness 1/4 in. The tube is cut to length, and then the ears are machined using a plasma cutting tool. The 1 in holes are drilled using a machine shop drill press.

For assembly, a press fit is used to seat the pins inside the case body. The pieces are pressed together using a hydraulic shop press. The universal joint is placed between the driveshafts, and the bushings are inserted and held in place with a C-clip.

15.6.7 Cost Estimation

As displayed in SolidWorks, the volume of the universal joint cast body is 3.48 in^3. With a density of 0.284 lbm/in^3 and a cost of \$0.79 per lb of 4140 steel, the material cost of each cast body is $C_{Material} = V\rho C = (3.48\ in^3)(0.284\ lbm/\ in^3)(\$0.79) = \$0.78$ per unit. The manufacturing cost to cast 1000 of these units is \$700. The total cost for a cast body is \$700/1000 + \$0.78 = \$1.48.

The volume of the universal joint bushing is 0.78 in^3. With a density of 0.284 lbm/in^3 and a cost of \$1.11 per lb of 4340 steel, the material cost of each cast body is $C_{Material} = V\rho C = (0.78\ in^3)(0.284\ lbm/in^3)(\$1.11) = \$0.25$ per unit. The manufacturing cost to cast 1000 of these units is \$650. The total cost for a cast body is \$650/1000 + \$0.25 = \$0.16.

The volume of each driveshaft is 65 in^3. With a density of 0.284 $lbm/\ in^3$ and a cost of \$0.79 per lb of 4140 steel, the material cost of each cast body is $C_{Material} = V\rho C = (65\ in^3)(0.284\ lbm/\ in^3)(\$0.79) = \$14.50$ per unit. The manufacturing cost to cast 1000 of these units is \$1200. The total cost for a cast body is \$1200/1000 + \$14.50 = \$15.70.

Total cost of the universal joint is \$1.48 + 4(\$0.16) = \$2.12. This is an excellent overhead cost. A universal joint is typically sold retail for \$9.99. Nearly 80% profit will be made from this component.

The automotive driveshaft system being designed requires one universal joint and two driveshafts. Two driveshafts at \$15.70 each and one universal joint at \$2.12 gives a total cost of \$33.52 for the completed system.

15.6.8 Codes and Standards

A factor of safety of 2.5 was designed into every component of the driveshaft and universal joint system. Automobiles are often pushed to (and frequently even past) their limits. Automotive drive trains must be robust enough to handle the anticipated abuse that they will experience. A factor of safety of 2.5 should be more than enough to safely handle the anticipated loads.

15.6.9 Assembly

A final assembly of components was created in SolidWorks in order to display the completed operating system. Each component is inserted, and the system is constructed into one unit and animated. The animation allows the system function to be seen and analyzed for any unexpected interferences. Creating a modeled assembly is useful, because it can be used to catch errors or interferences in the design process, rather than during prototyping. This certainly saves engineers a lot of time and money.

For purposes of animation, a handle was added to one of the driveshafts in order to physically move the system and display the motion. This handle is not part of the real system.

The assembly was physically created by first importing each component into an assembly file. The first step was to insert each axle pin into one of the four holes on the U-joint body using the "concentric-mate" command. Similarly, each bushing was mounted on an axle pin, this time by mating the exterior face of each pin to the interior face of the hollow bushing.

Each driveshaft ear-hole was first concentrically mated to the bushings. Then, the face of the driveshaft had to have a tangent-mate to the face of the bushing.

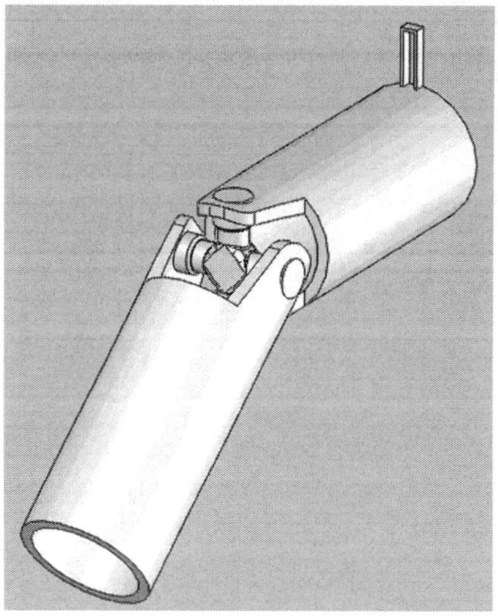

Figure 15.81

Universal joint assembly.

At this point, the system was assembled, but it could not yet be animated. In order to animate the motion of the joint, one of the parts needs to be fixed in space for a motion reference. In this case, it was necessary for all the parts to be moving. The solution to this problem was to create two shafts, mate them concentrically into each driveshaft, and fix them both in space. This was successful in allowing the desired parts to spin freely. These two shafts were hidden using an invisible command in the assembly toolbar (Figure 15.81).

15.6.10 FEA

One of the most useful tools that computer-aided drafting provides to an engineer is the ability to analyze the stresses present in a component or system. This way, the system can be designed and even tested before any prototyping has to take place, saving a lot of experimental time and money down the road.

An FEA was done on the assembled universal joint in NX-IDEAS. The calculated torsion stresses of 4726 in·lb were placed at the end of each arm and then animated in a color code that represented areas of high and low stresses. The program can even simulate deformation of the component, though it is exaggerated for purposes of animation. Figure 15.82 shows the results of the FEA on the partially assembled universal joint. The stress calculated from the FEA matched the stress calculated in Section 15.6.2 at around 400 kpsi.

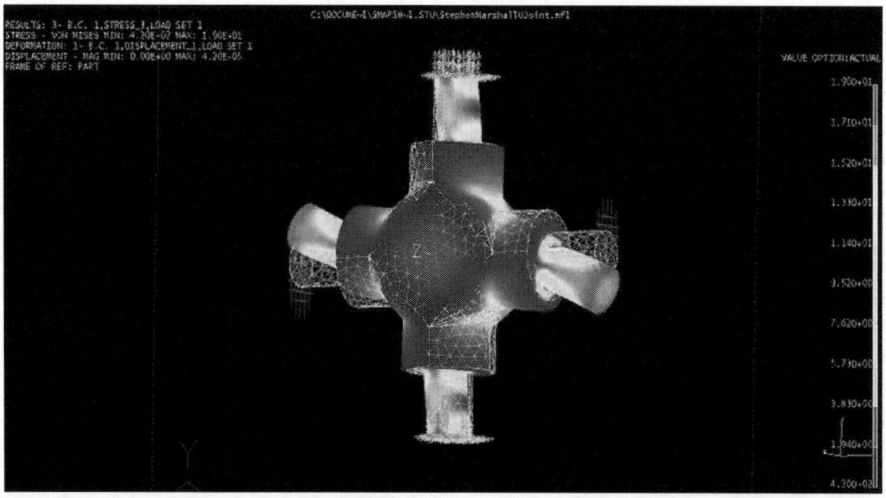

Figure 15.82

Finite element analysis of partially assembled universal joint.

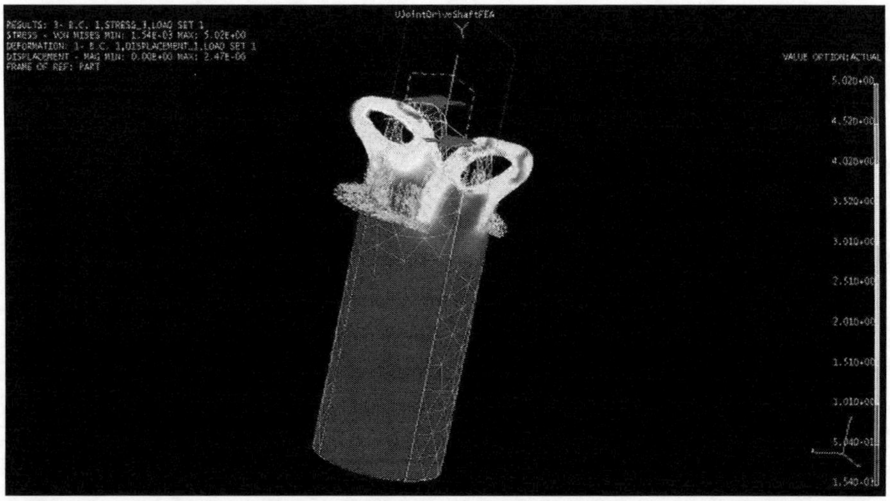

Figure 15.83

Finite element analysis of drive shaft.

A similar analysis was done on the driveshaft, placing the same 4726 in·lb of torque on the ears of the shaft. The FEA in Figure 15.83 showed stresses of about 200 kpsi, which closely matched the calculated stress on the shaft in the analysis section of this report.

15.7 CAD Example 7: Piston and Connecting Rod

Computer-aided design of piston and connecting rod for a four-cylinder, 2.0 L, four-stroke engine operating at 6000 rpm. The compression ratio is 10:1. The engine efficiency is 50%, and the combustion efficiency is 85%. The temperature inside the cylinder at the intake is 60°C. There is also 5% residual exhaust left over between the combustion and exhaust cycles.

The piston–connecting rod subsystem of an internal combustion engine is where the actual combustion occurs and where the power is developed. The piston head lies within the cylinder and is sealed by the piston rings. At the top of the cylinder, there are two valves. The intake valve injects a mixture of fuel and air, which is ignited by the spark plug. The combustion pushes the cylinder and connecting rod down. As the crankshaft spins, the counterweight carries the piston upward, which pushes all the combustion gas out of the exhaust valve.

The flow of forces in this system begins at the source of combustion. The expanding gases from the combustion process create a pressure, which transmits the force to the cylinder head. The cylinder head experiences a large stress from the high pressure. As the cylinder head moves downward, a force is imparted to the connecting rod. The largest stress the connecting rod experiences will be during midcycle, when the connecting rod is farthest from the cylinder. All components are fast moving and operate many cycles during their lifetime. The most critical part of the assembly are the piston rings. These components experience high stress and can overheat. Special consideration must be given to choosing a material that will best suit these extreme conditions.

The manufacturing of these components can be accomplished through CNC milling and turning. The cylinder head can be turned from a solid cylinder, with cuts being made on the underside of the shape. Since the connecting rod contains more detail, it will require more cuts. It can be milled from a solid piece of raw material with little waste.

Analysis and Design:

The calculations for the piston head are as follows.

We start by analyzing the piston cylinder assembly as a thermodynamic system. First, the displacement volume was found for one cylinder

$$V_d = \frac{V}{\text{cylinders}} = 0.5\ L$$

$$V_d = \frac{V_d}{1000} = 5\times10^{-4}\ \text{L}$$

where:

cylinders is the number of cylinders

V is the combined volume of all cylinders

V_d is the displacement volume

Next, the clearance volume was found for one cylinder.

$$V_c := \frac{V_d}{CR-1} = 0.056\ \text{L}$$

$$V_c = \frac{V_c}{1000} = 5.556\times10^{-5}\ \text{L}$$

where:

CR is the compression ratio

V_c is the clearance volume

Next, the bore, stroke, and rod length were determined. The bore is the diameter of the cylinder, the stroke is the length of the reciprocating motion, and the rod length determines how long the connecting rod should be.

$$B = \left(\frac{V_{dm}}{\frac{\pi}{4}\cdot 1.025}\right)^{\frac{1}{3}} \cdot 10$$

$$B = 0.085\ \text{m}$$

In this equation, B is the bore and V_{dm} is the displacement volume in milliliters. The stroke S is represented as

$$S = 1.025B$$

$$S = 0.087\ \text{m}$$

while the rod length is determined to be

$$RL = 2S$$

$$RL = 0.175\ \text{m}$$

Now, each state will be analyzed using thermodynamics to find mass, temperature, and pressures. The intake, State 1, has temperature T and pressure P_1, while the gas constant is R.

$$T_1 = 333K$$

$$P_1 = 101.10^3\ \text{Pa}$$

$$R = 0.287$$

$$V_1 = V_{dm} + V_{cm}$$

$$V_1 = 5.556 \times 10^{-4}\ \text{L}$$

V_1 is the total volume with the boundary condition. The mass of the fuel mixture is found using the ideal gas equation, solving for m_m, mass of the mixture m_m

$$m_m = \frac{P_1 \cdot V_1}{R \cdot T_1}$$

Mass of mixture $= 5.871 \times 10^{-4}$ kg.
The second state (State 2) is when the compression of the fuel begins.
Pressure and temperature were found, knowing that the process is isentropic.

$$k = 1.35$$

$$P_2 = \frac{P_1 \cdot (CR)^k}{1000} = 2.261 \times 10^3\ \text{Pa}$$

$$T_2 = T_1 \cdot (CR)^{k-1} = 745.494\ \text{K}$$

$$V_2 = \frac{m_m \cdot R \cdot T_2}{P_2} = 0.056\ \text{L}$$

where:

k is the isentropic constant
P_2 is the pressure at state 2
T_2 is the temperature at state 2
V_2 is the total volume displaced during the process

Now, the mass of the air fuel and exhaust are calculated.

$$m_a = \frac{15}{16} S$$

$$m_f = \frac{1}{16} S$$

$$m_e = 0.05 \cdot 5.871 \times 10^{-4}\ \text{kg} = 2.936 \times 10^{-5}\ \text{kg}$$

where:

m_a is the air mass
m_f is the fuel mass
m_e is the mass of the exhaust fumes

Now the total mass m_t that passes the boundaries can be found:

$$m_t = m_a + m_f + m_e$$

The third state is the combustion stage. The fuel–air mixture is ignited to provide power into the system. The heat added during a cycle must now be calculated to find T_3, which will allow us to find P_3.

$$\text{CV} = 0.812$$

$$Q_{hv} = 45{,}000$$

$$\mu_c = 0.75$$

$$T_3 := \frac{m_f \cdot Q_{hv} \cdot \mu_c}{m_m \cdot CV \cdot 100}$$

In this set of equations, CV is a constant, Q_{hv} is the heat added to the system, and μ_c is the chemical potential. The pressure can also be obtained.

$$P_3 = P_2 \cdot \left(\frac{T_3}{T_2}\right)$$

The fourth and final stage is exhaust. During this process, all the combustion gas is forced out of the piston. Using the isentropic relation, the temperature, pressure, and volume of state 4 can be obtained:

$$T_4 = T_3 \cdot \left(\frac{1}{CR}\right)^{k-1}$$

$$P_4 = P_3 \cdot \left(\frac{1}{CR}\right)^{k}$$

$$V_4 = \frac{m_m \cdot R \cdot T_4}{P_4} = 0.556\text{ L}$$

Next, the process is analyzed to find the work during each stage, in order to find net work and engine horsepower. First, work from Stage 3 to Stage 4:

$$W_{34} = \frac{m_m \cdot R \cdot (T_4 - T_3)}{1 - k} = 1.019\text{m}$$

Then, work from Stage 1 to Stage 2:

$$W_{12} = \frac{m_m \cdot R \cdot (T_2 - T_1)}{1 - k}$$

Net work produced

$$W_{\text{net}} = W_{12} + W_{34}$$

From this analysis, the thermal efficiency, total horsepower, and mean piston speed can be calculated.

$$Q_{\text{in}} = m_f \cdot Q_{hv} \cdot \mu_c$$

$$Q_{in} = 184.6 \text{ kJ}$$

Thermal efficiency

$$\mu_t = \frac{W_{\text{net}}}{Q_{in}}$$

Mean effective pressure

$$\text{MEP} = \frac{W_{\text{net}}}{(V_1 - V_2)}$$

Indicated power at 5000 rpm:
Indicated horsepower calculation:

$$N = \frac{6000}{60}$$

$$n = 2 \text{ (factor of safety, assumed)}$$

$$W_{\text{dot}} = \frac{W_{\text{net}} \cdot N}{n} \cdot \text{cylinders}$$

$$HP = W_{\text{dot}} \cdot 1.34$$

$$HP = 220 \text{ hp (indicated HP)}$$

Mean piston speed

$$U_p = 2 \cdot S \cdot N$$

Brake horsepower at 6000 rpm, as well as the total horsepower produced by the system, is found as follows. The torque of the engine is also calculated.

Brake horsepower calculation:

$$\mu_m = 0.5$$

$$W_{\text{dot}} = \mu_{\text{m}} \cdot W_{\text{dot}}$$

$$BHP = \mu_m \cdot W_{\text{dot}} \cdot 1.34$$

$$BHP = 115 \text{ hp}$$

Friction horsepower

$$FHP = HP - BHP$$

Torque at 6000 rpm

$$\tau = \frac{W_{dot}}{2 \cdot \pi \cdot N}$$

$$\tau = 177.2 \text{ ft.lb}$$

Now, the piston can be sized using the relations that were just calculated. The first dimension calculated is the total piston area

$$A_{\text{piston}} = \frac{\text{displacement}}{\text{stroke}} = 5.747 \times 10^{-3}\ \text{m}^2$$

Area of each cylinder

$$A_{\text{cylinder}} = \frac{A_{\text{piston}}}{4} = 1.437 \times 10^{-3}\ \text{m}^2$$

Diameter of piston head

$$d_{\text{cylinder}} = \text{bore}$$

$$d_{\text{piston}} = d_{\text{cylinder}} \cdot \left(\frac{624}{625}\right) = 0.085\,\text{m}$$

The material characteristics of the piston were considered, and an aluminum alloy was chosen. For T4 aluminum alloy 2024

$$S_y = 296 \times 10^3\ \text{Pa}$$

$$S_u = 446 \times 10^3\ \text{Pa}$$

$$v = 0.35$$

$$c = 0.25$$

Piston head thickness

$$t_h = \left(\frac{3 \cdot P_2 \cdot d_{\text{piston}}}{16 \cdot v \cdot S_y}\right)^{1/2}$$

where:

S_y is the yield strength
ν is Poisson's ratio

The maximum stress on the piston head is

$$\sigma_{\max} = \frac{c \cdot MEP \cdot d_{\text{piston}^2}}{t_h^2}$$

The piston ring needs to be analyzed and its dimensions determined.
The face width is

$$w = \frac{1}{50} \cdot \text{bore} = 1.7 \times 10^{-3}\,\text{m}$$

Face height

$$h = w$$

Clearance

$$c = d_{\text{cylinder}} - d_{\text{piston}}$$

Now, the piston wrist pin is to be analyzed (Figure 15.84)

$$W = A_{\text{cylinder}} \cdot \sigma_{\max}$$

$$W = 106.3$$

$$A = \frac{\pi(0.0249 - 0.0196)^2}{4}$$

$$\sigma = \frac{R_a}{A} = 4.818 \times 10^8$$

Calculations to find critical load of connecting rod:

$$C = 1$$

$$b = 0.02 \text{ m}$$

$$E = 69 \times 10^9 \text{Pa}$$

$$h = 0.01 \text{ m}$$

$$P_{cr} = 37 \text{ kN}$$

$$l = 0.175 \text{ m}$$

$$I = \frac{b.h^3}{12} = 1.667 \times 10^{-9} \text{m}^4$$

$$P_{cr} = \frac{C \cdot \pi^2 \cdot E \cdot I}{L^2}$$

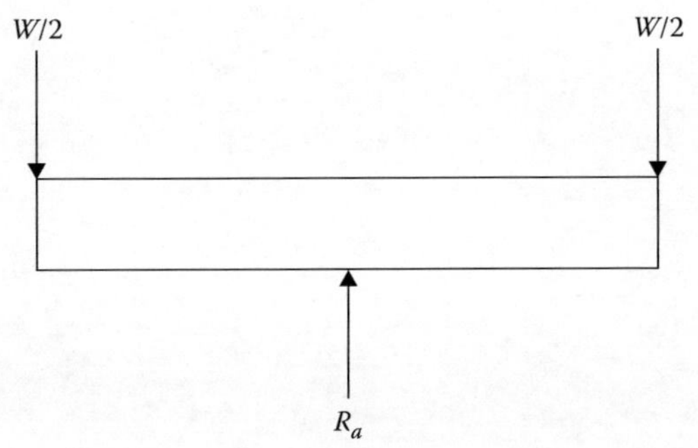

Figure 15.84

Free body diagram.

Figure 15.85

Cylinder head.

15.7.1 Solid Modeling

The first component designed was the cylinder head. It is a basic cylinder with grooves cut for the piston rings. The underside is hollowed out to allow the connecting rod to swing. The pin is also inserted (Figures 15.85 and 15.86).

This part was developed in SolidWorks. The process is rather simple. I created a cross section of the cylinder and revolved it around a center axis to create the main shape of the part. Next, the hole where the wrist pin is located was sketched and cut. Then, the piston ring grooves were cut using the "cut revolve" feature.

After the head is created, the piston needs piston rings to fill the grooves. These are easy to make, because the cross section and piston diameter are known. The cross section is sketched and revolved around the axis located at the same distance as the radius, and thus the ring is made (Figures 15.87 and 15.88).

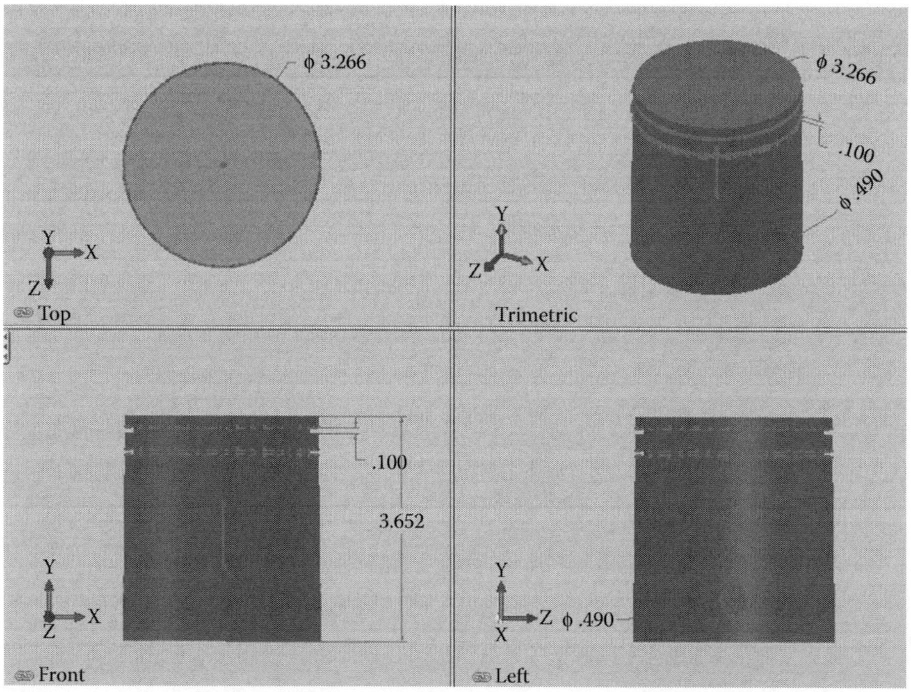

Figure 15.86

Dimensioned views of cylinder.

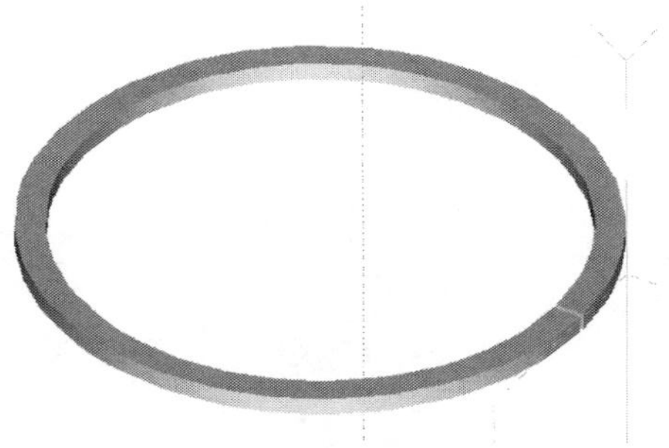

Figure 15.87

Piston ring.

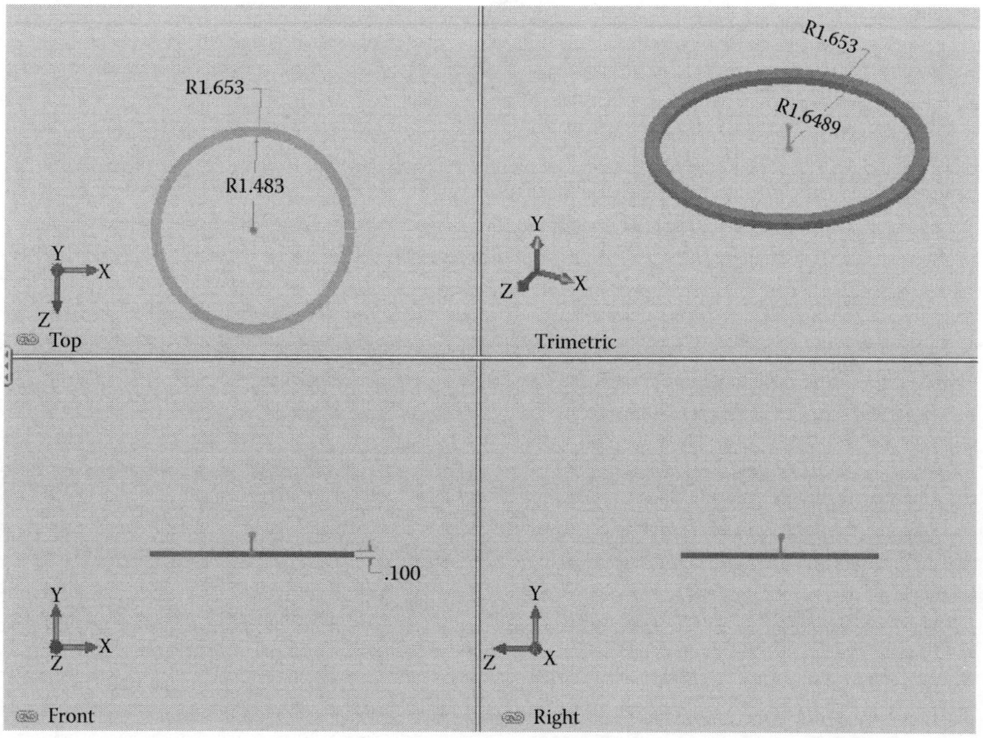

Figure 15.88

Dimensioned view of piston ring.

The next part to be generated was the connecting rod. The connecting rod was a little more complicated to make, but can be easily milled with blocks of a set size. This is because the part was made by first cutting a rectangular prism to the general shape. The piece was refined and smoothed until the final result was created.

Forming the general shape included sketching the rounded perimeter for the two holes on either end. The sketch was extruded as a cut to remove excess material from the rectangular prism. Since the shape was still solid, the wrist pin hole was cut next. The hole for the crankshaft was a little more complicated. A chamfer was used to cut a flat surface at a 35° angle. Then, a cylinder with the diameter of the crankshaft was used to cut the shape. The sides had a groove for removing excess material. This was done by selecting the face and sketching the groove shape into the plane. The groove was cut using the "extrude" feature. This leaves sharp edges in the groove; therefore, a fillet was used to smooth the edges of the interior groove (Figures 15.89 and 15.90).

The rod would not be effective if it could not hold on to the crankshaft, so the rod cap was made to be screwed to the connecting rod so that it would fit securely on the crankshaft. This was made in a similar fashion

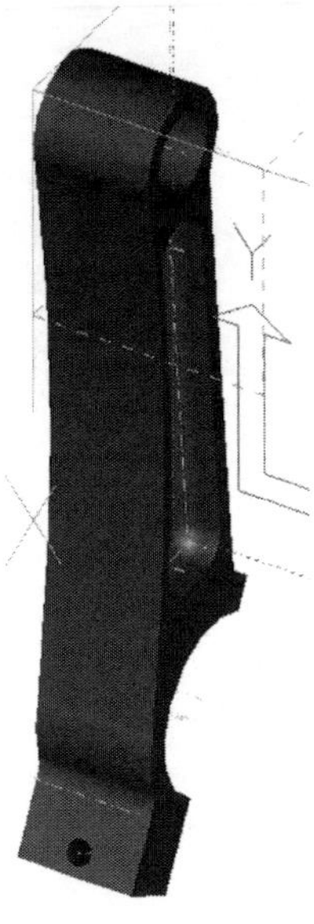

Figure 15.89

Connecting rod.

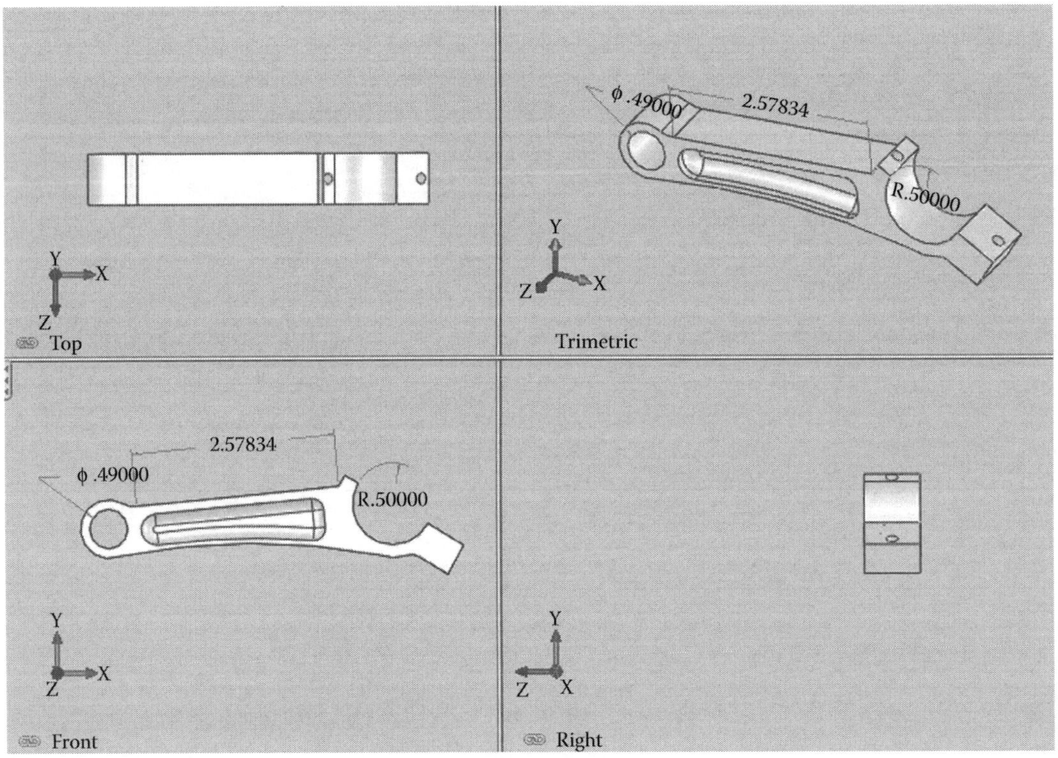

Figure 15.90

Dimensioned views of connecting rod.

to the connecting rod. Starting from a solid block, a sketch was made that matched the shape of the crankshaft end of the connecting rod. The sketch was extruded, and the part was made (Figures 15.91 and 15.92).

15.7.2 Tolerance Analysis

For the components designed, the tolerances must be precise, as an engine is a critical system. For example, the piston head must fit into the cylinder such that the fit is tight and secure. The tolerance specified by the

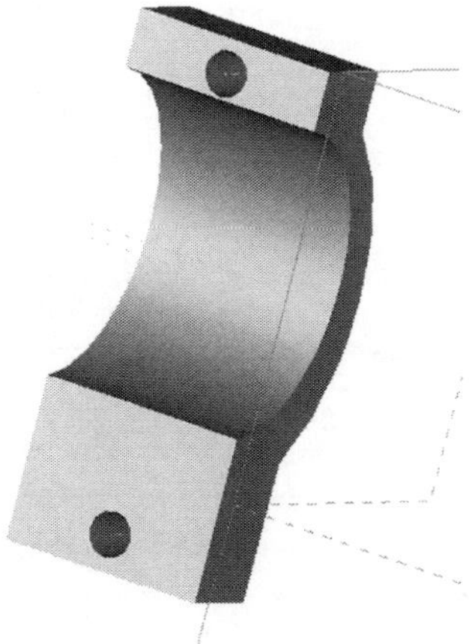

Figure 15.91

Connecting rod cap.

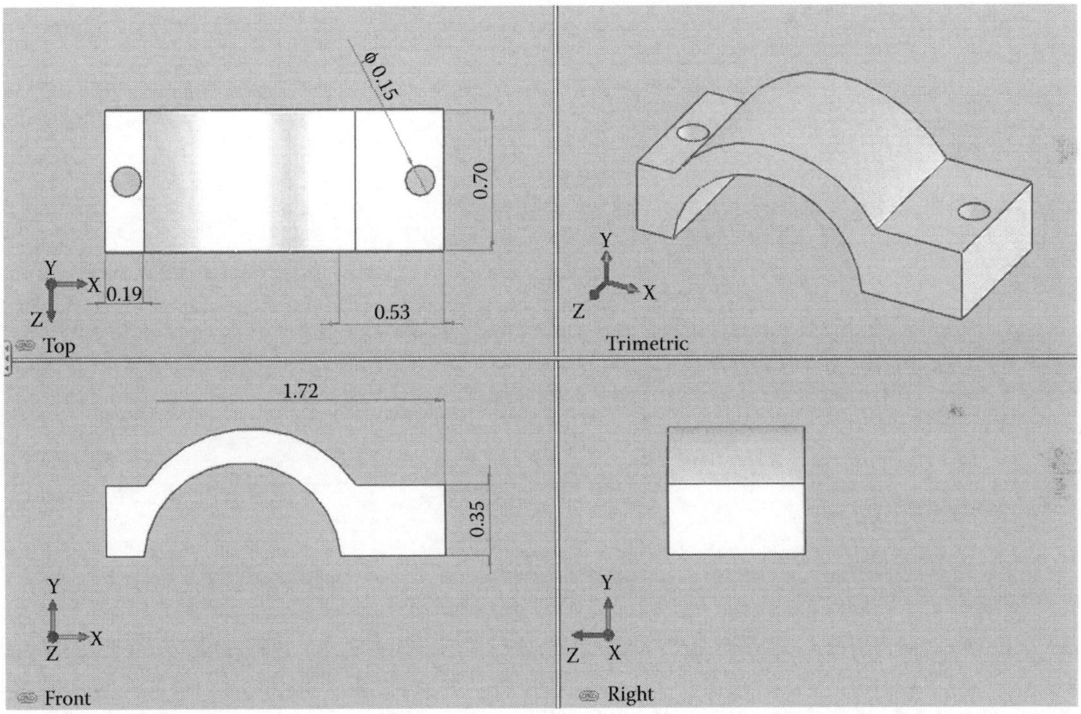

Figure 15.92

Dimensioned view of rod cap.

SAE ranges from ±0.01 to as low as ±0.0001 in. The tolerance cannot be too small, or the components will not move, which will generate massive amounts of heat and wear on the components.

A tolerance analysis was done using NX 9.0. This program performs a tolerance study for any dimension of the solid model. Table 15.23 shows the calculated tolerance for each component.

15.7.3 Manufacturing Process

The manufacturing process to develop the components could be done using the CNC system. The piston head could be machined from a solid cylinder of material using a lathe operation. Material would have to be removed from the interior using a rough bit, then a finishing bit to keep the material smooth. The piston ring grooves can be cut in a similar fashion to screw threads, but not angled. The piston rings themselves could be made by drawing raw material through a die and creating a long spool that can be

Table 15.23 Tolerance for Each Component

Piston ring	±0.01 in
Cylinder head	±0.001 in
Connecting rod	±0.01 in
Rod cap	±0.01 in

cut to lengths equal to the required circumference. The sections would then be shaped into the circular form required.

The connecting rod is more intricate and has more detail than the piston head. It would be manufactured by a CNC milling operation. The manufacturing would start with a block of raw material. All the curves and the general shape of the rod would be cut first. The grooves on the sides would then be cut, and the holes for the screws and for the wrist pin could be drilled.

15.7.4 Cost Estimation

To begin the cost analysis, the volume of each component was multiplied by the density of the material to find the weight.

$$W = V \cdot \rho$$

where:

- W is the weight
- V is the volume
- ρ is the density

The cost of raw steel is roughly $0.22 per pound.

$$C = S_{steel}\, W$$

The total weight for a four-cylinder engine is about 177 lb. Therefore, the cost of material is roughly $38.93. The cost of manufacturing and overhead is estimated as 150% of the material cost, while the cost of forging is estimated at 125% of the material cost.

$$C_T = C + 1.5C + 1.25C$$

The total cost of manufacturing, therefore, is almost $194.5 per unit.

15.7.5 Codes and Standards

The organizations that have developed codes applicable to this project are the Society of Automotive Engineers (SAE) and the American Society for Testing and Materials (ASTM). The SAE mission statement is as follows: "SAE International is a global body of scientists, engineers, and practitioners that advances self-propelled vehicle and system knowledge in a neutral forum for the benefit of society." SAE provides standards for materials, testing, and safety for the automotive and aerospace industry.

ASTM International, formerly known as the American Society for Testing and Materials (ASTM), is a globally recognized leader in the development and delivery of international voluntary consensus standards. Today, some 12,000 ASTM standards are used around the world to improve product quality, enhance safety, facilitate market access and trade, and build consumer confidence.

15.7.6 Assembly

The assembled system is an in-line four-cylinder engine. Figure 15.93 shows the pistons, connecting rod, and crankshaft. The two cylinders in the middle are in the bottom dead center position, while the two on the end are in the top center position.

15.7.7 FEA

FEA was performed for the connecting rod. The connecting rod is a slender member that experiences a compressive load; therefore, it is of interest to test the component for buckling. The geometry of the connecting

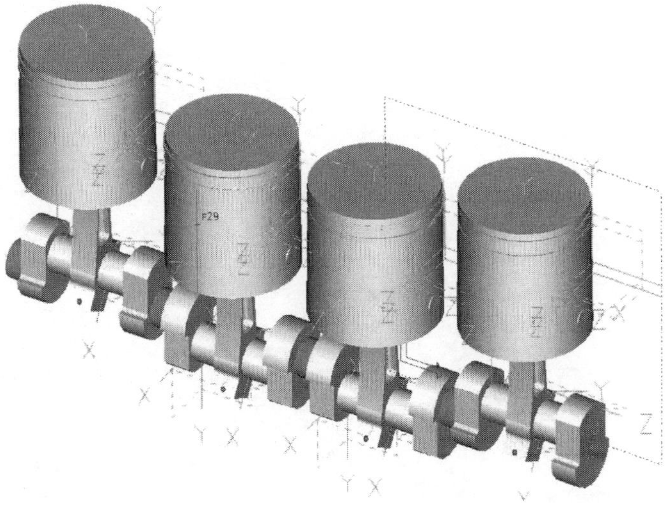

Figure 15.93

(See color insert) Assembled crankshaft.

rod makes it difficult to place a load on the pin end. To counter this problem, the pinned end was cut to form a flat surface. A load was placed on the top flat surface, and constraints were placed at the edges where the connecting rod links to the rod cap. Figure 15.94 shows that the rod will deflect but will not fail.

Higher stress is experienced along the short edge of the rod. The highest stress is produced at the top end, where the load is applied, as would be expected. The part would fail somewhere along the short end, but under the loading conditions of the engine, the part is safe.

The next component to be analyzed using 3-D FEA is the piston ring. The piston ring experiences a lot of wear around its outer edge. For the study, the ring was constrained on the inside, and the applied load was distributed along the outer face.

Figure 15.95 shows that the piston ring does not experience any high stresses. This is due to the material selection. The ring is made of a high-strength steel in order to withstand the constant abuse the ring faces.

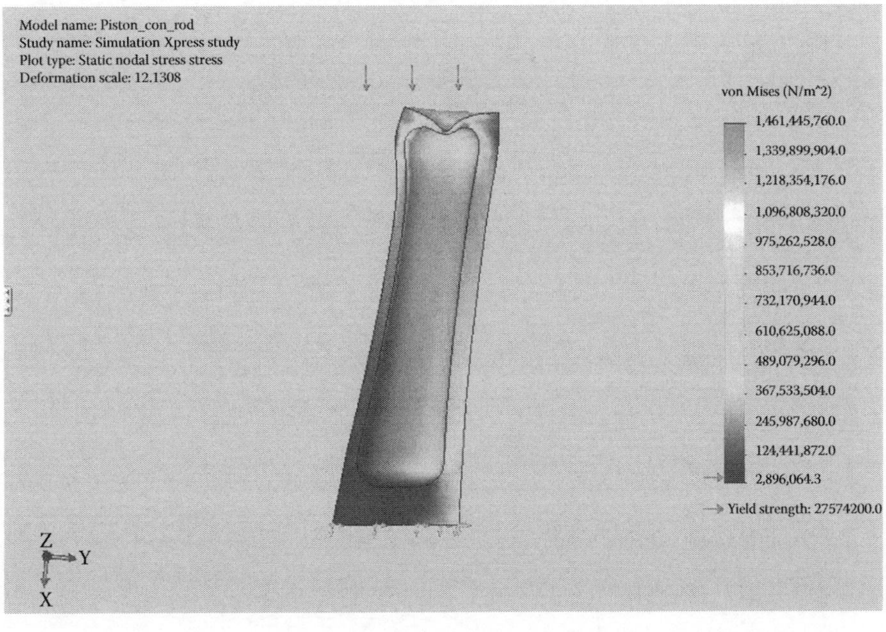

Figure 15.94

Connecting rod FEA.

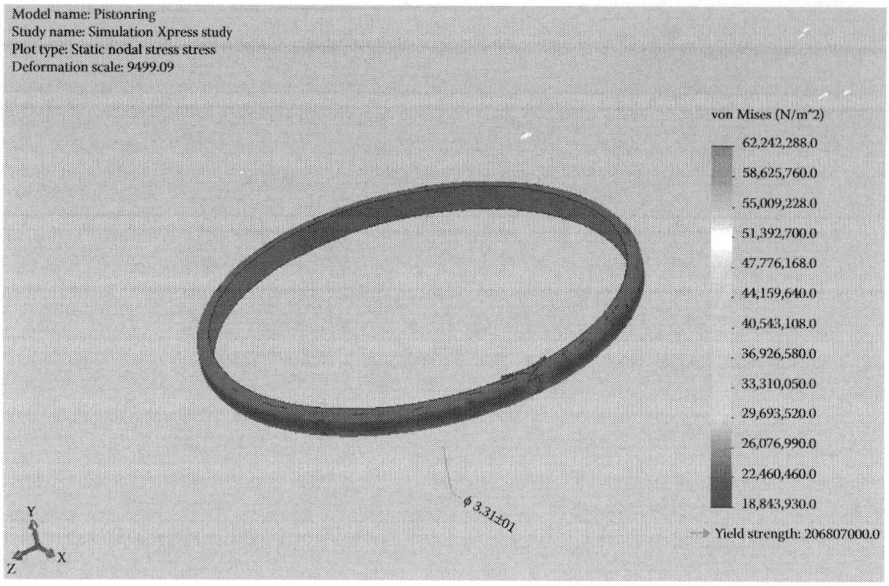

Figure 15.95

Piston ring FEA.

The last component to be analyzed using FEA was the piston head. This analysis was not done in 3-D. There were limitations with NX 9.0 that did not allow the part to be evaluated properly. Instead, a 2-D study was done on the top face of the part, because this is the point most likely to fail. The final result can be seen in Figure 15.96.

The part was constrained on the bottom face, while the top face was exposed to a distributed pressure load. The results show that the part experiences high stress near the middle. Under normal operating conditions, the piston head would survive repeated use.

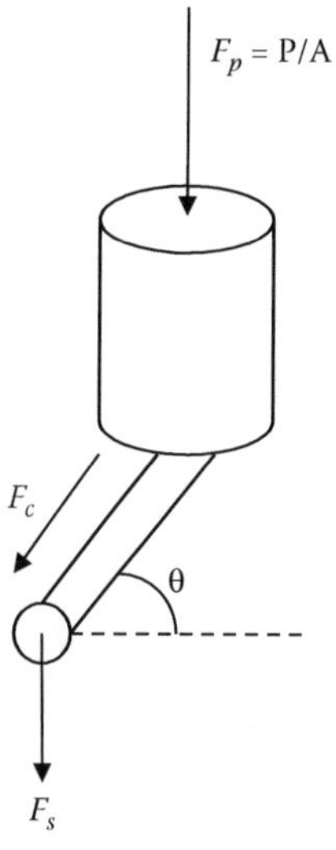

15.8 CAD Example 8: Universal Testing Machine

A universal testing press is designed with the aid of the computer software NX 9.0. Stresses and deformations are calculated and compared with the results of this software.

A solid model and assembly of a universal testing press was created using NX-9.0. After the solid model was prepared, high-stress concentrations in the model were analyzed in order to test for failure. The values were then checked using accepted engineering stress principles and calculations.

A universal testing press is used to test the compressive strength of materials. It is usually powered by a hydraulic motor, which, through the use of spur gears, applies a torque to two screw threads. The screw

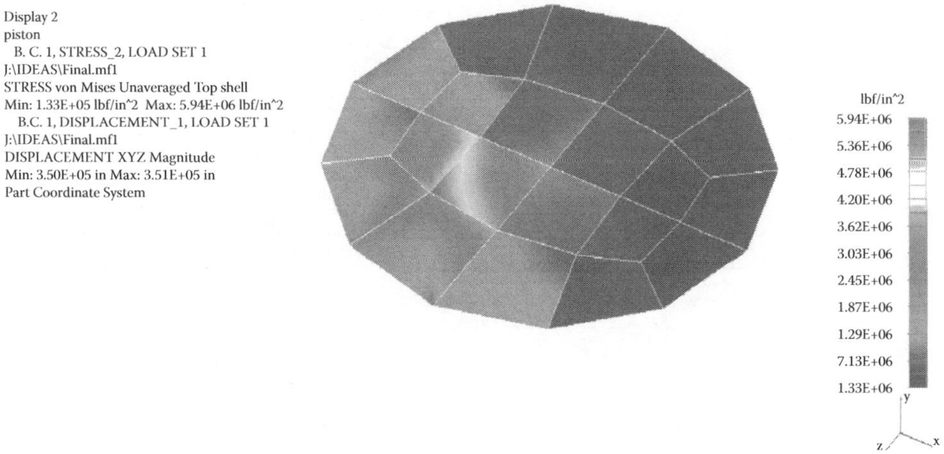

Figure 15.96

2D FEA of cylinder head.

threads rotate freely on thrust bearings, and the rotation of the screw threads serves as the driving force for the crosshead, which moves in the vertical direction. As the crosshead moves downward, it applies a compressive load to the object being tested.

The components of the universal testing press are as follows:

- One motor
- One 18-tooth spur gear
- Two 54-tooth spur gears
- Two screw threads
- Crosshead
- Thrust bearings
- Frame

15.8.1 Solid Modeling

A solid model was created using NX I-DEAS for each component of the universal testing press. The following is a breakdown of the individual components, showing how they appeared when designed in I-DEAS.

18-tooth spur gear:

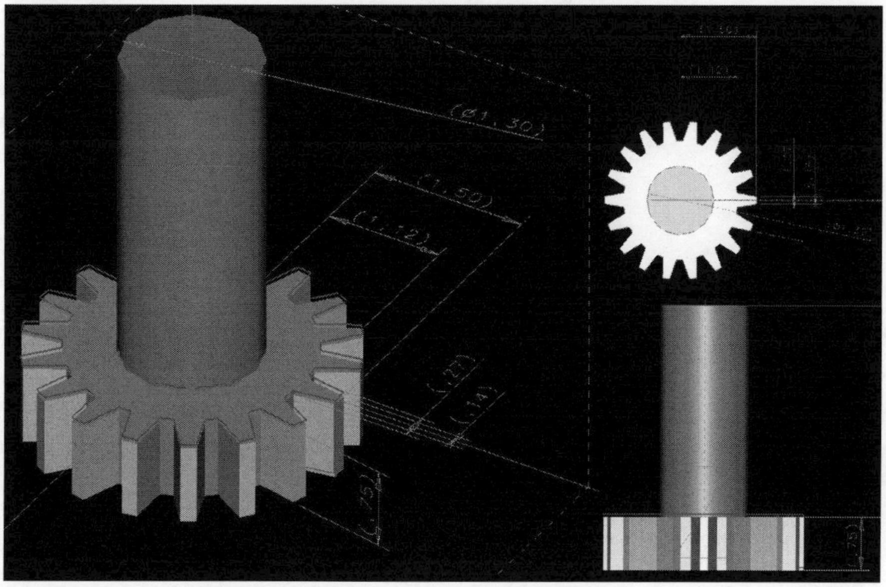

54-tooth spur gear:

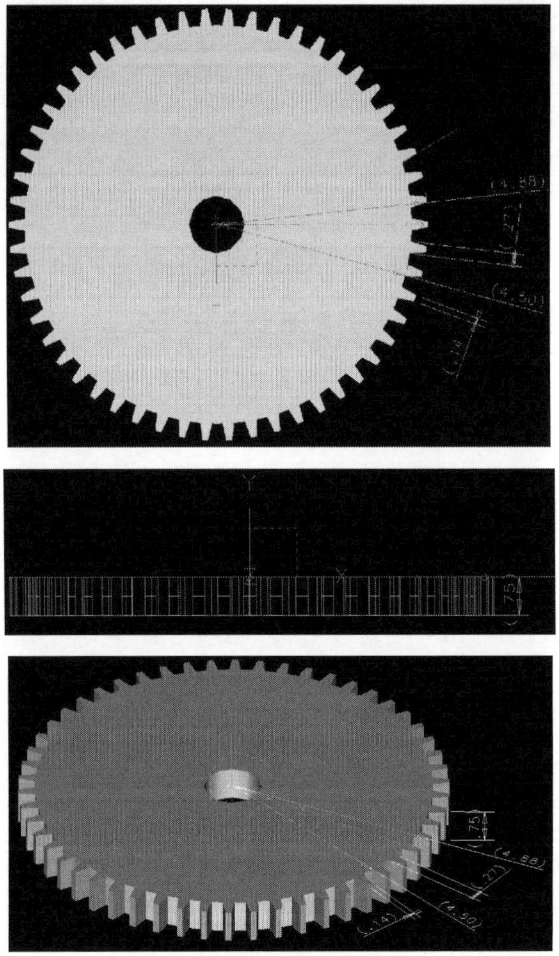

Screw thread:

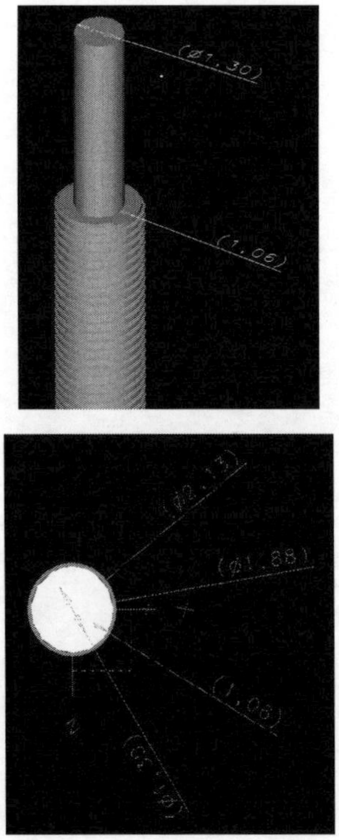

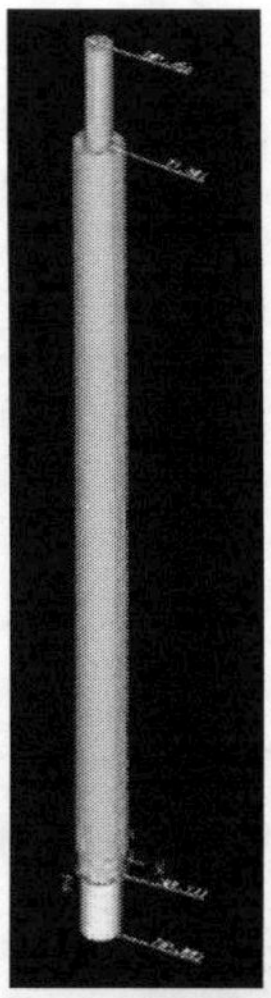

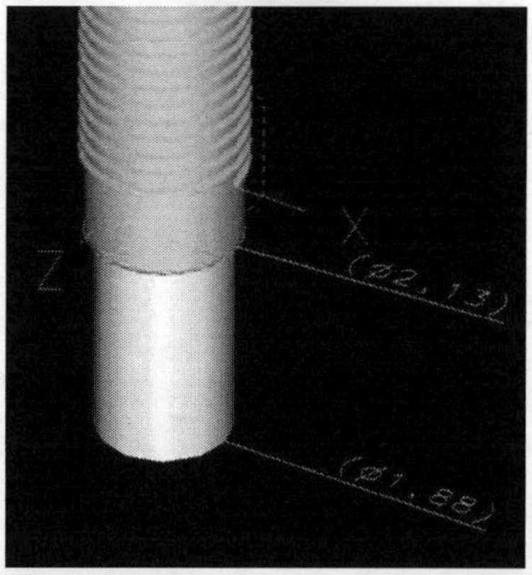

Crosshead:

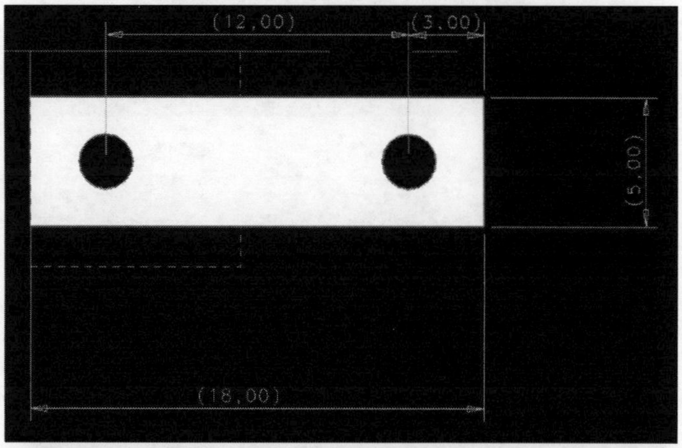

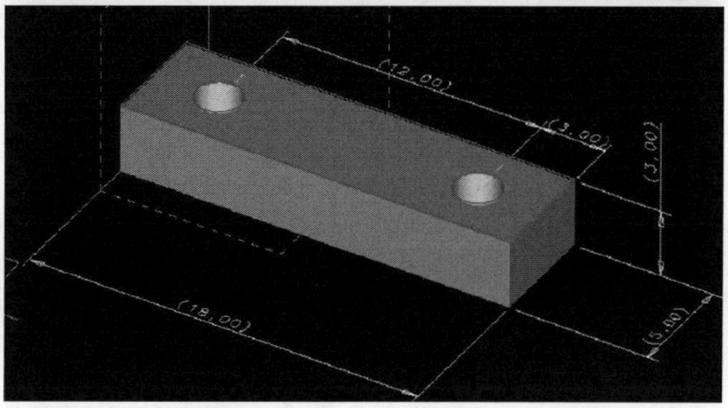

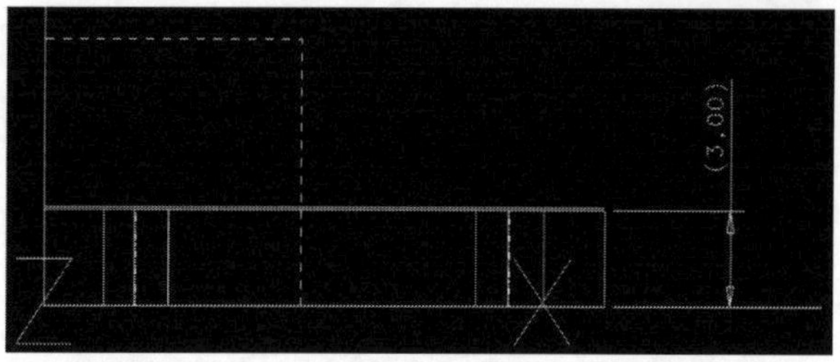
(3.00)

Frame:

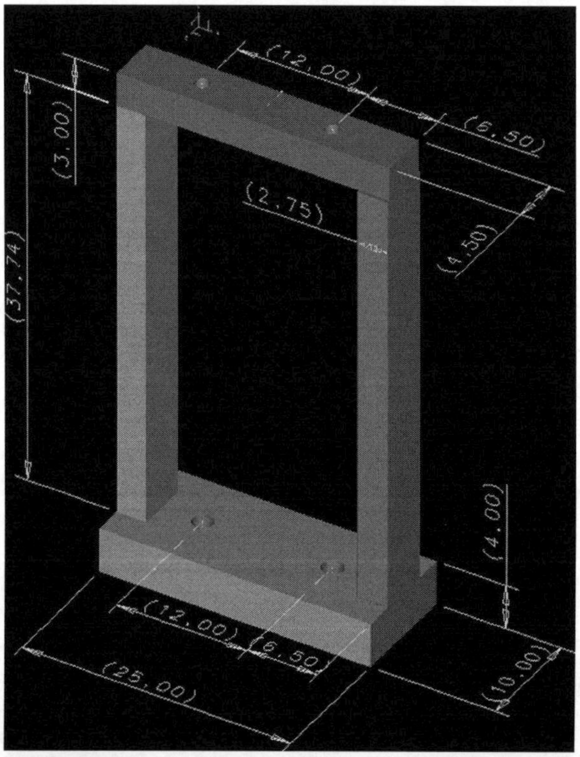
(12.00)
(6.50)
(3.00)
(2.75)
(4.50)
(37.74)
(4.00)
(12.00)
(6.50)
(25.00)
(10.00)

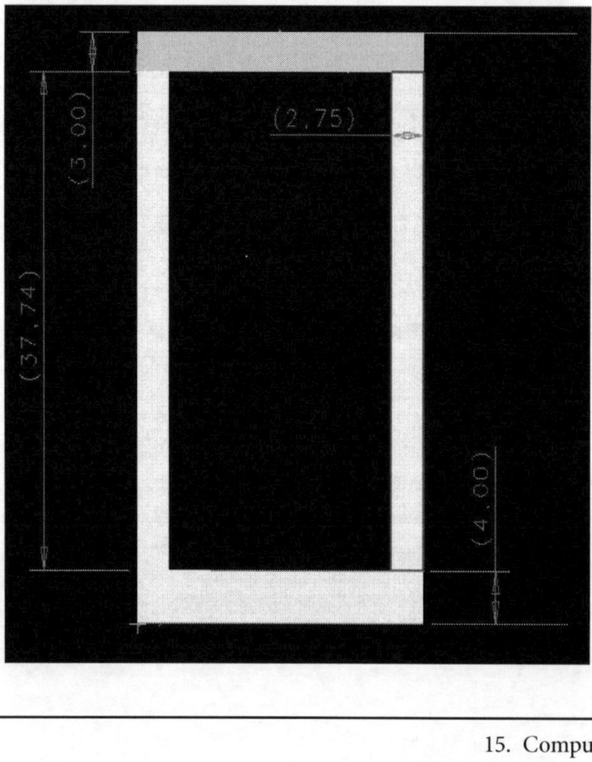
(3.00)
(2.75)
(37.74)
(4.00)

Thrust bearings:

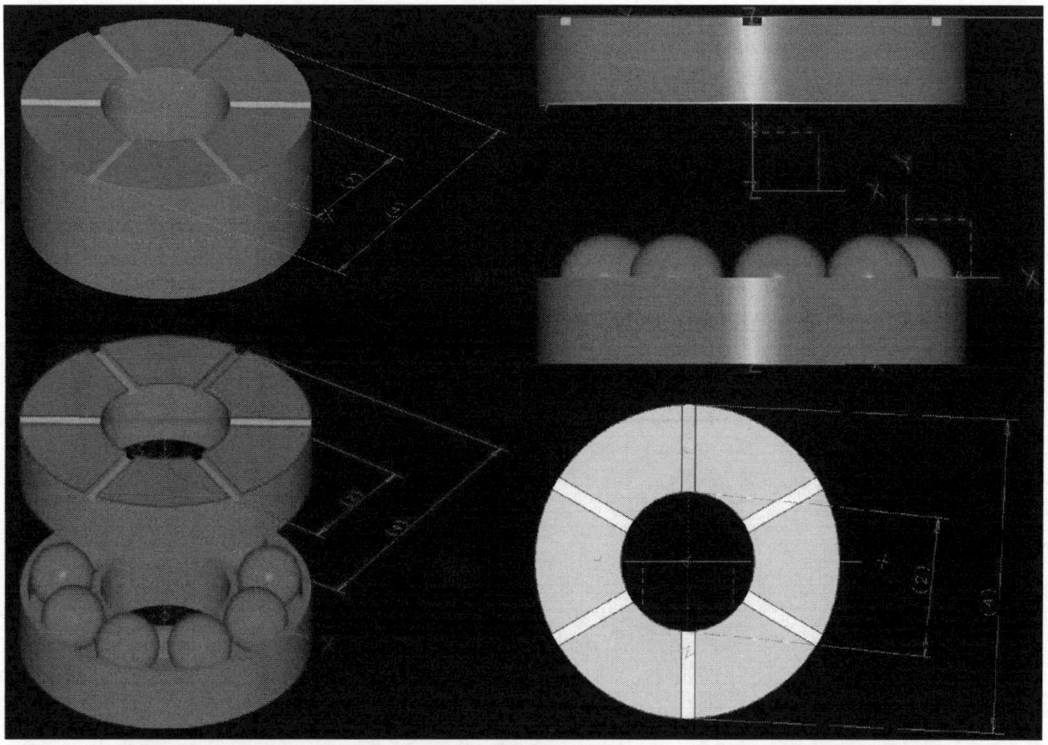

Overview of entire system:

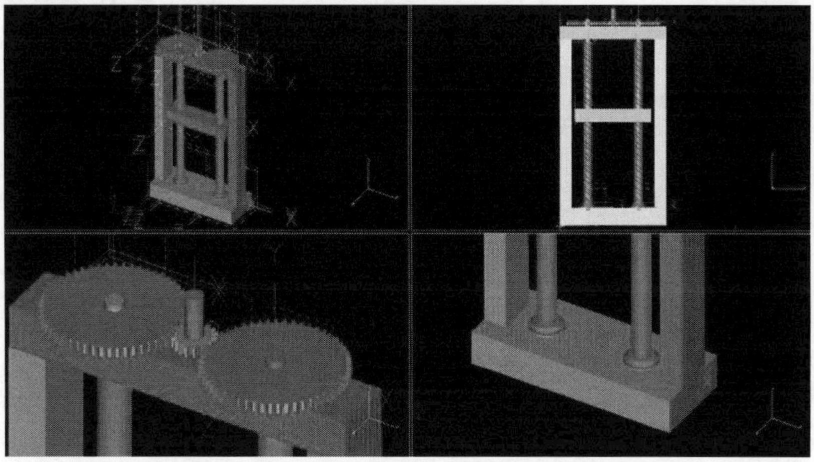

15.8.2 Manufacturing Process

15.8.2.1 Gears

The gears are made of high-grade steel containing a certain amount of carbon to make them even stronger. A band saw cuts off a piece of steel from a larger piece of steel that is just thick enough to make the gear. The blank is then mounted to a computer-controlled blade. As it spins, a turret carrying a series of tools moves in. One tool after another takes turns machining the blank into a specific shape. A shower of coolant prevents overheating when a hole is drilled into the center of the blank. Teeth on the inside of the gear are created by a cutter, whereas the teeth on the outside of the gear are created through the use of a gear hobber. The blank is constantly being rotated as the gear teeth are being cut, so as to prevent damage to the cutter blades.

Afterward, each gear is inspected by a probe that checks every gear tooth and ensures that the gear is of proper dimensions. Following the inspection, the gear is then heat treated in a furnace in order to strengthen the steel. When the furnace reaches a certain temperature, carbon is injected inside. The gears absorb this carbon, strengthening the gears further. The gears are then soaked in oil in order to harden the metal. After the heat treatment, the gears enter a computer-controlled grinder in order to correct any distortions created during the heat treatment.

15.8.2.2 Screw Threads

The screw threads are also made of high-grade steel, which contain a certain amount of carbon to make them even stronger. As for the gears, a band saw cuts off a piece of steel from a larger piece of steel that is just thick and long enough to make the screw thread. The blank is then mounted to a computer-controlled blade. As it spins, a turret carrying a series of tools moves in. One tool after another takes turns machining the blank into a specific shape, also giving it a specific finish. The blank is then mounted to a centerless cylindrical die mount, which allows for two or three round dies in order to create the finished thread.

Afterward, each screw thread is inspected by a probe that checks and ensures that the screw thread is of proper dimensions. Following the inspection, the screw threads are then heat treated in a furnace in order to strengthen the steel. When the furnace reaches a certain temperature, carbon is injected inside. The screw threads absorb this carbon, strengthening the gears further. The screw threads are then soaked in oil in order to harden the metal. After the heat treatment, the screw threads enter a computer-controlled grinder in order to correct any distortions created during the heat treatment.

15.8.2.3 Crosshead and Frame

The crosshead and frame are created by pouring molten steel into molds in order to form the general shape of each. Imperfections in the molded steel are corrected using a grinder. Holes are then drilled where appropriate using a computer-controlled drill bit. Coolant is added in order to prevent deformation of the drill bit.

The crosshead then undergoes screw threading by use of a computer-controlled cutter, which cuts the interior screw threads to precise dimensions. All the pieces undergo strengthening using the same heat-treating process used for the gears and screw threads, adding carbon to help strengthen the material, and correcting any distortions using computer-controlled machinery. The frame pieces are then assembled by welding them together.

15.8.2.4 Thrust Bearings

Thrust bearings are composed of a strong material in order to withstand the forces applied to them. They are made in a similar way to gears. The disks of the thrust bearings are made by shaping blanks with a computer-controlled blade. A turret with different tools then cuts and shapes each piece appropriately. After the pieces are shaped, they undergo an inspection process in order to ensure precision. The pieces are then heat treated in a furnace, and carbon is injected in order to help the material gain strength. Any slight distortion is corrected using further tools.

The balls of the thrust bearings are made using ball milling machines. They, too, are inspected, heat treated, and carburized, and then inspected and corrected once again.

15.8.3 Cost Estimation

Refer to Section 15.7.4 for cost calculations

The cost estimation was done as follows:

Cost of component is the sum of

- Material cost
- Manufacturing cost
- Material removal process—150% of material cost
- Forging cost—125% of material cost
- Foundry cost—135% of material cost
- Overhead cost—150% of material cost

Component	Material Cost	Manufacturing Cost	Overhead Cost	Subtotal
54-tooth spur gear				2 at $255.07 each
18-tooth spur gear				1 at $69.25 each
Crosshead	$31.672	$129.86	$47.51	$209.04
Frame	$211.68	$867.88	$317.52	$1397.08
Screw threads	$30.10	$123.39	$45.14	2 at $198.63 each
Thrust bearings				2 at $44.45 each
			Total =	$2671.66

15.8.4 Cost Calculations

$$\rho = 490\frac{\text{lbm}}{\text{ft}^3}$$

Cost of Steel

$$C = \frac{912}{\text{ton}} = 0.414\frac{1}{\text{lbm}}$$

15.8.4.1 Crosshead Cost Calculations

$$A = 5.183\,\text{in}^3 = 0.156\,\text{ft}^3$$

$$\text{Subtotal}_3 = \text{Cost} + \text{ManCost} + \text{Overhead} = 198.627$$

$$\text{Cost} = A \cdot \rho \cdot C = 31.672$$

$$\text{Overhead} = \text{Cost}(150\%) = 47.508$$

$$\text{ManCost} = \text{Cost}(150\% + 125\% + 135\%) = 129.856$$

15.8.4.2 Frame Cost Calculations

$$\text{Cost} = A \cdot \rho \cdot C = 211.679$$

$$\text{Subtotal}_1 = \text{Cost} + \text{ManCost} + \text{Overhead} = 209.036$$

$$\text{Overhead} = \text{Cost}(150\%) = 317.518$$

$$\text{ManCost} = \text{Cost}(150\% + 125\% + 135\%) = 867.884$$

$$\text{Cost} = A \cdot \rho \cdot C = 30.095$$

15.8.4.3 Screw Thread Cost Calculations

$$\text{Subtotal}_2 = \text{Cost} + \text{ManCost} + \text{Overhead} = 1397.081$$

$$A = \left[(10 \cdot 25 \cdot 4) + (4.5 \cdot 25 \cdot 3) + (37.74 \cdot 4.5 \cdot 2.75)\right]\,\text{in}^3 = 1.044\,\text{ft}^3$$

$$\text{Total} = \text{Subtotal}_1 + \text{Subtotal}_2 + 2\ \text{Subtotal}_3 + 255.072 + 69.25 + 2 \cdot 44.45 = 2671.661$$

$$\text{Overhead} = \text{Cost}(150\%) = 45.143$$

15.8.5 Assembly

15.8.5.1 18-Tooth Spur Gear

The 18-tooth spur gear was created by first creating a circle to a specified diameter. This circle was then extruded to a specified thickness. Using geometry and focus/reference points, the outline of a tooth was drawn on the shape. This tooth was extruded into a new part. Using the "circular pattern" command, the remainder of the 17 teeth were created. Finally, the teeth and the circular shape were joined together to form the gear. Figure 15.97 is a screenshot showing the assembly process of this component.

15.8.5.2 54-Tooth Spur Gear

The 54-tooth spur gears were created in the same manner as the 18-tooth spur gear, except that more teeth and a larger disk had to be incorporated into the design. Figure 15.98 is a screenshot showing the assembly process of this component.

15.8.5.3 Crosshead

The crosshead was created by first creating a square to proper dimensions. This square was then extruded to a specified thickness. Two circles were drawn on one of the faces at appropriate distances from the centerline of the piece. These circles were then extruded into two cylindrical cutouts. Using geometry and focus/reference points, portions of a screw thread were drawn into the cylinders themselves, and using the "revolve" feature (with appropriately set translation), screw threads were formed on the interior portion of the cutouts. Figure 15.99 is a screenshot showing the assembly process of this component.

$$A := \frac{2\pi \cdot (2.13\ \text{in})^2}{4} \cdot 36\ \text{in} = 0.148 \cdot \text{ft}^3$$

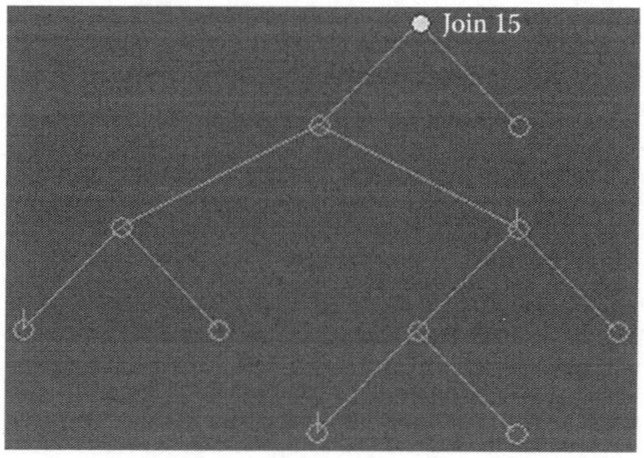

Figure 15.97

NX history of 18-tooth spur gear.

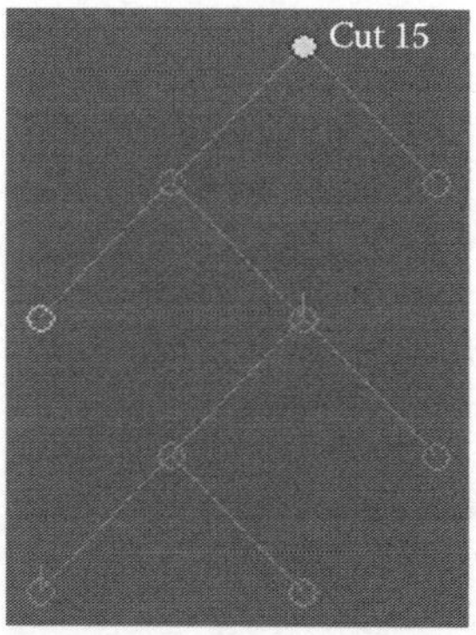

Figure 15.98

NX 9.0 history of 54-tooth spur gear.

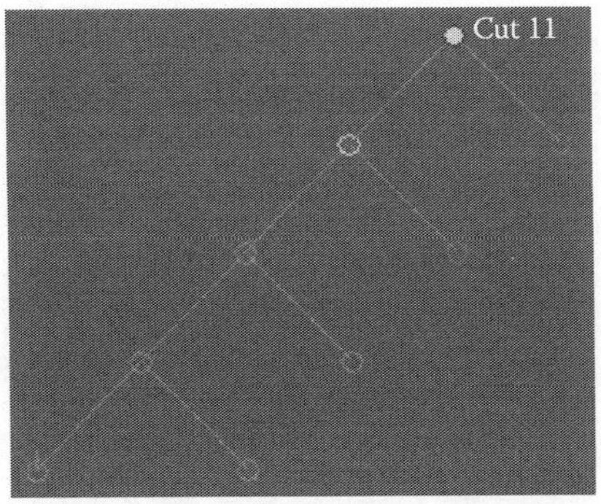

Figure 15.99

NX 9.0 history of crosshead.

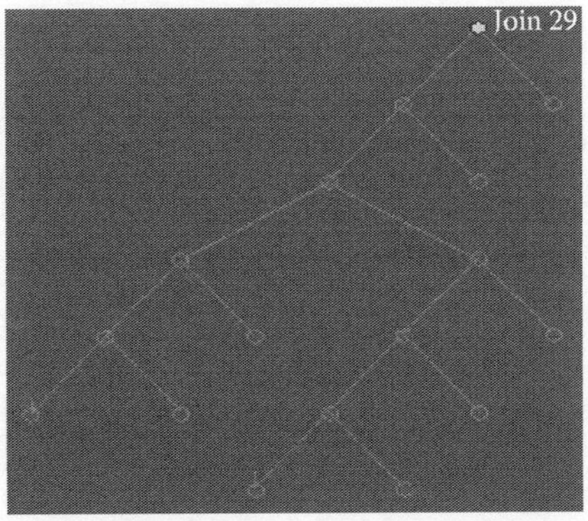

Figure 15.100

NX 9.0 history of frame.

15.8.5.4 Frame

The frame was created in a similar fashion to the crosshead. Squares were created to appropriate dimensions and then extruded to appropriate thicknesses. This was done first with the base, then with the support columns, and then with the overhead beam. The cylindrical cutouts in the overhead beam and the base were created in the same fashion as the cylindrical cutouts performed on the crosshead. Figure 15.100 is a screenshot showing the assembly process of this component.

15.8.5.5 Screw Threads

The screw threads were created by first creating and extruding a circle in order to form a column of appropriate dimensions. Using geometry and focus/reference points, a portion of a screw thread was sketched on the side of the column. Then, using the "revolve" feature with an appropriate pitch, the screw threads were created and added as a protrusion to the column. The maximum 50,000° of rotation that I-DEAS was able to handle was used. Figure 15.101 is a screenshot showing the assembly process of this component.

15.8.5.6 Thrust Bearings

The bottom plate of the thrust bearing was created first. This was done by creating concentric circles to appropriate dimensions and then extruding and cutting the different surfaces in order to produce a hollowed-out disk. The top plate was simply a copy of the bottom plate, which was then rotated about the z-axis in order to lie on top of the bottom plate with their hollow sides facing each other. The bearing balls were placed inside the hollowed-out portion of the two plates. These balls were created by rotating a semicircle of appropriate dimensions around an axis in order to create a sphere. This sphere was then duplicated into a pattern using the "circular patterns" command. Nine balls in total were produced and placed into the bearing. Figure 15.102a is a screenshot showing the assembly process of this component.

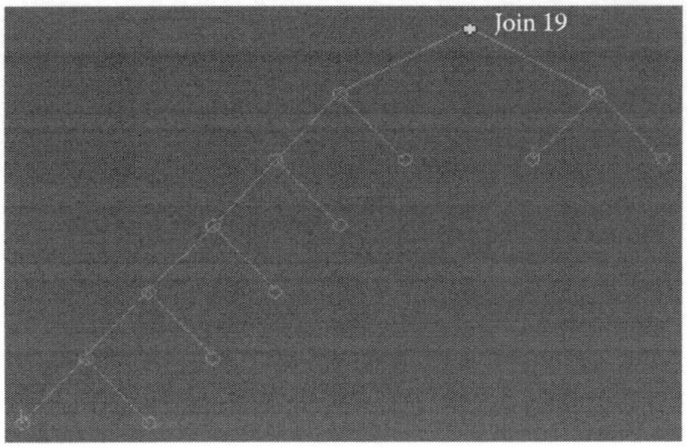

Figure 15.101

NX 9.0 history of screw threads.

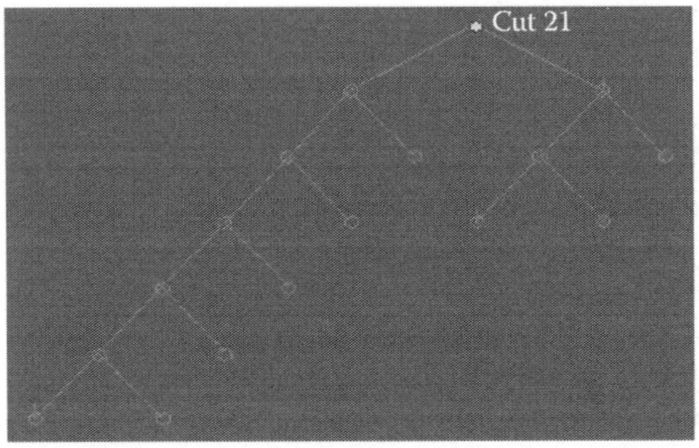

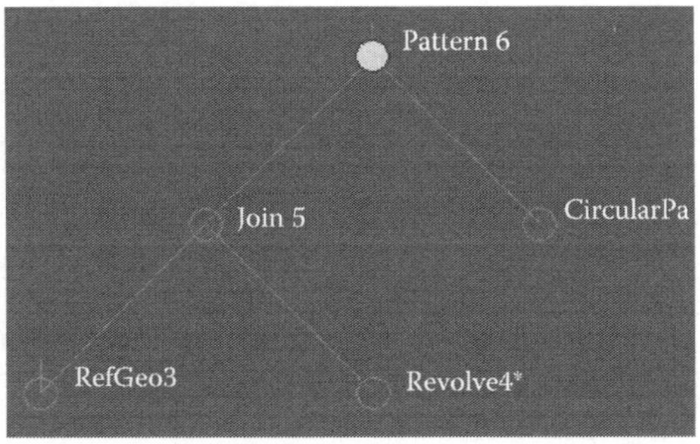

Figure 15.102

NX 9.0 history of thrust bearings.

15.8.6 FEA

15.8.6.1 Crosshead

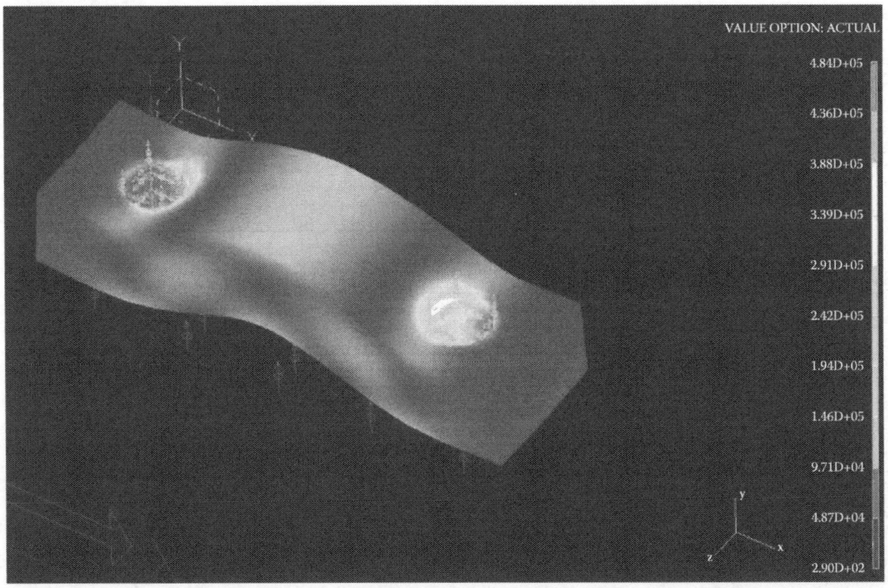

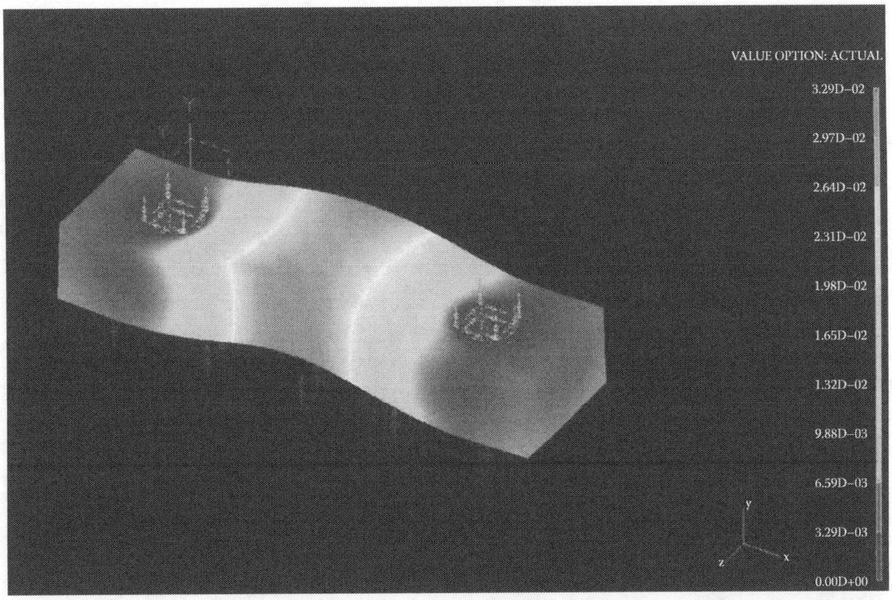

15.8.6.2 Gear Tooth

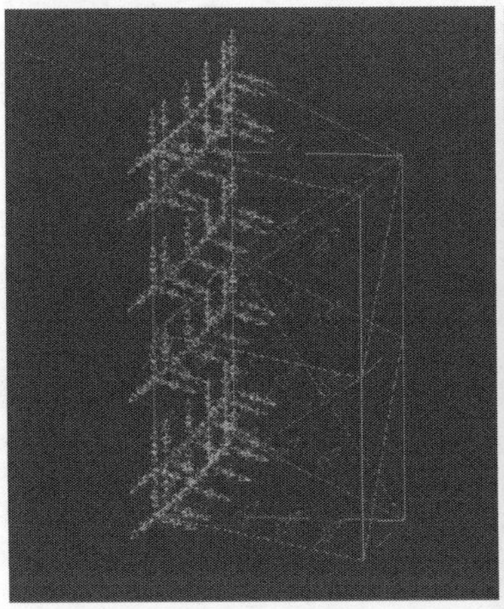

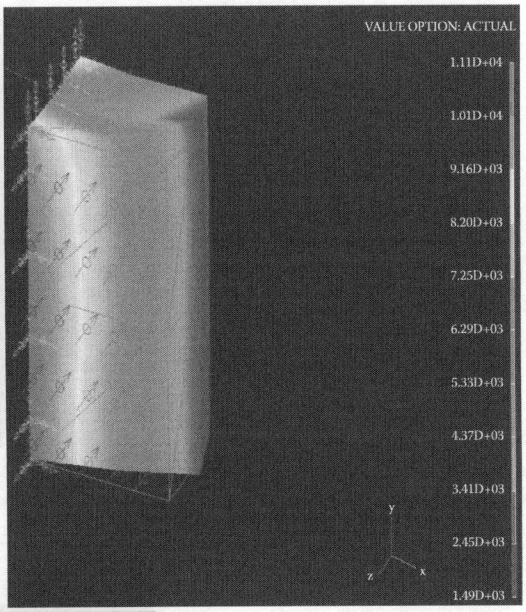

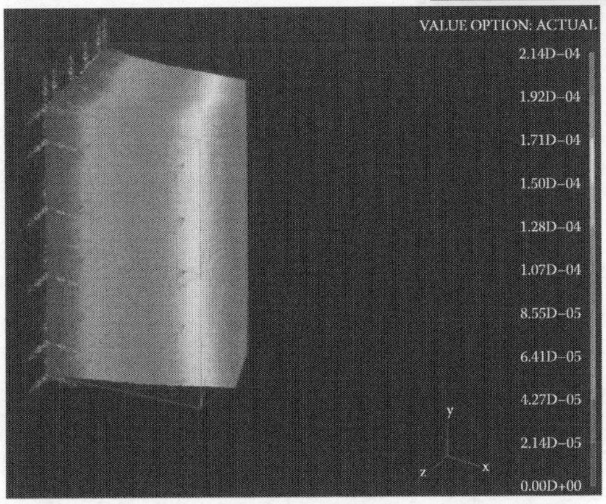

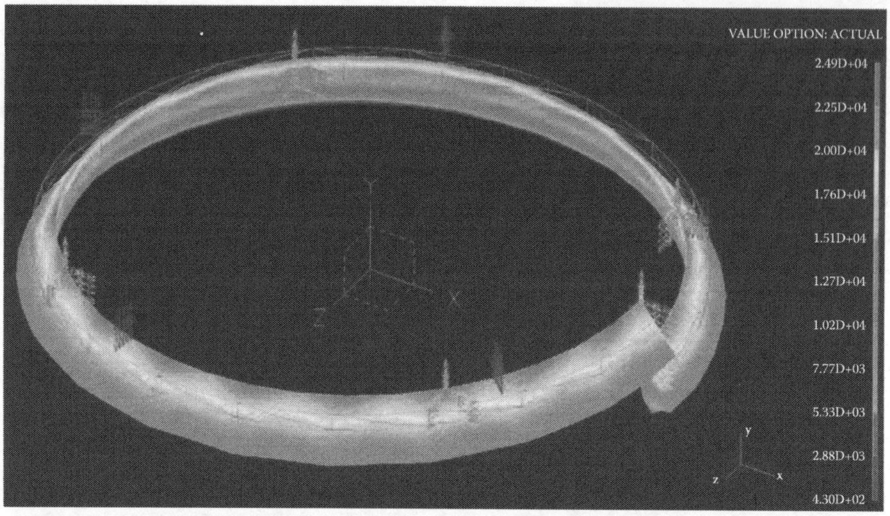

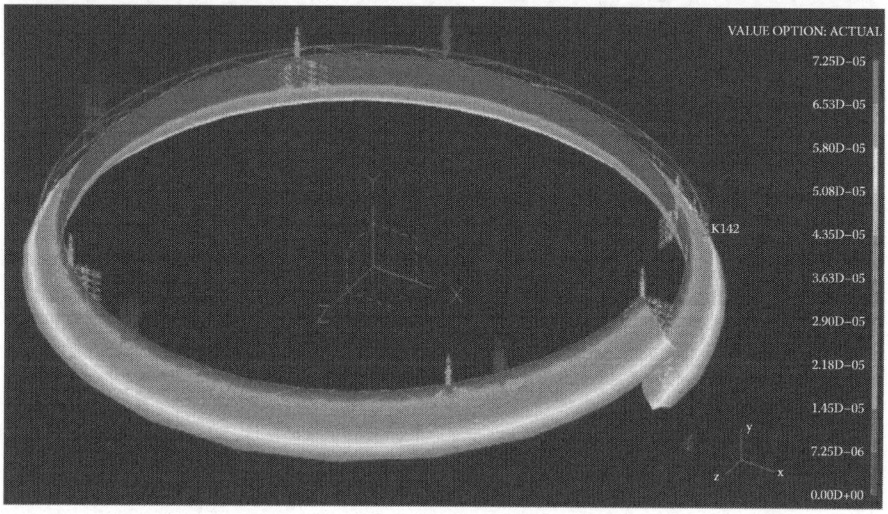

15.9 CAD Example 9: Crankshaft, Connecting Rod, and Flywheel

A crankshaft and its main bearing are analyzed and designed. The crankshaft, connecting rod, and flywheel of a six-cylinder in-line engine are to be analyzed. Each of these components undergoes a different force and reacts in a different way. This project requires that a crankshaft and main bearings be completely and fully designed using NX. Each force and component is separately drawn. After the drawing, each part must undergo FEA to check for failure. Lastly, the components must be assembled and move together in an animation.

15.9.1 Introduction

Crankshafts and their main bearings are typically used for automotive applications in larger engines. For this project, a 250 hp six-cylinder in-line engine is used to model the crankshaft. The crankshaft then regulates the movement of the pistons in the engine. This movement is accomplished by converting rotational motion to linear reciprocating motion. The conversion of movement is done through several pins, or "throws," which are several off-axis rotational bearings. As the crankshaft turns, the throws move the pistons up and down their cylinders. The crankshaft needs to be analyzed, as the rotation creates torque on the shaft. There are many cycles to this movement, so the crankshaft also needs to be checked for fatigue. The overall movement of the crankshaft allows the engine to perform properly in terms of intake and discharge fluids.

Additionally, connecting rods are placed between the crankshaft and the piston to control the movement. As a result, these connecting rods endure a compressive loading and can buckle under a heavy load. The main bearings on the crankshaft are also important for this application. The crankshaft rotates on these bearings. The purpose of these bearings is to keep the crankshaft stationary while minimizing the amount of friction caused by the contact of the crankshaft with its support and the friction caused by the rotation of the crankshaft. Also, there is a flywheel attached to the crankshaft to generate the movement. The diameter of this flywheel needs to be sized and its teeth checked for failure.

15.9.2 Analysis and Design

Crankshafts are typically made from a form of rolled steel. The pistons are usually made out of aluminum. The main bearings, however, can consist of three layers of different materials. The material that contacts the actual crankshaft is a layer of steel or copper, depending on the material of the shaft. The materials of the shaft and the bearing cannot be the same, as intensified friction could potentially cause a weld. The outer layer consists of an aluminum alloy, and the outermost layers are made from zinc plating. The purpose of the zinc is to make the bearing corrosion resistant. For this project, the crankshaft is composed of AISI 1050 steel. This steel delivers the necessary strength and durability to allow the crankshaft to perform properly and without failure. The bearings are made from the aforementioned materials. With these materials, many calculations were performed.

The first calculations performed were for the force and torque on the crankshaft:

$$F = \frac{2HP}{N \cdot s} \tag{15.35}$$

where:

HP is the horsepower
N is the rotational speed
s is the stroke length

The resulting force (F) is 2626 lbf.

$$T = F \cdot s \tag{15.36}$$

The resulting torque (T) was 656 lbf-in. Using these values, the crankpin could be sized with the following equations:

$$A(d) = \frac{\pi \cdot d^2}{4} \tag{15.37}$$

$$\tau_{\max} = \frac{S_y}{n} \tag{15.38}$$

$$\tau(d) = \frac{F}{A(d)} \tag{15.39}$$

where:

d is the diameter of the pin
S_y is the yield strength of the steel
n is the factor of safety
$\tau_{\max}$ is the maximum shear stress
A is the cross-sectional area for the pin

Solving these equations simultaneously gives a final diameter of 0.457 in. For safety and standard sizing purposes, a pin size of 0.625 in was ultimately used.

The pin size also allows the web thickness to be determined:

$$h = 0.4 \cdot d_{\text{pin}} \tag{15.40}$$

where d_{pin} is the pin diameter. With this calculation, it was found that the web needs to be at least 0.25 in thick. A thicker web was ultimately used for additional safety.

Next, the journal bearing can be designed. The following equations were used for the calculations:

$$J = \frac{\pi \cdot d^4}{32} \tag{15.41}$$

$$I = \frac{\pi \cdot d^4}{64} \tag{15.42}$$

$$\tau_{xy} = \frac{T \cdot r}{J} \tag{15.43}$$

$$\sigma_x = \frac{M \cdot r}{I} \tag{15.44}$$

$$\sigma = \sqrt{\sigma_x^2 + \tau_{xy}^2} \tag{15.45}$$

where:

- d is the diameter of the bearing
- r is the radius of the bearing
- J is the polar moment of inertia
- I is the second moment of inertia
- τ_{xy} is the shear stress
- σ_{xy} is the normal stress
- σ is a principal stress

With the factor of safety and design considerations in mind, a maximum principal stress was used to ultimately solve for the diameter of the pin. This diameter equated to 1.598 in. For reasons of standard sizing, a journal bearing of 1.75 in is used.

Subsequently, a flywheel needs to be designed for the shaft. The flywheel is designed based on the minimum and maximum velocities of the engine. These velocities are outlined in the following equations:

$$V\left(d_{fw}\right) = N \cdot \pi \cdot d_{fw} \tag{15.46}$$

$$V_{\max}\left(d_{fw}\right) = N_{\max} \cdot \pi \cdot d_{fw} \tag{15.47}$$

$$V_{\min}\left(d_{fw}\right) = N_{\min} \cdot \pi \cdot d_{fw} \tag{15.48}$$

where d_{fw} is the diameter of the flywheel (to be designed). The minimum and maximum rotational speeds are based on those of a typical automobile. A range of 1–7000 rpm was selected for these calculations. With these values known, the capable power of the flywheel can be found:

$$P = \frac{I}{r^2} \cdot V \cdot \left(V_{\max} - V_{\min}\right) \tag{15.49}$$

where:

- I is the second moment of area for the flywheel
- r is the radius of the flywheel

The final value for this power is compared with that of 85% of the rated horsepower of the engine, because the flywheel is expected to receive 85% of the power. Using these values, a diameter can be determined. The resulting calculation was 9.742 in. In conclusion, a flywheel of 10 in was designed and fitted onto the crankshaft.

Some components were also checked for possible failure. The stress on the journal bearing was analyzed, as was any possible bending of the web. Both calculations gave factors of safety well above 3, which is acceptable for this design. Ultimately, these calculations yielded the constraining dimensions to design each part, as described in Section 15.9.3.

15.9.3 Solid Modeling

To create the main bearing, a very simple process was performed. Two circles were drawn in 2-D, and were then extruded to the length of the bearing. The final result is shown in Figure 15.103.

Next, the crank throw was designed in two halves. One-half was created by drawing the basic 2-D cross section. Then, it was filleted in 2-D to add the curves in the cross section. Once extruded, a cut was made to create the indent at the crank pin. Then, two circles were drawn on the surface and extruded to produce the out-of-axis rotation involved in the movement. All the edges were filleted to create a smooth finish to the part. Once half of the crank throw was made, a rotation was set up to make it full. With the "copy switch" setting activated, the half throw was rotated 180° about the vertical axis. Then, the two halves were joined, creating one complete crank throw. The final result is shown in Figure 15.104.

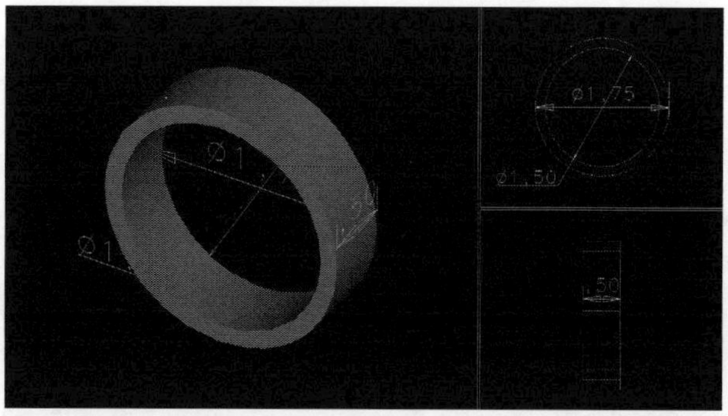

Figure 15.103

Bearing.

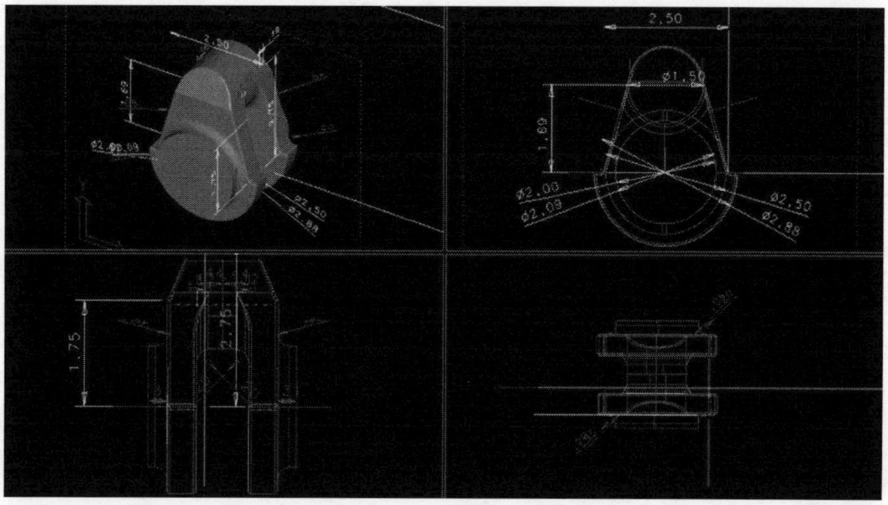

Figure 15.104

Crank throw.

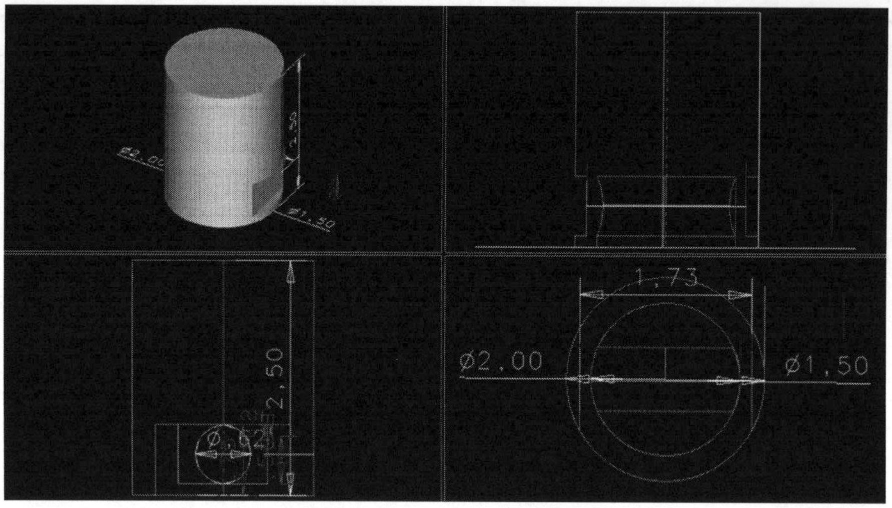

Figure 15.105

Piston.

A piston was drawn after the throw was completed. It started out as a simple cylinder. A circle was drawn for the cross section and then extruded. Once the cylinder was formed, three cuts were performed. One cut was made on the bottom surface to machine out the material. Then, two cuts were made on the sides. Once these cuts were made, a circle was drawn on one of the surfaces and extruded to the other surface. The function of this cylinder is for the piston to be properly pinned to the connecting rod. The final design of the piston is shown in Figure 15.105.

The next part to be electronically fabricated was the connecting rod. To create this part, two circles of different size were drawn. These circles were sized and dimensioned to be able to connect to the crankpin and the piston. The circles were then connected with tangent lines, and the cross section was extruded. The circles remained hollow, and the connection between them took form. Once the part was 3-D, the extra edges on each side were cut out. The sharp edges produced on the connecting rod by these cuts were filleted for aesthetic reasons. Additionally, two rectangular planes were drawn and extruded in order to connect the two halves of the larger circle with bolts. A hole was cut from each plane. The bolt for the hole will be taken from the built-in reference library in IDEAS. The final connecting rod, without the bolts, is shown in Figure 15.106.

The last components that needed to be drawn were for the flywheel. First, a tooth needed to be drawn. A simple square cross section was drawn and extruded. After extrusion, the sides of the tooth were cut

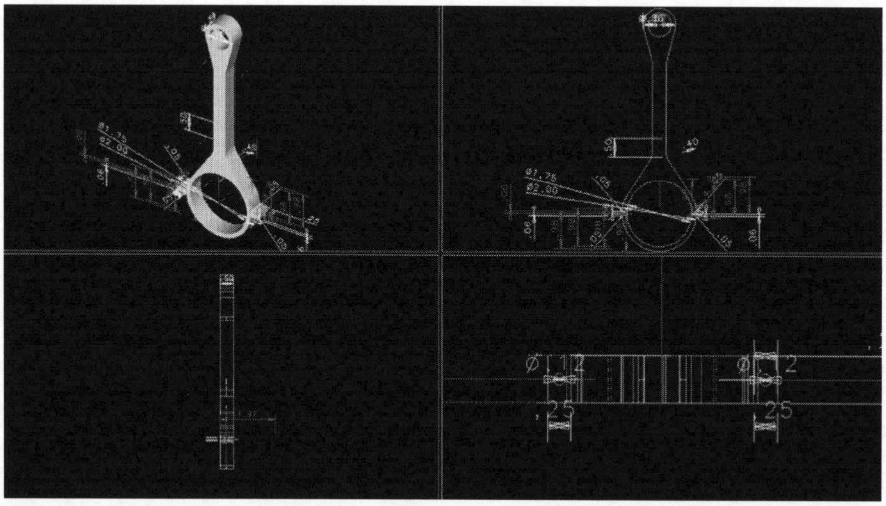

Figure 15.106

Connecting rod.

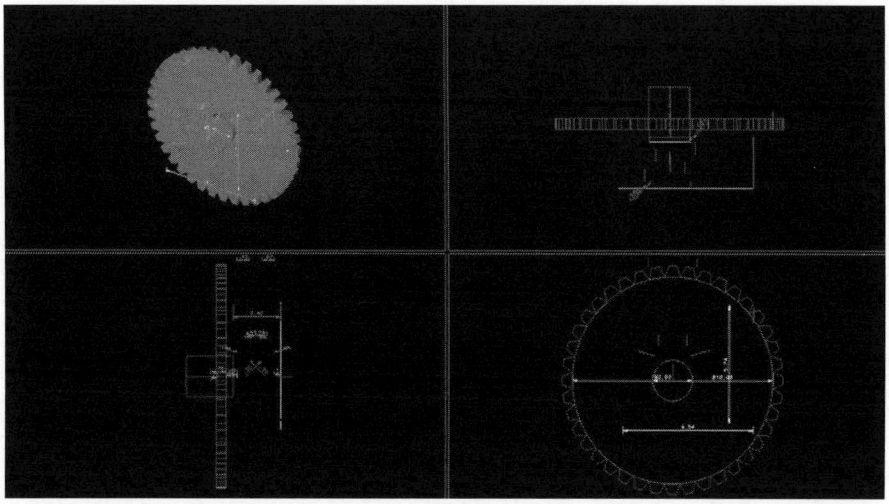

Figure 15.107

Flywheel.

in a V-shape, giving the tooth its definition. The flywheel was simply created by drawing and extruding a circle with the predefined dimensions, and the tooth that had just been designed was placed on the surface. Once properly placed, the tooth was replicated by rotation with the "copy" switch on. The rotation was performed about the radius for 360°, copying the tooth 40 times. The completed flywheel is shown in Figure 15.107.

All these parts were designed in consideration of the dimensions calculated in Section 15.9.2.

15.9.4 Tolerances and Allowances

For the crankshaft, tolerances and allowances are a crucial component of the design. The main bearing and crankpin must have the proper clearance fit to maintain the rotation involved in the shaft. For this crankshaft, a loose-running clearance fit will be used. The following equation is modeled for the clearance fit.

$$d_{max} = d + \delta \tag{15.50}$$

where:

d_{max} is the maximum size of the hole
d is the basic size of the hole
δ is the fundamental deviation

For these bearings, a fundamental deviation of 0.0001 in was used to permit movement and precision. This allows the manufacturing of the bearings to deviate by ±0.0001 in.

15.9.5 Manufacturing Process

A variety of manufacturing processes are involved in the fabrication of the crankshaft system. Crankshafts can either be forged from a bar of steel or machined from a billet. They can also be cast in steel. High-performance crankshafts are forged from steel, with a forging temperature around 2000°F. Then the steel is pressed and pressured to give the crankshaft its shape.

Bearings are also manufactured by several different means, depending on the type of bearing and its application. For a journal bearing, a CNC machine can be used to cut the bearing precisely. This allows the necessary clearances to be made. The journal bearing can then be lubricated and installed properly.

15.9.6 Cost Estimation

To determine the cost estimation of the project, the volume of each component needs to be established. With these volumes known, the total volume of the system can be found. This total volume needs to be separated into each component material. The following table shows the volumes.

Part	Volume (in³)	Quantity	Total (in³)
Steel Parts			
Connecting rod	1.3968	6	8.3809
Throw	11.0026	6	66.0156
Bolt	0.0079	12	0.0943
Flywheel	40.3694	1	40.3694
Tooth	0.1534	40	6.1360
		Total:	120.9961
Copper Parts			
Bearing	0.3191	6	1.9144
		Total:	1.9144
Aluminum Parts			
Piston	6.8652	6	41.1914
		Total:	41.1914

Once these volumes are known, the costs can be found, based on the density of the materials. These values are given in the following table:

Material	Cost per Kilogram ($/kg)	Density (g/cm³)
Steel	1.30	7.85
Aluminum	6.10	2.70
Copper	7.00	8.89
Cast iron	4.46	7.30

Now, with these values known, the total material mass required for the engine can be calculated, and the total cost of materials can be derived:

Material	Total Mass (g)	Total Cost
Steel	15564.6880	$44.36
Copper	278.8918	$2.37
Aluminum	1944.0091	$12.05
	Total Cost:	$58.78

Lastly, the manufacturing costs, as well as the overhead cost, can be calculated based on the total material cost:

Material cost	$58.78
Material removal process	$88.17
Forging cost	$73.48
Foundry cost	$79.36
Overhead cost	$88.17
Total cost	$387.97

15.9.7 Codes and Standards

The American Society of Mechanical Engineers (ASME) and SAE dictate many restrictions on the safety and construction of vehicle engines. Both societies have performed numerous and rigorous tests on vehicle components and produced various publications on each component that is applicable to engines. The codes are used to regulate the size of the crankshaft and pistons to ensure safety for engines of certain calibres. The materials, fuels, lubrication methods, and construction of the parts are also regulated. The main bearings of the crankshaft must be capable of the proper rotation for an extended life of the automobile. All these regulations were considered in the design and calculations of this crankshaft.

15.9.8 Assembly

Figures 15.108 through 15.111 show the fully assembled in-line six-cylinder engine in NX I-DEAS.

15.9.9 FEA

Using the NX software, an FEA of the crankshaft was generated. Figures 15.112 through 15.119 show the mesh and result of a 1-D analysis of the shaft. Additionally, a 3-D FEA was performed for each part.

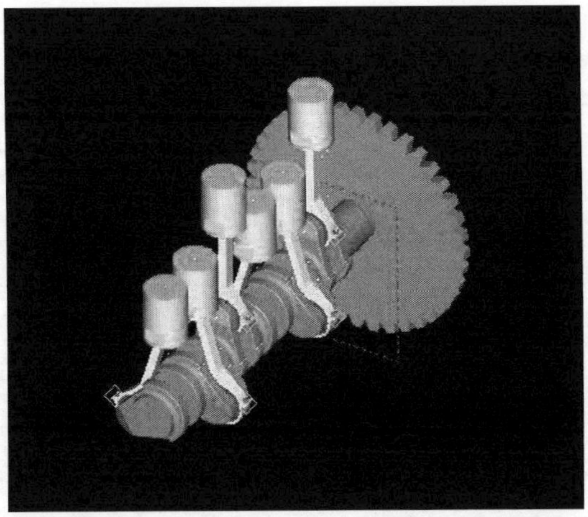

Figure 15.108

Isometric view of engine assembly.

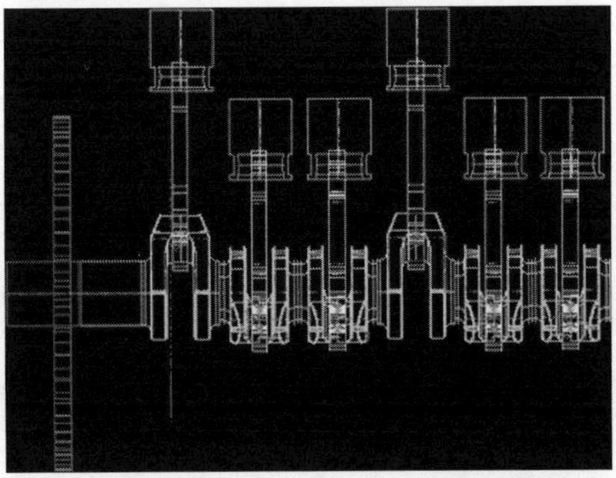

Figure 15.109

Side view of engine assembly.

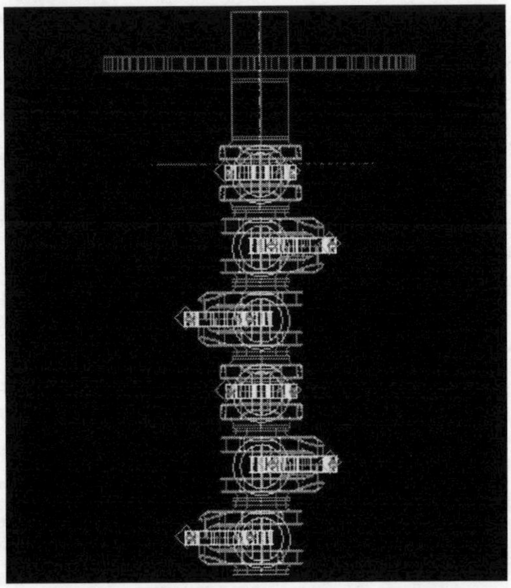

Figure 15.110

Top view of engine assembly.

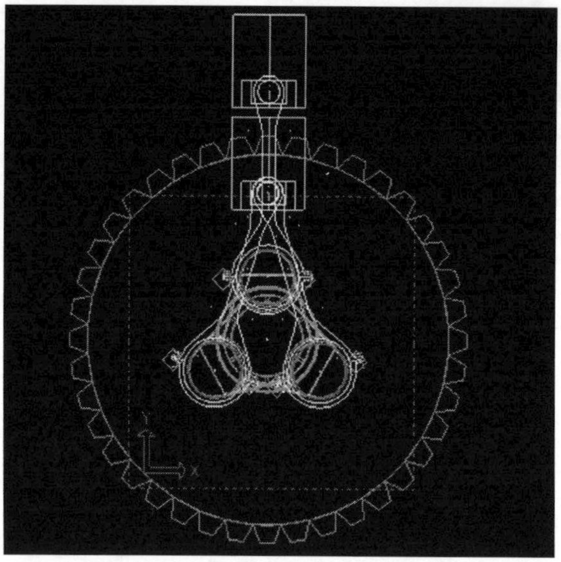

Figure 15.111

Front view.

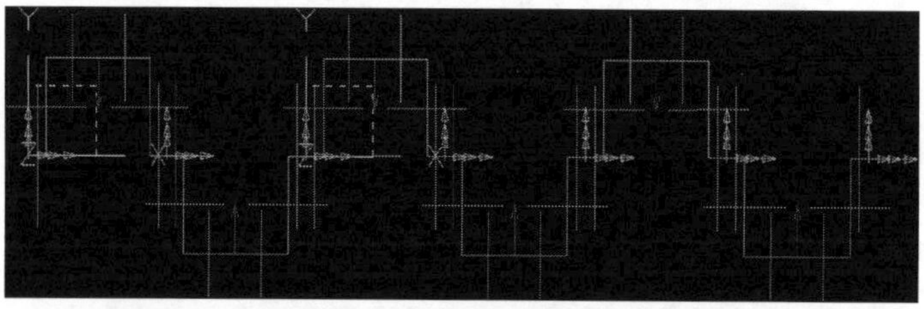

Figure 15.112

Nodal mesh for the crankshaft.

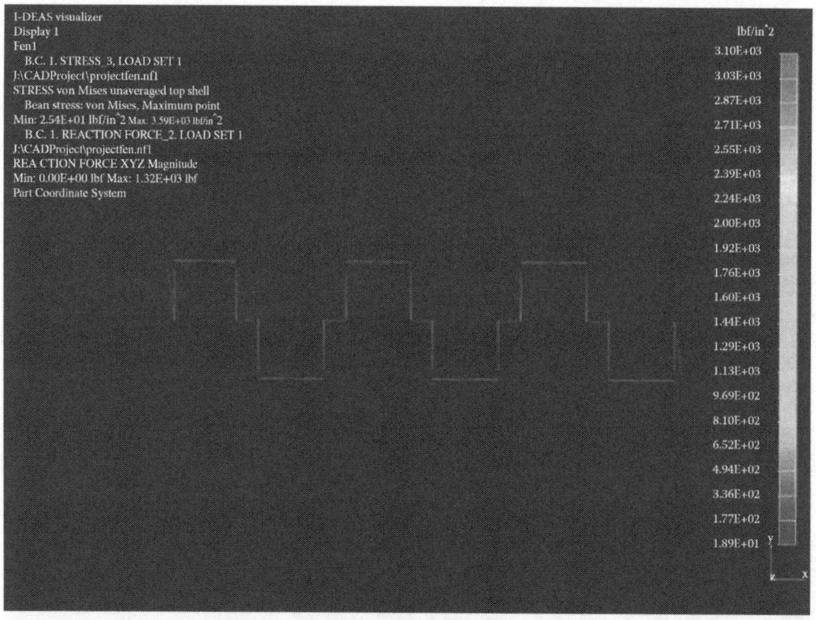

Figure 15.113

Simulated and processed stress and displacement results for the crankshaft.

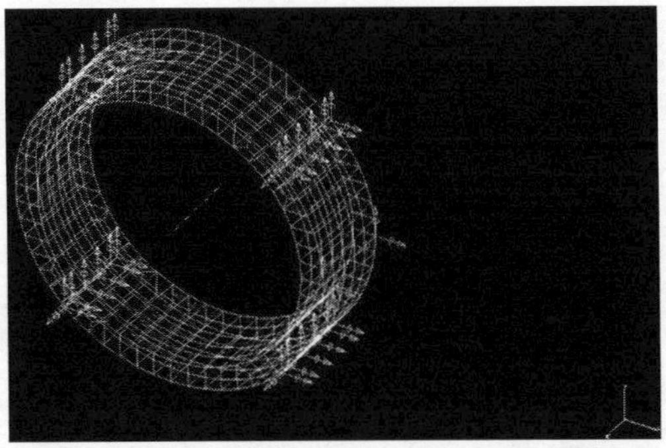

Figure 15.114

Nodal mesh for the main bearing.

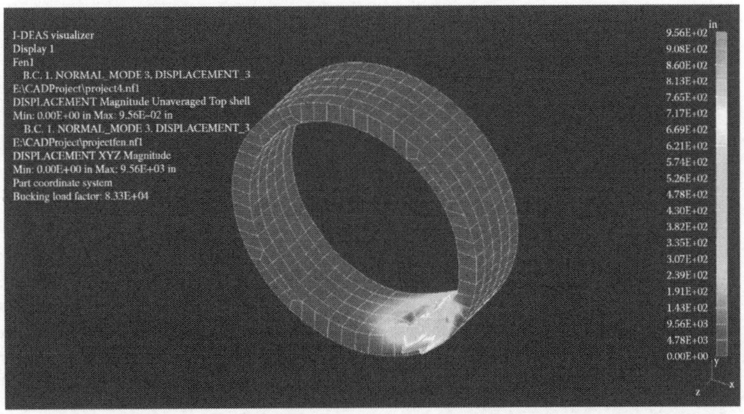

Figure 15.115

Simulated and processed stress and displacement results for the main bearing.

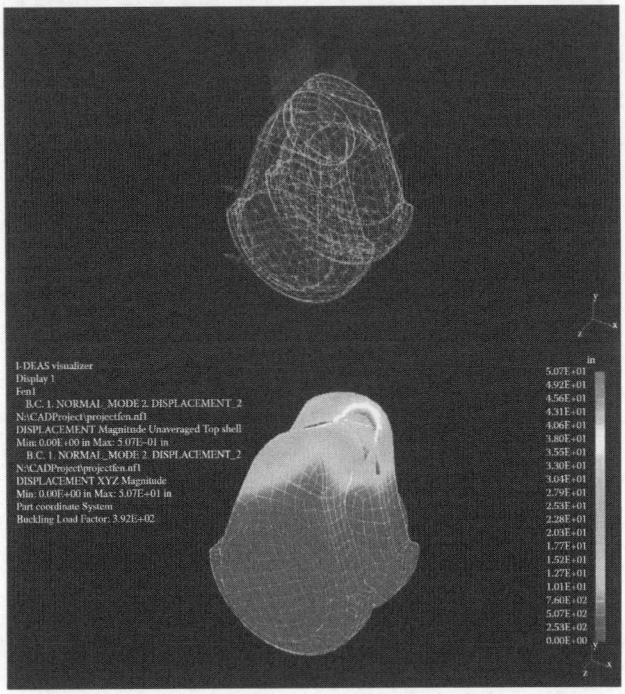

Figure 15.116

(See color insert) Nodal mesh for a crank throw.

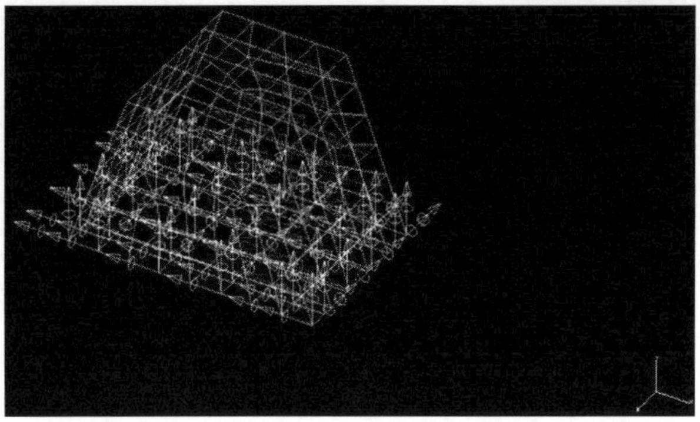

Figure 15.117

(See color insert) Nodal mesh for a flywheel tooth.

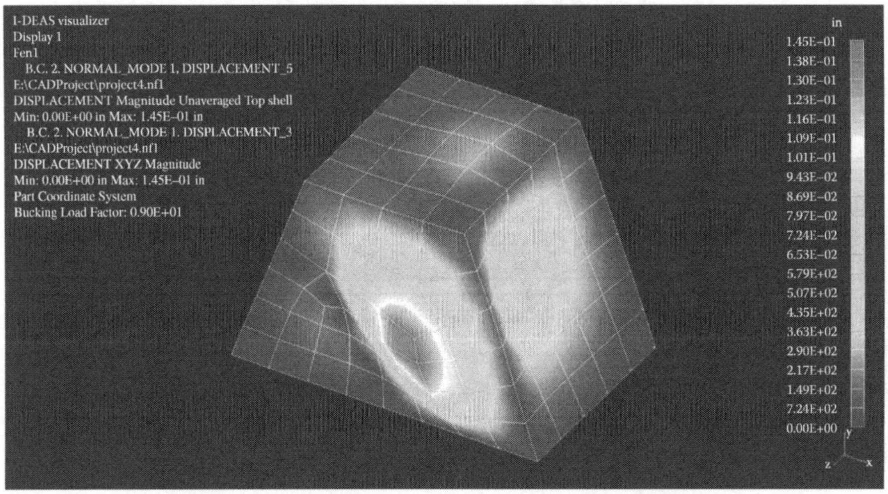

Figure 15.118

Simulated and processed stress and displacement results for a flywheel tooth.

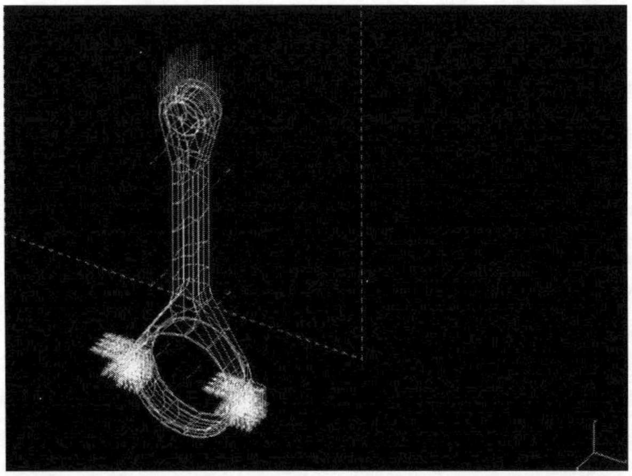

Figure 15.119

Nodal mesh for the connecting rod.

Each FEA provided the maximum displacement of each part, as well as the stress endured by the part. In all cases, the stress and displacement were minimal and would have no effect on the overall engine. In conclusion, the crankshaft and its various parts have been confirmed as safe and effective by 3-D FEA.

References

1. Ashby, M. F. (2011). *Materials and Environment.* Butterworth & Heinemann, Boston, USA
2. Groover, M. P. (2010). *Fundamental of Modern Manufacturing*, 4th edn. Wiley, New York.
3. Gonzalez, T. (1983). Remanufacturing technology. S.M. Thesis, MIT Department of Mechanical Engineering, Cambridge, MA.
4. Budynas, R. and Nisbett, J. (2008). *Mechanical Engineering Design*, 8th edn. McGraw-Hill, New York.

Bibliography

Brown Asharp Manufacturing Co.: Prodvidence, RI.

Chen, R., Navin-Chandra, D., and Prinz, F. (1993). Product design for recyclability: A cost benefit analysis model and its application. In *IEEE International Symposium on Electronics and the Environment* (pp. 178–183), Arlington, VA.

D'Amore, R. (1984). Product integrity in a remanufacturing operation. SAE Technical Paper Series, No. 840795.

Desai, A. and Mital, A. (2005). Incorporating work factors in design for disassembly in product design. *Journal of Manufacturing Technology Management* 16: 712–732.

Henstock, M. (1988). *Design for Recyclability.* The Institute of Metals, London, UK.

Holzwasser, H. (1983). Engineering for remanufacturing. SAE Technical Paper Series, No. 830549.

Jovane, F., Alting, L., Armillotta, A., Eversheim, W., Feldmann, K., Seliger, G., and Roth, N. (1993). A key issue in product life cycle: Disassembly. *Annals of the CIRP* 42-2: 651–658.

Kroll, E. and Carver, B. S. (1999). Disassembly analysis through time estimation and other metrics. *Robotics and Computer-Integrated Manufacturing* 15: 191–200.

Kroll, E. and Hanft, T. A. (1998). Quantitative evaluation of product disassembly for recycling. *Research in Engineering Design—Theory, Applications, and Concurrent Engineering* 10: 1–14.

Kutta, R. (1980). Remanufacture of durable products: A value-preserving technology. S.M. Thesis, MIT, Cambridge, MA.

Lund, R. (1983). Remanufacturing, United States experience and implications for developing nations. MIT Center for Policy Alternatives Report, CPA/83-17, Cambridge, MA.

McGlothlin, S. and Kroll, E. (1995). Systematic estimation of disassembly difficulties: Application to computer monitors. In *IEEE International Symposium on Electronics & the Environment* (pp. 83–88), *Jurnal Mekanikal, December.8,* Orlando, FL.

NX 9.0 Siemens. (2015). Siemens PLM Software.

Overby, C. (1980). Product recycling (remanufacturing) as a conservation option: An international dimension. In *Advances in Materials Technology in the Americas-1980, Vol. 1—Materials Recovery and Utilization* (pp. 113–119), 6th Inter-American Conf. Sanfransisco, August 1980.

Schrader, T. and Bawel, D. (1985). Specification criteria for remanufacturing. SAE Technical Paper Series, No. 840794.

Toyota Motor Company. (2005). Responses to automobiles recycling laws in Japan. http://www.toyota.co.jp/.

Warnecke, H. and Steinhilper, R. (1983). Designing products with built-in reconditioning features. In *Proceedings of the 6th Annual Conference on Design Engineering* (pp. 241–248), Birmingham, UK.

Whitney, D. E. (2004). *Mechanical Assemblies.* Oxford University Press, New York.

SECTION V
Environmentally Benign Manufacturing

16

Introduction to Environmentally Benign Manufacturing

A brief review of traditional manufacturing processes is presented to gain an understanding of the differences in approach between sustainable and environmentally benign manufacturing. A large number of processes are used in manufacturing operations, some of which can be categorized as follows:

1. Casting and foundry processes. Molten metal, plastic, or glass is cast into a mold and solidified into shape. The design considerations in casting can be noted as
   - **a.** Do not have thick sections at the bottom of a sand casting.
   - **b.** Do not use sharp corners. Sharp corners can cause cracking and tearing at corners.
   - **c.** If the fillet is too large, voids can occur.
   - **d.** Do not have two walls crossing, offset the crossing.
   - **e.** Keep all sections as uniform as possible.
   - **f.** Core diameters should be no less than the wall thickness.
   - **g.** If possible, hollow out solids.
   - **h.** Avoid large flats because they warp.
   - **i.** Include draft angles to allow pattern/casting to be removed.
   - **j.** Use simple part shapes.
   - **k.** Avoid the need for cores; however, if they are used, join cores to ensure accurate placement.
   - **l.** Avoid undercuts.
   - **m.** Avoid rapid wall thickness changes.
   - **n.** Consider machining requirements.
2. Deformation processes. Typical deformation processes are forging, rolling, extrusion, and wire drawing. In all these processes, metal is usually deformed either hot or cold to give it improved properties and to change its shape. Sheet metal forming is a special category in which deformation occurs in a two-dimensional stress state. The design considerations in deformation processes can be listed as

   Forging design considerations:
   - **a.** Use shapes that would occur as a result of die forging.
   - **b.** Avoid undercuts.
   - **c.** Provide tapers.
   - **d.** Avoid rapid thickness changes.

e. Parting line should be straight and halfway along the part.
f. Parts should have a rotational symmetry.

Extrusion design considerations:

a. Symmetric hollow sections are preferred.
b. Rounded tongues should be used.
c. Voids should be kept to a minimum.
d. Use fillets where possible.
e. Keep wall thickness uniform.
f. Use ribs to strengthen large flats.
g. Avoid undercuts.
h. Avoid rapid changes in thickness.
i. Avoid small or lateral holes.
j. Make parts as axisymmetric as possible.

3. Material removal or cutting (machining) processes. Using a sharp tool, material is removed from a workpiece by a variety of methods such as turning, milling, grinding, shaving, polishing, and lapping. The design considerations in metal-removal processes can be listed as:

Turning design considerations:

a. Avoid the need for complex tooling.
b. Provide adequate tool runout.
c. Avoid interior grooves.
d. Consider clamping issues.
e. Avoid excessive material removal.

Milling design considerations:

a. Design parts that only require vertical or horizontal operations.
b. Provide adequate tool runout.
c. Design parts to lie flat on the bed.
d. Avoid the need for complex tooling.
e. Avoid excessive material removal.

Drilling design considerations:

a. Allow drills to be used for blind holes.
b. Provide flats normal to the drill at the start of the hole.
c. Try and align holes on the opposite sides of a part to allow through-hole operations.

In general, it can be stated that the manufacturing process should always be considered when designing a part. When using a solid modeling (computer-aided design [CAD]) software package, use manufacturing-like operations. Design and manufacturing have a close relation. Sometimes, the material selected during the design dictates the manufacturing process and, at other times, the tolerance and allowance allowed in a design will force the selection of a particular manufacturing process. Both traditional manufacturing and environmentally conscious manufacturing (ECM) are linked to a design. A brief review of the relationship between design and manufacturing is presented before manufacturing operations are discussed.

16.1 Introduction to Green Manufacturing

> Green manufacturing is defined as the creation of manufacturing products that use materials and processes that minimize negative environmental impacts conserve energy, and natural resources, are safe for employees, communities, and consumers, and are economically sound. (Adapted from the Department of Commerce Definition)

It can be said that sustainable manufacturing is all about minimizing the diverse business risks inherent in any manufacturing operation while maximizing the new opportunities that arise from improving processes and products. The economic, environmental, and social aspects embraced by this concept are discussed below. This guides us in helping to improve the environmental performance of a manufacturing facility. The present philosophy in manufacturing operations is to maximize profit or minimize cost or maximize the production rate without thinking about its effect on the atmosphere by way of greenhouse gas (GHG) emissions and use of energy. Sustainable manufacturing reorients the objectives to minimizing the energy and the materials consumed. It is dictated by the depletion of resources in every aspect of production and global warming due to GHG emissions. It has long been established that GHG emission is directly related to energy consumption. GHG emissions will also depend on the kind of energy that is generated; for example, the energy that is generated using coal will create the largest GHG emissions. Renewable energy sources such as wind power, solar power, natural gas, and so on will reduce GHG emissions considerably. Hence, there is a focus on renewable energy generation. The concerns of manufacturers and retailers focus on the topics:

- How to achieve energy efficiency in factories and warehouses.
- How to comply with current environmental regulations and predict and plan for upcoming regulations.
- How a green agenda can help the industry cope with increased global competition and growing supply chain pressure.
- How to push recycling and waste management and implement cost-effective green manufacturing technology.
- How to design a sustainable environment, with an emphasis on water sustainability and a focus on facilities and sustainable packaging.

Research needs to be conducted to improve the energy efficiency and sustainability of manufacturing processes. Sustainability is defined in many ways, such as

- A reduction in the emission of global warming gases (carbon dioxide [CO_2], methane [CH_4], nitrous oxide [N_2O], chlorofluorocarbons [CFCs]); per capita, per gross domestic product (GDP), and per area/nation
- An increase in recyclability
- An improvement in the reuse of materials
- A lowering of energy consumption
- A reduction in pollution (air, water, land), which affects society directly
- The maintenance of the fair ecological footprint—"fair share"
- An improvement in exergy (available energy) or other thermodynamic measures

Several measures have been outlined in the literature for progressing green engineering. These measurements are defined in terms of return on investment (ROI), such as GHG return on investment (GROI), energy payback time, water payback time (or materials, consumables), and carbon footprint, including energy efficiency improvement. It is obvious that these measures have shown an improvement in several environmental and sustainability aspects of green manufacturing.

Attempts are afoot to reduce the waste from all input resources such as energy, materials, and water. To achieve these objectives, more research needs to be conducted on precision manufacturing technology, and sustainability considerations in design and manufacturing processes modeling need to be included. The manufacturing sector produces a large amount of waste in the form of chips and contaminants and appropriate attention should be paid to increase the safety of humans, society, and the world.

16.1.1 Manufacturing Job Creation

The United States is the second-largest producer of manufactured goods after China. In 2010, the sector was worth about 1 trillion dollars. In the United States, blue-collar workers received a wage of about $74,447 in 2009, according to the National Association of Manufacturers. The problem of sustainable manufacturing or green manufacturing needs to be tackled, keeping in mind the productivity and the job creation potential of manufacturing industries. We need a twenty-first-century job creation strategy in this regard (Figure 16.1).

16.1.2 Efficient Sustainable Manufacturing

Key sustainability concerns such as energy consumption, emissions, and raw material usage in the manufacture of a part should be simulated and optimized. The concept of recycling at end of life, reuse, reducing waste and cost is essential for sustainability. Over designing can be avoided by using CAD including finite element analysis software.

It has been observed that sustainable design can be achieved by the proper selection of materials, which should be of lowest cost, easily machinable, and environmentally friendly. With these objectives in mind, manufacturing engineers should design proper machine tools and cutting tools. Manufacturers the world over see sustainable design and manufacturing as an effective response to global competition

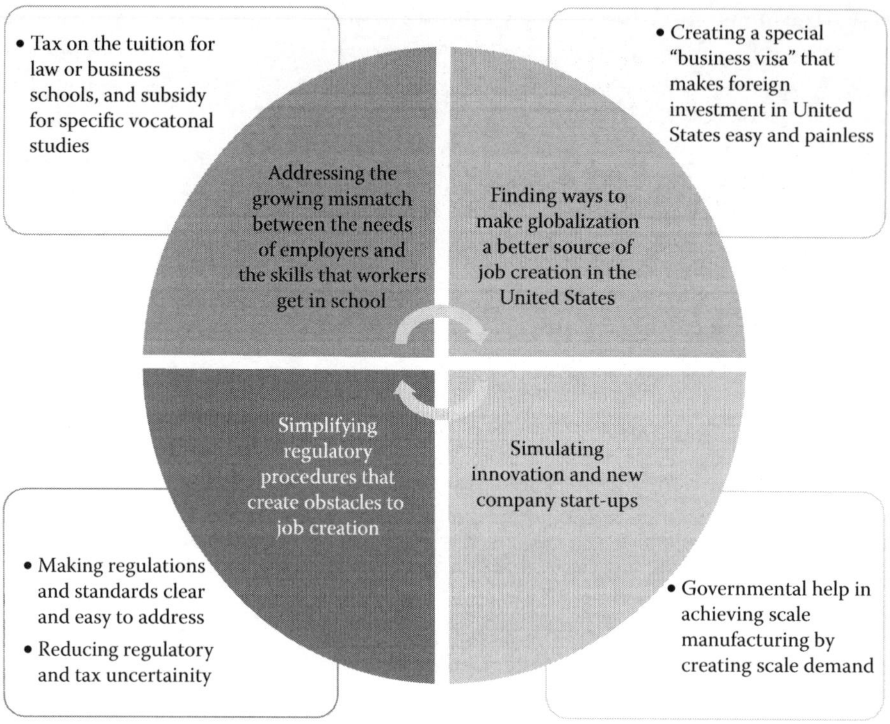

Figure 16.1

Manufacturing job creation.

and regulations. It requires a holistic view, from designing products that use less materials and energy to improving the machines and processes. A holistic view of the life cycle of a manufacturing system is shown in the following graph.

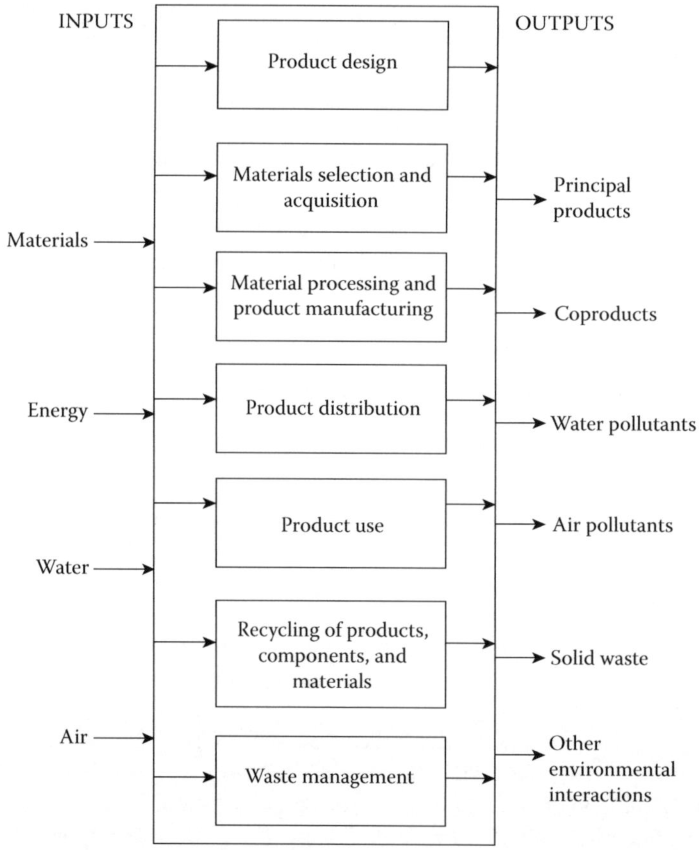

Our attention should start from the extraction of materials and their preparation using water, energy, and air. The product design should start with the selection of the material that has the least environmental impact and its distribution. The distribution of a manufactured product requires energy, which in turn emits CO_2. Therefore, the manufacturing center should be as close as possible to the market, thereby requiring less transportation and having less impact on the environment. The use phase of the product will again require power and energy consumption, which should be minimized. The reuse and recycle of the product at its end of life should be encouraged. Society will need to be encouraged to accept products that have been manufactured with recycled materials. The manufacturing industry should pay attention to all waste streams, including solid, liquid, and air pollutants. The pollution of air is not confined to the neighborhood where the manufacturing is taking place, but it can travel for thousands of miles. It does not obey the boundaries of nations or recognize the sovereignty of a nation.

Therefore, waste minimization should be the prime responsibility of the manufacturing industry. Waste minimization involves efforts to minimize resources and energy use during manufacturing. For the same product, using fewer materials will create less waste. It is obvious that waste minimization requires knowledge of the production process, tracking materials from their extraction to their return to the earth (cradle to grave), and even detailed knowledge of the composition of waste. Waste composition will affect the pollution of the earth, water, and air. The creation of waste may be reduced by selecting and using better manufacturing processes.

16.1.3 Waste Streams in Machining

The waste that is generated due to machining is a major environmental concern for society. The form and state of the waste streams generated and their transportation of waste differ with the process used and also vary within the process. Each type of waste, including solid, liquid, gaseous, and heat, has a different impact on the environment.

16.1.4 Life Cycle Analysis of Manufacturing Waste

Normally, the waste from manufacturing operations includes material, energy, cutting fluid, and tooling. The life cycle of a waste stream can be presented graphically as shown in Figure 16.2.

In a life cycle analysis (LCA), the input resources must balance the output including waste [1]. The inputs in any manufacturing operation, whether a turning, milling, grinding, drilling, forming, or casting process, are

1. Raw material
2. Energy
3. Cutting fluids
4. Toolings

However, the output of the manufacturing system has the useful machined part including waste such as a material stream (chips), worn-out cutting tools, waste cutting fluids used in cooling and lubricating, and

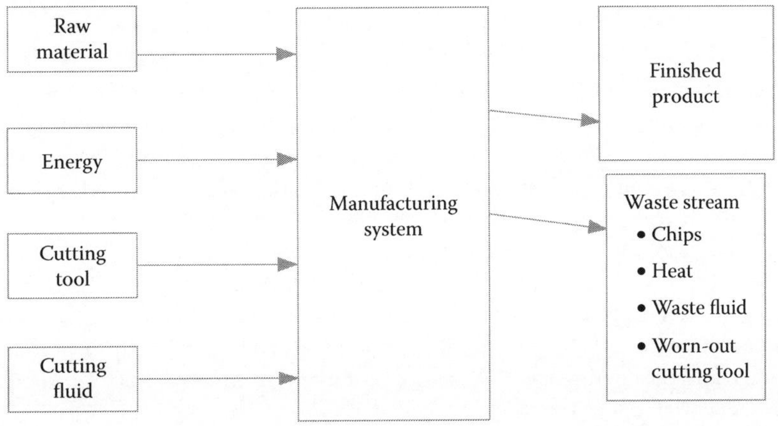

Figure 16.2

The LCA manufacturing system.

energy in the form of the heat that is generated during the machining operation. There may even be leakage of some energy. Therefore, the waste streams consist of material in the form of

- Chips
- Energy used and heat generated
- Fluid wastes
- Worn cutting tools

In an LCA, the mass balance of the material entering the machining operation should account for the materials lost. Wastage of material during manufacturing has serious environmental consequences due to the depletion of material and energy used during extraction and processing of material. There are also CO_2 emissions during processing and this should also be noted. Manufacturing operations need energy and its associated environmental effects in terms of toxic and hazardous gases need to be investigated. It should be noted that a manufacturing operation cannot avoid creating waste material during machining. However, if some amount of material, waste cutting fluid, and worn-out tools are recycled, this will reduce the environmental impact of the waste stream. The physical properties of the work material and the machining conditions affect the shape and size of the chips that are formed during the metal-removal process. It is more economically viable to recycle large volumes of chips rather than small volumes. Another constraint in recycling is the contamination of the stream due to the presence of dissimilar materials.

Obviously, the handling and disposal of chips are dependent on their shape, size, and texture. The shape of the chip influences the amount of cutting fluid that is carried away. Continuous chips are formed while cutting ductile metals such as iron, steel, copper, and aluminum. The chips are formed from the shearing action of a cutting tool on the work. This is called the *primary deformation zone*. Forces that are transmitted to the chips at the tool–chip interface are sufficient to slide the chips along the tool face. Sometimes during the machining of hard materials, the chips get fractured and segmented and these are called *discontinuous chips*. The machining of brittle materials such as cast iron and cast brass and ductile materials at extremely low speeds and high feeds causes the formation of discontinuous chips.

The volume (mass) of a chip that is formed during a machining operation should be easily measured or calculated as

$$\text{Mass}_{\text{raw material}} = \text{Mass}_{\text{product}} + \text{Mass}_{\text{chips}}$$

that is

$$\text{Mass}_{\text{chips}} = \text{Mass}_{\text{raw material}} - \text{Mass}_{\text{product}}$$

The mass of chips calculated can be used to calculate the environmental impact of a waste material stream.

16.1.5 Waste of Energy in Form of Heat

Energy can neither be created nor destroyed—it can only change from one form to another. With depleting sources, energy conservation is important for sustainable manufacturing. The machine tools that are used during a cutting operation require energy. The amount of energy that is required for any manufacturing process can be estimated using equations that are available in the literature as well as through the specific energy already tabulated for engineering materials. Specific energy (unit power) is used to compute the amount of energy that is required for the machining operation. Unit power is defined as the amount of power that is required to machine a unit volume of material. The amount of energy consumption (power) can be estimated by multiplying the specific energy (energy/unit volume of material) by the material removal rate (volume of material removed). The energy consumption has an impact on the heat that is generated. The heat that is generated and the energy that is used have serious environmental impacts in the form of a carbon footprint and subsequent global warming.

The heat that is generated during machining affects the tool wear and the surface finish (Figure 16.3). It may be assumed that the heat generated (H) during machining is equal to the energy consumed. The conversion of energy into heat occurs in two regions: the primary deformation zone or the shear zone and

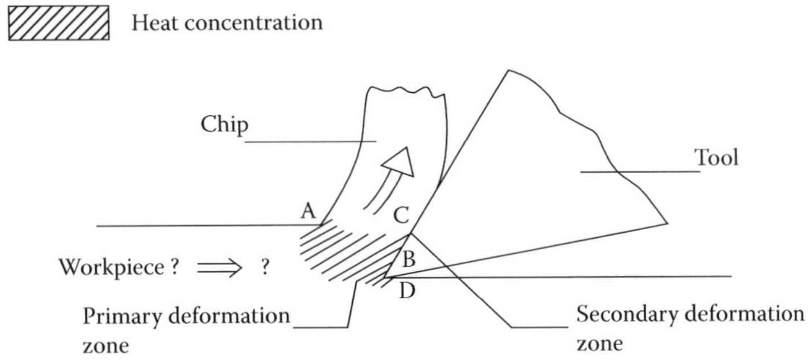

Figure 16.3

Heat generation in metal cutting.

the secondary deformation zone or the friction zone. The heat that is generated during machining causes a change in the temperature of the workpiece, the fluid used, and the tool. The heat that is generated is dissipated to the workpiece, the tool, and the chip. It has been experimentally established that virtually all of the heat that is generated in machining is dissipated to the chip and the workpiece. The rate of the heat flow to the tool forms a small portion of the total rate of the heat flow and may be neglected.

16.1.6 Cutting Fluid Heat Transfer

One important energy issue to be considered is the heat transfer between the fluid and the workpiece. The amount or rate of heat transfer is a direct measure of the cutting fluid performance. The cutting fluid heat transfer in machining operations depends on the nature of the cutting fluid that is used and the fluid application strategy. The transfer of heat between the fluid and the workpiece is assumed to be primarily due to convection.

16.1.7 Cutting Fluids Waste Stream

Many researchers have questioned the role of cutting fluids in machining operations. It is claimed that cutting fluids play many roles in metal machining. This section provides an insight into these claims and the reasons behind the use of cutting fluids. When properly applied, cutting fluids are said to increase productivity and quality and reduce costs by making it possible to apply higher cutting speeds, higher feed rates, and greater depths of cuts. Depending on the machining operation that is being performed, a cutting fluid has one or more of the following claimed functions:

- Cooling the tool, workpiece, and chip
- Lubricating (reducing friction and minimizing the erosion of the tool)
- Controlling the built-up edge (BUE) formation on the tool
- Flushing away chips
- Providing corrosive resistance to the workpiece

The relative importance of each function depends on the work material, the tool material, the cutting conditions, and the finish required. The selection of the fluid is governed by these functional requirements.

16.1.8 Types of Cutting Fluids

A variety of cutting fluids are available to satisfy the requirements of machining processes. Although there is no all-purpose cutting fluid, some offer considerable versatility while some are for specific applications. The preferred fluids have a long life and require very little changing or modifying. The four basic types of cutting fluids are as follows:

16.1.8.1 Straight Oils

A straight oil is a petroleum or vegetable oil that is used without dilution with water. Soluble oils may or may not be compounded with additives. Paraffin oils, naphthenic oils, and vegetable oils are some examples. It is claimed that straight oils provide excellent lubrication. For environmentally favorable requirements, vegetable oils are preferable due to their ease of biodegradation and disposal.

16.1.8.2 Soluble Oils

Soluble oils or emulsifiable oils are the largest type of cutting fluids used in machining operations. These are composed of an oil with emulsifiers and additives added and diluted with water. The dilution levels vary from 1% to 20% with 5% being the most common dilution level. Emulsifiers are chemical substances that cause the suspension of oil droplets in water to make a translucent solution in water. The predominant emulsifier is sodium sulfonate, which is used with fatty acid soaps, esters, and coupling agents that provide a white emulsion with no oil or cream separating out after mixing with water.

16.1.8.3 Synthetic Fluids

Synthetic fluids are water-based fluids and contain no mineral oil. They have a typical particulate size of 0.003 mm. Water provides excellent cooling properties. However, this compromises the lubricity and also causes corrosion. To inhibit rust formation, multiple rust inhibitors are added. Synthetic fluids are widely used in applications that require a long tank life and modest lubrication.

16.1.8.4 Semisynthetic Fluids

Semisynthetic fluids are similar to soluble oils in that they are emulsions and are similar to synthetic fluids in that they are water-based fluids. About 5%–20% of mineral oil is emulsified with water to produce a microemulsion. The particle size is in the range of 0.1–0.01 mm. This is small enough to transmit all incident light. These types of fluids are preferred largely due to their similarities with both soluble oils and synthetic fluids. Some of their major advantages are rapid heat dissipation, cleanliness of the system, resistance to rancidity, and bioresistance (due to the small particle size). Figure 16.4 outlines the various decisions and criteria involved in the selection of a cutting fluid.

16.1.8.5 Environmental Issues of Cutting Fluids

The advantages of using cutting fluids are offset by the associated environmental, health, and safety consequences. The waste streams emanating from the use of cutting fluids have an impact on the environment and on the operator. Fumes, odor, and smoke from the fluid result in both dermal and eye irritation. Some waste streams are also toxic and carcinogenic.

Mist is one of the waste streams that is generated from cutting fluids. Cutting fluids vaporize due to the heat that is generated during machining. The vapors subsequently condense to form mist droplets. The nonaqueous components of the cutting fluid, such as biocides, end up as a fine aerosol. Cutting fluid mists are

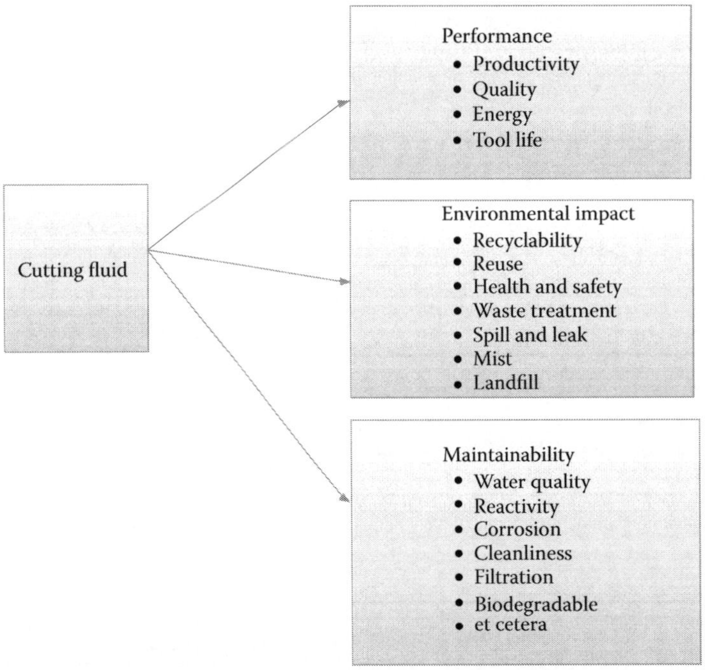

Figure 16.4

Cutting fluid selection criteria.

also formed by atomization mechanisms. As a cutting fluid stream impacts a stationary surface, the fluid may splash to form a mist (converting the mechanical energy in the stream to droplet surface energy). The drops that are formed by vaporization, condensation, or atomization are under the influence of buoyancy force, aerodynamic drag force, and gravity force.

The mist that is formed, if not collected, remains suspended in air and finally settles on a large area of the work floor. Most industries employ collection systems to ensure that their workers have minimum exposure to the fluid mist, which is in accordance with the regulations laid down by the Occupational Safety and Health Administration (OSHA). One such standard is the 5 mg/m^3 specification for airborne particles (largely due to fluid mist). However, mist collectors and machine enclosures are expensive to install and maintain.

16.1.8.6 Tooling Stream

Most manufacturing processes use some form of cutting tools and dies. These processes often require regular replacement of the tools. At the same time, there are processes that involve minimal or no tool losses. Most joining processes are of this type.

Worn tools contribute significantly to waste streams in the form of wear particles and a worn tool at the end of its useful life. The wear particles are usually carried away by the fluid. From an environmental perspective, the tools remaining at the end of their tool life are of importance as they are often disposed of and hence are a burden to the environment. It is thus necessary that tools have a long tool life so that the number of tools expended is minimized.

Tool life is one of the most important economic considerations in metal cutting. High grinding and replacement costs act as restrictions to the selection of process parameters that lead to shorter tool life. On the other hand, the use of low feeds and speeds not only results in an improved tool life but it also reduces the production rate. Tool failure occurs in two ways:

1. The gradual or progressive wearing away of the rake face (crater wear) or the flank (flank wear) or both of the cutting tool
2. Failures leading to the premature end of the tool

As the tool wears, its geometry changes, thus influencing the process performance in terms of the forces, the energy consumed, and the surface finish. Tool wear in machining operations occurs by adhesion, abrasion, and diffusion wear. The wear process is to a large extent dependent on the cutting temperatures. Abrasion is the common phenomenon in a low-temperature regime. At higher cutting speeds, which result in higher temperatures, the chemical solubility of the material determines the wear rate. At still higher cutting temperatures, diffusion wear occurs.

The LCA of a manufacturing system should take all four waste streams, including material, energy, cutting fluids, and worn-out cutting tools, into consideration. These topics will be discussed separately in more detail during the presentation of the manufacturing operations.

The forming processes create less waste than the metal-removal-processes and so the nature and the kind of manufacturing are important considerations in reducing waste. However, the creation of waste also arises during the supply chain as a company that is handling the product may insist on using a particular packing so that the product can fit into their packaging equipment. Waste management in particular focuses on processing waste after it has been created, concentrating on recycling, reuse, and waste-to-energy conversion.

Industries using sustainable, efficient, and better materials will generate less waste. These sustainable measures can make a clear and positive contribution to business in terms of ROIs. An environmentally friendly business would improve its stock price and increase its sales, and in turn improve its profit margins. By reducing the use of expensive and hazardous materials and waste, the business can build better community relations.

16.1.9 Critical Areas for Sustainable Manufacturing

The manufacturing industry must expand the boundaries of analysis by including social and environmental effects besides component production; eco-indicators (environmental impact) must go hand in hand with GDP; the global nature of manufacturing should include these sustainable considerations across geographical boundaries; greater emphasis should be placed on complex interactions between the various aspects of sustainability; exergy losses and entropy production at the global level must be considered; manufacturers must understand the effect of their business on such societal topics as (1) climate change, (2) ocean acidification, (3) ozone depletion, (4) nitrogen and phosphorus cycle overloading, (5) global freshwater withdrawals, (6) land system change, (7) biological diversity, (8) atmospheric aerosol loading, and (9) chemical pollution.

The expectations of society are obvious: reducing energy consumption; reducing waste; reducing material utilization; enhancing product durability; increasing operational safety; reducing toxic dispersion; reducing health hazards/improving health conditions; consistently improving manufacturing quality; improving recycling, reuse of scrap material, and remanufacturing; and maximizing sustainable sources of renewable energy.

In Europe, cleaner production is a preventive, company-specific environmental protection initiative. It is intended to maximize product output within the constraints of waste minimization and toxic emission [2]. By analyzing the flow of materials and energy in a company, the engineer tries to identify options to minimize waste and reduce emissions from manufacturing processes through some reduction strategies. The improvement in manufacturing technology has provided better choices in the materials and energy used, to reduce waste including wastewater generation and gaseous emissions. In Europe, industries participate in the 3P program of *pollution prevention pays*. The United Nations Environmental Program (UNEP) and the United Nations Development Organization (UNIDO) work on the same concept. UNEP and UNIDO believe that the 3P program will help developing countries to leapfrog from pollution to less pollution using new technologies. The concept of cleaner production is an outgrowth of this concept, which goes as far as to increase the resource efficiency of production in general. In the United States, the term *pollution prevention* is commonly used for cleaner production.

To improve sustainability in manufacturing, several international organizations have advocated ways to achieve sustainability. The basic elements of *sustainable* manufacturing can be listed as follows:

- Reducing energy consumption
- Reducing waste
- Reducing material utilization
- Enhancing product durability
- Increasing operational safety
- Reducing toxic dispersion
- Reducing health hazards/improving health conditions
- Consistently improving manufacturing quality
- Improving recycling, reuse, and remanufacturing
- Maximizing sustainable sources of renewable energy

The traditional mantra is high quality at low cost; however, in sustainable manufacturing, we are looking at waste reduction and GHG reduction, including recycling, reuse, and manufacturing. Various international organizations have presented ways and means of sustainable manufacturing, and these will be presented in the following sections.

16.1.10 Organization for Economic Cooperation and Development

16.1.10.1 Sustainable Manufacturing Toolkit

Companies across the world are likely to face increased costs of production due to compliance with local regulations coupled with the higher expectations of customers, investors, and communities at large. Industries need to manufacture economically and environmentally sustainable products to meet the needs of an environmentally conscious society. The Organization for Economic Cooperation and Development (OECD) has developed a sustainable manufacturing toolkit to measure the environmental performance of manufacturing factories. The start-up guide provides easy-to-read guidance to help the reader understand the basic issues and to start step-by-step measurement. There are seven step-by-step guidelines. These steps are divided into three categories as follows.

16.1.10.2 Prepare

- Step 1: Map your impact and set priorities
- Step 2: Choose indicators and understand data needs

The aim of the first step is to establish a general understanding of your positive and negative environmental impact by mapping your activities and determining which ones affect your performance the most.

There are many ways to reduce the environmental impact of your facility and improve its performance. Try to find out the local environmental regulations and the interest of customers and purchasers. If your product is likely to be sold in foreign markets, be aware of their environmental regulations, particularly EU countries that have very tough environmental regulations regarding the environmental affect of products sold in their market. Anticipating the needs of customers and investors is the surest way to attract external investors and is likely to help you to be a preferred supplier and brand. The OECD guidelines for preparing a company for sustainable manufacturing are enunciated as follows:

1. Identify any hazardous material that is being used for producing a part and replace it with less damaging alternatives.
2. Identify poor efficiency and heat loss in your facility. If there are any, identify opportunities for improvements. Try to involve staff in this action.
3. Plastics used for packaging should be replaced with recycled cardboard.

It is advisable to prioritize your environmental impact area as high, medium, and low.

Obviously, one should start with the high priority area, which is likely to have an immediate impact in the market with the potential for substantial gain. Undoubtedly, medium- and low-impact areas should follow. The company should be able to reassure its business customers that all environmental concerns have been addressed. Employees and the local community should receive assurances that the air quality and their exposure are satisfactory. Concerns about the GHG emissions from products should be dealt with immediately. However, there need to be some quantifiable measurements to show that you are succeeding. The OECD has proposed the following measurements for sustainability. Such measurements are used by businesses to monitor their performance.

16.1.10.3 Measure

- Step 3: Measure inputs used in production
- Step 4: Assess operations of your facility
- Step 5: Evaluate your products

The first set of indicators relates to the raw materials and intermediate products used in your production processes to make your products. Material selection would have the highest impact on the environmental performance. In fact, you can improve the environmental performance of your material inputs in a number of ways that will create win-win benefits and often save costs. For instance, in 2009, companies in Massachusetts participating in a program to reduce the use of toxic substances reported a range of benefits, including

- 55% reported management's increased attention to environmental practices.
- 51% reported improvements in workers' health and safety.
- 41% reported financial savings from reducing the use of toxic substances [3].

The key benefits from greening inputs are recognized as reduced expenditure on materials and reduced costs related to handling, storage, and treatment. It carries a lower regulatory compliance burden and reduced waste disposal requirements. Any materials and substances that go into making a product will have an environmental impact, so it is important to be prudent in material selection.

16.1.11 British Standard Institution

16.1.11.1 PAS 2050 (Publicly Available Specifications)

The British Standard Institution (BSI) has formulated the following steps to bring sustainability to manufacturing through publicly available specifications (PAS) 2050. These steps are useful for assessing the life cycle GHG emissions of goods and services. Climate change continues to be one of the greatest challenges facing nations, governments, businesses, and citizens and will influence the way that we live and work in future decades. Past and current actions, including the release of CO_2 and other GHGs through human activities such as the burning of fossil fuels, emissions from chemical processes, and other sources of anthropogenic GHGs, will have an effect on the future global climate. The GHG emissions associated with goods and services reflect the impact of processes, materials, and decisions occurring throughout the life cycle of these goods and services.

Life cycle GHG emissions are the emissions that are released as part of the processes of creating, modifying, transporting, storing, using, recycling, and disposing of such goods and services. PAS 2050 is focused on the single environmental issue of GHG emissions and their contribution to climate change. The range of possible environmental impacts from specific goods or services and the principles and techniques should include

1. Cradle-to-gate and cradle-to-grave GHG emissions assessment data as part of the life cycle GHG emissions assessment of goods and services.
2. Scope of GHGs emissions and removals to be included. Both emissions to the atmosphere and removals from the atmosphere shall be accounted for in the assessment of the overall GHG emissions of the product being assessed. The assessment of GHG emissions and removals during the life cycle of products shall include emissions and removals as identified in Annex A from processes including but not limited to
   a. Energy use (including energy sources, such as electricity, which are themselves created using processes that have GHG emissions associated with them)
      i. Combustion reactions
      ii. Loss to the atmosphere of refrigerants and other fugitive GHGs
      iii. Process operations
      iv. Service provision and delivery
      v. Land use and land use change
      vi. Livestock production and other agricultural processes
      vii. Waste management
3. Criteria for global warming potential (GWP) data. GHG emissions and removals shall be measured by mass and shall be converted into CO_2e using the latest Intergovernmental Panel on Climate Change (IPCC) 100-year GWP coefficients. It has been estimated that methane has a GWP coefficient of 25, that is, 1 kg of methane is equivalent to 25 kg of CO_2 in terms of its GWP.
4. Treatment of emissions and removals from land use changes the biogenic and fossil carbon sources.
5. Treatment of the impact of carbon storage on products and offsetting.
6. Requirements for the treatment of GHG emissions arising from specific processes.
7. Data requirements and accounting for emissions from renewable energy generation.

For the life cycle assessment, PAS has provided definitions of a system boundary, which includes all of its material life cycle processes.

The system boundary for an assessment identified as cradle to gate shall include the emissions and removals that have occurred up to and including the point where the product leaves the organization that is undertaking the assessment, for transfer to another party that is not the consumer. Cradle-to-gate GHG emissions assessment information shall be clearly identified as such so that it is not mistaken for a full assessment of the life cycle GHG emissions of a product.

16.1.11.2 Production Materials

The GHG emissions and removals arising from all processes used in the formation, extraction, or transformation of materials that are used in production (including farming, horticulture, fishing, and forestry) shall be included in the assessment, including all sources of energy consumption or direct GHG emissions associated with that formation, extraction, or transformation.

16.1.11.3 Energy

The GHG emissions and removals associated with the provisions and use of energy in the life cycle of the product shall be included in the emissions arising from the energy-supply system. Emissions from energy supply include

- Emissions at the point of consumption of the energy (e.g., emissions from the burning of coal and gas)
- Emissions from the provision of energy, including the generation of electricity and heat
- Emissions from transmission losses, transport fuels
- Upstream emissions (e.g., the mining and transport of fuel to the electricity generator or other combustion plant)

- Downstream emissions (e.g., the treatment of waste arising from the operation of nuclear electricity generation)
- Growing and processing of biomass for use as fuel

16.1.11.4 Capital Goods

The GHG emissions and removals from the production of capital goods used in the life cycle of product shall be excluded from the assessment unless provided for in the supplementary requirements in accordance with the principles set out in Section 16.3.1.

16.1.11.5 Manufacturing and Service Provision

The GHG emissions and removals arising from manufacturing and service provision that occur as part of the life cycle of the product, including emissions associated with the use of consumables, shall be included in the assessment of the GHG emissions of the life cycle of the product.

16.1.11.6 Transport

The GHG emissions and removals arising from road, air, water, rail, or other transport methods that form part of the life cycle of the product shall be included in the assessment of the GHG emissions.

16.1.11.7 Storage of Products

The GHG emissions and removals arising from storage shall be included in all assessments of the life cycle GHG emissions of the product, including

1. Storage of inputs, including raw material, at any points in the product life cycle.
2. Environmental controls (e.g., cooling, heating, humidity control, and other controls) related to a product at any point in the product life cycle for all operations, including environmental control, of factories in which the product may be stored.
3. Storage of products in the use phase.
4. Storage prior to reuse, recycling, or disposal activities.

16.1.11.8 Use Phase

The GHG emissions and removals arising from the use of products shall be included in the assessment of the life cycle GHG emissions of products, subject to the provision of cradle-to-gate assessments. When use-phase emissions and removal form part of an assessment carried out under the PAS, the detail of the basis on which the use phase for the product is assessed shall be recorded and retained.

16.1.11.9 Allocation to Coproducts

The preferred approach to the allocation of emissions and removals to a coproduct shall be, in order of preference:

1. Dividing the unit processes to be allocated into two or more subprocesses and collecting the input and output data related to these subprocesses or
2. Expanding the product system to include additional functions related to the coproducts where
   a. A product that is displaced by one or more of the coproducts of the processes being considered can be identified.
   b. The avoided GHG emissions associated with the displaced product represent the average emissions arising from the provision of the avoided product.

16.1.11.10 General Allocation Principle

Where waste results in GHG emissions (e.g., organic matter disposed of in a landfill), those emissions (CO_2 and non-CO_2) shall be allocated to the product system that gives rise to the waste. This allocation also applies to methane that is combusted without the generation of useful energy (i.e., flaring).

16.1.11.11 Use of Recycled Materials and Recycling

The method for assessing emissions arising from recycled or recyclable material shall be as specified in Annex D.

16.1.11.12 Treatment of Emissions Associated with Reuse

The following relation is a good tool to analytically measure GHG emissions when a product is reused. GHG emissions per instance of use shall be assessed on the basis of

$$\text{GHG emissions} = \frac{a+f}{b} + c + d + e$$

where:

- a is the total life cycle GHG emissions of the product, excluding use-phase emissions
- b is the anticipated number of reuse instances for a given product
- c is the emissions arising from an instance of refurbishment of the product to make it suitable for reuse (e.g., recovering and sterilizing a glass bottle)
- d is emissions arising from the use phase
- e is emissions arising from transport returning the product for reuse
- f is emissions arising from disposal

16.1.12 Greenhouse Gases

We noted earlier that water vapor (H_2O) and carbon dioxide (CO_2) are two of the major GHGs. Both of these molecules are naturally present on Earth, their amounts having evolved to steady levels as a total effect of plant and animal life. Several other gases contribute to the greenhouse effect. Most prominent among these are methane (CH_4), nitrous oxide (N_2O), ozone (O_3), and CFCs. Each gas is present in the atmosphere at a different concentration. In addition to knowing the concentration of the GHG, it is important to understand that each gas has a different relative greenhouse efficiency. In other words, some gases are better at absorbing solar radiation than others, and therefore have a greater overall impact on the greenhouse effect. As a measure of this efficiency, GHGs are often assigned a value for GWP. This value is simply a comparison of the efficiency of a gas relative to CO_2 over a time span of 100 years. Therefore, a gas with a GWP of 20 is 20 times more efficient at retaining heat than CO_2 over a 100 year time period.

Another important characteristic to examine when considering the potential damage of a GHG is the average time it remains in the atmosphere once it has been emitted. Policies made now that reduce industrial and other emissions of GHGs take years to result in changes in the atmosphere and its nature. In fact, scientists have stated that no matter how aggressively GHG emissions are restricted and reduced now, "the planet and the nation are certain to experience more than a century of climate change, due to the long lifetime of greenhouse gases already in the atmosphere and the momentum of the climate system" [4].

Of all the GHGs, carbon dioxide is present at the highest concentration by far. Based on 1990 concentrations, carbon dioxide is said to be responsible for almost 60% of the total greenhouse effect when efficiency and concentration are considered. Its concentration is increasing in the atmosphere due in large part to the extensive burning of coal and other fossil fuels for energy production. Another cause of the increase in CO_2 is the destruction of large areas of trees, which leads to a reduction in the use of carbon dioxide for photosynthesis.

Methane is present in the atmosphere at less than 1% of the levels of carbon dioxide; however, it is 25 times more efficient as a GHG. It contributes to just over 10% of the total greenhouse effect based on current concentrations. Its primary anthropogenic sources are the combustion of fossil fuel and the decomposition of organic materials associated with wetlands, rice paddies, and livestock manure.

Nitrous oxide also occurs in low concentrations relative to carbon dioxide, but it is 230 times more efficient as a GHG. These factors combine to make it a 6% contributor to the total greenhouse effect. The primary anthropogenic sources are fossil fuel combustion, fertilizers, and deforestation.

The only major GHGs that are not naturally occurring are CFCs. They come solely from anthropogenic sources such as the production or use or both of foams, aerosols, refrigerants, and solvents. They are present at an extremely low concentration in the atmosphere; however, they are 15,000 times more efficient as a GHG relative to carbon dioxide. As a result, they contribute to approximately 25% of the total greenhouse effect based on 1990 concentrations.

Table 16.1 lists the major GHGs, their emission levels and sources, and the approximate amount of time that they remain in the atmosphere once they have been emitted. The table also gives their approximate concentrations 100 years ago, today, and their projected concentration for the year 2030. Note that concentrations are given in parts per billion (ppb), referring to the proportion of molecules of the gas per billion molecules in the atmosphere. For example, the concentration of CO_2 today is 350,000 ppb, meaning that for every billion molecules in the atmosphere, there are 350,000 molecules of CO_2.

Table 16.1 Summary of Greenhouse Gases

Gas	Major Anthropogenic Sources	Amount Released per Year (millions of tons)	Average Time in the Atmosphere	Global Warming Potential (over 100 years)	Preindustrial Concentration (around 1860) (ppb)	Average Concentration Now (ppb)	Expected Concentration in 2030 (ppb)
CO_2	Burning of fossil fuels	5500	100 years	1	290,000	350,000	500,000
CH_4	Fossil fuel production, rice fields	500	10 years	21	850	1,700	2,300
N_2O	Fertilizers, deforestation, burning biomass	30	Days	310	0.001–7	0.001–50	0.001–50
CFCs	Aerosol sprays, refrigerants	1	60–100 years	1500–8100	0	about 3	2.4–6

16.1.13 Calculation of GHG Emissions of Products Presented by PAS

PAS requires that both emissions to the atmosphere and removals from the atmosphere be taken into account in calculating the total GHG emissions of a product over its life cycle. The following method is used to calculate the GHG emissions per functional unit of the product under assessment.

1. Determine the emissions and removals for each activity within the system boundary as primary activity data or secondary data, with emissions included as positive values and removals included as negative values.
2. Convert the primary activity data and the secondary data to GHG emissions and removals per functional unit of the product under assessment by multiplying the data by the emissions factor for each activity.
3. Convert the GHG emissions and removals data into units of CO_2 by multiplying the individual GHG emissions or removals figures by the relevant GWP.
4. Calculate the overall impact of carbon storage associated with the product.
5. Sum the CO_2 emissions and removals occurring in the life cycle of the product under assessment.

16.1.14 Global Warming Potential

The values of GWPs for GHGs that are to be used in calculations shall be in accordance with Table 16.2. Table 16.2 presents the data for global warming from different sources up to the Montreal protocol [5]. The GWP of various GHGs, chloroflorides, hydrofluorocarbons, fluorinated ethers, perfluoropolyethers, and perfluoropolymethylisopropyl ethers (PEPMIEs) are included in Table 16.2. It can be noticed that sulfur hexafluoride has the maximum GWP relative to carbon dioxide (CO_2), followed by trifluoromethyl sulfur pentafluoride (SF_5CF_3) with 17,700 GWP, and third in the list is nitrogen trifluoride (NF_3) with 17,200, and so on. The GWPs of these chemicals can cause tremendous injuries to all living beings on the earth and their use and applications need to be controlled or eliminated altogether.

16.1.15 Assessment of Emissions Arising from Recycled or Recyclable Material Inputs

The sustainable reuse of materials should be developed with the goal of reducing the environmental footprint in the reuse of materials. Energy and carbon footprint analyses are important considerations for green manufacturing. Where a product uses recycled or recyclable material, the emissions associated with this material are estimated by adding the emissions associated with recycled and virgin material and with the disposal of the material after use in accordance with following equation.

The emissions and removals arising per unit (E) from that material shall reflect the product-specific recycled content or the recycling rate or both based on the following calculation:

$$E = (1 - R_1)E_V + R_1 E_R + (1 - R_2)E_D$$

where:

R_1 = proportion of recycled material input

R_2 = proportion of material in the product that is recycled at end of life

Table 16.2 Recent (except for CH_4) Global Warming Potential (GWP) Relative to CO_2

Industrial Designation or Common Name	Chemical Formula	GWP for 100-year Time Horizon
Carbon dioxide	CO_2	1
Methane	CH_4	25
Nitrous oxide	N_2O	298
Substances Controlled by the Montreal Protocol		
CFC-11	CCl_3F	4,750
CFC-12	CCl_2F_2	10,900
CFC-13	$CClF_3$	14,400
CFC-113	CCl_2FCClF_2	6,130
CFC-114	$CClF_2CClF_2$	10,000
CFC-115	$CClF_2CF_3$	7,370
Halon-1301	$CBrF_3$	7,140
Halon-1211	$CBrClF_2$	1,890
Halon-2402	$CBrF_2CBrF_2$	1,640
Carbon tetrachloride	CCl_4	1,400
Methyl bromide	CH_3Br	5
Methyl chloroform	CH_3CCl_3	146
HCFC-22	$CHClF_2$	1,810
HCFC-123	$CHCl_2CF_3$	77
HCFC-124	$CHClFCF_3$	609
HCFC-141b	CH_3CCl_2F	725
HCFC-142b	CH_3CClF_2	Rev 2,310
HCFC-225ca	$CHCl_2CF_2CF_3$	122
HCFC-225cb	$CHClFCF_2CClF_2$	595
Hydrofluorocarbons		
HFC-23	CHF_3	14,800
HFC-32	CH_2F_2	675
HFC-125	CHF_2CF_3	3,500
HFC-134a	CH_2FCF_3	1,430
HFC-143a	CH_3CF_3	4,470
HFC-152a	CH_3CHF_2	124
HFC-227ea	CF_3CHFCF_3	3,220
HFC-236fa	$CF_3CH_2CF_3$	9,810
HFC-245fa	$CF_3CH_2CF_3$	1,030
HFC-365mfc	$CH_3CF_2CH_2CF_3$	794
HFC-43-10mee	$CF_3CHFCHFCF_2CF_3$	1,640
Perfluorinated Compounds		
Sulfur hexafluoride	SF_6	22,800
Nitrogen trifluoride	NF_3	17,200
PFC-14	CF_4	7,390
PFC-116	C_2F_6	12,200
PFC-218	C_3F_8	8,830
PFC-318	$c\text{-}C_4F_8$	10,300
PFC-3-1-10	C_4F_{10}	8,860
PFC-4-1-12	C_5F_{12}	9,160
PFC-5-1-14	C_6F_{14}	9,300
PFC-9-1-18	$C_{10}F_{18}$	>7,500
Trifluoromethyl sulfur pentafluoride	SF_5CF_3	17,700
Fluorinated Ethers		
HFE-125	CHF_2OCF_3	14900
HFE-134	CHF_2OCHF_2	6,320
HFE-143a	$CHF_3 OCF_3$	756
HCFE-235da2	$CHF_2OCHClCF_2$	350
HEF-245fa2	$CHF_2OCH_2 CF_2$	708
HFE-254cb2	$CH_2OCF_2 CHF_2$	659
HFE-254fa2	$CHF_2 OCH_2 CF_3$	359

Table 16.2 (Continued) Recent (except for CH_4) Global Warming Potential (GWP) Relative to CO_2

Industrial Designation or Common Name	Chemical Formula	GWP for 100-year Time Horizon
HFE-347mcc3	$CH_3OCF_2CF_2CF_3$	575
HFE-347pcf2	$CHF_2CF_2OCH_2CF_3$	580
HFE-356pcc3	$CH_3OCF_2CF_2CHF_2$	110
HFE-449s1(HFE-7100)	$C_4F_9OCH_3$	297
HFE-569sf2 (HFE-7200)	$C_4F_9OC_2H_5$	59
HFE-43-10-pccc124 (H-Galden 1040x)	$CHF_2OCF_2OC_2F_4OCHF_2$	1,870
HFE-236ca12 (HG-10)	$CH_2OCF_2OCHF_2$	2,800
HFE-338pcc13 (HG-01)	$CHF_2OCF_2CF_2OCHF_2$	1,500
Perfluoropolyethers		
PFPMIE	$CF_2OCF(CF_2)$	10,300
	$CF_2OCF_2OCF_3$	10,300
Hydrocarbons and Other Compounds—Direct Effects		
Dimethylether	CH_3OCH_3	1
Methylene chloride	CH_2Cl_2	8.7
Methyl chloride	CH_2Cl	13

E_R = emissions and removals arising from recycled material input per unit of material
E_V = emissions and removals arising from virgin material input per unit of material
E_D = emissions and removals arising from disposal of waste material per unit of material

This method is also called the *end-of-life approach* or *recyclability substitution* or the *0–100 output method.*

$$E = (1 - R_2)E_V + R_2 \cdot E_R + (1 - R_2)E_D$$

where:
R_2 = proportion of material in the product that is recycled at end of life
E_V = emissions and removals arising from virgin material input per unit of material
E_D = emissions and removals arising from disposal of waste material per unit of material
E_R = emissions and removals arising from recycled material input per unit of material

16.1.16 Single Release vs. Delayed Release

Where emissions from the use phase or the final disposal phase of a product are calculated according to a single release within 25 years of the formation of the product, the weighting factor (FW) to be applied to the GHG emissions that are released at that time shall reflect the number of years of delay in the emissions being released (i.e., the number of years between the formation of the product and the single release of the emissions) according to:

16.1.16.1 Single Release

$$\text{FW} = \frac{100 - (0.76xt_0)}{100}$$

where t_0 is the number of years between the formation of the product and the single release of the emissions.

16.1.16.2 Delayed Release

The weighted average time that the emissions are in the atmosphere can be calculated according to

$$\text{FW} = \frac{\sum_{i=1}^{100} x_i(100 - i)}{100}$$

where:
i = each year in which emissions occur
x = the proportion of the total emissions occurring in any year i

Example 16.1

If use-phase emissions were to be delayed for 10 years following the formation of the product, with total emissions being released evenly over the following 5 years, then the FW that represents the weighted average time that these emissions are present in the atmosphere would be

$$FW = \frac{(0.2x(100-11))+(0.2x(100-12))+(0.2x(100-13))+(0.2x(100-14))+(0.2x(100-15))}{100}$$

where:

0.2 is the average of 5 years

11, 12, ..., 15 represent the years after the 10th year

FW = 0.87

In this example, the total amount of use-phase emissions, expressed as CO_2e, released during the 100-year quantification period would be multiplied by a weighting factor of 0.87 to reflect the weighted average time that these emissions are present in the atmosphere during the 100-year quantification period.

16.2 Thermodynamics in Manufacturing Sustainability

The main purpose of manufacturing processes is to transform materials into useful products. In the course of these operations, energy resources are consumed and the usefulness of material resources is altered. Each of these effects can have significant consequences for the environment and for sustainability, particularly when the processes are practiced on a very large scale. We summarize thermodynamic data for three aspects of the manufacturing processes: (1) the energy requirements for the materials used in a manufacturing process, (2) the energy requirements for the manufacturing processes themselves, and (3) the efficiency of the material and energy transformations in manufacturing *processes.*

Thermodynamics is particularly well suited to analyze the magnitude of these effects as well as the efficiency of the transformations [6]. Thermodynamics is a science that primarily deals with energy. Energy can viewed as the capacity to cause change. The study of thermodynamics is concerned with the ways that energy is stored within a body and how energy transformations, which involve heat and work, may take place. The first law of thermodynamics is the law of conservation of energy. It is stated as "If a system passes through a cycle of processes so that it returns to its initial state the sum of heat and work effects will be zero[1]. Heat and work are mutually convertible but since energy can neither be created nor destroyed; the total energy associated with an energy conversion remains constant or no machine can produce energy without corresponding expenditure of energy, i.e. it is impossible to construct a perpetual motion machine of first kind." This is the very fundamental principle of sustainable manufacturing.

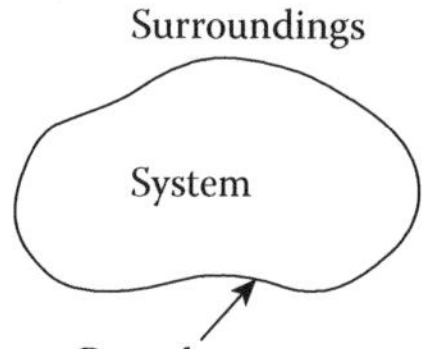

Systems may be considered closed or open, depending on whether a fixed mass or a fixed volume in space is chosen for study. A *closed system* consists of a fixed amount of mass and no mass may cross the system boundary. The closed system boundary may move. Examples of closed systems are sealed tanks and piston cylinder devices (note that the volume does not have to be fixed). However, energy in the form of heat and work may cross the boundaries of a closed system.

For a closed system, we have

$$\Delta E = Q + W(\text{closed system}) \tag{16.1}$$

where:

ΔE is the change in the energy content of the system

Q is the amount of heat transferred to the system from its surroundings

W is the amount of work that is done on the system by its surroundings

In many textbooks, we find the closed system equation defined as $\Delta E = Q - W$, where W is defined as the work done by the system on its surroundings. For practical application in manufacturing, we can write the aforementioned equation as

$$\Delta E = \Delta U + \Delta E_p + \Delta E_k = Q + W_{sh} - W_{pv} \tag{16.2}$$

This expression defines a change in the energy content (E) of the closed system in three parts: (i) changes in the internal energy (U), (ii) changes in the potential energy (E_p), and (iii) changes in the kinetic energy (E_k). The internal energy can be changed by applying force on the system and causing deformation (strain) of the material, caused by the application of force per unit area (stress). The potential energy is changed by shifting the system (material) location in a force field (e.g., gravitational, electrical, and magnetic fields). The kinetic energy is varied by increasing or decreasing the system velocity. Equation 16.2 splits the work into W_{pv} and W_{sh}, which we call "PV" work and "shaft work," respectively. PV work arises from the fact that every system, however small, has some volume. We can understand that changing the volume of work (material) changes the system's potential energy. The work is done on the system to reach a certain volume of the product.

An *open system*, or a *control volume*, has mass as well as energy crossing the boundary, called a *control surface*. Examples of open systems are pumps, compressors, turbines, valves, and heat exchangers.

Manufacturing systems can be idealized as open systems (Figure 16.5). Raw material and energies are input into the system and the finished product and waste are the output of the system. Therefore, manufacturing processes can be treated as thermodynamic processes. They play an important role in renewable energy. The calculation of the mass balances for each activity will be facilitated by the new option to characterize each input and output flow in terms of its specific properties, including the material content of specific elements or substances, and the new option to link specific inputs to specific outputs through the use of transfer coefficients. Also, waste can now be described as a physical flow with specific material properties, so that a specific mass balance can be maintained throughout the waste treatment or recycling processes or both. An input or output from an activity data set can only be characterized in units of one specific property, typically mass or energy. However, in different contexts it can be different properties that are of interest. To make an energy balance, of relevance is the energy content, the dry mass and water content for mass balances, the carbon content for carbon balances—maybe even different kinds of carbon according to their origin (fossil, biogene, carbonate), not to mention the many specific heavy metals. And for material flow analysis, the lifetime of a product may also be a property of interest. Additions to stock need to be characterized by these properties, in the same way as all other inputs and outputs. Seen from the waste-generating activity, the treatment of waste is a service input, while the wastes are physical outputs. To avoid describing the waste treatment service in isolation from the physical waste, we can use the simple balancing rule, that a positive input is identical to a negative output. From the perspective of the waste-generating activity, waste treatment can therefore be expressed as a negative service output, with a positive mass (or alternatively as a positive service input with a negative mass). Mass, energy, and monetary balances can also be maintained in a material flow accounting model for a product system in which the individual activities needed to produce

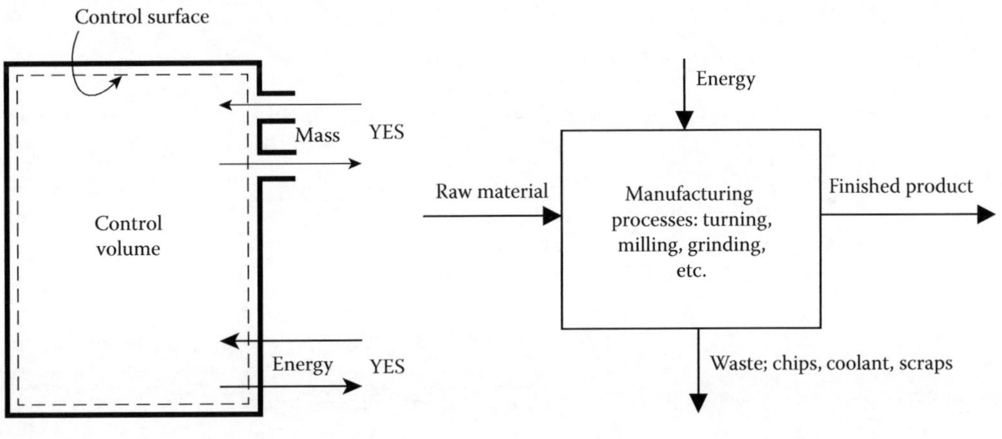

Figure 16.5

Manufacturing systems as open thermodynamic systems.

the product are linked. However, this requires that the individual activity data sets are maintained intact and are not allocated. The allocation (partitioning) of the inputs and outputs of an activity over the product outputs from the activity implies an artificial subdivision of the activity according to a specified allocation key. Most often, the applied allocation key is the revenue obtained from the product; however, the use of mass or energy content is also quite common in LCA practice, although allocation is generally dissuaded by both the ISO 14044 LCA standard and by economic modelers [7,8]. After allocation, the resulting artificial subactivities have only one product output, which is the purpose of the allocation, and all the other inputs and outputs are included in the allocated subactivities in proportion to the specified allocation key. If the allocation key is mass, the mass balance remains intact for the subactivities, but energy and elementary balances (e.g., the carbon balance) are skewed; with economic allocation, none of the physical balances remains intact in the subactivities except if by chance there is a physical parameter that follows the revenue from the products.

16.2.1 Method of Material and Energy Balance

The following steps are helpful in solving problems that combine material and energy balance calculations.

1. Since material balances have to be obeyed even if an energy balance is to be performed, the first thing to do is to solve the material balance part of the problem (i.e., choose a basis, make a flowchart, identify your unknown, perform a degrees-of-freedom analysis, write equations, and solve).
2. If an energy balance calculation is required in addition, there will be an extra unknown in the problem. This unknown may be a temperature, an amount of heat added, or some other variable that appears in the energy balance. The extra equation that you will use to solve for this additional variable is the energy balance. To begin, identify the "system" and write the appropriate form of the energy balance for it (i.e., closed, open, isolated). If no shaft work is taking place, then $W_{sh}=0$. Other simplifications may apply based on the given information.
3. For each species for which $\hat{H}$ or $\hat{U}$ are required, you will need to choose a reference state. If using data from an external source (i.e., tables, graphs), the reference state will be specified in that source. If you are calculating the internal energies or enthalpies, then you will need to decide what reference state to use. Often, it is convenient to choose one of the inlet or outlet conditions as the reference state because then the $\hat{H}$ or $\hat{U}$ for that stream becomes zero, thus simplifying the calculations.
4. Make a list of all the terms in the energy balance that you will need to calculate or look up to solve for the unknown of interest. This list may include internal energies, enthalpies, elevations, and mass flow rates (mass flow rates were presumably already calculated in Step 1).

This is shown in Figure 16.5 where the components of a manufacturing process system are represented as (a) the manufacturing process and (b) the energy conversion process for manufacturing, and (c) the materials production process and (d) the energy conversion process for the materials process. Each box represents an open, bulk flow process in steady state with enthalpy and entropy flows, heat transfer with the environmental reservoir at temperature TR, and shaft work or energy transfer between the processes.

16.3 Computational Tools for Green Manufacturing

Manufacturing has been identified as one of the major energy-consuming industries and should be analyzed in an ecodesign perspective, targeting the reduction of its environmental impact. It has been observed that manufacturing consumes about 30% of energy and produces about 60% of GHGs. Manufacturers are looking for guidelines to improve their products in an effective and low-cost manner. Several authors have been highlighting the importance of energy efficiency and the environmental impact of manufacturing processes in general, as many of the environmental concerns are directly related to the generation of energy, resulting in the depletion of natural energy resources. Most of these studies focus on the analysis of metal-removal processes, dealing with the influence of material removal and cutting fluids, in parallel to the impact of electricity consumption [6,9–11].

The environmental issues in metal-removal processes are as follows:

- Waste material creation
- Energy consumption

The large quantities of scraps and chips that are produced during material removal processes need to be reduced. This could be done by a better design or by designing with CAD software to minimize the volume of materials to be used in the manufacture of the part. Thus, as the volume of material is reduced so the energy requirements for the production and embodied energy will be reduced. Hence, there is need for better design and better machine and cutting tool design.

- Machine, material (embodied energy), temperature-controlled environment

Metal-removal processes, including forming processes, need lubricants and cutting fluids to lower the temperature during the metal-removal process, which in turn improves the life of the cutting tool as well as the quality of the product. However, these cutting fluids contain chemicals that have the potential to contribute to global warming by polluting the land, water, and air. Sometimes, they are injurious to the health of machinists and other workers on the factory floor. One way to reduce the use of cutting fluids is dry machining, which is more a theory than practice in the industries. Therefore, attempts should be made to reduce the use of

- Lubricants and hydraulic fluids
- Cutting fluids

Dry machining is one way to eliminate the use of lubricants and cutting fluids; however, it is difficult to practice.

In the case of chipless-shaping processes with no significant material waste or consumable usage, such as rolling, forging, and bending, the saving on the electrical consumption of the machine tool becomes the dominant factor to analyze, and pushes for a detailed analysis of the process sequence toward the identification of the main consumption drivers.

16.3.1 Life Cycle Assessment

An LCA is a tool that provides a methodology for quantifying a holistic view of the environmental impact of a product at all stages of its life, "cradle to grave." An LCA must include all essential phases: goal and scope definition, inventory analysis, impact assessment, and obviously interpretation of results.

An LCA is aimed at increasing the knowledge of the environmental impacts of manufacturing processes to develop and implement environmentally conscious processes, reduce environmental impacts, and take knowledge-based decisions. The application of an LCA encourages sustainability oriented decision-making based on the improvement of its operational applicability through increased knowledge and advanced methods. It includes decision-making problems integrating the broader sustainability, especially economic and performance considerations (eco-efficiency life cycle engineering), and operational applicability problems with their integration into existing industrial decision-making processes (design for environment [DFE]). Therefore, ECM involves manufacturing products so that their overall negative environmental effects are minimized. Understanding the life cycle of the product and its impact on the environment at each of its life stages help in making better decisions during the product stage itself. In addition, understanding the end-of-life stage of the product is critical since one of the largest impacts on the environment occurs at that stage.

Depending on the end-of-life strategy of the product, the design of the product can be realized to increase its recyclability, manufacturability, and disassembability and to minimize its environmental impact. When designing a product with environmental features, material selection should also be considered as a key element. Once the design decisions of a product are complete and the materials for its production are identified, the product's environmental attributes are pretty much set. However, in addition to the design and materials decisions, issues involving the selection of the energy source and cooling systems, and the handling of hazardous by-products and so on must be controlled during the manufacturing process to achieve a complete ECM concept. In this process, an LCA examines and quantifies the energy and the material used and wasted, and assesses the impact of the product on the environment. An LCA usually facilitates the systematic collection, analysis, and presentation of environmentally related data. The steps involved in an LCA, which are commonly repeated, are [5]: the identification of the goals and boundaries of an LCA, an analysis of an inventory to achieve a balance between material and energy in the system, an evaluation of the system's impact on the environment, and an assessment of the most promising system improvements to reduce the negative environmental impact.

The results of an LCA may provide the basis for the development of environmental laws, taxes, and regulations. The qualitative and quantitative characteristics of the product life cycle are taken into account by means of an LCA during the conceptual design of each new product. This enables designers to estimate the costs and benefits associated with the design attributes of the product, and the energy consumption, materials requirement, and end-of-life choices of the product. The scope of an LCA involves tracking all the materials and energy flows of a product from the retrieval of its raw materials out of the environment to the disposal of the product back into the environment. The complexity of the LCA problem grows when the product structure is large and complex and the number of factors to be considered increases. In practice, however, such a process could be extremely involved if the limits of the system are not clearly defined. Therefore, prior to the execution of an LCA, the associated goals and boundaries of the life cycle must be defined. Based on the observations made by Keolein and Menerey [12], an LCA can be treated as an optimization model by maximizing the added value and minimizing the resource consumption and waste dispersion activities. Emblemsvag and Bras [13] have developed activity-based cost functions by considering various uncertainty factors that are likely to occur in an LCA.

Ishii [14] developed the software called LINKER (developed using ToolBook under Microsoft Windows) by concentrating on advanced planning for product retirement and addressing the level to which a product should be disassembled. LINKER allows the user to evaluate a design at various stages of a product's life cycle. After entering the required data into the software's input stream, LINKER displays the disassembly times for components and fasteners, the compatibility index, and the retirement cost breakdown for each part. The knowledge gained during an LCA needs to be transferred into the initial design of a new product (Figure 16.1). It is actually possible to focus on a specific stage of the product's life such that the environmental impact is minimized.

Figure 16.6 presents the life cycle of a lamp. During the manufacturing phase, materials such as glass, steel, and copper are usually used along with mercury (Hg) and phosphorous. One can notice that reducing or eliminating the mercury and phosphorous from the input will improve the environmental impact of the lamp. During the manufacturing phase, a reduction in the solid waste in the form of scraps and chips is welcome. Air pollution reduction could be achieved by reducing the intake of dangerous substances in the making of the lamp. During the use phase, a reduction in energy consumption is always welcome. At the end of life, both possibilities of landfill and incineration exist. Therefore, an LCA should include calculations for both direct and indirect impacts, for example,

- Direct product and processes in the extraction and transportation of raw materials
- Direct energy, water, and fuel consumption as well as energy and heat losses through energy balancing
- Emissions such as the direct release of gases and indirect environmental impacts from processing

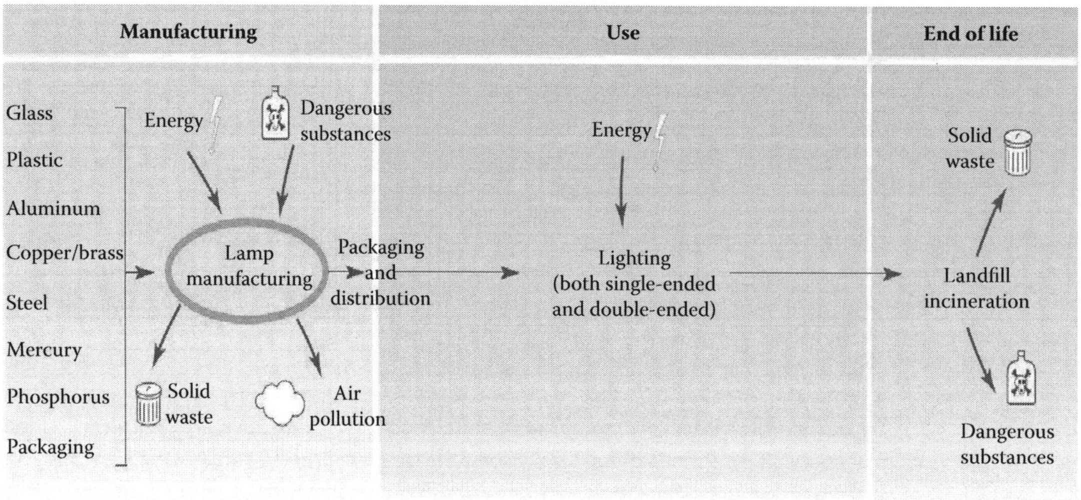

Figure 16.6

Life cycle analysis.

16.3.2 Environmentally Conscious Production (ECP)

In addition to environmentally friendly product designs resulting from DFE initiatives, issues involving production must also be addressed to have a complete concept of ECM. These issues include selecting energy sources that are necessary for production, designing cooling systems, and handling hazardous by-products. Currently, numerous production techniques, material handling systems, and energy sources are available and using an assessment tool to select among them may be valuable to improve the environmental features of the production system. Bock [15] has developed a tool to select a good material and process combination. Similar models have been developed to analyze how the selection of different manufacturing processes affect the environment [16–18]. Many companies monitor their waste generation from their manufacturing processes. Several techniques have been proposed for such a monitoring process. For example, weighting methods have been proposed to measure the chemical and toxic discharges of the different manufacturing methods used [19,20]. Pellerin has examined the effect of automation on the environmental indicators of a production. Major developments to date are waste-stream models for traditional machining techniques such as turning, tapping, milling, drilling, and nontraditional machining techniques such as electro discharge machining (EDM) processes. The waste streams identified with their associated process performance measures for traditional and EDM processes are depicted in Table 16.3.

The waste stream for both metal-removal processes and EDM includes solid, liquid, and gaseous substances and needs to be reduced. Table 16.3 also shows the measures for improving the processes by increasing the rate of materials and tool life. Reducing the heat generated or using the heat generated for some means will definitely be welcome. Gutowski et al. [21–23] has presented the energy requirements for the manufacture of materials in terms of megajoule per kilogram, as shown in Table 16.4.

Table 16.3 Waste Stream in Manufacturing

		Turning, Tapping, Milling, Drilling	Electro Discharge Machining (EDM)
Water Stream	Gas/aerosol	• Cutting fluid mist	• Electrolyte vapor
		• Dust (dry machining)	• Gaseous products of chemical reactions
	Solid	• Chips, worn tools	• Sludge
	Liquid	• Spent cutting fluids	• Used electrolyte
Process Performance Measures		• Material removal rates	• Material removal rates
		• Chip morphology	• Surface quality
		• Power consumption	
		• Heat generated	
		• Mist size and mass distribution	
		• Tool life/cutting fluid life	

Table 16.4 Typical Energy Costs of Engineering Materials

Material	Energy Cost	Made or Extracted From
Aluminum	227–342	Bauxite
Bricks	2–5	Clay
Cement	5–9	Clay and limestone
Copper	60–125	Sulfide ore
Glass	18–35	Sand, etc.
Iron	20–25	Iron ore
Limestone	0.07–0.1	Sedimentary rock
Nickel	230–70	Ore concentrate
Paper	25–50	Standing timber
Polyethylene	87–115	Crude oil
Polystyrene	62–108	Crude oil
Polyvinylchloride	85–107	Crude oil
Sand	0.08–0.1	Riverbed
Silicon	230–235	Silica
Steel	20–50	Iron
Sulfuric acid	2–3	Sulfur
Titanium	900–940	Ore concentrate
Water	0.001–0.01	Streams, reservoirs
Wood	3–7	Standing timber

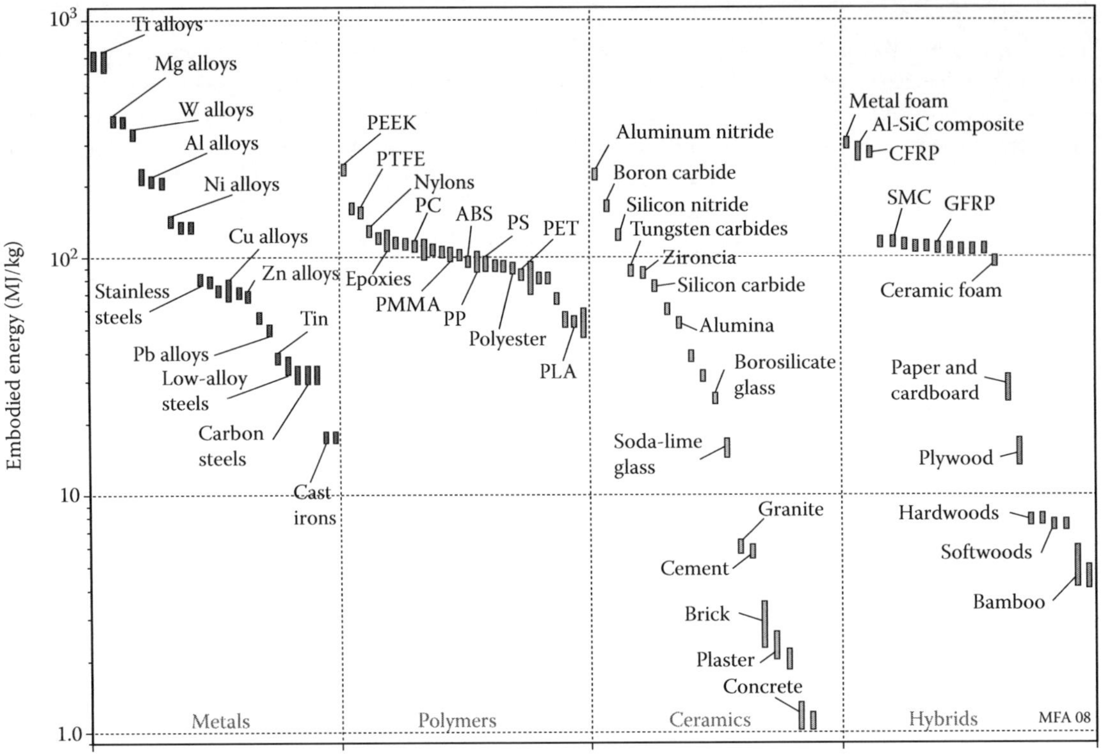

Figure 16.7

(See Color Insert) A bar chart of the embedded energies of materials per unit mass.

The embodied energy per unit mass (MJ/kg) for materials presented by Ashby [21], is shown in Figure 16.7. The embodied energy requirements for manufacturing a part can be calculated as follows.

Sample calculation

- 1 kg part made from 2 kg of aluminum stock 2024 of density 2.7 g/cm^3
- The production machining (turning) needs 14.2 kJ/cm^3 of energy per cubic centimeter
- Volume of material for the part = 1000 g/2.7 g/cm^3 = 370 cm^3
- Therefore, the energy needed to manufacture the part is (14.2 × 370 = 5.25 MJ) × 3 = 15.8 MJ

Using Ashby's chart, the embodied energy for 2 kg of material production is calculated as follows: The embodied energy is 284.5 MJ/kg, so for 2 kg the embodied energy is 2 kg × 284.5 MJ/kg = 569 MJ. Therefore, the total energy required for producing a part of 2 kg = 569 MJ + 15.8 = 584.5 MJ.

We can calculate the emission of toxic gases or GHGs to manufacture a 1 kg part as follows. Approximately 585 MJ of energy is equivalent to 0.06 KWh and the GHGs emitted in the atmosphere are emissions from a power station.

- 585 MJ of primary energy megawatt hour (MWh). This would produce the following GHG.
  –33.35 kg of CO_2
  –140 g of SO_2
  –0.6 g of Hg all for a 1 kg part

16.3.3 Estimation of Energy Consumption in Manufacturing Operations

Kordonowy [2] proposed that power should be measured at the different start-up stages of the machine tool; turning on the computer and monitor, fans, servos in charge of manipulating the clamps, spindle motor, coolant system, air compressor, mist/dust collector, coolant pump, and so on. The power consumed during

Table 16.5 Energy Breakdown of Machine Tool

Activities	Power Consumption (W)
Coolant system	
Air compressor	
Mist/dust collector	
Computer and monitor	
Fans	
Servos	
Spindle key	
Coolant pump	
Material/waste handling	
Unloaded motors	
x-, *y*-, *z*-axis translation	
Spindle *z*-axis translation	
Tool change	
Carousel rotation	
Machining	

machining should then be measured for a particular material removal rate. It is possible to collect the power consumption data in tabular fashion as shown in Table 16.5.

The power consumption depends on the voltage (V) and the current (I) during the idle (start-up) and the machining phase. Measurement equipment can be installed into a machine tool's electrical cabinet to measure and calculate the multiple electrical variables (voltage, current, power factor, frequency, harmonics, and so on).

The electricity consumption per line measured (P_{active}) in kilowatts, is calculated according to

$$P_{\text{active}} = V^{*} I^{*} P_f = \sqrt{3}.VI \tag{16.3}$$

where:

V = voltage

I = current

P_f = power factor, $\sqrt{3}$

The total electricity consumption of each system in the line can be calculated as

$$P_{\text{active system}} = P_{\text{active1}} + P_{\text{active2}} + P_{\text{active}n} \tag{16.4}$$

The total power consumption could be decomposed into constant and variable parts as

$$P_{\text{active,system}} = P_0 + kx\text{RMR} \tag{16.5}$$

where:

P_0 is the constant part corresponding to the total standby power (kW)

RMR is the rate of material removal, typically cubic centimeters per second (in^3/s)

k is a specific energy constant for a material, provided in Newton-meter per cubic millimeter or joules per cubic millimeter (in-lb/in^3)

Energy use is relatively high in a variety of machining and metal fabrication facilities. A variety of processes require energy, which can potentially be altered to reduce the amount of energy used. Electricity consumption is often estimated by multiplying the real power by the running time. The idle power (E_0) also needs to be added to the cutting tool power for estimating the total power consumption as demonstrated next.

The energy requirement (E) for machining can be calculated from Equation (16.5) as

$$E = (P_0 + kx\text{RMR})T_c = E_0 + E_{\text{cutting}} \tag{16.6}$$

where:

- T_c is the total cycle time for machining
- E_0 is the energy consumed by the machine tool before engaging the cutting tool
- E_{cutting} is the energy used for actual material removal in joules (in-lb)
- E_0 is the idle energy and depends on the equipment features that support the machine tool

It is desired that E_0 or P_0 should be reduced. The energy for cutting (E_{cutting}) depends on the cycle time. The cycle time (T_c) for different machining is calculated as

$$T_c = T_h + T_m + T_p \tag{16.7}$$

where:

- T_h is the part handling time
- T_m is the machining time
- T_p is the tool change time per part

The part handling time (T_h) can be estimated from predetermined time (PDT) standards provided in Sections II and III. The machining time (T_m) varies with the type of machining, such as

Turning:

$$T_m = \frac{\pi \cdot D \cdot L}{v \cdot f} \tag{16.8}$$

where:

- D = workpart diameter (mm [in.])
- L = length of part
- v = cutting speed (mm/min [in./min])
- f = feed (mm/rev [in./rev])

Milling

$$T_m = \frac{L + A}{f_r} = \frac{L + A}{N \cdot n_t \cdot f} \tag{16.9}$$

where:

- N = $v/\pi.D$
- n_t = number of teeth on milling cutter
- f = feed (chip load) in millimeters per tooth (in/tooth)
- L = length of part
- A = approach distance, which varies with slab or face milling

Drilling

$$T_m = \frac{t + A}{N \cdot f} \tag{16.10}$$

where:

- t = thickness of part being drilled (mm [in.])
- A = approach allowance that accounts for the drill point angle, representing the distance that the drill must travel before reaching the full diameter
- N = spindle (rev/min = v/πD)

The tool change time per is estimated as

$$T_p = \frac{T_t}{n_p} = T_t \frac{(\pi D L v^{1/n-1})}{f C^{1/n}} \tag{16.11}$$

where:
- D = diameter of drill (mm [in.])
- v = cutting speed (mm/min [in./min])

where C and n are parameters from the tool life equation.

16.3.4 Idle or Start-up Energy Estimation

Theoretically, the idle power is the difference between the maximum rated power of the motor and the maximum power consumption during full load. The voltage is constant during machining as well as during the idle stage; however, the current changes during machining operation. As an example, let us assume that the motor of a milling machine (X) has a rated power of 7.1 kW. The voltage of the machine is 220 V and the milling machine (X) needs 40 A in full load. Therefore,

Power consumption under full load $= \sqrt{3}(220)(40) = 9058$ W
Estimated idle power = 9058–7100 = 1957 W

This is quite a substantial amount, and neglecting the idle power in machining operations is fraught with risk. However, the foregoing description for idle power consumption can be used when the maximum power rating of the machine tool in use is known as well as the spindle motor working at 100%. The literature shows that it is a valid assumption for milling machines, but it may not be true for turning machines.

16.3.5 Environmental Burden of Cutting Tool (*Te*)

In a numerical control (NC) program, the term *Te* indicates the environmental burden due to the cutting tools. All cutting tools are measured by tool life; hence, the tool life is compared with the machining time to calculate the environmental burden from machining. Cutting tools are both disposable and regrindable. The environmental burden due to a cutting tool is calculated as follows by considering the aforementioned processes.

$$T_e = \frac{T_m}{\sum_{j=1}^{RGN+1} T_j}((TD_e + TP_e)xTW + RGNxRG_e) \tag{16.12}$$

where:
- T_m = machining time (s)
- T = tool life (s)
- TP_e = environmental burden of cutting tool production (kg-CO_2/kg)
- TD_e = environmental burden of cutting tool disposal (kg-CO_2/kg)
- TW = tool weight (kg)
- RGN = total number of regrinding processes
- RG_e = environmental burden of regrinding (kg-CO_2)

16.3.6 Environmental Burden of Metal Chips (*CHe*)

The environmental burden caused by the removal of material as chips in machining is indicated by *CHe*. Metal chip recycling, in which a chip compactor, a chip crusher, a centrifugal separator, and an arc furnace are used, generates an environmental burden. Here, the environmental burden is calculated based on chip weight as follows.

$$CH_e = (\mathrm{WPV} - \mathrm{PV})xMDxWD_e \tag{16.13}$$

- WPV = volume of workpart blank in cubic centimeters or cubic inches (cm^3)
- PV = product volume (cm^3)

MD = material density of workpart (kg/cm^3) or (lb/in^3)
WD_e = environmental burden of metal chip processing (kg-CO_2/kg)

Example 16.2

A carbon steel shaft of 5 cm in diameter and 1 m in length is being turned to reduce the diameter of the shaft to 2 cm. Calculate the environmental burden of the metal chips.

The workpart volume (WPV) can be calculated as WPV $= \pi/4(2.5)^2(100) = 1964$ cm^3 and the product volume (PV) is calculated as PV $= \pi/4(2)^2(100) = 314.2$ cm^3; therefore, the volume of chips removed is

Volume of chips removed = 1650 cm^3.
The density of the carbon steel, $\rho = 7800$ kg/m^3.
Hence, the mass of the chips removed is $1650/(100)^3 \times 7800 = 13$ kg.
The environmental burden of chip processing (WD_e) $= 13 \times 0.8 = 11$ kg of CO_2.

The embodied energy in recycling can also be calculated as
Recycling embodied in processing chips $= 13 \times 9.8 = 127.5$ MJ.

In the United States, most electricity is generated by burning coal (about 50%) and the CO_2 emitted into the atmosphere is calculated as CO_2 environmental burden $= (667 \times 0.0003) = 0.20$ kg and it should be added to the CO_2 emission calculated for the chips removed.

It can be observed that as the cutting speed increases, so the machining time decreases, but the life of the tool is adversely affected. However, cutting at a higher cutting speed will definitely reduce the energy footprint, which is one of the goals of green manufacturing.

The Toyota Company has analyzed the energy breakdown in machining operations as depicted in Figure 16.8. The figure shows that an actual machining operation consumed only 14.8% of energy and the remaining 85.2% was consumed by other factors.

The power used in manufacturing industries is consumed by a variety of peripheral equipment such as burners, compressors, heating systems, and lighting systems. The following list contains suggestions for improving energy efficiency

1. Exchange heated process exhaust to heat building air.
2. Design make-up air units around processing areas with exhaust to withdraw recently introduced outside air rather than air that has already been heated.
3. Destratify air with fans and use smaller furnaces or spot-heating units where necessary on the shop floor to conserve the load on the facility furnace unit(s).
4. Utilize spot heating near operator stations or critical processes in larger facilities with large thermal loads.
5. Redirect oven heating.

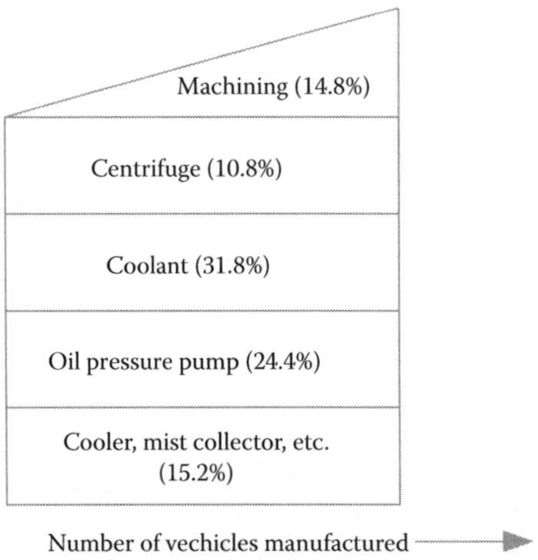

Figure 16.8

Energy use breakdown.

A range of opportunities in a variety of process areas exist for energy efficiency improvements. For example, thermal savings can be found by making sure your burners are operating as efficiently as possible and your burner controls are working properly. Electrical savings can be found in a variety of projects that use powered equipment effectively, such as compressed air auditing; welding control; motor, fan, and pump optimization; and lighting improvements. The Minnesota Pollution Control Agency [24] has published the following standards for savings on electrical energy in an industrial place.

16.3.7 Burners

Efficient burners should be capable of maintaining a stable flame throughout the firing range at low excess air. Many burners claim significant turndown capability, but if they cannot operate at low excess air, savings from turndown are smaller than suggested. A significant operating time above 7% O_2 suggests a large conservation opportunity, but even below this level, a 1% reduction in O_2 yields a 1% improvement in operating efficiency. Larger efficiency improvements occur at higher excess air levels. The use of an economizer can decrease the impact of excess air reduction by a small amount.

16.3.8 Burner Controls

In terms of improving burner controls, O_2 trim typically has the biggest impact. O_2 control is required by code to operate boilers at less than 3% O_2 and can achieve levels below 1%. Trim control does require a burner that can operate stably within the desired control range.

16.3.9 Compressed Air

Compressed air is an expensive and energy-intensive utility that is commonly used for air tools and control valves, as well as occasionally for drives, media blasting, cooling, and blow-offs. Common improvements include repairing leaks in piping and equipment, reducing the compressor output pressure, using a cold-air intake, setting up remote air receivers, correcting large system pressure drops, controlling the loading pattern of the compressed air system, and reducing inappropriate uses such as process cooling and cleaning. You may opt for more costly solutions, which can include the proper sizing and distribution of the compressed air system, improving sequencing controls, and installing variable-speed compressors to handle variable loads.

16.3.10 Heating System Best Practices

Heating system best practices include testing for leaks in steam traps and lines, repairing leaks, improving the thickness of insulation, repairing insulation as needed, improving condensate return, improving boiler feedwater quality to reduce blowdown, and cleaning heat exchange surfaces to improve heat transfer.

16.3.11 Lighting Improvements

Installing T-5 or T-8 fluorescence, reflectors, compact fluorescent lights (CFLs), occupancy sensors, and light-emitting diode (LED) configurations in your facility can be a low-cost, rapid payback opportunity.

16.3.12 Process Controls

Automating or refining controls can shut off an idle machine tool or move it to low-energy standby mode. By improving these controls, the Minnesota Technical Assistance Program (MnTAP) estimates that some facilities could save up to 9% of their total energy use.

16.3.13 Process Heat Recovery

Process heat recovery is possible wherever waste heat is released, although the best recovery opportunities will be in cases where the waste heat temperature and the total amount of heat are high. One difficulty with this technology is determining how to use the recovered heat; this will likely vary in each facility. Possible uses of waste heat include raw material preheating, process or possibly in boiler systems. The simplest heat recovery systems can use waste heat to preheat other operations directly, but it may also be feasible or necessary to upgrade waste heat with heat pumps or to store heat to allow for differences in supply and demand timing.

16.3.14 Process Motors, Pumps, and Fans

Developing a motor replacement plan can be useful in helping you to prepare for eventual motor failure by predetermining what type of motor you will use to replace a failed motor (National Electrical Manufacturers

Association [NEMA] premium, rewound, spare). In a failed motor situation, you will want to understand which other motors in the facility can be moved from less critical applications to cover for the failed equipment. Doing so can provide more time to install the best replacement rather than the fastest replacement. Properly sizing pumps, fans, and electric motors is important, as is matching supply and demand in variable flow applications. You should avoid throttling the flow with valves or dampers as this can be costly.

16.3.15 Coolant Alternatives

Machining coolants are an important component of metal working operations. Coolants improve machinability, increase productivity, and extend the tool life by cooling and lubricating the part and cutting tool. When performing these functions, coolants can quickly become contaminated with foreign materials, causing coolants to lose their effectiveness and develop foul odors. For this reason, most coolants are only used for a short time before being discarded.

The practice of discarding and replacing coolants is costly and wasteful. Developing and implementing a coolant maintenance program can help you minimize contamination, prolong the coolant life, and reduce costs. MnTAP's fact sheet, "Coolant Maintenance for Machining Operations," can help guide efforts to develop a maintenance program.

Machining and metal fabricating operations must comply with multiple environmental, health, and safety regulations. Information on air, *hazardous waste*, and wastewater regulations can be obtained through the following links.

16.3.16 NESHAP: National Emission Standards for Hazardous Air Pollution

The following industrial processes are regulated: (1) dry abrasive blasting, (2) dry grinding and polishing with machines, (3) machining, (4) spray painting, and (5) welding. Laws requiring the reduction of lead, mercury, hexavalent chromium, and chromium metals in specified products exist. All have known adverse effects on public health and the environment.

16.3.17 LCA of Environmental Burden of Machining Operations

Manufacturing operations revolve around machining operations and their environment; therefore, the emissions of GHG during machining need to be analyzed. Narita's detailed explanation of estimating the environmental burden due to machining is worth presenting here [25]. The need for ECM, that is, the relationship between manufacturing and its impact on the environment has been fully analyzed by Narita [25]. The environmental burden analyses for machine tool operations that have been developed include the difference among the depth of cuts, the feed rate, the spindle speed, and the tool path pattern. Furthermore, if the removal volume and the material type are the same, the environmental burden remains the same.

Narita's environmental burden analyzer for machine tool operations seems to be most suitable for sustainable manufacturing. The analyzer also helps in deciding about cutting conditions to achieve a minimum environmental burden.

16.3.18 Milling Machine Life Cycle Assessment

Generally, an LCA is used for estimating the environmental burden at each stage of a product or service: raw material production; manufacture; transportation; use, repair, and maintenance; and disposal or recycling. An LCA has four processes: goal and scope definition, inventory analysis, impact assessment, and interpretation, from which it realizes a holistic assessment of the environmental aspect and helps with making a more informed decision regarding product design modifications and business strategies (Figure 16.9).

Figure 16.9 shows an overview of the environmental burden analyzer for machine tool operations. The electric consumption of a machine tool's peripheral device and motors, a cutting tool's wear, coolant, lubricant, and metal chip quantities, and other factors are calculated. The electric consumption of air conditioning, light, automatic guided vehicle (AGV) transportation, product washing, and so on is also estimated. From these total electrical energy consumption estimates, the GHG emission is estimated for producing a part.

The emission intensity is the rate of an emission matter for an impact category. For example, the quantity of carbon dioxide emitted per joule of energy produced for global warming. The emission intensity data could be prepared according to an impact category such as global warming, acidification, toxicity to ecosystem, toxicity to human, eutrophication, and nuclear radiation. The data include the machine tool specification data, the cutting tool parameters, and the physical parameters of the cutting force for the estimation of a machining process. The cutting tool parameter corresponds to the tool's diameter, helical angle, rake angle, and number of teeth. Narita's analyzer can calculate the environmental burden in various cutting

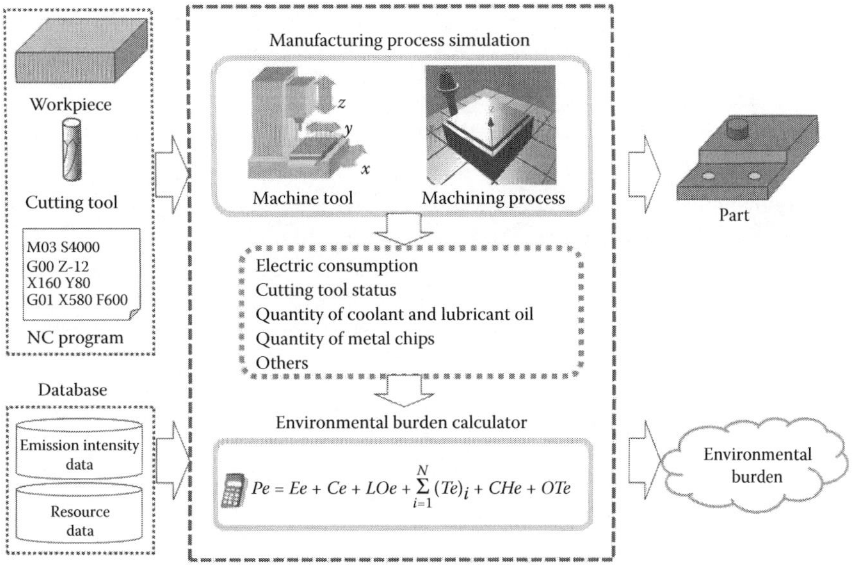

Figure 16.9

Environmental burden analyzer.

conditions, because a machining process has been properly evaluated. Generally, an impact category must be determined and the relevant emission factors of the impact category are selected for an LCA. Global warming is determined as an impact category and carbon dioxide (CO_2), methane (CH_4), and dinitrogen monoxide (N_2O) are selected as the relevant emission factors. The influences of halocarbon and sulfur hexafluoride (SF6) on global warming are very high and severe but they have been ignored in Narita's analyzer. All emissions are converted to their equivalent CO_2 emission by multiplying them by characterization factors, and then the total equivalent CO_2 emission is calculated as the environmental burden. The GWP of a 100-year impact (IPCC, 2007 [26]) is used as the characterization factors, as shown in Table 16.6. This is the usual practice in GWP estimations.

16.3.19 Narita's Algorithm of Environmental Burden Estimation

Equation 16.14 describes the total environmental burden due to a machine tool operation, which is obtained from the electric consumption of the machine tool, the coolant, the lubricant oil, the cutting tool's wear, the metal chip, and other factors.

$$Me = Ee + Ce + LOe + \sum_{i=1}^{N} (Te)_i + CHe \tag{16.14}$$

where:

- Me = environmental burden of a machining operation (kg-CO_2)
- Ee = environmental burden of a machine tool component (kg-CO_2)
- Ce = environmental burden of a coolant (kg-CO_2)
- LOe = environmental burden of a lubricant oil (kg-CO_2)
- Te = environmental burden of a cutting tool (kg-CO_2)
- CHe = environmental burden of a metal chip (kg-CO_2)
- N = number of cutting tools used in an NC program, which includes the number and variety of cutting tools in machining

Table 16.6 Characterization Factors of Global Warming

	CO_2	CH_4	N_2O
Global warming potential (GWP)	1	25	298

Source: Intergovernmental Panel on Climate Change (IPCC). (2007). UNEP.

Some other relevant factors could be added to Equation 16.14 as the situation demands.

The calculation algorithms of the factors *Ee*, *Ce*, *LOe*, *Te*, and *CHe* are introduced in detail as follows.

16.3.20 Electric Consumption of Machine Tool (*Ee*)

The factor *Ee* indicates the environmental burden due to the electric consumption in machining. The electric consumption model of a machine tool and the environmental burden due to the electric consumption of the machine tool include servo motors, spindle controller, spindle cooling system, compressor, coolant pump, lift-up chip conveyor, tool magazine motor, and oil, air, and mist compressor. The chip air blow compressor is obviously included in the electric consumption. Every machine tool consumes power during the standby period and this also needs to be included in the total energy consumption estimation. It is also possible that every machine tool may not possess all the aforementioned peripherals and obviously they should be excluded from the electric power consumption estimation.

$$\begin{aligned} Ee = k(&SME + SPE + NCE + SCE + CME + CPE + TCE1 + TCE2 \\ &+ ATCE + MGE + OAE + COE + CUE + SBE) \end{aligned} \tag{16.15}$$

where:

k = CO_2 emission intensity of electricity (kg-CO_2/kWh) (like 60 kg-CO_2/MWh for U.S. power grid)
SME = electricity consumption of servo motors (kWh)
SPE = electricity consumption of a spindle motor (kWh)
NCE = electricity consumption of an NC controller (kWh)
SCE = electricity consumption of a spindle cooling system (kWh)
CME = electricity consumption of a compressor (kWh)
CPE = electricity consumption of a coolant pump (kWh)
$TCE1$ = electricity consumption of a lift-up chip conveyor (kWh)
$TCE2$ = electricity consumption of a chip conveyor in a machine tool (kWh)
$ATCE$ = electricity consumption of an auto tool changer (ATC) (kWh)
MGE = electricity consumption of a tool magazine motor (kWh)
OAE = electricity consumption of an oil air compressor (kWh)
COE = electricity consumption of an oil mist compressor (kWh)
CUE = electricity consumption of a chip air blow compressor (kWh)
SBE = standby energy of a machine tool (kWh)

The standby energy consumption of a machine tool can be estimated by the machine tool specifications supplied by its manufacturer. In Equation 16.15, the electric consumption of peripheral devices such as an NC controller, a spindle cooling system, a compressor, a coolant pump, a lift-up chip conveyor, a chip conveyor in a machine tool, an ATC, and a tool magazine motor is calculated by their running times. But the electricity consumption of a coolant pump (*CPE*), lift-up chip conveyor (*TCE*), chip conveyor in a machine tool (*TCE2*), an oil mist compressor (*COE*), and a chip air blower (*CUE*) may not be used according to a machine tool operation with or without coolant usage. Based on the machine tool specification, we must decide whether or not to consider these electric consumptions. The electric consumption of the servo motors and the spindle motor also varies dynamically according to the machining process, for which Narita [25] developed a new analysis model.

The load torques of servo motors are calculated as follows:

$$TL\ servo = TU + TM \tag{16.16}$$

where:

$TL\ servo$ = load torque of servo motor (N · m)
TU = axis friction torque (N · m)
TM = application torque of ball screw (N · m)

Here, *TU* is the torque due to rubber sealing and cannot be obtained theoretically, thus its value needs to be decided experimentally.

The application torque of a ball screw (*TM*) is calculated as follows

$$TM = \frac{(\mu.M \mp f).l.\cos(\theta) \pm (M - f).l.\sin(\theta)}{2\pi\eta} \quad (16.17)$$

where:

- μ = friction coefficient of a slide way
- η = transmissibility of a ball screw system
- l = ball screw lead (m)
- M = moving part weight (table and workpiece) (N)
- f = cutting force in an axis (N)
- θ = gradient angle from horizontal plane (rad)

Equation 16.17 was developed by Fujimura et al. [27]. for the cutting force with normal and reverse rotations of the servo motor. The value of θ is 0 in the x- and y-axes, and $\pi/2$ in the z-axis. The cutting force in an axis (f) for machining operations is well established and can be obtained from the literature. The load torque of the spindle motor ($T_{Lspindle}$) is calculated from the cutting force model.

The calculated motor torque is converted to electric consumption as follows.

$$P = \frac{T_L \cdot n \cdot t}{9.55} \quad (16.18)$$

where:

- P = electric consumption (Wh)
- T_L = load torque (*TL servo* or *TL spindle*) (N · m)
- n = motor rotation speed (rpm)
- t = time (h)

The electric consumption of the compressors corresponding to *CPE*, *TCE1*, *TCE2*, *COE*, and *CUE* also varies according to a discharge pressure and the air dryer; hence, Equation 16.19 is adapted for the estimation.

$$Ec = CEn + CF + ADFr + ADFa \quad (16.19)$$

where:

- Ec = electric consumption related to a compressor (kW)
- CEn = electric motor power of the main motor for a compressor (kW)
- CF = rated electric motor power of a fan motor for a compressor (kW)
- $ADFr$ = rated power of a cooling machine of the air dryer (kW)
- $ADFa$ = rated power of a cooling fan of the air dryer (kW)

16.3.21 Environmental Burden of Coolant (*Ce*)

The factor *Ce* indicates the environmental burden due to the coolant used in a machining operation. There are two types of cutting fluid, hence two equations are proposed for *Ce*. A water-based coolant is generally used to enhance machining performance and is circulated in a machine tool by a coolant pump until the coolant has been replaced. The reduction of the dilution fluid (water) due to vapor should be considered in estimating the total coolant. Here, Equation 16.20 is adopted to calculate the environmental burden due to the coolant.

$$Ce = \frac{CUT}{C_L}\left\{(CP_e + CD_e)x(CC + AC) + WA_e x\left(WAQ + AWAQ\right)\right\} \quad (16.20)$$

where:

- CUT = coolant usage time in machining (s)
- C_L = mean interval of coolant update (s)
- CP_e = environmental burden of cutting fluid production (kg-CO_2/L)

CD_e = environmental burden of cutting fluid disposal ($kg\text{-}CO_2/L$)
CC = initial coolant quantity (L)
AC = additional supplement quantity of coolant (L)
WA_e = environmental burden of water distribution ($kg\text{-}CO_2/L$)
WAQ = initial quantity of water (L)
$AWAQ$ = additional supplement quantity of water (L)

16.3.22 Environmental Burden of Lubricant Oil (*LOe*)

The factor *LOe* indicates the environmental burden due to the lubricant oil used in machining. A lubricant oil is used for two main purposes. One is for a spindle and the other is for a slide way. Here, minute amounts of oil are supplied by a pump to the spindle and the slide way within a specific interval. Grease as a lubricant is not introduced here, but the same equation can be applied to calculate the environmental burden due to grease. The environmental burden due to the lubricant oil is calculated as follows:

$$LOe = \frac{SRT}{SI} xSVx(SP_e + SD_e) + \frac{LUT}{LI} xLVx(LP_e + LD_e) \tag{16.21}$$

where:
SRT = spindle runtime in machining (s)
SV = discharge rate of spindle lubricant oil (L)
SI = mean interval between discharges (s)
SP_e = environmental burden of spindle lubricant oil production ($kg\text{-}CO_2/L$)
SD_e = environmental burden of spindle lubricant oil disposal ($kg\text{-}CO_2/L$)
LUT = slide way runtime in machine tool (s)
LI = mean interval between supplies (s)
LV = lubricant oil quantity supplied to slide way (L)
LP_e = environmental burden of slide way lubricant oil production ($kg\text{-}CO_2/L$)
LD_e = environmental burden of slide way lubricant oil disposal ($kg\text{-}CO_2/L$)

16.3.23 Environmental Burden of Cutting Tool (Te)

The factor *Te* indicates the environmental burden due to cutting tools. All cutting tools are measured by their tool life, hence the tool life is compared with the machining time to calculate the environmental burden of a machining operation. The environmental burden due to a cutting tool is calculated as follows by considering the aforementioned processes.

$$Te = \frac{T_m}{\sum_{j=1}^{RGN+1} T_j}((TD_e + TP_e)xTW + RGNxRG_e) \tag{16.22}$$

where:
T_m = machining time (s)
T = tool life (s)
TP_e = environmental burden of cutting tool production ($kg\text{-}CO_2/kg$)
TD_e = environmental burden of cutting tool disposal ($kg\text{-}CO_2/kg$)
TW = tool weight (kg)
RGN = total number of regrinding processes
RG_e = environmental burden of regrinding ($kg\text{-}CO_2$)

16.3.24 Environmental Burden of Metal Chips (*CHe*)

The factor *CHe* indicates the environmental burden due to the metal chips in machining based on chip weight. Metal chips can be recycled by either of the processes as in a chip compactor, a chip crusher, or a centrifugal separator.

$$CHe = (WPV - PV)xMDxWD_e \tag{16.23}$$

where:

WPV = workpiece volume (cm^3)

PV = product volume (cm^3)

MD = material density of workpiece (kg/cm^3)

WD_e = environmental burden of metal chip processing (kg-CO_2/kg)

Narita [25] has presented a case study of a vertical machining center and to benefit from an understanding of the algorithm described in section 16.3.18. It is presented here in full.

Example 16.3: Environmental Burden of Manufacturing a Case Study

For this case study, a vertical machining center with a carbide end mill of 6 mm in diameter and two flutes is considered. The workpiece material is medium-carbon steel.

The parameters to calculate the electric consumption of the servo motor of the machine tool and the electric power of the peripheral devices of the machine tool are summarized in Tables 16.7 and 16.8, respectively, which have been obtained from the instruction manual of the machine tool. The friction torque of servo motors has also been determined by experiment in advance. Table 16.9 shows the other parameters regarding a machine tool operation. Table 16.10 shows the CO_2 emission intensities that are required to calculate the environmental burden of machining operations.

Table 16.7 Parameters Needed to Calculate the Electric Consumption of a Servo Motor

Table part weight (kg)	x: 903, y: 230, z: 512
Friction coefficient of slide way—all axes	0.01 (linear guide)
Ball screw lead (mm)	x, y: 20, z: 16
Transmissibility of ball screw system	0.95

Table 16.8 Electric Consumption and Power of Machine Tool

NC controller (kW)	0.16
Spindle cooling system (kW)	0.45
Compressor (kW)	1
Coolant pump (kW)	0.25
Lift-up chip conveyor(kW)	0.1
Chip conveyor in machine tool (kW)	0.6
ATC (Wh)	0.08
Tool magazine (Wh)	(1 round) 0.087
Vampire (standby) energy (kW)	0.64

Table 16.9 Other Parameters for Machine Tool Operation

Initial cutting fluid quantity (L)	8.75
Additional supplement of cutting fluid (L)	4.3
Initial dilution fluid quantity (L)	175
Additional supplement of dilution fluid (L)	82.25
Mean interval between replacements of coolant in pump (month)	5
Discharge rate of spindle lubricant oil (mL)	0.03
Mean interval between discharges for spindle lubrication (s)	480
Lubricant oil supplied to slide way (mL)	228
Mean interval between supplies (h)	2000
Tool life (s)	5400
Total number of regrinding processes	2
Material density of cutting tool (g/cm^3)	11.9
Coolant tank capacity of machine tool (L)	175
Material density of workpiece (g/cm^3)	7.1

Table 16.10 Equivalent CO_2 Emission Intensities

Electricity (kg-CO_2/kWh)	0.381
Cutting fluid production (kg-CO_2/L)	0.978
Cutting fluid disposal (kg-CO_2/L)	0.0029
Dilution liquid (water) (kg-CO_2/L)	0.189
Spindle and slide way lubricant oil production (kg-CO_2/L)	0.469
Spindle and slide way lubricant oil disposal (kg-CO_2/L)	0.0029
Cutting tool production (kg-CO_2/kg)	33.7
Cutting tool disposal (kg-CO_2/kg)	0.0135
Regrinding (kg-CO_2/number)	0.0184
Metal chip processing (kg-CO_2/kg)	0.322

Table 16.11 Cutting Conditions of NC Programs

	Program 1	Program 2
Spindle speed (rpm)	2500	7500
Feed rate (mm/min)	200	600

Narita's study shows different cutting conditions as well as different coolants. He shows that the tool life increases 1.5 times of the original one due to the coolant effect. The different cutting conditions adopted by Narita are given in Table 16.11.

The environmental burden due to the cutting tool is reduced by the coolant effect, but the environmental burden due to the electric consumption of peripheral devices is increased by the use of a coolant pump. The main reason for the increased environmental burden is peripheral devices, as shown in Figure 16.9.

The environmental burden due to peripheral devices must be reduced to reduce the total environmental burden. The environmental burden due to the peripheral devices is proportional to time; hence, high-speed milling might be a better replacement.

Narita established a very useful relation of equivalent carbon dioxide (CO_2) with cutting speeds. The approximate equation is obtained regarding the plotted data as follows.

$$y = 3.38x10^{-2}x^2 - 27.0x + 6.02.10^3 \quad (16.24)$$

where:

y = equivalent CO_2 emission

x = cutting speed

The minimum cutting speed is obtained by the iterative calculation and becomes 399.408 m/min (about 12702 rpm).

Example 16.4: Cutting Speed vs. CO_2 Emission

The cutting speeds for a host of tool materials are calculated using the Taylor tool-life equation, which is presented in the following table.

Tool Material	Nonsteel Cutting (m/min)	Steel Cutting (m/min)
Plain carbon tool steel	70	20
High-speed steel	120	70
Cemented carbide	900	500
Cermet	—	600
Coated carbide	—	700
Ceramic	—	3000

Calculate the CO_2 emission in the atmosphere.

$y = 3.38x10^{-2} x^2 - 27.0x + 6.02.10^3$, where y represents the CO_2 emission in grams and x is the cutting speed in meters per minute.

The results obtained for CO_2 emission during the cutting operations are calculated in the following table

	Cutting Nonsteel		Cutting Steel	
Tool Material	Cutting Speed (m/min)	CO_2 Emission (g)	Cutting Speed (m/min)	CO_2 Emission (g)
Plain carbon tool steel	70	4296	20	5494
High-speed steel	120	3267	70	4296
Cemented carbide	900	9098	500	970
Cermet	—	—	600	1988
Coated carbide	—	—	700	3682
Ceramic	—	—	3000	2.292×10^5

We can observe that at a particular speed, CO_2 is lowest. Another important point worth noting is that cutting tool costs are in increasing order from lowest at plain carbon tool steel to highest at ceramic.

16.3.25 Electrical Energy Estimation for Manufacturing Processes

Manufacturing processes are made up of a series of processing steps, which for high-production situations are usually automated. For some manufacturing processes, many steps can be integrated into a single piece of equipment. A modern milling machine, for example, can include a wide variety of functions including work handling, lubrication, chip removal, tool changing, and tool break detection, all in addition to the basic function of the machine tool, which is to cut metal by plastic deformation. The result is that these additional functions can often dominate energy requirements. In an article on the optimization of energy consumption and surface quality in finish turning, Yansong Guo et al. [28] developed equations for specific constant energy (SCE) and specific process energy (SPE) and showed experimentally how the total specific energy (TSE) is calculated. In general, there is a significant energy requirement to start up or stand by and maintain the equipment in a "ready" position. Once in the "ready" position, there is then an additional requirement that is proportional to the quantity of material being processed. This situation is modeled in equation below.

The SPE is calculated by dividing the variable power by the material removal rate R_{MR}, that is,

$SPE = F_C/R_{MR}$; where F_C is the cutting force and R_{MR} is the material removal rate.

$SCE = F_o/R_{MR}$; where F_o is the constant power requirement to start up and maintain the equipment in the ready position. The constant energy can be assigned to machine tool accessories such as pumps, computers, fans, and lighting.

The TSE is just the addition of both the SPE and the SCE.

$$TSE = SPE + SCE$$

However, the presentation up to this point has not included the concept of ECM or green manufacturing. Intensive energy consumption in manufacturing industries has drawn increasing attention due to its adverse environmental impact and obviously its depletion of natural resources. The energy consumed by manufacturing industries accounts for 30% of the total world energy and 36% of global CO_2 emissions. In addition, increasing energy and material costs are a heavy burden on the manufacturing sector all over world.

Yansong Guo et al. have developed an empirical relation for the TSE as follows.

$$TSE = C_0 v^{\alpha} f^{\beta} d^{\gamma} D^{\varphi} + C_1 / v \cdot f \cdot d$$

where v (m/min), f (mm/rev), and d (mm) are the cutting variables and D is the final workpiece diameter. They have estimated the values of C_0, C_1, α, β, γ, and φ for steel and these values are presented in Table 16.12.

Table 16.12 Fitted Constants for TSE

C_0	C_1	A	B	γ	φ
1.9205	85.442	0.486	−0.685	−0.8214	−0.8040

The overall R^2 coefficient is .995, which presents a good relation between the experimental and the theoretical values. For a given feed and depth of cut, the process power shows an allometric growth for increasing cutting speed. The cutting variables, *v*, *f*, and *d* can be optimized to minimize the TSE. The TSE equation with the fitted values can be written as

$$\text{TSE} = 1.905v^{0.486}f^{-0.685}d^{-0.8214}D_F^{-0.8040} + \frac{85.442}{v \cdot f \cdot d}$$

Example 16.5

In a turning operation on aluminum, the cutting conditions are: cutting speed (v) = 900 ft/min, feed (f) = 0.020 in/rev, and the depth of cut (d) = 0.250 in; the final diameter (D_F) = 1.25 in. (a) What is the TSE? (b) What is the energy required for cutting? (c) What is the environmental effect, that is, CO_2 emission?

From the previous displayed equation, we can calculate

$$\text{TSE} = 1.95(900)^{0.486}(0.020)^{-0.605}(0.25)^{-0.8214}(1.25)^{0.8040} + \frac{85.442}{(900).(0.020).(0.25)}$$

where:

TSE = 2138 in-lb/in^3

Total turning energy required = 9621 ft-lb

Green manufacturing or the sustainable manufacturing concept is now considered. Here, we are considering the minimum energy criterion along with the dry turning process. The use of an oil-based cooling/lubricant fluid is one of the most unsustainable aspects of a manufacturing operation [9]. The total energy (E) used in a turning operation can be evaluated from

- Energy consumed during the set-up of a turning operation (E_s)
- Energy consumed during turning (E_m)
- Energy consumed during tool change (E_c)
- Energy consumed during the making of a cutting tool (E_t)

The workpiece material energy that is consumed during the production of a workpiece is not considered here. Hence, the total energy E can be calculated as

$$E = E_s + E_m + E_c + E_t$$

The energy consumed during set-up (E_s) is evaluated from the power consumed by the machine and the total time taken for the tool and workpiece setup. It is generally assumed that the set-up is done when the spindle has not yet been turned on. The energy consumed during turning (E_m) is evaluated from the energy consumed for powering the machine modules and the energy for material removal, as modeled by Gutowski et al. [6]) in the following equation:

$$E_m = (P_0 + U \cdot \text{RMR})T_m$$

where:

P_0 = power consumed by machine modules (hp [W])

U = the specific energy requirements in cutting operations (in-lb/in^3 [N-m/mm^3]), which is a property of the material

RMR = material removal rate (in^3/s [mm^3/s])

These parameters are presented in Table 16.13. Using this measure, different work materials can be compared in terms of their power and energy requirements. The values in Table 16.13 are based on two assumptions: (1) the cutting tool is sharp and (2) the chip thickness before the cut t_0 = 0.25 mm (0.010 in). If these

Table 16.13 Specific Energy

Material	Brinell Hardness	Specific Energy (U) or Unit Power (P_u) N.m/mm³	in-lb/in³	Unit Horsepower $hp_{uhp/(in^3/min)}$
Carbon steel	150–200	1.6	240,000	0.6
	201–250	2.2	320,000	0.8
	251–300	2.8	400,000	1.0
Alloy steels	200–250	2.2	320,000	0.8
	251–300	2.8	400,000	1.0
	301–350	3.6	520,000	1.3
	351–400	4.4	640,000	1.6
Cast iron	125–175	1.1	160,000	0.4
	176–250	1.6	240,000	0.6
Stainless steel	150–250	2.8	400,000	1.0
Aluminum	50–100	0.7	100,000	0.25
Aluminum alloys	100–150	0.8	120,000	0.3
Brass	100–150	2.2	320,000	0.8
Bronze	100–150	2.2	320,000	0.8
Magnesium alloys	50–100	0.4	60,000	0.15

assumptions are not met, some adjustments must be made. For worn tools, the power required to perform the cut is obviously greater, and this is reflected in the higher values of specific energy and unit horsepower. It is suggested in the machinery handbook that the values in Table 16.13 should be multiplied by a factor of between 1.00 and 1.25, depending on the degree of dullness of the cutting tool. For cutting tools in a finishing operation, the factor is assumed to be 1.10, and for a roughing operation, the factor is 1.25.

Chip thickness before the cut (t_0) also affects the specific energy and unit horsepower values. The unit horsepower tends to increase as the chip thickness before the cut (t_0) is reduced. It is found that in a grinding operation where the chip thickness is extremely small compared with turning and milling, the specific energy requirement is very high. To account for a different chip thickness other than $t_0 = 0.25$ mm (0.010 in), a correction factor is applied. However, other factors also affect the specific energy requirements, such as the rake angle, cutting speed, and cutting fluid. If the rake angle or the cutting speed are increased or a cutting fluid is added, the specific energy or unit horsepower is slightly reduced. However, very few experimental or analytical data are available for such corrections.

T_m is the machining time in seconds. The energy consumed during tool change (E_c) is evaluated by multiplying the machine power by the tool change time. A turning tool is generally replaced when the machine is not running. The energy used during the making of a cutting tool (E_t) can be evaluated by dividing the energy footprint of the cutting tool by the number of cutting edges. This is calculated from the energy embodied in the cutting-tool material, the energy used during the cutting-tool manufacture, and the energy of a supplementary process such as coating. The quantity (E_t) is calculated by multiplying the energy per cutting edge (Y_E) by the number of cutting edges required to complete the manufacturing pass. The embodied energy per cutting edge is Y_E. Finally, a new equation for energy consumption in single-pass turning has been developed as shown in the following equation.

$$E = P_0 T_h + (P_0 + U.RMR)T_m + P_0 T_t\left(\frac{T_m}{T}\right) + Y_E\left(\frac{T_m}{T}\right)$$

This equation can be further modified by including the cutting time (T_m) and tool-life equations. Taylor's extended tool life is

$$vT^n f^m = C \tag{16.25}$$

or we can write

$$T = \frac{C^{1/n}}{v^{1/n} f^{m/n}}$$

and

$$T_m = \frac{\pi DL}{fv}$$

We can also write the equation for the material removal rate (RMR) as

$$\mathrm{RMR} = vfd$$

where:

f = feed (mm)
d = depth of cut (mm)
v = velocity (ft/min)

Of course, we can write the previous equation as

$$\mathrm{RMR} = vf\frac{(D_i - D_f)}{2}$$

and we finally get

$$E = P_0T_h + P_0\frac{\pi DL}{fv} + Uvf\frac{D_i - D_f}{2}\frac{\pi DL}{fv} + P_0T_t\left(\frac{\pi DL / fv}{C^{1/n} / (v^{1/n}f^{m/n})}\right) + y_E\left(\frac{\pi DL / fv}{C^{1/n} / (v^{1/n}f^{m/n})}\right)$$

The equation can be further simplified as

$$E = P_0T_h + \frac{P_0\pi DL}{fv} + \frac{U\pi L(D_i^2 - D_f^2)}{4} + (P_0T_t + y_E)\left(\frac{\pi DL / fv}{C^{1/n}/(v^{1/n}f^{m/n})}\right) \tag{16.26}$$

The objective is to minimize the total consumption of energy (E) during the machining process. By differentiating equation (11) with respect to v and equating to zero, we get

$$\frac{\partial E}{\partial v} = \frac{P_0\pi DL}{fv^2} + (P_0T_t + y_E)\left(\frac{1}{n} - 1\right)\frac{\pi DLf^{(m/n-1)}v^{(1/n-2)}}{C^{1/n}} = 0$$

It can further be simplified as

$$\frac{C^{1/n}}{f^{(m/n-1)}v^{(1/n-1)}} = \left(\frac{1}{n} - 1\right)\frac{(P_0T_t + y_E)}{P_0}$$

By substituting Taylor's tool-life equation into the preceding equation, the equation for optimizing tool life and minimizing total energy consumption is obtained as

$$T^* = \left(\frac{1}{n} - 1\right)\left(\frac{P_0T_t + y_E}{P_0}\right) \tag{16.27}$$

We can examine the equation and conclude that for minimizing the total energy consumption during turning, the following factors are important.

- The tool-life velocity exponent (i.e., the rate at which the tool life reduces as the cutting speed increases).
- The power used by the turning machine during a noncutting operation or standby power. It clearly shows that the design of a machine has to be improved to consume less power during the noncutting phase.
- The total time for tool change, which obviously affects the noncutting power consumption.

All these are key factors influencing the selection of the optimum tool life for achieving minimum energy consumption.

The values for the unit horsepower and the specific energy (U) for selected work materials using sharp cutting tools and chip thickness before the cut $t_0 = 0.25$ mm (0.01 in) are readily available in the literature and are given in Table 16.13.

16.3.26 Estimated Energy Footprint for Tooling

The energy that is used for manufacturing a cutting tool does not represent direct energy consumption in the machining process. However, it influences the total energy for the system (inputs to the machining process), which needs to be optimized. The energy for tool manufacturing mainly consists of energy embodied in the tool materials, sintering (and grinding for some tools), and coating process energy. Dahmus and Gutowski [19] present reference data for the energy footprint of tooling. Here, we use the energy footprint provided by Dahmus and Gutowski [13] in Table 16.14. They presented energy footprint data for two cases. Case 1 whereby the energy embodied in the tooling material was presented and Case 2 in which only the energy for tooling manufacture was considered. Here, the inserts used were assumed to be manufactured from a sintering process. Each insert was coated. The total weight of the inserts was estimated as 9.5 g. The ecoproperties of materials (Ashby [21]) are presented in Table 16.15.

16.3.27 Cutting Tool Examples

Example 16.6

A high-speed steel tool is used to turn a steel workpart that is 300 mm in length and 80 mm in diameter. The parameters in the Taylor equation are: $n=0.13$ and $C=75$ (m/min) for a feed of 0.4 mm/rev. The operator and the machine tool rate = \$30.00/h and the tooling cost per cutting edge = \$4.00. It takes 2.0 min to load and unload the workpart and 3.50 min to change tools. Determine (a) the cutting speed for the maximum production rate, (b) the tool life in minutes of cutting, and (c) the cycle time and cost per unit of product. (d) Find T & V for minimum cost. (e) T & V for minimum energy. Assume the tool weight is 9.5 g. (T is tool life in minutes, V is cutting speed in m/min.)

Solution

a. $C_0 = \$30/\text{h} = \$0.50/\text{min}$

$v_{max} = 75/[(1/0.13-1)(3.5)]^{13} = 75/[6.692 \times 3.5]^{27} = 49.8$ m/min

b. $T_{max} = (75/49.8)^{1/0.13} = (1.506)^{7.692} = 23.42$ min

c. $T_m = \pi DL/fv = \pi(80)(300)/(0.4 \times 49.8 \times 10^3) = 3.787$ min

$n_p = 23.42/3.787 = 6.184$ pc/tool life. Use $n_p = 6$ pc/tool life

$T_c = T_h + T_m + T_t/n_p = 2.0 + 3.787 + 3.5/6 = 6.37$ min/pc

$C_c = 0.50(6.37) + 4.00/6 = \$3.85/\text{pc}$

Example 16.7

Example 16.6 was solved to calculate the cutting speed for minimum cost.

$C_0 = \$30/\text{h} = \$0.50/\text{min}$

$v_{min} = 75[0.50/((1/0.13 - 1)(0.50 \times 3.5 + 4.00))]^{13} = 75[0.50/(6.692 \times 5.75)]^{13} = 42.6$ m/min

(b) $T_{min} = (75/42.6)^{1/0.13} = (1.76)^{7.692} = 76.96$ min

(c) $T_m = \pi DL/fv = \pi(80)(300)/(0.4 \times 42.6 \times 10^3) = 4.42$ min/pc

$n_p = 76.96/4.42 = 17.41$ pc/tool life. Use $n_p = 17$ pc/tool life

$T_c = T_h + T_m + T_t/n_p = 2.0 + 4.42 + 3.5/17 = 6.63$ min/pc

$C_c = 0.50(6.63) + 4.0/17 = \$3.55/\text{pc}$

Table 16.14 Energy to Produce an Insert Tool

	Case 1 (Material and Manufacturing)	Case 2 (Manufacturing Only)
Embodied tool material energy (MJ/kg)	400	–
Sintering and coatings (MJ per process per cutting insert)	1–2 (avg. 1.5)	1–2 (avg. 1.5)
Total energy per insert (MJ)	5.3	1.5

Table 16.15 Ecoproperties of Engineering Materials

Materials	Ecoproperties: Material						Ecoproperties: Processing				Recycling		
	a	b	c	d	e	f	g	h	I	j	k	l	m
Aluminum alloys	$33\times10^6-34\times10^6$	20×109 tons 22×109 tons	200–240	11–13	*125–375	740–820	*2.4–2.9	0.14–0.17	*.2.4–2.9	0.19–0.23	18–21	1.1–1.2	33–55
Magnesium alloys	$645\times10^3-655\times10^3$	$1\times10^9-1.1\times10^9$	356–394	22.4–24.8	*500–1500	-	*2.1–2.6	*0.13–0.16	3.8–4.6	*0.3–0.37	8.4–9.	1.8–1.95	9.2–11.4
Titanium alloys	$120\times10^5-122\times10^5$	$50\times10^6-52\times10^6$	600–700	*38–44	*470–1410	-	5.02–5.77	0.301–0.346	*4.21–5,7	0.377–0.456	228–281	*14.4–16.7	20–24%
Copper alloys	$15\times10^6-16\times10^6$	$470\times10^6-490\times10^6$	68–74	4.9–5.6	*150–450	1300–1500	*2,5–2.8	*0.15–0.17	*1.8–2.2	0.14–0.15	17–18.5	1.2–1.4	40–60&
Lead alloys	$3.3\times10^6-3.4\times10^6$	$66\times10^6-67\times10^6$	53–58	3.3–3.7	*175–525	610–670	*0.4–0.49	*0.024–0.03	81.6–2.7	*0.13–0.14	8.4–9.2	0.45–0.51	70–75%
Zinc die-casting alloys	$9.7\times10^6-1.0\times10^7$	$2.18\times10^8-2.21\times10^8$	70–75	3.7–4	*160–521	3000–3400	*1.09–1.32	*0.065–0.08	-	-	12.6–13.5	0.66–0.72	20–25%
Nickel-chromium alloys	$1.5\times10^6-1.6\times10^6$	$63\times10^6-65\times10^6$	127–140	*7.89–8.82	*134–512	4900–5500	*3.68–4.06	0.221–0.243	*2.52–2.78	*0.202–0.222	31.8–35	*1.97–2.21	22–26%
Nickel-based superalloys	$1.5\times10^6-1.6\times10^6$	$6.3\times10^7-6.5\times10^7$	135–150	*7.89–9.2	*134–484	4900–5500	*3.38–4.09	*0.20–0.24	*3.26–3.95	0.26–0.31	33.8–37.5	*1.97–2.3	22–26%
Low-carbon steel	$1.1\times10^9-1.2\times10^9$	$78\times10^9-79\times10^9$	29–35	2.2–2.8	*23–69	79–87	*4.2–4.6	*0.25–0.28	*2.2–2.6	*0.18–0.21	8.1–9.8	0.6–0.8	40–44%
Low-alloy steel	$1.1\times10^9-1.2\times10^9$	$78\times10^9-79\times10^9$	*32–48	*1.95–2.3	*37–111	100–120	*3.8–4.6	*0.23–0.28	*3.2–3.9	0.26–0.31	*9.0–11	*0.55–0.64	40–44%
Stainless steel	$30\times10^6-31\times10^6$	$*2.5\times10^9-2.9\times10^9$	*77–85	*4.7–5.4	*112–336	860–960	*3.9–4.4	*0.24–0.29	*3.0–3.7	*0.26–0.31	*22–24	*1.4–1.5	35–40
Cast iron, ductile (nodular)	$1.1\times10^9-1.2\times10^9$	$78\times10^9-79\times10^9$	16–18	1.0–1.1	*13–39	38–42	3.2–3.7	*0.19–0.22	*2.5–3.1	*0.2–0.24	4.9–5.5	0.29–0.32	60–80%

* (a) Annual world production (ton/year); (b) reserves (ton); (c) embodied energy, primary production (MJ/kg); (d) CO_2 footprint, primary production (kg/kg); (e) water usage (L/kg); (f) eco-indicator (millipoints/kg); (g) casting energy (MJ/kg); (h) casting CO_2 footprint (MJ/kg); (i) deformation processing energy (MJ/kg); (j) deformation processing CO_2 footprint (kg/kg); (k) embodied energy, recycling (MJ/kg); (l) CO_2 footprint, recycling (kg/kg); (m) recycle fraction in current supply (%). These values are collected from different sources including Ashby [21].

(e) T&V for minimum energy:

$$T = \left(\frac{1}{n} - 1\right)\left(\frac{PoTt + yE}{Po}\right)$$

where:

Po = power consumed by machine = 3.594 kW

Tt = tool change time = 3.5

yE = embodied energy = 400 × 9.5/100 + 1.5 = 5.3 MJ/per insert

n = 0.13

T = (1/0.13–1)(3.594*3.5 + 5.3 × 4)/3.594.=62.97 min for four inserts (tool life for minimum energy)

We can compare the tool life for minimum cost which is 76.96 min.

We can also calculate the velocity:

$$VT^n = C$$

where:

C = 75

n = 0.13

V = 75/(62.79)^0.13 = 43.786 m/min, which is close to V for minimum cost.

16.4 Eco-Manufacturing Processes: Material and Process Families and Eco-Data

Common material properties are general properties (cost, density), mechanical properties (strength, stiffness, toughness), thermal properties (conductivity, diffusivity, expansion, heat capacity), electrical properties (electrical conductivity, dielectric constant), and optical (refraction, absorption), magnetic, and chemical properties (corrosion resistance). Materials properties determine the suitability of a material based on design requirements. A successful product, one that performs well, is good value for money, and gives pleasure to the user, uses the best materials for the job, and fully exploits its potential and characteristics. Materials selection is not about choosing a material, but it is a profile of properties that best meets the needs of the design. Material and process are interdependent and grouped into families; each family has a characteristic profile (family likeness) that is useful to know when selecting which family to use for a design. In general, there are six families for materials [15]: metals (steels, cast irons, alloys, etc.), ceramics (alumina, silicon carbides), polymers (polyethylene, polypropylene, polyethylene terephthalate), glasses (soda glass, borosilicate glass), elastomers (isoprene, neoprene, butyl rubber, natural rubber), and hybrids (composites, foams) as shown in Figure 16.6.

Processes are also classified based on the design requirements (material, shape, dimensions, precision, and the number of parts to be made). The process families are shaping (casting, molding, deformation, machining, heat treatment), joining (fastening, welding, adhesives), and surface treatments (polishing, painting), as shown in Figure 16.6 [21].

The material and process selection based on some design requirements relies on the materials' mechanical, thermal, electrical, and chemical properties. The rational selection of materials to meet environmental objectives starts by identifying the phase of the product life that causes greatest concern: production, manufacture, use, or disposal. Dealing with all of these requires data for the obvious eco-attributes such as energy, CO_2 and other emissions, toxicity, ability to be recycled, and the like (see Table 16.1). Thus, if material production is the phase of concern, selection is based on minimizing the production energy or the associated emissions (e.g., CO_2 production). However, if it is the use phase that is of concern, selection is based instead on the light weight of the material, and its excellence as a thermal insulator or as an electrical conductor (while meeting other constraints on stiffness, strength, cost, and so on). Additional information on such eco-data (embodied energy and CO_2 footprint as shown in Figure 16.7 and Table 16.16) are needed for sustainable engineering and ecodesign.

16.4.1 Energy Requirements for Materials Used in Manufacturing Processes

The energy required to produce a kilogram of input material for a manufacturing process depends on the nature of the input material, the quality requirements, and the technology employed. Furthermore, as the

Table 16.16 General Characteristics of Metal Processes

Process	Economic Quantity	Materials (typical)	Optimum Size	Minimum Section (mm)	Holes Possible	Inserts Possible
Sand casting	Small/large	No limit	1–100 kg	3	Yes	Yes
Die-casting, gravity	Large	Al, Cu, Mg, Zn alloys	1–50 kg	3	Yes	Yes
Die-casting, pressure	Large	Al, Cu, Mg, Zn alloys	50 g–5 kg	1	Yes	Yes
Centrifugal casting	Large	No limit	30 mm–1 m diameter	3	Yes	Yes
Investment casting	Small/large	No limit	50 g–50 kg	1	Yes	No
Closed die forging	Large	No limit	3000 cm^2	3	Yes	No
Hot extrusion	Large	No limit	500 mm diameter	1	–	No
Hot rolling	Large	No limit	–	–	No	No
Cold rolling	Large	No limit	–	–	No	No
Drawing	Small/large	Al, Cu, Zn, mild steel	3 mm/6 m diameter	0.1	No	Yes
Spinning	One-off, large	Al, Cu, Zn, mild steel	6 mm/4.5 m diameter	0.1	No	Yes
Impact extrusion	Large	Al, Pb, Zn, Mg, Sn	6–100 mm diameter	0.1	–	No
Sintering	Large	Fe, W, bronze	80 g–4 kg	0.5	Yes	Yes
Machining	One-off, large	No limit	–	–	Yes	Yes

technology improves, with time this energy requirement decreases. Alternatively, as the required purity of the material increases, the energy requirement increases. This energy requirement is important because manufacturing processes differ widely in the efficiency of their material use. As a consequence, the loss of energy-intensive materials could be just as important, or even more important, than the direct energy requirements of the manufacturing process per se. Figure 16.2 gives the approximate electrical energy requirements for five materials. These inputs correspond to *embodied energy* (in Figure 16.1) integrated over the time it takes to make 1 kg of material. The first, electric arc furnace steel, is lowest because the quality of the input (mostly recycled steel) closely matches the quality requirements of the output. Aluminum is larger in part because of the requirements to break the strong Al_2O_3 bonds. The two silicon entries are large due to the high purity required [26], and the single-wall carbon nanotube (SWNT) is very large due to the vapor processes used, the low yields, and the newness of the technology. We can expect the nanotube energy to come down significantly, but even so, the trend is clear that large-scale production of SWNTs would be very energy intensive.

16.4.2 Life Cycle Analysis and Selection Strategies

The material life cycle is shown in Figure 16.2. Ore and feedstock, drawn from the earth's resources, are processed to give materials. These materials are manufactured into products that are used and then discarded at the end of their lives, a fraction perhaps entering a recycling loop, the rest committed to incineration or landfill. Energy and materials are consumed at each point in this cycle (phases), with an associated penalty of CO_2, SOx, NOx, and other emissions, heat, and gaseous, liquid and solid waste.

The steps for an LCA are

1. Define the goal and scope of the assessment: Why the assessment needs to be done? What is the subject and which part of its life is to be assessed?
2. Compile an inventory of relevant inputs and outputs: What resources are consumed (bill of materials)? What are the emissions generated?
3. Evaluate the potential impacts associated with those inputs and outputs.
4. Interpret the results of the inventory analysis and impact assessment phases in relation to the objectives of the study: What do the results mean? What needs to be done about them?

The study examines the energy and material flows in raw material acquisition; processing and manufacturing; distribution and storage (transport, refrigeration, etc.); use; maintenance and repair; and recycling options.

16.4.3 Strategy for Minimizing Eco-Impact

The first step is to develop a tool that is approximate but retains sufficient discrimination to differentiate between alternative choices. A spectrum of levels of analysis exists, ranging from a simple eco-screening against a list of banned or undesirable materials and processes to a full LCA, with overheads of time and cost.

The second step is to select a single measure of eco-stress. On one point, there is some international agreement: the Kyoto Protocol of 1997 committed the developed nations that signed up to it to progressively reduce their carbon emissions, meaning CO_2 [29]. At the national level, the focus is more on reducing energy consumption, but since energy consumption and CO_2 production are closely related, they are nearly equivalent. Thus, there is a certain logic in basing design decisions on energy consumption or CO_2 generation; they carry more conviction than the use of a more obscure indicator. We shall follow this route, using energy as our measure.

The third step is to separate the contributions of the phases (material, manufacturing, use, transportation, and disposal) of life because subsequent action depends on which is the dominant one with respect to energy consumption and CO_2 emissions (see Figure 16.6).

For selection to minimize the eco-impact, we must first ask: which phase of the life cycle of the product under consideration makes the largest impact on the environment? The answer guides material selection. To carry out an eco-audit or an LCA, we need the bill of material, the shaping or manufacturing process, the transportation used for the parts of the final product, the duty cycle during the use of the product, and also the eco-data for the energy and CO_2 footprints of the materials and the manufacturing process.

16.5 Manufacturing Processes

Metals can be processed in a variety of ways. These can be roughly classified into casting, forming, and machining. Table 16.16 gives the characteristics of the different processes for metals, although some may also apply to nonmetallic materials such as plastics and composites.

Machining processes are classified in various ways as will be explained. The first classification is based on the type of cutting tool that is used in machining.

- Single-point machining
  - Turning, boring, trepanning, planing
- Multiple point machining
  - Drilling, milling, reaming, sawing, broaching, grinding

The second type of classification is based on the dynamics of the tool and the workpiece, which are specified as follows.

- Tool stationary: turning, boring, and so on
- Tool moves: sawing, milling, drilling, broaching
- Workpiece moves: milling, boring, and so on
- Both move: milling, five-axis milling

However, all machining processes have the same issues, such as

- Power, forces
- Heat, tool materials, rate limits
- Temperature
- Surface finish

These are some of the issues that we discuss in types of machining operations. All materials have a property defined as specific energy, in W.s/mm^3 or hp-min/in^3. This is a well-established characteristic of materials and helps in estimating the power and energy requirement for machining. This in turn will help in estimating the carbon footprint or the environmental impact of the manufacturing process. It has been well established that the consumption of energy is related to GHG emissions. However, it would depend on the type of energy generation. In the U.S. grid, about 50% of electricity comes from coal and the rest comes from other renewable energy sources (Figure 16.10).

16.5.1 Carbon Dioxide (CO_2) Emission from Energy Generation

In the United States, 50% of energy comes from coal.

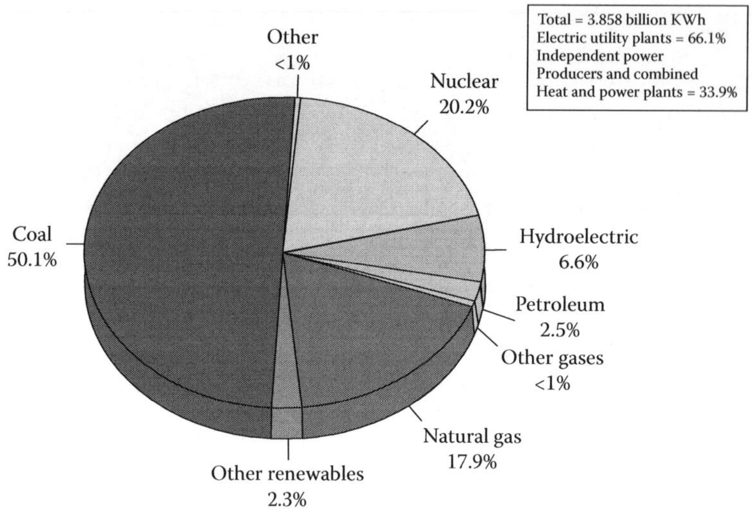

Figure 16.10

(See color insert) Components of U.S. net electric power generation, 2002. (Data from U.S. Energy Information Administration, DOE-STD-1151-2002, DOE, May 2002 (30).)

- 667 kg of CO_2/MWh
- 2.75 kg of SO_2/MWh
- 1.35 kg of NOx/MWh
- 12.3 g Hg/GWh
- and so on

Energy Generation Industry: The percentages of energy generated by different sources are presented in the following grid, which is about 30% efficient.

U.S. Electricity Composition by Source						
Hydro	Nuclear	Coal	Gas	Oil	Waste/Renewable	Other
7.1%	19.6%	50.7%	16.7%	3.1%	2.27%	0%

For every megajoule of electricity, we also get

- 171.94 g of CO_2
- 0.76 g of SO_2
- 0.31 g of NOx
- 6.24 g of CH4
- 0.0032 mg of Hg

(*Source*: EPA 2010 [31])

References

1. Scientific Applications International Corporation (SAIC). (2006). Life Cycle Assessment (LCA): Principles and Practice, EPA/6000R-06/060, May 2006.
2. Kordonowy, D. N. (2001). A power assessment of machining tools. BS thesis, Department of Mechanical Engineering, MIT, Cambridge, MA.
3. Toxic Use Reduction Institute. (2009). Toxic Use Reduction Act Program Assessment: Executive Summary, TURI Methods and Policy Report 26, University of Massachusetts Lowell, MA.
4. U.S. Global Change Research Program. (2000). *Climate Change Impacts on the United States*, Cambridge University Press: Cambridge, New York.

5. United Nations Environment Programme. (2009). *Handbook for the Montreal Protocol on Substances That Deplete the Ozone Layer*, 8th edn. UNEP.
6. Gutowski, T., Jeffrey, D., Thiriez, A. (2006). Electrical energy requirements for manufacturing processes. In *13th CIRP International Conference on Life Cycle Engineering*, Leuven, May 31–June 2 2006.
7. Alman, C. (1989). *The Craft of Economic Modeling*, 2nd edn. Ginn Press.
8. Suh, S. (2009). *Handbook of Input-Output Economics in Industrial Ecology*. Springer.
9. Pusavec, F., Deshpande, A., Yang, S., M'Saoubi, R., Kopac, J., Dillon Jr., O. W., and Jawahir, I. S. (2014). Sustainable machining of high-temperature nickel-Alloy – Inconel 718: Part 2 -Chip breakability and optimization. *Journal of Cleaner Production* 87(1): 941–952.
10. Rajemi, M. F., Mativenga, P. T., and Aramcharoen, A. (2010). Sustainable machining selection of optimum turning conditions based on minimum energy consideration. *Journal of Cleaner Production* 18: 1059–1065.
11. Dietmair, A. and Verl, A. (2008). Energy consumption modeling and optimization for production machines. *IEEE International Conference on Sustainable Energy Technologies*, 24–27 November, Singapore.
12. Keolein, G. A. and Menerey, D. (1994). Sustainable development by design: Review of life cycle design and related approaches. *Journal of the Air and Waste Management Association* 44 (5): 645–668.
13. Emblemsvag, J. and Bras, B. (1998). ISO 14000 and activity based life cycle assessment in environmentally conscious design and manufacturing: A comparison. *1998 ASME Design Engineering Technical Conference.*
14. Ishii, K. (1995). Life-cycle engineering design. *Journal of Vibration Acoustics* 117 (B): 42–47.
15. Santos, J. P., Oliveira, M., Almeida, F. G. et al. (2011). Improving the environmental performance of machine tools: Influence of technology and throughput on the electrical energy consumption of a press-brake. *Journal of Cleaner Production* 19: 356–364.
16. Dittrich, M. A., Gutowski, T. G., Cao, J., Roth, J. T., Xia, C., Kiridena, V., Ren, F., Henning, H. (2012). Exergy analysis of incremental sheet forming. *Production Engineering* 6 (2): 169–177.
17. Narita, H., Desmira, N., Fujimoto, H. (2008). Environmental burden analysis for machining operation using LCA method. In *Proceedings of the 41st CIRP Conference on Manufacturing Systems* (pp. 65–68), Tokyo, Japan.
18. Szargut, J., Morris, D. R., Steward, F. R. (1998). *Exergy Analysis of Thermal, Chemical and Metallurgical Processes*, Hemisphere Publishing Corporation, New York.
19. Dahmus, J. B. and Gutowski, T. G. (2004). An environmental analysis of machining. In *Proceedings of the 2004 ASME International Mechanical Engineering Congress and RD&D Exposition*, Anaheim, CA.
20. Taniguchi, M., Kakinuma, Y., Aoyama, T., Inasaki, I. (2006). Influences of downsized design for machine tools on the environmental impact. In *Proceedings of the MTTRF 2006 Annual Meeting* (pp. 1–4), San Francisco, CA.
21. Ashby, M. F. (2009). *Materials and the Environment: Eco-Informed Material Choice* (pp. 112, 114, 284–287, 290–291), Butterworth-Heinemann, Burlington, MA.
22. Haapala, K., Rivera, J. L., Sutherland, J. (2009). Reducing environmental impacts of steel product manufacturing. *Transactions North American Manufacturing Research Institution of the SME* 37: 419–426.
23. Fthenakis, V. M. and Kim, H. C. (2007). Greenhouse-gas emissions from solar electric- and nuclear power: A life-cycle study. *Energy Policy* 35 (4): 2549–2557.
24. Hondo, H. (2005). Life cycle GHG emission analysis of power generation systems: Japanese case. *Energy* 30 (11–12): 2042–2056.
25. Narita, H. (2012). Environmental burden analyzer for machine tool operations and its application. *Manufacturing System*. www.intechopen.com/books/manufacturing-systems.
26. Intergovernmental Panel on Climate Change (IPCC). (2007). UNEP.
27. Fujimura, Y. and Yasui, T. (1994). Machine tool and manufacturing system [in Japanese]. Kyoritsu Shipping Co. Ltd.
28. Pettersson, A. (2007). High-performance base fluids for environmentally adapted lubricants. *Tribology International* 40: 638–645.
29. United Nations. (1997). Kyoto Protocol, December. UN Climate Change Convention.
30. U.S. Energy Information Administration. (2002). DOE-STD-1151-2002, May, DOE.
31. Environmental Protection Agency (EPA). (2010). Emission standards. U.S. Government.

Bibliography

Abdalla, H. S. and Patel, S. (2006). The performance and oxidation stability of sustainable metalworking fluid derived from vegetable extracts. *Proceedings of the Institution of Mechanical Engineers, Part B: Engineering Manufacturing* 220: 2027–2040.

Alves, S. M. and Oliveira, J. F. G. (2008). Vegetable based cutting fluid-an environmental alternative to grinding process. In: *15th CIRP Proceedings of the International Conference on Life Cycle Engineering* (pp. 664–668), Sydney.

Fthenakis, V. M. and Kim, H. C. (2007). Greenhouse-gas emissions from solar electric- and nuclear power: A life-cycle study. *Energy Policy* 35 (4): 2549–2557. Brookhaven National Lab., NY.

Guo, Y., Duflou, J. R., Lauwers, B. (2014). Energy-based optimization of the material stock allowance for turning-grinding process sequence. *The International Journal of Advanced Manufacturing Technology* 75 (1–4): 503–513.

Guo, Y., Loenders, J., Duflou, J., Lauwers, B. (2012). Optimization of energy consumption and surface quality in finish turning. *Procedia CIRP* 1: 512–517.

Gutowski, T., Branham, M., Dahmus, J., Jones, A., Thiriez, A. Sekulic, D. (2009). Thermodynamics analysis of resources used in manufacturing processes. *Environmental Science & Technology* 43 (5): 1584–1590.

Gutowski, T., Dahmus, J., Thiriez, A. (2006). Electrical energy requirements for manufacturing processes. In *Proceedings of the 13th CIRP International Conference on Life Cycle Engineering* (pp. 623–627), Leuven, Belgium.

Hong, S. Y. and Broomer, M. (2000). Economical and ecological cryogenic machining of AISI 304 austenitic stainless steel. *Clean Products Processes* 2: 157–166.

Kalpakjian, S. and Schmids, R. (2010). *Manufacturing Processes for Engineering Materials*, 6th edn, Pearson Prentice Hall, Upper Saddle River, NJ.

Machinability Data Center. (1980). Machinability data center. In *Machining Data Handbook*, Metcut Research Associates, Cincinnati, OH.

Niggeschmidt, S., Helu, M., Diaz, N., Behmann, B., Lanza, G., Dornfeld, D. (2010). Integrating green and sustainability aspects into life cycle performance evaluation. In *Proceedings of the 16th CIRP International Conference on Life Cycle Engineering* (pp. 366–371), Heifei, China.

Rao, D. N. and Srikant, R. R. (2006). Influence of emulsifier content on cutting fluid properties. *Proceedings of the Institution of Mechanical Engineers, Part B: Engineering Manufacturing* 220: 1803–1806.

Shokrani, A., Dhokia, V., Newman, S. T. (2012). Environmentally conscious machining of difficult-to-machine materials with regard to cutting fluids. *International Journal of Machine Tools Manufacture* 57: 83–101.

Tönissen, S. (2009). Power demand of precision machine tools. MS Thesis, Department of Mechanical Engineering, University of California, Berkeley, CA.

U.S. Energy Information Administration, DOE 2002 and Klee and Graedel.

17

Turning Operation

17.1 Introduction to Single-Point Metal Cutting: Turning

Machining plastically deforms a material using a hard tool in order to obtain a desired physical shape and properties. With metal cutting, a wedge-shaped tool is used to remove material from the workpiece in the form of a *chip*. Two motions are required: the *primary motion*, such as the rotation of the workpiece in a lathe; and the *secondary motion*, such as the feed of a lathe cutting tool. Single-point tools are used for turning, shaping, planning, and so on, and multipoint tools are used for milling and so on. It is necessary to understand the forces acting on the tool and their effects on power requirement, tool life, and production cost. In the following tables of tool forces and formulae, specific power consumption, metal-removal rate, tool life, and so on are given (Figures 17.1 and 17.2).

Tool forces vary with cutting speed (v), feed rate (f), depth of cut (d), and rake angle (α). Force may be measured experimentally by using a *cutting tool dynamometer,* in which the tool is mounted on a flexible steel diaphragm and its deflections in three planes are measured by three electrical transducers. Three meters indicate the force, typically of 25 N, up to, say, 2000 N.

Symbols used: F_c = cutting force (in newtons), F_r = radial force (in newtons), F_f = feed force (in newtons), and resulting force on tool in Figure 17.3.

$$\text{horizontal plane} = \sqrt{F_f^2 + F_r^2}$$

17.1.1 Merchant's Circle for Tool Forces

Merchant's circle is a well-known construction for the analysis of cutting forces for a single-point tool. If the cutting and feed forces, the initial and final chip thickness, and the tool rake angle are known, then the other forces, friction, and shear angles can be found. In drawing the Merchant's force circle, we should know F_c = cutting force, F_f = feed force, t_0 = initial chip thickness, t_1 = final chip thickness, and α = tool rake angle. After drawing the force circle, we can obtain F_s = shear force, F_r = resultant force, F = friction force on tool face, F_{ns} = force normal to shear force, F_n = force normal to F, μ = coefficient of friction = F/F_n, θ = friction angle = $\tan^{-1}\mu$, and ϕ = shear angle.

The forces involved in cutting processes are largely by shear plane angle (ϕ) and tool rake angle (α) (Figure 17.4). The relations for the forces in cutting are established as follows:

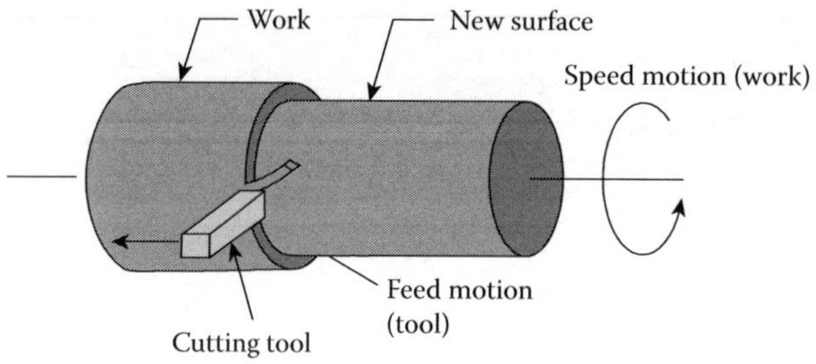

Figure 17.1

Turning operation.

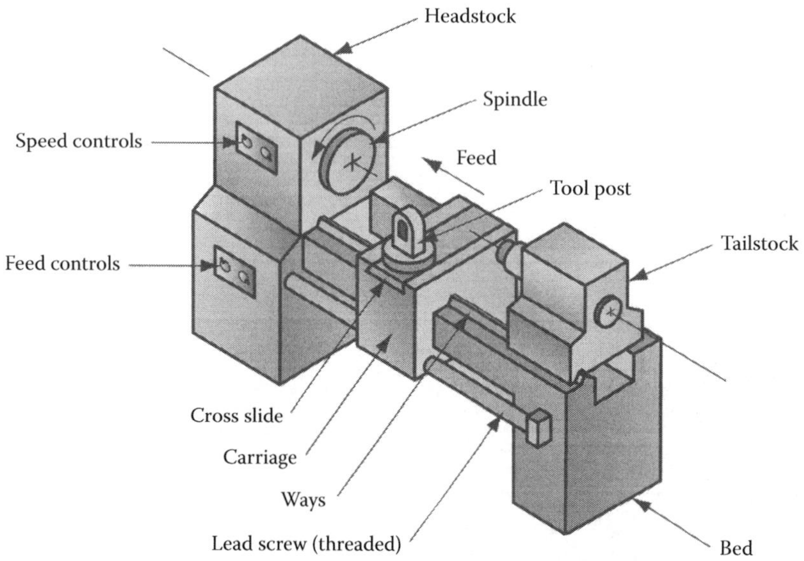

Figure 17.2

Basic turning machine tool.

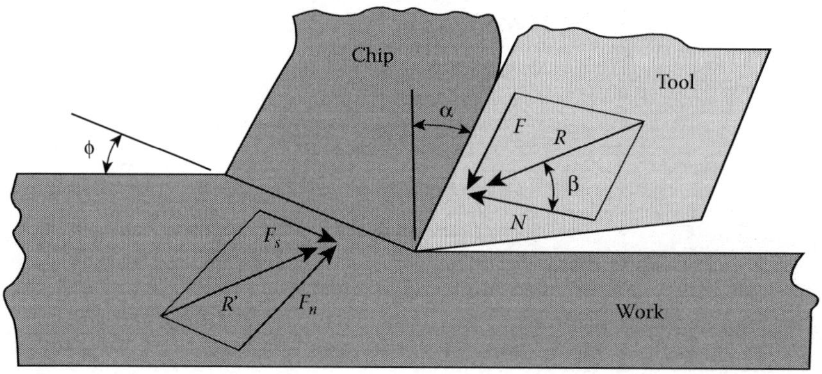

Figure 17.3

Forces during turning operation.

Friction force on tool face (F):

$$F = F_c \sin\alpha + F_t \cos\alpha \tag{17.1}$$

It should be recognized that in cutting operations, the cutting forces (F_c) and the feed force (F_t) are measured with help of dynamometer. All other forces in the force circle are derived from these.

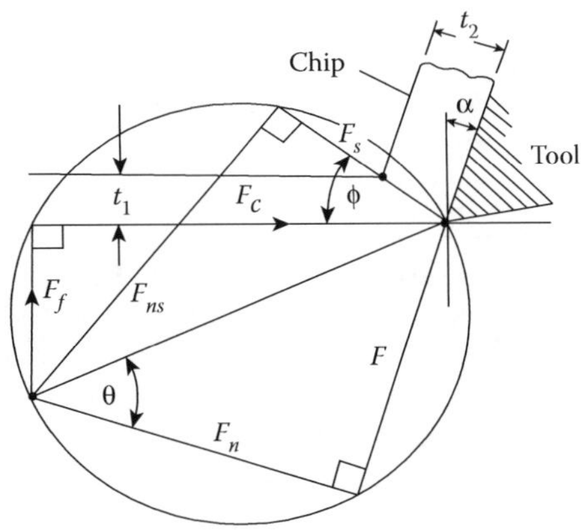

Figure 17.4

Merchant's force circle.

The normal force to friction (F_n):

$$F_n = F_c \cos\alpha - F_t \sin\alpha \tag{17.2}$$

The shear force on the shear plane (F_s):

$$F_s = F_c \cos\phi - F_t \sin\phi \tag{17.3}$$

The shear plane angle (ϕ) also plays an important role in machining operation.

The normal force to shear (F_{ns}):

$$F_{ns} = F_c \sin\phi + F_t \cos\phi \tag{17.4}$$

From Figure 17.4, the cutting force equation can also be derived as follows:

$$F_c = \frac{\cos(\beta - \alpha)}{\cos(\phi + \beta - \alpha)} F_s \tag{17.5}$$

The rake angle (α) varies from negative to positive values passing through 0. When $\alpha = 0$, $F = F_t$, from Equation (17.1) and $F_n = F_c$, from Equation (17.2).

The shear plane angle (ϕ) affects the shear plane area and that, in turn, affects the shear stress (τ) on the shear plane. The length of shear plane (l_s) can be represented in terms of initial chip thickness (t_1) as $l_s = t_1/\sin\phi$. The area of the shear plane is calculated as the length (l_s) × width of the plane (w). The width of the shear plane is equal to the width of the work or alternatively equivalent to the wedge-shaped cutting tool. Therefore, the shear plane area (A_s) is $A_s = (t_1/\sin\phi)w$. The shear stress (τ) is equal to

$$\tau = F_s/A_s \tag{17.6}$$

The thrust force (F_t) can also be written in a direct way as

$$F_t = \frac{\sin(\beta - \alpha)}{\cos(\phi + \beta - \alpha)} F_s \tag{17.7}$$

It is recognized that as the shear plane angle (ϕ) is increasing, the shear plane area (A_s) decreases and so the shear stress (τ) also reduces. This is good, as we will need less shear force to remove material. Analytically, we can state that increasing shear plane angle (ϕ) as much as possible would lower shear plane length (l_s), lower the shear plane area and, finally, reduce shear force for material removal. It will lower the cutting power and energy requirements and it also lowers the cutting temperature, which will in turn lower the cutting fluid requirements and possibly cause less damage to the product surface.

Using less cutting fluid is also good for the environment. The cutting fluid makes the machining place dirty and it is also injurious to the health of workers. The downstream release of used cutting fluid would pollute the air, water, and land. Less power and energy due to increased ϕ also helps in reducing greenhouse gas emissions.

In orthogonal cutting, the width of cutting tool is perpendicular to the direction of the velocity. The feed (f) is small compared with depth of cut (d) and most cutting happens in the direction of feed. Merchant derived a very important equation for the shear plane angle (ϕ). According to him, shear plane angle (ϕ) is most important factor in cutting as shear deformation occurs, that is, at this angle the shear stress is equal to shear strength of the work material. He hypothesized that the shear angle (ϕ) at which shear energy is minimized. At all other angles the shear stress is less than shear strength. Let us derive the Merchant's equation for shear plane angle (ϕ) at which energy is minimum. We know shear stress (τ) is represented by Equation (17.5). The Equation (17.5) is further modified as

$$\tau = \frac{F_c \cos\phi - F_t \sin\phi}{\left(t_1 w / \sin\phi\right)}$$

after replacing F_s by Equation (17.3). We can see from Figure 17.5, F_t can be represented by $F_t = F_c \tan(\beta - \alpha)$. Therefore,

$$\tau = \frac{F_c \cos\phi - Fcs \tan(\beta - \alpha)\sin\phi}{\left(t_1 w / \sin\phi\right)} = (F_c / t_1 w)x\left(\sin\phi \cdot \cos\phi - \tan\left(\beta - \alpha\right)\right) \tag{17.8}$$

To minimize energy, we can minimize the shear stress. To achieve this, the derivative $\partial\tau/\partial\phi = 0$, we get $\partial\tau/\partial\phi(\sin\phi \cos\phi - \tan(\beta - \alpha)\sin^2\phi) = 0$ as the terms before the parenthesis in equation is a constant. We get $-\sin^2\phi + \cos^2\phi - \tan(\beta - \alpha) \times 2\sin\phi.\cos\phi = 0$, which could be simplified to

$$\tan\left(2\phi\right) = 1/\tan\left(\beta - \alpha\right) = \cot\left(\beta - \alpha\right)$$

The $\tan(2\phi) = \cot(\pi/2 - 2\phi)$, so $\pi/2 - 2\phi = \beta - \alpha$ and the Merchant's equation can be written as follows:

$$\phi = 45 + \alpha/2 - \beta/2 \tag{17.9}$$

This result, the so-called Merchant equation, gives one a good sense for how the shear angle would vary with changes in rake angle and friction. However, because the actual shear strength of a material may vary, for example, with strain rate and temperature, rather than remain constant, this equation should be viewed as an approximate relationship rather than an exact one.

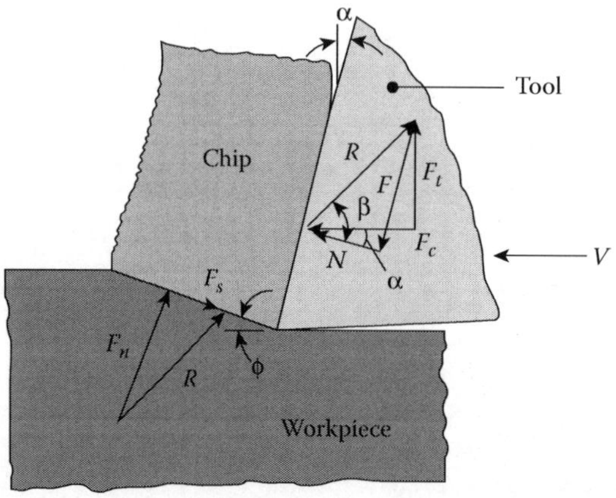

Figure 17.5

Forces in orthogonal cutting.

Example 17.1: Numerical Study of Effect of Rake Angle (α) on the Forces in Cutting

Let us illustrate this through an example. The shear strength of the material under processing is 50,000 psi. The chip thickness before the cut (t_1) = 015 in. and final chip thickness after cutting (t_2) = 03 in. The width of cut (w) = 0.15 in. The rake angle varies from −10° to 20° in steps of 5°. Estimate the variation of the forces in the rake angle range mentioned above. Let us remember that in case of orthogonal cutting feed (f) = chip thickness before cut (t_1) and depth of cut (d) is equivalent to width of cut (w). Cutting speed (V = 400 ft/min) does not change. The chip thickness ratio, $r = t_1/t_2$, decreases as rake angle (α) is reduced and represented by $r = 0.38 + 0.006\alpha$. For possible discrete values of rake angle, the chip thickness ratio is calculated as follows.

We can calculate the friction angle (β) and the shear plane angles for all possible values of rake angles as follows. The friction angle (β) and shear plane angle (ϕ) are calculated from equations as follows. The shear plane angle (ϕ) is represented as

$$\tan(\phi) = \frac{r\cos\alpha}{1 - r\sin\alpha} \tag{17.10}$$

which is derived as follows.

The chip thickness ratio is written as

$$r = t_1/t_2 = l_s \sin\phi / l_s \cos(\phi - \alpha) = \frac{\sin\phi}{\cos(\phi - \alpha)}$$

The equation can be rearranged to produce Equation (17.10). Similarly the friction angle (β) can be obtained from Merchant's equation (17.7) and written as

$$\beta(\alpha) := 2\left(45 + \frac{\alpha}{2}\phi(\alpha)\right) \tag{17.11}$$

The ϕ and β values obtained are given below.

A numerical study of variation in the values of rake angle (α) on shear plane angle and all the forces arising due to material removal processes are presented below obtained by spreadsheet (Table 17.1).

We can notice that as rake angle (α) increases, so the friction angle (β) and shear plane angle (ϕ) values increase. The increase in ϕ values shows a very significant effect on the values of shear force (F_s,), feed force or thrust force (F_t), friction force (F), and cutting force (F_c). All forces decrease as rake angle increases, but the shear plane angle (ϕ) increases. We can also notice that shear plane (A_s) area also reduce with increasing value of ϕ. The effect of varying rake angles on the forces in material removal are presented in the Table 17.2.

The values presented above can be calculated using Excel or any similar software. In this case they are obtained by using Mathcad. Another point of interest in the above table is that cutting force is very high in a negative rake angle compared with a positive rake angle. This illustrates that a negative rake angle may be good from a strength consideration of the cutting tool or life of the tool but from a power consumption perspective, it is not good. Power or energy consumption in the negative rake angle cutting tool is tremendously high compared with positive rake angle. From an environmentally

Table 17.1 Studies of Rake Angle vs. Cutting Angles

Rake Angle (α)	Chip Thickness Ratio (r)	Friction Angle (β)	Shear Plane Angle (ϕ)	Shear Plane Area (A_s)
−10	0.32	46.754	16.623	7.865×10^{-3}
−5	0.35	47.614	18.693	7.02×10^{-3}
0	0.38	48.386	20.807	6.334×10^{-3}
5	0.41	49.087	22.956	5.769×10^{-3}
10	0.44	49.732	25.134	5.297×10^{-3}
15	0.47	50.335	27.333	4.9×10^{-3}
20	0.5	50.914	29.543	4.563×10^{-3}

All angles are in degrees and shear plane area in in.^{2}.

Table 17.2 Rake Angles vs. Cutting Forces

Rake Angle (α)	Force on Shear Plane (F_s)	Thrust Force (F_t)	Friction Force (F)	Cutting Force (F_c)
−10	393.258	1150	1001	753.645
−5	351.015	870.215	808.946	664.998
0	316.707	666.586	666.586	592.105
5	288.44	514.533	558.871	531.192
10	264.868	398.602	475.824	479.579
15	245.016	308.623	410.777	435.322
20	228.158	237.719	359.16	396.987

All forces are calculated for chip thickness before cut $(t_1) = 0.015$ in. and width of cut $(w) = 0.15$ in. and chip thickness ratio are calculated from empirical relation: $r(\alpha) = 0.38 + 0.006\alpha$ [1]. The shear strength (S) of material is 50000 lb/in.[2].

conscious manufacturing or green manufacturing viewpoint, negative rake is inappropriate as larger energy consumption creates more greenhouse gases and CO_2.

Example 17.2: Let Us Compare the Power Consumption for −10° and 10° Rake Angle Cutting Tool from Table 17.2

$$P_C^{-10^\circ} = 753.45(400) = 3.0138 \times 10^5 \text{ ft} - \text{lb/min} \quad \text{or} \quad HP_C^{-10^\circ} = 9.133$$

$$P_C^{10^\circ} = 479.579 \times 400 = 191831.6 \text{ ft} - \text{lb/min} \quad \text{or} \quad HP_C^{10^\circ} = 5.813$$

The horsepower required for cutting in both cases varies widely and just from the power or energy viewpoint the negative rake angle is undesirable.

To check the carbon dioxide emission due to change of rake angle from −10° to 10°, we find

$$CO_2 \text{ emission due to } -10^\circ \text{ rake angle} = 4.55 \text{ kg}$$

$$CO_2 \text{ emission due to } 10^\circ \text{ rake angle} = 2.58 \text{ kg}$$

The Environmental Protection Agency (EPA) has presented in its 2014 report the following figures for carbon dioxide generation depending on forms of energy generation as in Table 17.3.

If we use the CO_2 emission data from EPA report 2014, the estimate values are

$$CO_2 \text{ gas emission from } -10^\circ \text{ rake angle} = 9.133(1/1.34)1001/1000 = 6.823 \text{ kg}$$

$$CO_2 \text{ gas emission from } 10^\circ \text{ rake angle} = 5.183(1/1.34)1001/1000 = 3.87 \text{ kg}$$

Results differ due to the emission data by different sources.

The analytical estimates for tool wear based on abrasion and diffusion proposed in [2,3] can be used to evaluate tool life. An interesting aspect of tool life analysis is that from an environmental

Table 17.3 Life Cycle Greenhouse Gas Emissions by Electricity Source

Technology	Description	50th Percentile (g CO_2/kWh$_e$)
Hydroelectric	Reservoir	4
Wind	Onshore	12
Nuclear	Various generation II reactor types	16
Biomass	Various	18
Solar thermal	Parabolic trough	22
Geothermal	Hot dry rock	45
Solar PV	Polycrystalline silicon	46
Natural gas	Various combined cycle turbines without scrubbing	469
Coal	Various generator types without scrubbing	1001

Source: Kalpakjian, S. and Schmid, S. R. *Manufacturing Processes for Engineering Materials*, 6th edn, 2010. Pearson Prentice Hall, Upper Saddle River, NJ.

perspective, the significant waste stream generated from tool wear is not the portion of the tool which is worn during the process but rather the remaining portion of the tool which is disposed of after its useful life. This is in contrast to the tribological view of tool wear, which concentrates on the particles of the tool removed. Wear generally occurs by mechanical and chemical means, mechanical wear occurring by both abrasion and adhesion and chemical wear occurring by the processes of diffusion and dissolution. It is assumed that by carefully selecting appropriate cutting conditions, modes such as deformation, chipping, corrosion, fatigue, microfracture, and coating delamination will not be responsible for the end of tool life. Instead, tool life will be determined by the rate of gradual wear of the tool surfaces that are in contact with the workpiece. For this analysis, tool life is defined as the process time at which the flank surface wears to a depth of 0.3 mm [4]. An abrasive wear model can be combined with a chemical-dissolution wear model to provide an algorithm that predicts the wear rates of a hard coating throughout the speed range of application of high-speed steel and cemented tungsten-carbide tooling [2]. For materials of moderate chemical stability, the wear due to mechanical activation is modeled by the abrasive wear theory developed in [5], where the wear rate is modeled as proportional to the tool/workpiece hardness and the cutting velocity, and is expressed as:

17.1.2 Cutting Power

Let D = work diameter (mm), d = depth of cut (mm), and N = number of revolutions per minute; $V = \pi DN$ in mm/m; the Power $(P) \approx u_s \times \text{MRR}$; where u_s is specific energy and MRR is material removal rate (MRR). μ_s values can be taken from Table 17.4.

Power is $P = F_C{}^*V$ and MRR = Vfd, where f is feed and d is depth of cut. For orthogonal cutting in turning operation, feed f = chip thickness before cut and depth of cut d = width of cut w; hence, MRR can be written as, d(volume)/dt = Vt_0w = A*V, where A is the cross-sectional area of the undeformed chip.

The cutting force can be estimated as $F_C \approx u_s \times A$. This approximation of the cutting force is in the cutting direction. It is obvious that higher MRR is always desired but limitations on power, cutting force, vibration, and chatter, including the heat generated, will force the user to reduce the MRR.

The total process energy is the integral of the process power over the total processing time, where the total processing time can be expressed as

$$t_{\text{process}} = \frac{\text{Vol}}{Vt_0w} \tag{17.12}$$

From Equations (17.11) and (17.12), the total process energy (TPE) can be expressed as

$$TPE = \frac{\cos(\beta - \alpha)}{\cos(\beta + \varphi - \alpha)}\left(\frac{\tau \text{Vol}}{\sin\varphi}\right) \tag{17.13}$$

Equation (17.13) shows that the ideal process energy is independent of operating parameters such as the tool speed, feed, and depth of cut. Instead, the machining energy is dependent on setup parameters, such as the choice of cutting fluid and tool rake angle, and part design parameters such as material selection and the volume of material removed.

Process power becomes a dominant factor in the selection of a machine tool to perform specific operations. Minimization of the process power reduces the size of the machine tool components required and, ultimately, the capital investment. However, reduction in process power increases the associated time required to perform the process, resulting in higher operating costs.

Table 17.4 Typical Values of P_s

Material	Specific Power Consumption, P_s
Plain carbon steel	34
Alloy steel	71
Cast iron	24
Aluminum alloy	12
Brass	25

In customary US units, power is defined as horsepower (HP), ft-lb/min and as

$$HP = \frac{F_c V}{33000}$$

The *unit power, Pu*, or *unit horsepower*, HP_u, are given as $P_u = P_C$/MRR.
The MRR can be calculated as vt_0w, and unit power, P_u, can be also written as

$$P_U = \frac{F_C}{t_0 w} \text{ in-lb / in.}^3 \left(\text{N-m / mm}^3\right)$$

The unit power (P_U) is also called as *specific energy U*.

The power required at the machine tool motor, called gross power, is estimated by dividing the power at the tool by mechanical efficiency, η, of the machine tool as

$$P_g = \frac{P_c}{h}$$

And the gross horsepower is represented as

$$HP_g = \frac{HP_c}{\eta}$$

Example 17.3

Estimate the total process energy for a round shaft of 2.5 in. reduced to 2 in. by a plain carbon steel-cutting tool of 10° rake angle (α) and shear strength of workpiece material is 50,000 psi. The chip thickness before cut is. The chip thickness before the cut (t_1) = 0.15 in. and final chip thickness after cutting (t_2) = 0.30 in. The width of cut (w) = 0.15 in. Let us remember that in case of orthogonal cutting feed (f) = chip thickness before cut (t_1) and depth of cut (d) is equivalent to width of cut (w). Cutting speed (V = 400ft/min) does not change. The shaft length is 10 in.

The chip thickness ratio $r = t_1/t_2$, decreases as rake angle (α) is reduced and represented by

$$r = 0.38 + 0.006\alpha$$

$$r = 0.38 + 0.006(10) = 0.44$$

The shear plane angle (ϕ) is calculated as 25.134° and the friction angle (β) = 49.73°. The volume of material removed is Vol = π/4(2.5² − 2²)(10) = 17.68 in.³. Hence, the total process energy is estimated as

$$\text{TPE} = \frac{\cos(\beta-\alpha)}{\cos(\beta+\varphi-\alpha)}\left(\frac{\tau \text{ Vol}}{\sin\varphi}\right) = \frac{\cos(49.73-10)}{\cos(25.134+49.73-10)}\left[\frac{50000(5.625)}{\sin(25.134)}\right] = 1000 \text{ lb-ft}$$

17.1.3 Life Cycle Analysis (LCA) of Turning Operation

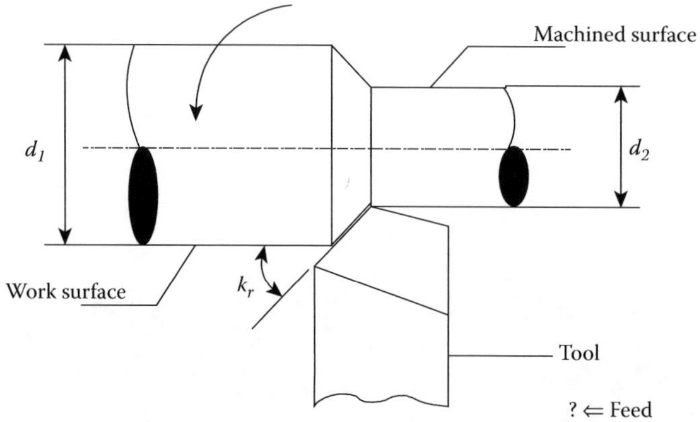

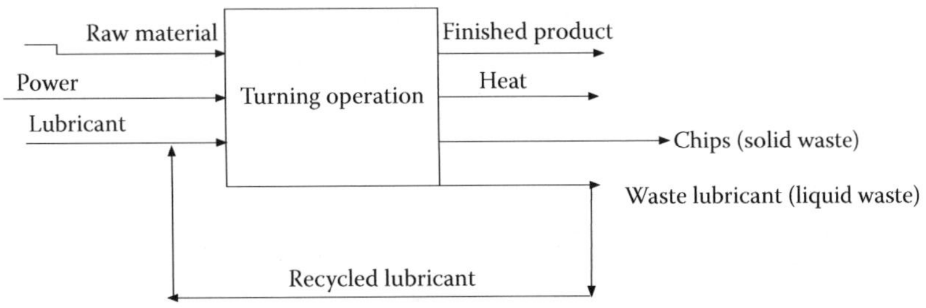

Figure 17.6

Mass balance of turning operation.

17.1.3.1 Material Balance

In mass-reducing processes, such as, turning, the desired geometry is produced by removing material from the initial piece of raw material (Figure 17.6). We will consider Example 17.3 for material balance. We can write mass balance equation as

$$M_i = M_f + \Delta M$$

where:

M_i = initial volume (mass) of material
M_f = final volume of material
ΔM = volume of material removed
M_i = 49.1 in.^{3}
M_f = 31.42 in.^{3}
ΔM = 17.68 in.^{3}

Power used (total process energy) = 1000 lb-ft
The energy used in processing the rod of length 10 in. from 2.5 in. to 2.0 in. is equivalent to

Oil equivalent = 1000 lb-ft (32.4*10^{-12}) × 2000 = 7*10^{-5} lb (data from Ashby [6])

This much oil has produced 7(10^{-5}) (4/2) = 14(10^{-5}) kg of CO_2
If we manufacture 1,000 pieces of this product, the total CO_2 emission = 1 kg of CO_2
All the relevant (data is from Ashby [6]).

Some energy is lost in heat, which is generated due to friction between cutting tool and the work material surface. This is dissipated by a cooling liquid in manufacturing. However, some part of the lubricant is recycled and some is lost in mist.

All the input resources used must be balanced with output and accounted for. Analytically, the temperature during machining could be estimated by Cook's equation as

$$\Delta T = \frac{0.4U}{\rho C}\left(\frac{vt_0}{K}\right)$$

where:

ΔT = mean temperature rise at tool-chip interface, C° (F°)
U = specific energy in the operation, N-m/mm^3 (in-lb/in^3)
V = cutting speed, m/s (in/s)
t_0 = chip thickness before cut, m(in)
ρC = volumetric specific heat of the work material, J/mm^3-C (in-lb/in^3-F)
K = thermal conductivity of the work material, m^2/s(in^2/s)

17.1.3.1.1 Tool Life The values of C and n for different cutting-tool materials, determined experimentally, are given in Table 17.5.

$$T = \left(\frac{C}{v}\right)^{1/n} (\text{min})$$

Table 17.5 Values of C and n

Tool Material	C	n	Wear Land Width (mm)	
			Roughing	Finishing
High-speed steel	60–400	0.08–0.15	1.5	0.25–0.38
Cemented carbide	200–330	0.16–0.5	0.75	0.25–0.38
Ceramic	330–600	0.40–0.6	0.25–0.38	0.25–0.38

Figure 17.7 shows the face, flank, and wear land of a single-point cutting tool. Figure 17.8 shows the normal wedge angle and feed of a cutting tool. Figure 17.9 shows the tool life plotted against cutting speed for high-speed steel, carbide, and ceramic tools.

17.1.3.1.2 Tool Characteristics

Force vs. cutting speed: F_c is constant over normal range cutting speed; however, thrust force (F_f) increases slowly with cutting speed (Figure 17.9).

Force vs. depth of cut: The cutting force (F_c), increases with depth of cut but the friction force (F_f) increases at a slow rate with depth of cut (Figure 17.10). It is possible that the cutting force (F_c) increases above and beyond the shear yield strength of the cutting tool and then the cutting tool yields or fails. However, it is suggested that a machinist should reach to maximum depth of cut to avoid another cut or second pass. This will reduce energy consumption during machining. This is main objective of *sustainable manufacturing.*

Force vs. rake angle: F_c and F_f fall slowly with rake angle (Figure 17.11).

Force vs. feed rate: F_c increases linearly with feed rate and F_f increases in a curve with decreasing rate. A numerical study of rake angles vs. forces has been presented in Table 17.2. The table shows that all forces involved in cutting decreases with increase in rake angles from negative to positive. There are rapid reduction in thrust force and friction forces (Figure 17.12).

It is always recommended to increase feed to increase rate of material removal rate (MRR) but this produces a bad surface. Therefore, it could be concluded that improvements in sustainability and reductions in energy consumption may not produce good surface finishes or a good-quality product (Figure 17.13).

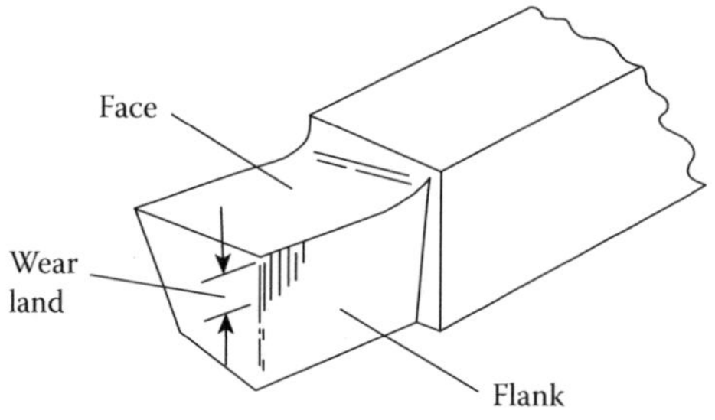

Figure 17.7

Tool diagram.

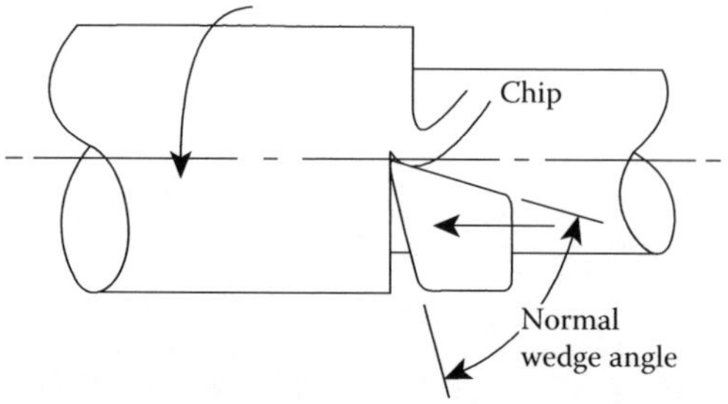

Figure 17.8

Cutting angle.

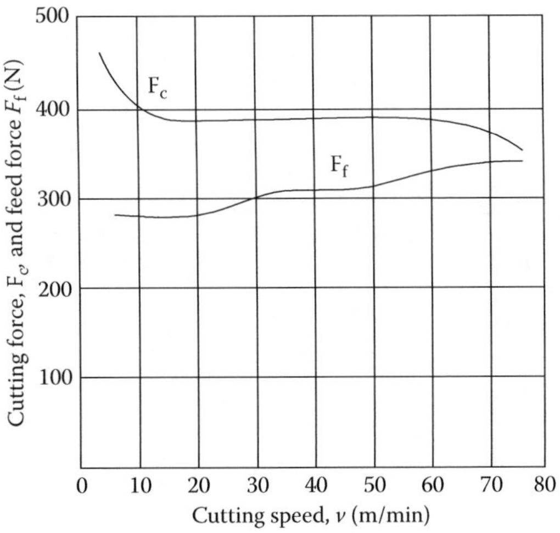

Figure 17.9

Force vs. cutting speed.

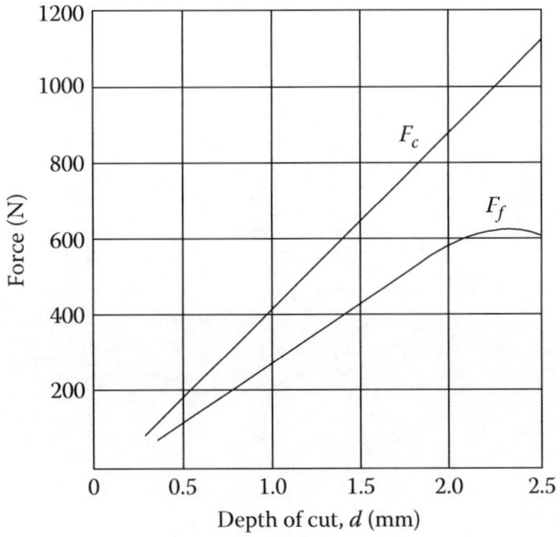

Figure 17.10

Force vs. depth of cut.

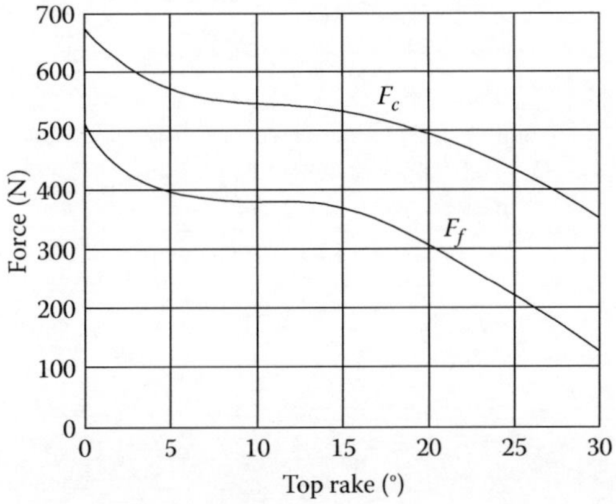

Figure 17.11

Force vs. rake angle.

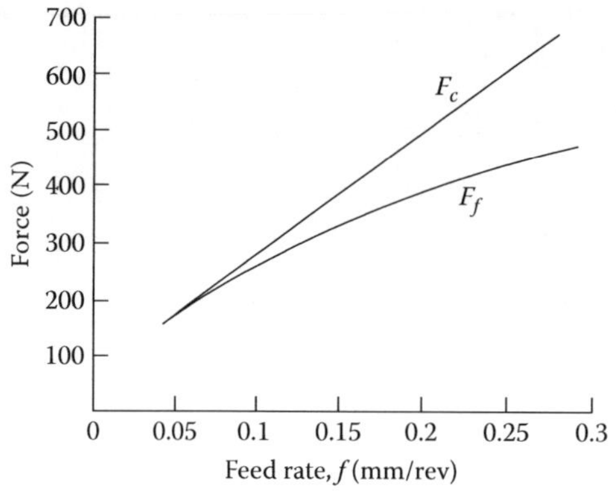

Figure 17.12

Force vs. feed rate.

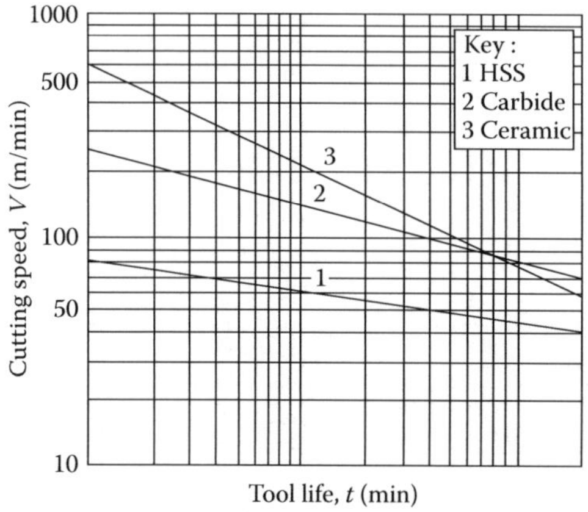

Figure 17.13

Cutting speed vs. tool life.

Figure 17.11 shows the effect of cutting speed on tool life. It is noticed here that at higher speed, tool life is shorter and, obviously, at lower speed tool life is longer. Table 17.5 presents values of C and n for high-speed steel, carbide and ceramic tools. The wear land width for rough and finish cutting are also presented. Remarkably, the wear land width for the ceramic tool is the same for both rough and finish cutting. Cutting speeds of various cutting-tool materials are presented in Table 17.6.

Table 17.6 presents the cutting speed (V) for various cutting-tool materials in a turning operation. The table shows that tungsten carbide has the highest cutting speed for all work materials and high-speed steel has the lowest.

Table 17.6 Turning Speeds

Material	High-Speed Steel (m/min)	Super-High-Speed Steel (m/min)	Stellite (m/min)	Tungsten Carbide (m/min)
Aluminum alloys	70–100	90–120	>200	>350
Brass, free cutting	70–100	90–120	170–250	350–500
Bronze	40–70	50–80	70–150	150–250
Gray cast iron	35–50	45–60	60–90	90–120
Copper	35–70	50–90	70–150	100–300
Magnesium alloys	85–135	110–150	85–135	85–135
Monel metal	15–20	18–25	25–45	50–80
Mild steel	35–50	45–60	70–120	–
High-tensile steel	5–10	7–12	20–35	–
Stainless steel	10–15	12–18	30–50	–
Thermosetting plastic	35–50	45–60	70–120	100–200

Table 17.7 Turning of Plastics- Depth of Cut, Feed, and Cutting Speed

Material	Condition	Depth of Cut (mm)	Feed (mm rev^{-1})	HSS (m/min)	Brazed Carbide (m/in)	Throw-Away Carbide Tip (m/min)
Thermoplastics, polyethylene, polypropylene, TFE fluorocarbon	Extruded, molded or cast	4	0.25	50	145	160
High-impact styrene, modified acrylic	Extruded, molded or cast	4	0.25	53	160	175
Nylon, acetals and polycarbonate	–	4	0.25	50	160	175
Polystyrene	Molded or extruded	4	0.25	18	50	65
Soft grades of thermosetting plastic	Cast, molded or filled	4	0.25	50	160	175
Hard grades of thermosetting plastic	Cast, molded or filled	4	0.25	48	145	160

Table 17.8 Typical Standard Times for Capstan and Turret Lathe Operations

Operation	Time (s)	Operation	Time (s)
Change speed	3	Engage feed	1.5
Change feed	3	Feed to bar stop	3.5
Index tool post	3.5	Chuck in, 3-jaw chuck	4.5

17.1.4 Turning of Plastics

All lathes have some specified operation times for change of speed and feed, setting up chuck, and so on (Table 17.7). Some of these specified standard times for both capstan and turret lathes are given in Table 17.8.

17.1.5 Typical Standard Times for Capstan and Turret Lathe Operations

A snapshot of standard times for operations like changing speed, feed, and indexing is provided in the Table 26. Table 26 also shows standard times for engaging feed, feed to bar stop, and chuck in.

Example 17.4

Let us compare the machining time of a 2 in. diameter cylindrical part of 10 in. length with the setting-up time given above in Table 17.6 for a turret lathe. The feed (f) = 0.015 in. The velocity (v) = 400 ft/min. The machining time:

$$T_m = \frac{\pi D_o L}{fv} = \frac{\pi(10)(2)}{0.015(400 \times 12)} = 0.87\,\text{min}$$

From Table 17.7, we can note that the total setting-up time for the turret lathe is 19 s and the ratio of setting-up time to machining time is 0.36, which is quite significant.

17.2 Lathe Tool Nomenclature and Setting

There are many types of lathe tool, the principal ones being bar turning; turning and facing, parting-off, facing, boring, and screw cutting. Some are made from a bar of tool steel, others with high-speed steel tips welded carbon steel shanks and some with tungsten-carbide tips brazed to a steel shank. A tool holder with interchangeable tips can also be used (Figure 17.14).

17.2.1 Lathe Components

Of the many standard and special types of turning machines that have been built, the most important, versatile and widely recognized is the engine lathe. The standard engine lathe is not a high-production machine but it can be readily tooled up for many one-piece or short-run jobs. It is also possible to modify the basic machine for many higher-production applications. The modern engine lathe provides a wide range of speeds and feeds which allow optimum settings for almost any operation. There have been advances in headstock design to provide greater strength and rigidity. This allows the use of high-horsepower motors so that heavy cuts with carbide tools are practical. To utilize this high power without losing accuracy, new lathes incorporate heavier beds, wider hardened ways and deeper-sectioned carriages.

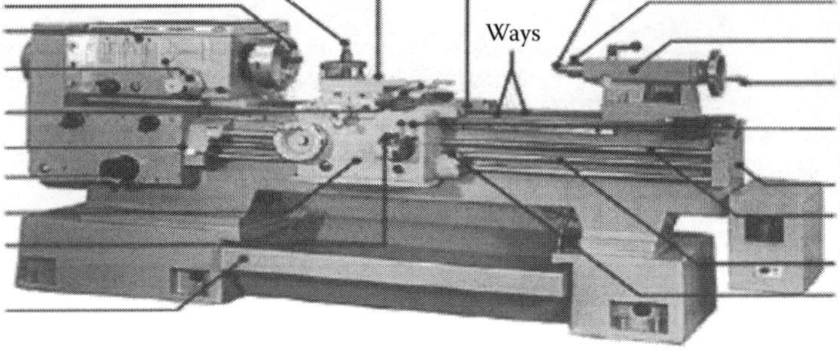

Figure 17.14

General view of typical lathe machine.

17.2.2 Typical Lathe Components

17.2.2.1 Headstock

The headstock is the powered end and is always at the operator's left. This contains the speed-changing gears and the revolving, driving spindle, to which any one of several types of work holders is attached. The center of the spindle is hollow so that long bars may be put through it for machining.

17.2.2.2 Tailstock

The tailstock is nonrotating but on hardened ways, it can be moved to the left or right to adjust to the length of the work. It can also be offset for cutting small-angle tapers.

17.2.2.3 Carriage

The carriage can be moved left or right either by hand wheel or power feed. This provides the motion along the z-axis. During this travel turning cuts are made.

17.2.2.4 Apron

The apron attached to the front of the carriage holds most of the control levers. These include the levers that engage and reverse the feed lengthwise (z-axis) or crosswise (x-axis), and the lever that engages the threading gears.

17.2.2.5 Cross Slide

The cross slide is mounted on the carriage and can be moved in and out (x-axis) perpendicular to the carriage motion. This is the part that moves when facing cuts are made with power feed, or at any time a cut must be made *square* with the z-axis. This, or the compound, is also used to set the depth of cut when turning. The cross slide can be moved by its hand wheel or by power feed.

17.2.2.6 Compound Rest

The compound rest, or compound for short, is mounted on the carriage. It can be moved in and out by its hand wheel for facing or for setting the depth of cut. It can also be rotated 360° and fed by its hand wheel at any angle. The compound does not have any power feed but it always moves longitudinally with the cross slide and the carriage.

17.2.2.7 Tool Post

The tool post is mounted on the compound rest. This can be any of several varieties but in its simplest form is merely a slotted cylinder that can be moved left or right in the T-slot in the compound and clamped in place. It can also be rotated so as to present the cutter to the work at whatever angle is best for the job.

17.2.2.8 Bed

The bed of the lathe is its backbone. It must be rigid enough to resist deflection in any direction under load. The bed is made of cast iron or a steel weldment, in a box or I-beam shape, and it is supported on legs, a cabinet or a bench.

17.2.2.9 Ways

The ways of the lathe are the flat or V-shaped surfaces on which the carriage and the tailstock are moved left and right. Each has its separate pair of ways, often one flat surface for stability and one V-way for guidance in a perfectly straight line. These ways are hardened and scraped or ground to close tolerances. The basic accuracy of movement of the carriage depends on the ways.

17.2.2.10 Size

The size of a lathe is specified by two or three dimensions:

- The largest diameter workpiece that will clear the bed of the lathe. The center is the headstock spindle center.
- The largest diameter workpiece that will clear the cross slide is sometimes also specified.
- The longest workpiece that can be held on centers between the headstock and the tailstock.

17.2.3 Turret Lathe

The standard engine lathe is versatile but it is not a high-production machine. When production requirements are great, more automated turning machines must be used. The turret lathe represents the first step from the engine lathe toward the high-production turning machines. The turret lathe is similar to the engine lathe except that tool-holding turrets replace the tailstock and the tool post compound assembly. These machines possess special features that adapt them to production. The *skill of the worker* is transferred to machines, making it possible for inexperienced operators to reproduce identical parts. In contrast, the engine lathe requires a skilled operator and requires more time to produce parts that are dimensionally the same.

The principal characteristic of turret lathes is that the tools for consecutive operations are set up for use in the proper sequence. Although skill is required to set and adjust the tools properly, once they are correct, less skill is required to operate the turret lathe. Many parts can be produced before adjustments are necessary. These machines are normally used for small to medium-sized production runs where the engine lathe is too slow but the additional production rate desired does not warrant a special machine.

17.2.4 Square and Hex Turrets

A square turret is mounted on the top of the cross slide and is capable of holding four tools. If several different tools are required, they are set up in sequence and can be quickly indexed and locked in correct working position. The slide is provided with positive stops or feed trips so that cuts can be duplicated. Likewise, the longitudinal position of the entire assembly may be controlled by positive stops on the left side of the apron. Cuts may be taken with square turret tools and with tools mounted on the hexagon turret simultaneously.

An outstanding feature is the turret in place of the tailstock. This turret, mounted on either the sliding ram, the saddle, or on the back of the structure carries four to 18 tool stations. The tools are preset for the various operations. The tools are mounted in proper sequence on the various faces of the turret so that as the turret indexes between machining operations, the proper tools are engaged into position. For each tool there is a stop screw or electric/electronic transducer, which controls the distance the tool will feed and cut. When this distance is reached, an automatic trip lever stops further movement of the tool by disengaging the drive clutch.

Like the engine lathe, the modern turret lathe provides fast spindle speeds, wide speed and feed ranges, high power and great rigidity. The machine is operated in the high end of its speed range more than the engine lathe, partly because the tools placed in the turret often work on small diameters on the workpiece, but also because the operator is more production conscious (Figure 17.15).

17.2.5 Horizontal Turret Lathes

Horizontal turret lathes are made in two general designs and are known as the ram and saddle types. The ram-type turret lathe has the turret mounted on a slide or ram that moves back and forth on a saddle clamped to the lathe bed. The saddle-type turret lathe has the turret mounted directly on a saddle, which moves back and forth with the turret.

17.2.6 Vertical Turret Lathes

A vertical turret lathe resembles a vertical boring mill, but it has the characteristic turret arrangement for holding the tools. It consists of a rotating chuck or table in the horizontal position, with the turret mounted

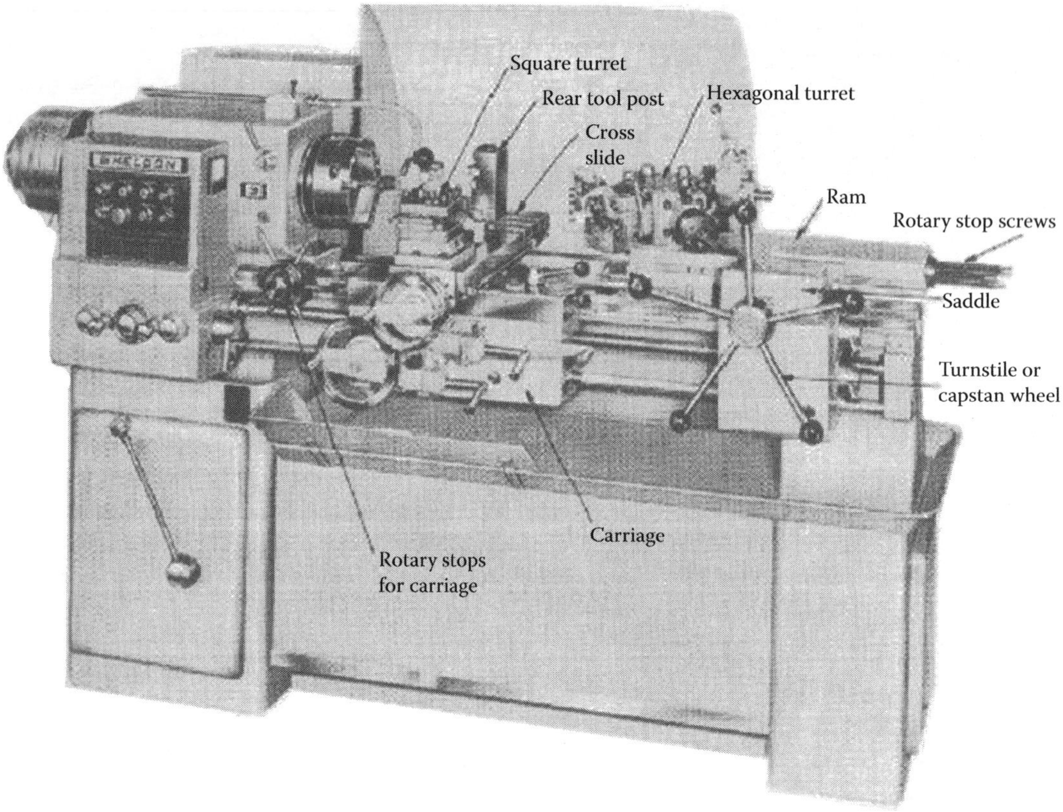

Figure 17.15

A horizontal turret lathe.

above on a cross rail. In addition, there is at least one side head provided with a square turret for holding tools. All tools mounted on the turret or side head have their respective stops set so the length of cuts can be the same in successive machining cycles. It is, in effect, the same as a turret lathe standing on the headstock end, and it has all the features necessary for the production of duplicate parts. This machine was developed to facilitate mounting, holding and machining of large-diameter heavy parts. Only chucking work is done on this kind of machine.

A vertical turret lathe is provided with two cutter heads: the swiveling main turret head and the side head. The turret and side heads function in the same manner as the hexagonal and square turrets on a horizontal lathe. To provide for angle cuts both the ram and turret heads may be swiveled 30° right or left of center.

The machine can be provided with a control that permits automatic operation of each head, including rate and direction of feed, change in spindle feed, indexing of turret, starting, and stopping. Once a cycle of operations is preset and tools are properly adjusted, the operator need only load, unload, and start the machine. Production rate is increased over those manually operated machines because they operate almost continuously and make changes from one operation to another without hesitation or fatigue. By reducing the handling time and making the cycle automatic, an operator can attend more than one machine.

The turret lathe normally has a jawed chuck to hold the workpiece; however, a collet may be more suitable when producing parts from bar stock. A turning machine equipped with a collet and a turret is called a screw machine, but it is actually a special turret lathe. The special features of screw machines are aimed primarily at reducing idle time on the parts being machined, thereby increasing productivity.

17.2.7 Computer-Controlled Lathes

In the most advanced lathes, movement and control of the machine and its components are actuated by computer numerical controls (CNC). These lathes are usually equipped with one or more turrets. Each turret is equipped with a variety of tools and performs several operations on different surfaces of the workpiece. These machines are highly automated, the operations are repetitive, and they maintain the desired accuracy. They are suitable for low to medium volumes of production.

17.2.8 Tool Features

For cutting to take place the tool must have a *front clearance angle*, which must not be so large that the tool is weakened. There must also be a *top rake angle* to increase the effectiveness of cutting. The value of this angle depends on the material being cut. Typical values are given in Figures 17.16 and 17.17.

For efficient removal of materials of specified strength, the rake angles of the cutting tools are given below in Table 17.9.

Table 17.9 is useful in designing or selecting a cutting tool for efficient removal of material and productivity. We can see that as the strength of material increases, so the negative rake angle also increases. As was

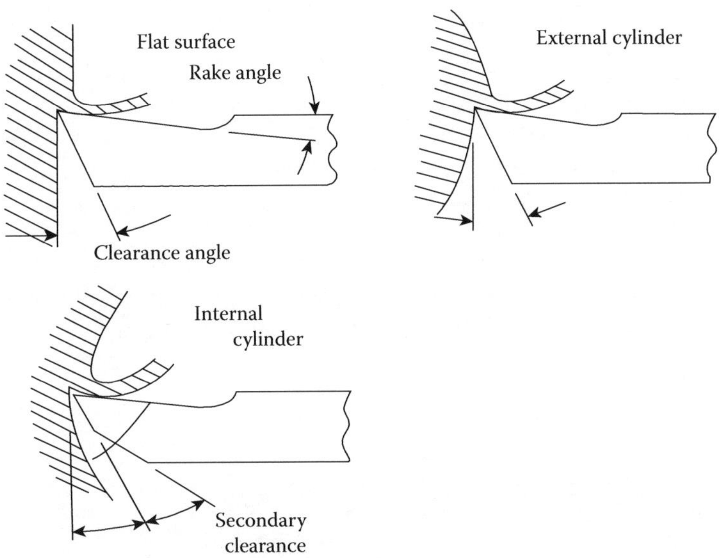

Figure 17.16

Cutting angle vs. surface.

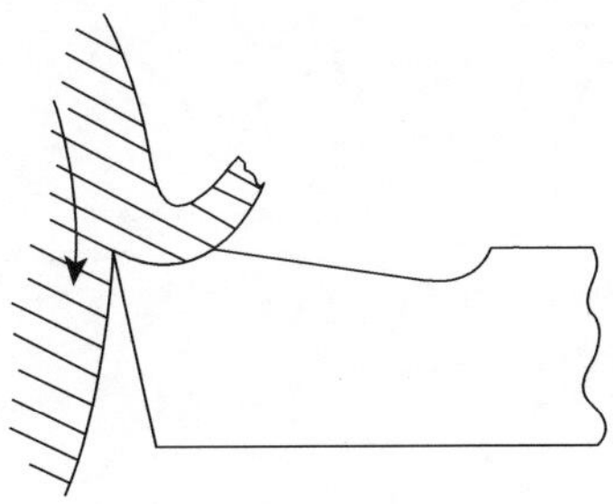

Figure 17.17

Chip breaker.

Table 17.9 Rake Angle for Different Materials of Workpart

Workpiece Material	Tensile Strength (N mm^{-2})	Tool Rake Angle (°)
High-tensile steel	1550	-8
Nickel-chrome steel	1000–1150	-5
Steel	750	-3
Steel forging	450–600	-2
Brass and bronze	–	0
Cast iron	–	2
Mild steel	–	7
Free-cutting mild steel	–	10
Light alloys	–	12

presented earlier with negative rake angle increasing, the cutting tool becomes stronger, but it takes more power or energy to remove materials.

Another feature is the *chip breaker*, which breaks long, dangerous and inconvenient streamers of *swarf* into chips shown in Figure 17.13.

17.2.9 Plan Approach Angle

To reduce the load on the tool for a given depth of cut, the cutting edge can be angled to increase its length (Figures 17.18 and 17.19). Note the direction of chip flow; if the angle is too large there is a danger of chatter, shown in Figure 17.11.

17.2.10 Other Features

In addition to front clearance and top rake, there is also side clearance and side rake. A small nose radius improves cutting and reduces wear. All these angles are shown in Figure 17.12. They are presented as given here with symbols: φ = top rake angle, α = front clearance angle, β = wedge angle, δ = plan relief or trail angle, ε = plan approach angle, θ = true rake angle, γ = true wedge angle, λ = side clearance angle, and ψ = side rake angle.

17.2.11 Tool Setting

The tool must not be set too high or too low or inclined at an angle. The effects are shown in Figure 17.20. The following precautions should be followed for smooth and efficient manufacturing:

- **i.** Above center: tool tends to rub
- **ii.** Below center: work tends to climb over tool

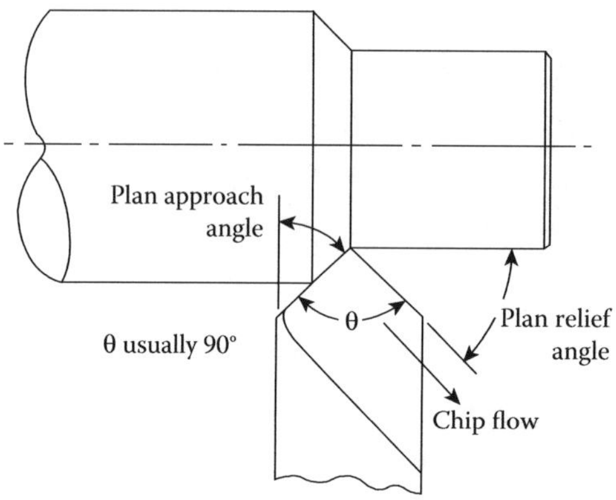

Figure 17.18

Chip flow.

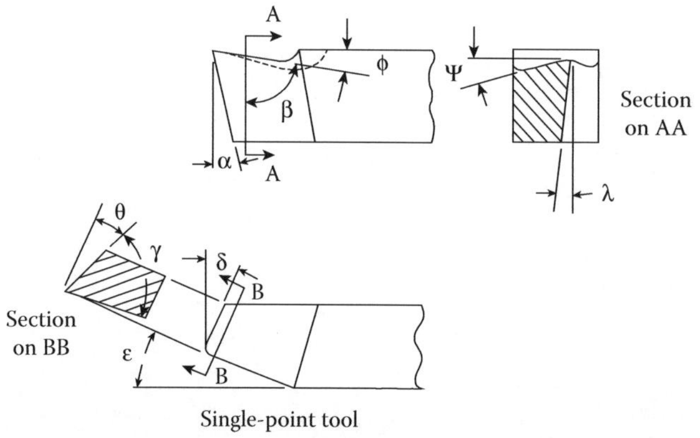

Figure 17.19

Tool geometry.

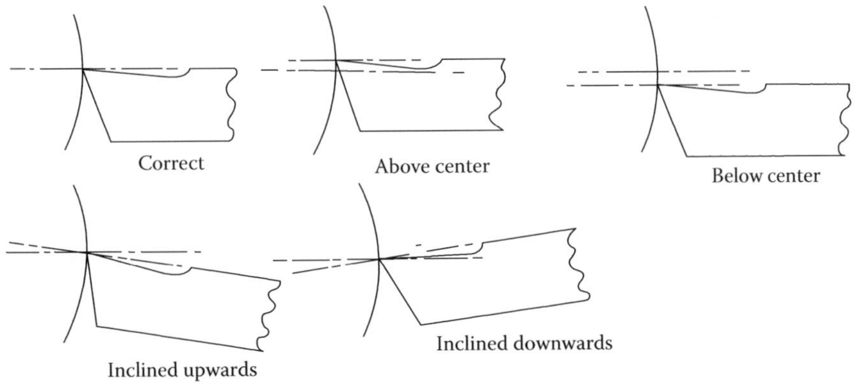

Figure 17.20

Tool settings.

iii. Inclined upwards: tool rubs
iv. Inclined downwards: work tends to drag tool in

17.2.11.1 Parting-Off Tool

This is used for *parting-off* the workpiece from bar stock held in a chuck. Note that there is *body clearance* on both sides as well as *side clearance*. The tool is weak and must be used with care. It must be set on or slightly above center. This prescription is not in agreement with normal settings precautions. If set even slightly below center, the work will climb onto the tool before parting-off (Figure 17.21).

17.2.12 Environmental Burden of Cutting Tool (T_e)

The environmental burden (T_e) due to the cutting tools is estimated here. All cutting tools are governed by tool life; hence a tool life is compared with machining time to calculate the environmental burden in a machining [7]. Cutting tools, especially for turning, are often renewed by regrinding. The environmental burden due to the cutting tool is calculated as follows by considering the aforementioned processes (Table 17.10).

$$T_e = \frac{T_m}{\sum_{j=1}^{RGN+1} TL_j}((TD_e + TP_e) \times TW + RGN \times RG_e) \tag{17.14}$$

where:

T_m = Machining time [s]
TL = Tool life [s]

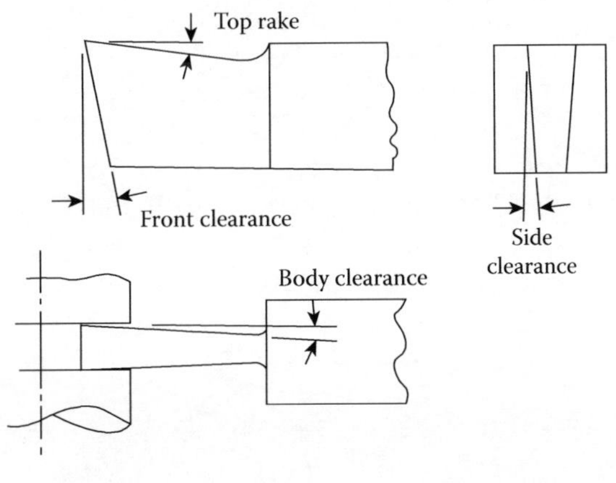

Figure 17.21

Parting-off tool.

Table 17.10 Suggested Cutting Point Angles for Turning Tools

Material		HSS		Carbide Brazed		Carbide Disposable	
		B Rk	S Rk	B Rk	S Rk	B Rk	S Rk
Martensitic	160 HB	0°	10°/15°	0°	6°	0°/–5°	–5°
Martensitic	300 HB	0°	15°	0°	6°	–5°	–5°
Free machining grades	160 HB	5°	8°	0°	6°	–5°	–5°
Wrought ferritic and cast plain Cr	160 HB	5°	8°	0°	6°	0°	5°
Wrought and cast austenitic	165 HB	0°	10°/15°	0°	6°	0°	5°

B Rk, back rake angle; S Rk, side rake angle. Nose radius = 0.5-3 mm. Use as large as possible radius consistent with part being turned.

TP_e = Environmental burden of cutting tool production [kg-CO_2/kg]
TD_e = Environmental burden of cutting tool disposal [kg-CO_2/kg]
TW = Tool weight [kg]
RGN = Total number of regrinding processes
RG_e = Environmental burden of regrinding [kg-CO_2]

Example 17.5

A square shoulder carbide-tipped turning tool is used to remove material from a shaft of 2.5 in. diameter and length of 5 in. The final diameter of the shaft is to be 2 in. The overall length of the tool is 4 in. and shank width and height are 5/8 in. The turning is to be regrinded five times during its life. Calculate the environmental burden of regrinding. The cutting speed is 100 ft./min and feed is 0.002 in./rev. The depth of cut is assumed to be 0.25 in. The tool life is 10 min.

Environmental burden calculation of the turning tool [5]:

Machining time

$$(T_m) = L/Nf$$

$$N = \frac{12(100)}{\pi(2.5)} = 1563 \text{ rpm}$$

and

$$T_m = \frac{5}{153 \times .002} = 16.5 \text{ min}$$

As tool life is 10 min, it needs regrinding twice.

The tool weight (TW) = 4 × 5/8 × 5/8 × 0.283 = 0.50 lb (the tool shank if plain carbon steel, and density (ρ) = 0.283 lb/in.^{3})

The environmental burden of cutting tool production (TP_e) and environmental burden of cutting tool disposal (recycling) (TD_e) = 1.06 × 1/4 = 0.15 lb of CO_2. (Here 1.06 is CO_2 for both together.)

Now we can calculate environmental burden of a carbide-tipped square shoulder turning tool.

$$T_e = \frac{16.5}{\sum_1^3 10}((0.15) \times 0.5 + 2 \times 4)$$

$$T_e = \frac{16.5}{20}(0.08 + 8) = 6.7 \text{ lb of } CO_2$$

Here environmental burden of regrinding (RG_e) = 4 lb/lb (Figure 17.22).

Figure 17.22

Orthogonal single-point cutting tool.

17.2.13 Waste Streams after Machining

One complication in developing the process model is quantifying the mass loss from the process. Primary and catalytic mass losses may occur in many forms in the gaseous, liquid and solid states. Primary mass loss consists of chips generated during the machining process, while catalytic mass losses consist of the waste streams generated by the cutting fluid evaporated from the cutting front, fluid coated on the chips and the workpiece, and expended tools. Obviously, equal amounts of these materials do not have an equivalent environmental impact. Therefore, a weighting scheme needs to be created in order to evaluate the relative environmental impact of solid, liquid and vapor waste streams for each process.

17.2.14 Toxicity Ratings for Some of the Materials

Graedel and Allenby [8] have provided toxicity ratings of various elements and it would be useful to use them in both design and manufacturing considerations. It is obvious that high-toxicity materials such as beryllium, arsenic, cadmium, and mercury including lead should be avoided. Mercury is particularly used in various electronic parts and, being highly toxic, has high potential for environmental damage. In old US houses, lead was used widely in plumbing and other spheres and it has been recently replaced with other low toxicity materials. If we look at the list of moderate toxic materials, most of them are used in manufacturing fields either as alloying elements in cutting tools or in some other way. However, the effect of chromium, cobalt, nickel, and copper would be difficult to achieve by some other means.

High toxicity: beryllium, arsenic, cadmium, mercury, lead
Moderate toxicity: lithium, boron, chromium, cobalt, nickel, copper, bismuth
Low toxicity: aluminum, silicon, titanium, iron, zinc, bromine, silver, tin, tungsten, gold

17.2.15 Classes of Supply for Some of the Elements

We are interested in sustainable manufacture and it is proper to examine the available supply of materials in nature. As we know, every day materials are being depleted and their regeneration is not possible, and so we should be prudent in using them for sustainable manufacturing. Graedel has listed the supply of materials in nature. It has been presented in categories of infinite supply, ample supply, adequate supply, potentially limited supply, and potentially highly limited supply. Ashby has presented this list in terms of reserve and depletion each year in his book [8].

Infinite supply: bromine, calcium, chlorine, krypton, magnesium, silicon
Ample supply: aluminum (gallium), carbon, iron, potassium, sulfur, titanium
Adequate supply: lithium, phosphorus
Potentially limited supply: cobalt*, chromium†, nickel*, lead (arsenic, bismuth), platinum†, zirconium
Potentially highly limited supply: silver, gold, copper, mercury, tin, zinc (cadmium)

17.2.16 Dry vs. Wet Machining

Just two decades ago, cutting fluids accounted for less than 3% of the cost of most machining processes. Fluids were so cheap that few machine shops gave them much thought. Times have changed. Today, cutting fluids account for up to 15% of a shop's production costs, and machine shop owners constantly worry about fluids. Cutting fluids, especially those containing oil, have become a huge liability. Not only does the EPA

* Supply is adequate but virtually all from South Africa and Zimbabwe. This geographical distribution makes supplies potentially subject to cartel control.

† Maintenance of supplies will require mining seafloor nodules.

regulate the disposal of such mixtures, but many states and localities also have classified them as hazardous wastes and impose even stricter controls if they contain oil and certain alloys. Because many high-speed machining operations and fluid nozzles create airborne mists, governmental bodies also limit the amount of cutting fluid mist allowed into the air. The EPA has proposed even stricter standards for controlling such airborne particulate and the Occupational Safety and Health Administration (OSHA) is considering an advisory committee's recommendation to lower the permissible exposure limit of fluid mist. The cost of maintenance, record keeping and compliance with current and proposed regulations is rapidly raising the price of cutting fluids [2,3].

Consequently, many machine shops are considering eliminating the costs and headaches associated with cutting fluids altogether by cutting dry. The decision to cut wet or dry must be made on a case-by-case basis. A lubricious fluid often will prove beneficial in low-speed jobs, hard-to-machine materials, difficult applications, and when surface finish requirements are demanding. A fluid with high cooling capacity can enhance performance in high-speed jobs, easy-to-machine materials, simple operations and jobs prone to edge-buildup problems or having tight dimensional tolerances.

Many times, though, the extra performance capabilities that a cutting fluid offers is not worth the extra expense incurred, and in a growing number of applications, cutting fluids are simply unnecessary or downright detrimental. Modern cutting tools can run hotter than their predecessors and sometimes compressed air can be used to carry hot chips away from the cutting zone.

17.3 Minimum Quantity Lubrication (MQL) during Turning

MQL reduced the cutting temperature, tool wear and surface roughness as compared to dry machining. The reduction of about 5%–12% at average cutting temperature using MQL by vegetable oil as compared to using dry machining was observed depending upon the levels of the process parameters (cutting speed, feed rate). Cutting forces decreased by about 5%–15% using MQL by vegetable oil. The experimental studies of sunflower and canola oils-based cutting fluids, including different percentages (8% and 12%) of extreme pressure additive and two commercial cutting fluids (semisynthetic and mineral based) in turning of AISI 304L stainless steel with respect to surface roughness, cutting force, feed force and tool wear as reported in literature. Experiments were also conducted at dry cutting conditions that caused rapid tool wear and fracture.

The higher tool life in vegetable-based cutting fluids was due to the fatty acid content. Canola-based cutting fluids showed better performance than sunflower-based cutting fluids because of different length of carbon chains. Canola oil has three carbons more in formulae and a longer carbon chain outstanding in high cutting temperature, thus improving the surface protection. Moreover, the high viscosity of canola oil had a tendency to resist flow. This high viscosity provides more effective lubricating at the tool-chip interface, thus reducing the friction between the tool and workpiece and removing heat developed at the interface easily. A high percentage of extreme pressure additive in vegetable-based cutting fluids showed the higher surface roughness values. Results indicated that 8% of extreme pressure additive including canola-based cutting fluid performed better than the rest. It is suggested that mineral and semisynthetic cutting fluids can be replaced by vegetable-based cutting fluids in turning. Krahenbuhl [9] used vegetable-based cutting fluids (groundnut oil, coconut oil, palm kernel oil and shea butter oil) during turning of mild steel, aluminum and copper and measured cutting force. Although, it was found that the effects of vegetable-based cutting fluids were material dependent, groundnut oil showed the best performance among the four vegetable-based cutting fluids investigated. In another study, coconut oil outperformed the other two cutting fluids (soluble oil and straight cutting oil) in terms of reducing the cutting force and temperature.

References

1. Groover, M. P. (2012). *Fundamentals of Modern Manufacturing: Materials, Processes, and Systems*, 5th edn. John Wiley.
2. Batzer, S. A. and Sutherland, J. W. (1998). The dry cure for coolant ills. *Cutting Tool Engineering* 50 (4): 34–44.
3. Haan, D. M., Batzer, S. A. Olson, W. W., and Sutherland, J. W. (1997). An experimental study of cutting fluid effects in drilling. *Journal of Materials Processing Technology* 71 (2): 305–313.
4. Keolein, G. A. and Menerey, D. (1994). Sustainable Development by Design: Review of Life Cycle Design and Related Approaches. CSS Publication # CSS94-01, *Journal of the Air and Waste Management Association* 44 (5): 645–668.

5. Narita, H., Desmira, N., and Fujimoto, H. (2012). Environmental Burden Analyzer for Machine Tool Operations and Its Application. www.intechopen.com/books/manufacturing-systems.
6. Ashby, M. F. (2009). *Materials and the Environment: Eco-informed Material Choice*, Butterworth-Heinemann/Elsevier, Burlington, MA.
7. Taniguchi, M., Kakinuma, Y., Aoyama, T., and Inasaki, I. (2006). Influences of downsized design for machine tools on the environmental impact. In *Proceedings of the MTTRF 2006 Annual Meeting* (pp. 1–4).
8. Graedel, T. and Allenby, B. (2002). *Industrial Ecology*, 2nd edn. Prentice Hall.
9. Krahenbuhl, U. (2002). Vegetable oil-based coolants improve cutting performance (Cutting Fluids). Tooling & Production, Nelson.

Bibliography

Abdalla, H. S. and Patel, S. (2006). The performance and oxidation stability of sustainable metalworking fluid derived from vegetable extracts. *Proceedings of the Institution of Mechanical Engineers Part B: Journal of Engineering Manufacture* 220: 2027–2040.

Abou-El-Hossein, K. A. (2008). Cutting fluid efficiency in end milling of AISI 304 stainless steel. *Industrial Lubrication and Tribology* 60 (3): 115–120.

Alves, S. M. and Oliveira, J. F. G. (2008). Vegetable based cutting fluid-an environmental alternative to grinding process. In *Proceedings of the 15th CIRP International Conference on Life Cycle Engineering* (pp. 664–668), Sydney.

Baniszewski, B. (2005). An environmental impact analysis of grinding, S.B. thesis, Department of Mechanical Engineering, Massachusetts Institute of Technology, Cambridge, MA.

Cetin, M. H., Ozcelik, B., Kuram, E., and Demirbas, E. (2011). Evaluation of vegetable based cutting fluids with extreme pressure and cutting parameters in turning of AISI 304L by Taguchi method. *Journal of Cleaner Production* 19: 2049–2056.

Dahmus, J. B. and Gutowski, T. G. (2004). An environmental analysis of machining. In *Proceedings 2004 ASME International Mechanical Engineering Congress and RD&D Exposition* (pp. 13–19), Anaheim, CA.

Dittrich, M. A., Gutowski, T. G., Cao, J., Roth, J. T., Xia, C., Kiridena, V., Ren, F., and Henning, H. (2012). Exergy analysis of incremental sheet forming. *Production Engineering Research Development* 6: 169–177.

Fox, N. J. and Stachowiak, G. W. (2007). Vegetable oil-based lubricants—A review of oxidation. *Tribology International* 40: 1035–1046.

Fthenakis, V. M. and Kim, H. C. (2007). Greenhouse-gas emissions from solar electric- and nuclear power: A life-cycle study. *Energy Policy* 35 (4): 2549–2557.

Gutowski, T., Branham, M., Dahmus, J., Jones, A., Thiriez, A., and Sekulic, D. (2009). Thermodynamics Analysis of Resources Used in Manufacturing Processes. *Environmental Science and Technology* 43 (5): 1584–1590.

Gutowski, T., Dahmus, J., and Thiriez, A. (2006). Electrical energy requirements for manufacturing processes. In *Proceedings of the 13th CIRP International Conference on Life Cycle Engineering* (pp. 623–627), Cambridge, MA.

Gutowski, T., Jeffrey, D., and Thiriez, A. (2006). Electrical Energy Requirements for Manufacturing Processes. In *Proceedings of the13th CIRP International Conference on Life Cycle Engineering*, Leuven, May 31st–June 22nd, Cambridge, MA.

Haapala, K., Rivera, J. L., and Sutherland, J. (2009). Reducing environmental impacts of steel product manufacturing. *Transactions North American Manufacturing Research Institution of the SME* 37: 419–426.

Hoffman, E. G. (1984). *Fundamentals of Tool Design*, 2nd edn. Society of Manufacturing Engineers, Dearborn, MI.

Hondo, H. (2005). Life cycle GHG emission analysis of power generation systems: Japanese case. *Energy* 30 (11–12): 2042–2056.

Hong, S. Y. and Broomer, M. (2000). Economical and ecological cryogenic machining of AISI 304 austenitic stainless steel. *Clean Products and Processes* 2: 157–166.

Jan, E. and Bras, B. (1998). ISO 14000 and activity based life cycle assessment in environmentally conscious design and manufacturing: A comparison. 1998 ASME design engineering technical conference, American Society of Mechanical Engineers, Atlanta, GA.

Johnson, W. (1956). The pressure for the cold extrusion of lubricated rod through square dies of moderate reduction at slow speeds. *Journal of the Institute of Metals* 85: 403.

Kalpakjian, S. and Schmid, S. R. (2010). *Manufacturing Processes for Engineering Materials*, 6th edn. Pearson Prentice Hall, Upper Saddle River, NJ.

Khan, M. M. A. and Dhar, N. R. (2006). Performance evaluation of minimum quantity lubrication by vegetable oil in terms of cutting force, cutting zone temperature, tool wear, job dimension and surface finish in turning AISI-1060 steel. *Journal of Zhejiang University–Science A: Applied Physics & Engineering* 7 (11): 1790–1799.

Kordonowy, D. N. (2001). A power assessment of machining tools. B.S. Thesis, MIT, Department of Mechanical Engineering, Cambridge, MA.

Kurtz, M. (2007). *Environmental Conscious Manufacturing*, John Wiley.

Lange, K. (1985). *Handbook of Metal Forming*, McGraw-Hill, New York.

Life Cycle Assessment (LCA): Principles and Practice, EPA/6000R-06/060, May 2006.

Machining Data Handbook, Machinability data center, Metcut Research Associates, 1980.

Narita, H., Desmira, N., and Fujimoto, H. (2008). Environmental burden analysis for machining operation using LCA method. In *Proceedings of the 41st CIRP Conference on Manufacturing Systems* (pp. 65–68), Springer vesley.

Niggeschmidt, S., Helu, M., Diaz, N., Behmann, B., Lanza, G., and Dornfeld, D. (2010). Integrating green and sustainability aspects into life cycle performance evaluation. In *Proceedings of the 16th CIRP International Conferences on Life Cycle Engineering* (pp. 366–371), Heifei, China.

Pettersson, A. (2007). High-performance base fluids for environmentally adapted lubricants. *Tribology International* 40: 638–645.

Rangarajan, A. and Dornfeld, D. (2004). Efficient tool paths and part orientation for face milling, *CIRP Annals* 53 (1): 73–76.

Rao, D. N. and Srikant, R. R. (2006). Influence of emulsifier content on cutting fluid properties. In *Proceedings of the Institute of Mechanical Engineering Part B: Journal of Engineering Manufacture* 220: 1803–1806.

Santos, J. P., Oliveira, M., and Almeida, F. G., et al. (2011). Improving the environmental performance of machine-tools: Influence of technology and throughput on the electrical energy consumption of a press-brake. *Journal of Cleaner Production* 19: 356–364.

Schey, J. A. (2000). *Introduction to Manufacturing Processes*, 3rd edn. McGraw Hill, New York.

Shokrani, A., Dhokia, V., and Newman, S. T. (2012). Environmentally conscious machining of difficult-to-machine materials with regard to cutting fluids. *International Journal of Machine Tools and Manufacture* 57: 83–101.

Srikant, R. R., Rao, D. N., and Rao, P. N. (2009). Influence of emulsifier content in cutting fluids on cutting forces, cutting temperatures, tool wear, and surface roughness. *Proceedings of the Institute of Mechanical Engineering Part J: Journal of Engineering Tribology* 223: 203–209.

Szargut, J., Morris, D. R., and Steward, F. R. (1988). *Exergy Analysis of Thermal, Chemical and Metallurgical Processes*, Hemisphere Publishing Corporation, New York.

Tlusty, G. (2011). *Manufacturing Processes and Equipment*, Prentice Hall, Upper Saddle River, NJ.

Tönissen, S. (2009). Power demand of precision machine tools. MS Thesis, Department of Mechanical Engineering, University of California, Berkeley, Berkeley, CA.

Yansong, G., Loenders, J., Duflou, J., and Lauwers, B. (2012). Optimization of energy consumption and surface quality in finish turning. In *Proceedings of 5th CIRP Conference on High Performance Cutting 2012* (pp. 551–556).

18

Drilling Operations

18.1 Introduction to Drilling and Reaming

All drilling machines operate on the same basic principle. The spindle turns the cutting tool, which is advanced either by hand or automatically into a workpiece that is mounted on a table or held in a drill press vise. Many types and sizes of drilling machines are used in manufacturing [1–3]. They range in size from a simple bench-mounted sensitive drill press to large multiple-spindle machines that are able to drive many drills at the same time.

18.1.1 Simple Drill Press

A simple drill press, shown in Figure 18.1, can be floor mounted or it can have a shorter main post and can be mounted on a bench. The motions of this machine are very simple. The table of a floor model can be raised or lowered and rotated around the machine column. The spindle rotates and can be raised and lowered, with a stroke of 4–8 in. Stops can be set to limit and regulate the depth.

A twist is a manually or machine-rotated tool with cutting edges to produce circular holes in metals, plastics, wood, and so on. It consists of a hardened steel bar, usually with two helical grooves or "flutes" ending in two angled cutting edges. The flutes permit many regrinds and assist in the removal of cuttings. Drills vary in size from a fraction of a millimeter to over 10 cm (Figure 18.2). As with a lathe turning tool, the cutting edges must have top rake and clearance.

18.1.2 Radial Drilling Machine

A radial drilling machine or a radial arm press is a geared drill head that is mounted on an arm assembly that can be moved around to the extent of its arm reach. The most important components are the arm, column, and the drill head. The drill head of a radial drilling machine can be moved, adjusted in height, and rotated. Aside from its compact design, the radial drill press is capable of positioning its drill head to the workpiece through this radial arm mechanism.

This is probably one of the reasons why more machinists prefer using this type of drilling machine. In fact, the radial drilling machine is considered the most versatile type of drill press. The tasks that a radial drilling machine can do include boring holes, countersinking, and grinding off small particles in masonry work.

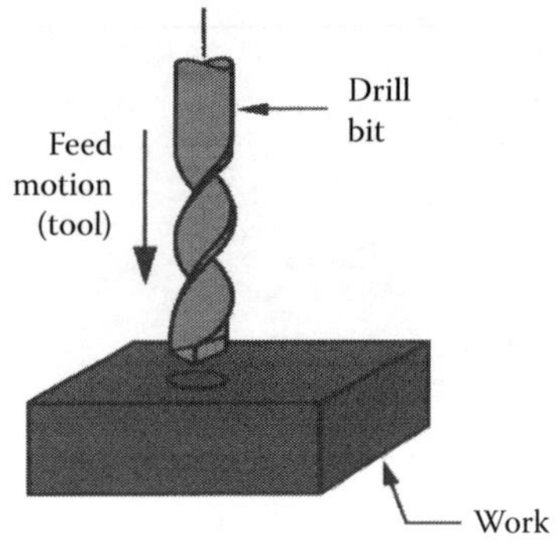

Figure 18.1

Drilling operation.

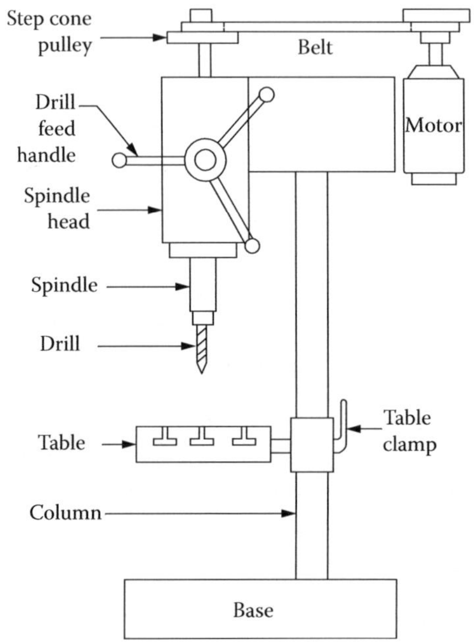

Figure 18.2

Drilling machine.

Although some drill presses are floor mounted, the most common setup of radial arm drill presses are those that are mounted on workbenches or tables. With this kind of setup, it is easier to mount the drill and the workpieces. There is no need to reposition workpieces because the arm can extend as far as its length will allow. Moreover, it is easier to maneuver large workpieces with a radial drilling machine (Figure 18.3).

Some of the major parts of the radial arm drilling machine are [4,5]

Column: the part of the radial arm drill press that holds the radial arm, which can be moved around according to its length

Arm raise: adjusts the vertical height of the radial arm along the column

On/off button: the switch that activates and deactivates the drill press

Arm clamp: secures the column and the arm in place

Table: the area where the workpieces are fed and worked on

Base: the radial arm drill press part that supports the column and the table

Spindle: the rotated part of the drill press that holds the drill chuck used in holding the cutting tool

Drill head: the part of the drill press that penetrates through the material or workpiece and drills a specific hole size

Radial arm: holds and supports the drill head assembly and can be moved around to the extent of its length

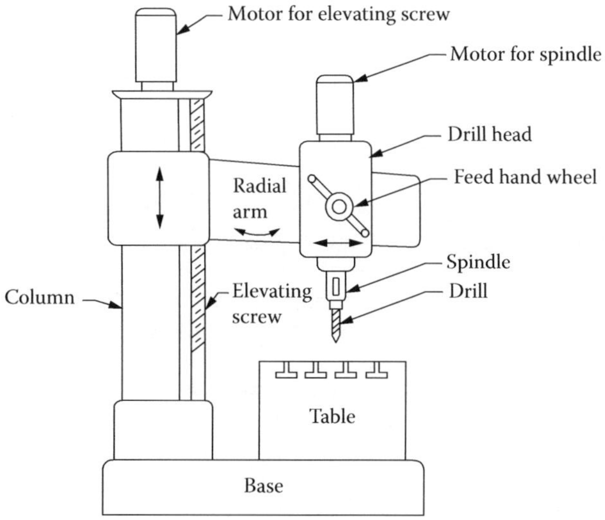

Figure 18.3

Radial drilling machine with components.

There are a number of advantages to using the radial arm drill press. One of these advantages is the amount of area that it can cover, which is only dependent on the length of the arm. Another advantage is the considerable size of work that it can handle since the arm can actually swivel out of the working area allowing cranes and derricks to place workpieces on the table. Finally, less effort is required during the drilling process because the arm assembly seemingly is doing all that is needed for specific tasks to be completed.

18.2 Drilling Tools

Drill bits are cutting tools that are used to create cylindrical holes, almost always of circular cross section. Drill bits come in many sizes and have many uses. Bits are usually connected to a mechanism, often simply referred to as a *drill*, which rotates them and provides *torque* and axial force to create a hole.

The *shank* is the part of the drill bit that is grasped by the chuck of a drill. The cutting edges of the drill bit are at one end, and the shank is at the other. Drill bits come in standard sizes. A comprehensive drill bit and tap size chart is shown in Table 18.5.

The term *drill* may refer to either a drilling machine or a drill bit that is used in a drilling machine. In this chapter, for clarity, drill bit or bit is used throughout to refer to a bit for use in a drilling machine, and drill always refers to a drilling machine.

Drill bit geometry has several characteristics:

- The *spiral* (or rate of twist) in the drill bit controls the rate of chip removal. A fast spiral (high twist rate or "compact flute") drill bit is used in high feed rate applications under low spindle speeds, where the removal of a large volume of chips is required. Low spiral (low twist rate or "elongated flute") drill bits are used in cutting applications where high cutting speeds are traditionally used, and where material, such as aluminum or copper, has a tendency to fall on the bit or otherwise clog the hole.
- The *point angle*, or the angle formed at the tip of the bit, is determined by the material that the bit will be operating on. Harder materials require a larger point angle, and softer materials require a sharper angle. The correct point angle for the hardness of the material controls wandering, chatter, hole shape, wear rate, and other characteristics.
- The *lip angle* determines the amount of support provided to the cutting edge. A greater lip angle will cause the bit to cut more aggressively under the same amount of point pressure as a bit with a smaller lip angle. Both conditions can cause binding, wear, and eventual catastrophic failure of the tool. The proper amount of lip clearance is determined by the point angle. A very acute point angle has more web surface area presented to the work at any one time, requiring an aggressive lip angle,

where a flat bit is extremely sensitive to small changes in the lip angle due to the small surface area supporting the cutting edges.
- The *length* of a bit determines how long a hole can be drilled, and it also determines the stiffness of the bit and the accuracy of the resultant hole. Twist drill bits are available in standard lengths, referred to as stub length or screw-machine length (short), the extremely common jobber length (medium), and taper length or long series (long).

Most drill bits for consumer use have straight shanks. For heavy-duty drilling in industry, bits with tapered shanks are sometimes used.

The diameter-to-length ratio of the drill bit is usually between 1:1 and 1:10. Much higher ratios are possible (e.g., "aircraft length" twist bits and pressured-oil gun drill bits), but the higher the ratio, the greater the technical challenge of producing good work.

The best geometry to use depends on the properties of the material that is being drilled.

18.3 Drill Helix and Point Angles

The helix angle is usually a standard size, but "quick" and "slow" helix angles are used for particular materials. The helix angle can be calculated as

$$\delta = \tan^{-1}(\pi D/L)$$

where:
- D is the diameter of the drill
- L is the pitch length of the helix

For example, a drill of 1 in. (925.5 mm) in diameter with a helix angle of 38.15° is found to have a pitch length (L) of 4 in. (101 mm). This pitch length may then be used to find the helix angle at the point on the drill where $D = 0.5$ in. (12.7 mm). If the helix angle is calculated now, it is equal to 21.5°.

It is sometimes necessary, for example, for brass and thin material, to grind a short length of straight flute, as shown in Figure 18.4. It is also sometimes necessary to thin down the web or core. The point angle was traditionally about 120° (included angle), but other angles are now used to suit the material. The lip clearance also varies (see Table 18.6).

Core drills: Core drills have three of four flutes and are used for opening out existing holes, for example, core holes in castings (Figure 18.4).

18.3.1 Cutting Parameter Calculation in Drilling

Drills are attached to the spindle of a drilling machine and the spindle or drill rotational speed (N) may be calculated as

$$N = \frac{v}{\pi D}$$

where D is the drill diameter (mm [in.]). The feed (f) in drilling (mm/rev [in./rev]) is proportional to the drill diameter; higher feeds are recommended for larger tool diameters. The uncut chip thickness is roughly half the feed. The feed rate (f_r) is calculated the same as in turning:

$$f_r = fN \text{ (mm/min [in./min])}$$

We drill either a through hole or a blind hole. In the case of a through hole, the drill exits opposite the work; however, in the case of a blind hole, the drilled hole length is less than the thickness of the part. The machining time (T_m) is calculated as

$$T_m = \frac{t + A}{f_r}$$

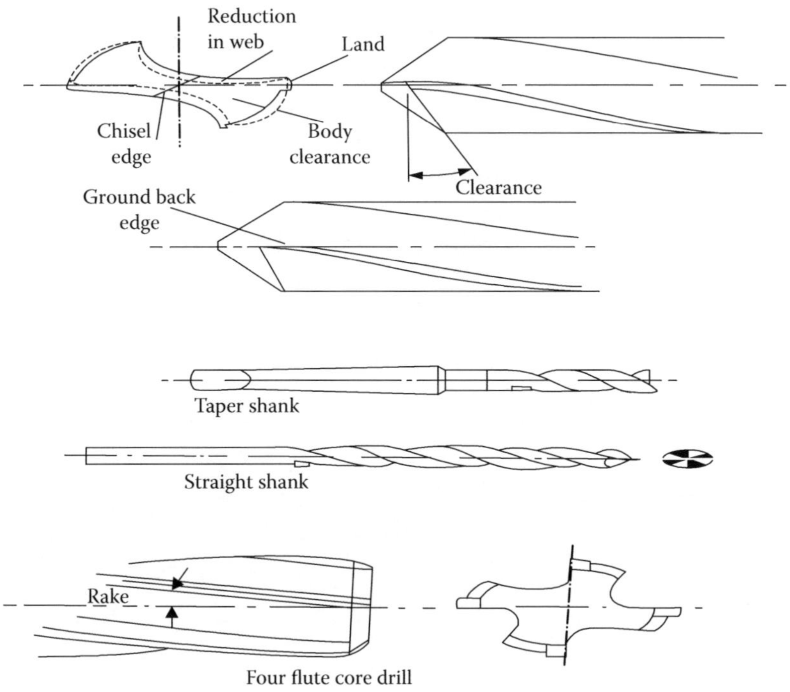

Figure 18.4

Core drills geometry.

where:

- t is the thickness of the workpiece
- A is an approach distance that accounts for the drill point angle

This is calculated as

$$A = 0.5D\tan\left(90 - \frac{\theta}{2}\right)$$

where θ is the drill point angle.
For a blind hole, if d is the hole depth, then

$$T_m = \frac{d + A}{f_r}$$

The rate of metal removal in drilling can be calculated as

$$R_{MR} = \frac{\pi D^2 f_r}{4} \ (\text{mm}^3/\text{min}\,[\text{in.}^3/\text{min}])$$

18.4 Reamers

A reamer is used to finish a hole accurately with a good surface finish. It is a periphery cutting tool, unlike the drill, which is end cutting. Rake and clearance are required as shown in Figure 18.5; note that a reamer must be ground on the clearance face otherwise the size will be lost. Flutes may be straight or helical (usually left-handed). A hand reamer requires a long slow taper, but machine reamers have a short 45° lead. The hole is drilled only slightly smaller than the reamer diameter, with an allowance of about 0.015 mm/mm, but it depends on the material. Taper reamers are used for finishing holes for taper pins.

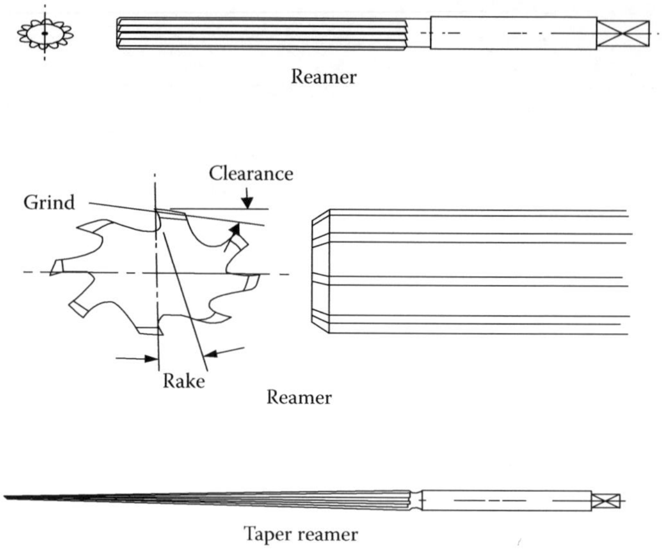

Figure 18.5

Types of reamers.

18.4.1 Drilling Parameters

Drilling feeds and drilling speeds are shown in Tables 18.1 and 18.2, respectively. Table 18.3 provides information on drilling plastics. Cutting lubricants for drilling, reaming, and tapping, and tapping drill sizes for metric coarse threads are presented in Tables 18.4 and 18.5, respectively.

Plastics are increasingly used in the manufacturing of parts especially in the aircraft industry. Manufacturers are increasingly using plastics in the manufacture of household products and this practice is being accepted by a large population for a variety of reasons including cost, weight, environment, and so on. Table 18.3 presents the feed (mm/rev) for various drills.

Table 18.1 Drilling Feeds

Drill Diameter (mm)	Hard Materials[a]	Soft Materials[b]
1.5	0.05	0.05
3.0	0.05	0.07
6.0	0.07	0.10
9.0	0.10	0.15
12.0	0.12	0.20
19.0	0.18	0.30
25.0	0.22	0.35

[a] Steels above 0.3%C and alloy steels.

[b] Gray cast iron, steels below 0.3%C, brass, bronze, aluminum alloys, and so on.

Table 18.2 High-Speed Drill Speeds

Material	Speed[a] (m/s)
Cast iron	0.4–0.6
Mild steel	0.3–0.5
60/40 brass	0.8–1.0
Medium-carbon steel	0.2–0.3

[a] Speed $= \pi DN/60{,}000$ m/s, where D = diameter (mm), N = number of revolutions per minute.

Table 18.3 Drilling Plastics, Cutting Speeds, and Feeds

Material	Conditon	Cutting Speed (m/min)	Feed (mm/rev) for Nominal Hold Diameter (mm) of							
			1.5	3.0	6.0	12.0	20.0	25.0	30.0	50.0
Polyethylene, polypropylene, TFE-flourocarbon	Extruded, molded, or cast	33	0.12	0.25	0.30	0.38	0.46	0.50	0.64	0.76
High-impact styrene, modified acrylic	Extruded, molded, or cast	33	0.05	0.10	0	0.15	0.15	0.20	0.20	0.25
Nylon, acetals, polycarbonate	Molded	33	0.05	0.12	0.1	0.20	0.25	0.30	0.38	0.38
Polystyrene	Molded or extruded	66	0.03	0.05	0	0.10	0.13	0.15	0.18	0.20
Soft grades of thermosetting plastic	Cast, molded, or filled	50	0.08	0.13	0	0.20	0.25	0.30	0.38	0.38
Hard grades of thermosetting plastic	Cast, molded, or filled	33	0.05	0.13	0.15	0.20	0.25	0.30	0.38	0.38

Table 18.4 Cutting Lubricants for Drilling, Reaming, and Tapping

Material	Drilling	Reaming	Tapping
Mild steel (hot and cold rolled)	Soluble oil, mineral oil, lard oil	Mineral lard oil	Soluble oil, lard oil
Tool steel (carbon and high speed)	Soluble oil, lard oil with sulfur	Lard oil	Sulfur-base oil, mineral lard oil
Alloy steel	Soluble oil, mineral oil	Lard oil	Sulfur-base oil, mineral lard oil
Brass and bronze	Dry, lard oil, paraffin mixture	Soluble oil	Soluble oil, lard oil
Copper	Soluble oil	Soluble oil	Soluble oil, lard oil
Aluminum	Paraffin, lard oil	Mineral lard oil	Soluble oil, mineral lard oil
Monel metal	Lard oil, sulfur-base oil	Mineral lard oil, sulfur-base oil	Mineral lard oil, sulfur-base oil
Malleable iron	Soluble oil	Soluble oil	Soluble oil
Cast iron	Dry	Dry	Dry, lard oil for nickel cast iron

Table 18.5 Tapping Drill Sizes for Metric Coarse Threads

Nominal Diameter (mm)	Thread Pitch (mm)	Tap Drill Size (mm)	Nominal Diameter (mm)	Thread Pitch (mm)	Tap Drill Size (mm)
1.6	0.35	1.20	20.0	2.50	17.5
2.0	0.40	1.60	24.0	3.0	21.0
2.5	0.45	2.05	30.0	3.50	26.5
3.0	0.50	2.50	36.0	4.00	32.0
3.5	0.60	2.90	42.0	4.50	37.5
4.0	0.70	3.30	48.0	5.00	43.0
5.0	0.80	4.20	56.0	5.50	50.5
6.0	1.0	5.30	64.0	6.00	58.0
8.0	1.25	6.80	72.0	6.00	66.0
10.0	1.50	8.50	80.0	6.00	74.0
12.0	1.75	10.20	90.0	6.00	84.0
14.0	2.00	12.00	100.0	6.00	94.0
16.0	2.00	14.00			

18.5 Drill Bit Materials

Many different materials are used for or on drill bits, depending on the required application [6]. Many hard materials, such as carbides, are much more brittle than steel, and are far more subject to breaking, particularly if the drill is not held at a very constant angle to the workpiece; for example, when it is hand held.

18.5.1 Steel

- Soft *low-carbon* steel bits are inexpensive, but they do not hold an edge well and require frequent sharpening. They are only used for drilling wood; even working with hardwoods rather than softwoods can noticeably shorten their lifespan.
- Bits made from *high-carbon* steel are more durable than low-carbon steel bits due to the properties conferred by hardening and tempering the material. Overheating (e.g., by frictional heating while drilling) causes them to lose temper, resulting in a soft cutting edge. These bits can be used on wood or metal.
- *High-speed* steel (HSS) is a form of tool steel; HSS bits are hard and much more resistant to heat than high-carbon steel. They can be used to drill metal, hardwood, and most other materials at greater cutting speeds than carbon steel bits, and have largely replaced carbon steels.
- *Cobalt steel alloys* are variations on HSS, but contain more cobalt. They hold their hardness at much higher temperatures, and are used to drill stainless steel and other hard materials. The main disadvantage of cobalt steels is that they are more brittle than standard HSS.

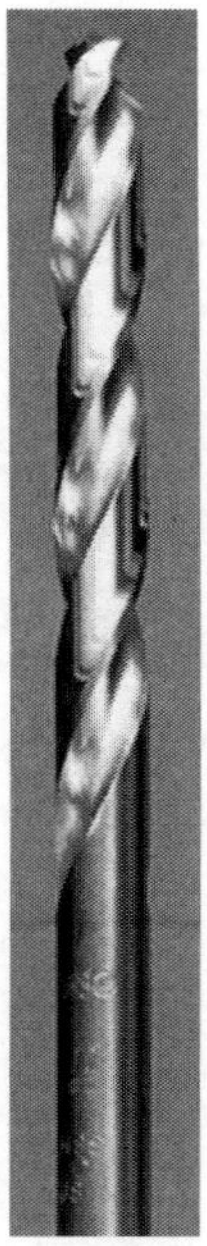

Figure 18.6

Titanium nitride-coated twist bit.

Other materials used for drill bits are expensive and are used for special purposes.

- *Tungsten carbide* and other carbides are extremely hard, and can drill virtually all materials while holding an edge longer than other bits. The material is expensive and much more brittle than steel. Consequently, they are mainly used for drill bit tips: small pieces of hard material fixed or brazed onto the tip of a bit made of less hard metal. However, it is becoming common in job shops to use solid carbide bits. In very small sizes, it is difficult to fit carbide tips; in some industries, most notably polychlorinated biphenyl (PCB) manufacturing, requiring many holes with diameters less than 1 mm, solid carbide bits are used.
- *Polycrystalline diamond* (PCD) is among the hardest of all tool materials and is therefore extremely resistant to wear. It consists of a layer of diamond particles, typically about 0.5 mm (0.020 in.) thick, bonded as a sintered mass to a tungsten-carbide support. Bits are fabricated using this material either by brazing small segments to the tip of the tool to form the cutting edges, or by sintering PCD into a vein in the tungsten carbide "nib." The nib can later be brazed to a carbide shaft; it can then be ground to complex geometries that would otherwise cause braze failure in the smaller "segments." PCD bits are typically used in the automotive, aerospace, and other industries to drill abrasive aluminum alloys, carbon-fiber-reinforced plastics, and other abrasive materials, and in applications where machine downtime to replace or sharpen worn bits is exceptionally costly (Figure 18.7).

Coatings on drills:

- Black oxide is an inexpensive black coating. A black oxide coating provides heat resistance and lubricity, as well as corrosion resistance. Coating increases the life of HSS bits.
- Titanium nitride (TiN) is a very hard ceramic material that can be used to coat a HSS bit (usually a twist bit), extending the cutting life by three or more times. However, when the bit is sharpened, the new edge will not have the benefits of the coating.
- Titanium aluminum nitride (TiAlN) is a similar coating that can extend the tool life by five or more times.
- Titanium carbon nitride (TiCN) is another coating that is also superior to TiN (Figure 18.6).
- Diamond powder is used as an abrasive, most often for cutting tile, stone, and other very hard materials. Large amounts of heat are generated by friction, and diamond-coated bits often have to be water cooled to prevent damage to the bit or the workpiece.

Figure 18.7

Diamond-coated 2 mm bits, used for drilling materials such as glass.

Table 18.6 Drill Angles

Material	Helix Angle	Point Angle (°)	Lip Clearance (°)
Aluminum alloy	Quick	140	12–15
Magnesium alloy	Standard	100	12–15
Brass	Slow	130	10–12
Copper	Quick	125	12–15
Bakelite	Slow	30	12–15
Manganese steel	Slow	130	7–10

- Zirconium nitride has been used as a drill bit coating for some tools under the Craftsman brand name.

18.6 Lubricants for Drilling

Lubricants are used in drilling to reduce the temperature during drilling, which is normally very high and damages the quality of the hole and the life of drills. However, it is the supposed greatest polluter of workplace and downstream land, water, and air [7,8]. Although attempts in manufacturing research are focused on reducing the use of cutting fluids, the present state-of-the-art technologies do not seem to assure that cutting fluids will be entirely eliminated. The new trend concerning cutting fluids in machining is to replace hazardous components with environmentally friendly compounds. These new compounds must not only show the same properties as cutting fluids but they must also improve the machining performance, such as productivity. The most suitable lubricants for drilling, reaming and tapping different materials are presented in Table 18.4.

Tapping is used to provide internal screw threads on an existing hole. Threaded components are widely used as fasteners in assembly and for the transmission of motion in machinery. Table 18.5 presents the nominal diameter of drilled holes from 1.6 to 100 mm with the tap drill size and thread pitch.

The point angle and lip clearance angles are peculiar to drills and are specified in Table 18.6 for various materials. These angles help in the design of a better drill. The point angles in particular help in calculating the approach distance during drilling operations.

References

1. Groover, M. P. (2012). *Fundamentals of Modern Manufacturing: Materials, Processes, and Systems*, 5th edn. Wiley, Hoboken, NJ.
2. Kalpakjian, S. and Schmids, R. (2010). *Manufacturing Processes for Engineering Materials*, 6th edn. Pearson Prentice Hall, Upper Saddle River, NJ.
3. Schey, J. A. (2000). *Introduction to Manufacturing Processes*, 3rd edn. McGraw Hill, New York.

4. Tlusty, G. (2011). *Manufacturing Processes and Equipment*. Prentice Hall, Englewood Cliffs, NJ.
5. Hoffman, E. G. (1984). *Fundamentals of Tool Design*, 2nd edn. Society of Manufacturing Engineers, Dearborn, MI.
6. Machinability Data Center, and Metcut Research Associates. (1980). *Machining Data Handbook*. Machinability Data Center, Metcut Research Associates, Cincinnati, OH.
7. Batzer, S. A. and Sutherland, J. W. (1998). The dry cure for coolant ills. *Cutting Tool Engineering* 50 (4): 34–44.
8. Haan, D. M., Batzer, S. A., Olson, W. W., and Sutherland, J. W. (1997). An experimental study of cutting fluid effects in drilling. *Journal of Materials Processing Technology* 71 (2): 305–313.

Bibliography

Abdalla, H. S. and Patel, S. (2006). The performance and oxidation stability of sustainable metalworking fluid derived from vegetable extracts. *Proceedings of the Institution of Mechanical Engineers, Part B: Journal of Engineering Manufacture* 220 (12): 2027–2040.

Alves, S. M. and Oliveira, J. F. G. (2008). Vegetable based cutting fluid-an environmental alternative to grinding process. In *15th CIRP International Conference on Life Cycle Engineering* (pp. 664–668), Sydney.

Ashby, M. F. (2009). *Materials and the Environment: Eco-informed Material Choice* (pp. 112, 114, 284–287, 290–291). Butterworth-Heinemann, Burlington, MA.

Cetin, M. H., Ozcelik, B., Kuram, E., and Demirbas, E. (2011). Evaluation of vegetable based cutting fluids with extreme pressure and cutting parameters in turning of AISI 304L by Taguchi method. *Journal of Clean Production* 19 (17–18): 2049–2056.

Dahmus, J. B. and Gutowski, T. G. (2004). An environmental analysis of machining. In *Proceedings of ASME International Mechanical Engineering Congress and RD&D Expo*, Anaheim, CA.

Fox, N. J. and Stachowiak, G. W. (2007). Vegetable oil-based lubricants—A review of oxidation. *Tribology International* 40 (7): 1035–1046.

Gregory, A. K. and Menerey, D. (1994). Sustainable development by design: Review of life cycle design and related approaches. *Journal of the Air and Waste Management Association* 44 (5): 645–668, CSS Publication # CSS94-01.

Guo, Y., Loenders, J., Duflou, J., and Lauwers, B. (2012). Optimization of energy consumption and surface quality in finish turning. In *5th CIRP Conference on High Performance Cutting*, Zürich, Switzerland.

Gutowski, T., Branham, M., Dahmus, J., Jones, A., Thiriez, A., and Sekulic, D. (2009). Thermodynamics analysis of resources used in manufacturing processes. *Environment Science and Technology* 43 (5): 1584–1590.

Gutowski, T., Dahmus, J., and Thiriez, A. (2006). Electrical energy requirements for manufacturing processes. In *Proceedings of 13th CIRP International Conference on Life Cycle Engineering* (pp. 623–627), Leuven, Belgium.

Gutowski, T., Jeffrey D., and Thiriez, A. (2006). Electrical energy requirements for manufacturing processes. In *13th CIRP International Conference on Life Cycle Engineering*, Leuven, Belgium, May 31–June 2.

Haapala, K., Rivera, J. L., and Sutherland, J. (2009). Reducing environmental impacts of steel product manufacturing. *Transactions of the North American Manufacturing Research Institution of the SME* 37: 419–426.

Hirohisa, N. (2012). *Environmental Burden Analyzer for Machine Tool Operations and Its Application*. InTech, Rijeka. www.intechopen.com/books/manufacturing-systems.

Hong, S. Y. and Broomer, M. (2000). Economical and ecological cryogenic machining of AISI 304 austenitic stainless steel. *Clean Products and Processes* 2: 157–166.

Jan, E. and Bras, B. (1998). ISO 14000 and activity based life cycle assessment in environmentally conscious design and manufacturing: A comparison. In *1998 ASME Design Engineering Technical Conference*. American Society of Mechanical Engineers, Atlanta, GA.

Johnson, W. (1956). The pressure for the cold extrusion of lubricated rod through square dies of moderate reduction at slow speeds. *Journal of the Institute of Metals* 85: 403–408.

Khan, M. M. A. and Dhar, N. R. (2006). Performance evaluation of minimum quantity lubrication by vegetable oil in terms of cutting force, cutting zone temperature, tool wear, job dimension and surface finish in turning AISI-1060 steel. *Journal of Zhejiang University Science A* 7 (11): 1790–1799.

Kordonowy, D. N. (2001). A power assessment of machining tools. BS thesis, Department of Mechanical Engineering, MIT, Cambridge, MA.

Krahenbuhl, U. (2002). Vegetable oil-based coolants improve cutting performance (cutting fluids). In *Tooling & Production*. Nelson Publishing.

Kurtz, M. (2007). *Environmental Conscious Manufacturing*. Wiley, Hoboken, NJ.

Lange, K. (1985). *Handbook of Metal Forming*. McGraw-Hill, New York.

Life Cycle Assessment (LCA): Principles and Practice. (2006). EPA/6000R-06/060.

Narita, H., Desmira, N., and Fujimoto, H. (2008). Environmental burden analysis for machining operation using LCA method. In *Proceedings of 41st CIRP Conference on Manufacturing Systems* (pp. 65–68), Tokyo, Japan.

Niggeschmidt, S., Helu, M., Diaz, N., Behmann, B., Lanza, G. and Dornfeld, D. (2010). Integrating green and sustainability aspects into life cycle performance evaluation. In *Proceedings of 16th CIRP International Conference on Life Cycle Engineering*, Heifei, Israel.

Pettersson, A. (2007). High-performance base fluids for environmentally adapted lubricants. *Tribology International* 40 (4): 638–645.

Rangarajan, A. and Dornfeld, D. (2004). Efficient tool paths and part orientation for face milling. *CIRP Annals* 53 (1): 73–76.

Rao, D. N. and Srikant, R. R. (2006). Influence of emulsifier content on cutting fluid properties. *Proceedings of the Institution of Mechanical Engineers, Part B: Journal of Engineering Manufacture* 220: 1803–1806.

Santos, J. P., Oliveira, M., Almeida, F. G., et al. (2011). Improving the environmental performance of machine-tools: Influence of technology and throughput on the electrical energy consumption of a press-brake. *Journal of Cleaner Production* 19: 356–364.

Shokrani, A., Dhokia, V., and Newman, S. T. (2012). Environmentally conscious machining of difficult-to-machine materials with regard to cutting fluids. *International Journal of Machine Tools and Manufacture* 57: 83–101.

Srikant, R. R., Rao, D. N., and Rao, P. N. (2009). Influence of emulsifier content in cutting fluids on cutting forces, cutting temperatures, tool wear, and surface roughness. *Proceedings of the Institution of Mechanical Engineers, Part J: Journal of Engineering Tribology* 223: 203–209.

Szargut, J., Morris, D. R., and Steward, F. R. (1988). *Exergy Analysis of Thermal, Chemical and Metallurgical Processes*. Hemisphere, Washington, DC.

Tönissen, S. (2009). Power demand of precision machine tools. MS thesis, Department of Mechanical Engineering, University of California, Berkeley, CA.

19

Milling

19.1 Introduction to Milling Operation

Milling machines produce mainly flat surfaces by means of a rotating cutter with multiple cutting edges. The two main types of machine are the horizontal and the vertical spindle. Milling cutters usually have teeth cut on the periphery and/or at the end of a disk or cylinder. Alternatively, *inserted-tooth* cutters with replaceable teeth may be used. In horizontal milling, *up-cutting* is the usual practice, but *down-cutting* may also be used. The types of cutter are listed in the following (Figure 19.1).

There are multiple cutting edges on a milling cutter. However, most of the cutting tool parameters are common as in a single-point cutting tool. The rake angle and clearance are common and have the same meaning and significance as that of a single-point cutting tool. When the rotation of the milling cutter is clockwise and work is fed from left to right, it is called *up milling*, and when the milling cutter is rotating counter-clockwise and work is fed from left to right, it is called *down milling*, as shown below. However, these two operations are performed only in horizontal milling (Figures 19.2 and 19.3).

A milling machine is used to transform pieces of metal into multiple geometric shapes [1–3].

19.1.1 Up Milling

Up milling is the operation when the milling tool is rotating in the opposite direction as that of the feed. This means that the teeth of the tool are pulling the chips out. It is scooping the metal starting at the bottom of the depth of cut scooping upwards. In up milling, the chips start small and have a very low force at contact. This low force can cause a pull of the table, which can be dangerous if the work piece is not securely fastened or if the force is great enough to pull the table off the ground.

19.1.2 Down Milling

Down milling is the operation when the tool is rotating in the same direction as that of the feed. This means that the teeth of the tool are pushing down on the work piece. It scoops the metal starting at the top of the depth of cut and goes all the way to the bottom. The chips are generally bigger because of the high forces at contact. The high forces and high impact secure the work piece to the table.

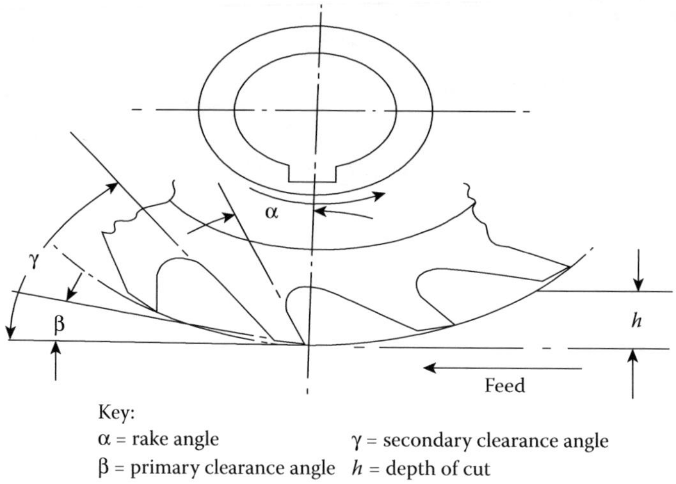

Figure 19.1

Milling parameters.

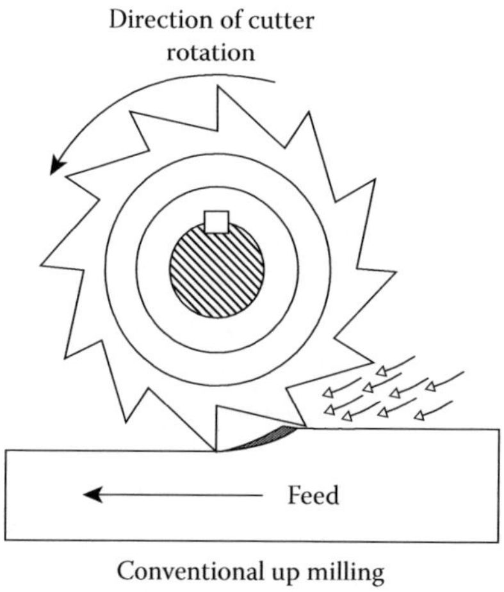

Figure 19.2

Horizontal up milling cut.

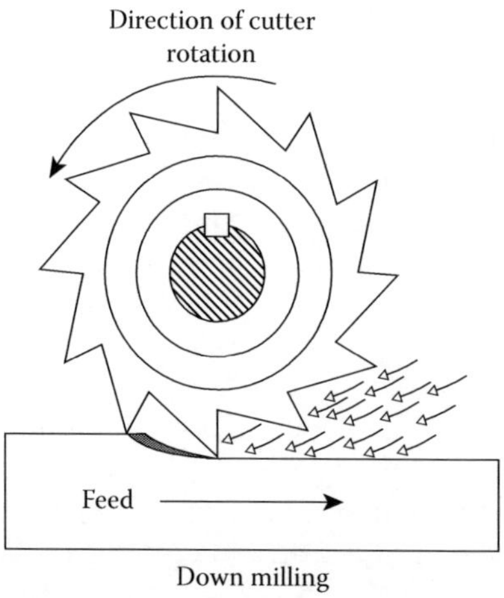

Figure 19.3

Horizontal down milling cut.

19.1.3 Up vs. Down Milling

Down milling tends to be the safest choice because it secures the work piece to the table instead of trying to pull it off as in up milling. Also the chips in down milling are less likely to fly off and hit people while operating the mill. The downside to down milling is that it takes higher forces and impact which equate to higher energy, costing more money. For large manufacturers, up milling is the choice, but for many small industrial operations, down milling is chosen most of the time due to its safety.

There are two major classes of milling process:

- In *face milling*, the cutting action occurs primarily at the end corners of the milling cutter. Face milling is used to cut flat surfaces (faces) into the workpiece, or to cut flat-bottomed cavities. In *face milling*, the axis of the cutter is perpendicular to the surface being crated and metal is removed by cutting edges on both the end and outside periphery of the cutter. There are various forms of face milling, such as; (a) *conventional face milling*, where the diameter of the cutter is greater than the workpart width, so the cutter overhangs on both sides, (b) *end milling*, in which the cutter diameter is less than the work width, so slot is cut into the part. (c) *pocket milling*, another form of end milling used to cut shallow pockets into flat parts; (d) *surface countering*, in which a ball-nose cutter rather than square end-cutter is fed back and forth across the work along a curvilinear path to cut a three-dimensional surface form.
- In *peripheral milling*, the cutting action occurs primarily along the circumference of the cutter, so that the cross section of the milled surface ends up receiving the shape of the cutter. In this case, the blades of the cutter can be seen as scooping out material from the work piece. Peripheral milling is well suited to the cutting of deep slots, threads, and gear teeth. Peripheral milling is also called *plain milling*; the axis of tool is parallel to the surface being milled, and material is removed by cutting edges on the outside periphery of the milling cutter, hence the name *peripheral milling*. Various types of plain milling are performed for different purposes as (a) *slab milling*, here the cutter extends beyond the workpiece on both sides; (b) *slot milling*, here, the width of the milling cutter is less than the workpiece width, creating a slot in the work; (c) *side milling*, where the milling cutter mills the sides of the workpiece; (d) *straddle milling* is the same as side milling, in which milling takes place on both sides of the work; and (e) *form milling*, in which milling teeth creates a special profile of a slot in the work. It transfers the profile on the milling teeth to the work (Figure 19.4).

Milling is the metal-removal process of using rotary *cutters* to remove material from a workpiece advancing (or *feeding*) in a direction at an angle with the axis of the tool. It covers a wide variety of different operations and machines, on scales from small individual parts to large, heavy-duty gang-milling operations. It

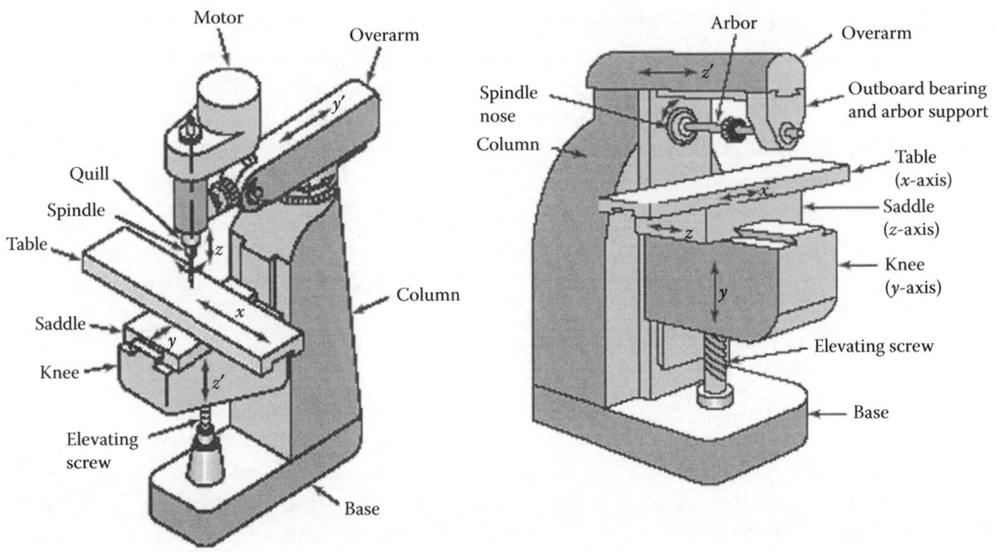

Figure 19.4

Milling machines: (a) vertical milling, (b) horizontal milling.

is one of the most commonly used processes in industry and machine shops today for machining parts to precise sizes and shapes. The original class of machine tools for milling was the *milling machine* (often called a *mill*). After the advent of *computer numerical control (CNC)*, milling machines evolved into *machining centers* (milling machines with automatic tool changers, tool magazines, or carousels, CNC control, coolant systems, and enclosures), generally classified as *vertical machining centers (VMCs)* and *horizontal machining centers (HMCs)*. The integration of milling into *turning* environments and of turning into milling environments, begun with live tooling for lathes and the occasional use of mills for turning operations, led to a new class of machine tools, *multitasking machines (MTMs)*, which are purpose-built to provide for a default machining strategy of using any combination of milling and turning within the same work envelope.

A vertical Milling machine holds a work piece (the metal piece that you are shaping) in place while a milling tool, much like a drill bit, spins and cuts the metal. The tool and the work piece come into contact by moving the table that the work piece is attached to by two wheels, one moving it forward and back and the other moving it left and right. The moving of the work piece is called "the feed of the machining process." When the tool comes in contact with the metal work piece, chips are formed and the metal is cut into shape as you move the work piece along the rotating tool. Chips are the pieces of metal that form when the tool cuts or scoops metal away from the work piece. In the *vertical mill*, the spindle axis is vertically oriented. Milling cutters are held in the spindle and rotate on its axis. The spindle can generally be extended (or the table can be raised/lowered, giving the same effect), allowing plunge cuts and drilling. The important components of vertical milling machines are (1) milling cutter, (2) spindle, (3) top slide or overarm, (4) column, (5) table, (6) y-axis slide, (7) knee, and (8) base.

There are two subcategories of vertical mills: the bed mill and the turret mill.

- A *turret mill* has a stationary spindle, and the table is moved both perpendicular and parallel to the spindle axis to accomplish cutting. The most common example of this type is the Bridgeport, described below. Turret mills often have a quill which allows the milling cutter to be raised and lowered in a manner similar to a drill press. This type of machine provides two methods of cutting in the vertical (z) direction: by raising or lowering the quill and by moving the knee.
- In the *bed mill*, however, the table moves only perpendicular to the spindle's axis, while the spindle itself moves parallel to its own axis.

Turret mills are generally considered by some to be more versatile of the two designs. However, turret mills are only practical as long as the machine remains relatively small. As machine size increases, moving the knee up and down requires considerable effort and it also becomes difficult to reach the quill feed handle (if equipped). Therefore, larger milling machines are usually of the bed type.

A third type also exists—a lighter machine, called a mill-drill, which is a close relative of the vertical mill and quite popular with hobbyists. A mill-drill is similar in basic configuration to a small drill press, but equipped with an x-y table. They also typically use more powerful motors than a comparably sized drill press, with potentiometer-controlled speed, and generally have more heavy-duty spindle bearings than a drill press to deal with the lateral loading on the spindle that is created by a milling operation. A mill-drill also typically raises and lowers the entire head, including motor, often on a dovetailed vertical, where a drill press motor remains stationary, while the arbor raises and lowers within a driving collar. Other differences that separate a mill-drill from a drill press may be a fine-tuning adjustment for the z-axis, a more precise depth stop, the capability to lock the x-, y-, or z-axis, and often a system of tilting the head or the entire vertical to allow angled cutting. Aside from size and precision, the principal difference between these hobby-type machines and larger true vertical mills is that the x-y table is at a fixed elevation; the z-axis is controlled in basically the same fashion as that of drill press, where a larger vertical or knee mill has a vertically fixed milling head, and changes the x-y table elevation.

A horizontal mill has the same sort of x-y table, but the cutters are mounted on a horizontal arbor (Figure 19.4b) across the table. Many horizontal mills also feature a built-in rotary table that allows milling at various angles; this feature is called a *universal table*. While endmills and the other types of tools available to a vertical mill may be used in a horizontal mill, their real advantage lies in arbor-mounted cutters, called side and face mills, which have a cross section rather like a circular saw, but are generally wider and smaller in diameter. Because the cutters have good support from the arbor and have a larger cross-sectional area than an end mill, quite heavy cuts can be taken, enabling rapid material removal rates. These are used to mill grooves and slots. Plain mills are used to shape flat surfaces. Several cutters may be ganged together on the arbor to mill a complex shape of slots and planes. Special cutters can also cut grooves, bevels, radii, or indeed any section desired. These specialty cutters tend to be expensive. Simplex mills have one spindle and duplex mills have two. It is also easier to cut gears on a horizontal mill. Some horizontal milling machines are equipped with a power-take-off provision on the table. This allows the table feed to be synchronized to a rotary fixture, enabling the milling of spiral features such as hypoid gears. The important components of horizontal milling machines are (1) base, (2) column, (3) knee, (4) and, (5) table (x-axis slide is integral), (6) overarm, and (7) arbor (attached to the spindle).

The choice between vertical and horizontal spindle orientation in milling machine design usually hinges on the shape and size of a workpiece and the number of sides of the workpiece that require machining. Work in which the spindle's axial movement is normal to one plane, with an endmill as the cutter, lends itself to a vertical mill, where the operator can stand before the machine and have easy access to the cutting action by looking down upon it. Thus, vertical mills are most favored for diesinking work (machining a mold into a block of metal). Heavier and longer workpieces lend themselves to placement on the table of a horizontal mill. Through the years, various different versions of milling operations have developed, depending on its applications. This prompted the development of various other versions of milling machines, as provided in the table below. In addition to horizontal vs. vertical, other distinctions are also important:

Criterion	Example Classification Scheme	Comments
Spindle axis orientation	Vertical vs. horizontal; turret vs. non-turret	Among vertical mills, "Bridgeport-style" is a whole class of mills inspired by the Bridgeport original, rather like the IBM PC spawned the industry of IBM-compatible PCs by other brands.
Control	Manual; mechanically automated via cams; digitally automated viaNC/CNC	In the CNC era, a very basic distinction is manual vs. CNC. Among manual machines, a worthwhile distinction is non-DRO-equipped vs. DRO-equipped.
Control (specifically among CNC machines)	Number of axes (e.g., three, four, or more)	Within this scheme, also: • Pallet-changing vs. non-pallet-changing • Full-auto tool changing vs. semi-auto or manual tool changing
Purpose	General purpose vs. special purpose, or single purpose	
Purpose	Tool room machine vs. production machine	Overlaps with above.
Purpose	"Plain" vs. "universal"	A distinction whose meaning evolved over decades as technology progressed, and overlaps with other purpose classifications above. Not relevant to today's CNC mills. Regarding manual mills, the common theme is that "plain" mills were production machines with fewer axes than "universal" mills; for example, whereas a plain mill had no indexing head and a non-rotating table, a universal mill would have those. Thus, it was suited to universal service, that is, a wider range of possible tool paths. Machine tool builders no longer use the "plain"-vs.-"universal" labeling.
Size	Micro, mini, benchtop, standing on floor, large, very large, gigantic	
Power source	Line-shaft-drive vs. individual electric motor drive	Most line-shaft-drive machines, ubiquitous circa 1880–1930, have been scrapped by now.
	Hand-crank-power vs. electric	Hand-cranked not used in industry but suitable for hobbyist micromills.

19.2 Variants of Milling Machines and Operations

Bed mill: This refers to any milling machine where the spindle is on a *pendant* that moves up and down to move the cutter into the work, while the table sits on a stout *bed* that rests on the floor. These are generally more rigid than a knee mill. Gantry mills can be included in this bed mill category.

- *Box mill or column mill:* This is a very basic hobbyist bench-mounted milling machine that features a head riding up and down on a column or box way.

- *C-frame mills*: These are larger, industrial production mills. They feature a knee and fixed spindle head that is only mobile vertically. They are typically much more powerful than a turret mill, featuring a separate hydraulic motor for integral hydraulic power feeds in all directions, and a 20–50 horsepower motor. Backlash eliminators are almost always standard equipment. They use large NMTB 40 or 50 tooling. The tables on C-frame mills are usually 18″ by 68″ or larger, to allow multiple parts to be machined at the same time.
- *Floor mills*: These have a row of rotary tables, and a horizontal pendant spindle mounted on a set of tracks that runs parallel to the table row. These mills have predominantly been converted to CNC, but some can still be found (if one can even find a used machine available) under manual control. The spindle carriage moves to each individual table, performs the machining operations, and moves to the next table while the previous table is being set up for the next operation. Unlike other mills, floor mills have movable floor units. A crane drops massive rotary tables, *x-y* tables, into position for machining, allowing large and complex custom milling operations.
- *Gantry mill*: The milling head rides over two rails (often steel shafts) which lie at each side of the work surface.
- *Horizontal boring mills*: These are large, accurate bed horizontal mills that incorporate many features from various machine tools. They are predominantly used to create large manufacturing jigs, or to modify large, high-precision parts. They have a spindle stroke of several (usually between four and six) feet, and many are equipped with a tailstock to perform very long boring operations without losing accuracy as the bore increases in depth. A typical bed has *x* and *y* travel, and is between three and four feet square with a rotary table or a larger rectangle without a table. The pendant usually provides between four and eight feet of vertical movement. Some mills have a large (30″ or more) integral facing head. Right angle rotary tables and vertical milling attachments are available for further flexibility.
- *Jig borers*: These are vertical mills that are built to bore holes, and very light slot or face milling. They are typically bed mills with a long spindle throw. The beds are more accurate, and the handwheels are graduated down to .0001″ for precise hole placement.
- *Knee mill* or *knee-and-column mill* refers to any milling machine whose *x-y* table rides up and down the column on a vertically adjustable knee. This includes Bridge ports.
- *Planer-style mills*: These are large mills built in the same configuration as planers, except with a milling spindle instead of a planing head. This term is becoming outdated as planers themselves are largely a thing of the past.
- *Ram-type mill*: This can refer to any mill that has a cutting head mounted on a sliding ram. The spindle can be oriented either vertically or horizontally. In practice, most mills with rams also involve swiveling ability, whether or not it is called "turret" mounting. The Bridgeport configuration can be classified as a vertical-head ram-type mill. Van Norman specialized in ram-type mills through most of the twentieth century. Since the wide dissemination of CNC machines, ram-type mills are still made in the Bridgeport configuration (with either manual or CNC control), but the less common variations (such as were built by Van Norman, Index, and others) have died out, their work being done now by either Bridgeport-form mills or machining centers.
- *Turret mills*: These are more commonly referred to as Bridgeport-type milling machines. The spindle can be aligned in many different positions for a very versatile, if somewhat less rigid, machine.
- *Gang milling* refers to the use of two or more milling cutters mounted on the same *arbor* (i.e., ganged) in a horizontal-milling setup. All of the cutters may perform the same type of operation, or each cutter may perform a different type of operation. For example, if several workpieces need a slot, a flat surface, and an angular *groove*, a good method to cut these (within a non-CNC context) would be gang milling. All the completed workpieces would be the same, and the milling time per piece would be minimized.

Gang milling was especially important before the CNC era, because for duplicate part production, it was a substantial efficiency improvement over manual-milling one feature at an operation, then changing machines (or changing setup of the same machine) to cut the next op. Today, CNC mills with automatic tool change and 4- or 5-axis control obviate gang-milling practice to a large extent.

19.3 Computer Numerical Control

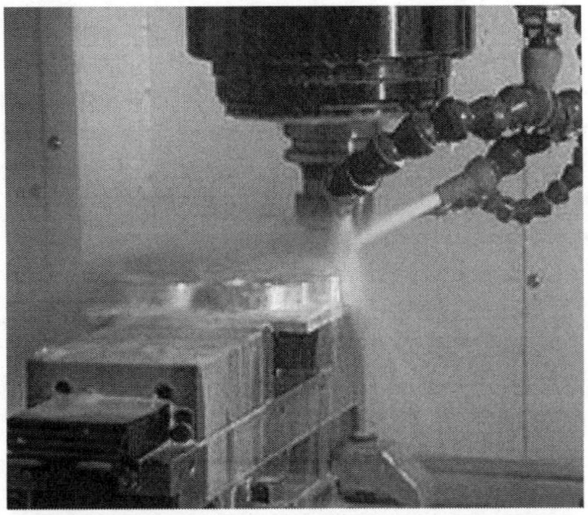

Most CNC milling machines (also called *machining centers*) are computer-controlled vertical mills with the ability to move the spindle vertically along the *z*-axis. This extra degree of freedom permits their use in die sinking, engraving applications, and 2.5D surfaces such as relief sculptures. When combined with the use of *conical* tools or a *ball-nose cutter*, it also significantly improves milling precision without impacting speed, providing a cost-efficient alternative to most flat-surface hand-engraving work (Figure 19.5).

The most advanced CNC milling machines, the multiaxis machine, add two more axes in addition to the three normal axes (*xyz*) [4]. Horizontal milling machines also have a C or Q axis, allowing the horizontally mounted workpiece to be rotated, essentially allowing asymmetric and eccentric turning. The fifth axis (B axis) controls the tilt of the tool itself. When all of these axes are used in conjunction with each other, extremely complicated geometries, even organic geometries such as a human head can be made with relative ease with these machines. But the skill to program such geometries is beyond that of most operators. Therefore, 5-axis milling machines are practically always programmed with CAM. A set of instructions (called a program) is used to guide the machine for desired operations.

19.4 Milling Cutters

The milling cutter is a rotary *cutting tool*, often with multiple cutting points. As opposed to *drilling*, where the tool is advanced along its rotation axis, the cutter in milling is usually moved perpendicular to its axis so that cutting occurs on the circumference of the cutter. As the milling cutter enters the workpiece, the cutting edges (flutes or teeth) of the tool repeatedly cut into and exit from the material, shaving off *chips* from the workpiece with each pass. The cutting action is shear deformation; the metal is pushed off the workpiece in tiny clumps that hang together to more or less extent (depending on the metal type) to form chips. The milling process removes material by performing many separate, small cuts. This is accomplished by using

Figure 19.5

Five-axis machining center with rotating table and computer interface.

a cutter with many teeth, spinning the cutter at high speed, or advancing the material through the cutter slowly; most often it is some combination of these three approaches. The *speeds and feeds* used are varied to suit a combination of variables. The speed at which the piece advances through the cutter is called *feed rate*, or just *feed*; it is most often measured in length of material per full revolution of the cutter.

Many different types of cutting tools are used in the milling process. Milling cutters such as *endmills* may have cutting surfaces across their entire end surface, so that they can be drilled into the workpiece (plunging). Milling cutters may also have extended cutting surfaces on their sides to allow for peripheral milling. Tools optimized for face milling tend to have only small cutters at their end corners.

The cutting surfaces of a milling cutter are generally made of a hard and temperature-resistant material, so that they *wear* slowly. A low-cost cutter may have surfaces made of high-speed *steel* but more expensive but slower-wearing materials include *cemented carbide*. Thin film coatings may be applied to decrease friction or further increase hardness (Figure 19.6).

Figure 19.5 shows revolution ridges on a surface milled by the side of the cutter, showing the position of the cutter for each cutting pass and how it corresponds with the ridges (cutter rotation axis is perpendicular to the image plane).

As material passes through the cutting area of a milling machine, the blades of the cutter take piece of material at regular intervals. Surfaces cut by the side of the cutter (as in peripheral milling), therefore, always contain regular ridges. The distance between ridges and the height of the ridges depend on the feed rate, the number of cutting surfaces and the cutter diameter. With a narrow cutter and rapid feed rate, these revolution ridges can be significant variations in the surface height (Figure 19.7).

The face milling process can, in principle, produce very flat surfaces. However, in practice, the result always shows visible *trochoidal* marks following the motion of points on the cutter's end face. These *revolution marks* give the characteristic finish of a face milled surface. Revolution marks can have significant roughness, and it depends on factors such as flatness of the cutter's end face and the degree of perpendicularity between the cutter's rotation axis and feed direction. Often, a final pass with a slow feed rate is used to compensate for a poor milling setup in order to reduce the roughness of revolution marks. In a precise face milling operation, the revolution marks will only be microscopic scratches due to imperfections in the cutting edge (Figure 19.8).

As we can see from various forms of milling operations discussed above, we need a large number and variety of milling cutters for a particular type of milling operation. The milling cutters for various operations, including their size, applications, and arrangement of teeth, are presented in Table 19.1.

The general recommendations for milling operations are presented in Table 19.2. The table shows the starting condition of speed and feed as well as the range of speed and feed for a combination of material and milling cutter.

19.4.1 Milling Parameters

Power for peripheral milling operation is calculated as

$$P = kvzbCf^{x}\left(\frac{d}{r}\right)^{y}$$

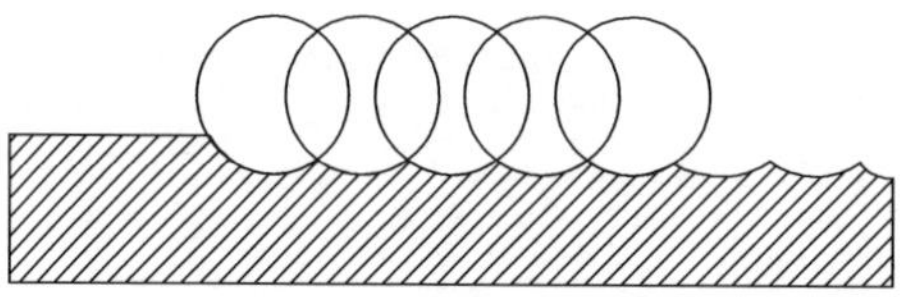

Figure 19.6

Milled surface.

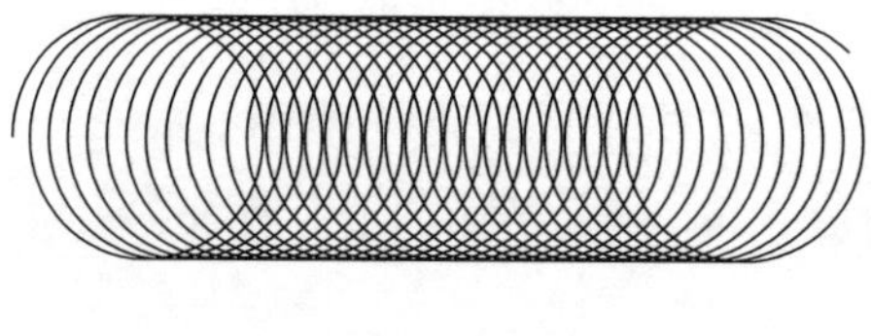

Figure 19.7

Trochoidal marks, characteristic of face milling.

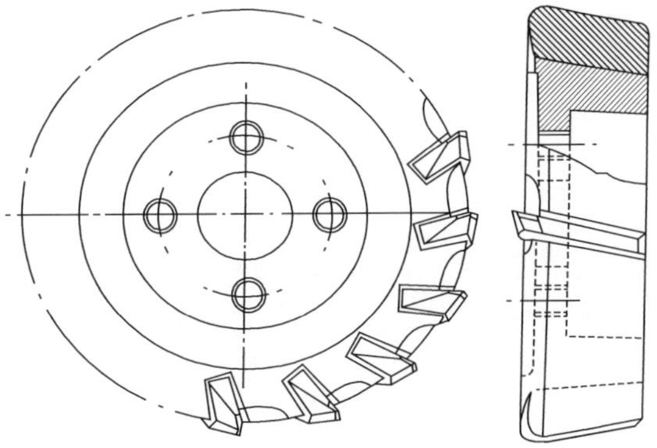

Figure 19.8

Inserted tooth cutter.

where:

P = power (watts)
v = cutting speed (m/s^{-1})
z = number of teeth
b = chip width (mm)
C = constant
f = feed per tooth (mm)
d = depth of cut (mm)
r = radius of cutter (mm)
x, y = indices
k = constant

The x, y, and k parameters for steel and cast iron are given in the table below:

Material	x	y	k
Steels	0.85	0.925	0.164
Cast iron	0.70	0.85	0.169

The constant C values for various materials in milling power equation above are given in Table 19.3. Cutting speed in milling is calculated as

$$\text{Cutting Speed } v = \pi DN/1000 \text{ (m/min)}$$

where:

D = cutter diameter (mm)
N = number of revolutions per minute

American Society for Testing Materials (ASTM) and other standardizing associations have standardized the cutting-tool materials as a guide to users in the selection of a carbide grade for a particular application. Four systems of classification of cemented carbide are in use today.

a. The US unofficial "C" classification system, which is based on performance
b. The British Hard Metal Association (BHMA) system, which is based on properties
c. The International Standard Organization (ISO) system (R513-*Application of carbides for machining by chip removal*)
d. The Russian system, which is based on composition

The "C" classification has eight numbers C-1 to C-8 which cover the grades of carbide manufactured for cutting tool applications. The BHMA designation consists of three numbers. The ISO system is akin to the "C" system and classifies all cutting grade carbides into three categories. P (steel-cutting grades), K (cast-iron-cutting grades), and M (an intermediate category for ductile irons, hard steels, and high-temperature

Table 19.1 Types of Milling Cutter

Type	Arrangement of Teeth	Application	Size	Appearance
Cylindrical (slab or rolling)	Helical teeth on periphery	Flat surfaces parallel to cutter axis	Up to 160 × 160 mm	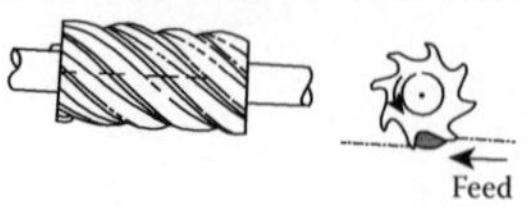
Side and face	On periphery and both sides	Steps and slots	Up to 200 mm diameter, 32 mm wide	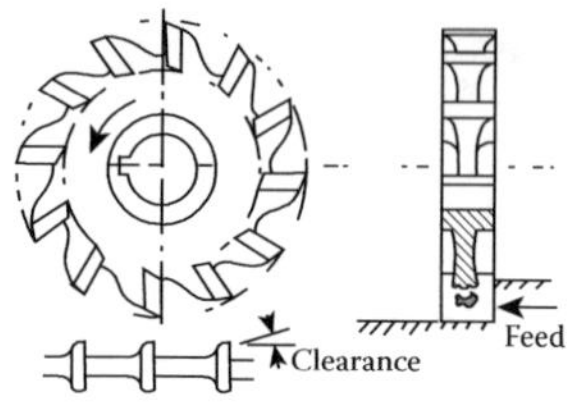
Straddle ganged	On periphery and both sides	Cutting two steps	Up to 200 mm diameter, 32 mm wide	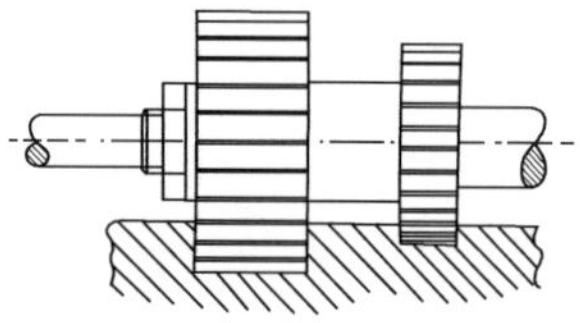
Side and face staggered tooth	Teeth on periphery, face teeth on alternate sides	Deep slots	Up to 200 mm diameter, 32 mm wide	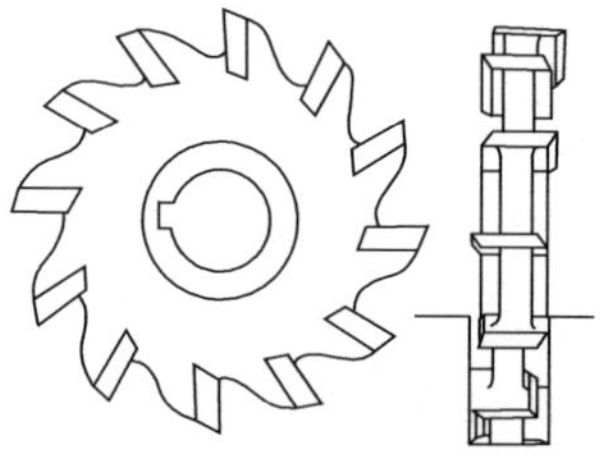
Single angle	Teeth on conical surface and flat face	Angled surfaces and chamfers	60°–85° in 5° steps	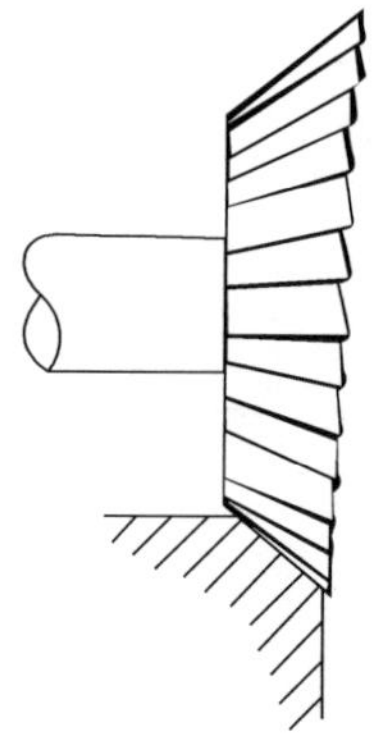
Double angle	Teeth on two conical faces	Vee slots	45°, 60°, 90°	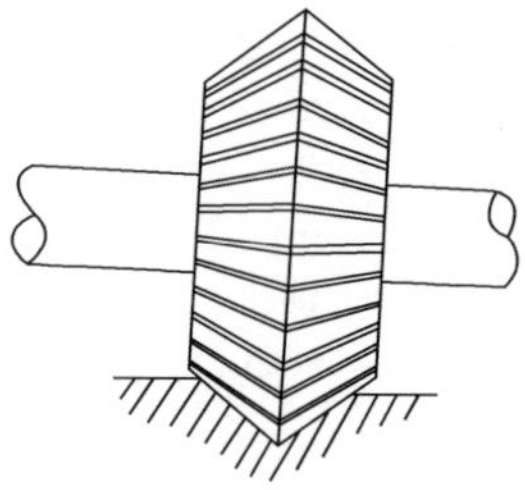
Rounding	Concave quarter circle and flat face	Corner radius on edge	1.5–20 mm radius	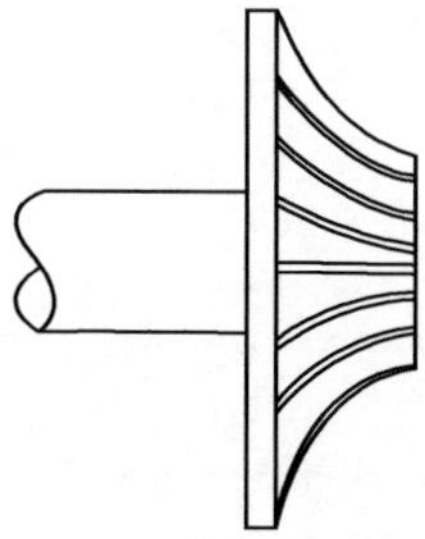

Table 19.1 (Continued) Types of Milling Cutter

Type	Arrangement of Teeth	Application	Size	Appearance
Involute gear cutter	Teeth on two involute curves	Involute gears	Large range	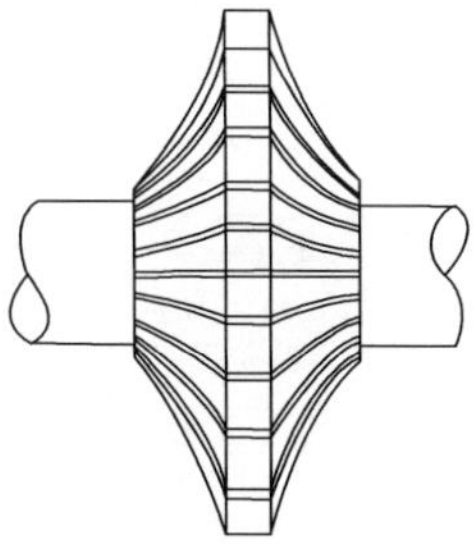
End mill	Helical teeth at one end and circumferential	Light work, slots, profiling, facing narrow surfaces	≤ 50 mm	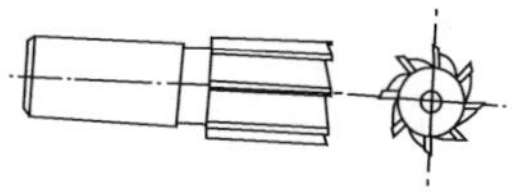
Tee slot	Circumferential and both sides	Tee slots in machine table	For bolts up to 24 mm diameter	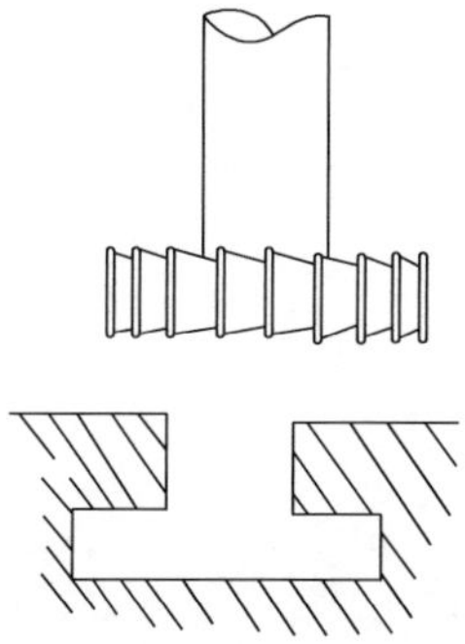
Dovetail	On conical surface and one end face	Dovetail machine slides	38 mm diameter, 45° and 60°	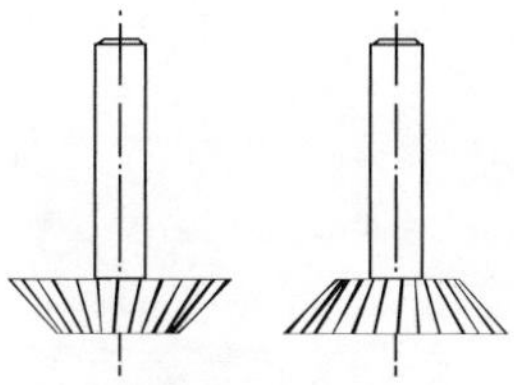
Shell end mill	Circumferential and one end	Larger work than end mill	40–60 mm diameter	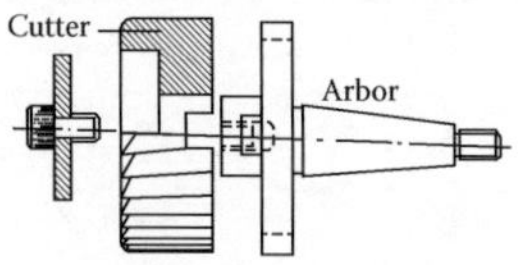
Slitting saw (slot)	Circumferential teeth	Cutting off or slitting. Screw slotting	60–400 mm diameter	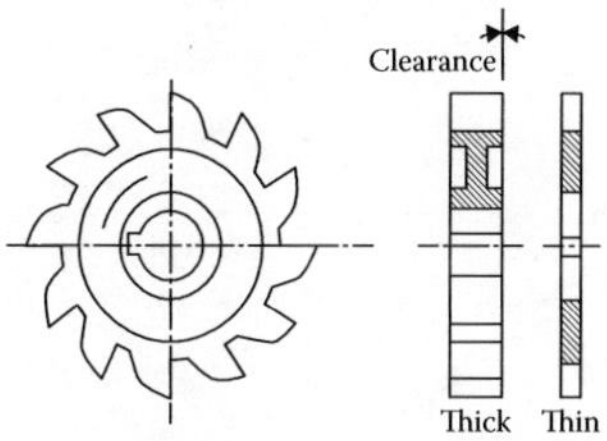
Concave-convex	Curved teeth on periphery	Radiusing	1.5–20 mm radius	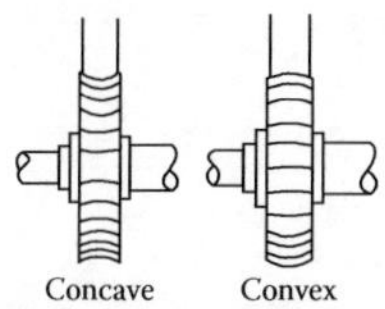

Source: Hoffman, E. G. *Fundamentals of Tool Design*, 2nd edn. Society of Manufacturing Engineers, Dearborn, MI, 1984 [5].

Table 19.2 General Recommendation of Milling Operation

Material	Cutting Tool	General Purpose Starting Conditions: Feed mm/tooth (in./tooth)	General Purpose Starting Conditions: Spread m/min (ft/min)	Range of Conditions: Feed mm/tooth (in./tooth)	Range of Conditions: Spread m/min (ft/min)
Low-carbon and carbon-free machining steels	Uncoated carbide, coated carbide, cermets	0.13–0.20 (0.005–0.008)	120–180 (400–600)	0.085–0.38 (0.003–0.015)	90–425 (300–1400)
Alloy Steels					
Soft	Uncoated, coated, cermets	0.10–0.18 (0.004–0.007)	90–170 (300–550)	0.08–0.30 (0.003–0.012)	60–370 (200–1200)
Hard	Cermets, PcBN	0.10–0.15 (0.004–0.006)	180–210 (600–700)	0.08–0.25 (0.003–0.010)	75–460 (250–1500)
Cast Iron, Gray					
Soft	Uncoated, coated, cermets, SiN	0.10–10.20 (0.004–0.008)	120–760 (400–2500)	0.08–0.38 (0.003–0.015)	90–1370 (300–4500)
Hard	Cermets SiN, PcBN	0.10–0.20 (0.004–0.008)	120–210 (400–700)	0.08–0.38 (0.003–0.015)	90–460 (300–1500)
Stainless steel, Austenitic	Uncoated, coated, cermets	0.13–0.18 (0.005–0.007)	120–370 (400–1200)	0.08–0.38 (0.003–0.015)	90–500 (300–1800)
High-temperature alloys Nickel based	Uncoated, coated, cermets, SiN, PcBN	0.10–0.18 (0.004–0.007)	30–370 (100–1200)	0.08–0.38 (0.003–0.015)	30–550 (90–1800)
Titanium alloys	Uncoated, coated, cermets	0.13–0.15 (0.005–0.006)	50–60 (175–200)	0.08–0.38 (0.003–0.015)	40–140 (125–450)
Aluminum Alloys					
Free machining	Uncoated, coated, PCD	0.13–0.23 (0.005–0.009)	610–900 (2000–3000)	0.08–0.46 (0.003–0.018)	300–3000 (1000–10,000)
High silicon	PCD	0.13 (0.005)	610 (2000)	0.08–0.38 (0.003–0.015)	370–910 (1200–3000)
Copper alloys	Uncoated, coated, PCD	0.13–0.23 (0.005–0.009)	300–760 (1000–2500)	0.08–0.46 (0.003–0.018)	90–1070 (300–3500)
Plastics	Uncoated, coated, PCD	0.13–0.23 (0.005–0.009)	270–460 (900–1500)	0.08–0.46 (0.003–0.018)	90–1370 (300–4500)

Source: Tlusty, G. *Manufacturing Processes and Equipment*. Prentice Hall, Englewood Cliffs, NJ, 2011.

Note: Depths-of-cut, *d*, usually are in the range of 1–8 mm (0.04–0.3 in.). PcBN: polycrystalline cubic-boron nitride. PCD: polycrystalline diamond. See also Table 23.4 for range of cutting speeds within tool material groups.

Table 19.3 Values of x, y, k, and C Are Given in the Tables

Material	C*
Free machining carbon steel	980 (120 BHN) 1190 (180 BHN)
Carbon-steels	1620 (125 BHN) 2240 (225BHN)
Nickel-chrome steels	1460 (125 BHN) 2200 (270 BHN)
Nickel-molybdenum and chrome-molybdenum steels	1600 (150 BHN) 1960 (280 BHN)
Chrome-vanadium steels	1820 (170 BHN) 2380 (190 BHN)
Flake graphite cast iron	635 (100 BHN) 1330 (263 BHN)
Nodular cast irons	1110 (annealed) 1240 (as cast)

*BHN numbers are hardness grades.

alloys). In Table 19.4, milling cutter materials speed (m/min) for a feed rate of 0.2 mm/tooth is shown. In Table 19.5, speed for cutting-tool material for a select list of materials is presented both for indexable and brazed cutters.

Table 19.5 shows cutting speeds for both indexable and brazed cutters for various metals being milled.

19.4.2 Table Feed Rates

In milling operation, the feed rate depends on three parameters; *f*, feed/tooth, *N*, rpm of cutter, and the number of teeth on the milling cutter. Milling is a multitooth operation and is hence used for bulk removal of materials.

Table 19.4 Milling Cutting Speeds at a Feed Rate of 0.2 mm per Tooth

	Indexable Inserts ISO Carbide Grade				Brazed Cutters ISO Carbide Grade			
Metal Being Cut	P10	P30	P40	K20	P10	P30	P40	K20
Mild steel	150	130	100	20	200	170	130	–
Carbon steel 0.7%	120	90	75	–	150	90	75	–
Steel castings	60	45	50	–	80	75	50	–
Stainless steel	100	100	100	–	125	125	115	–
Gray cast iron	150	130	110	–	150	130	110	–
Aluminum alloy	–	–	–		–	–	–	600

Table 19.5 Milling Cutting Speeds for High-Speed Steel Cutters

Material Being Cut	Cutting Speed (m/min)
Alloy steel	10
Cast iron	20
Low-carbon steel	28
Bronze	35
Hard brass	45
Copper	60
Aluminum alloy	100

Table 19.6 Typical Values of Feed per Tooth (mm)

Material Being Cut	Face Mills HSS Carbide	Side and Face Mills HSS Carbide	End Mills HSS Carbide	Saws HSS Carbide
Aluminum	0.55 0.50	0.33 0.30	0.28 0.25	0.13 0.13
Alloy, brass, bronze		0.18 0.18		
Copper	0.30 0.30	0.18 0.18	0.15 0.15	0.07 0.07
Cast iron	0.40 0.50	0.22 0.30	0.20 0.25	0.10 0.13
Low-carbon steel	0.25 0.35	0.15 0.20	0.13 0.18	0.07 0.10
Alloy steel	0.20 0.35	0.13 0.20	0.10 0.18	0.05 0.10

$$\text{Feed Rate } f_t = fzN \text{ (mm/min)}$$

where:

f = feed/tooth (mm)
z = number of teeth
N = number of revolutions per minute of cutter

The feed (mm/min) specifications for various milling operations are provided in Table 19.6. These values should be lowered for finishing and increased for rough milling. The metal-removal rate during milling is provided in Table 19.7, which is the inverse of unit power. Sometimes these data could be used directly without any calculation. An example is presented for the material removal rate, cutting forces, and the environmental impact of the milling operation.

Example 19.1

An aluminum workpiece is being milled by the face milling operation. The number of teeth on cutter is four and the diameter of cutter is 2 in. The face milling cutter is a TiN-coated tool, and its surface speed is 2500 ft/min. Let the width of cut be 1 in. depth of cut = 0.1 in., and feed per tooth = 0.1 in. Calculate (a) material removal rate, (b) cutting force per tooth, and (c) the environmental impact of chips.

Table 19.7 Material Removal Rate in Milling

Material Being Cut	Metal-Removal Rate (mm^3/kW-min)
Mild steel	18,900
Alloy steel	10,500
Cast steel	12,600
Gray cast iron	12,600
Stainless steel	8400
Copper	18,900
Aluminum	4200
Magnesium	4200
Titanium	10,500

The rotational rpm (*N*) for the spindle is

$$N = \frac{V_S}{\pi D} = \frac{2500}{2\pi} = 398 \text{ rpm}$$

Now, we can calculate *Vw*, workpiece velocity,

$$f = \frac{Vw}{z \cdot N} \Rightarrow Vw = 159.2\,(\text{in./min})$$

Material-removal rate, MRR = *Vw*w*d* = 15.92 (in.^{3}/min), and Power requirement, *P* = Us*MRR. The specific energies for different materials are shown in Table 19.7. For Al, Us = 100,000 in.-lb/in.^{3} and so power requirement, *P* = 100,000(15.92)/33,000 = 48.24 [hp]. Cutting force/tooth, *F* ~ us**d***f* = 1000 (lbf).

19.5 Environmental Effect of Chips Removed (CH_e) in Milling (7–13)

CH_e means the environmental burden due to the metal chips in machining based on chip weight. Metal chips can be recycled by one of these processes: in a chip compactor, a chip crusher, or a centrifugal separator. However, if an arc furnace is used it generates heat and therefore an environmental burden.

Let the size of the rectangular work piece be length 2 in, width 1.5 in, and depth 0.25 in. Hence, the volume of workpiece = 2 × 1.5 × .25 = 0.75 in^3. The volume of product volume = 2 × 1.5 × 0.15 = 0.45 in^3. Now, we can estimate the environmental impact of chips = CH_e = (0.75–0.45) × 0.098 × (1/4.54) × 12 × 4.54 = 0.35 lb of CO_2 emitted into the atmosphere:

$$CH_e = (WPV - PV) * MD * WD_e \tag{19.1}$$

where

WPV: workpiece volume (cm^3)
PV: product volume (cm^3)
MD: material density of workpiece (kg/cm^3)
WDe: environmental burden of metal chip processing [kg-CO_2/kg]

We can also calculate the embodied energy in the chips:

$$\text{Embodied energy} = 0.3 \times 220 \times 0.738 \times 10^6 = 48.7 \times 10^6 \text{ ft.lb}$$

Chips produced 0.35 lb of CO_2 and needed 48.7 × 10^6 ft.lb of energy to produce it.

19.5.1 LCA Analysis of Milling Machine (8,9,14,15)

For LCA analysis, the following specifications of a milling machine have been selected. The specifications are presented in Table 19.8. In general, the LCA of machine tools is divided into four categories shown in Figure 19.9.

The GHG emissions and removals arising from all processes used in the formation, extraction, or transformation of materials used in production shall be included in the assessment, including all sources of energy consumption or direct GHG emissions associated with that formation, extraction, or transformation. For convenience of computation, we assume that all parts are made of gray cast iron, low-carbon steel, and aluminum alloys. Most of the machine tool structure and frame is made of cast iron and it is approximately about 70% of the total weight. Some parts, including the spindle, table, head, and quill, are of steel and these together are approximately 25% of the total weight. Some parts, including monitor, enclosures of the work table, etc. may be made of particular glass and some of synthetic rubber. The weights and percentage of different materials are presented in Table 19.9.

The total CO_2 emission in air is 5043 kg and the total embodied energy during extraction phase is 79,320 MJ.

Some other materials category is associated with some unknown materials. This is assumed here as 2%.

a. GHG emissions and energy consumption during the extraction phase:
The total CO_2 emission in air is 5043 kg and the total embodied energy during extraction phase is 79,320 MJ.

b. Manufacturing of machine tool:
Major portions of materials are gray cast iron and low-carbon steel, approximately 95% of the total weight. Gray cast iron needs foundry process for manufacturing of the milling machine structure

Table 19.8 Specifications of Milling Machine for LCA

Items	Specifications
Longitudinal traverse (x)	800 mm
Vertical traverse (y)	400 mm
Cross traverse (z)	225 mm
Spindle motor power	30–37 kW
Spindle speed range	4000–6000 rpm
Digital CNC	Flat screen
Coolant system	0.15 HP
Feed (longitudinal and cross feed)	0–1.3 mpm (0–50 ipm)
Vertical feed	0.8 mpm (31.5 ipm)
Input voltage	230
Power capacity	4 kVA
Working surface	1525 × 325 mm
Approximate net weight	6390 lb (3000 kg)
Maximum weight over the table	880 lb (400 kg)

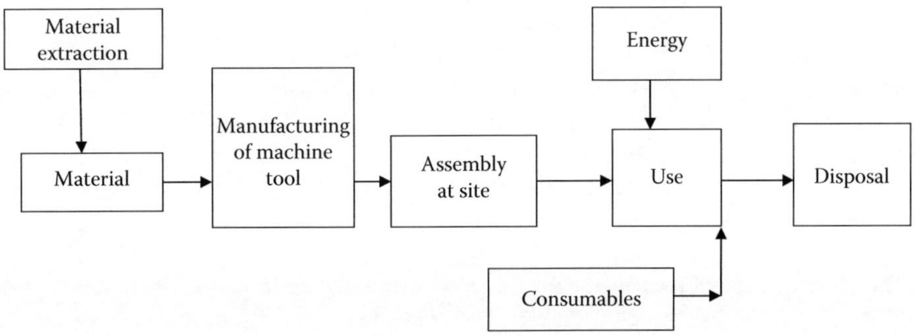

Figure 19.9

LCA for milling machine.

Table 19.9 Total Weights, GHG Emissions, and Embodied Energy during Extraction

Material	Percentages	Weight	GHG (CO_2) Emission (kg/kg)	Embodied Energy (MJ/kg)
Gray cast iron	70%	2100 kg	1.1 kg/kg × 2100 = 2310 kg	18MJ/kg × 2100 = 37,800 MJ
Steel	25%	750 kg	2.8 kg/kg × 750 = 2100 kg	38 MJ/kg × 750 = 28,500 MJ
Synthetic rubber	0.5%	15 kg	10 (kg/kg) × 15 = 150 kg	168 MJ/kg × 15 = 2520 MJ
Glass (ceramic)	2.5%	75 kg	2.2 kg/kg × 75 = 165 kg	40 MJ/kg × 75 = 3000 MJ
Other (thermoplastics)	2%	60 kg	5.3 kg/kg × 60 = 318 kg	125 MJ/kg × 60 = 7500 MJ
Total			5043 kg	79,320 MJ

Table 19.10 Estimation of CO_2 and Processing Energy of Milling Machine

Material	Percentages	Weight	GHG (CO_2) Emission (kg/kg)	Processing Energy (MJ/kg)
Gray cast iron	70%	2100 kg	0.29 kg/kg × 2100 = 609 kg	3.7MJ/kg × 100 = 7770 MJ
Steel	25%	750 kg	0.31 kg/kg × 750 = 233 kg	3.9 MJ/kg × 750 = 2925 MJ
Synthetic rubber	0.5%	15 kg	0.646(kg/kg) × 15 = 9.70 kg	8.08 MJ/kg × 5 = 121.2 MJ
Glass (ceramic)	2.5%	75 kg	0.72 kg/kg × 75 = 165 kg	8.98 MJ/kg × 75 = 694 MJ
Other (thermoplastics)	2%	60 kg	0.898 kg/kg × 60 = 54 kg	11.2 MJ/kg × 60 = 672 MJ
Total			1071 kg	12,182 MJ

and frame. However, spindle, table, and other steel parts like arbor will need turning, milling, and drilling. During the assembly stage of milling, a machine will need welding, soldering, and fastening.

The energy consumption and GHG emissions in the atmosphere can be estimated as presented in Table 19.10. All data are taken from Ashby [16].

c. Transportation of milling machine parts:
Manufacturing is truly a global phenomenon. It is estimated that Boeing makes its planes by assembling parts from 70 countries. Most of the automobiles produced in the United States gets its parts from several manufacturers and assembled at their plants. In this situation, the parts and equipment manufactured at different sites have to be transported to the plant site and assembled and shipped back to the market. During transportation, a lot of energy is consumed and harmful GHG emissions occur.

In the LCA of the milling machine, we have two processes of casting and machining of milling machine parts.

Assumptions:

Casting is carried out 2000 km from the plant site, and machining is carried out 1500 km and transported to the assembly and USE site.

These parts will be transported on diesel light goods vehicle. The energy and CO_2 costs of transport is provided as 0.90 MJ/ton.km and 0.064 kg CO_2/ton.km, respectively [16]. Therefore,

Energy used during transportation of:

Castings: $0.9 \times 2 \times 2000 = 3600$ MJ

Machined parts: $0.9 \times 0.75 \times 1500 = 1013$ MJ

CO_2 emission during transportation:

Casting: $0.064 \times 2 \times 2100 = 269$ kg

Machined parts: $0.064 \times 0.75 \times 1500 = 72$ kg

d. Assembly on site: Assembly and finishing operations are carried out on site. The operations on site are mainly welding, soldering, and fastening and paintings. Due to lack of data, this part of energy and CO_2 cost are not presented here.

e. USE phase: Milling machines need electricity and air supply for its operations. The power consumed will depend on the hours of use.

Assumption:

The economic life of the milling machine is assumed to be 10 years. The machine is used in three shifts of 8 h per day. The total number of hours in the life time of milling machine $= 3 \times 8 \times 30 \times 10 = 7200$ h.

Machines have a downtime for scheduled maintenance and changeover losses due to cutting tool changes, part load/unload, and so on. Sometimes due to failure, there is unplanned downtime. It is assumed that a machine is effectively used for 80% time and the rest 20% is for maintenance and failure. Therefore, a milling machine is used for $0.8\times7200=5760$ h in its total economic lifespan.

19.5.2 Energy

The GHG emissions and removals associated with the provisions and use of energy in the life cycle of the milling machine shall be included in the emissions arising from the energy-supply system. Emissions from energy supply include

- Emissions at the point of consumption of the energy (e.g., emissions from the burning of coal and gas).
- Emissions from provision of energy, including the generation of electricity and heat.
- Emissions from transmission losses, transport fuels.
- Upstream emissions (e.g., the mining and transport of fuel to the electricity generator or other combustion plant).
- Downstream emissions (e.g., the treatment of waste arising from the operation of nuclear electricity generation).

The power at the spindle (real power) provided in the data in Table 19.7 is 37 kW.

The power consumed during the life time of a milling machine $=37\times5760=213{,}120.0$ kW.

If all this power is generated by oil; CO_2 emitted $=0.54\times37\times5760=11{,}084.8$ kg.

It shows that during its entire economic life time, a milling machine consumes 213,120 kW of electricity and contributes 11,085 kg of CO_2 in the atmosphere.

This electricity consumption has been divided into *x*/*y*/*z* modules, head/spindle unit, tool changer, coolant system, and hydraulics and pneumatics. The air will also need power to compress it. Air compressors of different sizes are available in the market and a 1.2 kW compressor is selected in this example. The power consumed by the compressor is estimated as

$$\text{Power compressor}=1.2\times5760=6912.0\ \text{kW}$$

f. Consumables: Milling machines need hydraulic oil, lubricant grease, and cutting fluid. These are oil-based and can be considered to have the same ecoproperties. Normally cutting fluids have 96% water and 4% oil. Cutting tools and water can also be included in the consumable category. Milling cutters are coated tungsten-carbide type, and tungsten can be used to estimate ecoproperties.

19.5.3 Assumptions

Lubricant oil: It is assumed that the lubricant is changed daily after 24 h and its weight is 5 kg. Therefore, in 10 years, the total lubricant used $=30\times5\times10=1500$ kg of lubricant.

Milling cutter: Tungsten carbide inserts tools are estimated as follows: It is assumed that each cutter is of 2 kg and replaced in every ship daily. Therefore, tungsten milling cutter $=2\times3\times30\times10=1800$ kg. Now the energy and CO_2 costs of consumables are estimated as

	Embodied Energy	CO_2 Emission
Tungsten milling cutter	346 MJ/kg $\times$ 1800 = 622,800 MJ	22 kg/kg $\times$ 1800 = 39,600 kg
Lubricating oil		

- Disposal: The milling machine tool is easily disassembled and sorted out into different categories of materials. Table 19.10 shows that about 95% of milling machine consists of metals gray cast iron and steel. All metals will be recycled and all other materials such as rubber and glass will be sent to a landfill.

  All the data regarding energy and CO_2 have been taken from Ashby [16]. Lubricant oil and water mixed with cutting fluid are toxic and should be sent to a special waste incinerator. The total environmental impact estimated through LCA is presented here.

19.5.4 Waste-Stream Creation in Milling

The mechanics of the machining process has a significant influence on all waste streams generated through the machining speed, the volume of material removed, the geometry of the chip formed, the cutting forces generated, and the process energy utilized.

19.5.4.1 Oblique Machining

Oblique cutting is characterized by the rotation of the cutting tool by an inclination angle 2 about the y-axis. There are two main reference frames of interest: (1) the axes orthogonal to the workpiece and (2) the axes rotated with the tool about the y-axis (Figure 19.5a). The oblique model can serve as a first-order estimate of the mechanics of milling processes. Unlike the orthogonal case, the chip velocity Vc is no longer contained in the x-y plane. The chip flows along the rake face at an angle measured in this plane from the x'-axis (Figure 19.5b). The total process energy for oblique cutting ($\text{TPE}_{\text{oblique}}$) as

$$\text{TPE}_{\text{oblique}} = \frac{\left(\cos(\beta_n - \gamma_n)\cos\eta_S\cos\lambda + \cos(\phi_n + \beta_n - \gamma_n)\sin\eta_S\sin\lambda\right)}{\cos(\phi_n + \beta_n - \gamma_n)}\frac{\tau_S Vol}{\sin(\phi_n)\cos\lambda}$$

The process energy for oblique machining is also dependent on the setup parameters (i.e., the tool rake angle, the tool oblique angle, and cutting fluid selection), as well as material selection and part geometry.

19.6 Environmentally Conscious Machining (12,17–20)

The need for environmentally conscious manufacturing; that is, the relationship between manufacturing and its impact on environment, has not been fully analyzed. Machining is a major manufacturing activity, and an analyzer developed by Narita [21] enables the estimation of the environmental burden caused by machine tool operations. The machine tool operations have a big potential influence on the environment, and the environmental burden analyzer is developed on the basis of LCA.

The environmental burden analyses for machine tool operations developed not only evaluate the difference among dry, wet, and MQL (minimal quantities of lubricant) machining operations, but also include difference among the depth of cuts, feed rate, spindle speed, and the tool path pattern. Furthermore, if removal volume and material type are the same, the environmental burden remains the same.

Narita's environmental burden analyzer for machine tool operations seems to be most suitable for sustainable manufacturing. The novelty about Narita's method is that it helps in deciding cutting conditions to achieve minimum environmental burden by using the analyzer developed. Toyata has presented the energy consumption breakdown as in Figure 19.10.

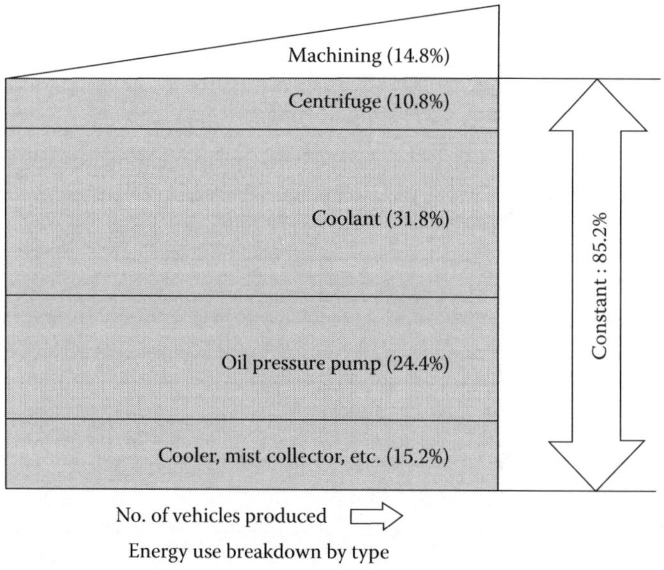

Figure 19.10

Energy consumption breakdown.

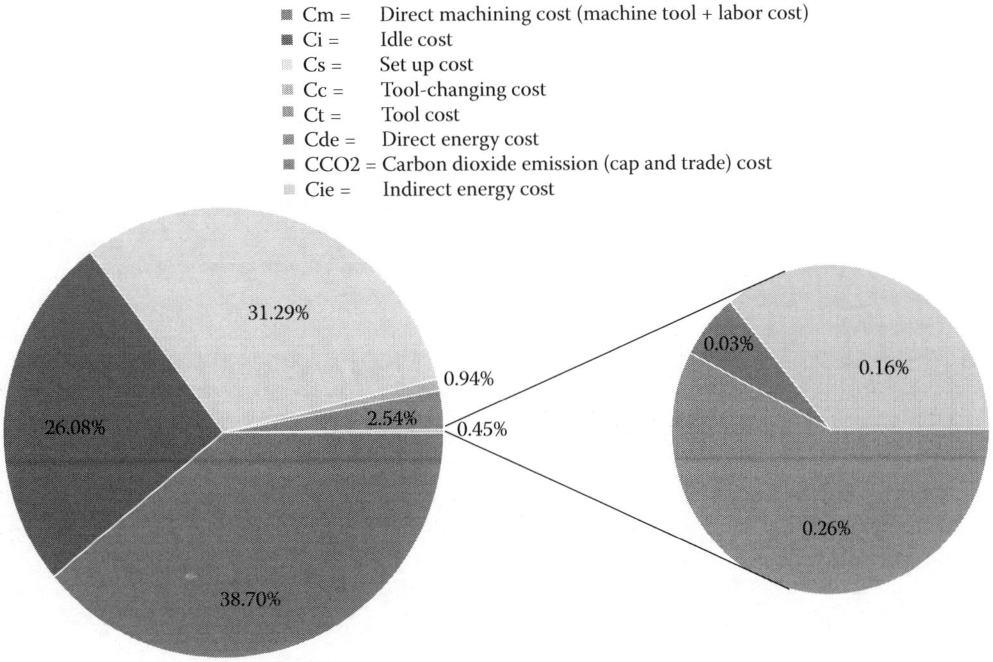

Figure 19.11

Relative contribution of cost component.

It is seen from Figure 19.11 that the idle cost is quite substantial in CNC and there should be some device to reduce it. Some European countries are trying to devise a software which can help turn and off the CNC as and when required.

References

1. Groover, M. P. (2012). *Fundamentals of Modern Manufacturing: Materials, Processes, and Systems*, 5th edn. Wiley, Hoboken, NJ.
2. Kalpakjian, S. and Schmids, R. (2010). *Manufacturing Processes for Engineering Materials*, 6th edn. Pearson Prentice Hall, Upper Saddle River, NJ.
3. Schey, J. A. (2000). *Introduction to Manufacturing Processes*, 3rd edn. McGraw Hill, New York.
4. Kordonowy, D. N. (2001). A power assessment of machining tools. B.S. Thesis, MIT, Department of Mechanical Engineering, Cambridge, MA.
5. Hoffman, E. G. (1984). *Fundamentals of Tool Design*, 2nd edn. Society of Manufacturing Engineers, Dearborn, MI.
6. Tlusty, G. (2011). *Manufacturing Processes and Equipment.* Prentice Hall, Englewood Cliffs, NJ.
7. Kurtz, M. (2007). *Environmental Conscious Manufacturing.* Wiley, Hoboken, NJ.
8. Life Cycle Assessment (LCA): Principles and Practice. (2006). EPA/6000R-06/060.
9. Gregory, A. K. and Menerey, D. (1994). Sustainable development by design: Review of life cycle design and related approaches. *Journal of the Air and Waste Management Association* 44 (5): 645–668.
10. Dahmus, J. and Gutowski, T. (2004). An environmental analysis of machining. In *ASME International Mechanical Engineering Congress and RD&D Expo*, Anaheim, CA.
11. Santos, J. P., Oliveira, M., Almeida, F. G. et al. (2011). Improving the environmental performance of machine-tools: Influence of technology and throughput on the electrical energy consumption of a press-brake. *Journal of Cleaner Production* 19: 356–364.
12. Baniszewski, B. (2005). An environmental impact analysis of grinding. S.B. thesis, Department of Mechanical Engineering, MIT, Cambridge, MA.
13. Hirohisa, N. (2012). *Environmental Burden Analyzer for Machine Tool Operations and its Application.* InTech, Rijeka. www.intechopen.com/books/manufacturing-systems.

14. Jan, E. and Bras, B. (1998). ISO 14000 and activity based life cycle assessment in environmentally conscious design and manufacturing: A comparison. In *1998 ASME Design Engineering Technical Conference*. American Society of Mechanical Engineers, Atlanta, GA.
15. Niggeschmidt, S., Helu, M., Diaz, N., Behmann, B., Lanza, G., and Dornfeld, D. (2010). Integrating green and sustainability aspects into life cycle performance evaluation. In *Proceedings of 16th CIRP International Conference on Life Cycle Engineering.*
16. Ashby, M. F. (2009). *Materials and the Environment: Eco-informed Material Choice* (pp. 112, 114, 284–287, 290–291). Butterworth-Heinemann, Burlington, MA.
17. Gutowski, T., Branham, M., Dahmus, J., Jones, Thiriez, A., and Sekulic, D. (2009). Thermodynamics analysis of resources used in manufacturing processes, *Environment Science and Technology* 43 (5): 1584–1590.
18. Dittrich, M. A., Gutowski, T. G., Cao J., et al. (2012). Exergy analysis of incremental sheet forming. *Production Engineering* 6 (2): 169–177.
19. Hong, S. Y. and Broomer, M. (2000). Economical and ecological cryogenic machining of AISI 304 austenitic stainless steel. *Clean Products and Processes* 2: 157–166.
20. Shokrani, A., Dhokia, V., and Newman, S. T. (2012). Environmentally conscious machining of difficult-to-machine materials with regard to cutting fluids. *International Journal of Machine Tools and Manufacture* 57: 83–101.
21. Narita, H., Desmira, N., and Fujimoto, H. (2008). Environmental burden analysis for machining operation using LCA method. In *Proceedings of 41st CIRP Conference on Manufacturing Systems* (pp. 65–68), Tokyo, Japan.

Bibliography

Abdalla, H. S. and Patel, S. (2006). The performance and oxidation stability of sustainable metalworking fluid derived from vegetable extracts. *Proceedings of the Institution of Mechanical Engineers, Part B: Journal of Engineering Manufacture* 220 (12): 2027–2040.

Abou-El-Hossein, K. A. (2008). Cutting fluid efficiency in end milling of AISI 304 stainless steel. *Industrial Lubrication and Tribology* 60 (3): 115–120.

Alves, S. M. and Oliveira, J. F. G. (2008). Vegetable based cutting fluid-an environmental alternative to grinding process. In *15th CIRP International Conference on Life Cycle Engineering* (pp. 664–668), Sydney.

Batzer, S. A. and Sutherland, J. W. (1998). The dry cure for coolant ills. *Cutting Tool Engineering* 50 (4): 34–44.

Cetin, M. H., Ozcelik, B., Kuram, E., and Demirbas, E. (2011). Evaluation of vegetable based cutting fluids with extreme pressure and cutting parameters in turning of AISI 304L by Taguchi method. *Journal of Clean Production* 19 (17–18): 2049–2056.

Dahmus, J. B. and Gutowski, T. G. (2004). An environmental analysis of machining. In *Proceedings of ASME International Mechanical Engineering Congress and RD&D Expo*, Anaheim, CA.

Fox, N. J. and Stachowiak G. W. (2007). Vegetable oil-based lubricants—A review of oxidation. *Tribology International* 40 (7): 1035–1046.

Fthenakis, V. M. and Kim, H. C. (2007). Greenhouse-gas emissions from solar electric- and nuclear power: A life-cycle study. *Energy Policy* 35 (4): 2549–2557.

Guo, Y., Loenders, J., Duflou, J., and Lauwers, B. (2012). Optimization of energy consumption and surface quality in finish turning. In *5th CIRP Conference on High Performance Cutting*, Zürich, Switzerland.

Gutowski, T., Dahmus, J., and Thiriez, A. (2006). Electrical energy requirements for manufacturing processes. In *Proceedings of 13th CIRP International Conference on Life Cycle Engineering* (pp. 623–627), Leuven, Belgium.

Gutowski, T., Jeffrey, D., and Thiriez, A. (2006). Electrical energy requirements for manufacturing processes. In *13th CIRP International Conference on Life Cycle Engineering*, Leuven, Belgium, May 31st–June 2.

Haan, D. M., Batzer, S. A., Olson, W. W., and Sutherland, J. W. (1997). An experimental study of cutting fluid effects in drilling. *Journal of Materials Processing Technology* 71 (2): 305–313.

Haapala, K., Rivera, J. L., and Sutherland, J. (2009). Reducing environmental impacts of steel product manufacturing. *Transactions of the North American Manufacturing Research Institution of the SME* 37: 419–426.

Hondo, H. (2005). Life cycle GHG emission analysis of power generation systems: Japanese case. *Energy* 30 (11–12): 2042–2056.

Khan, M. M. A. and Dhar, N. R. (2006). Performance evaluation of minimum quantity lubrication by vegetable oil in terms of cutting force, cutting zone temperature, tool wear, job dimension and surface finish in turning AISI-1060 steel. *Journal of Zhejiang University Science A* 7 (11): 1790–1799.

Krahenbuhl, U. (2002). Vegetable oil-based coolants improve cutting performance (Cutting Fluids). In *Tooling & Production*. Nelson Publishing.

Lange, K. (1985). *Handbook of Metal Forming*. McGraw-Hill, New York.

Machinability Data Center, and Metcut Research Associates. (1980). *Machining Data Handbook*. Machinability Data Center, Metcut Research Associates, Cincinnati, OH.

Pettersson, A. (2007). High-performance base fluids for environmentally adapted lubricants. *Tribology International* 40 (4): 638–645.

Rangarajan, A. and Dornfeld, D. (2004). Efficient tool paths and part orientation for face milling. *CIRP Annals* 53 (1): 73–76.

Rao, D. N. and Srikant, R. R. (2006). Influence of emulsifier content on cutting fluid properties. *Proceedings of the Institution of Mechanical Engineers, Part B: Journal of Engineering Manufacture* 220: 1803–1806.

Srikant, R. R., Rao, D. N., and Rao, P. N. (2009). Influence of emulsifier content in cutting fluids on cutting forces, cutting temperatures, tool wear, and surface roughness. *Proceedings of the Institution of Mechanical Engineers, Part J: Journal of Engineering Tribology* 223: 203–209.

Szargut, J., Morris, D. R., and Steward, F. R. (1988). *Exergy Analysis of Thermal, Chemical and Metallurgical Processes*. Hemisphere, Washington, DC.

Taniguchi, M., Kakinuma, Y., Aoyama, T., and Inasaki, I. (2006). Influences of downsized design for machine tools on the environmental impact. In *Proceedings of MTTRF Annual Meeting* (pp. 1–4).

Tönissen, S. (2009). Power demand of precision machine tools. MS Thesis, Department of Mechanical Engineering, University of California, Berkeley, CA.

20

Grinding

20.1 Introduction to Grinding

The main abrasive machining operations are surface grinding, cylindrical grinding, internal grinding, centerless grinding, creep-feed grinding, honing, and lapping [1]. A number of other, specialty grinding machines are also used in manufacturing. Steels and cast iron are most commonly ground, but aluminum, brass, bronze, copper, titanium, nickel alloys, nylon, carbides, and ceramics are also ground. Grinding is one of the few effective ways of shaping hard, brittle materials like ceramics while they are cold. Surface grinding is the most common abrasive machining operation, and it involves grinding flat surfaces. Work pieces are held to the work table of the grinder via a magnetic chuck. If the work piece is not magnetic, it is held by vises, vacuum chucks, double-sided adhesive tape, or special fixtures [2]. A grinding wheel, mounted on the horizontal spindle of the grinder, rotating at 1500–3000 m/min, is used to remove material. A typical horizontal spindle surface grinder is shown in Figure 20.1.

Three typical surface grinding operations are depicted in Figure 20.2. In transverse grinding, the work table reciprocates longitudinally, bringing part of the work piece in front of the grinding wheel, then moves laterally to feed the part under the wheel. It reciprocates again to bring a new section of the part in the range of the wheel until the entire surface has been ground. In plunge grinding, the grinding wheel is moved radially into the work piece, grinding a groove into the surface. Grinders with vertical spindles can allow numerous pieces such as ball bearings to be ground in one setup by bringing the face of the grinding wheel down flat against a special rotary fixture.

20.1.1 Grinding Machines

20.1.1.1 Types of Grinding

There are many forms of grinding, but the four major industrial grinding processes are as follows:

20.1.1.1.1 Cylindrical Grinding

In cylindrical grinding, the workpiece rotates about a fixed axis and the surfaces machined are concentric to that axis of rotation. Cylindrical grinding produces an external surface that may be either straight, tapered, or contoured. The basic components of a cylindrical grinder include a wheel head, which incorporates the spindle and drive motor; a cross slide, that moves the wheel head to and from the workpiece; a headstock, which locates, holds, and drives the workpiece; and a tailstock, which holds the other end of the work.

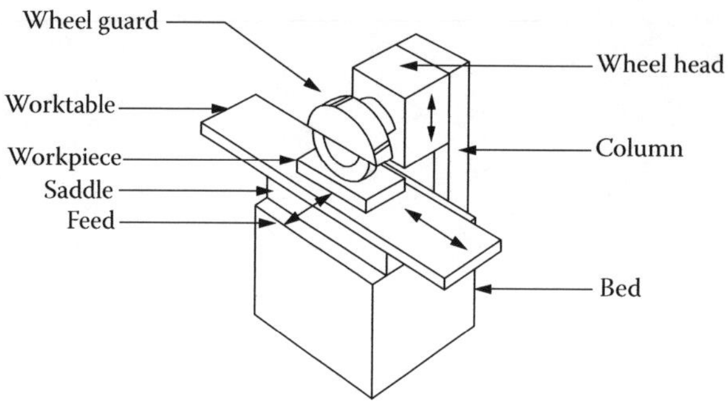

Figure 20.1

Horizontal spindle surface grinding machine.

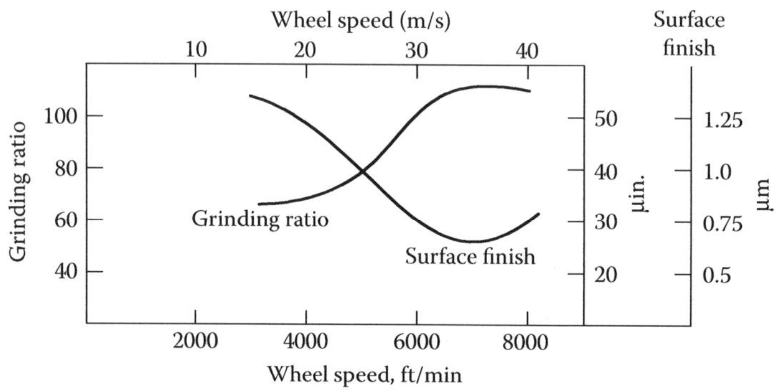

Figure 20.2

Wheel speed vs. grinding ratio and surface finish.

20.1.1.1.2 Internal Grinding

Internal diameter or I.D. grinders finish the inside of a previously drilled, reamed, or bored hole, using small grinding wheels at high RPM. The principal elements of an internal grinding machine are the workhead, which holds the work and has its own drive, and the wheel head, which is the internal grinding spindle. In addition to the rotary motions of work and wheel, an internal grinder has a traverse movement to bring the wheel to and from the work zone, and a reciprocating spindle movement for both the wheel's approach to the work surface and for the feed movement of the wheel during grinding.

20.1.1.1.3 Centerless Grinding

In centerless grinding, the workpiece rotates between a grinding wheel and a regulating drive wheel. The work is supported from below by a fixed work-rest blade. The two basic modes of centerless grinding are *thru-feed* and *in-feed* or *plunge* mode. In the thru-feed mode, the work proceeds in the axial direction through the slowly narrowing gap between the grinding wheel and the regulating wheel. Work is advanced by the axial force exerted on it by the rotating surface of the regulating wheel. This is a highly productive form of grinding in that a number of workpieces can be ground simultaneously and in a continuous stream. The in-feed mode is used for work with projecting heads that would prohibit thru-feeding; the work is placed on the work-rest blade while one wheel is retracted and fed to an end stop. The wheel is then brought back, reducing the gap between the wheels, grinding the work.

20.1.1.1.4 Surface Grinding

Surface grinding produces flat, angular, or contoured surfaces by feeding work in a horizontal plane beneath a rotating grinding wheel. Work is most often magnetically attached to the table, and may be ground by either a traversing or rotating movement of the table. Most surface grinding machines use a horizontal spindle which adjusts up and down allowing either the edge or the face of the wheel to contact the work.

Workpiece surfaces produced by grinding are influenced by the following factors:

- Workpiece material—harder materials allow finer finishes
- Type of wheel—fine grains yield finer finishes
- Dressing procedure: improperly dressed wheels will mar the work surface
- Feed rate: finer finishes are obtained with slower feed rates
- Machine rigidity: older, worn machines yield a poor quality finish
- Wheel condition: clogged wheels cannot produce a good finish
- Lubricant cleanliness: coolant filtration removes waste that could damage workpiece surface

Grinding operation produces a smooth flat surface and as the speed of the wheel is increased, surface roughness gets progressively lower, that is, surface finish getting better with an increase of speed of the grinding wheel. It is shown in Figure 20.2. It also shows that as speed increases, the grinding ratio increases.

The grinding ratio (GR) is a term used to indicate the slope of wheel wear curve and estimated as follows:

$$GR = \frac{V_w}{V_g}$$

V_w and V_g are the volume of work material removed and corresponding volume of the grinding wheel that is worn, respectively. Figure 20.2 shows that as speed increases grinding ratio increases, as the volume of material from the work is likely to increase.

Operating (surface) speeds are usually between 20 m/s and 30 m/s (4000–6000 ft/min). There is usually an optimum speed at which the G ratio is highest. The material removal rate increases with the force imposed on the wheel, but beyond a certain limit, the G ratio drops sharply. Speeds up to 90 m/s are achieved in high-speed grinding with specifically constructed wheels. High rotational speeds impose a huge stress on the wheel and all grinding wheels must be inspected for flaw prior to installation (Figure 20.3).

Grinding machines produce flat, cylindrical, and other surfaces by means of high-speed rotating abrasive wheels. Grinding is a means of giving more accurate finish to a part already machined, but is also a machining process in its own right. The main types of machine are the *flat grinding machine* for flat surfaces and the *cylindrical grinding machine* for cylindrical surfaces. More complex shapes are produced by shaped wheels called *contour grinding wheels*. *Bench* and *pedestal* grinders are used for tool sharpening and so on.

20.1.2 Grinding Wheels

Typical materials for wheels are bonded abrasive powders such as aluminum oxide (Al_2O_3), silicon carbide (SiC), and diamond dust. The contour grinding wheels for various purposes are shown below (Figures 20.4 through 20.6).

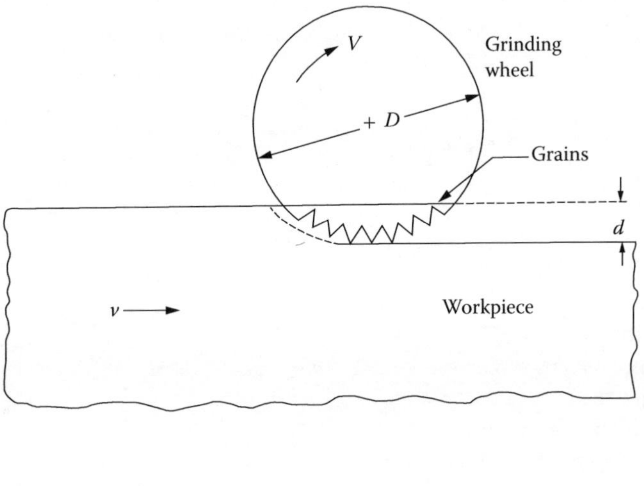

Figure 20.3

Surface grinding.

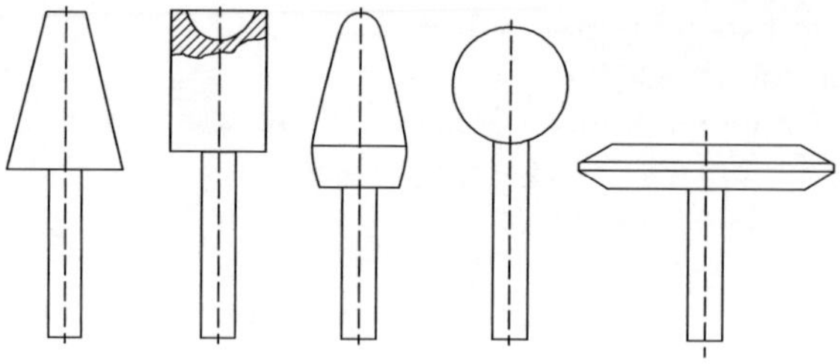

Figure 20.4

Contour grinding wheels.

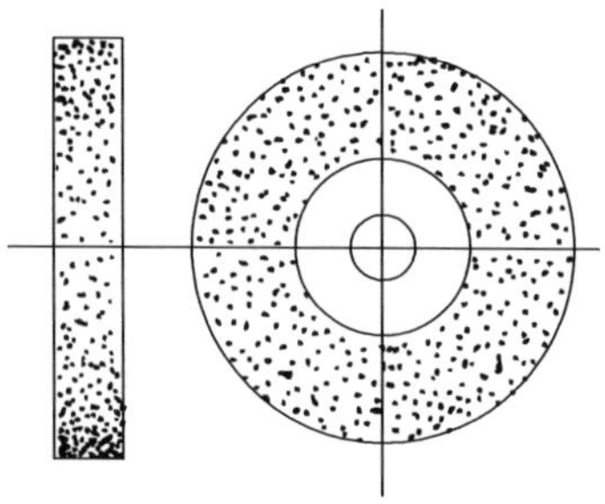

Figure 20.5

Standard grinding wheel.

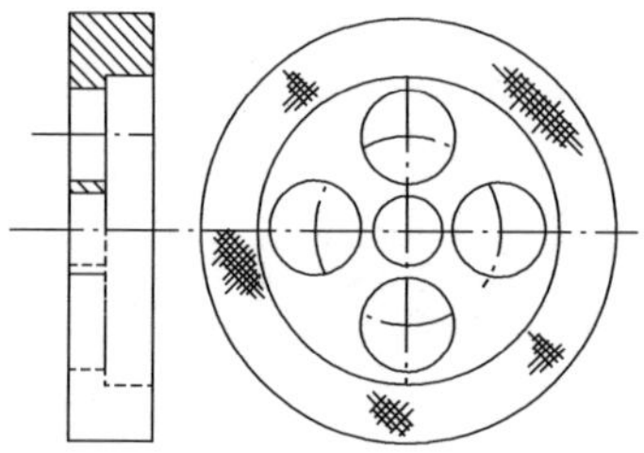

Figure 20.6

Steel wheel coated with abrasive.

20.2 Grinding Process Calculations (Cylindrical Grinding)

Very fine chips are removed in grinding and so are always used for fine finishing operations.

t = chip thickness (mm), f = feed or depth of cut (mm), p = pitch of grains (mm) b = width of cut (mm), P = power (watts), v = wheel peripheral velocity (mm/s) u = work peripheral velocity (mm/s), d = wheel diameter (mm), D = work diameter (mm) F = tangential force on wheel (newtons)

The chip thickness in grinding is calculated as follows:

$$\text{Chip thickness } t = \frac{2pu}{v}\sqrt{\frac{(D+d)f}{Dd}}$$

where:

Minus sign is used for internal grinding

- t = chip thickness (mm)
- f = feed or depth of cut (mm)
- p = pitch of grains (mm)
- b = width of cut (mm)
- u = work peripheral velocity (mm/s)
- d = wheel diameter (mm)
- D = work diameter (mm)

The power needed for grinding is estimated as

$$\text{Power } P = \frac{Fv}{1000}$$

where:

P = power (watts)
v = wheel peripheral velocity (mm/s)
F = tangential force on wheel (newtons)

$$\text{Energy per unit Volume Removed } E = \frac{p}{bfv}\left(\text{J/mm}^3\right)$$

This is the specific energy of chip formation in grinding operation. The specific energy in grinding is about 10×10^6 in-lb/in^3 (7000 kg/mm^3) as the chip size is extremely small compared to other metal-removal processes. Grinding is similar to the milling operation as removal of material happens either on the periphery or on the face of the grinding wheel, similar to peripheral and face milling. However, peripheral grinding is much more common than face grinding. In grinding, there is a lot of speed compared to milling. The chips that are removed are much smaller than milling operation. The material removal rate (R_{MR}) can be calculated as

$$R_{MR} = bfv \ (\text{mm}^3/\text{s})$$

The temperature at the work surface is extremely high and it has been experimentally determined that surface temperature is dependent on energy per surface area ground, almost similar to specific energy. It has been estimated that surface temperature T_s is related to grinding parameters as follows:

$$T_s = K_2 f^{0.75}\left(\frac{r_g CV}{v_w}\right) d^{0.25}$$

where:

V = wheel speed
v_w is work speed
d = depth of cut
f = feed
r_g = grain aspect ratio
C = number of active grits per square inches on the grinding wheel
K_2 is a constant of proportionality (Figure 20.7)

In contrast to metal-cutting operation, the individual cutting edges are *randomly distributed* and more or less *randomly oriented*. The depth of engagement (the undeformed chip thickness) is not equal for all

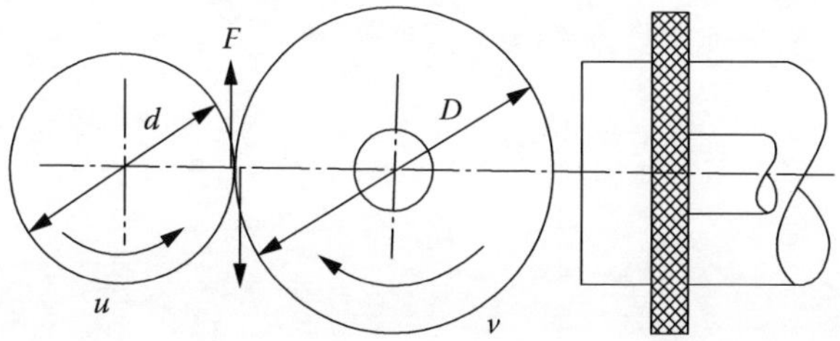

Figure 20.7

Grinding diagram.

abrasive grains. These factors change the character of the process substantially. Because only a fraction of grit-workpiece encounters result in actual material removal, and because there are many sources of friction, abrasive machining is inefficient; the energy required for removing a unit is tremendously high. Most of the energy is converted into heat, only some of which is removed with chip; much heat remains in the workpiece with a number of undesirable consequences. Discoloration of the surface is due to oxidation and the color id is indicative of temperatures reached. The surface of heat-treated steel may reach high-enough temperature to cause a transformation to austenite. After the grit has passed, sudden chilling causes the formation of martensite and associated cracks. A *deep blue* color is indicative of such *burning*.

20.3 Specific Energy Requirements in Surface Grinding

The specific energy required for grinding is very high compared to grinding. As we know, specific energy (U) is estimated as

$$U = \frac{F_c}{t_0 w} \text{(N-m/mm}^3\text{)(in-lb/in}^3\text{)}$$

where:

F_c = cutting force
w = width of cut
t_0 = thickness of chip before cut

This equation is similar to the above specific energy equation shown above.

In grinding, the depth of engagement (the undeformed chip thickness) is small compared to the metal-cutting process and the specific energy required for removing the unit volume may be up to *10 times higher* than in cutting. Table 20.1 shows the specific energy in grinding some materials used in manufacturing.

Example 20.1

Find the total grinding time to remove 0.015 in. of material from a steel block of 2×4 in. surface area. The grinding wheel diameter is 6-in and the table speed is 100 ft/min. The downfeed is 0.002 in using a A80HB wheel. The reversal of the wheel takes 1 s at the end of stroke. Determine the power required.

Solution

The wheel speed of 5000 ft/min (between 4000 and 6000 ft/min) is selected. It is usual to grind in the length direction. To obtain a uniform surface finish, the wheel must clear the workpiece at the end of the stroke; therefore, the total stroke is 6 + 4 = 10 in.

The surface speed is calculated as

$$V = 5000 = \frac{\pi DN}{12} = \frac{\pi(6)N}{12}, N = 3183 \text{ rpm}$$

Table 20.1 Specific Energy Requirements for Surface Grinding

Material	Hardness (HB)	Specific Energy (U) HP.min/in³	W.s/mm³
1020 steel	110	4.6	13
Cast iron	215	3.9	11
Titanium alloy	300	4.7	13
Superalloy	340	5.0	14
Tool steel (T15)	67 HRC	5.5	15
Aluminum	150	2.2	6

Source: *Machining Data Handbook*, Metcut Research Associate, 5th edn.

The number of passes (n) = 0.015/0.000 = 7 passes

Thus, time/pass = stroke/table speed = 10/(100 × 12/60)=0.5 s/pass

Time for every pass = 1 + 0.5 = 1.5 s

Total time for 7 passes = 7 × 1.5 = 10.5 s

Total volume of material removed = 2 × 4 × 0.015 = 0.1 in^3

Therefore, material removal rate (Z_w) = 0.1/10.50 = 0.6 in^3/min

Power (HP) required = 0.6 × 4.6 = 2.63 hp

If the efficiency of the grinding machine is 0.90, the gross power needed would be Power (HP) = 2.921 hp

20.4 Detailed Description of Coolants and Coolant Recycling Systems in Grinding

Grinding fluids improve the efficiency of grinding operations by reducing the wear on grinding wheels and by lowering power consumption. The fluids absorb heat created during grinding and reduce friction, improving the parts' surface finish and dimensional accuracy. Grinding fluids also remove chips from the surface of the part being worked [3–11].

Water-based emulsions are typically used for general grinding, while oils are most often used for thread grinding. The fluid is applied in a fast steam under pressure through nozzles, or applied as a mist to reach inaccessible areas and provide better visibility of the workpiece.

As grinding fluids remove heat from the part, they can increase temperature significantly, particularly if they are water-based. Grinding fluids which drip off the part are collected and circulated through refrigerating systems, allowing them to continue to cool the part when they are reapplied. After machining, cutting fluids must be removed from parts, either mechanically or via solvents. Water-based emulsions are commonly used for grinding fluids since they are easy to remove. In flood cooling, a stream of coolant is applied at a rate of about 10 L (2.64 gallons) per minute, at pressures anywhere from 700 to 14,000 kPa [11–14]. Mist cooling is usually applied at air pressures of 70–600 kPa. Since mist cooling deposits less coolant at once, it is not as useful in situations requiring a large cooling capacity, but it usually works well for grinding since only small amounts of material are removed at a time. Mist cooling requires venting in order to prevent machine operators from inhaling toxic fluid particles. Fumes from cutting fluids can cause respiratory problems and severe skin irritation; so many factories make use of mist collectors to reduce risk to their machinists. As grinding fluids are recirculated through a grinder, they slowly accumulate tiny bits of metal chips. If a grinder is left standing unused long enough for the chips to settle, many of the chips may be scooped out. This method of cleaning the fluid does not remove all the impurities, and eventually the grinding fluid will need to be more thoroughly cleaned or replaced. In order to cut down on costs on account of buying new grinding fluid and paying for safe disposal of old fluids, high-production grinding facilities recycle their own grinding fluid or pay to have it recycled. Numerous machines exist, which cleanse high volumes of grinding fluids via centrifuging and filtering, extending fluid life almost indefinitely. The XYBEX® series of coolant recycling systems, made by Master Chemical Corporation, allow closed-loop coolant recycling inside an operating production plant. There are several variations of XYBEX-like system 600, 750, 1000, 2001, and so on. They are portable and low-cost coolant recycling systems. Here, series 2001 is presented for illustration.

Master Chemical's XYBEX System 2001 is a complete, full-capability single-product coolant recycling system for a medium-to-large-sized metalworking plant operating with machine tools fitted with their own coolant sumps (Figure 20.8).

The XYBEX System 2001 will provide a makeup coolant and will recycle a sufficient quantity of fluid daily to support a total of up to 15,000 U.S. gallons (56,700 l) of coolant or more in the machines.

20.4.1 Features

- 300 gallon (1140 L) dirty coolant holding tank
- 450 gallon (1700 L) clean coolant holding tank
- UNIMIX™ automatic coolant proportioning pump
- High-speed, disc-bowl, liquid-from-liquid centrifuge to clean coolant to less than 0.5% emulsified tramp oil and 2–5 μm of particulate matter

20.4.2 Automatic Operation

The coolant recycling system, shown in Figure 20.8, is a sump pump, which pumps dirty coolant, chips, and fines through a filter to remove solids (Master Chemical Corporation, 2002). Once the fluid has been

Figure 20.8

XYBEX 2001 coolant recycling system.

filtered, the sump pump returns it to the machine tool for reuse. A side stream of filtered fluid can be purified of tramp oils and other extraneous fluids by the XYBEX centrifuge even while the machine tool is running. Tramp oils are extraneous oils, usually from machine lubrication, which become mixed with the coolant [15]. They tend to float on top of the coolant where they produce harmful bacteria. If tramp oils are not filtered out, bactericides containing toxic agents such as chlorine, bromine, and formaldehyde must be added to grinding fluids to keep bacteria levels down. If tramp oils are present at the machining surface, smoke containing toxic heavy metals can result. Once tramp oils are centrifuged out, the XYBEX system may then be moved to another machine.

References

1. Groover, M. P. (2012). *Fundamentals of Modern Manufacturing: Materials, Processes, and Systems*, 5th edn. Wiley, Hoboken, NJ.
2. Kalpakjian, S. and Schmids, R. (2010). *Manufacturing Processes for Engineering Materials*, 6th edn. Pearson Prentice Hall, Upper Saddle River, NJ.
3. Rao, D. N. and Srikant, R. R. (2006). Influence of emulsifier content on cutting fluid properties. *Proceedings of the Institution of Mechanical Engineers, Part B: Journal of Engineering Manufacture* 220: 1803–1806.
4. Srikant, R. R., Rao, D. N., and Rao, P. N. (2009). Influence of emulsifier content in cutting fluids on cutting forces, cutting temperatures, tool wear, and surface roughness. *Proceedings of the Institution of Mechanical Engineers, Part J: Journal of Engineering Tribology* 223: 203–209.
5. Hong, S. Y. and Broomer, M. (2000). Economical and ecological cryogenic machining of AISI 304 austenitic stainless steel. *Clean Products and Processes* 2: 157–166.
6. Shokrani, A., Dhokia, V., and Newman, S. T. (2012). Environmentally conscious machining of difficult-to-machine materials with regard to cutting fluids. *International Journal of Machine Tools and Manufacture* 57: 83–101.
7. Abou-El-Hossein, K. A. (2008). Cutting fluid efficiency in end milling of AISI 304 stainless steel. *Industrial Lubrication and Tribology* 60 (3): 115–120.
8. Fox, N. J. and Stachowiak, G. W. (2007). Vegetable oil-based lubricants—A review of oxidation. *Tribology International* 40: 1035–1046.
9. Abdalla, H. S. and Patel, S. (2006). The performance and oxidation stability of sustainable metalworking fluid derived from vegetable extracts. *Proceedings of the Institution of Mechanical Engineers, Part B: Journal of Engineering Manufacture* 220: 2027–2040.
10. Krahenbuhl, U. (2002). Vegetable oil-based coolants improve cutting performance (Cutting Fluids). In *Tooling & Production*. Nelson Publishing.
11. Khan, M. M. A. and Dhar, N. R. (2006). Performance evaluation of minimum quantity lubrication by vegetable oil in terms of cutting force, cutting zone temperature, tool wear, job dimension and surface finish in turning AISI-1060 steel. *Journal of Zhejiang University Science A* 7 (11): 1790–1799.

12. Pettersson, A. (2007). High-performance base fluids for environmentally adapted lubricants. *Tribology International* 40 (4): 638–645.
13. Cetin, M. H., Ozcelik, B., Kuram, E., and Demirbas, E. (2011). Evaluation of vegetable based cutting fluids with extreme pressure and cutting parameters in turning of AISI 304L by Taguchi method. *Journal of Clean Production* 19 (17–18): 2049–2056.
14. Alves, S. M. and Oliveira, J. F. G. (2008). Vegetable based cutting fluid-an environmental alternative to grinding process. In *15th CIRP International Conference on Life Cycle Engineering* (pp. 664–668), Sydney.
15. Garrett by Honeywell. (1996). White paper: Burst and containment. www.turbobygarrett.com.

Bibliography

Ashby, M. F. (2009). *Materials and the Environment: Eco-informed Material Choice.* (pp. 112, 114, 284–287, 290–291) Butterworth-Heinemann, Burlington, MA.

Baniszewski, B. (2005). An environmental impact analysis of grinding. S.B. thesis, Department of Mechanical Engineering, MIT, Cambridge, MA.

Batzer, S. A. and Sutherland, J. W. (1998). The dry cure for coolant ills. *Cutting Tool Engineering* 50 (4): 34–44.

Dahmus, J. B. and Gutowski, T. G. (2004). An environmental analysis of machining. In *Proceedings of ASME International Mechanical Engineering Congress and RD&D Expo*, Anaheim, CA.

Dittrich, M. A., Gutowski, T. G., Cao, J. et al. (2012). Exergy analysis of incremental sheet forming. *Production Engineering* 6 (2): 169–177.

Fthenakis, V. M. and Kim, H. C. (2007). Greenhouse-gas emissions from solar electric- and nuclear power: A life-cycle study. *Energy Policy* 35 (4): 2549–2557.

Gregory, A. K. and Menerey, D. (1994). Sustainable development by design: Review of life cycle design and related approaches. *Journal of the Air and Waste Management Association* 44 (5): 645–668.

Guo, Y., Loenders, J., Duflou, J., and Lauwers, B. (2012). Optiimization of energy consumption and surface quality in finish turning. In *5th CIRP Conference on High Performance Cutting*, Zürich, Switzerland.

Gutowski, T., Branham, M., Dahmus, J., Jones, A., Thiriez, A., and Sekulic, D. (2009). Thermodynamics analysis of resources used in manufacturing processes. *Environmental Science and Technology* 43 (5): 1584–1590.

Gutowski, T., Dahmus, J., and Thiriez, A. (2006). Electrical energy requirements for manufacturing processes. In *Proceedings of 13th CIRP Intlernational Conference on Life Cycle Engineering* (pp. 623–627).

Gutowski, T., Jeffrey, D., and Thiriez, A. (2006). Electrical energy requirements for manufacturing processes. In *13th CIRP International Conference on Life Cycle Engineering*, Leuven, Belgium, May 31st – June 2.

Haan, D. M., Batzer, S. A., Olson, W. W., and Sutherland, J. W. (1997). An experimental study of cutting fluid effects in drilling. *Journal of Materials Processing Technology* 71 (2): 305–313.

Haapala, K., Rivera, J. L., and Sutherland, J. W. (2009). Reducing environmental impacts of steel product manufacturing. *Transactions of the North American Manufacturing Research Institution of the SME* 37: 419–426.

Hirohisa, N. (2012). *Environmental Burden Analyzer for Machine Tool Operations and Its Application.* Intech, Rijeka. www.intechopen.com/books/manufacturing-systems.

Hoffman, E. G. (1984). *Fundamentals of Tool Design*, 2nd edn. Society of Manufacturing Engineers, Dearborn, MI.

Hondo, H. (2005). Life cycle GHG emission analysis of power generation systems: Japanese case. *Energy* 30 (11–12): 2042–2056.

Jan, E. and Bras, B. (1998). ISO 14000 and activity based life cycle assessment in environmentally conscious design and manufacturing: A comparison. In *1998 ASME Design Engineering Technical Conference.* American Society of Mechanical Engineers, Atlanta, GA.

Johnson, W. (1956). The pressure for the cold extrusion of lubricated rod through square dies of moderate reduction at slow speeds. *Journal of the Institute of Metals* 85: 403–408.

Kordonowy, D. N. (2001). A power assessment of machining tools. B.S. Thesis, MIT, Department of Mechanical Engineering, Cambridge, MA.

Kurtz, M. (2007). *Environmental Conscious Manufacturing.* Wiley, Hoboken, NJ.

Lange, K. (1985). *Handbook of Metal Forming.* McGraw-Hill, New York.

Life Cycle Assessment (LCA): Principles and practice. (2006). EPA/6000R-06/060.

Machinability Data Center, and Metcut Research Associates. (1980). *Machining Data Handbook*. Machinability Data Center, Metcut Research Associates, Cincinnati, OH.

Narita, H., Desmira, N., and Fujimoto, H. (2008). Environmental burden analysis for machining operation using LCA method. In *Proceedings of 41st CIRP Conference on Manufacturing Systems* (pp. 65–68), Tokyo, Japan.

Niggeschmidt, S., Helu, M., Diaz, N., Behmann, B., Lanza, G., and Dornfeld, D. (2010). Integrating green and sustainability aspects into life cycle performance evaluation. In *Proceedings of 16th CIRP International Conference on Life Cycle Engineering*.

Rangarajan, A. and Dornfeld, D. (2004). Efficient tool paths and part orientation for face milling. *CIRP Annals* 53 (1): 73–76.

Santos, J. P., Oliveira, M., Almeida, F. G., et al. (2011). Improving the environmental performance of machine-tools: Influence of technology and throughput on the electrical energy consumption of a press-brake. *Journal of Cleaner Production* 19: 356–364.

Szargut, J., Morris, D. R., and Steward, F. R. (1988). *Exergy Analysis of Thermal, Chemical and Metallurgical Processes*. Hemisphere, Washington, DC.

Taniguchi, M., Kakinuma, Y., Aoyama, T., and Inasaki, I. (2006). Influences of downsized design for machine tools on the environmental impact. In *Proceedings of MTTRF Annual Meeting* (pp. 1–4).

Tlusty, G. (2011). *Manufacturing Processes and Equipment*. Prentice Hall, Englewood Cliffs, NJ.

Tönissen, S. (2009). Power demand of precision machine tools. MS Thesis, Department of Mechanical Engineering, University of California, Berkeley, CA.

21

Cutting Tools and Sustainability Considerations

21.1 Introduction to Cutting-Tool Materials

21.1.1 Carbon Steels

Their use is restricted to the cutting of soft materials and wood. Performance is poor above 250°C.

21.1.2 High-Speed Steels

These are used extensively, particularly for multipoint tools. They have been replaced to a large extent by carbides for single-point tools. Their main application is for form tools and complex shapes, for example for gear-cutting and broaching. They are also used for twist drills, reamers and so on.

21.1.3 Carbides

These consist of powdered carbides of tungsten, titanium, tantalum, niobium, and so on, with powdered cobalt as binder. They are produced by pressing the powder in dies and sintering at high temperature. They are then ground to the final shape. They are generally used as tips and can operate up to 1000°C.

21.1.4 Laminated Carbide

These consist of a hard thin layer of titanium carbide bonded to a tungsten-carbide body. The surface has a very high strength at high temperature, whilst the body has high thermal conductivity and is thus an efficient remover of heat.

21.1.5 Diamonds

These are the hardest of all cutting materials with low thermal expansion and good conductivity. They are twice as good as carbides under compression. A good finish can be obtained with nonferrous metals and finish polishing can be eliminated. Diamonds are particularly good for cutting aluminum and magnesium alloys, copper, brass, and zinc. They have a long life (Tables 21.1 and 21.2).

21.1.6 Carbide and Ceramic Tools

Carbides are graded according to series (see Table 21.3) and by a number from 01 (hardest) to 50 (toughest), for example, P01 to K40 (Tables 21.4 and 21.5).

Table 21.1 Carbon Steels (for Softer Metals and Wood; Poor Performance above 250°C) and Characteristics of Steel Tools

Composition	Characteristics	Applications
Plain carbon steel, 0.2% Mn, water hardening 0.7%–0.8%C	High toughness, low hardness	Shear blades, chisels, turning mandrels
0.9%C	General purpose. Best combination of toughness and hardness	Large taps and reamers
1.0–1.4%C	High hardness, keen edge, low, shock resistance	Taps, screw dies, twist drills, mills for soft metals, files
Carbon steel + vanadium, 0.8–1.0%C, 0.2%Mn, 0.2%Va	Water hardening, takes keen edge, more shock resistant than plain carbon steel	Screw taps and dies, twist drills, reamers, broaches
Chrome steel, 0.9%C, 0.2%Mn, 0.5%Cr	Water hardening, good abrasion resistance, takes high compression	Drawing dies, wood planes, chisels
High manganese steel, 0.55–0.8%C, 0.6–0.8%Mn	Oil hardening, tougher but less hard, high shock resistance	Bending form dies, hammers, tool shanks

Table 21.2 High-Speed Steels

	Composition (%)						
Type	C	W	Cr	Va	Co	Mo	Characteristics
Super	0.8	18/22	4.5	1.5	10/12	–	Highest temperature of HSS. Very hard but not so tough. Most expensive. For materials with tensile strength >1225 MPa
General purpose	0.75	18	4.15	1.2	–	–	Tougher than super and cheaper, for materials over >1225 MPa tensile strength
General purpose tungsten/molybdenum	1.2	57	4.3	2.8	6	5.5	Better impact resistance and cheaper than general purpose HSS. High wear resistance
High vanadium	1.55	12.5	4.75	5.0	5	–	Best abrasion resistance. Used for highly abrasive materials

Table 21.3 Carbides

Series	Material Machined	Carbides
P	Steel, steel castings	W, Ta, Tt, Ni with Co binder
M	Cast iron, nonferrous, plastic	W with Co binder
K	Heat-resistant steels, stainless steels	W with Co binder

Table 21.4 Sintered Carbide Tool-Cutting Conditions and Positive Rake

	Cutting Speed (m/min)		Top Rake (°)	
Material Being Cut	Rough	Fine	Clean Metal	Rough Scaly Metal
Steel				
Low-medium carbon	75	120–210	8	0–4
Medium-high carbon	60	90–180	4	0–3
Nickel carbon	30	75–120	0.4	0
Cast Iron				
200 Brinell hardness	45–60	90–120		3.5–8
White Heart	3–6	6–4		0
Copper	150–240	240–360		13–16
Brass	120–240	240–360		0–3.5
Bronze and gun metal	120–180	240–300		0–3.5
Aluminum alloy	90–150	180–225		13–16
Plastics	120–180	240–300		0
Glass	9–15	15–21		3.5

Table 21.5 Ceramic Tools (Sintered Aluminum Oxide with Grain Refiners and Binder)

Material Being Cut	Cutting Speed (m/min)
Cast iron	60–610
Steel	90–450
Aluminum	>610

21.2 General Information on Metal Cutting

21.2.1 Cutting Speed and Feed Rates (Tables 21.6 and 21.7)

21.2.2 Surface Finish

Different processes produce different degrees of finish on machined surfaces (Figure 21.1). These are graded from N1 with an average height of roughness of 0.025 μm, up to N12 roughness 50 μm. The manner in which a machined surface is indicated is as shown in Table 21.8.

$$\text{Average height of roughness, } h_a = \frac{a+b+c+\cdots}{L}$$

where:

a, b, c, and so on = area on graph
L = length of surface

Table 21.6 Cutting Speeds and Feed Rates

	Cutting Speed									
	High-Speed Steel Turning				Carbide, Turning		Stellite, Turning		Feed Rate	
Material of Workpiece	R	F	R & T	D	R	F	R	F	Rough	Fine
Mild steel	40	66	7.5–15	30	90	180	50	75	0.625–2.0	0.125–0.75
Cast steel	15	24	3.5	12	45	100	24	33	0.5–1.25	0.125–0.50
Stainless steel	15	18	3.0	12	27	45	22	25	0.5–1.00	0.075–0.175
Gray cast iron	18	27	3.5	13	60	100	23	45	0.4–2.5	0.20–1.00
Aluminum	90	50	15	72	240	360	120	180	0.1–0.5	0.075–0.25
Brass	75	100	18	60	180	270	90	150	0.375–2.0	0.20–1.25
Phosphor bronze	18	36	4.5	13	120	180	30	50	0.375–0.75	0.125–0.50

Table 21.7 Material Constants: Power Used and Volume Removed in Metal Cutting

Material	k_L	k_D	k_M	Material	k_L	k_D	k_M
Aluminum	700	0.11	0.9	Mild steel	1200	0.36	2.7
Brass	1250	0.084	1.6	Tool steel	3000	0.40	7.0
Cast iron	900	0.07	1.9				

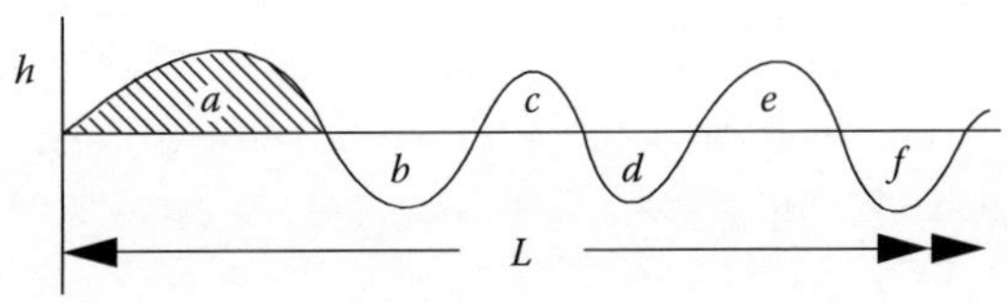

Figure 21.1

Surface graph.

Table 21.8 Roughness Grades

Roughness grade	N1	N2	N3	N4	N5	N6	N7	N8	N9	N10	N11	N12
h_a (μm)	0.025	0.05	0.1	0.2	0.4	0.8	1.6	3.2	6.3	12.3	25	50

Finishing processes		Surface indication
Mill Bore Turn Grind	Ream Broach Lap Hone, etc.	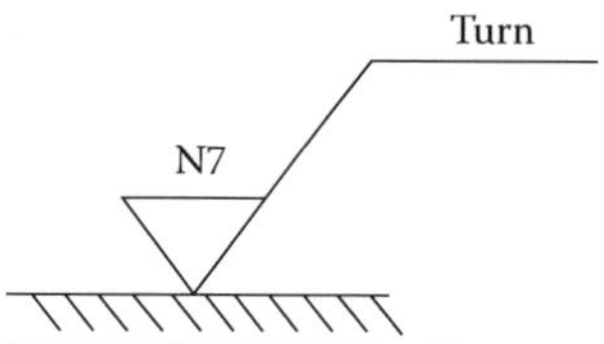

Table 21.9 Machining Properties of Thermoplastics

			Milling		Turning		Drilling	
Material	Rake Angle	Clearance Angle	Cutting Speed	Feed	Cutting Speed	Feed	Cutting Speed	Feed
Nylon	0/-10	20-/30	2.5–5.0	0.1–0.25	2.0–5.0	0.1–0.38	5	≤4
PTFE	0/-5	20-/30	1.0–2.5	0.05–0.25	1.25–5.0	0.1–0.38	5	≤4
Polystyrene	0/-5	20–30	1.5–5.0	0.05–0.25	0.5–10	0.1–0.38	5	≤4
Rigid PVC	0/-10	20–30	1.5–5.0	0.25–0.75	2.5–30	0.05–0.13	5	≤4

Table 21.10 Negative Rake Cutting

Material	Roughing Speed	Finishing Speed	Feeds
Steel 0.15%C	230	300	Milling: 0.2–0.4
0.4%C	160	210	Milling: 0.2–0.4
0.8%C	120	135	Milling: 0.2–0.4
Steel castings	90	105	Turning: 0.25–0.5
Phosphor bronze and gun metal	300	420	Turning: 0.25–0.5
Copper	450	540	Turning: 0.25–0.5
Brass	600	900	Turning: 0.25–0.5
Aluminum and alloys	900	1200	

21.3 Machining Properties of Thermoplastics

Machining parameters for some thermoplastics are presented in Table 21.9. These values are given in a range and need to be selected suitably depending on the situation (Table 21.10).

21.4 Calculation of Machining Cost

The *total-time cost per work piece* is made up of *machine-time cost, nonproductive time cost*, and *total cost. Machining-time cost* is for actual machining and includes overheads and wages. *Nonproductive time cost* covers *setting-up* and *loading- and unloading time cost. Tool cost* combines *tool change time cost* and actual *tool cost.* The former is the cost of changing the cutting edge, and the latter is the cost of cutting plus resharpening. When *total cost* is plotted against *cutting speed*, an optimum speed for minimum cost is found (Figure 21.2).

The total cost of manufacturing a product consists of the costs of materials, tooling, and labor, fixed costs, and capital costs. Several factors are involved in each cost category. Manufacturing cost can be minimized by analyzing the product design to determine whether the part size and shape are optimal and whether the materials selected are the least costly ones that possess the desired properties and characteristics. The possibility of substituting one material with the other is an important consideration in minimizing cost.

In the past, several books contained the machining optimization based on minimization of cost and technological considerations. For example, the total cost, C_c, can be written as

$$C_c = C_0 T_h + \frac{C_0 \pi D L}{f v} + \frac{(C_0 T_t + C_t)\left(\pi D L v^{1/n - 1}\right)}{f C^{1/n}} \tag{21.1}$$

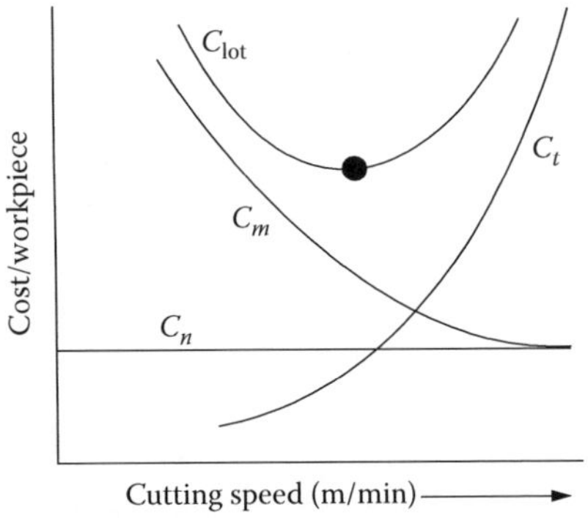

Figure 21.2

Cost work piece vs. cutting speed.

where:

- C_0 = the cost rate (e.g., \$/min) for the operator and the machine. Sometimes costs for the operator and the machine can differ and this aspect should be taken into consideration.
- T_h = part handling time. This is the time the operator takes in loading the part on the machine at the beginning of the manufacturing cycle and unloading the part after machining is over.
- D = average diameter = $(D_0 + D_f)/2$.
- D_0 = diameter of the part at beginning and D_f = diameter of the part at end of turning.
- T_t = is tool change time at the end of tool life. It should be converted to tool change time per part.

This is calculated differently for throwaway tool and regrindable cutting tool. The tool change time per part = $T_t/n_p \cdot n_p$ represents the number of parts manufactured in one tool life (the number of pieces cut with one cutting edge until the tool is changed).

The number of parts (n_p) manufactured in one tool life is calculated as

$$n_p = \frac{T}{T_m}$$

where:

- T = tool life, min/tool
- T_m = machining time per part, min/pc

Both T and T_m are functions of speed; therefore, the n_p is presented as

$$n_p = \frac{fC^{1/n}}{\pi DLv^{1/n-1}}$$

This has been included in Equation 21.1.

C_t is the tooling cost and it differs in the disposable and regrindable cutting tools. For disposable (throw away) tools, the tool cost is determined as

$$C_t = \frac{P_t}{n_e}$$

where:

- P_t = price of the throw away cutting tool
- n_e = number of cutting edges per cutting tool

In case of regrindable cutting tools, the cost/part includes purchase price and the grinding cost. It is presented as

$$C_t = P_t / n_e + T_g C_g$$

where:

- T_g = time to grind or regrind the tool, min
- C_g = grinder's rate, \$/min

In Equation 21.1, n and C are tool-life parameters as outlined in Taylor's equation. In this equation, material constant is neglected as it is independent of the cutting speed. The objective is to determine cutting conditions for specified workpiece materials and part geometry. A better relationship than Equation 21.1 has been suggested by Jonsson [4], including the scrap rate and average down time per part. However, it is not used in this book due to lack of proper data. The minimum cost of the part can be determined by differentiating Equation 21.1 w.r.t. cutting velocity and equating it to zero and solving for v_{min}

$$v_{min} = C\left(\frac{n}{1-n}\cdot\frac{C_0}{C_0T_t + C_t}\right) \tag{21.2}$$

As we can see, the tooling cost, C_t, is the major component that differentiates between disposable and regrindable tool, and this can change the optimum cutting parameters in both types of tool. The corresponding tool life can be calculated using Taylor's tool-life equation, $vT^n = C$. Plugging v_{min} from Equation 21.3 into Taylor's equation, we get

$$T_{min} = \left(\frac{1}{n} - 1\right)\left(\frac{C_0T_t + C_t}{C_0}\right) \tag{21.3}$$

It is clear from Equation 21.3 that tool life will be higher for regrindable tools, as they have a higher C_t than disposable tools. This also illustrates along with Equation 21.2 that cutting velocity is slower and the cycle time is longer for regrindable tools than disposable tools.

21.5 Sustainable Machining Economics: Selection of Optimum Turning Conditions Based on Minimum Energy Considerations

Energy consumption as driven by consumption demand is a key contributor to carbon dioxide emissions and climate change. Hence, reducing energy usage is an essential consideration in sustainable manufacturing. Reducing the energy footprint leads to a reduction of carbon dioxide emissions. The link between the carbon dioxide emission and energy footprints is established by a carbon emission signature (CES) as presented by Jeswiet and Kara [2].

Energy used in manufacturing has to be reduced in order to cut down carbon emissions derived from energy generation. For machining processes, the energy requirement decreases as the material removal rate increases [3,4]. Reducing energy consumption leads to sustainable manufacturing. It is essential to optimize and improve manufacturing productivity while simultaneously mitigating the effect of manufacture on the environment. In the past, in manufacturing, metal-cutting operations have been mainly optimized on the basis of economic and technological considerations without the environmental dimension. However, the presentation up to this stage did not include the concept of environmentally conscious manufacturing or green manufacturing.

Here, we are considering the minimum energy criterion along with the dry turning process. The use of oil-based cooling/lubricant fluid is one of the most unsustainable aspects of the manufacturing operation (Pusvec et al., 2010). The total energy E used in the turning operation can be evaluated from the energy consumed by the machine during

- Energy during setup of turning operation, E_s
- Energy consumed during turning, E_m
- Energy consumed during tool change, E_c
- Energy consumed during making of cutting tool, E_t

The workpiece material is fixed and the energy consumed during production of workpiece is not considered here. Hence, the total energy E can be calculated as

$$E = E_s + E_m + E_c + E_t \tag{21.4}$$

The energy consumed during setup, E_s is evaluated from the power consumed by the machine and the total time taken for tool and workpiece setup. It is generally assumed that setup is done when the spindle has not yet been turned on. The energy consumed during turning, E_m, is evaluated from the energy consumed for powering the machine modules and energy for material removal, as modeled by Gutowski et al. 2006 in Equation 21.5:

$$E_m = (P_0 + U.RMR)T_m \tag{21.5}$$

where:

P_0 is idle power consumed by machine modules, hp (W)

U is the specific energy requirements in cutting operations in-lb/in³ (N-m/mm³), which is a property of material

RMR is the metal-removal rate, in³/s (mm³/s)

These parameters are presented in Table 21.11. Using this measure, different work materials can be compared in terms of their power and energy requirements. The values in Table 21.11 are based on two assumptions: (1) cutting tool is sharp and (2) the chip thickness before the cut t_0=0.25 mm (0.010 in.). If these assumptions are not met, some adjustments must be made. For worn-out tools, the power required to perform the cut is obviously greater, and this is reflected in higher values of specific energy and unit horsepower. It is suggested in the machinery handbook that the values in Table 21.11 should be multiplied by a factor between 1.00 and 1.25 depending upon the degree of dullness of the cutting tool. For cutting tools in finishing operation, the factor is assumed to be 1.10 and for roughing operation, the factor is 1.25.

Chip thickness before the cut t_0 also affects the specific energy and unit horsepower values. The unit horsepower tends to increase as chip thickness before cut t_0 is reduced. It is found that in grinding operation where chip thickness is extremely small compared to turning and milling, the specific energy requirement is very high. T_m is the machining time in seconds. The energy consumed during tool change, E_c, is evaluated by multiplying machine power and tool change time. In turning, the tool is generally replaced when the machine is not running. The energy used during the making of the cutting tool, E_t, is evaluated by dividing the energy footprint of the cutting tool by the number of cutting edges. This is calculated from the energy embodied in the cutting-tool material, the energy used during cutting-tool manufacture, and the energy of the supplementary process such as coating. The quantity E_t is calculated by multiplying the energy per cutting edge y_E by the number of cutting edges required to complete the manufacturing pass. Finally, a new equation for energy consumption in single-pass turning is developed:

$$E = P_0T_h + (P_0 + U.RMR)T_m + P_0T_t\left(\frac{T_m}{T}\right) + y_E\left(\frac{T_m}{T}\right) \tag{21.6}$$

Table 21.11 Specific Energy

Material	Brinell Hardness	Specific Energy U or Unit Power P_u		Unit Horsepower
		N · m/mm³	in-lb/in³	$HP_{u \cdot hp/(in^3/min)}$
Carbon steel	150–200	1.6	240,000	0.6
	201–250	2.2	320,000	0.8
	251–300	2.8	400,000	1.0
Alloy steels	200–250	2.2	320,000	0.8
	251–300	2.8	400,000	1.0
	301–350	3.6	520,000	1.3
	351–400	4.4	640,000	1.6
Cast iron	125–175	1.1	160,000	0.4
	176–250	1.6	240,000	0.6
Stainless steel	150–250	2.8	400,000	1.0
Aluminum	50–100	0.7	100,000	0.25
Aluminum alloys	100–150	0.8	120,000	0.3
Brass	100–150	2.2	320,000	0.8
Bronze	100–150	2.2	320,000	0.8
Magnesium alloys	50–100	0.4	60,000	0.15

The total energy equation can be further modified by including cutting time (T_m) and tool-life equations. Taylor's extended tool life is

$$vT^n f^m = C \tag{21.7}$$

or can be written as

$$T = \frac{C^{1/n}}{v^{1/n} f^{m/n}}$$

and

$$T_m = \frac{\pi DL}{fv} \tag{21.8}$$

The equation for the material removal rate (RMR) can also be written as

$$RMR = vfd$$

where:
- f is the feed (mm)
- d is depth of cut (mm)
- v is velocity (ft/min)

Of course, the above equation can be written as

$$RMR = vf\frac{(D_i - D_f)}{2} \tag{21.9}$$

Plugging Equations 21.8 and 21.9 into Equation 21.6, we get

$$E = P_0T_h + P_0\frac{\pi DL}{fv} + Uvf\frac{D_i - D_f}{2}\frac{\pi DL}{fv} + P_0T_t\left(\frac{\pi DL/fv}{C^{1/n}/(v^{1/n}f^{m/n})}\right) + y_E\left(\frac{\pi DL/fv}{C^{1/n}/(v^{1/n}f^{m/n})}\right) \tag{21.10}$$

The above equation can be rewritten as

$$E = P_0T_h + \frac{P_0\pi DL}{fv} + \frac{U\pi L(D_i^2 - D_f^2)}{4} + (P_0T_t + y_E)\left(\frac{\pi DL / fv}{C^{1/n} / (v^{1/n}f^{m/n})}\right) \tag{21.11}$$

The objective is to minimize the total consumption of energy E during the machining process. By differentiating the total energy equation w.r.t. v and equating to zero, we get

$$\frac{\partial E}{\partial v} = \frac{P_0\pi DL}{fv^2} + (P_0T_t + y_E)\left(\frac{1}{n} - 1\right)\frac{\pi DLf^{(m/n-1)}v^{(1/n-2)}}{C^{1/n}} = 0 \tag{21.12}$$

It can further be simplified as

$$\frac{C^{1/n}}{f^{(m/n-1)}v^{(1/n-1)}} = \left(\frac{1}{n} - 1\right)\frac{(P_0T_t + y_E)}{P_0} \tag{21.13}$$

By substituting Taylor tool-life equation with the above equation, the equation for optimum tool life for minimizing the total energy consumption during machining is obtained as

$$T_{\min E} = \left(\frac{1}{n} - 1\right)\left(\frac{P_0T_t + y_E}{P_0}\right) \tag{21.14}$$

We can examine the equation and conclude that for minimization of total energy consumption during turning, the following factors are important:

- Tool-life velocity exponent (i.e., rate at which tool life reduces as the cutting speed increases).
- The power used by a turning machine during noncutting operation. It clearly indicates that the design of the machine has to improve to consume less power during the noncutting phase.
- Total time for tool change, which obviously affects the noncutting power consumption.

All the above are key factors influencing the selection of optimum tool life for achieving minimum energy consumption.

The values of unit horsepower and specific energy (U) for selected work materials using sharp cutting tools and chip thickness before the cut $t_0=0.25$ mm (0.01 in.) are given in Table 21.11.

21.5.1 Estimated Energy Footprint for Tooling

The energy for manufacture of the cutting tool does not represent direct energy consumption in the machining process. However, it influences the total energy for the system (inputs to the machining process which needs to be optimized. Energy for tool manufacture mainly consists of energy embodied in tool materials, sintering (and grinding for some tools) and coating process energy. Dahmus and Gutowski [5] presents reference data for energy footprint of tooling. Here, we use the energy footprint provided by Dahmus and Gutowski in Table 21.12. They presented energy footprint data for two cases. Case 1 whereby the energy embodied in the tooling material is presented and case 2 in which only the energy for tooling manufacture was considered. Here the inserts used were assumed to be manufactured from a sintering process. Each insert was coated. The total weight of the insert was estimated as 9.5 g (Table 21.13).

Example 21.1

A high-speed steel (HSS) tool is used to turn a steel workpart that is 300 mm long and 80 mm in diameter. The parameters in the Taylor equation are: $n=0.13$ and $C=75$ (m/min) for a feed of 0.4 mm/rev. The operator and machine tool rate = \$30.00/h, and the tooling cost per cutting edge = \$4.00. It takes 2.0 min to load and unload the workpart and 3.50 min to change tools. Determine (a) cutting speed for maximum production rate, (b) tool life in min of cutting, (c) cycle time and cost per unit of product, (d) T&V for minimum cost, and (e) T&V for minimum energy. Assume tool weight as 9.5 g.

Solution

(a) $C_0=\$30/\text{h}=\$0.50/\text{min}$

$v_{max}=75/[(1/0.13-1)(3.5)]^{0.13}=75/[6.692\times 3.5]^{0.27}=49.8$ m/min

(b) $T_{max}=(75/49.8)^{1/0.13}=(1.506)^{7.692}=23.42$ min

(c) $T_m=\pi DL/fv=\pi\,(80)(300)/(0.4\times 49.8\times 10^3)=3.787$ min

$n_p=23.42/3.787=6.184$ pc/tool life Use $n_p=6$ pc/tool life

$T_c=T_h+T_m+T_t/n_p=2.0+3.787+3.5/6=6.37$ min/pc

$C_c=0.50(6.37)+4.00/6=\$3.85/\text{pc}$

Example 21.2

Solve previous Problem 21.1 except that in part (a), determine cutting speed for minimum cost.

Solution

(a) $C_0=\$30/\text{h}=\$0.50/\text{min}$

$v_{min}=75[0.50/((1/0.13-1)(0.50\times 3.5+4.00))]^{0.13}=75[0.50/(6.692\times 5.75)]^{0.13}=42.6$ m/min

(b) $T_{min}=(75/42.6)^{1/0.13}=(1.76)^{7.692}=76.96$ min

Table 21.12 Energy to Produce an Insert Tool

	Case 1 (Material and Manufacturing)	Case 2 (Manufacturing Only)
Embodied tool material energy (MJ/kg)	400	–
Sintering and coatings (MJ per process per cutting insert)	1–2 (avg. 1.5)	1–2 (avg. 1.5)
Total energy per insert (MJ)	5.3	1.5

Table 21.13 Ecoproperties of Engineering Materials

Materials	Ecoproperties: Material						Ecoproperties: Processing				Recycling		
	A	B	C	d	e	F	g	H	i	j	k	l	m
Aluminum alloys	$33\times10^6-34\times10^6$	20×109 22×109	200–240	11–13	*125–375	740–820	*2.4–2.9	0.14–0.17	*.2.4–2.9	0.19–0.23	18–21	1.1–1.2	33–55
Magnesium alloys	$645\times10^3-655\times10^3$	$1\times10^9-1.1\times10^9$	356–394	22.4–24.8	*500–1500	–	*2.1–2.6	*0.13–0.16	3.8–4.6	*0.3–0.37	8.4–9.	1.8–1.95	9.2–11.4
Titanium alloys	$120\times10^5-122\times10^5$	$50\times10^6-52\times10^6$	600–700	*38–44	*470–1410	–	5.02–5.77	0.301–0.346	*4.21–5,7	0.377–0.456	228–281	*14.4–16.7	20–24%
Copper alloys	$15\times10^6-16\times10^6$	$470\times10^6-490\times10^6$	68–74	4.9–5.6	*150–450	1300–1500	*2,5–2.8	*0.15–0.17	*1.8–2.2	0.14–0.15	17–18.5	1.2–1.4	40–60&
Lead alloys	$3.3\times10^6-3.4\times10^6$	$66\times10^6-67\times10^6$	53–58	3.3–3.7	*175–525	610–670	*0.4–0.49	*0.024–0.03	81.6–2.7	*0.13–0.14	8.4–9.2	0.45–0.51	70–75%
Zinc die-casting alloys	$9.7\times10^6-1.0\times10^7$	$2.18\times10^8-2.21\times10^8$	70–75	3.7–4	*160–521	3000–3400	*1.09–1.32	*0.065–0.08	–	–	12.6–13.5	0.66–0.72	20–25%
Nickel-chromium alloys	$1.5\times10^6-1.6\times10^6$	$63\times10^6-65\times10^6$	127–140	*7.89–8.82	*134–512	4900–5500	*3.68–4.06	0.221–0.243	*2.52–2.78	*0.202–0.222	31.8–35	*1.97–2.21	22–26%
Nickel-based superalloys	$1.5\times10^6-1.6\times10^6$	$6.3\times10^7-6.5\times10^7$	135–150	*7.89–9.2	*134–484	4900–5500	*3.38–4.09	*0.20–0.24	*3.26–3.95	0.26–0.31	33.8–37.5	*1.97–2.3	22–26%
Low-carbon steel	$1.1\times10^9-1.2\times10^9$	$78\times10^9-79\times10^9$	29–35	2.2–2.8	*23–69	79–87	*4.2–4.6	*0.25–0.28	*2.2–2.6	*0.18–0.21	8.1–9.8	0.6–0.8	40–44%
Low-alloy steel	$1.1\times10^9-1.2\times10^9$	$78\times10^9-79\times10^9$	*32–48	*1.95–2.3	*37–111	100–120	*3.8–4.6	*0.23–0.28	*3.2–3.9	0.26–0.31	*9.0–11	*0.55–0.64	40–44%
Stainless steel	$30\times10^6-31\times10^6$	$*2.5\times10^9-2.9\times10^9$	*77–85	*4.7–5.4	*112–336	860–960	*3.9–4.4	*0.24–0.29	*3.0–3.7	*0.26–0.31	*22–24	*1.4–1.5	35–40
Cast iron, ductile (nodular)	$1.1\times10^9-1.2\times10^9$	$78\times10^9-79\times10^9$	16–18	1.0–1.1	*13–39	38–42	3.2–3.7	*0.19–0.22	*2.5–3.1	*0.2–0.24	4.9–5.5	0.29–0.32	60–80%

*A, annual world production, tons/year; B, reserves, tons; C, embodied energy, primary production, MJ/kg; D, CO_2 footprint, primary production, kg/kg; E, water usage, L/kg; F, ecoindicator, millipoints/kg; G, casting energy, MJ/kg; H casting CO_2 footprint, MJ/kg; I, deformation processing energy, MJ/kg; J, deformation processing CO_2 footprint, kg/kg; K, embodied energy, recycling, MJ/kg; L, CO_2 footprint, recycling, kg/kg; M, recycle fraction in current supply, %.

(c) $T_m = \pi DL/fv = \pi(80)(300)/(0.4 \times 42.6 \times 10^3) = 4.42$ min/pc
$n_p = 76.96/4.42 = 17.41$ pc/tool life Use $n_p = 17$ pc/tool life
$T_c = T_h + T_m + T_t/n_p = 2.0 + 4.42 + 3.5/17 = 6.63$ min/pc
$C_c = 0.50(6.63) + 4.0/17 = \$3.55$/pc

(d) T&V for minimum energy:

$$T = \left(\frac{1}{n} - 1\right)\left(\frac{PoTt + yE}{Po}\right)$$

Po = Power consumed by machine = 3.594 kW
Tt = tool change time = 3.5
yE = embodied energy
$yE = 400 \times 9.5/100 + 1.5 = 5.3$ MJ/per insert
$n = 0.13$
$T = (1/0.13 - 1)(3.594 * 3.5 + 5.3 \times 4)/3.594 = 62.97$ min for four inserts

We can compare the tool life for minimum cost, which is 76.96 min. We can also calculate the velocity:

$$VT^n = C,\ C = 75,\ \text{and}\ n = 0.13$$

$V = 75/(62.79)^{0.13} = 43.786$ m/min, which is close to V for minimum cost.

21.6 Sustainability Considerations in Metal Cutting Tools

Eco-indicators of cutting tools are classified into three categories, namely, material-, process-, and geometry related. Factors related to eco-indicators consist of density, price, embodied energy, CO_2 footprint, NOX, SOX, water usage, material-processing energy, and recycle fraction of tool materials.

Process-related eco-indicators consist of material removal rate, cutting velocity, feed rate, spindle speed, and surface coating. Geometry-related eco-indicators consist of special geometric features of cutting tool that make the tool's performance robust in terms of process-related eco-indicators. It is obvious that at present the cutting tool design does not take into account these environmental factors or eco-indicators. To design an ideal cutting tool needs considerations of material-, process-, and geometry-related eco-indicators. In literature, we find that researchers have concluded that cutting tools, coolants, and peripheral devices (chip removal devices, coolant circulation pump, tool changers, etc.) have a huge effect on sustainability. It has also been identified that constant spindle speed is good for reducing overall power consumption during material removal process. It is also obvious that near dry and cryogenic machining are good for sustainability. However, the environmental burden of cutting-tool materials has received less attention in literature. Here, we present some of the factors that need to be considered for sustainability in cutting-tool materials. Various types of cutting-tool materials are used for production processes. To name a few, they are HSS, cemented carbides, cBN, diamond, etc. Sometimes, the tool surface is coated by using single or multiple layers of ceramics or diamond to make the cutting tools more productive and robust during cutting. The most common types of cutting tools materials used in industry are presented in Table 21.14.

21.6.1 Cemented Carbides

Cemented carbides or sintered carbides are the most popular cutting-tool materials for production operations. They are composed of hard metallic particles (about 75%–95% by volume) bonded together in metallic matrix. The main ingredients of most commercial tools are tungsten carbide (WC) and a cobalt binder. Titanium carbide (TiC) or tungsten titanium carbide (WTiC) and, to a lesser extent, tantalum and columbium carbides (TaC and CbC), suffer less cratering and wear in cutting steel because of lower friction forces and temperatures and less chip welding. For such advantages, TiC tools with nickel-molybdenum binder are popular, although not as strong and as resistant to chipping and breaking as are WC tools. Cemented carbides are made and sold under a number of trade names, such as Kennametal and Carboloy (Table 21.14).

Table 21.14 Typical Properties of Tungsten-Carbide-Based Cutting-Tool Hard Metals

ISO Application Code	Composition (%)				Density	Hardness	Traverse Rupture Strength (N/mm^2)
	WC	TiC	TaC	CO			
P_{01}	50	35	7	6	8.5	1900	1100
P_{05}	78	16		6	11.4	1820	1300
P_{10}	69	15	8	8	11.5	1740	1400
P_{15}	78	12	3	7	11.7	1660	1500
P_{20}	79	8	5	8	12.1	1580	1600
P_{25}	82	6	4	8	12.9	1530	1700
P_{30}	84	5	2	9	13.3	1490	1850
P_{40}	85	5		10	13.4	1420	1950
P_{50}	78	3	3	16	13.1	1250	2300
M_{10}	85	5	4	6	13.4	1590	1800
M_{20}	82	5	5	8	13.3	1540	1900
M_{30}	86	4		10	13.6	1440	2000
M_{40}	84	4	2	10	14.0	1380	2100
K_{01}	97			3	15.2	1850	1450
K_{05}	95		1	4	15.0	1790	1550
K_{10}	92		2	6	14.9	1730	1700
K_{20}	94			6	14.8	1650	1950
K_{30}	91			9	14.4	1400	2250
K_{40}	89			11	14.1	1320	2500

Source: Machinability Data Center, and Metcut Research Associates. *Machining Data Handbook*. Machinability Data Center, Metcut Research Associates, Cincinnati, OH, 1980 [6].

21.6.2 Coating

In the final stage of carbide manufacture, coatings of various kinds are applied mainly to cutting tools, where for cutting steel in particular it is advantageous to give the rank and clearance surface characteristics that are quite different from those of the body of the insert. Coatings of titanium carbide, nitride, or carbonitride; of aluminum oxide; and of other refractory compounds are applied to a variety of hard metal substrates by chemical or physical vapor deposition (CVD or PVD) by newer plasma method.

21.6.3 ISO Standard 513

This is summarized in Table 21.15. It divides all machining grades into three color-coded groups: straight tungsten carbides (letter K, color red) for cutting gray cast iron, nonferrous metals, and nonmetallics; highly alloyed grades (letter P, color blue) for machining steel; and less alloyed grades (letter M, color yellow), generally with less TiC than the corresponding P series, which are multipurpose and may be used on steels, nickel-based superalloys, ductile cast irons, and so on. Each grade within a group is also given a number to represent its position in a range from maximum hardness to maximum toughness (shock resistance). Technical application is described for grades at more or less regular numerical intervals. Although coated grades scarcely existed when ISO standard was prepared, it is extremely easy to classify coated as uncoated carbides- or carbonitrides, ceramics, and superhard materials according to this system (Table 21.15).

21.6.4 High-Speed Steel

Before 1870, all turning tools were forged from plain carbon steel consisting of about 1% carbon, 0.2% manganese, and the rest iron. These tools had low hot-hardness at elevated temperature during manufacturing and could only be used for machines at low speeds of about 15 f.p.m. This tool also required water quenching for hardening, which frequently resulted in cracks of the tool. To increase the abrasive resistance of the HSS tool, Taylor started using 0.3% vanadium in his tool. By 1910, most HSS tools consisted of 18% tungsten, chromium of about 4%, and vanadium of about 1%, and thus the famous 18-4-1 HSS tool came into being. The cutting speed was subsequently raised to 100 f.p.m. and all necessary changes in machine tool designs were made. One can realize the tremendous increment in the rate of production or productivity and associated reduction in the cost of manufacturing. Current HSS tools enable steel to be machined at speeds that are

Table 21.15 ISO Classification of Hard Metals (Cemented Carbides and Carbonitrides) by application

Main Types of Chip Removal		Group of Applications			Direction of Decrease in Characteristics	
Symbol and Color	Broad Categories of Material to Be Machined	Designation (Grades)	Specific Material to Be Machined	Use and Working Condition	Of Cut	Of Carbide
P, blue	Ferrous with long chips	P_{01}	Steel, steel castings	Finish turning and boring; high cutting speeds, small chip sections, accurate dimensions, fine finish, vibration-free operations		
		P_{10}	Steel, steel castings	Turning, copying, threading, milling, high cutting speeds, small or medium chip sections		
		P_{20}	Steel, steel castings, ductile cast iron with long chips	Turning, copying, milling, medium cutting speeds and chip sections, planning with, small chip sections		
		P_{30}	Steel, steel castings, ductile cast iron with long chips	Turning, milling, planning; medium or large chip sections, unfavorable machining conditions.		
		P_{40}	Steel, steel castings, with sand inclusions and cavities.			
		P_{50}	Steel, steel castings of medium or low tensile strength, with sand inclusions and cavities			
M, yellow	Ferrous metals with long or short chips, and nonferrous metals	M_{10}	Steel, steel castings, manganese steel, gray cast iron, alloy cast iron	Turning; medium or high cutting speeds, small or medium chip sections		
		M_{20}	Steel, steel castings, austenitic or manganese steel, gray cast iron	Turning, milling; medium cutting speeds and chip sections		
		M_{30}	Steel, steel castings, austenitic steel, gray cast iron, high-temperature–resistant alloys	Turning, milling; planning; medium cutting speeds' medium or large chip sections		
		M_{40}	Mild, free cutting steel, low-tensile steel, nonferrous metals and light alloys	Turning, parting off; particularly on automatic machines		
K, red	Ferrous metals with short chips, nonferrous metals and nonmetallic materials	K_{01}	Very hard gray cast iron, chilled castings over 85 Shore, high-silicon aluminum alloys, hardened steel, highly abrasive plastics, hard cardboard, ceramics	Turning, finish turning, boring, milling, scrapping		
		K_{10}	Gray cast iron over 220 Brinell, malleable cast iron with short chips, hardened steel, silicon-aluminum and copper alloys, plastics, glass, hard rubber, hard cardboard, porcelain, stone	Turning, milling, drilling, boring, broaching, scrapping		
		K_{20}	Gray cast iron up to 220 Brinnel, nonferrous metals, copper, brass, aluminum	Turning, milling,, boring, broaching, demanding very tough carbide		
		K_{30}	Low-hardness gray cast iron, low tensile strength steel, compressed wood	Turning, milling, planning, slotting, unfavorable conditions, and possibility of large cutting angles		
		K_{40}	Softwood or hard wood, nonferrous metals	Turning, milling, planning, slotting, unfavorable conditions. And possibility of large cutting angles		

Table 21.16 Composition of Some Commercially Available High-Speed Steel

AISI Designation	Weight Percent					
	C	W	Cr	V	Mo	Co
M1	0.8	1.75	3.75	1.15	8.75	
M2, class 1	0.85	6.25	4.00	2.00	5.00	
M3, class 1	1.05	6.25	4.00	2.50	5.75	
M4	1.30	5.50	4.00	4.00	4.75	
M6	0.80	4.25	4.00	1.50	5.00	12.00
M7	1.02	1.75	3.75	2.00	8.25	
M8	0.80	5.00	4.00	1.50	5.00	(1.25 Cb)
M10, class 1	0.89	0.70	4.00	2.00	8.00	
M15	1.50	6.50	4.00	5.00	8.50	5.00
M30	0.80	1.80	4.00	1.20	8.25	5.00
M34	0.90	1.75	3.75	2.10	8.75	8.25
M36	0.85	6.00	4.00	2.00	5.00	8.25
M41	1.10	6.35	4.25	2.00	3.75	5.25
M42	1.08	1.60	3.75	1.15	9.60	8.25
M43	1.20	2.70	3.75	1.60	8.00	8.20
M44	1.15	5.25	4.25	2.00	6.50	11.75
M45	1.27	8.25	4.20	1.60	5.20	5.50
M46	1.24	2.10	4.00	3.20	8.25	8.25
T1	0.73	8.00	4.00	1.00		
T2	0.85	18.00	4.00	2.00		
T3	1.05	18.00	4.00	3.00	0.60	
T4	0.75	18.00	4.00	1.00	0.60	5.00
T5	0.80	18.00	4.25	2.00	0.90	8.00
T6	0.80	20.50	4.25	1.60	0.90	12.25
T7	0.75	14.00	4.00	2.00		
T8	0.80	14.00	4.00	2.00	0.90	5.00
T9	1.20	18.00	4.00	4.00		
T15	1.55	12.50	4.50	5.00	0.60	5.00

often in excess of 300 f.p.m. A major development in HSS design came by use of composition such as 1.4% C, 4% Cr, 4% V, 9% W, 4% Mo, 12% Co. This helped to increase hardness and reasonable toughness. Table 21.16 gives the composition of commercially available HSSs.

The American Iron and Steel Institute (AISI) has introduced symbols for the two major classes of HSS: the symbol T designates a tungsten-base steel and the symbol M designates a molybdenum-HSS.

21.6.5 Cast Alloy Steels

About 1915 nonferrous high-temperature alloys containing significant amounts of cobalt, chromium, and tungsten were developed. These materials were not heat-treatable but were used as cast and are, therefore, sometimes known as cast alloy tools. These tools are not as hard as HSS at room temperature but they do retain their hardness at high temperatures and hence are occasionally used for such special application as form tools. Table 21.17 gives the composition of a number of cast alloy steel materials. The room temperature hardness of these materials is generally in the R_c range 60–65.

The *eco-indicators* of cutting-tool materials are presented:

Density: Density of material can be considered as an eco-indicator as it affects the volume of material being used. Low density means less material for a given volume and for sustainability, less resources (i.e., material) are always consumed.

Price: Price of material can be considered as an eco-indicator because low price means enough supply, perhaps huge reserve and low-cost-end-products due to low cutting tool cost.

Embodied Energy: Embodied energy perhaps has been recognized as the most important eco-indicator. It is because less embodied energy means less CO_2 emissions and good for the atmosphere. Embodied

Table 21.17 Composition of Some Commercially Available Cast Alloy Tool Materials

Trade Designation	Weight Percent			
	Co	Cr	W	C
Blackalloy 525	44	24	20	2
BlackalloyTx90	42	24	22	2
Crobalt 1	48	30	14	2
Crobalt 2	40	33	18	2.5
Crobalt 3	40	32.5	17.5	3
Haynes Stellite R star J	43	32.5	17.5 (1.2 b, 4 V)	2.5 (4 Ni)
Haynes Stellite R3	50	31	12.5	2.4
Haynes Stellite R19	53	31	10.5	1.8
Haynes Stellite R98M2	38	30	18.5	2
Tantung G	47	30	15	3
Tantung 144	45	28	18	3

Table 21.18 Eco-Indicators of Two Selected Tool Materials

No	Eco-Indicator	HSS (Tool Type: Tungsten Alloy) (AISI T1)	Tungsten Carbide (with 90.22% Cobalt)
1	Density (kg/m^3)	[8590, 8760]	[14,400, 14,700]
2	Price ($/cm^3)	[0.223, 0.251]	[0.65, 0.73]
3	Embodied energy (KJ/cm^3)	[755.061, 855.852]	[8590, 9723.6]
4	CO_2 footprint (g/cm^3)	[47.503, 53.5236]	[[645.109, 727.08]
5	NOX (g/cm^3)	[0.287, 0.323]	[2.577, 2.908]
6	SOX (g/cm^3)	[0.49, 0.553]	[5.163, 5.817]
7	Water usage (L/cm^3)	[4.982, 5.615]	[1.237, –]
8	Fraction (%)	[52.3, 57.8]	[29.4, 32.5]

energy provides an estimation of energy required to manufacture unit mass of material from its primary source (e.g., ores extraction). Obviously, low embodied energy is desirable.

CO_2 Footprint: CO_2 footprint provides an estimation of carbon dioxide generated due to the production of unit mass of material from its primary source (e.g., ores). Low CO_2 in turn means less greenhouse gas (GHG) and it is good from the sustainability point of view.

NOX and SOX: NOX (oxides of Nitrogen) and SOX (oxides of Sulfur) are supposed to be more harmful than CO_2. They are responsible for acid rain and very harmful to health. They also affect the natural water sources. Therefore, the quantity of NOX and SOX to be used for unit production of material is an important eco-indicator. Obviously, low NOX and SOX are desirable.

Water Usage: For sustainability it is desired that less water is used in any production endeavor. The amount of fresh water needed to produce a unit mass of material from its primary source is also considered as an eco-indicator. Low water usage is desirable.

Material-Processing Energy: For any production activity, energy is needed for casting, forming, forging, machining, grinding, or polishing processes. The energy needed to process a unit mass or volume of material is supposed to be one of the most important eco-indicators as it affects cost and carbon footprint as well. Low-energy consumption is always advocated.

Recycle Fraction: Recycle fraction is the percentage of recycled material in the total worldwide supply of material. A large percentage of recycled material is encouraged from the point of view of sustainability. It is hard to get exact data for recycle percentage in literature.

21.7 Representation of Eco-Indicators of Tool Materials

Data related to the eco-indicators of tool material can be gathered from various sources. Most of the indicators are estimated by numerical ranges. For example, see the database developed by Granta Design. Since the volume of a cutting tool remains the same irrespective of the materials used, the unit of the eco-indicators should be in unit/unit volume of material. For example, if the density of a tool material is [14,000, 15,000] kg/m^3 and its CO_2 footprint is [70, 72] kg-CO_2/kg, then its CO_2 footprint should be represented by [14,000 × 70, 15,000 × 72] kg-CO_2/m^3 = [0.98, 1.08] kg-CO_2/cm^3. Table 21.18 shows a comparison between two typical tool materials, HSS and tungsten carbide.

As seen from Table 21.18, HSS is twice as much lighter, three times cheaper, ten times energy efficient, ten times better in terms of CO_2 footprint, NOX and SOX than that of tungsten carbide. From the point of view of water usage, tungsten carbide is better than HSS. The data of water usage for tungsten carbide shown in Table 21.18 is only for pure tungsten carbide. When alloying materials like cobalt is added, water usage increases. For this reason, a dash mark "–" is used, as shown in Table 21.18. The recycle fraction for tungsten carbide is the recycle fraction of pure form. For tungsten carbide with alloying material, the recycle fraction might be further low. However, this figure shown in Table 21.18 shows HSS is at least twice as much better than tungsten carbide in terms of recycle fraction. It is observed that recycle materials sometimes need more Embodied Energy and can produce more CO_2 footprint, NOX, SOX, and water usage. Therefore, if the recycle fraction is large, then the values of the eco-indicator of tool materials presented above should be modified.

21.8 Environmentally Friendly Machining: Model for Environmental Impact of Machining Process

The analytical model integrates aspects of the process mechanics, wear characteristics, and lubricant flows. The quantifiable dimensions in this analysis include energy utilization, process rate, workpiece primary mass flow, and secondary flow of process catalysts. The generation of multiple waste streams can be compared by examining factors such as toxicity and flammability. The sensitivity of environmental factors to variations in operating parameters such as depth of cut, speed, feed, and tool rake angle are examined. The prioritization of environmental factors was evaluated for both high-rate transfer line and flexible job shop environments through utility analysis. This model can serve as a framework for decision-making in environmentally conscious manufacturing, including the design of parts for machining, process planning, and the selection of operating parameters.

Environmentally conscious manufacturing has received increasing interest in recent years due to the adoption of the pollution-prevention legislation by several countries. There is a growing consumer demand for "green" products in many countries, especially in Europe, and there is a restriction on imports and production processes. Until recently, industrial efforts at alleviating the environmental impact of their products and processes have focused on two areas: the development of recycling processes, especially for metals and plastics, and chemical substitutions for selected manufacturing processes such as cleaning, degreasing, and plating. In the area of recycling, government and industry efforts have resulted in the recycling of 17% of all solid wastes produced. In the area of chemical substitution, recent studies have focused on the development of substitute chemicals and processes for specific processing applications, such as metal plating and finishing, development of environment friendly solvents in electronics manufacturing, and cleaning and flux less soldering. While these few selected waste streams have been the subject of studies, little analytical work has been done on the total life cycle trade-offs between waste streams in manufacturing processes, the impact of product design on waste streams, and linkages between environmental impact and traditional industrial requirements of production rate. The influence of product design on recyclability factors was considered in the machining process. A framework for analyzing the environmental impact of machining processes is presented in which the three principal methods for minimizing environmental impact are the modification of existing processes, replacement with alternative processes, and the development of new manufacturing process. The trade-offs in part design, process planning, and process are needed. A set of quantifiable dimensions need to be analyzed, including process time, energy utilization, primary flow of workpiece and secondary flows of process catalysts. Machining process catalysts may include lubricants, coolants, tooling, and other expendable materials which aid in the transformation of the workpiece from the initial to the final states. Evaluation of these dimensions requires the development of a process model which incorporates the analysis of energy and mass loss, process rate, and generation of secondary waste streams in vapor, liquid, and solid forms. More importantly, the interdependencies between these aspects of machining must be captured in order to analyze trade-offs between different dimensions due to operating, process planning, and design decisions. The objective should be to present an integrated analysis of the environmental impact of machining processes. The analysis combines the effects of chip-formation mechanics for orthogonal- and oblique-cutting conditions, tool life due to flank wear, and the generation of cutting-fluid waste streams in both liquid and vapor forms. Through analysis, the effects of changes in operating parameters (speed, feed, depth of cut, tool angle, etc.) on process energy consumption, process rate, and mass flows of waste streams can be estimated. Waste-stream flows can be compared through a weighting analysis which incorporates the effects of toxicity, flammability, reactivity, and disposability.

21.9 Model Configuration

In order to develop the relationships between manufacturing parameters and environmental impact, four general aspects of the machining process are analyzed:

- The material removal mechanics
- Tool life
- Scrap production
- Cutting-fluid flow

The inputs to the process include the type of operation (drilling, milling, turning, etc.), the operating parameters, the initial- and final-workpiece geometries, the workpiece material, the tool geometry, and the tool material. The outputs are the process rate, the process energy, the machining forces, and the total volume of material removed. The process rate is used to estimate tool life through a wear analysis. Tool life and the volume of material removed are evaluated in the material scrap model to estimate the generation of solid waste streams (chips, used tools, swarf, and foam). The process rate, along with the cutting fluid properties and flow rate, is used to estimate the quantities of cutting fluid released in liquid (through chip coating, workpiece coating, and biodegradation) or vapor forms.

The challenge for manufacturing is to be able to manufacture a sustainable and environmentally friendly product without increasing unit costs. This is especially important in metal cutting processes where liquid cutting fluid is used to prolong tool life. Traditionally, cutting fluids have been used to remove heat from the work piece due to the high thermal capacity and conductivity of the cutting fluid. However, numerous health problems like chronic skin conditions have been linked to prolonged exposure to the hazardous chemicals and the bacterial constituents found in cutting fluids.

Cutting fluids are designed on the basis of the following criteria:

• Optimal cooling and lubrication	• Low foaming properties
• Acceptability to operator	• Machine tool compatibility
• Long life	• Workpiece compatibility
• Corrosion protection	• Water supply compatibility
• Not gummy or tacky	• Cost effective

Note that some cutting fluids can only be used for specific machining processes.

The focus on lubricants has shifted from biodegradability to renewability over the years and owing to a change in thinking on the environment. Several aspects of an environmentally adapted lubricant are as follows:

- Biodegradability
- Toxicity
- Renewability
- Bioaccumulability and biomagnifications
- Life cycle assessment (LCA)
- Energy saving and fuel economy

Removal of cutting fluid is also an absolute necessity. Improving environmental performance and lowering the cost of manufacturing (in turn increasing profit) simultaneously will require process modification and recycling of cutting fluids. To account for total volume of cutting fluid used in the process, a complete mass balance analysis is required. Sometimes dry machining has been advocated, but it has not been possible to implement this. However, minimal use of cutting fluid, mist reduction, and recycling of cutting fluid should be encouraged. Such integrated strategy of reduce/recycle would reduce the cost of manufacturing and reduce workers' exposure to cutting fluid mist in the manufacturing environment, including greater recycling or reduced disposal cost. This will greatly further the goal of green manufacturing.

Green manufacturing is very expensive and appropriate regulatory processes must be put in place to ensure its safe environmental disposal. Recently the concept of dry machining was introduced in an attempt to reduce reliance on cutting fluid and an improvement to the working environment. Reducing unit costs low was also an objective, as cutting fluids increase the cost of manufacturing. To date, these goals have not been met, even with improvements made to cutting-tool materials. Dry machining involving air has low convective heat removal rate and is inadequate for dissipating intense heat generation in metal-removal processes.

The cutting fluid used during machining not only acts as a coolant and lubricant, but also results in a better surface finish and longer tool life. The fluid also acts as a rust preventative and assists in swarf removal. Therefore, it can be visualized that the absence of cutting fluid would have a negative impact on the quality of the machined product and accelerated tool wear. The residual stresses within the component being machined and dimensional errors would be added consequences. The cutting fluids used in metal-removal processes are complex and vary significantly, depending on the manufacturing operations they are used in. By traditional definitions, they are grouped in four categories: straight oils, soluble oils, semi-synthetics, and synthetics. Straight oils consist of a petroleum or vegetable oil base with or without specialty additives. The remaining three types of cutting fluids are water-soluble and classified by the ratio of water to mineral oil in their concentrated form. The concentration is usually diluted in 80% to 95% water when used in process. Synthetic cutting fluids contain no mineral oil. Water-soluble cutting fluids contain varying amounts of specialty additives, including lubricants, corrosion inhibitors, emulsifiers, chelating agents, pH buffers, defoamers, and biocides. The following table lists various fluids and their advantages (Table 21.19).

Machining coolants are an important component of metal working operations. Coolants improve machinability, increase productivity, and extend tool life by cooling and lubricating the part and cutting tool. When performing these functions, coolants can quickly become contaminated with foreign materials, causing coolants to lose effectiveness and develop foul odors. As a result, many coolants are used only for a short time and then discarded. The practice of discarding and replenishing coolants can be costly and wasteful. Developing and implementing a coolant maintenance program can help minimize contamination, prolong coolant life, and reduce cost.

21.9.1 Coolant Selection

Coolants affect the overall performance of machining operations, so proper selection is important. Using an unsuitable coolant may lead to premature tool failure or produce an undesirable finish on parts and may shorten the useful life of the coolant. When selecting a coolant, it should be determined if it is suitable for the type of metal and machining operation being used and if it will produce the desired finish quality.

Cutting fluids should have the following properties to fulfill their functions properly:

- Good lubricating properties
- High cooling capacity
- Low viscosity to provide free flow of cutting fluid
- Chemically stable
- Noncorrosive
- High flash point to reduce fire risks
- Allergy free
- Less evaporative
- Low cost

Table 21.19 Cutting Fluid Applications

Group	Description	Advantages
Soluble oil	Oil, emulsifier and 2%–10% water	Good coolant, poor lubricant
Clear soluble oil	As above, with more emulsifier	Good coolant, poor lubricant
Water-based fluids	Solution of sodium nitride and triethanolamine	Good coolant, poor lubricant
EP soluble oils	Soluble oils with EP additives, e.g., sulfur and/or chlorine	Fairly good lubricant
Straight oils	Mineral or fatty oils (lard, sperm, olive, neat's foot, rape seed, etc.) alone or compounded	Good lubricant, often unstable
Sulfurized EP oils	Straight oils with sulfur, zinc oxide or other additives (0.2%–0.8%S)	Average coolant, good lubricant, pressure-resistant, prevents welding of chip on tool
Sulfochlorinated EP oils	Mineral and fatty oil blends with sulfur and chlorine additives	More efficient than sulfurized oils, for most arduous conditions, highly resistant to welding of chip on tool
Chlorinated materials	Carbon tetrachloride and trichloroethylene alone or blended with oils	Very good EP fluid, highly dangerous to use
Gases and vapors	Air, oil mist, CO_2	Limited cooling power, chip dispersed

The selection of appropriate cutting fluid is very important because it could affect machining performance (tool life, cutting forces, surface roughness, power consumption, etc.) and the selection depends on some parameters such as workpiece material used, cutting-tool material, and the type of machining process. For instance, cutting fluids containing sulfur and chlorine additives should not be used with nickel-based alloys and titanium, respectively. Cutting fluids with high lubricity ability are generally used in low-speed machining such as screw cutting, broaching, and gear-cutting and on difficult-to-cut materials, whereas cutting fluids with high cooling ability are generally used in high-speed machining.

Biodegradable fluids—subject to breakdown by bacteria and living organisms—should be considered if coolant spills are likely, or if spent coolant will be discharged to a wastewater treatment system or land spread. The most stable, nonbiodegradable coolants give most machine shops the best performance over the longest period. Disposal of these coolants generally includes separating the oil for fuel blending.

Facilities should use a minimum number of coolant types. Many metalworking facilities require no more than two types of coolants for their operations, one for machining and one for grinding. The use of a large number of coolant types requires extra storage space, adds to inventory and maintenance needs, increases the chance for mix-ups and cross-contamination, and requires more staff expertise.

21.9.2 Contaminant Removal and Prevention

Coolants are most frequently contaminated by metal chips and fines, tramp oil, bacteria, and dissolved salts. Chips and fines are removed from coolants to eliminate surfaces for bacteria to grow on. Free oils are removed to eliminate a food source for bacteria and surface films that prevent oxygen from dissolving into the coolant. They also can deplete oil soluble components of the coolant and oils reduce the cooling capability of the metalworking fluid. Dissolved salts are removed because they affect tool and coolant life, foaming characteristics, filtering efficiency, and emulsion stability.

By removing contaminants, machining facilities can reduce costs by prolonging the effective life of coolants. The table at the end of this fact sheet describes the equipment and techniques used to control and remove the four most frequent contaminants. A good first step is teaching machine operators not to use coolant sumps as trash receptacles. Waste food, cigarettes, and other trash can cause bacterial contamination and feed bacterial growth.

21.9.3 Coolant Monitoring and Control

When a coolant is in use, heat and contaminants can change its composition and performance. Routine monitoring of coolants for quality is necessary to optimize performance and prevent problems. Monitoring pH and maintaining concentration should be included in a coolant maintenance program.

21.9.3.1 pH Control

A change in pH can quickly deteriorate coolant quality, corrode metals, and irritate the skin. Proper coolant pH operating ranges are specified by the manufacturer. For most coolants, this range is 8.5 to 9.0. Within the proper pH range, corrosion and bacterial growth are minimized. Drops below the recommended limit may be a symptom of a bigger problem and are most likely caused by excessive bacteria or excessive diluting. These problems must be corrected first. If needed, pH levels can be increased to the recommended range by a careful addition of an alkaline chemical like dilute caustic soda. Adding chemicals to coolants on a frequent, ongoing basis can cause problems, such as the buildup of dissolved solids, and indicates that other problems have not been solved. The use of low-cost tools for monitoring pH levels such as portable pH meters and pH paper can help determine when pH needs adjusting. For specific recommendations about maintaining coolant pH, chemical or coolant suppliers must be contacted.

21.9.4 Coolant Concentration

Maintaining a proper coolant-to-water ratio is critical because coolants are designed to perform best at specific concentrations. Overly diluted coolant may reduce tool life, and concentrated coolant results in using more coolant than necessary. The same is true for coolant additives. Coolant suppliers can provide information on proper coolant concentrations for specific applications.

Refractometers can provide accurate measurements to determine the coolant-to-water ratio and are fairly inexpensive and simple to operate. Titration kits, supplied by some coolant manufacturers, are another way to maintain proper concentration. Coolant metering or feed pumps can also improve the rate at which coolants are added.

In addition to monitoring pH and coolant concentration, a comprehensive coolant maintenance program should include monitoring tramp oil concentration, suspended solids, alkalinity, and spot corrosion of parts and bacteria count.

21.9.5 Contaminant Control and Removal

Contaminant	Control Method	Contaminant Removal Process
Metal chips and fines	Bag and cartridge filters	Over time, small chips and metal fines accumulate and settle in coolant sumps. Coolants should be filtered routinely to remove these contaminants. Usually a 50-μm bag filter is used, followed by a 5- or 10-μm cartridge filter.
	Centrifuges	With high-speed centripetal force, centrifuges can separate very fine, suspended particles from a coolant.
	Hydrocyclones	Hydrocyclones help concentrate solids from a coolant. With coolant entering at high speed, the conical shape of a hydrocyclone draws a clean coolant flow upward and forces a heavy solids flow downward. Solids are then removed by filtration.
	Screens and conveyors	Placing metal screens and drag conveyors at coolant sumps will collect the majority of metal chips and turnings.
Tramp oil	Belt and disk skimmers	Used to skim off tramp oils floating on the surface of the coolant, skimmers are designed to fit on top of settling tanks or accessible machine sumps.
	Coalescers	Coalescers contain a series of plates that allow tiny oil droplets to cling together causing the oil to rise at a faster rate, increasing the amount of oil removed.
	Prevent and fix leaks	Repair leaks to prevent tramp oil from contaminating coolant.
	Settling tanks	A settling tank provides a calm environment and ample time for tramp oil to rise to the surface of the coolant. Tramp oil can then be removed using a skimmer. Baffles within a tank help localize tramp oil for more efficient removal.
Bacteria (anaerobic)	Aeration	Aeration is the simplest method to control the growth of anaerobic bacteria. Aeration creates an oxygen-rich environment, preventing anaerobic conditions and the growth of anaerobic bacteria.
		Nonaerated coolant sumps provide ideal conditions for the growth of anaerobic bacteria: water and no oxygen. Anaerobic bacteria break down sulfur-containing compounds in coolants, and generate acidic hydrogen sulfide (H_2S) gas, recognized by its "rotten egg" smell. If bacterial growth is not controlled in the coolant, the H_2S can impede lubricating qualities and can create a corrosive environment that could damage tools and parts.
		Aeration is accomplished by recirculating the coolant in the sump when the machine is not in operation, or by adding a stirring mechanism or an air bubbler to the sump.
	UV light, ozone, heat pasteurization, and chemical biocides	Many other methods are available to control bacterial growth problems. When selecting one, determine if it is compatible with the coolant used and if it will control the bacterial strains present. Coolant suppliers can help determine which method to use for specific applications.
		Add chemical biocides only when needed. Excess biocide can cause coolant pH to fluctuate and can have an adverse affect on human health, such as aggravated dermatitis. As a last resort for controlling bacterial growth, use an EPA-approved chemical biocide.
	Sump cleaning	Routine sump cleaning controls bacterial growth. When a coolant is replaced, the sump should be chemically or steam-cleaned to remove any residual bacteria that could contaminate the new coolant.
Dissolved salts	Water treatment: deionization, reverse osmosis, or distillation	City or well water usually contains significant amounts of calcium and magnesium, making it hard water. It may also contain other minerals and possibly some suspended solids. These contaminants affect tool and emulsion stability. As coolant life is extended, these contaminants build up. Hard water minerals can cause scaling on equipment and upset the chemical balance of the coolant. Suspended solids provide surfaces for bacterial growth and can end up as abrasive solids in the cutting zone. Water softening is not a useful treatment method because it introduces sodium salts, which can be corrosive when combined with chlorides and sulfates in the coolants. Using purified water to make coolants generally improves performance and life. If foaming problems occur, try blending purified water with city water to create an intermediate water quality.

Suppliers of coolants and industrial chemicals can provide recommendations on what types of coolant to use in specific machining operations. Suppliers can also assist with management information that will extend the usefulness of coolants.

The MnTAP [7] reference list, Coolant Maintenance Equipment and Supplies, provides information about sources for the equipment identified in this fact sheet.

21.9.6 Cutting Fluid

Environmental concerns, market forces and legislative requirements make imperative a search for new solutions that minimize environmental impact [8]. To minimize the adverse environmental effects associated with the use of cutting fluids, the best solution is to remove the hazardous components from their formulation. But we focused on alternative methods rather than formulation. The effective way of minimizing adverse environmental effects from the use of cutting fluids is to minimize the volumes used and replace mineral-based cutting fluids with environmentally friendly cutting fluids such as vegetable-based cutting fluids. From an environmental point of view, the best method is dry machining. Dry machining not only reduces the contamination of water and air but also reduces danger to health of the operator. However, dry machining is not efficient for many cutting operations.

21.9.7 Dry Machining

Dry machining means that no cutting fluid is used during process. For economic as well as environmental reasons, machining process is carried out without any cutting fluid, but dry machining has some disadvantages. During dry machining process, temperature of the cutting tool is very high and this induces excessive tool wear, thus decreasing tool life. Also, the chips generated at machining cannot wash away and these chips cause deterioration on the machined surface. The problems of cutting fluid contamination and disposal are not seen in dry machining. Dry machining does not induce the pollution of the atmosphere or water resources. Contrary to dry machining, in wet machining (machining with cutting fluids by any means flooding and minimum quantity lubricant), environment, water sources, and soil become polluted during disposal of the cutting fluid. Dry machining will also diminish the manufacturing costs. In some cases, for instance, in interrupted machining process like milling, dry cutting gives a longer tool life than machining with cutting fluid. In milling, the cutting tool does not cut continuously and the use of cutting fluids increases thermal shock effect. Hence, dry machining is better suited for milling operations. In addition, machining with ceramic tools must be conducted in dry conditions due to the thermal shock. In drilling, especially gun drilling, the most important function of the cutting fluid is chip removal, and dry cutting may induce drill breakage. Dry cutting shows positive effects in some workpiece materials such as AISI 316 stainless steel [9]. Tool life for dry cutting was 3.5 times higher than those obtained when abundant fluid was utilized. It was found that the method of fluid application did not influence tool life. However, in some workpiece materials, dry machining presents many problems. For instance, aluminum is a soft material and dry machining of aluminum induces BUE. This influences the surface quality of the workpiece. Higher friction between tool and workpiece in dry machining can increase the temperature in cutting region. This high temperature will cause dimensional inaccuracies at the workpiece and excessive tool wear. So, the disadvantages of dry machining have to be compensated. Improving cutting tool properties by producing better tool materials with lower friction coefficient and high heat resistance, coatings, or tool geometries is being explored by researchers in order to compensate the effects of the elimination of cutting fluids in machining. Some attempts have been made by researchers in order to make the application of dry machining feasible. One of these attempts is to use tool materials with coating. In dry machining, advanced cutting-tool materials and selection of cutting parameters are inevitable; however, these cutting tools are very expensive and increase the machining costs [10]. Cutting fluid is used to both cool and lubricate the cutting surfaces. It is extremely important in situations (like creating threads with a tap) where there is a large surface area in constant contact and little airflow to help remove the heat. Cutting fluid can help to extend tool life, reduce warping of the work piece, and improve the surface finish. In addition, it can 'damp down' the dust particles from certain machining operations.

There are three common ways to apply cutting fluid (Figure 21.3):

- Squeeze bottle (e.g., Tap Magic)
- Brushed on (e.g., machine oil)
- Aerosol can (e.g., WD-40)

21.9.8 Quenching the Workpiece

Some processes, such as grinding, heat up the work piece very quickly, making it difficult to handle. A bucket of water or oil (depending on the material) is used to quickly cool off the piece. However, this may

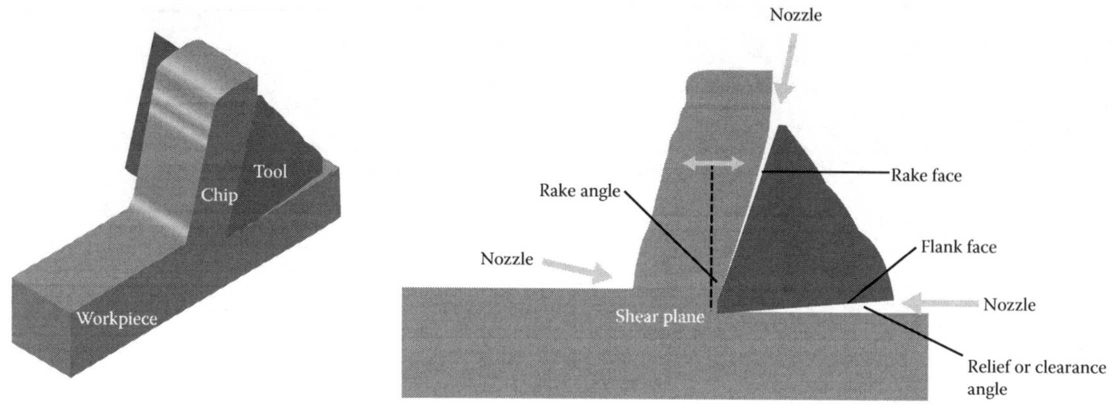

Figure 21.3

(See color insert) Application of cutting fluid.

introduce residual stresses and should not be used for precision parts. The cutting fluid is assumed to diverge into four paths during the machining process: (1) a liquid waste stream created through fluid coating on the chips generated during the machining process, (2) a liquid waste stream resulting from cutting-fluid coating of the workpiece, (3) a vapor waste stream generated through cutting-fluid diffusion into the surrounding environment, and (4) lubricant flow collected and recirculated through the system. The total mass flow of the cutting fluid can be expressed as:

$$V_{\text{fluid}} = \text{Fluid}_{\text{chip}} + \text{Fluid workpiece} + \text{Fluid}_{\text{evaporated}} + \text{Fluid}_{\text{recirculated}} \tag{21.15}$$

Nominally, the recirculating portion of the cutting-fluid stream (including splash and leakage) is recovered and reused. A significant portion of the chip-coated fluid may also be recovered through centrifugal or steam-injection methods; however, in this analysis, all chip-coated fluid is assumed to be unrecoverable. Workpiece-coated fluid is removed and disposed of through secondary cleaning processes. The evaporated cutting fluid is lost to the environment. For the effect of chip coating, chip geometry with height h and thickness b must be known. Assuming that the dominant mechanism for fluid coating is surface tension, the geometry of the coating liquid can be modeled as cylindrical, with curvature in the direction of the smallest length scale of the chip surface. For each chip surface, the cross-sectional area of the coating layer based on the force balance principle is estimated as

$$A = \frac{2\sigma}{\rho g}$$

where σ and ρ are stress and density of the cutting fluid.

The total volume of cutting fluid coating the chips formed during a machining process is dependent on the aggregate length of the chip (or chips) formed:

$$\text{Vol}_{\text{fluid}} = 4AL = \frac{8\sigma}{\rho g} L$$

The volume of cutting fluid coating the workpiece can be determined in a similar fashion by using the major length of the workpiece as L.

The fluid lost through evaporation is calculated using an energy balance. Assuming that the energy generated at the shear plane and rake face is directed toward heating and evaporation of a portion of the cutting-fluid waste stream, the rate of fluid mass loss due to evaporation can be estimated as

$$m \cdot E_{vap} = \frac{f_p V}{C_p(T_{vap} - T_\infty + H)}$$

The thermal properties of the cutting fluid can be derived through a mass fraction average of the constituent components.

21.10 Environmental Impact of Waste Streams

Given the mass flows for each component waste stream, an important issue in the analysis of waste streams is the determination of the relative environmental impact of dissimilar substances. The workpiece scrap may be composed of metals, ceramics, polymers, or composite materials, whilst the waste stream of expended tools can be high-speed steel, ceramic, or ceramic-metal composites. Perhaps the most significant waste is the fluid stream. Lubricants used in the machining process are divided typically into three categories: water-based, oil-based, and synthetics. Water-based lubricants can be categorized further into emulsions, suspensions, gels, or pastes.

One way to establish a quantitative index for comparing cutting fluids is by using experimental data developed for toxicity and flammability measurements. This model evaluates the relative toxicity and flammability for each constituent of the cutting fluid mixture, as well as the solid waste streams comprised of expended tools and scrap material. Animal test data on the median lethal concentration (LCs) and lethal dosage (LDs) are good indicators of the degree of toxicity of a substance, whilst the flash point of a substance is a good indicator of its relative chemical stability, which is related to factors such as transportation and storage, fire risk, and chemical degradation.

References

1. Johnson, W. (1956). The pressure for the cold extrusion of lubricated rod through square dies of moderate reduction at slow speeds. *Journal of the Institute of Metals* 85: 403–408.
2. Jeswiet, J. and Kara, S. (2008). Carbon emissions and CESTM in manufacturing. *CIRP Annals of Manufacturing Technology* 57 (1): 17–20.
3. Rajemi, M. F., Mativenga, P. T., and Aramcharoen, A. (2010). Sustainable machining: Selection of optimum turning conditions based on minimum energy considerations. *Journal of Cleaner Production* 18: 1059–1065.
4. Gutowski, T., Branham, M., Dahmus, J., Jones, A., Thiriez, A., and Sekulic, D. (2009). Thermodynamics analysis of resources used in manufacturing processes. *Environment Science and Technology* 43 (5): 1584–1590.
5. Dahmus, J. and Gutowski, T. (2004). An environmental analysis of machining. In *ASME International Mechanical Engineering Congress and RD&D Expo*, Anaheim, CA.
6. Machinability Data Center, and Metcut Research Associates. (1980). *Machining Data Handbook*. Machinability Data Center, Metcut Research Associates, Cincinnati, OH.
7. Narita, H., Desmira, N., and Fujimoto, H. (2008). Environmental burden analysis for machining operation using LCA method. In *Proceedings of 41st CIRP Conference on Manufacturing Systems* (pp. 65–68), Tokyo, Japan.
8. Haapala, K., Rivera, J. L., and Sutherland, J. (2009). Reducing environmental impacts of steel product manufacturing. *Transactions of the North American Manufacturing Research Institution of the SME* 37: 419–426.
9. Hondo, H. (2005). Life cycle GHG emission analysis of power generation systems: Japanese case. *Energy* 30 (11–12): 2042–2056.
10. Kurtz, M. (2007). *Environmental Conscious Manufacturing*. Wiley, Hoboken, NJ.

Bibliography

Abdalla, H. S. and Patel, S. (2006). The performance and oxidation stability of sustainable metalworking fluid derived from vegetable extracts. *Proceedings of the Institution of Mechanical Engineers, Part B: Journal of Engineering Manufacture* 220: 2027–2040.

Abou-El-Hossein, K. A. (2008). Cutting fluid efficiency in end milling of AISI 304 stainless steel. *Industrial Lubrication and Tribology* 60 (3): 115–120.

Alves, S. M., and Oliveira, J. F. G. (2008). Vegetable based cutting fluid-an environmental alternative to grinding process. In *15th CIRP International Conference on Life Cycle Engineering* (pp. 664–668), Sydney.

Ashby, M. F. (2009). *Materials and the Environment: Eco-informed Material Choice.* (pp. 112, 114, 284–287, 290–291). Butterworth-Heinemann, Burlington, MA.

Baniszewski, B. (2005). An environmental impact analysis of grinding. S.B. Thesis, Department of Mechanical Engineering, MIT, Cambridge, MA.

Batzer, S. A. and Sutherland, J. W. (1998). The dry cure for coolant ills. *Cutting Tool Engineering* 50 (4): 34–44.
Cetin, M. H., Ozcelik, B., Kuram, E. and Demirbas, E. (2011). Evaluation of vegetable based cutting fluids with extreme pressure and cutting parameters in turning of AISI 304L by Taguchi method. *Journal of Clean Production* 19 (17–18): 2049–2056.
Dahmus, J. B. and Gutowski, T. G. (2004). An environmental analysis of machining. In *Proceedings of ASME International Mechanical Engineering Congress and RD&D Expo*, Anaheim, CA.
Dittrich, M. A., Gutowski, T. G., Cao J., et al. (2012). Exergy analysis of incremental sheet forming. *Production Engineering* 6 (2): 169–177.
Fox, N. J. and Stachowiak, G. W. (2007). Vegetable oil-based lubricants—A review of oxidation. *Tribology International* 40 (7): 1035–1046.
Fthenakis, V. M. and Kim, H. C. (2007). Greenhouse-gas emissions from solar electric- and nuclear power: A life-cycle study. *Energy Policy* 35 (4): 2549–2557.
Gregory, A. K. and Menerey, D. (1994). Sustainable development by design: Review of life cycle design and related approaches. *Journal of the Air and Waste Management Association* 44 (5): 645–668, CSS Publication # CSS94-01.
Groover, M. P. (2012). *Fundamentals of Modern Manufacturing: Materials, Processes, and Systems*, 5th edn. Wiley, Hoboken, NJ.
Guo, Y., Loenders, J., Duflou, J., and Lauwers, B. (2012). Optimization of energy consumption and surface quality in finish turning. In *5th CIRP Conference on High Performance Cutting*, Zürich, Switzerland.
Gutowski, T., Dahmus, J., and Thiriez, A. (2006). Electrical energy requirements for manufacturing processes. In *Proceedings of 13th CIRP International Conference on Life Cycle Engineering* (pp. 623–627), Leuven, Belgium.
Gutowski, T., Jeffrey D., and Thiriez, A. (2006). Electrical energy requirements for manufacturing processes. In *13th CIRP International Conference on Life Cycle Engineering*, Leuven, Belgium, May 31–June 2.
Haan, D. M., Batzer, S. A., Olson, W. W., and Sutherland, J. W. (1997). An experimental study of cutting fluid effects in drilling. *Journal of Materials Processing Technology* 71 (2): 305–313.
Hirohisa, N. (2012). *Environmental Burden Analyzer for Machine Tool Operations and Its Application*. Intech, Rijeka. www.intechopen.com/books/manufacturing-systems.
Hoffman, E. G. (1984). *Fundamentals of Tool Design*, 2nd edn. Society of Manufacturing Engineers, Dearborn, MI.
Hong, S. Y. and Broomer, M. (2000). Economical and ecological cryogenic machining of AISI 304 austenitic stainless steel. *Clean Products and Processes* 2: 157–166.
Jan, E. and Bras, B. (1998). ISO 14000 and activity based life cycle assessment in environmentally conscious design and manufacturing: A comparison. In *1998 ASME Design Engineering Technical Conference*. American Society of Mechanical Engineers, Atlanta, GA.
Kalpakjian, S. and Schmids, R. (2010). *Manufacturing Processes for Engineering Materials*, 6th edn. Pearson Prentice Hall, Upper Saddle River, NJ.
Khan, M. M. A. and Dhar, N. R. (2006). Performance evaluation of minimum quantity lubrication by vegetable oil in terms of cutting force, cutting zone temperature, tool wear, job dimension and surface finish in turning AISI-1060 steel. *Journal of Zhejiang University Science A*7 (11): 1790–1799.
Kordonowy, D. N. (2001). A power assessment of machining tools. B.S. Thesis, MIT, Department of Mechanical Engineering, Cambridge, MA.
Krahenbuhl, U. (2002). Vegetable oil-based coolants improve cutting performance (cutting fluids). In *Tooling & Production*. Nelson.
Lange K. (1985). *Handbook of Metal Forming*. McGraw-Hill, New York.
Life Cycle Assessment (LCA): Principles and Practice. (2006). EPA/6000R-06/060.
Niggeschmidt, S., Helu, M., Diaz, N., Behmann, B., Lanza, G., and Dornfeld, D. (2010). Integrating green and sustainability aspects into life cycle performance evaluation. In *Proceedings of 16th CIRP International Conference on Life Cycle Engineering*, Windsor, Canada.
Pettersson, A. (2007). High-performance base fluids for environmentally adapted lubricants. *Tribology International* 40 (4): 638–645.
Rangarajan, A. and Dornfeld, D. (2004). Efficient tool paths and part orientation for face milling. *CIRP Annals* 53 (1): 73–76.
Rao, D. N. and Srikant, R. R. (2006). Influence of emulsifier content on cutting fluid properties. *Proceedings of the Institution of Mechanical Engineers, Part B: Journal of Engineering Manufacture* 220: 1803–1806.

Santos, J. P., Oliveira, M., Almeida, F. G., et al. (2011). Improving the environmental performance of machine-tools: Influence of technology and throughput on the electrical energy consumption of a press-brake. *Journal of Cleaner Production* 19: 356–364.

Schey, J. A. (2000). *Introduction to Manufacturing Processes*, 3rd edn. McGraw Hill, New York.

Shokrani, A., Dhokia, V. and Newmanm S. T. (2012). Environmentally conscious machining of difficult-to-machine materials with regard to cutting fluids. *International Journal of Machine Tools and Manufacture* 57: 83–101.

Srikant, R. R., Rao, D. N. and Rao, P. N. (2009). Influence of emulsifier content in cutting fluids on cutting forces, cutting temperatures, tool wear, and surface roughness. *Proceedings of the Institution of Mechanical Engineers, Part J: Journal of Engineering Tribology* 223: 203–209.

Szargut, J., Morris, D. R., and Steward, F. R. (1988). *Exergy Analysis of Thermal, Chemical and Metallurgical Processes*. Hemisphere, Washington, DC.

Taniguchi, M., Kakinuma, Y., Aoyama, T., and Inasaki, I. (2006). Influences of downsized design for machine tools on the environmental impact. In *Proceedings of MTTRF Annual Meeting* (pp. 1–4).

Tlusty, G. (2011). *Manufacturing Processes and Equipment.* Prentice Hall, Englewood Cliffs, NJ.

Tönissen, S. (2009). Power demand of precision machine tools. MS Thesis, Department of Mechanical Engineering, University of California, Berkeley, CA.

22

Foundry Processes

22.1 Introduction to Solidification Manufacturing Processes

Some industries use manufacturing processes in which the starting material is in liquid form and a product is created through a solidification process in a mold. Casting is one such important process that has been used for more than 2500 years.

Foundry work involves casting molten metal into a mold. Casting can be done manually by static casting or automatically by injection, dye, continuous, spin or spray casting [1–4]. Casting is the forming of metal or plastic parts by introducing a liquid material to a suitably shaped cavity (mold), allowing it to solidify, and then removing it from the mold. Further processing is usually required. As a production process, it is carried out in foundry factories by foundrymen. The casting processes are used for creating a complex geometry of parts that would be difficult to manufacture by metal-removal processes. The term *net shape* refers to manufacturing that requires no further operations to obtain the required shape and geometry. Casting is supposed to be one such process. Parts with a very large geometry and weight can easily be manufactured by casting. The only requirement is that the metal should be heated to liquid state and poured into a mold. A typical process includes preparing a mold for casting, melting and pouring the metal into the mold, and removing and finishing the casting. Foundries and *die-casting* operations specialize in melting a wide variety of metals and alloys and casting them into the desired shapes. Often, these facilities also engage in heat treating to change the metallurgical and mechanical properties of a cast part. Castings are used in 90% of all finished manufactured parts. The metal casting sector includes establishments that pour molten ferrous metals (iron and steel) or nonferrous metals under high pressure into molds to manufacture castings. A ferrous metal casting includes those castings made with gray iron, ductile iron, malleable iron, and steel. Each type of iron contains different elements that affect its characteristics. Nonferrous castings are predominantly aluminum, but they might also be brass, bronze, zinc, magnesium, and titanium. More than 90% of all manufactured goods in the United States contain cast metal components. These include engine blocks, transmission housings, and suspension parts for cars and trucks; the undercarriages of farm and construction equipment; and pipes and valves for plumbing fixtures and boilers.

22.1.1 Environmental Issues and Efficiency of Metal Casting

Working in a foundry presents a range of work health and safety risks, including

- Explosions and burns from molten metal and other hot materials
- Respiratory effects from exposure to gases, vapors, fumes, and dusts
- Skin effects from contact with corrosive or sensitizing chemicals
- Eye damage from light radiation, metal fragments, dust, and chemical splashes
- Heat stress, heatstroke, and fatigue from hot working conditions
- Slips, trips, and falls
- Joint and muscle sprains and strains
- Physical injuries from machinery and equipment, for example, by entanglement or crushing
- Health effects from machinery and equipment, for example, caused by vibration and noise [5]

In Minnesota, the metal casting industry uses 2,700,000 MM Btu and generates volatile organic compounds (VOCs), other organics, air emissions, and solid and *hazardous waste*. Pollution prevention and energy efficiency opportunities exist within many metal casting facilities and can help reduce the environmental impact and costs. Specific opportunities depend largely on the type of operation, the metal and nonmetal inputs, the processes, and the equipment. Foundry work can involve working with *lead*. A lead process in a foundry means foundry processes involving:

- Melting or casting lead alloys containing more than 1% by weight of lead metal in which the temperature of the molten material exceeds 450°C.
- Dry machine grinding, dicing, buffing, or cutting using power tools lead alloys containing more than 1% by weight of lead metal.
- Lead risk work means work that is carried out in a lead process that is likely to cause the blood lead level of a worker carrying out the work to exceed 10 μg/dL (0.48 μmol/L) in a female of reproductive age and 30 μg/dL (1.45 μmol/L) in any other case.

Exposure to lead must be kept to the lowest level practicable and control measures should be continually reviewed to ensure that they are effective at minimizing exposure as far as is reasonably practicable. Lead is a poison that accumulates in the body and can never be fully removed. It enters the body by ingestion or inhalation. The symptoms of lead exposure include headaches, tiredness, irritability, constipation, nausea, stomach pains, and anemia. The signs and symptoms of toxicity will not appear unless the exposure is very high. Other illnesses can trigger the release of lead that has accumulated in the bones into the blood. In excess, it leads to kidney, nerve, and brain damage. Lead is extremely hazardous to the unborn child and to breastfeeding infants and it is recommended that women intending to become pregnant be excluded from lead-exposed work.

22.1.2 Latest Environmental Statistics Provided by the Environmental Protection Agency for Metal Casting Sector

Energy use: 157 trillion Btu

Emissions of criteria air pollutants: 75,000 tons

Releases of chemicals reported to the Toxics Release Inventory (TRI): 49.6 million lb; air emissions: 3.8 million lb; water discharges: 68,500 lb; waste disposals: 45.7 million lb; recycling, energy recovery, or treatment: 127.5 million lb

Hazardous waste generated: 30,000 tons

22.1.3 Energy Use

In 2002, the metal casting sector consumed 157 trillion Btu. The major furnaces that casting operations use are cupola (used primarily for ferrous metal casting), electric, reverberators, and crucible furnaces. The heating and melting of these *various metals* consume large amounts of energy, accounting for 72% of the sector's total energy use, according to the U.S. Department of Energy (DOE) estimates. Mold and core making account for 7% of the sector's energy use, and finishing accounts for 6%. During molding, foundries use energy for transporting materials, mechanical mixing, and making mold cores.

The foundry sector is heavily dependent on natural gas and purchased electricity, making up 48% and 34%, respectively, of the sector's fuel inputs for energy in 2002. Coke, the primary fuel for cupola furnaces, was the third largest energy source at 15%. A DOE report on the sector identified several energy-saving opportunities. Casting operations using iron induction can automate the furnace temperature and power controls to prevent overshooting temperature settings, and can minimize the time that the lid is open while melting or holding iron. Operations using a cupola can dehumidify blast air to reduce coke consumption and can cover coke storage areas to prevent water from being introduced into the furnace.

Air emissions are a primary environmental concern in the sector, and include criteria air pollutants (CAPs), greenhouse gases (GHGs), and a number of chemicals reported to the Environmental Protection Agency's (EPA) TRI. In general, the "toxic chemicals" tracked by the TRI are found in the raw materials and fuels that are used. CAPs and GHGs are also generated from the on-site combustion of fuels. The TRI of toxic chemicals includes all but six of the hazardous air pollutants (HAPs) that are regulated under the Clean Air Act (CAA). Air pollution has a significant environmental impact, particularly pollution from ferrous metal casting. Because aluminum, which is used in nonferrous operations, melts at a lower temperature than ferrous metals, nonferrous casting usually results in lower air emissions. The sector's air emissions result from the various operations in a facility, including metal melting, mold making, handling foundry sand, and die-casting. The majority of metal emissions come from the metal melting operations, while most organic emissions are from handling the binder that holds sand together to produce the cores and molds. Once the binder is combined with the sand, there may be additional organic, particulate, and carbon monoxide (CO) emissions from pouring the molten metal into the casting and from breaking apart the cast. The handling of foundry sand results primarily in particulate emissions.

22.1.4 Air Emission Reported to Toxics Release Inventory

considering the toxicity of air emissions, the EPA's Risk-Screening Environmental Indicators (RSEI) model assigns every TRI chemical a relative toxicity weight, then multiplies the pounds of media-specific releases (e.g., pounds of mercury released to air) by a chemical-specific toxicity weight to calculate a relative toxicity score. RSEI methodological considerations are discussed in greater detail in the data guide, which explains the underlying assumptions and important limitations of the RSEI.

Summing the toxicity scores for all of the air emissions reported to the TRI by the sector produces the trend illustrated in the following graph. The sector's toxicity score declined 82% from 1996 to 2005 [10]. Three chemicals—manganese, chromium, and diisocyanates—accounted for 81% of the sector's total toxicity score. Manganese and chromium emissions result from melting; furnaces that melt metal emit dust, metallic particles, and metal oxide fumes, along with the products of combusted fuel. Diisocyanates, associated with binding materials, are emitted as a result of exposure to air. The apparent spike in 1996 was exacerbated by diisocyanates emissions reported by one facility, which, in subsequent years, reported no diisocyanates emissions. The sector's reported emissions of all three chemicals have declined since 1996. During the same period, regulations led to the increased use of pollution control equipment. The technology related to the binding process has also improved; changes in binder ingredients and processing, for example, have promoted reductions in VOC emissions.

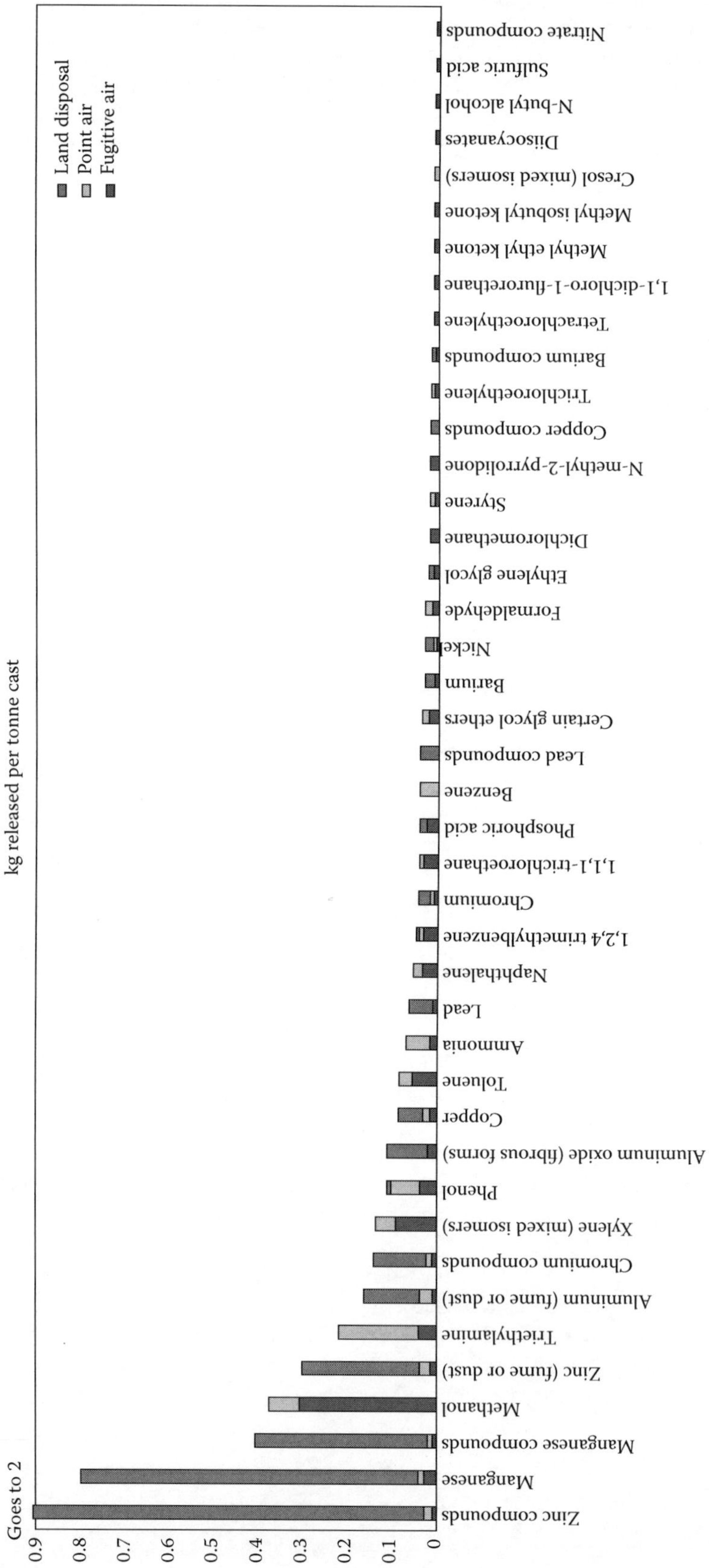
kg released per tonne cast
0
0.1
0.2
0.3
0.4
0.5
0.6
0.7
0.8
0.9
Goes to 2
Land disposal
Point air
Fugitive air
Zinc compounds
Manganese
Manganese compounds
Methanol
Zinc (fume or dust)
Triethylamine
Aluminum (fume or dust)
Chromium compounds
Xylene (mixed isomers)
Phenol
Aluminum oxide (fibrous forms)
Copper
Toluene
Ammonia
Lead
Naphthalene
1,2,4 trimethylbenzene
Chromium
1,1,1-trichloroethane
Phosphoric acid
Benzene
Lead compounds
Certain glycol ethers
Barium
Nickel
Formaldehyde
Ethylene glycol
Dichloromethane
Styrene
N-methyl-2-pyrrolidone
Copper compounds
Trichloroethylene
Barium compounds
Tetrachloroethylene
1,1-dichloro-1-flurorethane
Methyl ethyl ketone
Methyl isobutyl ketone
Cresol (mixed isomers)
Diisocyanates
N-butyl alcohol
Sulfuric acid
Nitrate compounds

22.2 Sand Casting

In sand casting, the mold is made in a "molding box" using special sand and a wooden "pattern." Holes are produced by inserting previously made "cores" of baked sand. Molten metal is poured into runners until it appears in the risers. The casting is cleaned by chipping, grinding, and sandblasting. Practically any metal can be cast (Figures 22.1 and 22.2).

22.2.1 Environmentally Conscious Sand Casting

Environmental issues are related to every aspect of casting, including energy use, water usage, toxic air emission, and sand itself (Figure 22.3). The foundry process involves sand collection and preparation, mold making, metals extraction, metal preparation and melting, and finally casting. At each and every step, the process requires energy and water, and emits toxic and hazardous gases. The environmental impact of casting processes is great and can be very harmful to air, water, and soil. Some data on cast iron is provided by Gutowski [7, 8] as follows.

Cast Iron Example (Cupola)

Stage	MJ/kg
Mold preparation	3.0
Metal preparation	6.7
Casting	0.7
Finishing	1.2
Total at foundry	11.6
Electricity losses	6.0
Total	~18

The typical components produced by sand casting are automobile engine blocks and ship propellers (Figure 22.4).

Different types of sand are used in sand casting:

- Petro-bond: This is a mixture of quality sand and oil or synthetic resin.
- Green sand: A mixture of sand, clay, water, and sometimes other additives. It is called green sand because it is reusable. The right amount of water has to be added to prevent porosity.

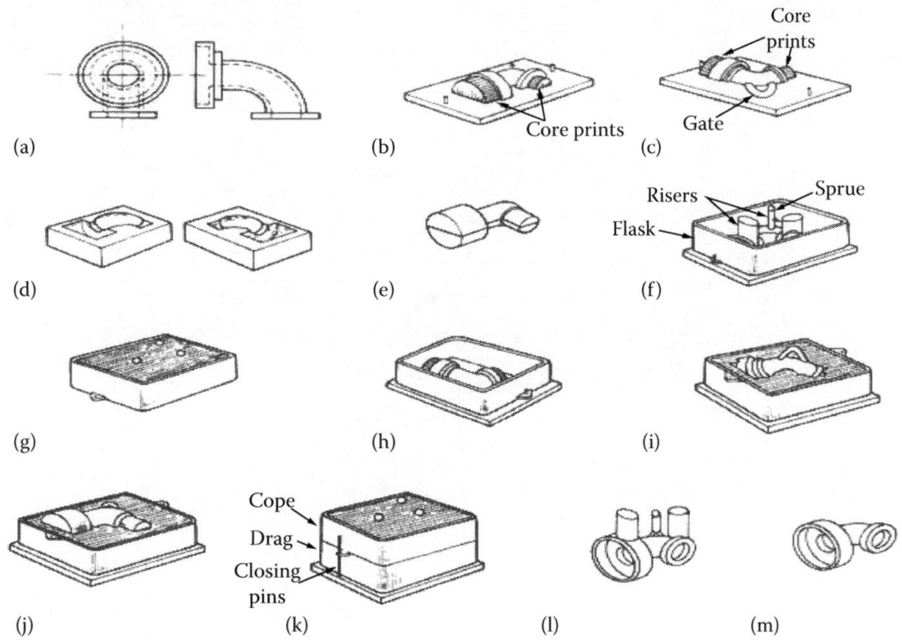

Figure 22.1

Typical sand casting products: (a) mechanical drawing of part, (b) cope pattern plate, (c) drag pattern plate, (d) core boxes, (e) core halves pasted together, (f) cope ready for sand, (g) cope after ramming with sand and removing pattern, sprue, and risers, (h) drag ready for sand, (i) drag after removing pattern, (j) drag with core set in place, (k) cope and drag assembled ready for pouring, (l) casting as removed from mold: heat treated, and (m) casting ready for shipment.

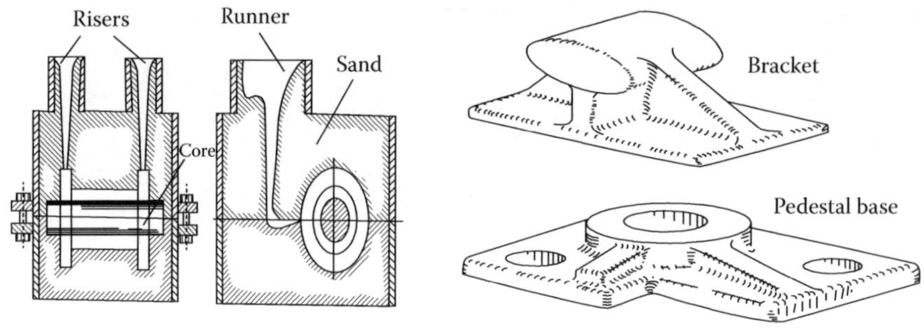

Figure 22.2

Sand casting mold and products.

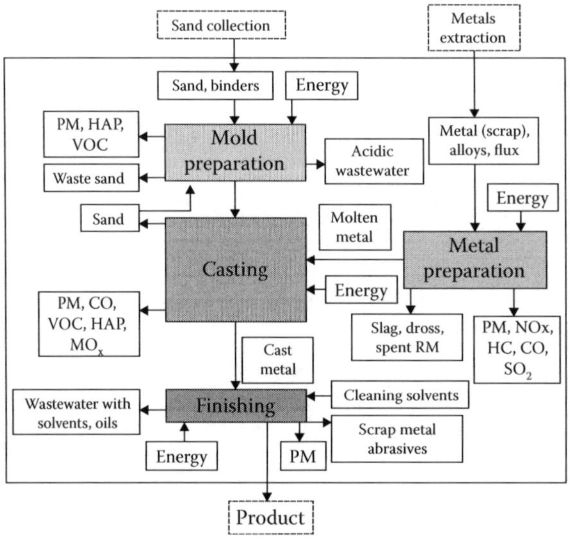

Figure 22.3

Environmental impact aspects of casting.

Figure 22.4

Sand casting of crankshaft. A crankshaft for an auto engine, which has been produced by sand casting.

- Sand mixed with water glass ($NaO.nSiO_2.mH_2O$) can be hardened with CO_2 gas through the chemical reaction:

$$Na_2O.nSiO_2.(mn + x)H_2O + CO_2 = Na_2CO_3.xH_2O + n(SiO_2.mH_2O)$$

This transforms the sand into a solid mold, which can be used after the cope and drag are removed. A parting powder similar to talcum powder is put on the pattern to make it easier to remove. More complicated castings can be made by producing a pattern from polystyrene foam, which is left in place when the molten metal is added. When the metal is poured into the mold, the heat vaporizes the foam a short distance away from the metal surface, allowing the metal to fill the mold cavity.

- Advantages
    - Low capital investment means that short production runs are viable.
    - Use of sand cores allows fairly complex shapes to be cast.
    - Large components can be produced.
- Disadvantages
    - The process has a high unit cost, as it is labor intensive and time consuming.
    - The sand mold leaves a poor surface finish, which often requires further processing.
    - Cannot make thin sections.

22.2.2 Shell Molding

The basic process of permanent mold casting is similar to that for sand casting, except that a permanent metallic mold is used. This mold is made of two parts, containing the necessary gates and risers to ensure the correct flow of metal. The metal molds give a better surface finish than sand casting, and selective cooling rates give control over the scale and morphology of the microstructure formed. This is a form of sand casting that is done using very fine sand mixed with synthetic resin. The pattern is made of machined and polished iron. The sand mixture is blown into a box containing the pattern, which is heated to produce a hard, thin (6–10 mm) mold that is split and removed from the pattern and then glued together. It is a high-speed process, producing highly accurate castings; however, the energy consumed is greater than that for sand casting.

22.3 Investment Casting (Lost Wax Casting)

Wax patterns are made from a permanent metal mold. The wax patterns are coated with ceramic slurry, which is hardened and baked so that the wax is melted out. The cavity is filled with a molten metal to give a precision casting. Any metal can be cast using this process. Investment casting is used to make precision parts with a good surface finish. It is used to make turbine blades from a single crystal Ni-based superalloy. As the molds are made from ceramic, metals with high melting points can be cast in this way.

- A pattern of the required component is formed out of wax usually through injection molding into a metal mold. A runner and a gate are included within the wax pattern. Preformed ceramic cores can also be added, so that hollow castings can be made.
- Many wax patterns are then connected together to form a tree, and this is then dipped into a ceramic slurry, which sets through drying. Different layers of ceramic are created around the wax patterns, increasing in coarseness.
- The wax patterns are then melted out, ready for the molten metal to be poured in to create the castings.
- The ceramic mold is broken to reveal the castings, and any ceramic cores added at the start can be chemically etched out to leave a hollow structure.

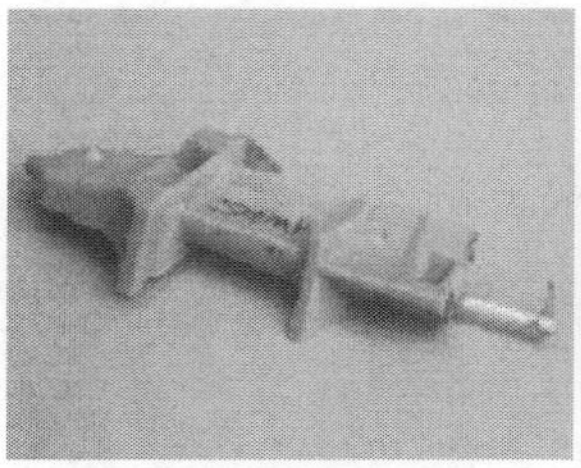
Investment wax

Investment mold

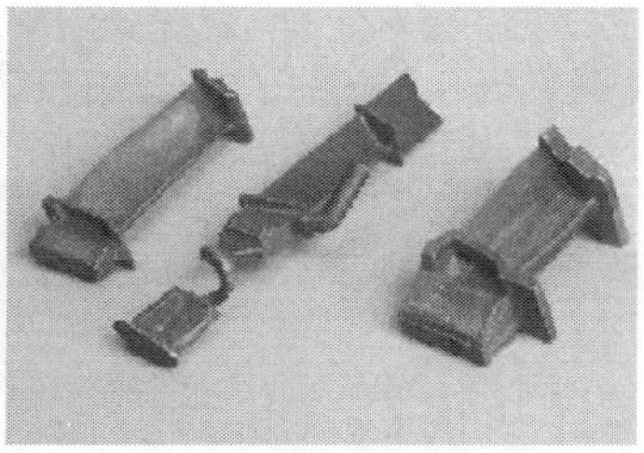
Investment blades

- Advantages
  - Metals with a high melting temperature can be cast due to the use of ceramic molds.
  - Complex shapes can be formed by using ceramic liners in the original wax patterns.
  - Good surface finish can be obtained using fine ceramic material.
- Disadvantages
  - Expensive as the mold cannot be reused.
  - Time consuming (the drying time for ceramic is roughly 24 h) (Figure 22.5).

22.4 Die-Casting

Die-casting is an automated process that forms castings under high pressure in a metal mold. The typical items produced by die-casting are toys and small precision parts such as sprockets and gears. Aluminum, copper, and zinc alloys are commonly used in die-casting; however, ferrous alloys are harder to cast in this way as the molds are made from hardened steel.

There are various methods of die-casting:

- Gravity fed: Similar to permanent mold casting, the material is poured into the top of the mold and is forced downward by gravity.

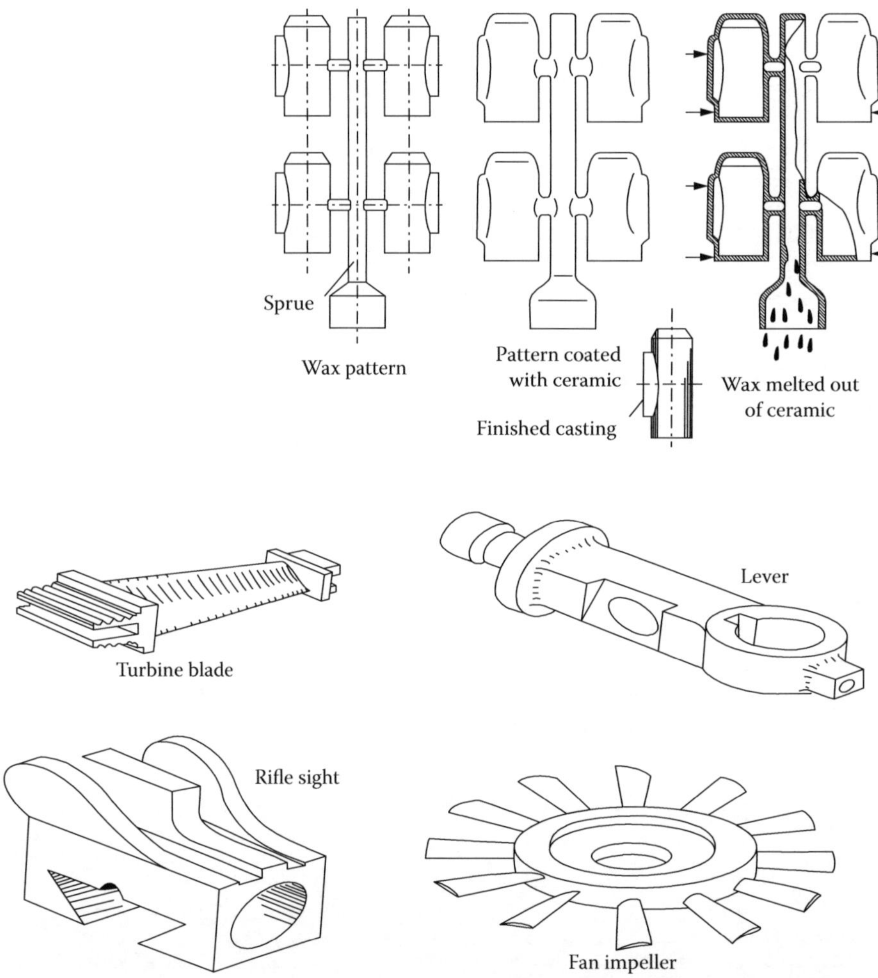

Figure 22.5

Types of investment casting and products.

- Pressure: The metal is forced into the mold under pressure.
- Cold chamber: Molten metal is poured into the system and then forced into the mold.
- Hot chamber: The metal is heated within the system and then forced from the crucible into the die.
- Squeeze: Similar to injection molding, a given amount of material is forced into the mold.

The mold is formed from pieces of steel that are dowelled together. Molten metal is fed into the mold by gravity or pressure and, when solid, it is ejected by pins. Aluminum, copper, manganese, and zinc alloy are suitable for casting by this method (Figures 22.6 and 22.7).

- Advantages
    - Very low unit cost.
    - High definition surface finish.
    - Excellent dimensional accuracy.
    - Cool metal mold gives fast solidification, leading to a fine-grain structure.
    - Can produce thin sections.

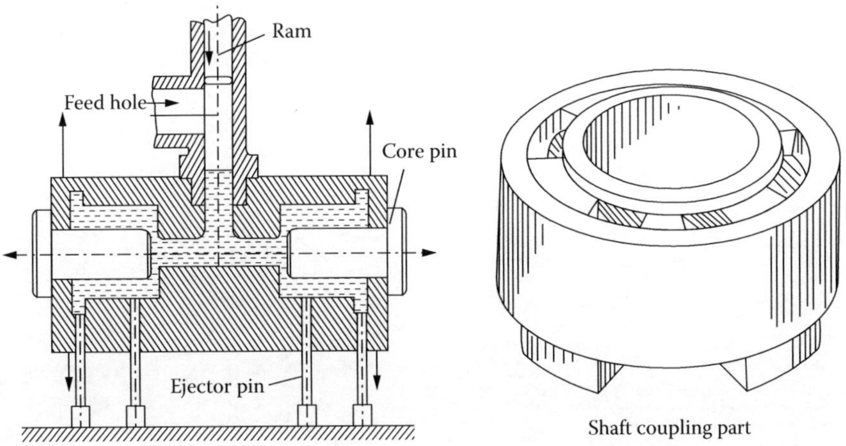

Figure 22.6

Die-casting diagram.

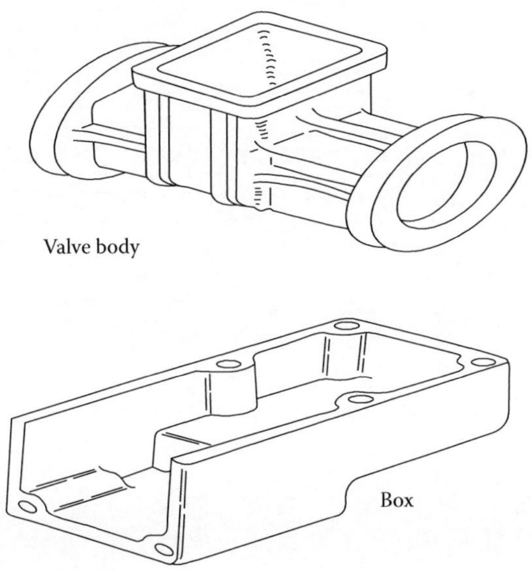

Figure 22.7

Die-casting products.

- Disadvantages
  - A large capital investment is required to set up a die-casting process.
  - It is difficult to control the microstructure of the solid.
  - The alloys used must have a low melting point, often at the expense of other properties, such as strength and stiffness.
  - Cannot be used for complex shapes, as the casting cannot be ejected from the mold.
  - Cannot be used for large castings.

22.5 Centrifugal Casting

Cylindrical or circular components such as piston rings, cylinder liners, pipes, and so on, may be cast in a rotating mold. Centrifugal pressure gives a fine-grain casting. Any metal may be cast using this process (Figures 22.8 and 22.9).

22.6 Continuous Casting

Continuous casting is used to produce very large quantities of metals in simple shapes, by casting and rolling the metal continuously in one process. The slabs are then generally further processed by methods such as rolling and extrusion or another casting method to make more useful items.

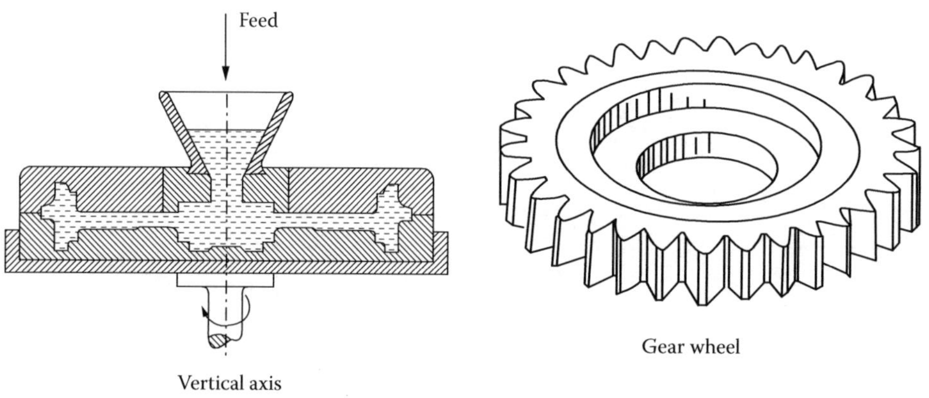

Figure 22.8

Centrifugal casting.

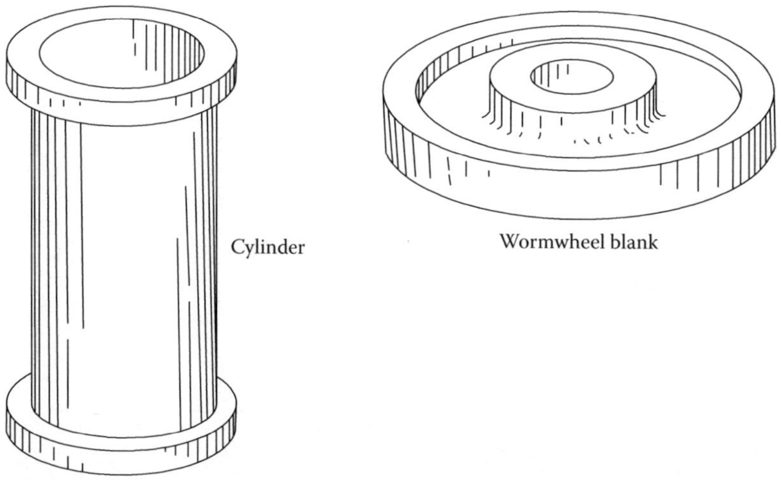

Figure 22.9

Centrifugal casting products.

- Advantages
  - High yield of casting for a given volume of liquid, mainly thanks to the lack of a contraction pipe.
  - Good surface finish.
  - Extremely low unit cost due to the very high volume of metal that can be cast.
- Disadvantages
  - A large capital investment is required to set up the process.
  - Only simple shapes can be cast and they must have a constant cross section.

22.6.1 Heat Transfer in Casting

Heat transfer during solidification is steady state. The thermal conductivity of material used for a mold plays a vital part in the heat transfer and solidification of molten metal. The thermal conductivity of some materials used in casting and molding is given:

Thermal conductivity k of various materials for parts and molds (W/m°K)
Copper: 394; Sand: 0.61
Aluminum: 222; poly(methyl methacrylate) (PMMA): 0.20
Iron: 29; polyvinyl chloride (PVC): 0.16

22.7 Solidification Time Estimation for Sand Casting

The solidification time is the time required for the casting to solidify after poring. Obviously, this time will depend on the size and the shape of the casting. *Chvorinov's rule* is a mathematical relationship that was first expressed by Nicolas Chvorinov in 1940, which relates the solidification time for a simple casting to the volume and surface area of the casting. In simple terms, the rule establishes that under otherwise identical conditions, a casting with a large surface area and a small volume will cool more rapidly than a casting with a small surface area and a large volume. The relationship can be written as [2]

$$t = B\left(\frac{V}{A}\right)^n$$

where:

- t is the solidification time
- V is the volume of the casting
- A is the surface area of the casting that contacts the mold
- n is a constant
- B is the mold constant

The mold constant B depends on the properties of the metal, such as density, heat capacity, heat of fusion, and superheat, and the properties of the mold, such as initial temperature, density, thermal conductivity, heat capacity, and wall thickness. The SI units of the mold constant B are seconds per meter squared. According to Askeland and Fulay [9], the constant n is usually 2; however, Degarmo [9] claims it is between 1.5 and 2. The mold constant of Chvorinov's rule, B, can be calculated using the following formula:

$$B = \left[\frac{\rho_m L}{(T_m - T_0)}\right]^2 \left[\frac{\pi}{4kpc}\right]\left[1 + \frac{(c_m \Delta T_s)^2}{L}\right]$$

where:

- T_m = melting or freezing temperature of the liquid (K)
- T_0 = initial temperature of the mold (K)
- ΔT_s = $T_{pour} - T_m$ = superheat (K)
- L = latent heat of fusion (J/kg)
- k = thermal conductivity of the mold (W/m/K)
- ρ = density of the mold (kg/m³)

c = specific heat of the mold (J/kg/K)
ρ_m = density of the metal (kg/m^3)
c_m = specific heat of the metal (J/kg/K)

It is most useful in determining if a *riser* will solidify before the casting, because if the riser solidifies first then defects such as shrinkage or porosity can form.

Usually Chvorinov's empirical rule is used for solidification time estimation. This rule is presented as follows:

$$t_s = B\left(\frac{V}{A}\right)^2$$

where:

t_s = solidification time (min)
V = volume of casting (cm^3 [in.^{3}])
A = surface area of the casting (cm^2 [in.^{2}])
B = mold constant (min/cm^2 [min/in.^{2}])

B is determined experimentally for different mold materials. For aluminum and copper it is usually 0.2 s/mm^2. It can also be observed that to lower the solidification time (t_s), the (V/A) ratio must be lower or the surface area (A) should be proportionately higher than the volume (V).

Example 22.1

Determine the solidification time for casting a magnesium part of $10 \times 10 \times 3$ cm if the metal is poured at its melting point.

$$V = 100 \times 100 \times 30 = 3 \times 10^5 \text{ mm}^3$$

$$A = 2(100 \times 100) + 4(100 \times 30) = 32{,}000 \text{ mm}^2$$

$$t_s = 0.2(3 \times 10^5/32{,}000)^2 = 17.6 \text{ s}$$

References

1. Groover, M. P. (2012). *Fundamentals of Modern Manufacturing: Materials, Processes, and Systems*, 5th edn. Wiley, Hoboken, NJ.
2. Kalpakjian, S. and Schmids, R. (2010). *Manufacturing Processes for Engineering Materials*, 6th edn. Pearson Prentice Hall, Upper Saddle River, NJ.
3. Schey, J. A. (2000). *Introduction to Manufacturing Processes*, 3rd edn. McGraw Hill, New York.
4. Tlusty, G. (2011). *Manufacturing Processes and Equipment.* Prentice Hall, Englewood Cliffs, NJ.
5. U.S. EPA. (2008). Sector Strategies Performance Reports: Metal Casting and Iron and Steel Operations. 2008 Sector Performance Report.
6. EPA. (2006). Life Cycle Assessment (LCA): Principles and Practice. EPA/6000R-06/060.
7. Gutowski, T., Jeffrey, D., and Thiriez, A. (2006). Electrical energy requirements for manufacturing processes. In *13th CIRP International Conference on Life Cycle Engineering*, Leuven, Belgium, May 31–June 2.
8. Dahmus, J. and Gutowski, T. (2004). An environmental analysis of machining. In *ASME International Mechanical Engineering Congress and RD&D Expo*, Anaheim, CA.
9. Askeland, D. and Fulay, P. (2010). *The Science and Engineering of Materials*, 4th edn. Cengage Learning.
10. Black, J. T. and Kohser, R. A. (2011). *DeGarmo's Materials and Processes in Manufacturing*, 11th edn. Wiley.

Bibliography

Abdalla, H. S. and Patel, S. (2006). The performance and oxidation stability of sustainable metalworking fluid derived from vegetable extracts. *Proceedings of the Institution of Mechanical Engineers, Part B: Journal of Engineering Manufacture* 220: 2027–2040.

Abou-El-Hossein, K. A. (2008). Cutting fluid efficiency in end milling of AISI 304 stainless steel. *Industrial Lubrication and Tribology* 60 (3): 115–120.

Alves, S. M. and Oliveira, J. F. G. (2008). Vegetable based cutting fluid-an environmental alternative to grinding process. In *15th CIRP International Conference on Life Cycle Engineering* (pp. 664–668), Sydney.

Ashby, M. F. (2009). *Materials and the Environment: Eco-informed Material Choice* (pp. 112, 114, 284–287, 290–291). Butterworth-Heinemann, Burlington, MA.

Baniszewski, B. (2005). An environmental impact analysis of grinding. SB thesis, Department of Mechanical Engineering, MIT, Cambridge, MA.

Batzer, S. A. and Sutherland, J. W. (1998). The dry cure for coolant ills. *Cutting Tool Engineering* 50 (4): 34–44.

Cetin, M. H., Ozcelik, B., Kuram, E., and Demirbas, E. (2011). Evaluation of vegetable based cutting fluids with extreme pressure and cutting parameters in turning of AISI 304L by Taguchi method. *Journal of Clean Production* 19 (17–18): 2049–2056.

Dahmus, J. B. and Gutowski, T. G. (2004). An environmental analysis of machining. In *Proceedings of ASME International Mechanical Engineering Congress and RD&D Expo*, Anaheim, CA.

Dittrich, M. A., Gutowski, T. G., Cao, J., Roth, J. T., Xia, C., Kiridena, V., Ren, F., Henning, H. (2012). Exergy analysis of incremental sheet forming. *Production Engineering* 6 (2): 169–177.

Fox, N. J. and Stachowiak, G. W. (2007). Vegetable oil-based lubricants—A review of oxidation. *Tribology International* 40 (7): 1035–1046.

Fthenakis, V. M. and Kim, H. C. (2007). Greenhouse-gas emissions from solar electric- and nuclear power: A life-cycle study. *Energy Policy* 35 (4): 2549–2557.

Gregory, A. K. and Menerey, D. (1994). Sustainable development by design: Review of life cycle design and related approaches. *Journal of the Air and Waste Management Association* 44 (5): 645–668.

Guo, Y., Loenders, J., Duflou, J., and Lauwers, B. (2012). Optimization of energy consumption and surface quality in finish turning. In *5th CIRP Conference on High Performance Cutting*, Zürich, Switzerland.

Gutowski, T., Branham, M., Dahmus, J., Jones, A., Thiriez, A., and Sekulic, D. (2009). Thermodynamics analysis of resources used in manufacturing processes. *Environment Science and Technology* 43 (5): 1584–1590.

Gutowski, T., Dahmus, J., and Thiriez, A. (2006). Electrical energy requirements for manufacturing processes. In *Proceedings of 13th CIRP International Conference on Life Cycle Engineering* (pp. 623–627), Leuven, Belgium.

Gutowski, T., Jeffrey, D., and Thiriez, A. (2006). Electrical energy requirements for manufacturing processes. In *13th CIRP International Conference on Life Cycle Engineering*, Leuven, Belgium, May 31–June 2.

Haan, D. M., Batzer, S. A., Olson, W. W., and Sutherland, J. W. (1997). An experimental study of cutting fluid effects in drilling. *Journal of Materials Processing Technology* 71 (2): 305–313.

Haapala, K., Rivera, J. L., and Sutherland, J. W. (2009). Reducing environmental impacts of steel product manufacturing. *Transactions of the North American Manufacturing Research Institution of the SME* 37: 419–426.

Hoffman, E. G. (1984). *Fundamentals of Tool Design*, 2nd edn. Society of Manufacturing Engineers, Dearborn, MI.

Hondo, H. (2005). Life cycle GHG emission analysis of power generation systems: Japanese case. *Energy* 30 (11–12): 2042–2056.

Hong, S. Y. and Broomer, M. (2000). Economical and ecological cryogenic machining of AISI 304 austenitic stainless steel. *Clean Products and Processes* 2: 157–166.

Jan, E. and Bras, B. (1998). ISO 14000 and activity based life cycle assessment in environmentally conscious design and manufacturing: A comparison. In *1998 ASME Design Engineering Technical Conference.* American Society of Mechanical Engineers, Atlanta, GA.

Johnson, W. (1956). The pressure for the cold extrusion of lubricated rod through square dies of moderate reduction at slow speeds. *Journal of the Institute of Metals* 85.

Khan, M. M. A. and Dhar, N. R. (2006). Performance evaluation of minimum quantity lubrication by vegetable oil in terms of cutting force, cutting zone temperature, tool wear, job dimension and surface finish in turning AISI-1060 steel. *Journal of Zhejiang University Science A* 7 (11): 1790–1799.

Kordonowy, D. N. (2001). A power assessment of machining tools. BS thesis, Department of Mechanical Engineering, MIT, Cambridge, MA.

Kurtz, M. (2007). *Environmental Conscious Manufacturing*. Wiley, Hoboken, NJ.

Lange, K. (1985). *Handbook of Metal Forming*. McGraw-Hill, New York.

Machinability Data Center and Metcut Research Associates. (1980). *Machining Data Handbook*. Machinability Data Center, Metcut Research Associates, Cincinnati, OH.

Narita, H., Desmira, N., and Fujimoto, H. (2008). Environmental burden analysis for machining operation using LCA method. In *Proceedings of 41st CIRP Conference on Manufacturing Systems* (pp. 65–68), Tokyo, Japan.

Niggeschmidt, S., Helu, M., Diaz, N., Behmann, B., Lanza, G., and Dornfeld, D. (2010). Integrating green and sustainability aspects into life cycle performance evaluation. In *Proceedings of 16th CIRP International Conference on Life Cycle Engineering*.

Pettersson, A. (2007). High-performance base fluids for environmentally adapted lubricants. *Tribology International* 40 (4): 638–645.

Rangarajan, A. and Dornfeld, D. (2004). Efficient tool paths and part orientation for face milling. *CIRP Annals* 53 (1): 73–76.

Rao, D. N. and Srikant, R. R. (2006). Influence of emulsifier content on cutting fluid properties. *Proceedings of the Institution of Mechanical Engineers, Part B: Journal of Engineering Manufacture* 220: 1803–1806.

Santos, J. P., Oliveira, M., and Almeida, F. G. et al. (2011). Improving the environmental performance of machine-tools: Influence of technology and throughput on the electrical energy consumption of a press-brake. *Journal of Cleaner Production* 19: 356–364.

Shokrani, A., Dhokia, V., and Newman, S. T. (2012). Environmentally conscious machining of difficult-to-machine materials with regard to cutting fluids. *International Journal of Machine Tools and Manufacture* 57: 83–101.

Srikant, R. R., Rao, D. N., and Rao, P. N. (2009). Influence of emulsifier content in cutting fluids on cutting forces, cutting temperatures, tool wear, and surface roughness. *Proceedings of the Institution of Mechanical Engineers, Part J: Journal of Engineering Tribology* 223: 203–209.

Szargut, J., Morris, D. R., and Steward, F. R. (1988). *Exergy Analysis of Thermal, Chemical and Metallurgical Processes*. Hemisphere, Washington, DC.

Taniguchi, M., Kakinuma, Y., Aoyama, T., and Inasaki, I. (2006). Influences of downsized design for machine tools on the environmental impact. In *Proceedings of MTTRF Annual Meeting* (pp. 1–4).

Tönissen, S. (2009). Power demand of precision machine tools. MS thesis, Department of Mechanical Engineering, University of California, Berkeley, CA.

23

Metal-Forming Processes

23.1 Introduction to Forming Processes

The metal-forming sector produces the basic metal components that form the building blocks for other manufacturing industries. This sector can be divided into three distinct subsectors:

Forgings: This involves production of high-strength parts by pressing and squeezing the metal at high pressure. Forgings are used in a huge variety of safety critical and demanding engineering applications such as automotive transmission shafts and jet-engine blades.

Sheet metal: This involves production of components by pressing and cutting flat, thin pieces of metal. Various fabrications, ductwork, and automotive body parts are examples of sheet metal products.

Fasteners: This involves production of fasteners such as bolts, and is similar in some respects to forging.

The components manufactured by the metal-forming industry usually require some form of finishing. Finishing processes include heat treatment, welding, surface cleaning, and coating. Finishing may be carried out on the same site as the metal-forming process or subcontracted out to another site.

23.2 Forging

Forging is a manufacturing process involving the shaping of metal using localized compressive forces. The blows are delivered with a hammer (often a power hammer) or a die. Forging is often classified according to the temperature at which it is performed: cold forging (a type of cold working), warm forging, or hot forging (a type of hot working). Forming processes are classified into hot working and cold working. Hot working is the operation carried out at a temperature above recrystallization temperature. The benefits and drawbacks of cold and hot working can be seen in the table given below.

In general, hot working is used in the beginning and intermediate stages, while cold working is used at the end to capitalize on the surface finish and tolerance advantages.

Property	Cold Work	Hot Work
Surface finish	Good	Poor due to oxidation or decarburization
Dimensional tolerance	Good	Poor due to expansion during heating and contraction during cooling
Strain hardening	Yes	No
Energy for deformation	High	Low
Strength	Increased	Nearly the same
Ductility	Reduced	Nearly the same
Chemical inhomogeneities	Not reduced	Reduced due to diffusion at high temperature
Uniformity in properties	Anisotropy	Anisotropy

Forged parts can range in weight from less than a kilogram to hundreds of metric tons. They are widely used in mechanisms and machines where a component requires high strength; such forgings usually require further processing (such as machining) to achieve a finished part. Some examples include aluminum/steel frames of doors and windows, coins, springs, elevator doors, cables and wires, and almost all sheet metal. Forgings are high-strength parts produced when the metal is pressed, pounded, or squeezed at high pressure.

23.2.1 Preheating

In the hot forging process, the metal is preheated in a furnace to the desired temperature (up to 1260°C) before it is worked. Furnaces are usually gas fired, but may in some cases be electrically heated. Depending on the size of the piece to be forged and the required temperature, preheating can take many hours and represents the largest use of energy at hot forging sites.

The heated metal is worked to the desired shape using presses or hammers. Presses and hammers are typically driven by hydraulic, pneumatic, or mechanical flywheel systems and typically account for 20% of the electrical consumption at a forging site. The dies used on presses are often heated, using either electricity or gas. Die heating may be carried out *in situ* or in a separate die heater.

23.2.2 Heat Treatment

Once the metal has been worked to the desired shape, it undergoes heat treatment. This entails heating the piece in a gas or electric furnace to a specific temperature and time profile to impart the required properties, for example, softening, normalizing, stress relieving, and hardening. Products may undergo a number of heat treatments depending on the material and the required properties.

23.2.3 Quenching

Depending on the heat-treatment process employed, the product may be rapidly cooled in a controlled manner by quenching. In most cases, quenching is carried out in an oil bath at room temperature. In a limited number of cases, quenching is carried out in a heated water bath.

23.2.4 Tempering

Steel products may be tempered to decrease hardness and increase toughness to produce the desired combination of mechanical properties. Tempering involves heating the steel to a temperature below the transformation range and holding for a suitable time at the temperature, followed by cooling at a suitable rate.

23.2.5 Machining

Machining is not considered part of the core forging process; however, forging sites often include machine shops to provide rough machined or fully finished forgings. Machining operations include finishing lathes, hole boring, and milling.

23.2.6 Surface Treatment

Surface treatments such as painting or galvanizing are not considered part of the core forging process and are usually carried out at another site. There are a number of different methods used to make forgings. Three of the most common methods are described below.

Open-die forging refers to the working of metal using dies that do not laterally confine the work piece. The metal is typically worked to the desired shape between flat-faced dies as shown in Figure 23.2. Open-die forging comprises many process variations, permitting an extremely broad range of shapes and sizes of up to 30 m in length and ranging from a few kilograms to many tons in weight.

Most forgeable ferrous and nonferrous alloys can be open-die forged, including some less common materials like age-hardening superalloys and corrosion-resistant refractory alloys.

Closed-die forging, also known as impression-die forging, refers to the working of metal between two or more dies containing impressions of the part shape, as shown in Figure 23.3.

Metal flow is restricted by the die contours and, as a result, the process generally yields more complex shapes than open-die forging processes. This process forms forged parts that range in weight from a few grams to 25 tons. Metals and alloys, such as carbon and alloy steels, tool steels, aluminum and copper alloys, and certain titanium alloys, can be forged by the impression-die processes.

Seamless rolled ring forging is a specific process used to produce ring-shaped forgings. The process starts by punching a hole in a thick piece of metal using an open-die forge, forming a hollow donut shape. This donut is heated above the recrystallization temperature and placed over the idler or mandrel roll. Under high pressure, the idler roll moves toward the drive roll that continuously rotates to reduce the wall thickness, thereby increasing the diameters of the resulting ring, as shown in Figure 23.7. This process can be used to form rings with outside diameters that vary in size from a few centimeters to over 10 m and weights ranging from a half a kilogram up to over 160,000 kg. Cold forging is similar to hot forging described earlier, except that the work piece is not preheated before working. Cold forging encompasses many processes such as bending, cold drawing, cold heading, coining, and extrusion to yield a diverse range of part shapes.

Cold forgings are frequently used in automotive steering and suspension parts, antilock-braking systems, hardware, and other applications where high strength, close tolerances, and volume production make them an economical choice. Metals range from lower-alloy and carbon steels to 300 and 400 series stainless steels, selected aluminum alloys, brass, and bronze.

23.3 Fasteners

Fasteners, such as bolts, nuts, and screws, are used across industries including aerospace and automotive engineering and construction industries. Fastener manufacture is similar in some respects to forging. Fasteners, such as bolts, may be manufactured using either a "hot" or a "cold" production process. The majority of UK fastener manufacturers use cold processing techniques in which products up to 12–15 mm in diameter are forged in continuous forging machines (transfer headers and bolt-makers). Wire is fed at the front of the machine, and a finished or semifinished product results after cold forging. In some cases, heat may be used on continuous forging machines, and this is often provided by induction heating. The wire drawing follows:

- Blanks of strong ductile metals like steels are rolled between threaded dies.
- Large threads are hot rolled followed by finishing, and smaller threads are straight cold rolled to desired finish.
- Cold rolling attributes more strength and toughness to the threaded parts.
- Widely used for mass production of wire.

Hot processing tends to be limited to larger-diameter products (>15–30 mm diameter). Cold process fastener manufacturing is similar to that described below, except that the material is not heated before being worked and is often lubricated for pressing.

23.3.1 Preheating

In hot processing, the metal is preheated to the desired temperature (up to 1000°C) before pressing. A variety of heating methods are used in the fasteners subsector, including gas-fired furnaces, electrically heated furnaces, and electrical induction coils.

23.3.2 Pressing

The work piece is inserted into the press to form the general shape of the bolt head. A second pressing finalizes the head shape, and this is then trimmed to remove excess material.

23.3.3 Thread Rolling

With the general shape of the bolt formed, a thread rolling machine is used to "roll" the thread at the other end to the head.

23.3.4 Heat Treatment and Quenching

Irrespective of whether the bolts are cold or hot formed, they may require hardening by heat treatment. This is usually a bulk process where a large batch of machine-finished bolts is first degreased, separated onto a conveyer, heated in an oven to 900°C, quenched in an oil bath, and then tempered at 400°C to toughen the metal. Many smaller fastener manufacturing sites in the United Kingdom do not have in-house heat-treatment facilities.

23.3.5 Tempering

Steel products may be tempered to decrease hardness and increase toughness. Tempering involves heating the steel to a temperature below the transformation range and holding for a suitable time at the temperature, followed by cooling at a suitable rate.

23.3.6 Washing

Bolts often need to be washed in water before dispatch. The processes described above may be automated and integrated to varying degrees at different sites and on different production lines within a site. For example, the sequential stages of forming the bolt shape and thread may be carried out on separate machines or on a single integrated machine.

Painting is not considered to be part of the core sheet metal manufacturing process and has not been investigated as part of this project. However, a number of sheet metal sites incorporate paint shops, which can be significant energy users.

23.4 Hand Forging and Drop Forging

Forging is the forming of metal parts by hammering, pressing, or bending to the required shape, usually at red heat. *Hand forging* involves the use of an anvil and special hammers, chisels, and swages. A *drop forging machine* uses pneumatic or hydraulic pressure to compress hot metal blanks between hard steel dies.

Forging is the process where (heated) metal is beaten with a heavy hammer to give it the required shape. For example, ancient sword-making used flat hammers to beat on a heated strip of metal kept on a flat piece of iron called an anvil (you may have seen this action in many movies). However, forging is used to make many more complex shapes, and to let the metal form into such shapes, the hammer and the supporting pieces are cut into the reverse of the required shape—in other words, they form the forging dies.

Drop hammers are of two kinds: (a) gravity die drop hammers and (b) power drop hammers. In gravity drop hammers the ram falls under gravity, whereas in power drop hammers pressurized air power or steam power is used accelerate the ram. The force of the blow is determined by the height of drop and weight of the ram.

Figure 23.1 shows an *open-die forging* process. If we desire to hammer down the stock to a well-defined shape, then it is customary to set a *closed-die forging process*; in this process, the hammerhead and the anvil are basically hardened dies with the inverse of the shape we want. Figure 23.2 shows a schematic of the dies and stock in the closed-die forging process. Usually, the stock volume is a little in excess of the part volume; this ensures that the entire die cavity gets filled properly. However, the excess material flows out through the gap between the dies; this excess is called *flash* and must later be machined away (this operation is called *trimming*). If the stock and the final part are very different in shape, then the forging is done in several stages. Figure 23.2 shows a common forged part—a connecting rod. This part is used in almost all petrol engines. The part is made from bar stock in four stages, marked in the figure as (b) edging, (c) blocking, (d) finishing, and (e) trimming.

Engineering components and parts such as the connecting rod, landing gear, axle, and lever are made through the forging process (Figures 23.3 through 23.5).

In some modern applications, the volume of the stock and the precision of the last-stage dies are controlled very tightly. This method is sometimes called precision forging. One form of precision forging is used to make coins. If you look closely at a newly manufactured coin, you can see very high-quality details in the forging; usually, good-quality coins may need between three and five stages of forging. This is because in coin-making, no lubricant or oil can be applied to the die surface; it may get trapped in the tiny cavities of the design and lead to poor feature definition in the forged coin.

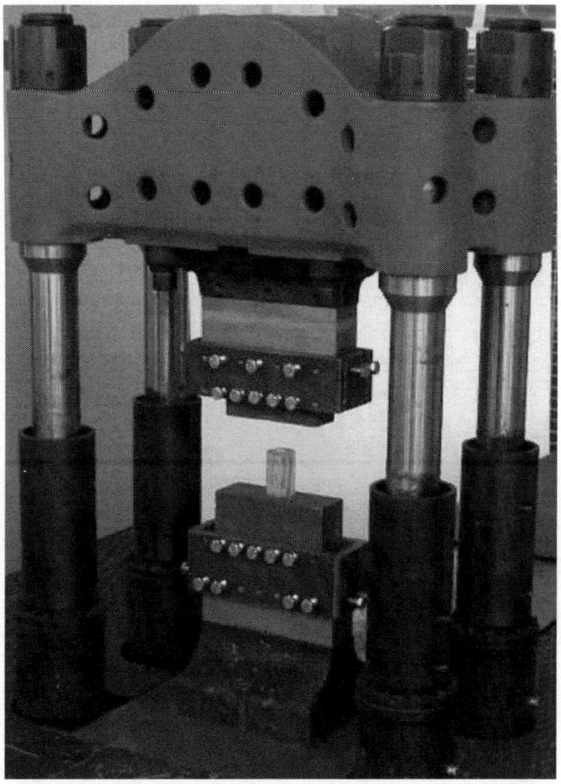

Figure 23.1

A simple forging press; note the red-hot bar stock in place.

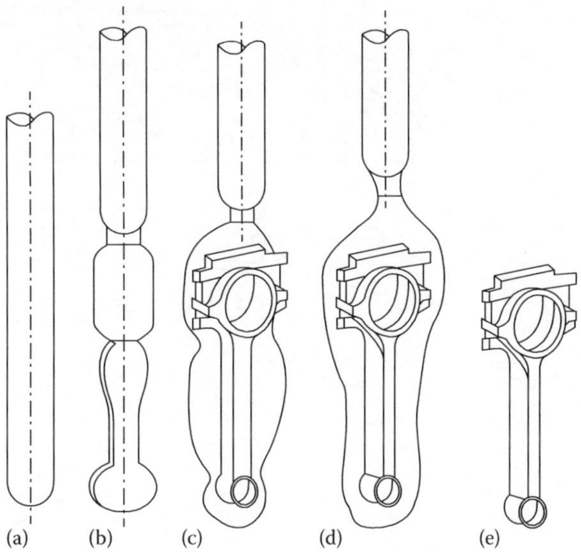

Figure 23.2

Stages of connecting rod forging: (a) blank, (b) edging, (c) blocking, (d) finishing, and (e) trimming.

23.4.1 Forging Parameters Calculation

23.4.1.1 Open-Die Forging

Under ideal conditions, there is no friction between work and die, and then there is homogeneous deformation of workpart. In this process, the initial height, h_0, of the work is reduced to final height, h_f, in stages. The strain on the workpart can be estimated by

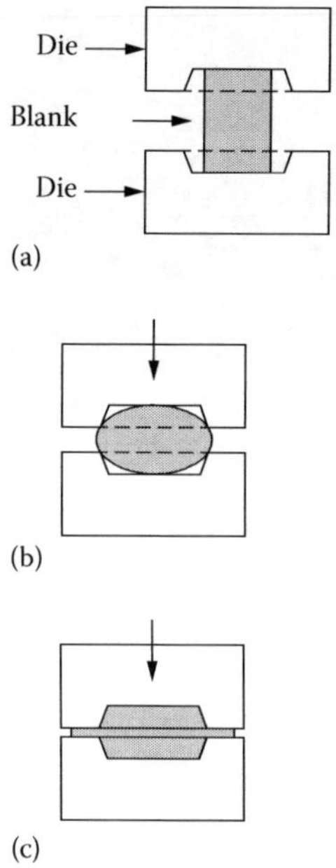

Figure 23.3

(a–c) Stages of closed-die forging.

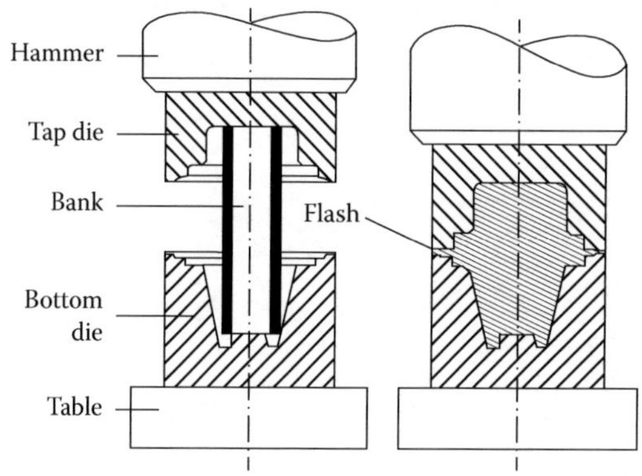

Figure 23.4

Drop forging machine.

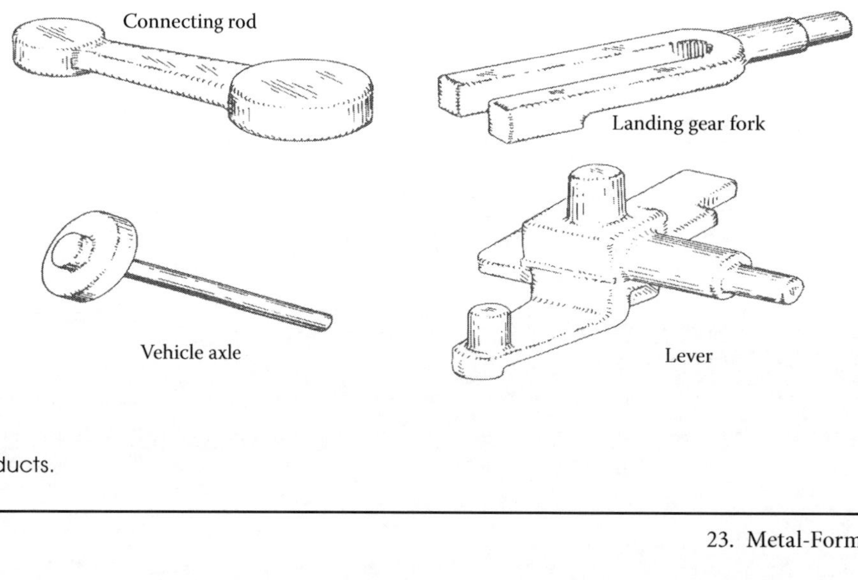

Figure 23.5

Forging products.

$$\varepsilon = \ln\frac{h_0}{h}$$

where h is the intermediate height at a particular stage.

The part also experiences flow stress, Y_F, corresponding to strain and is determined as

$$Y_F = K\varepsilon^n$$

where:

- K is the strength coefficient, MPa (lb/in.^{2})
- n is the strain hardening exponent

We can visualize that Y_F is akin to stress (σ) in elastic deformation, and when $n=1$, it is elastic. However, metal-forming operations are carried out at the plastic stage, and K and n values are different values due to strain hardening. Y_F is also defined as the instantaneous stress value required to keep the metal flowing, and so is called *flow stress*. The upsetting force, F, required to produce deformation is estimated as $F=Y_FA$, where A is the cross-sectional area of the workpart in mm^2 (in.^{2}). The unit of force is naturally in pounds (N). The height is progressively reduced in stages, but the area (A) continuously increases. The upsetting force, F, also increases due to strain hardening. However, in hot working, $n=0$, and flow stress (Y_F) is constant. However, real upset forging does not occur as neatly as explained above. Barreling occurs, that is, the cross section in the middle of the work bulges, and due to this barreling the force equation is modified as $F=K_FY_FA$. K_F is called the barreling shape factor and is always greater than 1 to take barreling into account. The barreling shape factor (K_F) is estimated as

$$K_F = 1 + \frac{0.4\mu D}{h}$$

where:

- D is the workpart diameter to be calculated at every stage
- μ is the coefficient of friction between the work and the die surface

As explained, ideally there is no friction between the surfaces, but there is always some friction in the real world. We can also understand that $0 \le n \le 1$; n is zero for the purely elastic case and 1 for the purely plastic stage (hot working), and for in-between metal deformation processes, n is less than 1 and greater than zero.

23.4.1.2 Impression-Die Forging

In impression-die forging the die contains the inverse of the desired shape, and after forging this impression is transferred to the workpart. Automotive parts like the connecting rod are forged in a closed-die or impression-die forging operation in stages, as shown in Figure 23.4. There is some flash, which is removed by trimming as shown above in the last stage of closed forging.

23.4.1.3 Flashless Forging

This is also known as precision forging. Coining is closed-die or impression-die forging where the coin details are impressed in inverse to the die on the top and bottom of the die and are transferred to the workpart during the forging operation. In this operation, to prevent any flash, flashless or precision die forging is used. As we can observe, in precision forging there is need for very close tolerance between the volume of die cavity and work volume. In short, no subsequent machining is required in precision forging. It can also be called a near net-shape process.

23.5 Heading

In the fastener industry, upset forging is used to manufacture the heads of nails, bolts, and other such parts. Both open-die and closed-die forging operations are used to create heads.

23.6 Quality of Forged Parts

Usually forged parts are much stronger/tougher than cast or machined parts made from the same material. This is because the hammering process arranges the microstructure of the metal such that the crystal grains get aligned along the part profile. This leads to an increase in strength. It is therefore common to use forging to make parts that can carry very high stresses during their use, rather than casting.

23.7 Energy Cost in Forging Plants

Energy cost represents 10%–12% of the cost of production for a forging/heat-treating plant. Energy cost is expected to rise at a higher rate than other costs for the plant. Energy cost is controllable in a short term through assessment, analysis, and cost-effective energy conservation measures.

23.8 Trends in the Forging Industry

The forging industry is moving toward high-capacity continuous furnace operations in countries such as India and China. The newer units are considerably more efficient due to better furnace design and energy-saving systems. In some cases induction is replacing fuel-fired systems for two reasons: one being energy use per pound and the other being lack of availability of clean fuel such as natural gas. For fuel-fired systems, use of recuperators and a regenerative system helps reduce energy use by 25%–40%. Use of improved controls such as pulse firing helps in continuous as well as batch operations. The energy use observed is in the range of 1.6 MM Btu/ton to 1.8 MM Btu/ton with "good" operating practices.

23.9 Major Energy User Systems for Forging Plants

- Electric motors
- Pumps (various fluids)
- Air compressors
- Induction heaters
- Fans and blowers
- Atmosphere generators

23.10 Observations on Energy Use Pattern for the Forging Industry

In most forging industry plants, furnaces or heating systems such as induction heating use the largest amount of heat. A large amount of total plant energy, between 65% and 80%, is used for reheating and heat treating of the production parts—commonly known as process heating operations. In some plants, boiler energy use can be between 20% and 30% of the fuel energy used in the plant. The process heating equipment (furnaces, ovens, etc.) has relatively low efficiency, between 20% and 60%, compared to higher than 75% for many boilers. Waste heat from furnaces and ovens represents the largest amount of energy wasted in the plant.

23.11 Health Hazards of Forging Operations

In the iron and steel and other manufacturing industries, foundries and forges produce a lot of pollutants in the environment—both working and ambient. In these processes, metals are extracted and produced from ores by various metallurgical processes, and processes for molding, melting and casting, and so on, are accompanied by evolution of heat, noise, dust fines, fly ash, and oxides of nitrogen, sulfur, and metals. Particulate matters are generated in large quantities when preparing mold core sands and molds melting metals, pouring metal, knocking out poured molds, and loading and unloading raw materials. Here metals are given a specific shape by metal castings for various engineering purposes. Gaseous matters like gases, vapors, fumes, and smoke are produced during melting and pouring operations. The major pollutants are emitted from various work areas in the foundry, that is, pattern shop, sand preparation, molding and core making, mold drying and ladle heating, cupola, electric arc furnace, pouring and mold cooling, knockout, fettling, and heat treatment. In addition, various air pollutants and noise pollutants (Davis, 2002) are produced from forge shops and other manufacturing industrial units. Pollutants are also emitted in sintering, pelletization, rolling mills, coke-oven plants, refractories, and so on, in steelmaking and by-products manufacturing. These pollutants cause various health hazards in human beings.

23.12 Rolling

Rolling is done in either hot stage or cold stage. The deformation of metal is achieved in plastic range. Both hot rolling and cold rolling are quite ancient processes and used extensively in engineering manufacturing.

23.12.1 Rolling Processes

23.12.1.1 Hot Rolling

Hot rolling is a metalworking process that occurs above the recrystallization temperature of the material. After the grains deform during processing, they recrystallize, which maintains an equiaxed microstructure and prevents the metal from work hardening. The starting material usually consists of large pieces of metal like semifinished casting products such as slabs, blooms, and billets. If these products are from a continuous casting operation, they are usually fed directly into the rolling mill at the proper temperature. In smaller operations the material starts at room temperature and must be heated. This is done in a gas- or oil-fired soaking pit for larger workpieces; for smaller workpieces induction heating is used. As the material is worked, the temperature must be monitored to make sure it remains above the recrystallization temperature. To maintain a safety factor, a *finishing temperature* is defined above the recrystallization temperature; this is usually 50°C–100°C (90°F–180°F) above the recrystallization temperature. If the temperature drops below this, the material must be reheated before further hot rolling. Hot-rolled metals generally have little directionality in their mechanical properties and deformation-induced residual stresses. However, in certain instances nonmetallic inclusions impart some directionality, and workpieces less than 20 mm (0.79 in.) thick often have some directional properties. Also, nonuniform cooling induces a lot of residual stresses; this usually occurs in shapes that have a nonuniform cross section, such as I-beams. While the finished product is of good quality, the surface is covered with mill scale, which is an oxide that forms at high temperatures. It is usually removed through pickling or the smooth clean surface process, revealing a smooth surface. Dimensional tolerances are usually 2%–5% of the overall dimension.

Hot-rolled mild steel seems to have a wider tolerance for amount of carbon than cold rolled steel, making it a bit more problematic for a blacksmith to use. Also among similar metals, hot-rolled steel seems to be typically less costly.

Hot rolling is used mainly to produce sheet metal or simple cross sections, such as rail tracks. Other typical uses for hot-rolled metal include truck frames, automotive wheels, pipe and tubular, water heaters, agricultural equipment, strappings, stampings, compressor shells, railcar components, wheel rims, metal buildings, railroad-hopper cars, doors, shelving, discs, guard rails, and automotive clutch plates [1].

23.12.1.2 Cold Rolling

Cold rolling occurs with the metal below its recrystallization temperature (usually at room temperature); this which increases the strength by strain hardening up to 20%. It also improves the surface finish and holds tighter tolerances. Commonly cold rolled products include sheets, strips, bars, and rods; these products are usually smaller than the corresponding hot-rolled products. Because of the smaller size of the workpieces and their greater strength, as compared to hot-rolled stock, four-high or cluster mills are used [2]. Cold rolling cannot reduce the thickness of a workpiece as much as hot rolling in a single pass. Cold rolled sheets and strips come in various conditions: *full-hard*, *half-hard*, *quarter-hard*, and *skin-rolled*. Full-hard rolling reduces the thickness by 50%, while the others involve less reduction. Skin-rolling, also known as a *skin-pass*, involves the least amount of reduction: 0.5%–1%. It is used to produce a smooth surface, with a uniform thickness, and to reduce the yield point phenomenon (by preventing bands from forming in later processing). It locks dislocations at the surface and thereby reduces the possibility of formation of Lüders bands. To avoid the formation of Lüders bands, it is necessary to create substantial density of unpinned dislocations in ferrite matrix. It is also used to break up the spangles in galvanized steel. Skin-rolled stock is usually used in subsequent cold-working processes where good ductility is required.

Other shapes can be cold rolled if the cross section is relatively uniform and the transverse dimension is relatively small. Cold rolling shapes requires a series of shaping operations, usually along the lines of sizing, breakdown, roughing, semiroughing, semifinishing, and finishing. Typical uses for cold rolled steel include furniture, desks, filing cabinets, shelves, tables, chairs, motorcycles, exhaust pipes, computer cabinet and hardware, all home appliances and components, lighting fixtures, hinges, tubing, steel drums, lawn movers, electronic cabinetry, water heaters, metal containers, and a variety of construction-related products.

Rolling is a process in which the metal is squeezed between two hard rollers. The purpose is to change the thickness (and since volume is conserved, the length is increased). It involves the rolling of slabs, strips, sheets, and plates. Thickness of the part is reduced by an amount called *draft* and is calculated as $d = t_0 - t_f$, where t_0 and t_f are the initial and final thickness, respectively. *Reduction* is defined as $r = d/t_0$; it can also be written as $r = 1 - (t_f/t_0)$.

Spreading of the workpart also occurs during rolling, that is, work width increases along with length. However, volume of work remains constant and so the relation $t_0 w_0 L_0 = t_f w_f L$; as discussed earlier, $t_0 > t_f$, $w_f > w_0$, and $L_f > L_0$.

Sometimes, volume rates of material flow before and after are also used as $t_0 w_0 v_0 = t_f w_f v_f$, where v_0 and v_f are the entering and exiting velocities of the work under the rollers. Whenever there is a change of dimension, there is strain, stress, and related force on the rollers.

Strain is calculated as $\varepsilon = \ln(t_0/t_f)$, and stress is calculated as $\overline{Y}_F = (\overline{K}\varepsilon^n)/(1+n)$. It is called average flow stress due to the physics of rollers as stress varies under the roller in the contact length. The force on the roller to accomplish the rolling action is calculated as

$$F = \overline{Y_F} wL$$

where:

- L is the contact length between the work and roller and is estimated as $L = \sqrt{R(t_0 - t_f)}$
- R is the roll radius
- wL is termed as the roll–work contact area

The torque required in rolling is estimated as $T = 0.5FL$. The power of each roll can be estimated as $2\pi NT$, where N is the rotational speed of the roll. The power for two rolls is estimated as $P = 2\pi\text{FNL}$ (J/s or W [in.-lb/min]).

The rolling process may not be able to reduce the initial thickness (t_0) to the desired final thickness (t_f) in a particular pass or stage, and the maximum draft in the pass is calculated as $d_{max} = \mu^2 R$, where μ is the coefficient of friction between roller and work. If $d_{max} < d$, rolling is possible to accomplish the draft in that stage if not more than one pass is required for rolling. An interesting fact that emerges is that for rolling to be feasible, friction (μ) is essential between the roller as at zero coefficient of friction the draft would be zero and it would be difficult to accomplish rolling.

Figure 23.6 shows a schematic of rolling mills. The main use of rolling is in plants where the metal is made. For example, in steelmaking plants, liquid iron is first formed in a blast furnace by reducing the iron oxide. After further processing of the liquid metal, including converting the iron to steel, it is cast by a process called *continuous casting* into raw stock shapes. These are very large pieces of steel (several tons each), with typical cross sections including rectangles (blooms, billets, slabs), circles (rounds), or I-sections (beams). These pieces are too large to be directly used—they are rolled in rolling mills that squeeze them into much smaller, but usable shapes. These usable shapes are the raw stock for almost all types of manufacturing that uses steel.

Rolling mills are categorized as *hot-rolling* or *cold-rolling* mills; in hot rolling, the metal is heated to just below its melting point before being fed into the rollers. This is useful, for example, if the initial billet is in a brittle form, for example, cast iron; hot-rolled steel cools down with finer grains in the crystalline microstructure and is stronger and less brittle (*wrought iron*). Rolling mills can also use a variety of roller shapes to get different cross sections of rolled bars. Typical process flows are shown in Figure 23.6.

In each stage of rolling, the raw stock is reduced in thickness by a small percentage; therefore, typical rolling mills have several stages, where each stage successively reduces the thickness of the stock until the desired cross section is achieved (see Figure 23.7). A very important use of the rolling process is in the making of screws and bolts—the threads of the screw are made by rolling a cylindrical stock between two dies that form the thread-shapes on the stock. A single rolling machine of this type can produce tens of screws per second (which is why they are so cheap).

23.13 Rolling

In a rolling mill, red-hot ingots of steel or other metals are passed through successive pairs of specially shaped rollers to produce flat bars, sheets, I-sections, T-sections, channels, angles, or other section bars. Final cold rolling may be carried out to give a better finish.

Universal beams, universal columns, joists, angles, and channels are made to British Standards BS 4: Part 1 and BS 4848: Part 4.

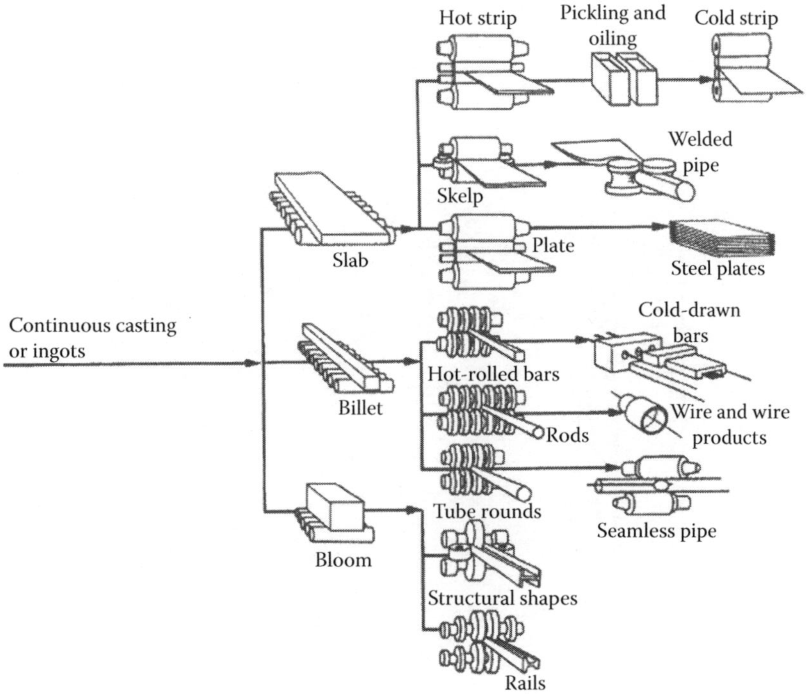

Figure 23.6

Different types of flat- and shaped-rolling processes.

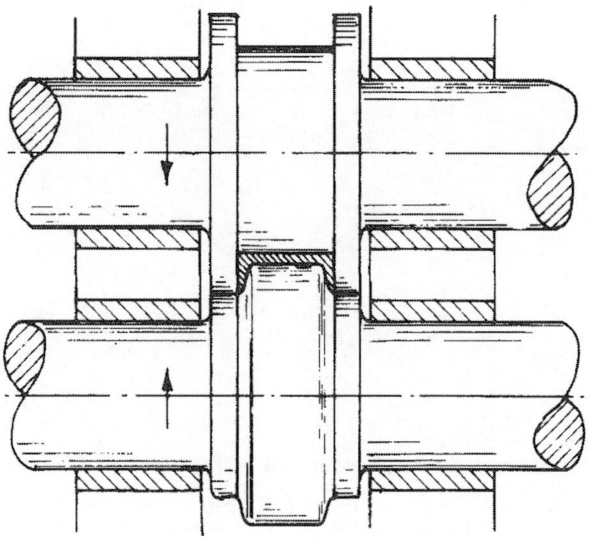

Figure 23.7

Rolling mill.

23.13.1 Rolled Sections

Rolled sections are made to standard sizes. Several I-sections, channel, angles, and others are rolled (Figure 23.8). The standard sizes for rolled sections are provided:

Universal beams are normally $D \times B = 127$ mm × 76 mm to 914 mm × 419 mm. t and T are several sizes in each case. Normally these are designed dimensions, and standard sections are selected from the available stocks of manufacturers (Figure 23.9).

Universal columns are normally $D \times B = 152$ mm × 152 mm to 356 mm × 406 mm. t and T are several sizes in each case.

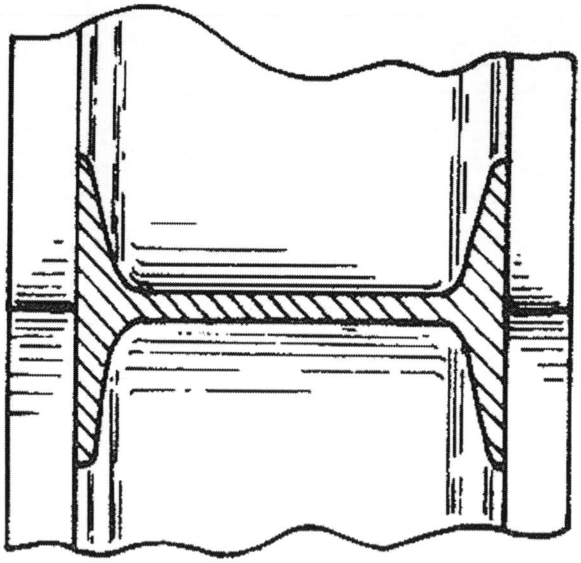

Figure 23.8

Rolls for I-section.

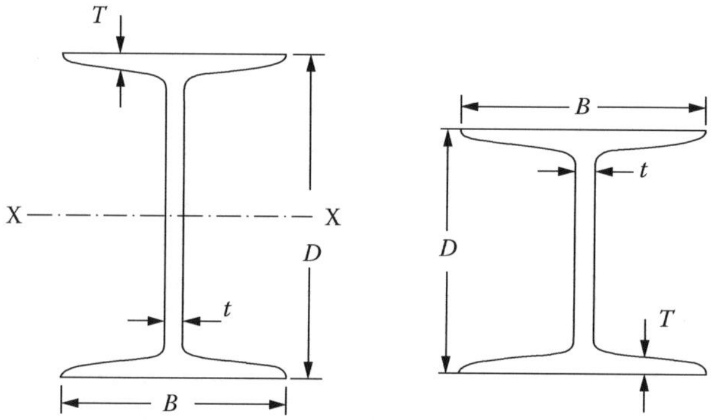

Figure 23.9

Beams, columns, and joists.

Joists are normally from 76 mm × 76 mm to 254 mm × 203 mm.

Equal angles normally have D × B from 25 mm × 25 mm to 200 mm × 200 mm. There are several values of *t* in each case (Figure 23.10).

Unequal angles normally have *D* × B from 40 mm × 25 mm to 200 mm × 150 mm (Figure 23.11).

Channels normally have *D* × B from 76 mm × 38 mm to 432 mm × 102 mm. There is one value of *t* in each case (Figures 23.12 and 23.13).

23.14 Extrusion

Extrusion is a process in which metal is forced, or squeezed, out through a hole (die). The process is similar to squeezing toothpaste out of the tube. This process is used mostly for metals that are ductile, including copper, steel, aluminum, and magnesium; it is also used for some plastics and rubbers. Common examples of parts made by extrusion are the aluminum frames of whiteboards, and door- and window-frames. Usually, long strips of the required cross section are extruded and sold as raw stock. Any hardware store will have between ten and hundred different cross-section bars—almost all are made by extrusion. The process can be used to make hollow as well as solid cross sections, as seen from the example parts in this figure.

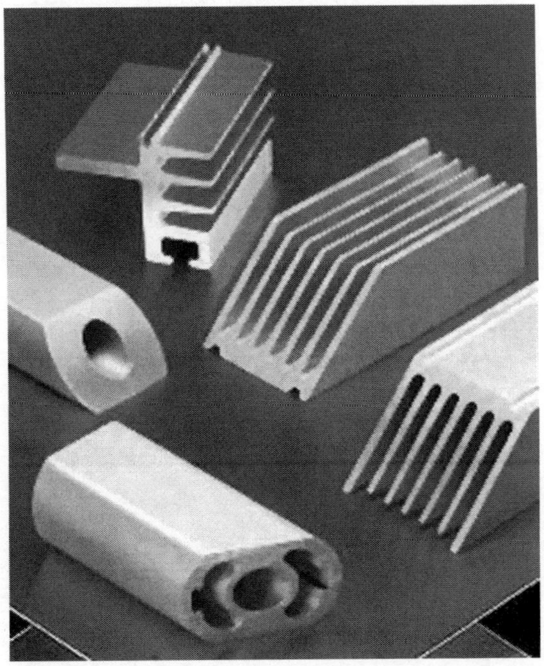

The extrusion press has a pressure chamber into which the raw stock is loaded; the die is made of hardened steel, with a hole that is the shape of the required cross section. The metal is then squeezed out of the die hole using a high-pressure hydraulic piston. The schematic below shows the basic process, and the images show two examples of aluminum extrusion dies.

23.14.1 Hot Extrusion

A piece of red-hot bar or billet is placed in a cylinder and forced through a specially shaped die by a piston to produce long bars. Hollow sections can be made by placing a mandrel in the die orifice (Figure 23.14).

23.14.2 Extrusion Parameters

The extrusion ratio or reduction ratio is a useful parameter for estimating the strain and pressure to be applied to the ram for extrusion. Extrusion ratio (r) estimation involves the original cross-sectional area (A_0) and the final cross-sectional area (A_f) after extrusion, and the ratio is calculated as $r = A_0/A_f$. The resulting

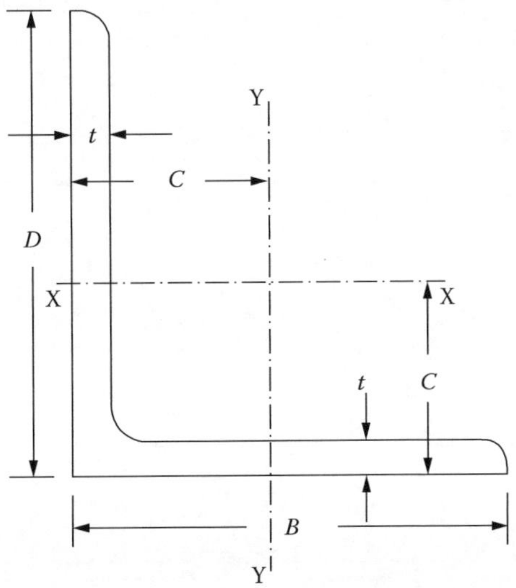

Figure 23.10

Joist diagram.

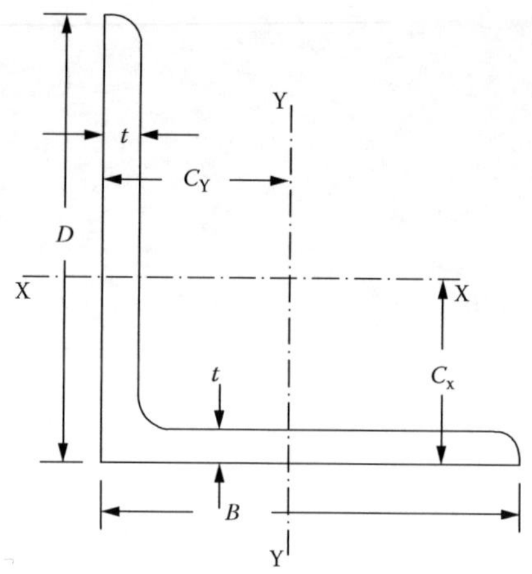

Figure 23.11

Unequal angle diagram.

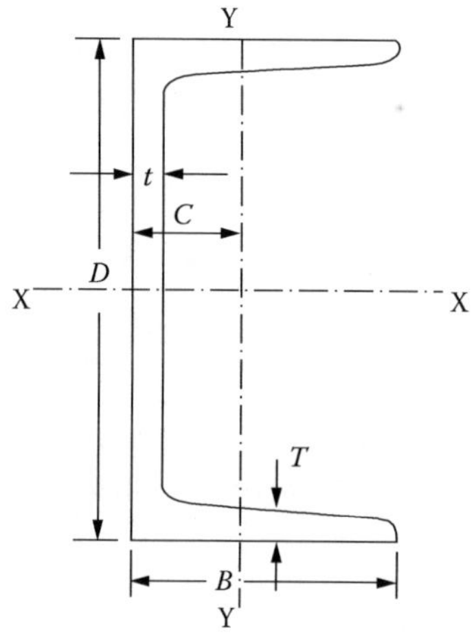

Figure 23.12

Channel diagram.

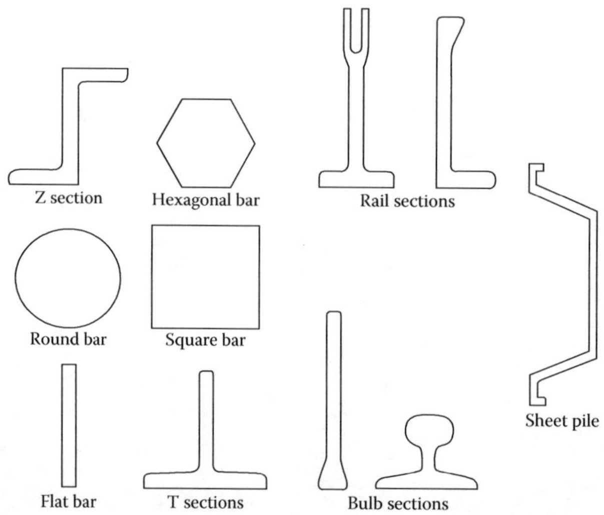

Figure 23.13

Miscellaneous rolled sections.

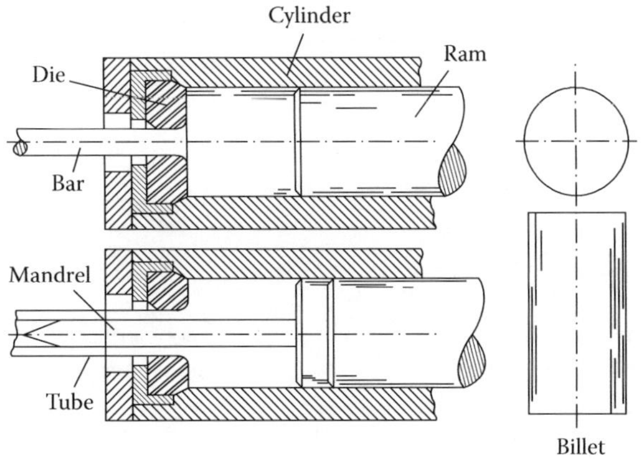

Figure 23.14

Hot extrusion.

strain (ε) is calculated as $\varepsilon = \ln(r) = \ln(A_0/A_f)$. The pressure ($p$) applied to the ram to deform the billet to the extruded section through the die opening needs the average flow stress $\overline{Y_F}$. It is calculated as follows:

$$p = \overline{Y_F}\varepsilon$$

where

$$\overline{Y_F} = \frac{K\varepsilon^n}{(1+n)}$$

All the estimations above are given for an ideal frictionless condition, which is not practical, and therefore the strain in both indirect and direct extrusion needs to be modified. Johnson has worked extensively on plastic deformation and presented the following empirical relation for strain during deformation due to friction. In the case of indirect extrusion, only the friction between the billet and the die is considered.

$\varepsilon_x = \mathrm{a} + b\ln(r)$ (indirect extrusion).
a and b values depend on the die angles, but typical values are $a = 0.8$ and $b = 1.2$–1.5.
$p = \overline{Y_F}\varepsilon_x$ (pressure estimation in indirect extrusion).

However, direct extrusion involves additional friction of the container wall and the billet, and pressure is estimated as follows.

$$p = \overline{Y_F}\left(\varepsilon_x + \frac{2L}{D_0}\right) \text{ (pressure estimation in direct extrusion)}$$

In direct extrusion, $2L/D_0$ accounts for additional friction of the container and the billet. Therefore, strain in the case of direct extrusion is always higher than that for indirect extrusion and so the pressure needed for extrusion is more. L is the billet length remaining to be extruded, and D_0 is the original diameter of the billet.

The force needed on the ram is calculated as

$$F = pA_0$$

The ram force is in N (lb), and the power required for the extrusion is

$$P = Fv$$

where:

P = J/s (in.-lb/min)
v = ram velocity, m/s (in./min)

We can see that to reduce the power requirement for extrusion or for environmentally benign manufacturing or environmentally conscious manufacturing (ECP), we must reduce the friction, and so lubrication should be used. However, we should be careful in the use of lubrication and its disposal (Figure 23.15).

23.14.3 Cold Extrusion

Soft metals such as aluminum and copper can be extruded cold. Practically all metals may be extruded (Figure 23.16).

23.14.4 Impact Extrusion

A metal which is plastic when cold may be extruded by the impact of a high-velocity punch. The metal of the blank flows up the sides of the punch to produce a cylinder. The process is used for manufacturing toothpaste tubes, ignition coil cans, and so on (Figure 23.17).

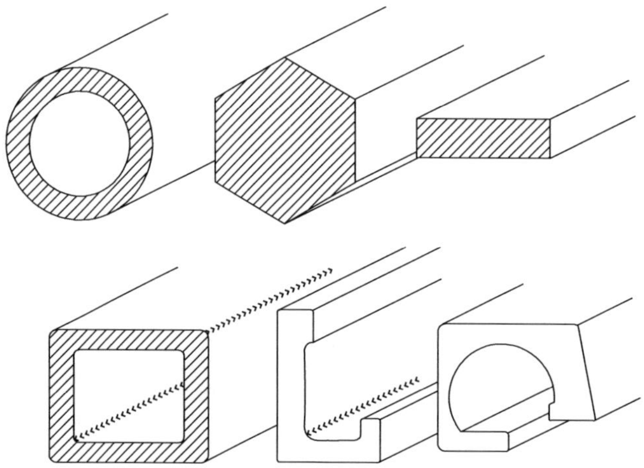

Figure 23.15

Hot extruded sections.

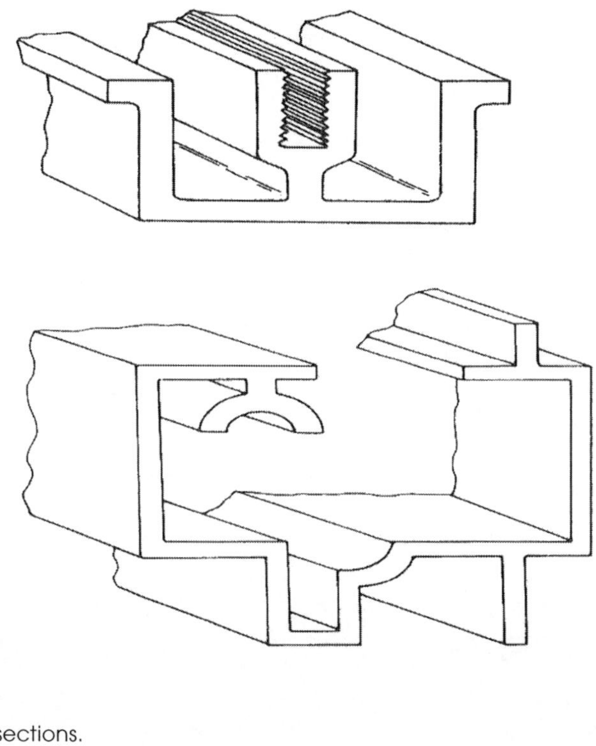

Figure 23.16

Aluminum architectural sections.

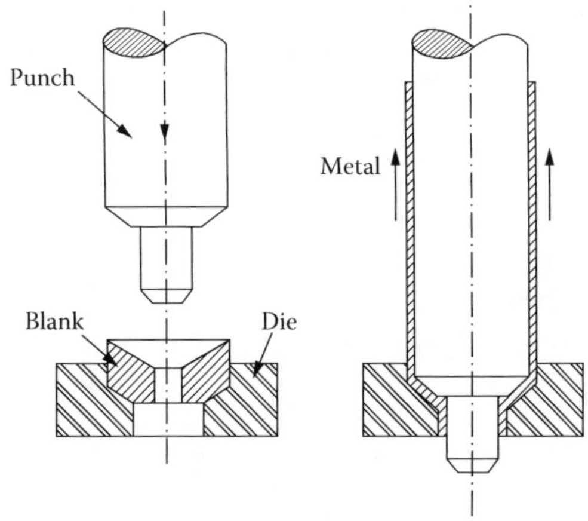

Figure 23.17

Impact extrusion.

23.15 Drawing Processes

23.15.1 Drawing

Drawing is the process most commonly used to make wires from round bars; this process is very similar to extrusion, except that instead of applying pressure from the back end, in drawing, the wire is pulled from the side where it emerges from the circular die. A schematic is shown below. Drawing is the forming of flat metal blanks into box and cup-like shapes by pressing them with a shaped punch into a die. The process is used for cartridge cases, boxes, electrical fittings, and so on. Dies are made of specially hardened tool steels or tungsten carbide. Diamond dies are used for drawing very fine wires. Both extrusion and drawing may be hot (i.e., the stock is heated to a high temperature for processing) or cold (i.e., the stock is not heated).

23.15.2 Drawing Force Estimation

The force of drawing is usually estimated as $F = \text{stress} \times \text{area}$. In the case of plastic deformation, the ideal stress resulting from the operation is estimated as follows:

$$\sigma = \overline{Y_F}\varepsilon$$

where ε is the true strain and is calculated as

$$\varepsilon = \ln\left(\frac{A_0}{A_f}\right)$$

where A_0 and A_f are the starting and final cross-sectional areas of the drawn part. Area reduction (r) during drawing is calculated as

$$r = \frac{A_0 - A_f}{A_0}$$

With inclusion of reduction ratio in the true strain equation, we can write

$$\varepsilon = \ln\left(\frac{A_0}{A_f}\right) = \ln\left(\frac{1}{1-r}\right)$$

The stress (σ) can be rewritten in terms of reduction ratio as

$$\sigma = \overline{Y_F}\ln\left(\frac{A_0}{A_f}\right) = \overline{Y_F}\ln\left(\frac{1}{1-r}\right)$$

Because friction is present and no drawing operation is frictionless, the actual stress is higher than that estimated by the above equation. Schey [3] has presented a modified draw stress (σ_d) as follows:

$$\sigma_d = \overline{Y_F}\left(1+\frac{\mu}{\tan\alpha}\right)\phi\ln\left(\frac{1}{1-r}\right)$$

where:

- μ is the coefficient of friction between the work and die
- α is the die angle (half-angle)
- ϕ is a modification factor for inhomogeneous deformation and is estimated as follows:

$$\phi = 0.88 \pm 0.12\frac{D}{L}$$

where:

- D is the average diameter of the work during drawing
- L is the contact length of the work with draw die in mm (in)
- D and L are estimated as follows:

$$D = \frac{D_0 + D_f}{2} \text{ and } L = \frac{D_0 - D_f}{2\sin\alpha}$$

The resultant draw force (F) is calculated as

$$F = A_f \sigma_d$$

23.16 Health Hazards of Foundries and Forges

A huge amount of dust, suspended particle matter (SPM), respirable SPM (RSPM), and gaseous pollutants is released into the environment—both workplace and ambient—from various foundries and forges of iron and steel industries and other manufacturing industries. These pollutants cause health hazards. Many health hazards are reported in foundry workers and forge shop workers of industries. In these processes, metals are extracted and produced from ores by various metallurgical processes, and processes for molding, melting and castings and so on are accompanied by evolution of heat, noise, dust fines, fly ash, oxides of nitrogen, sulfur, and metals. Particulate matters are generated in large quantities when preparing mold core sands and molds melting metals, pouring metal, knocking out poured molds, and loading and unloading raw materials. Here metals are given a specific shape by metal castings for various engineering purposes. Gaseous matters like gases, vapors, fumes, and smoke are produced during melting and pouring operations. The major pollutants are emitted from various work areas in the foundry, that is, pattern shop, sand preparation, molding and core making, mold drying and ladle heating, cupola, electric arc furnace, pouring and mold cooling, knockout, fettling, and heat treatment. In addition, various air pollutants and noise pollutants are produced from forge shops and other manufacturing industrial units. Pollutants are also emitted in sintering, pelletization, rolling mills, coke-oven plants, refractories, and so on in steelmaking and by-products manufacturing. These pollutants cause various health hazards in human beings.

Many foundry and forge workers are exposed to common air pollutants in their occupations, for example, smoke, dust, SPM, RSPM, carbon monoxide, sulfur dioxide, oxides of nitrogen (Nox), hydrocarbons, and heavy metals like Pb, Cd, Cr, As, and Ni. Their prolonged exposure causes various health hazards. Heavy metals cause acute and chronic poisoning. Some disastrous episodes have focused attention upon air pollution as a health hazard. Some fugitive gaseous and dust emissions may cause primary occupational health problems to workers engaged in foundries; gaseous and dust emissions from foundries pose potential health risk to populations residing in the surrounding areas. Diseases such as black lung, metal fume fever, silicosis, and pneumoconiosis are all occupational maladies which are attributed to inhalation of one or other type of fine dust particles.

Some general health hazards arise as a result of contact between the pollutants and the body. These hazards are as follows:

- Eye irritation.
- Headache.
- Nose and throat irritation.

- Irritability of respiratory tract.
- Gases like hydrogen sulfide, ammonia, and mercaptans cause odor nuisance even at low concentrations.
- High temperature can cause fatigue and dehydration.
- Chronic pulmonary diseases like bronchitis and asthma are aggravated by a high concentration of SO_2, NO_2, particulate matter, and photochemical smog.
- Carbon monoxide combines with the hemoglobin in the blood and consequently increases stress on those suffering from cardiovascular and pulmonary diseases.
- Dust particles cause respiratory disease. Diseases like silicosis and asbestosis result from specific dust.
- Carcinogenic agents like PAHs, Cr(VI), and Cd cause cancer.
- Hydrogen fluoride causes diseases of bone (fluorosis) and mottling of teeth.
- Certain heavy metals like lead, cadmium, mercury, chromium, nickel, and manganese enter into the body during inhalation, by skin absorption, and through the food chain. They cause acute and chronic poisoning.

23.16.1 Specific Health Hazards of Foundries and Forges

Apart from the general health hazards described above, some specific health hazards are prevalent among workers in foundries and forges of iron and steel industries and other manufacturing industries. Dust, SPM, noise, and gaseous pollutants pose a potential threat to the health of workers in industries and populations residing in the surrounding areas. Dust also absorbs gases and in this combination proves to be a more serious health hazard due to synergism. It has recently been demonstrated that SO_2 absorbed on submicroscopic particles penetrates deep into the lungs, and this is a greater danger to health. Silicosis and siderosis are common diseases in foundry men and forge shop workers. Most of the cases of silicosis, however, have arisen in the manufacture of silica-containing materials and in foundry workers. The pottery industry has a bad reputation for silicosis. Grinders in the cutlery industry using sandstone wheels died in large numbers from silicosis. Pneumoconiosis and black lung diseases are caused due to coal dust. Asbestosis is common in asbestos workers in industries. Direct IR radiation poses a risk to sight. Contact with hot metal or hot water may result in severe burns. Workers exposed to gamma rays and related ionizing radiations suffer from several hazards. Explosion and fire hazards occur during handling of liquid metal and the presence of flammable chemicals and liquid fuel. Iron foundry slag may be highly reactive if calcium carbide is used to desulfurize the iron.

23.16.2 Health Hazards due to Heavy Metals

Heavy metal pollution either directly or indirectly entering the food chain is becoming an increasing threat to health. Mercury metal is toxic for living beings. It enters the body through the food chain, skin, or inhaled air and causes mercury poisoning. The symptoms of mercury poisoning are tremor, gingivitis, and loss of teeth, together with a condition called "erethism," which includes excessive shyness [4]. Organic mercurials are used as bactericides and fungicides, and are used in the paint, paper, and alkali industries. Methyl mercury is more poisonous than mercury itself, which is used as a catalyst in a factory in Minamata. Tetraethyl lead (TEL) used as an antiknock agent in gasoline and TML pose potential health hazard. The harmful effects include blocking of spindle fiber mechanism in cell division, uncoupling of oxidative phosphorylation, and altered rapid eye movement phase of sleep leading to insomnia. Other metals with toxic effects known to occur as pollutants in some localized areas are Be, Pb, Cu, Zn, Cd, Fe, Ni, and so on. Lead causes *plumbism* in human beings. Vanadium levels in the environment are rising as a consequence of the burning of vanadium-containing fossil fuels and mining and processing to meet the growing needs for the metal in industries. Both acute and chronic effects of occupational exposure to V compounds are manifested in the respiratory tract by irritation, including bronchitis and pneumonia. Some studies on health hazards of dust, fume, gaseous pollutants, noise, heavy metals, and so on, from foundries and forges of the iron and steel and other manufacturing industries have reported, for example, risk of bladder cancer in foundry workers.

In epidemiological studies, a meta-analysis of foundry workers was done. Summary risk estimates (SRE) were calculated from 40 systematically extracted results. Weekly increased risks were overall with an SRE of 1.11. Exposure response findings showed significantly increased risk of about 1.6–1.7 after 20 or more years of employment. Occupation-specific SREs showed a 40%–50% increased risk among molders, casters, and unskilled foundry workers. There was limited evidence that bladder cancer risk correlated with lung cancer risk, which is a more established risk among foundry workers. Mortality from lung cancer disease in steel industry workers had been reported due to chromium and PAH. Silicosis and cardiovascular diseases have been reported. Public health assessment in forges and foundries was done. Health hazards from emissions of propane burners are seen in forges and foundries of industries. Sometimes propane burners cause explosions. Noise from forges and foundries can cause loss of hearing and disorders in the central nervous system.

23.17 Sheet Metal Work

23.17.1 Sheet Metal

Sheet metal is a flat, thin piece of metal that can be formed into a variety of shapes by applying force to the metal and modifying its geometry. The applied force stresses the metal beyond its yield strength, causing the material to plastically deform, but not to fail. There are a number of different sheet metal-forming processes:

- Bending
- Spinning
- Deep drawing
- Stretch forming
- Incremental sheet forming

Sheet metal sites principally use electricity. Natural gas is typically used for space heating as the vast majority of sheet metal processes are carried out cold.

Normally, sheet metal thickness is between 0.4 mm (1/64 in.) and 6 mm (1/4 in.); above 6 mm (1/4 in.), it is called a plate in industries. Commercially, sheet metal has several applications in the automotive, aircraft, locomotive, and construction industries.

In all forming operations, processing is done either in the cold or the hot stage. Most sheet metal operations are at room temperature, except when the sheet is thick. The sheet metal used in this forming operation should be malleable and drawable. The most commonly used steel is low-carbon steel (0%–0.15% C). The steel sheets are manufactured by rolling and have good formability. The sheet metal operations are stamping, cutting, and bending. All these operations are performed on a press (called a stamping press to distinguish from a forging press). The tool for stamping, cutting, or drawing is usually called a *punch-and-die.*

23.17.2 Punching Characteristics

Punching is the most cost-effective process of making holes in a strip or sheet metal for average to high fabrication.

- It is able to create multiple shaped holes.
- Punches and dies are usually fabricated from conventional tool steel or carbides.
- It creates a burnished region roll-over and a die break on the sidewall of the resulting hole [5].
- It is quick.

23.17.3 Press Work

A press is used for a wide range of processes such as punching, piercing, blanking, notching, bending, drawing, and folding. It may be operated by means of a crank connected to a heavy flywheel using hydraulic power. Formulae are given for various processes. Most punch presses are mechanically operated, but simple punches are often hand powered. Major components of this mechanical press are the frame, motor, ram, die posts, bolster, and bed. The punch is mounted into the ram, and the die is mounted to the bolster plate. The scrap material drops through as the workpiece is advanced for the next hole. A large computer-controlled punch press is called a CNC turret punch. It houses punches and their corresponding dies in a revolving indexed turret. These machines use pneumatic, hydraulic, or electrical power to press the shape with enough force to shear the metal (Figure 23.18).

23.17.4 Press Tool Theory

In pressworking, mostly three operations are performed: shearing, blanking, and punching. Shearing involves simply cutting along a straight line with a punch. Blanking separates out a useful part from the sheet metal by cutting along a closed outline (usually circular). This process is carried out in one step. It is opposite to making holes in a sheet metal part for heat exchanger punching, where the stamped pieces are trash.

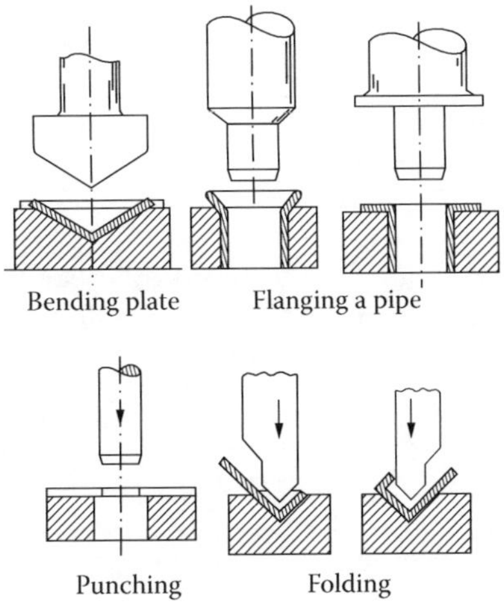

Figure 23.18

Press work.

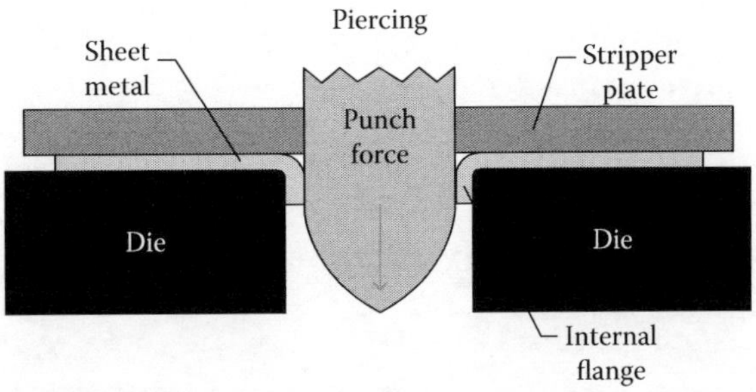

The punch diameter should be less than the die opening for the part to slip out of the punch press. The clearance (C_l) as shown in Figure 23.19 is calculated as

$$C_l = A_c t$$

where

A_c is the clearance allowance (values provided by Groover [6]) (Table 23.1)

t is the thickness

Sheet metals are grouped into three categories of cold rolled steel, and aluminum alloys. It is advisable to select the values of any other metals from the table.

Groover [6] has compiled these A_c values from Hoffman [7]. The estimated clearance value (C_l) can be applied for proper blanking and punching operations to find proper punch and die dimensions for a round blank of diameter D. The punch diameter D_p is calculated as $D_p = D - 2C_l$, and the die diameter D_d as $D_d = D_p + 2C_l$.

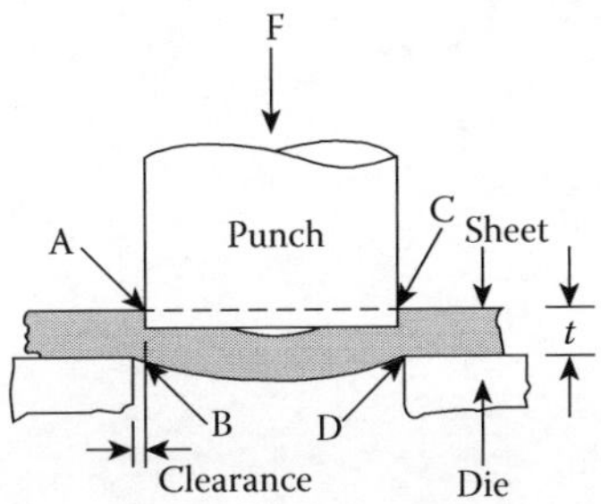

Figure 23.19

Bending operation.

Table 23.1 Clearance Allowance Metal Group

Sheet Metal Group	Clearance Allowance (A_c)
1100S and 5052S aluminum alloys, all tempers	0.045
2024ST and 6061ST aluminum alloys; brass, all tempers; soft cold rolled steel, soft stainless steel	0.060
Cold rolled steel, half-hard; stainless steel, half-hard and full-hard	0.075

In order for the blank to falls through the die, the die opening should have an angular clearance of 0.25°–1.5°. It is likely that if clearance is less than estimated above, this will cause excessive force and work may not be accomplished, whereas if the part is too large, it will have big burrs or very rough edges (Figure 23.20).

23.17.5 Punching Process

In a punching press, the maximum shear force required for bending is calculated as

$$\text{Maximum shear force } F_{max} = \tau_u t_p$$

where:

F_{max} = maximum shear force
τu = ultimate shear stress
t = material thickness
x = penetration
p = perimeter of the profile or hole being cut

Therefore, work done (W) during punching is calculated as

$$\text{Work done W} = F_{max} x$$

and the penetration ratio (c) is estimated as

$$\text{Penetration Ratio } c = \frac{x}{t}$$

The punch force required to punch a piece of sheet metal can be estimated from the tensile strength consideration as follows:

$$F_{max} = 0.7(UTS)tp$$

where:

t is the sheet metal thickness
p is the total length sheared (perimeter of the shape)
UTS is the ultimate tensile strength of the material

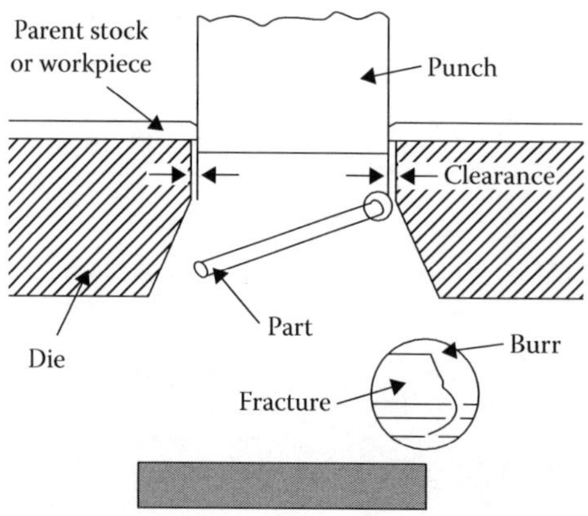

Figure 23.20

Blanking process.

Die and punch shapes affect the force during the punching process. The punch force increases during the process as the entire thickness of the material is sheared at once. A beveled punch helps in the shearing of thicker materials by reducing the force at the beginning of the stroke. However, beveling a punch will distort the shape because of lateral forces that develop. Compound dies allow multiple shaping to occur. Compound dies will generally slow down the process and are typically more expensive than other dies. Progressive dies may be used in high-production operations. Different punching operations and dies may be used at different stages of the operation on the same machine (Figures 23.21 through 23.23).

23.17.6 Shearing Process

$$\text{Shearing force } F = \frac{F_{max}}{\left(1 + \frac{h}{x}\right)}$$

where h is shear.

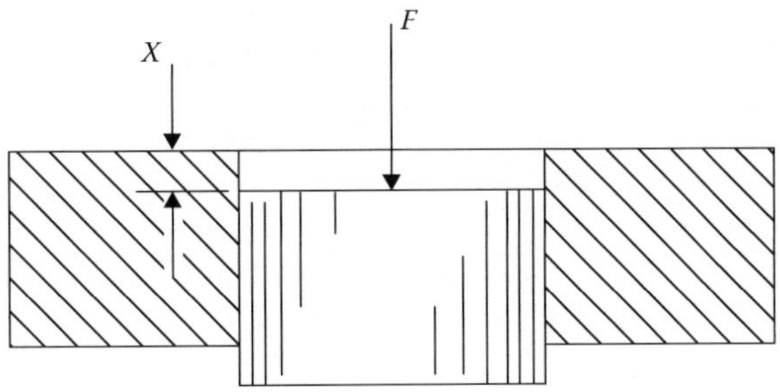

Figure 23.21

Punching diagram.

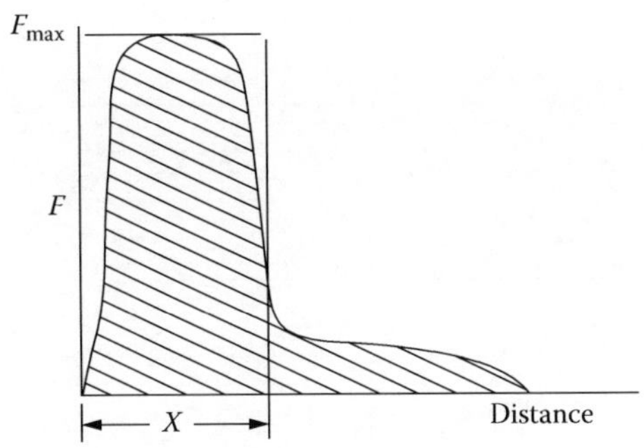

Figure 23.22

Maximum shear force graph.

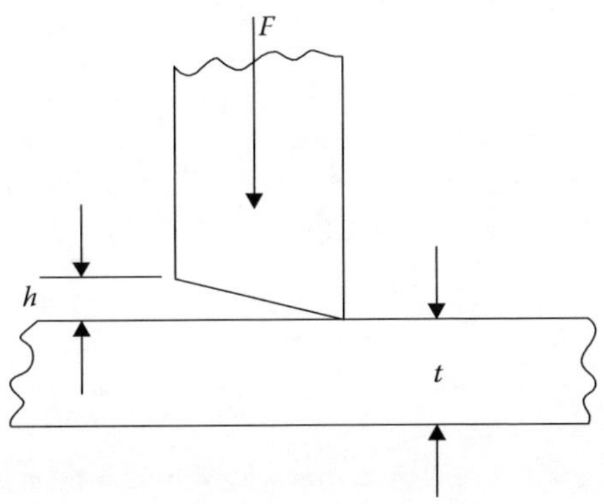

Figure 23.23

Shearing diagram.

23.17.7 Plastics

Punching in plastics fabrication usually refers to the removal of scrap plastic from the desired article. For example, in extrusion blow molding it is common to use punching dies to remove tails and molding flash (scrap plastics) and handle slugs from bottles or other molded containers. In shuttle machinery, the containers are usually trimmed in the machines, and finished containers leave the blow molding machine. Other blow molding equipment, such as rotary wheel machinery, requires the use of downstream trimming. Types of downstream trimming equipment include detabbers for tail removal, rotary or reciprocating punch trimmers, and spin trimmers.

23.17.8 Drawing Process

In a drawing operation, the diameter of the blank (D) required is calculated as follows for drawing a cup of diameter d and height h.

$$\text{Blank Diameter } D = \sqrt{d^2 + 4dh}$$

The force required for drawing is calculated from tensile strength as

$$\text{Required Force } F = \pi d t \sigma_u$$

where σ_u is ultimate tensile stress (Figure 23.24).

One of the measures of deep drawing prevalent in sheet metal industries is drawing ratio (dr); it is calculated as $dr = D/D_p$, where D_p is the punch diameter calculated earlier (Figures 23.25 and 23.26).

23.17.9 Bending Operation in Sheet Metal Industries

Large numbers of parts are made by the bending operation in sheet metal industries. These are found in the automotive, aircraft, and construction sectors. Normally, in bending, the sheet on the inside of the neutral axis is compressed and the outside is stretched. In this operation, we again need a punch and die. Depending on the bending type, V-bending or edge-bending, the die varies. For V-bending, a V-die is required, while

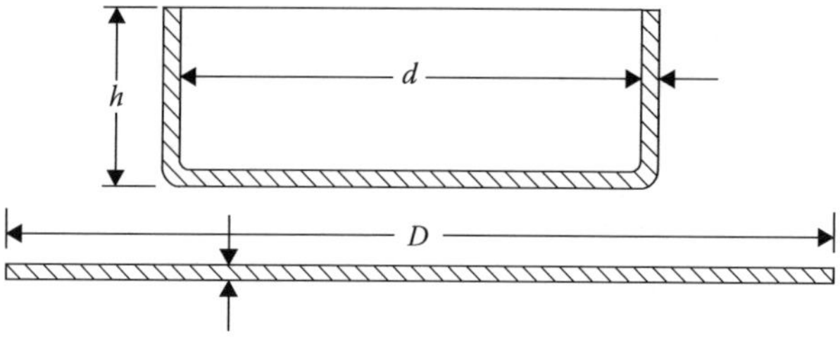

Figure 23.24

Drawing diagram.

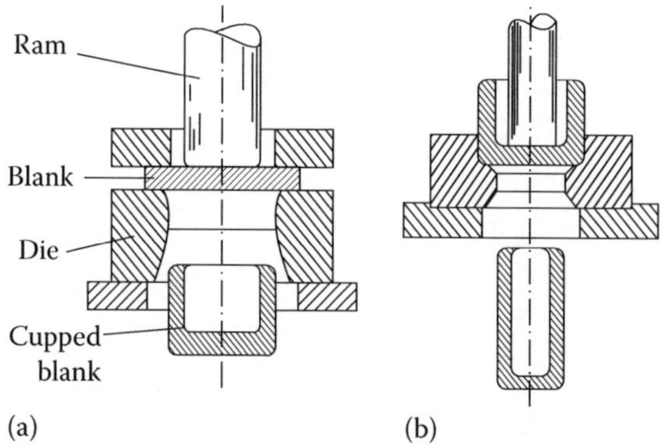

Figure 23.25

Deep drawing: (a) first stage, (b) second stage.

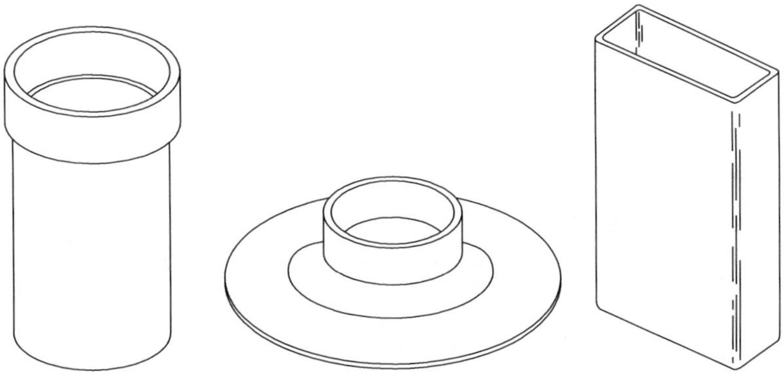

Figure 23.26

Drawing products.

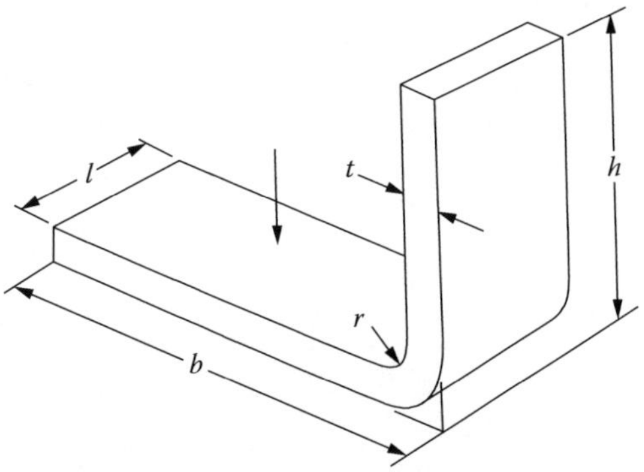

Figure 23.27

Bending diagram.

for edge-bending, the part placed as cantilever on the die punch is used to bend the part at the edges. A pressure pad is used to hold the part on the die.

In sheet metal work, allowance must be made for bends depending on the thickness of the material, the radius of the bend, and the bend angle (Figure 23.27).

23.17.9.1 Allowance for Right Angle Bend

Lengths a and b are reduced by an "allowance" c, and

$$c = r + t - \frac{\pi}{4}\left(r + \frac{t}{2}\right)$$

When $r = 2t$ (as is often the case), $c = 1.037t$.

23.17.9.2 Allowance for Bend with Outside Angle θ (Figure 23.28)

$$c = (r + t)\tan\frac{\theta}{2} - \frac{\pi\theta}{360}\left(r + \frac{t}{2}\right), (\theta \text{ in degrees})$$

$$\text{When } r = 2t, c = \left(3\tan\frac{\theta}{2} - 0.02180\right)t$$

23.17.10 Bending Process

$$\text{Bending Force } F_b = \tau_u L t$$

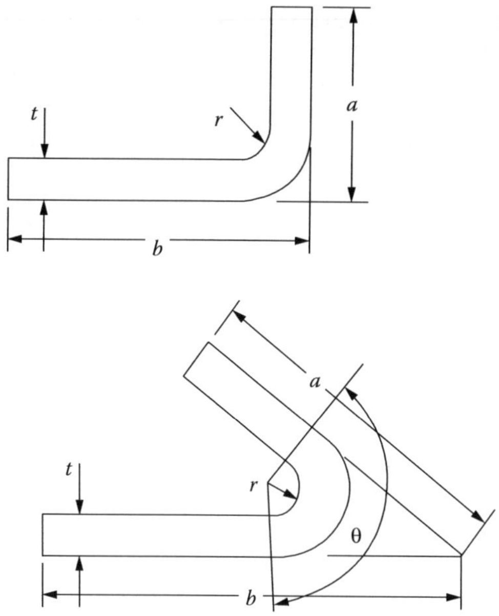

Figure 23.28

Bending examples.

$$\text{Planishing Force } F_t = \sigma_y Lb$$

where σ_y = yield stress.

$$\text{Initial length of strip } L_i = h - t - 2r + b + \frac{\pi}{2}\left(r + \frac{t}{2}\right)$$

After bending a part, the punch is removed, and the elastic energy in the part deformed will cause it to recover partially toward the original shape, and this is called *springback*. The springback depends on the yield strength as well as the elastic strength of the material. It is likely to increase the bend radius as well as the included angle (Figure 23.29).

23.17.11 Design Considerations for Bending Operations

Blank design: Material scrap is a major concern in stamping, so it is important that blanks be designed so that they can be easily nested in order to reduce scrap.

Bending: Bending creates large distortions of the stress concentration within the sheet metal. At times, these concentrations can become large enough to create wrinkling, buckling, or even fracture. A common practice to reduce these effects is the incorporation of carefully located notches in the design of the part. These notches can help avoid tearing and wrinkling in right angle bending operations. It is also important to avoid placing holes too close to a bend, since holes already create areas of higher stress concentration.

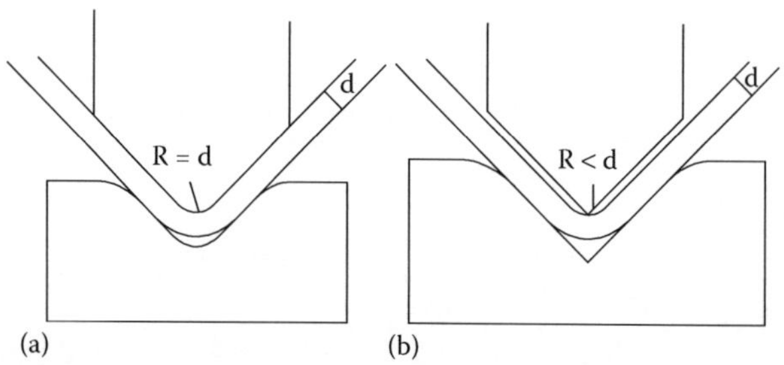

Figure 23.29

Springback in bending: (a) bottoming, (b) coining.

Progressive die operations: Tooling costs can become somewhat high in stamping operations, so it is important to design the part with as few features as possible or with features that require few stamping stations. If two features are placed too closely together, there might be little clearance for simultaneous punches, thus adding unnecessary tooling costs to the process.

Springback: Plastic deformation of a material is always followed by elastic recovery once the load causing the deformation is removed. A common fix for this phenomenon is simply overbending the part, so it springs back to the desired angle.

23.17.12 Cost

Tooling and equipment costs for stamping operations can be somewhat high, whereas labor costs are typically very low. However, stamping dies can cost reasonably less than tooling used in molds, forging, and casting because of the repetitive nature of combined forming operations that stamping allows. Many of the products created with those methods could be easily redesigned for stamping instead. One of the advantages of switching to stamping operations is that higher production rates can be achieved, resulting in lower manufacturing costs. The cost of secondary operation or down-the-chain operation can also be reduced by stamping, since many stamped parts require no machining. Scrap production from stamping operations is typically on the order of magnitude of 10%–25% of the original material, which is on the low end for most manufacturing processes, with some having scrap production as high as 60%. Once the scrap rate is reduced, the energy consumed and embodiment energy are reduced, resulting in less GHG emissions.

Machine tools have been identified as one of the main energy-consuming systems in the manufacturing industry, and they should be targeted for the reduction of their environmental impact. Machine tool manufacturers are looking for guidelines to improve their products in an effective and low-cost manner. The energy efficiency in machine tools will have great positive impact on the environment directly with regard to the generation of energy, resulting in the depletion of natural material resources. During the use phase machine tools have high environmental cost associated with manufacturing. However, the total energy requirement for operation is not constant; instead it depends on constant and variable factors. While the constant part is used to insure the active response of system parts, such as the main driving system and the integrated electronics or the auxiliary subsystems (positioning servo motors, ventilators, or chillers), and is independent of whether or not a part is being produced, the variable part corresponds to the energy needed to produce a workpiece and is proportional to the amount of material removed or being processed. The total electricity requirement for machine tool operation as a function of the rate of material processed has been modeled for a set of manufacturing operations involving material removal or deposition [8]. In general, redesign strategies for machine tools should aim to improve energy efficiency by conversion of hydraulic systems to all-electric systems and to improve the rate at which the physical mechanism can perform the desired operation, that is, improve machine usage.

Some electric models of the press-brake are now commercially available, and although initially restricted to a lower range of bending capacity, some technologies are already enabling solutions for a higher loading range.

References

1. Machinability Data Center and Metcut Research Associates. (1980). *Machining Data Handbook*. Machinability Data Center, Metcut Research Associates, Cincinnati, OH.
2. www.oecd.org/innovation/green/toolkit OECD Sustainable Manufacturing Toolkit, 2015.
3. Schey, J. A. (2000). *Introduction to Manufacturing Processes*, 3rd edn. McGraw Hill, New York.
4. Vimy, M. J. (1998). *Your Toxic Teeth: Mercury Poisoning from Fillings*. Prevident Management.
5. Kordonowy, D. N. (2001). A power assessment of machining tools. B.S. Thesis, MIT, Department of Mechanical Engineering, Cambridge, MA.
6. Groover, M. P. (2012). *Fundamentals of Modern Manufacturing: Materials, Processes, and Systems*, 5th edn. Wiley, Hoboken, NJ.
7. Hoffman, E. G. (1984). *Fundamentals of Tool Design*, 2nd edn. Society of Manufacturing Engineers, Dearborn, MI.
8. Gutowski, T., Jeffrey, D., and Thiriez, A. (2006). Electrical energy requirements for manufacturing processes. In *13th CIRP International Conference on Life Cycle Engineering*, Leuven, Belgium, May 31–June 2.

Bibliography

Abdalla, H. S. and Patel, S. (2006). The performance and oxidation stability of sustainable metalworking fluid derived from vegetable extracts. *Proceedings of the Institution of Mechanical Engineers, Part B: Journal of Engineering Manufacture* 220: 2027–2040.

Abou-El-Hossein, K. A. (2008). Cutting fluid efficiency in end milling of AISI 304 stainless steel. *Industrial Lubrication and Tribology* 60 (3): 115–120.

Alves, S. M. and Oliveira, J. F. G. (2008). Vegetable based cutting fluid-an environmental alternative to grinding process. In *15th CIRP International Conference on Life Cycle Engineering* (pp. 664–668), Sydney.

Ashby, M. F. (2009). *Materials and the Environment: Eco-informed Material Choice* (pp. 112, 114, 284–287, 290–291). Butterworth-Heinemann, Burlington, MA.

Baniszewski, B. (2005). An environmental impact analysis of grinding, S.B. Thesis, Department of Mechanical Engineering, MIT, Cambridge, MA.

Batzer, S. A. and Sutherland, J. W. (1998). The dry cure for coolant ills. *Cutting Tool Engineering* 50 (4): 34–44.

Cetin, M. H., Ozcelik, B., Kuram, E., and Demirbas, E. (2011). Evaluation of vegetable based cutting fluids with extreme pressure and cutting parameters in turning of AISI 304L by Taguchi method. *Journal of Clean Production* 19 (17–18): 2049–2056.

Dahmus, J. B. and Gutowski, T. G. (2004). An environmental analysis of machining. In *Proceedings of ASME International Mechanical Engineering Congress and RD&D Expo*, Anaheim, CA.

Dittrich, M. A., Gutowski, T. G., Cao, J., Roth, J. T., Xia, C., Kiridena, V., Ren, F., and Henning, H. (2012). Exergy analysis of incremental sheet forming. *Production Engineering* 6 (2): 169–177.

Fox, N. J. and Stachowiak, G. W. (2007). Vegetable oil-based lubricants—a review of oxidation. *Tribology International* 40 (7): 1035–1046.

Fthenakis, V. M. and Kim, H. C. (2007). Greenhouse-gas emissions from solar electric- and nuclear power: A life-cycle study. *Energy Policy* 35 (4): 2549–2557.

Gregory, K. A. and Menerey, D. (1994). Sustainable development by design: Review of life cycle design and related approaches. *Journal of the Air and Waste Management Association* 44 (5): 645–668.

Guo, Y., Loenders, J., Duflou, J., and Lauwers, B. (2012). Optimization of energy consumption and surface quality in finish turning. In *5th CIRP Conference on High Performance Cutting*, Zürich, Switzerland.

Gutowski, T., Branham, M., Dahmus, J., Jones, A., Thiriez, A., and Sekulic, D. (2009). Thermodynamics analysis of resources used in manufacturing processes. *Environment Science and Technology* 43 (5): 1584–1590.

Haan, D. M., Batzer, S. A., Olson, W. W., and Sutherland, J. W. (1997). An experimental study of cutting fluid effects in drilling. *Journal of Materials Processing Technology* 71 (2): 305–313.

Haapala, K., Rivera, J. L., and Sutherland, J. (2009). Reducing environmental impacts of steel product manufacturing. *Transactions of the North American Manufacturing Research Institution of the SME* 37: 419–426.

Hirohisa, N. (2012). *Environmental Burden Analyzer for Machine Tool Operations and its Application*. Intech, Rijeka. www.intechopen.com/books/manufacturing-systems.

Hondo, H. (2005). Life cycle GHG emission analysis of power generation systems: Japanese case. *Energy* 30 (11–12): 2042–2056.

Hong, S. Y. and Broomer, M. (2000). Economical and ecological cryogenic machining of AISI 304 austenitic stainless steel. *Clean Products and Processes* 2: 157–166.

Horne, R., Grant, T., and Verghese, K. (2009). *Life Cycle Assessment (LCA): Principles and Practice*, EPA/6000R-06/060, May 2006. Collingwood, Australia.

Jan, E. and Bras, B. (1998). ISO 14000 and activity based life cycle assessment in environmentally conscious design and manufacturing: A comparison. In *1998 ASME Design Engineering Technical Conference*. American Society of Mechanical Engineers, Atlanta, GA.

Johnson, W. (1956). The pressure for the cold extrusion of lubricated rod through square dies of moderate reduction at slow speeds. *Journal of the Institute of Metals* 85.

Kalpakjian, S. and Schmids, R. (2010). *Manufacturing Processes for Engineering Materials*, 6th edn. Pearson Prentice Hall, Upper Saddle River, NJ.

Khan, M. M. A. and Dhar, N. R. (2006). Performance evaluation of minimum quantity lubrication by vegetable oil in terms of cutting force, cutting zone temperature, tool wear, job dimension and surface finish in turning AISI-1060 steel. *Journal of Zhejiang University Science A* 7 (11): 1790–1799.

Krahenbuhl, U. (2002). Vegetable oil-based coolants improve cutting performance (Cutting Fluids). In *Tooling & Production*. Nelson.

Lange, K. (1985). *Handbook of Metal Forming*. McGraw-Hill, New York.

Narita, H., Desmira, N., and Fujimoto, H. (2008). Environmental burden analysis for machining operation using LCA method. In *Proceedings of 41st CIRP Conference on Manufacturing Systems* (pp. 65–68), Tokyo, Japan.

Niggeschmidt, S., Helu, M., Diaz, N., Behmann, B., Lanza, G., and Dornfeld, D. (2010). Integrating green and sustainability aspects into life cycle performance evaluation. In *Proceedings of 16th CIRP International Conference on Life Cycle Engineering.*

Pettersson, A. (2007). High-performance base fluids for environmentally adapted lubricants. *Tribology International* 40 (4): 638–645.

Rangarajan, A. and Dornfeld, D. (2004). Efficient tool paths and part orientation for face milling. *CIRP Annals* 53 (1): 73–76.

Rao, D. N. and Srikant, R. R. (2006). Influence of emulsifier content on cutting fluid properties. *Proceedings of the Institution of Mechanical Engineers, Part B: Journal of Engineering Manufacture* 220: 1803–1806.

Santos, J. P., Oliveira, M., and Almeida, F. G. et al. (2011). Improving the environmental performance of machine-tools: Influence of technology and throughput on the electrical energy consumption of a press-brake. *Journal of Cleaner Production* 19: 356–364.

Shokrani, A., Dhokia, V., and Newman, S. T. (2012). Environmentally conscious machining of difficult-to-machine materials with regard to cutting fluids. *International Journal of Machine Tools and Manufacture* 57: 83–101.

Srikant, R. R., Rao, D. N., and Rao, P. N. (2009). Influence of emulsifier content in cutting fluids on cutting forces, cutting temperatures, tool wear, and surface roughness. *Proceedings of the Institution of Mechanical Engineers, Part J: Journal of Engineering Tribology* 223: 203–209.

Szargut, J., Morris, D. R., and Steward, F. R. (1988). *Exergy Analysis of Thermal, Chemical and Metallurgical Processes.* Hemisphere, Washington, DC.

Taniguchi, M., Kakinuma, Y., Aoyama, T., and Inasaki, I. (2006). Influences of downsized design for machine tools on the environmental impact. In *Proceedings of MTTRF Annual Meeting* (pp. 1–4).

Tönissen, S. (2009). Power demand of precision machine tools. MS Thesis, Department of Mechanical Engineering, University of California, Berkeley, CA.

www.oecd.org/innovation/green/toolkit OECD Sustainable Manufacturing Toolkit, 2015.

Section VI
Regulations for Sustainable Manufacturing

24

International Regulations for Green Design and Manufacturing

24.1 Introduction to International Regulations for Green Manufacturing

The environmental crisis has several idiosyncrasies that necessitate implementation of a new vision of environmental protection—a vision transcending national boundaries and embracing the global collective. A cursory examination of the legal environmental systems of some developing countries lends credibility to the argument that nation states, particularly those of the developing world, cannot be relied upon to pursue environmental objectives beneficial to the global commons. It has been observed that the enforcement of environmental standards tended to be lax in developing countries. A survey of air and water pollution laws revealed that the failure of the legal system rose out of a lack of financial resources dedicated to legal enforcement, multiple agencies performing overlapping functions, and a general failure to implement laws.

The enactment of far-reaching environmental legislation can also be observed in India. Yet, a variety of problems remained; namely, uncertainty about the legislative authority of the central government to implement environmental statutes, enforcement problems, lack of a private right of action, and insufficient resources to implement environmental laws. The trend in Africa is similar to that which has been observed elsewhere in the developing world. It has been observed that there was a proliferation of pollution control legislation that created conflict in asserting jurisdiction. This resulted in difficult implementation of environmental protection regulations. Environmental expenditures are normally borne by the state, and it would appear that the will of the developing world is undermined by competing interests for limited financial resources. In most instances, developing countries are only motivated to adopt environmental measures when external financial and technical assistance is provided. Developing countries appear to be more concerned with pursuing developmental objectives that are driven by the forces of industrialization. Further, while the developed world certainly bears responsibility for many current environmental problems, developing countries' present and future contributions to the decimation of the environment cannot be underestimated. The evidence clearly points to a significant increase in developing nations' contribution to environmental problems affecting the global commons. Deterioration of the environment appears to hold grave consequences for the international economy, the security and health of humankind, and the stability of habitation patterns. A purely domestic focus on environmental solutions is therefore untenable. It has been observed that from rather hesitant and tentative beginnings, environmental law has progressed rapidly under a combined stimulus of ever more powerful means of inflicting irreversible environmental damage and an ever-increasing awareness of the fragility of the global environment. Furthermore, principles such as intergenerational equity, sustainable development, and common heritage are attaining a level of prominence. It should be part of thinking every aspect of life, such as living space, transportation, consumption, air quality, and education, including training.

Another dimension to the uneasy fit between the protection of the environment and sovereign rights is the depletion of the stratospheric ozone layer. The stratospheric ozone layer, by its very nature, cannot be defined either as part of the territory of a particular state or its property. The fact that transboundary pollution affected not only neighboring states but also entire regions and even the entire international community must be recognized by all countries. The realization that acid rain may be responsible for the decimation of much of the aquatic ecosystems in Europe prompted the negotiation of the Convention on Long-Range Transboundary Air Pollution. Common but differentiated responsibilities should be pursued to redress the historical imbalances. This has been clearly contemplated by the Framework Convention on Climate Change, where Article 4 states "All Parties, taking their common but differentiated responsibilities ... shall," where the use of the word "shall" implies an obligation on the part of member states. The framework convention prescribes different responsibilities for dealing with climate change and this was manifested in the Kyoto Protocol to the framework convention, where developed countries accepted certain specific restrictions on greenhouse gas emissions but to which developing nations were not required to adhere.

The preamble to the Montreal Protocol requires special provisions for meeting the needs of developing countries. This includes the setting up of the Multilateral Fund for the Implementation of the Montreal Protocol (MFIMP) and the exemption of developing countries from making contributions to the MFIMP, as per Article 10(6) of the amended Montreal Protocol. Without some effort to recognize the plight of developing countries and provide appropriate incentives, little would be accomplished. Furthermore, not only has an international agreement provided for environmentally suitable technology, but another has provided for the transfer of technology to gain commercial returns as an incentive for ensuring participation of developing nations. Additionally, equity seems to demand preferential treatment of developing countries in the pursuit of effective solutions to environmental problems. This preferential treatment includes different technical standards and time frames for compliance for developed and developing countries in order to provide recognition for the interaction between environmental protection and economic development.

Even if it is possible to repair an environmental harm, the costs might be staggering, or at least beyond the means of any one nation. Also, the causal link between a harmful activity and damage to the environment may be difficult to establish. For example, in respect of acid rain, how can one firmly establish the causal link between the emission of air pollutants in India and the decimation of aquatic ecosystems in Pakistan? In the area of depletion of the stratospheric ozone layer, it is impossible to pinpoint whose chlorofluorocarbons (CFCs) are causing damage to the stratospheric ozone layer at any given time.

While worldwide concern with the environment is growing, developed and developing countries must find common ground in the pursuit of environmental objectives. Nevertheless, the problems confronting developing nations are somewhat different. Indoor air pollution remains a more immediate threat to the health and well-being of many people in the developing world. There are also serious problems associated with water and sanitation in developing countries. The domination of the developed world in the process of agenda setting, agreement articulation, and international institutions ensures that these issues are not given any prominence on the international agenda for environmental protection. By adopting a comprehensive approach to defining the environment and dealing with environmental issues on an all-embracing basis, the objections of agenda manipulation and lack of sympathy for the most virulent problems affecting developing countries would be undermined and provide much-needed optimism among developing nations that their issues are also important.

All over the world environmental regulations are being enforced in manufacturing sector. Therefore, one of the first steps for a manufacturing company is to ensure that it is operating in compliance with the local environmental regulations. In the manufacturing business, consideration of environmental issues and laws has become an integral part of doing business in the United States and the international or global market. To be globally competitive and sometimes just to stay in business, a company must comply with environmental regulations and consider its options to operate in a more comprehensive environmentally conscious manner. This consciousness must start from the design stage itself, and if a country does not permit certain materials (toxic or otherwise), they should not be used. Most of the regulations are bound by environmental footprint indicators. The term footprint has become popular in the environmental lexicon, but it means different things to different people, as well as in different countries. A variety of different boundaries commonly used for footprint are shown in Figure 24.1.

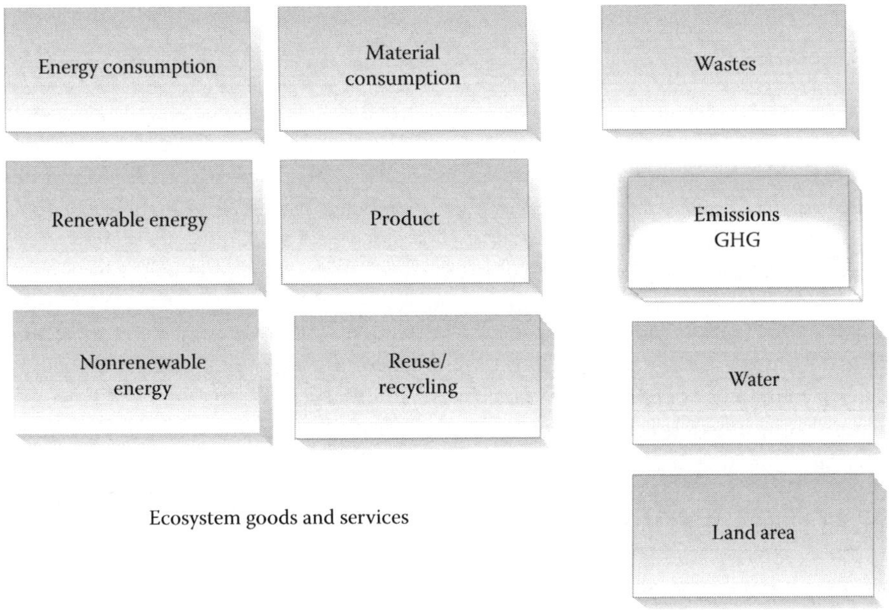

Figure 24.1

Boundaries of international regulations.

It is likely that some international regulations may include only the energy footprint on nonrenewable energy sources such as coal, oil, and so on, some others include renewable sources of energy. A material footprint may analyze total mass throughput, focus on consumption of input materials, or focus on reuse and recycling of materials as well. Most of the regulations have strict definitions of waste, which in turn may include solids, liquids, and/or airborne emissions. It should be noted that carbon footprint focuses on just one type of airborne emissions, greenhouse gases (GHG).

24.1.1 Green Manufacturing Technologies: Requirements and Response

1. Need for green manufacturing
   a. Customers are increasingly concerned about environmental quality
   b. Every country in the world is concerned about environmental protection and so there is the creation of an environmental protection agency (EPA)
   c. Reasons for environment protection consciousness:
      i. Energy crisis as fuel supply is running out
      ii. Global warming due to pollution of atmosphere
      iii. Depletion of ozone layer
      iv. Loss of forest and greeneries on the earth

Therefore, there is need for green design and manufacturing to sustain the growth of the economy. The green manufacturing concept starts at the design stage, so we have the term green design. This is a multidisciplinary approach that can help in reducing energy consumption along with less material-intensive design. Advanced and improved material removal processes and other fabrication techniques can reduce waste and hazardous materials. We can see that if we have a Pb-less welding material, there will be less lead (Pb) in the atmosphere. There is a need for manufacturing or fabrication techniques that requires less energy to produce a part or component, compared with others. It is possible to imagine that material removal processes such as turning will require less energy to turn raw material into a component than foundry processes. It may not be out of place to mention that fabrication process such as forging may need less energy compared with turning or milling.

It has been observed through the years some products have become green or eco-friendly. The Canon copier is one such product which has become greener in every stages of design. It is suggested that to make the product green or environmentally friendly, its energy consumption during manufacturing should be as low as possible, materials used in parts and packages should toxic-free, and it should be recyclable as far as possible. The case study of Canon copiers shows landmarks in product development through the years.

In 1993, lead-free optical glass was used, 1995 saw the elimination of PBB and PBDE in covers for all products, 1996 saw energy-efficient on-demand printing technology, 2002 saw the introduction of recycled materials in covers, 2004 saw the introduction of the first RoHS-compliant copier, the iRC 6800, and 2004 saw a new iR generation with on-demand technology up to 45 pages per minute, as well as a new iR generation with sleep mode use of 1.5 W.

The on-demand printing technology has been utilized since 1990. With this technology, a ceramic heater localizes the heating to a specific area through a fixing film during printing. Surplus energy consumption is avoided and energy is utilized efficiently. In 2002, Canon moved to IH fixing technology. This technology employs an electromagnetic induction heater in which an eddy current is generated when the magnetic field passes through metal coils, directly heating the fixing sleeve. Efficiency results because heating is unnecessary during standby time. Energy consumption compared with the previous model is about 18%. It is estimated that CO_2 emissions reduced by 3.73 million tons, with a tremendous effect on environmental pollution. The energy costs saved millions of dollars.

Some Japanese have shown great ingenuity in material selection for green technology. Lead-free implementation examples include the following. Matsushita Electric (Panasonic) completed a switchover to Pb-free solder for printed circuit boards (PCBs) in March 2003; 12,000 models have been manufactured in 22 facilities in Japan and 79 overseas facilities of the Matsushita group. Approximately 80% of Sony's products use Pb-free solders and Pb-free products include LCD televisions, mobile phones, laptop PCs, and digital camcorders. Conversion plans to be completed by the end of March 2005. All of Toshiba's new hard disk drives have already implemented Pb-free soldering and halogen-free PCBs.

As was stated earlier, the recycling of materials and reuse of parts are important measurements for green technology and the recycled materials used in toy manufacturing is one good example. The recycled materials used in toys include the following. Plastic drink containers made from every pound of toys can help to keep at least 10 milk jugs out of dumps. Aluminum cans, industrial aluminum scrap, plastic containers and a percentage of reused steel can be used for bike pedals, grips, handlebar, and frames. The utilization of recycled materials in product design enhances the recyclability of components at the end of useful life. Thermoplastics are used for almost all toy components. They are melted at approximately 235°F, molded under pressure and harden as they cool in cold mold. In principle, these plastics can be melted are remolded an infinite number of times.

24.2 Environmental Regulations Worldwide

24.2.1 European Union Directive 2002/96/EC

The responsible treatment of electronic health care products at the end of their useful life is a critical issue worldwide. An important driver for increased and improved environmental regulations has been the influence of civil society the world over. The broad regulatory directives of the European Union and US government aimed at curbing use of toxic materials and waste have been very successful. The European Union (EU) adopted the Waste Electrical and Electronic Equipment (WEEE) directives in 2005 and it is strictly applied and implemented. These directives have been very successful in recycling, reuse and end-of-life disposal of WEEEs. WEEE directives have also identified certain hazardous and toxic materials that cannot be used in WEEE and it is necessary for the designers worldwide to know that the lead cannot be used in most electrical and electronic equipment. Mercury, cadmium, hexavalent, chromium, polybrominated biphenyls, and polybrominated diphenyl ethers face similar restrictions with the Restriction of Hazardous Substances (RoHS) directive.

1. WEEE: Waste Electrical and Electronic Equipment (WEEE) Directives.

   "Producers or third parties acting on their behalf, in accordance with Community legislation, set up systems to provide for treatment of WEEE using best available treatment, recovery and recycling techniques."

   The European Union has legislated very clearly about the recycling and reuse of electronic and electrical equipment. This law is effective since August 13, 2005. The legislation also provides for the following:

   - The legislation extends the manufacturer's responsibility toward end-of-line disposal.
   - Legislation sets minimum standard for collection targets, such as with recycling, reuse, and so on.
   - It encourages measures for green design using recycled or reused parts, components, and materials. Environmental aspects of green design and manufacturing are identified with reference to the following stages of the life cycle of the product:

- Smart raw material selection and use; as we can see, right at the design stage we need to select raw material while thinking of the manufacturing process, which can save energy and less material wastage.
- Manufacturing or fabrication; as mentioned above, material selection will lead to an energy-efficient fabrication process.
- Packaging, transport, and distribution; all the energy and hazardous material use saved during design and manufacturing should not be jeopardized during packaging, transport or at the distribution stage.
- Installation and maintenance; installation of the fabrication or manufacturing plant should not waste any energy or cause pollution during fabrication stage.
- Use at customer end; at the customer end, the product should not create any pollution or need large amounts of energy. It should be an energy-efficient process.
- End of life, meaning state of an energy-using product (EuP) having reached the end of its first use until its final disposal.

• Legislation to reduce disposal, hazardous impacts on the atmosphere, and resource loss (Figure 24.2).

It is obvious that the EU has set up quite a high standard for recover, reuse, and recycle. It has not been determined how successful this has been so far (Table 24.1).

24.2.2 Environmental Laws Highlights of Various Countries

One of the most important pieces of legislation of the European Union regards the restriction of hazardous substance (RoHS) (Figure 24.3). The legislation set up in 2002/95/EC is most comprehensive as far as restriction on hazardous material is concerned (Table 24.2). To reduce the hazardous materials at source, the legislation states:

> RoHS: European Union Member states shall ensure that, from 1 July 2006, new electrical and electronic equipment put on the market does not contain lead (Pb), mercury (Hg), cadmium (Cd), hexavalent chromium (Cr6+), polybrominated biphenyls (PBB) or polybrominated diphenyl ethers (PEDE).

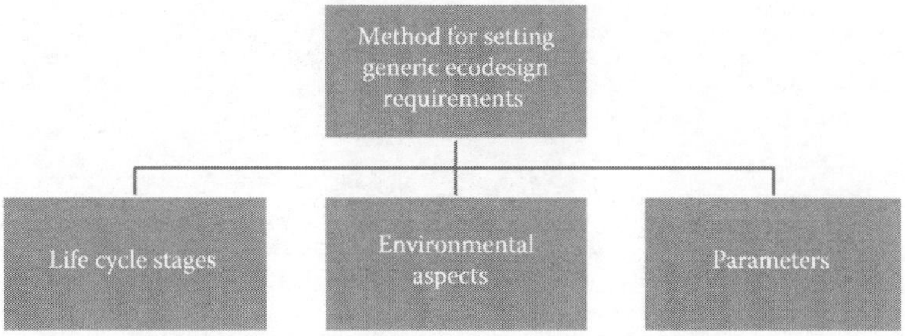

Figure 24.2

Eco-profile.

Table 24.1 Categories of Electrical and Electronic Equipment Covered by WEEE Legislation

Categories	Minimum Recovery Target by Weight (%)	Minimum Reuse and Recycling Target by Weight (%)
Large household appliances	80	75
Small household appliances	70	50
IT and telecommunication equipment	75	65
Consumer equipment	75	65
Lighting equipment	70	50
Electrical and electronic tools	70	50
Toys, leisure, and sports equipment	70	50
Medical devices	No target found	No target found[a]
Monitoring and control instruments	70	50
Automatic dispensers	80	75

[a] It was supposed to be set by December 2008.

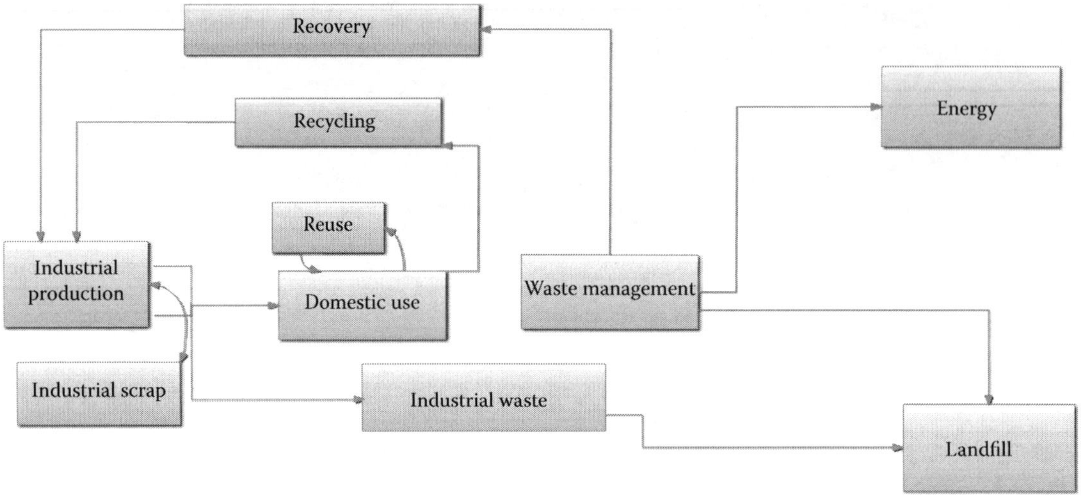

Figure 24.3

Flow of materials in society.

Table 24.2 End-of-Life Targets

Countries	Recycling (%)	Waste Combustion (%)	Landfill (%)
United States	25	15	60
United Kingdom	5	10	85
Switzerland	42	48	10
Sweden	35	38	27
Norway	15	18	67
The Netherlands	38	42	20
Japan	5	74	21
Germany	31	25	44
France	12	43	45
Finland	30	4	66
Denmark	25	55	20
Canada	28	4	68
Belgium	3	54	43
Austria	25	10	65

Table 24.3 Permissible Toxic Elements

Materials	Permissible Quantity (ppm)
Lead (Pb)	1000
Mercury (Hg)	1000
Cadmium (Cd)	100
Hexavalent chromium (Cr^{6+})	1000
Polynominated biphhenyls (PBB)	1000
Polubrominated ethers (PVDE)	1000

The legislation also provides the guidelines for the permissible quantity of above mentioned hazardous substance in electrical and electronic components. These are by parts-per-million (ppm) (Table 24.3).

The RoHS objective is to reduce the above mentioned hazardous materials in WEEE, as well as assist the recycling efforts set forth by WEEE documents. RoHS legislation outlines maximum enforcements requirements on member states of the European Union. It sets forth stringent requirements such as:

- Restricting hazardous substances that may have adverse effects on human health
- Restricting hazardous substances that may have adverse effects on the environment

The International Electrotechnical Commission (IEC) TC 111 working group 3 released the committee draft IEC 62321/1CD on 6/24/2005 for determining the levels of six regulated substances (lead, cadmium,

Table 24.4 List of Banned Substances by Application

#	Banned/Restricted Substance	Applications
1	Cadmium	Batteries, paints, pigments; additives in plastics (especially polyvinyl chloride (PVC) used in cable assemblies; phosphorescent coatings; detector/devices/LEDs
2	Mercury	Switches, pigments, paints, polyurethane materials (high glass PU windows); lamps, bulbs/lightings (displays, scanners, projectors)
3	Hexavalent chromium	Metal finishes for corrosion protection (chasses, fasteners); aluminum conversion coatings, alloys; pigments, paints
4	Leads	Solder and interconnects, batteries, paints, pigments, discrete components, sealing glasses, PVC cables (UV/heat stabilizer), metal parts, chasses, washers
5	Polybrominated biphenyls (PBBs)	Used as flame retardants (plastics, housings, cables, connectors, fans, components, paints)
6	Polybrominated diphenyl ethers (PBDEs)	Same as PBBs

mercury, hexavalent chromium, polybrominated biphenyls, and polybrominated biphenyls ether) in electrotechnical products. The law forbids the use of these banned substances particularly in the applications as illustrated in Table 24.4.

The IEC also recommends performing a tear-down analysis to find a lead-free product to ensure RoHS compliance. It wants to ensure that there is no contamination during the assembly process and no surprises from suppliers. It also wants to keep documentation of the tear-down analysis process for a statistically valid sample of homogenous materials. It also recommends that only one tear-down per sample is needed.

The European Union's Directive 2005/32/EC presents a framework for the setting of ecodesign requirements for energy-using products (EuP). EuP means a product that, when it comes to market, is dependent on the energy input (electricity, fossil fuels and renewable energy sources) to work as intended, or a product for the generation, transfer and measurement of such energy.

The EuP directive first appeared on July 6, 2005, and needed to be transposed to a law by August 11, 2007. Based on Article 95 of the framework, integration of various environmental aspects, including energy efficiency, hazardous substance control, water consumption, noise emission, and so on, into the design of an energy-using product is highly recommended. The EuP measure is the result of the integration of the WEEE and energy efficiency requirements (EER) proposals and it is supposed to be the first integrated product policy (IPP) model legislation. The EuP legislation covers all products that

1. Have a significant environmental impact within European Community.
2. Present significant potential for improvement in terms of its environmental impact without entailing excessive costs, taking into account in particular:
    a. The absence of other relevant European Community legislation or failure of market forces to address the issue properly.
    b. A wide disparity in the environmental performance of EuPs available on the market with equivalent functionality.
3. Represent a significant volume of sales and trade, indicatively more than 200,000 units a year within the European Community. Examples include refrigerators, freezers, radio-control toy car, robot dog, electronic pets, and so on.

The 2005/32/EC directive also defines very clearly the term ecodesign. It states that ecodesign means the integration of environmental aspects into product design with the aim of improving the environmental performance (parameters) of the EuP throughout its entire life cycle. These parameters are

1. Weight and volume of the product
2. Use of materials
3. Consumption of energy, water and other resources
4. Use of hazardous materials
5. Quantity and nature of consumables
6. Ease of reuse and recycling
7. Extension of lifetime
8. Emission of air, water, soil

The European environmental laws focus on the design aspects of the product. It is a paradigm shift in our approach to achieving sustainable development; each builds on the concept of life cycle estimation. Over 80% of all product-related environmental impacts are determined during the design phase. The life cycle thinking incorporates life cycle assessment (LCA), design for environment (DfE), product service systems (PSS) and integrated product policy (IPP) in the green design. The EuP six "RE" philosophy is outlined below.

1. Rethink: The product and its function, for example, how the product may be used more efficiently.
2. Reduce: Energy and material consumption throughout the product's life cycle.
3. Replace: Harmful substances with more environmentally friendly alternatives.
4. Recycle: Select materials that can be recycled and build the product in a way that it can be easily disassembled for recycling.
5. Reuse: Design the products so parts can be reused.
6. Repair: Make the product easy to repair so that the product does not yet need to be replaced.

Automotive manufacturing has increased in the last 20 years, reaching about 58 million units (excluding commercial vehicles) in 2000 [1]. At present, approximately 75%–80% of end-of-life vehicles in terms of weight, mostly metallic fractions, both ferrous and nonferrous, are being recycled. However, the remaining 20%–25% in weight, consisting mainly of heterogeneous mix of materials such as resins, rubber, glass, textile, and so on, is still being disposed of [2]. According to an EU directive [3], the disposal of vehicles is a major source of hazardous waste and toxic emissions. About 25% of a vehicle's weight is classified as hazardous waste. In Europe, about 12 million tons of vehicles reach their end of life every year, and 25% of them are disposed to the landfill [4].

Now, the automotive manufacturers are facing the impact of vehicles on the environment, consumption of energy and resources, waste generation, greenhouse gases, hazardous substance emissions, and disposal. These are burdens created by vehicle production and use [1]. To reduce the environmental impact of end-of-life vehicles, laws in the European Union.

Japan, the United States, and Australia require manufacturers to take back their products at the end of their useful life and recycle them [5]. In order to enhance the recycling rate of the vehicle, it is important to disassemble every part of the vehicle, depending on the materials. From end-of-life vehicles, dismantling companies first remove the oil, engine, transmission, tires, battery, catalytic converter, and other parts, which are commonly recycled or reused. The dismantling process is the process or procedure of breaking down an end-of-life vehicle into its individual components and material-specific elements [6]. Shredding companies then sort out the ferrous and nonferrous metals and resin from the remaining vehicle bodies. At shredder facilities, hulks are inspected prior to shredding to ensure that potentially hazardous components such as batteries, gas tanks, and fluids have been removed. Hulks (and other collected materials) are then shredded into fist-sized pieces using large hammer mills. The post-shredder process separates the stream of materials into the basic streams, ferrous material and nonferrous material. By using air separation, the nonferrous materials are separated into metal (containing aluminum, brass, bronze, copper, lead, magnesium, nickel, stainless steel, and zinc) and nonmetal, known as the ASR (containing plastic, glass, rubber, carpet, foam, textile, etc.) fraction. The ferrous metal fraction is sent to the steel smelters for recycling. The metal fraction is typically sent to another specialized facility to separate the stream into its individual metals by a variety of means. The ASR is disposed into landfill [7]. ASR often contains hazardous substances such as lead, cadmium, and PCBs. Therefore, some countries have classified ASR as hazardous waste and have established legislative controls. ASR represents about 20%–25% of the end-of-life vehicle weight [1].

However, this may be impossible if the disassembly and recycling costs are very high and unrealistic. Because of this, the disassembly effort should be determined in the vehicle development phase, where its properties are fixed. Based on that, the disassembability or the degree of easiness of disassembly, becomes important. The environmental impacts of products can be minimized only if the products can be disassembled and recycled easily [8] and the cost-effectiveness of recycling will be increased if disassembly is made easier [9].

24.2.3 End of-Life Vehicles (ELV) Regulations

In September 2000, after much consultation, the European Commission produced an end-of-life vehicles (ELV) directive. The EU ELV directive is the legislation enacted by the European Commission to minimize pollution resulting from vehicles that have reached the end of their useful life [4]. The EU directive states that the vehicle manufacturers and material and equipment manufacturers must meet the following objectives [3]:

1. Endeavour to reduce the use of hazardous substances when designing vehicles
2. Design and produce vehicles that facilitate the dismantling, reuse, recovery and recycling of end-of-life vehicles
3. Increase the use of recycled materials in vehicle manufacture
4. Ensure that components of vehicles placed on the market after July 1 2003 do not contain mercury, hexavalent chromium, cadmium, or lead

According to the EU directive [3], the requirements for dismantling, reuse and recycling of end-of-life vehicles and their components should be integrated in the design and production stages of new vehicles and producers should ensure that vehicles are designed and manufactured in such a way as to allow the quantified targets for reuse, recycling and recovery to be achieved. Various countries have adopted laws for recycling of automotive parts and reuse as much as possible. However, the European Union regulations are broad-based and more eco-friendly.

24.2.3.1 Japan

On 1 January 2005, the end-of-life vehicle recycling law was fully enforced in Japan. This law is intended to promote environmental conservation and the effective use of resources by implementing measures to ensure the proper and smooth recycling of end-of-life vehicles. Under the law, automobile manufacturers are obliged to collect and properly dispose of the following three items: CFCs that are used as air conditioner refrigerants and would destroy the ozone layer and contribute to global warming if emitted into the air; airbags that are difficult to dispose of because of their explosive nature; and automobile shredder residue (ASR) that remains after the collection of useful materials from end-of-life vehicles. The target of the Japan law for recycling is to establish recycling system with the goal of attaining 95% recycle rate by 2015.

24.2.3.2 United States

In the United States, there is no specific legislation regarding the management of end-of-life vehicles. All materials, either waste, materials or recycled materials are considered as a solid waste, so the recycling industry has received much less interest.

Currently, there is no shortage in waste-disposal sites because of abundant land. This situation can make the costs of waste disposal low. Furthermore, there is no standard waste legislation for the whole of the US. Every state has its own legislation, so the target and implementation varies from state to state.

However, Ford, Daimler Chrysler and General Motors have provided a special program to study how to improve the recyclability rate and methods to decrease the current ASR burden [7]. The US Environmental Protection Agency (EPA) is trying to promote the recycling concept among the vehicle manufacturers.

24.2.3.3 Australia

According to the Australian Department of the Environment [7], there is no specific end-of-life vehicle directive in Australia, but some progress has been made toward encouraging end-of-life vehicle recycling through informal encouragement of recyclers and dismantlers.

A joint project between the Department of Environment and the Auto Parts Recyclers Association of Australia (APRAA) has produced guide booklets on waste oil recycling, which were sent to recyclers and dismantlers throughout Australia during 2003. They are also encouraging recyclers to prepare their own environmental action plan, highlighting the efficiency benefits and improved reputation amongst councils, environmental protection agencies, customers and staff of any operators taking actions to protect the environment. There is also a current project being undertaken by the plastics and chemicals industries. The European Union (EU) target is to produce 22% of all electricity from renewable sources by 2010. By 2001, 15% of electricity was produced from renewable sources. However, further efforts are needed to ensure that EU will meet its indicative target of 22% by 2010.

24.3 New International Standards for Life Cycle Assessment: ISO 14040 and ISO 14044

The International Organization of Standardization (ISO) is the world's largest standard setting organization, with membership consisting of national standard bodies from 143 countries. ISO 14000 is a series of international standards on environmental management that provides a framework for implementing corporate environmental management systems (EMS).

24.3.1 Background, Aims, and Scope

The development of the international standards for life cycle assessment (ISO 14040:1997, ISO 14041:1999, ISO 14042:2000, ISO 14043:2000) was an important step to consolidate procedures and methods of LCA. Their contribution to the general acceptance of LCA by the international community was crucial. Currently, the process of the revision of this first generation of LCA standards is close to completion. The new international standards of LCA, that is, new ISO 14040 and ISO 14044, explain the outline and the technical changes.

A consensus has been achieved on the following four key objectives:

1. Increase readability by compiling only two documents/merging different documents/reorganizing the current standards, but
   a. Keep the technical contents (only improvements are acceptable)
   b. Keep the consensus/balance
   c. Keep the requirements
2. Address applications of LCA (life cycle thinking, relations to eco-label, DFE, life cycle managements etc.)
   a. Integrate the different application of life cycle in TC 207 into SC 5 documents
   b. Include the identification of significant environmental aspects (regarding the products of an organization) as additional application of LCA
3. Inclusion of economic and social aspects are beyond the scope of TC207, but links should be addressed
4. Give guidance/training for application in industry, government and so on, especially in developing countries
   a. Translate the LCA language for experts coming from other fields
   b. Facilitate the use of the LCA standards
   c. Collect case studies using ISO standards showing their applicability

The new standard 14044 contains all the technical requirements and guidance for Environmental management life cycle assessment while the new 14040 contains the framework and guidance standard.

24.3.2 General Outline of the New Standards

The general outline of the two new standards and the relation to the current standard is shown in Figure 24.4.

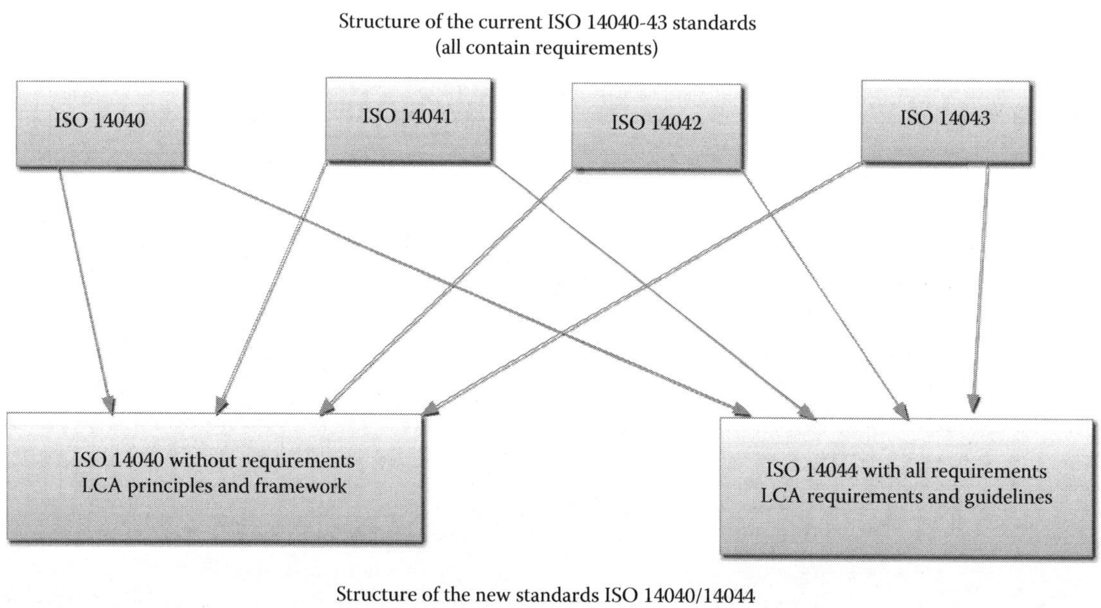

Figure 24.4

ISO regulations.

As defined by the scope of revision, the content of the current standards is transferred to the two new standards. Basically, all technical requirements were transferred to the new ISO 14044, making it the core reference document for the practitioner of LCA. The new ISO 14044 aims to provide a description of LCA principles and framework that is readable and accessible not only for LCA practitioners, but also a broader target audience. The revised 14040 will contain a single, formal requirement of compliance with the new ISO 14044 standard.

In order to arrive at the outlined structure in an effective and efficient way, the working group adopted a revision strategy of three steps: (1) rearrange, (2) change, and (3) enhance.

This strategy basically meant that, first, the technical content of the original 14040-43 standards is to be rearranged according to new work item proposals (NWIPs), and that potential changes are to be addressed as a second step.

Based on this strategy, the roadmap for the revision was developed (Figure 24.2). The starting point was the existing documents, and so on, the standards plus the so-called *should* and *shall* documents, respectively. The latter were compiled by Denmark as an exercise to separate all the existing requirements (*shall* document) and the guidance (*should* document). The work was started with the main technical content (excel.scope, normative references, definitions, annexes). As indicated by Figure 10.4, the so-called *backbone* documents of 14040 and 14044 were generated, followed by an allocation of the remaining guidance to general guidance (→14040) and specific guidance (→14044). As the next step, errors and inconsistencies were addressed and, finally, the text was edited for readability, including scope, references, definitions, and annexes. The next two sections summarize the resulting formal changes. They show the formal changes and technical changes.

24.3.3 Main Formal Changes of the New Standards

A first important formal change is the clarification concerning the compliance with standards. Currently, the missing references between the standards lead to an ambiguity of compliance claims (today compliance with 14040 does not necessarily include compliance with 14041-43). Linking both standards with the only requirement in the new 14040 to comply with 14044 eliminates this situation. Therefore, a compliance claim to the new ISO 14040 does unambiguously include compliance with new ISO 14044.

New standards for LCA have reduced the number of standards and pages containing requirements. The new annexes are much smaller than the original 1044. The new documents also include a set of definitions both in ISO 14040 and ISO 14044 specific to LCA. All these changes are intended to increase the readability and accessibility of the standards.

24.3.4 Main Technical Changes in the New Standards

Generally, the main technical content of the current standards was reconfirmed to be still valid. Many important issues of fundamental importance, such as allocation requirements for comparative assertions or the phases of LCA, were not changed. Therefore, the new LCA standards will not generate a comprehensive need to adapt LCA practices, if based on the current standards. The modified technical content is in line with the previous requirements and serves mainly as a clarification of the technical content, and as a correction of errors and inconsistencies. It includes, for example, the addition of several definitions (e.g., the product, processes, etc.), the addition of principles for LCA, clarifications concerning LCA intended to be used in comparative assertions intended to be disclosed to the public, clarifications concerning the system boundary, clarifications concerning the critical review panel and the addition of an annex about applications.

24.3.5 Definitions

The new document ISO 14040 and ISO 14044 adds some new terms such as *product* and *processes*, based on the definitions ISO 9000 and ISO 14021, respectively. Among the relevant revisions of definitions are the *waste* and *system boundary*. The definition of waste was modified by relating to the Basel Convention on the Control of Transboundary Movements of Hazardous Wastes and Their Disposal, including a clarification that is not limited to hazardous waste for the application within the standards. In the current standards the term *system boundaries* was used for both the interface between product systems and the interface between a product system and the environment. The new standards clarify that "system boundary" only relates to the issue of which unit processes are part of the product system, that is, as part of the inventory analysis.

24.3.6 Principles of LCA

Both the current and new ISO 14000 have the title *principles and framework* but the current version does not include any principles. To remove this inconsistency, the following principles were added to the new ISO 14040.

- Life cycle perspective: LCA considers the entire life cycle of a product, from new material extraction and acquisition, through to energy and material production and manufacturing, to use and end-of-life treatment and final disposal. Through such a systematic overview and perspective, the shifting of a potential environmental burden between life cycle stages or individual processes can be identified and possibly avoided.
- Environmental focus: LCA addresses the environmental aspects and impacts of a product system. Economic and social aspects and impacts are, typically, outside the scope of LCA. Other tools may be combined with LCA for more extensive assessments.
- Relative approach and functional unit: LCA is a relative approach, which is structured around a functional unit. This functional unit defines what is being studied. All subsequent analyses are then relative to that functional unit as all the inputs and outputs in the life cycle interpretation (LCI) and consequently the life cycle impact assessment (LCIA) profile is related to the functional unit.
- Iterative approach: LCA is an iterative technique. The individual phases of an LCA use results of the other phases. The iterative approach within and between the phases contributes to the comprehensiveness and consistency of the study and reported results.
- Transparency: Due to the inherent complexity with an LCA, transparency is an important guiding principle in executing LCAs in order to ensure a proper interpretation of the result.
- Comprehensiveness: LCA considers all attributes or aspects of natural environment, human health and resources. By considering all attributes and aspects within one study in a cross-media perspective, potential trade-offs can be identified and assessed.
- Priority of specific approach: Decisions within an LCA are preferably based on natural science. If this is not possible, other scientific approaches (e.g., from social and economic sciences) can be used or international conventions can be referred to. If neither a scientific basis exists nor a justification based on other scientific approaches or international conventions is possible, then, as appropriate, decisions may be based on value choices.

It is explained that these principles are fundamental and should be used as guidance for decisions relating to both the planning and conducting of an LCA. As a general requirement in ISO 14044, it is stated that LCA is to be conducted in accordance with the principles and the framework described in ISO 14040. The current standards of LCA have a specific set of requirements for LCA intended to be used for a comparative assertion intended to be disclosed to the public.

24.3.7 Life Cycle Impact Assessment

The general procedure and outline for LCA were reconfirmed in the revision of standards. Also, the limitations and restrictions were confirmed to be still valid; for example, there is the statement that there is no scientific basis for reducing LCA results to a single overall score or number or the requirement that weighting is not intended to be disclosed to the public. For weighting, there was the clarification that weighting steps are based on value choices and are not scientifically based, where the current text relates this statement to natural science.

A new section was added between the mandatory and optional elements of LCA that addresses resulting data after characterization. It clarifies that after characterization and before further optional elements, the inputs and outputs of the product system are represented by, for example,

- A set of impact category indicator results
- A set of inventory results that are elementary flows but have not been assented to impact categories
- A set of data that do not represent elementary flows

24.3.8 Life Cycle Interpretation

The final objective of this phase, according to the current standards, is to draw conclusions and to make recommendations reflecting the identification of significant issues and an evaluation element. A clause added to the new standard is the requirement to identify limitations when drawing conclusions and making recommendations. Methodological and study limitations were also added to the list of items that need to be checked for consistency when assessing preliminary conclusions. Therefore, the term *limitations* is also added to the relevant section heading and to the figure describing the interpretation phase. Another inconsistency that has been removed was the missing link between the requirements for conclusions and recommendations. The new standards clarify that conclusions shall be drawn from the study, that recommendations shall be based on the final conclusions of the study, and this shall reflect a logical and reasonable consequence of the conclusion.

24.3.9 Critical Review and Reporting

For critical review of LCA, several clarifications were added to the new standards. The section for internal critical review and external critical review were merged. In the scope definition phase, it was added that it is not only whether and how a critical review is conducted and the type of critical review, as well as who conducts the review that needs to be defined, but also that the level of expertise of the reviewers need to be justified. Finally, it was clarified that review panels, especially for the review by interested parties, consist of at least three members.

24.3.10 Innovative Annex on Application of LCA

The new ISO 14040 includes an entirely new informative annex on application of LCA. In this annex, applications of LCA in the field of other environmental management systems and tools are addressed:

- Environmental management systems and environmental performance evaluation (ISO 14001, ISO 14004, ISO 14031, and ISO 14032), for example; identification of significant environmental aspects of the products and services of an organization
- Environmental labels and declarations (ISO 14020, ISO 14021 and ISO 14025)
- Integration of environmental aspects into product design and development (design for environment) (ISO 14062)
- Inclusion of environmental aspects in product standards (ISO Guide 64)
- Environmental communication (ISO 14063)
- Quantification, monitoring, and reporting of entity and project emissions and removals and validation, verification, and certification of greenhouse gas emissions (ISO 14064)

The variety of potential further applications in private and public organizations is addressed as well. These techniques, methods and tools are not based on the LCA technique as such, but the life cycle approach, principles, and framework can be usefully applied to them.

There is no one-size-fit-all solution as to how LCA can be applied within the decision-making context in industries. Each company has to solve and decide on each case depending, among other things, on the size and culture of the organization, its products, the strategy, the internal systems, tools and procedures, and external drivers. LCA can be used for a broad spectrum of applications. The individual use, adaptation, and practice of LCA for all potential applications are based on ISO 14040 and ISO 14044.

In addition, the LCA technique with proper justification could be applied in studies that are not LCA or LCI studies. Examples are

1. Cradle-to-gate studies
2. Gate-to-gate studies
3. Specific parts of the life cycle (e.g., waste management, components of a product)

Sometimes it may be appropriate to determine indicator results of each unit process or of each stage of a life cycle individually and to calculate the indicator results of the whole product system by adding up the indicator results of different units processes or stages.

ISO 14040 and ISO 14044 represent the widest possible consensus. It is recommended that new standards serve as core reference documents for the users and practitioners of LCA.

24.3.11 References

ISO 9000: Quality management systems—Fundamentals and vocabulary

ISO 14001: Environmental management systems—Requirements with guidance for use

ISO 14004: Environmental management systems—General guidance on principles, systems and supporting techniques

ISO 14020: Environmental labels and declarations—General principles

ISO 14021: Environmental labels and declaration—Self-declared environmental claims (Type II environmental labeling)

ISO 14025: Environmental labels and declaration —Type III environmental declarations-Principles and procedures

ISO 14031: Environmental management—Environmental performance evaluation: Guidelines

ISO 14032: Environmental management—Environmental performance evaluation: Example of environmental performance evaluation (EPE)

ISO 14040: Environmental Management—Life Cycle Assessment: Principles and Framework

ISO 14041: Environmental Management—Life Cycle Assessment: Goal and Scope Definition and Inventory Analysis

ISO 14042: Environmental Management—Life Cycle Assessment: Life Cycle Impact Assessment

ISO 14043: Environmental Management—Life Cycle Assessment: Life Cycle Interpretation

ISO 14062: Environmental Management—Integrating environmental aspects into product design and development

ISO 14063: Environmental communication—Guidelines and examples

ISO 14064: ISO 14064—Part 1, Specification for the quantification, monitoring and reporting of entity emissions and removal; ISO 14064-2, Greenhouse gases—Part 2—Specification for the quantification, monitoring and reporting of project emissions and removals; ISO 14064-3, Greenhouse gases—Part 3—Specification and guidance for validation, verification and certification

ISO Guide 64: Guide for the inclusion of environmental aspects in product standards

24.4 Environmental Protection Agency (EPA)

In the US Code (USC), more than a dozen major environmental laws are codified. The Environmental Protection Agency (EPA) is the premier federal agency for environmental regulations is the US. The federal register publishes all the regulations of environmental protection as part of the Code of Federal Regulations (CFR). The CFR is divided into 50 titles, and most environmental regulations are contained in Title 40. Title 40 is revised every July 1. The practicing engineers mostly need access to chapter 1 of CFR 40, or 40 CFR. It provides regulations and responsibilities of businesses, manufacturers, governments, and institutions affected by these laws. States may vary in the implementation of federal environmental laws and may in some cases be more stringent than federal laws.

24.4.1 Major Environmental Regulations Affecting Manufacturing Sector

There are dozens of environmental laws affecting a company, but the following three major pollution control laws and the resulting regulations are most important. These three laws have broad implication for regulated industries.

24.4.2 Clean Air Act (CAA)

The Clean Air Act is the comprehensive federal law that regulates air emissions from area, stationary, and mobile source throughout the United States. It aims to protect air quality and promote human health and public welfare. The stationary sources of pollution such as power plants and industrial plants, and mobile sources such as cars, trucks, trains, or planes are regulated under CAA laws. CAA authorized the EPA to establish the National Ambient Air Quality Standards (NAAQS) to protect public health and environment. The setting up of maximum pollutant standards was coupled with directing the states to develop their own standards states implementation plans (SIP) applicable to appropriate industrial sources in the state. The original 1977 CAA was amended in 1990 to include acid rain, stratospheric ozone depletion, and air toxics such as hazardous air pollutants (HAP). The 1990 amendments to CAA include the following:

- Establishing an air permit system throughout the United States, which regulated specific sources of pollutants with a single permit that addressed all types of pollutants based on level of pollutant discharge into the air
- Providing the EPA with authority to enforce the permit conditions through fines and penalties without always having to go to the court
- Providing for public participation in the process of how the CAA is carried out through making public comments and the opportunity to challenge any aspect of law, rules, or permits developed or issued
- Recognizing that interstate and international transport air pollution affects air quality in the United States and needs to be addressed when considering a SIP

- Implementing market-based approaches that provide the opportunity to buy or sell air pollution credits that achieve net reduction and economic incentives for reducing pollution (e.g., cleaner fuels)
- Tying deadlines to SIPs and other activities that were thought to be more realistic than previously described in the CAA

The CAA regulates the following common pollutants by permit system and also need to be calculated for the permits.

1. Nitrogen dioxide: A smog-forming chemical, commonly arising from burning of fuels. It is an ingredient of acid rain.
2. Carbon monoxide: Sources are commonly from burning of various types of fuels with extreme health effects in concentration.
3. Sulfur dioxide: An ingredient of acid rain. It commonly results from burning of oil, coal, and industrial processes.
4. Particular matter: Many sources generate dust, smoke, and soot and are regulated by particular size, commonly referred to as PM-10.
5. Lead: This is being regulated out of products but many sources remain.
6. Ozone: A major smog ingredient, ozone is created by the chemical reaction of nitrogen dioxide and volatile organic compounds (VOC).

Many VOC chemicals are also HAPs. VOCs are not criteria pollutants but are heavily regulated because they contribute to the chemical reaction that creates smog. The harmful impact of global warming requires countries to quickly change to clean energy and low-carbon technologies to decrease greenhouse gas (GHG) emissions of NO_2, CO, CO_2, SO_2, and so on. These gases have different levels of global warming potential. The EPA's Clean Air Markets program is called *allowance trading* or *cap-and-trade*. This relies on the use of a comprehensive air quality simulation model for advanced assessment. Cap-and-trade is an attempt to control large amounts of emissions from a group of sources, whether stationary or mobile, at costs that are lower than if sources were regulated individually. The approach first sets overall cap, or maximum amount of emissions per compliance period, that will achieve the desired environmental effects. Total emission allowances provided cannot exceed the cap. Trading experience in the past has helped this policy in implementation. However, without international treaties, a cap-and-trade program alone would not be sufficient to meet the challenge of climate change and global warming.

24.4.3 Clean Water Act (CWA)

The Federal Water Pollution Control Act was enacted in 1972 for the purpose of surface water quality protection in the United States. Later, the Clean Water Act (CWA) of 1987 emphasized the regional water quality management planning that is a requirement of section 208. Public participation is part of the section 208 process and allows collaboration with all stakeholders. The regional water quality management plans normally include the identification of existing and proposed wastewater facilities, provide general planning and guidance nonpoint source pollution control, and ensure that control alternatives over the discharge can be maintained.

24.4.4 Resources Conservation and Recovery Act (RCRA)

The Resources Conservation and Recovery Act (RCRA) was enacted in 1976 as an amendment to the 1965 Solid Waste Disposal Act. The initial focus of RCRA was on the establishment of a solid waste management system with laws for disposal of municipal and industrial waste. In a 1984 amendment of RCRA, the new provisions required phasing out land disposal of hazardous waste. The 1986 amendment to RCRA enabled the EPA to address environmental problems that could result from underground tanks storing petroleum and other hazardous substances. The EPA's RCRA manual provides an in-depth summary of the law. In all, there are 10 subtitles in the RCRA act as it has been amended through 1996.

Solid waste is defined in RCRA as garbage, refuse, sludge from waste treatment plants and water supply plants, sludge from pollution control facilities, nonhazardous wastes, and other discarded materials that include solid, semisolids, liquid, or contained gaseous materials resulting from industrial, commercial, mining, agricultural, and community activities. The EPA guidelines for municipal solid waste management provide for a hierarchical approach, with the highest-priority management practices being source reduction, which emphasizes prevention of waste through all phases of product manufacturing and use.

The hierarchical priorities of legislation are recycling and reuse, combustion, and, finally, landfill. The recycling to recover and reuse can bring great cost savings to industries; however, it depends on the market condition. If recycled materials are cheaper and accepted by the society, it can help reduce the pressure on landfill. The second priority is combustion or incineration of waste that is hard or difficult to recycle. It can help reduce the need for landfilling but it would increase the pollution of air. The landfill requires large amount of land and water and nobody wants it in the backyard. It is a big problem for municipalities. RPA established technical criteria to address seven aspects of municipal solid waste landfills:

1. Location: avoid water resources contamination, unstable substances, and other hazards such as attracting birds near airports
2. Design: including a landfill liner and leachate collection system
3. Operation: including control of disease vectors, daily cover, controlling explosive gases, and a prohibition on burning
4. Groundwater monitoring: monitoring wells to observe any potential failures in the liner system resulting in rainwater leakage or leachate
5. Corrective action: to be taken in the event of a failure in the liner and leachate collection system
6. Closure and post closure: criteria for end-of-life management of the landfill facility
7. Financial assistance: plans to address the financial responsibilities of the landfill owner and operator during closure and postclosure phases, including future potential failure in the liner or leachate collection

The solid waste programs listed in subtitle D of RCRA are largely managed by the states that must obtain authorization for solid waste permit provisions through the EPA.

Hazardous waste is dangerous to human health as well as to the environment and it is listed in subtitle C of RCRA, outlining cradle-to-grave regulatory system for such wastes. The regulations are very comprehensive regarding the generation of hazardous waste, transportation, and final treatment, storage, or disposal facilities (TSDFs).

The first step in hazardous waste management is to identify the hazardous waste. It is identified as hazardous waste if it contains toxic chemicals that are corrosive and reactive. There are four lists located in 40 CFR, part 261. It contains waste from common manufacturing processes such as spent solvents and electroplating and metal finishing waste. The K list has waste from 13 specific industries. They are the coking process, explosive manufacturing; ink formulation, petroleum manufacturing, inorganic chemicals manufacturing, pesticides manufacturing, organic chemical, inorganic pigment manufacturing, iron and steel manufacturing, primary aluminum production, secondary lead processing, veterinary pharmaceutical manufacturing, and wood preservation. The P and U lists are codified in 40 CFR and consist of chemicals of pure or commercial grade or as constituents of formulations. The P list chemicals are acutely toxic and may be fatal or seriously harmful in low doses. The U list chemicals are generally toxic but may also exhibit other hazardous characteristics such as ignitability, reactivity, or corrosivity.

Hazardous waste that is transported is subject to the packaging and labeling requirements adopted by EPA from the Department of Transportation (DOT) regulations.

Reuse, recycling, and reclamation of hazardous waste is preferable to disposal in a hazardous waste landfill under RCRA guidelines. Most hazardous waste being recycled is still fully regulated by RCRA rules. Hazardous waste such as batteries, pesticides, hazardous waste lamps, and hazardous waste thermometers are listed under the universal waste rules. Mercury is a common hazardous material in lamps, as well thermometers.

The third major sector in the cradle-to-grave system is the TSDF. TSDFs are permitted facilities that may engage in treating the waste to alter the hazardous properties of the material, storage of the hazardous waste for some temporary period, and final disposal by some means on land or water where the material will remain after the TSDF discontinues its operation.

24.5 British Standard Institution (BSI): PAS 2050 (Publicly Available Specifications)

The specifications for assessment of the life cycle greenhouse gas emissions of goods and services are presented in PAS 2050. PAS should not be regarded as British Standards, European Standards, or International Standards.

24.5.1 Introduction

Climate change continues to be one of the greatest challenges facing nations, governments, business, and citizens, and it will influence the way we live and work in future decades. Past and current actions, including the release of carbon dioxide (CO_2) and other greenhouse gases through human activities, such as the burning of fossil fuels, emissions from chemical processes, and other sources of anthropogenic greenhouse gases will have an effect on future global climate.

While greenhouse gas (GHG) emissions are often viewed at global, national, corporate, or organizational levels, emissions within these grouping can arise from supply chains within business, between businesses, and between nations. The GHG emissions associated with goods and services reflect the impact of processes, materials and decisions occurring throughout the life cycle of these goods and services.

PAS 2050 was developed in response to a broad community and industry desire for a consistent method for assessing the life cycle GHG emissions of goods and services. Life cycle GHG emissions are the emissions that are released as part of the processes of creating, modifying, transporting, storing, using, providing, recycling, or disposing of such goods and services. PAS 2050 offers organizations a method to deliver improved understanding of the GHG emissions arising from their supply chains, but the primary objective of this PAS is to provide a common basis for GHG emission quantification that will inform and enable meaningful GHG emission reduction programs.

A new edition of PAS includes a set of principles governing the development of supplementary requirements for the application of PAS 2050 to particular product types. These principles are intended to ensure that such supplementary requirements are not in conflict with the requirements of PAS. PAS puts great emphasis on proper recording of the processes and outcomes. Where an organization implementing this PAS chooses to disclose all or part of the results of an assessment of GHG emissions, all relevant supporting information should also be made available.

This PAS is focused on a single environmental issue (i.e., GHG emissions and their contribution to climate change) but this is only one of a range of possible environmental impacts from specific goods or services. The relative importance of those impacts can vary significantly from product to product, and it is important to be aware that decisions taken on the basis of a *single issue* assessment could be detrimental to other environmental impacts potentially arising from the provision and use of the same product.

24.5.2 Background, Benefits, and Context of PAS 2050

PAS 2050 builds on existing life cycle assessment methods established through BS EN ISO 14040 and BS EN ISO 14044 by giving requirements specifically for the assessment of GHG emissions within the life cycle of goods and services. These requirements further clarify the implementation of these standards in relation to the assessment of GHG emissions of goods and services, and they establish particular principles and techniques, including:

1. Cradle-to-gate and cradle-to-grave GHG emissions assessment data as part of the life cycle GHG emissions assessment of good and services.
2. Scope of greenhouse gas emissions and removals to be included.

    Both emissions to the atmosphere and removals from the atmosphere shall be accounted for in the assessment of the overall GHG emissions of the product being assessed. The assessment shall include the gases listed in Annex A arising from both fossil and biogenic sources for all products, with the exception of human food and animal feed products. This permitted exception avoids the need to calculate the CO_2 emissions and removals caused by the consumption and digestion of food and feed and from human and animal waste.

    The assessment of GHG emissions and removals during the life cycle of products shall include emissions and removals as identified in Annex A from processes including but not limited to

    - Energy use (including energy sources, such as electricity, that were themselves created using processes that have GHG emissions associated with them)
    - Combustion reactions
    - Loss to atmosphere of refrigerants and other fugitive GHGs
    - Process operations
    - Service provision and delivery
    - Land use and land use change
    - Livestock production and other agricultural processes
    - Waste management

3. Criteria for global warming potential (GWP) data; GHG emissions and removals shall be measured by mass and shall be converted into CO_2e using the latest IPCC 100-year global warming potential (GWP) coefficients (Annex A), except where otherwise specified. It has been estimated that methane has a GWP coefficient of 25, and 1 kg of methane is equivalent to 25 kg of CO_2 in terms of its GWP.
4. Treatment of emissions and removal from land use change and biogenic and fossil carbon sources.
5. Treatment of the impact of carbon storage in products and offsetting.
6. Requirements for the treatment of GHG emissions arising from specific processes.
7. Data requirements and accounting for emissions from renewable energy generation.

This PAS benefits organization, businesses, and other stakeholders by providing a clear and consistent method for the assessment of the life cycle GHG emissions associated with goods and services. Specifically, PAS provides the following benefits:

1. For organization that supply goods and services, this PAS
   a. Allows internal assessment of the existing life cycle GHG emissions of goods and services
   b. Facilitates the evaluation of the alternative product configurations, sourcing and manufacturing methods, raw material choices and supplier selection on the basis of life cycle GHG emissions associated with goods and is to be used as a basis for comparison of services
   c. Provides a benchmark for programs aimed at reducing GHG emissions
   d. Allows for the quantification, management, and potential comparison of GHG emissions from goods or services using a common, recognized, and standardized approach to life cycle GHG emissions assessment
   e. Supports reporting (e.g., on corporate responsibility)
2. For consumers of goods and services, PAS provides a common basis for understanding the assessment of life cycle GHG emissions when making purchasing decisions and using goods and services

24.5.3 Emissions and Removals: Scope of GHG Emissions and Removals

Both emissions to the atmosphere and removals from the atmosphere shall be accounted for in the assessment of the overall GHG emissions of the product being assessed. This assessment shall include the gases listed in Annex A arising from both fossil and biogenic sources for all products, with the exception of human food and animal feed products.

For food and feed, emissions and removals arising from biogenic sources that become part of the product may be excluded.

24.5.4 Product Life Cycle Processes to Be Included

The assessment of GHG emissions and removals during the life cycle of products shall include emissions and removals shall include

1. Energy use (including energy sources, such as electricity, that were themselves created using processes that have GHG emissions associated with them)
2. Combustion processes
3. Chemical reactions
4. Loss to atmosphere of refrigerants and other fugitive GHG
5. Process operations
6. Service provision and delivery
7. Land use and land use change
8. Livestock production and other agricultural processes
9. Waste management

The assessment of the GHG emissions and removals arising from the life cycle of products shall include the GHG emissions and removals occurring during the 100-year period following the formation of the product (i.e., the 100-year assessment period).

24.5.5 Global Warming Potential (GWP)

GHG emissions and removal shall be measured by mass and shall be converted into CO_2 using the latest IPCC 100-year GWP coefficients (Annex A), except where otherwise specified.

For example, methane has a GWP coefficient of 25, and 1 kg of methane is equivalent to 25 kg of CO_2 in terms of its GWP.

24.5.6 System Boundary

The system boundary shall be clearly defined for each product under assessment and shall include all of its material life cycle processes. While supplementary requirements specifying a system boundary value have been developed for the product being assessed in accordance with principles set out in 4.3, they should be used.

24.5.7 Cradle-to-Gate System Boundary

The system boundary for an assessment identified as cradle-to-gate shall include the emissions and removal identified in 5.1 that have occurred up to and including the point where the product leaves the organization undertaking the assessment for transfer to another party that is not the consumer. For products that use recycled material as an input, the emissions and removals associated with processing of that material shall be included as defined by Annex D.

Cradle-to-gate GHG emissions assessment information shall be clearly identified as such so as not to be mistaken for full assessment of the life cycle GHG emissions of a product.

24.5.8 Material Contribution and Threshold

Calculations carried out in accordance with this PAS shall include all emissions and removals within the system boundary that have the potential to make a material contribution to the assessment of GHG emissions of the product (see 3.31).

For GHG emissions, and removals arising from the life cycle of product, the assessment shall include

- All sources of emissions and processes for removal anticipated to make a material contribution to the life cycle GHG emissions of the functional unit
- At least 95% of the anticipated life cycle GHG emissions and removals associated with the functional unit

24.5.9 Elements of Production System

The life cycle elements covered in 6.4.2–6.4.10 should be included in the system boundary for the assessment of the life cycle GHG emissions and removals associated with the products under assessment.

24.5.10 Production Materials

The GHG emissions and removals arising from all processes used in the formation, extraction, or transformation of materials used in production (including farming, horticulture, fishing, and forestry) shall be included in the assessment, including all sources of energy consumption or direct GHG emissions associated with that formation, extraction, or transformation. The GHG emissions and removals arising from materials used in production shall include, for example those from

- Development of material sources (e.g., surveying, prospecting)
- Mining or extracting raw materials (solids, liquids and gases such as iron, oil and natural gas), including emissions from any machinery used
- Consumables used in sourcing production materials
- Fertilizers (e.g., N_2O emissions arising from the application of nitrogen fertilizer and emissions arising from the production of the fertilizer)
- Waste generated at each stage of the extraction and preprocessing of production materials
- Direct land use change (e.g., draining of peatland or removal of forest)
- Energy-intensive atmospheric growing conditions (e.g., heated farmhouse)
- Emissions from crop production (e.g., methane from the rice cultivation) and livestock (e.g., methane from cattle)

24.5.11 Energy

The GHG emissions and removals associated with the provisions and use of energy in the life cycle of the product shall be included in the emissions arising from the energy-supply system.

Emissions from energy supply include

- Emissions at the point of consumption of the energy (e.g., emissions from the burning of coal and gas)
- Emissions from provision of energy, including the generation of electricity and heat
- Emissions from transmission losses, transport fuels
- Upstream emissions (e.g., the mining and transport of fuel to the electricity generator or other combustion plant)
- Downstream emissions (e.g., the treatment of waste arising from the operation of nuclear electricity generation)
- And the growing and processing of biomass for use as fuel

24.5.12 Capital Goods

The GHG emissions and removals from the production of capital goods used in the life cycle of product shall be excluded from the assessment unless provided for in the supplementary requirements in accordance with the principles set out in 4.3.

24.5.13 Manufacturing and Service Provision

The GHG emissions and removals arising from manufacturing and service provision that occur as part of the life cycle of the product, including emissions associated with use of consumables, shall be included in the assessment of the GHG emissions of the life cycle of the product.

Where a process is used for prototyping a new product, the emissions associated with the prototyping activities shall be allocated to the resulting product(s) and co-product(s) of the process.

24.5.14 Operation of Premises

The GHG emissions and removals arising from the operation of premises, including those from factories, warehouses, central supply centers, offices, retail outlets, and so on, shall be included in the assessment of the GHG emissions of the life cycle of the product.

24.5.15 Transport

The GHG emissions and removals arising from road, air, water, rail, or other transport methods that form part of the life cycle of the product shall be included in the assessment of the GHG emissions of the life cycle of the product taking account of the requirements in 8.6.

24.5.16 Storage of Products

The GHG emissions and removals arising from storage shall be included in all assessment of the life cycle GHG emissions of the product, including

- **a.** Storage of inputs, including raw material, at any points in the product life cycle
- **b.** Environmental controls (e.g., cooling, heating, humidity control, and other controls) related to a product at any point in the product life cycle (6.4.6) for all operation, including environmental control, of factories in which product may be stored
- c. Storage of products in the use phase
- **d.** Storage prior to reuse, recycling, or disposal activities

24.5.17 Use Phase

24.5.17.1 General Principle

The GHG emissions and removals arising from the use of products shall be included in the assessment of the life cycle GHG emissions of products, subject to the provision of 6.2 for cradle-to-gate assessments. The emission factor associated with energy used in the use phase of products shall be determined in accordance with 6.4.3.

24.5.17.2 Basis of Use Profile

Determination of the use profile for the use phase of products shall be based on a hierarchy of boundary definitions in the following order of preference:

1. Supplementary requirements in accordance with the principles set out in 4.3, which include a use phase for the product being assessed

2. Published international standards that specify a use phase for the product being assessed
3. Published national guidelines that specify a use phase for the product being assessed
4. Published industry guidelines that specify a use phase for the product being assessed

When no method for determining the use phase of products has been established in accordance with points 1–4 of this clause, the approach taken in determining the use phase of the products shall be established by the organization carrying out the assessment of GHG emissions for the product and recorded.

24.5.17.3 Time Period for Use-Phase GHG Emission

1. The emissions and removals during use phase of the product during 100-year assessment period shall be included.
2. The method used for calculating the weighted average impact of the product over a period of more than 1 year shall be that provided in Annex E.

24.5.17.4 Recording Basis of Use-Phase Calculations for Products

When use-phase emissions and removals from part of an assessment carried out under this PAS, the detail of the basis on which the use phase for the product is assessed shall be recorded and retained (4.4 and Annex B).

24.5.18 Data

General: The data recorded in relation to a product shall include all GHG emissions and removals occurring within the system boundary of the product.

24.5.19 Allocation of Emissions

24.5.19.1 General Requirements

Allocation to coproducts: The preferred approach to allocation of emissions and removals to coproduct shall be, in order of preference:

1. Dividing the unit processes to be allocated into two or more subprocesses and collecting the input and output data related to these subprocesses; or
2. Expanding the product system to include additional functions related to the co products where:
   a. A product that is displaced by one or more of the coproducts of the processes being considered can be identified, and
   b. The avoided GHG emissions associated with the displaced product represent the average emissions arising from the provision of the avoided product

24.5.19.2 Recording Allocation Assumption

The applied approach to the allocation of emissions to coproduct shall be recorded by the organization implementing this PAS.

24.5.19.3 Emissions from Waste

General allocation principle: Where waste results in GHG emissions (e.g., organic matter disposed of in a landfill, those emissions (CO_2 and non-CO_2) shall be allocated to the product system that gives rise to the waste. This allocation also applies to methane combusted without the generation of useful energy (i.e., flaring).

Waste combustion with energy recovery: Where waste or fuel derived from waste is combusted to generate useful electricity and/or heat, GHG emissions shall be allocated to the generation of the energy. GHG removals shall also be allocated to the energy generation system.

Use of recycled materials and recycling: This method for assessing emissions arising from recycled or recyclable material shall be as specified in Annex D.

24.5.19.4 Treatment of Emissions Associated with Reuse

Where a product is reused, GHG emissions per instance of use shall be assessed on the basis of:

$$\text{GHG emissions} = \frac{a+f}{b} + c + d + e$$

where:

- a is the total life cycle GHG emissions of the product, excluding use-phase emissions
- b is the anticipated number of reuse instances for a given product
- c is emissions arising from an instance of refurbishment of the product to make it suitable for reuse (e.g., recovering and sterilizing a glass bottle)
- d is emissions arising from use phase
- e is emissions arising from transport returning the product for reuse
- f is emissions arising from disposal

24.5.19.5 Emissions from Energy Production Using CHP

Where energy production CHP is expected to a larger system (e.g., export of electricity to a national electricity network), the avoided GHG emissions arising from the exported energy shall be allocated in accordance with 8.1.1.

24.5.20 Emissions from Transport

24.5.20.1 Calculation of GHG Emissions of Products

This PAS requires that both emissions to the atmosphere and removals from the atmosphere to be taken into account in calculating the total GHG emissions of a product over its life cycle. The following method shall be used to calculate the GHG emissions per functional unit of the product under assessment. Determine the emissions and removals for each activity within the system boundary as primary activity data or secondary data, with emissions included as positive values and removals included as negative values.

1. Convert the primary activity data and secondary data to GHG emissions and removals per functional unit of a product under assessment by multiplying the data by emissions factor for each activity.
2. Convert the GHG emissions and removal data into units of CO_2 by multiplying the individual GHG emissions or removals figures by the relevant GWP.
3. Calculate the overall impact of carbon storage associated with the product in accordance with 5.5.1 expressed as $CO_2\,e$ (5.3) and recorded in accordance with 5.5.2.
4. Sum the $CO_2\,e$ emissions and removals occurring in the life cycle of the product under assessment.

24.5.21 Annex A (Normative)

Global Warming Potential: The values of global warming potentials for GHGs to be used in calculations shall be in accordance with Table 24.A.1 (IPCC 2007, Table 2.14, see clause 2). Note attention is drawn to the requirement that the GWP actually used in calculations is the latest available from the IPCC (see 5.1.1). It is the responsibility of the organization undertaking GHG emissions assessment to confirm the currency of GWP values given in Table 24.A.1 before use.

24.5.22 Annex B (Normative)

- Recording of supporting information B.1 General to provide for analysis and verification when required. Information supporting greenhouse gas assessments for this PAS shall be recorded and made available when requested.
- System boundary: description of the system boundary that has been used for the assessment of the GHG emissions of the product under assessment.
- Carbon storage: a full description of the basis on which the impact of any carbon storage was calculated including the emissions profile of the product.
- Land use change: data sources and identification of the options taken in the event of limited knowledge of sources and timing of land use change.
- Exclusions on the basis of materiality: where the decision to exclude an emission source assessed as being below the 1% threshold has been made, the assessment details leading to that decision.
- Production materials: identification of all materials required to meet the specification of the product under assessment.
- Energy sources: all sources of energy contributing to the product under assessment together with related emissions factors where these are publicly available.

Table 24.A.1 Direct (Except for CH_4) Global Warming Potentials (GWP) Relative to CO_2

Industrial Designation or Common Name	Chemical Formula	GWP for 100-Year Time Horizon
Carbon dioxide	CO_2	1
Methane	CH_4	25
Nitrous oxide	N_2O	298
Substances Controlled by the Montreal Protocol		
CFC-11	CCl_3F	4,750
CFC-12	CCl_2F2	10,900
CFC-13	$CClF_3$	14,400
CFC-113	CCl_2FCClF_2	6,130
CFC-114	$CClF_2CClF_2$	10,000
CFC-115	$CClF_2CF_3$	7,370
Halon-1301	$CBrF_3$	7,140
Halon-1211	$CBrClF_2$	1,890
Halon-2402	$CBrF_2CBrF_2$	1,64
Carbon tetrachloride	CCl_4	1,400
Methyl bromide	CH_3Br	5
Methyl chloroform	CH_3CCl_3	146
HCFC-22	$CHClF_2$	1,810
HCFC-123	$CHCl_2CF_3$	77
HCFC-124	$CHClFCF_3$	609
HCFC-141b	CH_3CCl_2F	725
HCFC-142b	CH_3CClF_2	Rev 2,310
HCFC-225ca	$CHCl_2CF_2CF_3$	122
HCFC-225cb	$CHClFCF_2CClF_2$	595
Hydrofluorocarbons		
HFC-23	CHF_3	14,800
HFC-32	CH_2F_2	675
HFC-125	CHF_2CF_3	3,500
HFC-134a	CH_2FCF_3	1,430
HFC-143a	CH_3CF3	4,470
HFC-152a	CH_3CHF_2	124
HFC-227ea	CF_3CHFCF_3	3,220
HFC-236fa	$CF_3CH_2CF_3$	9,810
HFC-245fa	$CF_3CH_2CF_3$	1,030
HFC-365mfc	$CH_3CF_2CH_2CF_3$	794
HFC-43-10mee	$CF_3CHFCHFCF_2CF_3$	1,640
Perfluorinated Compounds		
Sulfur hexafluoride	SF_6	22,800
Nitrogen trifluoride	NF_3	17,200
PFC-14	CF_4	7,390
PFC-116	C_2F_6	12,200
PFC-218	C_3F_8	8,830
PFC-318	$c\text{-}C_4F_8$	10,300
PFC-3-1-10	C_4F_{10}	8,860
PFC-4-1-12	C_5F_{12}	9,160
PFC-5-1-14	C_6F_{14}	9,300
PFC-9-1-18	$C_{10}F_{18}$	>7,500
Trifluoromethyl sulfur pentafluoride	SF_5CF_3	17,700
Fluorinated Ethers		
HFE-125	CHF_2OCF_3	14900
HFE-134	CHF_2OCHF_2	6,320
HFE-143a	CHF_3 OCF3	756

(Continued)

Table 24.A.1 (Continued) Direct (Except for CH_4) Global Warming Potentials (GWP) Relative to CO_2

Industrial Designation or Common Name	Chemical Formula	GWP for 100-Year Time Horizon
HCFE-235da2	$CHF_2OCHClCF_2$	350
HEF-245fa2	$CHF_2OCH_2CF_2$	708
HFE-254cb2	$CH_2OCF_2CHF_2$	659
HFE-254fa2	$CHF_2OCH_2CF_3$	359
HFE-347mcc3	$CH_3OCF_2CF_2CF_3$	575
HFE-347pcf2	$CHF_2CF_2OCH_2CF_3$	580
HFE-356pcc3	$CH_3OCF_2CF_2CHF_2$	110
HFE-449s1(HFE-7100)	$C_4F_9OCH_3$	297
HFE-569sf2(HFE-7200)	$C_4F_9OC_2H_5$	59
HFE-43-10-pccc124(H-Galden 1040x)	$CHF_2OCF_2OC_2F_4OCHF_2$	1,870
HFE-236ca12(HG-10)	$CH_2OCF_2OCHF_2$	2,800
HFE-338pcc13(HG-01)	$CHF_2OCF_2CF_2OCHF_2$	1,500
Perfluoropolyethers		
PFPMIE	$CF_2OCF(CF_2)$	10,300
	$CF_2OCF_2OCF_3$	10,300
Hydrocarbons and Other Compounds: Direct Effects		
Dimethylether	CH_3OCH_3	1
Methylene chloride	CH_2Cl_2	8.7
Methyl chloride	CH_2Cl	13

- Use phase analysis: where use phase emissions form part of the assessment, full description of the use profile used.
- Allocation of emissions: the basis on which allocations have been made including any assumptions that have contributed to those decisions.
- Recycling: the option selected for the treatment of recycled or recyclable material including any assumptions made about end-of-life treatment in cradle-to-gate assessments.

24.5.23 Annex C (Normative)

Default land use change values for selected countries.

24.5.24 Annex D (Normative)

Assessment of emissions arising from recycled or recyclable material inputs.

24.5.24.1 D.1 General

Where a product uses recycled or recyclable material, the emissions associated with recycled or recyclable material, the emissions associated with recycled and virgin material and with disposal of material after use shall be assessed in accordance with D.2–D.6.

24.5.24.2 D.2 Recycled Content Method

Note: this method is also called the cut-off method and the 100-0 method.

If the recycled material does not maintain the same inherent properties as the virgin material input, the emissions and removals arising per unit (E) from that material shall reflect the product-specific recycled content and/or recycling rate based on the following calculation:

$$E = (1 - R_1)E_V + R_1E_R + (1 - R_2)E_D$$

where:

R_1 = proportion of recycled material input
R_2 = proportion of material in the product that is recycled at end of life
E_R = emissions and removals arising from recycled material input per unit of material
E_V = emissions and removals arising from virgin material inputs per unit of material
E_D = emissions and removals arising from disposal of waste material per unit of material

24.5.24.3 D.3 Closed-Loop Approximation Method

If the recycled material maintains the same inherent properties as the virgin material input, the emissions and removals arising per unit (E) from material shall reflect the product-specific recycling rate based on the calculation given in this clause (closed-loop approximation method)

Note 1: If the product system includes many recycled material inputs and outputs, it can be difficult to correctly perform the closed-loop approximation method even when the same inherent properties are maintained. In this case companies should use the recycled content method.

Note 2: This method is also called the end-of-life approach, recyclability substitution, and/or the 0–100 output method.

$$E = (1 - R_2)E_V + R_2 \cdot E_R + (1 - R_2)E_D$$

where:

R_2 = proportion of material in the product that is recycled at end of life
E_V = emissions and removals arising from virgin material input, per unit of material
E_D = emissions and removals arising from disposal of waste material, per unit of material
E_R = emissions and removals arising from recycled material input, per unit of material

24.5.24.4 D.4 Recycling Data

24.5.24.4.1 D.4.1 Material Input with System Average Recycle Content and Recycling Rate Where the life cycle of a product includes a material input that contains the system average proportion of recycled content and is recycled at the system average recycling rate for that product category, the calculation in D.2 shall reflect the system average recycle content and recycle rate.

Note: It is assumed that materials are recycled in a steady state system. This might not be the case for some materials where the total stock of material in use is increasing or decreasing over time.

24.5.24.4.2 D.4.2 Material Input with a Product-Specific Recycle Content and/or Recycling Rate Where the life cycle of a product includes a material input with a specific proportion of recycled content and/or the material in the product has a recycle rate that is different from the average recycle rate for that product category, the emissions and removals arising from that material shall reflect the product-specific recycle content and/or recycling rate.

24.5.24.4.3 D.4.3 Demonstration of Product-Specific Recycled Content and/or Recycling Rate

24.5.24.5 D.5 Other Types of Recycling

24.5.24.6 D.6 Recording the Basis of Treatment of Recycling

24.5.25 Annex E (Normative)

Calculation of the weighted average impact of delayed emissions arising from the use and final disposal phase of products.

24.5.25.1 General Application

Where the impact of delayed emissions is to be assessed, the emissions arising during the use phase or during final disposal of the product over a period of more than one year shall be taken into account using one of the methods provided in E2 or E3.

Note 1: The formula presented in this annex represent a simplification of approach outlined in IPCC 2007, Table 2.14 (footnote(a)), chapter 2 (see clause 2). Full implementation of the IPCC approach gives a more precise result.

Note 2: The approach presented in IPCC 2007 Table 2.14 (footnote (a)), chapter 2 (see Clause 2) is applicable to CO_2 emissions only, while the approximation presented here is applied to GWP data, assessed under this PAS. As a result, the approximation is less accurate where the overall $CO_2\,e$ emissions of the product include a significant non-CO_2 component.

24.5.25.2 Specific Case-Delayed Single Release

Where emissions from the use phase or the final disposal phase of product calculated according to occur as a single release within 25 years of the formation of the product, weighting factor, FW, to be applied to the GHG emissions released at that time shall reflect the number of years of delay in the emissions being released (i.e., the number of years between formation of the product and the single release of the emissions) according to

$$\mathrm{FW} = \frac{100 - (0.76 \times t_0)}{100}$$

where t_0 is the number of years between formation of the product and single release of the emissions.

24.5.25.3 General Case: Delayed Release

The weighted average time the emissions are in the atmosphere shall be calculated according to

$$FW = \frac{\sum_{i=1}^{100} x_i (100 - i)}{100}$$

where:

i = each year in which emissions occur

x = the proportion of total emissions occurring in any year i

Example: If use-phase emissions were to be delayed for 10 years following formation of the product, with total emissions being released evenly over the following five years, then the weighting factor that represents the weighted average time these emissions are present in the atmosphere would be

$$FW = \frac{\left(0.2 \times (100 - 11)\right) + \left(0.2 \times (100 - 12)\right) + \left(0.2 \times (100 - 13)\right) + \left(0.2 \times (100 - 14)\right) + \left(0.2 \times (100 - 15)\right)}{100}$$

$$FW = 0.87$$

In this example, the total amount of use-phase emissions, expressed as $CO_2\,e$, released during the 100-year quantification period would be multiplied by a weighting factor 0.87 to reflect the weighted average time these emissions are present in the atmosphere during the 100-year quantification period.

24.5.25.4 Recording the Basis of Carbon Storage Assessment

Where the assessment of the life cycle GHG emissions of a product is accompanied by assessment of the impact of delayed release of the stored carbon, the data sources from which impact of stored carbon was calculated, together with the carbon storage profile of the product over 100-year assessment period, shall be recorded and retained (see 4.4).

24.6 Environmental Regulations in India

In the Directive Principles of State Policy, Article 48-A was inserted which enjoins the State to make endeavor for protection and improvement of the environment and for safeguarding the forest and wildlife of the country. It shall be the duty of every citizen of India to protect and improve the natural environment, including forests, lakes, rivers, and wildlife and to have compassion for living creatures. The environmental regulations in India have evolved in number of steps through number of years. A list of these regulations and laws are outlined in Table 24.5.

India being a federal structure for governance, the laws and regulations are applied in different ways in different states of India. The environmental protections regulations for water, air, and land have developed in different stages and are presented in brief as below.

1. The Water (Prevention and Control of Pollution) Act, 1974
   - Provide information to the SPCB
   - Provide access to the SPCB for taking samples
   - Allow entry to the SPCB to ascertain that the provisions of the Act are being compiled with.

   Responsibilities:
   - Obtain "Consent to Establish"
   - Obtain "Consent to Operate"

Table 24.5 Environmental Regulations in India

Year	Environmental Regulations
1974	Water (Prevention & Control of Pollution Act) Amendments, 1988
1975	The Water (Prevention & Control of Pollution) Rules
1977	The Water (Prevention & Control of Pollution) Cess Act
1978	The Water (Prevention & Control of Pollution) Cess Rules
1981	The Air (Prevention & Control of Pollution) Act, Amendments, 1987
1982/1983	The Air (Prevention & Control of Pollution) Rules
1986	The Environment (Protection) Act, Amendments (1989, 1990, 1993, 1996, 1997, 1998, 1999, 2000, 2001)
1986	The Environmental (Protection) Rules
1989	The Hazardous Wastes (Management and Handling) Rules, Amendments, 2000, Draft Amendments 2002
1989	Manufacture, Storage and Import of Hazardous Chemical Rules, Amendments, 1994, 2000
1991	The Public Liability Insurance Act/Rules, 1992
1992	E (P) Act Notification—"Environment Statement"
1994	E (P) Act Notification—"Environmental Clearance"
1995	The National Environment Tribunal Act
1997	Amendments in the Environment Clearance, Notification—"Public Hearing" made mandatory
1997	Prohibition on the Handling of Azo dyes
1997	The National Environment Appellate Authority Act
1998	The Bio-Medical Waste (M&H), Rules
1999	Notification for making 100% Utilization of Fly-ash made mandatory
1999	Regulation on recycling of waste oil and nonferrous scrape
2000	Municipal Solid Waste (M&H) Rules
2000	Ozone Depleting Substance (R&C) Rules
2000	Noise pollution (regulations and control)
2001	Batteries (M&H) Rules

- Apply for renewal of the "Consent to Operate" before the expiry of validity period
- Consent to be deemed as granted automatically and unconditionally after 4 months from the date of application already given or refused before this period
- Refusal of "Consent" to be recorded in writing
- Pay Water Cess as indicated in the assessment order
- Affix water meters of the prescribed standards
- Provide access to SPCB
- Pay interest in case of delay in paying the Water Cess
- Pay penalty for nonpayment of Cess
- Industry is entitled to 25% rebate if meeting certain conditions

2. The Air (Prevention & Control of Pollution) Act, 1981
   - Comply with the conditions in the "Consent to Establish" or "Consent to Operate"
   - Not to discharge air pollutant(s) in excess of the prescribed standards
   - Furnish information to the SPCB of any accident or unforeseen act or event
   - Allow entry to the SPCB to ascertain that provisions of the Act are being complied with
   - Provide information to enable SPCB to implement the Act
   - Provide access to the SPCB for taking samples
   - Comply with the directions issued in writing by the SPCB
   - Obtain "Consent to Establish"
   - Obtain "Consent to Operate"
   - Apply for the renewal of "Consent to Operate" before expiry of the validity period
   - Consent to be deemed as granted after 4 months from the date of receipt of application if no communication from the SPCB is received
   - A prior "Notice of Inspection" to be served by the SPCB
   - Industry to ensure that specified emission sampling procedure is being followed by the SPCB
   - Opportunity to file objections with the SPCB within 15 days from the date of service of notice
   - PCB to record reasons in writing in case it does not provide an opportunity to the industry to file objections

In 1986, the federal government established the Environment (Protection) Act, which is a comprehensive set of laws and their applications in all the states and regions of India.

3. Environment (Protection) Act, 1986
    - Comply with the directions issued by the Central Government. The direction may include
        - Closure, prohibition or regulation of any industry, or
        - Stoppage or regulation of the supply of electricity, water or any other service
    - Prevent discharges or emissions excess of the prescribed standards
    - Furnish information of any accidental or unforeseen event
    - Allow entry and inspection to ascertain compliance
    - Allow samples to be taken
    - Submit an "Environmental Statement" every year before 30 September to the SPCB
    - Obtain prior "Environmental Clearances" from MoEF, in case of a new project or for modernization/expansion of the existing project

The Hazardous Waste Rules of 1999 and amended in 2000 is most comprehensive regarding hazardous waste treatment and disposal. It is comparable to EPA regulations in the United States.

4. The Hazardous Waste (Management and Handling) Rules, 1989, Amendments 2000
    - Check whether the waste(s) generated covered in Schedule 1 and 2 of the amendment rules, 2000
    - If covered, apply in the Prescribed Format to obtain an "Authorization" for proper treatment and disposal of hazardous waste(s) and comply with the conditions specified in the authorization
    - Take steps, wherever feasible for reduction, recovery and recycling of wastes
    - Ensure proper collection, reception, treatment, storage and disposal of hazardous wastes
    - Apply for renewal of authorization before expiry of the validity period
    - Maintain records of hazardous wastes handling (Form 3)
    - Submit "Annual Returns" to the SPCB (Form 4)
    - Report all accidents to the SPCB
    - Labeling, packaging, transportation of HW as per Motor Vehicle Act, 1988 and Rules 1989

The Hazardous Waste (Management and Handling) Rules were again amended in 2002 in accordance with international regulations of hazardous waste. The amendments included regulations regarding registration of reprocessors of nonferrous scrap and waste oil. They also included regulations and laws regarding manufacture, storage, and import of hazardous chemical. They require that all containers of hazardous waste must be labeled with specific information about the chemical in the container. The records of the imported hazardous chemicals must be kept and provided to the authorities. The regulations also provide relief in case of death or injury or damage to property from an accident on the principle of no fault. The regulations are complete with provision of contribution to Environmental Relief Fund and pay the amount of an award as specified by the collector. The following industrial projects are required to obtain environmental clearance from the government:

- Nuclear power plant and related projects, such as heavy water plants, nuclear fuel complex, rare earths
- River valley projects including hydro power, major irrigation projects and a combination, including flood control
- Ports harbors, airports (except minor ports and harbors)
- Petroleum refineries including crude and product pipelines
- Chemical fertilizers
- Exploration for oil and gas and their production transportation and storage
- Synthetic rubber
- Primary metallurgical industries
- Electric arc furnaces (mini steel plants)
- Viscose staple fiber and filament yarn
- Storage batteries integrated with the manufacture of oxides of lead and lead antimony alloy
- Thermal power plants
- Pulp paper and newsprint

- Cement
- Any project proposed to be located within 25 km of
  - Reserved forests
  - Ecologically sensitive areas which may include national parks, sanctuaries
  - Biosphere reserves
- Critically polluted areas
- Or within 50 km of interstate boundary shall require environmental clearance from the central government

The following project categories, irrespective of the investment:

- Pesticides
- Bulk drugs and pharmaceuticals
- Asbestos and asbestos products
- All tourism projects between 200–500 m of a high tide line and at locations with an elevation of more than 1000 m with investments of more than Rs. 5 crores
- Mining projects (with leases of more than 5 ha)
- Highway projects except projects relating to improvement work
- Petrochemical complexes
- Tarred roads in the Himalayas and forest areas
- Distilleries
- Raw skins and hides
- Dyes
- Foundries
- Chlor-alkali industry
- Hydrocyanic acid
- Electroplating
- Meta-aminophenol
- Small-scale industrial units in the project categories mentioned above
- With investments less than Rs. 1 crore that are on the reserved list
- Are exempted from environmental clearance

The regulations cover biomedical waste disposal, including the installation of pollution control systems such as incinerators and autoclaves and meeting the prescribed limits of hazardous chemicals. The regulations regarding the recycling of waste materials include waste oils, lead-acid batteries, and nonferrous waste. Manufacturers are to take the initiative to collect spent batteries. They must maintain records of hazardous waste produced and transportation of hazardous chemicals according to Motor Vehicle Act/Rules.

DEFINITIONS. In this Act, unless the context otherwise requires,

a. "environment" includes water, air and land and the inter relationship which exists among and between water, air and land, and human beings, other living creatures, plants, micro-organism and property;
b. "environmental pollutant" means any solid, liquid or gaseous substance present in such concentration as may be, or tend to be, injurious to environment;
c. "environmental pollution" means the presence in the environment of any environmental pollutant;
d. "handling," in relation to any substance, means the manufacture, processing, treatment, package, storage, transportation, use, collection, destruction, conversion, offering for sale, transfer or the like of such substance;
e. "hazardous substance" means any substance or preparation which, by reason of its chemical or physico-chemical properties or handling, is liable to cause harm to human beings, other living creatures, plant, micro-organism, property or the environment;
f. "occupier," in relation to any factory or premises, means a person who has, control over the affairs of the factory or the premises and includes in relation to any substance, the person in possession of the substance;
g. "prescribed" means prescribed by rules made under this Act.

24.7 Environmental Protection Regulations of Japan

Japan established its Environment Agency and since then environmental situation has undergone substantial changes. During the high economic growth notable achievements have been made in controlling pollution. However, it has been reported that in Japan air pollution by nitrogen oxides in major urban areas and water pollution caused by household effluent and waste disposal have continued. During high economic growth Japan's share of global warming, depletion of ozone layer, transboundary movement of acid rain, and hazardous waste have been substantial. In Japan, the Basic Environment Law, which set out basic principles and directions for formulating environmental policies, was enacted in 1993. The plan systematically clarifies the measures to be taken by the national and local governments, as well as actions to be carried out by citizens, business and private organizations. In its first chapter, the Basic Environment Law sets out three basic principles for environmental conservation and the responsibilities of each sector of the society-including the national and local governments, corporations, and citizens-in living up to these principles. The second chapter gives a list of basic policies for environmental conservation, including the formulation of the Basic Environment Plan, promotion of environmental impact assessment, new policy measures such as economic measures to clear obstacles from environmental conservation policies, and measures to deal with global environmental problems. The third chapter stipulates councils needed to develop the policies listed in Chapter 2. A summary of the Basic Environment Law is presented here.

24.7.1 Chapter 1 General Provisions

- Article 1: Purpose
- Article 2: Definition
- Article 3: Enjoyment and Future Succession of Environmental Blessing
- Article 4: Creation of Society Ensuring Sustainable Development with Reduced Environmental Load
- Article 5: Active Promotion of Global Environmental Conservation through International Cooperation
- Article 6: Responsibility of State
- Article 7: Responsibility of Local Governments
- Article 8: Responsibility of Corporations
- Article 9: Responsibility of Citizens
- Article 10: Environment Day
- Article 11: Legislative Measures, and so on
- Article 12: Annual Report
- Article 13: Prevention of Air Pollution and the Like by Radioactive Substances

24.7.2 Chapter 2 Basic Policies for Environmental Conservation

Chapter 2 sets the guidelines for policy formulation. This chapter has eight sections:

- Section 1: Guidelines of Policy Formulation
  - Article 14: Guidelines for Policy Formulation
- Section 2: Basic Environment Plan
  - Article 15: Basic Environment Plan
- Section 3: Environmental Quality Standards
  - Article 16: Environmental Quality Standards
- Section 4: Environmental Pollution Control in specified Areas
  - Article 17: Formulation of Regional Environmental Pollution Control Program
  - Article 18: Promotion of Attainment of Regional Environmental Pollution Control
- Section 5: Implementation of Policies for Environmental Conservation by the State
  - Article 19: Consideration in Formulation of Policies by the State
  - Article 20: Promotion of Environmental Impact Assessment
  - Article 21: Regulations to Prevent Interference with Environmental Conservation
  - Article 22: Economic Measures to Prevent Interference with Environmental Conservation
  - Article 23: Promotion of Construction of Facilities and Other Projects for Environmental Conservation
  - Article 24: Promotion of Use of Products Contributing to Reduction of Environmental Load

- Article 25: Education and Learning on Environmental Conservation
- Article 26: Measure to Promote Voluntary Activities by Private Organization
- Article 27: Provision of Information
- Article 28: Implementation of Researches
- Article 29: Improvement in Systems for Monitoring and Others
- Article 30: Promotion of Science and Technology
- Article 31: Settlement of Environmental Pollution Disputes and Relief of Damage
- Section 6: International Cooperation for Global Environmental Conservation, **and so on**
  - Article 32: International Cooperation for Global Environmental Conservation, and so on
  - Article 33: Ensuring International Cooperation for Monitoring, Observation and Others
  - Article 34: Measures to Promote Activities by Local Governments and Private Organizations
  - Article 35: Consideration in Implementation of International Cooperation and Others
- Section 7: Implementation of Policies by Local Governments
  - Article 36: Implementation of Policies by Local Governments
- Section 8: Cost Bearing and Financial Measures
  - Article 37: Cost Bearing by Causers
  - Article 38: Cost Bearing by Beneficiaries
  - Article 39: Financial Measures for Local Governments
  - Article 40: Cooperation between the State and Local Governments

24.7.3 Chapter 3 Environment Council, and So On

- Section 1: Environment Council
  - Article 41: Central Government Council
  - Article 42: Organization of the Central Environment Council
  - Article 43: Prefectural Environment Council
  - Article 44: Municipal Environment Council
- Section 2: Conference on Environmental Pollution Control
  - Article 45: Establishment and Mandates
  - Article 46: Organization, and so on

The US–Japan Agreement on Cooperation in Environmental Protection was signed in 1975 and the US EPA and Japanese Ministry of Environment have been in constant over the implementation of environmental laws. The agreement established a Joint Planning and Coordination Committee (JPCC) which meets periodically to discuss major environmental policy issues, and to coordinate and review activities and accomplishments under the Agreement. The periodic JPCC meeting takes place alternate between the U.S. and Japan. The areas of cooperation stipulated in the agreement include:

- Air pollution control (mobile and stationary sources)
- Water pollution (including municipal and industrial wastewater treatment, marine pollution control, agricultural runoff and pesticide control)
- Control and disposal of toxic substances
- Endocrine disrupting chemicals
- Global climate change

24.8 Environmental Laws in China

China's environmental regulatory apparatus is in the process of attempting to develop a comprehensive framework for the regulation of soil pollution. The current scheme consisting of a few obscure notices and generally ignored soil standards is clearly inadequate to the task. A new soil pollution prevention law has reportedly been drafted and received expert feedback. It has now been submitted to the "relevant departments for comments." It is not clear whether the reference to "relevant departments" refers to the various internal departments of the Ministry of Environmental Protection or other ministries which may have some stake in the issue. Both steps are preludes to the eventual passage of the law by the Standing Committee of the National People's Congress. Combined with the draft Guidelines for Risk Assessment of Contaminated Sites that was released last year for comment, this news indicates that the efforts to develop a soil pollution legal framework are moving forward.

China recently released some findings from its survey of air and water pollution sources. The most striking fact was the significant impact of agricultural nonpoint sources on water pollution levels. Air and water (at least surface water) pollution which are easily observed generate more attention than pollution of the soil, yet once you stop the sources of air and surface water pollution, those resources return to pre-pollution states fairly quickly. The same cannot be said for soil and sediment contamination. Stopping the source of pollutants to the soil does not return the soil to the status quo ante; without active remediation the soil will remain contaminated for decades, even centuries, depending on the pollutant. Soil contamination is a problem because it reduces crop yields, taints the food supply, and creates a source of pollution of ground and surface water.

China initiated its first soil pollution survey in 2006 with a budget of 1 billion Yuan. The survey is designed "to assess soil quality across the country by analyzing the amount of heavy metals, pesticide residue and organic pollutants in the soil." The results of the survey were supposed to have been released at the end of 2009, but they were not. It is unclear whether they will be released this year. Nevertheless, some experts have "estimated one fifth, or about 20 million hectares, of China's arable land had been polluted" [10].

Soil pollution is caused by both industrial discharges (through direct discharges to the soil and through deposition of industrial air pollutants (such as mercury) on the soil) and agricultural activities. Indeed, as with water pollution, it is becoming clearer that China's agricultural sector is a major contributor to soil contamination. A new study published by *Science* magazine underscores the seriousness of the soil pollution problem.

In some select areas "soils are approaching pH values at which potentially toxic metals such as Al and manganese (Mn) could be mobilized," the authors write, concluding that "overall, anthropogenic acidification driven by [nitrogen] fertilization is at least 10 to 100 times greater than that associated with acid rain."

"Quantified economy-wide emissions targets" and NAMAs are, of course, well-established concepts under the UNFCCC and perfectly illustrate the gaping divide between what is required of developed and what is required of developing countries [2]. China's announced target of a 40%–45% reduction in its carbon intensity per unit of GDP by 2020 (over the 2005 level) is more in the nature of a quantified economy-wide emissions target than a NAMA. A Chinese NAMA would be something along the lines of "we will close X MW of small, inefficient coal-fired electric generation capacity by 2015."

There is a BASIC meeting scheduled for the third week in January when I suspect this issue will be hashed out among this new climate bloc consisting of Brazil, South Africa, India, and China. In the meantime, what do you think will happen?

China amended its Renewable Energy Law (Chinese version) on December 26, 2009. Chinese spokespersons have emphasized the fact that the amendments require the state grid companies to purchase all power produced by renewable energy sources, but they (and the news agencies that have run with the story) fail to point out that the prior version of the law required the same thing. In fact, the actual amendments to the law are extremely minor.

The changes include somewhat more national oversight of the preparation of sub-national renewable energy development and utilization plans, and a more explicit recitation of what should be considered in the preparation of such plans (Articles 8 & 9).

As relates to the purchase of renewable power by the grids, an argument could be made that the amendments actually weaken, or at least complicate, this obligation. The law as originally enacted included this simple and clear requirement at Article 14:

> Grid enterprises shall enter into grid connection agreement with renewable power generation enterprises that have legally obtained administrative license or for which filing has been made, and buy the grid-connected power produced with renewable energy within the coverage of their power grid, and provide grid-connection service for the generation of power with renewable energy.

Oriental Outlook Weekly published an interview with Cao Rongxiang, climate change researcher at the CPC Central Committee. He is the author of the forthcoming book, Climate Change and the Failure of Democracy, and he provides a primer on China's core negotiating principles and the major issues it anticipates at Copenhagen.

The 10 items of the core negotiating principles are

1. One target
   A global temperature rise below 2°C, or atmospheric carbon dioxide concentration of less than 450 ppm.

2. Two main channels

   The 1992 United Nations Framework Convention on Climate Change and the 1997 Kyoto Protocol.

3. Three principles
   a. "Common but differentiated responsibilities."
   b. "Fair."
   c. "Sustainable development."

   The "sustainable development" principle addresses the developing countries' concerns that efforts to address climate change will not interfere with their economic development and poverty eradication efforts.

4. Four consensus
   a. Developed countries must commit to quantified mid-term emission reduction obligations because of their historical emissions and current high per capita emissions.
   b. Developing countries should undertake efforts to reduce their greenhouse gas emissions and adapt to climate change in as aggressive manner as their national conditions and funding and technology transfers from developed countries permit.
   c. Developed countries should provide financial support to developing countries.
   d. Effective institutional arrangements should be devised to promote technology transfers to developing countries to enhance their capacity to address climate change.

As Cao Rongxiang notes, there really is not a consensus surrounding these last four items (but who wants to spoil a catchy phrase), indeed, there is not a consensus about any of the 10 points set forth above. Therefore, he believes the negotiations will be very intense.

He predicts the major controversies will center around five items:

1. Whether the parties proceed inside or outside the framework established by the "two main channels." He spends a lot of time criticizing the US for its desire to scrap some of the basic features of these prior international agreements (which are basically set forth in his three principles and four consensus).
2. Appropriate emission reduction commitments from developed countries.
3. The amount of financial assistance provided by developed countries to developing ones.
4. The transfer of technology from developed countries to developing ones.
5. Whether the focus of future actions should on adaptation or mitigation. Cao Rongxiang states that many small developing countries need help to adapt to rising sea levels, impacts on agricultural production, and so on, while developed countries tend to focus on capacity-building, the use of new energy sources, and so on to slow down the temperature increases.

He believes that the two most important issues for China are to hold firm on negotiating within the existing framework and to pursue technology transfers over financial assistance because lack of technology is China's most pressing need.

There is nothing new here, but it does make for a handy review of where China stands at the outset of negotiations.

24.9 Environmental Laws of Canada

The Canadian Environmental Assessment Act, 2012 (CEAA 2012) and its regulations establish the legislative basis for the federal practice of environmental assessment in most regions of Canada. The first Canadian Environmental Assessment Act was passed in 1992 by Government of Canada. The Act requires federal departments, including Environment Canada, agencies, and Crown corporations to conduct environmental assessments for proposed projects where the federal government is the proponent or where the project involves federal funding, permits, or licensing. The original version of the act was repealed in 2012 and re-written. The new law came into effect on July 6, 2012. The present act is comprehensive and looks into all aspects of environment.

24.9.1 Purpose of CEAA 2012

- **Protect components of the environment** that are within federal legislative authority from significant adverse environmental effects caused by a designated project;

- Ensure that designated projects **are considered and carried out in a careful and precautionary manner** in order to avoid significant adverse environmental effects when a federal authority is exercising a power or performing a duty or function required for the project to proceed;
- Promote cooperation and **coordination between federal and provincial governments**;
- Promote communication and cooperation with **Aboriginal peoples**;
- Ensure that opportunities are provided for **meaningful public participation**;
- Ensure that environmental assessments are **completed in a timely manner**;
- Ensure that proposed projects on federal lands or that are outside Canada and carried out or financially supported by a federal authority, **are considered in a careful and precautionary manner** in order to avoid significant adverse environmental effects;
- Encourage federal authorities to take actions in a manner that **promotes sustainable development** in order to achieve or maintain a healthy environment and a healthy economy; and
- Encourage further studies of the **cumulative effects of physical activities in a region** and the consideration of the study results in environmental assessments.

When the Canadian Environmental Assessment Agency (the Agency) is the responsible authority for a designated project that is described in the regulations designating physical activities, upon acceptance of a project description, an analysis is undertaken by the Agency to decide if a federal environmental assessment is required. This step does not apply to designated projects regulated by the National Energy Board and the Canadian Nuclear Safety Commission for which conducting an environmental assessment is mandatory when such a project is designated.

A responsible authority ensures that an environmental assessment of a designated project is conducted in accordance with CEAA 2012, including ensuring the public is provided with an opportunity to participate in the environmental assessment.

Under CEAA 2012, responsible authorities can be the Canadian Nuclear Safety Commission, the National Energy Board or the Agency.

The following factors must be considered:

- Environmental effects, including environmental effects caused by accidents and malfunctions, and cumulative environmental effects
- Significance of those environmental effects
- Public comments
- Mitigation measures and follow-up program requirements
- Purpose of the designated project
- Alternative means of carrying out the designated project
- Changes to the project caused by the environment
- Results of any relevant regional study
- Any other relevant matter

There are two types of environmental assessment conducted under the Canadian Environmental Assessment Act, 2012 (CEAA 2012):

1. Environmental assessment by a responsible authority.
2. Environmental assessment by a review panel.

An environmental assessment by a responsible authority is conducted by the Agency, the National Energy Board or the Canadian Nuclear Safety Commission. Information on the process for environmental assessments conducted by the Agency is provided below. Information with respect to environmental assessments conducted by the National Energy Board or Canadian Nuclear Safety Commission may be found on their websites.

An *environmental assessment by review panel* is conducted by a panel of individuals appointed by the Minister of the Environment and supported by the Agency.

Both types of assessments can be conducted by the federal government alone or in cooperation with another jurisdiction, such as a province.

An environmental assessment conducted by the Agency must be completed within *365 days*. This timeline starts when a notice of the commencement of the environmental assessment is posted on the Registry

Internet site and ends when the Minister of the Environment makes a decision as to whether the designated project is likely to cause significant adverse environmental effects.

The Minister of the Environment may refer a project to a review panel within 60 days of the notice of commencement of an environmental assessment. An environmental assessment by a review panel needs to be completed within *24 months*. This timeline starts when the proposed project is referred to a review panel and ends when the Minister of the Environment issues the environmental assessment decision statement.

For every environmental assessment,

- The Minister may extend the time limits by an additional 3 months, to facilitate cooperation with another jurisdiction or to take into account other circumstances specific to the project.
- Upon recommendation of the Minister of the Environment, the Governor in Council may also extend the time limit (in addition to the 3 month extension granted by the Minister).
- The period that is taken by the proponent to respond to a request from the Agency or a review panel (conduct studies, prepare environmental impact statement, collect further information, etc.), is not counted in the timelines.

24.9.2 Pollution Control Act

The Canadian Environmental Protection Act of 1999 is "An Act respecting pollution prevention and the protection of the environment and human health in order to contribute to sustainable development."

The goal of the renewed Canadian Environmental Protection Act (CEPA 1999) is to contribute to sustainable development through pollution prevention and to protect the environment, human life and health from the risks associated with toxic substances. It covers a diversity of activities that can affect human health and the environment, and acts to address any pollution issues not covered by other federal laws. As such, CEPA 1999 is a "catch all" piece of legislation that ensures potentially toxic substances are not inadvertently exempt from federal oversight as a result of unforeseen legislative loopholes. CEPA 1999 also recognizes the contribution of pollution prevention and the management and control of toxic substances and hazardous waste to reducing threats to Canada's ecosystems and biological diversity. It acknowledges, for the first time, the need to virtually eliminate the most persistent toxic substances that remain in the environment for extended periods of time before breaking down and bioaccumulative toxic substances that accumulate within living organisms.

24.10 Brazil Environmental Law

The National Environment Policy Act (Law 6938/81) established the National Environmental System (SISNAMA), which consists of federal, state and municipal government agencies and entities that are responsible for the protection and control of environmental quality. The Brazilian Constitution empowers state and local governments to legislate on environmental matters in a complementary manner. The government has established the basic requirements that are applicable across the board, while state and local agencies introduce the specific requirements and standards that may be even more rigorous than the federal rules, except in areas assigned to exclusive federal jurisdiction such as nuclear and off shore activities. Here it seems, Brazilian Government has taken different routes than most other countries in framing the environmental laws, and they have given more powers to the state and local governments. The establishment of environmental regulations at the federal level takes place through either the National Congress or the National Environment Council (CONAMA), which is the core agency of Brazil's National Environment System (SISNAMA).

The first state to enact laws on environmental licensing and pollution control is Rio de Janeiro. The state level agency is called the State Environmental Engineering Foundation (FEEMA).

The Environmental Impact Assessment (EIA) for environmental licensing purposes, including "any power generation plant with capacity of over 10 MW," is very broad. Later in 1997, CONAMA established clear jurisdiction between federal and state for licensing purposes. It was a more advanced version that the previous one from 1986 and leaves no ambiguity in the framing of laws. The Environmental Impact Assessment (EIA) is quite broad and more sophisticated than other countries environmental laws. For licensing purposes, the Federal Environmental Agency, CONAMA has established clear three stages at which the environmental concerns would be addressed. These three stages are; (i) Preliminary License (LP), required during the planning stage and issued on the basis of an environmental assessment study, that is, an Environmental Impact Assessment (EIA) or similar type of study, (ii) Installation License (LI), required prior to construction through submitting the Environmental Control and Management Plan (EMCP), detailing all the environmental control systems and monitoring programs, (iii) Operating License (LO), issued after the completion of construction and inspection of the facilities, complying with all the requirements in the Installation License, prior to start-up of plant operations. It could be safely observed that these three stages are quite broad and comprehensive. They have also put time limits on the licenses awarded at various stages and the licensee cannot dragging the project far beyond the technological limits. Sometimes, if the licensee wishes to expand the company, they have to go through all three stages again. All federal, state, and municipal legislation are comparable the best international standards.

Federal CONAMA established the National Air Quality Program (PRONAR) and sets the Air Quality Policy criteria, National Water Resource Management System, and projects for final treatment and disposal of solid wastes.

24.10.1 Air Quality Policy Criteria

The Air Quality Policy criteria set the standard for air quality, as presented in Table 24.6, which also compares them to international standards. This table sets the national maximum limits for atmospheric emissions from new stationary sources burning oil or coal. These standards for Air Emissions Standards proposed by CONOMA, 1990, are given in Table 24.7. Table 24.8 shows a comparison of Brazilian and World Bank Air Emission Standards.

Brazil also has legislation for damage to the environment or any portion of it and establishes the principles that bind the responsible obliged to repair such damages. It stipulates the value of damages and requiring environmental reclamation. It also establishes that owners, companies and managers/employees responsible for damages to the environment or any of its part shall be sued for criminal action, irrespective of civil liabilities and fines.

Table 24.6 Ambient Air Quality Standards

		CONAMA (11)		World Bank
		Primary (μg/m³)	Secondary(μg/m³)	Guidelines (12)[a]
	24 H	240[a]	150[a]	230[d]
	Annual	80[b]	60[b]	80[d]
	24 H	150[a]	100[a]	
	Annual	60[c]	40[c]	
	24 H	150[a]	150[a]	150
	Annual	50[c]	50[c]	50
	24 H	365[a]	100[a]	150
	Annual	80[c]	40[c]	80
	1 H	40[a] (35 ppm)	40[a] (35 ppm)	
	8 H	10[a] (9 ppm)	10[a] (9 ppm)	
O_3	1 H	160[a]	160[a]	
	1 H	320[a]	190[a]	
	24 H			150
	Annual	100[c]	100[a]	100

Source: Data for World Bank guidelines from *Pollution Abatement Handbook*, World Bank, 1998.

Note: These limits are to be used only for carrying out an environmental assessment in the absence of local ambient standards. They do not substitute in any way for local ambient air quality standards.

Primary standards are meant to prevent harm to public health. Secondary standards are thresholds above which there might be harm to the environment. The primary air quality standards are applicable until states designate air quality classes within their territory.

[a] Not to exceed once per year; [b] annual geometric mean; [c] annual arithmetic mean; [d] measurement of PM_{10} is preferable to measurement of TPM as recommended by World Bank Guidelines.

Table 24.7 Air Emission Standards (CONAMA 1990)

Pollutants	Classes of Areas			
	Class I (Conservation Units)	Other Class I (Resort Areas)	Class II	Class III
Sources with Less than 70 MW				
SO_2	NA	2000 g/10^6 kcal	5000 g/10^6 kcal	5000 g/10^6 kcal
Total particulate matter	NA	120 g/10^6 kcal	350 g/10^6 Kcal – oil 1,500 g/10^6 kcal-coal	350 g/10^6 kcal-coal 1500 g/10^6 Kcal-coal
Smoke (opacity)	NA	20% Ringelman 01	20% Ringelman 01	20% Ringelman 01
Sources with More than 70 MW				
SO_2	NA	NA	2000 g/10^6 kcal	2000 g/10^6 kcal
Total particulate matter	NA	NA	120 g/10^6 kcal-oil 800 g/10^6 kcal-coal	120 g/10^6 kcal-oil 800 g/10^6 kcal-coal
Smoke (opacity)	NA	NA	20% Ringelman 01	20% Ringelman 01

Source: CONOMA. Nacional do Meio Ambiente. CONOMA.MMA, 2014.

Table 24.8 Comparison of Air Emission Standards: Brazil and World Bank

Bases	CONAMA	World Bank
TPM	120 g/10^6 kcal-oil	50 mg/Nm^3
SO_2	2000 g/10^6 kcal	0.20 tons per day (tpd) per Mwe of capacity for the first 500 Mwe plus 0.10 (tpd) for each additional Mwe of capacity over 500 Mwe concentration should not exceed 2000 mg/Nm^3, with a maximum emission level of 500 tpd. Construction of two or more plants in the same airhed to circumvent this cap is not acceptable.
Smoke (opacity)	20% Ringelman	NR

Source: Data for World Bank guidelines from *Pollution Abatement Handbook*, World Bank, 1998.

Notes: NA, not allowed; NR, not regulated.

24.10.2 National Water Resource Management System

CONAMA established Brazil's water resources policy and created the National Water Resource Management System in 1997. Two types of standards are introduced by this law. One sets water quality standards for water bodies by their proposed use, classifying inland waters into seven types. The standards for class II waters are for waterways for commercial shipping, scenic harmony and secondary contact recreation. The standard for Class II is given in Table 24.9. The effluent discharge standards are presented in Table 24.10, which shows the effluent discharge standards in milligrams per liter.

24.10.3 Solid Waste

The next set of legislation established by CONAMA is related to solid waste treatment and disposal. This is subject to approval by the competent state and municipal authorities. The solid waste-disposal legislation establishes three classes of residues: Class I, hazardous; Class II, noninert; and Class II, inert. Using NBR 10004 for technical support, it establishes solid waste inventories to be submitted to the environmental authorities for both new and existing activities. These classes of hazardous (Class I), noninert (Class II), and inert (Class III) are established according to their potential risks to the environment and public health, ensuring proper handling and disposal of these solid wastes. The requirements for differentiating between noninert (Class II) and inert (Class III) wastes are established, applicable only to wastes in a solid state. It also established the requirements for the design and operation of hazardous wastes landfills. It stipulates that solid wastes may be dumped in landfills or at other disposal sites provided that they do not affect neighboring water bodies.

CONAMA also define nationwide noise emissions evaluation criteria, according to ABNT standards 10.d 10.152.

24.10.4 Noise Emissions

Federal CONAMA resolution N001/90 established noise levels limits acceptable to citizens at large at night and day separately. These limits are presented in Table 24.11.

If the noise goes above the limits in the table above, it triggers the noise impact evaluation. The CONAMA established the National Silence Program and empowers the State and Municipal Government to impose stricter limits on noise emissions.

Table 24.9 Class II Bodies of Fresh Water: Water Quality Standards

Parameter	Concentration/Description (mg/L)
pH	6.0 < pH<9.0
Oil and greases	Virtually absent
DBO (oxygen demand)	<3
Aluminum	<0.1
NH_4	<0.02
Arsenic	<0.05
Barium	<1.0
Benzene	<0.01
Baron	<0.75
Cadmium	<0.001
Lead	<0.03
Cyanide	<0.01
Copper	<0.02
Hexavalent chromium	<0.05
Chloride	<250
Tin	<2.0
Phenol	<0.001
Soluble iron (Fe^{2+})	<0.3
Fluoride	<1.4
Manganese	<0.1
Mercury	<0.0002
Nickel	<0.025
Silver	<0.0011
Selenium	<0.01
Sulfite	<0.002
Phosphate	<0.025
Nitrate	<10.0
Trichorethylene	<0.003
Vanadium	<0.18
Tetrachlorinated carbn	<0.01
DDT	<0.005
PCBs	<0.1 μg/L
Zinc	<0.01

24.10.5 Storage of Hazardous Materials

Brazil's National Environmental Policy Act stipulates the best practices of adequate storage for hazardous materials in order to protect workers and environment. It also establishes Hazardous materials Management Guidelines, applicable to facilities and activities involving the shipment, production, handling, storage, and disposal of hazardous materials.

It stipulates through a Federal Decree N 96.044/88 the strict guidelines for transportation of hazardous materials. These guidelines regulate the road shipments of hazardous materials and establish that

- Any vehicle carrying a hazardous substance shall avoid thoroughfares running through densely populated areas or protected areas encompassing water sources, water reservoirs, or forest or ecological reserves or areas lying close to any of these.
- The issuer shall annually inform the National Highway Department (DNER) of the flow of hazardous materials shipped on a regular basis.
- Should the origin or destination of hazardous cargo require the use of a restricted roadway, the shipper shall prove this fact to the authority with jurisdiction over said roadway, whenever requested to do so.
- The shipping contact should assign responsibility for emergency expenses and costs, in order to provide technical support in case of accident.

The local transportation rules of traffic Department must also be observed.

Table 24.10 Effluent Discharge Standards

Parameter	CONAMA Limits	World Bank Limits
pH	5.0 < pH < 9.0	6.0 < pH < 9.0
Temperature	<40.0°C (not to generate a variation of more than 3°C)	
Setting matter	1.0 in 1 h Imhoff test	50
Oil and grease (mineral)	20	10
DBO (oxygen demand)	60	50
Residual chlorine	NR	0.2
Total phosphates	NR (but not to exceed 0.025 outside the mixing zone)	NR
Nitrates	10	NR
NH_4	5.0	10.0
Cadmium	0.2	0.1
Lead	0.5	0.1
Copper	1.0	0.5
Hexavalent chromium	0.5	0.1
Chromium	2.0	0.5
Phenol	0.5	0.5
Soluble iron (Fe^{2+})	15.0	3.5
Fluoride	10.0	20.0
Soluble manganese (Mn^{2+})	1.0	NR
Nickel	2.0	0.5
Sulfite	1.0	NR
Zinc	5.0	1

Note: NR, not regulated.

Table 24.11 Noise Level Limits (Maximum Acceptable L_{eq}* per Period)

Area Classification	Day (db(A))	Night (db(A))
Brazilian NBR 10.151		
Urban residential area	50	45
Industrial area	70	60
World Bank		
Residential/institutional/educational	55	45
Industrial/commercial	70	70
US Export-Import Bank		
Residential	60	55
Commercial	65	60
Industrial	75	75

Note: L_{eq}*, equivalent noise level, defined as constant noise levels with same acoustic energy as level of the real noise levels nonstable, varying during time of measurement.

Brazilian Ministry of Labor has regulations regarding occupational health and safety almost similar to US OSHA regulations. Brazil has developed a substantial body of pollution control laws and regulations, similar in many respects to those of the United States. The environmental legislation in Brazil could be considered sufficient to give judicial support to public policies and private actions in the field of environmental protection. In essence, the problem is political rather than legal, raising questions as to the role of the economic interests of Brazil's elite, and casting doubt on the merits of passing new laws without a change in attitude.

References

1. Ramlogan, R. (2001–2002). Promoting compliance with international environmental agreements. *Vermont Journal of Environmental Law* 3.
2. Findley, R. W. (1988). Pollution control in Brazil. *Ecology Law Quarterly* 15 (1): 1–68.

3. – (2012). Canadian Environmental Assessment Act. Laws.justice.gc.ca. http://laws.justice.gc.ca/en/C-15.2/. Retrieved 2012-10-19.
4. Kanari, N., Pineau, J. L., and Shallari, S. (2002). End-of-life vehicle recycling in the European Union. *JOM* 55 (8): 15–19.
5. European Commission, 2000. Directive 2000/53/EC of the European Parliament and of the Council on end-of-life vehicles, EU.
6. Gilmore Hankey Limited Company. (2006). A study to examine the benefits of the end-of-life vehicles directive and the costs and benefits of a revision of the 2015 targets for recycling, re-use and recovery under the ELV directive. www.ghkint.com, UK.
7. Australian Department of the Environment and Heritage. (2002). Environmental impact of end-of-life vehicles: An information paper. Adelaide, Australia.
8. Staudinger, J. and Keoleian, A. (2001). Management of end-of-life vehicles (ELVs) in the US. Centre for Sustainable Systems. Report No. CSS01-01, March 2001, University of Michigan.
9. The United State Environmental Protection Agency (EPA). (2007). Report on environment. EPA.
10. Falk, R. and Wee, J. (2014). China's new environmental protection law. Morrison & Foerster LLP, October.
11. CONOMA. (2014). Nacional do Meio Ambiente. CONOMA/MMA.
12. World Bank. (1998). *Pollution Prevention and Abatement Handbook 1998: Toward Cleaner Production*. The World Bank Group, Washington, DC.

Index

D

E

H

I

J

K

L

M

N

O

P

Q

R

S

T